SIXTH EDITION

Architecture: Drafting and Design

DONALD E. HEPLER
PAUL R. WALLACH
DANA J. HEPLER

GLENCOE
McGraw-Hill

New York, New York
Columbus, Ohio
Mission Hills, California
Peoria, Illinois

Cover Design: NSG Design
Cover CAD Drawing: Versacad

Library of Congress Cataloging-in-Publication Data

Hepler, Donald E.
Architecture: drafting and design / Donald E. Hepler, Paul R. Wallach, Dana J. Hepler. — 6th ed.
p. cm.
ISBN 0-07-028322-2
1. Architectural drawing. 2. Architectural design. I. Wallach, Paul Ross. II. Hepler, Dana J. III. Title.
NA2700.H4 1990
720'.28'4—dc20 90-2818
CIP

Architecture: Drafting and Design, Sixth Edition
Imprint 1996

ISBN 0-07-028322-2

7 8 9 10 11 12 13 14 15 VH/LP 00 99 98 97 96

Contents

Part Three
Basic Architectural Plans *131*

Part Four
Specialized Architectural Plans *281*

The design of space structures is one of the architectural challenges of the nineties. *(NASA)*

ABOUT THE AUTHORS

Donald E. Hepler completed his undergraduate work at California State College, California, Pennsylvania, and his graduate work at the University of Pittsburgh. He has been an architectural designer and drafter for several architectural firms, has served as an officer with the United States Army Corps of Engineers, and has taught architecture, design, and drafting at both the secondary and the college level. He is currently marketing manager, Glencoe/McGraw-Hill.

Paul R. Wallach received his undergraduate education at the University of California at Santa Barbara and did his graduate work at California State University, Los Angeles. He has acquired extensive experience in the drafting, designing, and construction phases of architecture and has taught architecture and engineering drawing for many years in Europe and California at the secondary school and post-secondary levels. He currently devotes full time to technical writing and consulting.

Dana J. Hepler received his bachelor's degree from Ohio State University. He is an ASLA licensed landscape architect and a member of the Construction Specifications Institute. He has been associated with several of the largest architectural firms in the world as both designer and construction manager. He is currently Director of Planning, Westwood Companies and Director, Environetics Inc., Kingsport, New York.

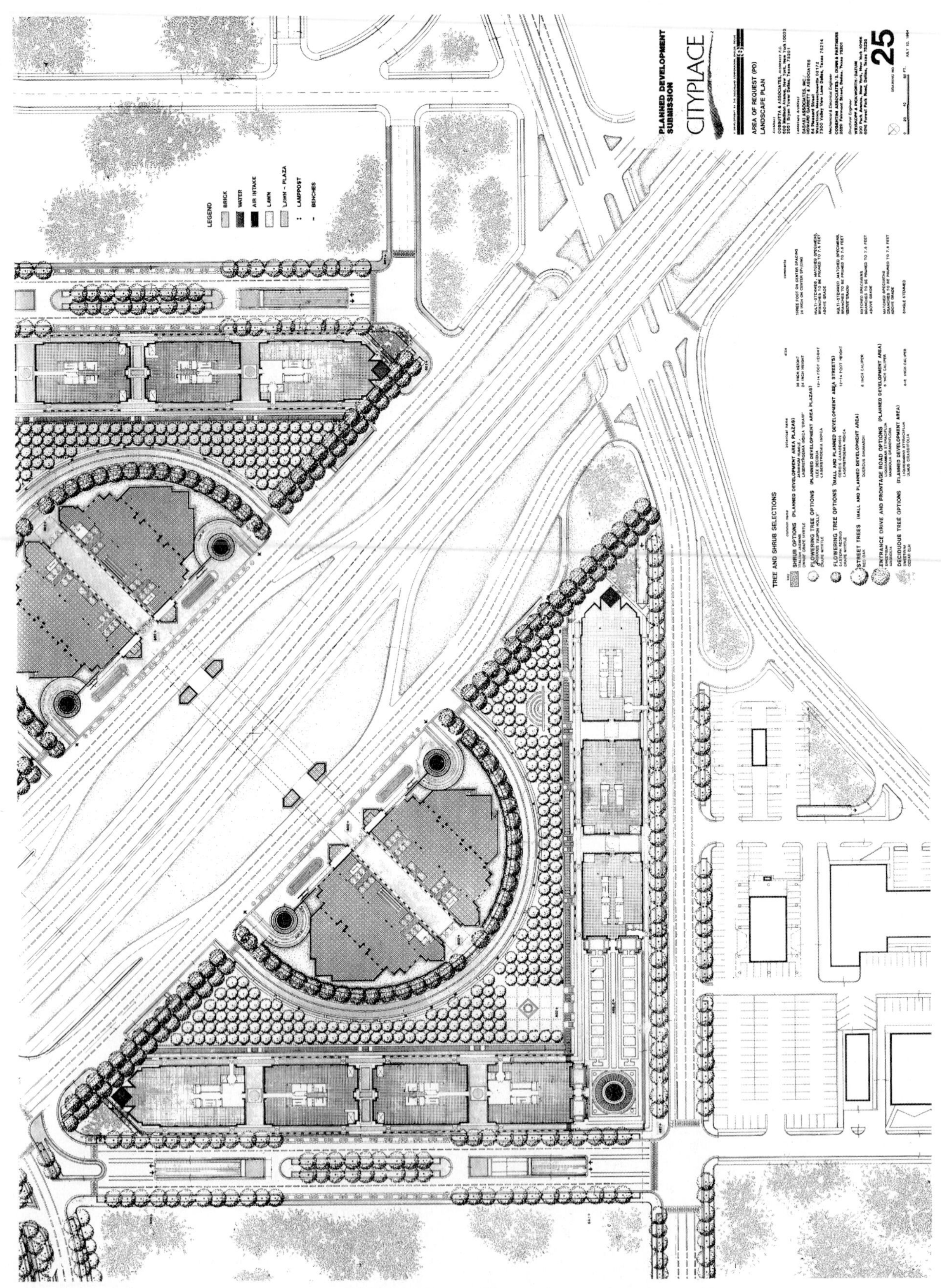

The design of Southland Corporation's City Place in Dallas, Texas required over 10,000 drawings to complete. *(Southland)*

Preface

Architecture: Drafting and Design is intended for use in a first course in architectural drafting. Since a study of basic drafting normally precedes a course in architecture, only the principles and practices essentially related to architectural drafting are presented. However, orthographic principles and design practices as applied to architectural drawings are reviewed.

Architecture: Drafting and Design has been revised and expanded to include the latest technological information, methodology, and standards relating to drafting, design, and construction. Portions completely revised in this edition include the units on computer-aided drafting (CAD), energy, structural calculations, solar planning, and site development. In addition, the design factors and procedures unit has been extensively expanded to include the most current design sequences and techniques used in contemporary architectural practice. Unit 52, Construction Systems, has also been extensively expanded to include structural steel, skeleton frame, heavy timber, and concrete and masonry systems of construction.

The sixth edition is divided into six parts:

Part One, Preliminary Design and Planning, exposes the student to the elements, sequence, and principles of architectural design. It also covers the preliminary design considerations necessary for effective planning, such as solar orientation, energy planning, density planning, and ecological planning.

Part Two, Area Planning, covers the basic elements of planning the areas of a structure and the functional techniques used to integrate these areas into a composite, effective architectural plan.

Part Three, Basic Architectural Plans, includes the basic drafting techniques and design procedures used in preparing architectural floor plans, elevations, and pictorial drawings. Information about scales (including metric) and CAD is also introduced in this part.

Part Four, Specialized Architectural Plans, covers the basic types of construction systems and shows how to prepare the many specialized architectural drawings that are necessary for a complete and detailed description of a basic architectural design. These include site development, sectional, foundation, framing, electrical, HVAC, plumbing, and modular drawings.

Part Five, Architectural Support Services, covers the activities in which an architectural drafter may participate but which do not directly involve the drafting function; these include the preparation of architectural models, schedules, and specifications. Part Five also contains an introduction to the related legal and financial aspects of architectural planning.

Part Six, the Appendix, includes reference materials which can be used in the preparation of the types of architectural drawings covered in the first five parts. The major section, Section 23, in the Appendix, is Basic Engineering Calculations, which includes instructional and reference material on general architectural, geometric, and structural design calculations. The Appendix also includes career information and complete coverage of architectural terms and abbreviations used in architectural drawings and documents.

Architecture: Drafting and Design is organized to be used consecutively from Part One through Part Five, with Part Six functioning as an architectural reference source. However, other sequences of study may be more suitable for students with different needs. When the basic emphasis is placed on developing fundamental architectural drafting skills and techniques, Part Three may be studied first. Classes that are specifically oriented to the construction phase of architecture may find Part Four a logical point of departure.

Coverage of specific topics is not restricted to one unit or section. Topics are covered to whatever depth and with whatever emphasis are needed to meet the goals of each unit. Thus the same topic may be presented from a different perspective and with different levels of concentration in several different units.

All the illustrations have been specifically selected and/or prepared to reinforce and amplify the principles and procedures described in the text. Whenever possible, each principle and practice has been reduced to its most elementary form and, for easy comprehension, has been directly related to a familiar environment.

Reinforcement is continuous throughout the text through the constant use of multiple views of the same subject. The plan, elevation, and pictorial interpretation of components are included at every opportunity in order to strengthen understanding of the functions of different architectural views and types of drawings. The progression within each section and unit is from the simple to the complex and from the familiar to the abstract.

The exercises that appear at the end of each unit are organized to provide the maximum amount of flexibility and reinforcement of the material covered. Most units include exercises that range from the very simplest, which can be completed in a few minutes, to the more complex, which require considerable time and application of the principles of architectural drafting and design. Exercises that require original design work are marked with a special symbol . Completion of these exercises by the student will result in the creation of a complete set of related architectural plans and documents.

Since communication in the field of architectural drafting and design depends largely on understanding the vocabulary of architecture, new terms, abbreviations, and symbols are defined when they first appear and are reinforced throughout the remainder of the text, and in the Glossary.

The authors express sincere thanks to Wendy Talcott for the design contributions of Home Planners, Inc., Farmington Hills, Michigan; to Vardy Vincent, Lee Kwoler, William Wagoner, Brenda Guthrie, David Umfress, Walter McKey, Jay Helsel, Cecil Jensen, Byron Urbanic, Tony Iorlano, Arthur Hidalgo, Robert Weiss, Hugh Phares, Stuart Soman, Frank Mahan, and Norman Ouellette for their technical reviews and suggestions; and to Diane Kingston for her rendering.

Donald E. Hepler
Paul R. Wallach
Dana J. Hepler

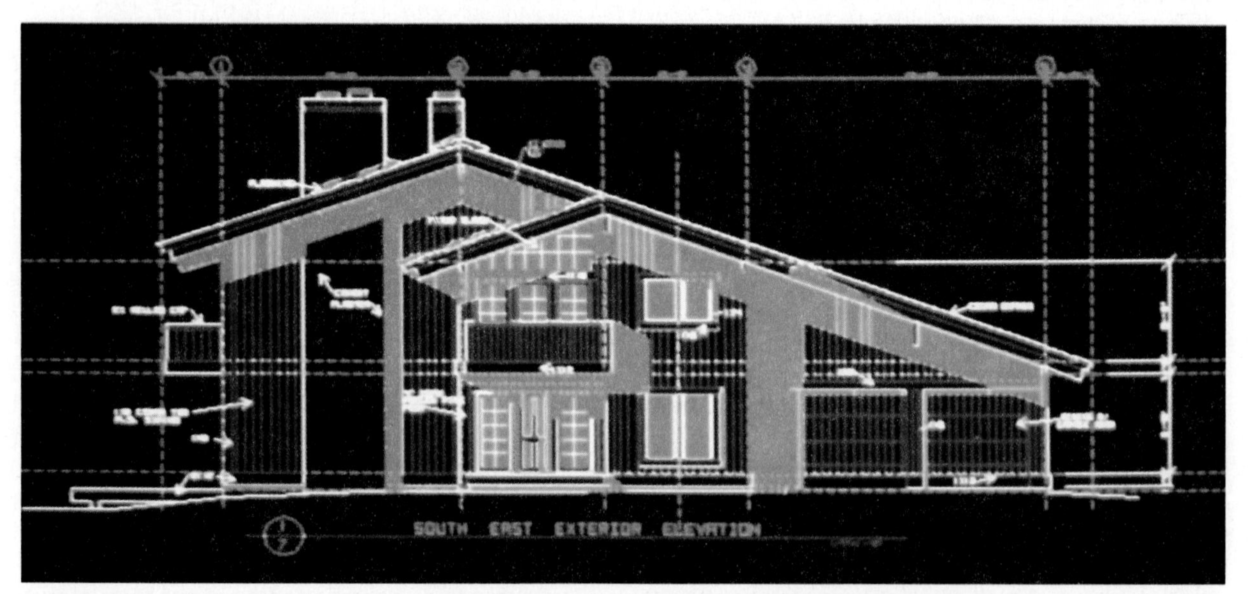

Architectural drafting and design began with a stick and has evolved into computer-aided drafting and design systems; such as the one used to produce this drawing. *(Versacad)*

INTRODUCTION

HISTORY

About 10,000 years ago human beings gave up the nomadic life, turned to agriculture, and began settling in fixed locations for longer periods. Before this time there had been little need for drawing or planning in the making of dwellings, since shelter usually consisted of natural caves or portable tents made of animal skins. But as people became more sedentary, they began constructing permanent tents, adobe huts, modified caves, or lean-to shelters, using existing natural materials.

The concentration of more permanent dwellings near fertile areas gave rise to villages. And village life created the need for more planning to serve the developing social structure. And so, with the planning and construction of the first dwellings, fortifications, and public areas, the art and science of architecture began. Likewise, the field of architectural drafting began when early people first drew the outline of a shelter or a village plan in the sand or dirt and planned ahead for the use of existing materials.

Centuries passed and civilization developed, changing the form, function, and complexity of architecture and construction. As materials, processes, and structures evolved, and as human needs expanded and lifestyles became more sophisticated, the role of the architectural designer became more critical in the development of the environments for human growth and potential. Then as structures became more complex, more complete, accurate, and detailed drawings became necessary.

In planning a structure a designer uses the cumulative knowledge of centuries. The history of architectural design is directly related to progress in other areas of learning. For example, architecture has relied heavily upon advances in science and mathematics. From these advances have come new building materials and building methods. New engineering developments and new building materials have brought about more changes in architectural design in the last several decades than had occurred in all the earlier history of architecture. Yet, many of the basic principles of modern architecture, such as bearing-wall construction and skeleton-frame construction, have been known for centuries. Even today, architectural structures are divided into two basic types: the bearing wall and the skeleton frame.

Bearing-Wall Construction

Bearing walls are solid and support both themselves and the roof of a structure. A log cabin is an example of bearing-wall construction. Most early architecture used the bearing wall for support. In fact, one of the first major problems in architectural drafting and design was how to provide openings in supporting walls without sacrificing the needed support. One of the first solutions to this problem was the development of the *post and lintel* (Fig. 1). In this type of construction, posts large enough to support the lintel (upper horizontal beam), wall, and roof above are used. The ancient Greeks used *post-and-lintel* construction to erect many of their most beautiful buildings (Fig. 2).

Since most ancient peoples used stone as their primary building material, they were limited by the great weight of the stone in their application of post-and-lintel construction. Furthermore, stone post-and-lintel construction could not support wide openings. Therefore, many posts (columns) were placed close together to provide the needed support. The Greeks and Romans developed many styles of columns and gave names to them. The various styles of column designs were known as *orders*. The orders of architecture developed by the Greeks are known as the *Doric*, the *Ionic*, and the *Corinthian* orders. Later, the Romans developed the *Composite* and the *Tuscan* orders as shown in Fig. 3.

Since the climate of Greece was well suited to open-air construction, the Greeks used the post-and-lintel technique to great advantage. The Parthenon is a classic example of the Greeks' use of the post and lintel.

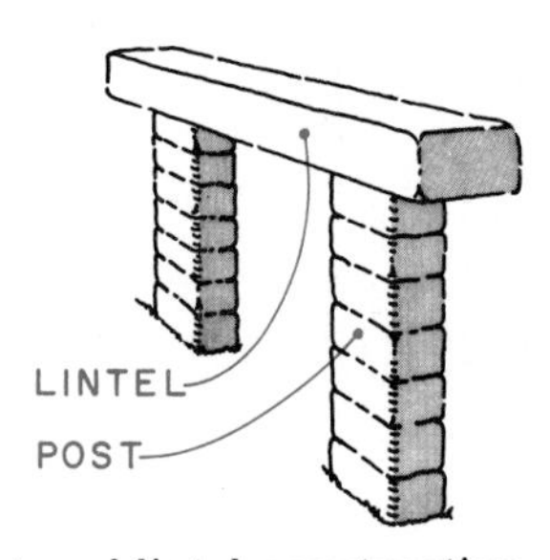

Fig. 1 Post-and-lintel construction.

Fig. 2 Ancient Greeks used post-and-lintel construction *(Celotex Corp.)*

Oriental architects also made effective use of the post and lintel. They were able to construct buildings with larger openings under the lintel because they used lighter materials, such as wood. The use of lighter materials resulted in the development of a style of architecture that was very light and graceful. The oriental post-and-lintel designs were also used extensively for gates and entrances.

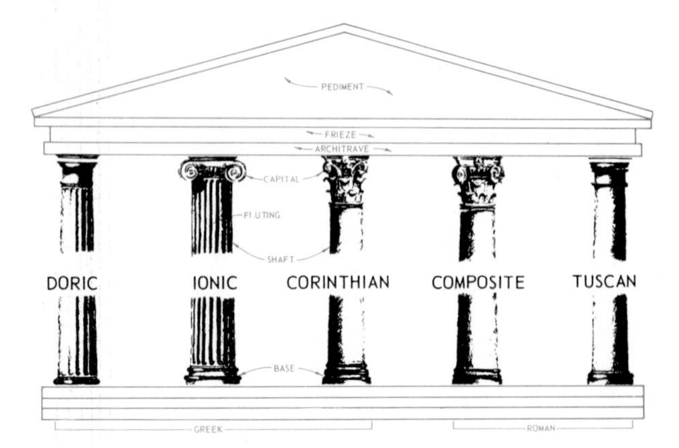

Fig. 3 Column orders in architecture.

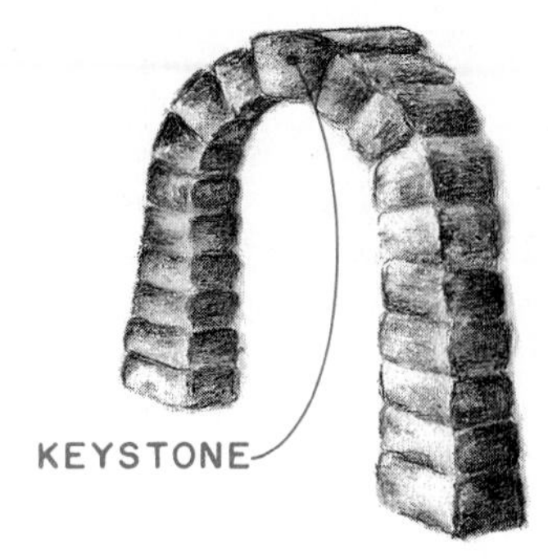

Fig. 4 A keystone supports both sides of an arch.

The Arch

The Romans began a new trend in the design of wall openings when they developed the *arch.* The arch (Fig. 4) is different from the post and lintel because it can *span* (extend over) greater areas without support. It is easier to erect because it is made from many smaller, lighter pieces of stone. (In contrast, post-and-lintel construction is limited in span by the size of the stone used for the lintel.) The principle of the arch is that each stone is supported by leaning on the *keystone* in the center. The keystone is shaped like a wedge and locks the other stones in place.

The Vault

The simple arch led to the development of the *vault* (Fig. 5). The vault is simply a series of arches that forms a continuous covering. This development allowed the use of the arch as a passageway rather than as just an opening in a wall. The *cross vault* (Fig. 6) is the intersection of two *barrel vaults.* The barrel vault and the cross vault were popular Roman construction devices. The intersections make additional strength possible with less foundation support. The Romans combined the use of the vault and the column very extensively in their architecture.

Fig. 5 The barrel vault is a series of arches.

Fig. 6 A cross vault is the intersection of two barrel vaults.

The Dome

The *dome* is a further refinement of the arch. A dome is made of arches so arranged that the bases make a circle and the tops meet in the middle of the ceiling (Fig. 7). The Romans felt that the dome gave a feeling of power. Therefore, they used domes often in religious and governmental structures, as shown in Fig. 8.

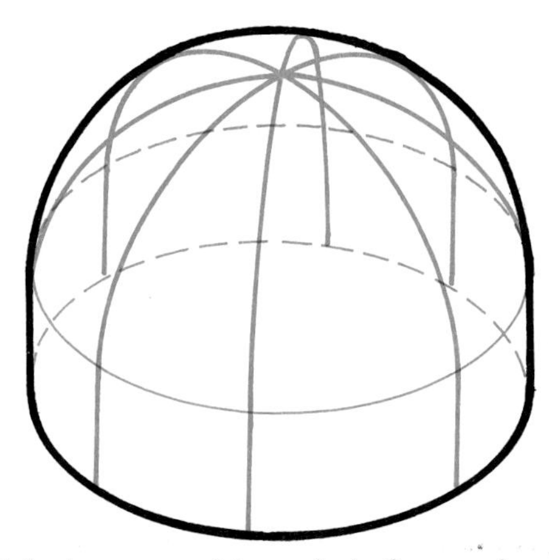

Fig. 7 Arches spaced in a circle form a dome.

Fig. 8 Domes became an early symbol of piety. *(Celatex Corp.)*

The Gothic Arch

Gothic architecture originated in France. It spread throughout western Europe between 1160 and 1530. Another variation of the arch, the *pointed arch* was developed in Gothic architecture. The pointed arch (Gothic arch) became very popular in Gothic cathedrals because it created a sense of reaching and aspiring by its emphasis on vertical lines. Construction of the pointed arch posed the same problem as did conventional arches, that of spreading at the bottom.

To support the arch at the bottom, a new device known as a *buttress*, or pilaster, was developed (Fig. 9). Buttresses were gradually moved up the walls, until their higher placement resulted in the development of the *flying buttress*. A flying buttress helps to support the sides of a wall without adding additional weight to the wall. Therefore thinner walls and more windows are possible.

Early builders knew little about the theory of structures, so many early buildings were built by trial and error. That is, if a building didn't collapse, the design would be duplicated by another builder. Then another designer or builder would become more daring, and so on, until the limits of the structural system were discovered

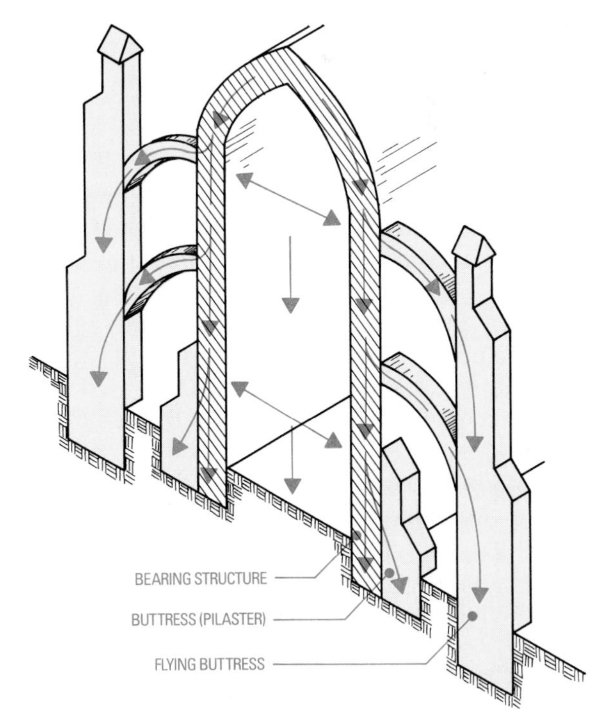

Fig. 9 Methods of supporting arches.

by the collapse of a building. This practice continued over many centuries until a mass of information was gathered which allowed builders to predict the structural success of a building with reasonable accuracy. But of course this approach cannot be tolerated in today's fast-changing world of new architectural materials and methods.

Technological Advances

Bearing-wall construction is still used in modern architecture. But new building materials, such as reinforced and prestressed concrete, enable the architect to span greater areas. This allows a greater flexibility in design, and therefore greater variety. The development of new materials and new building methods has also meant greatly improved structures. Modern buildings are larger, lighter, and safer than ever before. They are also more functional because they serve more uses.

New Materials

Advances in architecture throughout history have depended on the building materials at hand. As recently as American colonial times, builders had only wood, stone, and ceramic materials to work with. Early American architecture reflects the use of these materials. But a great change came with the development of steel, aluminum, structural glass, prestressed concrete, wood laminates, plastics, and other new synthetics. Now, buildings can be designed in sizes and shapes never before possible.

Many new materials are really old materials used in new ways or in new forms. Sometimes, they are old materials manufactured in a different way. For example, glass is not a new material. But the development of structural glass, glass blocks, corrugated glass, thermal glass, and plate glass in larger sizes has given the architect much greater freedom in the use of this material.

Wood is also one of the oldest materials used in construction. Yet, the development of new structural wood forms, plywoods, and laminates has revolutionized the use of wood in building. The manufacture of stressed-skin panels, boxed beams, curved panels, folded roof plates, and laminated beams has given builders new ways to build larger structures without sacrificing the warmth and beauty of wood.

Among the truly new architectural materials is plastic. The development of vinyl and laminated plastic has provided the architect with a wide range of new materials. The material that has contributed most to architectural change is steel. Without the use of steel, construction of most of our large high-rise buildings would be impossible. Even smaller structures can now be built on locations and in shapes that were impossible without the structural stability of steel, as shown in Fig. 10.

The manufacture of aluminum into lightweight, durable sheets and structural shapes has also given greater variety to design. But an old material, concrete, actually changed the basic nature of structural design. New uses of concrete are found in factory-made reinforced and prestressed structural shapes. These shapes are used for floors, roofs, and walls. They have provided the architect with still other tools for structural design.

Today's architects have the opportunity to design the framework of a building with steel but to use a variety of other materials as well. They can use large glass sheets for walls, prestressed concrete for floors, aluminum for casements, plastics for skylights, and wood for cabinets. A wide variety of still other materials makes possible different combinations. Thus the development of each new architectural material may make another one more useful.

New Construction Methods

The development of new materials is usually not possible without the development of new construction methods. For example, large glass panels could not have been used in the eighteenth century even if they had been available,

Fig. 10 Structural use of steel. *(Bethelehem Steel Corp.)*

because no large-span lintel-support system had been developed. Only when both new materials and new methods exist is the architect free to design with complete flexibility.

Present-day structures are usually a combination of old and new. In a modern building, examples of the old post-and-lintel method may be used together with skeleton-frame, curtain-wall, or cantilevered construction.

Skeleton Frame

One of the first methods developed to employ modern materials makes use of the *skeleton frame*. This kind of construction has an open frame to which a wall covering is attached. The frame provides the primary support, and the covering provides the needed shelter. The skeleton frame became popular with the development of framing materials and wall coverings that are light, strong, and usable in a variety of ways. The skeleton frame is now commonly used for light construction, as shown in Fig. 11. When steel is used for the skeleton, the skeleton frame is known as *steel-cage* construction as shown in Fig. 12.

The use of the skeleton frame, as opposed to bearing-wall construction, has given architects new opportunities. Lighter and stronger materials allow much more flexibility in design in such areas as cantilever construction. In cantilever construction the loads can be supported at locations other than at the extreme ends. As shown in Fig. 13, cantilevering allows a larger use of space without structural support on the end. Cantilever construction is uniquely suited to steel because unsupported steel beams can be extended farther without sagging than any other material. Before the development of steel girders, large cantilevered overhangs, such as the ones shown in Fig. 14, were not possible.

Since loads in steel-cage construction are not supported by the outside wall, *curtain walls* are possible. In the type of building known as *cur-*

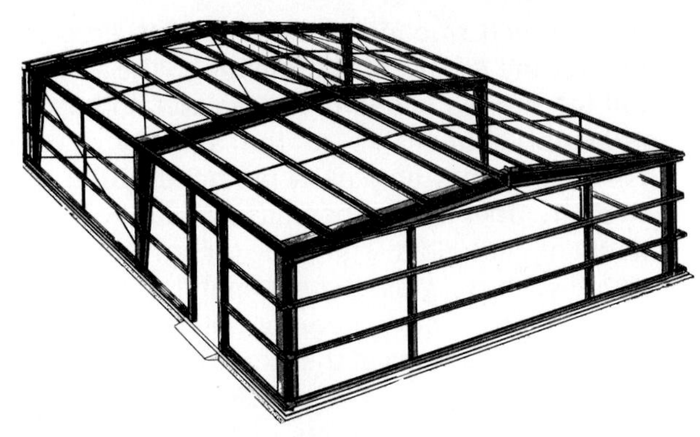

Fig. 12 Steel cage construction.

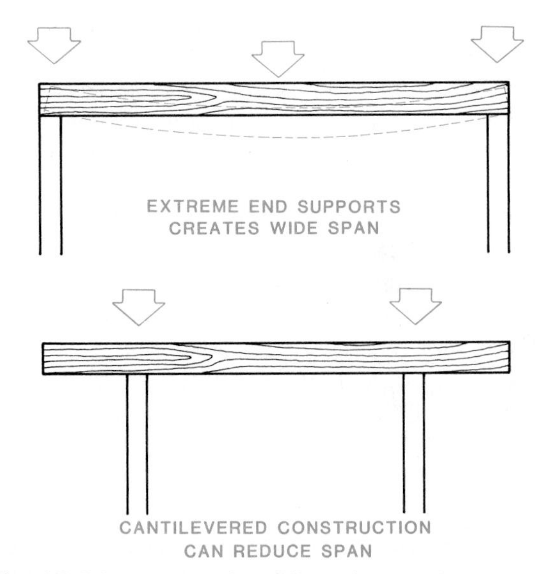

Fig. 13 Advantages of cantilevering.

Fig. 11 Wood skeleton-frame construction. *(Kauffman and Broad)*

Fig. 14 Cantilevered decks.

tain-wall construction, a steel cage is erected, forming the shape of the building. The curtain wall, or skin, is added last. This curtain has no structural relationship to the stability of the building; it acts only as a protection from the weather. Therefore, the curtain wall can be made of materials with little or no structural value, such as glass, sheet metal, or plastic. This is not possible in bearing-wall construction, where large portions of the exterior surface must be used to support the building, thus leaving only relatively small areas of the wall for windows and doors.

Shapes

For centuries, architectural development has been restricted by the use and overuse of the square and the cube (which are based on right angles) as the basis for most structures. Architects are now using other shapes, such as the triangle, octagon, pyramid, pentagon, circle, and sphere (Fig. 15). This has come about with the development of materials that are stronger and lighter and have a variety of uses. New construction methods also enable architects to design buildings that are completely *functional* (able to fulfill all needs) without reference to any basic geometric form. Many forms are now possible, and even the basic shapes of floor plans can be designed to meet a variety of needs.

Right-angle structures dominate today's buildings because of their efficient use of space. However, the early Egyptians recognized that the triangle provided the strongest rigidity with the smallest number of members. Today we are using new triangular principles to provide structural stability to buildings. In contrast to conventional curtain-wall construction, in which the wall is applied to a steel frame, some buildings have an exposed structural steel frame, eliminating all vertical columns from the skin to the core.

Sizes

New technology uses knowledge gained from advances in science. One of the most striking results has been the use of new materials and new methods to design and build structures of size greater than ever before (Fig. 16). The Sears Tower in Chicago is now the tallest building in the world. But as technology develops even more, buildings can increase to sizes previously thought impossible.

Who can say what will be possible? The idea of building a geodesic dome over central Manhattan, in New York City, as shown in Fig. 17, certainly seems impossible at the moment. But remember that the landing of human beings on the moon, and flights to and landings on distant planets, also seemed impossible not many years ago.

Location

Today, architects not only design buildings of enormous size but can also choose locations for buildings that were unthought of years ago. Further advances in transportation and architectural engineering will make even more difficult locations not only possible but workable.

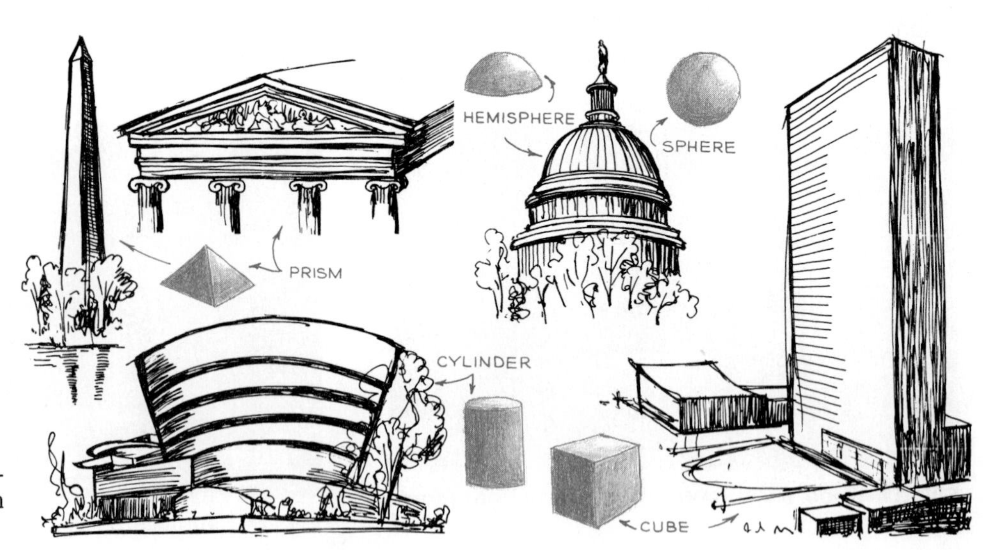

Fig. 15 Basic geometric shapes used in architecture.

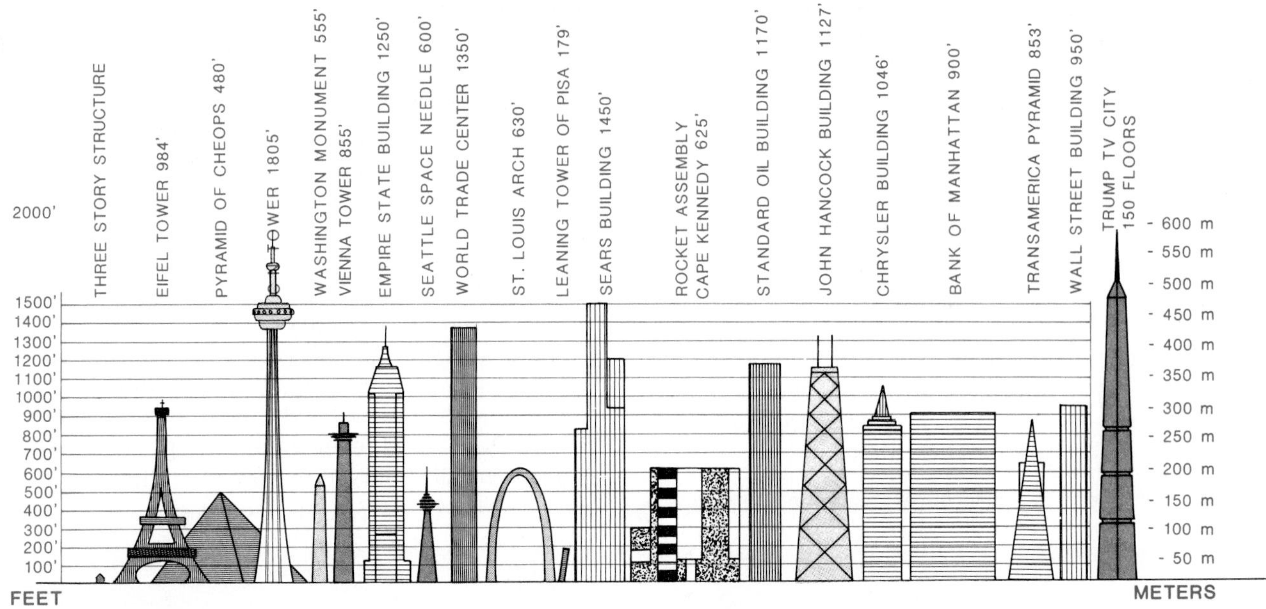

Fig. 16 Comparison of the largest structures in the world.

Components

One of the most significant advances of the past several years is the design and construction of architectural *components*. Components are preconstructed parts of a building; they are parts or sections made in advance. Using components allows the builder to construct parts away from the building site. This does away with much on-the-job construction work. Typical components include preassembled wall sections, windows, and trusses and molded bathrooms.

Fig. 17 Proposed Manhattan geodesic dome.

As more components are developed, construction changes from *on-site*, piece-by-piece building to the assembly of component parts on the site. The development of component systems does not necessarily change the nature of design, but it does change the way architects may design.

Designing with the use of components means the designer must adhere strictly to certain sizes in creating an architectural plan. Sizes of the components are standard, or uniform, just as an automobile is designed with many different, interchangeable parts.

THE FUTURE

The future of architecture will certainly be influenced by the development of new materials, new construction methods, and sociological changes dealing with the way people live in a given society. With the development of new materials and methods of construction, the architect is freed from the restrictions of traditional materials and methods. The architect becomes a coordinator of the activities of the structural engineer, the electrical engineer, the acoustical engineer, the sociologist, the interior designer, and so forth.

The architect's plan is in a sense a blend of all such activities and a refining of the relationship between art and technology.

Designing a structure that will last is a great challenge. The architect, the designer, and the builder must keep abreast of technological changes and advancements in architectural engineering and building design. This is true whether they are designing a residence, a large building (Fig. 18), redeveloping a neighborhood or city, or planning a completely different kind of structure. In fact, not all future buildings will be designed for land base use. Space stations above the earth and underwater structures as shown in Fig. 19 are now structurally feasible. The computer now plays a major role in architectural design. Drafters can now quickly and accurately generate all the working drawings needed for a structure. In any case, anyone working in the field of architectural design must understand people, their habits, their needs, and their activities. Such a person must also be capable of working with shapes, materials, colors, and proportions in order to design aesthetically pleasing and structurally sound buildings. The primary purpose of this text is to help provide some of the skills and knowledge necessary to achieve this end.

Fig. 18 World Trade Center. *(The Port Authority of New York)*

Fig. 19 Large underwater structures are now structurally feasible. *(McDermott International, Inc.)*

PART ONE

Preliminary Design and Planning

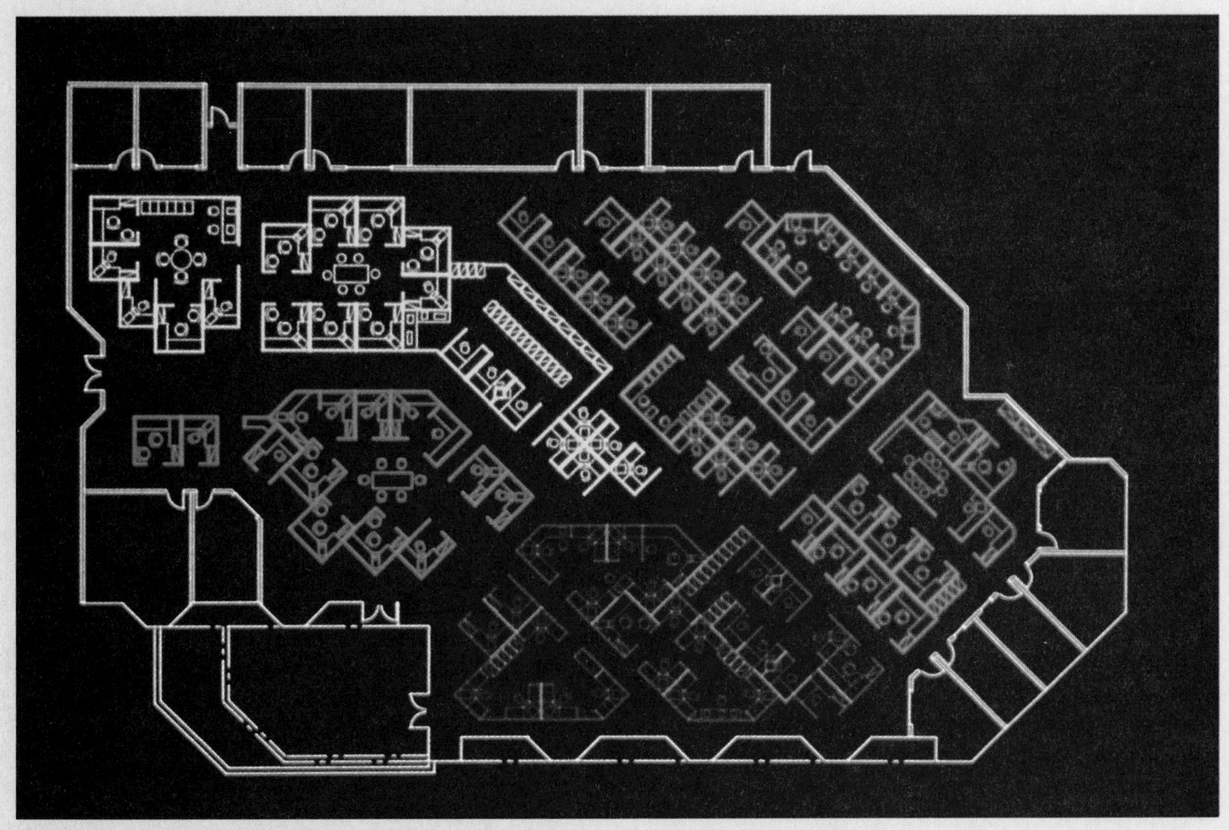

(Versacad)

Part One covers activities and considerations that must be undertaken prior to the formulation of any architectural design. Included here is a coverage of architectural styles, aesthetic design factors, solar orientation, passive energy, and ecological and density planning. A basic understanding of the principles and practices of these areas is an essential step before specific architectural design and drafting activities.

SECTION 1

FUNDAMENTALS OF DESIGNING AND PLANNING

Before one begins to design, there are many aesthetic and procedural factors to be considered. This section provides background information on architectural styles and types. It also covers the aesthetic factors of design that must be considered prior to the development of an architectural plan.

UNIT 1

ARCHITECTURAL STYLES AND TYPES

SPECIFIC ARCHITECTURAL STYLES

Our architectural heritage is largely derived from European and early American architecture. Nevertheless, specific architectural styles have developed through the years as a result of technological advances and the demands of our culture. Styles of the past reflect the culture of the past. Styles of the present reflect our current living habits and needs. Styles of the future will be largely determined by advances in technology and changes in our living styles and patterns.

Regardless of the basis for identification, there are few homes, past or present, which are totally pure examples of a specific architectural style. This is because the development of our architectural styles has been greatly affected by the changes from one culture to another, from one time period to another, and from one part of the world to another.

These transitions have usually taken place simultaneously. For example, the European influence was brought to this country by many different ethnic groups and continued to change and evolve as we passed through the colonial, Georgian, Federal, and romantic periods into the contemporary period. It is difficult to precisely identify any specific architectural style purely with one country of origin or time period. Nevertheless, there are common and dominant characteristic features which do help us communicate architectural style preferences.

European Styles

The English, French, Italians, and Spanish have provided the most significant influence on our architecture. Since most of the early settlers of this country emigrated from England and France, the European styles that dominated early American residential architecture were naturally those from England and France.

English Architecture English styles include old English (Cotswold), Tudor, Elizabethan (Fig. 1-1), and Georgian. English styles can be

Fig. 1-1 Elizabethan style. *(Schulz Homes, Inc.)*

distinguished from each other by some very specific differences in features. However, there are also many features common to all of these styles. For example, English styles all have relatively high-pitched roofs, massive chimneys, light leaded windows, and masonry siding. But the Cotswold style is found in small cottages, while the Elizabethan is distinguished by its half-timber construction, the Tudor by its multiple gables, and the Georgian by its classic box form. Within this framework of a general style, English styles can range from the very simple to the very lavish.

French Architecture Since the French styles were brought to this country much later than the English styles—from 1700 to 1800—their impact on colonial residential architecture was far less pronounced. However, some French styles, such as Regency, mansard, provincial, and château, were accepted and used in many areas. French provincial architecture was brought to this continent when the French settled Quebec. French provincial architecture can be identified by the *mansard roof* (Fig. 1-2A). This roof design was developed by the French architect François Mansard. On the French provincial home, this roof is high-pitched, with steep slopes and with rounded dormer windows projecting from the sides.

Southern European Architecture Spanish architecture was brought to this country by Spanish colonials who settled the southwest. Spanish architecture is characterized by low-pitched roofs of ceramic tile and by stucco exterior walls. A distinguishing feature of almost every Spanish home is a courtyard patio. Two-story Spanish homes contain open balconies enclosed in grillwork. One-story Spanish homes were the forerunners of the present ranch-style homes that were developed in southern California.

Italian architecture is very similar to Spanish architecture. Distinguishing features are the use of columns and arches at a *loggia* entrance, and windows or balconies opening onto a loggia. A loggia is an open passage covered by a roof. The use of classical moldings around first-floor windows also helps to distinguish the Italian style from the Spanish.

Southern European styles are also classified under the heading *Mediterranean*.

Early American

The early colonists came to the New World from many different cultures and were familiar with many different styles of architecture.

The label *early American styles* is something of a misnomer, since all styles that found their way to America during our early development can also be labeled *early American*. There is probably as much overlapping (or more) of characteristics among early American styles as there is in the European styles, since all of these styles have a northern European base.

New England Colonial The colonists who settled the New England coastal areas were influenced largely by English styles of architecture. Lack of materials, time, and equipment greatly simplified their adaptation of these styles. One of the most popular of the New England styles was the *Cape Cod* (Fig. 1-2B). This is a one-and-one-half-story gabled-roof house with dormers. It has a central front entrance, a large central chimney, and exterior walls of clapboard or bevel siding. Double-hung windows are fixed

Fig. 1-2A French mansard style. *(Home Planners, Inc.)*

Fig. 1-2B New England colonial style. *(Home Planners, Inc.)*

with shutters, and the floor plan is generally symmetrical. Cold New England winters also influenced the development of many design features, such as shutters, small window areas, and enclosed breezeways.

Dutch Colonial Gambrel roofs characterized many small farm buildings in Germany. A *gambrel roof* is a double-pitched roof with projecting overhangs. Many of the Germans who settled in New York and Pennsylvania made the gambrel roof a part of the Dutch colonial style of architecture.

Mid-Atlantic Colonial The availability of brick, a seasonal climate, and the influence of the architecture of Thomas Jefferson led to the development of the mid-Atlantic style of architecture. The style was formal, massive, and ornate. In colonial days, buildings from Virginia to New Jersey were designed in this manner. It was an adaptation of many urban English designs.

Southern Colonial When the early settlers migrated to the South, warmer climates and outdoor living activities led them to develop the southern colonial style of architecture. As the house became the center of plantation living, the size was increased and a second story was added. Two-story columns were used to support the front-roof overhang and the symmetrical gable roof (Fig. 1-2C).

Later American Styles After the colonial period, architectural styles either evolved from existing styles or climatic needs, or were copied from southern European styles as part of the classical revival movement. Styles that developed during this period include the Federal, Victorian, classical revival, Gothic revival, and several Spanish western styles, including the western ranch, western adobe, Monterey, and Spanish Mediterranean.

Fig. 1-2C Southern colonial style. *(Home Planners, Inc.)*

Ranch Style

As settlers moved west, they adapted architectural styles to meet their needs. The availability of space at ground level eliminated the need for second floors. Since the needed space was spread horizontally rather than vertically, a rambling plan resulted (Fig. 1-3). The Spanish and Mexican influence also led to the popularization of the western *ranch*, which used a U-shaped plan with a patio in the center, as shown in Fig. 1-4.

Contemporary Styles

The styles of today are based on, and often dominated by, the materials and technology of today. Consequently, contemporary homes (Fig. 1-5) tend to accent large, open areas with less structural restriction. Lines therefore become simpler and bolder and are usually less cluttered. Proponents of the contemporary style claim that any deviation from any pure historical style will always includes some compromise with architectural style authenticity. For example, colonial homes were designed and built with small window lights because large-sized glass could not be manufactured in those times. Today, glass size is not a design restriction; however, to design a truly pure colonial house, the designer should use small window lights. To many designers, this is a waste of an opportunity offered by an advance in technology. So the designer must decide how much liberty should be taken in the cause of architectural authenticity and how many contemporary features can or should be incorporated into the design.

FUNCTIONALISM

Louis Sullivan wrote, "Our architecture reflects us as truly as a mirror." Modern architecture is now reflecting our freedom, our functionalism, and our technological advances. In the modern art and science of architecture, architects are working to achieve even more functionalism, freedom, technological refinement, and relationships among the parts of a structure and be-

Fig. 1-3 Ranch style. *(Home Planners, Inc.)*

Fig. 1-4 Rambling ranch style. *(Home Planners, Inc.)*

Fig. 1-5 A contemporary style. *(Home Planners, Inc.)*

tween a structure and its environment. *Functionalism* is the quality of being useful, of serving a purpose other than adding beauty or aesthetic value. Louis Sullivan and Frank Lloyd Wright's "form follows function" idea has now been accepted by most modern architects. Few items can find their way into an architectural design without performing some specific function, or job. This is the line of distinction between architecture and sculpture. Architecture performs a function; a piece of sculpture does not.

Functionalism in architecture has led to the application of simplicity in design. Simplicity and functionalism complement each other.

Frank Lloyd Wright is considered one of the greatest American architects. He believed that architecture should be *organic*—that the materials, function, form, and surroundings in nature should be completely coordinated. All of these things should be in agreement and harmony.

Wright showed his genius by continually developing new styles and trends in architecture. He believed that even the basic shapes of floor plans should be more diversified with the development of new and more flexible building materials. Wright's Falling Waters (Fig. 14 in the Introduction), designed and built more than 40 years ago, still maintains a functional, contemporary look. Today, relating structures to the environment through the use of materials and structural shapes characterizes contemporary architecture. New technological developments allow designers to study the relationship of design elements to the environment with a level of accuracy and flexibility once impractical. For example, the three images of the Frank Lloyd Wright structure shown in Fig. 1-6 illustrate the capacity of computer graphics software to simulate precisely the effects of sunlight and artificial light on an architectural design.

Fig. 1-6 The use of Intergraphs Model View software to model sunlight and artificial light effects. *(Intergraph Corp.)*

INTERIOR DESIGN

The total architectural style of a structure must be one of the first considerations in developing the interior design style. For a style to be truly authentic, periods should be matched internally and externally. That is, the internal construction and furnishing should be as authentic as the exterior architecture. This means that the elements of design, both inside and outside, must be matched to get a desired consistent style. For example, the interior shown in Fig. 1-7 is consistent in style with the exterior shown in Fig. 1-5. The interior shown in Fig. 1-8 is consistent with the exterior shown in Fig. 1-1.

The role of the architect as a coordinator in these activities will increase. The relationship between art and technology will be refined to enable all types of buildings to be technically appropriate and aesthetically acceptable.

Fig. 1-7 Contemporary interior decor. *(California Redwood Assoc.)*

Fig. 1-8 Traditional interior decor. *(Benjamin Moore Co.)*

Exercises

1. Name three early American styles.
2. List the characteristics of the Mediterranean styles.
3. Find examples of several different styles of architecture in your area. Photograph the structures and list their distinguishing characteristics.
4. List several architectural pioneers and state their contribution to architectural style.
5. Find three buildings in other units in this text that would blend in style with the interiors shown in Figs. 1-7 and 1-8.

UNIT 2

DESIGN FACTORS AND PROCEDURES

Design activities may be formal or informal. *Informal design* occurs when a product is made by the designer without the use of a plan. *Formal design* involves the complete preparation of a set of working drawings. The working drawings are then used in constructing the product. Architectural design is nearly always of the formal type.

Ideas in the creative stage may be recorded by sketching basic images. These sketches are then revised until the ideas are crystallized and given final form. First sketches rarely produce a finished design. Usually, many revisions are necessary.

A basic idea, regardless of how creative and imaginative, is useless unless the design can be built successfully. Designing involves the transfer of basic sketches into architectural working drawings. Every useful building must perform a specific function. Every part of the structure should also be designed to perform a specific function. Today's buildings must be not only functional but also aesthetically conceived, and for both purposes the elements and principles of design must be applied.

ELEMENTS OF DESIGN

The *elements of design* are the tools of the designer. They are the ingredients of every successful design. The basic elements are line, form, color, space, light, and material.

Fig. 2-1 Curved lines dominate this design. *(Benjamin Moore Co.)*

Line

The element of *line* is used to produce a sense of movement within an object or to produce a greater sense of length or height. Lines enclose space and provide the outline or contour of forms. *Straight lines* are either vertical, horizontal, or diagonal. *Curved lines* have an infinite number of directional variations; they are not limited in the direction they can take. Curved lines dominate the design shown in Fig. 2-1. The straight lines found in Fig. 2-2 create a vertical emphasis.

A vertical line creates the illusion of an increase in height because the eye moves upward to follow the line. A horizontal line creates the illusion of an increase in width as the eye moves horizontally.

In duplicating the various positions of the human body, straight vertical lines create a feeling of strength, simplicity, and alertness. Horizontal lines suggest relaxation and repose. Diagonal lines create a feeling of restlessness or transition. Curved lines indicate soft, graceful, and flowing movements.

As in any art form, the combination of straight and curved lines in patterns is further combined with the other elements of design to create the most pleasing total design configuration.

Fig. 2-2 Vertical straight-line emphasis. *(California Redwood Assoc.)*

Form

Lines joined together produce *form* and create the shape of an area. Straight lines joined together produce rectangles, squares, and other geometric shapes. Curved lines form circles, ovals, and ellipses. The proportion of these forms or shapes is an important factor in design. Circles and ovals convey a feeling of completeness. Squares and rectangles produce a feeling of mathematical precision (Fig. 2-3) and should be used accordingly. The form of an object may be closed and solid or closed and volume-containing. The form of the structure, however, should always be determined by its function.

Fig. 2-3 Rectangles provide a feeling of precision. *(Western Wood Products Assoc.)*

Color

Color either is an integral part of an architectural material or else must be added to create the desired effect. Color in architecture serves to distinguish items, strengthen interest, and reduce eye contact. Color makes a considerable difference in the final appearance of any design, as illustrated in Figs. 2-4 and 2-5. These illustrations show the same view of a room with and without color.

Fig. 2-4 A room without color. *(Sol Vista Homes)*

Fig. 2-5 The room in Fig. 2-4 with color added. *(Sol Vista Homes)*

The Color Spectrum Colors in the spectrum are divided into primary, secondary, and tertiary colors. *Primary* colors are those colors that cannot be made from any other color. The primary colors are red, yellow, and blue.

A *secondary* color is made from equal mixtures of two primaries. Green is a combination of yellow and blue. Violet is a combination of blue and red; orange is a combination of red and yellow.

A *tertiary* color is the combination of a primary color and a neighboring secondary color, as shown on the color wheel in Fig. 2-6. The tertiary colors are *red-orange, yellow-orange, yellow-green, blue-green, blue-violet,* and *red-violet.*

A *neutral* shows no color in the ordinary sense of the word. The neutrals are white, gray, and black. The three primary colors, if mixed in equal strengths, will produce black. When colors cancel each other out in this manner, they are neutralized.

Color Quality For greater accuracy in describing a color's exact appearance, colorists

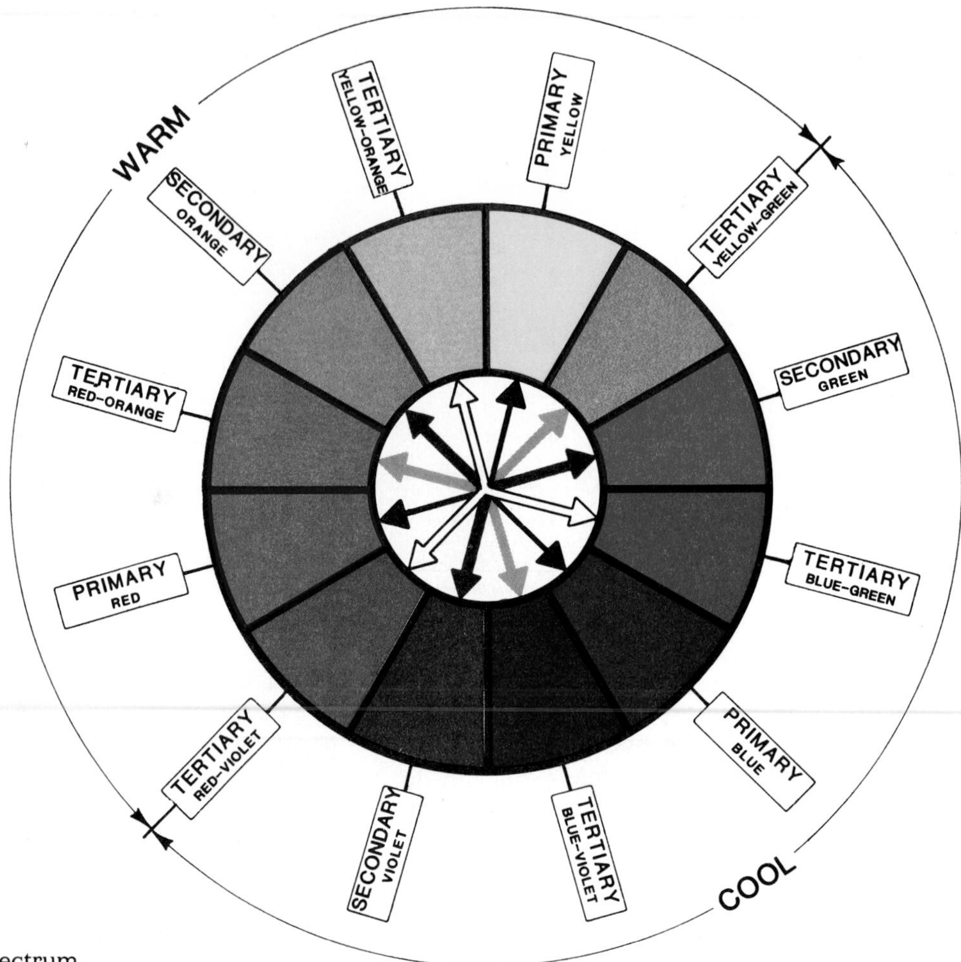

Fig. 2-6 The color spectrum.

distinguish three qualities: hue, value, and intensity.

The *hue* of a color is its basic consistent identity. A color hue may be identified as being yellow, yellow-green, blue, blue-green, and so forth. Even when a color is made lighter or darker, the hue remains the same.

The *value* of a color (see Fig. 2-7) refers to the lightness or darkness of the hue. A *tint* is lighter (or higher) in value than the normal value of a color. A tint of a hue will make a room look larger in area. Varying the value properties can dramatically change the mood of a room. A tint is produced by adding white to the base color.

A *shade* is darker (or lower) in value than the normal value of the color. A great many degrees of value can be obtained. A shade is produced by adding black to the normal color. A dark shade will make a room smaller and fuller.

A *tone* is produced by adding gray or some value to the normal color. Each color on a color wheel has a value that can be expressed as equivalent to the degree of gray included.

The *intensity* (strength) of a color is its degree of purity (or brightness), that is, its freedom from

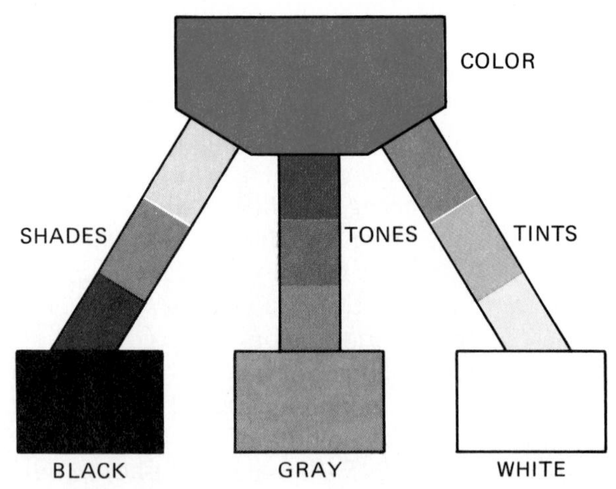

Fig. 2-7 Color values.

Fig. 2-8 Warm colors. *(California Redwood Assoc.)*

neutralizing factors. This quality is also referred to in color terminology as *chroma*. A color entirely free of neutral elements is called a *saturated* color. The intensity of a color can be changed without changing the color's value by mixing the color with a gray of the same value.

Harmonies are groups of colors which relate to each other in a predictable manner. The basic harmonies are complementary, analogous, split-complementary, monochromatic, and triadic, as shown on the color wheel in Fig. 2-6.

Color Effects The use of color can have a very strong effect on the atmosphere of a building. The perceived level of formality, temperature, and mood are all influenced by the color design. Colors such as red, yellow, and orange create a feeling of warmth, informality, cheer, and exuberance (Fig. 2-8). Colors such as blue and green create a feeling of quiet, formality, and coolness.

Color is also used to change the apparent visual dimensions of a building. It is used to make rooms appear higher or longer, lower or shorter. Bold colors, such as red, create the illusion of advancement, while pale colors (pastels) tend to recede.

Light and Shadow

Light reflects from the surfaces of forms. *Shadows* appear in the area that light cannot reach. Light and shadow both give a sense of depth to any structure. The effective designer plans the relationship of light and dark areas accordingly (Fig. 2-9). The designer must, therefore, consider which surfaces reflect light instead of absorbing light, and which surfaces refract (bend) light as it passes through the material. The designer must also remember that with continued exposure to light, visual sensitivity decreases. Thus we become adapted to degrees of darkness, or lightness, after extended exposure.

Space

Space surrounds form and is contained within it. The design can create a feeling of space. Architectural design is the art of defining space and space relationships in a manner that makes use of all other elements of design in a functional and aesthetic manner.

Materials

Materials are the raw substances with which designers create. Materials possess their own color, form, dimension, degree of hardness, and texture. The hardness of the material cannot be altered. However, to some degree, the color, form, and dimension of materials can be al-

Fig. 2-9 Dark and light relationships. *(Armstrong Cork Co.)*

tered. Texture is the unique and most significant factor in the selection of appropriate materials. *Texture* refers to the surface finish of an object—its roughness, smoothness, coarseness, or fineness. Surfaces of materials such as concrete, stone, and brick are rough and dull, and suggest strength and informality. Smoother surfaces, such as those of glass, aluminum, and plastics, create a feeling of luxury and formality. The designer must be careful not to include too many different textures of a similar nature. For this reason, masonries (brick and stone, for example) are not usually combined in areas close together. Textures, such as the wood and stone shown in Fig. 2-10, are more pleasing when combined and contrasted with other surfaces.

Rough surfaces reduce the apparent height of a ceiling or distance of a wall and make colors appear darker. Smooth surfaces increase the apparent height of a ceiling or wall and reflect more light, thus making colors appear brighter.

Fig. 2-10 Contrasting textures. *(Schulz Homes, Inc.)*

PRINCIPLES OF DESIGN

The basic *principles of design* are the guidelines for using the elements of design to create aesthetically functional buildings. The basic principles of design are balance, variety, emphasis, unity, opposition, proportion, rhythm, subordination, transition, and repetition.

Balance

Balance is the achievement of equilibrium in design. Buildings are *formally balanced* if they are symmetrical. They are *informally balanced* if there is variety, yet a harmonious relationship in the distribution of space, form, line, color, light, and shade. The building shown in Fig. 2-11 is informally (asymmetrically) balanced. The building shown in Fig. 1-2A is formally (symmetrically) balanced.

Rhythm

When lines, planes, and surface treatments are repeated in a regular sequence (order or arrangement), a sense of rhythm is achieved. Rhythm is used to create motion and carry the viewer's eyes to various parts of the space. This may be accomplished by the repetition of lines, colors, and patterns.

Emphasis

The principle of *emphasis* (domination) is used by the designer to draw attention to an area or subject. Emphasis is achieved through the use of color, form, texture, or line.

In architectural design, some emphasis or focal point should be designed into each elevation and interior space. Directing attention to the *point of emphasis* (focal point) is accomplished by the arrangement of features, the use of contrasting colors, line direction, light variations, space relationships, or material changes.

Proportion

The proportional dimensions (scale) of a building are important. The early Greeks found that rectangular proportions in the ratio of 2 to 3, 3 to 5, and 5 to 8 were more pleasing than others. For example a room or a rug with dimensions of 9′ x 15′ or 10′ x 16′ will have the proportions 3 to 5 and 5 to 8.

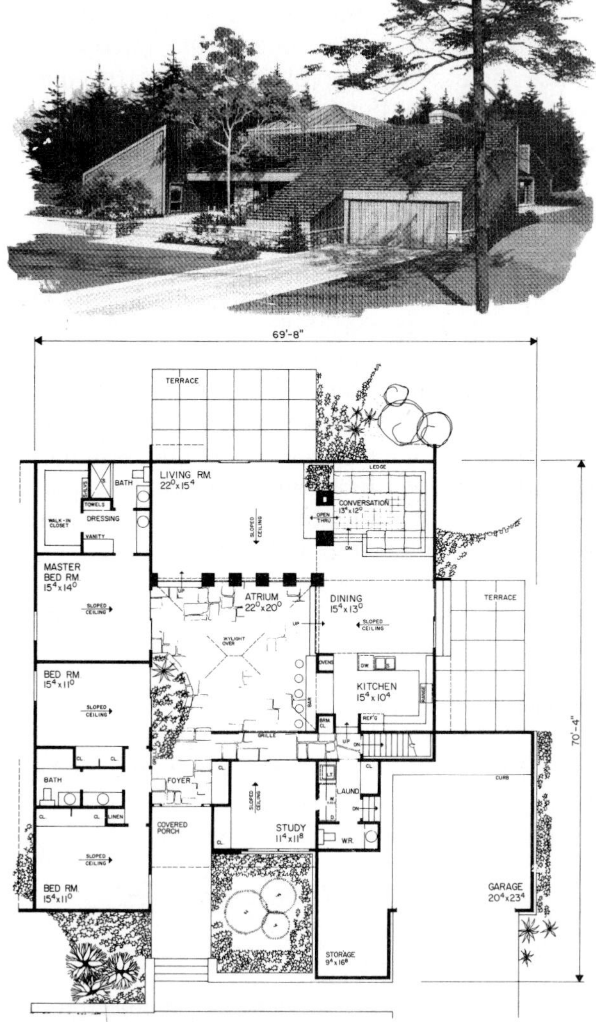

Fig. 2-11 Informally balanced design. *(Home Planners, Inc.)*

The scale between interior space, furniture, and accessories should be harmonious. Bulky components in small rooms should be avoided. Small components in large rooms should not be used. Figure 2-12 shows several proportional systems used in two-dimensional design. Areas will appear completely different depending on the division of space within the area, as shown in Fig. 2-13.

Unity

Unity is the expression of the sense of wholeness in the design. Every structure should appear complete. No parts should appear as appendages or afterthoughts. Designers achieve unity through the use of consistent line and color, even though the building is composed of many different parts. Unity, or harmony, as the name implies, is the joining together of the basic elements of good design to form one harmonious, unified whole. Unity (or harmony) is achieved through the utilization of any or all of the elements of design, for there is potential for unity within each.

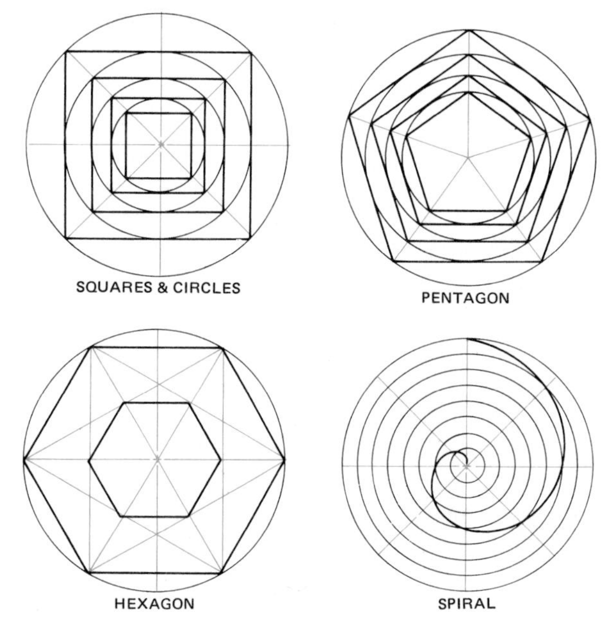

Fig. 2-12 Proportional systems.

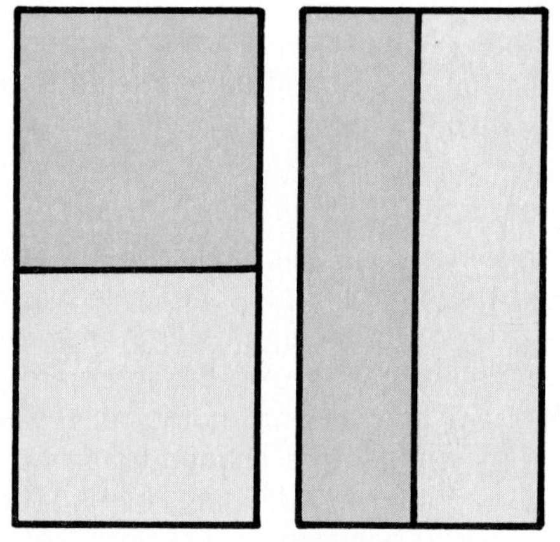

Fig. 2-13 Identical rectangles appear different when subdivided.

Variety

Without variety, any area can become dull and tiresome to the eye of the observer. Too much rhythm, too much repetition, too much unity ruin a sense of variety or contrast. Likewise, too little of any of the elements of design will also result in a lack of variety. Light, shadow, and color are used extensively to achieve variety.

Repetition

Unity is often achieved through *repetition*. Vertical lines, spaces, and textures are repeated throughout the design to tie the structure together aesthetically and to achieve unity.

Opposition

Opposites in design add interest. *Opposition* involves contrasting elements such as short and long, thick and thin, straight and curved, black and white. Opposite forms, colors, and lines in a design, when used effectively with the other principles of design, achieve balance, emphasis, and variety.

Subordination

When emphasis is achieved through some design feature, other features naturally become subordinate—lesser in emphasis or importance. Subordination can be related in design to lines, shapes, or color.

Transition

The change from one color to another, or from a curved to a straight line, if done while maintaining the unity of the design, is known as *transition*. Transition may involve the intersection of molding from one wall to another in the same room, or may apply to a change from one floor surface to another in adjoining rooms. The designer's task in achieving successful transition in all aspects of the design contributes to the harmony of different elements of design without sacrificing unity. And thus using the elements of design (line, form, material, color, light, space, texture) and applying the principles of design (balance, variety, emphasis, unity, opposition, proportion, rhythm, subordination, transition, and repetition), combined with innovative engineering, the modern designer creates aesthetically pleasing environments which enhance contemporary lifestyles and provide functional convenience.

CREATIVITY

Creativity in architecture involves the ability to create mental images of arrangements and forms not yet seen. Creative imagination is the ability to present new patterns, use new objects, and invent new configurations. Thus, creativity and imagination both relate to the forces that cause isolated and unrelated factors to come together into arrangements of cohesive unity and beauty.

FUNCTIONAL DESIGN

Any basic idea, no matter how creative or imaginative, is useless unless the design can be implemented successfully and function as planned. Architectural design involves not only how a structure appears but how it functions. Thus, architectural design begins with an assessment of human needs. Remember that *form follows function*. However, functional success alone does not guarantee that a design will be aesthetically pleasing. The task of the competent designer is to combine functional efficiency and aesthetics in a unified design. The designer must manipulate the elements of design successfully through the effective application of the principles of design.

No design can exist in isolation. It must always be related to all situations that influence it. Thus, creating a successful design involves manipulating the entire environment.

CHANGING PATTERNS AND TASTES

Not only must the architectural designer blend the basic elements of design to create a good functional plan, but this must be done according to changing styles and tastes. There are periods when people prefer open planning. At other times, complete privacy is of primary concern. Individual and public tastes constantly change, but the designer must not be caught in a "fad trap." An effective, creative designer will recognize the difference between trends and fads. But the designer must be constantly on the alert, always looking for the link between present and future. The contemporary designer must recall past experiences and apply old ideas to new situations in combinations of endless variety.

THE DESIGN PROCESS

The architectural design process involves the relationships between many personal, social, economic, and technical variables and the crea-

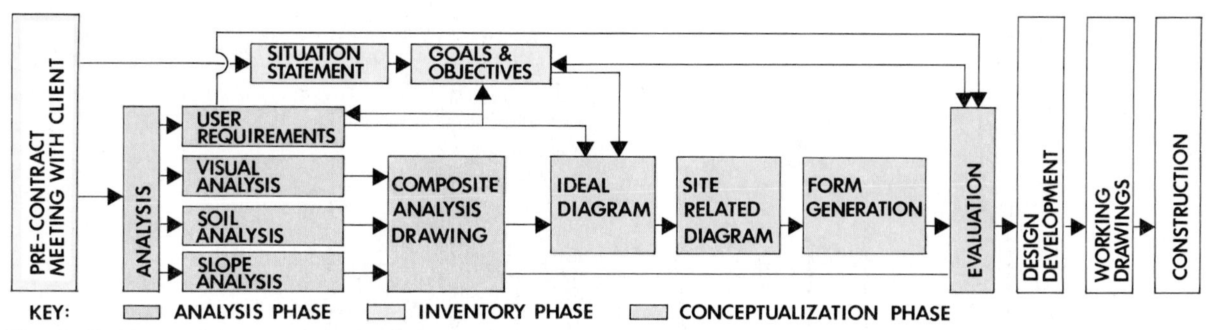

Fig. 2-14 The architectural design process.

tion of perhaps hundreds or thousands of detailed working drawings. The process is therefore too complex and potentially chaotic to be left to chance or unorganized thought. Thus, to effectively apply the principles and elements of design to an architectural project, established design sequences and procedures must be followed. Only then can the elements of design be combined with the technical requirements of the project to produce a fully functional and aesthetically pleasing design. This process is a logical sequence of thought and action which begins with an inventory and analysis of the project. This process continues through the completion of working drawings as shown in Fig. 2-14.

Initial Contact

The success of any design depends on the degree to which the finished project meets the needs of the client. It is therefore essential to establish a firm understanding of these needs from the beginning. During the first meeting between client and designer, the scope of the project and the budget and schedules are described. During this stage the designer must be a keen listener to extract a description of all specific wants and needs. This includes not only physical needs but also the social habits, special interests, and ideas of the client.

Inventory and Goal Setting

To ensure that the designer, client, and all concerned agree on the purpose and theme of the project, the designer next prepares a *situation statement.* This statement identifies and records the client's needs and requirements and any special problems. This statement firmly establishes and documents the mutual understanding between client and designer, as illustrated in Fig. 2-15.

SITUATION

MR. AND MRS. JOHN SMITH HAVE ACQUIRED A FIVE ACRE PARCEL OF WOODED LAND. THEY WANT THE SITE DEVELOPED AS A RESIDENCE AND ALSO FOR HOME OFFICE USE. THE SMITH FAMILY OF FOUR HAVE STRONG FEELINGS FOR ENVIRONMENTAL PRESERVATION AND ARE VERY FITNESS ORIENTED. THEY WANT A HIGH DEGREE OF PRIVACY AND NEED THE CAPACITY TO ENTERTAIN WEEKEND GUESTS.

Fig. 2-15 Design situation statement.

Once all parties agree with this statement, the designer then prepares a very specific set of goals and objectives for every phase of the project based on the situation statement. This is a most important step, since constant reference is made to these goals in the creation and evolution of the final design. Without recorded specific goals the designer can easily lose focus. Furthermore, without recorded design goals the functional success of the final design cannot be accurately evaluated. Figure 2-16 shows a sample set of goals and objectives.

Analysis

The analysis stage involves an organized and sequenced analysis of user wants and needs, site features, soil conditions, slope studies, and visual observations. All of these separate analyses lead logically to the development of a comprehensive analysis.

User Analysis In a user analysis each goal is further refined into descriptions of specific usage, space requirements, and the relationships between areas. The user analysis clari-

MAJOR GOALS

DESIGN A RESIDENCE FOR THE EXISTING SITE WITH GOOD VISUAL PROFILES. AESTHETIC APPEAL AND EMPHASIS ON FUNCTIONAL, NON-DESTRUCTIVE USE OF ALL SITE FEATURES, INCLUDING MAXIMUM USE OF SOLAR ENERGY. PLAN WORKING LIVING AND RECREATION AREAS TO CONFORM TO SPACE AND PRIORITY NEEDS. DEVELOP PROGRESSIVE REALIZATION OF ALL FACILITIES SO THAT ALL ARE NOT VISIBLE FROM ONE VANTAGE POINT.

OBJECTIVES

1. PROVIDE STIMULATING, CASUAL ATMOSPHERE FOR FAMILY AND GUESTS.
2. LOCATE PRIVATE AND PUBLIC AREAS IN A SEQUENCE TO AVOID USER CONFLICTS.
3. POSITION BUILDINGS AND FACILITIES FOR MINIMUM ENVIRONMENTAL IMPACT.
4. ORIENT STRUCTURES FOR MAXIMUM SOLAR USE.
5. BUILDING FOOTPRINT TO BE HOMOGENOUS WITH EXISTING SITE LANDFORM.
6. RELATE INTERIOR LIVING AREAS TO EXTERIOR SPACE FUNCTIONALLY.
7. RESIDENCE TO BE NOT COMPLETELY VISIBLE FROM ACCESS ROAD.
8. PLAN CIRCULATION PATTERNS FOR BOTH VEHICULAR AND PEDESTRIAN TRAFFIC.
9. PLAN FACILITIES FOR BASKETBALL, SWIMMING, SUNBATHING, JOGGING, AND WHIRLPOOL.
10. PROVIDE COURTYARD FOR SEASONAL USE.
11. PLAN FOR BOTH INTERIOR AND EXTERIOR DINING FACILITIES FOR 8-10 GUESTS.
12. PROVIDE MR. SMITH, A PUBLISHER, WITH A HOME OFFICE FOR EVENING AND WEEKEND USE.
13. PROVIDE MRS. SMITH, AN ACCOUNTANT, WITH AN ACCESSIBLE OFFICE TO MEET CLIENTS DAILY.
14. USE NATURAL CONTEMPORARY LINES AND MATERIALS CONSISTENT WITH SITE.
15. DESIGN A FIREPLACE FOR LIVING AREA AND MASTER BEDROOM SUITE.
16. DESIGN PROGRESSIVE REALIZATION FOR VEHICULAR APPROACHING TRAFFIC.
17. CENTRAL KITCHEN AND GREENHOUSE IS REQUESTED BY THE SMITHS.
18. KEEP TOTAL COST WITHIN LIMITS ESTABLISHED BY CLIENTS. PRIORITIZE FEATURES TO CUT IF DESIGN EXCEEDS BUDGET.

Fig. 2-16 Design goals and objectives.

SPACE ELEMENTS	PRIMARY USERS	MIN. SIZE	NOTES AND RELATIONSHIPS
LIVING ROOM	8-16 ADULTS	16'x 22'	POOL VIEW - FIREPLACE ACCESS TO FOYER & DR.
DINING ROOM	6-12 ADULTS	14'x 20'	ACCESS TO KIT & LR.
STUDY #1	MR. SMITH	14'x 20'	PRIVATE - QUIET.
STUDY #2	MRS. SMITH	12'x 14'	PRIVATE - CLIENT ACCESSIBLE - JOINT OFFICE W MR. S ?
ENTRY	FAMILY - GUESTS	8'x 10'	VISIBLE FROM DRIVE - BAFFLE FROM STREET.
PARKING	2 FAMILY CARS 6 GUEST CARS	9'x 10' STALLS	ACCESS TO MAIN. ENTRY.
DECKS OR TERRACE	FAMILY - GUESTS	16'x 20'	OVERLOOK POOL - NEXT TO LIVING AREA.
COURTYARD	FAMILY - GUESTS	200 SQ FT	FOR CASUAL ENTERTAINMENT NEXT TO KIT.
KITCHEN	FAMILY	12'x 16'	ACCESS TO DECK LR. & LAUNDRY.
GARAGE	FAMILY	20'x 24'	ACCESS TO KIT - CONVERT TO SHOP.
SERVICE PICKUP	SERVICE PERS.	40 SQ. FT.	SCREEN FROM LIVING AREAS.
MASTER BEDROOM	2 ADULTS	16'x 24'	MORNING SUN - KING BED - ACCESS TO POOL - SUITE W BATH - QUIET AREA.
BEDROOMS	2 CHILDREN	16'x 18'	PLAN FOR TEEN GROWTH - AWAY FROM LIVING & MASTER BR.
BATHS	CHILDREN & GUESTS	2-8'x 10'	ACCESS FROM CHILDRENS ROOMS & GUESTS.
GUEST BEDROOM	GUESTS	12'x 16'	BATH ACCESS - OR CONVERTIBLE STUDY ?
SITE CONSIDERATIONS	ALL	ENTIRE SITE	SOLVE SITTING WATER PROBLEM. USE ROCK FORMATIONS & ADD FOLIAGE FOR VISUAL APPEAL.
SOLAR CONSIDERATIONS	ALL	BLDGS & SITE	USE PASSIVE TECHNIQUES - CARE IN ORIENTATION OF FACILITIES.
RECREATION FACILITIES	ALL	COURTS, POOL,	ORIENT W SUN & SCREEN FROM RESIDENCE &

Fig. 2-17 User analysis.

fies problems by enabling the designer to break down each design element into parts of a manageable size. The user analysis is usually prepared in chart form (see Fig. 2-17) for ease of evaluation, verification, and discussion with the client. Since this analysis has great influence on the conceptualization of the design, no area should be omitted. For if the user analysis is inadequate or contains erroneous information, the final design will also be inadequate and/or erroneous.

Site Analysis Any building designed without an analysis of the site will be artificially imposed on the site. The development of every architectural project should therefore be analyzed and tailored to take advantage of the positive features, while minimizing the effect of the negative features, of the site. Thus, the site analysis aids the designer in making proper design decisions and also helps ensure appropriate land use. Three types of site analysis techniques are used to develop a final site analysis drawing: soil analysis, slope analysis, and visual analysis. Each of these covers distinct factors affecting the potential use of different areas of the site.

However, before a soil, slope, or visual analysis is started, a base map of the site must be prepared. This base map includes all fixed items that are a permanent part of the site. This may include a USGS (United States Geological Survey) topographic map; the outline and location of property lines, adjacent streets, existing structures, walkways, paths, terraces, utility lines, and easements; setback limits; and a compass direction. Base maps are usually prepared to a scale of 1″ = 10′, 1″ = 20′, 1″ = 30′, or ⅛″ = 1′-0″. An accurate base map is essential, since many copies of this map will be used in the design process for analysis drawings, design development drawings, and site-plan working drawings. Figure 2-18 shows a typical base map.

Soil Analysis Soil is composed of gaseous, water, organic, and rock constituents. Variations in the percentage of these ingredients determine the physical characteristics of the soil and its capacity to support the weight of structures (its bearing capacity). Local soil information can be obtained through USDA (United States Department of Agriculture) county soil surveys or an analysis of test borings from the site. In general, coarse-grained soils, because of their drainage and bearing capacity, are preferred for building but are not ideal for planting areas. Conversely, fine-grained soils with high organic content are preferred for planting areas but not for building. To aid in the analysis process, the USDA classifies soils for building as follows:

1. *Excellent:* Course-grained soils—no clays, no organic matter

2. *Good to Fair:* Fine, sandy soils (minimum organic and clay content)

3. *Poor:* Fine-grained silts and clays (moderate organic content)

4. *No development:* Organic soils (high clay and peat content)

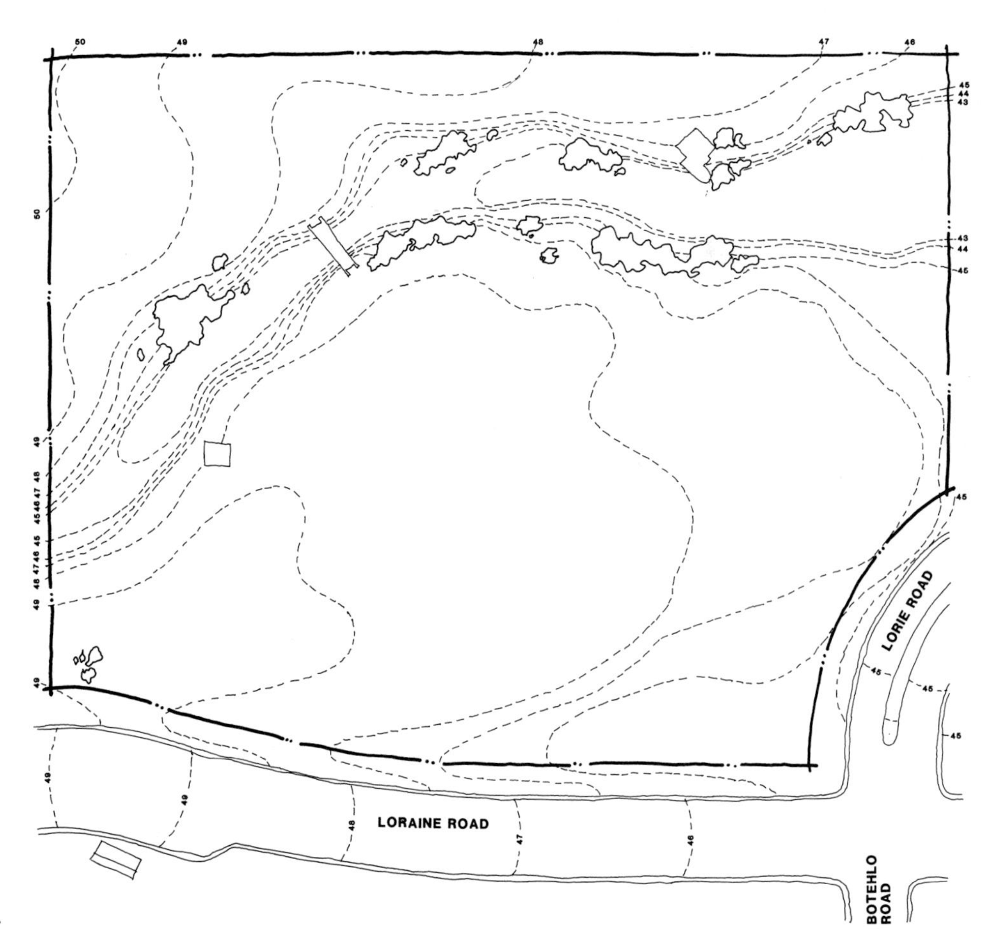

Fig. 2-18 Base map.

To prepare a soil analysis drawing, follow these steps:

1. Obtain a soil classification for the site from a county soil survey or from private borings.

2. Draw areas on the base map representing the different soil types, as shown in Fig. 2-19.

3. Note the bearing capacity, and depth to bedrock, for each soil category. This information is given in kilopounds (kips, or K) in the USDA survey book (1 K = 1000 lb for 1 ft^2 of soil).

4. Provide a legend showing the categories of soil types and describe the soil characteristics of each type. On the drawing, note developmental potential and constraints.

5. Color-code each soil capacity type in the legend and on the drawing.

Slope Analysis The slope of a particular site greatly affects the type of building that can or should be designed for it. The slope percentage may also determine what locations are acceptable, preferred, difficult, or impossible for building. The cost of building may also be acutely affected by excessive slope angles. To complete a slope analysis drawing, refer to Fig. 2-20 and follow these eight steps, as shown in Fig. 2-22A.

1. To the base map add contour lines derived from a USGS map of the area. If the site is very hilly, additional contours may be needed to provide a more detailed description of the slope of the site. The existing contour lines should be dashed, since the finished contour grade lines will later be drawn solid.

2. Identify four classifications of slopes on the drawing:

0% to 5% —excellent
5% to 10%—good/fair
10% to 25%—poor
over 25% —no development

3. Identify each slope category by the use of colors or tones to show the degree of develop-

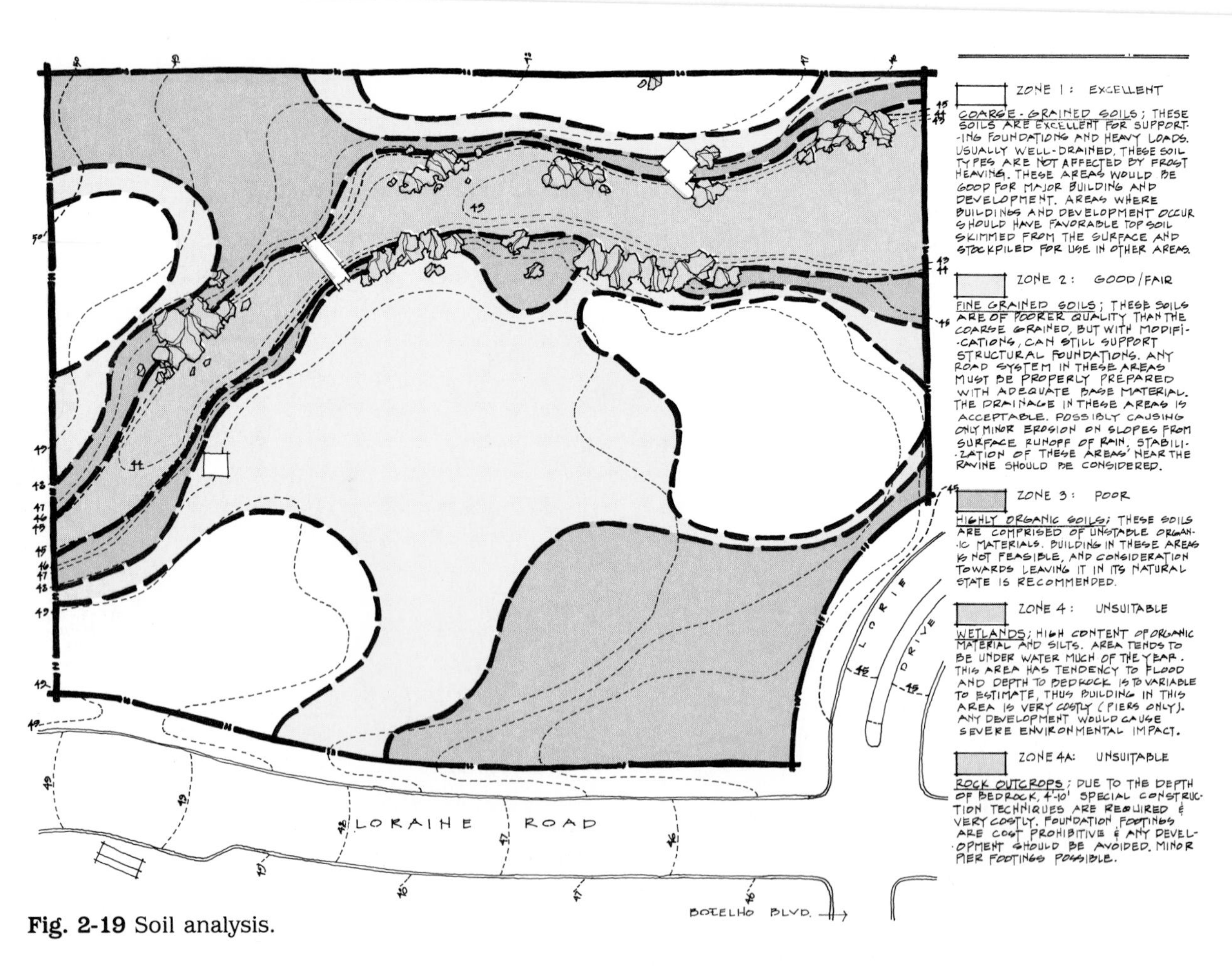

Fig. 2-19 Soil analysis.

ment potential of each section. Generally, light colors are used for areas suitable for development, whereas dark colors are used for less suitable areas. Interpreting the slope classifications from the contour lines can be done by the use of a "tic strip." A tic strip is a scale the designer can make which identifies the percent slope when placed perpendicular to the contour lines.

4. For each slope classification find the corresponding perpendicular distance between contours.

5. Provide a color-keyed legend of slope categories.

6. Note both potential and constraints for development for each slope category.

7. Note economic assets and/or limitations for each slope category.

8. Note erosion or drainage problems (if any) for each category.

Visual Analysis The architectural designer must be able to discover and analyze the aesthetic and environmental qualities of a site and visualize its potential. Because visual observations and aesthetic qualities are often subjective and elusive, some organized method of recording and analyzing is important. By following the steps listed here, a visual analysis drawing (Fig. 2-21) can be prepared to provide input for the future design conceptualization phase.

1. On the base map locate the direction of the best views from each important viewer position. Show the viewer position with an *x* and a line and arrow pointing in the direction of the view. Label the nature of each view and rate its value good, fair, or poor. Also make recommendations for the treatment of each view ("enhance" or "screen," for example).

2. Identify existing structures on the base map and describe their condition as good, fair, poor, unsound, or hazardous. Note suggestions to enhance, remove, or rehabilitate, and note any possible uses such as storage.

3. Draw the outline and location of all existing and significant plant material such as large shrubs and trees. Label the type, and indicate

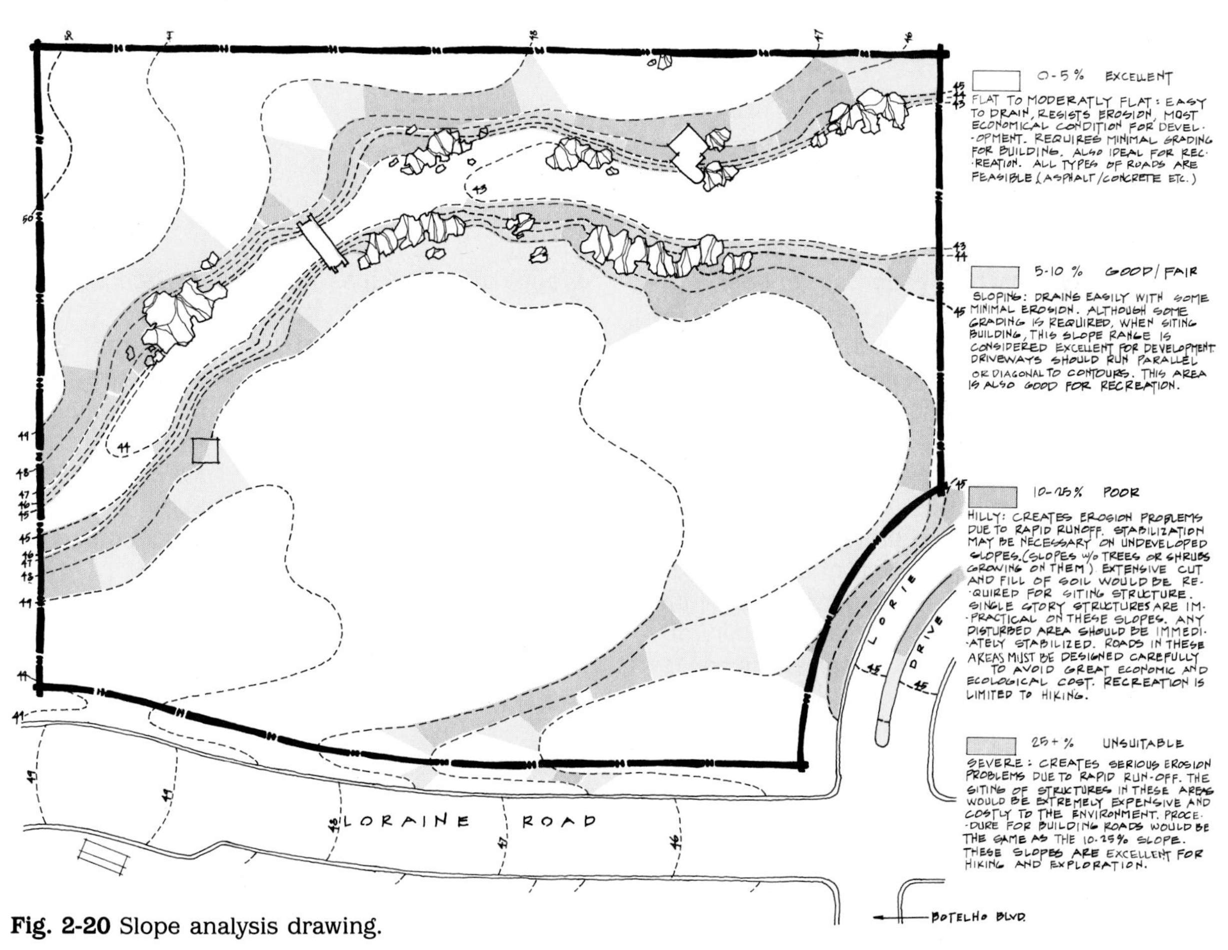

Fig. 2-20 Slope analysis drawing.

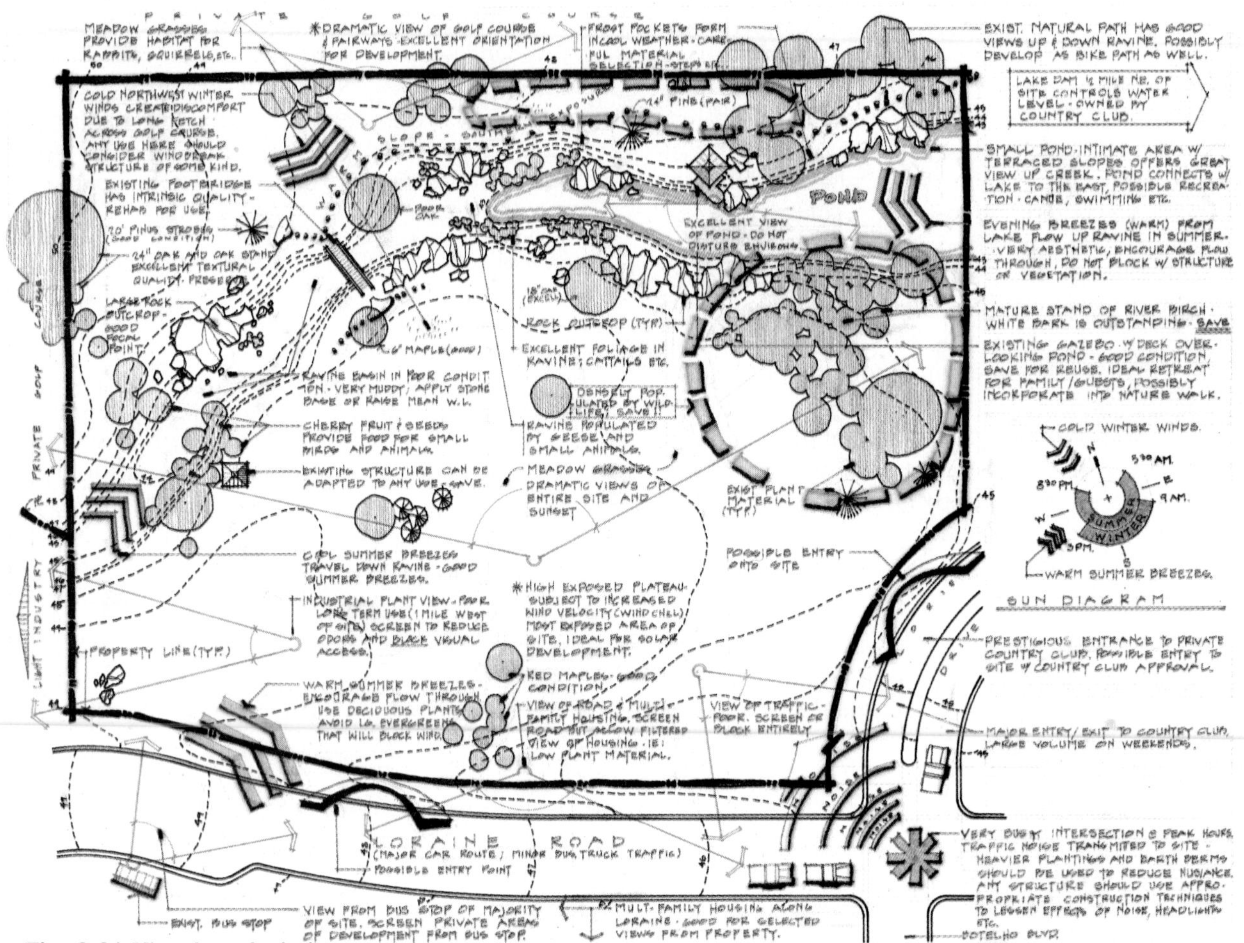

Fig. 2-21 Visual analysis input.

the condition of each as good, fair, or poor. Also locate, draw, and indicate large stands of ground vegetation to be saved.

4. Identify any wildlife population and habitat areas to be saved. Indicate animal food and water sources.

5. With directional arrows, show the direction of prevailing winter winds. Also show the direction of prevailing summer breezes.

6. Find and label the source of any desirable fragrances and/or undesirable odors. For the latter indicate possible solutions, such as minimizing with aromatic vegetation, screening, or removal of the source.

7. Locate and label exposed open space, semienclosed public space, and private space.

8. Add side notes to the base map to include the average annual rainfall of the area, average high and low temperatures, depth of local frost, and average sun days per year. All of this data is available from the USDA, from architectural reference books such as *Time Saving Standards*, or from local weather bureaus.

Composite Analysis Once the soil, slope, and visual analysis drawings are completed, the information needs to be combined into a composite analysis drawing. The major purpose of the composite analysis drawing is to determine the best location zones for the placement of structures on the site. Location zones are divided into four categories:

1. Excellent development potential
2. Good or fair development potential
3. Poor development potential
4. No development potential

To prepare a composite analysis drawing, the following steps should be followed:

1. Place the soil analysis drawing directly over the slope analysis and align the property lines with the base map and tape the base map to the drawing board.

2. Attach tracing paper over the slope and soil drawings and trace a line around each distinct area.

3. Determine which development zone each of the areas outlined represents. For example, if a 0 to 5 percent slope area overlaps with a coarse-grained soil area the zone is labeled "1, excellent potential." If a poor, clay soil area overlaps with a 20 percent slope area, the zone is labeled "3, poor." If any area, in any category, falls within a no development zone, obviously the zone is "4." Other combinations, such as excellent potential in one category and poor in another, become gray areas requiring judgment trade-offs in the design decision-making process. Label each zone on an overlay drawing.

4. Place the overlay drawing over the visual analysis drawing and repeat the same outlining of areas covered in step 3 to complete the composite analysis drawing as shown in Fig. 2-22. Apply your own judgment concerning priorities when overlapping areas conflict. For example, if a wildlife habitat area falls directly over an excellent building area, according to the soil and slope analysis, should the wildlife be removed or should a slightly less favorable building site be considered? Here the client's goals, objectives, and priorities must be considered; the client should be consulted before you proceed.

Conceptualization

Idealized Diagram The ideal diagram is a series of study sketches usually on inexpensive tracing paper ("trash" or "bum wad"). The diagram represents the *ideal special* relationships of the major user elements from the user analysis. Ideal diagrams are freehand, bubble-like sketches that sort out the major user elements and show how they go together. The *bubbles* are not used to relate the sizes of the areas, *only* their spatial relationship to other elements, as shown in Fig. 2-23. Designers do as many studies or sketches as necessary until they

ZONE 1: EXCELLENT
THIS AREA IS A PRIME DEVELOP-MENT LOCATION. SOILS ARE COMPATIBLE WITH CONSTRUCTION & HAVE EXCELLENT DRAINAGE CHARACTERISTICS. A RELATIVELY FLAT AREA (0-5%) SLOPE W/ MINIMAL EROSION PROBLEMS, & FEW RESTRICTIONS. SOLAR ARRANGEMENT OF STRUCTURES VERY PRACTICAL. AREA GOOD FOR ALL TYPES OF RECREATION, ROADS, PATHS & ALL STRUCTURES
ZONE 2: GOOD/FAIR
THIS GENTLE SLOPING (0-10%) AREA REPRESENTS GOOD DEV-OPMENT POTENTIAL. THE SOILS HERE ARE COMPATIBLE W/CON-STRUCTION ESPECIALLY WHEN POST & BEAM TECHNIQUE IS USED. THIS AREA HAS GOOD DRAINAGE & EROSION IS MINIMAL, EXCEPT NEAR THE RAVINE. THIS AREA IS A PRIME AREA FOR SITING STRUCTURES ACCORDING TO SOLAR ORIENTATION DUE TO ITS SOUTHERN EXPOSURE.
ZONE 3: POOR [RESTRICTED DEVELOPMENT]
DUE TO SLOPE (10-25%), SOIL CONSID-ERATIONS, AND ECONOMIC COSTS, CONSTRUCTION IN THIS ZONE SHOULD BE LIMITED. STRUCTURES MUST BE CUSTOM DESIGNED. DESIGN IS LIMITED BECAUSE OF HIGH RUNOFF AND RISKS OF EXTREME EROSION PROBLEMS. THIS ZONE SHOULD BE PRESERVED WHENEVER POSSIBLE; REMOVAL OF VEGETATION COULD RESULT IN IRRE-VERSIBLE DAMAGE TO THE ECOTONE. EXCELLENT USE COULD BE MADE W/ DEVELOPMENT OF HIKING PATHS.
ZONE 4: NO DEVELOPMENT
ORGANIC SOIL, POOR DRAINAGE, AND IT'S GENERALLY UNDER WATER, MAKE THIS ZONE UNFEASIBLE FOR DEVEL-OPMENT. NIGHT TIME CIRCULATION COULD BE DANGEROUS BECAUSE OF FOG AND FROST POCKETS. IT WOULD BE BEST TO LEAVE THIS AREA UN-DEVELOPED, EXCEPT FOR MINOR CIRCULATION PATHS. LEAVE THIS ZONE AS NATURAL AS POSSIBLE, ALL VEGETATION SHOULD BE PRE-SERVED TO PROTECT THE ECOTONE.
ZONE 4a: NO DEVELOPMENT
THESE AREAS HAVE ESTABLISHED WOODED STANDS AND HEAVILY POPULATED BY WILDLIFE AND TEND TO BE FEEDING GROUNDS-SAVE.
LORAINE ROAD
LORIE DRIVE
BOTELHO BLVD.

Fig. 2-22 Composite analysis.

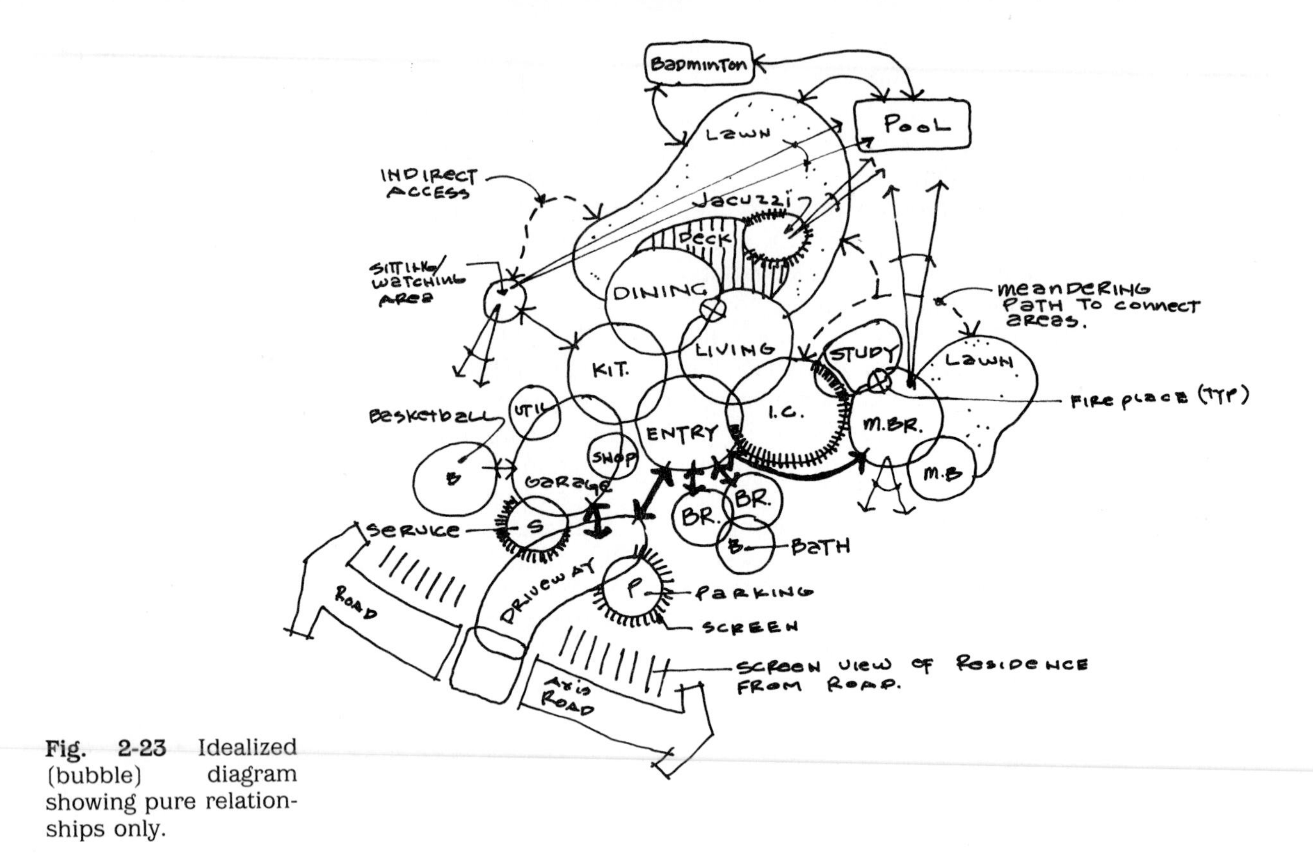

Fig. 2-23 Idealized (bubble) diagram showing pure relationships only.

achieve the one study sketch that provides in the best way possible the ideal relationship between these elements.

Site-Related Diagram The *site-related diagram* applies the idealized diagram to the site and introduces the size requirements from the user analysis. Frequently called the push-pull–bend-twist phase, the effort is concentrated on "fitting" all the various elements of the user analysis onto the site while maintaining their most ideal relationships. The relative *scale* of the elements is first introduced at this phase of the design process. The success of this phase relates to the designer's ability to concentrate on the site constraints and opportunities from the site analysis while applying the ideal relationships of the ideal diagram to the site. The format for developing site-related diagrams should be freehand sketches on tracing paper overlaid on top of the composite site-analysis drawing. At this point the design begins to take physical form. Several site-related studies should be completed which integrate the design with the site (see Fig. 2-24). Serious consideration must be given at this stage to the constraints and opportunities offered by the characteristics of the site as shown in the site-related drawing shown in Fig. 2-25.

Conceptual Design From all the site-related diagrams and sketches, one is chosen that responds best to the information on the site analysis and in the user analysis chart. From the site-related diagram the designer now begins to generate the form of the design. Drawings are now refined into a loose graphic format for evaluation, as shown in the plan drawing of Fig. 2-26. This plan may also include some section drawings and detail sketches both for clarity and for a test of some basic ideas.

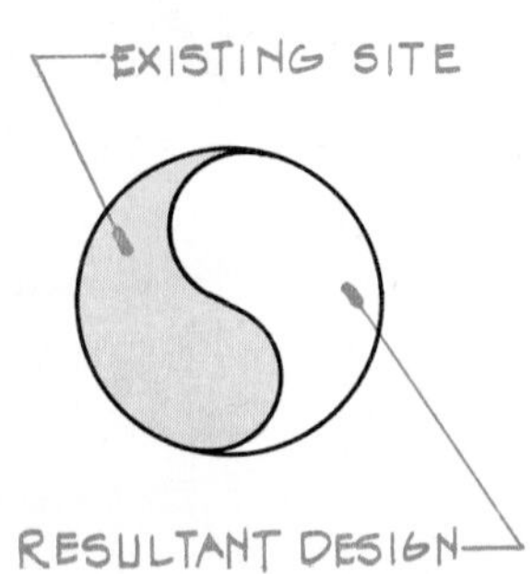

Fig. 2-24 Site and design must be integrated.

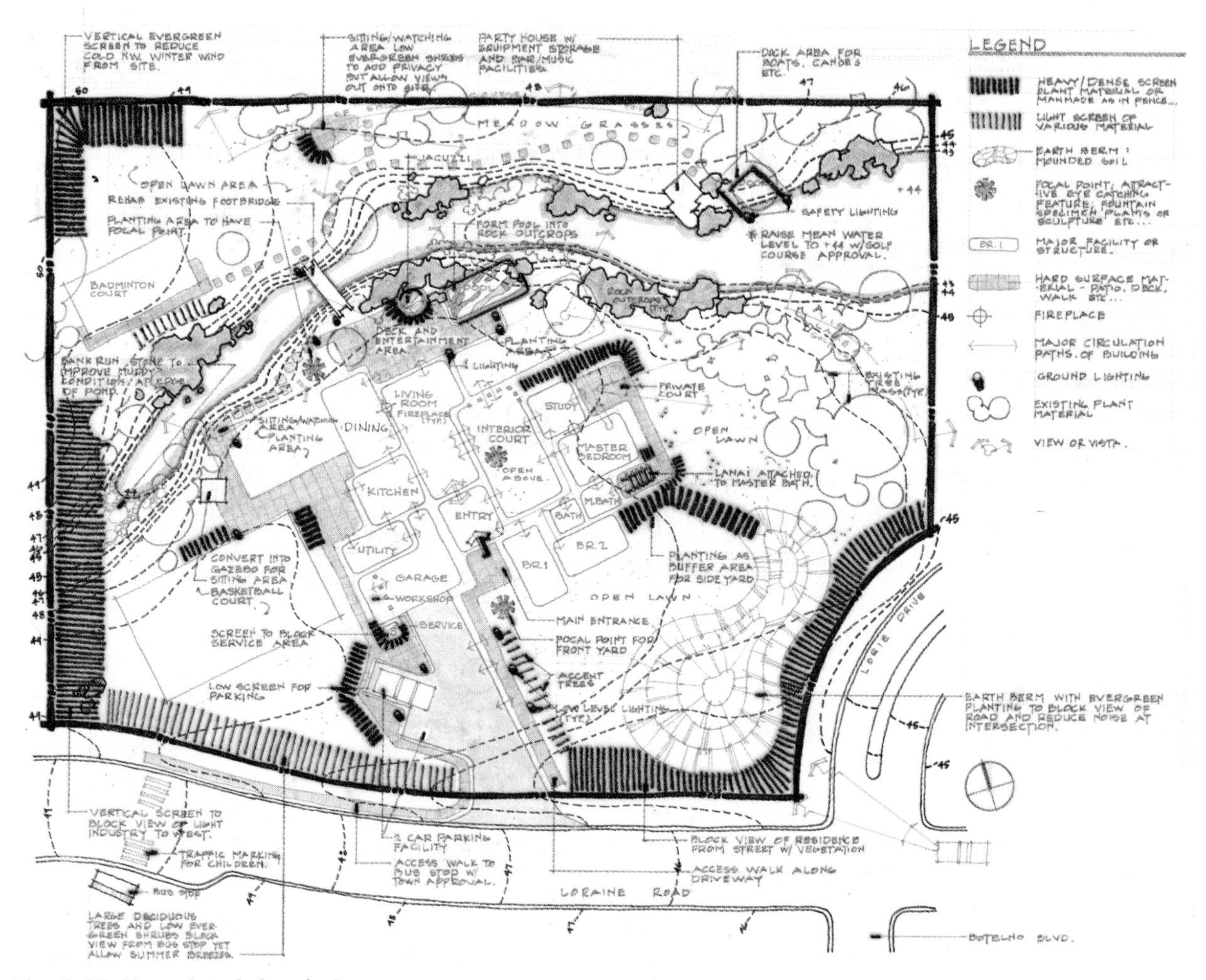

Fig. 2-25 Site-related drawing.

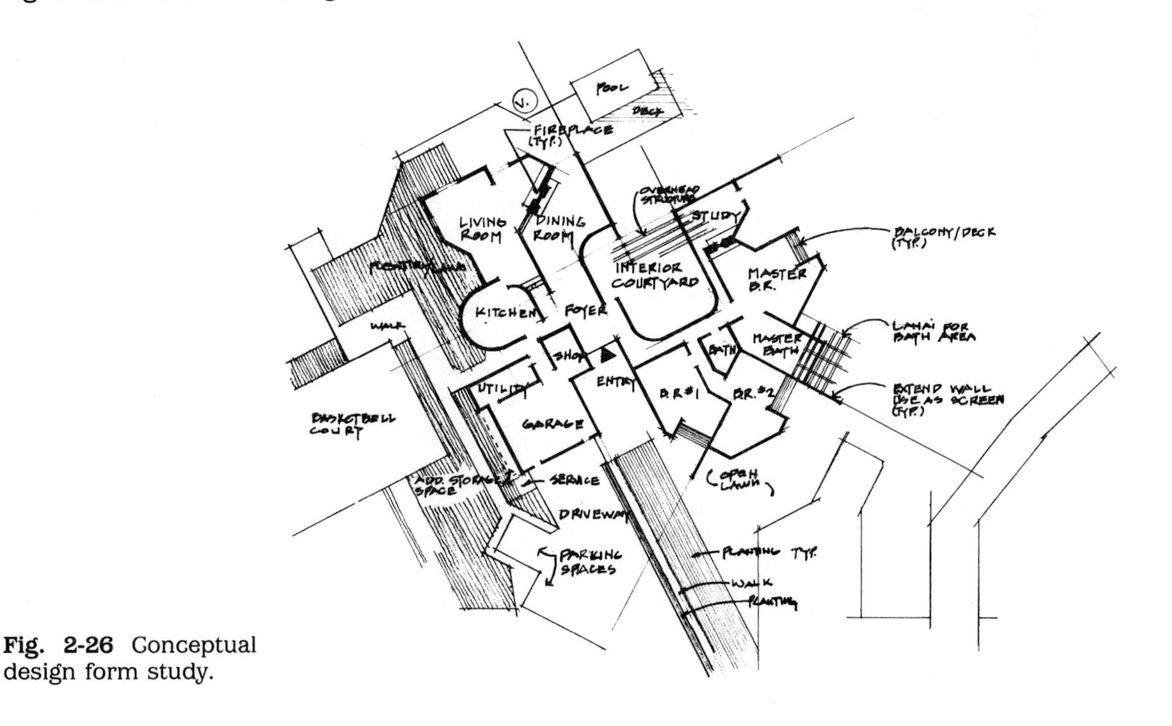

Fig. 2-26 Conceptual design form study.

Evaluation

Evaluation is needed to determine the degree of excellence of a design. This is a very difficult phase because it is subjective and based on judgment, which varies from one individual to another. However, whether the product is a design or anything else, its quality will eventually be evaluated and judged by clients or users. Self-evaluation of a design is critical and necessary. This requires checking the quality to see if it measures up to the predetermined goals and objectives and the user analysis requirements. The conceptual design must be evaluated and altered prior to the beginning of the final design-development phase to avoid the possible need to redesign at a later time, perhaps after much time and money have been spent on the project. It is easier and more time-efficient to redesign some elements at this stage than later in the process. By the time the evaluation is complete and the necessary changes have been made, the major design concepts will have been established and only design detail adjustments should be made. Many details and sizes will not yet have been established, but the position of the structures and the relationship between the design elements should not change significantly after this step is completed.

To evaluate a design the contents of the conceptual design must be compared with each specific goal and objective found in the user analysis. If a goal has not been accomplished in the design, then that part of the design must be altered to achieve the desired result. A well-developed design will have few discrepancies between user-analysis goals and the conceptual design. However, if many omissions are found, care must be taken to avoid correcting one omission and causing another in the process. At times it may be better to redo the conceptual design or analysis diagram than to try to "save" a design by numerous changes.

Design Development

After the necessary changes have been made in the conceptual design as a result of the evaluation and client feedback, the final phase, design development, can begin. During this design-development phase details are added to the site-related diagram in progressive sketches. Sketches are redone until the outlines of the design parts fit together without overlapping. During this stage awkward offsets are eliminated. Once the design is "smoothed out" in this manner a single line drawing to scale is prepared, as shown in Fig. 2-27, and key design details that are important to the final execution of the design are sketched for future reference. In addition to these schematic drawings and pictorial sketches, three-dimensional conceptual models may also be developed to aid in interpreting the conceptual design for the client and/or user. Once the plans of this stage are approved by all concerned, the preparation of working drawings can begin. Working drawings and documents are prepared to further refine the basic design concepts into very exact plans which can be used for bidding, budgeting, and construction purposes. They will be covered in Parts Three through Five.

ERGONOMIC PLANNING

Buildings are for people, and so buildings must be biotechnically (ergonomically) planned. This means the design must match the size, shape, reach, and mobility of all occupants.

Human Dimensions

Figure 2-28 shows the dimensions and reach limits of the average adult. The top numbers refer to male dimensions and the bottom numbers to female dimensions. To design functionally scaled architectural facilities, these dimensions should be used in planning or locating the facilities used by occupants on a regular basis. Human dimensions are especially critical in planning the size and position of cabinets, shelves, work counters, traffic areas, door openings, windows, and kitchen, bath, and laundry facilities. When buildings (such as schools) are designed primarily for children, obviously the scale must be adjusted.

Planning for the Handicapped

In planning buildings for general public use, ergonomic planning must be extended to include special provisions for the handicapped. This means planning facilities for the physically handicapped as well as for those with hearing or visual impairment. Design requirements which prohibit the use of architectural barriers or in-

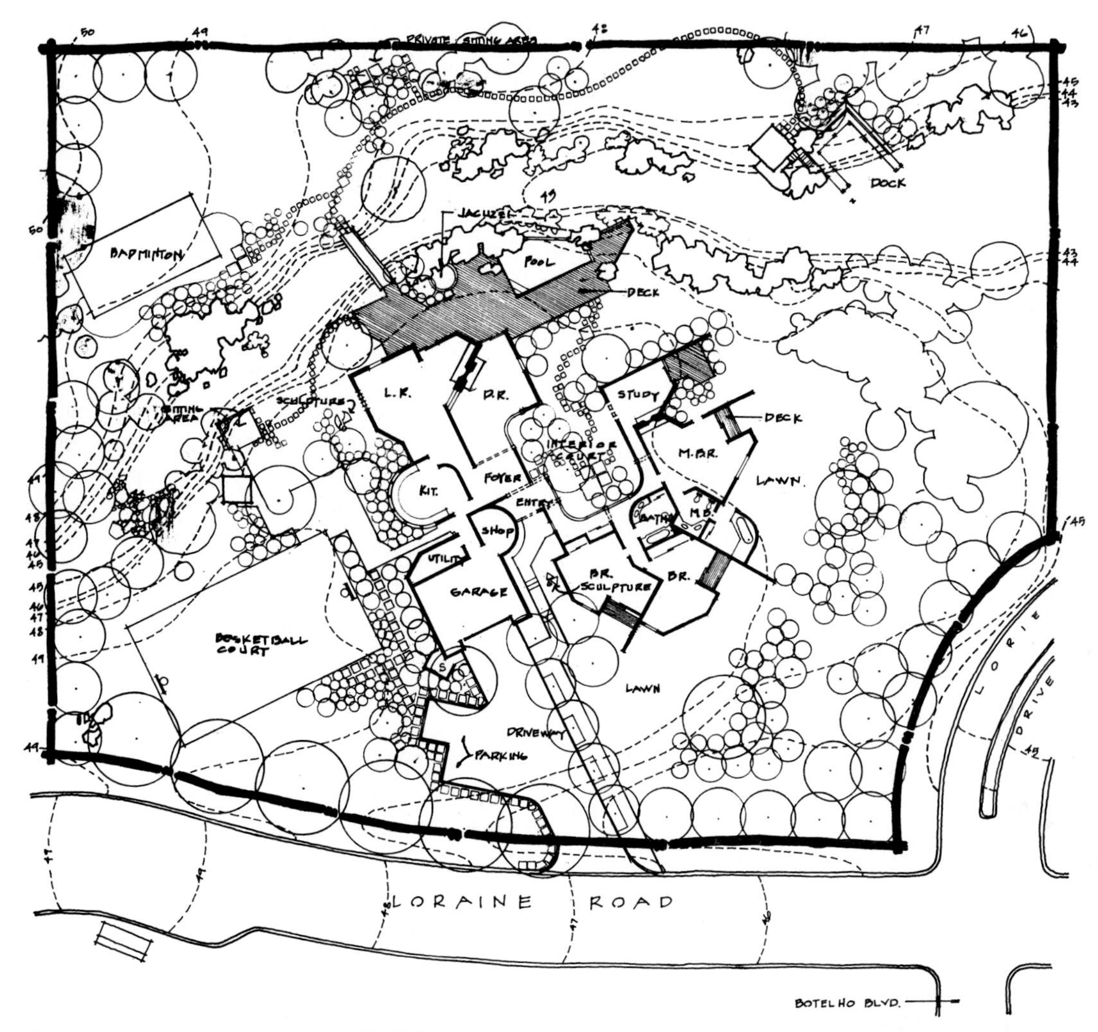

Fig. 2-27 Scaled single line (Scale: 1″-20′).

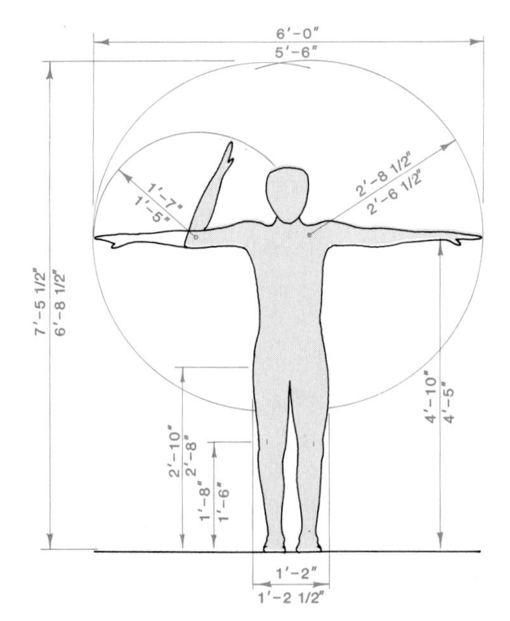

Fig. 2-28 Dimensions of the human figure.

fringements on the comfort and safety of the handicapped are found in every building code.

Design features that affect those with mobility handicaps usually relate to providing for wheelchair use. Since wheelchairs require the greatest amount of space, plans that accommodate wheelchairs will easily function for other design conditions. Figure 2-29 shows the dimensions and turning radius of a standard wheelchair. The following design guidelines apply to planning for adequate handicapped mobility, convenience, and safety in public buildings.

1. At least one entrance must be accessible to wheelchair traffic and provide access to the entire building. A passenger loading zone at least 4′ × 20′ must also be provided in this area.

2. At least one ramp (or elevator) must be provided as an alternative to stairs. The mini-

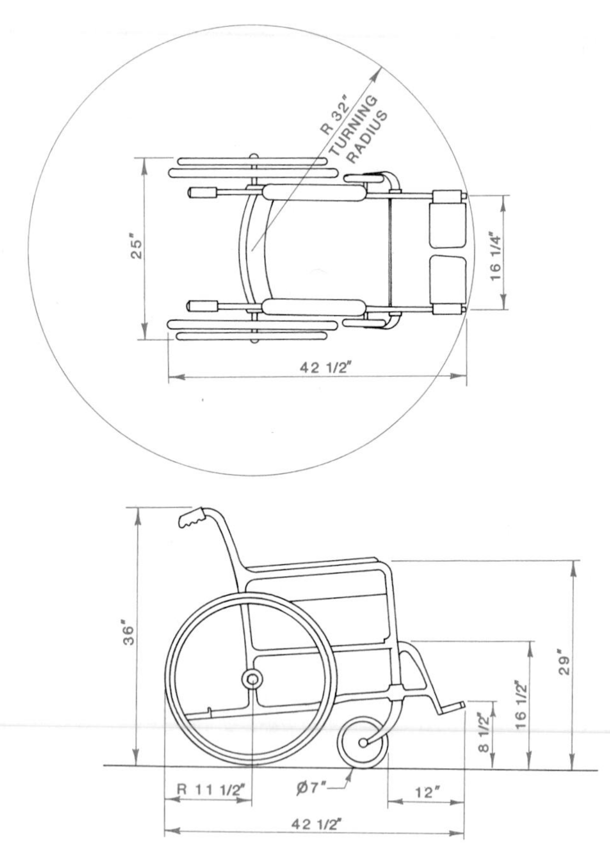

Fig. 2-29 Wheelchair dimensions.

mum width of ramps is 3′, with a maximum slope of 1:12 and a maximum rise of 2′-6″. Handrails must be provided for ramps longer than 6′-0″, and at least 6′-0″ of level area must be provided at the top of each ramp. Nonskid surfaces must be used on all ramps.

3. Doors must open at least 90° and be at least 3′-0″ wide, with threshold heights no more than ¾″ on exterior doors and ½″ on interior doors.

4. Halls must be at least 3′-0″ wide, with 5′-0″ provided in all turning areas. Halls must provide access to all areas of the building without the need to pass through other rooms.

5. Parking facilities must be provided in the parking area nearest the building and marked with the international handicapped access symbol (Fig. 2-30). This area must be out of the

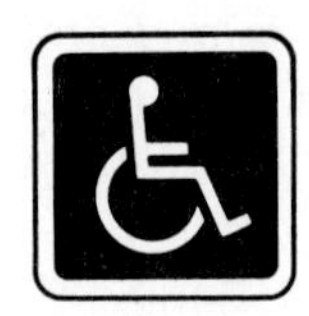
Fig. 2-30 International handicapped access symbol.

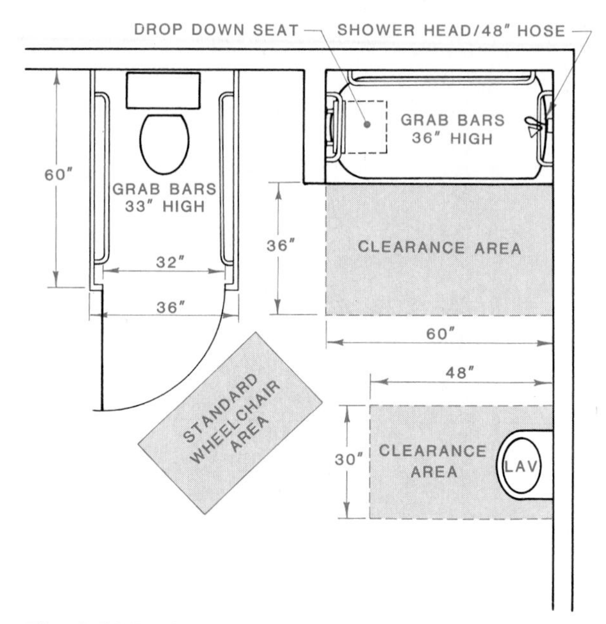

Fig. 2-31 Bath area dimensions required for wheelchairs.

main traffic flow and connected to the building by a ramp if the level changes. Parking slots must have at least 4′ of clearance on each side.

6. Walkways must be at least 4′ wide, with no less than 6′-8″ headroom clearances. Walks must be level and ramps must be used when it is necessary to change level. Walks must be free of obstructions and be surfaced with nonskid material.

7. Lavatory facilities for the handicapped must include at least 2′-6″ × 4′-0″ clear floor areas. Water closet seat tops must be 1′-6″ from the floor. A 3′-0″ knee-room height must be provided under sinks and drinking fountains. Grab bars must be provided near water closets, sinks, and bath areas, as shown in Fig. 2-31.

8. Public-phone amplifiers, high-frequency alarms, and eye-level warning lights to augment audio alarms should be provided for the hearing-impaired.

9. Braille signs, level-change warning surfaces, and restrictions on wall protrusions over 4″ must be provided for the visually impaired.

10. There should be at least three treads in a series of stairs. Treads and risers should be uniform and treads should have a minimum width of 11″ with round nosings. A landing should be planned for stair systems which contain more than 16 risers.

11. Doors should be provided with handles that do not require a tight grip to turn.

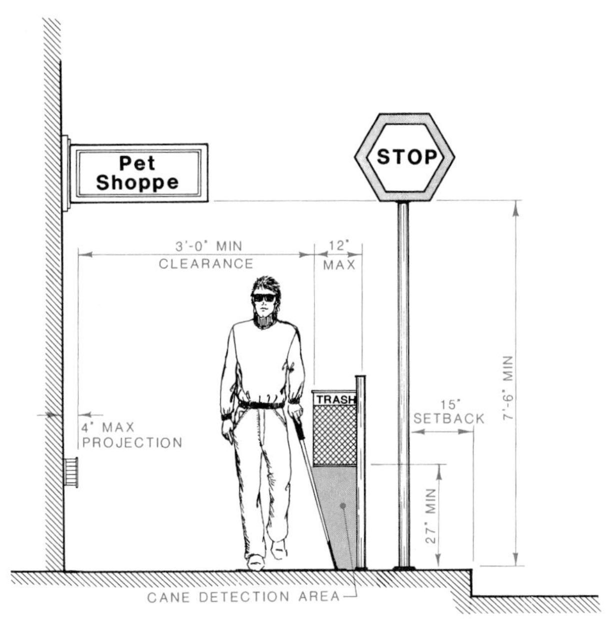

Fig. 2-32 Wall projection requirements.

12. Thresholds should not be higher than ½″ on swinging doors and ¾″ on sliding doors.

13. Appliance cooking controls should be placed in front of the burners.

14. A clear 28″, 31″, and 36″ floor space should be provided under selected base cabinets or next to appliances.

15. Ovens should have side-hinged doors and dishwashers should be front loading.

16. Refrigerators should contain a vertical freezer compartment.

17. Cabinet pulls should be recessed.

18. Texture changes using raised strips, grooves, rough, or cushioned surfaces should be used to warn the visually impaired of an impending danger area including ramp approaches.

19. Wall projections, if located between 27″ and 80″ from the floor, cannot extend more than 4″ from a wall. Objects mounted below 27″ from the floor may project any amount. But freestanding objects between 27″ and 80″ may only project 12″ from their support as shown in Fig. 2-32.

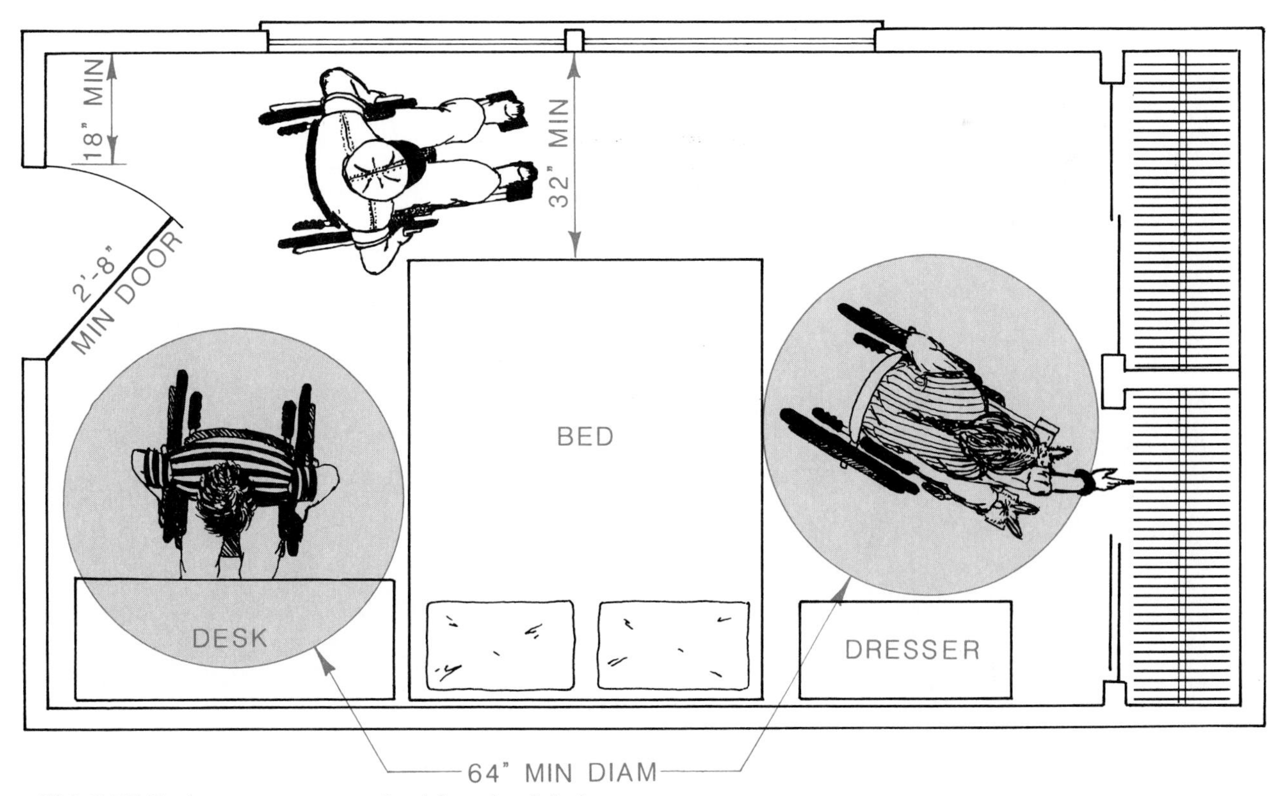

Fig. 2-33 Bedroom space required for wheelchairs.

20. Protruding signs must be at least 7′-6″ from the floor.

21. Emergency warning alarms should be both visible and audible.

22. Door kickplates should cover the bottom 10″ of a door used by wheelchair users.

23. Floors should have nonslip surfaces, even when wet, or be covered with carpeting with a pile thickness of no more than ½″.

24. Lighting should be free of glare or deep shadows.

25. Bedrooms should be designed to allow wheelchair maneuverability and access to the bed, storage table surfaces, and doors as shown in Fig. 2-33.

26. Design storage facilities should allow easy reach from a wheelchair as shown in Fig. 2-34.

27. Washbasins should be mounted no closer to the floor than 29″ and should extend a minimum of 17″ from the wall to provide adequate knee space as shown in Fig. 2-35.

28. A minimum floor space area of 2′-6″ × 4′-0″ must be provided around lavatories for wheelchair access as shown in Fig. 2-36.

29. Handicapped parking spaces must be at least 8′-0″ wide and have an access area of at least 5′-0″ wide as shown in Fig. 2-37.

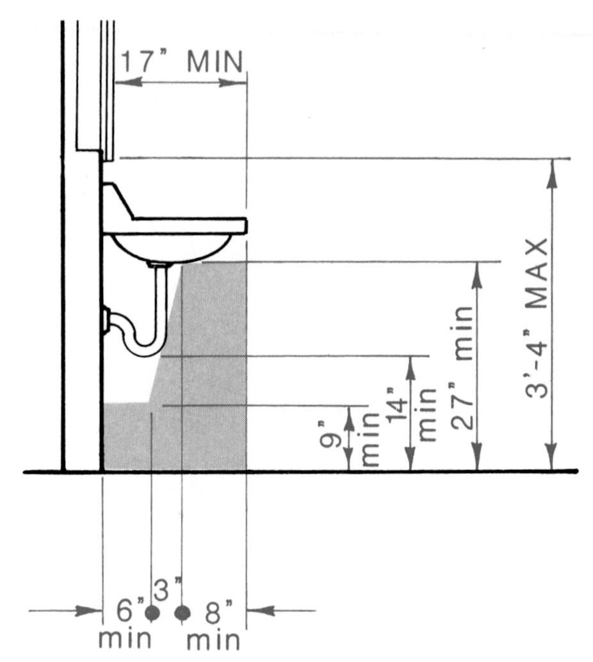

Fig. 2-35 Wheelchair kneespace dimensions for lavatories.

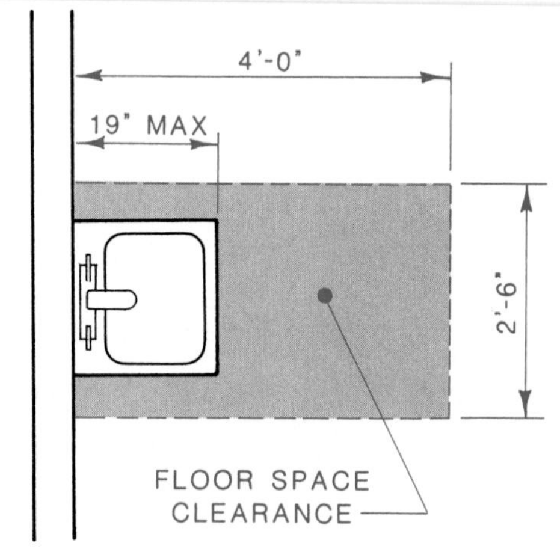

Fig. 2-36 Floor space clearance around lavatories for wheelchairs.

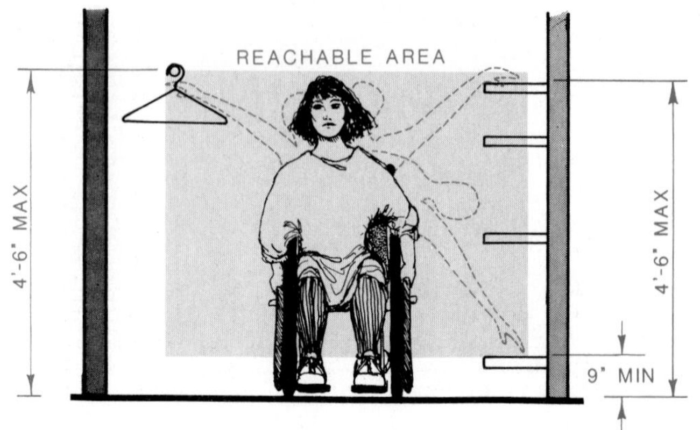

Fig. 2-34 Wheelchair reachability requirements.

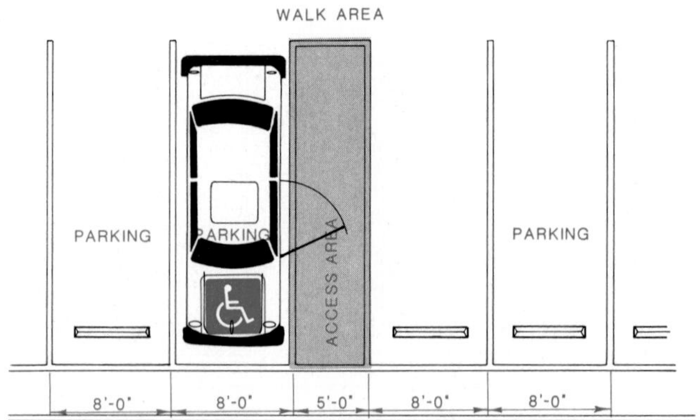

Fig. 2-37 Handicapped parking access area dimensions.

Safety Factors in Design

The designer must ensure that no design feature creates a public health or safety risk. Precautions include specifying appropriate mechanical equipment, such as gas furnaces and appliances, oil burners, electrical equipment, wiring devices, and machinery; and they also involve the avoidance of hazardous materials (such as asbestos) and of accident-causing materials or processes, such as extra smooth floors, thin glass, or unstable ceiling coverings.

Safe design also involves the safe arrangement of traffic areas to provide adequate vehicular turning angles, fire lanes, lane widths, slope angles, snow clearance, exit signs, handrails, and floor runoff. Naturally, safety in design also implies that a building will be structurally sound and will adequately support all anticipated live and dead leads. Air for an environmentally safe building should be electronically filtered to provide for the elimination of harmful pollutants.

Designing for New Technology

Before the design process is begun for any building, an assessment must be made to determine what special technology needs are anticipated. So-called smart buildings now contain such built-in equipment and accommodations as TV cables, static-proof surfaces, high- and low-voltage circuits, computer hookups, and even computer floors. If the need is anticipated, magnetic or radio-wave interferences can be blocked and the environment can be controlled to enable sensitive electronic and mechanical equipment and devices to operate effectively. Through the use of computers, many building control systems can effectively regulate the function of appliances, lighting, alarms, safety devices, communication systems, heating, and air conditioning. The system shown in Fig. 2-38 controls many of these functions through the use of a computer touchscreen. Other systems use infrared sensors for security surveillance and electronically synthesized voice commands for communications. Some systems can also monitor smoke, gas, and electrical consumption, while some include closed-circuit video cameras interfaced with a computer to record images for immediate use and future reference. High technology features which should be considered in the architectural design process include the ability to

1. Monitor smoke, gas, sound, and movement and sound alarm when established levels are exceeded.

2. Monitor and adjust heat, cooling, and humidity levels for each room.

3. Provide video phone intercom communication between entrances and selected rooms including synthesized voice response to visitors.

4. Open and close, lock and unlock doors, windows, vents, gates, and vents from a central control center.

5. Turn on or off appliances, audio, or VCR from a central control center directly or on timed sequences.

6. Monitor and/or time the opening and closing of solar energy devices including window shades and drapes.

Fig. 2-38 Computer touchscreen building control system. *(Unity Systems Inc.)*

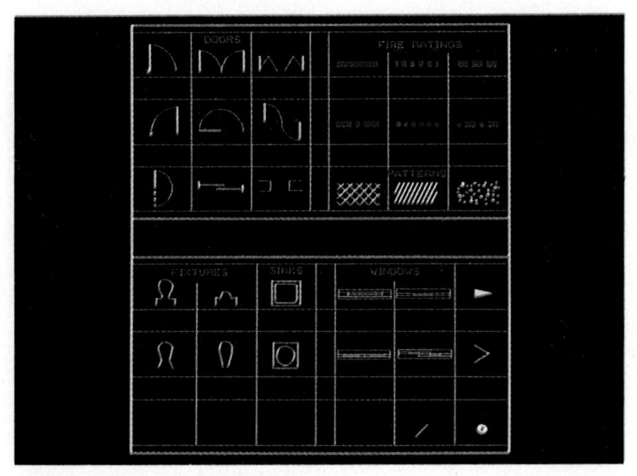

Fig. 2-39 Computer stored architectural symbol library. *(Intergraph Corp.)*

7. Control lighting circuits from a central control center directly or on timed sequences.

8. Program combination of controls to activate systems on a timed basis for night, morning, midday, evening, work week, weekend, or vacation modes of operation.

9. Systems should be capable of activation via outside and inside phone commands.

Designing with New Technology

Contemporary designers must not only design *for* new technological needs, but must design *with* new technological tools. New developments in computer graphics now allow architectural designs to be developed faster, more accurately, and with greater flexibility than ever before. For example, a complete architectural symbol library containing material, component, and fixture symbols (Fig. 2-39) can be stored in a computer's memory. These symbols can be placed on any computer-generated drawing with a variety of input devices such as a stylus or mouse. In addition, geometric elements such as points, lines, cubes, arcs, circles, and spheres can be added to a drawing by touching the desired location with an input device. A combination of these symbols and elements was used to create the computer-generated floor plan shown in Fig. 2-40. These same elements were used to create the elevation drawings shown in Fig. 2-41. Many elevation design variations can be stored in the computer and used to develop design alternatives with many variations. Different materials and components can also be changed with a few keyboard strokes or stylus moves.

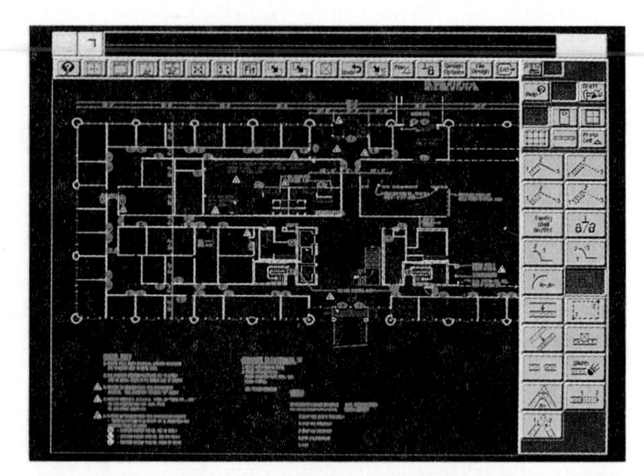

Fig. 2-40 Computer generated floor plan. *(Intergraph Corp.)*

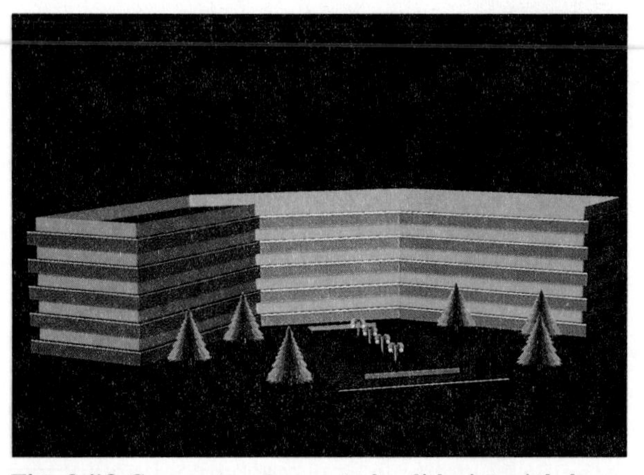

Fig. 2-42 Computer generated solid pictorial drawing. *(Versacad)*

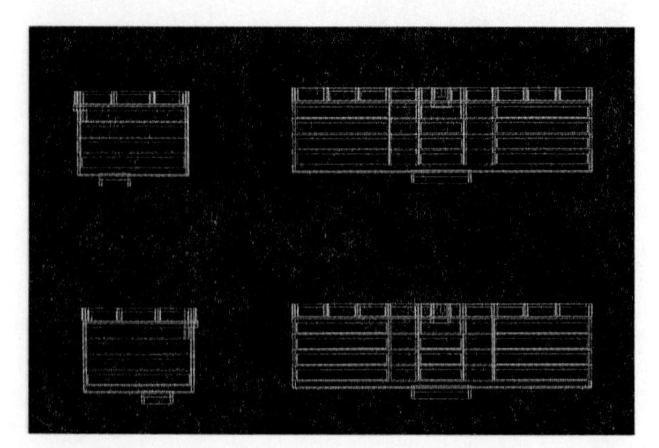

Fig. 2-41 Computer generated elevation drawing. *(Integraph Corp.)*

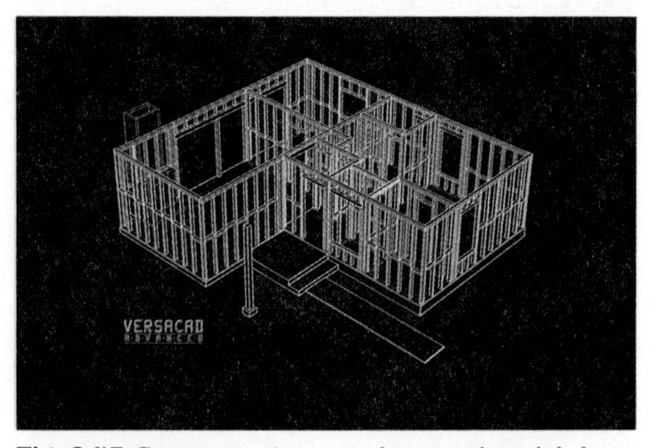

Fig. 2-43 Computer generated open pictorial drawing. *(Versacad)*

Computer generated pictorial drawings can create images that will show precisely how a proposed building design will fit into an existing environment. Many programs can also create shadows, reflections, textures and highlights by computing the suns position at any time. Pictorial drawings may be computer generated as solid objects as shown in Fig. 2-42, or open (wire-frame) as shown in Fig. 2-43. Wire frame drawings allow viewers to look through a building and view the interior from any angle. Sophisticated programs can also create "walk-through" images that simulate the process of walking through and around a building. This is accomplished by specifying a path and viewing positions on the path.

Very detailed sectional drawings as shown in Fig. 2-44 can also be computer generated using section lining symbols and automatic dimensioning. Computer generated base maps can be used conveniently as the basis for many site development drawings as shown in Fig. 2-45.

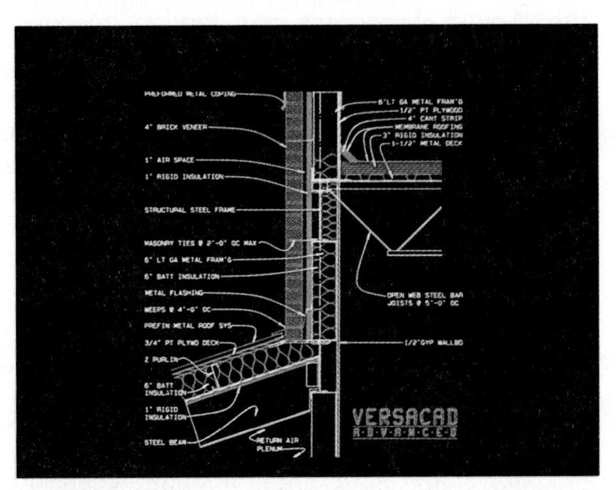

Fig. 2-44 Computer generated sectional drawing. *(Versacad)*

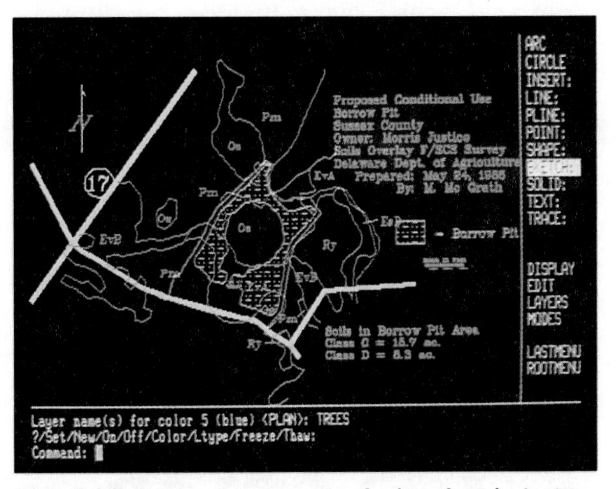

Fig. 2-45 Computer generated site drawing. *(Versacad)*

Exercises

1. **Sketch the front elevation of the house shown in Fig. 2-15. Convert the front-elevation design to a formally balanced elevation.**
2. **Sketch the house shown in Fig. 1-2A to provide more emphasis on one phase of the design.**
3. **Sketch the wall shown in Fig. 2-4 to improve the patterns of light and shadow and to provide more unity of texture and line in the design.**
4. **List the major color you would use to decorate each room in your home. List two supporting colors you would use for contrast or variety.**
5. **List each element of design and describe your preference for applying each to a residence of your own design.**
6. **Follow the design process and produce a concept plan for a building of your own design. Use an existing site or create one for this exercise.**
7. **Define these terms: technical design, creative design, aesthetic, informal design, formal design, function, form, space, light, shadow, texture, line, color, beauty, repetition, variety, emphasis, informal balance, formal balance, ergonomics, primary color, secondary color, tertiary color, color value, hue, tint, harmony, site analysis, slope diagram, conceptual design, kips.**

SECTION 2

Environmental Design Factors

General environmental conditions related to building design differs greatly from one part of the country to another. Specific conditions may also vary significantly within a geographical area. These include building-site characteristics, density requirements, energy sources, ecological problems, and special occupant needs. The designer must therefore carefully study and use these factors to ensure the development of fully functional architectural plans. The designer must also do the utmost to protect, and indeed improve, the environment in the process.

UNIT 3

ENERGY ORIENTATION AND SOLAR PLANNING

Throughout history, the environment has been used in conjunction with the need for shelter. Local resources and climatic conditions have always affected heating, cooling, and lighting needs. For example, early Native Americans built adobe houses under overhanging cliffs (Fig. 3-1). The cliffs provided shade and protection from the sun's rays during the day. At night, the material used in these houses released the heat accumulated during the day and warmed the area. Even later, when other heat sources were developed, people conserved fuel, whether it was wood, dung, twisted grass, or blubber. But this all changed when fossil fuel became popular. Inexpensive fossil fuel appeared to be an endless energy source. Fuel was used with no thought of saving it. That practice affected the design and construction of buildings. Consequently, architectural designs and materials that previously had been chosen to gain heating and cooling benefits from natural sources soon became *obsolete*. By relying almost exclusively on fossil fuels for energy, designers controlled the inside environments of many buildings artificially. There was little regard for energy efficiency or for taking advantage of nature.

Now, after years of unchecked use and misuse of energy resources at considerable cost to the environment, the end of cheap fossil fuels can be seen. Today, considering the finite supply of fossil fuels, there is a dire need to return to the use of energy-efficient principles in building design.

Sources of energy and methods to control building environments now available include fossil fuels, solar power, nuclear power, hydroelectric power, wind power, geothermal power, and the combustion of a variety of natural mate-

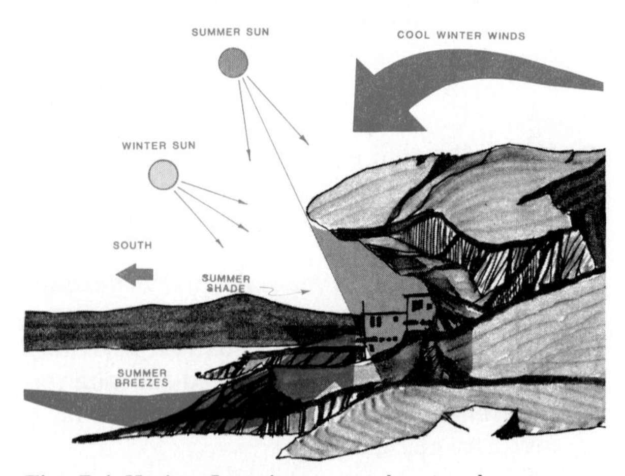

Fig. 3-1 Native Americans used natural resources for heating and cooling.

rials. These last include reclaimed waste materials, wood, and other organic materials. All of these sources and methods, except passive solar, require special equipment or devices to effectively heat and cool buildings.

Carefully designed, constructed, and sited buildings can use the power of the sun with or without devices to provide environmental control. When only the features and orientation of a building are used to gain and control the sun's energy, the designer is using the power of the sun *passively*. When mechanical or electrical devices are added to collect, store, distribute, and control the sun's energy, the designer is *actively* capturing the power of the sun. Thus there are two types of solar design for buildings: *passive* and *active*. However, any system, from fossil fuel to active solar, can function more effectively if the building is designed using basic passive solar design principles. Therefore, passive solar design is covered here, while active solar design is found in Unit 63.

PASSIVE SOLAR SYSTEMS

Designers and builders are rediscovering the natural cooling, heating, and lighting provided in many climates. This reborn sensitivity, combined with new technology, is fostering a redirection in architectural design. It is no longer necessary to sacrifice building quality and comfort to design buildings that place less demand on limited energy resources. Even in the mildest climates, effective solar planning can often cut heating and cooling costs significantly.

The process of solar heating or cooling of a building is based on four steps: (1) collecting, (2) storing, (3) distributing, and (4) controlling. The four steps occur in all systems, active or passive. However, the equipment, materials, and devices used differ greatly among systems.

Passive Solar Design Principles

The earth's annual revolution around the sun and daily rotation on its axis determines how much solar energy is available at any time on any location on earth. Both the daily and seasonal paths of the sun over a building site are the first consideration in passive solar planning.

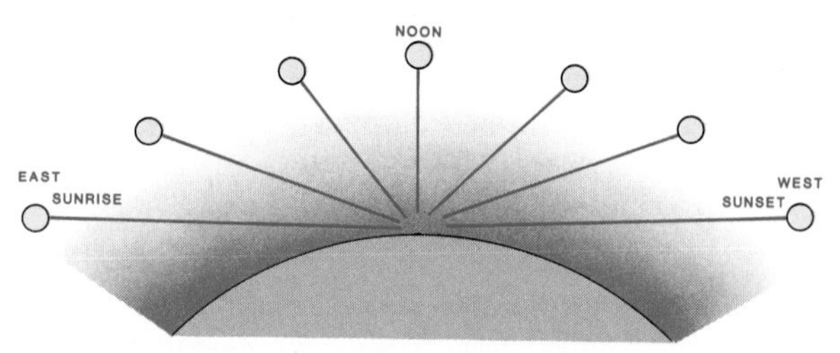

Fig. 3-2 Sun's rays must travel different distances through the atmosphere.

The amount of solar radiation reaching a site also depends on the thickness and density of the atmosphere. When the sun is directly overhead, the sun's rays travel the shortest distance through the atmosphere. As the sun moves closer to the horizon, the amount of atmosphere the rays travel through increases greatly. This greater amount of atmosphere decreases the amount of solar radiation reaching the site. This means winter, early morning, and evening rays must pass through more atmosphere than summer and midday rays. This atmospheric effect is shown in Fig. 3-2.

The slope of a site also affects the amount of usable solar radiation. This is because the sun's rays when striking a surface perpendicularly are more concentrated than when they intersect the surface at an angle. Therefore solar radiation striking a south-facing slope, as shown in Fig. 3-3, will be more concentrated than rays striking nearly flat terrain.

In addition to the atmospheric relationship of the sun to the earth's surface, two other principles of solar physics are used in passive solar planning. One is the *greenhouse effect*. The other is the natural law of rising warm air.

A car parked in direct sunlight with the windows closed illustrates the *greenhouse effect*. The interior of the car becomes heated because sunlight enters through the windows. The heat is absorbed by the interior surfaces of the car and is trapped inside the car as stored heat. In a sim-

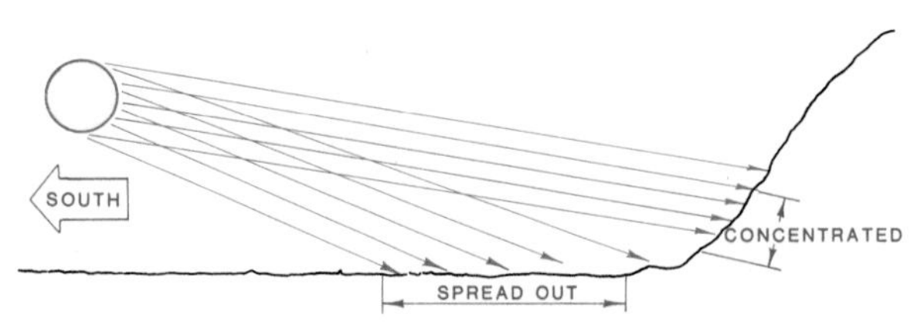

Fig. 3-3 South-facing slopes receive more concentrated solar energy.

ilar manner, heat from the sun enters a building through windows and is stored in a *thermal mass* so that the heat can be used later when the sun's heat is not available. A *thermal mass* is any material that will absorb heat from the sun and later radiate the heat back into the air. The seats in the car act as a thermal mass. Walls, floors, and fireplaces can all work in this manner in a building designed for maximum solar effectiveness. In passive solar systems the thermal mass functions as both the storage and the distribution system. Storing and using, or dissipating, the trapped heat to either lower or raise the temperature of a building is one of the most important features of passive solar design. Figure 3-4 shows an example of the use of the greenhouse effect to store heat in the winter and expel it in the summer.

Heated air will always rise until trapped. Therefore, recirculating heated air from high places to cooler lower areas helps heat a building's living levels. Likewise, expelling warm air that accumulates and would otherwise gradually move downward will help reduce living-level temperatures. Figure 3-5 show how convection currents are used to draw out unwanted warm air by

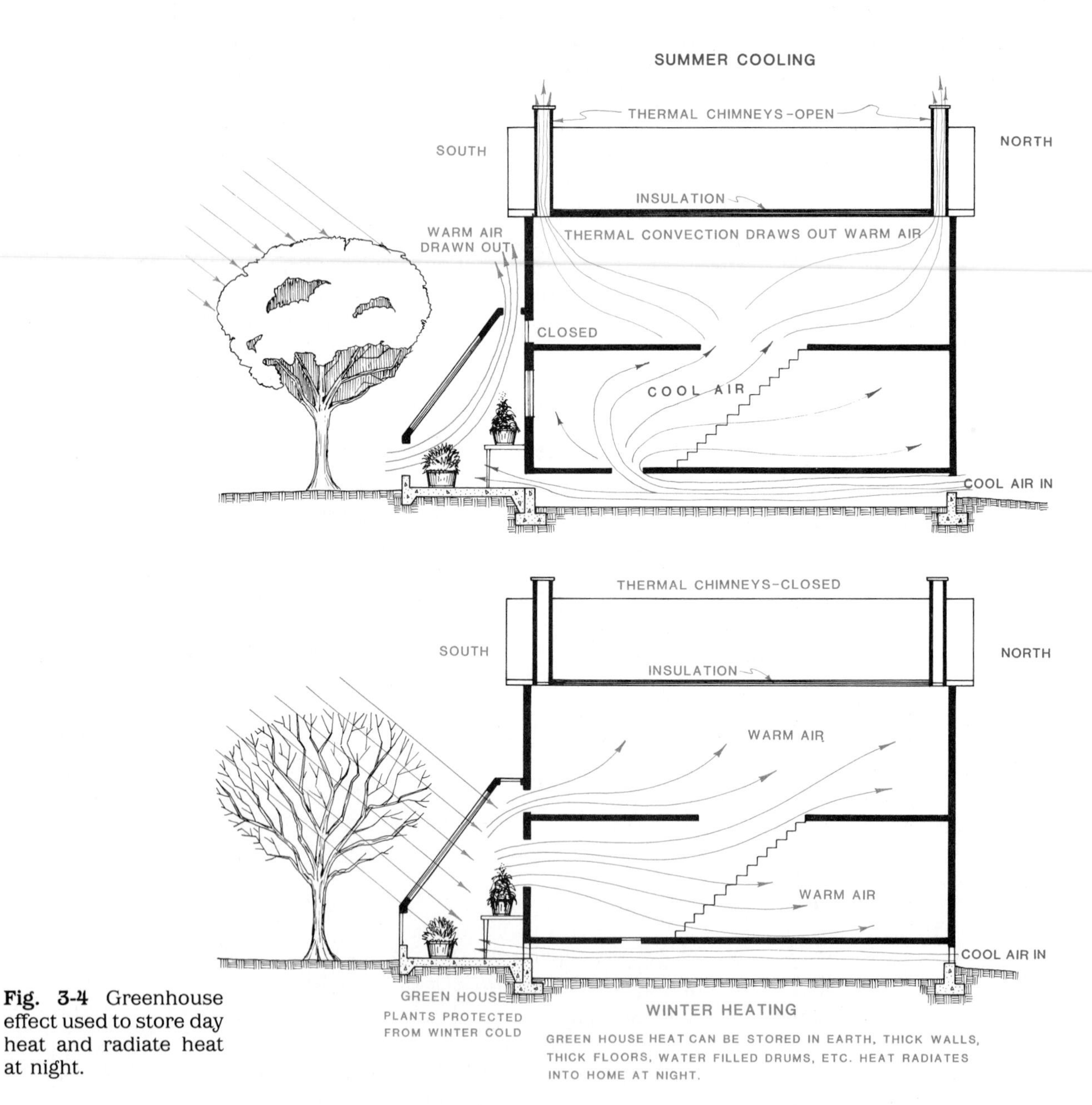

Fig. 3-4 Greenhouse effect used to store day heat and radiate heat at night.

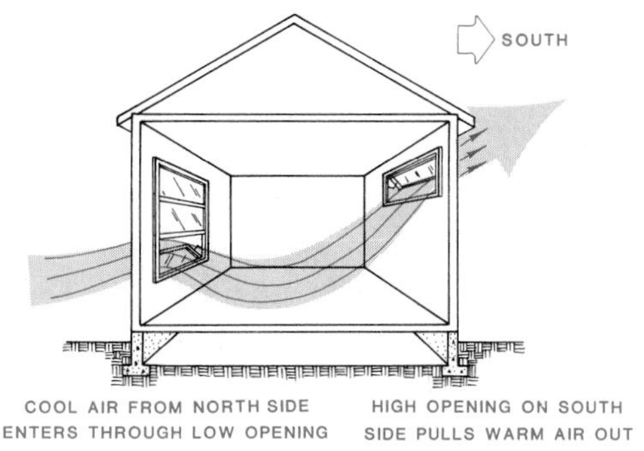

Fig. 3-5 Convection draws out warm air.

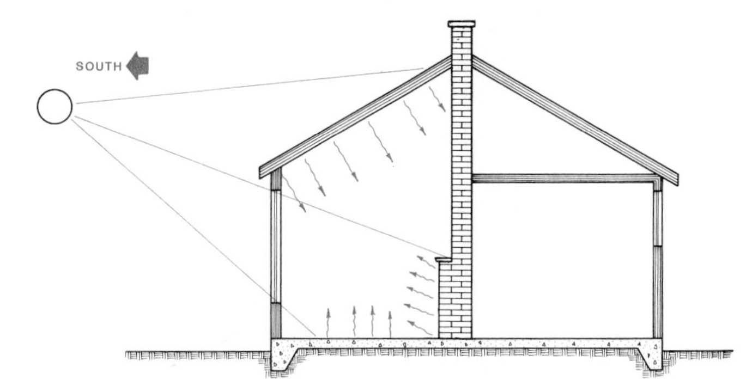

Fig. 3-7 Direct-gain method of heating.

passively manipulating the natural flow of rising warm air. In this example the level and placement of windows provide natural convection and ventilation to both circulate and exhaust warm air. Figure 3-6 shows the effect of ceiling heights and exhausts in the control of high, heated air.

Passive Solar Methods

There are two passive solar methods which make use of environmental elements without additional technical assistance from devices such as solar panels and heat pumps. These two methods are the *direct-gain method* and the *indirect-gain method*. Both methods are designed to take full advantage of the sun to provide heat when and where it is needed and to block the sun's heat when and where it is not wanted.

Direct-Gain Method With the direct-gain method the inside of a building is heated by the sun's rays directly as they pass through large glass areas, as shown in Fig. 3-7. To maximize the amount of winter heat directly entering a building, large south-facing glass areas are used. Once the winter rays enter a building through windows, they are absorbed and the heat is stored in thermal-mass objects to be used later. These objects include floors, walls, and furniture. When the sun is not shining, the stored heat in the thermal masses is slowly released, keeping the inside temperature higher. Large thermal masses such as masonry floors, walls, and fireplaces store more heat than wood and other fibrous materials. Therefore they hold heat longer for later use.

Some direct-gain walls are designed with reflective insulating units to provide illumination and heat in cold weather yet reflect most of the sun's heat in hot seasons.

Indirect-Gain Method The indirect-gain method uses a thermal mass placed between the sun and the inside of a building. This thermal mass is heated directly by the sun. When heat is needed, the thermal mass is exposed to the inside of the building and indirectly heats the inside air. The advantage of this method is in the opportunity it offers to control the amount of heat emitted to the interior by opening or closing the separations between the thermal mass and the inside. The Trombe wall shown in Fig. 3-8 is a type of indirect-gain system. In this system the temperature is controlled by directing varying amounts of the rising warm air to the inside or to the exterior of the building, depending on the comfort level needed.

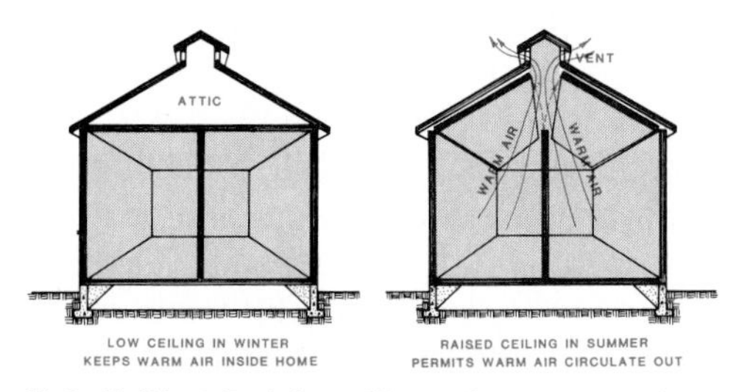

Fig. 3-6 Ceiling height affects heat retention and exhaust.

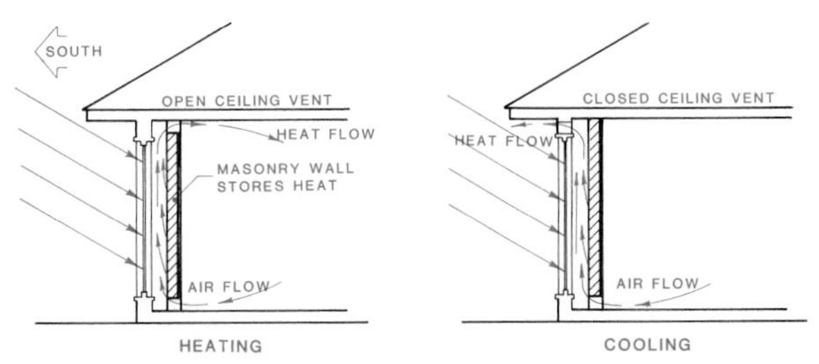

Fig. 3-8 The Trombe wall is a type of indirect-gain system.

ORIENTATION

To take full advantage of the principles of passive solar design, a building must be positioned to maximize the desirable features and minimize the negative aspects of the environment. This is accomplished through effective *orientation*. The orientation of a building is the relationship of a building to its environment. In determining the most appropriate orientation of a building on a site, the following factors must be carefully considered: sun heat, sunlight, existing vegetation, desirable and undesirable views, objectionable noise, velocity and direction of prevailing winds or breezes, land forms and shapes, lot and structure size, relationship of the site to the neighborhood, and local building codes.

Solar Orientation

The first step in effective solar planning is the correct orientation of the building to the sun. It is usually desirable to collect the maximum amount of the sun's heat in the winter and also provide protection from the sun in the summer. Fortunately, the sun's angle changes from summer to winter, as shown in Fig. 3-9. In the northern hemisphere, the south and west sides of a structure are warmer than the east and north sides. The south side of a building is therefore the warmest side, because of the nearly constant exposure to the sun during its intense periods of radiation. Ideally, a building should be oriented to absorb this southern-exposure heat in winter and to repel the excess heat in summer. Figure 3-10 shows the compass location of the sun during both winter and summer days and nights. This chart can be used to locate and orient a structure to ensure that the areas requiring the most solar exposure will be correctly positioned in reference to the sun. Keep in mind that effective solar orientation should not only provide the greatest heating or cooling effect but should also be planned to provide the greatest amount of natural sunlight where needed.

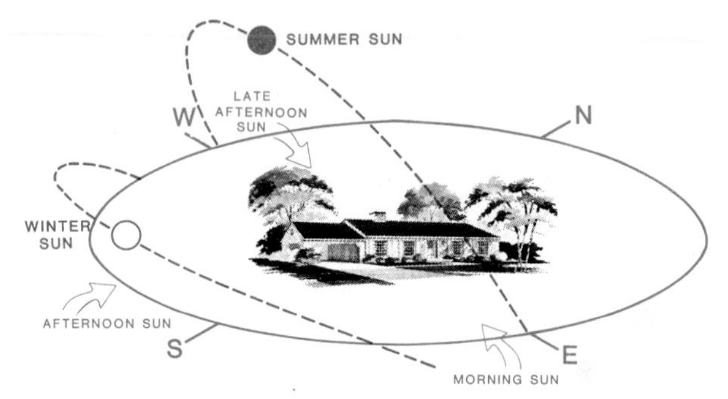

Fig. 3-9 Summer sun angles provide more heat.

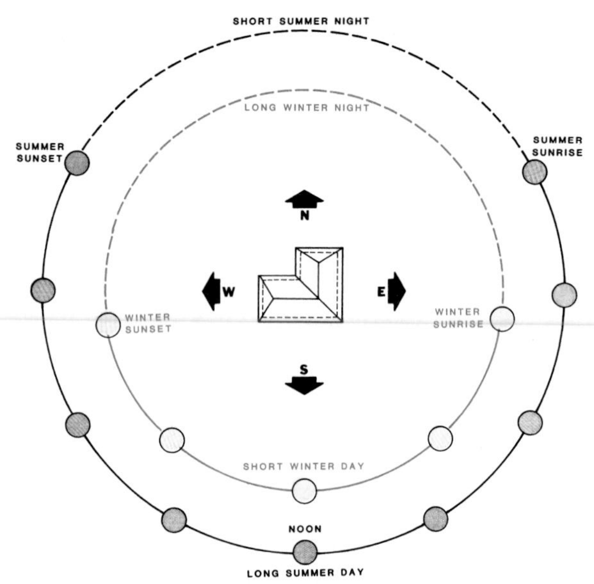

Fig. 3-10 Compass location of the sun in summer and winter.

Integration of Structure with Landform

Every structure should be designed as an integral part of the site, regardless of the shape or size of the terrain. Buildings should not appear as appendages to the land but as a functional part of the landscape. For the indoor and outdoor areas to effectively function as part of the same plan, the building and the lot must be designed together. In order to achieve the maximum compatibility of a building with a site the conditions of the building site must be studied for climatic factors, such as air temperature, type and amount of precipitation, humidity, wind speed, wind direction, and available sunlight. The designer must also consider the specific physical characteristics of the site, such as hills, valleys, fences, other buildings, and trees. Physical obstructions such as these may affect wind patterns and the amount and direction of available sunlight in different seasons. Large bodies of water may also affect air temperature and air movements. Surrounding pavement areas and buildings can raise or lower temperatures, because concrete and asphalt collect and store the sun's heat. When the characteristics of a site are known, the designer can then begin orienting and designing a structure.

Don't forget to consider the view options in orienting the house. Orientation of specific areas toward the best view, or away from an objectionable view, usually means careful planning of the position of the various areas of the building.

A particular plan may be compatible with one lot and site and yet appear totally out of place in another location. Sloping sites offer a variety of conditions. Bilevel and trilevel houses are well adapted to an almost endless combination of site contours.

Frank Lloyd Wright probably did more than any architect to popularize the integration of the structure and the site. The design of a building to completely integrate with its surroundings he labeled *organic architecture*.

The Lot

The size of the lot affects the flexibility of choice in locating structures. For planning purposes, lots are divided into three areas, according to function: the private area, the public area, and the service area (Fig. 3-11). The *private area* includes the house and outdoor living space. The *public area* is the area of the lot that can be viewed by the public. This area is usually located at the front of the house and should provide off-street parking and access to the main entrance. The *service area* of the lot should be adjacent to the service area of the house.

The placement of the house on the lot determines the relative size and relationship of the three areas. Making the features of the lot an integral part of the total organic design is as important in the design process as the basic floor plan of a structure.

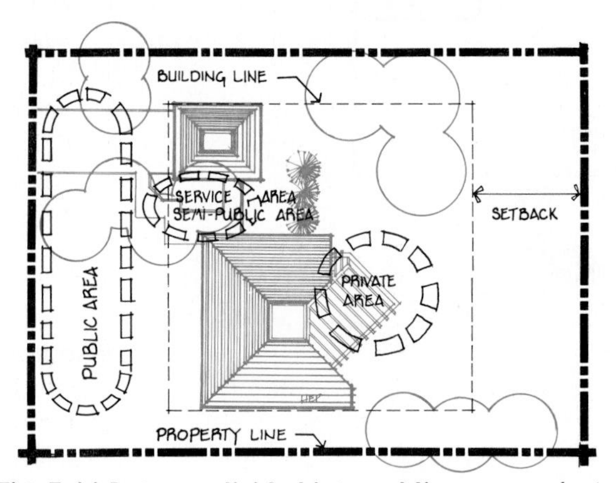

Fig. 3-11 Lots are divided into public areas, private areas, and service areas.

Zoning Code Considerations

An architectural plan should not be started until local building and zoning codes have been thoroughly checked. Some codes require that a structure be placed no closer than 10′ from the property line. Others may require 50′ or more. To determine the allowable space for building, draw lines within and parallel to the property line to represent the code minimum distance. The area within the inside lines represents the building areas, as shown in Fig. 3-12. Zoning codes also restrict the distance from a building to the street. These distances are known as *setbacks*.

Zoning codes often limit the percentage of the lot that can be covered by a structure. Some states also have laws which restrict a new structure from shading the solar panels of an existing building (see Fig. 3-13). A careful review of local

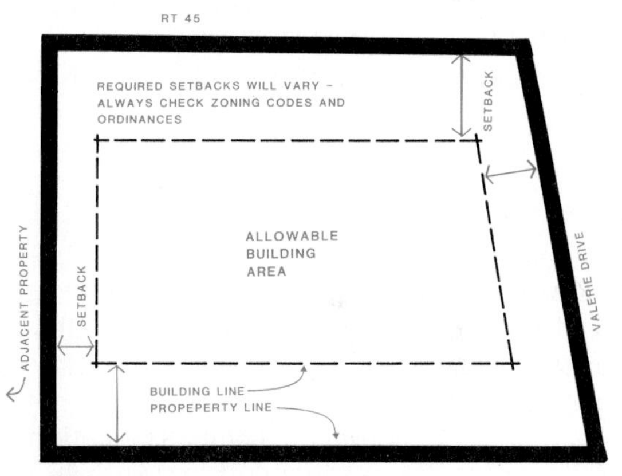

Fig. 3-12 Building codes restrict the placement of a structure on a lot.

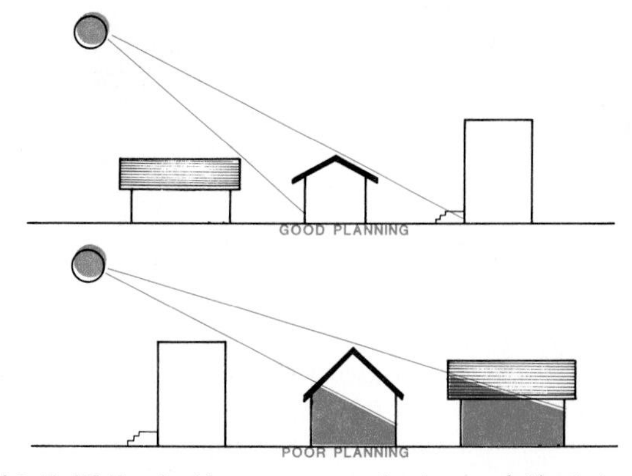

Fig. 3-13 Zoning laws may protect an existing structure's sun exposure.

zoning codes is always necessary before any building design is started.

Room Locations

A room should be located to absorb the heat of the sun or to be baffled from the heat of the sun, depending on the function of the room, season, and on the time of day the room is likely to be used. The location of each room should also make maximum use of the light from the sun.

Generally, sunshine should be available in the kitchen during the early morning and should reach the living areas by afternoon. To accomplish this, kitchen and dining areas should be placed on the south or east side of the house. Living areas placed on the south or west side are desirable because they receive the late-day rays of the sun. The north side is the most appropriate side for placing sleeping areas, since it provides the greatest darkness in the morning and evening and is also the coolest side. North light is also consistent and diffused and has little glare. Figure 3-14 summarizes the basic guidelines for room placement.

To take full advantage of the sun, outdoor living areas requiring sun in the morning should be located on the east side of the floor plan. Those requiring the sun in the evening should be placed on the west side. Remember that the midday sun in the northern hemisphere shines from the south, which means that the north side of the building receives no direct sunlight. A separate entry hall on any side can also reduce heat transfer considerably, because it always protects the main house from direct contact with the outdoor temperature (see Fig. 3-15).

Overhang Protection

The angle of the sun differs in summer and in winter. Therefore, roof overhangs can be designed with a length and angle that will shade windows in summer and allow the sun to enter during the winter. The length, height, and angle of a roof overhang affect the amount of winter or summer sun allowed to enter a window. The relationship of the edge of the overhang to the height of the window also has great importance. Figure 3-16 shows the computation of a desira-

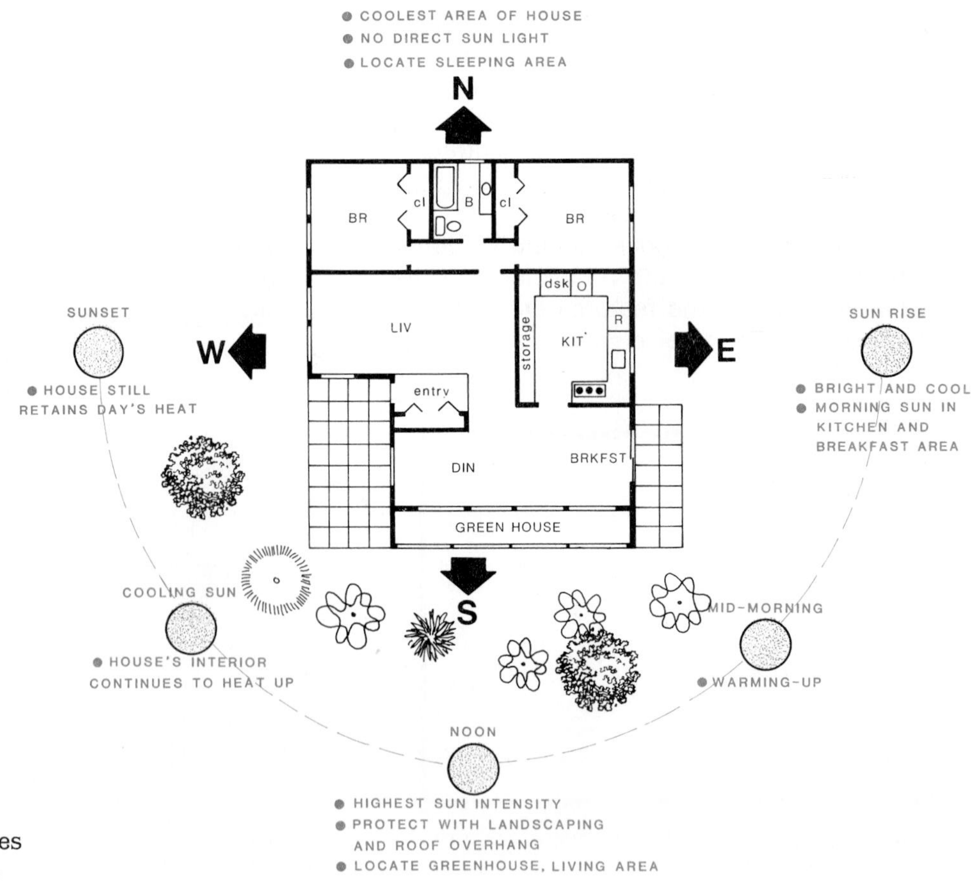

Fig. 3-14 Guidelines for room placement.

ble roof overhang using the latitude and the window height. Figure 3-17 shows various effective overhang baffling systems.

Overhangs and baffles should be designed to allow the maximum amount of sunlight and heat to penetrate the inside in winter. Conversely, the maximum amount of sun heat should be shielded from entering the interior during a summer midday. This should be accomplished without blocking out diffused natural light. Keep in mind that overhangs block out only direct-gain radiation and some atmospheric diffused radiation. Little reflected radiation is shielded in winter, as shown in Fig. 3-18.

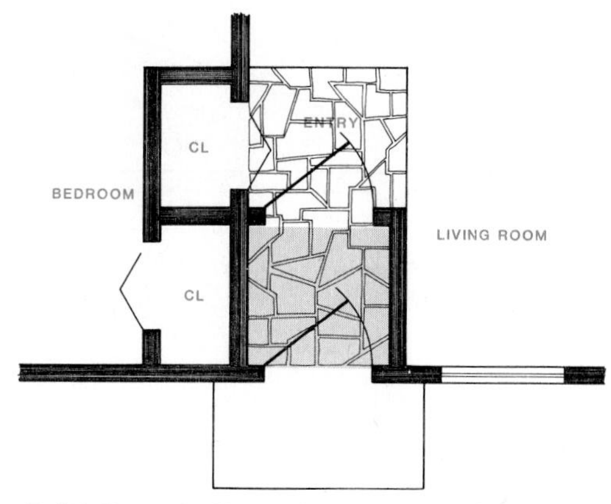

Fig. 3-15 Entry hall used to prevent heat loss.

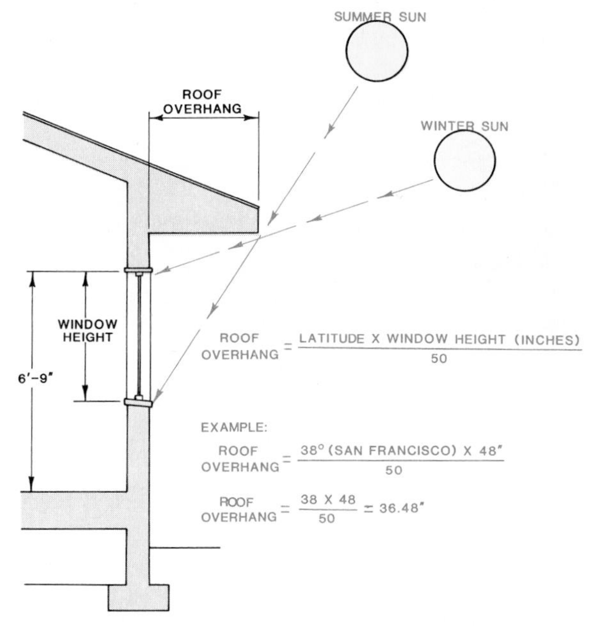

Fig. 3-16 Computation of roof overhang.

Vegetation

Foliage of all types is a great aid in heat, light, wind, humidity, and noise control. Trees, shrubs, and ground-cover foliage, when effectively used, can also baffle undesirable views and enhance attractive scenes.

Deciduous trees (trees which lose their leaves in winter) maximize summer cooling and winter heating by providing shade in the summer and permitting the sun's warmth to penetrate the building in winter (see Fig. 3-19). However, because of their dense structure, coniferous (evergreen) trees and shrubs are most effective in blocking or redirecting north or northwest storm winds, thus further insulating a building during all seasons. Low, dense trees with large circumferences provide better shade than tall, narrow trees. But vegetation should not be used as a

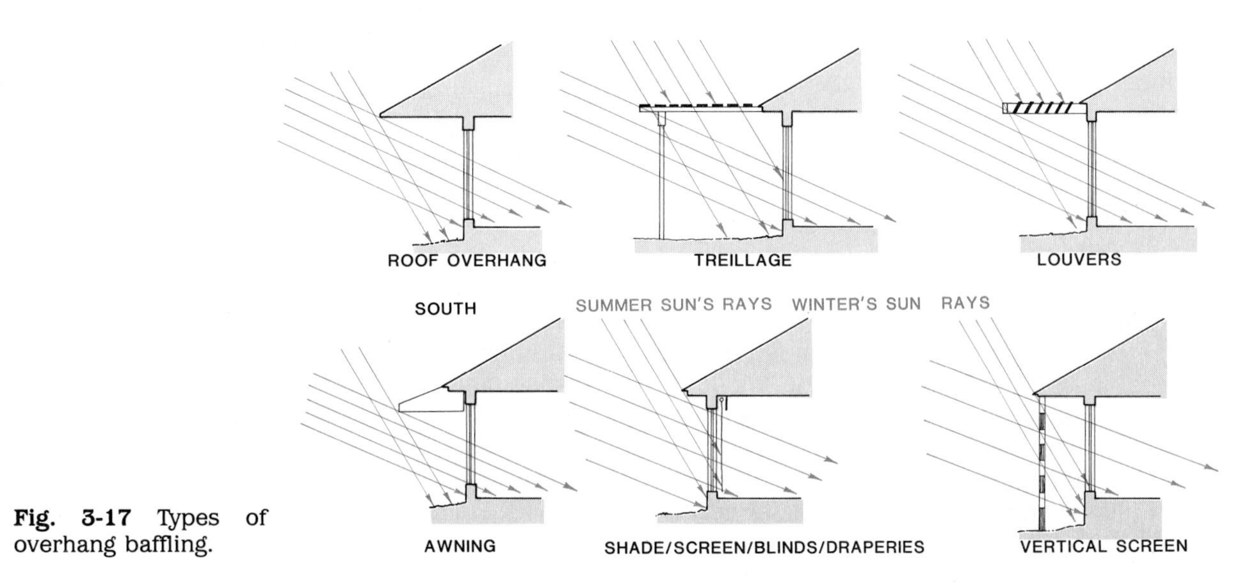

Fig. 3-17 Types of overhang baffling.

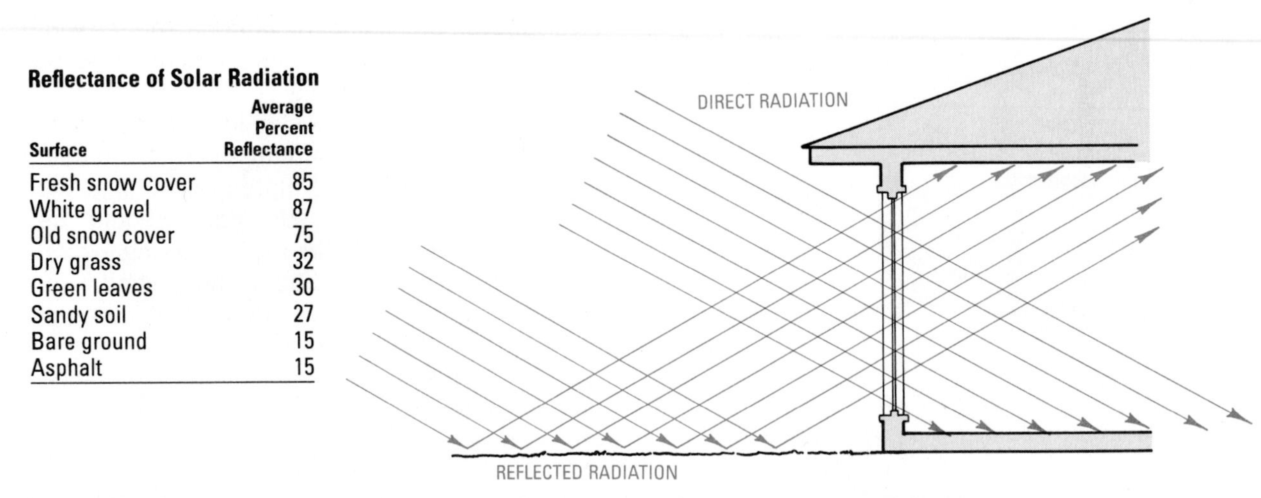

Reflectance of Solar Radiation

Surface	Average Percent Reflectance
Fresh snow cover	85
White gravel	87
Old snow cover	75
Dry grass	32
Green leaves	30
Sandy soil	27
Bare ground	15
Asphalt	15

Fig. 3-18 Windows receive both direct and reflected radiation.

Fig. 3-19 Use of deciduous trees to control sun exposure.

Fig. 3-20 Trees provide effective windbreaks.

substitute for appropriate orientation to control a building's environment.

Noise Control

In orienting a structure, carefully study the direction of any objectionable noises. If possible, orient the living areas to provide the greatest amount of noise baffling. However, noise protection must be evaluated in comparison with other orientation priorities.

Wind Control

One of the functions of effective orientation is wind control. While the sun can provide natural energy to a structure, wind can easily take it away. Existing heat can be lost very rapidly when cold air is forced into buildings through minute crevices, usually around doors and windows. Heat also escapes by *windchill* loss through walls, windows, roofs, and foundations. The windchill effect is the loss of internal stored heat. Building orientation, vegetation, structural barriers, and the natural terrain can be used to control or minimize the negative effects of prevailing winds.

Once the wind patterns are known, buildings should be sited to take full advantage of (or offer full protection from) the cooling effect of prevailing winds. Protection can be provided by locating buildings in sheltered valleys or opposite the windward side of hills. Locating buildings to take full advantage of the baffling effect of existing wooded areas, as shown in Fig. 3-20, is a most effective way of reducing wind velocity. If no wooded areas exist or if vegetation is young or is not available or practical, construction baffles such as detached fences (Fig. 3-21) or attached fences or walls (Fig. 3-22) may be necessary. Another method that can be used in conjunction with other methods is the orientation of buildings to protect as much building surface as possible from perpendicular wind contact. Figure 3-23 shows the use of building angles and narrow-side orientation to avoid wind impact. When the site conditions allow, low roof angles combined with earth berms (Fig. 3-24) can also

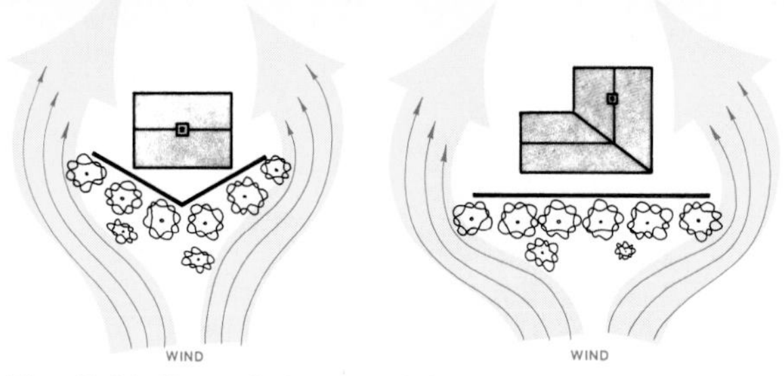

Fig. 3-21 Use of detached fences to deflect wind.

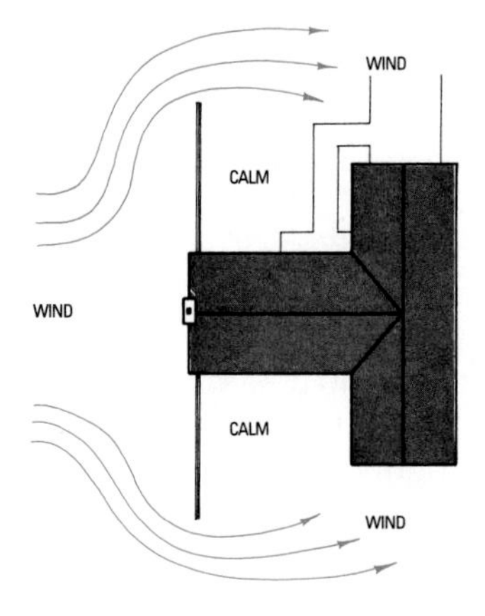

Fig. 3-22 Attached fences or walls used to deflect wind.

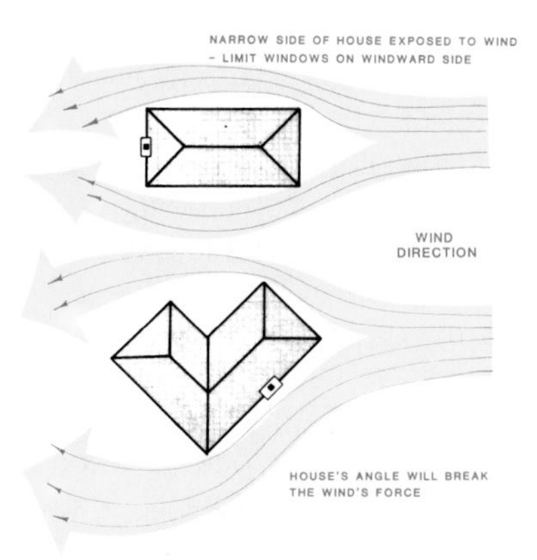

Fig. 3-23 Building position can help reduce wind effect.

Fig. 3-24 Wind deflected by earth berm.

deflect prevailing winds upward and over the building.

Nature aids the site designer by usually providing desirable summer breezes from one direction and winter winds from the opposite side. Nature also provides a consistent source of gentle breezes near large bodies of water, since cool air will always move in to replace rising warm air, as shown in Fig. 3-25. Outdoor and indoor areas should be located to avoid or take advantage of these air movements.

In planning urban buildings or isolated clusters of buildings, care must be taken to avoid the Monroe effect. This effect is caused by high-velocity winds striking the upper floors of high-rise buildings and being forced downward and back against lower buildings, so that they cause turbulence on the surface (Fig. 3-26). This effect is also caused, on a smaller scale, by winds trapped in structural offsets, courtyards, and patios.

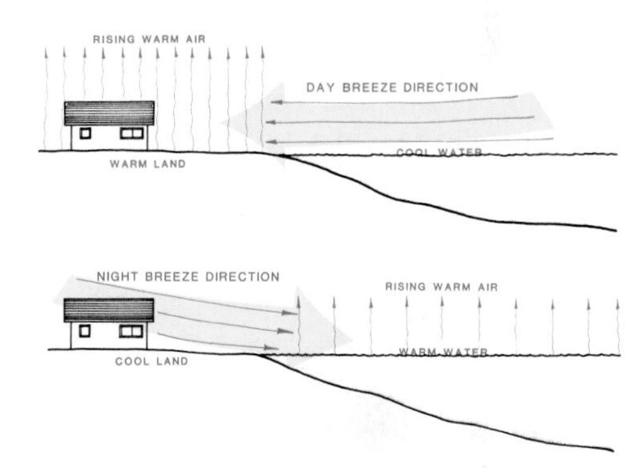

Fig. 3-25 Effect of large bodies of water on air movement.

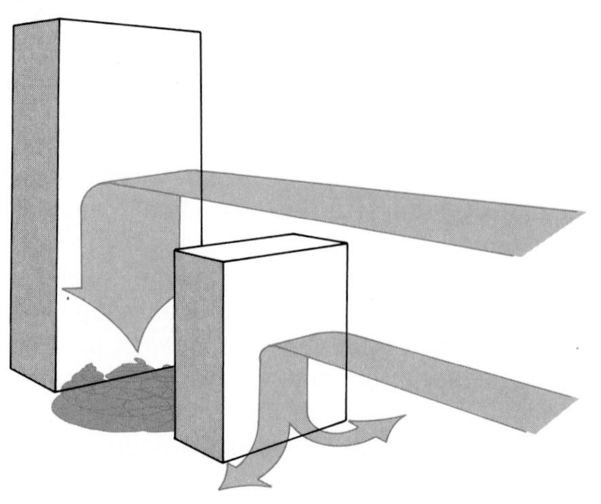

Fig. 3-26 Turbulence caused by trapping of wind currents.

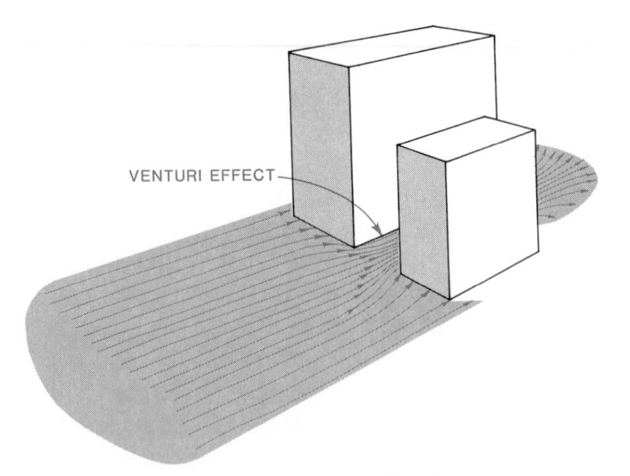

Fig. 3-27 Example of the venturi effect.

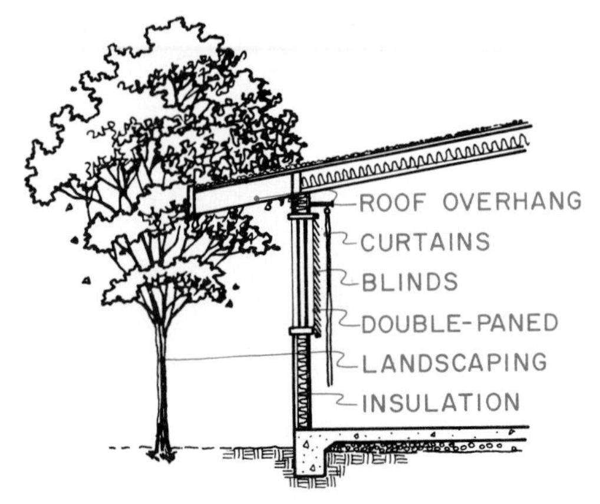

Fig. 3-28 Common deterrents to heat transfer through walls.

In urban situations the *venturi effect* also complicates wind control. The venturi (wind-tunnel) effect is created as large amounts of moving air are forced into narrow openings. The reduced area through which the wind must pass creates a partial vacuum, and the air picks up speed as it is pulled through the opening, as shown in Fig. 3-27. The venturi effect can be partially controlled by avoiding the alignment of streets or buildings with the direction of prevailing winds.

BUILDING MATERIALS

Building materials greatly affect the energy planning of a structure. Some surface materials, such as the adobe used by Native Americans, effectively collect the sun's heat during the day and radiate that heat during the night.

Collecting the heat, however, is quite useless if the heat is allowed to escape through the roof, windows, doors, vents, and crannies. Insulating attics, floors, and walls; plugging crannies with insulation or weather stripping; weather-stripping doors; and double-glazing windows—all these prevent much unnecessary heat loss. Heat is lost twice as fast through glass as through walls. Double glazing creates a buffer zone of dead-air space to reduce heat loss through windows. Open interiors and high ceilings also encourage ventilation and cooler temperatures. Low ceilings and closed floor plans tend to increase temperatures. However, it is possible to circulate warm air in the summer and hold heat in the winter through the use of a variety of devices.

Devices and features that contribute to better energy efficiency include attic exhaust fans and vents; insulation on pipes and ducts, in walls and floors; energy-saving windows; sun control for windows; weather stripping; caulking of joints; flue heat-recovery devices; and energy-saving thermostats. Figure 3-28 summarizes some of the common devices and features used to prevent heat transfer.

PASSIVE SOLAR APPLICATIONS

Not all passively planned buildings use all of the principles of passive solar design. However, it should be the goal of the designer to include as many passive solar features as the design situation allows or requires. For example, Fig. 3-29 shows a passive solar–designed residence using an atrium skylight to admit the sun's heat when desired and to block it when it is not wanted. The solarium can be opened or closed from either the foyer, the master bedroom, the breakfast room, or the terrace. Therefore, the opening or closing of any or all of the solarium doors or windows can adjust the amount of heat admitted from the solarium. The occupant also has control not only of the heat gain from the solarium through use of the skylight panels, but also of the distribution of the solarium heat into other parts of the house. In this plan the atrium skylight substitutes for south-facing windows and helps create a greenhouse effect in either the solarium or the studio.

On warm days, warm air is released outside through the use of exhaust fans as shown in Fig.

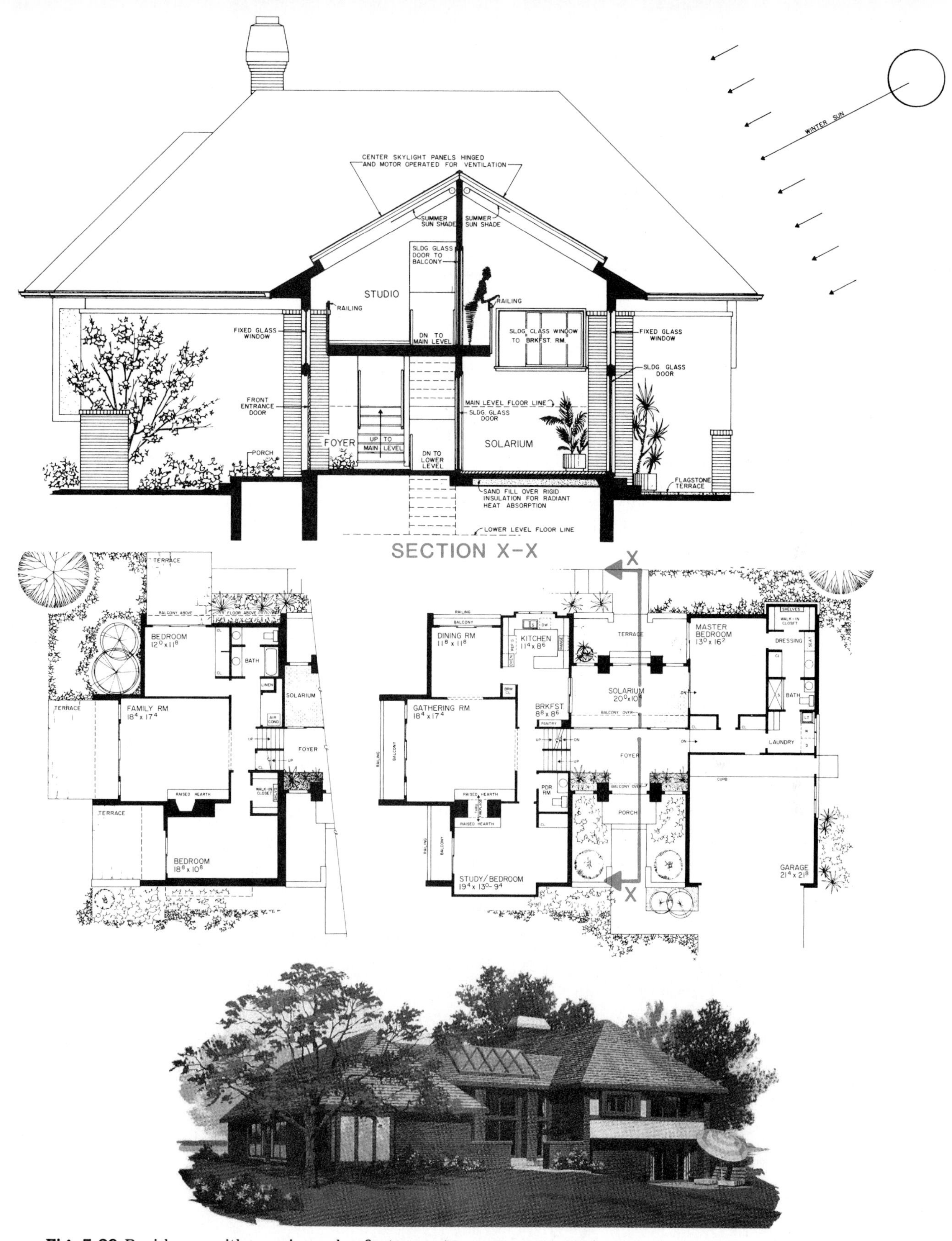

CENTER SKYLIGHT PANELS HINGED
AND MOTOR OPERATED FOR VENTILATION
WINTER SUN
SUMMER SUN SHADE
SUMMER SUN SHADE
SLDG. GLASS DOOR TO BALCONY
STUDIO
RAILING
RAILING
FIXED GLASS WINDOW
SLDG. GLASS WINDOW TO BRKFST. RM.
FIXED GLASS WINDOW
DN TO MAIN LEVEL
SLDG. GLASS DOOR
FRONT ENTRANCE DOOR
MAIN LEVEL FLOOR LINE
SLDG. GLASS DOOR
FOYER
UP TO MAIN LEVEL
SOLARIUM
PORCH
DN TO LOWER LEVEL
FLAGSTONE TERRACE
SAND FILL OVER RIGID INSULATION FOR RADIANT HEAT ABSORPTION
LOWER LEVEL FLOOR LINE
SECTION X-X
TERRACE
BALCONY ABOVE
FLOOR ABOVE
BEDROOM 12⁰ x 11⁸
BATH
SOLARIUM
TERRACE
FAMILY RM. 18⁴ x 17⁴
FOYER
RAISED HEARTH
WALK-IN CLOSET
TERRACE
BEDROOM 18⁸ x 10⁸
RAILING
BALCONY
DINING RM. 11⁸ x 11⁸
KITCHEN 11⁴ x 8⁶
TERRACE
MASTER BEDROOM 13⁰ x 16²
DRESSING
SOLARIUM 20⁰ x 10
BATH
GATHERING RM. 18⁴ x 17⁴
BRKFST. 8⁸ x 8⁶
BALCONY OVER
LAUNDRY
FOYER
RAISED HEARTH
PDR RM.
BALCONY OVER
RAISED HEARTH
PORCH
GARAGE 21⁴ x 21⁸
STUDY/BEDROOM 19⁴ x 13⁰-9⁴
X
X

Fig. 3-29 Residence with passive solar features. *(Home Planners, Inc.)*

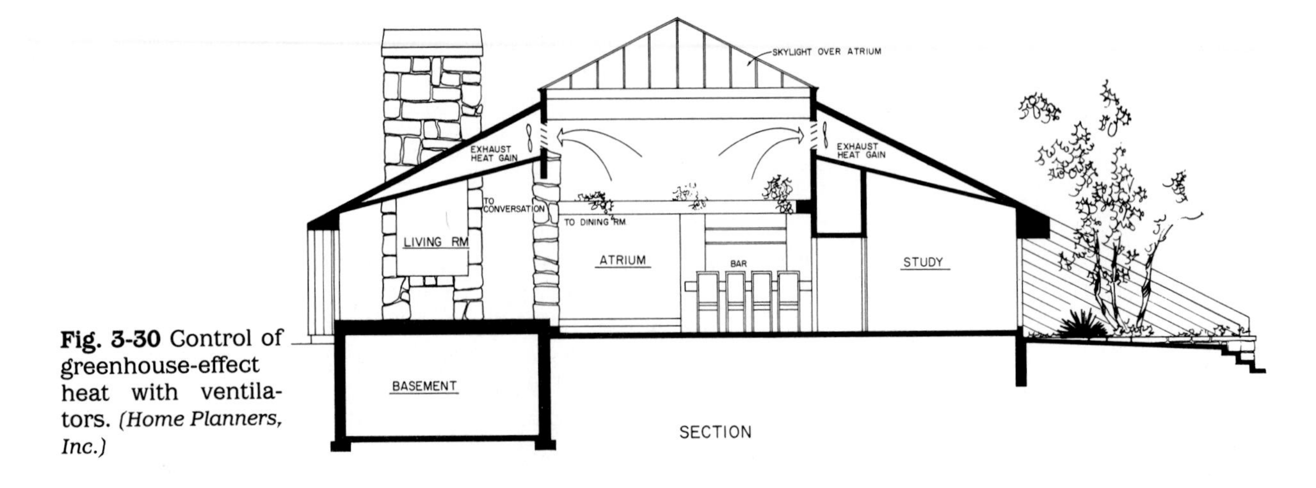

Fig. 3-30 Control of greenhouse-effect heat with ventilators. *(Home Planners, Inc.)*

3-30. This plan is based on the natural rise of warm air. Thus, anything that can be done either to accelerate the rise of warm air when the warm air is not wanted on lower levels or to trap and redirect the hot air when desired is effective in controlling heat. In this plan, vents and windows also provide optional natural air convection and circulation. The use of heavy brick or concrete construction delays the entry of daytime heat into the house and also has good thermal storage capabilities. Figure 3-31 shows the optional site orientation of the solarium for buildings facing different compass directions.

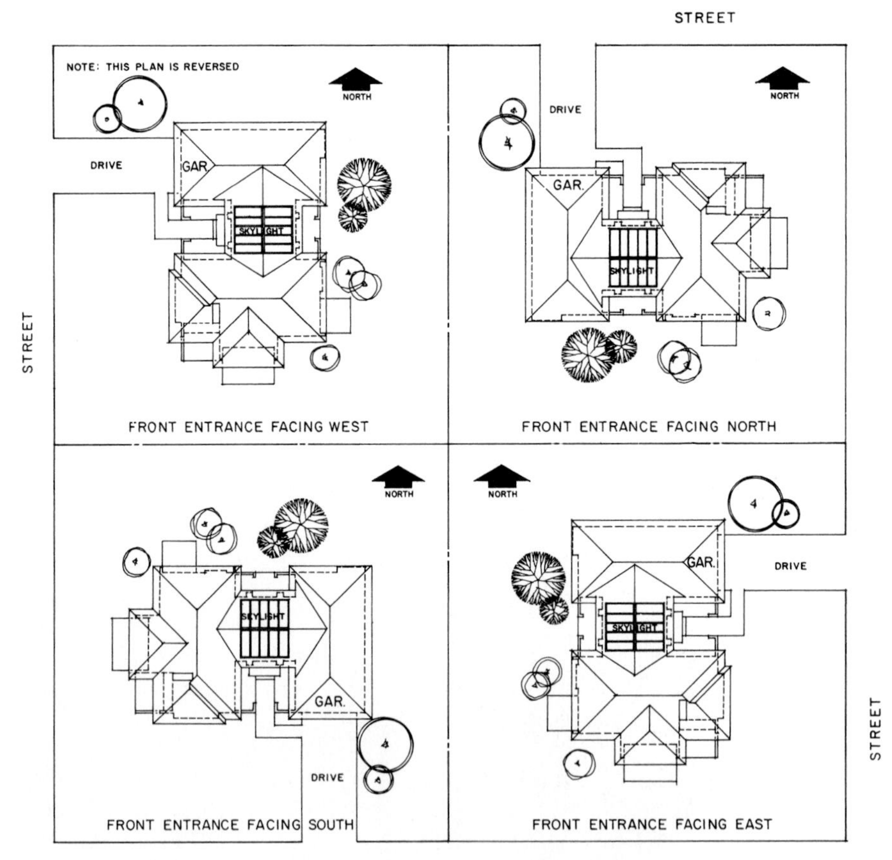

Fig. 3-31 Optional site orientation to maximize solarium or greenhouse effect. *(Home Planners, Inc.)*

EARTH-SHELTERED HOMES

The earth's characteristic moderate temperature is maximized in the design of earth-sheltered buildings (Fig. 3-32). Regardless of how high or low the outer temperature is, the soil just a short distance below the surface remains at a comfortable and constant temperature.

The thought of living partly underground may seem oppressive. However, with effective planning and proper orientation, adequate natural light can be achieved. The underground location avoids the problems of wind resistance, extreme temperature conditions, and winter storm winds. Underground structures are most effective with solar collectors placed above ground to supply energy.

Construction costs for earth-sheltered homes can be less than those for conventional types of construction if experienced builders are employed. The major care in construction must be waterproofing, which can be accomplished with paint, sealants, membrane blankets, and proper drainage. However, the structure of an earth-sheltered home needs to be heavier than that of conventional buildings, to support the heavy soil loads.

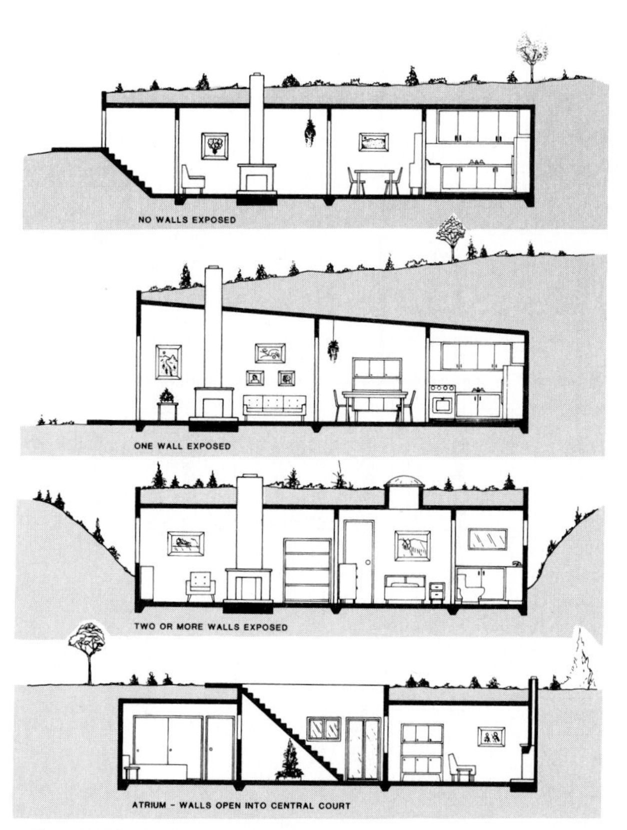

Fig. 3-32 Types of earth-sheltered homes.

The best site location for an earth-sheltered home is on a gentle downward slope. The exposed walls should make maximum use of glass to capture as much light and heat as possible. These glass areas should also be oriented away from prevailing winds but should face the best possible view.

The location of rooms in earth-sheltered homes must be carefully planned. Seldom-used rooms or rooms not requiring windows must be located in ground-locked areas so that windowed areas can be reserved for the rooms that require the most light. However, skylights can be used in rear areas if channeled to the surface and effectively sealed and drained.

Figure 3-33 shows an earth-sheltered home with the entire living area, two bedrooms, and the vestibule exposed to a front window wall. Because this is an open plan, light does extend to the family room, the study, and the kitchen, which are located in the rear. The skylight in this plan provides additional light for this area, as shown in the pictorial drawing.

Although sloping lots are better for earth-sheltered designs, earth berms can be created to provide similar protection. Figure 3-34 shows an earth-sheltered house with berms on two rear sides and sod added to the roof.

Underground structures have the following advantages: low maintenance, low utility costs, safety from fire and wind, security from vandalism, lower insurance rates, long-life construction, smaller heating units, and less use of heating and air-conditioning units.

But there are some disadvantages to underground structures, such as geographical restriction, humidity-control concerns, view control, the risk that builders will not be familiar with the needed construction techniques, difficulty in borrowing money for construction, and the need to control groundwater. In fact, groundwater problems account for most of the maintenance problems in earth-sheltered homes. But these can be solved with careful study of drainage patterns and effective waterproofing.

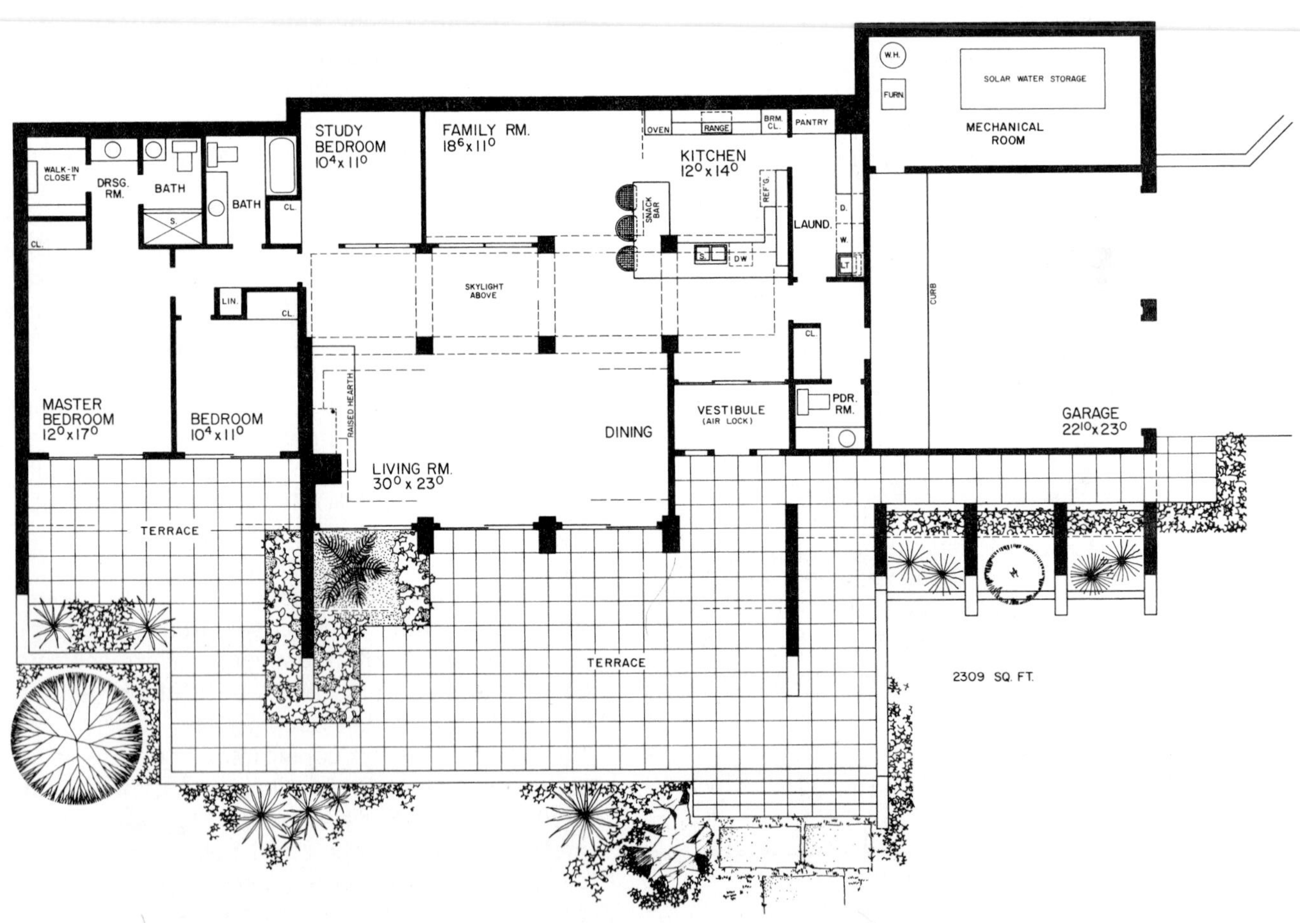

Fig. 3-33 Example of effective earth-sheltered design. *(Home Planners, Inc.)*

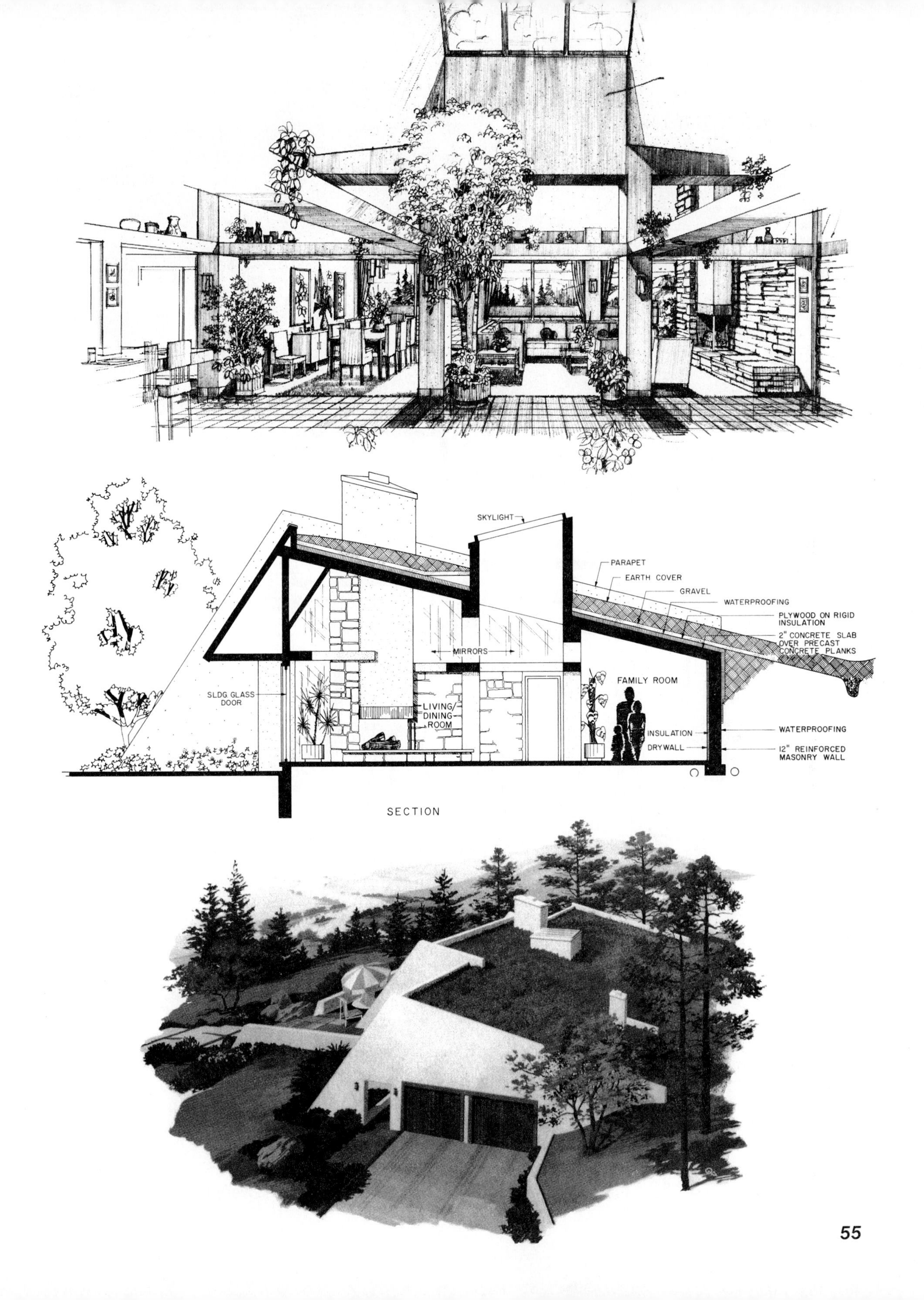
SKYLIGHT
PARAPET
EARTH COVER
GRAVEL
WATERPROOFING
PLYWOOD ON RIGID INSULATION
2" CONCRETE SLAB OVER PRECAST CONCRETE PLANKS
MIRRORS
FAMILY ROOM
SLDG. GLASS DOOR
LIVING/ DINING ROOM
INSULATION
DRYWALL
WATERPROOFING
12" REINFORCED MASONRY WALL
SECTION

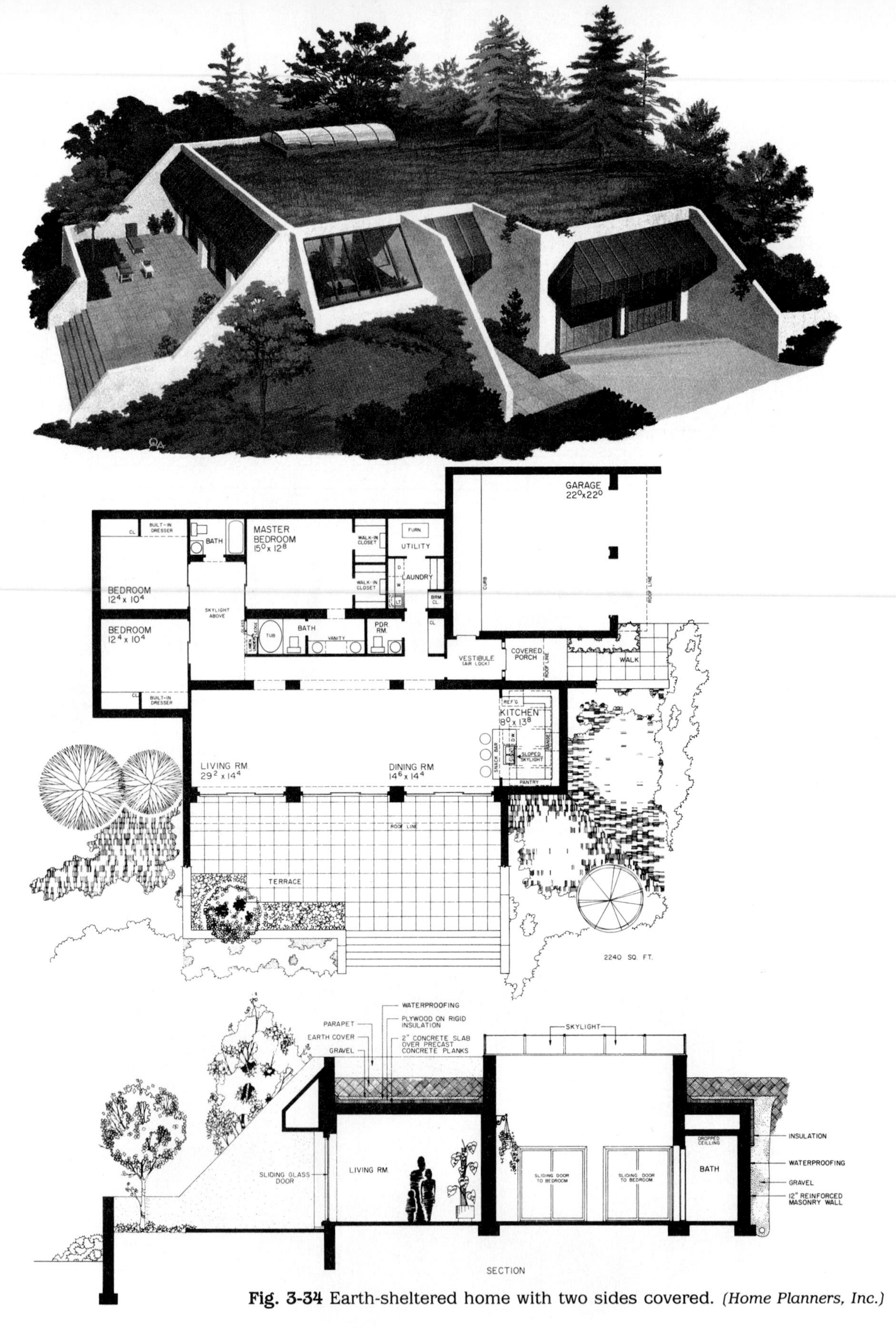

Fig. 3-34 Earth-sheltered home with two sides covered. *(Home Planners, Inc.)*

Exercises

1. **Sketch a 75′ x 110′ property. Sketch the floor plan shown in Fig. 3-29 to the same scale and place it on the property in the most desirable location.**
2. **Sketch the lot layout shown in Fig. 3-35. Sketch a floor plan of a house and garage on this lot in the most desirable position.**
3. **Sketch an active solar system for the house shown in Fig. 3-14. Label and sketch the position of collectors, storage facilities, distribution channels, and control devices.**
4. **List the passive and/or active solar features you would include in a residence of your own design.**
5. **Define the following terms: solar orientation, site, lot, overhang, wind baffle, building line, public area, private area, service area, organic architecture, earth-sheltered home, active solar planning, passive solar planning, solar collector, natural convection, thermal mass, southern exposure, double glazing, deciduous tree, Trombe wall, berm.**

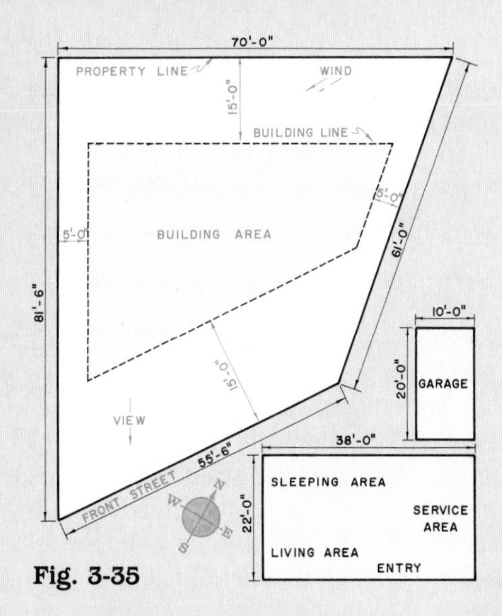

Fig. 3-35

UNIT 4

DENSITY PLANNING

Density planning is not new. People have always tended to cluster into communities for protection and to share common facilities. For centuries people have chosen to live in clusters of expanded families even when unlimited land has been available. However, the differences in size and population distribution between various communities offer different advantages and disadvantages. Today, the development of transportation has enabled people to live in semi-isolation without sacrificing conveniences or safety (Fig. 4-1). Yet most people work with others and prefer to live in close proximity to others while striving for maximum privacy in their homes. This desire to live close to others, to maintain and enjoy common conveniences, and

Fig. 4-1 Semi-isolated site. *(Potlach Corp.)*

yet to have a high degree of privacy makes density planning extremely difficult and complex.

Density, in architectural terms, is the relationship of the number of residential structures and people to a given amount of space. The density of an area is the number of people or families per acre or square mile. For example, a town may have a density of 10 families per acre (hectare) or 50 families per square mile, or mi^2 (square kilometer).

AVERAGE DENSITY

The *average density* of an area is the ratio of inhabitants to a geographic area. In large geographic areas such as counties or cities, the density patterns may vary greatly among different parts of the area. For that reason, the average density of a larger geographic area is not as significant as the average density of smaller areas. The average density of the area shown in Figure 4-2 at A is the same as the average density of the area shown in Fig. 4-2 at B. However, the *density patterns* of the two areas are significantly different. The number of people living in a large geographic area cannot always be controlled, but planning the most effective density patterns can provide the best possible use of the available land. Density patterns must always be planned for the maximum number of people who may eventually use the area.

PLANNING PHILOSOPHIES

There are several basic approaches to density planning. The first and most common is to restrict the size of each building lot through local *zoning* ordinances. This method automatically restricts the number of families allowed to occupy a specific area. This method also spreads the density pattern equally. Zoning laws also

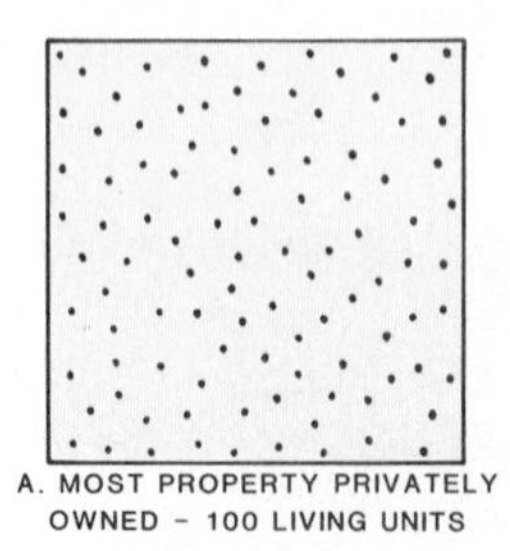

A. MOST PROPERTY PRIVATELY OWNED - 100 LIVING UNITS

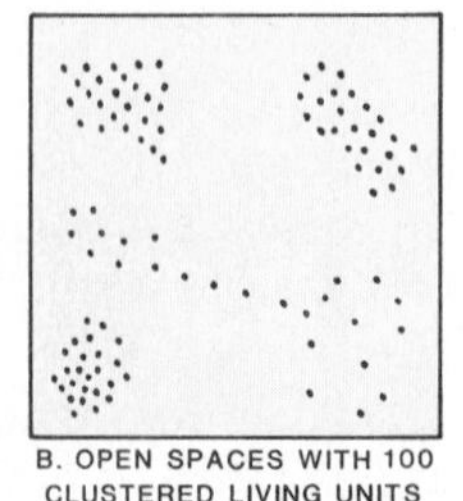

B. OPEN SPACES WITH 100 CLUSTERED LIVING UNITS

Fig. 4-2 The average density is the same for both areas.

indicate which areas may be used for *industrial*, *commercial*, or *residential* construction. Within residential-zoned areas, there may also be restrictions concerning the number of multiple-family dwellings or the size and capacity of apartment buildings. These restrictions are primarily designed to avoid overcrowding of local school, transportation, and recreational facilities.

The second approach involves clustering residents into fewer structures, such as high-rise apartments or row houses. Transportation, shopping centers, schools, theaters, golf courses, swimming pools, tennis courts, and other support facilities are then planned adjacent to housing units.

The third approach is actually a combination of plans. It involves zoning part of the area for single-family residences, specifying other areas for row houses, and reserving some areas for high-rise apartments. The amount of space planned for each type of structure depends on the average density desired.

Redevelopment

Often, undesirable density patterns develop as a result of population growth or shifts. This occurs especially in older communities. When undesirable density patterns develop, the most appropriate corrective action usually is *redevelopment*. Redevelopment can be either short-term or long-range. *Short-term redevelopment* involves the complete razing of buildings and substituting new housing units, parks, and shopping centers in their place. *Long-range redevelopment* is the changing of the density pattern of an area over a long period of time, according to a predetermined, phased building plan and time schedule.

Figure 4-3A and B shows an example of an urban redevelopment plan. In this plan a terraced promenade with landscaped plazas and arcades connects the residential clusters with the main street. Notice also how the dwelling units are jogged to provide each unit with a view. The view is San Francisco Bay. Figures 40-12A and B show opposite views of this area.

Neighborhood Planning

The smallest residential unit involving density planning is the *neighborhood*. The next larger unit, the *community*, is a combination of neighborhoods. *Regions* are a combination of interrelated communities, neighborhoods, and cities.

Fig. 4-3A Urban development plan. See Figs. 40-12A and B for other views.

Fig. 4-3B Terraced promenade of plan shown in Fig. 4-3A. *(Backen, Arrigoni and Ross Inc.)*

A *neighborhood* is a series of homes, whether arranged vertically, as in a high-rise apartment; connected, as in row houses; located on several acres, as in some rural areas; or on small lots, as in most cities. In planning neighborhoods, the designer must consider the following characteristics of people who will live there: their typical age, marital status, number of children, lifestyle, and economic level. The following features of the community must be matched to the needs of these residents: availability of parks and playgrounds, consistency of home design, nearness of shopping areas, and preservation of the natural landscape (Fig. 4-4). The architect must therefore be certain that the needs of the homeowners are consistent with the characteristics of the neighborhood, but keep in mind that a residence isolated from a neighborhood must be planned to include its

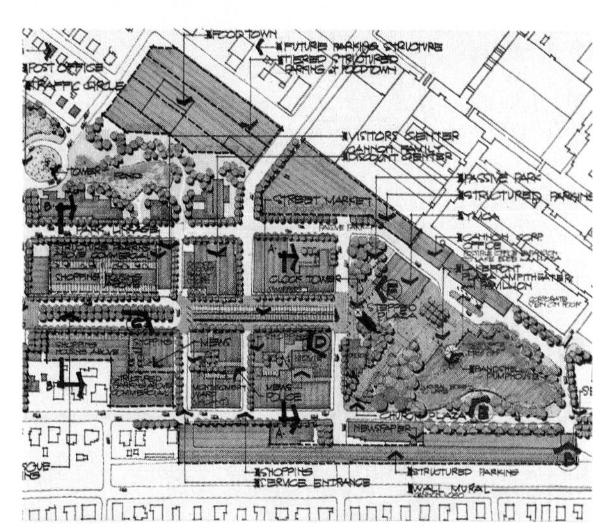

Fig. 4-4 Well-planned community features. *(Cannon Mills Co.)*

own facilities for transportation, recreation, security, and maintenance.

Relationship of Neighborhood to Community

Not only must a house be related to the site and the site related to the neighborhood, but the neighborhood must be related to the community. Features of the community that must be considered include the availability and quality of schools, theaters, playgrounds, houses of worship, parking facilities, shopping centers, highway access, automobile service, and police and fire protection; and traffic patterns, too, must be considered.

Cluster Planning

Figure 4-5 shows alternatives in planning a 1-acre neighborhood. Figure 4-6 shows how the same area can be planned for 4, 6, 8, and 16 houses per acre. Plan A is designed for 4 houses per acre, plan B for 6 houses per acre, plan C for 8 houses per acre, and plan D for 16 houses per acre. The houses used in each of these plans must be designed carefully to relate to adjacent property.

Designing areas with heavy population concentrations is more difficult than designing areas that are thinly populated. However, by using the best combination of plans, by effective orientation, and by designing traffic patterns, as shown in Fig. 4-7, the designer can achieve maximum efficiency in city planning.

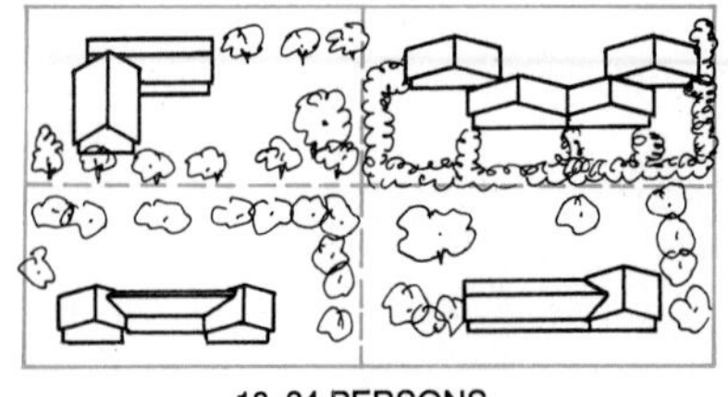

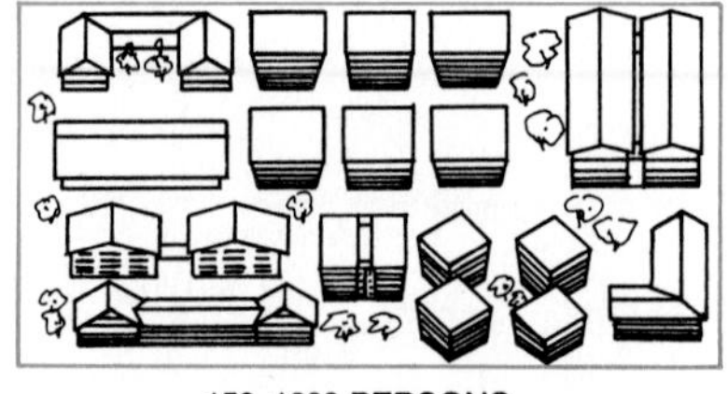

Fig. 4-5 Alternatives in neighborhood planning.

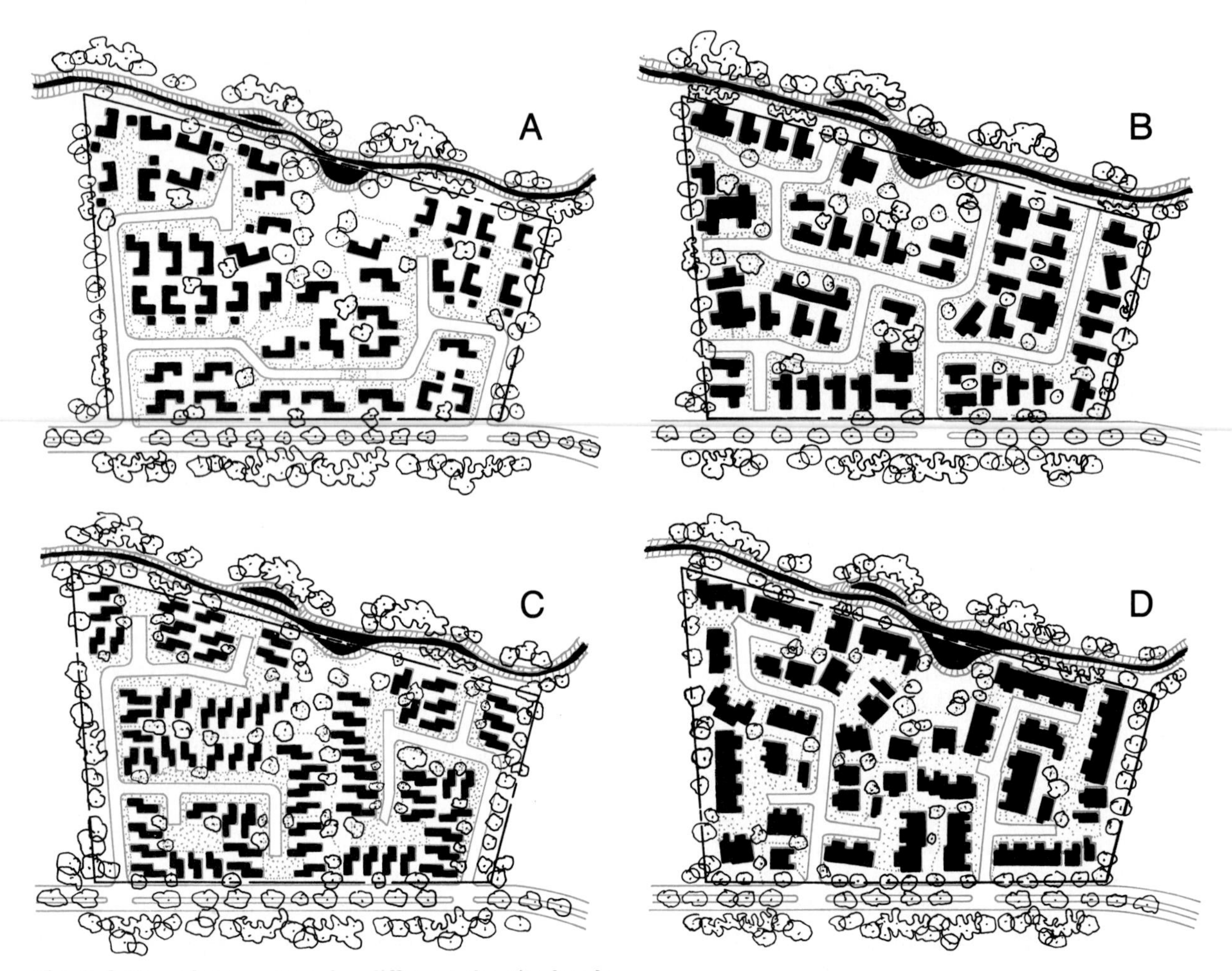

Fig. 4-6 Use of same area for different density levels.

COMMUNITY PLANNING

Most cities and towns were not planned; they developed without a plan. Thus, most architectural activity in community planning relates to redevelopment and/or efforts to control and direct future growth according to a long-range master plan. City plans are defined according to the geometric form produced on a map of the area. Each form has its distinct advantages and disadvantages, depending on the terrain, density, and future growth patterns of the area, as forecast according to the area's cultural and economic needs (Fig. 4-8).

Architectural plans in the future must make full use of technological advances in construction and transportation. The plans also must take into account the expected rate of growth of cities and towns, and the effect of such growth.

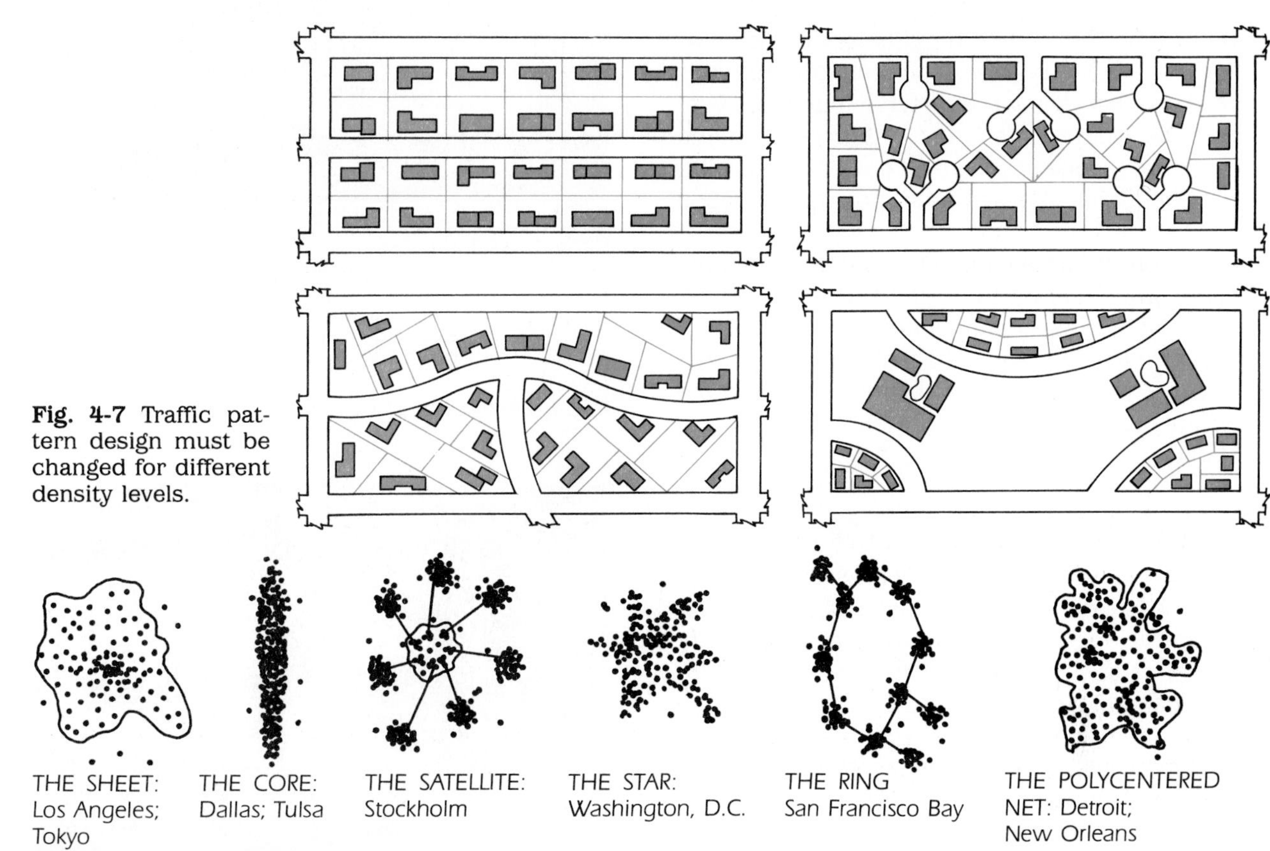

Fig. 4-7 Traffic pattern design must be changed for different density levels.

Fig. 4-8 Urban landform types.

Exercises

1. **What are the local zoning ordinances in your community? How can they be improved?**
2. **Redesign a square-mile or square-kilometer area around your home to provide better density balance.**

3. **Describe the density pattern you prefer for an area in which you wish to locate a house of your design.**
4. **Define these terms: *density, average density, density patterns, zoning.***

UNIT 5

ECOLOGICAL AND ENVIRONMENTAL PLANNING

There are now nearly three times as many people on the earth as there were less than a century ago. This population increase, combined with the shift from an *agrarian* (farming) society to an industrial one over the last century, has led to the creation of environmental problems previously unknown. The ever-increasing material needs of our technological economy have created enormous pollution problems that must be solved if humanity is to survive. Designers must plan in a way that eliminates or reduces pollutants. The preservation of natural ecological balances must be a prime requirement in the creation of every design. Architectural creations must be designed to preserve our supply of clean air, pure water, and fertile land. At the same time, noise levels must be controlled without sacrificing the aesthetic qualities of good design.

In short, modern buildings must be both functional and aesthetically pleasing, and something more. The contemporary architect or designer must be sure that structures do not interfere with or create problems relating to the environment. That is what is meant by *ecological planning*.

Fig. 5-1A Results of excessive foliage removal.

Fig. 5-1B Naturally preserved site. *(Julius Shulman)*

LAND POLLUTION

Land is polluted by the discharge of solid and liquid wastes on land surfaces or by the removal of topsoil, vegetation, or trees from large tracts of land. Notice the difference between the site development shown in Fig. 5-1A and the one shown in Fig. 5-1B. Figure 5-2 shows a site development plan that complements the topography. Land pollutants originate from industrial, agricultural, and residential waste and garbage. When pollutants exist in excessive quantities, they create health hazards, contribute to soil erosion, cause unpleasant odors, and overwork sewage-treatment plants.

Architecturally related ways of reducing land pollution include recycling waste material, *compacting* (condensing) waste material to reduce volume, providing for sanitary landfills, and providing for minimum removal of vegetation, especially trees.

Landfill practice involves placing solid waste material in low-lying areas, compacting it into layers (about 10′ thick), and covering it with clean soil. This is called *sanitary landfill*. Thorough landfill projects also include the planting of ground cover—trees and shrubs—on the filled area. Such practice restores the land to its original condition (or better), both ecologically and aesthetically.

AIR POLLUTION

Air is polluted by the discharge of industrial wastes. Some pollutants are nontoxic and are not suspended in the air for long periods of time. They present an annoyance but usually do not endanger life. Other materials, such as sulfur oxides and organic gases composed of hydrocarbons, oxides of nitrogen, and carbon monoxide, are extremely dangerous to both animal and plant life. These pollutants are emitted primarily as by-products of manufacturing plants, heating devices, and the exhaust systems of vehicles.

To avoid ecological problems, designers must reduce pollution by planning heavy traffic patterns away from heavily populated areas. Designers must also plan or specify electronic air filters or other solid-waste removal systems to eliminate particles before they become airborne. Designs must include features that promote energy conservation. Conserving energy not only diminishes pollution by decreasing fuel consumption but also helps reduce the operating costs of buildings.

WATER POLLUTION

Water is polluted by sewage, industrial chemicals, and agricultural wastes dumped into bodies of water. These wastes include pathogens

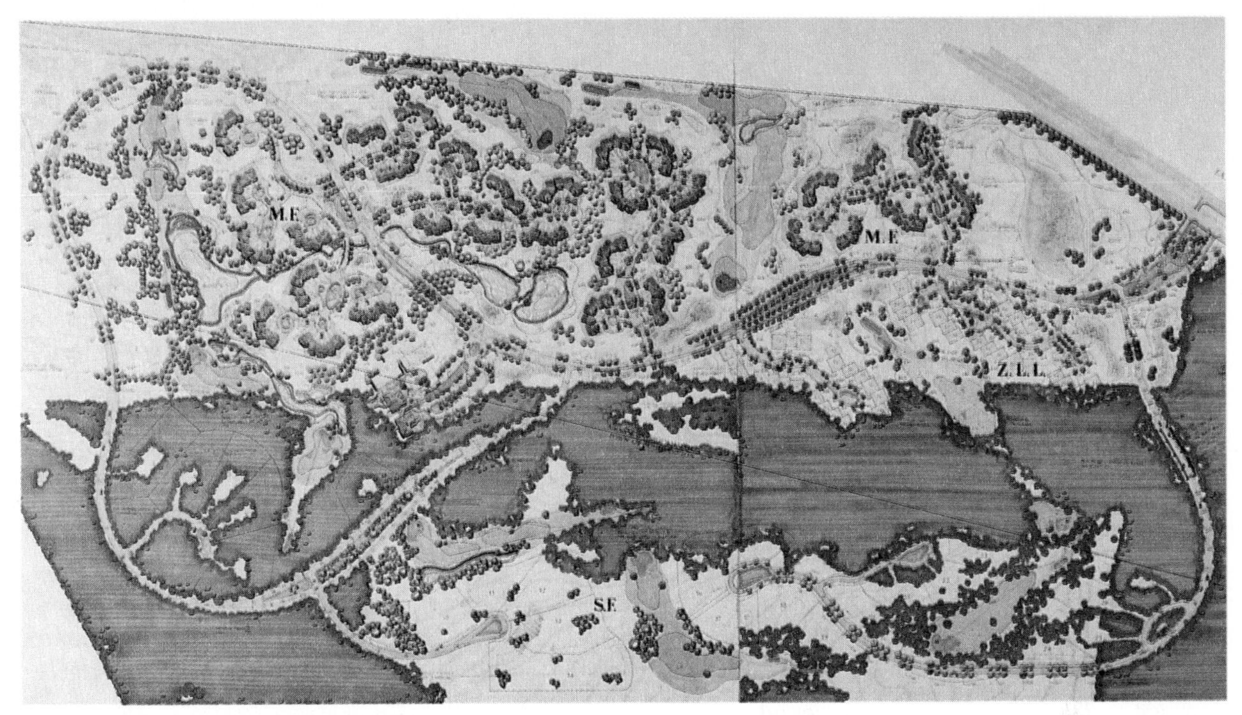

Fig. 5-2 Site plan that protects the environment.

(such as disease-causing bacteria), unstable organic solids, mineral compounds, plant nutrients, and agricultural insecticides. Water pollution results in the destruction of marine life. It presents very serious potential health hazards to animal and human life.

Effective architectural planning can help reduce water pollution through the design of sewage-treatment systems in conjunction with each new construction. Proper density planning, to levels low enough that water sources are not overused for fresh water or waste disposal, also helps reduce water pollution. Plans must also provide for the removal of industrial wastes without sending them in raw form into waterways. Plans must also eliminate excessive runoff of topsoil into rivers and streams.

VISUAL POLLUTION

Many air, water, and land pollutants, such as unsanitary garbage dumps and smog-producing agents, are not only unhealthy but also visually undesirable. Other sources of visual pollution, such as junkyards, exposed utility lines, public litter, barren land, and large billboards, may not create health or safety hazards. They are aesthetically objectionable, however, and must be avoided in the architectural design process. The best safeguards against visual pollution consist of following closely the basic principles of design, not only for structures but for the entire landscape, and enacting and enforcing stringent zoning regulations.

SOUND POLLUTION

Sound levels are measured in decibels, as shown in Fig. 5-3. Exposure to excessive decibel levels, over 80, or to even moderate levels, over 60, for long periods creates stress and can result in neurosis, irritability, and hearing loss in many people. Excessive noise can also create hazardous environments by eliminating people's ability to identify and discriminate between sounds, especially those that warn of danger.

The architect has a number of ways to reduce noise to acceptable levels. Among them are effective floor planning, orientation, landscaping to provide ample noise-buffer space, using acoustical wall panels, making maximum use of buffer foliage, and taking care to control patterns of vehicular traffic. Insulation also helps reduce excessive noise from the outside. Rooms

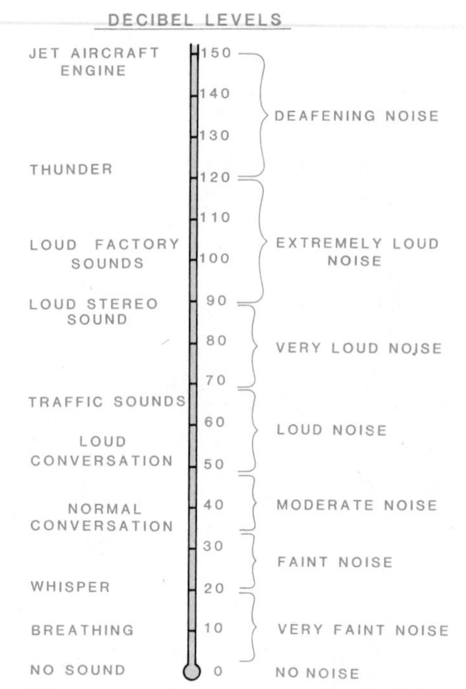

Fig. 5-3 Decibel levels of common sounds.

requiring quiet, such as bedrooms, should be located on the side away from the major source of noise. Fabrics such as carpets, drapes, and upholstery also reduce noise by absorbing sound within a room. Figure 5-4A through E summarizes the causes, effects, and cures for or prevention of ecological problems. Most of the preventative measures can be accomplished through effective architectural design.

For example, effective landscape planning which preserves existing vegetation alone can help maintain acceptable oxygen-nitrogen cycles, encourage wildlife habitats, preserve and enhance the natural beauty of the site, provide visual screening, curtail airborne wastes, provide summer solar screening, encourage ground water retention, and reduce noise.

Many pollution problems can be solved or prevented by architectural design professionals who become involved with local planning boards that initiate stringent environmentally related ordinances.

Causes	Effects	Cures or prevention
Excessive sewage	Destruction of marine life	Eliminate dumping
Excessive industrial wastes	Health hazards to humans and animals	Curtail use of insecticides
Agricultural insecticides	Destruction to marine plants	More sewer treatment facilities
Watershed removal	Poor quality of water	Less population congestion
Human wastes	Crop destruction	Water quality control
Construction too near water		
High population density		
Diversion of freshwater		

Fig. 5-4A Water-pollution problems.

Causes	Effects	Cures or prevention
Manufacturing industries exhausts	Excessive energy consumption	Develop nonpolluting energy sources
Transportation exhausts	Crowded roads	Less vehicular traffic
Heating exhausts	Serious health hazards	Use of low fuel burning vehicles
Removal of oxygen-producing plants	Offensive odors	Lower heat levels
Overpopulation of small areas	Reduced visual ceilings	Additional landscape plantings
		Curtailment of airborne wastes
		Mass transit improvement

Fig. 5-4B Air-pollution problems.

Causes	Effects	Cures or prevention
Industrial wastes	Poisons introduced into life cycles	Recycle materials
Sewage dumps	Soil erosion	Proper land fills
Agricultural insecticides	Losing topsoil	Improve landscape planning
Residential garbage wastes	Overtaxed sewer treatment plants	Manufacture biodegradable items
Landscape removal	Additional costs to taxpayers	Eliminate manufacture of disposable items
Overgrazing		Curtail use of insecticides
Overdevelopment		Hold land reserves for open spaces
Nuclear wastes		Preserve oxygen producing trees and vegetation

Fig. 5-4C Land-pollution problems.

Causes	Effects	Cures or prevention
Factories	Stress	Regulate loud noises
Vehicles	Sleep impairment	Muffle sounds
Aircraft	Relaxation impairment	Building orientation
Power tools	Psychological disorders	Vegetation barriers
Radios, TVs		Fences
Barking dogs		Zoning restrictions

Fig. 5-4D Noise-pollution problems.

Causes	Effects	Cures or prevention
Poor municipal planning	Barren land	Long-range city planning
Poor structure design	Offensive structures	Organic design of structures
Exposed utility wires	Elimination of desirable views	Underground utility wires
Landscape removal	Visual confusion	Landscape preservation
Site destruction	Lower property value	Limitation of public advertising spaces
Waste landfills	Public safety hazard	Trash receptacles
Excessive advertising signs		
Littering		

Fig. 5-4E Visual-pollution problems.

The shadow study shown in Fig. 5-6 and the reflecting insulating units used in the Century Building in San Antonio (Fig. 5-5) are examples of preplanning to prevent future environmental problems.

Fig. 5-5 Greenhouse effect in San Antonio's Century Building provides illumination and heat in cool weather yet reflects 90 percent of the sun's heat during hot seasons.

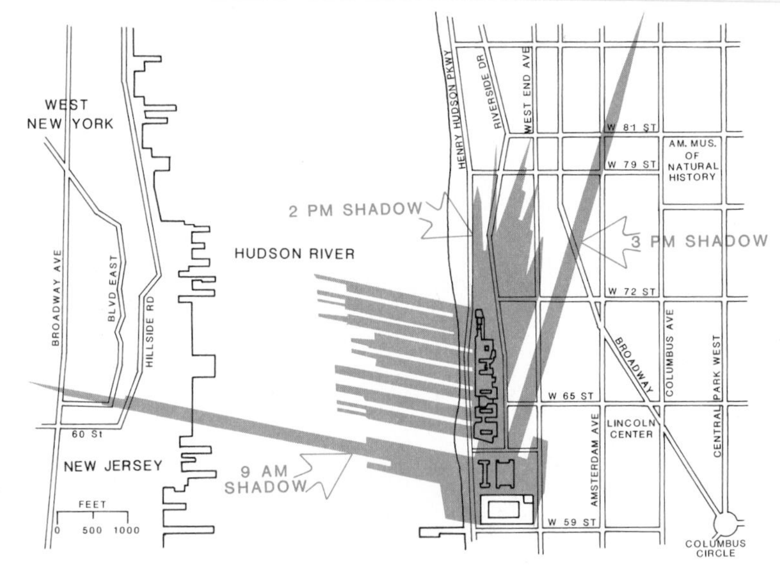

Fig. 5-6 Trump City shadow study for Manhattan's upper west side.

Exercises

1. **Find buildings in your area that are designed to control pollution.**
2. **Find buildings in your area that emit pollutants into the air or water or on the land. List ways of correcting these conditions.**
3. **List diseases caused by air, water, and land pollution.**

4. **List the ecological factors to be considered in planning a residence of your own design.**
5. **Define these terms: *air pollution, solar energy, electronic filter, water pollution, sanitary landfill, erosion, industrial waste.***

PART TWO

Area Planning

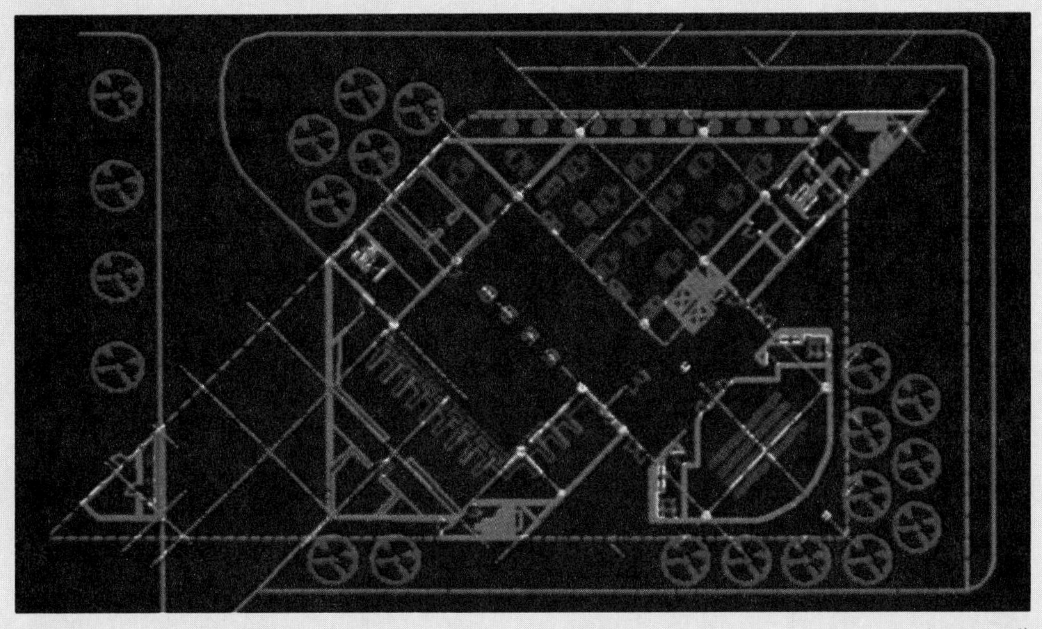

(Versacad)

In creating any architectural design, the designer must progress logically step by step through the design process as covered in Unit 2. One of the key steps in this process is to divide the functions of buildings into specific areas. For example, a school would be divided into such areas as those for administration, classrooms, service, physical activity, and so forth. A hospital would likewise be divided into such areas as those for reception, emergency services, food service, maintenance, patient rooms, laboratory functions, and so forth. In the same manner, a house can be divided into three major functional areas, for planning purposes: the living area, the service area, and the sleeping area. Areas are further subdivided into rooms so that all the rooms in an area will relate to its basic function. A designer must always be sure to become familiar with the functions and relationships of the areas of a building regardless of the type of structure being designed. Part Two presents the principles and practices involved in planning the three basic areas of a residential structure.

SECTION 3

Living Area

Your first impression of a home is probably the image you retain of the living area. In fact, this is the only area of the home that most strangers observe. The living area is just what the name states, the area where most of the living occurs. It is here the family entertains, relaxes, dines, listens to music, watches television, enjoys hobbies, and participates in other recreational activities.

The total living area is divided into smaller areas (rooms) which are designed to perform specific living functions. The subdivisions of most living areas may include the living room, dining room, recreation or game room, family room, patio, entrance foyer, den or study, and guest lavatories. Other specialized rooms, such as the library, music room, or sewing room, are often included as part of the living area of large houses that have the space to devote to such specialized functions. In smaller homes, many of the standard rooms combine two or more functions. For example, the living room and dining room are often combined. In extremely small homes, the living room constitutes the entire living area and provides all the facilities normally assigned to other rooms in the living area. Although the subdivisions of the living area are called rooms, they are not always separated by a partition or a wall. Nevertheless, they perform the function of a room, whether there is a complete separation, a partial separation, or no separation.

When room are completely separated by partitions and doors, the plan is known as a closed plan. When partitions do not divide the rooms of an area, the arrangement is called an open plan.

In most two-story dwellings, the living area is normally located on the first floor. However, in split-level homes or one-story homes with functional basements, part of the living area may be located on the lower level.

UNIT 6

LIVING ROOMS

The *living room* is the center of the living area in most homes. In small homes the living room may represent the entire living area. Hence, the function, location, decor, size, and shape of the living room are extremely important and affect the design, functioning, and appearance of the other living-area rooms.

FUNCTION

The living room is designed to perform many functions. The exact function depends on the living habits of the occupants. In the home, the living room is often the entertainment center, the recreation center, the library, the music room, the TV center, the reception room, the social room, the study, and occasionally the dining center. If the living room is to perform all or some of these functions, then it should be designed accordingly. The shape, size, location, decor, and facilities of the room should be planned to provide for each activity. Figure 6-1 shows the design of a media-centered room.

Any of the facilities normally associated with the living room can be eliminated if a separate special-purpose room exists for that activity. For example, if television viewing is restricted to a recreation room, then planning for TV in the living room can be eliminated. If a den or study is provided for reading and for storing books, facilities for the use of large numbers of books in the

Fig. 6-1 A media-oriented room. *(Feincraft Inc.)*

living room can be eliminated. Regardless of the exact activities anticipated, the living room should always be planned as a functional, integral part of the home. The living room is planned for the comfort and convenience of the family and guests.

LOCATION

The living room should be centrally located. It should be adjacent to the outside entrance, but the entrance should not lead directly into the living room. In smaller residences, the entrance may open into the living room, but whenever possible this arrangement is to be avoided. The living room should not be a traffic access to the sleeping and service area of the house. Since the living room and dining room function together, the living room should be adjacent to the dining room. Figure 6-2 shows the central location of a living room and its proximity to other rooms of the living area.

Open Plan

In an open-plan living area, the living room, dining room, and entrance may be part of one open area. The living room may be separated from

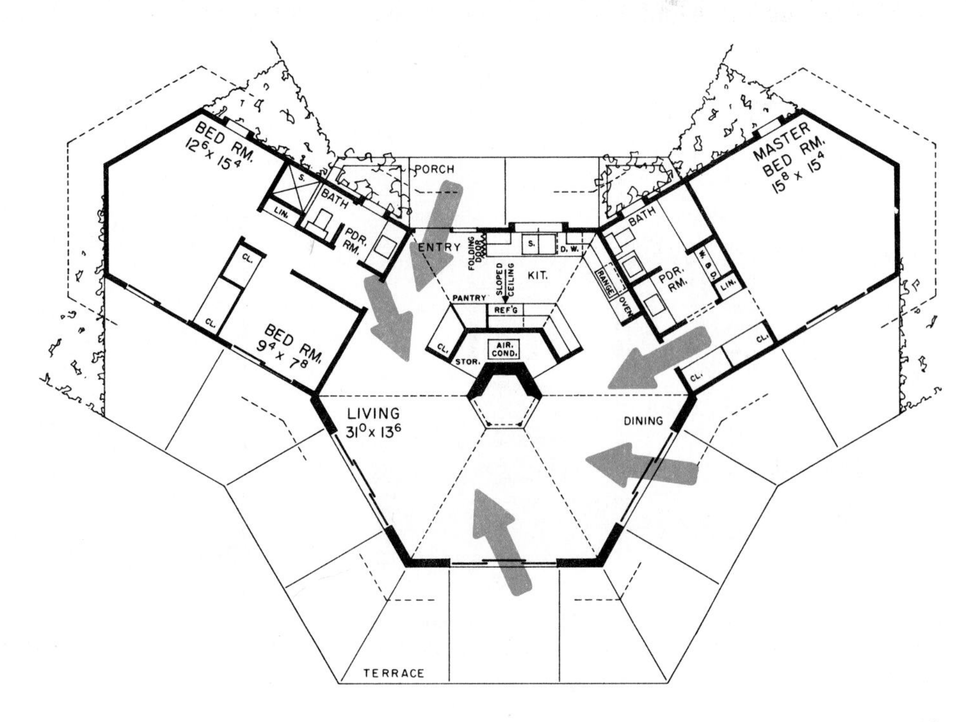

Fig. 6-2 A centrally located living room.

other rooms by means of a divider without doors, such as a storage wall. In Fig. 6-3A and B, the living room is separated from the dining room by a fireplace. In Fig. 6-3B, the living room is separated from the rest of the house by an atrium. Often a separation is accomplished by placing the living room on a different level. Separation may also be achieved by the use of area rugs or furniture placement. Of course, these features do not separate the rooms visually, but they do effect a functional separation.

When an open plan is desired and, yet, the designer wants to provide some means of closing off the room completely, sliding doors or folding doors can be used.

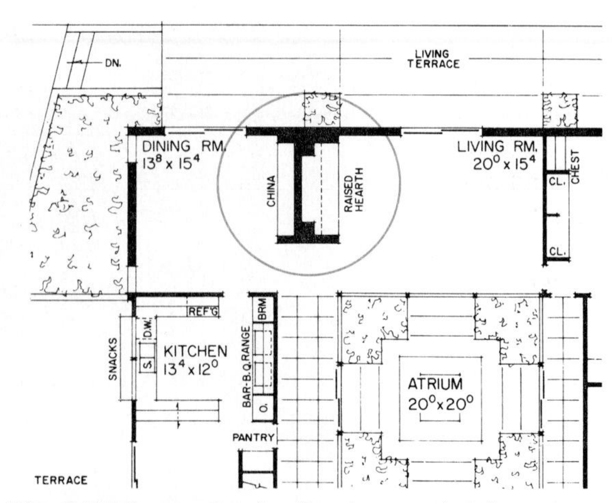

Fig. 6-3B Separation by fireplace and atrium. *(Home Planners, Inc.)*

Closed Plan

In a closed plan, the living room is completely closed from the other rooms by means of walls. Access is through doors, arches, or relatively small openings in partitions (Fig. 6-4). Closed plans are found most frequently in traditional or period-type homes.

DECOR

There is no one way to design and decorate a room. The decor depends primarily on the tastes, habits, and personalities of the people who will use the room. If the residents' tastes are modern, the wall, ceiling, and floor treatments should be consistent with the clean, functional lines of contemporary architecture and contemporary furniture. If the residents prefer colonial or period-style architecture, then this theme should be reflected in the decor of the room.

The living room should appear inviting, comfortable, and spacious. This appearance can be accomplished by an effective use of color and lighting techniques and by the tasteful selection of wall, ceiling, and floor-covering materials. The selection and placement of functional, well-designed furniture also helps the appearance. Decorating a room is much like selecting clothing. The color, style, and materials should be selected to minimize faults and to emphasize good points. The use of mirrors and floor-to-ceiling drapes along with proper furniture placement can create a spacious effect in a relatively small room.

Fig. 6-3A A fireplace can be used to separate the living room from the dining room without isolation. *(Home Planners, Inc.)*

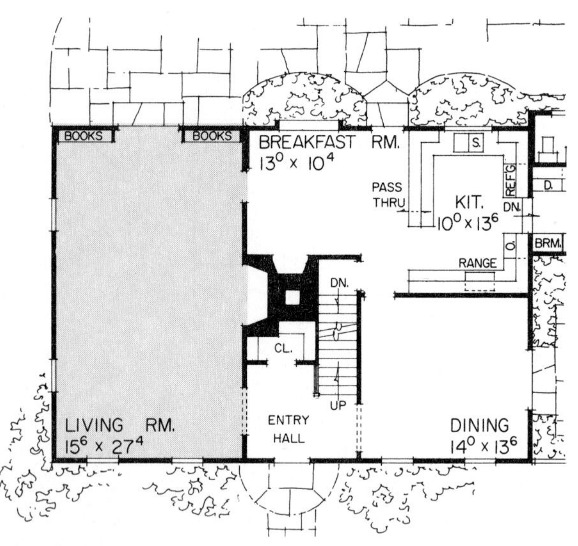

Fig. 6-4 Closed-plan living room. *(Home Planners, Inc.)*

Walls

The design and placement of doors, windows, and chimneys along the walls of the living room can change the entire appearance of the room. The kind of wall-covering material used can also affect the appearance. Wall coverings are selected from a variety of materials, including plaster, gypsum wallboard, wood paneling, brick, stone, and glass. Sometimes, furniture is built into the walls. Fireplaces, windows, doors, or openings to other areas should be designed as integral parts of the room. They should not appear as afterthoughts.

Orientation

The living room should be oriented to take full advantage of the position of the sun and the most attractive view. Since the living room is used primarily in the afternoon and evening, it should be located to take advantage of the afternoon sun.

Windows

When a window is placed in a living-room wall, it should become an integral part of that wall. The view from the window or windows becomes part of the living-room decor, especially when landscape features are near and readily observable. When planning windows, consider also the various seasonal changes in landscape features.

Although the primary function of a window is twofold, to admit light and to provide a pleasant view of the landscape, there are many conditions under which only the admission of light is desirable. If the view from the window is unpleasant or is restricted by other buildings, translucent glass, which primarily admits light, can be incorporated into the plan.

Fireplace

The primary function of a fireplace is to provide heat, but it is also a permanent decorative feature. The fireplace and accompanying masonry should maintain a clean, simple line consistent with the decor of the room and of the wall where they are placed. The fireplace and chimney masonry can cover an entire wall. Consequently, the fireplace can become the focal point of the room. The fireplaces shown in Fig. 6-3A and B are used as the major separation between the living room and the dining room.

Floors

The living-room floor should reinforce and blend with the color scheme, textures, and overall style of the living room. Exposed hardwood flooring, room-size carpeting, wall-to-wall carpeting, throw rugs, and sometimes polished flagstone are appropriate for living-room use.

Ceilings

Most conventional ceilings are flat surfaces covered with plaster or gypsum board. New building materials, such as laminated beams and arches, and new construction methods now enable architects to design ceilings that conserve building materials and utilize previously wasted space. Figure 6-5A shows types of open-beam ceiling treatments, and Fig. 6-5B shows a comparison of open-beam and conventional ceiling space.

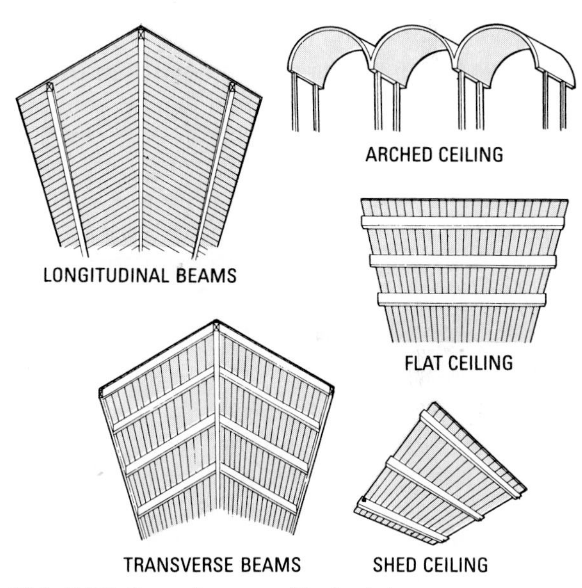

Fig. 6-5A Open-beam ceilings styles.

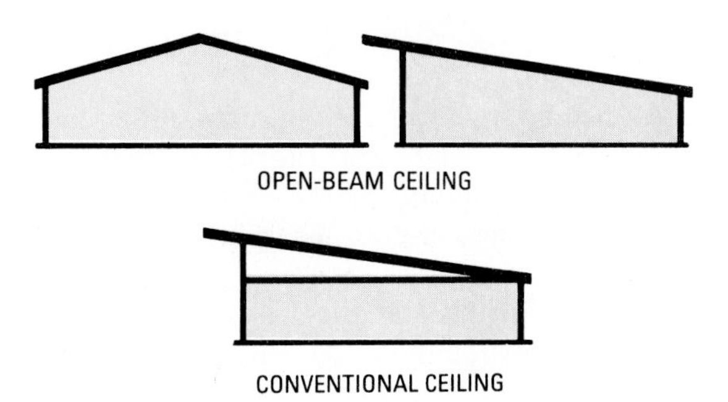

Fig. 6-5B Two basic types of open-beam ceilings.

Lighting

Living-room lighting is divided into three types: *general lighting, decorative lighting,* and *local lighting.* General lighting is designed to illuminate the entire room through the use of ceiling fixtures, wall spots, or cove lighting. Local lighting is provided for a specific purpose, such as reading, drawing, sewing, or decorative purposes. Local lighting can be supplied by table lamps, wall lamps, pole lamps, or floor lamps.

Furniture

Furniture for the living room may or may not reflect the motif and architectural style of the home. Most designers, when working directly with a client, attempt to match the exterior style of the house to the interior furniture style preference of the client.

A special effort should be made to have built-in furniture maintain lines consistent with the remaining wall treatment. Notice how the built-in bookshelves in Fig. 6-6 eliminate the need for other pieces of furniture in their end of the room. The built-in bookshelves and cabinets blend functionally into the total decor of the room. The furniture for the living room is chosen to fit the living needs of the residents. The size, shape, and layout of the room should be designed to accommodate the furniture. This top design is a result of establishing the size and shape of the room without considering the size and number of pieces of furniture to be used.

Fig. 6-6 Built-in bookshelves conserve floor space. *(Home Planners, Inc.)*

SIZE AND SHAPE

One of the most difficult aspects of planning the size and shape of a living room, or any other room, is to provide sufficient wall space for the effective placement of furniture. Continuous wall space is needed for the placement of many articles of furniture, especially musical equipment, bookcases, chairs, and couches. The placement of fireplaces, doors, or openings to other rooms should be planned to conserve as much wall space as possible for furniture placement.

Rectangular rooms are generally easier to plan and to place furniture in than are square rooms. However, the designer must be careful not to establish a proportion that will break the living room into several conversational areas.

Living rooms vary greatly in size. A room 12 feet by 18 feet (12′ X 18′), or 3.7 meters by 5.5 meters (3.7 m X 5.5 m), would be considered a small or minimum-sized living room. A living room of average size would be approximately 16′ X 20′ (4.9 m X 6.1 m), and a very large or optimum-sized living room would be 20′ X 26′ (6.1 m X 7.9 m) or more.

Exercises

1. **Sketch an open-plan living room. Indicate the position of windows, fireplace, foyer, entrance, and dining room.**
2. **Sketch a closed-plan living room. Show the position of adjacent rooms.**
3. **Sketch one wall of the living room you designed for Problem 2. Use Fig. 6-4 as a guide.**

4. **List the furniture you would include in the living room of the house of your design. Cut out samples of this furniture from catalogs or newspapers.**
5. **Define the following terms: *closed plan, open plan, decor, living area, living room, local lighting, general lighting.***

UNIT 7

DINING ROOMS

The dining facilities designed for a residence depend greatly on the dining habits of the occupants. The dining room may be large and formal, or the dining area may consist of a dining alcove. It may also be a breakfast nook in the kitchen. Large homes may contain dining facilities in all these areas.

FUNCTION

The function of a dining area is to provide a place for the family to gather for breakfast, lunch, or dinner in both casual and formal situations. When possible, a separate dining area potentially capable of seating from 8 to 12 persons for dinner should be provided in addition to breakfast or dinette facilities.

LOCATION

Dining facilities may be located in many different areas, depending on the capacity needed and the type of plan. In the closed plan, a separate dining room is usually provided. In an open plan, many different dining locations are possible (Fig. 7-1). Open-area dining facilities are sometimes provided in the kitchen or the living room.

Regardless of the exact position of the dining area, it must be placed adjacent to the kitchen. The ideal dining location is one that requires few steps from the kitchen to the dining table. However, the preparation of food and other kitchen activities should be baffled from direct view from the dining area.

If dining facilities are not located in the living room, they should be located next to it. Family and guests normally enter the dining room from the living room and use both rooms jointly.

The nearness of the dining room to the kitchen, and to the living room, requires that it be placed between the kitchen and the living area. The dining room in the closed plan shown in Fig. 7-2 is located in this manner.

The area between the living room and the dining room may be entirely open, partially baffled,

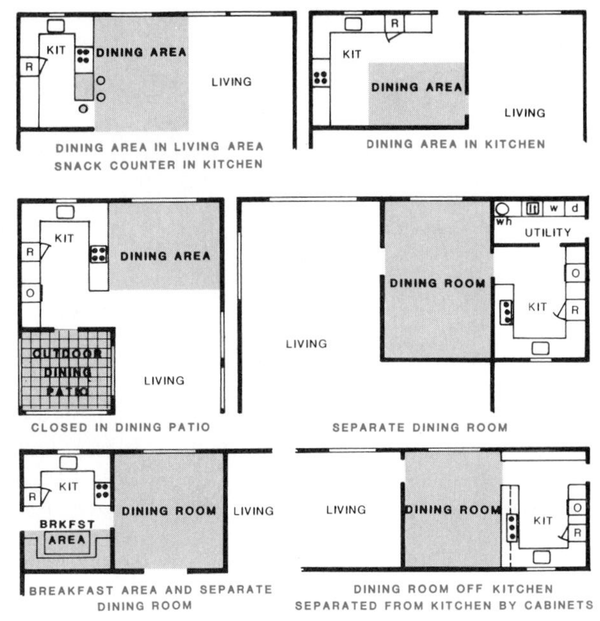

Fig. 7-1 Plans showing the location of dining facilities in many different areas.

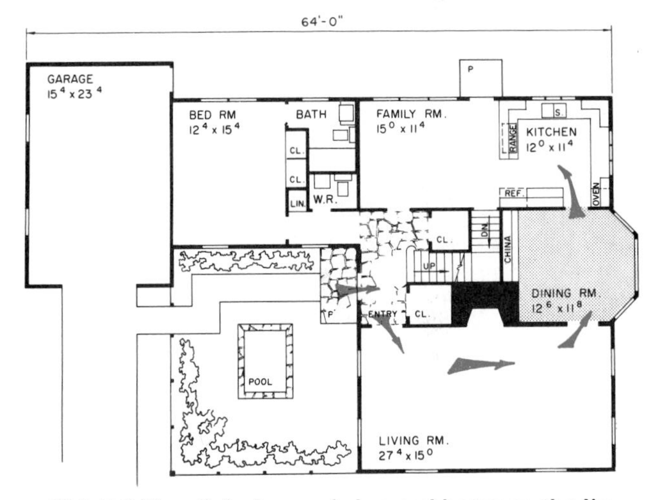

Fig. 7-2 The dining room is located between the living room and the kitchen. *(Home Planners, Inc.)*

or completely closed off. Sometimes, the separation of the dining room and the living room is accomplished by different floor levels or by dividing the rooms with common half walls, as shown in Fig. 7-3. Another method of separating the dining area in an open plan is through the use of partial partitions, as shown in Fig. 7-4. Compare this semi-isolated arrangement with the completely open dining area shown in Fig. 7-5.

There is often a need for dining facilities on or adjacent to the patio. The porch or patio should

Fig. 7-3 Separation with half walls. *(Home Planners, Inc.)*

Fig. 7-4 A partial wall without a door makes this an open plan. *(Home Planners, Inc.)*

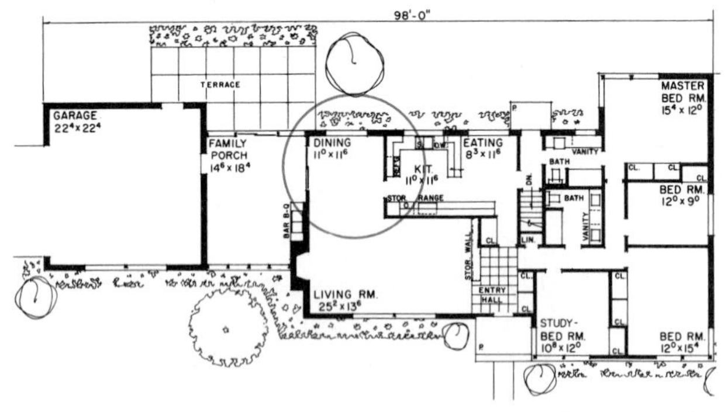

Fig. 7-5 An open dining-plan. *(Home Planners, Inc.)*

be near the kitchen and directly accessible to it. Locating the patio or dining porch directly outside the dining room or kitchen wall provides maximum use of the facilities. This minimizes the inconvenience of using outside dining facilities.

DECOR

The decor of the dining room should blend with the rest of the house. Floor, wall, and ceiling treatment should be the same in the dining area as in the living area. If a dining porch or a dining patio is used, its decor must also be considered part of the dining-room decor. This is because the outside dining area is viewed from the inside.

If semi-isolation is desired, partial divider walls can be used effectively. These dividers may be planter walls, glass walls, half walls of brick or stone, paneled walls, fireplaces, or grillwork.

Lighting

Controlled lighting can greatly enhance the decor of the dining room. General illumination that can be subdued or intensified can provide the right atmosphere for almost any occasion. Lighting is controlled by a rheostat, which is commonly known as a *dimmer switch*. In addition to general illumination, local lighting should be provided for the table either by a direct ceiling spotlight or by a hanging lamp (Fig. 7-6). A hanging lamp can be adjusted down for local dining lighting and up for general illumination when the dining facilities are not in use.

SIZE AND SHAPE

The size and shape of the dining area are determined by the size of the family, the size and

Fig. 7-6 The use of local lighting over the dining table. *(Home Planners, Inc.)*

amount of furniture, and the clearances and traffic areas between pieces of furniture. The dining area should be planned for the largest group that will dine in it regularly. There is little advantage in having a dining-room table that expands, if the room is not large enough to accommodate the expansion. One advantage of the open plan is that the dining facilities can be expanded in an unlimited manner into the living area, as shown in Fig. 7-7. Thus, the living area temporarily becomes part of the dining area.

Furniture

The dining room should be planned to accommodate the furniture. Dining-room furniture may include an expandable table, side chairs, armchairs, buffet, server or serving cart, china closet, and serving bar. In most situations, a rectangular dining room will accommodate the furniture better than a square room. Figure 7-8 shows a typical furniture placement for a dining room.

Regardless of the furniture arrangement, a minimum space of 2′ (610 millimeters, or mm) should be allowed between a chair and the wall or other furniture when the chair is pulled to the out position. This allowance will permit serving traffic behind chairs and will permit entrance to and exit from the table without difficulty. A distance of 27 inches, or 27″ (690 mm), per person should be allowed at the table. This spacing is accomplished by allowing 27″ (690 mm) from the centerline of one chair to the centerline of another, as shown in Fig. 7-9.

A dining room that would accommodate the minimum amount of furniture—a table, four chairs, and a buffet—would be approximately 10′ × 12′ (3 m × 3.7 m). A minimum-sized din-

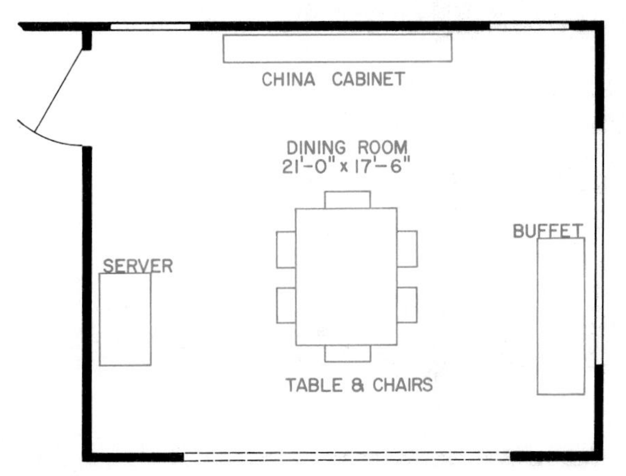

Fig. 7-8 Typical furniture placement in a dining room.

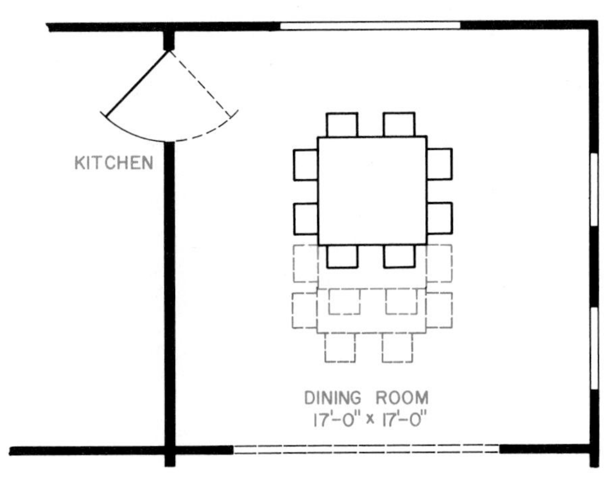

Fig. 7-7 A dining area planned for expansion.

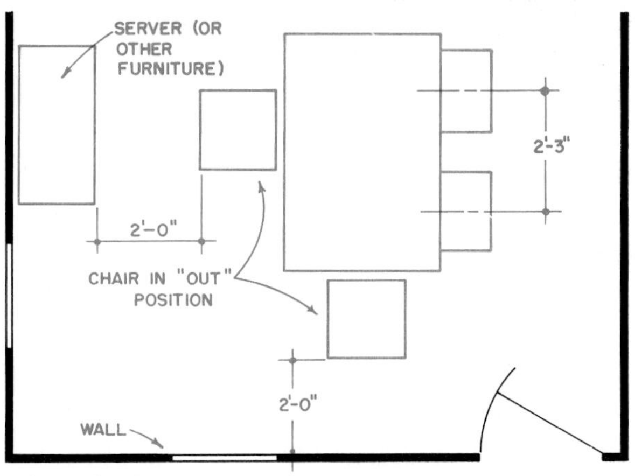

Fig. 7-9 Dining-room clearances.

dining room that would accommodate a dining table, six or eight chairs, a buffet, a china closet, and a server would be approximately 12′ × 15′ (3.7 m × 4.6 m). An optimum-sized dining room would be 14′ × 18′ (4.3 m × 5.5 m) or larger.

Exercises

1. **Sketch a dining room to include the following furniture: dining table to accommodate six, buffet, china closet. Indicate the relationship to the living room, and provide access to a patio.**
2. **Sketch a plan for an informal dining area directly adjacent to the kitchen.**
3. **Sketch an open-plan dining area. Show the relationship of this area to the living room.**
4. **Redesign the dining area of your own home.**

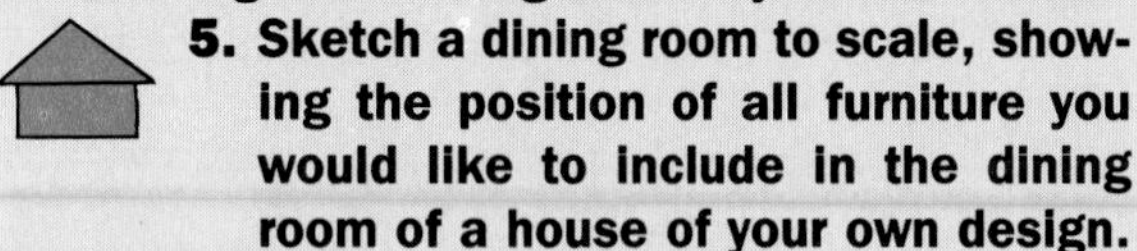

5. **Sketch a dining room to scale, showing the position of all furniture you would like to include in the dining room of a house of your own design.**
6. **Draw a floor plan of the dining area shown in Fig. 7-10.**

Fig. 7-10 *(Home Planners, Inc.)*

7. **Define the following terms: *buffet, china closet, server, rheostat, dining porch, dining patio, formal dining, casual dining.***

UNIT 8

FAMILY ROOMS

The trend toward more informal living because of more leisure time has influenced the popularity of the family room. Today, the majority of homes are designed to include a family room.

FUNCTION

The purpose of the family room is to provide facilities for family-centered activities. It is designed for the entire family, children and adults alike.

Only in extremely large residences is there sufficient space for a separate sewing room, children's playroom, hobby room, or music room. The modern family room often performs the functions of all those rooms.

LOCATION

Activities in the family room often result in the accumulation of hobby materials and clutter. Thus, the family room is often located in an area accessible from, but not visible from, the rest of the living area.

It is quite common to locate the family room adjacent to the kitchen, as shown in Fig. 8-1.

Fig. 8-1 A family room located adjacent to the kitchen. *(Olsen-Spencer Assoc.)*

This location revives the idea of the old country kitchen in which most family activities were centered.

When the family room is located adjacent to the living room or dining room, it becomes an extension of those rooms for social affairs. In this location, the family room is often separated from the other rooms by folding doors, screens, or sliding doors. The family room shown in Fig. 8-2 is located next to the kitchen and yet is accessible from the living room when the folding door is open.

Another popular location for the family room is between the service area and the living area. The family room shown in Fig. 8-3 is located between the garage, the kitchen, and the entrance. This location is especially appropriate when some service functions, such as home-workshop facilities, are assigned to the family room.

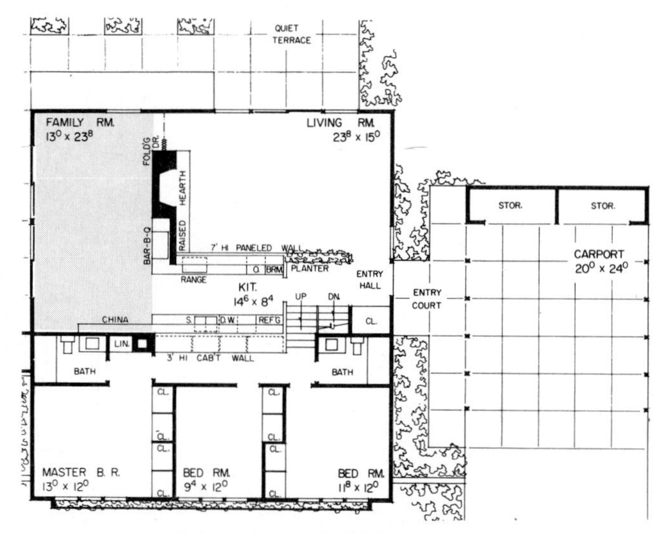

Fig. 8-2 A family room accessible from the kitchen and living room. *(Home Planners, Inc.)*

DECOR

The family room is also known as the *activities room* or *multiactivities room.* Decoration of this room should provide a vibrant atmosphere. Ease of maintenance should be one of the chief considerations in decorating the family room. Family-room furniture should be informal and suited to all members of the family. The use of plastics, leather, and wood provides great flexibility in color and style and promotes easy maintenance.

Floors should be resilient—able to keep original shape or condition despite hard use. Linoleum or tile made of asphalt, rubber, or vinyl will best resist the abuse normally given a family-room floor. If rugs are used, they should be the kind that will stand up under rough treatment. They should also be washable.

Soft, easily damaged materials such as wallpaper and plaster should be avoided for the family room. Materials such as tile and paneling are most functional. Chalkboards, bulletin boards, built-in cupboards, and toy-storage cabinets should be used when appropriate. Work areas that fold into the wall when not in use conserve space and may perform a dual function if the cover wall can also be used as a chalkboard or a bulletin board.

Since a variety of hobby and game materials will be used in the family room, sufficient space must be provided for the storage of these materials. Figure 8-4 shows the use of built-in storage facilities, including cabinets, closets, and drawer storage.

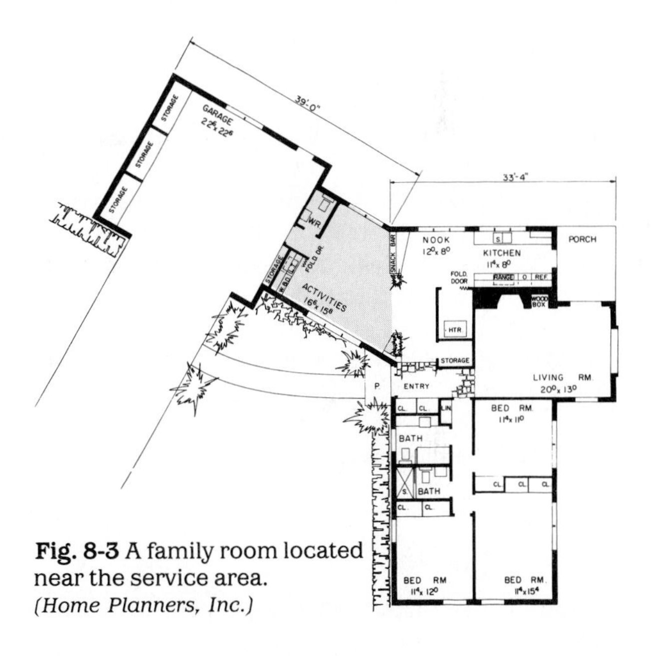

Fig. 8-3 A family room located near the service area. *(Home Planners, Inc.)*

Acoustical ceilings are recommended to keep the noise of the various activities from spreading to other parts of the house. This feature is especially important if the family room is located on a lower level.

SIZE AND SHAPE

The size and shape of the family room depends directly on the equipment needed for the activi-

Fig. 8-4 Plan for adequate storage in the family room.

ties the family will pursue in this room. The room may vary from a minimum-sized room, of approximately 150 sq. ft. (150 ft^2), to the more optimum-sized family room, of 300 ft^2 or more. Most family-room requirements lie somewhere between the two extremes.

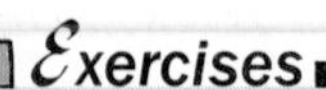

Exercises

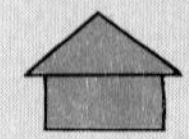

1. **Sketch a family room you would like to include in a home of your own design. Include the location of all furniture and facilities (scale ½″ = 1′-0″).**
2. **Design a family room primarily for children's activities.**
3. **Design a family room that doubles as a guest bedroom.**
4. **Define the following terms: *sewing room, children's playroom, hobby room, music room, family room, multiactivities, linoleum, asphalt tile, rubber tile, vinyl tile, acoustical ceiling.***

UNIT 9

RECREATION ROOMS

The *recreation room* (game room, playroom) is exactly what the name states. It is a room for play and recreation. It includes facilities for participation in recreational activities.

FUNCTION

The design of the recreation room depends on the number and arrangement of the facilities needed for the various pursuits. Activities for which many recreation rooms are designed include billiards, chess, checkers, table tennis, darts, television watching, eating, and dancing.

The function of the recreation room often overlaps that of the family room. Overlapping occurs when a multipurpose room is designed to provide for recreational activities such as table tennis and billiards and also includes facilities for more sedentary family activities such as sewing, knitting, model building, and other hobbies.

LOCATION

The recreation room is frequently located in the basement in order to use space that would otherwise be wasted. Basement recreation rooms often provide more space for the use of large equipment, such as table-tennis tables, billiard tables, and shuffleboard. A basement recreation-room fireplace can be located directly beneath the living-room fireplace on the upper level.

When the recreation room is located on the ground level, its function can be expanded to the patio or terrace. Regardless of the level, the recreation room should be located away from the quiet areas of the house.

DECOR

Designers take more liberties in decorating the recreation room than with any other room. They do so primarily because the active, informal atmosphere that characterizes the recreation room lends itself readily to unconventional furniture, fixtures, and color schemes. Bright, warm colors can reflect a party mood. Furnishings and accessories can be used to accent a variety of central themes. Regardless of the central theme, recreation-room furniture should be comfortable and easy to maintain. The same rules apply to recreation-room walls, floors, and ceilings as apply to those of the family room.

SIZE AND SHAPE

The size and shape of the recreation room depend on whether the room occupies an area on the main level or whether it occupies basement space. If basement space is used, the only restrictions on the size are the other facilities that will also occupy space in the basement, such as the laundry, the workshop, or the garage.

Exercises

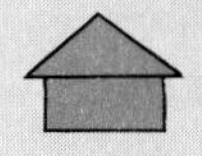

1. Sketch a plan of a recreation room you would include in a house of your own design.

2. Sketch a plan for a recreation room, including facilities for billiards, chess, shuffleboard, and television watching.

3. Define the following terms: *game room, playroom, recreation room.*

UNIT 10

PORCHES AND DECKS

A *porch* is a covered platform leading into an entrance of a building. Porches are commonly enclosed by glass, screen, or post and railings. A porch is not the same as a patio. The porch is attached structurally to the house, whereas a patio is placed directly on the ground. Balconies and decks are actually elevated porches.

FUNCTION

Porches serve a variety of functions. Some are used for dining and some for entertaining and relaxing. Others are furnished and function like patios for outdoor living. Still others provide an additional shelter for the entrance to a house or patio. The primary function of a porch depends on the structure and purpose of the building to which it is attached.

Verandas

Southern colonial homes were designed with large porches, or *verandas*, extending around several sides of the home. Outdoor plantation life centered on the veranda, which was very large.

Balconies

A *balcony* is a porch suspended from an upper level of a structure. It usually has no access from the outside. Balconies often provide an extension to the living area or a private extension to a bedroom.

The house shown in Fig. 10-1 is distinguished by several types of balconies. The upper balcony is supported by cantilevered beams and provides an extension that covers the porch below. The porch in turn shelters a patio below it. Hillside lots lend themselves to vertical plans and provide maximum flexibility in using outdoor living facilities.

Spanish- and Italian-style architecture is characterized by numerous balconies. The return of the balcony to popularity has been influenced and accelerated by new developments in building materials. The principle of cantileverage, or suspension in space, can also be used to a greater extent with steel construction. An example is shown in the balcony in Fig. 10-2.

Stoop

The *stoop* is a projection from a building, similar to a porch. However, a stoop does not provide sufficient space for any activities. It provides only shelter and an access to the entrance of the building.

The Modern Porch

Only in the last several years has the porch been functionally designed and effectively utilized for outdoor living. The classic front porch and back porch that characterized most homes built in this country during the 1920s and 1930s were designed and used merely as places in which to sit. Little effort was made to use the porch for any other activities. A porch for a modern home should be designed for the specific activities anticipated for it. The form of the porch should be determined by its function.

LOCATION

Since the porch is an integral part of the total house design, it must be located where it will function best. A porch can be made consistent with the rest of the house by extending the lines of the roof to provide sufficient *overhang*, or projection.

A dining porch should be located adjacent to either the dining room or the kitchen. The dining porch shown in Fig. 10-3 can be approached from the dining room or from the living-area porch.

The porch should be located to provide maximum flexibility. A porch that can function for dining and other living activities is desirable. The primary functions of the porch should be considered when orienting the porch with the sun. If much daytime use is anticipated and direct sunlight is desirable, a southern exposure should be planned. If little sun is wanted during the day, a northern exposure would be preferable. If morning sun is desirable, an eastern exposure would be best, and for the afternoon sun, a western exposure.

DECOR

The porch should be designed as an integral and functional part of the total structure. A blending of roof styles and major lines of the porch roof

Fig. 10-1 The upper balcony of this house provides protection for the patio below. *(Western Wood Products Assoc.)*

Fig. 10-2 Steel members make large cantilevered distances possible.

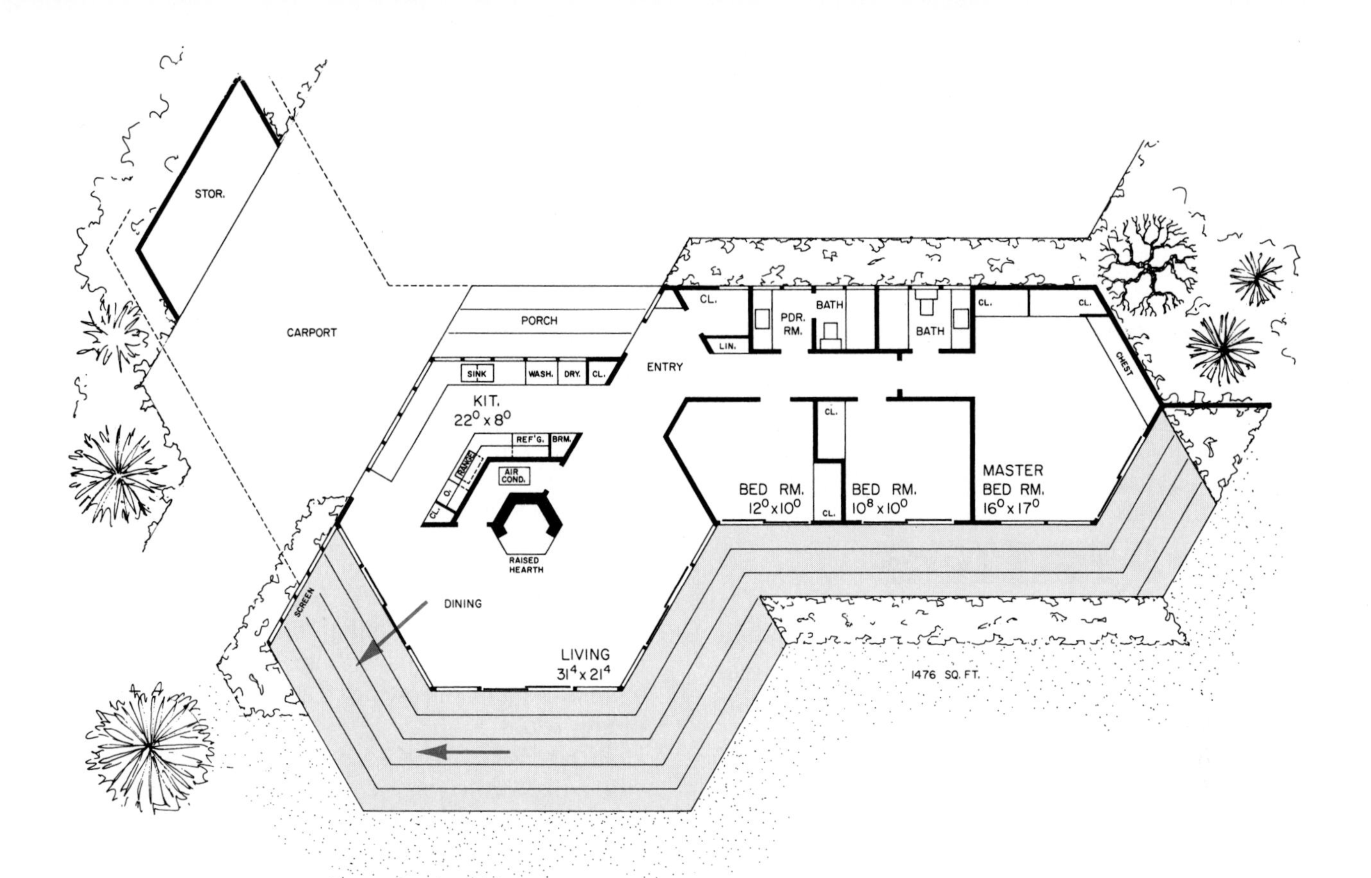

Fig. 10-3 Location of a dining porch. *(Home Planners, Inc.)*

and house roof is especially important (Fig. 10-4). A similar consistency should characterize the vertical columns or support members of the porch. Figure 10-4 shows some deck relationships that ensure uniformity in design.

Various materials and methods can be used for deck railings, depending on the degree of privacy or sun and wind protection needed. The sides of porches can provide adequate ventilation and also offer semiprivacy and safety. Railings on elevated porches should be designed at a height above 3′ (915 mm) to discourage the use of the top rail as a place to sit.

Fig. 10-4 Frank Lloyd Wright always sought to relate all building components structurally and aesthetically to the site. This is an example of structurally and aesthetically related decks.

Porch furniture should withstand deterioration in any kind of weather. The covering material should be waterproof, stain-resistant, and washable. Protection from wind and rain should be planned.

SIZE AND SHAPE

Porches range in size from the very large veranda to rather modest-sized stoops, which provide only shelter and a landing surface for the main entrance. A porch approximately 6′ X 8′ (1.8 m X 2.4 m) is considered minimum-sized. An 8′ X 12′ (2.4 m X 3.7 m) porch is about average. Porches larger than 12′ X 18′ (3.7 m X 5.5 m) are considered rather large.

The shape of the porch depends greatly upon how the porch can be integrated into the overall design of the house.

Exercises

1. **Add a porch to the plan shown in Fig. 8-2.**
2. **Add a porch to a floor plan of your own design.**
3. **From catalogs, newspapers, and magazines, cut out pictures of porch furniture you would choose for your own porch.**
4. **Define the following terms: *veranda, balcony, cantilever, stoop.***

UNIT 11

PATIOS

A *patio* is a covered surface adjacent or directly accessible to the house. The word *patio* comes from the Spanish word for courtyard. Courtyard living was an important aspect of Spanish culture, and courtyard design was an important part of early Spanish architecture.

FUNCTION

The patio at various times may perform outdoors all the functions that the living room, dining room, recreation room, kitchen, and family room perform indoors.

The patio is often referred to by other names, such as *loggia, breezeway,* and *terrace.*

Patios can be divided into three main types according to function: *living patios, play patios,* and *quiet patios.* The home shown in Fig. 11-1 contains all three kinds of patios.

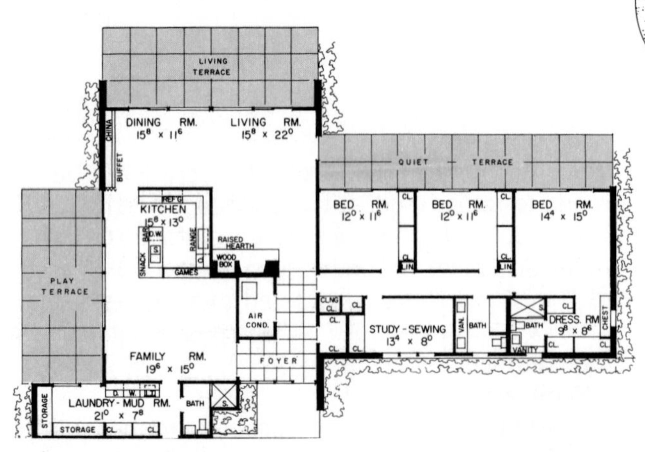

Fig. 11-1 This plan includes three kinds of patios. *(Home Planners, Inc.)*

LOCATION

Patios should be located adjacent to the area of the home to which they relate. They should also be somewhat secluded from the street or from neighboring residences.

Living Patio Living patios should be located close to the living room or the dining room. When dining is anticipated on the patio, access should be provided from the kitchen or dining room.

Play Patio It is often advantageous to provide a play patio for use by children and for physical activities not normally associated with the living terrace. The play terrace sometimes doubles as the service terrace and can conveniently be placed adjacent to the service area. Notice the location of the play terrace in Fig. 11-1. It is related directly to the service area and also to the family room in the living area.

Quiet Patio The quiet patio can actually become an extension of the bedroom. It can be used for relaxation or sleeping. A quiet terrace

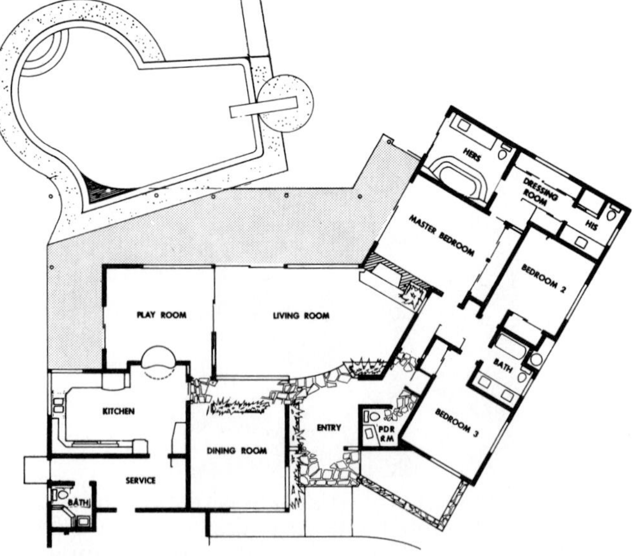

Fig. 11-2 Continuous patio. *(Southern California Gas Co.)*

should be secluded from the normal traffic of the home.

Often the design of the house will allow these separately functioning patios to be combined in one large, continuous patio. That type of patio is shown in Fig. 11-2. Here the playroom, living room, master bedroom, and kitchen all have access to the patio.

Placement

Patios can be conveniently placed at the end of a building, between corners of a house, or wrapped around the side of the house, as shown in Fig. 11-2. Or they may be placed in the center of a U-shaped house or in a courtyard. The courtyard patio shown in Fig. 11-3 offers complete privacy from all sides.

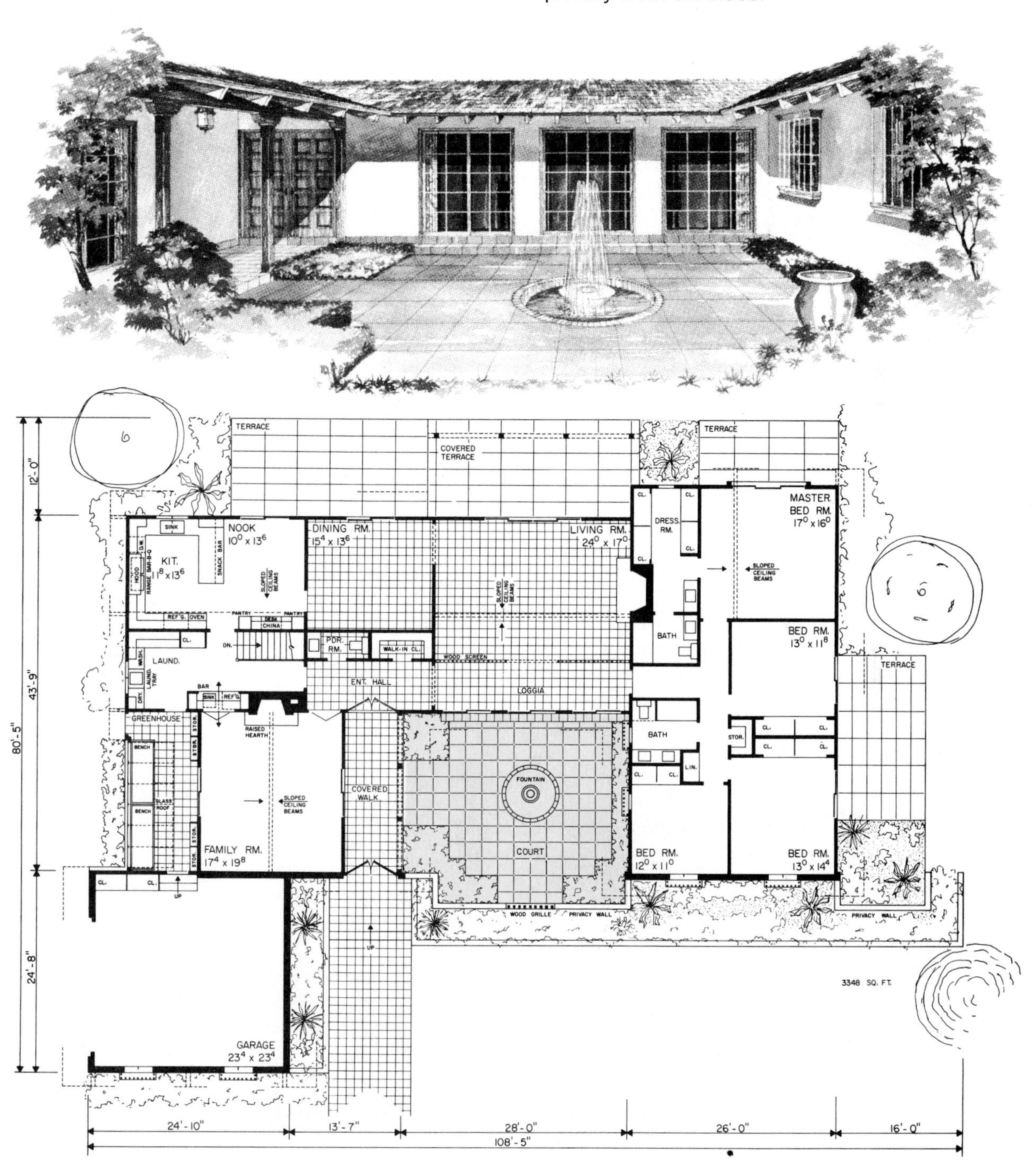

Fig. 11-3 A courtyard patio. *(Home Planners, Inc.)*

Separate Patios

In addition to the preceding locations, the patio is often located completely apart from the house. When a wooded area, a particular view, or a terrain feature is of interest, the patio can be placed away from the house. When the patio is located in this manner, it should be readily accessible.

Orientation

When the patio is placed on the north side of the house, the house itself can be used to shade the patio. If sunlight is desired, the patio should be located on the south side of the house. The planner should take full advantage of the most pleasing view and should restrict the view of undesirable sights.

DECOR

The materials used in the deck, cover, baffles, and furniture of the patio should be consistent with the lines and materials used in the rest of the home. Patios should not appear to be designed as an afterthought but should appear and function as an integral part of the total design.

Patio Deck

The *deck* (floor) of the patio should be constructed from materials that are permanent and maintenance-free. Flagstone, redwood, concrete, and brick are among the best materials for use on patio decks. Wood slats such as those shown in Fig. 11-4 provide for drainage between the slats and also create a warm appearance. However, they do require maintenance.

Fig. 11-4 A wood-slat patio deck. *(Julius Shulman)*

Brick-surface patio decks are very popular because bricks can be placed in a variety of arrangements to adapt to practically any shape or space. The area between the bricks may be filled with concrete, gravel, sand, or grass. A concrete deck is effective when a smooth, unbroken surface is desired. Patios where bouncing-ball games are played, or where poolside cover is desired, can use concrete advantageously.

Patio Cover

Patios need not be covered if the house is oriented to shade the patio during the times of the day when shade is normally desired. Since a patio is designed to provide outdoor living, too much cover can defeat the purpose of the patio. Coverings can be an extension of the roof structure, as shown in Fig. 11-5. They may be graded or tilted to allow light to enter when the sun is high and to block the sun's rays when the sun is lower. The graded effect can be obtained by placing louvers spaced straight or slanted to admit the high sun and block the low sun, as shown in Fig. 11-6.

Plastic, glass fiber, and other translucent materials used to cover patios admit sunlight and yet provide protection from the direct rays of the

Fig. 11-5 A covered patio. *(Home Planners, Inc.)*

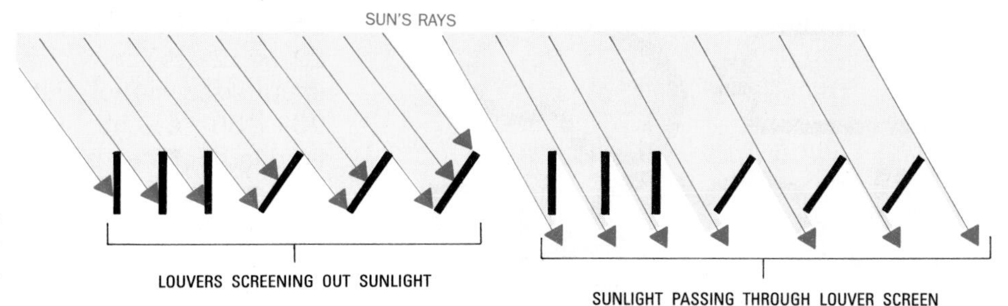

Fig. 11-6 The angle and spacing of louvers is important in proper sun screening.

sun. Translucent covers also provide shelter from rain. When translucent covering is used, it is often desirable to have only part of the patio covered. This arrangement provides sun for part of the patio and shade for other parts. Balconies can also be used effectively to provide shelter for a patio.

Walls and Baffles

Patios are designed for outdoor living, but outdoor living need not be public living. Some privacy is always desirable. Solid walls can often be used effectively to baffle the patio from a street view, from wind, and from the low rays of the sun. Baffling devices include solid fences, slatted fences, concrete blocks, post-and-rail, brick or stone walls, and hedges or other shrubbery.

A solid baffle wall is often undesirable because it restricts the view, eliminates the circulation of air, and makes the patio appear smaller.

In mild climates, completely enclosing a patio by solid walls can help make the patio function as another room. In such an enclosed patio, some opening should be provided to allow light and air to enter. The grillwork openings often provide an effective and aesthetically pleasing solution to this problem.

Day and Night Decor

To be totally effective, the patio should be designed for both daytime and nighttime use. Correct use of general and local lighting can make the patio useful for many hours each day. If the walls between the inside areas of the house and the patio are designed correctly, much light from the inside can be utilized on the patio.

If carefully designed and located the patio area can provide visual enhancement to the living area not available when deciduous trees are in leaf.

SIZE AND SHAPE

Patios may range from the size of a small garden terrace to the spaciousness of the courtyard patio shown in Fig. 11-7. The primary function will largely determine the size. A Japanese-garden terrace, for instance, has no furniture and is designed primarily to provide a baffle and a beautiful view. The courtyard patio is designed for many uses.

Activities should be governed by the amount of space needed for equipment. Equipment and furnishings normally used on patios include picnic tables and benches, lounge chairs, serving carts, game apparatus, and barbecue pits. The placement of these items and the storage of games, apparatus, and fixtures should determine the size of the patio. Patios vary more in length than in width, since patios may extend

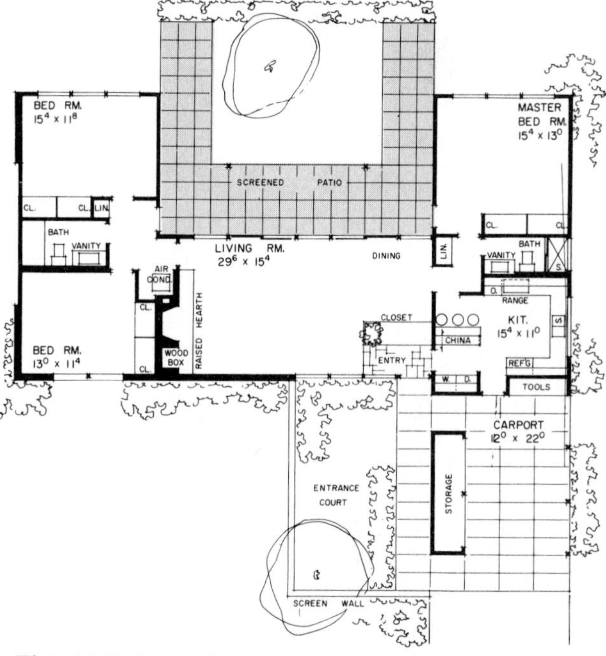

Fig. 11-7 A spacious courtyard patio. *(Home Planners, Inc.)*

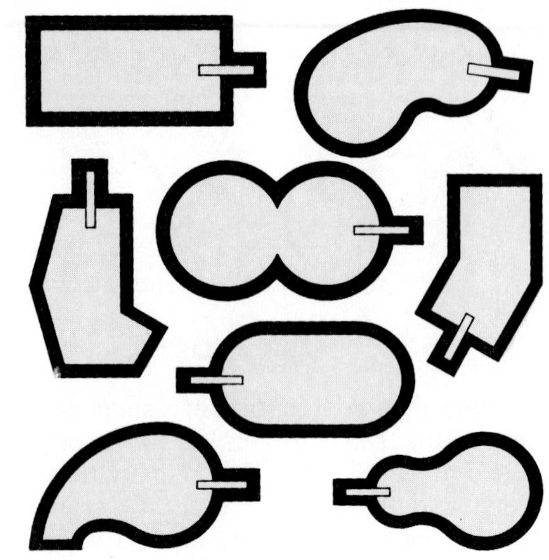

Fig. 11-8 Pool shapes.

over the entire length of the house. A patio 12′ × 12′ (3.7 m × 3.7 m) is considered a minimum-sized patio. Patios with dimensions of 20′ × 30′ (6.0 m × 9.1 m) or more are considered large. When a pool is designed for a home, it becomes an integral part of the patio. Many pool shapes now available allow the designer to blend the pool into the size and shape of the patio (Fig. 11-8).

When designing and locating a pool, the location of the filter, heater (if used), and electrical, plumbing, and filter lines must be planned. Since the filter runs continuously, it should be located as far from the patio as possible without the use of excessively long plumbing, electrical, and filter supply lines. Figure 11-9 shows the location of this equipment in relationship to the pool and patio.

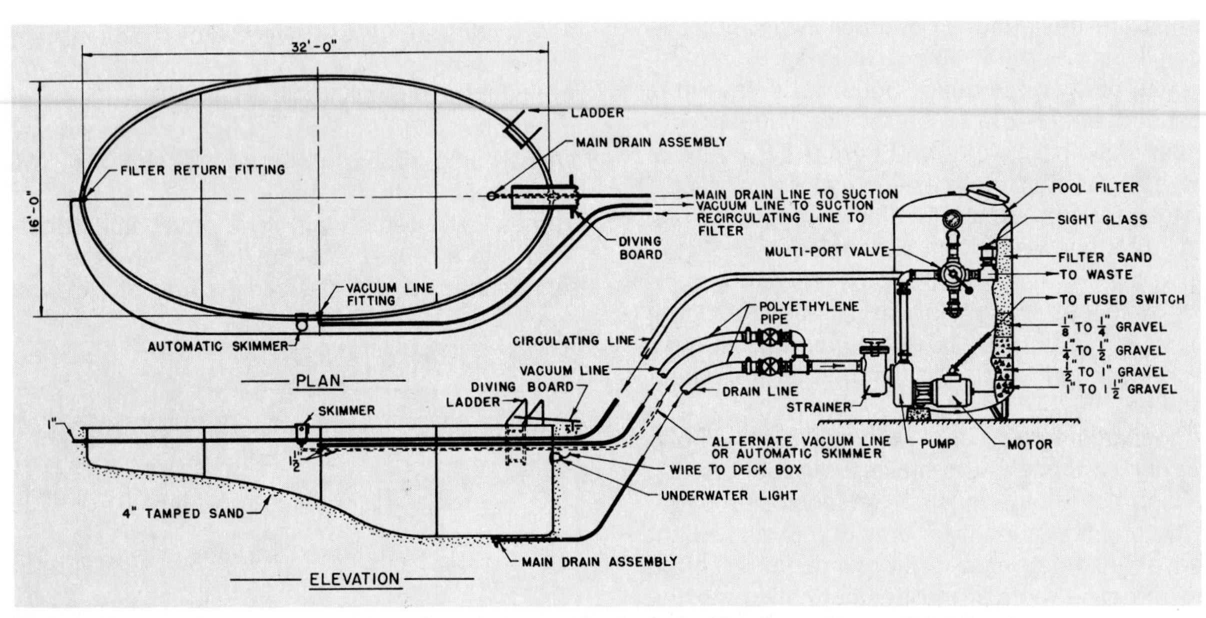

Fig. 11-9 Filtering-system equipment location must be included in plans. *(Lancer Pool Corp.)*

Exercises

1. Choose an appropriate pool shape and incorporate it in the patio design shown in Fig. 11-1.

2. Plan a patio for a house of your own design. Sketch the basic scheme and the facilities.

3. Define the following terms: *patio, loggia, breezeway, terrace, play patio, quiet patio, living patio, flagstone, redwood, concrete, patio deck, patio baffles.*

UNIT 12

LANAIS

Lanai is the Hawaiian word for porch. However, the word *lanai* is now also used to refer to a covered exterior passageway.

FUNCTION

Large lanais are often used as patios, although their main function is to provide shelter for the traffic accesses on the exterior of a building. Lanais are actually exterior hallways (Fig. 12-1A).

Lanais that are located parallel to exterior walls are usually created by extending the roof overhang to cover a traffic area as shown in Fig. 12-1B. A typical lanai plan eliminates the need for more costly interior halls. Lanais are used extensively in warmer climates.

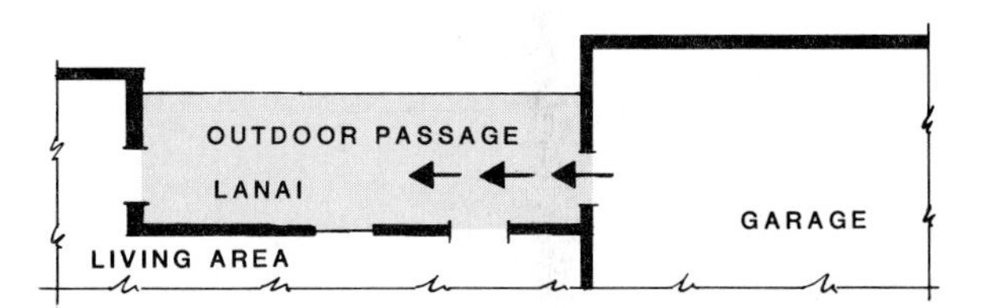

Fig. 12-1A Lanais connecting living areas.

Fig. 12-1B Lanais created by a large roof overhang.

LOCATION

In residence planning, a lanai can be used most effectively to connect opposite areas of a home. Lanais are commonly located between the garage and the kitchen, the patio and the kitchen or the living area, and the living area and the service area. U-shaped buildings are especially suitable for lanais because it is natural to connect the extremes of the U.

When lanais are carefully located, they can also function as sheltered access from inside areas to outside facilities such as patios, pools, outdoor cooking areas, or courtyards, as shown in Fig. 12-2. A covered or partially covered patio is also considered a lanai when it doubles as a major access from one area of a structure to another. The patio shown in Fig. 12-3 functions in this manner. A lanai can also be semi-enclosed and provide not only traffic access but also privacy and sun and wind shielding. When lanais are used to connect the building with the street, they actually function as marquees.

DECOR

The lanai should be a consistent, integral part of the design of the structure. The lanai cover may be an extension of the roof overhang or may be supported by columns, as shown in Fig. 12-4. If glass is placed between the columns, the lanai becomes an interior hallway rather than an exterior one. This separation is sometimes the only difference between a lanai and an interior hall.

It is often desirable to design and locate the lanai to provide access from one end of an extremely long building to the other end, as shown in Fig. 12-4. The lines of this kind of lanai strengthen and reinforce the basic horizontal and vertical lines of the building. The columns supporting the roof overhang in Fig. 12-4 also provide a visual boundary without blocking the view.

If a lanai is to be utilized extensively at night, effective lighting must be provided. Light from within can be used when drapes are open, but additional lighting fixtures are used for the times when drapes are closed.

Fig. 12-2 Lanais integrated with courtyard. *(Home Planners, Inc.)*

Fig. 12-3 Covered entrance lanais. *(Home Planners, Inc.)*

Fig. 12-4 A large commercial building lanai. *(Libby-Owens-Ford)*

SIZE AND SHAPE

Lanais may extend the full length of a building and may be designed for maximum traffic loads. They may be as small as the area under a 2′ or 3′ (610 or 915 mm) roof overhang. However, a lanai at least 4′ (1220 mm) wide is desirable. The length and type of cover is limited only by the location of areas to be covered.

Exercises

1. Draw the outline of a lanai you would plan for a home of your own design.

2. Sketch a floor plan of your own home. Add a lanai to connect two of the areas, such as the sleeping and living areas.

3. Define the following terms: *lanai cover, roof overhang, exterior hallway, traffic load, baffle, translucent.*

UNIT 13

TRAFFIC AREAS AND PATTERNS

When an architect plans a commercial structure both vehicular and pedestrian traffic volume and patterns must be considered. Traffic inside and outside the building must be planned for. Traffic areas for employees, visitors, and deliveries into and out of the building must be allocated using a minimum amount of space. In designing commercial traffic areas, the maximum volume of pedestrian traffic must be considered.

Planning the traffic areas of residence is not as complex, because of the small number of people involved. Nevertheless, the same basic principle of efficient space allocation prevails. The traffic areas of the home provide passage from one room or area to another. The main traffic areas of a residence include the halls, entrance foyers, stairs, lanais, and areas of rooms that are part of the traffic pattern.

TRAFFIC PATTERNS

Traffic patterns of a residence should be carefully considered in the design of the room layout. A minimum amount of space should be devoted to traffic areas. Extremely long halls and corridors should be avoided. They are difficult to light and provide no living space. Traffic patterns that require passage through one room to get to another should also be avoided, especially in the sleeping area.

The traffic pattern shown in the plan in Fig. 13-1 is efficient and functional. It contains a minimum amount of wasted hall space without creating a boxed-in appearance. It provides access to each of the areas without passing

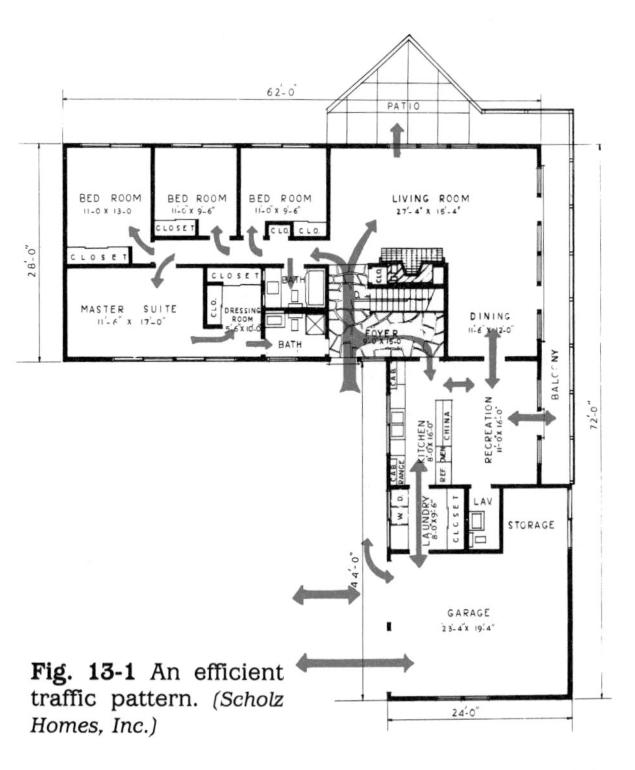

Fig. 13-1 An efficient traffic pattern. *(Scholz Homes, Inc.)*

through other areas. The arrows clearly show that the sleeping area, living area, and service area are accessible from the entrance without passage through other areas. In this plan the service entrance provides access to the kitchen from the carport and other parts of the service area.

One method of determining the effectiveness of the traffic pattern of a house is to imagine yourself moving through the house by placing your pencil on the floor plan and tracing your route through the house as you perform your daily routine. If you trace through a whole day's activities, including those of other members of the household, you will be able to see graphically where the heaviest traffic occurs and whether the traffic areas have been planned effectively. Figure 13-2 shows the difference between a poorly designed traffic pattern and a well-designed traffic pattern.

HALLS

Halls are the highways and streets of the home. They provide a controlled path that connects the various areas of the house. Halls should be planned to eliminate or keep to a minimum the passage of traffic through rooms. Long, dark, tunnel-like halls should be avoided. Halls should be well lighted, light in color and texture, and planned with the decor of the whole house in mind. The hall shown in Fig. 13-3 is extremely long; however, it is broken by level, by open partitions, and by light variations.

One method of channeling hall traffic without the use of solid walls is with the use of dividers. Planters, half walls, louvered walls, and even furniture can be used as dividers. Figure 13-3 shows the use of furniture components in dividing the living area from the hall. This arrangement enables both the hall and the living room to share ventilation, light, and heat.

Another method of designing halls and corridors as an integral part of the area design is with the use of movable partitions. The Japanese scheme of placing these partitions (shoji) between the living area and a hall is effective. In some Japanese homes, this hall actually becomes a lanai when the partition between the living area and the hall is closed and the outside wall is opened.

Figure 13-4 shows some of the basic principles of efficient hall design.

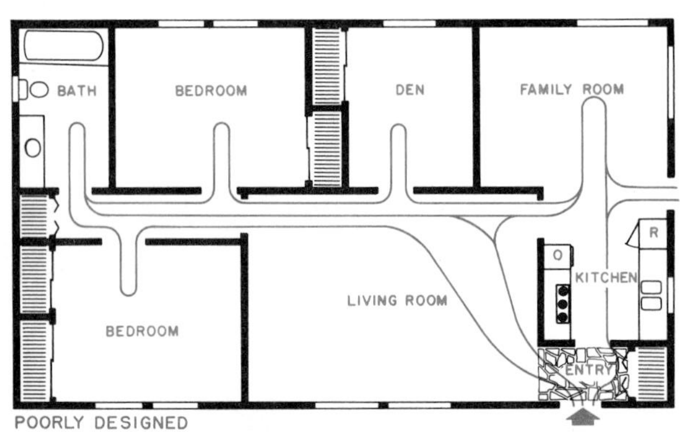

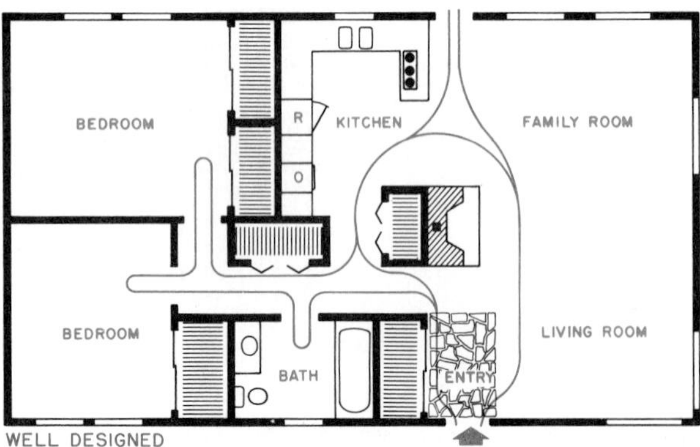

Fig. 13-2 The difference between a poorly designed and a well-designed traffic pattern.

Fig. 13-3 The use of furniture components in separating traffic areas. *(United States Plywood Corp.)*

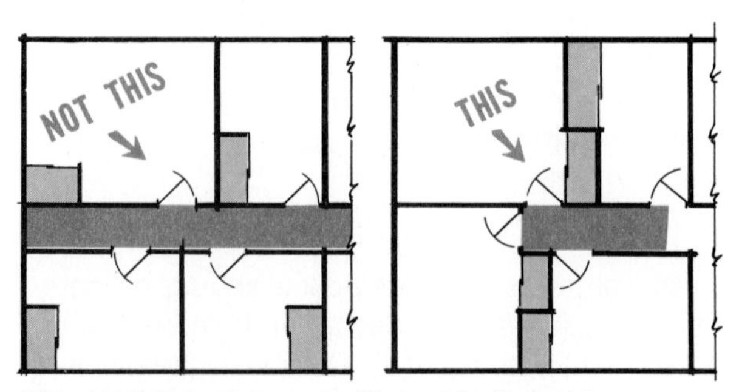

Fig. 13-4 Principles of efficient hall design.

STAIRS

Stairs are inclined hallways. They provide access from one level to another. Stairs may lead directly from one area to another without a change of direction, they may turn 90 degrees (90°) by means of a landing, or they may turn 180° by means of landings. Figure 13-5 shows the basic types of stairs.

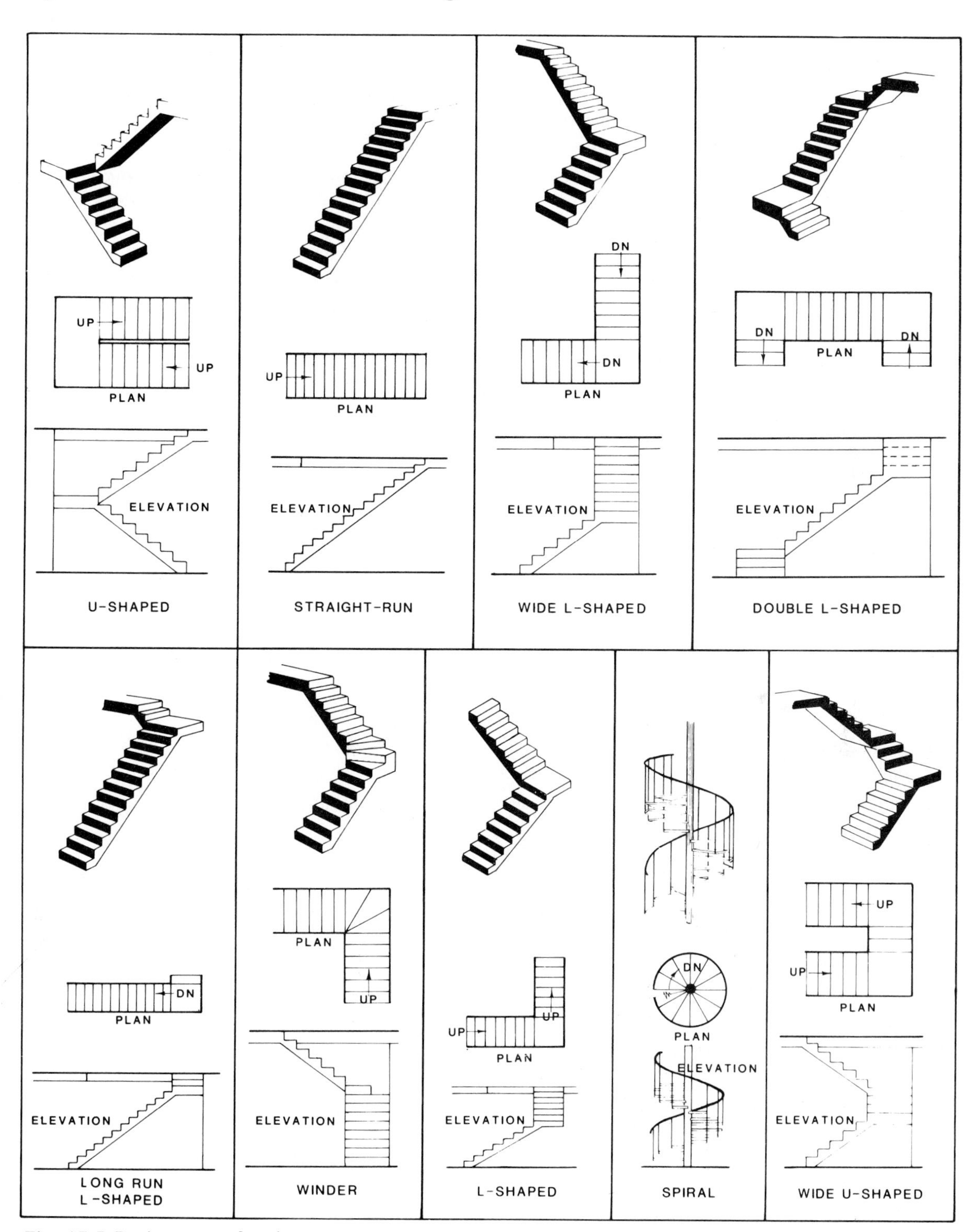

Fig. 13-5 Basic types of stairs.

With the use of newer, stronger building materials and new techniques, there is no longer any reason for enclosing stairs in walls that restrict light and ventilation. Stairs can now be supported by many different devices.

Windows should be placed to provide natural light for stairs. Because stairwells should be lighted at all times when in use, natural light is the most energy-efficient. If it is difficult to provide natural light, three-way switches should be provided at the top and bottom of the stairwell to control the stair lighting. See Unit 61, "Drawing Electrical Plans."

SPACE REQUIREMENTS

There are many variables to consider in designing stairs. The tread width, the riser width, the width of the stairwell opening, and the headroom all help to determine the total length of the stairwell.

The *tread* is the horizontal part of the stair, the part upon which you walk. The average width of the tread is 10″ (250 mm). The *riser* is the vertical part of the stair. The average riser height is 7¼″ (180 mm). Figure 13-6 shows the importance of correct tread and riser design.

The overall width of the stairs is the length or distance across the treads. A minimum of 3′ (915 mm) should be allowed for the total stair width. However, a width of 3′-6″ (1070 mm) or even 4′ (1220 mm) is preferred (Fig. 13-7).

Headroom is the vertical distance between the top of each tread and the top of the stairwell ceiling. A minimum headroom distance of 6′-6″ (2 m) should be allowed. However, distances of 7′ (2.1 m) are more desirable (Fig. 13-8).

Landing dimensions will probably be determined by the size of the stairs and the space for the stairwell. More clearance must be allowed where a door opens on a landing (Fig. 13-9). Landings should be located at the center between levels to eliminate long runs.

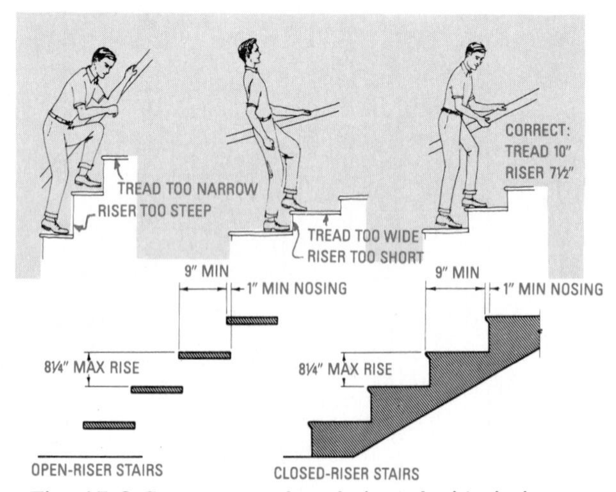

Fig. 13-6 Correct tread and riser design is important.

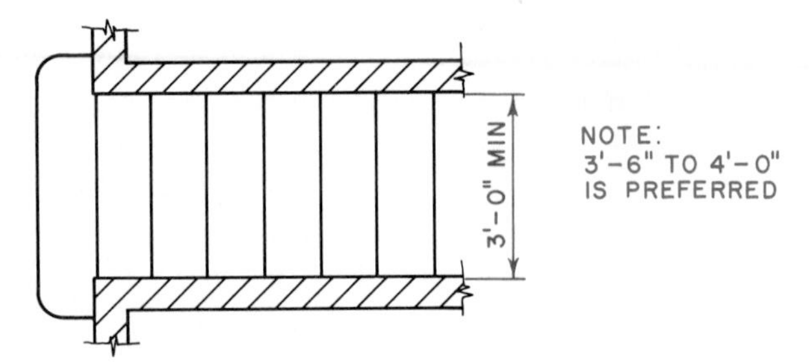

Fig. 13-7 Minimum width of stairs.

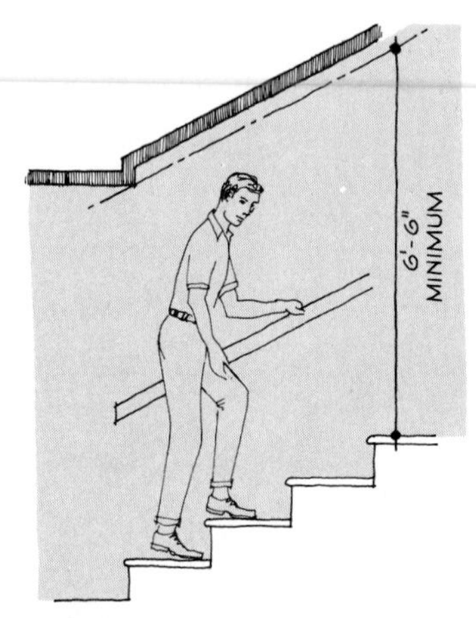

Fig. 13-8 Minimum headroom clearance.

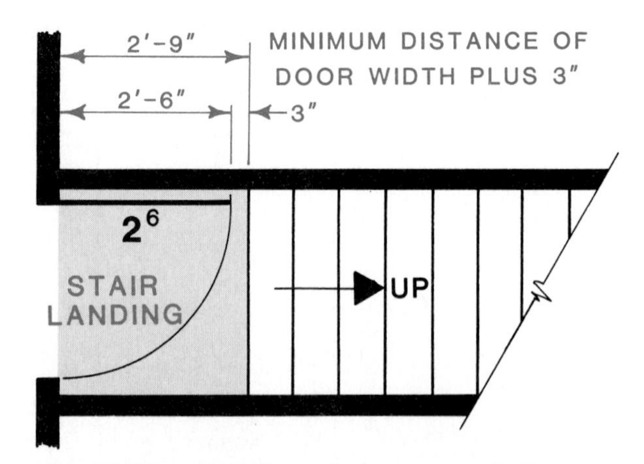

Fig. 13-9 Landing dimensions may need allowance for a door.

Exercises

1. **Sketch the floor plan of a home of your design. Plan the most efficient traffic pattern by tracing the route of your daily routine.**
2. **Define the following terms: *traffic pattern, halls, corridors, main traffic areas, movable partitions, minimum stair width, center-supported stairs, tread, riser, minimum stair width, headroom, landing.***

UNIT 14

ENTRANCES

Entrances are divided into several different types: the main entrance (Fig. 14-1), the service entrance, and the special-purpose entrance. The entrance is composed of an outside waiting area (porch, marquee, lanai), a separation (door), and an inside waiting area (foyer, entrance hall).

Fig. 14-1 Contemporary main entrance. *(Scholz Homes, Inc.)*

FUNCTION

Entrances provide for and control the flow of traffic into and out of a building. Different types of entrances have somewhat different functions.

Main Entrance

The *main entrance* provides access to the house. It is the one through which guests are welcomed and from which all major traffic patterns radiate. The main entrance should be readily identifiable. It should provide shelter to anyone awaiting entrance.

Some provision should be made in the main-entrance wall for the viewing of callers from the inside. This can be accomplished through the use of side panels, lights (panes) in the door or windows (Fig. 14-2) which face the side of the entrance.

The main entrance should be planned to create a desirable first impression. A direct view of other areas of the house from the foyer should be baffled but not sealed off. Figure 14-1 shows an atrium foyer that channels traffic without creating a closed area. Also, a direct view of exte-

Fig. 14-2 Side windows provide a view of the entrance from the inside. *(Home Planners, Inc.)*

rior parking areas should be baffled from view.

The entrance foyer should include a closet for the storage of outdoor clothing and guest's wraps. This foyer closet should have a capacity that will accommodate both family and guests.

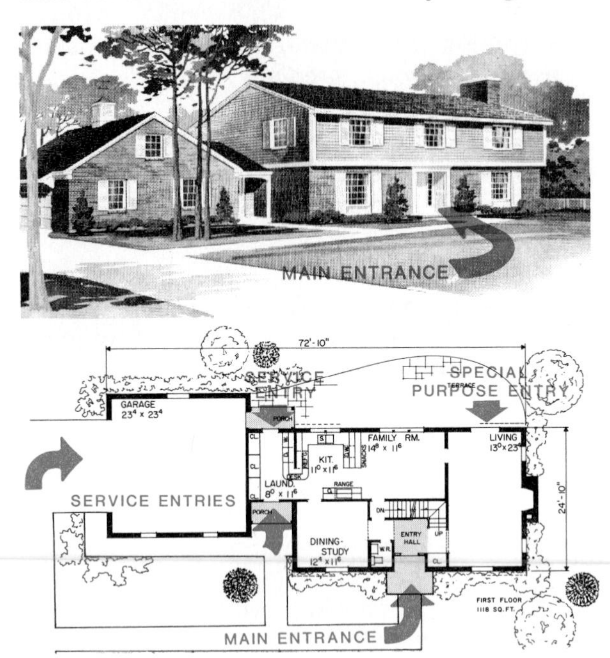

Fig. 14-3 The basic types of entrances. *(Scholz Homes, Inc.)*

Service Entrance

The *service entrance* provides access to the house through which supplies can be delivered to the service areas without going through other parts of the house. It should also provide access to parts of the service area (garage, laundry, workshop) for which the main entrance is inappropriate and inconvenient.

Special-Purpose Entrances

Special-purpose entrances and exits do not provide for outside traffic. Instead they provide for movement from the inside living area of the house to the outside living areas. A sliding door from the living area to the patio is a special-purpose entrance. It is not an entrance through which street, drive, or sidewalk traffic would have access. Figure 14-3 shows the difference between special-purpose entrances, the main entrance, and the service entrance. Figure 14-4 shows the use of numerous special-purpose entrances.

LOCATION

The main entrance should be centrally located to provide easy access to each area. It should be

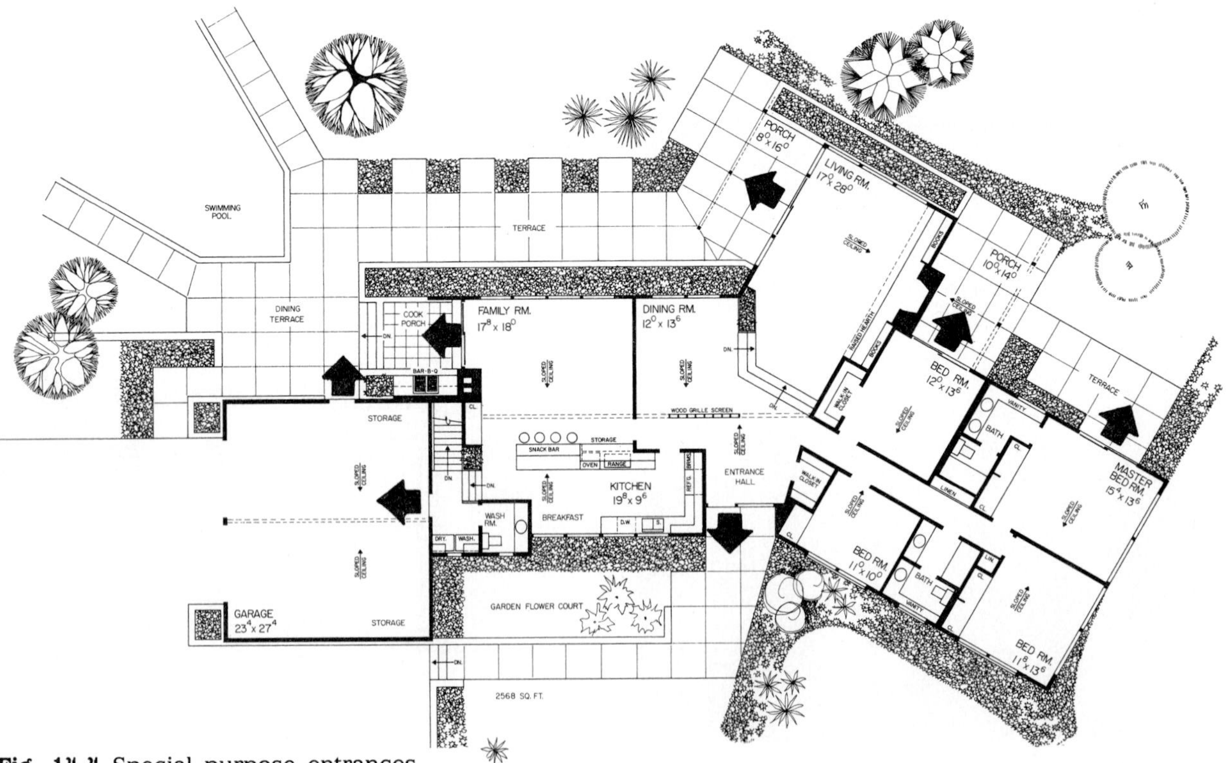

Fig. 14-4 Special purpose entrances. *(Home Planners, Inc.)*

Fig. 14-5 Left: an entrance with lines related to the lines of the structure. Right: an entrance with lines unrelated to the lines of the structure.

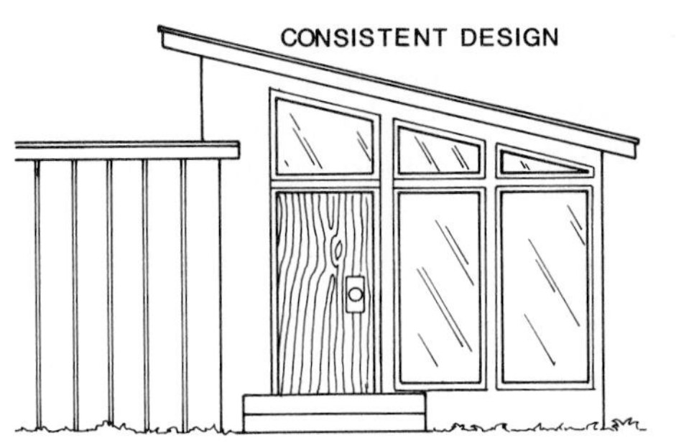

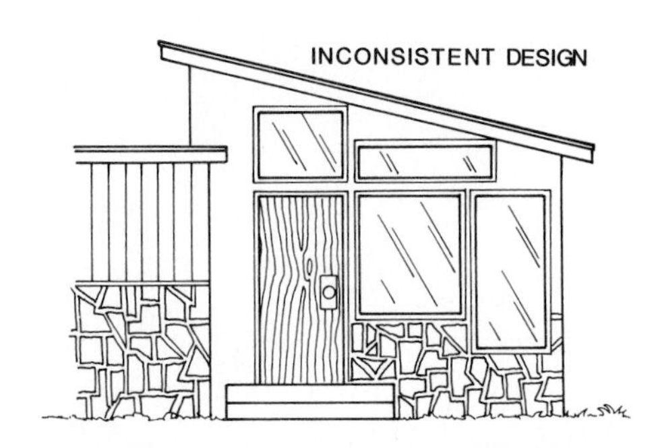

conveniently accessible from driveways, sidewalks, or street.

The service entrance should be located close to the drive and garage. It should be placed near the kitchen or food-storage areas.

Special-purpose entrances and exits are often located between the bedroom and the quiet patio, between the living room and the living patio, and between the dining room or kitchen and the dining patio. Figure 14-3 shows the functional placement of all these entrances.

DECOR

The entrance should create a desirable first impression. It should be easily identifiable yet an integral part of the architectural style.

Consistency of Style

The total design of the entrance should be consistent with the overall design of the house. The design of the door, the side panel, and the deck and cover should be directly related to the lines of the house. The lines of the entrances shown at the left in Fig 14-5 are designed as integral parts of the exterior. The lines of the entrance shown at the right in Fig. 14-5 are unrelated to the major building lines of the structure.

The entrance shown in Fig. 14-6A is a good example of entrance design involving all the principles of location, style consistency, lighting utilization, and size and shape effectiveness. Figure 14-6B shows a close view of an entrance with double swinging entry door.

Fig. 14-6A Entrance lines related to the remainder of the home. *(Scholz Homes, Inc.)*

Fig. 14-6B A close view of related entrance lines. *(Scholz Homes, Inc.)*

Open Planning

The view from the main entrance to the living area should be baffled without creating a boxed-in appearance. The foyer should not appear as a dead end. The extensive use of glass, effective lighting, and carefully placed baffle walls can create an open and inviting impression. This is accomplished in the entrance shown in Fig. 14-7 by the use of window walls, double doors, roof-overhang extension, and baffle walls that extend

Fig. 14-7 Open-plan entrance and foyer. *(Western Wood Products Assoc.)*

the length of the foyer. Open planning between the entrance foyer and the living areas can also be accomplished by the use of louvered walls or planter walls. These provide a break in the line of sight but not a complete separation. Sinking or elevating the foyer or entrance approach also provides the desired separation without isolation.

Flooring

The outside portion of the entrance should be weather-resistant stone, brick, or concrete. If a porch is used outside the entrance, a wood deck will suffice. The foyer deck should be easily maintained and be resistant to mud, water, and dirt brought in from the outside. Asphalt, vinyl or rubber tile, stone, flagstone, marble, and terrazzo are most frequently used for the foyer deck. The use of a different material in the foyer area helps to define the area when no other separation exists.

Foyer Walls

Paneling, masonry, murals, and glass are used extensively for entrance foyer walls. The walls of the exterior portion of the entrance should be consistent with the other materials used on the exterior of the house.

Lighting

An entrance must be designed to function day and night. General lighting, spot lighting, and all-night lighting are effective for this purpose. Lighting can be used to accent distinguishing features or to illuminate the pattern of a wall, which actually provides more light by reflection and helps to identify and accentuate the entrance at night. Natural lighting, as shown in Fig. 14-8, is also effective in lighting entrance areas during daylight hours.

SIZE AND SHAPE

The size and shape of the areas inside and outside the entrance depend on the budget and the type of plan. Foyers are not bounded by solid walls in the open plan. In formal or closed plans the foyer may be partially or fully closed off.

The Outside

The outside covered portion of the entrance should be large enough to shelter several people at one time. Sufficient space should be allowed on all sides, exclusive of the amount of space needed to open storm doors that open to the outside. Outside shelter areas range in size from the minimum arrangement to the more generous size shown in Fig. 14-9.

Fig. 14-8 Natural lighting used in a foyer. *(Western Wood Products Assoc.)*

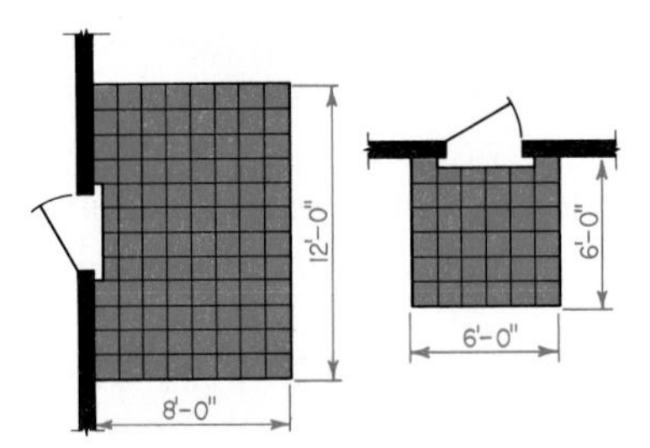

Fig. 14-9 The entrance area on the left has optimum dimensions. The entrance on the right has minimum dimensions.

The Inside

The inside of the entrance foyer should be sufficiently large to allow several people to enter at the same time, remove their coats, and store them in the closet. A 6′ X 6′ (1.8 m X 1.8 m) foyer is considered minimum for this function. A foyer 8′ X 10′ (2.4 m X 3.0 m) is average, but a more desirable size is 8′ X 15′ (2.4 m X 4.6 m).

A foyer arrangement must allow for the swing of the door, something that must be taken into consideration in determining the size of the foyer. If the foyer is too shallow, passage will be blocked when the door is open, and only one person can enter at a time.

Foyers are normally rectangular because they lead to several areas of the home. They do not need much depth in any one direction. The ideal entry includes:

1. Adequate room to handle traffic flow
2. Access to all three areas of a home
3. A closet
4. Bath access for guests
5. Consistent decor
6. Outside weather protection
7. Effective lighting day and night

Exercises

1. Redesign the entrance shown in Fig. 14-3B, adding sufficient shelter space that will be consistent with the main lines of the house.

2. Plan a foyer for a house you have designed.

3. Define the following terms: *main entrance, service entrance, foyer, special-purpose entrances, open planning.*

UNIT 15

DENS AND STUDIES

The den or study can be designed for many different purposes, depending on the living habits of its occupants.

FUNCTION

The den may function basically as a reading room, writing room, hobby room, or professional office. For the teacher, writer, or lawyer, the study may be basically a reading and writing room. For the engineer, architect, drafter, or artist, the den or study may function primarily as a studio and may include such facilities as those shown in the study in Fig. 15-1.

The den or study often doubles as a guest room. Quite often the children's bedroom can provide facilities normally included in a study, such as desk, bookcase, and hobby space.

LOCATION

The den is often considered part of the sleeping area, since it may require placement in a quiet part of the house. It also may function primarily in the living area, especially if the study is used as a professional office by a physician or an insurance agent whose clients call at home. Fig-

Fig. 15-1 A studio-study. *(Haas Cabinet Co.)*

ure 15-2 shows a professional study or office located near the main entrance hall and accessible from the main entrance and also through a side entrance directly from the garage.

If a study doubles as a public office, it should be located in an accessible area. However, if it is only for private use, then it can be located in the basement or attic, using otherwise wasted space.

DECOR AND LIGHTING

The decor of the study should reflect the main activity and should allow for well-diffused general lighting and glareproof local lighting. Windows positioned above eye level admit the maximum amount of light without exposing distracting eye-level images from the outside.

SIZE AND SHAPE

The size and shape of the den, study, or office will vary greatly with its function. Size and shape will depend on whether one or two persons expect to use the room privately or whether it should provide a meeting place for business clients. Studies range in size from just enough space for a desk and chair in a small corner to a large amount of space with a diversity of furnishings, depending on the major functions for which the room is planned.

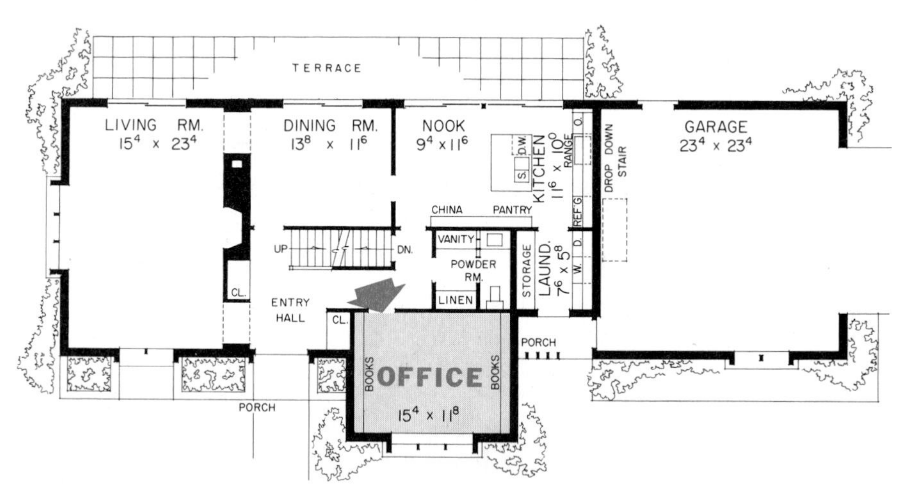

Fig. 15-2 A professional office should be accessible from the main entrance.

Exercises

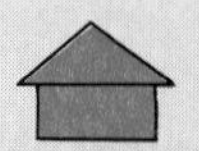

1. **Sketch a plan for a den in a home of your own design.**
2. **Sketch a plan for a den for your own home.**
3. **Sketch a plan for a den which will double as a guest bedroom.**
4. **Define the following terms: *den, study, living area, sleeping area, guest room, professional office, central theme.***

SECTION 4

SERVICE AREA

The service area includes the kitchen, laundry, garage, workshops, storage centers, and utility room. Since a great number of different activities take place in the service area, it should be designed for the greatest efficiency. The service area should include facilities for the maintenance and servicing of the other areas of the home. The functioning of the living and the sleeping areas is greatly dependent upon the efficiency of the service area.

UNIT 16

KITCHENS

A well planned kitchen is efficient, attractive, and easy to maintain. To design an efficient kitchen, the designer must consider the function, basic shape, decor, size, and location of equipment.

FUNCTION

The preparation of food is the basic function of the kitchen. However, the kitchen may also be used as a dining area and as a laundry.

The proper placement of appliances, storage cabinets, and furniture is important in planning efficient kitchens. Locating appliances in an efficient pattern eliminates much wasted motion. An efficient kitchen is divided into three areas: the storage and mixing center, the preparation and cleaning center, and the cooking center (Fig. 16-1).

Fig. 16-1 Efficient kitchens are divided into three activity areas. *(Tappan Co.)*

Storage and Mixing Center

The refrigerator is the major appliance in the storage and mixing center. The refrigerator may be freestanding, built-in, or suspended from a wall. The storage and mixing center also includes cabinets for the storage of utensils and ingredients used in cooking and baking, as well as a countertop work area.

Preparation and Cleaning Center

The sink is the major appliance in the preparation and cleaning center. Sinks are available in one- and two-bowl models with a variety of cabinet arrangements and countertop and drainboard areas. The preparation and cleaning center may also include a waste-disposal unit, an automatic dishwasher, a waste compactor, and cabinets for storing brushes, towels, and cleaning supplies.

Cooking Center

The range and oven are the major appliances in the cooking center. The range and oven may be combined into one appliance, or the burners may be installed in the counter-top while the oven is built into a cabinet. The cooking center should also include counter-top work space, as well as storage space for minor appliances and cooking utensils that will be used in the area. The cooking center must have an adequate supply of electrical outlets for the many minor appliances used in cooking. Figure 16-2 shows the size requirements for the storage or installation of many minor appliances that may be located in the various centers.

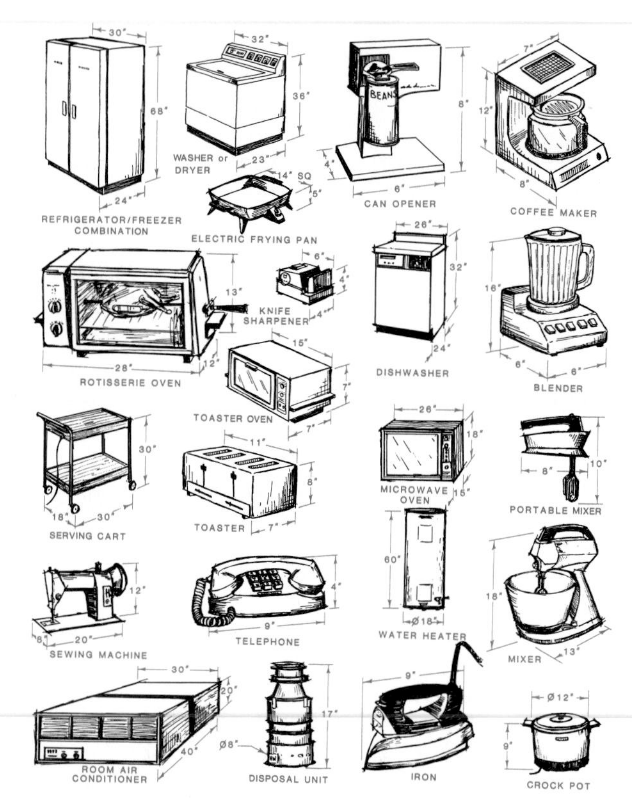

Fig. 16-2 Sizes of common appliances.

Fig. 16-3 The length of the work triangle should be between 12′ (3.7m) and 22′ (6.7m). *(Hotpoint Div. General Electric Corp.)*

Work Triangle

If you draw a line connecting the three centers of the kitchen, a triangle is formed (Fig. 16-3). This is called the *work triangle*. The perimeter of an efficient kitchen work triangle should be between 12′ and 22′ (3.7 and 6.7 m).

BASIC SHAPES

The position of the three areas on the work triangle may vary greatly. However, the most efficient arrangements usually fall into the following categories.

U-Shaped Kitchen The U-shaped kitchen is a very efficient arrangement. The sink is located at the bottom of the U, and the range and the refrigerator are at the opposite ends. In this arrangement, traffic passing through the kitchen

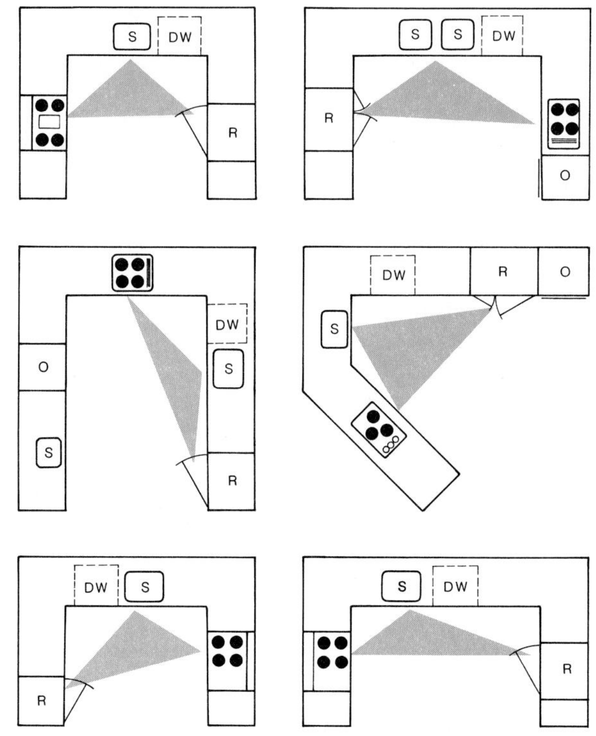

Fig. 16-4 Six arrangements for a U-shaped kitchen.

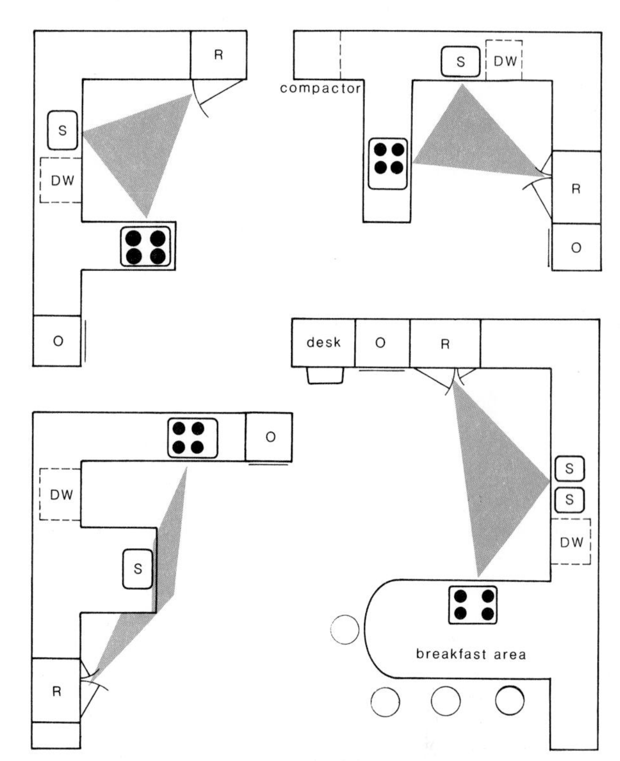

Fig. 16-5 Four arrangements for a peninsula kitchen.

is completely separated from the work triangle. The open space in the U between the sides may be 4′ or 5′ (1.2 or 1.5 m). This arrangement produces a very efficient but small kitchen. Figure 16-4 shows various U-shaped-kitchen layouts and the resulting work triangles (Fig. 16-4).

Peninsula Kitchen The peninsula kitchen is similar to the U kitchen. However, one end of the U is not enclosed with a wall. The cooking center is often located in this peninsula, and the peninsula is often used to join the kitchen to the dining room or family room. Figure 16-5 shows various arrangements of peninsula kitchens and the resulting work triangles.

L-Shaped Kitchen The L-shaped kitchen (Fig. 16-6) has continuous counters and appliances and equipment on two adjoining walls. The work triangle is not in the traffic pattern. The remaining space is often used for other kitchen facilities, such as dining or laundry facilities. If the walls of an L-shaped kitchen are too long, the compact efficiency of the kitchen is destroyed. Figure 16-7 shows several L-shaped kitchens and the work triangles that result from these arrangements.

Fig. 16-6 L-shaped kitchens permit a large area of open floor space. *(Home Planners Inc.)*

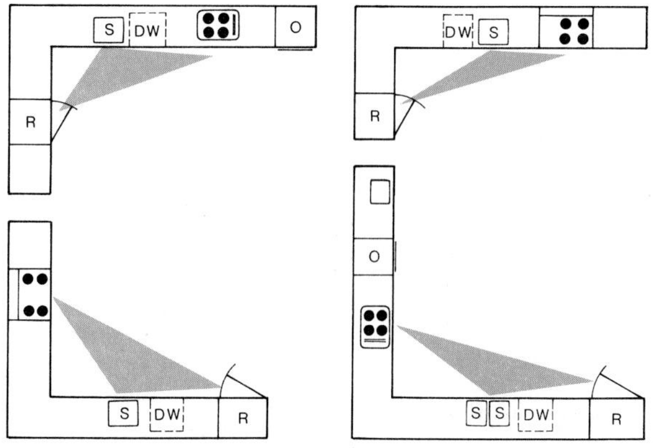

Fig. 16-7 Four arrangements for an L-shaped kitchen.

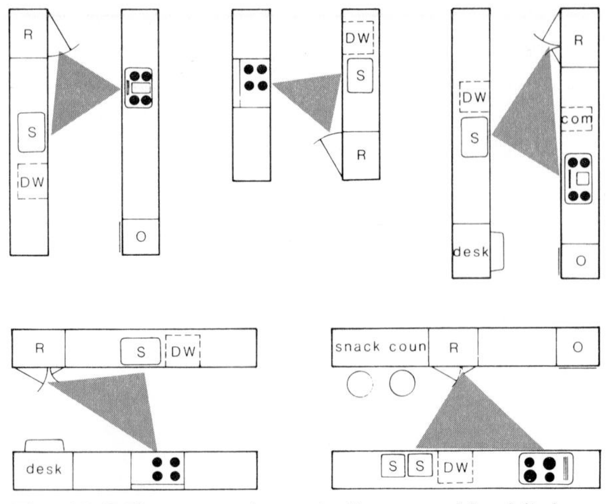

Fig. 16-8 Five arrangements for a corridor kitchen.

Corridor Kitchen The two-wall corridor kitchens shown in Fig. 16-8 are very efficient arrangements for long, narrow rooms. A corridor kitchen is unsatisfactory, however, if considerable traffic passes through the work triangle. A corridor kitchen produces a very efficient work triangle.

One-Wall Kitchen A one-wall kitchen is an excellent plan for small apartments, cabins, or houses in which little space is available. The work centers are located in open line and produce an efficient arrangement (Fig. 16-9). However, in planning the one-wall kitchen, the designer must be careful to avoid having the wall too long and must provide adequate storage facilities. Figure 16-10 shows several one-wall-kitchen arrangements.

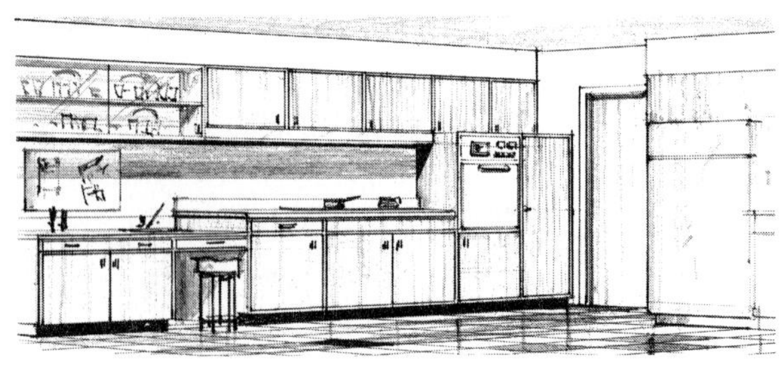

Fig. 16-9 A one-wall kitchen, *(Home Planners, Inc.)*

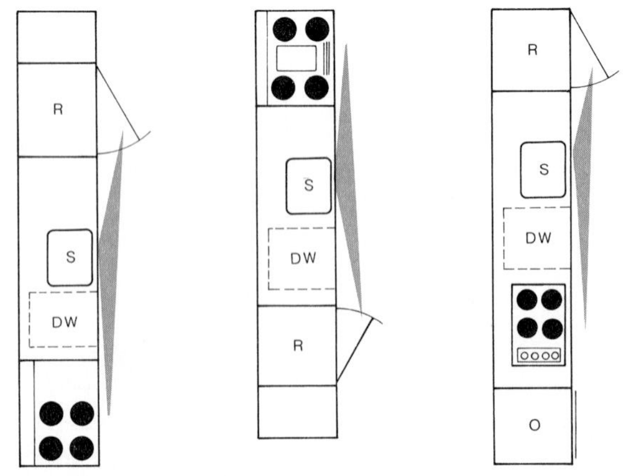

Fig. 16-10 Three arrangements for a one-wall kitchen.

Island Kitchen The island, which serves as a separator for the different parts of the kitchen, usually has a range top or sink, or both, and is accessible on all sides. Other facilities that are sometimes located in the island are the mixing center, work table (Fig. 16-11A), buffet counter, extra sink (Fig. 16-11B), and snack center. Figure 16-11C shows examples of other island facilities. The island design is convenient when two or more persons work in the kitchen at the same time.

Fig. 16-11A Worktable island. *(Home Planners, Inc.)*

Fig. 16-11B A sink island kitchen. *(Elkay Mfg. Co.)*

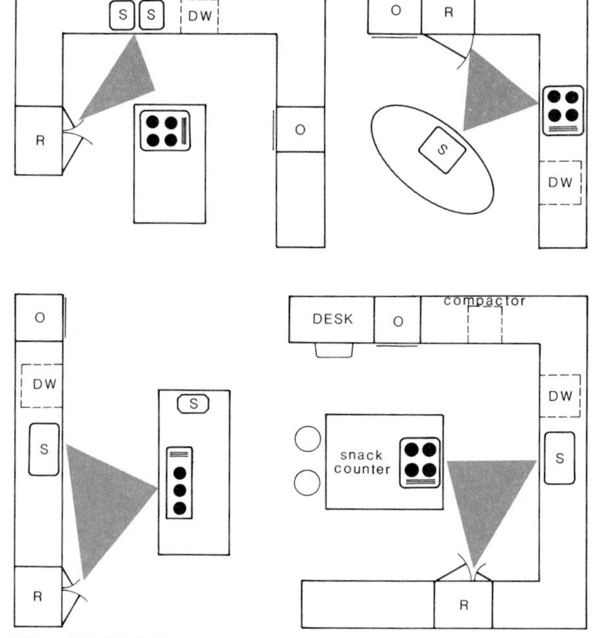

Fig. 16-11C Four island-kitchen arrangements.

Family Kitchen The family kitchen is an open kitchen using any of the basic plans. Its function is to provide a meeting place for the entire family in addition to providing for the normal kitchen functions. Family kitchens are normally divided into two sections. One section is for food preparation, which includes the three work centers; the other section includes a dining area and family-room facilities.

Family kitchens must be rather large to accommodate these facilities. An average size for a family kitchen is 15′ × 15′ or 225 sq. ft. (4.6 m) square. Figure 16-12 shows several possible arrangements for family kitchens.

DECOR

Even though kitchen appliances are of contemporary design, some homemakers prefer to decorate kitchens with a period or colonial motif. The design of the cabinets, floors, walls, and accessory furniture must therefore be accented to give the desired effect. Compare the colonial kitchen shown in Fig. 16-13 with the modern kitchen shown in Fig. 16-14. The kitchen shown in Fig. 16-13 is a colonial version of the contemporary kitchen shown in Fig. 16-14.

Regardless of the style, kitchen walls, floors, countertops, and cabinets should require a minimum amount of maintenance. Materials that are relatively maintenance-free include stainless steel, stain-resistant plastic, ceramic tile, washable wall coverings, washable paint, vinyl, and laminated plastic countertops.

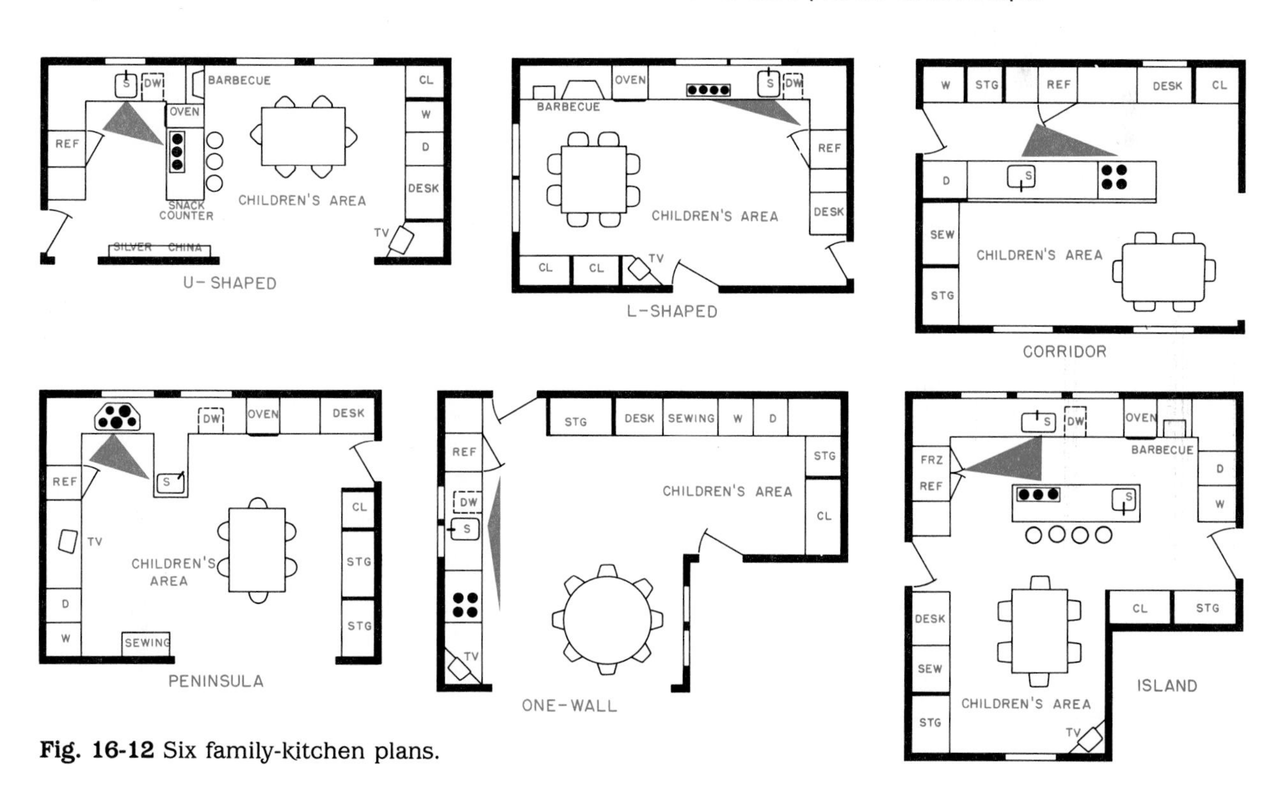

Fig. 16-12 Six family-kitchen plans.

Fig. 16-13 A colonial kitchen decor. *(Consoweld Corp.)*

Fig. 16-14 A contemporary kitchen decor. *(Consoweld Corp.)*

LOCATION

Since the kitchen is the core of the service area, it should be located near the service entrance and near the waste-disposal area. The children's play area should also be visible from the kitchen, and the kitchen must be adjacent to the dining area and outdoor eating areas.

KITCHEN PLANNING GUIDES

The following guides for kitchen planning provide a review of the more important factors to consider in designing efficient and functional kitchens:

1. The traffic lane is clear of the work triangle.

2. The work areas include all necessary appliances and facilities.

3. The kitchen is located adjacent to the dining area.

4. The kitchen should be located near the children's play area.

5. The kitchen is cheerful and pleasant.

6. The centers include (a) the storage center, (b) the preparation and cleaning center, and (c) the cooking center.

7. The work triangle measures between 12′ and 21′.

8. An adequate number of electrical outlets are provided for each work center.

9. Adequate storage facilities are available in each work center (Fig. 16-15A through D).

10. Shadowless and glareless light is provided and is concentrated on each work center.

11. Adequate counter space is provided for meal preparation.

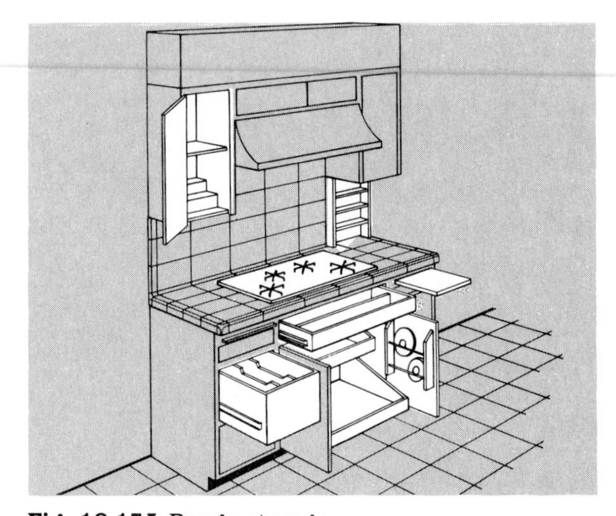

Fig. 16-15A Range storage.

Fig. 16-15B Sink storage.

12. Ventilation is adequate.

13. The oven and range are separated from the refrigerator by at least one cabinet.

14. Doors on appliances swing away from the work-triangle (Fig. 16-16).

15. Lapboard heights are 26″ (660 mm).

16. Working heights for counters are 36″ (915 mm).

17. Working heights for table are 30″ (760 mm).

18. The combination of base cabinets, wall cabinets, and appliances provides a consistent standard unit without gaps or awkward depressions or extensions.

19. Cabinet and appliance locations have been planned using standard manufacturer's modular dimensions (Figs. 16-17 and 16-18).

20. A counter is provided adjacent to the refrigerator to place items while loading or unloading (Fig. 16-19).

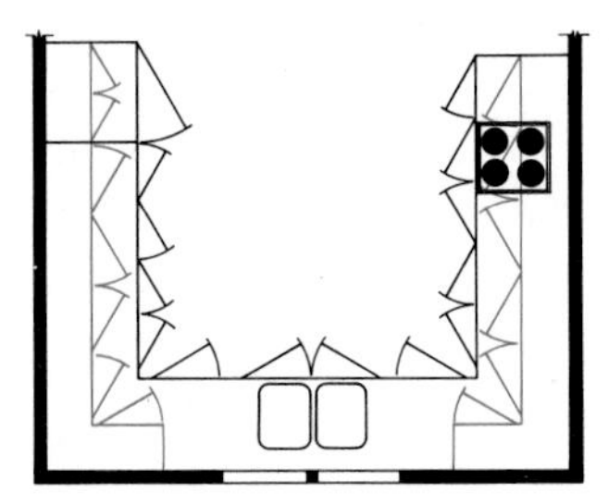

Fig. 16-16 Cabinet doors should open away from the work area.

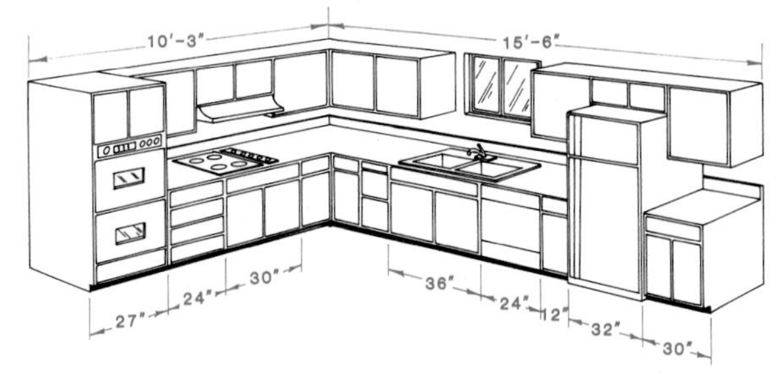

Fig. 16-17 Design standards for modular kitchen units. *(William Wagoner)*

Fig. 16-15C Corner storage.

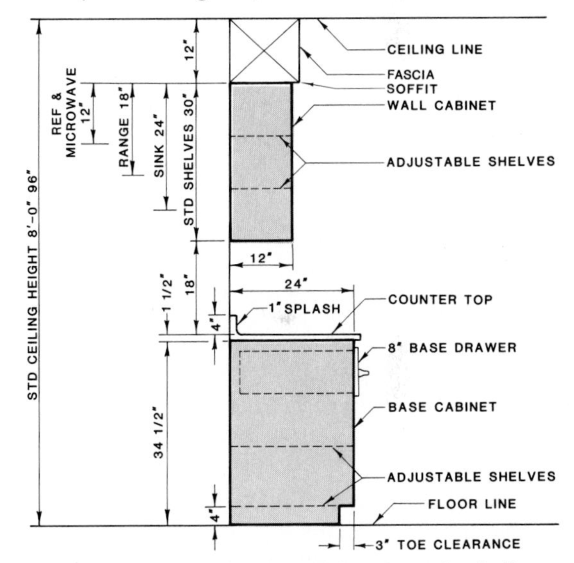

Fig. 16-18 Wall and base cabinet standard dimensions.

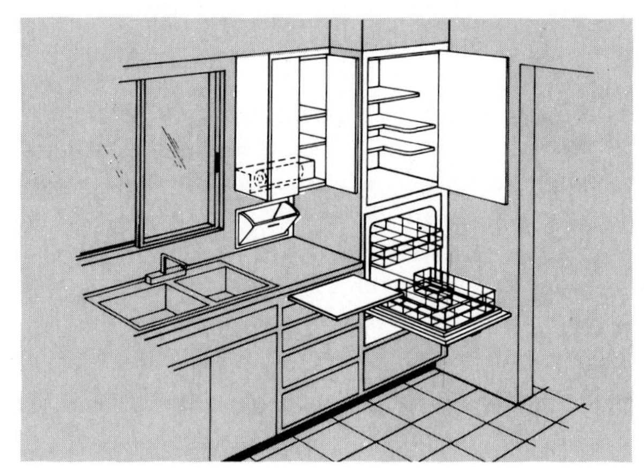

Fig. 16-15D Wall cabinet storage.

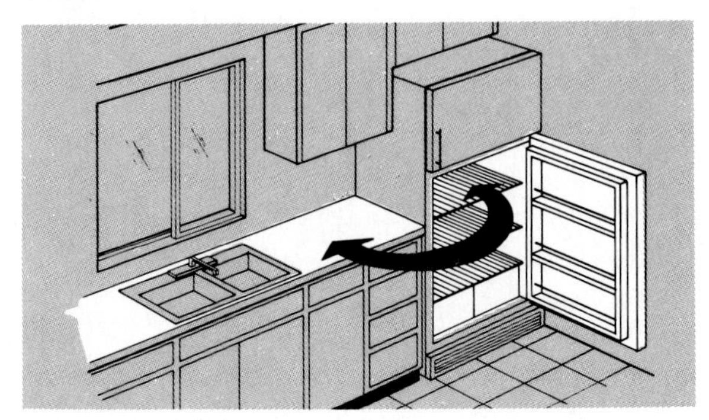

Fig. 16-19 Refrigerator storage. *(Southern California Gas Co.)*

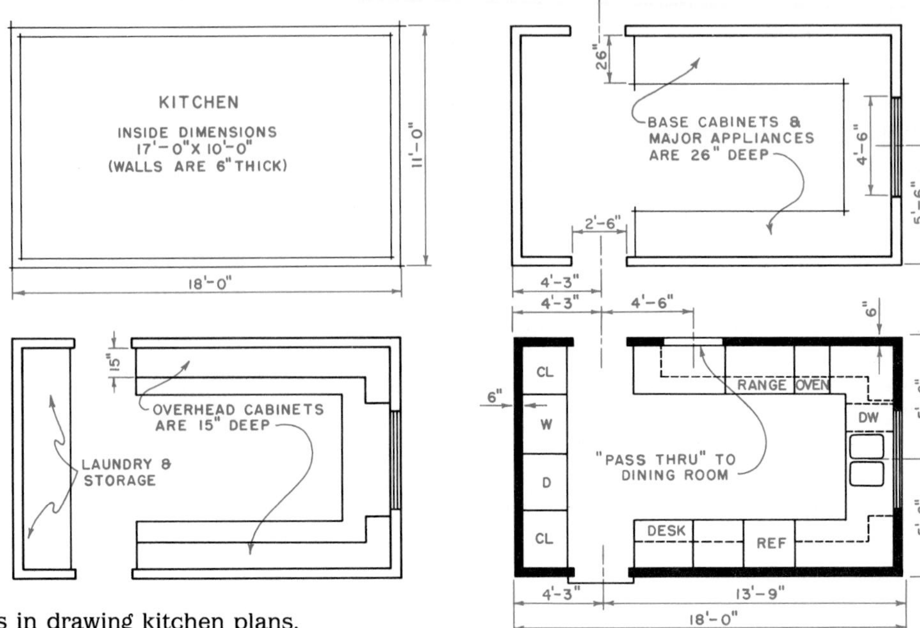

Fig. 16-20 Basic steps in drawing kitchen plans.

DRAWING KITCHENS

In planning and drawing kitchen floor plans, use template planning techniques and procedures as outlined in Unit 31, "Functional Room Planning." However, once basic dimensions are established or if room dimensions are predetermined, follow the steps outlined in Fig. 16-20 in drawing a kitchen floor plan.

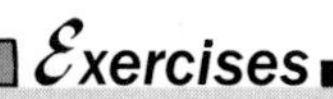

Exercises

1. **Sketch a floor plan of one of the U-shaped kitchens shown in Fig. 16-4. Show the position of the dining area in relation to this kitchen, using the scale ½" = 1′-0″.**
2. **Remodel the kitchen shown in Fig. 16-21. Change door arrangements and dining facilities as needed.**

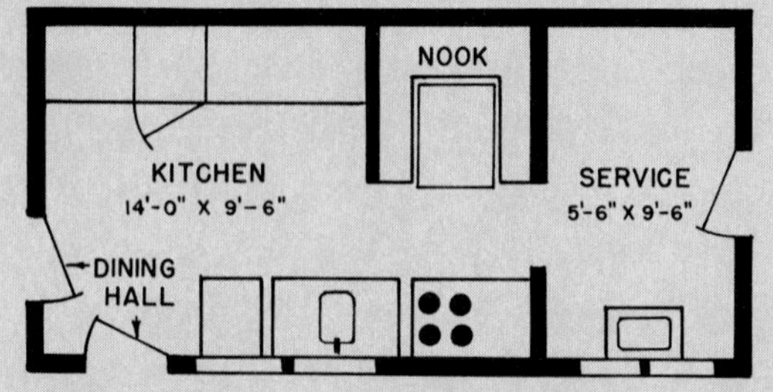

Fig. 16-21

3. **Sketch a floor plan of the kitchen in your own home. Prepare a revised sketch to show how you would propose to redesign this kitchen. Make an attempt to reduce the size of the work triangle.**

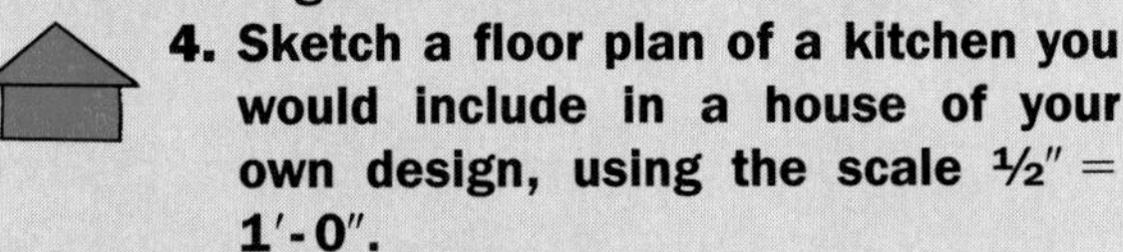

4. **Sketch a floor plan of a kitchen you would include in a house of your own design, using the scale ½" = 1′-0″.**
5. **Define the following terms: *work triangle, U shape, peninsula, L shape, corridor, island, family kitchen, storage and mixing center, planning and preparation center, cooking center, major appliances, minor appliances, base cabinet, wall cabinet, countertop, service area.***

UNIT 17

UTILITY ROOMS

The *utility room* may include facilities for washing, drying, ironing, sewing, and storing household cleaning equipment. It may contain heating and air-conditioning equipment and/or even pantry shelves for storing groceries. Other names for this room are *service room, all-purpose room,* and *laundry room.*

If the utility room is used for heating and air-conditioning, space must be planned for the furnace, heating and air-conditioning ducts, hot-water heater, and any related equipment such as humidifiers or air purifiers.

SHAPE AND SIZE

The shapes and sizes of utility rooms differ, as shown in Fig. 17-1. The average floor space required for appliances, counter, and storage area is 100 square feet, or ft^2 (10 m^2). However, this size may vary according to the budget or needs of the household.

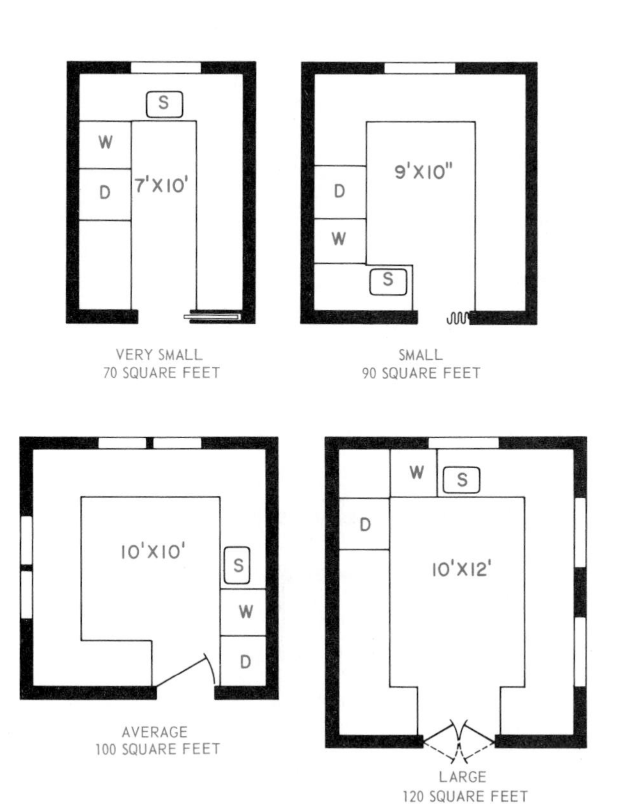

Fig. 17-1 The size of a utility room varies according to the budget and needs of the family.

STYLE AND DECOR

Style and decor in a utility room depend on the function of the appliances, which are themselves an important factor in the appearance of the room. Simplicity, straight line, and continuous counter spaces produce an orderly effect and permit work to progress easily. Such features also make the room easy to clean.

An important part of the decor is the color of the paint used for walls and cabinet finishes. Colors should harmonize with the colors used on the appliances. All finishes should be washable. The walls may be lined with sound-absorbing tiles or wood paneling.

The lighting in a utility room should be carefully planned so that it would be 48" (1220 mm) above the equipment used for washing, ironing, and sewing. However, the lighting fixtures placed above the preparation area and laundry sinks can be farther from the work-top area.

THE LAUNDRY AREA

The laundry area is only one part of the utility room, but it is usually the most important center. To make laundry work as easy as possible, the appliances and working spaces in a laundry area should be located in the order in which they will be used. Such an arrangement will save time and effort. There are four steps in the process of laundering. The equipment needed for each of these steps should be grouped so that the person doing the laundry can proceed from one stage to the next in an orderly and efficient way (Fig. 17-2).

Receiving and Preparing The first step in laundering—receiving and preparing the items—requires hampers or bins, as well as counters on which to collect and sort the articles. Near this equipment there should be storage facilities for laundry products such as detergents, bleaches and stain removers.

Washing The next step, the actual washing, takes place in the area containing the washing machine and laundry tubs, trays, or sinks.

Drying The equipment needed for this stage of the work includes a dryer, indoor drying lines, and space to store clothespins. Either a

Fig. 17-2 The appliances and working spaces should be arranged in the order in which they are used.

220-volt (220-V) outlet or gas access is necessary for all dryers.

Ironing and Storage For the last part of the process, the required equipment consists of an iron and a board, a counter for folding, a rack on which to hang finished ironing, and facilities for sewing and mending. If a sewing machine is included, it may be portable, or it may fold into a counter or wall.

Location

Separate Laundry Area The location of the laundry area in a utility room is desirable because all laundry functions, including repairs, will then be centered in one place. A further advantage of the separate room is that laundering is kept well apart from the preparation of foods. Figure 17-3 shows three optional arrangements for a separate laundry room.

Space is not always available for a utility room, however, and the laundry appliances and space for washing and drying may need to be located in some other area. Wherever it is placed, the equipment in the laundry unit should be arranged in the order in which the work must be done.

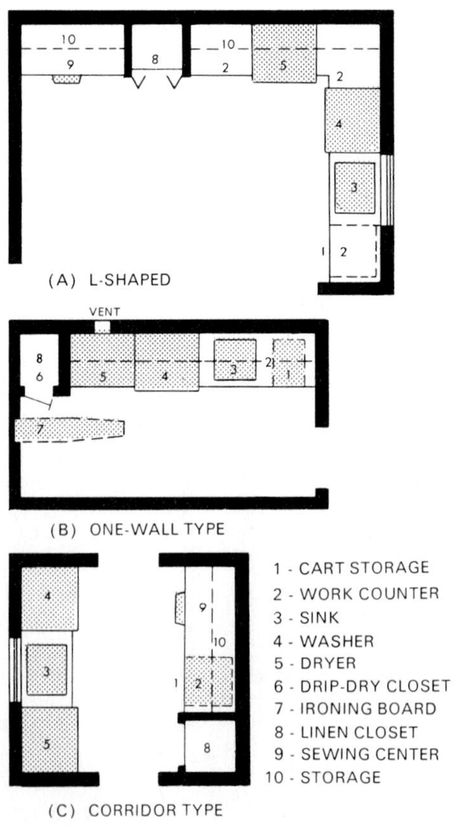

Fig. 17-3 Types of laundry arrangements.

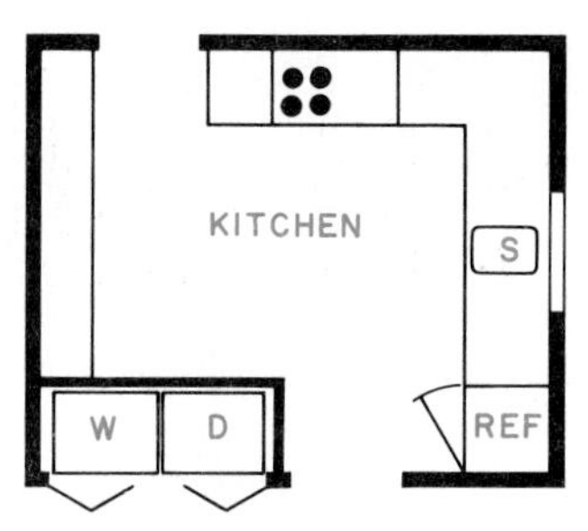

Fig. 17-4 Laundry facilities in a closet.

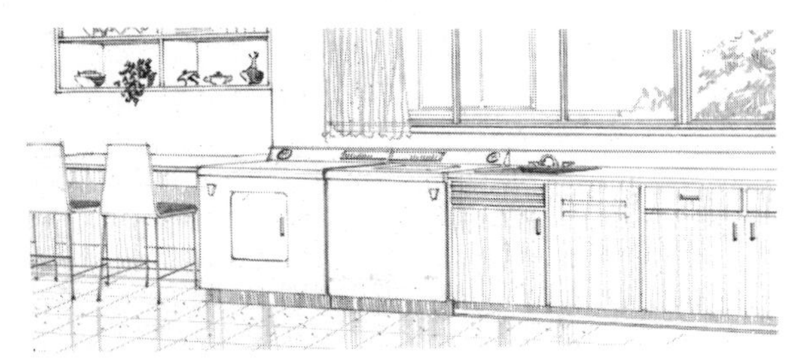

Fig. 17-5 Laundry area in a family room. *(Home Planners, Inc.)*

The Kitchen Placing the laundry unit in or near the kitchen has some advantages. The unit is in a central location and is near a service entrance. Plumbing facilities are near, and counters may be used for folding. However, the additional noise may be a disadvantage.

Other Locations Laundry appliances may be located in a closet (Fig. 17-4), on a service porch, in a basement, in a family room (17-5) or in a garage or carport. The service porch, basement, garage, or carport provides less expensive floor space than other parts of the house.

Exercises

1. Sketch a complete laundry floor plan from one of the plans shown in Fig. 17-1. Show positions of appliances and equipment on a total floor-plan sketch. Scale ½″ = 1′-0″.

2. Design a utility room with a complete laundry within an area of 10′ × 10′ (100 sq ft.).

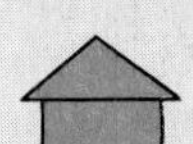

3. Design a utility room for the house you are designing.

4. Define the following terms: *utility room, laundry area, hamper, water heater, appliance, fixture.*

UNIT 18

GARAGES AND CARPORTS

Storage of an automobile occupies a large percentage of the available space of a house or property. Garages and carports must therefore be designed with the greatest care to ensure maximum utilization of space.

GARAGE

A *garage* is an enclosed structure designed primarily to shelter an automobile. It may be used for many secondary purposes—as a workshop, as a laundry room, or for storage space. A garage may be connected with the house (it is then an *integral* garage), or it may be a separate building (in which case it is *detached*). Figure 18-1 shows several possible garage locations.

CARPORT

A *carport* is a garage with one or more of the exterior walls removed. It may consist of a free-

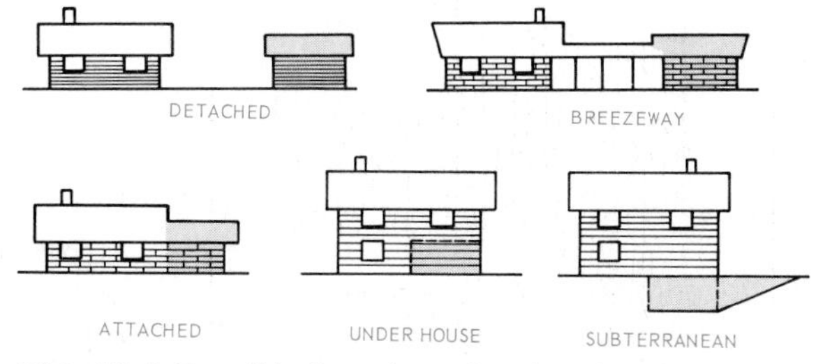

Fig. 18-1 Possible locations for the garage.

Fig. 18-2 A carport provides only overhead protection. *(Home Planners, Inc.)*

standing roof completely separate from the house, or it may be built against the existing walls of the house (Fig. 18-2). Carports are most acceptable in mild climates where complete protection from cold weather is not needed. A carport offers protection primarily from sun and moisture.

The garage and the carport both have distinct advantages. The garage is more secure and provides more shelter. However, carports lend themselves to open-planning techniques and are less expensive to build than garages.

DESIGN

The lines of the garage or carport should be consistent with the major building lines of the house. The style of the garage should be consistent with the style of architecture used in the house. The garage or carport must never appear as an afterthought. Often a patio, porch, or breezeway is planned between the garage and the house to integrate a detached garage with the house. A covered walkway from the garage or carport to the house should be provided if the garage is detached.

The garage floor must be solid and easily maintained . A concrete slab 3″ or 4″ (75 or 100 mm) thick with steel mesh provides the best deck for a garage or carport. The garage floor must have adequate drainage either to the outside or through drains located inside the garage (Fig. 18-3). A vapor barrier consisting of waterproof materials under the slab should be provided. The driveway should be of asphalt or concrete construction, preferably with welded-wire fabric to maintain rigidity (Fig. 18-4).

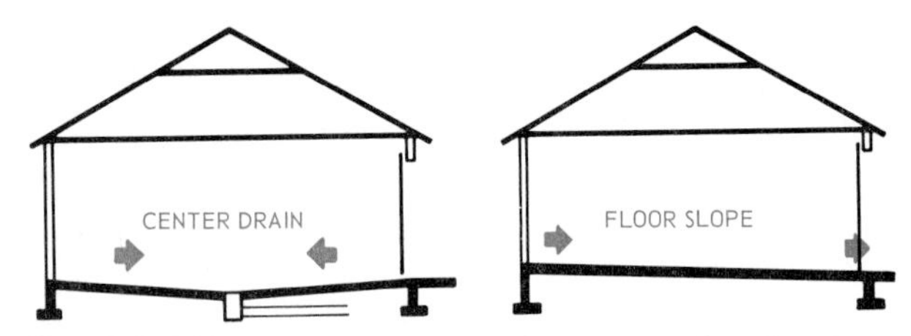

Fig. 18-3 Proper drainage is important for the garage.

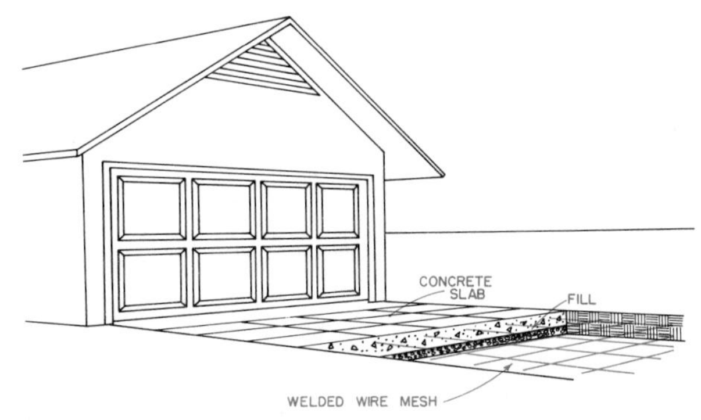

Fig. 18-4 Steel-wire mesh in the concrete will help keep the pavement and garage floor from cracking.

The design of the garage door greatly affects the appearance of the garage. Several types of garage doors are available. These include the two-leaf swinging, overhead, four-leaf swinging, and sectional roll-up doors (Fig. 18-5). Electronic devices are available for opening the door of the garage from the car.

SIZE

The size and number of automobiles and the additional facilities needed for storage or workshop use should determine the size of the garage (Fig. 18-6).

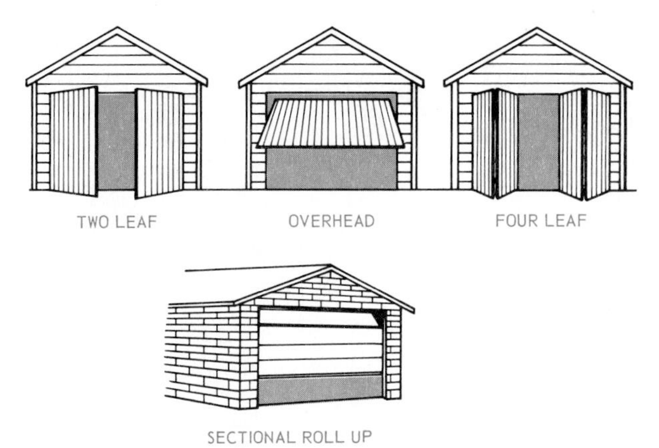

Fig. 18-5 Four common types of garage doors.

The dimensions of a single-car garage range between 11′ × 19′ (3.4 m × 5.8 m) and 13′ × 25′ (4.0 m × 7.6 m). A garage that is 16′ × 25′ (4.9 m × 7.6 m) is more desirable if space is needed for benches, mowers, tools, and the storage of children's vehicles. A full double garage is 25′ × 25′ (7.6 m × 7.6 m).

A two-car garage does not cost twice as much as a one-car garage. However, if the second half is added at a later date, the cost will more than double.

STORAGE

Storage is often an additional function of most garages. The storage space over the hood of the car should be utilized effectively. Figure 18-7 shows how storage facilities and even living space can be created by using otherwise wasted space. Cabinets should be elevated from the floor several inches to avoid moisture and to facilitate cleaning the garage floor. Garden-tool cabinets can be designed to open from the outside of the garage.

DRIVEWAY

A driveway can be planned for purposes other than providing access to the garage and temporary parking space for guests. By adding a wider space to an apron at the door of the garage, an area can be provided for car washing and polishing and for a hard, level surface for children's games. Aprons are often needed to provide space for turning the car in order to eliminate backing out onto a main street (Fig. 18-8).

The driveway should be accessible to all entrances, and the garage should provide easy access to the service area of the home. Suffi-

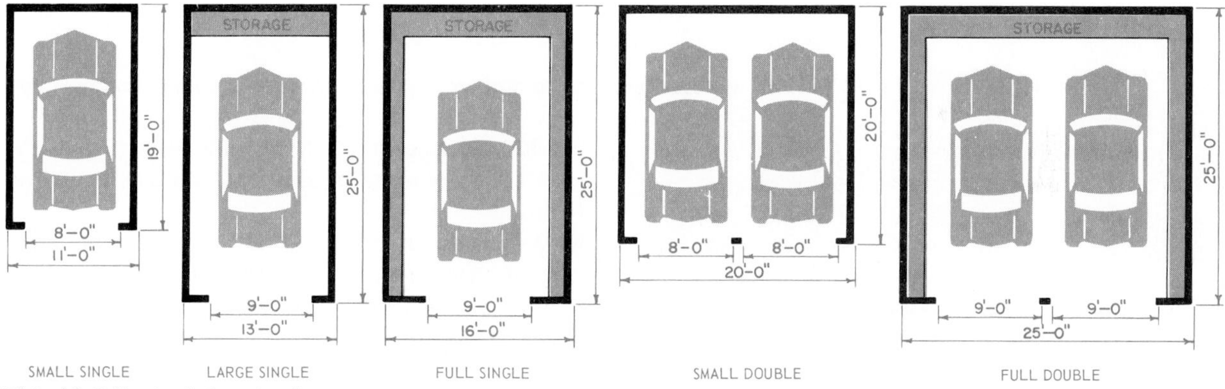

Fig. 18-6 Typical garage sizes.

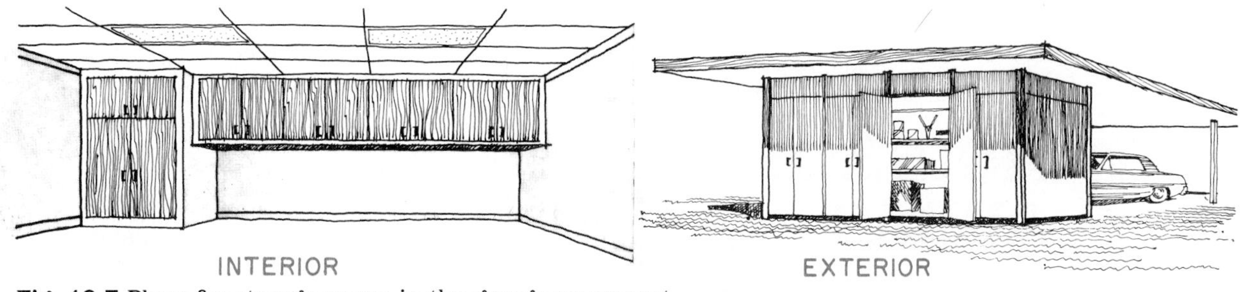

Fig. 18-7 Plans for storage space in the garage or carport.

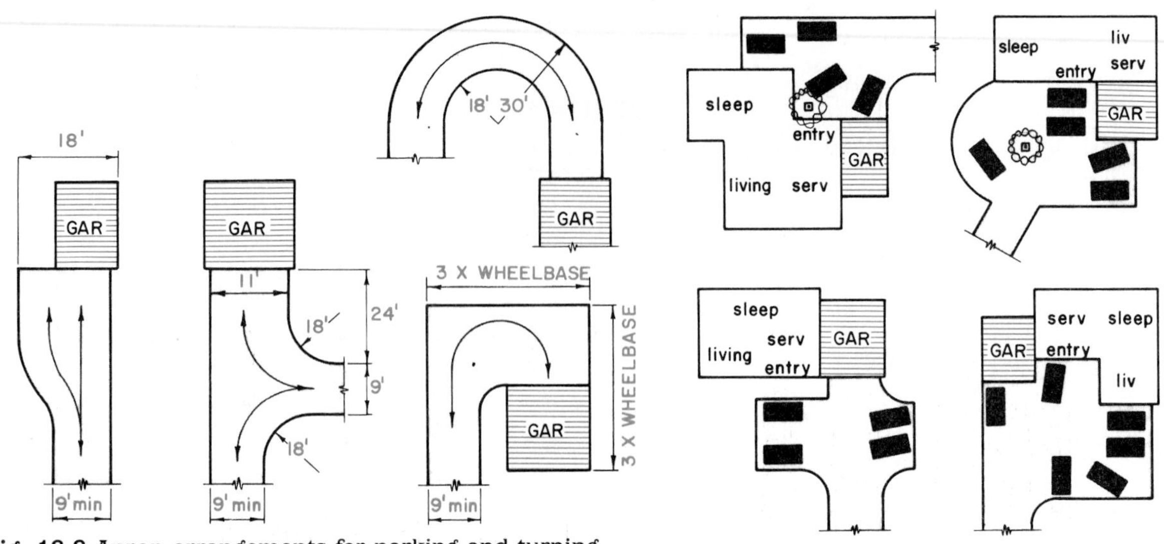

Fig. 18-8 Apron arrangements for parking and turning.

cient space in the driveway should be provided for parking of guests' cars.

Driveways should be designed at least several feet wider than the track of the car (approximately 5′-0″, or 1.5 m). However, slightly wider driveways are desirable (Fig. 18-9) for access and pedestrian traffic (approximately 7′ to 9′, or 2.1 to 2.7 m, wide).

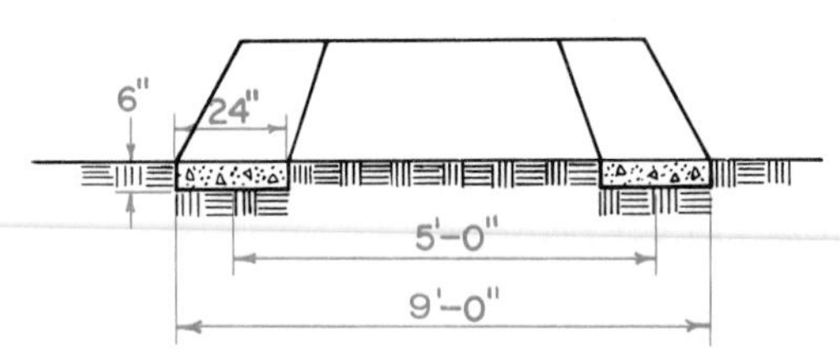

Fig. 18-9 Minimum track driveway width.

Exercises

1. **Sketch the floor plan from Fig. 13-2, and design a garage, apron, and driveway.**
2. **Design a full double garage for the house of your design, and draw in storage, laundry, and workbench.**
3. **Sketch a two-car garage plan. Show the following storage facilities: storage wall, outside storage, boat slung from ceiling, laundry area, gardening equipment, storage over the hood of the cars.**
4. **Define these architectural terms: *garage, carport, breezeway, subterranean, apron, integral garage, detached garage.***

UNIT 19

WORKSHOP AREAS

The home work area is designed for activities ranging from hobbies to home-maintenance work. The home work area may be located in part of the garage, in the basement, in a separate room, or in an adjacent building (Fig. 19-1).

LAYOUT

Power tools, hand tools, workbench space, and storage should be systematically planned. A workbench complete with vise is needed in every home work area. The average workbench is 36″ (915 mm) high. A movable workbench is appro-

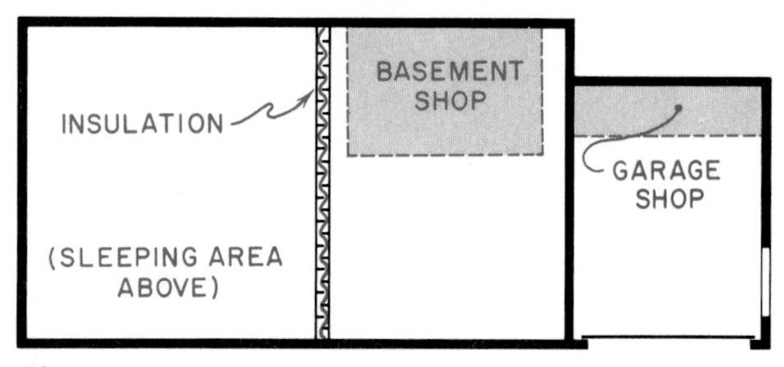

Fig. 19-1 The home workshop may be located in the garage, the basement, or a separate building.

priate when large projects are to be constructed. A *peninsula workbench* provides three working sides and storage compartments on three sides. A dropleaf workbench is excellent for work areas where a minimum amount of space is available.

Hand Tools

Some hand tools are basic to all types of hobbies or home-maintenance work. These basic tools include a claw hammer, carpenter's square, files, hand drills, screwdrivers, planes, pliers, chisels, scales, wrenches, saws, a brace and bit, mallets, and clamps.

Power Tools

Some of the more common power tools used in home workshops include electric drills, saber saws, routers, band saws, circular saws, radial-arm saws, jointers, belt sanders, lathes, and drill presses. Placement of equipment should be planned to provide the maximum amount of work space. Figure 19-2 suggests clearances necessary for safe and efficient machine operation.

Multipurpose machines are machines that can perform a variety of operations. Multipurpose equipment is popular for use in the home workshop since the purchase of only one piece of equipment is necessary and the amount of space needed is relatively small, compared with the amount of space needed for a variety of machines.

Tools and equipment needed for working with large materials should be placed where the material can be easily handled. Separate-drive motors can be used to drive more than one piece of power equipment, in order to conserve motors. Separate 110 V and 220 V electrical circuits for lights and power tools should be included in the plans for the home workshop area.

STORAGE FACILITIES

Maximum storage facilities in the home work area are essential. Hand tools may be stored in cabinets that keep them dust-free and safe, or hung on perforated hardboard, as shown in Fig. 19-3. Tools too small to be hung should be kept in special-purpose drawers. Any inflammable finishing material, such as turpentine or oil paint, should be stored in metal cabinets.

SIZE AND SHAPE

The size of the work area depends on the size and number of power tools and equipment, the amount of workbench area, and the amount of tool and material storage facilities provided. The

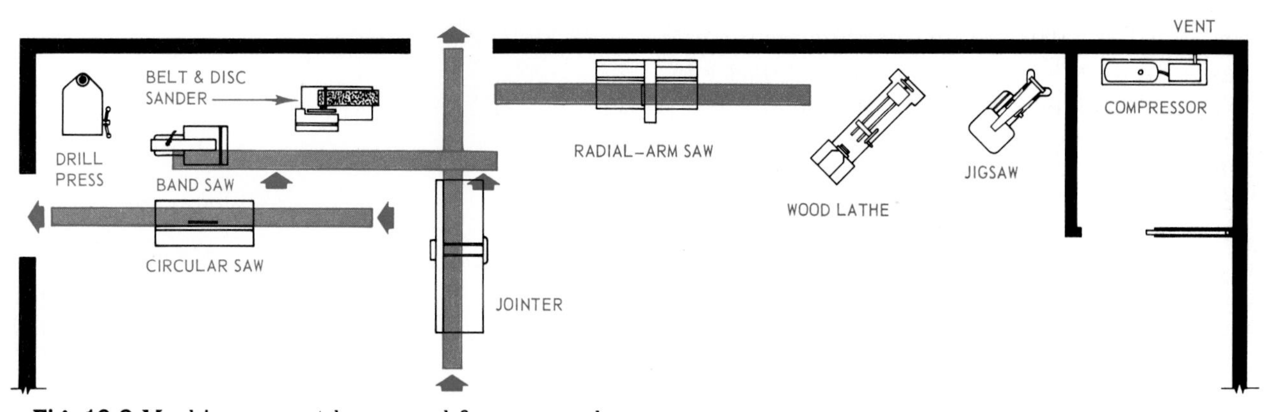

Fig. 19-2 Machinery must be spaced for proper clearances.

Fig. 19-3 Facilities must be planned for storage of tools and supplies. *(Better Homes and Gardens)*

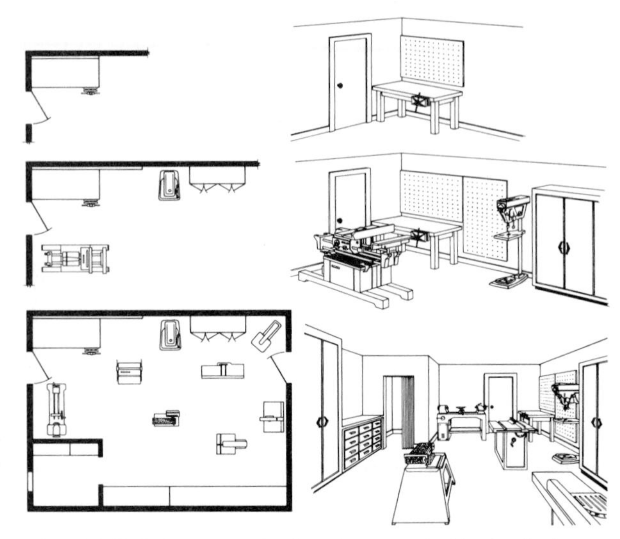

Fig. 19-4 A three-stage development of a workshop.

size of the work area should be planned for maximum expansion, even though only a workbench or a few tools may be available when the area is first occupied. Therefore, space for the maximum amount of facilities should be planned and located when the area is designed. As new equipment is added, it will fit appropriately into the basic plan (Fig. 19-4). The designer must anticipate the type and number of materials for which storage space will be needed and design the storage space accordingly.

DECOR

The work area should be as maintenance-free as possible. Glossy paint or tile retards an accumulation of shop dust on the walls. Exhaust fans eliminate much of the dust and the gases produced in the shop. The shop floor should be of concrete or linoleum for easy maintenance. Abrasive strips around machines will eliminate the possibility of slipping. Do not locate noisy equipment near the children's sleeping area. Interior walls and ceilings should be soundproofed by offsetting studs and adding adequate insulation to produce a sound barrier (Fig. 19-5).

Light and color are most important factors in designing the work area. Pastel colors, which reduce eyestrain, should be used for the general color scheme of the shop. Extremely light colors that produce glare, and extremely dark colors that reduce effective illumination, should be avoided. Adopting one of the major paint manufacturers' color systems for color coding will help to create a pleasant atmosphere in the shop and will also help to provide the most efficient and safest working conditions. General lighting should be provided in the shop to a level of 100 footcandles (1076 lux, or lx) on machines and worktable tops.

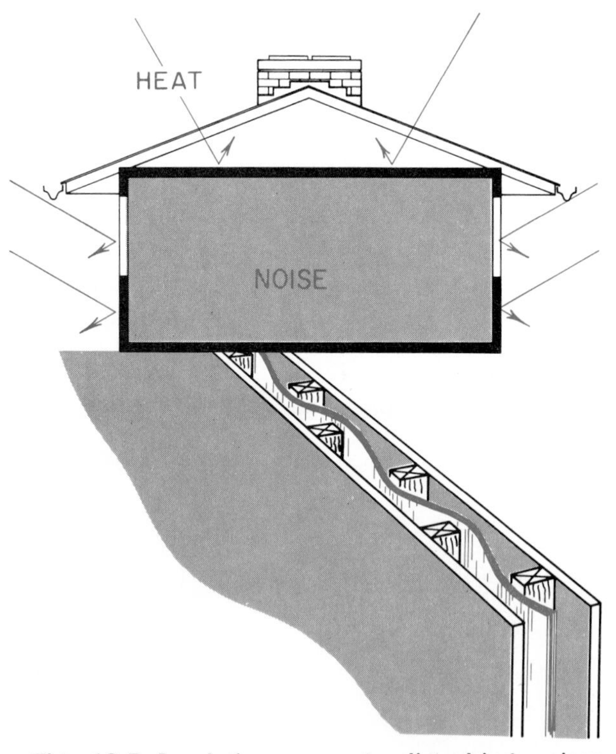

Fig. 19-5 Insulation prevents disturbing noises from entering other parts of the house.

Exercises

1. **Design a work area in a double garage for an activity other than woodworking; for example ceramics, jewelry, metalworking, or automotive repairs. Show what tools and work areas are necessary.**

2. **Design a work area for the house of your choice. Use a basement or garage location.**
3. **Define the following architectural terms: *dehumidifier, workbench, perforated hardboard, inflammable, hand tools, power tools, multipurpose tools.***

UNIT 20

STORAGE AREAS

Storage areas should be provided for general storage and for specific storage within each room (Fig. 20-1). Areas that would otherwise be considered wasted space should be used as general storage areas. Parts of the basement, attic, or garage often fall into this category. Effective storage planning is necessary to provide storage facilities within each room that will create the least amount of inconvenience in securing the stored articles. Articles that are used daily or weekly should be stored in or near the room where they will be used. Articles that are used only seasonally should be placed in more permanent general storage areas.

STORAGE FACILITIES

Storage facilities and equipment, including furniture used for storage within the various rooms

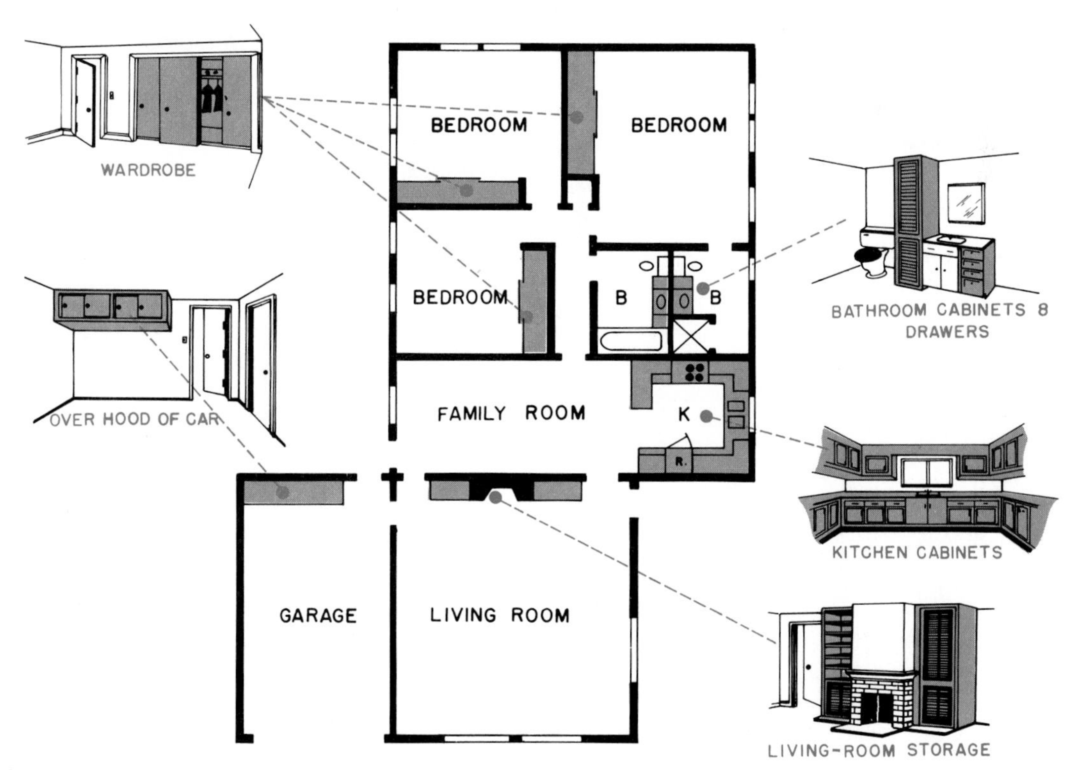

Fig. 20-1 Locations of storage areas.

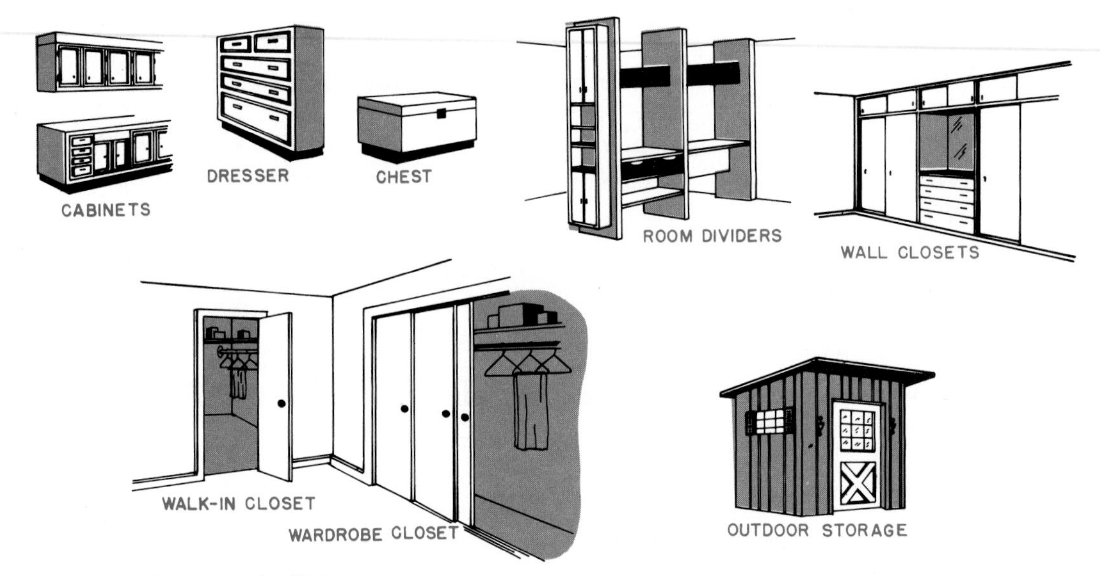

Fig. 20-2 Types of storage facilities.

of the house, are divided into the following categories (Fig. 20-2):

Wardrobe Closets A *wardrobe closet* is a shallow clothes closet built into the wall. The minimum depth for the wardrobe is 24″ (610 mm). If this closet is more than 30″ (760 mm) deep, you will be unable to reach the back of the closet. Swinging or sliding doors should expose all parts of the closet to your reach. A disadvantage of the wardrobe closet is the amount of wall space needed for the doors (Fig. 20-3).

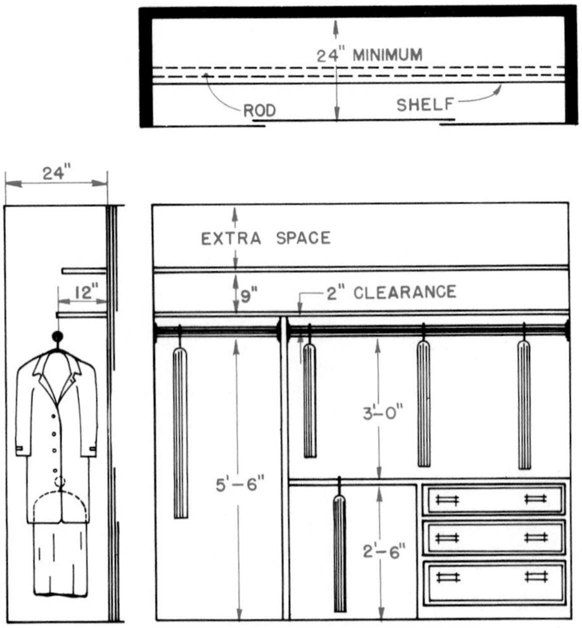

Fig. 20-3 Dimensions for wardrobe closets.

Walk-In Closets *Walk-in closets* are closets large enough to walk into. The area needed for this type of closet is an area equal to the amount of space needed to hang clothes plus enough space to walk and turn. Although some area is wasted in the passage, the use of the walk-in closet does provide more wall area for furniture placement, since only one door is needed (Fig. 20-4).

Wall Closets A *wall closet* is a shallow closet in the wall holding cupboards, shelves, and drawers. Wall closets are normally 18″ (460 mm) deep, since this size provides access to all stored items without using an excessive amount

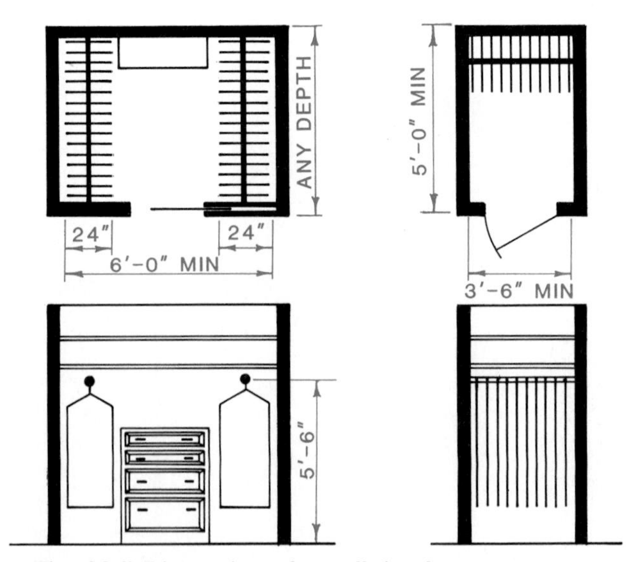

Fig. 20-4 Dimensions for walk-in closets.

of floor area (Fig. 20-5). Figure 20-6 is an example of effective wall storage closets.

Protruding closets that create an offset in a room should be avoided (Fig. 20-7). Often, by filling the entire wall between two bedrooms with closet space, it is possible to design a square or rectangular room without the use of offsets. Doors on closets should be sufficiently wide to allow easy accessibility. Swing-out doors have the advantage of providing extra storage space on the back of the door. However, space must be allowed for the swing. For this reason, sliding doors are usually preferred. All closets, except very shallow linen closets, should be provided with lighting.

Chests and Dressers Chests and dressers are freestanding pieces of furniture used for storage, generally in the bedroom. They are available in a variety of sizes, usually with shelves and drawers.

Room Dividers A room divider often doubles as a storage area especially when a protruding closet divides several areas. Room dividers often extend from the floor to the ceiling but may also be only several feet high. Many room dividers include shelves and drawers that open from both sides (Fig. 20-8).

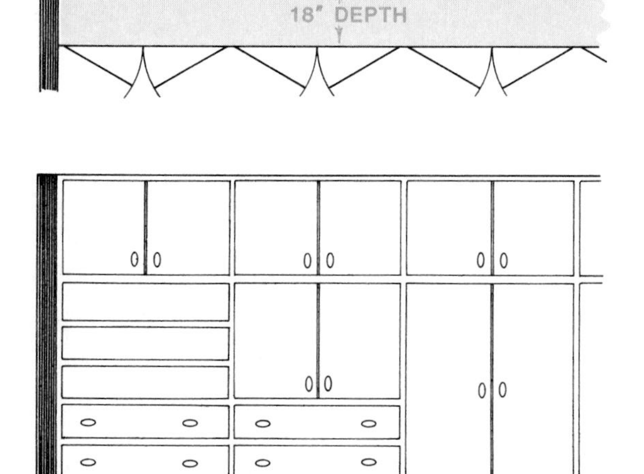

Fig. 20-5 Wall storage uses a minimum of floor space.

Fig. 20-6 Built-in closets and drawer storage. *(Home Planners, Inc.)*

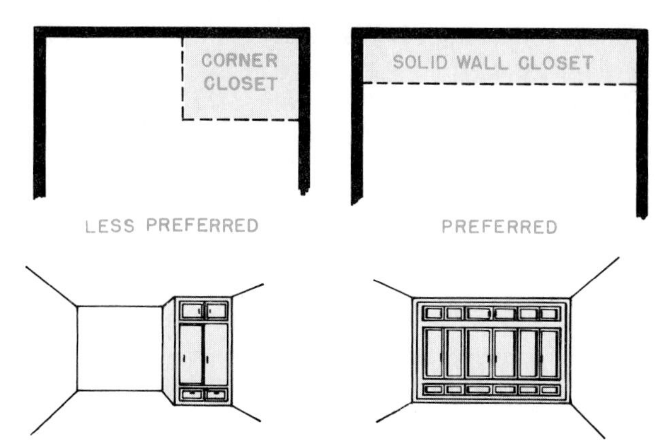

Fig. 20-7 Avoid closets that create offsets.

Fig. 20-8 Room dividers can be used for storage. *(Home Planners, Inc.)*

LOCATION

Different types of storage facilities are necessary for different areas of the home depending on the type of article to be stored. The most appropriate types of storage facilities for each room in the house are as follows:

Living room: room divider, built-in wall cabinets (Fig. 20-9), bookcases, window seats

Dining area: room divider, built-in wall closet (Fig. 20-10)

Family room: built-in wall storage, window seats (Fig. 20-11)

Recreation room: built-in wall storage

Porches: storage under porch stairs, window seats

Patios: sides of barbecue, auxiliary building

Outside: storage areas built into the side of the house

Halls: built-in wall closets, ends of blind halls

Entrance: room divider, wardrobe, walk-in closet

Den: built-in wall closet, window seats, bookcases (Fig. 20-12)

Kitchen: wall and floor cabinets, room divider, wall closets (Fig. 20-13A through D)

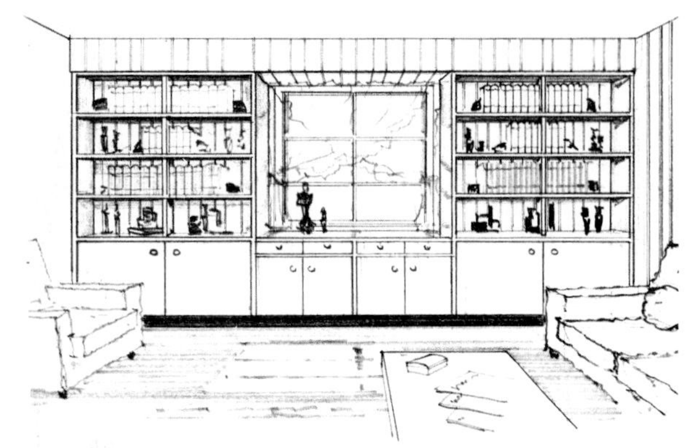

Fig. 20-9 Built-in wall cabinets. *(Home Planners, Inc.)*

Fig. 20-10 Dining-area storage. *(Home Planners, Inc.)*

Fig. 20-11 Family-room cabinet storage. *(Home Planners, Inc.)*

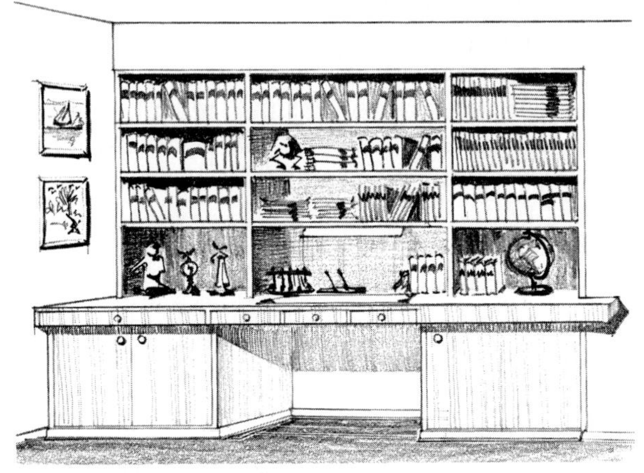

Fig. 20-12 Den or study storage. *(Home Planners, Inc.)*

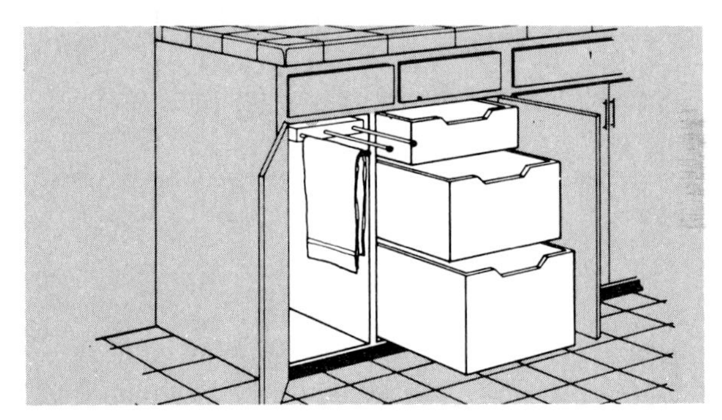

Fig. 20-13A Kitchen bin storage. *(California Gas Co.)*

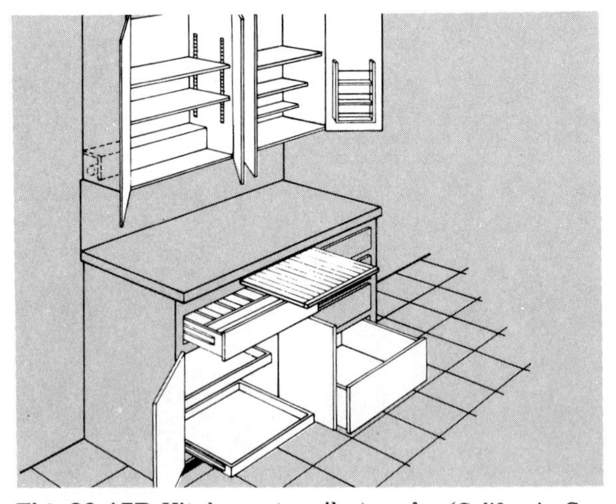

Fig. 20-13B Kitchen utensil storage. *(California Gas Co.)*

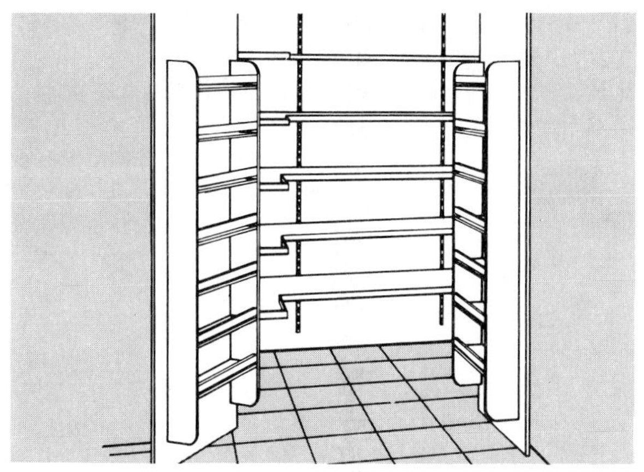

Fig. 20-13C Pantry storage. *(California Gas Co.)*

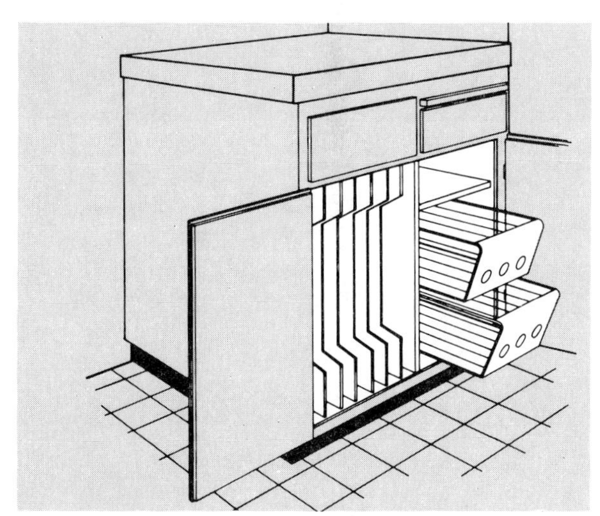

Fig. 20-13D Kitchen vertical storage. *(California Gas Co.)*

Utility room: cabinets on floor and walls
Garage: cabinets over hood of car, wall closets along sides, added construction on the outside of the garage (Fig. 20-14A through D)

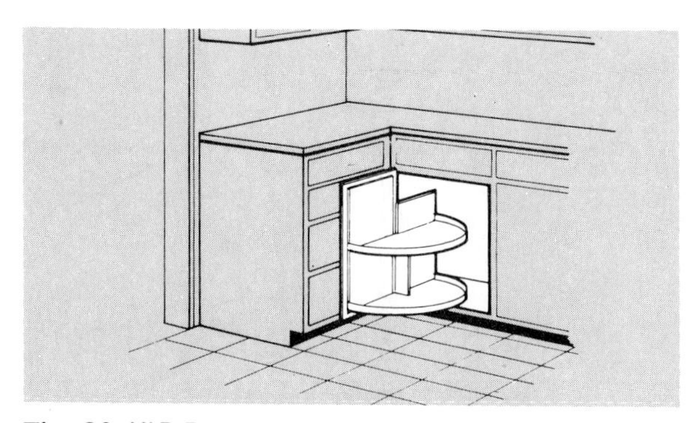

Fig. 20-14A Door storage. *(California Gas Co.)*

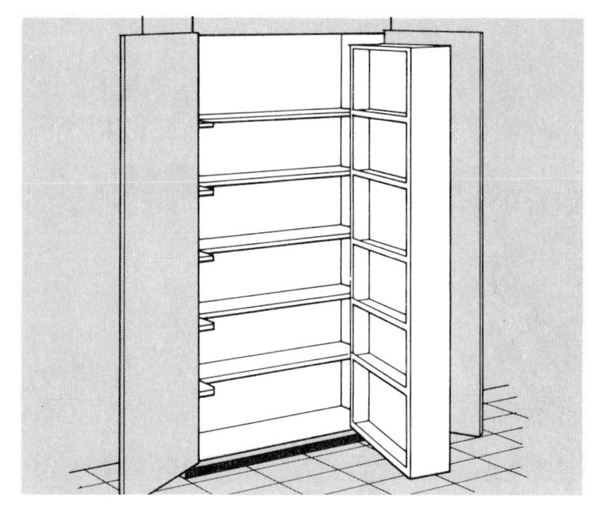

Fig. 20-14B Closet storage. *(California Gas Co.)*

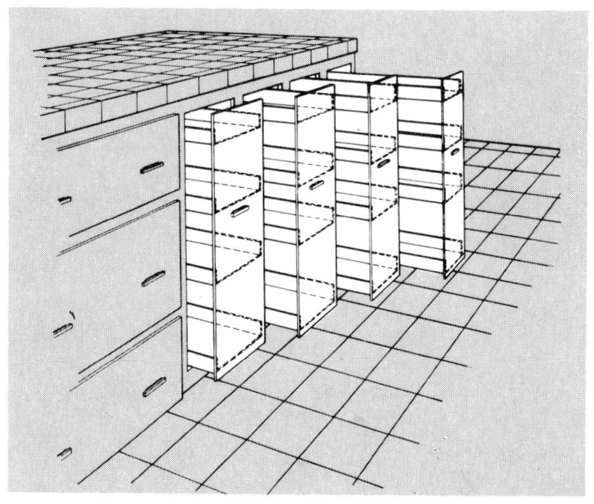

Fig. 20-14C Stock-bin storage. *(California Gas Co.)*

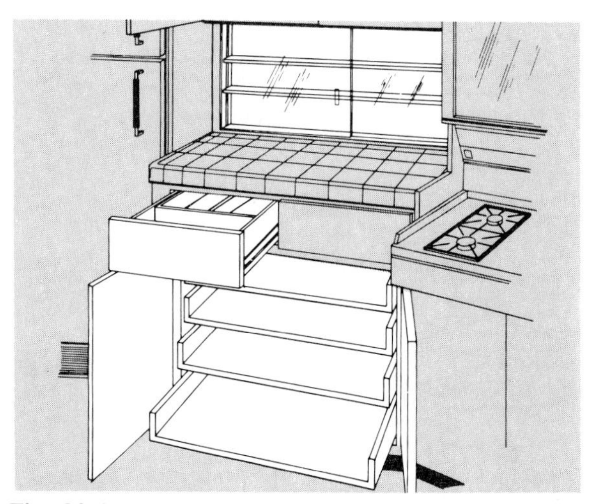

Fig. 20-14D Drawer storage. *(California Gas Co.)*

Work area: open tool board, wall closets, cabinets

Bedroom: walk-in closet; wardrobe closet; storage under bed, at foot of bed, at head of bed; built-in cabinets and shelves; dressers; chests; window seats

Bathroom: cabinets on floor and ceiling, room dividers

Additional storage facilities are shown in Fig. 20-15.

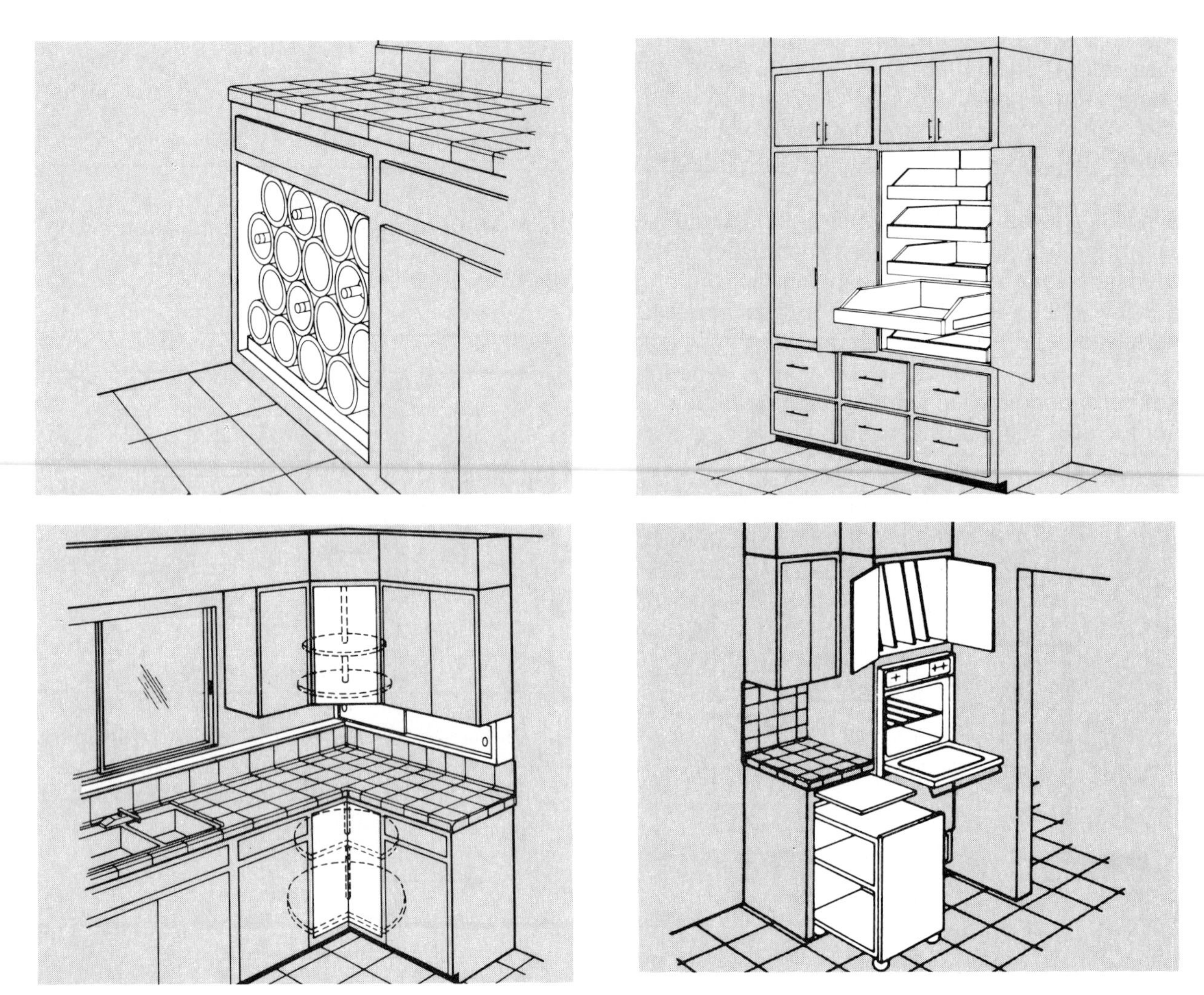

Fig. 20-15 Special storage facilities.

Exercises

1. **Add storage facilities to the first and second floor of the house shown in fig. 69-1. Sketch your solution and label each storage area.**

2. **Add storage facilities to the house of your design.**
3. **Define these terms: *window seat, built-in, wardrobe, auxiliary building, chest, room divider.***

SECTION 5

Sleeping Area

One-third of our time is spent in sleeping. Because of its importance, the sleeping area should be planned to provide facilities for maximum comfort and relaxation. The sleeping area is usually located in a quiet part of the house and contains bedrooms, baths, dressing areas, and nurseries.

UNIT 21

BEDROOMS

Houses are usually classified by size according to the number of bedrooms; for example, a three-bedroom home, or a four-bedroom home. In a home there are bedrooms, master bedrooms, and nursery rooms, according to the size of the family.

FUNCTION

The primary function of a bedroom is to provide facilities for sleeping. Some bedrooms may also provide facilities for writing, reading, sewing, listening to music, or relaxing.

NUMBER OF BEDROOMS

Ideally, each member of the family should have his or her own private bedroom. A family with no children may require only one bedroom. However, two bedrooms are usually desirable, in order to provide one for guest use (Fig. 21-1). Three-bedroom homes are most popular because they provide a minimum of accommodation for a family with one boy and one girl. As a family enlarges, boys can share one bedroom and girls can share the other.

SIZES AND SHAPES

The size and shape of a bedroom depend upon the amount of furniture needed. A minimum-sized bedroom would accommodate a single bed, bedside table, and dresser. In contrast, a complete master bedroom might include a double bed or twin beds, bedside stands, dresser, chest of drawers, lounge chair, dressing area, and adjacent master bath.

Space Requirements

The type and style of furniture included in the bedroom should be chosen before the size of the bedroom is established. Since the bed or beds require the most space, care must be taken to match the room size to the type of bed desired.

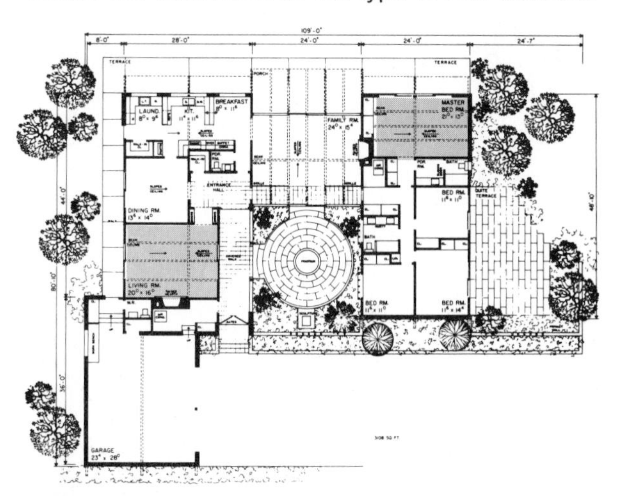

Fig. 21-1 The sleeping area should be away from the activity area of the home.

The size of the furniture should determine the size of the room, and not the reverse. Average bedroom furniture sizes are shown in Fig. 21-2. The wall space needed for twin beds is 8′-6″ (2.6 m). Figure 21-3 shows bed sizes and spacing requirements.

A minimum bedroom would be 100 ft^2, an average bedroom 100 to 200 ft^2, and a large bedroom over 200 ft^2 (19 m^2).

Wall Space

Since wall space is critical in the placement of furniture in the bedroom, the designer must plan for maximum wall space. One method of conserving wall space for bedroom furniture placement is to use high windows. High strip or ribbon windows allow furniture to be placed underneath, and they also provide some privacy for the bedroom. But many codes require an escape size window in each room.

Bedroom Doors

Unless it has doors leading to a patio, the bedroom will normally have only one access door. Entrance doors, closet doors, and windows should be grouped to conserve wall space whenever possible. Separating the doors slightly from one another and from the windows will spread out the amount of unusable wall space by eliminating long stretches of unused wall space that can be used for furniture placement. Pocket and bifolding doors for closets and for entrance doors help to conserve valuable wall space in bedrooms. If swinging doors are used, the door should always swing into the bedroom against an adjacent wall, and not into the hall (Fig. 21-4).

Dressing Areas

Dressing areas are sometimes separate rooms or an alcove or a part of the room separated by a divider (Fig. 21-5).

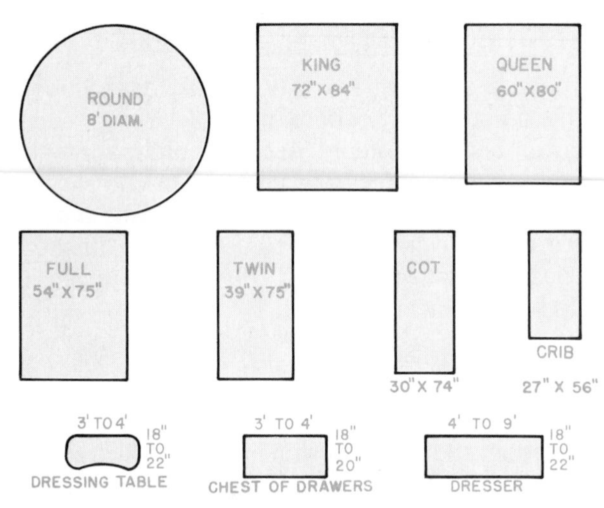

Fig. 21-2 Typical bedroom furniture sizes.

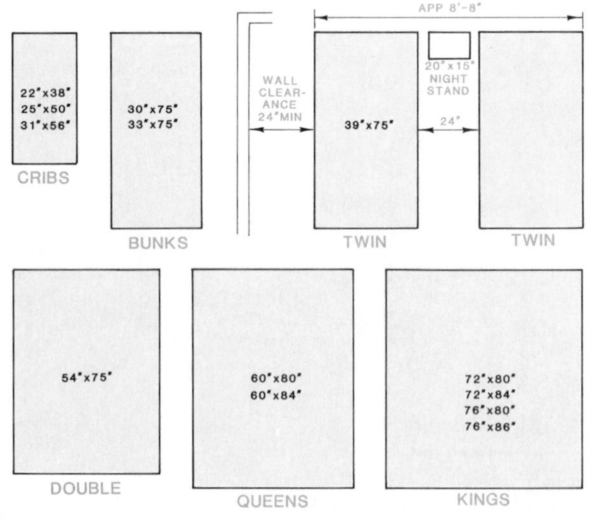

Fig. 21-3 Wall and floor space is needed for bedroom furniture.

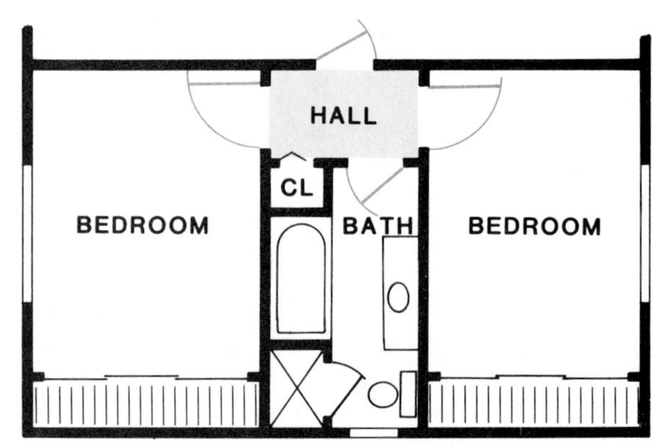

Fig. 21-4 Bedroom doors should not open into halls.

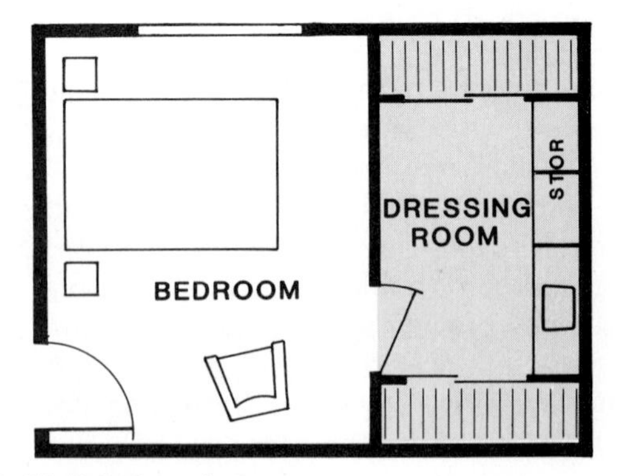

Fig. 21-5 A dressing area.

NOISE CONTROL

Since noise contributes to fatigue, it is important to plan for the elimination of as much noise as possible from the bedroom area (Fig. 21-6). The following guides for noise control will help you design bedrooms that are quiet and restful:

1. The bedroom should be in the quiet part of the house, away from major street noises.

2. Carpeting or porous wall and ceiling panels help to absorb many noises.

3. Rooms above a bedroom should be carpeted.

4. Floor-to-ceiling draperies help to reduce noise.

5. Acoustical tile in the ceiling is effective in reducing noise.

6. Trees and shrubbery outside the bedroom help deaden sounds.

7. The use of double-glazed insulating glass in windows and sliding doors helps to reduce outside noise.

8. The windows of an air-conditioned room should be kept closed during hot weather. This eliminates noise and aids in keeping the bedroom free from dust and pollen.

9. Air is a good insulator; therefore, closets provide additional buffers which eliminate noise coming from other rooms.

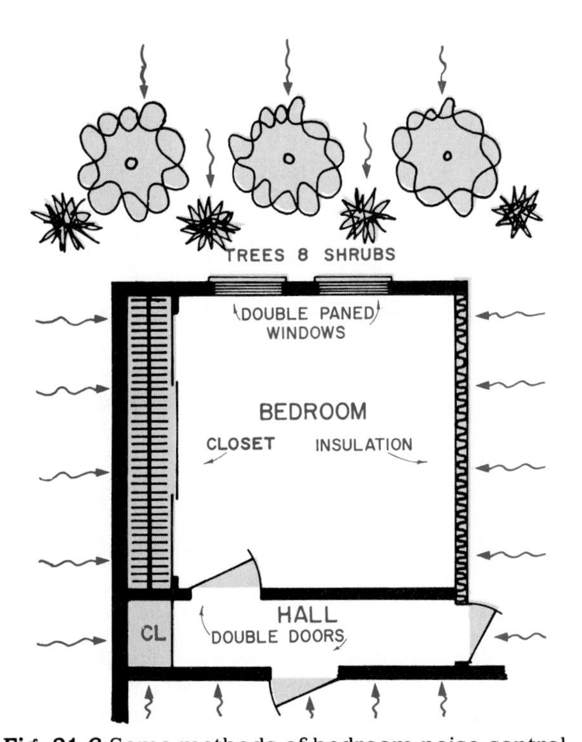

Fig. 21-6 Some methods of bedroom noise control.

10. In extreme cases when complete soundproofing is desired, fibrous materials in the walls may be used, and studs may be offset to provide a sound buffer.

11. Placing rubber pads under appliances such as refrigerators, dishwashers, washers, and dryers often eliminates vibration and noise throughout the house.

STORAGE SPACE

Storage space in the bedroom is needed primarily for clothing and personal accessories. Storage areas should be easy to reach and easy to maintain. Walk-in closets or wardrobe closets should be built in for hanging clothes (Fig. 21-7). Care should be taken to eliminate offset closets. Balancing offset closets from one room to an adjacent bedroom helps solve this problem. Providing built-in storage facilities also helps in overcoming awkward offsets, as shown in Fig. 21-8. Except for the storage space pro-

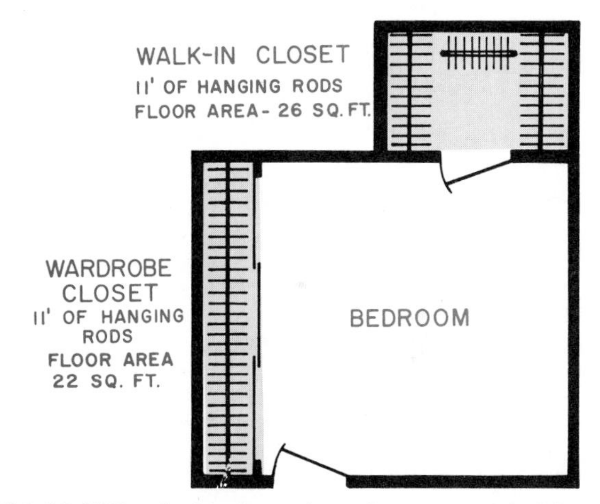

Fig. 21-7 Hanging rods are best for storage clothing.

Fig. 21-8 Built-in storage helps eliminate offsets.

vided in dressers, chests, vanities, and dressing tables, most storage space should be provided in the closet. Double rooms (Fig. 21-9) allow for maximum storage, yet provide flexibility and privacy as the family expands. Adequate storage facilities in a bedroom are absolutely essential.

VENTILATION

Proper ventilation is necessary and is conducive to sound rest and sleep. Central air-conditioning and humidity control provide constant levels of temperature and humidity and are an efficient method of providing ventilation and air circulation. When air-conditioning is available, the windows and doors may remain closed. Without air-conditioning, windows and doors must provide the ventilation. Bedrooms should have cross ventilation. However, the draft must not pass over the bed (Fig. 21-10). High ribbon windows provide light, privacy, and cross ventilation without causing a draft on the bed. Jalousie windows are also effective, since they direct air upward.

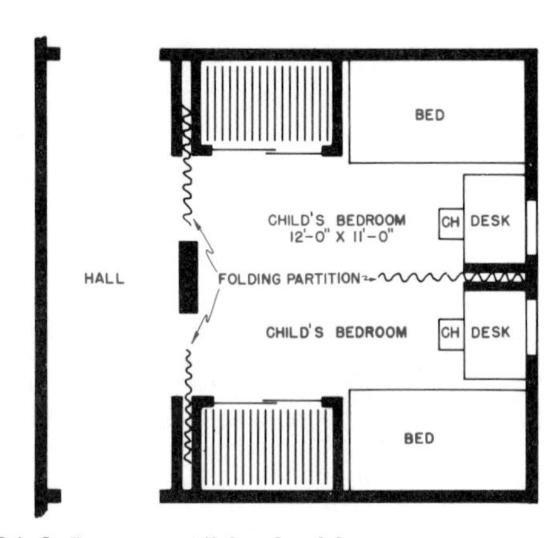

Fig. 21-9 A convertible double room.

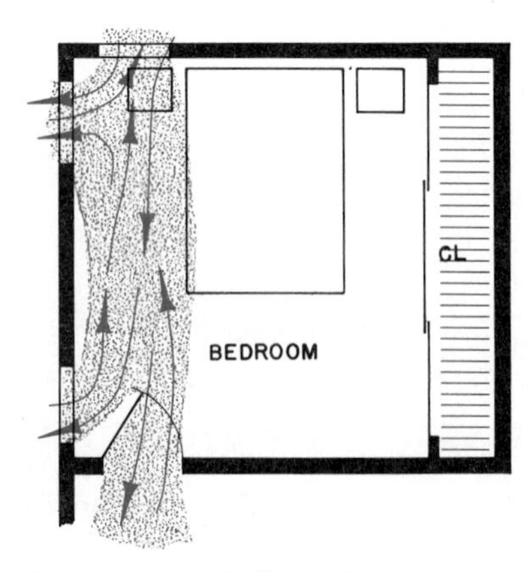

Fig. 21-10 Cross ventilation should not pass over the bed.

NURSERIES

Children's bedrooms and nurseries need special facilities. They must be planned to be comfortable, quiet, and sufficiently flexible to change as the child grows and matures (Fig. 21-11). Storage shelves and rods in closets should be adjustable so that they may be raised as the child becomes taller. Light switches should be placed low, with a delay switch which allows the light to stay on for some time after the switch has been turned off.

Chalkboards and bulletin boards help make the child's room usable. Adequate facilities for study and some hobby activities should be provided, such as a desk and worktable as shown in Fig. 21-12. Storage space for books, models, and athletic equipment is also desirable.

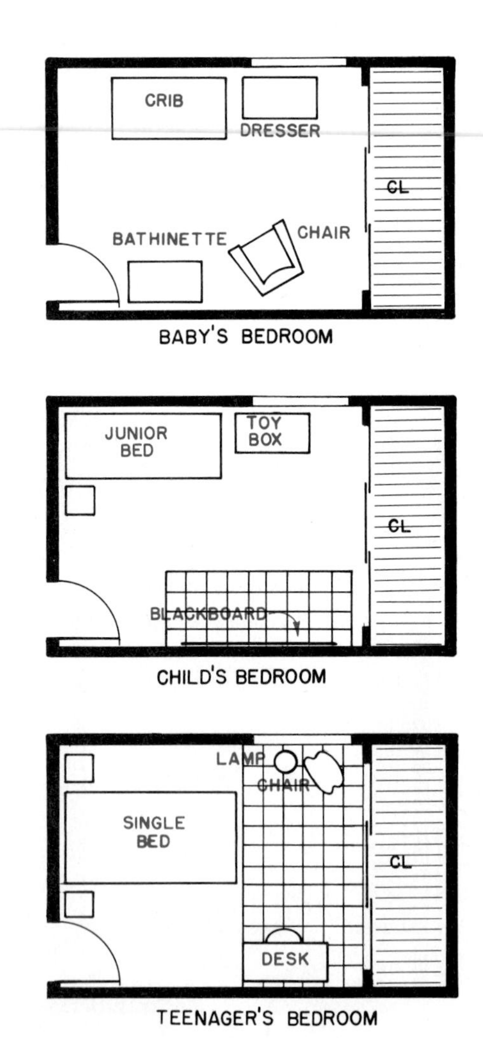

Fig. 21-11 Bedroom furnishings must change as children grow older.

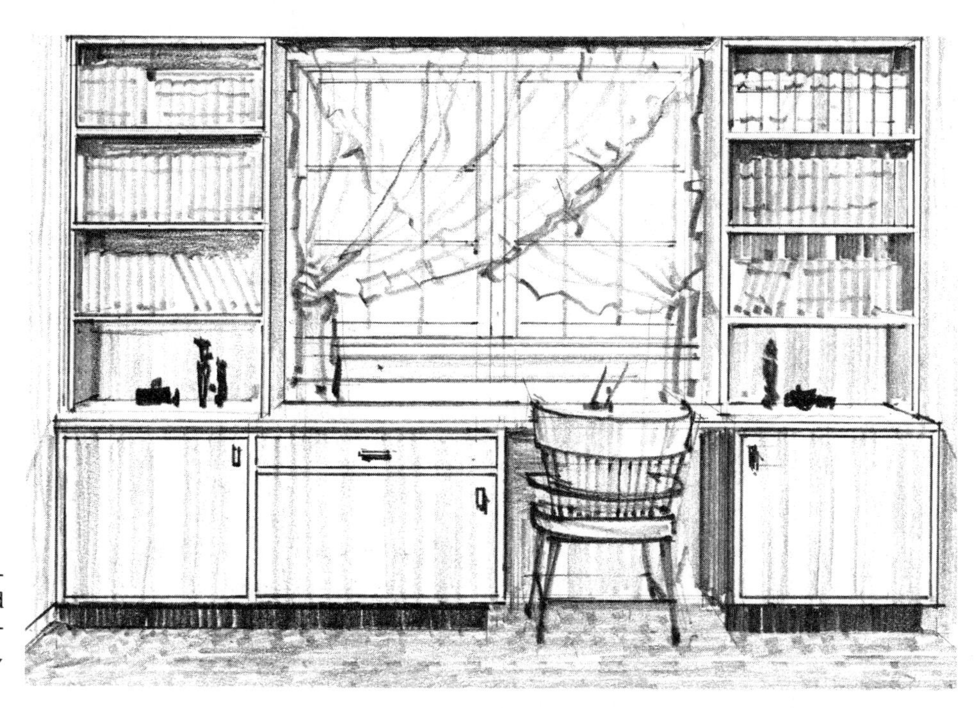

Fig. 21-12 Study facilities must be provided for school-age children. *(Home Planners, Inc.)*

DECOR

Bedrooms should be decorated in quiet, restful tones. Matching or contrasting bedspreads, draperies, and carpets help accent the color scheme. Uncluttered furniture with simple lines also helps to develop a restful atmosphere in the bedroom.

Exercises

1. **Design a bedroom, 100 ft^2 in size, for a very young child.**
2. **Design a bedroom, 150 ft^2 in size, for a teenager.**
3. **Design a master bedroom that is 200 ft^2 in size.**

4. **Design the bedroom areas for the home of your choice.**
5. **Define these architectural terms: *alcove, insulation, acoustical tile, cross ventilation, delay switch, room offset, sound buffer, humidity control.***

UNIT 22

BATHS

The design of the bathroom requires careful planning in the placement of fixtures. The bath must be planned to be functional, attractive, and easily maintained.

FUNCTION

In addition to the normal functions of the bath, facilities may also be included for dressing, exercising, sunning, and laundering (Fig. 22-1). Designing the bath involves the appropriate placing of fixtures; providing for adequate ventilation, lighting, and heating; and planning efficient runs for plumbing pipes.

Ideally, it would be advisable to provide a bath for each bedroom, as in Fig. 22-2. Usually, this

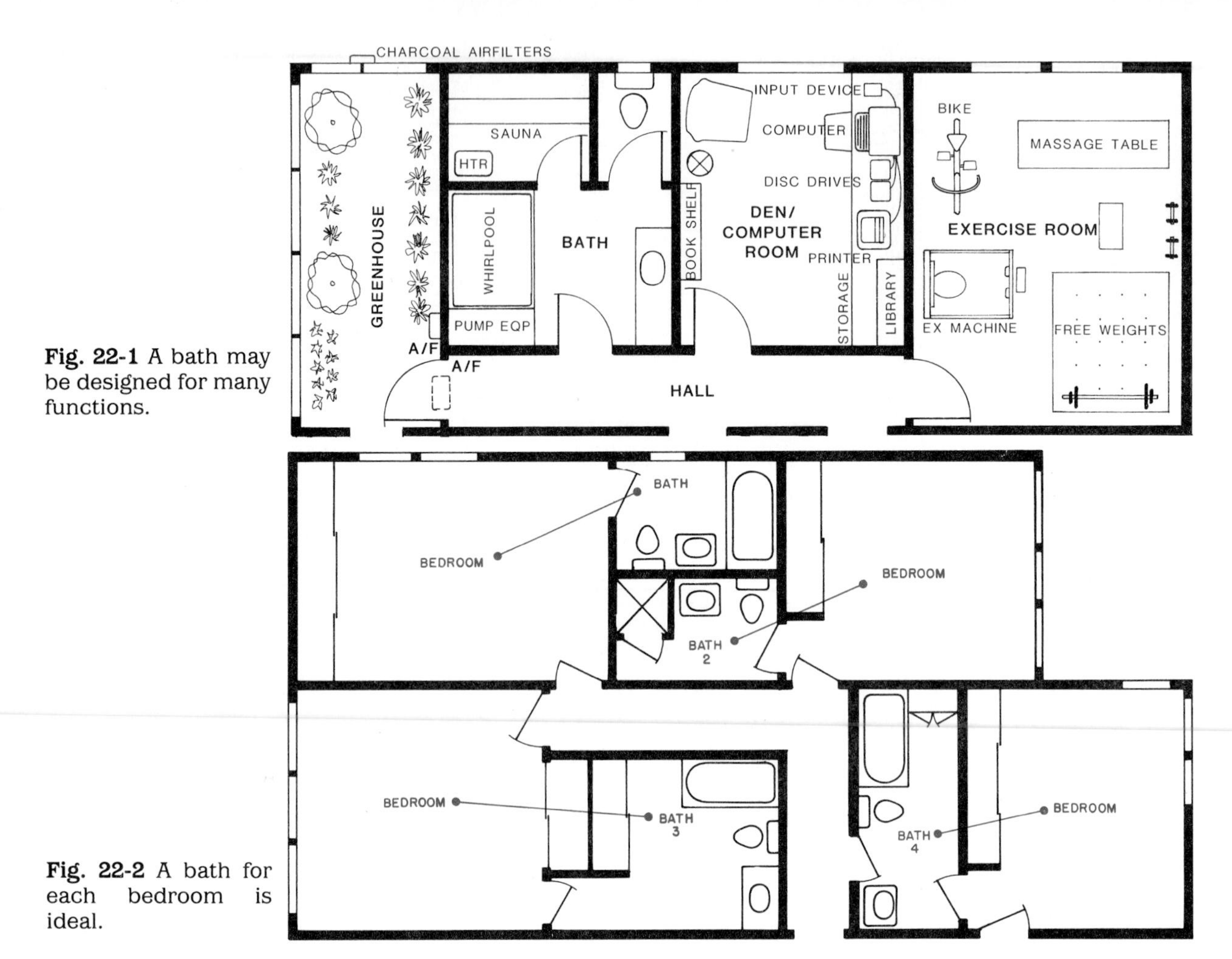

Fig. 22-1 A bath may be designed for many functions.

Fig. 22-2 A bath for each bedroom is ideal.

provision is not possible, and a central bath is designed to meet the needs of the entire family (Fig. 22-3). A bath for general use and a bath adjacent to the master bedroom is a desirable compromise (Fig. 22-4). When it is impossible to have a bath with the master bedroom, the general bath should be accessible from all bedrooms in the sleeping area. A bath may also function as a dressing room. Figure 22-5 shows a combination bath and dressing room with space for clothing storage.

Fixtures

The three *basic fixtures* included in most bathrooms are a *lavatory, a water closet,* and *a tub* or *shower.* The efficiency of the bath is greatly dependent upon the effectiveness of the arrangement of these three fixtures. Mirrors should be located a distance from the tub to prevent fogging. Sinks should be well lighted and free from traffic. If sinks are placed 18″ (460 mm) from other fixtures, they may share common vents. The water closet needs a minimum of

Fig. 22-3 A central bath serves all bedrooms.

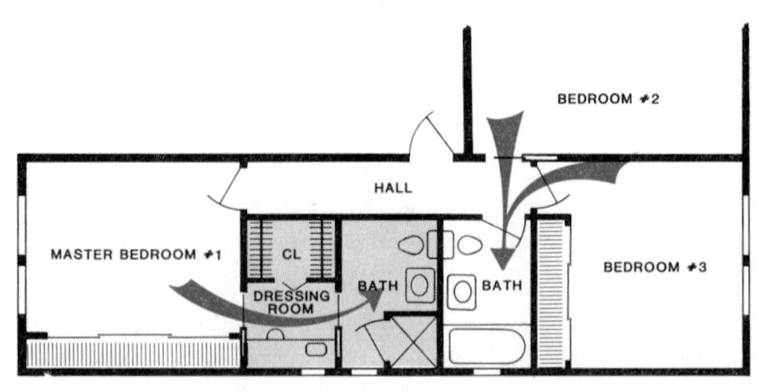

Fig. 22-4 A bath for the master bedroom and a second bath for other bedrooms

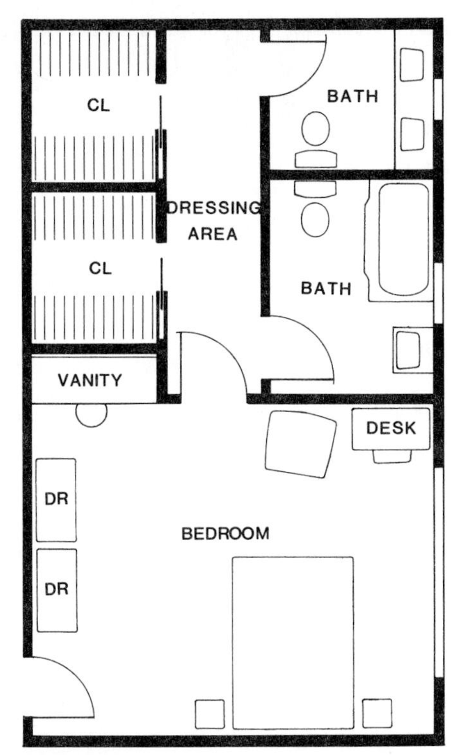

Fig. 22-5 A bedroom with a compartment bath and dressing area.

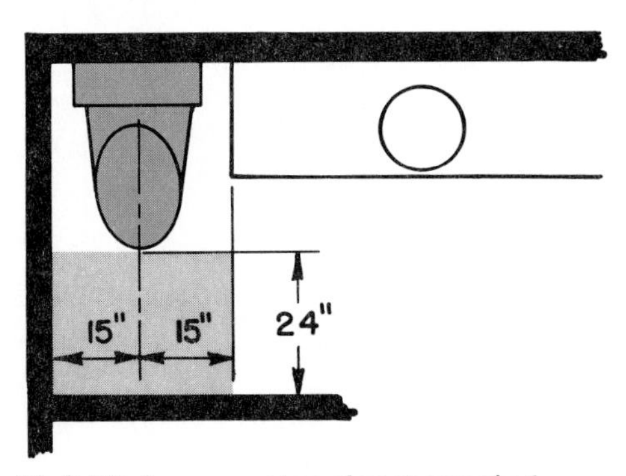

Fig. 22-6 Minimum water-closet spacing.

15″ (380 mm) from the center to the sidewall or to other fixtures (Fig. 22-6). Tubs and showers are available in a great variety of sizes and shapes. Square, rectangular, or sunken-pool tubs allow flexibility in fixture arrangement.

Ventilation

Baths should have either natural ventilation from a window or forced ventilation from an exhaust fan. Care should be taken to place windows in a position where they will not cause a draft on the tub or interfere with privacy. A bath can be designed without windows; however, a light source combined with a ventilating fan must be installed, and both must be controlled by a single switch.

Lighting

Lighting should be relatively shadowless in the area used for grooming. Shadowless general lighting can be achieved by the use of fluorescent tubes on the ceiling, covered with glass or plastic panels. Skylights can also be used for general illumination if the bath is without outside walls.

Heating

Heating in the bath is most important to prevent chills. In addition to the conventional heating outlets, an electric heater or heat lamp is often used to provide instant heat. It is advisable to have the source of heat under the window to eliminate drafts. All heaters should be properly ventilated.

Plumbing Lines

The plumbing lines that carry water to and from the fixtures should be concealed and minimized as much as possible. When two bathrooms are placed side by side, placing the fixtures back to back on opposite sides of the plumbing wall results in a reduction of the length of plumbing lines (Fig. 22-7). In multiple-story dwellings, the length of plumbing lines can be reduced and a common plumbing wall used if the baths are placed directly above one another. When a bath is placed on a second floor, a plumbing wall must be provided through the first floor for the soil and water pipes.

Layout

There are two basic types of bathroom layouts: the compartment and the open plan. In the *com-*

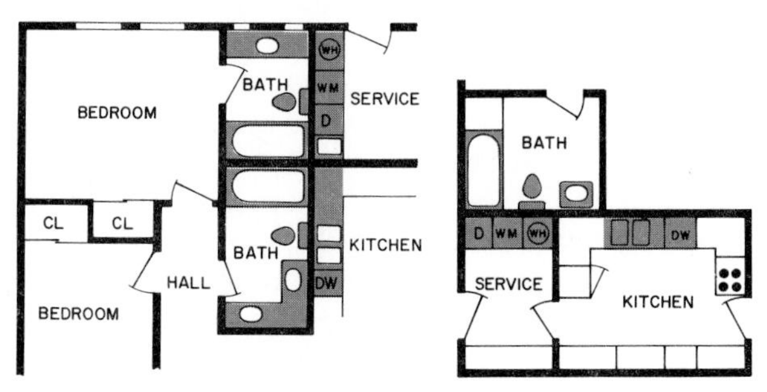

Fig. 22-7 Fixture arrangements that keep plumbing lines to a minimum.

Fig. 22-8A A compartment bath. *(Consoweld Corp.)*

partment plan, partitions (sliding doors, glass dividers, louvers, or even plants) are used to divide the bath into several compartments, one housing the water closet, another the lavatory area, and the third the bathing area (Fig. 22-8A). In the *open plan*, all bath fixtures are completely or partially visible.

A bath designed for, or used in part by, children should include a low or tilt-down mirror, benches for reaching the lavatory, low towel racks, and shelves for bath toys.

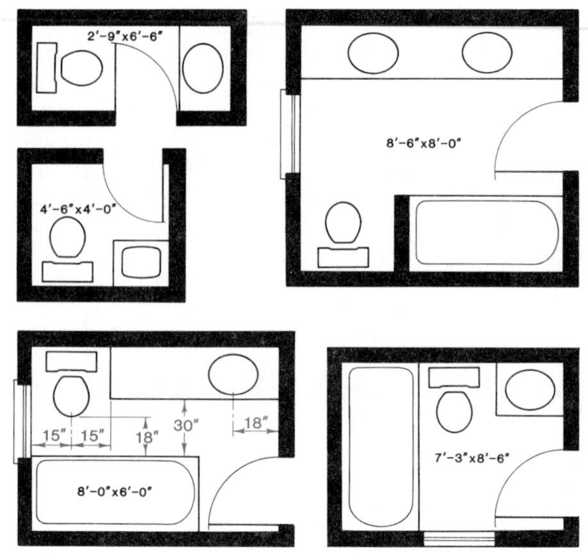

Fig. 22-8B Bath shapes and clearances.

SIZE AND SHAPE

The size and shape of the bath are influenced by the spacing of basic fixtures, the number of auxiliary functions requiring additional equipment, the arrangement or compartmentalization of areas, and the relationship to other rooms in the house. Figure 22-8B shows a variety of bath shapes and arrangements.

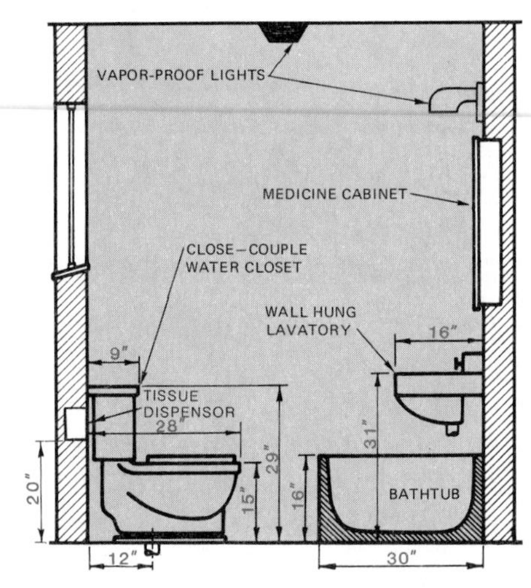

Fig. 22-9 Typical fixture sizes.

Furniture

Typical fixture sizes, as shown in Fig. 22-9, greatly influence the ultimate size of the bath. Figure 22-10 shows minimum-sized lavatory. Figure 22-11 shows small baths. Figure 22-12 shows average baths, and Fig. 22-13 shows large baths. Regardless of the size, these baths contain the three basic fixtures: sink, tub or shower, and water closet.

The sizes given here refer to complete baths and not to *half baths*, which include only a lavatory and water closet. Half baths are used in conjunction with the living area and therefore are not designed for bathing.

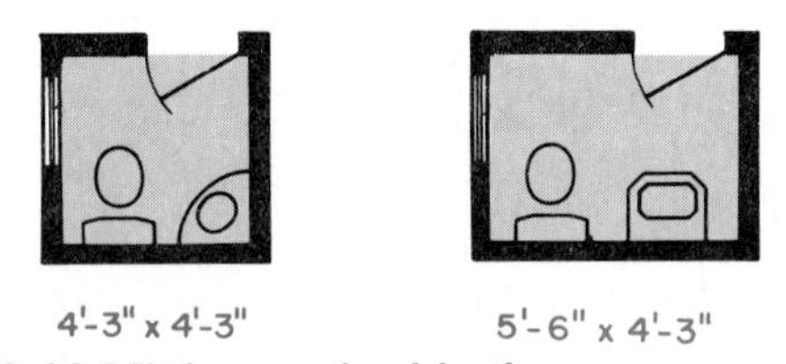

Fig. 22-10 Minimum-sized baths.

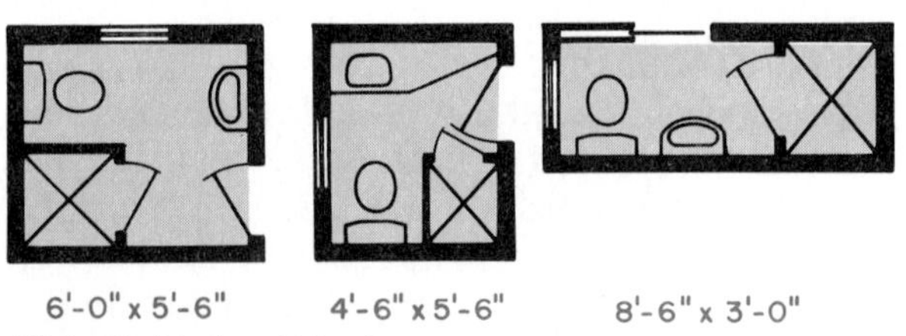

Fig. 22-11 Small baths.

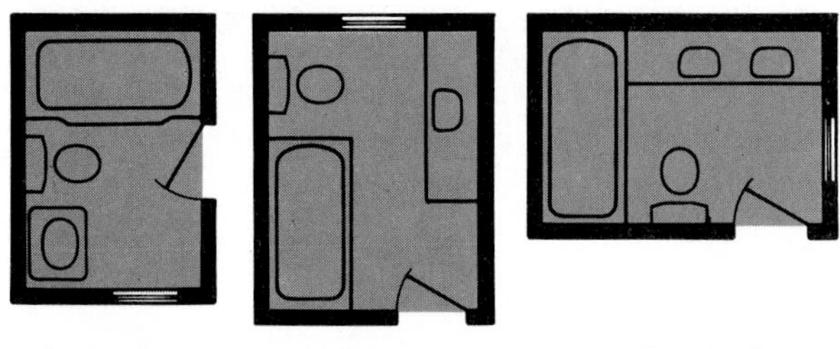

5'-0"x 7'-6" 6'-0" x 8'-0" 8'-0" x 5'-6"

Fig. 22-12 Average-sized baths.

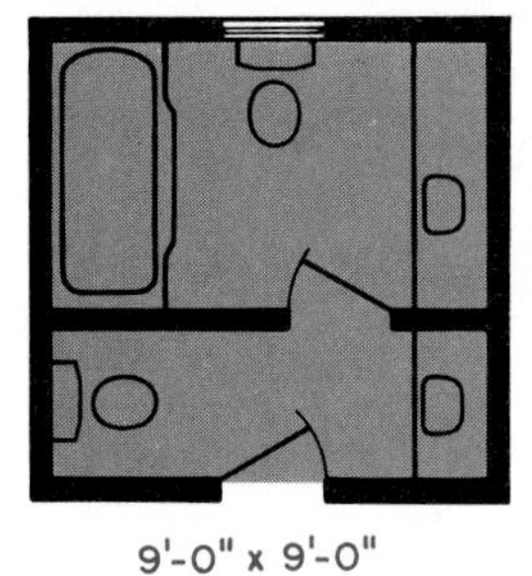

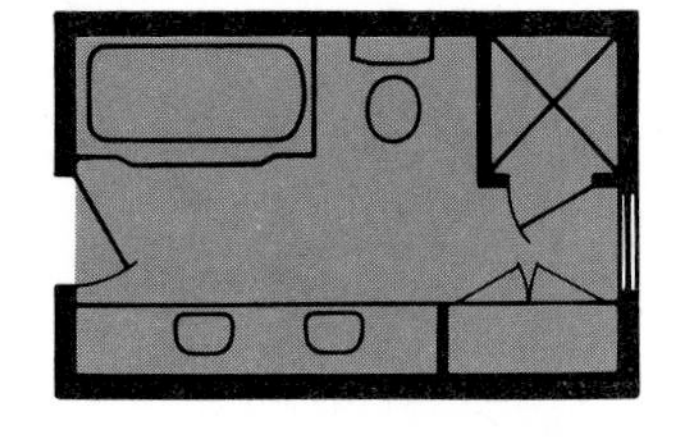

9'-0" x 9'-0" 11'-0" x 7'-0"

Fig. 22-13 Large baths.

Accessories

In addition to the three basic fixtures, the following accessories are often included in a bath designed for optimum use: exhaust fan, sunlamp, heat lamp, instant wall heater, medicine cabinet, extra mirrors, magnifying mirror, extra counter space, dressing table, whirlpool bath, foot-pedal control for water, single-mixing, one-control faucets, facility for linen storage, clothes hamper, bidet. Figure 22-14 shows a bath with many of these extra features.

Fig. 22-14 A bath designed with many extra features.

DECOR

The bath should be decorated and designed to provide the maximum amount of light and color. Materials used in the bath should be water-resistant, easily maintained, and easily sanitized. Tiles, linoleum, marble, plastic laminate, and glass are excellent materials for bathroom use. If wallpaper or wood paneling is used, it should be waterproof. If plastered or drywall construction is exposed, a gloss or semigloss paint should be used on the surface.

Fixtures and accessories should match in color. Fixtures are now available in a variety of colors. Matching countertops and cabinets are also available.

New materials and components are now available which enable the designer to plan bathrooms with modular units that range from one-piece molded showers and tubs to entire bath modules. In these units, plumbing and electrical wiring are connected after the unit is installed. Today's bathroom need not be strictly functional and sterile in decor. Bathrooms can be planned and furnished in a variety of styles. Figures 22-15 through Figure 22-17 show examples of a variety of bathroom decors and motifs.

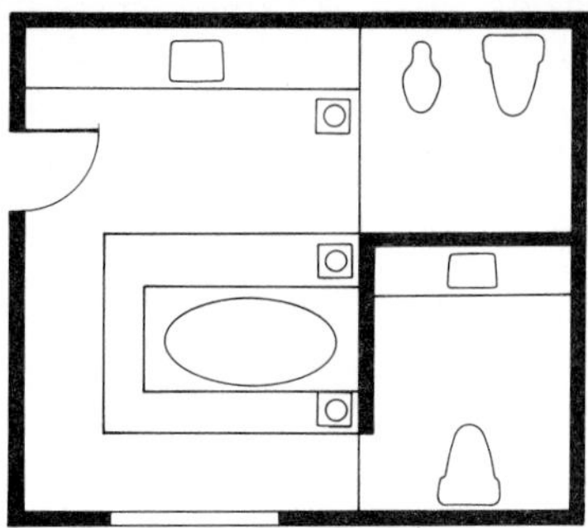

Fig. 22-15 Roman decor. *(American Standard, Inc.)*

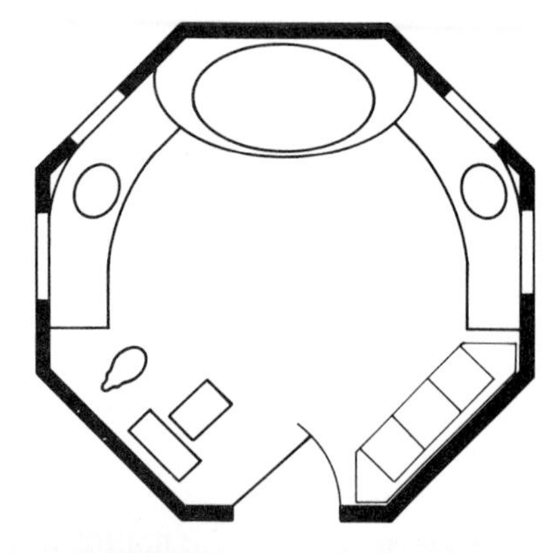

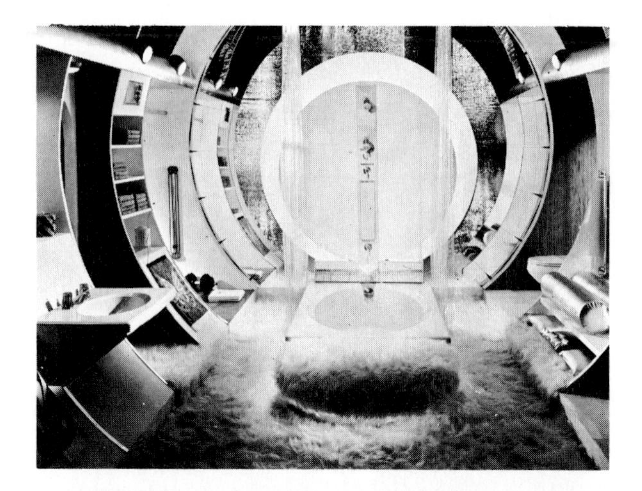

Fig. 22-16 Oriental decor. *(American Standard, Inc.)*

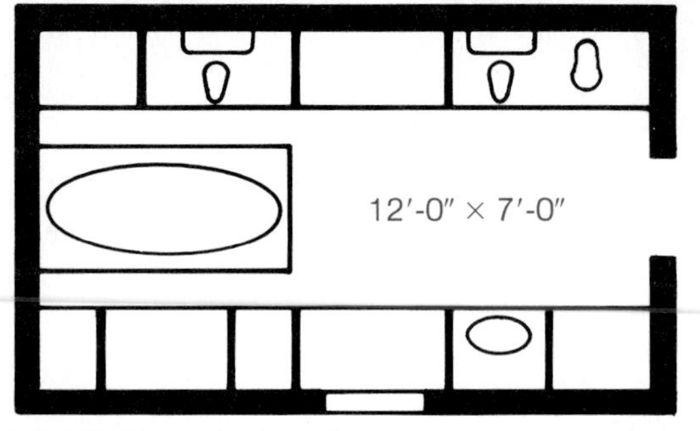

Fig. 22-17 Bath designed around space-age theme. *(American Standard, Inc.)*

Exercises

1. **Make a plan for adding fixtures to Fig. 22-18.**
2. **Draw a plan for remodeling the bath in Fig. 22-19, making the room more efficient.**
3. **Draw a plan for remodeling the bath in Fig. 22-20, making the room more efficient.**

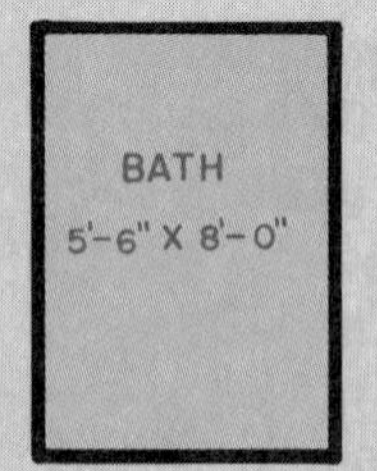

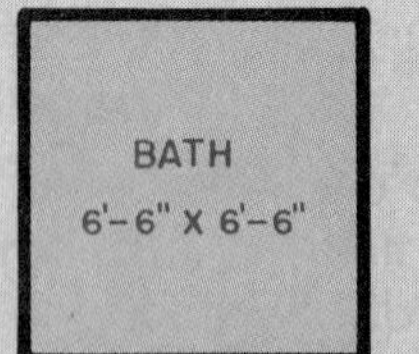

BATH
5'-5" X 5'-5"

Fig. 22-18

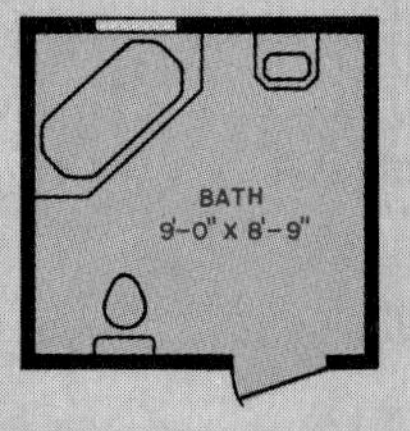

Fig. 22-19

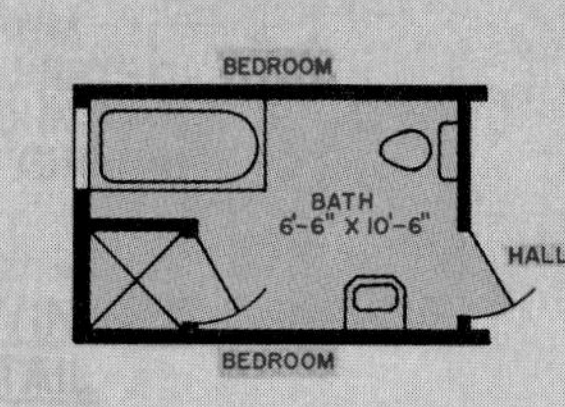

Fig. 22-20

4. **Draw the plans for the bath areas in the home of your design.**
5. **Define these architectural terms: *water closet, lavatory, fixture, open bath, sunken tub, shower stall, square tub, rectangular tub.***

PART THREE

Basic Architectural Plans

(Olympic Stains)

The general design of a structure is interpreted through several basic architectural plans. These include floor plans, elevations, and pictorial drawings. Floor plans show the arrangement of the internal parts of the design. Elevations graphically describe the exterior design. Pictorial drawings are prepared to show how the structure will appear when complete. In Part Three the preparations of these basic architectural drawings are explained in detail.

SECTION 6

Drafting Techniques

Most of the drafting skills and techniques used in architectural work are similar to those you have learned in mechanical drawing courses. However, there are some drafting procedures that are somewhat different. These involve the use of line techniques, templates, lettering practices, timesaving devices, and dimensioning practices. The differences are primarily due to the large size of most architectural drawings and to the great speed with which architectural plans must be prepared. For these reasons architectural drawings contain many abbreviated techniques. Methods of preparing architectural drawings now range from the use of preliminary sketches to very detailed computer-generated drawings. Section 6 covers this entire spectrum.

UNIT 23

ARCHITECTURAL LINE CONVENTIONS

Architects use various line weights to emphasize or deemphasize areas of a drawing. Architectural drafters also use different line conventions. Architectural line weights are standardized in order to make possible the consistent interpretation of architectural drawings. Figure 23-1 shows some of the common types of lines and line conventions used on architectural drawings. This figure is often called the alphabet of lines. You should learn the name of the line, the grade of the pencil used to make the line, and the technique used to draw the line.

ALPHABET OF LINES

Hard pencils (3H, 4H), as shown in Fig. 23-2, are used for architectural layout work. Medium pencils are used for most final lines (2H, H, F), and soft pencils are used for lettering, cutting-plane lines, and shading pictorial drawings (HB, B, 2B). Figure 23-2 shows a comparison of the various types of pencils and the lines they produce. Figure 23-3A shows the three basic types of drafting pencils, which can be sharpened to several types of points (Fig. 23-3B) depending on the type of line desired. Regardless of the type used, care must be taken to produce an even point. When uneven (chisel) points are produced, uneven lines will result, as illustrated in Fig. 23-3C.

Floor-Plan Lines

Figure 23-4 shows some of the common lines used on architectural floor plans. All lines are drawn dark. Different pencil grades are used to vary the width of lines. The lines are described as follows:

Object, or *visible, lines* are used to show the main outline of the building, including exterior walls, interior partitions, porches, patios, driveways, and walls. These lines should be drawn wide to stand out on the drawing.

Dimension lines are thin unbroken lines upon which building dimensions are placed.

Extension lines extend from the visible lines to permit dimensioning. They are drawn thin to eliminate confusion with the object outlines.

Hidden lines are used to show areas that are not visible on the surface but which exist behind the plane of projection. Hidden lines are also

NAME OF LINES	LINE SYMBOLS	LINE WIDTH	PENCIL
1. OBJECT LINES		THICK	H,F
2. HIDDEN LINES		MEDIUM	2H,H
3. CENTER LINES		THIN	2H,3H,4H
4. LONG BREAK LINES		THIN	2H,3H,4H
5. SHORT BREAK LINES		THICK	H,F
6. PHANTOM LINES		THIN	2H,3H,4H
7. STITCH LINES		THIN	2H,3H,4H
8. BORDER LINES		VERY THICK	F,HB
9. EXTENSION LINES		THIN	2H,3H,4H
10. DIMENSION LINES			
11. LEADER LINES		THIN	2H,3H,4H
12. CUTTING PLANE LINES		VERY THICK	F,HB
13. SECTION LINES		THIN	2H,3H,4H
14. LAYOUT LINES		VERY THIN LIGHT	4H
15. GUIDE LINES			
16. LETTERING	ARCHITECTURAL	THICK	H,F

Fig. 23-1 Architectural line weights; the alphabet of lines.

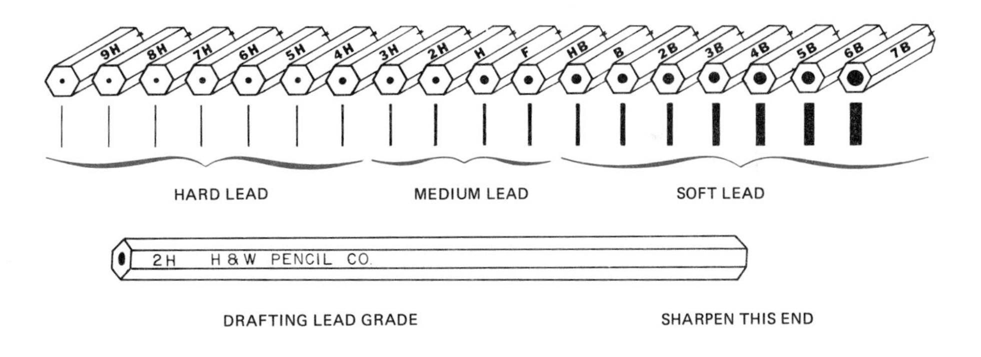

Fig. 23-2 Degrees of hardness of lead, cross sections of pencils, and their matching lines.

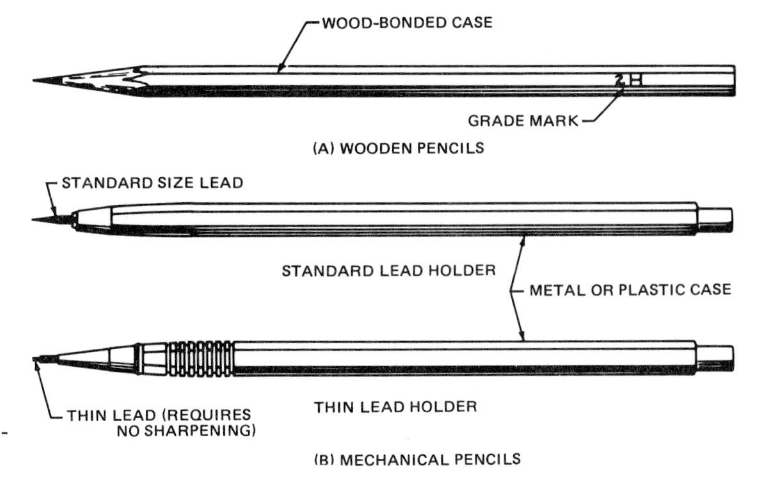

Fig. 23-3A Types of drafting pencils.

Fig. 23-3B Pencil point shapes.

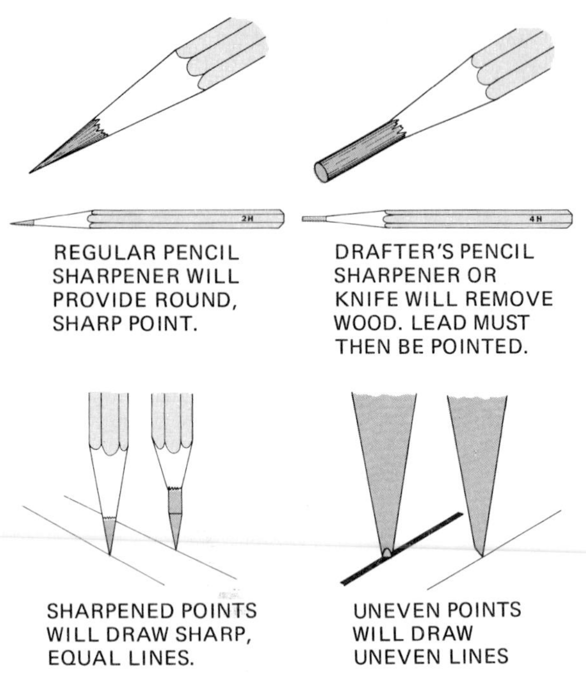

2H
4H
REGULAR PENCIL SHARPENER WILL PROVIDE ROUND, SHARP POINT.
DRAFTER'S PENCIL SHARPENER OR KNIFE WILL REMOVE WOOD. LEAD MUST THEN BE POINTED.
SHARPENED POINTS WILL DRAW SHARP, EQUAL LINES.
UNEVEN POINTS WILL DRAW UNEVEN LINES

Fig. 23-3C Results of pencil pointing.

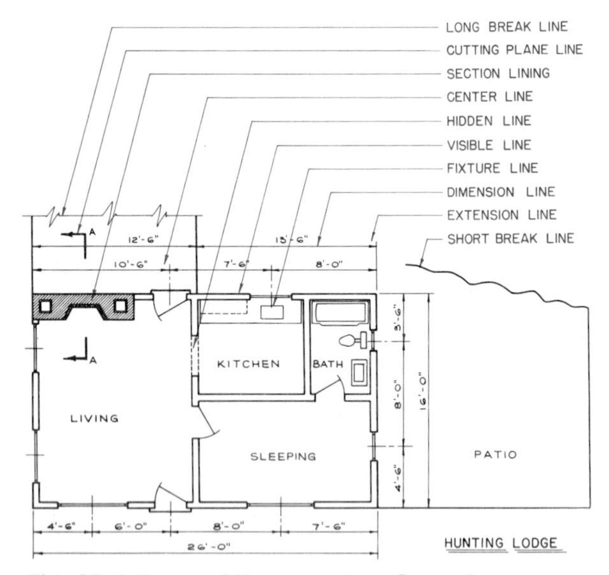

Fig. 23-4 Types of lines used on floor plans.

used in floor plans to show objects *above* the floor section, such as wall cabinets, arches, and beams. Hidden lines are drawn very thin.

Centerlines denote the centers of symmetrical objects such as exterior doors and windows. These lines are usually necessary for dimensioning purposes. Centerlines are drawn thin.

Cutting-plane lines are very wide lines used to denote an area to be sectioned. In this case, the only part of the cutting-plane line drawn is the extreme ends of the line. This is because the cutting-plane line would interfere with other lines on this drawing.

Break lines are used when an area cannot or should not be drawn entirely. A ruled line with freehand breaks is used for long, straight breaks. The long break line is thin. A wavy, uneven freehand line is used for smaller, irregular breaks. The short break line is wide.

Phantom lines are used to indicate alternate positions of moving parts, adjacent positions of related parts, and repeated detail. The phantom line is thin.

Fixture lines outline the shape of kitchen, laundry, and bathroom fixtures, or built-in furniture. These lines are thin to eliminate confusion with object lines.

Leaders are used to connect a note or dimension to part of the building. They are drawn thin and sometimes are curved to eliminate confusion with other lines.

Section lines are used to draw the section lining in sectional drawings. A different material symbol pattern is used for each building material. The section lining is drawn thin.

Elevation Lines

Figure 23-5 shows the application of the lines used on architectural elevation drawings. The technique and weight of each of the lines are exactly the same as those for the lines used on floor plans except that they are drawn on a vertical plane.

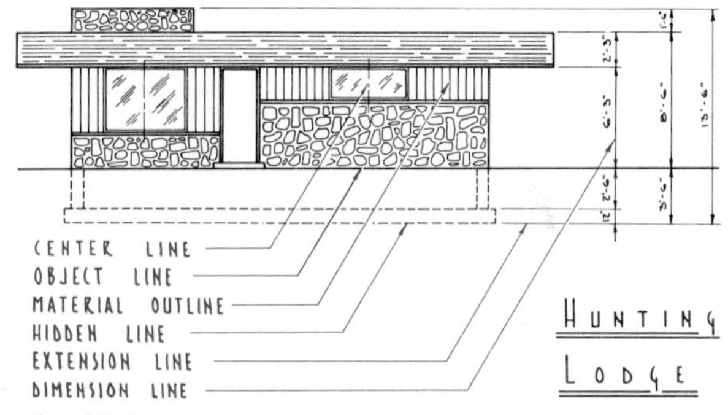

Fig. 23-5 Types of lines used on elevation drawings.

EFFECT OF THE DRAWING MEDIUM

Since the type of medium on which the line is drawn will greatly affect the line weight, different pencils may be necessary. Weather conditions such as temperature and humidity also greatly affect the line quality. During periods of high humidity, harder pencils must be employed. The drafter gradually gets the "feel" of the various pencil grades and their marking qualities on paper, vellum, and film.

Exercises

1. **Identify the types of lines indicated by the letters in Fig. 23-6.**
2. **List the grade of pencil you would use to draw each of the lines shown in Fig. 23-4.**
3. **Practice drawing each of the lines shown in Fig. 23-1, using a T square, triangle, or line tasks on a CAD system.**
4. **Practice lettering and drawing lines with different pencil grades on scrap paper, vellum, and film.**
5. **Define these terms: *line weights, alphabet of lines, hard lead, soft lead, object lines, dimension lines, extension lines, hidden lines, centerlines, cutting-plane lines, break lines, phantom lines, fixture lines, section lines, elevation lines.***

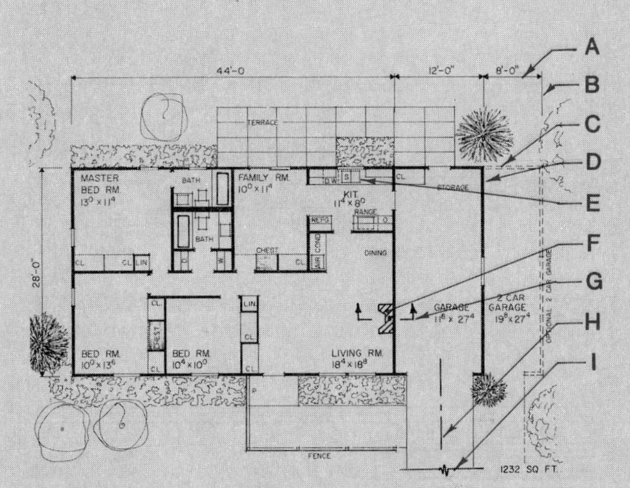

Fig. 23-6

UNIT 24

ARCHITECT'S SCALES

In ancient times, simple structures were built without detailed architectural plans and even without established dimensions. The outline of the structure and the position of each room could be determined experimentally by "pacing off" approximate distances. The builder could then erect the structure, using existing materials, by adjusting sizes and dimensions as necessary during the building process. In this case, the builder played the role of the architect, designer, contractor, carpenter, mason, and perhaps, the manufacturer of materials and components. Today, design requirements are so demanding, and materials so diverse, that a complete dimensioned set of drawings is absolutely necessary to ensure proper execution of the design as conceived by the designer. In the preparation of these drawings, the modern designer must use reduced-size scales. The ability to use architects' scales accurately is required not only in preparing drawings but in checking existing architectural plans and details. The architect's scale is used not only in preparing drawings, but also in a variety of related architectural jobs such as bidding, estimating, specification writing, and model building. The architect's scale is a measuring device, not a drawing instrument. *Never* use a scale as a straightedge for drawing.

REDUCED SCALE

The architect's scale is used to reduce a structure's sizes so that it can be drawn smaller than

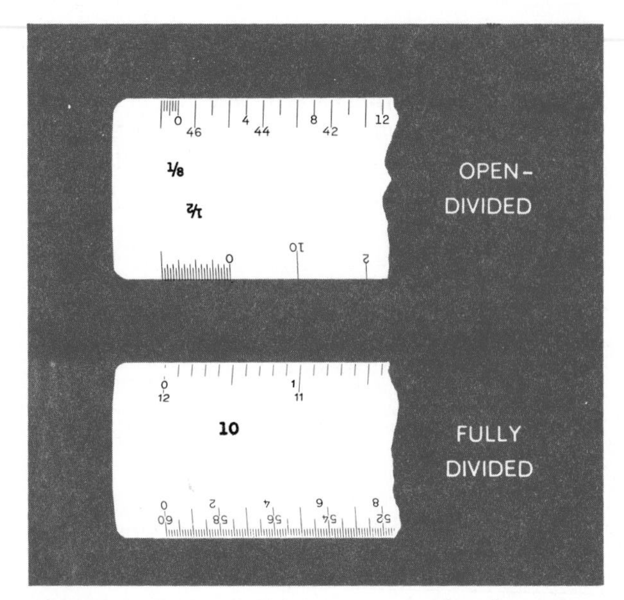

Fig. 24-1 Kinds of divisions on architect's scales.

actual size on paper. The architect's scale is also used to enlarge a small detail for clarity and to dimension it accurately.

Divisions

Architect's scales are either open-divided or fully divided. In *fully divided scales,* each main unit on the scale is fully subdivided into smaller units along the full length scale. On *open-divided scales*, only the main units of the scale are *graduated* (marked off) all along the scale. There is a fully subdivided unit at the start of each scale, as shown in Fig. 24-1. The main function of an architect's scale is to enable the architect, designer, or drafter to plan accurately and make drawings in relation to the actual size of the structure. For example, when a drawing is prepared to a reduced scale of 1/4″ = 1′-0″, a line that is drawn 1/4″ long is thought of by the drafter as 1′, not as 1/4″ (Fig. 24-2A).

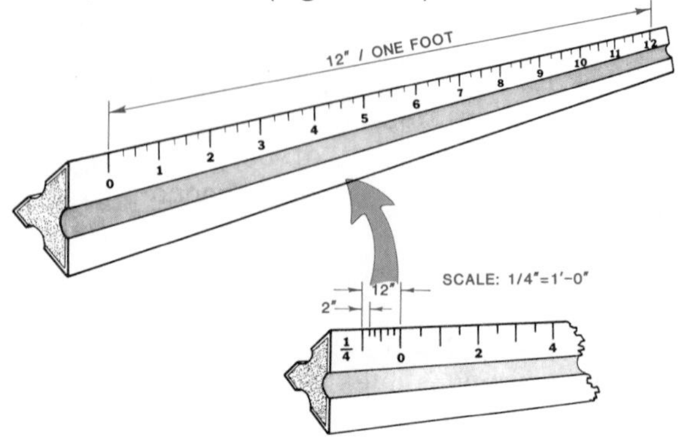

Fig. 24-2A On a drawing 1/4″ may represent 1′-0″

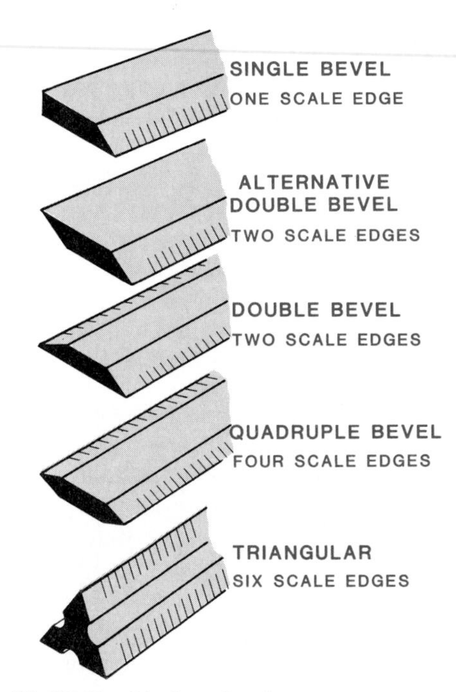

Fig. 24-2B Drafting scale shapes.

TYPES

Architect's scales are of either the bevel or the triangular style (Fig. 24-2B). You will notice that the triangular scale has 6 sides and accommodates 11 different scales. These scales are a full-size scale of 12″ graduated 16 parts to an inch, and 10 other open-divided scales which include the scales of 3/32, 1/8, 3/16, 1/4, 3/8, 1/2, 3/4, 1, 1 1/2, and 3. Two scales are located on each face. One scale reads from left to right. The opposite scale, which is half as large, reads from right to left. For example, the 1/4″ scale and the 1/8″ scale are placed on the same face. Similarly, the 3/4″ scale and the 3/8″ scale are placed on the same face but are read from opposite directions. Be sure you are reading in the correct direction when using an open-divided scale. Otherwise, your measurement could be wrong, since the second row of numbers read from the opposite side as seen in Fig. 24-3.

The architect's scale can be used to make the divisions of the scale equal 1′ or 1″. For example, on the 1/2″ scale shown in Fig. 24-4, 1/2″ represents 1″. The same scale in Fig. 24-5 is shown representing 1′-0″. Therefore 1/2″ can equal 1″ or 1′-0″. It can also represent any unit of measurement such as yards or miles.

Since buildings are large, most major architectural drawings use a scale that relates the parts of an inch to a foot. Architectural details

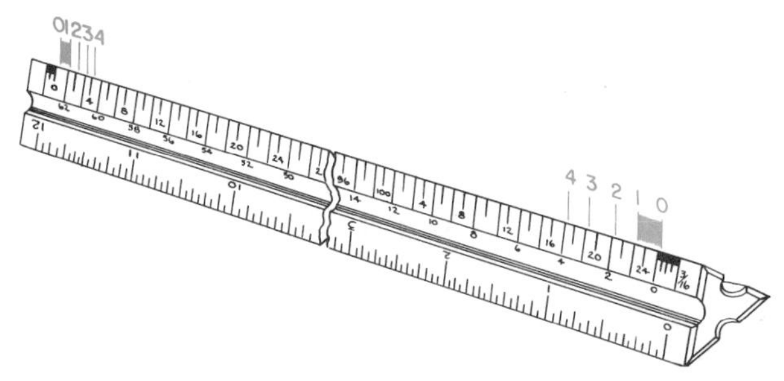

Fig. 24-3 The scale that reads from right to left is twice as large as the scale that reads from left to right.

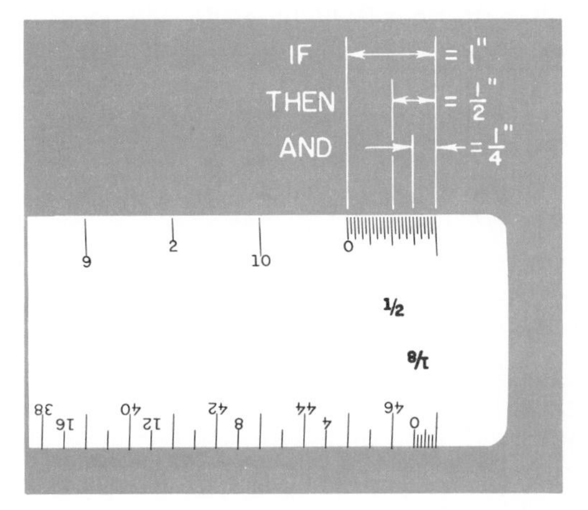

Fig. 24-4 If ½" equals 1", then the divisions represent fractions of an inch.

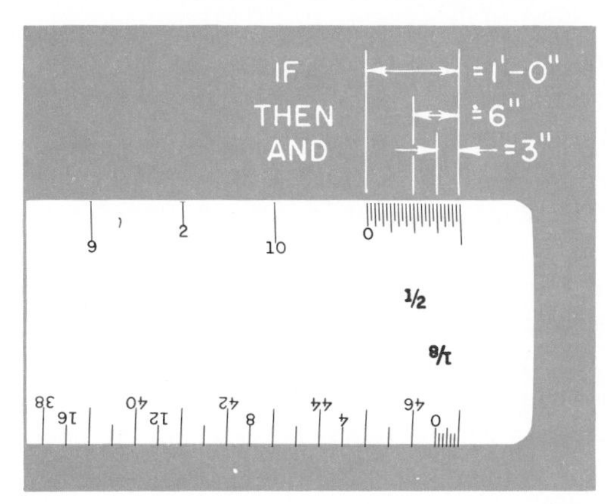

Fig. 24-5 If ½" equals 1-0", then the divisions represent fractions of a foot, or inches.

such as cabinet construction and joints often use the parts of an inch to represent 1". In either case, on open-divided scales the divided section at the end of the scale is not a part of the numerical scale. When measuring with the scale, start with the zero line, not with the fully divided section. Always start with the number of feet you wish to measure and then add the additional inches in the subdivided area. For example, in Fig. 24-6 the distance of 4'-11" is established by measuring from the line 4 to 0 for feet. Then, measure on the subdivided area 11" past 0. On this scale, each of the lines in the subdivided parts equals 1". On smaller scales, these lines may equal only 2". On larger scales, they may equal a fractional part of an inch. Figure 24-7 shows a further application of the use of the architect's scale. You will notice that the dimensioned distance of 8'-0" extends from the 8 to the 0 on the scale, and the 6" wall is shown as one-half of the subdivided foot on the end. Likewise, you can read the distance of 2'-0" on the ¼" scale shown in this illustration.

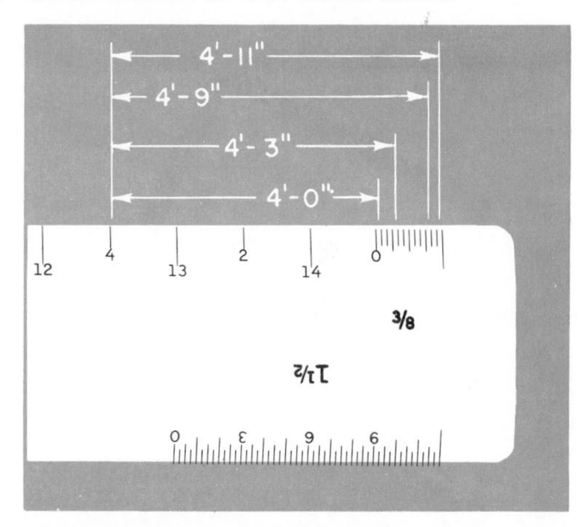

Fig. 24-6 Subdivisions at the end of an open-divided scale are used for an inch measurement.

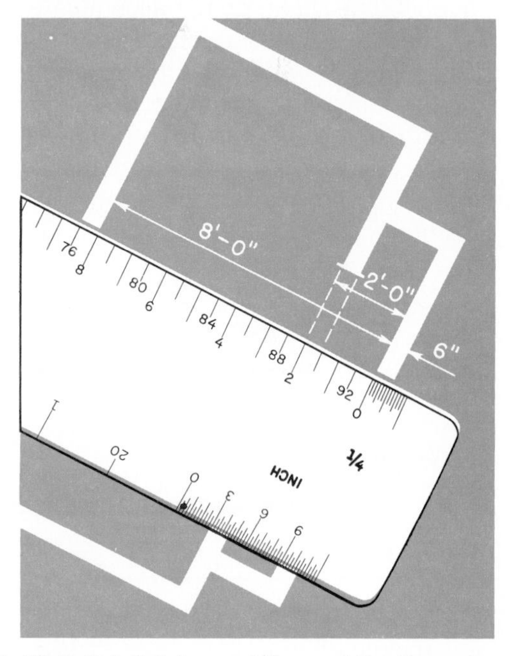

Fig. 24-7 Subdivisions of the architect's scale can be used to indicate overall dimensions and subdimensions.

Scale Selection

The selection of the proper scale is sometimes difficult. If the structure to be drawn is extremely large, a small scale must be used. Small structures can be drawn to a larger scale, since they will not take up as much space on the drawing sheet. Most floor plans, elevations, and foundation plans of residences are drawn to ¼″ scale, whereas construction details pertaining to these drawings are often drawn to ½″, ¾″, or even 1″ = 1′-0″. Remember that as the scale changes, not only does the length of each line increase or decrease but also the width of the various wall thicknesses increases or decreases. The actual appearance of a typical corner wall drawn to 1/16″ = 1′-0″, ⅛″ = 1′-0″, ¼″ = 1′-0″, and ½″ = 1′-0″ is shown in Fig. 24-8. You can see that the wall drawn to the scale of 1/16″ = 1′-0″ is small and that a great amount of detail would be impossible. The ½″ = 1′-0″ wall would probably cover too large an area on the drawing if the building were very large. Therefore, the ¼″ and ⅛″ scales are the most popular for drawing floor plans and elevations.

USE OF THE SCALE

The architect's scale is only as accurate as its user. In using the scale, do not accumulate distances. That is, always lay out overall dimensions first (Fig. 24-9). If the width and length are correct, only minor errors in subdimensions may occur. Furthermore, if your overall dimensions are correct, you will find it easier to check your subdimensions, because if one is off, another will also be incorrect.

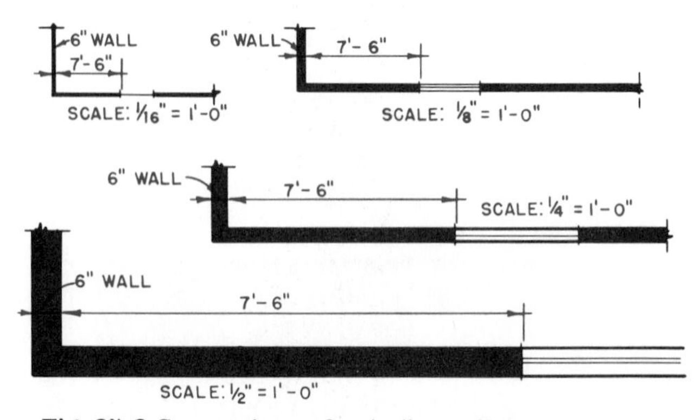

Fig. 24-8 Comparison of a similar wall drawn to several different scales.

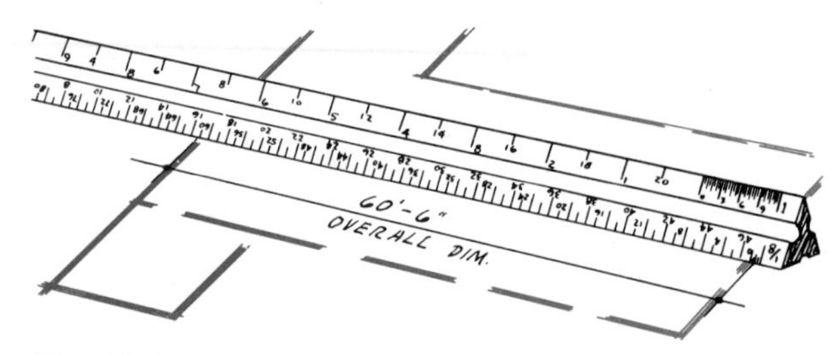

Fig. 24-9 Establish overall dimensions first.

Figure 24-10 shows the comparative distances used to measure 1′-9″ as it appears on various architect's scales. All these scales represent 1′-9″ as a reduced size. This same ratio would exist if we related the scales to 1″ rather than 1′. Using the scale of 1″ = 1′-0″, a distance of 1¾″ would have the same line length as 1′-9″. The 3/32″, 3/16″, ⅛″, ¼″, ⅜″, ½″, and ¾″ scales represent a distance smaller than full scale (1″-1′-0″). The scale 1½″ = 1″ will increase the drawing size by 50%. Likewise, the scale 3″ = 1″ will increase the drawing size 3 times. The scale 3″ = 1′-0″ will decrease the drawing size 4 times. Figure 24-11 shows the same comparison, using a distance of 5′-6″ on the foot-equivalent scale. If an inch-equivalent scale is used, the distance shown will be 5½″.

Civil Engineer's Scale

The civil engineer's scale is often used for plot plans, surveys, site plans, subdivision layouts, and landscape plans. Each scale divides the inch into decimal parts. These parts are 10, 20, 30, 40, 50, and 60 parts per inch (Fig. 24-12). Each one of these units can represent any distance, such as an inch, a foot, a yard, or a mile, depending on the final drawing size. The civil engineer's scale can also be used to draw floor plans. The scale ¼″ = 1′-0″ (a 1:48 ratio) is the same ratio as 1″ = 4′-0″ (also a 1:48 ratio) (Fig. 24-13). A civil engineer's scale of 1″ = 10′ is normally used for plot plans. If a site is very large, a scale of 1″-20′, or 1″ = 30′ may be needed to allow the plan to fit the sheet size.

Full Architect's Scale

The full side of the architect's scale is useful in dividing any area into an equal number of parts by following these steps:

Scale the distance available, as shown in Fig. 24-14. Next, place the zero point of the architect's scale on one side line. Then count off the correct number of spaces, using any convenient

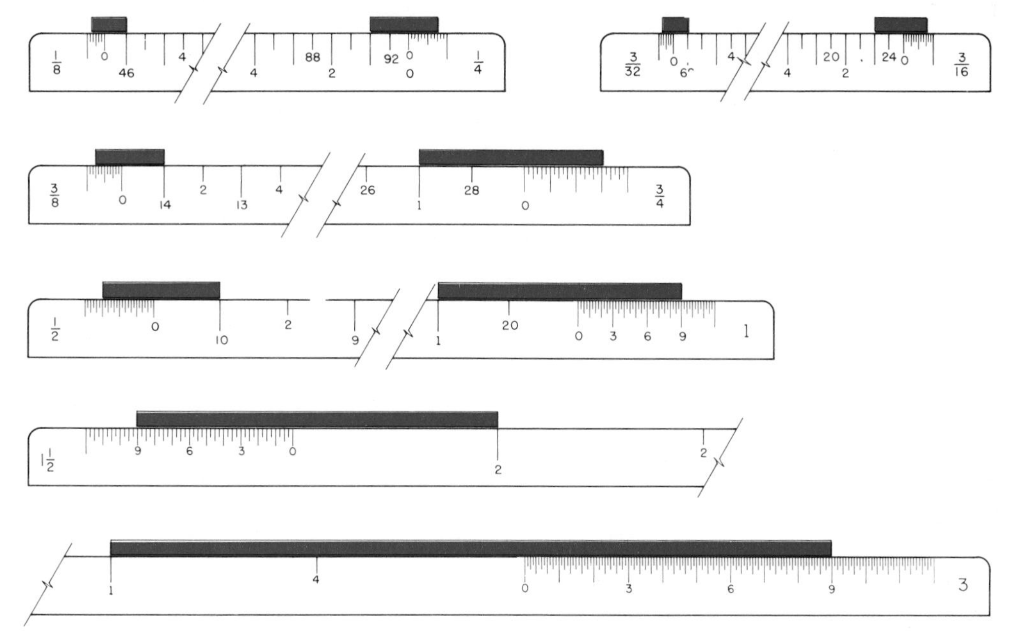

Fig. 24-10 The distance 1′-9″ as it appears on several architect's scales.

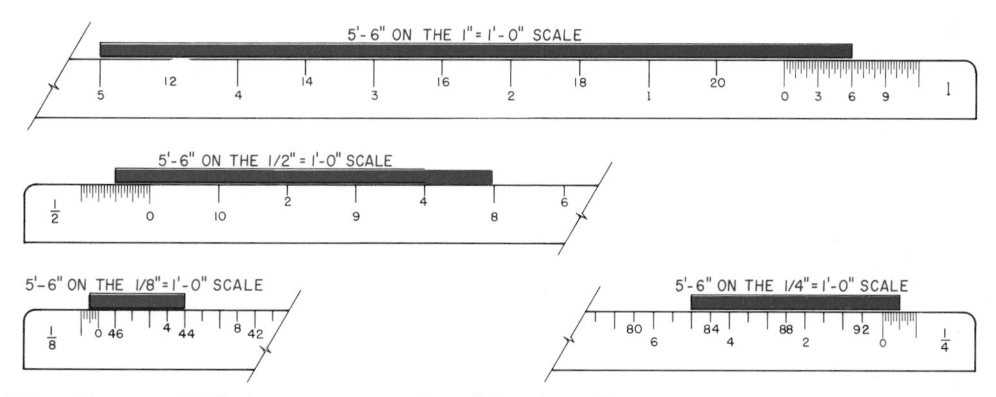

Fig. 24-11 The distance 5′-6″ shown on several architect's scales.

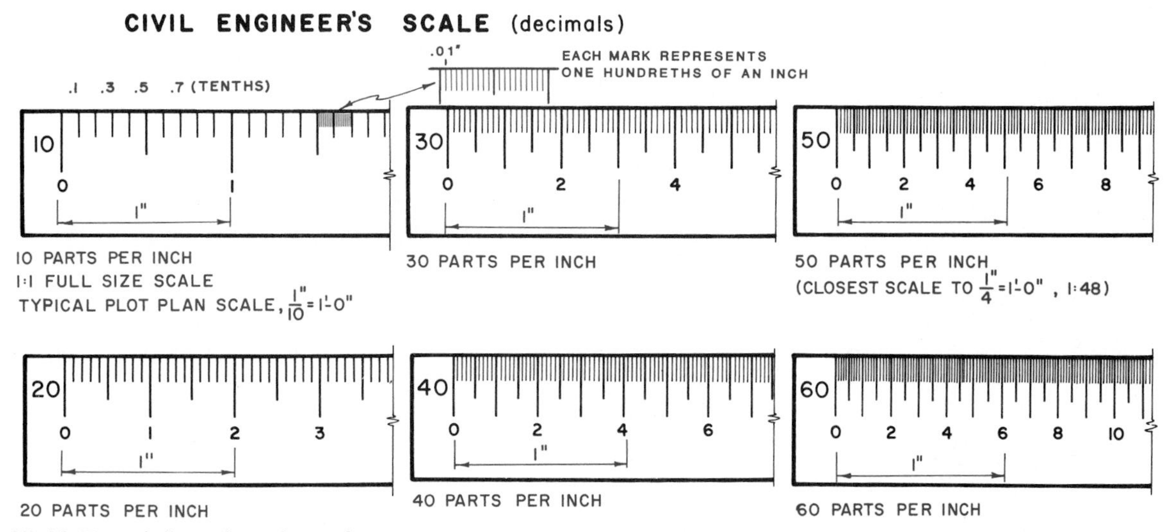

Fig. 24-12 The civil engineer's scale.

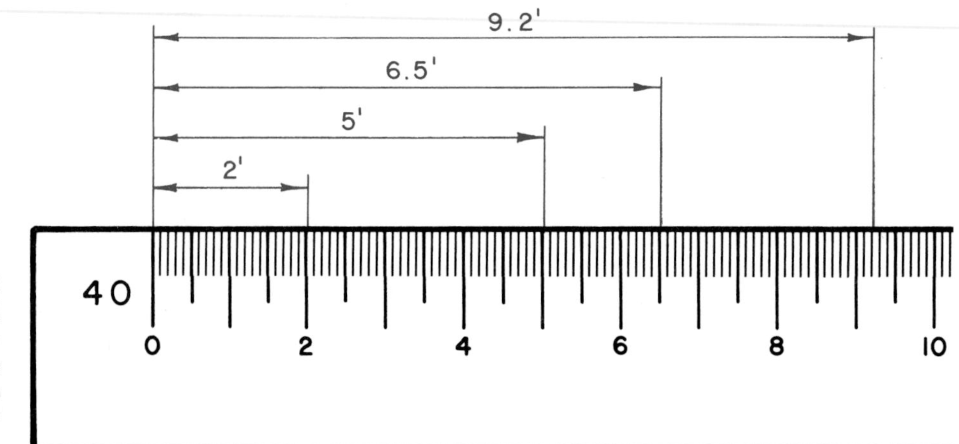

Fig. 24-13 Use of civil engineer's decimal dimensions for architectural drawing. The scale of 1″ = 4′-0″ is the same as ¼″ = 1′-0″.

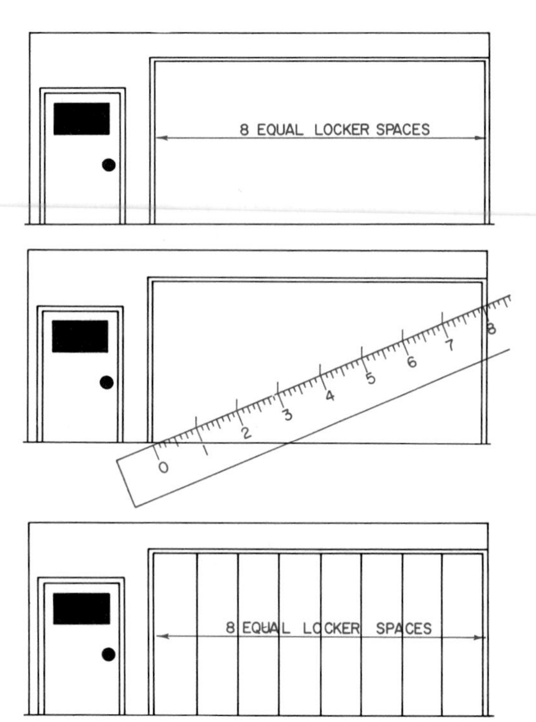

Fig. 24-14 Dividing an area into an equal number of parts.

unit, such as 1″ or ½″. Place the last unit mark on the line opposite the zero-point line. In Fig. 24-14, the 8″ mark is used since 1″ divisions are the most convenient. Mark each division and draw the dividing lines. Measure the horizontal distance between the lines to find the actual spacing; then multiply by the number of spaces to check your work.

DRAWING APPLICATIONS

Architectural drawings range widely in scale because of the wide range of objects or areas represented. For example, a construction detail may represent an area several feet wide, while a site plan may represent several acres. Before a drawing is started, determine the actual size of the area to be covered. Then select a scale that will provide the greatest detail and yet fit appropriately on the drawing sheet. Figure 24-15 shows the normal range of scales used on typical architectural drawings.

Drawing type	U.S. customary scales (feet/inches)	ISO Metric scales (millimeters)
Site plans	⅛″ = 1′-0″ or 1″ = 10′	1:100 or 1:150
Floor plans	¼″ = 1′-0″ or ⅛″ = 1′-0″	1:50 or 1:00
Foundation plans	¼″ = 1′-0″	1:50
Exterior elevations	¼″ = 1′-0″ or ⅛″ = 1′-0″	1:50 or 1:100
Interior elevations	½″ = 1′-0″	1:20
Construction details	1½″ = 1′-0″ thru ¾″ = 1′-0″	1:5 through 1:10
Cabinet details	½″ = 1′-0″	1:20

Fig. 24-15 Scales used for different drawing types.

Exercises

A 1/4" = 1'-0"
B 1/8" = 1'-0"
C 1/2" = 1'-0"
D 1/4" = 1'-0"
E 1/8" = 1'-0"
F 1/2" = 1'-0"
G 3/16" = 1'-0"
H 3/32" = 1'-0"

Fig. 24-16

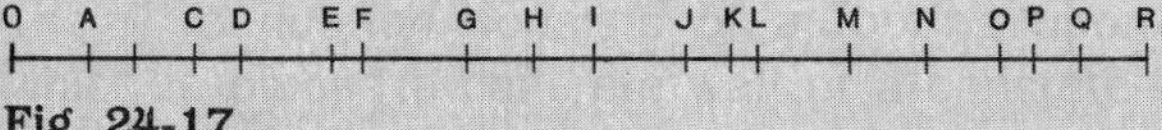

Fig. 24-17

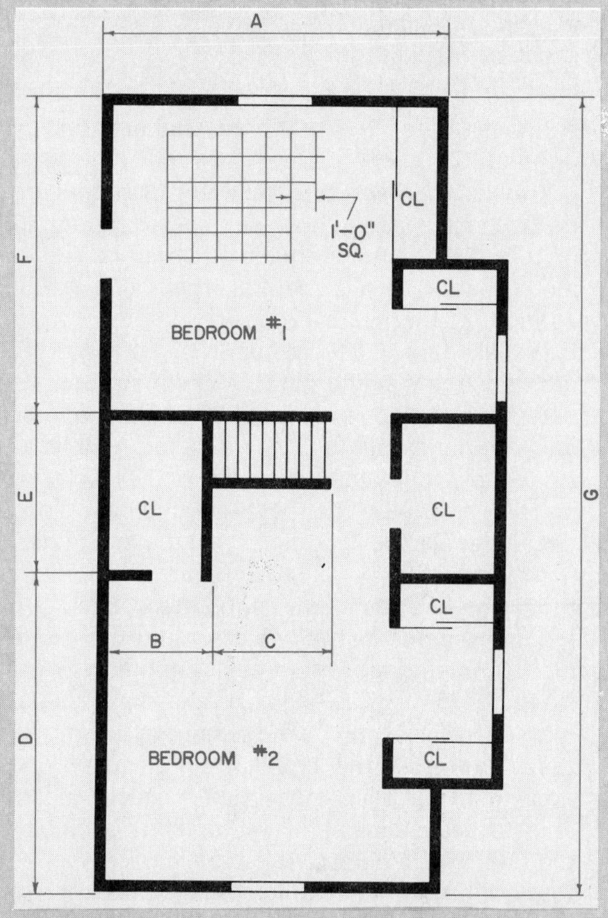

Fig. 24-18

1. **Measure the distances indicated on the horizontal lines, using the scale indicated in Fig. 24-16.**
2. **Measure the distances between the following letters in Fig. 24-17, using the 1/4" = 1'-0" scale: AB, AL, DE, EJ, KO, ST, FT, CK, EP.**
3. **Measure the distances shown in Problem 2, using the 1/8" = 1" scale. Each measurement is to originate from point 0 on the scale.**
4. **Answer the following questions concerning the plan shown in Fig. 24-18:**
   - ***a.*** **What is the overall length of the building?**
   - ***b.*** **What are the dimensions of bedroom 1, including the closets?**
   - ***c.*** **What is the length of the stairwell opening?**
   - ***d.*** **What are the dimensions of bedroom 2?**
   - ***e.*** **Determine the length of dimensions A through G.**
5. **Measure the distances in Fig. 25-16 using a scale of 1" = 1'-0", 1" = 10'-0", and 1" = 30'-0".**
6. **Define these terms: *open-divided, fully divided, architect's scale, decimal scale, triangular scale, reduced scale, division, inch-equivalent scale, foot-equivalent scale, full scale.***

UNIT 25

METRIC SCALES

The basic units of measure in the metric system are the *meter* (m) for distance, the *kilogram* (kg) for mass (weight), and the *liter* (L) for volume. Since most measurements used on architectural drawings are distances, only meters or millimeters are used.

PREFIXES

The meter is a base unit of one. To eliminate the use of many zeros, prefixes are used to change

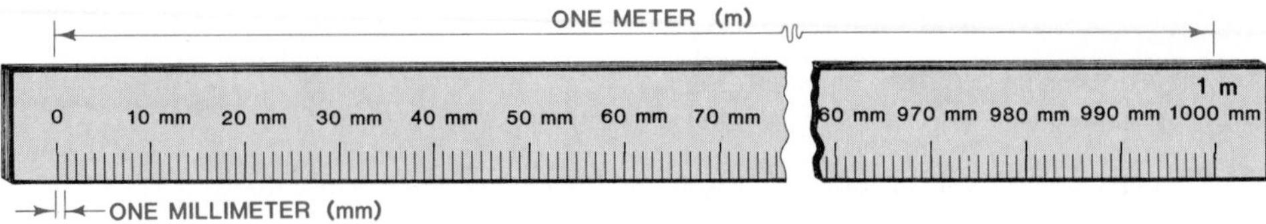

Fig. 25-1 A millimeter is one one-thousandth of a meter.

the base (meter) to larger or smaller amounts by units of 10.

Prefixes that represent multiples of meters are deka-, hecto-, and kilo-. A dekameter equals 10 meters. A hectometer equals 100 meters. A kilometer equals 1000 meters. The most useful multiple of the meter is the kilometer.

Prefixes that represent subdivisions of meters are deci-, centi-, and milli-. A decimeter equals one-tenth (0.1) of a meter. A centimeter equals one one-hundredth (0.01) of a meter. A millimeter equals one one-thousandth (0.001) of a meter (Fig. 25-1). The most useful subdivisions of a meter are the centimeter and the millimeter.

The numbers on a meter scale mark every tenth line and represent centimeters. Each single line represents millimeters. Note that there are 10 millimeters within each centimeter.

Figure 25-2 gives many of the most useful metric prefixes and shows the relationship of these prefixes to the meter. The prefixes may be applied to all base metric units except mass. This is because the base unit for mass is a multiple unit, the kilogram. The prefixes are applied to the *gram* for mass units. This consistent use of prefixes for distance, mass, and volume makes the metric system much easier to use than our customary system. The number of metric base units is fewer, making it easier to remember.

There are prefixes which extend the range upward to 10^{18}(exa) and downward to 10^{-18}(atto). These very large and very small units are used primarily for scientific notations in areas such as astronomy and microbiology.

	Prefix	Symbol	+ Meter =	
$1000 = 10^3$	kilo	k	kilometer	km
$100 = 10^2$	hecto	h	hectometer	hm
$10 = 10^1$	deka	da	dekameter	dam
$0.1 = 10^{-1}$	deci	d	decimeter	dm
$0.01 = 10^{-2}$	centi	c	centimeter	cm
$0.001 = 10^{-3}$	milli	m	millimeter	mm

Fig. 25-2 Prefixes change the base by increments of 10.

In the United States, there is a disagreement as to how to spell meter. All English-speaking countries using the metric system except the United States have adopted the spelling *metre*. Both spellings, meter and metre, are correct, and until one spelling becomes more popular in the United States than the other you should know that meter and metre mean the same thing. This is also true for liter and litre.

METRIC DIMENSIONS

Linear metric sizes used on basic architectural drawings such as floor plans and elevations are expressed in meters and decimal parts of a meter. Dimensions on these plans are usually carried to three decimal places. Figure 25-3

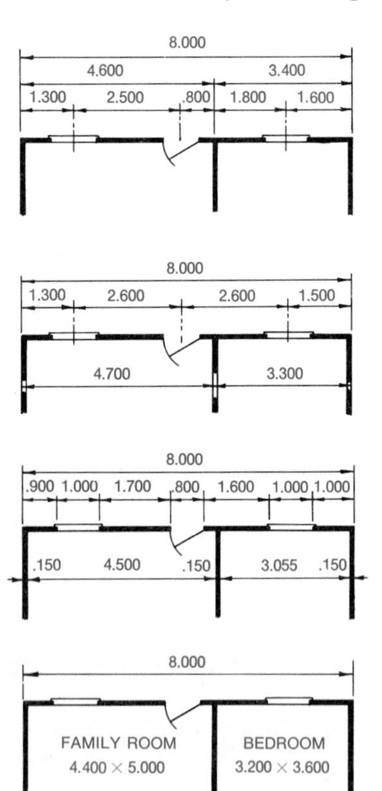

Fig. 25-3 Examples of dimensions in meters to three decimal places.

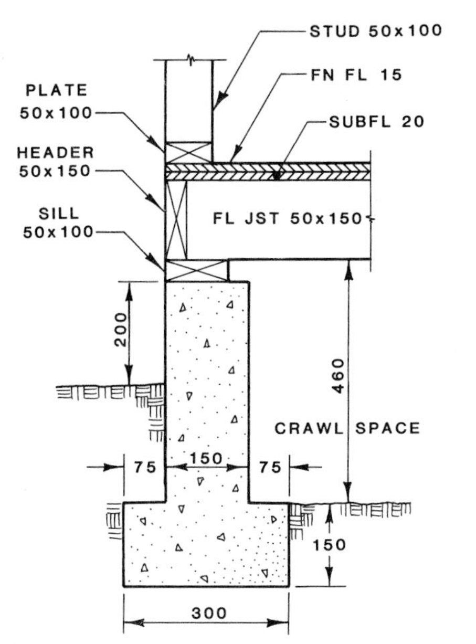

Fig. 25-4 The millimeter is used for detail dimensioning.

shows examples of meter dimensions carried to three decimal places. Small-detail drawings usually use millimeters, which eliminate the use of decimal points, as shown in Fig. 25-4. Note that the dimension 4.5 m could also be read as 4500 mm.

METRIC DRAWING RATIOS

Metric scales such as those shown in Fig. 25-5 are used in the same manner as the architect's scale to prepare reduced-size drawings. Metric scales, however, use ratios in increments of 10 rather than the fractional ratios of 12 used in architect's scales. Just as with fractional scales, the ratio chosen depends on the size of the drawing compared with the full size of the ob-

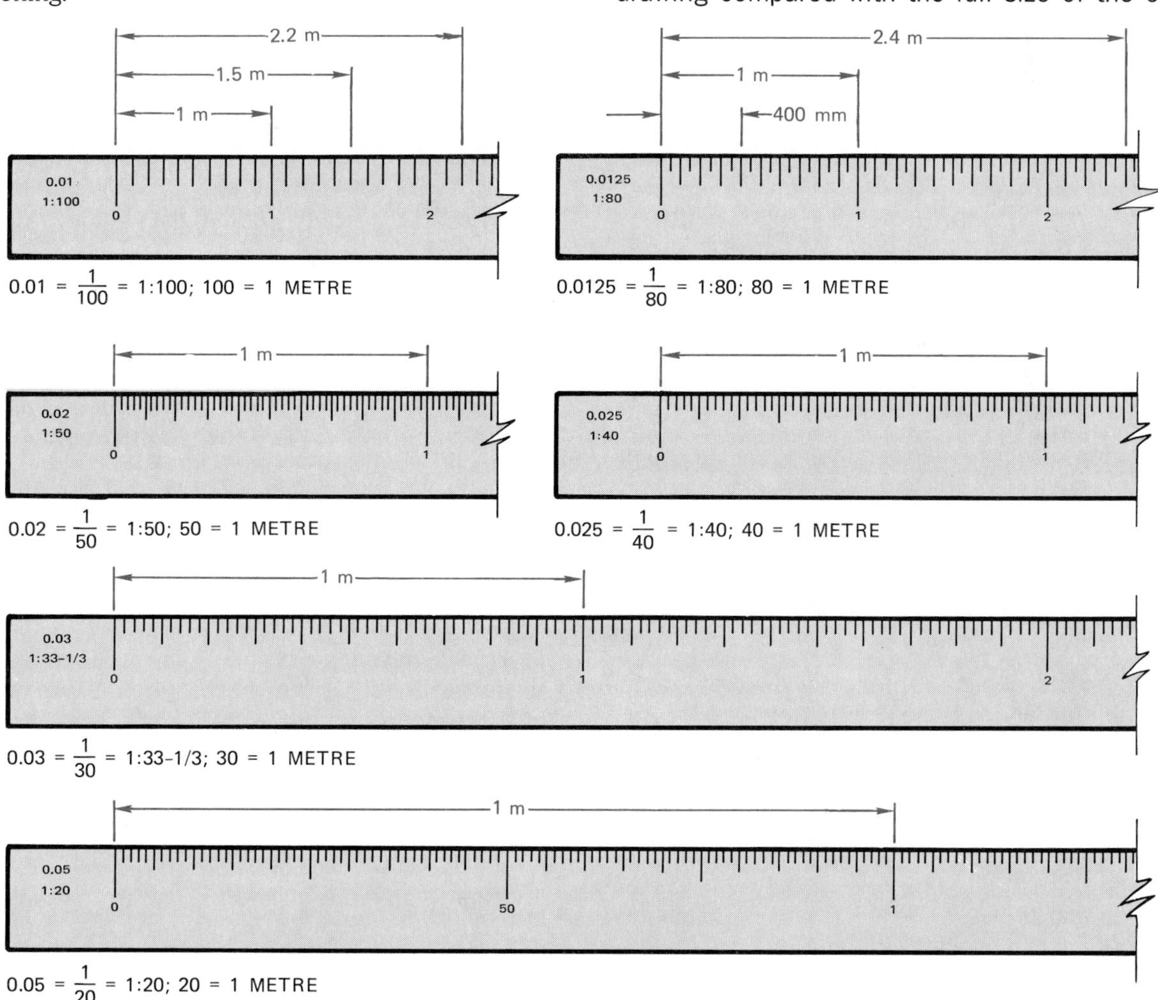

Fig. 25-5 Typical metric ratios.

ject. Figure 25-6 shows some common metric ratios and the various types of architectural drawings for which they are used. It is important to prepare *all* drawings in a set using metric ratios or to prepare all drawings in a set using the customary fractional system. Do not mix metric and customary units. If approximate conversion from one system to the other is necessary, refer to the appendix. When very accurate conversion from customary to metric units is necessary, consult a handbook or use *Metric Practice Guide*, ANSI/ASTM E 380-76.

Use	Ratio	Comparison to 1 meter
CITY MAP	1 : 2500	(0.4 mm equals 1 m)
	1 : 1250	(0.8 mm equals 1 m)
PLAT PLANS	1 : 500	(2 mm equals 1 m)
	1 : 200	(5 mm equals 1 m)
PLOT PLANS	1 : 100	(10 mm equals 1 m)
	1 : 80	(12.5 mm equals 1 m)
FLOOR PLANS	1 : 75	(13.3 mm equals 1 m)
	1 : 50	(20 mm equals 1 m)
	1 : 40	(25 mm equals 1 m)
DETAILS	1 : 20	(50 mm equals 1 m)
	1 : 10	(100 mm equals 1 m)
	1 : 5	(200 mm equals 1 m)

Fig. 25-6 Architectural use of metric ratios.

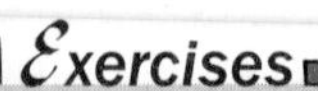

Exercises

1. Measure common objects, such as your book, desk, and room, using a metric scale. Record your results. Compare your measurements with customary measurements.

2. Define the following terms: *meter, kilogram, liter, millimeter, kilometer, prefix, centimeter, metric system, customary system, ratio.*

UNIT 26

DRAFTING INSTRUMENTS

Architectural designers need an assortment of instruments and supplies, including a large, adjustable drawing table or board. An overhead adjustable lamp is required, in addition to general illumination. The variety of pens, pencils, and instruments needed by designers are described in this unit.

T SQUARE

The *T square* is used primarily as a guide for drawing horizontal lines and for guiding the triangle when drawing vertical and inclined lines. The T square is also the most useful instrument for drawing extremely long lines that deviate from the horizontal plane. Common T-square lengths for use in architectural drafting are 18″, 24″, 36″, and 42″.

T squares must be held tightly against the edge of the drawing board, and triangles must be held firmly against the T square to ensure accurate horizontal and vertical lines. Since only one end of the T square is held against the drawing board, some sag may occur when extremely long T squares are not held securely.

Horizontal Lines

Horizontal lines are always drawn with the aid of instruments such as the T square, parallel slide, or drafting machine. In drawing horizontal lines with the T square, hold the head of the T square firmly against the left working edge of the drawing board (if you are right-handed). This procedure keeps the blade in a horizontal position to draw horizontal lines from left to right. Figure 26-1A shows the correct method of drawing horizontal and vertical lines, using the T square. Figure 26-1B shows the left-handed method.

Vertical Lines

Triangles are used with the T square for drawing vertical or inclined lines. The 8″, 45° triangle and the 10″, 30°-60° triangle are preferred for architectural work. Figure 26-1A and B shows the correct method of drawing vertical lines, using the T square and triangle.

Triangles are used to draw vertical and inclined lines with either a T square or a parallel slide. A variety of possible combinations produce numerous angles, as shown in Fig. 26-2A and B. The 45° triangle is used to draw miter (45°) lines that turn angles of buildings, as shown in Fig. 26-3. Triangles are also used to draw various symbols. Figure 26-3 also shows the use of the 30°-60° triangle in drawing a door symbol.

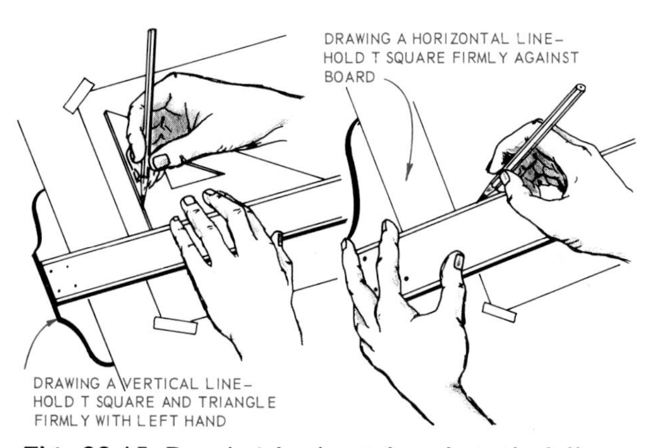

Fig. 26-1A Drawing horizontal and vertical lines with a T square and triangle.

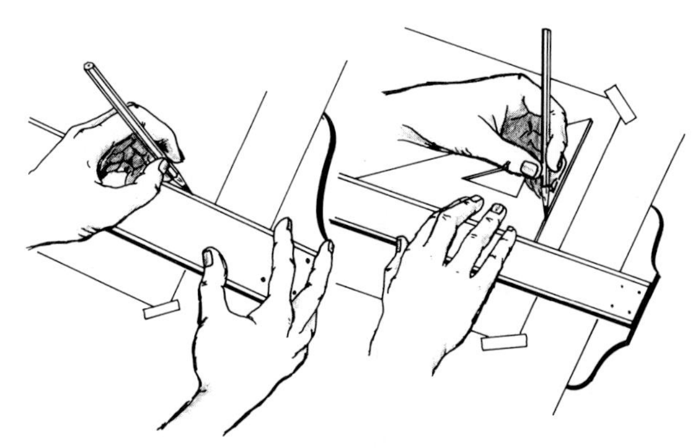

Fig. 26-1B Left-handed method of using T square and triangle.

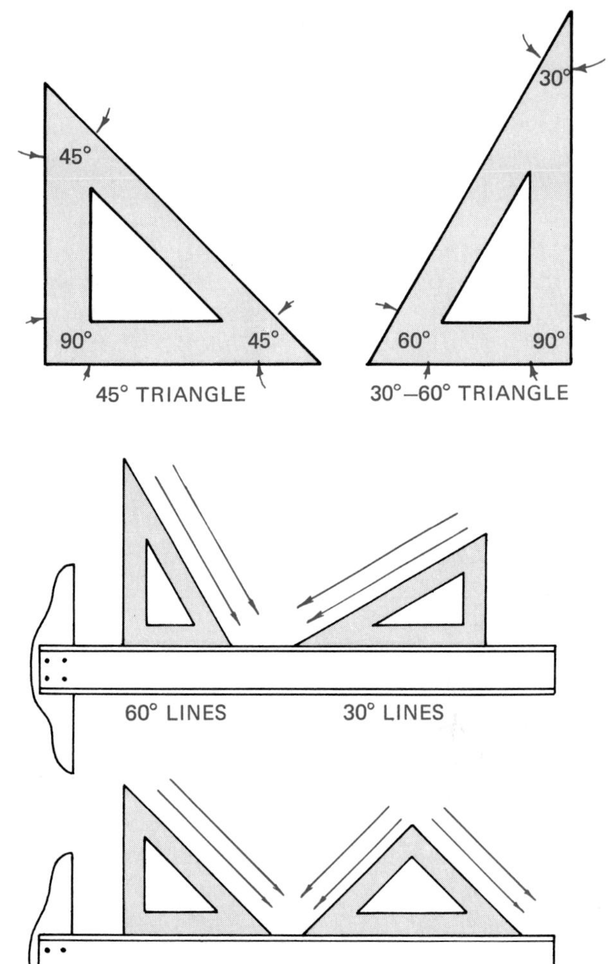

Fig. 26-2A Use of a T square and triangles to draw angles.

Triangles (and inverted T squares) are often used to project perspective lines to vanishing points, as shown in Fig. 26-4. The adjustable triangle is used to draw angles that cannot be laid out by combining the 45° and 30°-60° triangles. Figure 26-5 shows an architectural application of the use of an adjustable triangle.

PARALLEL SLIDE

The *parallel slide* (parallel rule) performs the same function as the T square. It is used as a

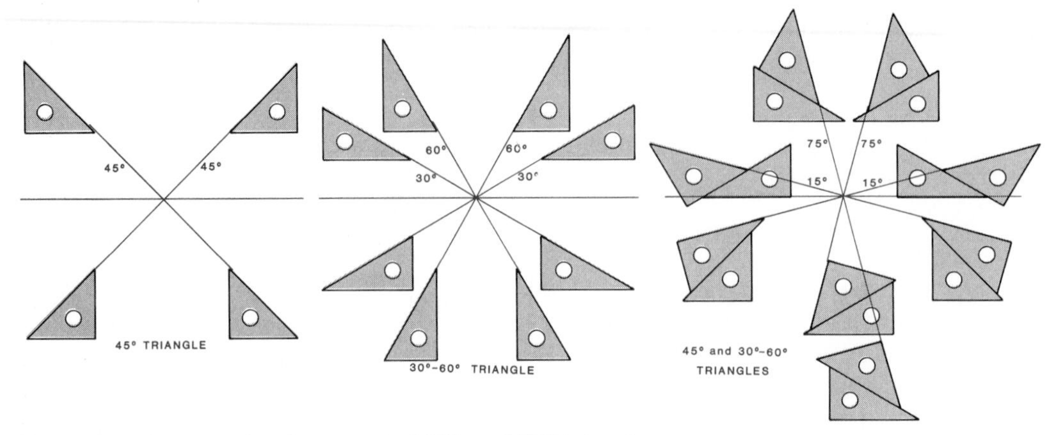

Fig. 26-2B Angles that may be drawn using 45° and 30° triangles.

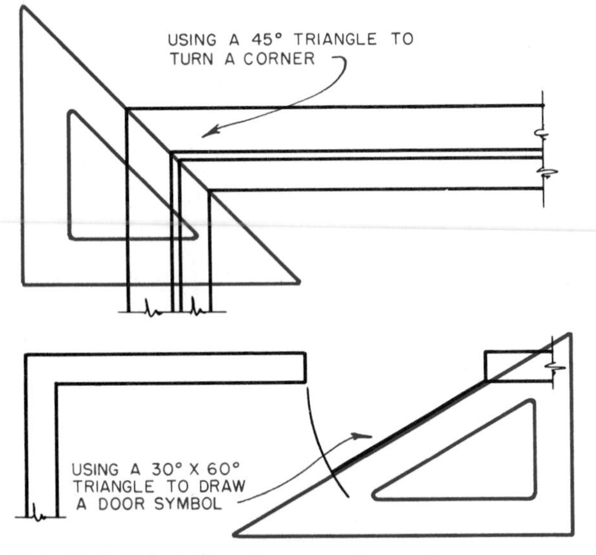

Fig. 26-3 Using triangles on a floor plan.

guide for drawing horizontal lines and as a base for aligning triangles in drawing vertical lines.

Extremely long lines are common in many architectural drawings such as floor plans and elevations. Since most of these lines should be drawn continuously, the parallel slide is used extensively by architectural drafters.

The parallel slide is anchored at both sides of the drawing board, as shown in Fig. 26-6. This attachment eliminates the possibility of sag at one end, which is a common objection to the use of the T square.

In using the parallel slide, the drawing board can be tilted to a very steep angle without causing the slide to fall to the bottom of the board. If the parallel slide is adjusted correctly, it will stay in the exact position in which it is placed.

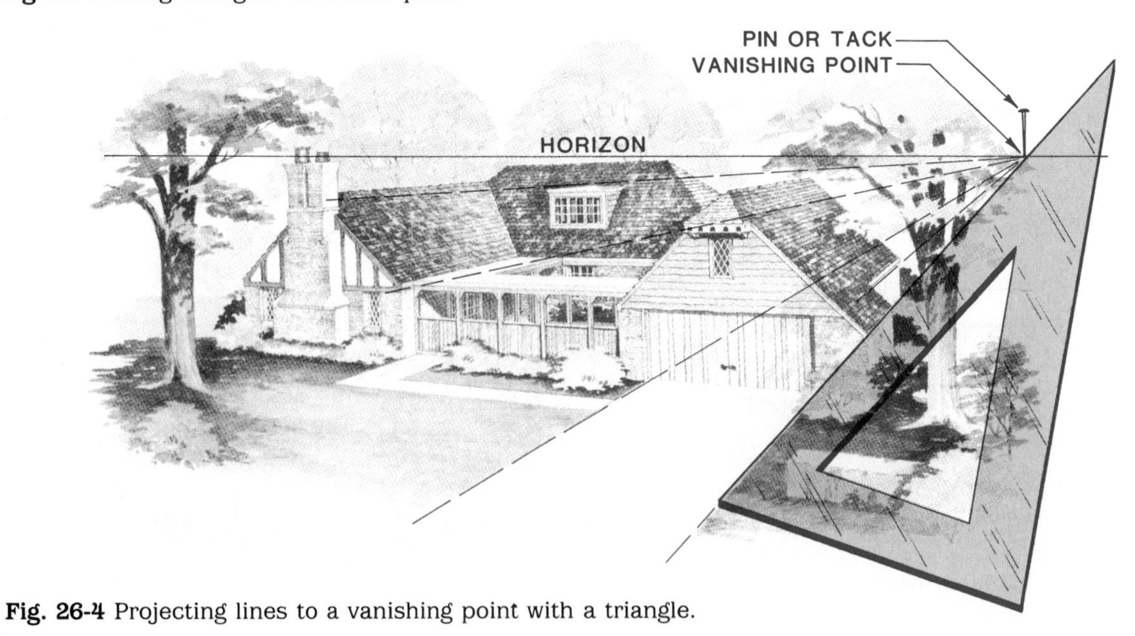

Fig. 26-4 Projecting lines to a vanishing point with a triangle.

Fig. 26-5 An adjustable triangle is used for angles of any number of degrees.

DRAWING HORIZONTAL LINES

DRAWING VERTICAL LINES

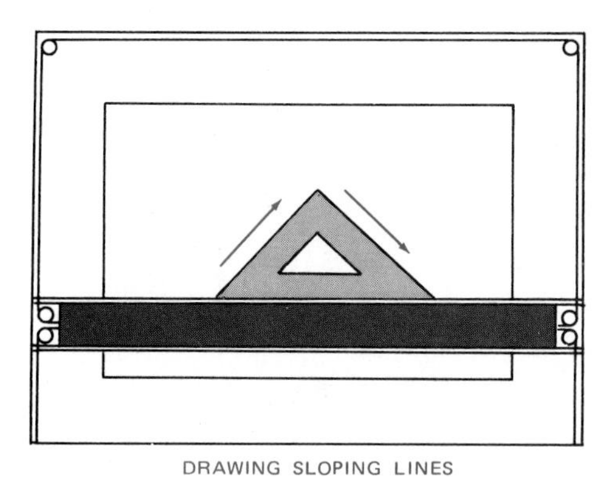

DRAWING SLOPING LINES

Fig. 26-6 A parallel slide used on a drawing board.

DRAFTING MACHINE

Using a drafting machine eliminates the need for the architect's scale, triangle, protractor, T square, or parallel slide. A *drafting machine* consists of a head to which two scales are attached (Fig. 26-7). These *scales* (arms) of the drafting machine are graduated like other architect's scales. They are usually made of aluminum or plastic. The two scales are attached to the head of the drafting machine perpendicular to each other. The horizontal scale performs the function of a T square or parallel slide in drawing horizontal lines. The vertical scale performs the function of a triangle in drawing vertical lines.

The head of the drafting machine can be rotated so that either of the scales can be used to draw lines at any angle. When the indexing thumbpiece, shown in Fig. 26-8, is depressed and then released, the protractor head of the drafting machine will lock into position every 15°. Figure 26-9 shows the intervals at which the

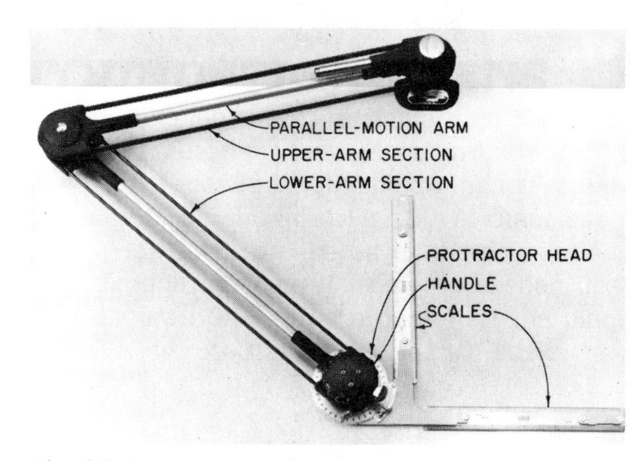

Fig. 26-7 A drafting machine. *(Vemco)*

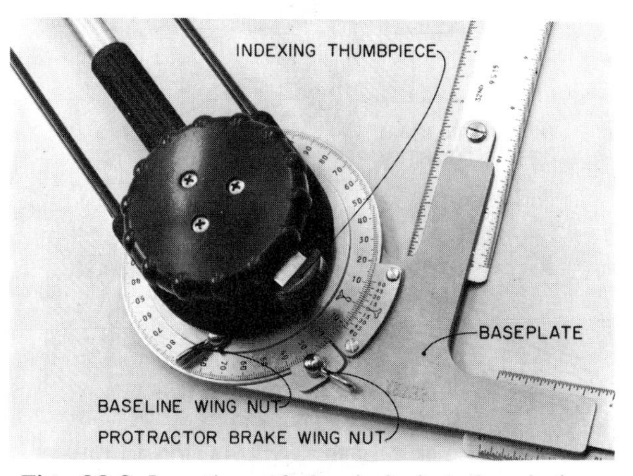

Fig. 26-8 Location of the indexing thumbpiece. *(Vemco)*

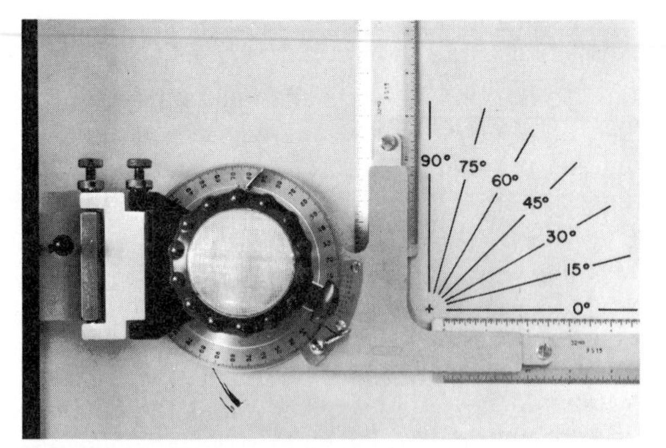

Fig. 26-9 Angles at which the scales will lock. *(Vemco)*

scales will index from a horizontal line. If the indexing thumbpiece remains depressed, the protractor head can be aligned to any degree. The protractor brake wing nut is used to lock the head in position. If accuracy in minutes is desired, the *vernier scale* (Fig. 26-10A) is used to set the protractor head at the desired angle. In this case, the vernier clamp is used to lock the head in the exact position when the desired setting is achieved.

Because of the large size of most architectural drawings, the drafting machine is often not satisfactory for large floor plans and elevations which require horizontal and vertical lines. For this reason the drafting machine is most often used for smaller architectural detail drawings.

Track drafting machines, shown in Fig. 26-10B, are smoother and faster than elbow-type machines. However, the operation of the head is identical. The head of a track machine is mounted on a vertical track which is attached to a horizontal track.

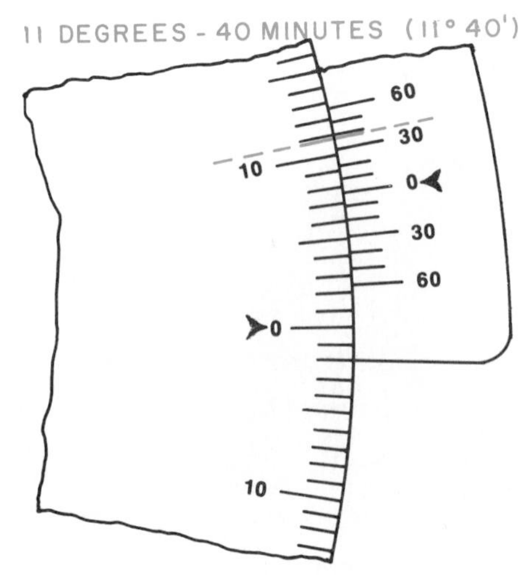

Fig. 26-10A Use of the vernier scale.

Fig. 26-10B Track drafting machine. *(Vemco)*

COMPASS

A *compass* is used in architectural work to draw circles, arcs, radii, and parts of many symbols. Small circles are drawn with a bow pencil compass. To use the bow, set it to the desired radius, and hold the stem between the thumb and forefinger. Rotate the compass with a forward clockwise motion and forward inclination as shown in Fig. 26-11.

Large circles on architectural drawings, such as those used to show the radius of driveways, walks, patios, and stage outlines, are drawn with a large *beam compass*, as shown in Fig. 26-12.

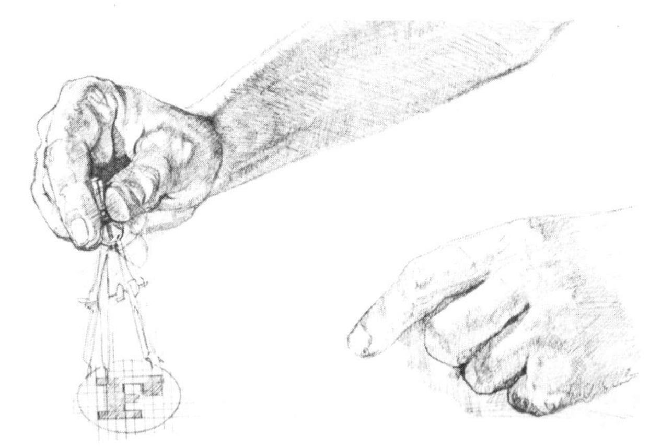

Fig. 26-11 Method of holding a compass. *(Fuller Instrument Co.)*

Fig. 26-12 A large beam compass.

Figure 26-13 shows the use of the compass in drawing door symbols. Very small circles on architectural drawings are drawn with either a template or a drop-bow compass. Figure 26-14 shows the correct method of sharpening and setting compass leads.

Curved Lines

Many architectural drawings contain irregular lines that must be repeated. *Flexible rules,* such as those shown in Fig. 26-15, are used to repeat irregular curves that have no true radius or series of radii and cannot be drawn with a compass. Curved lines that are not part of an arc can also be drawn with a *french* (irregular) *curve,* as shown in Fig. 26-16.

DIVIDERS

Dividing an area into an equal number of parts is a common task performed by architectural drafters. In addition to the architect's scale (see Unit

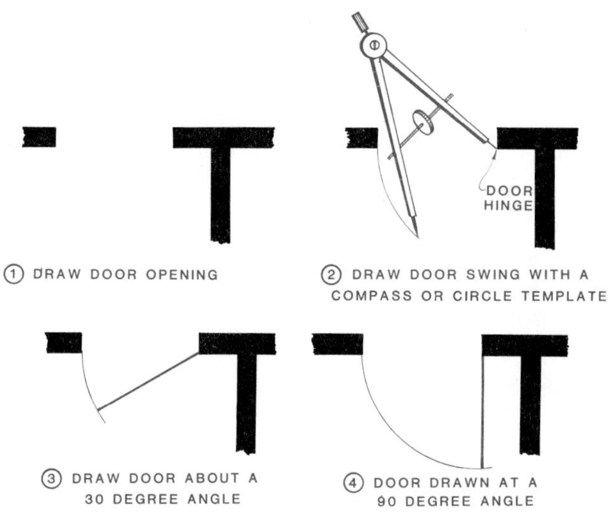

DOOR HINGE
① DRAW DOOR OPENING
② DRAW DOOR SWING WITH A COMPASS OR CIRCLE TEMPLATE
③ DRAW DOOR ABOUT A 30 DEGREE ANGLE
④ DOOR DRAWN AT A 90 DEGREE ANGLE

Fig. 26-13 Drawing door symbols.

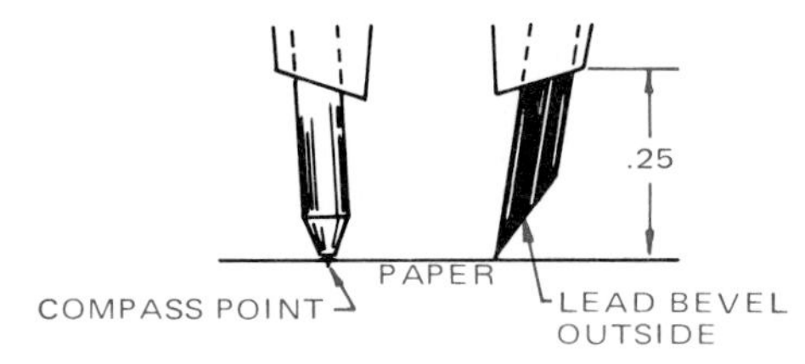

LEAD BEVEL OUTSIDE

Fig. 26-14 Sharpening and setting compass leads.

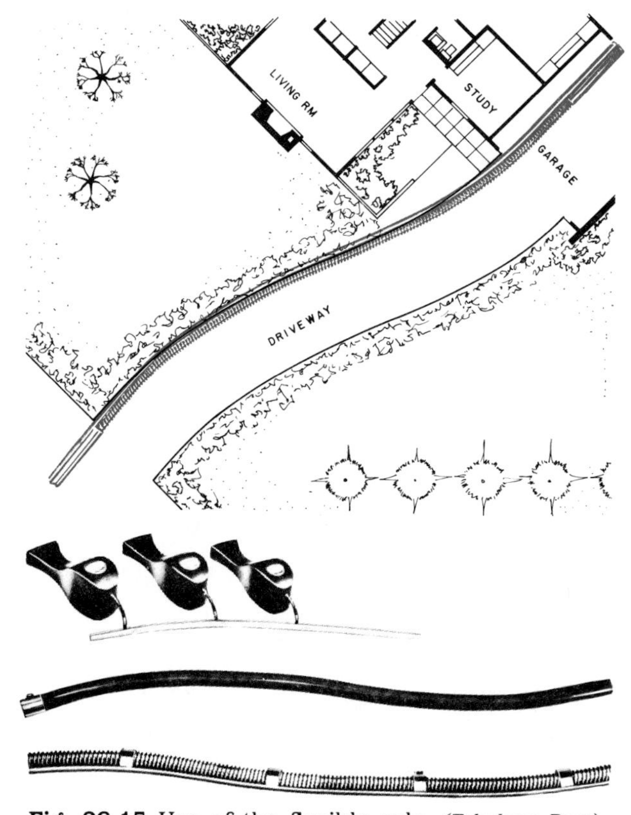

LIVING RM

Fig. 26-15 Use of the flexible rule. *(Teledyne Post)*

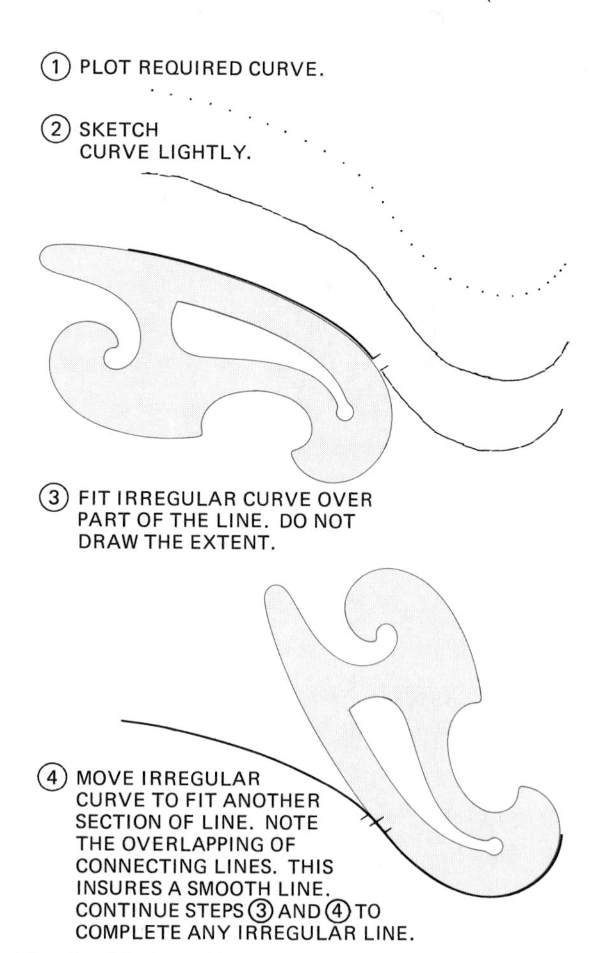

① PLOT REQUIRED CURVE.
② SKETCH CURVE LIGHTLY.
③ FIT IRREGULAR CURVE OVER PART OF THE LINE. DO NOT DRAW THE EXTENT.
④ MOVE IRREGULAR CURVE TO FIT ANOTHER SECTION OF LINE. NOTE THE OVERLAPPING OF CONNECTING LINES. THIS INSURES A SMOOTH LINE. CONTINUE STEPS ③ AND ④ TO COMPLETE ANY IRREGULAR LINE.

Fig. 26-16 Use of irregular curve.

24), the *dividers* (Fig. 26-17) are used for this purpose. To divide an area equally by the trial-and-error method, first adjust the dividers until they appear to represent the desired division of the area. Then place one point at the end of the area and step off the distance with the dividers. If the divisions turn out to be too short, increase the opening on the dividers. Repeat the process until the line is equally divided. If the divisions are too long, decrease the setting. Figure 26-18 shows the use of dividers in dividing an area into an equal number of parts.

Dividers are also used frequently to transfer dimensions and to enlarge or reduce the size of a drawing. Figure 26-19 shows the use of dividers to double the size of a floor plan. This work is done by setting the dividers to the distances on the plan and then stepping off the distance twice on the new plan.

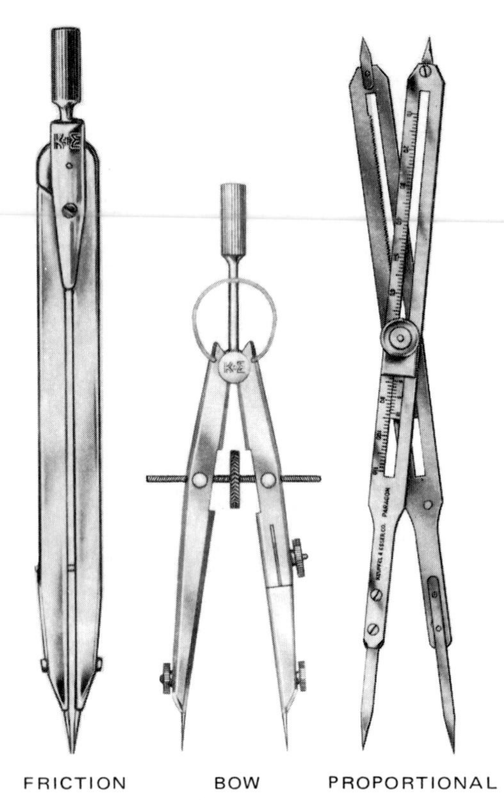

Fig. 26-17 Types of dividers. *(K&E)*

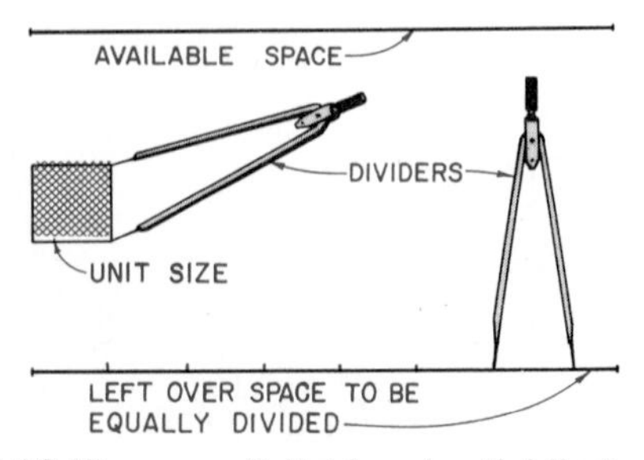

Fig. 26-18 The use of dividers in dividing areas.

CORRECTION EQUIPMENT

The designer employs a variety of erasers. *Basic erasers* are used for general purposes. *Gum erasers* are used for light lines. *Kneaded erasers* pick up loose graphite by dabbing, and dry cleaner bags remove smudges. Powder, sprinkled on the drawing, enables drafting instruments to move freely and also keeps the drawing and instruments clean.

Electric erasers are very fast and do not damage the surface of the drawing paper, since a very light touch can be used to eradicate lines.

A *drafting brush* is used periodically to remove eraser and graphite particles and to keep them from being redistributed on the drawing.

Erasing shields are thin pieces of metal or plastic with a variety of different-shaped openings. A line to be erased is exposed through an appropriate opening without disturbing nearby lines that are to remain on the drawing.

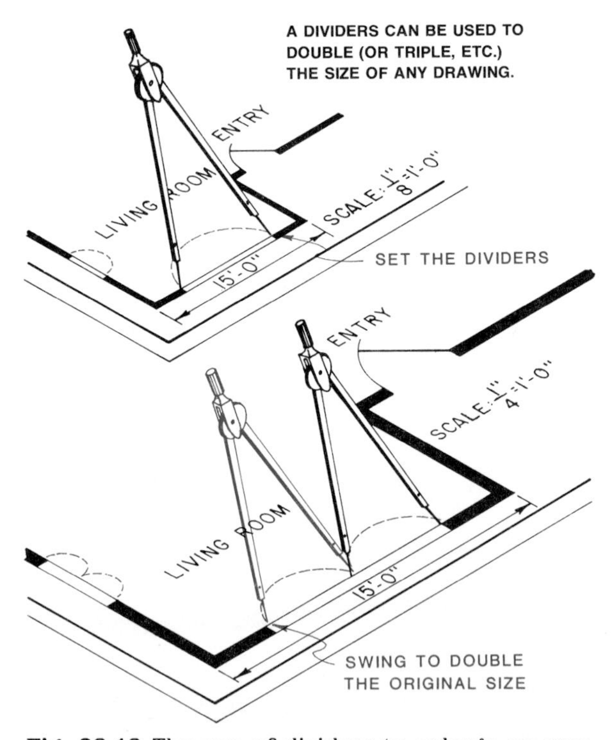

Fig. 26-19 The use of dividers to enlarge an area.

INKING PENS

Architectural drafters use ink drawings for much presentation work. Two basic types of inking pens are used. The *technical fountain pen* has a line of fixed width. As a result, a different technical pen is needed for each width represented on the drawing.

DRAWING MEDIA

Most architectural drawings are prepared on vellum or a good-quality tracing paper, although most preliminary design work and progressive sketches are done on extremely thin tracing paper (bum wad or trash) because these drawings are eventually discarded. The size of the drawing surface should be determined at the beginning of a project by selecting a drawing format at least one size larger than the largest drawing in the set. Figure 26-20 shows the standard sizes of papers or vellum used for architectural drawings.

Drawing paper and vellum sizes

Customary inches	Metric mm
8″ × 10″	
8″ × 11″	
*8.5″ × 11″ (A size)	210 × 297 mm (A4)
*9″ × 12″	
11″ × 14″	297 × 420 mm (A3)
*11 × 17″ (B size)	
*12″ × 18″	
14″ × 17″	
15″ × 20″	
*17″ × 22″ (C size)	420 × 594 mm (A2)
*18″ × 24″	
19″ × 24″	
21″ × 27″	
*22″ × 34″ (D size)	594 × 841 mm (A1)
*24″ × 36″	
*34″ × 44″ (E size)	841 × 1189 mm (A0)
*36″ × 48″	

*Most commonly used.

Fig. 26-20 American standard paper sizes.

Exercises

1. **Using a T square and triangle, parallel slide and triangle, or drafting machine, draw the floor plan shown in Fig. 33-12.**
2. **Using drafting instruments and an architect's scale, draw the elevation shown in Fig. 38-1 to the scale 1/4″ = 1′-0″. Add horizontal dimensions.**
3. **Define the following terms: *T square, 45° triangle, 60° triangle, vertical lines, horizontal lines, parallel slide, drafting machine, flexible rule, dividers, compass, drop-bow compass, track drafter, protractor head.***

UNIT 27

COMPUTER-AIDED DRAFTING

Computer-aided drafting involves the preparation of architectural or engineering drawings through the use of a computer. The use of computers in engineering and architecture is not new. Computers have been used for many years for structural calculation work. But only within the last few years has the computer been used extensively to prepare final working drawings.

INTRODUCTION TO CAD

A computer-aided drafting (CAD) system is a very sophisticated drafting tool. A CAD system can be compared to a drafting machine in the same way a slide rule can be compared to an electronic calculator. Both devices do the job well, but the CAD system and the calculator perform the same tasks faster, more accurately, and more consistently. The CAD system can also perform many more functions than manual drafting equipment, such as the preparation of schedules, specifications, budgets, and structural calculations.

CAD Advantages

There are both advantages and disadvantages to using CAD systems. However, the advantages of preparing architectural drawings using CAD greatly outweigh the disadvantages. Some of the greatest advantages include:

Speed
Accuracy
Information storage
Drawing quality
Repetitive task elimination
Cost savings
Automatic overlaying
Revision simplicity
Model simulation
Support document preparation
Structural calculations
Information retrieval

CAD systems do not create drawings—drafters produce drawings. The speed, accuracy, and consistency of CAD systems, however, enables drafters to produce more detail drawings for each design in less time. This results in the production of many detail drawings which eliminate many construction site errors. Often the greatest advantage of using CAD is not in the preparation of original drawings but in the speed, ease, and accuracy of the revision phase of the design process.

What CAD Cannot Do

Although CAD systems perform many tasks faster, more accurately, and more consistently than manual systems, nevertheless there are some important functions CAD systems *cannot do*. For example:

Computers cannot think
Computers cannot design
Computers cannot make decisions
Computers do not automatically eliminate all human errors

But actually computers and humans are very compatible. Each excels in what the other can not do well. Humans can think, create and reason—computers cannot. But humans are slow, inaccurate, and inconsistent, while computers are extremely fast, accurate, and consistent. If used appropriately, the computer, therefore, can be an effective tool to aid human creativity in architectural drafting and design. But to use a CAD system effectively, a thorough understanding of the principles and practices of architectural drafting and design is therefore essential.

System Components

CAD systems consist of hardware and software components. The physical devices used in a CAD system are known as *hardware*. The program which the drafter uses to operate the system is known as software. CAD hardware components include the central control system, input devices, memory and storage devices and output devices as shown in Fig. 27-1. Software, in disk or magnetic tape form, contains the specialized information which enables the operator to create the desired line, symbol, or numeral on a display device.

System Types

Computer systems are classified by the type of data handled, type of function performed, and memory storage capacity.

Data Type The two types of systems classified by data handled are digital computers and analog computers. An analog computer measures data continuously. But digital computers measure by discrete digital steps. Analog computers are used extensively for complex mathematical computations; digital computers are used widely for CAD systems.

Functional Types Computer systems function either passively or interactively. In passive systems the program runs from beginning to end without intervention or change. Interactive systems require human intervention to function. True CAD systems are interactive since the operator gives the computer a command, then waits for the computer to complete the task before entering a new command.

In robust interactive systems the computer also provides the operator with options necessary to proceed. Many systems also notify the

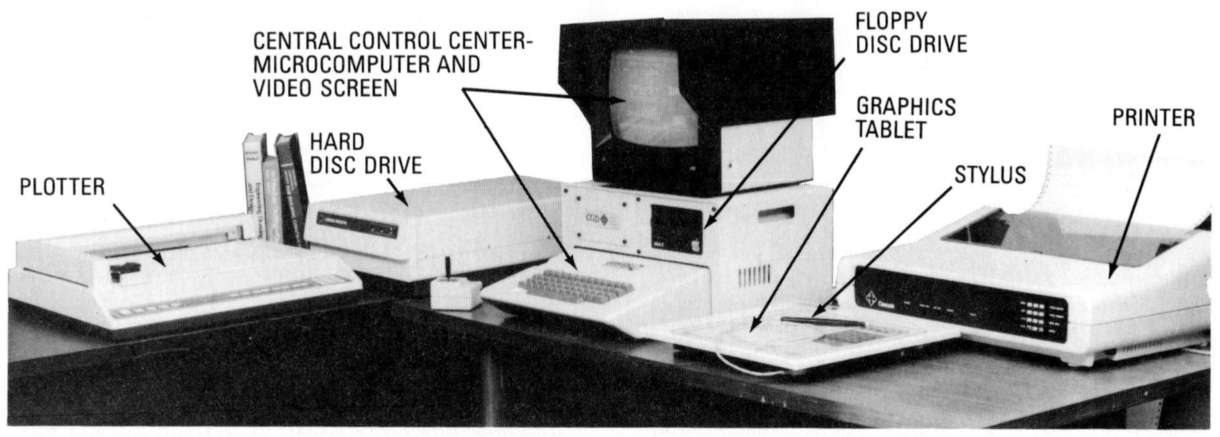

Fig. 27-1 Component of a complete CAD workstation. *(Cascade Computer Graphics)*

operator if a task requested is not valid. These instructions from the computer are known as *prompts.* Drafting on a CAD system is a series of interactions between computer and operator.

Capacity Types Digital computers are classified into three categories depending on the capacity of the computer memory. These include large mainframe computers, minicomputers, and smaller microcomputers. CAD programs are available for all three types.

How CAD Systems Work

CAD systems can be used to draw any object that can be drawn manually. The operator communicates with the computer through the use of an input device, usually a keyboard, mouse or stylus. When the operator requests a specific function such as a line, arc, or numeral, the computer memory switches to that function. The operator then indicates specifically, through a keyboard, grid system, or stylus pick, where the line, arc, or numeral is to be placed on the display screen. Once the function is completed on the screen the next individual task is performed in the same manner. Each task is performed in sequence until the drawing is complete.

CENTRAL CONTROL CENTER

At the heart of all CAD systems is the central control center, sometimes referred to as a central processing unit (CPU). This center consists of the computer and a display monitor. Here all operator input is digitized into the computer. That is, the information is translated into electric pulses and magnetic currents which constitute a language the computer understands. The computer has the capacity to modify the data and provide for the recall of any information in numerical or graphic form. Once information is stored in the computer memory, it can be recalled and viewed at any time on a display device (monitor) such as a cathode-ray tube (CRT).

INPUT

All information entered into a CAD system by an operator is known as *input.* Input information consists of either temporary or permanent data. At any time, the operator can make a variety of changes, deletions or additions to the drawing using this data. For example, libraries of architectural symbols or details can be stored in memory files and coded for recall and use on any part of the drawing, or on future drawings.

Menus

CAD system menus are listings of all tasks contained in the software program. There are two types of menus, the master menu and auxiliary menus. The *master menu* shows the major headings of all available functions in the program such as lines, arcs, dimensions, text, curves, and so forth. Figure 27-2 shows a typical CAD master menu.

Auxiliary menus show all the specific tasks available under each master menu heading. Numbers assigned to each function on the master menu refer to the heading number of each auxiliary menu. For example, Fig. 27-3 shows an auxiliary menu for circles and arcs. Note that this is task 54 on the master menu shown in Fig.

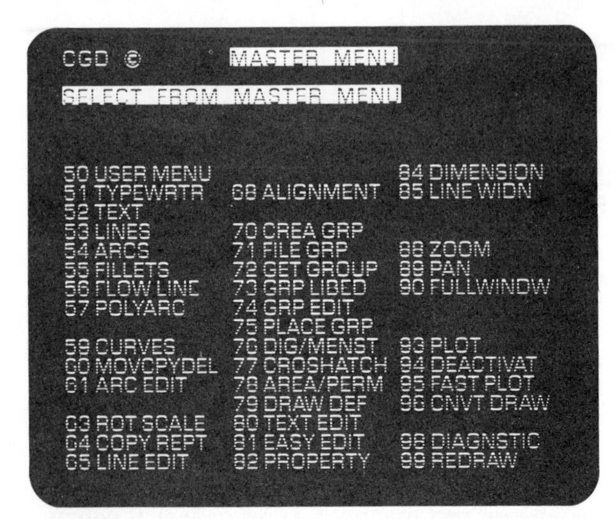

Fig. 27-2 CAD master menu. *(Cascade Computer Graphics)*

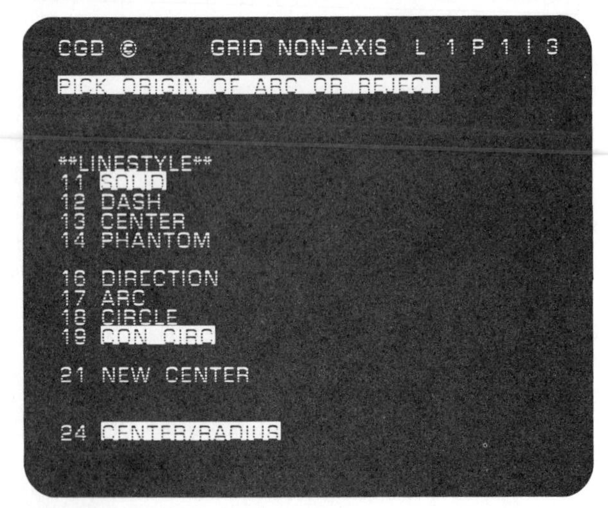

Fig. 27-3 CAD auxiliary menu. *(Cascade Computer Graphics)*

27-2. The auxiliary menu for arc (task 54) shows *all* the types of arcs available in the program. Similar auxiliary menus are provided for other major tasks listed on the master menu.

Input Devices

A wide variety of input devices can be used to provide instructions to the computer. These vary depending on the nature of the drawing and/or the preference of the operator. Input devices include:

A *menu board* (Fig. 27-4) includes a series of numerical, graphic, and instructional options that can be activated with the touch of a stylus.

An *alphanumeric keyboard* is used to enter data into the computer by typing code numbers and letters corresponding to graphic tasks in the program.

A *graphic tablet* is an electronic drafting board. The operator draws by touching points on the tablet with a stylus. A menu board and graphic tablet are often combined on one surface as shown in Fig. 27-5.

A *stylus* is used in conjunction with a graphic table (Fig. 27-5) to locate specific points on a drawing.

A *mouse* is a mechanism that is used to control the cursor on the monitor by sliding it along the surface of the graphic tablet.

A *joystick* is used to control the position of a cursor on a display screen. Tilting the stick up or down, right or left, moves the cursor in a corresponding direction on the screen (Fig. 27-5).

Thumb wheels are rarely used with CAD systems since accurately controlling the position of the cursor is difficult.

Light pens (Fig. 27-6) are used directly on CRT glass similar to the way a stylus is used on a graphic tablet. However, because of the lack of accuracy, light pens are used only for general design functions.

Digitizers (Fig. 27-7) are used to transfer graphic data from a drawing into numbered coordinates that can be recalled graphically.

Voice input devices enable an operator to activate CAD tasks by speaking the task number or title into a microphone. Lines can be drawn through voice input by stating the coordinates of each point in succession.

An *optical scanner* uses a camera to digitize a drawing for storage or viewing on a monitor.

Input Coordinate Grid System

Computers convert digitized electrical impulses into graphic form by plotting x and y cartesian coordinates as shown in Fig. 27-8. On a two-dimensional system, lines are created by locating the x-y coordinates (intersections) of each end of a line. Figure 27-9 shows a line segment on a coordinate grid system. The coordinates of the line end at point 1 are $x = .5$, $y = 1$. At point 2 the coordinates are $x = 5.5$, $y = 4.5$.

Figure 27-10 shows how a drawing is started by placing the stylus on the coordinate of one end of a line, then the other end. In this illustration the left-end coordinates are $x = 5$, $y = 4$. The coordinates of the right end of the line are $x = 22$, $y = 14$. Since quadrant I is used in this

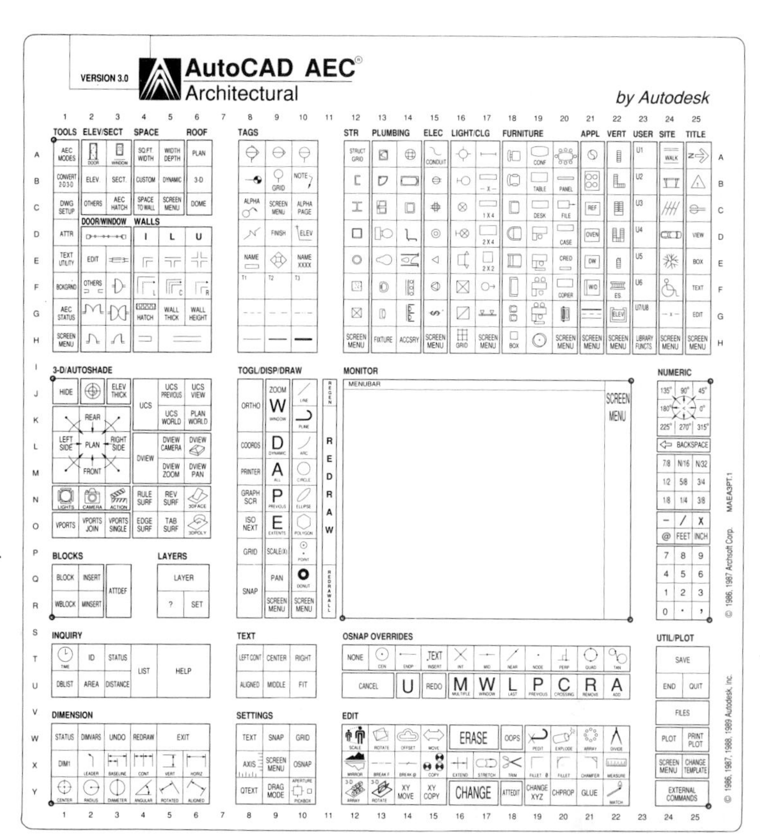

Fig. 27-4 Examples of a specialized architectural menu. *(Autocad)*

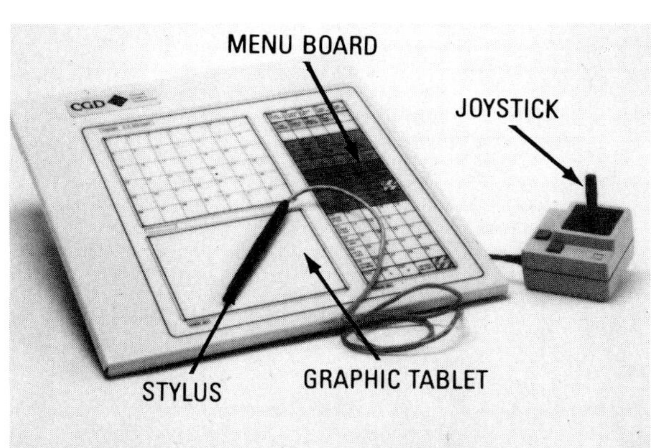

Fig. 27-5 Stylus and joystick with graphic tablet and menu board. *(Cascade Computer Graphics)*

Fig. 27-7 Digitizing a drawing. *(Synercom Tech. Inc.)*

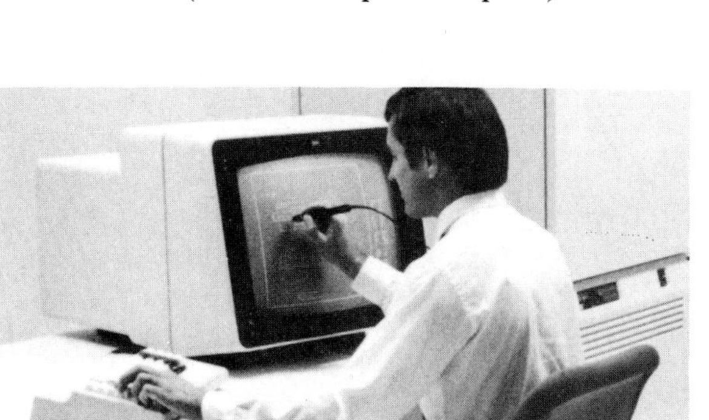

Fig. 27-6 Use of light pen. *(IBM Corp.)*

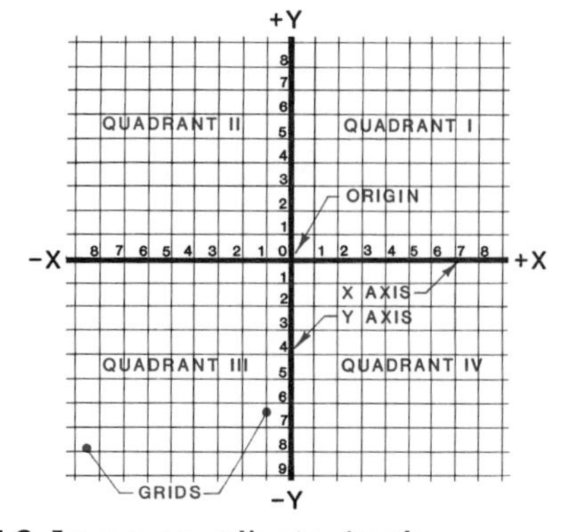

Fig. 27-8 An x-y coordinate graph.

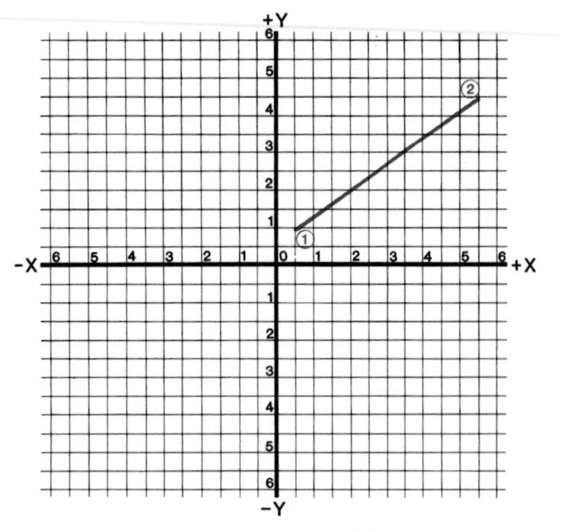

Fig. 27-9 Line segment on a grid system.

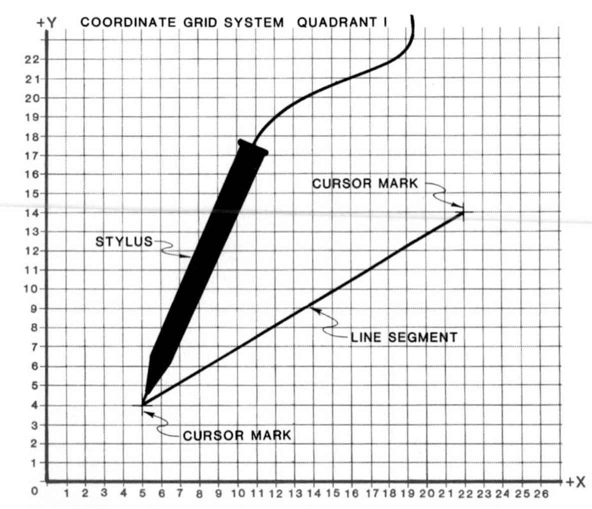

Fig. 27-10 Plotting coordinates with a stylus.

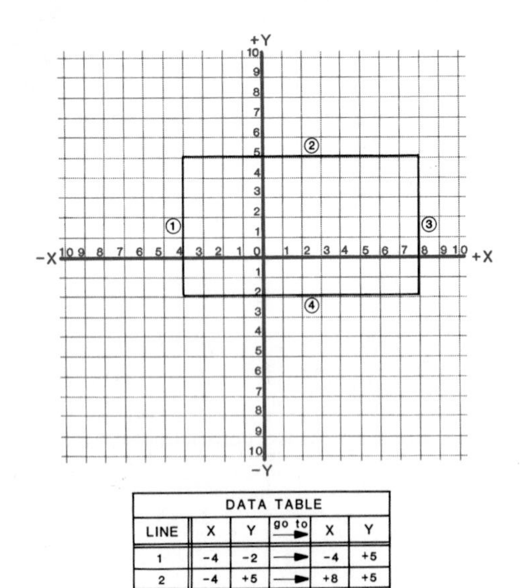

DATA TABLE					
LINE	X	Y	go to	X	Y
1	-4	-2	→	-4	+5
2	-4	+5	→	+8	+5
3	+8	+5	→	+8	-2
4	+8	-2	→	-4	-2

Fig. 27-11 Sequence of plotting polygrams.

case all positions are positive. Polygrams, as shown in Fig. 27-11, are constructed by repeating this process in the sequence shown in the data table. Match the sequence in the data table with the development of the four lines on the drawing to ensure that you understand this concept.

Although the development of curves requires the use of more closely spaced coordinate points, the process is the same. Figure 27-12 shows a curve and related data table containing coordinates for the development of a 20-point curve. The use of more points would produce a smoother curve. The use of fewer points would produce a flatter, segmented curve.

Figure 27-13 summarizes the sequence of using a stylus and graphic table in conjunction with a menu board to generate lines. In this application the stylus is first used to pick the task from the menu. Then the stylus is used to locate the cursor on the selected coordinate points on the screen.

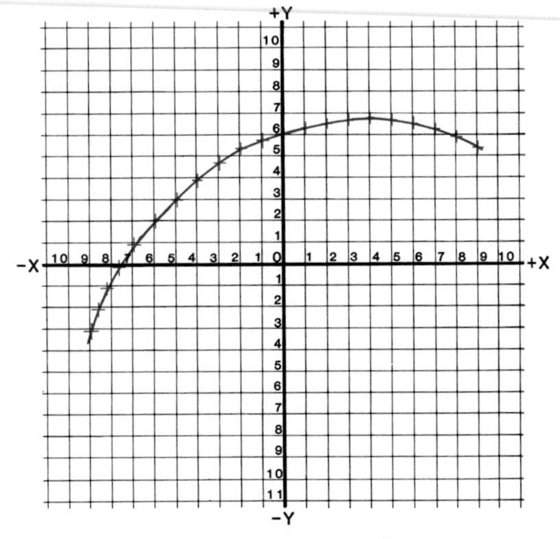

DATA TABLE					
SEGM	X	Y	go to	X	Y
1	-9	-3	→	-8.7	-2
2	-8.7	-2	→	-8.2	-1
3	-8.2	-1	→	-7.7	0
4	-7.7	0	→	-7	1
5	-7	1	→	-6	2.1
6	-6	2.1	→	-5	3.1
7	-5	3.1	→	-4	4
8	-4	4	→	-3	4.8
9	-3	4.8	→	-2	5.5
10	-2	5.5	→	-1	5.8
11	-1	5.8	→	0	6.2
12	0	6	→	1	6.4
13	1	6.4	→	2	6.6
14	2	6.6	→	3	6.8
15	3	6.8	→	4	6.9
16	4	6.9	→	5	6.8
17	5	6.8	→	6	6.6
18	6	6.6	→	7	6.4
19	7	6.4	→	8	6
20	8	6	→	9	5.5

Fig. 27-12 Plotting curve coordinates.

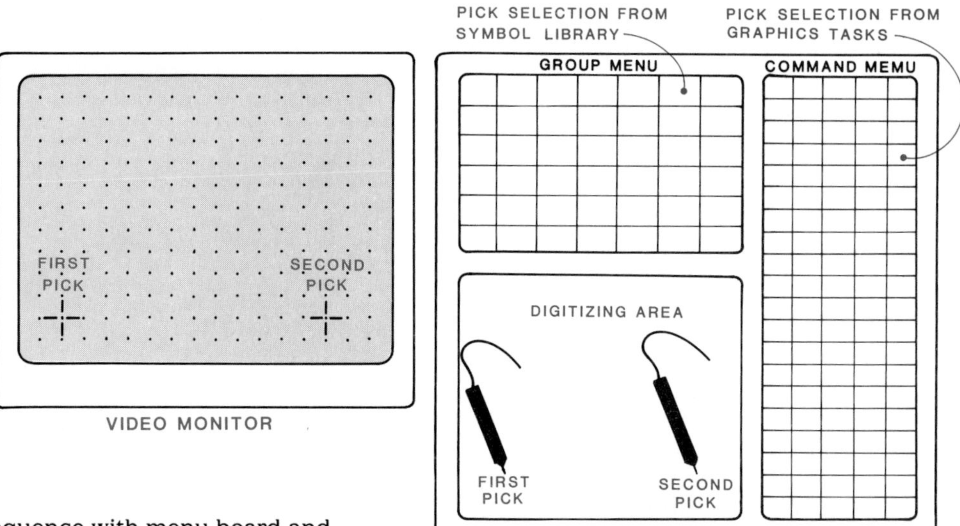

Fig. 27-13 Drawing sequence with menu board and graphic tablet.

An alphanumeric keyboard can also be used to create lines by using the keys corresponding to the *x-y* points, once the original point is established. For example, point *B* is established by striking left bracket, 30 (*x* axis), comma, 0 (*y* axis), and right bracket as shown in Fig. 27-14.

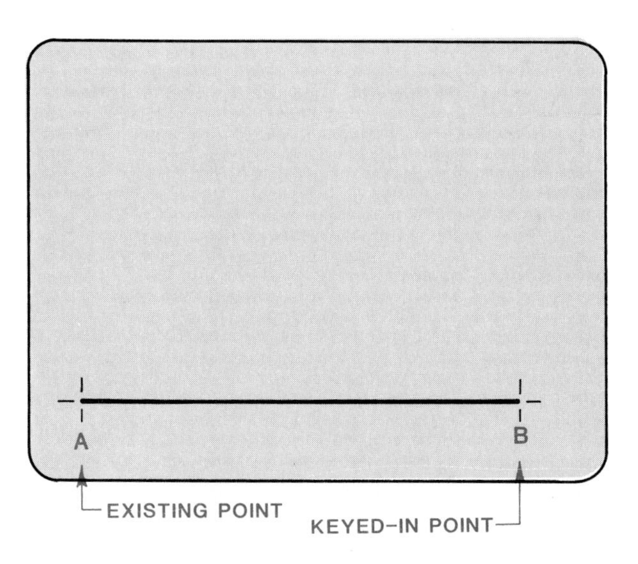

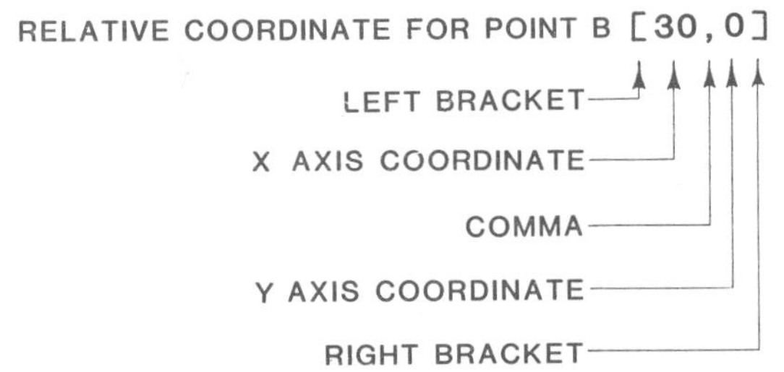

Fig. 27-14 Keyboarding x-y coordinates.

MEMORY AND STORAGE

A wide variety of graphic and numerical information can be stored in the computer's memory for recall when needed.

There are two types of computer memories, permanent and temporary. Permanent memory is known as read-only memory (ROM). Temporary memory is called random-access memory (RAM). The movement of data to and from the RAM is controlled by the central processing unit. When information is stored in the RAM it is labeled with a library number. This number (code) is then used to instruct the computer to display the stored drawing, detail or symbol on the monitor when needed. Both permanent and temporary information can be stored in the computer using magnetic tape, hard disk, optical disk, or floppy disks.

OUTPUT

Although an entire drawing can be viewed on a monitor, the completed work is not usable until a hard-copy print is produced. The hard copy (drawing on paper) produced on a CAD system is known as the *output*. Hard-copy output is produced on either a printer or a plotter.

Printer

A printer is an output device that closely resembles the size and quality of the image shown on the monitor display screen. Therefore, printers are used only to produce hard-copy drawings for checking purposes and are also used to print schedules, specifications, and other alphanumeric documents that do not require a high degree of line quality or dimensional accuracy.

Plotter

Pen plotters produce drawings that are superior in quality to most manually inked drawings. A plotter can produce lines of various weights and colors on the same drawing. There are two basic types of plotters: the *flatbed plotter* (Fig. 27-1) and the *drum plotter* (Fig. 27-15).

If different layers (levels) are used in the preparation of a drawing, a different pen width and/or color can be assigned to all lines used on a given level. For example, a plotter can produce a drawing with the base floor plan plotted in black, plumbing lines in blue, electrical lines in red, HVAC equipment in green, furniture in brown, and so forth.

The line quality of a finished plotted drawing is directly related to the quality of the plotter and not to the resolution of the CRT screen. A ragged line on a low-resolution screen can become a dense and distinct line on a high-quality plotter.

Fig. 27-15 Drum plotter. *(Hewlett-Packard)*

Fig. 27-16 Feeding commands to the computer. *(Cascade Computer Graphics)*

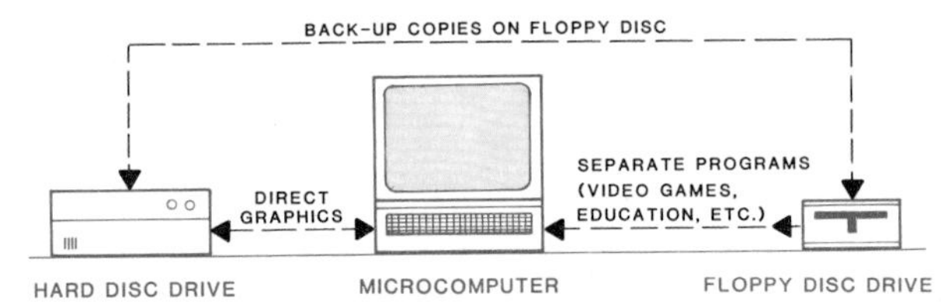

Fig. 27-17 Relationship of computer and disk drives.

CAD SYSTEM PROCEDURES

Each CAD system operates in a similar manner, but some operational differences do occur among systems. In a typical robust system, as shown in Fig. 27-16, input commands are fed into a hard disk that contains the software program. The graphic information is sent back to the computer which in turn displays the information on the monitor. The graphic data shown on the screen can then be stored on the hard disk for later recall. In this system the graphic information is passed between the computer and the hard-disk device as shown in Fig. 27-17. In this system, the floppy-disk drive is bypassed during the graphics work.

The floppy-disk drive is used when the operator wishes to store graphics from the hard disk to a floppy disk for a backup copy. The information on the floppy disk can be transferred back into the hard disk at any time. Different computer programs can be used directly from the floppy disk into the computer and displayed on the monitor, bypassing the hard-disk drive.

When the drawing is completed on the monitor, a fast plot on a printer is usually made before plotting the final drawing on the plotter. The printer produces a quick but rough copy that can be used to check the drawing for errors or changes. When the drawing checks out satisfactory, the finished drawing is then generated on the plotter.

Procedure Steps

Although different CAD systems require different sequences of operation most good user-friendly systems operate approximately as follows:

1. Switch on the system.

2. Enter the operator's initials.

3. The primary menu will appear on the video monitor (Fig. 27-18).

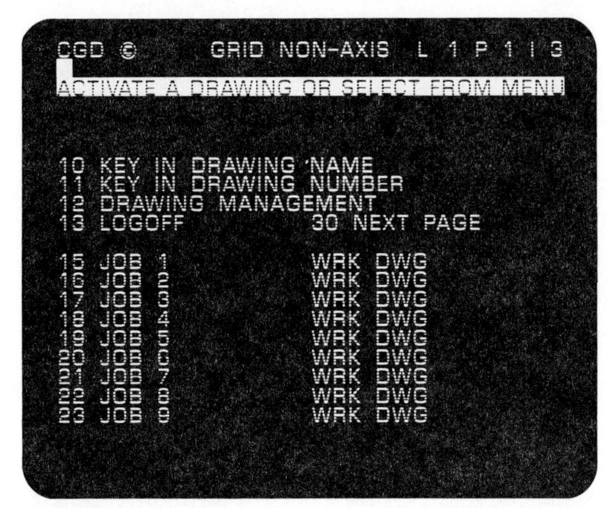

Fig. 27-18 Primary menu. *(Cascade Computer Graphics)*

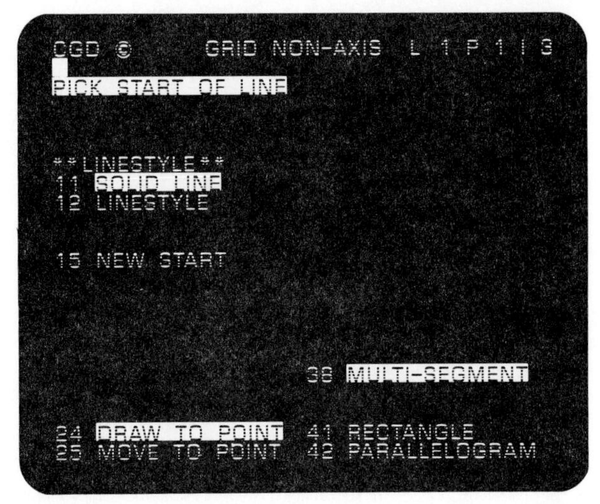

Fig. 27-19 Auxiliary menu for line task. *(Cascade Computer Graphics)*

4. Start the new drawing by entering *10* and then the drawing's name. Note that all information that is to be entered (keyed-in, typed) will be in italics.

5. Select the drawing's parameters. Each of the following items for the drawing's parameters will automatically appear in order:

U.S. Customary or metric units of measure

Scale size

Drawing's format (A, B, C, D, E) or metric paper size

Locate the origin *0* for the coordinate grid system

6. After the drawing's parameters have been selected, the master menu will appear on the screen, or it can be called up by keying-in the master menu code, which is ESC-M in this system.

7. Usually the first task for drawing is lines which is task 53 in the system illustrated. By entering *53*, the line task is activated by the computer. For information and instruction on the line task, type *ESC A* to display the auxiliary menu for lines (Fig. 27-19). The options most often used for line work are automatically set so the operator will not have to enter their code numbers. These commands are highlighted on the auxiliary menu. They are called *defaults*. The default commands task 53 in the system used are:

28 SOLID LINE A solid line is used for the line work.

24 DRAW TO POINT A line will appear between two keyed-in points.

12 AT MIDPOINT automatically connects at mid-point.

13 TANGENT to arc line drawn tangent to arc.

28-37 LINE STYLES Options for the alphabet of lines, on the second auxiliary menu, can be selected.

15 NEW START Used to start a new segment that is not connected to the existing geometric form.

25 MOVE TO POINT Used to create a gap in a line.

43 RECTANGLE A rectangle can be drawn by entering the lower left corner and then the upper right corner.

39 DASH A = .500

40 DASH B = .125

41 GAP .125

44 ANGLE select any angle

If there is confusion at any time during a graphics task, always return to the auxiliary menu, *ESC A*, and read the prompt at the top of the screen. These instructions are included on all friendly systems. The operator can then follow the prompt's instructions to complete the graphics task.

CREATING A DRAWING

Once the drawing format is established a dimensioned sketch similar to the one shown in Fig. 27-20 is first prepared. Then a sequence similar to the following is used to produce a CAD drawing. Although task numbers vary among sys-

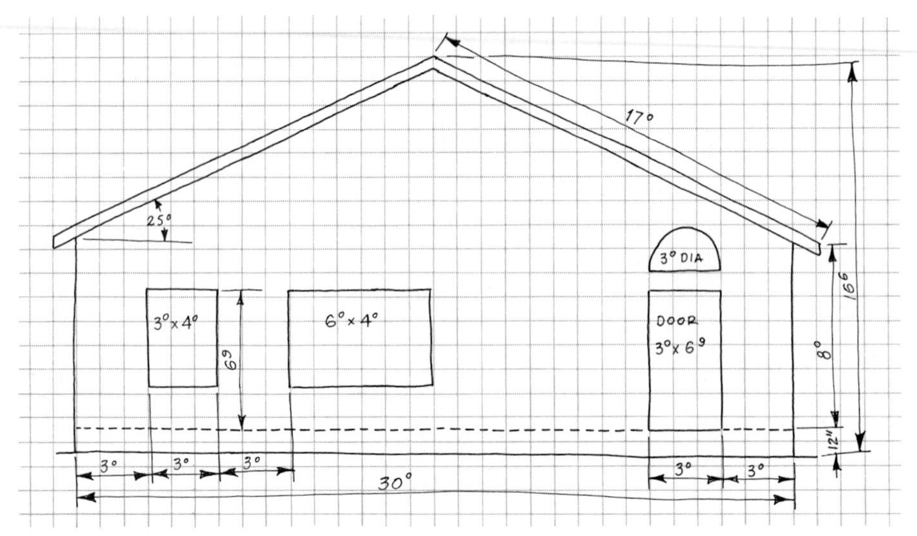

Fig. 27-20 Preliminary sketch.

tems, drawing procedures are very similar. For specific procedural instructions, always refer to the documentation manual for the system in use.

1. The drawing grid sizes should then be adjusted with task *79,* Drawing Definitions (Fig. 27-21). Enter *11,* change grids and follow the prompt which will ask to select the grid size. A convenient size grid for this elevation is a half-inch (.5″). If your drawing scale is ¼″ = 1′0″, then each grid will represent 2′. The grids may be set smaller if necessary, but if too small, the grids will clutter up the video screen. Most grid systems can be modified when picking points on the grid system as outlined in the following procedures and shown in Fig. 27-22.

00 grid pick—the point will snap to closest grid point.

01 free pick—the point will remain at the chosen spot.

Options that can be used with grid picks 00 and 01 are:

**03* nonaxis lock—allows drawing lines at any angle.

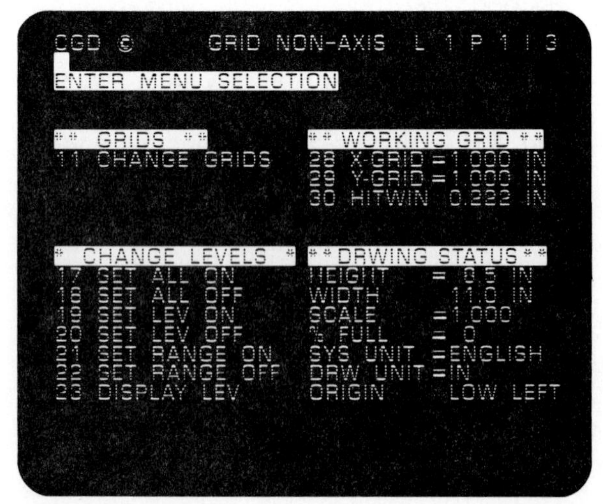

Fig. 27-21 Auxiliary menu for grids. *(Cascade Computer Graphics)*

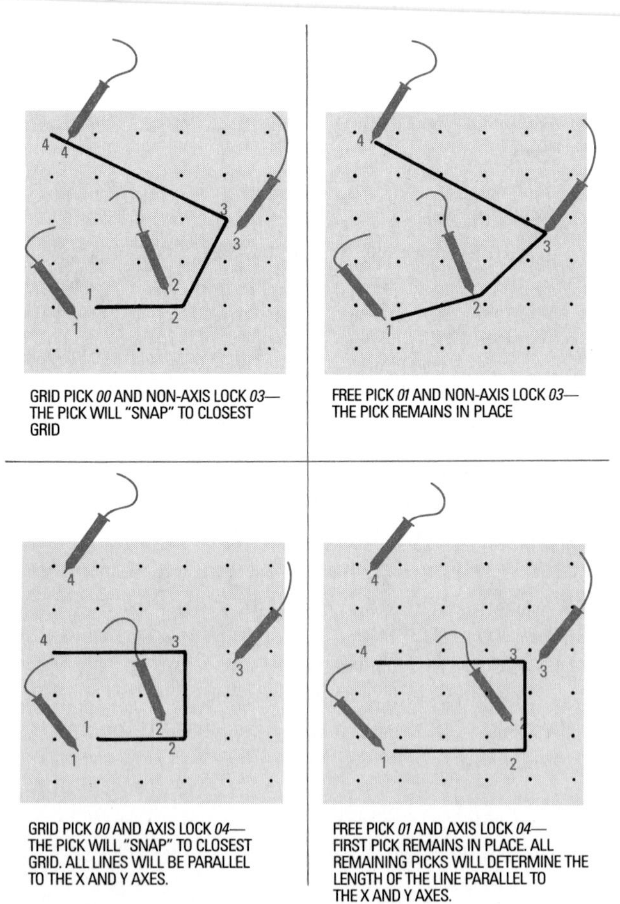

GRID PICK 00 AND NON-AXIS LOCK 03—
THE PICK WILL "SNAP" TO CLOSEST GRID
FREE PICK 01 AND NON-AXIS LOCK 03—
THE PICK REMAINS IN PLACE
GRID PICK 00 AND AXIS LOCK 04—
THE PICK WILL "SNAP" TO CLOSEST GRID. ALL LINES WILL BE PARALLEL TO THE X AND Y AXES.
FREE PICK 01 AND AXIS LOCK 04—
FIRST PICK REMAINS IN PLACE. ALL REMAINING PICKS WILL DETERMINE THE LENGTH OF THE LINE PARALLEL TO THE X AND Y AXES.

Fig. 27-22 Controlling stylus pick with grid positions.

*04 axis lock—for drawing straight lines on the *x* or *y* axis when in free pick (01). This is the only certain way to obtain straight lines that are horizontal or vertical when not on a grid line (00).

2. Enter task *53*, lines. Use defaults. If accuracy must be closer than the grid system allows, then each line and angle may be entered by typing in the specific coordinates for each line.

3. Enter the lower left corner of the elevation drawing with the stylus or enter the specific coordinates to start the line work (Fig. 27-23).

4. Enter 30′-0″ for the length of the elevation. The first line will quickly appear (Fig. 27-24).

5. Enter 9′-0″ for the height from the ground line to the ceiling line (Fig. 27-25).

6. Enter the length of the ceiling line, 30′-0″ (Fig. 27-26).

7. Enter the left side of the elevation to complete the rectangular outline of the wall (Fig. 27-27).

8. Enter *15* new start. Enter the peak of the roof (Fig. 27-28).

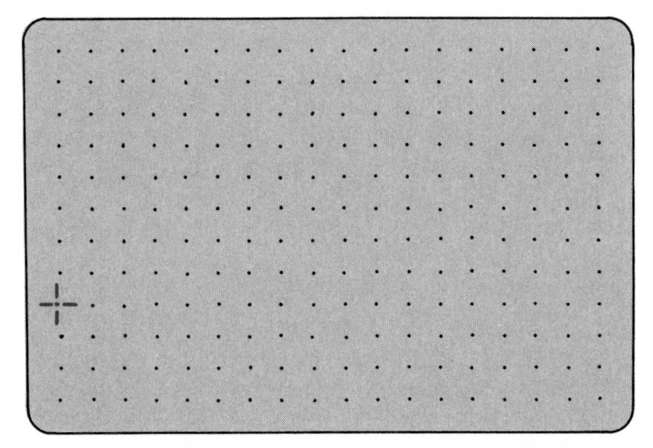

Fig. 27-23 Select lower left corner of elevation.

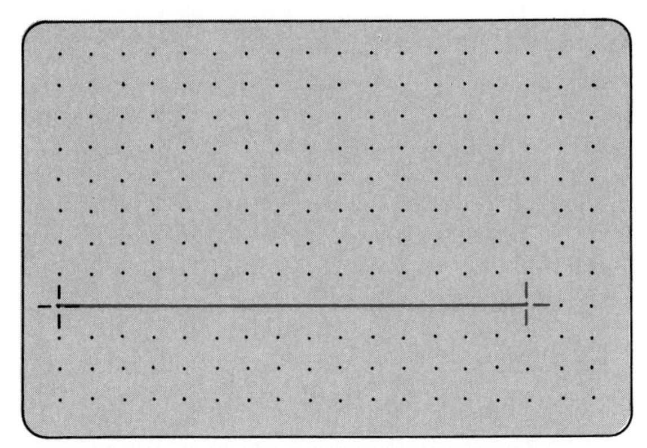

Fig. 27-24 Key in 30′-0″ length coordinates.

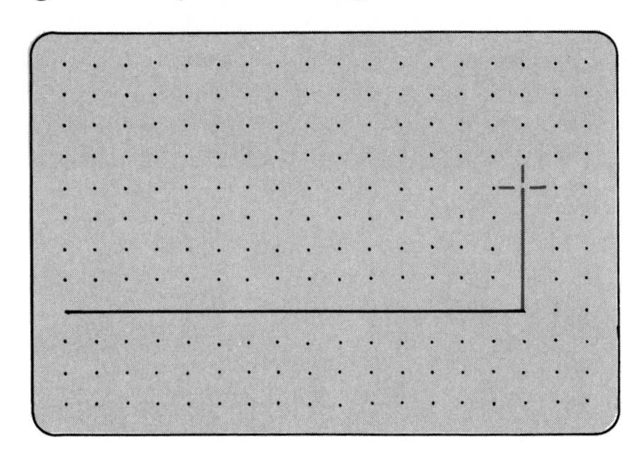

Fig. 27-25 Enter 9′-0″ ceiling height.

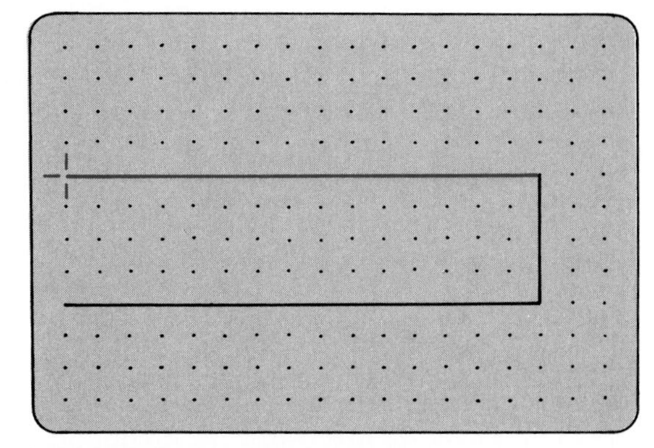

Fig. 27-26 Enter ceiling length.

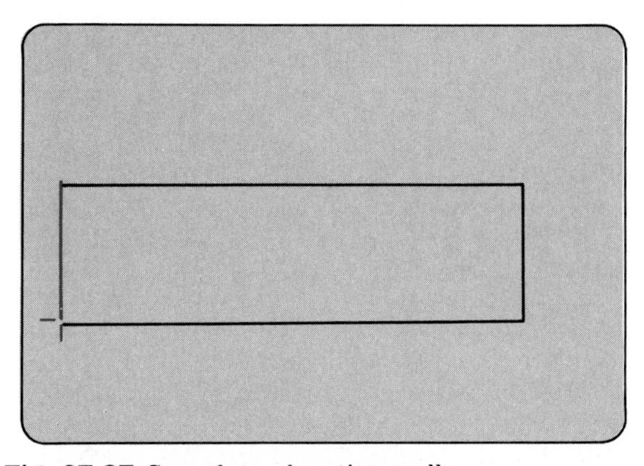

Fig. 27-27 Complete elevation walls.

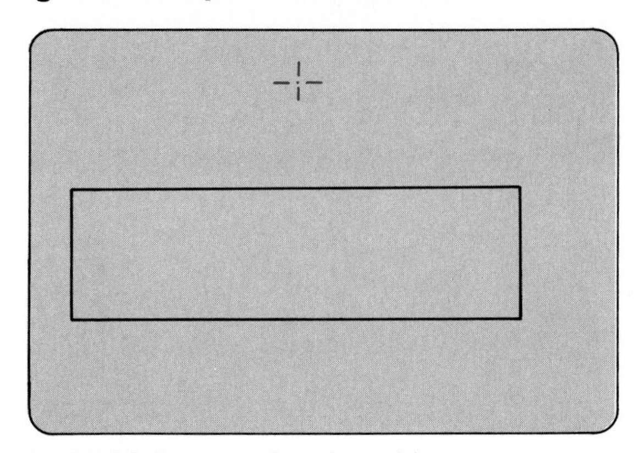

Fig. 27-28 Enter roof peak position.

9. Enter the bottom edge of the roof rafter by picking the point with a stylus or typing in the polar coordinates: length, angle *D*. The example is : *16,205 D*. The roof's pitch angle from the horizontal is 25°. The computer measures all angles counterclockwise, therefore 205° is entered (Fig. 27-29).

10. Repeat step 9 for the opposite of the roof (Fig. 27-30).

11. Enter the 6″ depth for the roof rafter's depth (Fig. 27-31).

12. Enter *15* for each new start and enter the windows and door (Fig. 27-32). Check original sketch for dimensions.

13. Enter *30* short dash. Enter the dashed line for the floor line (Fig. 27-33). The floor line is 12″ above the ground line (See Fig. 27-20).

14. To draw the circular window, enter task *54*, arcs (see Fig. 27-3). Use default 11, solid line. Enter *16* arc. Pick three points, in order, as shown in Fig. 27-34.

15. The result is the arc shown in Fig. 27-35. Complete the bottom of the window by entering in the bottom line with task *53*, lines.

16. Enter task *52* text. Enter *29*‸ to obtain ¼″ letters. It is important to type in ‸ after the command so the computer will know that "29" is a command and not text to be placed on the draw-

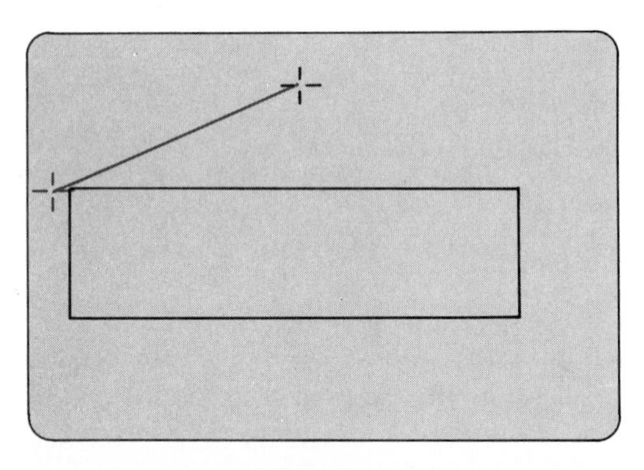

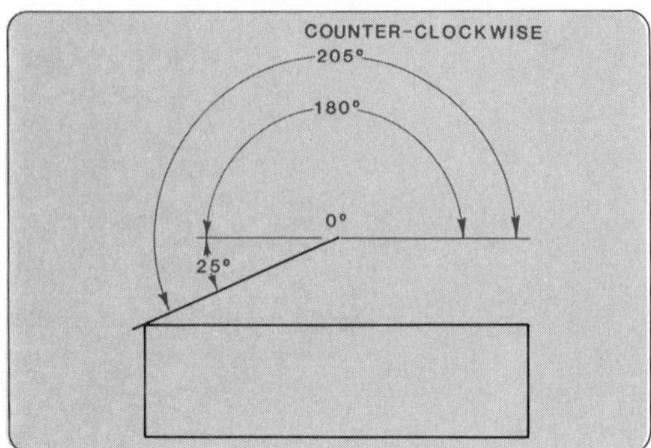

Fig. 27-29 Pick roof bottom edge and angle.

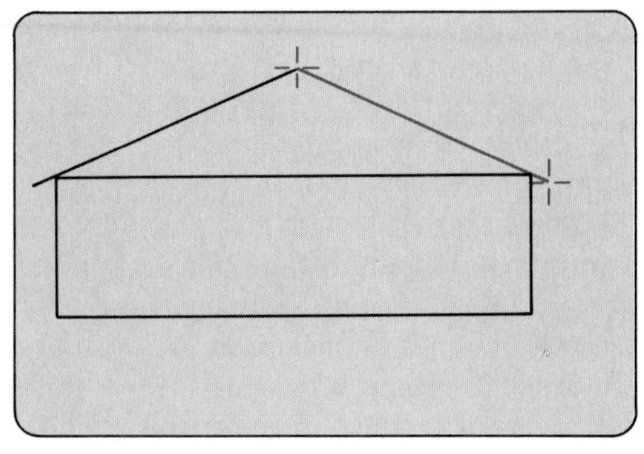

Fig. 27-30 Repeat previous step for right side.

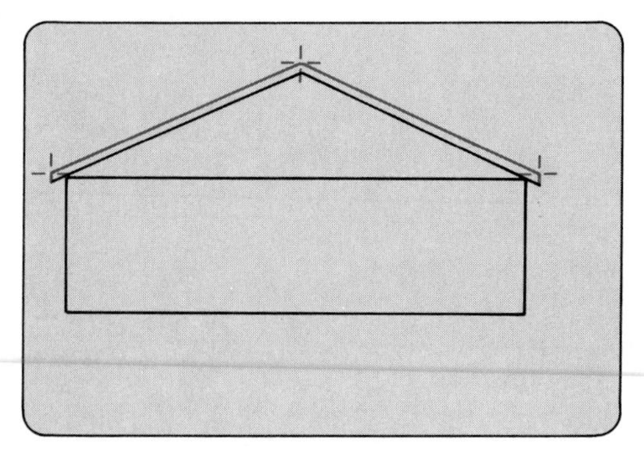

Fig. 27-31 Enter rafter 6″ depth.

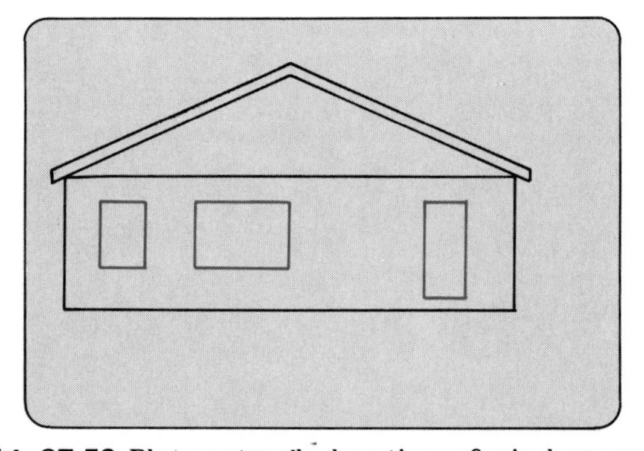

Fig. 27-32 Plot rectangle location of windows and doors.

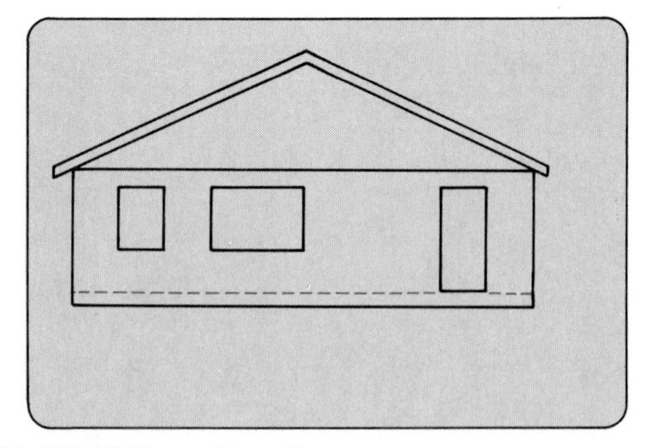

Fig. 27-33 Draw floor line.

ing. Type *FRONT ELEVATION*, then *return*. Typing "return" tells the computer this part of the text is finished. Enter *ESC G*, graphics, and touch the location on the monitor's screen where the text is to be located as shown in Fig. 27-36.

17. Enter task *65*^, line edit. Remember to enter ^ to get out of the text mode, or else the computer will use the 65 as text material. Enter *30*, erase segment. Touch the line with the cursor to erase it as shown in Fig. 27-37. Add trim, chimney, and so forth to finish the elevation (Fig. 27-38).

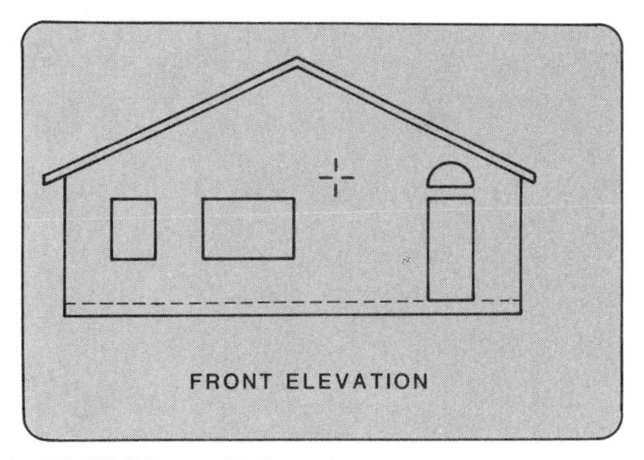

Fig. 27-37 Line edit function.

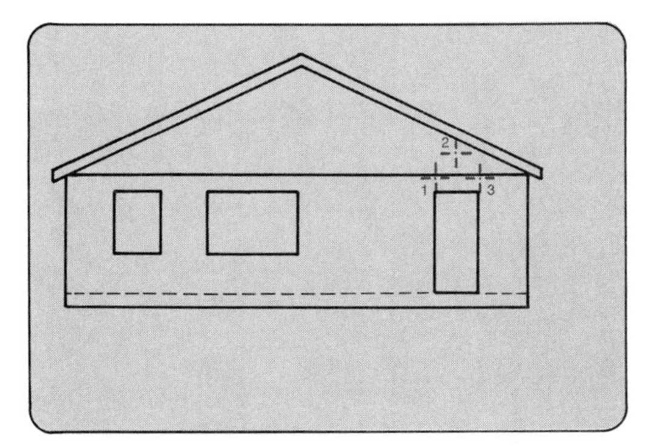

Fig. 27-34 Pick end points and radius of arcs.

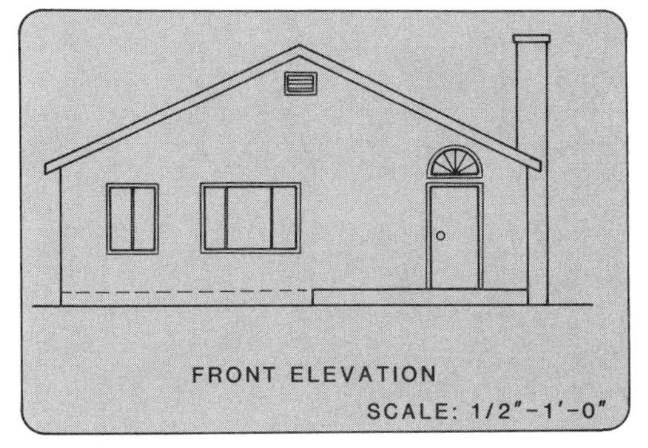

Fig. 27-38 Add trim details.

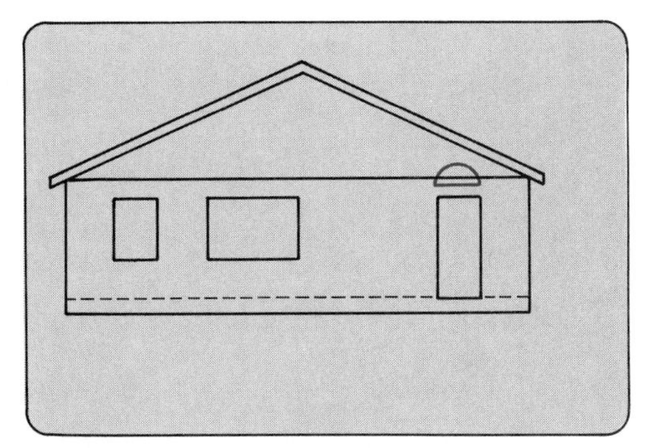

Fig. 27-35 Complete arc task.

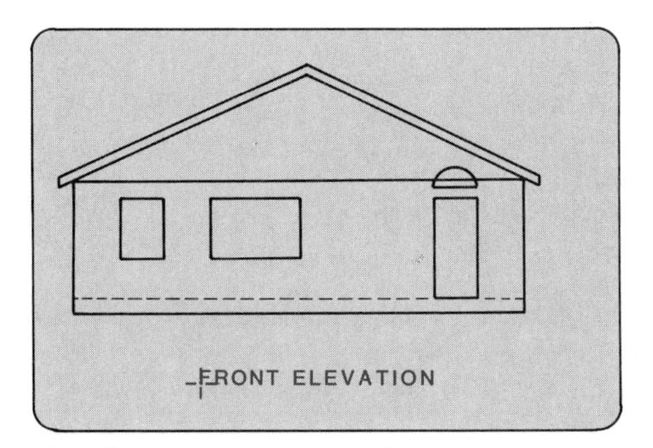

Fig. 27-36 Complete text entries.

This completes the front elevation. Remember by entering *ESC G* (graphics), *ESC M* (master menu), or *ESC A* (auxiliary menu), the required information needed to obtain the finished graphics will be provided. With practice, the operator will begin to remember the task numbers and will not need to go to the menus for instruction. It is at this time that CAD will surpass conventional methods of drafting.

This sample elevation drawing uses only a few of the many graphics tasks available in most CAD software programs. With practice, the other tasks can be learned and eventually the operator's efficiency, speed, and accuracy will surpass all levels of traditional drafting techniques.

CAD APPLICATIONS

Practically every type of architectural drawing can be produced on a CAD system, from simple single-line floor plans to complex three-dimensional perspective models. The specific applications of CAD tasks to each type of drawing is covered in the appropriate unit.

Fig. 27-39 Steps in drawing a floor plan on a CAD system.

When drawing floor plans on a CAD system, follow the ten steps shown in Fig. 27-39:

1. Using the line task, draw a single line outline of the floor plan perimeter.

2. Repeat the same line task procedure for all interior walls.

3. Use the CAD line widen task to extend the wall thickness to 6″.

4. Call up window symbols from the library using GET GROUP task. If no symbol library exists, draw one symbol, file in a group library, and call-up for use in each location. Or you may draw the symbol and, using the move, copy repeat functions to move a symbol to each location.

5. Repeat the same procedure to add door symbols to the drawing. Remember the wall lines must first be drawn before the symbol is added.

6. Use the line edit task to remove extraneous lines.

7. Call-up both fixture symbols and locate as shown.

8. Repeat the same procedure for kitchen fixture symbols.

9. Use the typewriting function to add all labels and notes.

10. Dimension the drawing using the automatic dimensioning task.

ARCHITECTURAL TASK APPLICATIONS

There are dozens of basic tasks and hundreds of task combinations which can be accessed with commands on a menu. Figure 27-40 shows an architectural master menu which contains symbols for architectural features such as doors, windows, fixtures, and appliances. The menu contains tasks which the operator can select to create lines, dimensions, geometric features, and text. The menu also includes many functions such as grid, zoom, and pan which make drawing with a computer faster, easier, and more accurate. In addition, there are many editing tasks which enable designers to rotate, repeat, erase, mirror, and stretch parts of a drawing.

Once the command mode is selected from a master (main) menu, most CAD systems display an auxiliary menu which shows the options available in the mode chosen. Some auxiliary menus display directly on the monitor with the drawing, as in Fig. 27-40. Some are displayed on a separate screen, split screen, or alternately on the drawing screen.

Basic Line Tasks

Line tasks are the most commonly used function. Most systems include object, layout, dashed, phantom, center, dimension, and cutting plane line options. The line widening task is especially useful in widening a line to create a floor plan wall without first drawing two separate lines as shown in Fig. 27-41.

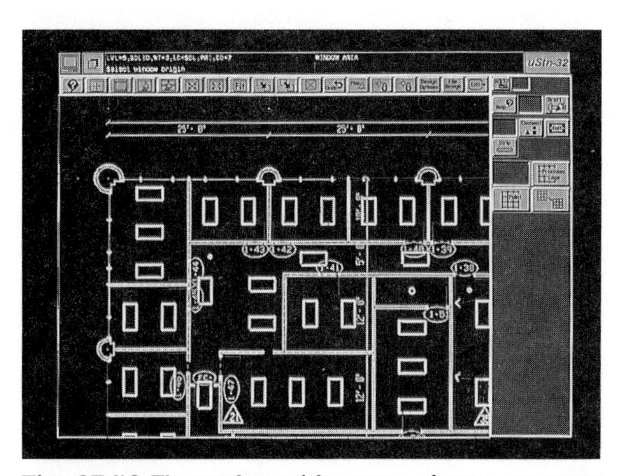

Fig. 27-40 Floor plan with menu shown on same screen. *(Intergraph Corp.)*

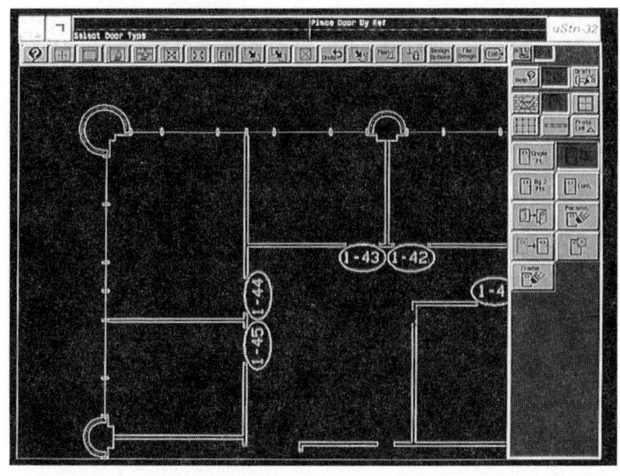

Fig. 27-41 CAD floor plan showing line widening of walls. *(Intergraph Corp.)*

Symbol Placement

Symbol libraries are either preprogrammed symbols embedded into a CAD program or created by an operator and stored for call-up at a future date. Most architectural programs allow symbols, such as doors or windows, to be inserted into walls automatically without the use of other commands to break and modify the wall. In addition to individual or component libraries, entire systems such as the stair assembly shown in Fig. 27-42, can be created, stored, and added to any drawing as one entity.

Geometric Forms

To avoid drawing common geometric forms using individual lines, arcs, and circles, many forms can be located on a drawing as one complete entity. Forms normally found in Architectural CAD programs include rectangles, polygons, circles, arcs, and ellipses. Note the many common geometric forms used in Fig. 27-43. These were located in each position without the need to draw each individual line.

Editing Functions

Many of the advantages in using a CAD system involves the use of editing functions. Often an original drawing can be manually prepared in the same or less time than with a CAD system. But revisions, which are usually numerous, can be completed on a CAD system in much less time than manual changes can be made. The editing function makes a CAD system a design tool and not just a computer system used to draw faster and create more accurate lines. Editing functions most commonly used in architectural work include copy, repeat, mirror, delete, stretch, rotate, scale, align, and break.

Drafting Assistance

Architectural CAD programs contain many features which enable drafters to compensate for the lack of space on a computer screen compared to a large drawing table. The zoom, pan (window), and layering commands are especially useful in architectural work. The zoom command allows a drafter to enlarge (zoom-in) a very small part of a drawing. Alternatively, the entire drawing, or any portion, can be viewed by zooming-out. The zoom feature is most helpful when working on large floor plans, elevations, and site plans as shown in Fig. 27-44.

The layering function is widely used in architectural work to separate drawing elements with a common base. For example, the electrical, plumbing, HVAC, or furniture layouts can be placed on the same base floor plan without redrawing or duplicating the base. Layers can be removed or combined to show the location of space conflicts during the design process.

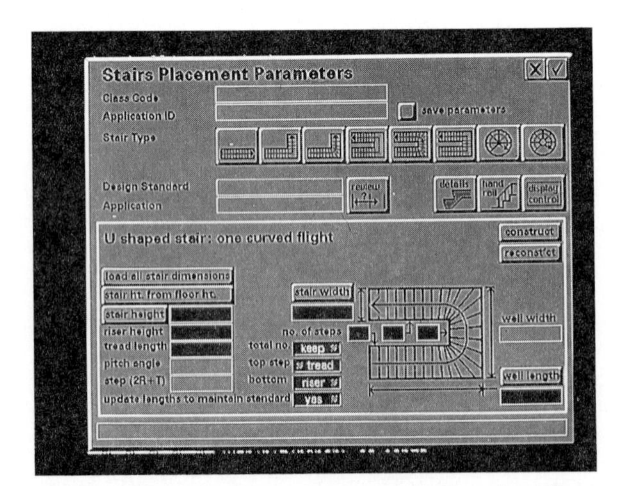

Fig. 27-42 Total stair system to insert onto a floor plan as a unit. *(Intergraph Corp.)*

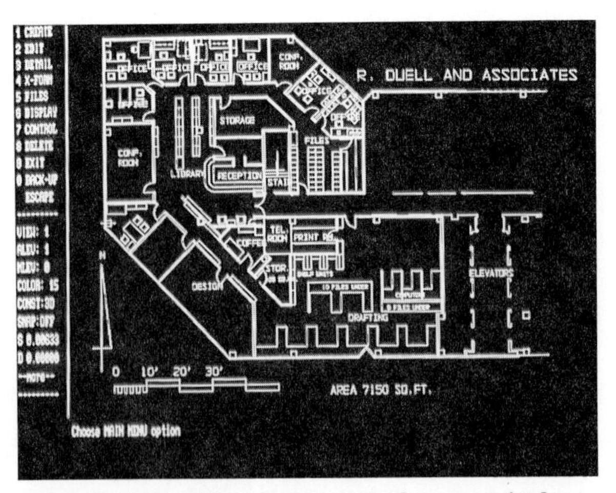

Fig. 27-43 Multiple positioning of geometric forms on a floor plan. *(Intergraph Corp.)*

Three Dimensional CAD

In the past, architectural pictorial drawings on models were usually prepared only after the design was finalized. This is because rendering and model building is very time consuming and expensive. Now 3D CAD systems, which use *x*, *y*, and *z* cartesian coordinates, can be part of the design process. Pictorial CAD programs are of three types; wire frame, surface and solid modeling. Wire frame pictorial models can be drawn with all object lines shown as in Fig. 27-45, with hidden lines removed, or with hidden lines dashed as in Fig. 75-3A. Surface models show a covering over a wire frame surface and solid models show the building as a solid mass (Fig. 2-42) which can be shaded as shown in Fig. 1-6.

The majority of architectural firms use CAD some of the time. However, 80 percent of all drawings produced in the United States in 1990 were prepared by conventional drafting methods (Fig. 27-46). So both CAD and manual systems must be learned. Just as the use of an electronic calculator does not eliminate the need to learn math principles, the use of a CAD system does not exempt novice designers or drafters from learning the basic principles of architectural drafting and design.

Remember, CAD systems do not operate on their own. CAD as a drafting and design tool is only as effective as the users knowledge of drafting and design principles and practices.

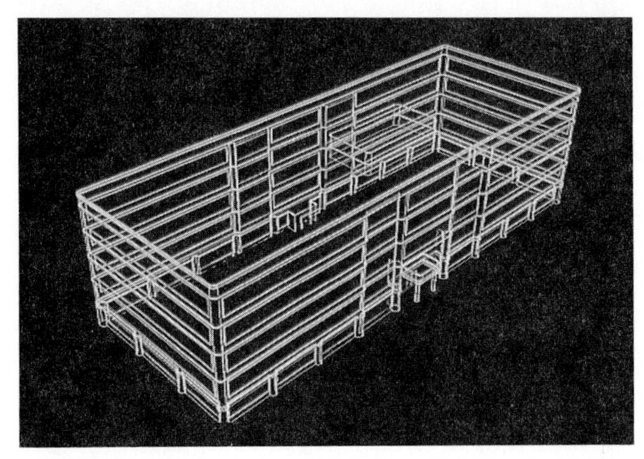

Fig. 27-45 Wire frame pictorial model drawing. *(Intergraph Corp.)*

ARCHITECTURAL TASK DEVELOPMENT

Although the completion of CAD tasks is similar in all drafting fields, the application of many tasks to architectural work differs in many situations. CAD software programs also vary widely in their capacity automatically to perform many drafting tasks. While command labels and drawing procedures differ somewhat among CAD systems the end result is the same.

Figure 27-46A through G shows the sequence of completing the most common architectural tasks available in most CAD software. A description of each task is shown at the left followed by the computer command used to complete each task. The computer display screen in columns three and four shows the first two steps used in beginning each task. The far right screen shows the completed task. Although additional steps are usually needed to complete a task, these steps are normally repetitions or adaptations of step two. And remember, the movement of the cursor as shown in these examples can be controlled by many input devices such as a keyboard, mouse, graphic tablet and stylus, or pressure sensitive tablet.

Once the application of these basic tasks is understood, learning to use any CAD system can proceed in a logical order.

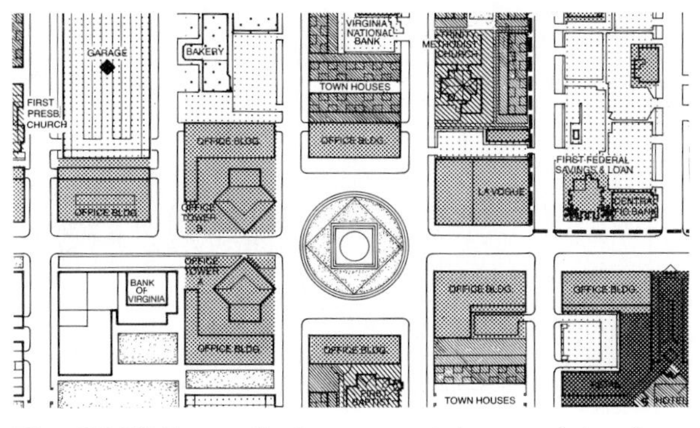

Fig. 27-44 Zoom features must be used to draw large plans. *(Veratec)*

Fig. 27-46 CAD and conventional drafting systems. *(A.M. Bruning)*

TASK	COMMAND	STEP 1	STEP 2	COMPLETED COMMAND
To draw a straight line from one point to another point	Lines	 Position cursor at start of line	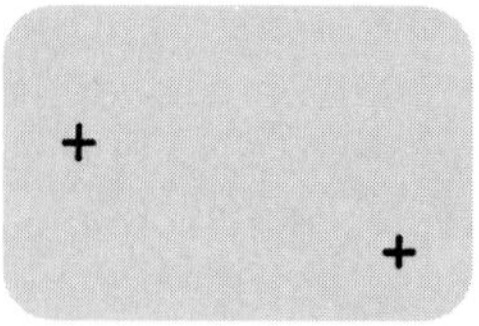 Position cursor at end of line	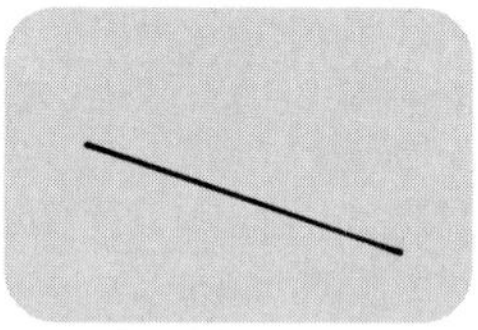 Completed line
To draw a straight line by entering (typing) cartesian coordinates	Lines	Type x-y coordinates for start of line	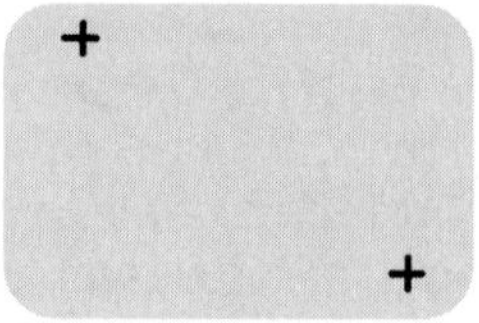 Type x-y coordinates for end of line	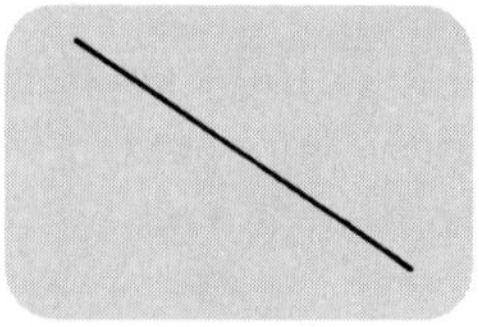 Completed line
To draw a line from a point to a tangency point	Lines	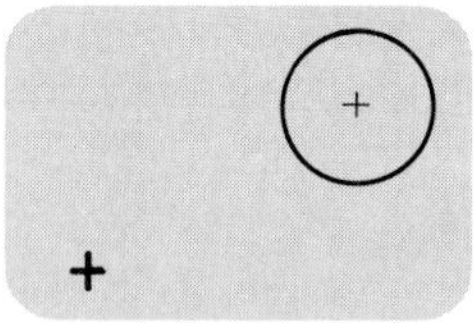 Draw arc/circle and locate end of line with cursor	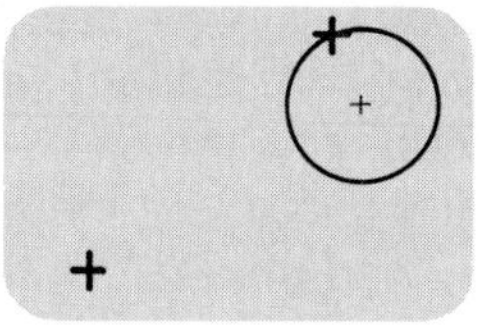 Locate cursor on arc approximately at tangency point	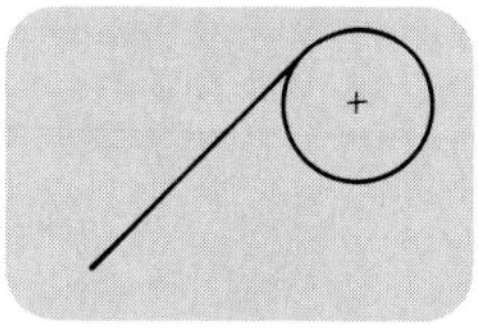 Line automatically lines up at tangency point
To draw a circle with the center and radius	Circles	 Position cursor at circle's center	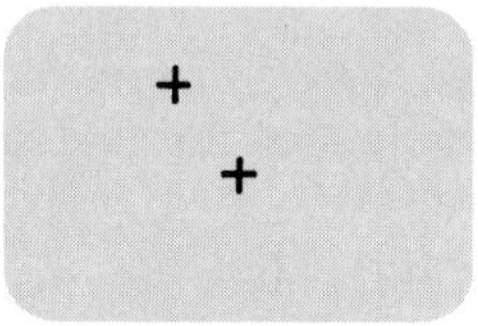 Position cursor at circle's radius (or type radius)	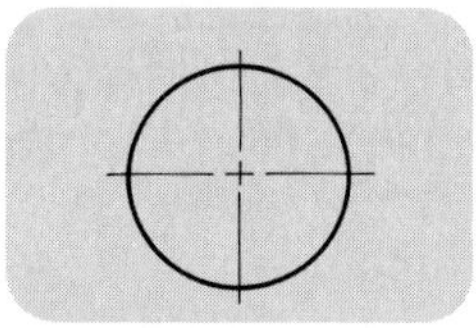 Completed circle
To draw a circle with the diameter	Circles	 Position cursor at one end of diameter	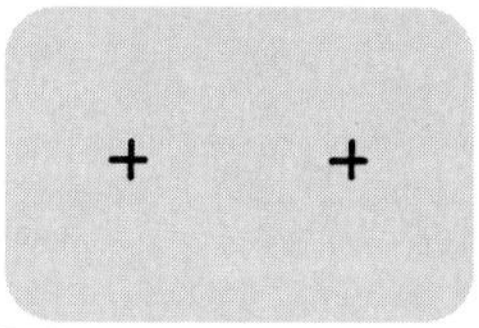 Position cursor at other end of the diameter	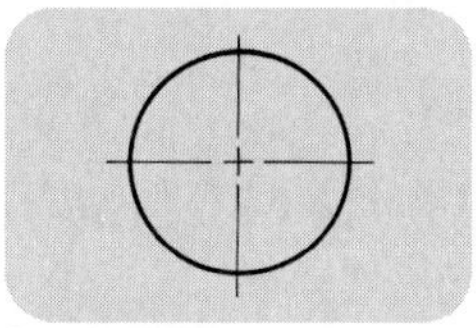 Completed circle
To draw an arc by center and angle (or radius)	Arcs	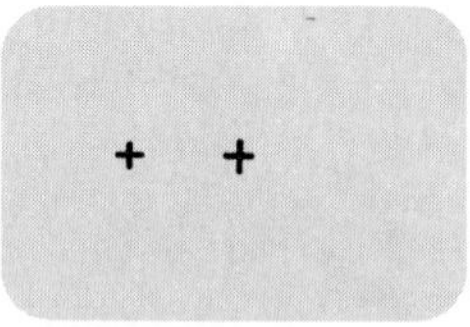 Locate arc's center and start	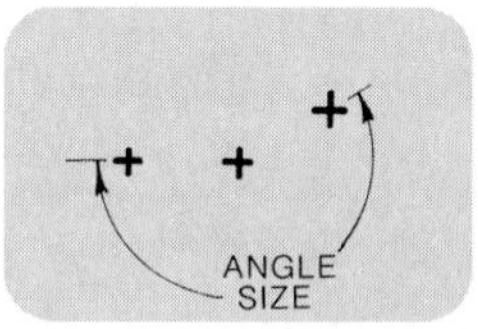 ANGLE SIZE Locate end of arc or type angle size	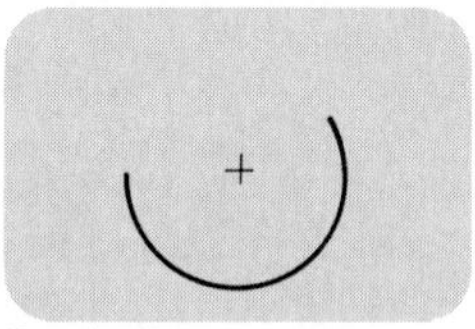 Completed arc
To draw an arc with three points (two ends and point somewhere in between)	Arcs	 With cursor locate arc's start	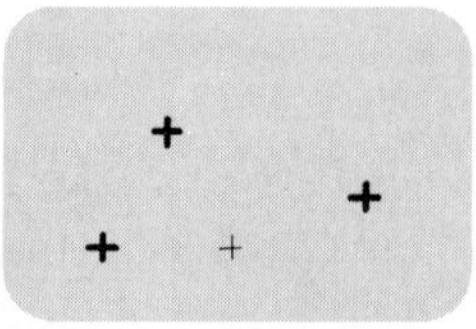 With cursor locate arc's end and a point between	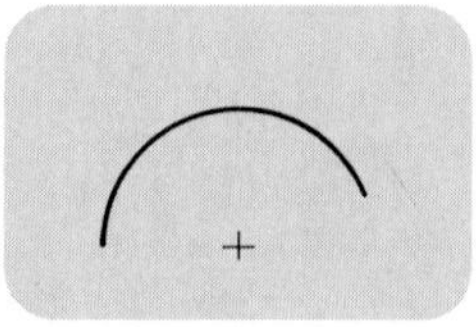 Completed arc

Fig. 27-47A Drawing Task procedures

TASK	COMMAND	STEP 1	STEP 2	COMPLETED TASK
To place text on a drawing	Text	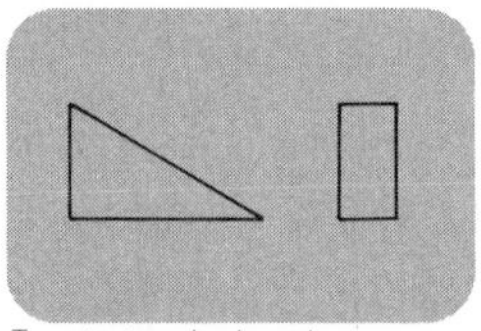 Type text on keyboard	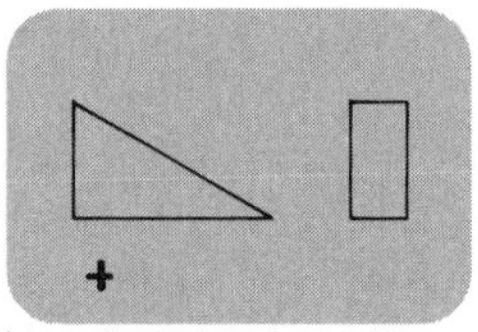 Locate beginning of text with the cursor	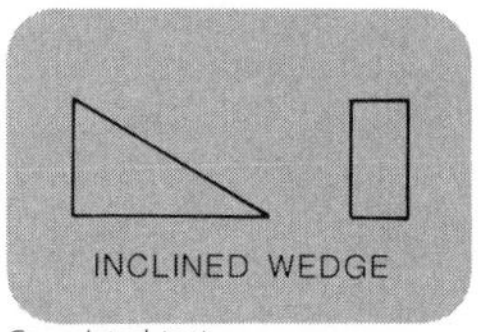 INCLINED WEDGE Completed text
To draw a rectangle	Rectangle	 Enter lower left corner with cursor	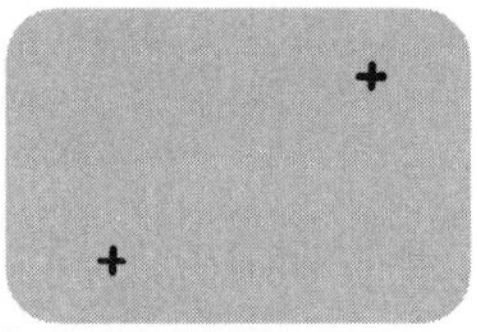 Enter upper right corner with cursor	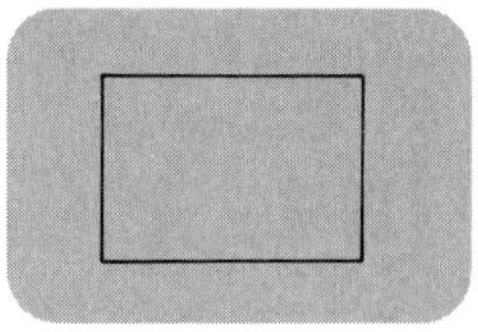 Completed rectangle
To draw any type of polygon	Polygon	 Enter polygon's center with cursor	 Type number of sides and inscribed diameter	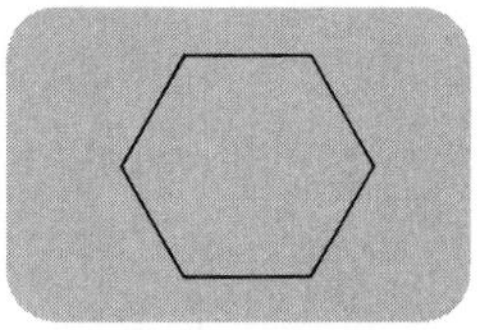 Completed polygon (hexagon)
To draw an ellipse by plotting a rectangle (major and minor axes)	Ellipse	 Enter rectangle's corners (major and minor ellipse diameter)	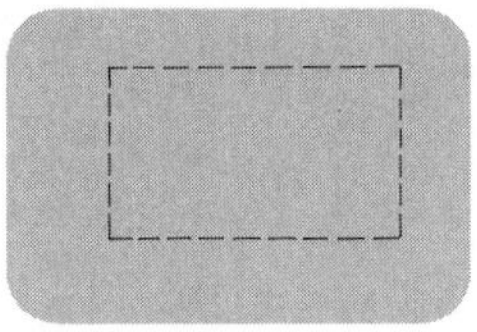 Digitized rectangle	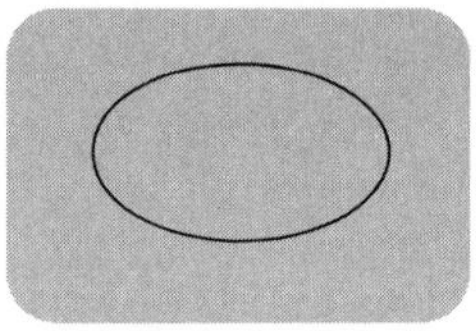 Completed ellipse
To draw an irregular curve from a series of plotted points	Curve	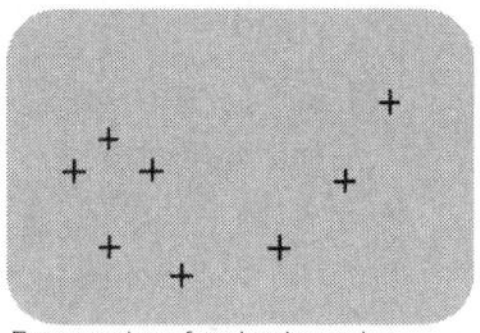 Enter points for the irregular curve	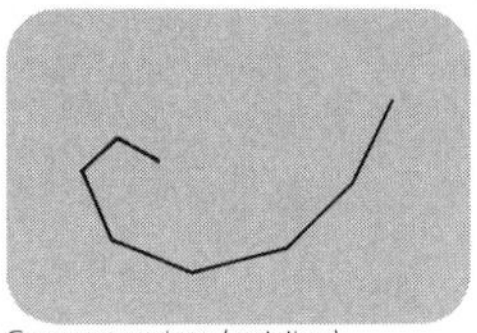 Connect points (polyline)	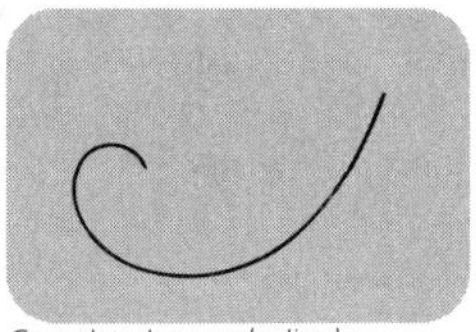 Completed curve (spline)
To use layout and projections lines to help with the completion of a drawing	Construction lines	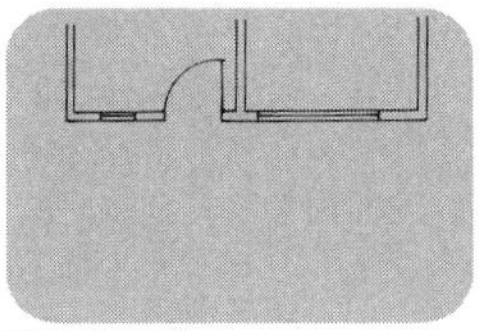 Draw floor plan	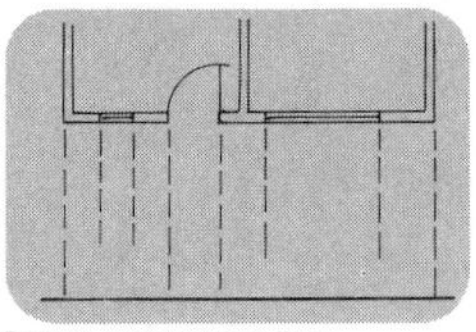 Project construction lines	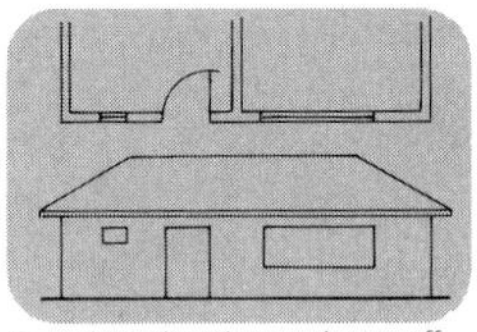 Complete elevation and turn off construction lines
To draw cross hatching lines for sectional drawings	Cross-hatching	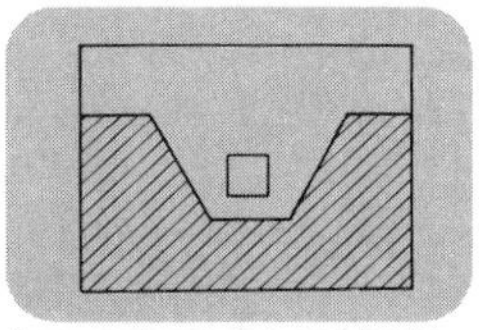 Draw sectional view and locate sectioned area with cursor	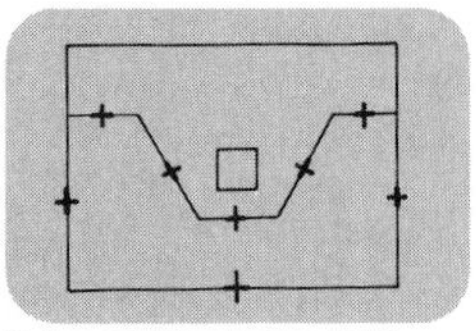 Define additional areas to be sectioned	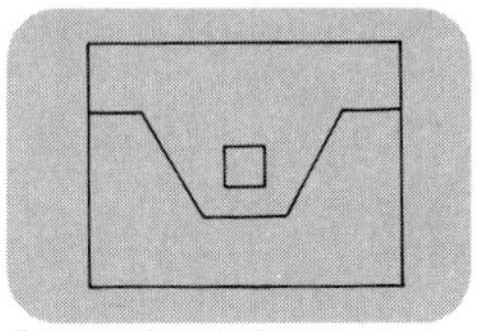 Completed sectioning

Fig. 27-47B Drawing Task procedures

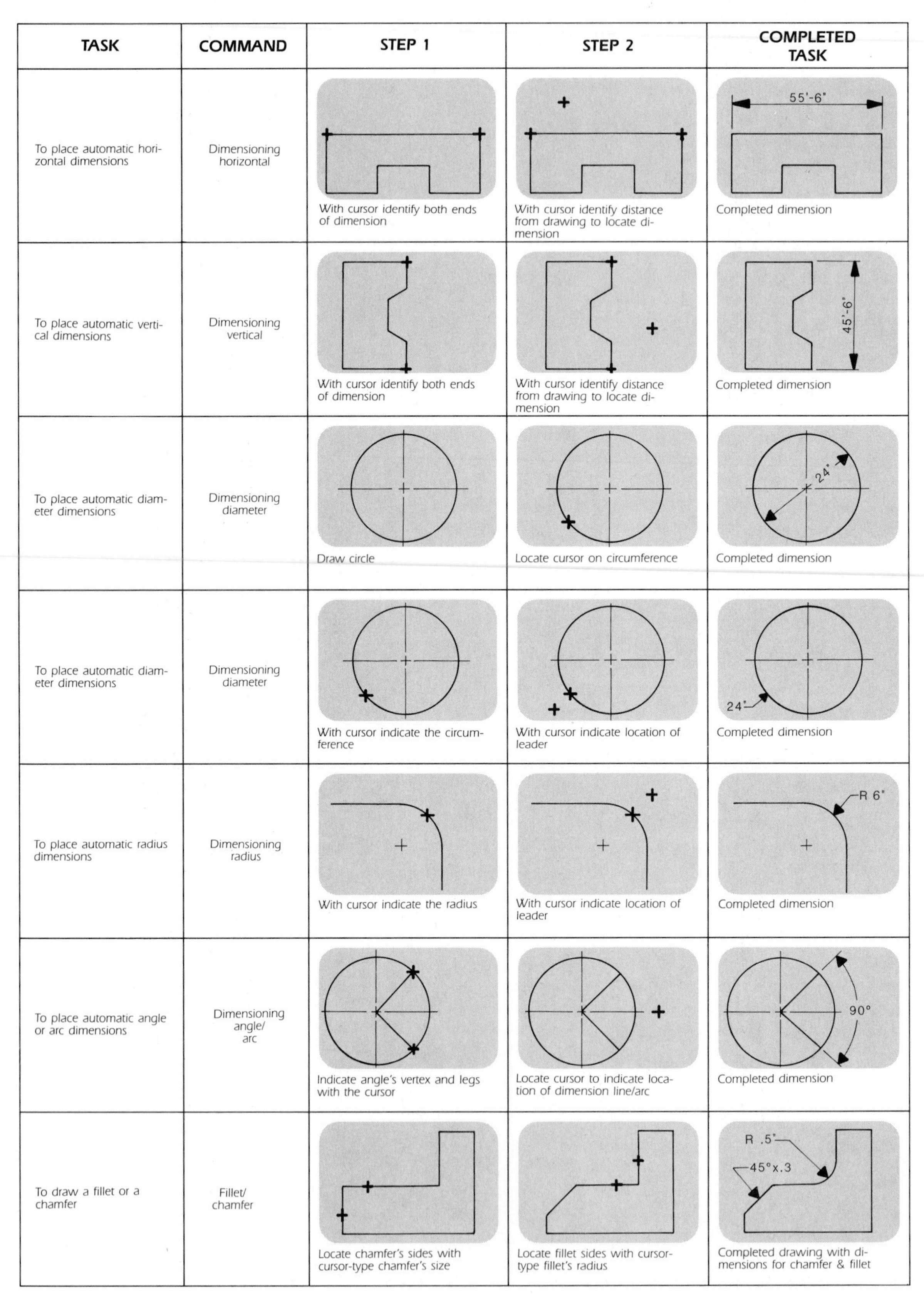

TASK	COMMAND	STEP 1	STEP 2	COMPLETED TASK
To place automatic horizontal dimensions	Dimensioning horizontal	With cursor identify both ends of dimension	With cursor identify distance from drawing to locate dimension	Completed dimension
To place automatic vertical dimensions	Dimensioning vertical	With cursor identify both ends of dimension	With cursor identify distance from drawing to locate dimension	Completed dimension
To place automatic diameter dimensions	Dimensioning diameter	Draw circle	Locate cursor on circumference	Completed dimension
To place automatic diameter dimensions	Dimensioning diameter	With cursor indicate the circumference	With cursor indicate location of leader	Completed dimension
To place automatic radius dimensions	Dimensioning radius	With cursor indicate the radius	With cursor indicate location of leader	Completed dimension
To place automatic angle or arc dimensions	Dimensioning angle/arc	Indicate angle's vertex and legs with the cursor	Locate cursor to indicate location of dimension line/arc	Completed dimension
To draw a fillet or a chamfer	Fillet/chamfer	Locate chamfer's sides with cursor-type chamfer's size	Locate fillet sides with cursor-type fillet's radius	Completed drawing with dimensions for chamfer & fillet

Fig. 27-47C Drawing Task procedures

TASK	COMMAND	STEP 1	STEP 2	COMPLETED TASK
To erase an entire item (grouped into a single entity)	Delete	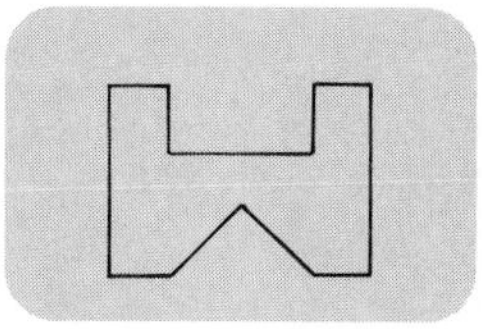 A single item drawing (grouped)	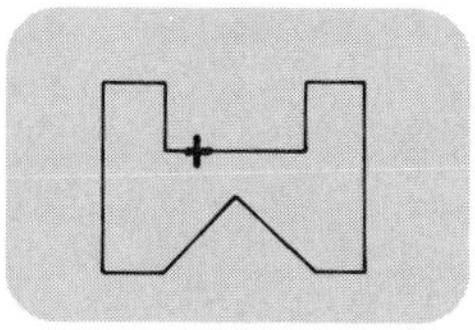 Place the cursor anywhere touching the drawing	Completed task
To erase a segment of a drawing	Delete window	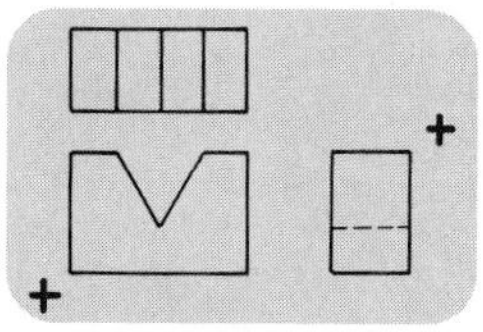 Create a rectangular window with cursor	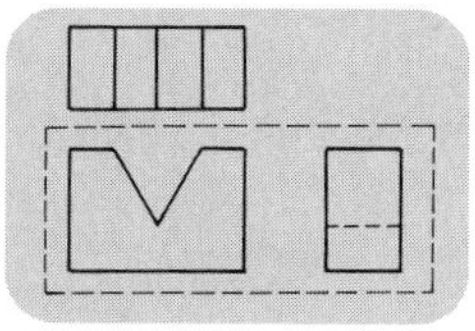 Area to be deleted	 Deletion completed
A command that will erase the last operation	Delete last	 Developing a drawing	 Completion of the last operation	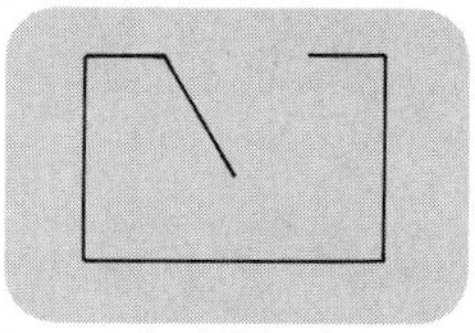 Last operation is deleted
To rotate an object about its point of origin	Rotation	 Completed drawing (grouped)	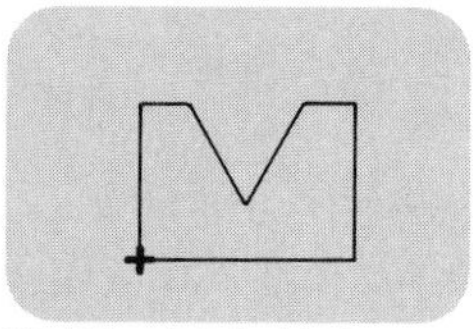 Establish point of origin for revolvement. Type angle of revolvement	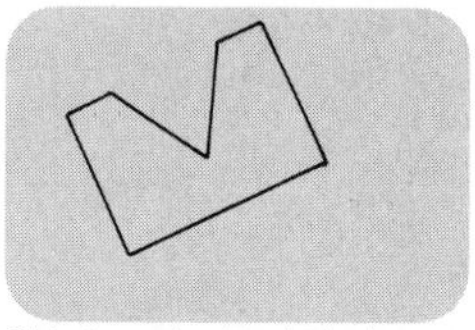 Objects revolved to desired position
To move a drawing to a different position	Move	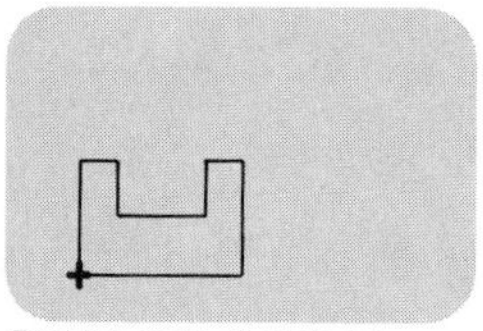 Establish point of origin on drawing	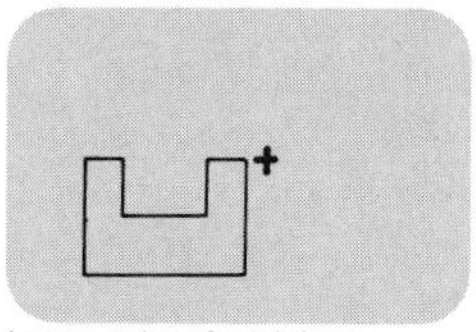 Locate point of origin's new position with cursor	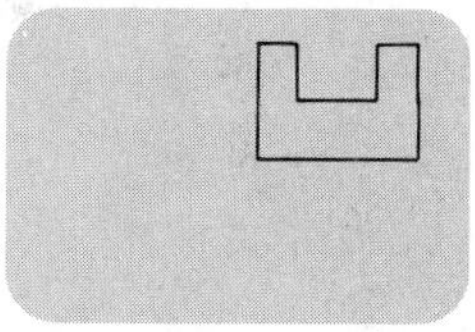
To mirror (flip-flop) a drawing	Mirror	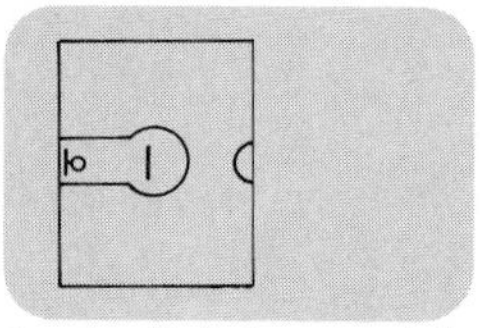 Completed half of a symmetrical drawing	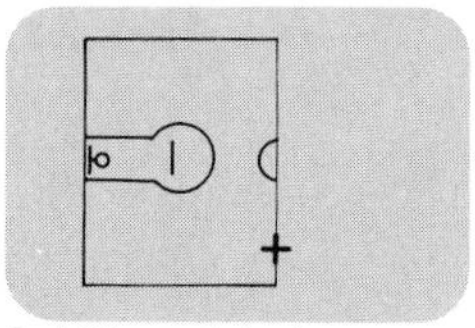 Establish an axis for the revolved position	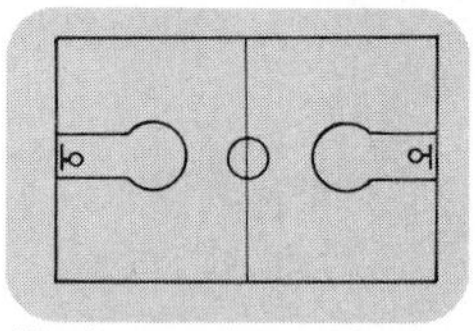 Mirroring operation completed
To duplicate a drawing singly and/or multiple	Copy/ repeat	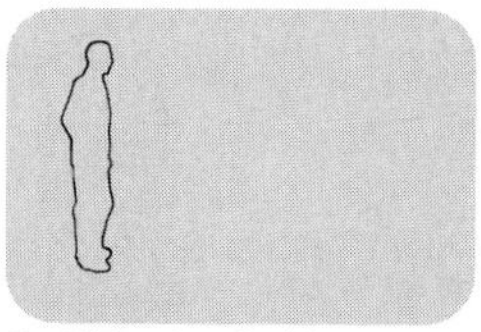 Establish point of origin on drawing	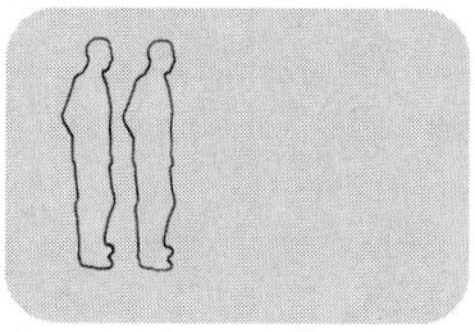 Copy drawing once. Type spacing	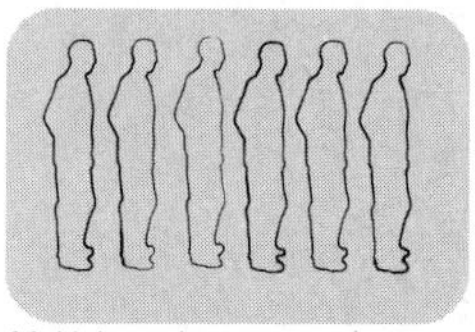 Multiple copies-type spacing and number of copies

Fig. 27-47D Editing Task procedures

TASK	COMMAND	STEP 1	STEP 2	COMPLETED TASK
To elongate or shorten a feature	Stretch	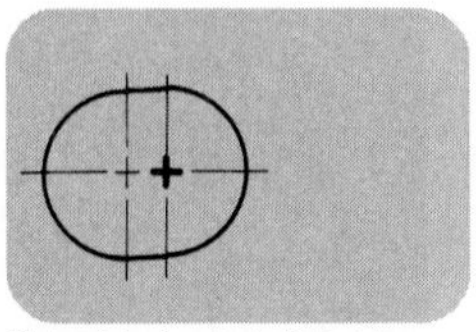 Pick origin for start of stretch	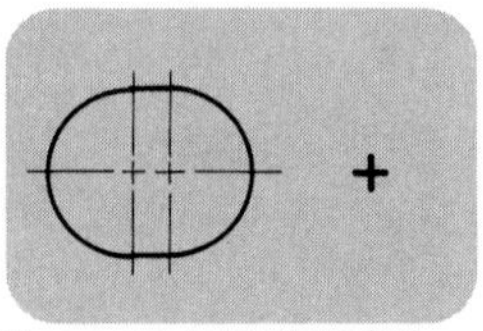 Move cursor to lengthened position	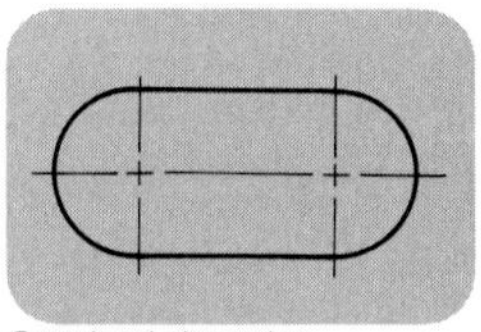 Completed elongation
To draw parallel lines to any desired width	Widen lines	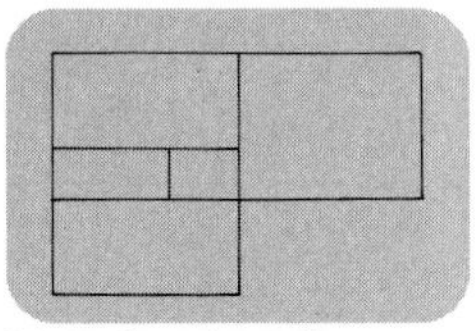 Existing floor plan outline	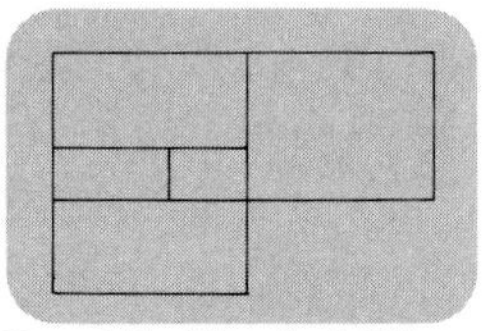 Type width of lines (wall thickness) define side to widen	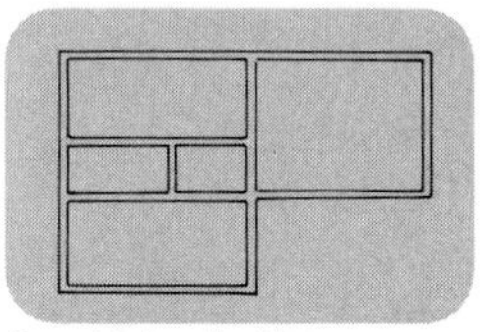 Completed wall widen
To change the working drawing's scaled size	Scale	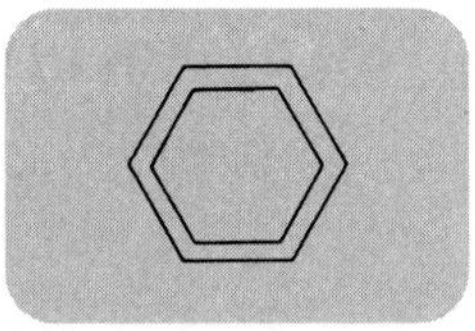 Origin drawing scale	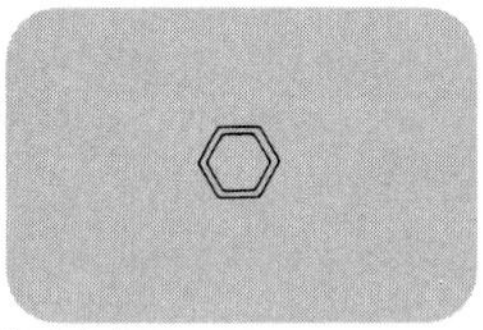 Scaled down	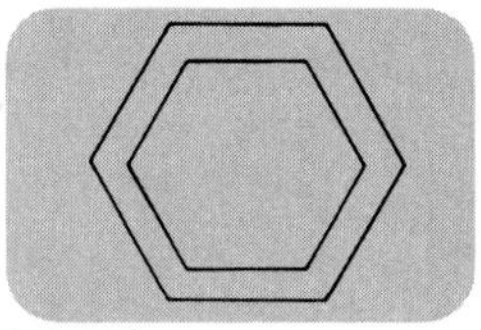 Scaled up
To line up objects along a horizontal or vertical axis	Align	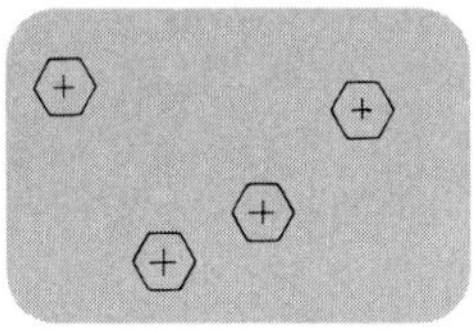 Determine the origin point for all items to align	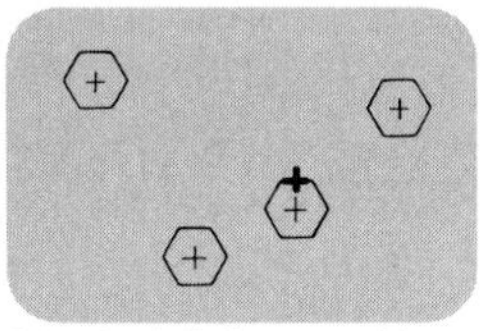 Select level of alignment with the cursor	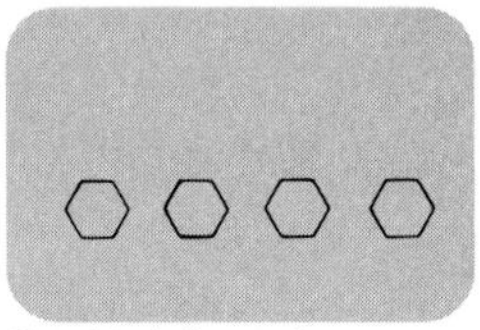 Completed alignment
To repeat object in a circular pattern	Array polar	Determine the origin point for the object to array	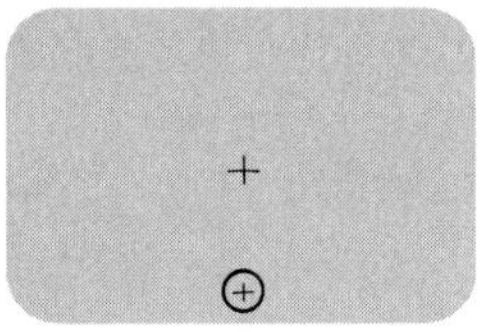 Select center of circular pattern. Type number of repeats	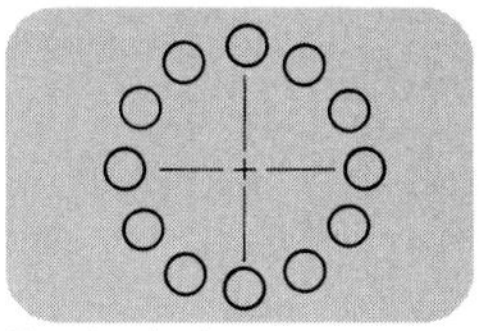 Completed polar array
To magnify or reduce in size selected portions of the working drawing	Zoom-in	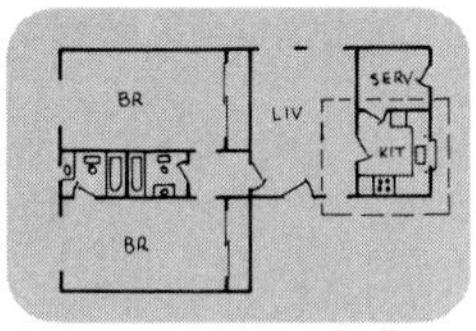 Delineate area for zooming-in or out	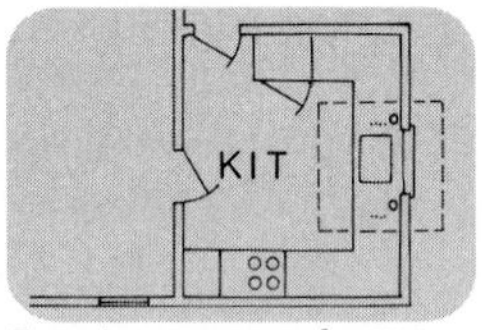 Zoom-in as necessary for magnification	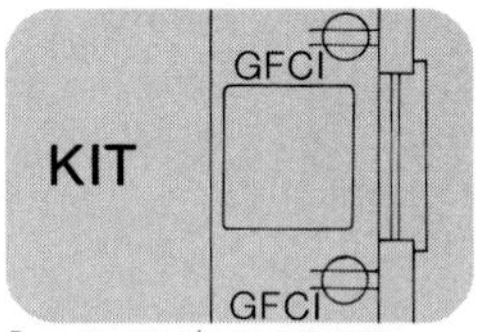 Repeat zoom-in as necessary
To move the working drawing across the monitor	Pan	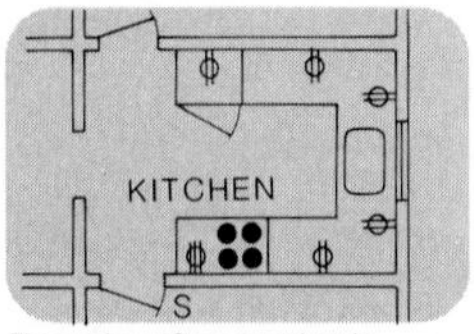 Floor plan after zooming-in	 Results of panning up	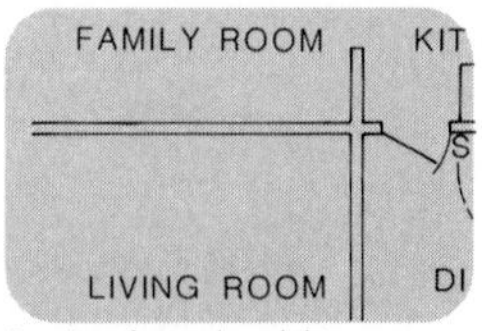 Results of panning right

Fig. 27-47E Editing Task procedures

TASK	COMMAND	STEP 1	STEP 2	COMPLETED TASK
To erase a segment of a drawing	Erase segment	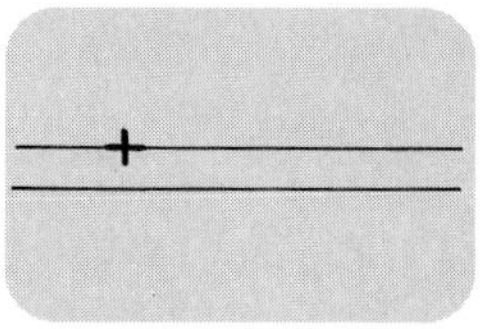 Place cursor at start of segment to be removed	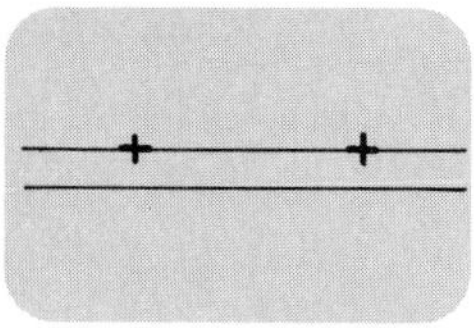 Place cursor at end of segment to be removed	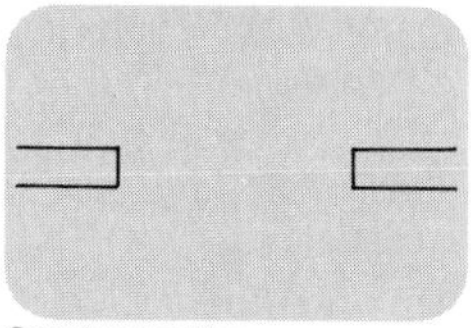 Completed editing
To shorten a line segment	Shorten segment	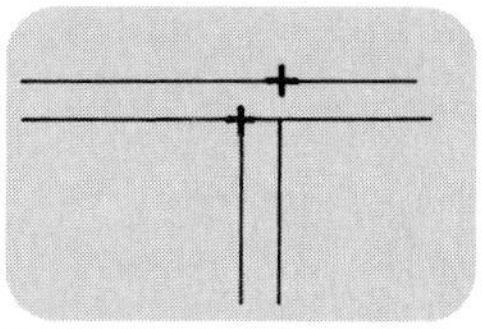 Place cursor at cutoff points	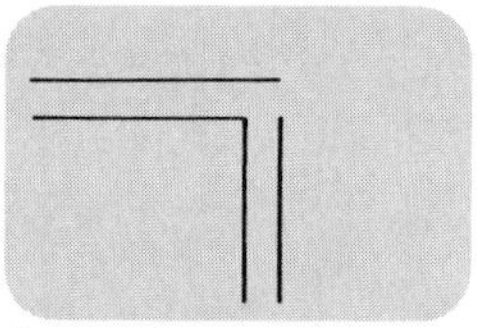 Use extend command to lengthen line	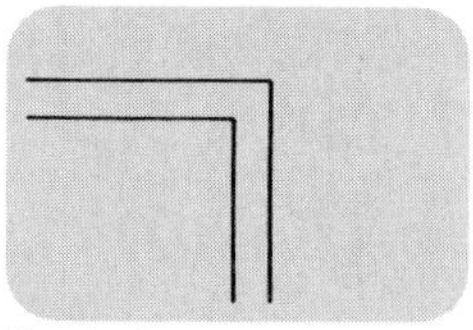 Completed editing
To extend a line segments	Extend segment	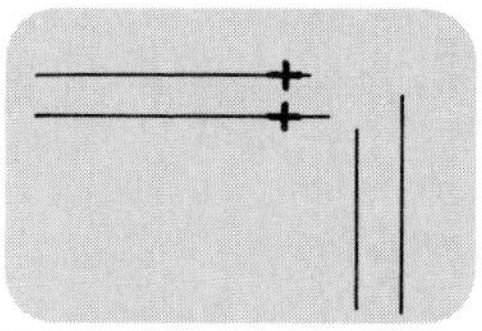 Place cursor on lines to extend	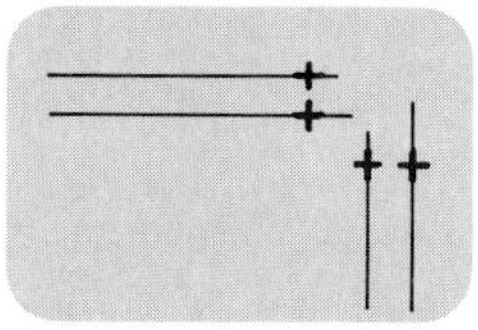 Place cursor on lines to extend	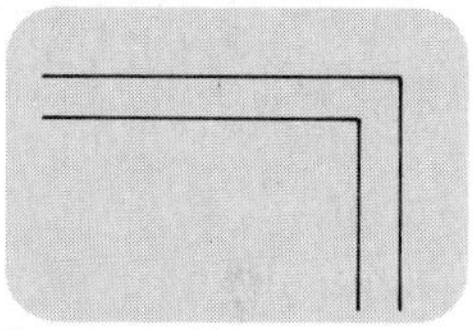 Completed editing
To create a gap or multiple gaps	Gap	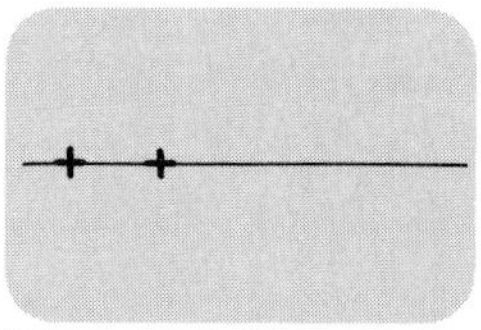 Define gap with cursor	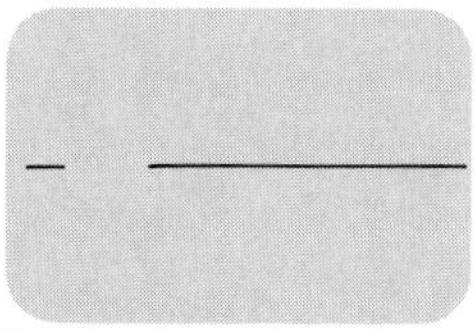 Type gap size, spacing and number of gaps	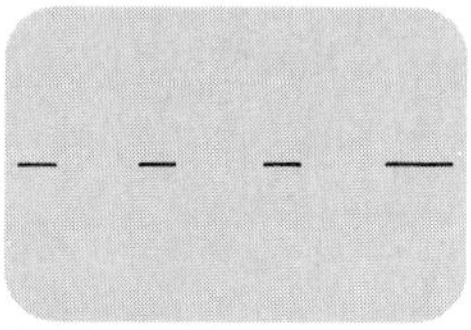 Completed editing
To create a break in any object that is not a blocked drawing	Break	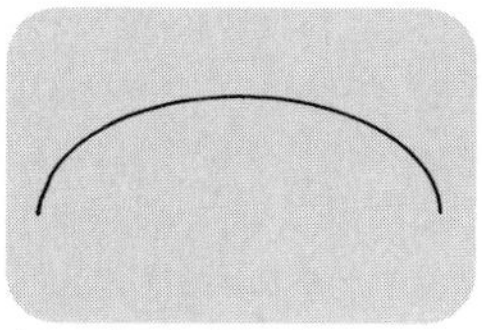 Any unblocked drawing	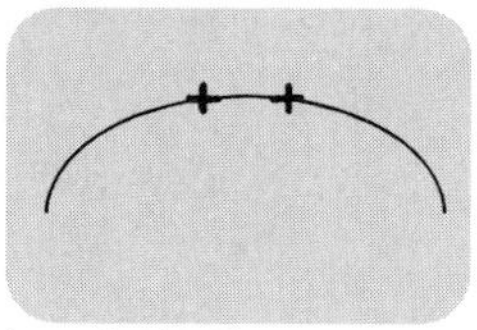 Locate break with cursor	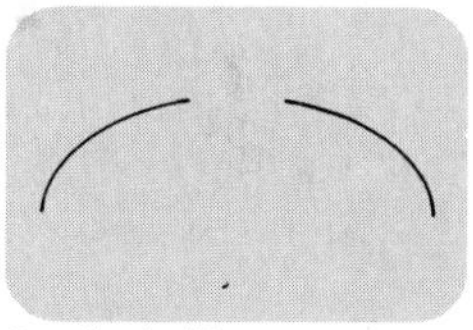 Completed editing
To create a sharp corner junction	Join	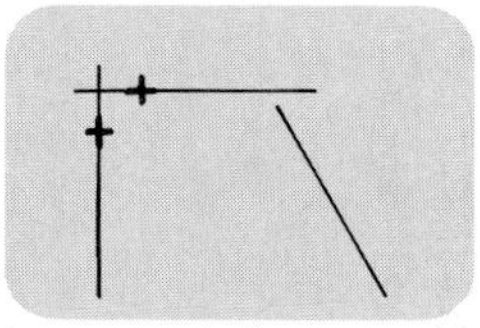 Locate cursor for first corner	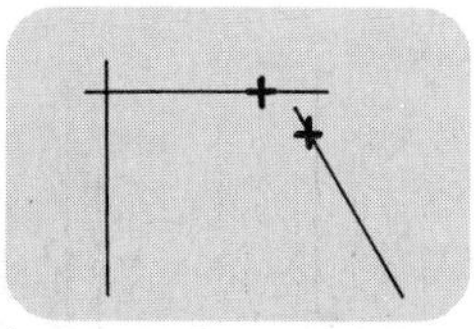 Locate cursor for second corner	 Completed editing
To square (90°) the joining of two lines	Orthogonalize	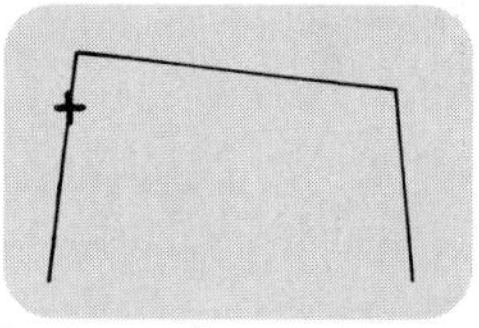 Locate cursor for one corner	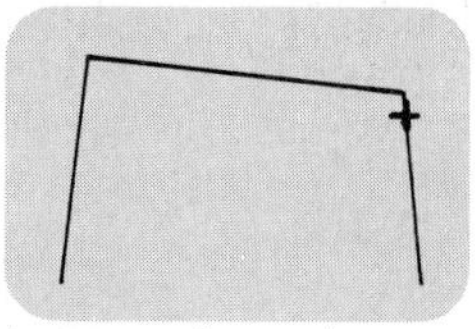 Locate cursor for second corner	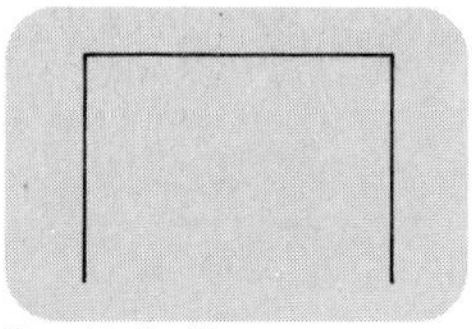 Completed editing

Fig. 27-47F Editing Task procedures

TASK	COMMAND	STEP 1	STEP 2	COMPLETED TASK
To select the required line convention for working drawings	Line styles	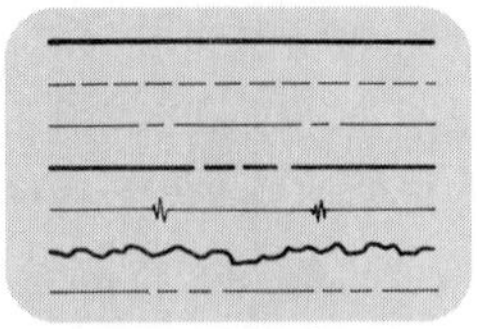 Call up the line styles from the command menu	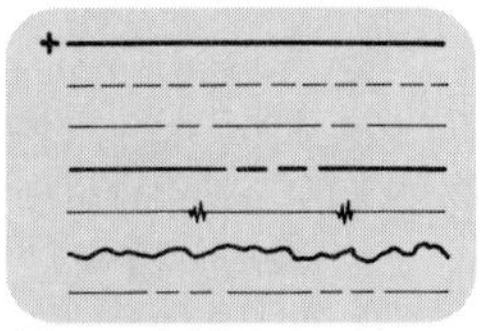 Select needed line convention	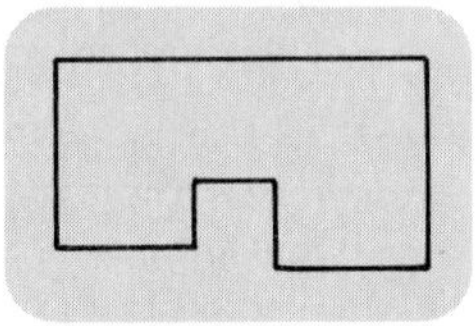 Continue drawing until next selection
The program keeps a file with data on all items entered into the working drawing and may automatically display or print schedules or bill of materials	Bill off materials	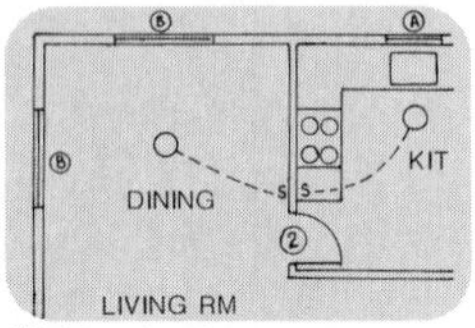 Working drawing	 Automatic window schedules	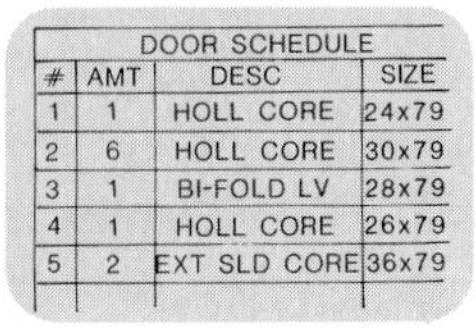 Automatic door schedules
To place segments of a working drawing on separate "layers" that can be displayed separately or combined. Usually over 100 layers per drawing is available on the CAD software	Layer	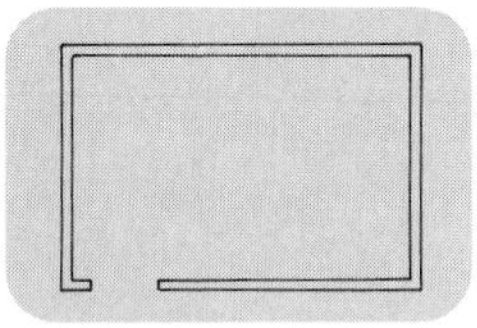 Draw layer one	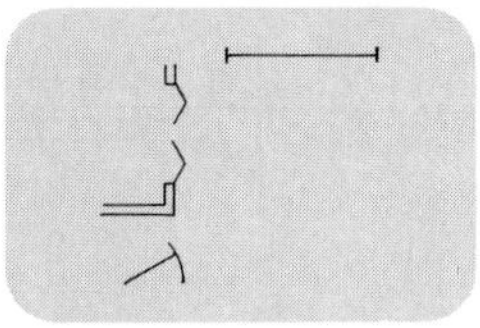 Separate layer two	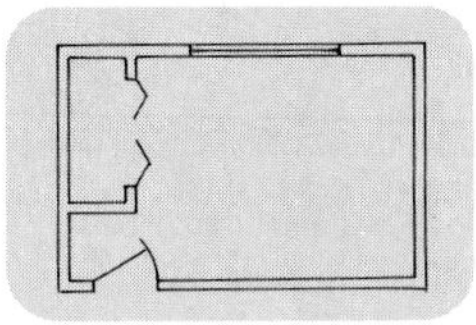 Combined layers
To select, from a large choice of library symbols, a symbol and place it on a working drawing	Library	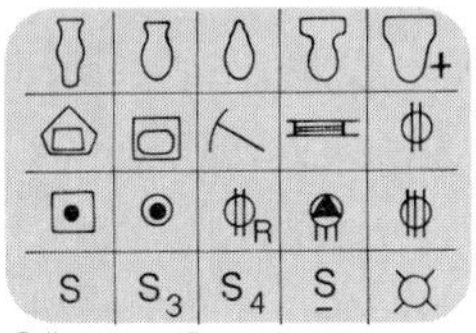 Call up specific symbol library and select	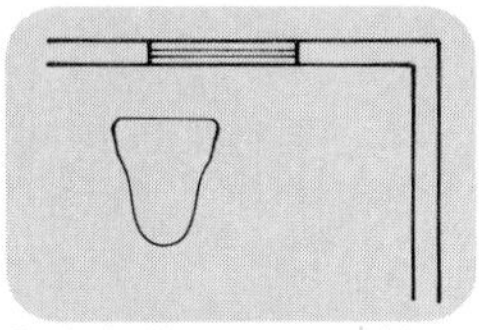 Symbol will appear on working drawing	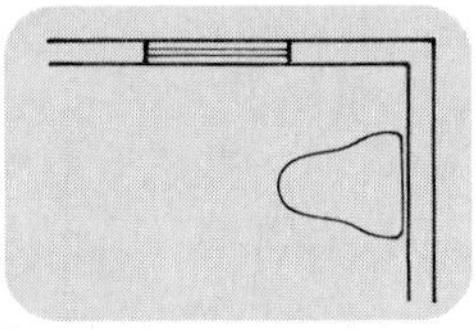 Rotate and move as needed
A function that sets a network of cartesian grid points to aid with the design of drawings	Grid	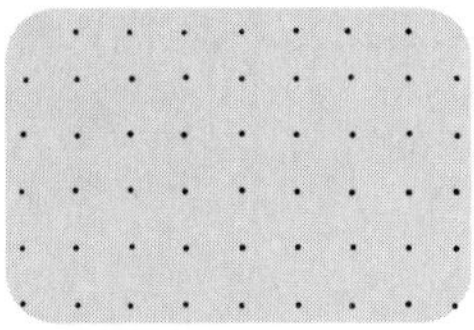 Type grid size to accommodate drawing format	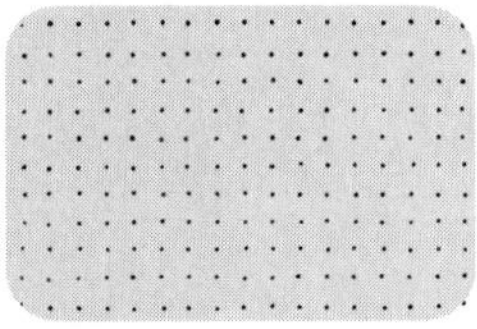 Reduce grid size when zooming-in enlargens the drawing & existing grids	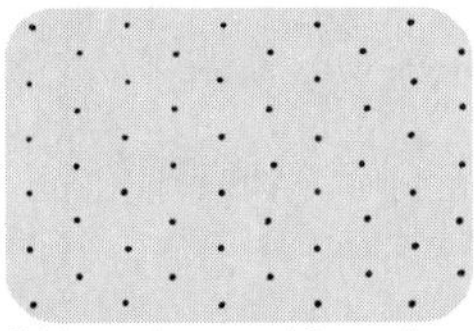 Grids may be rotated for auxiliary projection and for isometric drawings (30°)
A function that moves the cursor to the nearest grid point	Snap on	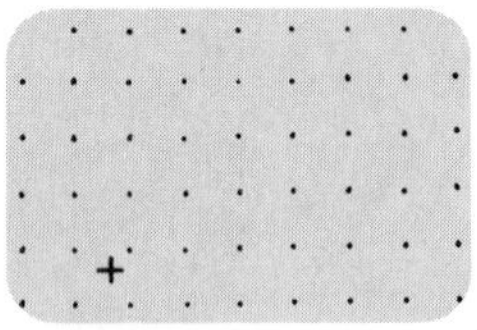 Pick first point	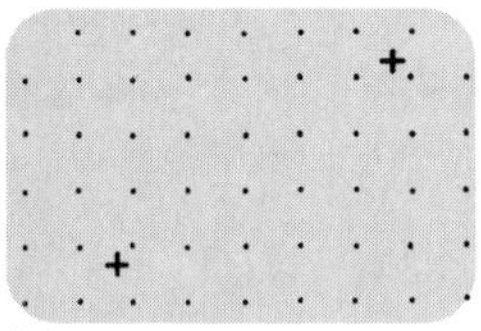 Pick second point	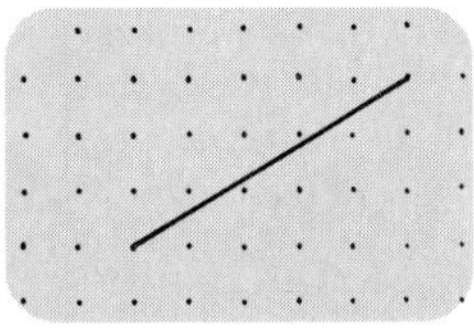 Points "snap" to the grids
A function that allows the selected point to remain where located regardless of grids	Snap off	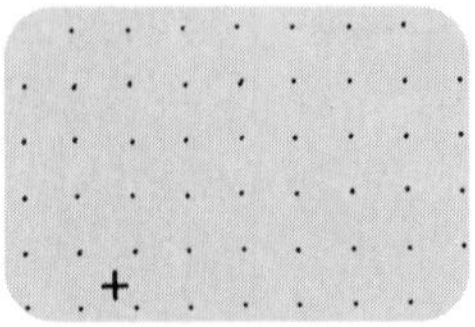 Pick first point	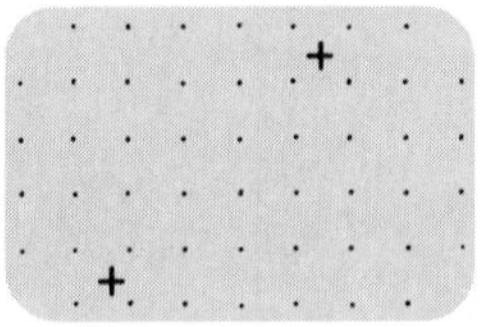 Pick second point	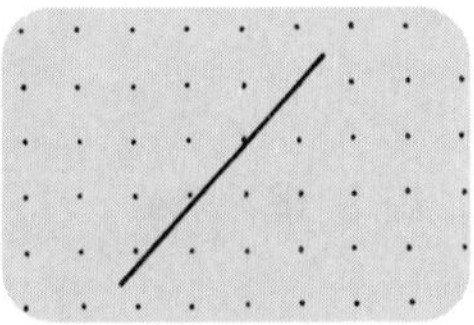 Points remain where positioned

Fig. 27-47G Editing Task procedures

ATTRIBUTES

In addition to aiding the preparation of drawings, computers are also used extensively in the preparation of construction documents. Bills of materials, specifications, schedules, construction costs and material estimates can all be completed by interfacing computer spreadsheet programs with CAD drawings. This is accomplished by assigning codes to building components and materials shown on each drawing.

Computers are also used in developing bid estimates based on square footage, material and labor cost formulas. For example, the *Dodge Estimator* described in Unit 78 contains a data base of all building material and labor costs for every zip code area of the United States. The computer generates bid estimates when this information is combined with specific information relating to the size and type of construction.

PREPARING FOR CAD

The following exercises provide opportunities to practice CAD procedure before using a CAD system. Plotting coordinates and completing data tables offers experience in using cartesian coordinates with paper and pencil before using a keyboard or stylus.

Exercises

1. **List the steps in the development of a CAD drawing in Fig. 27-48 using the graphics tablet for input.**
2. **Digitize the points in a data table for Fig. 27-48.**
3. **List the steps in the development of a CAD drawing shown in Fig. 27-49. Use the alphanumeric keyboard method for input.**
4. **Digitize each segment of Fig. 27-49 for a data table (roof, wall, door, two windows).**

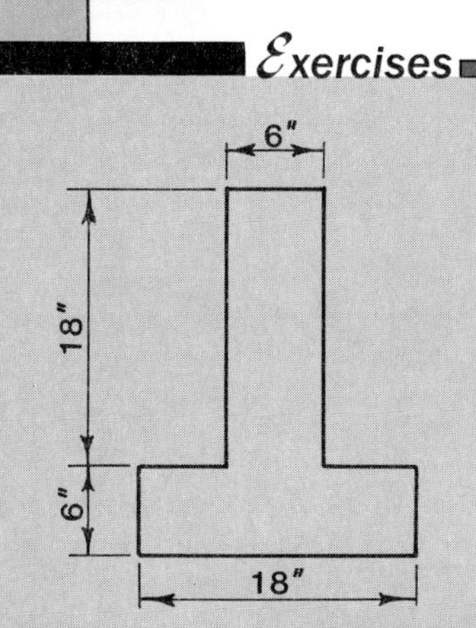

Fig. 27-48 Digitize points and draw this T-foundation detail.

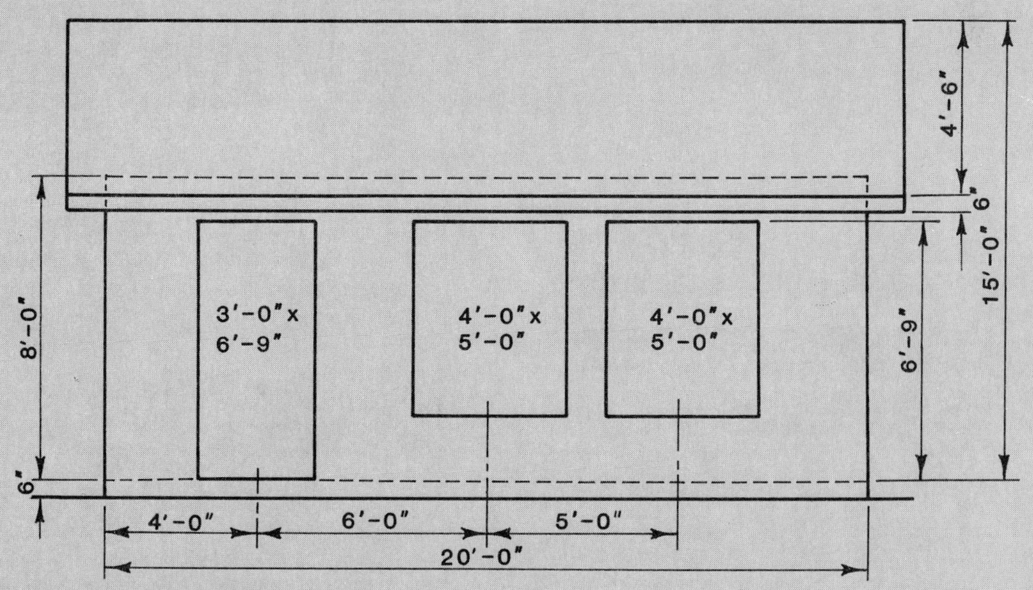

Fig. 27-49 Digitize points and draw this elevation.

5. **Complete the *x-y* coordinate points for the plot plan in Fig. 27-50.**
6. **Layout the *x-y* coordinate points in Fig. 27-51 on 1/4" grid paper.**
7. **Define the following terms: *software, hardware, digitizer, plotter, input, output, printer, printout, microcomputer, floppy disk, hard disk, disk drive, menu, stylus, graphics tablet, plotter, joystick, task, enter, key in, prompt, auxiliary menu, command, mode, monitor, hard copy.***
8. **Layout coordinate points on 1/4" grid system (scale: 1" = 1'-0") in Fig. 27-52.**

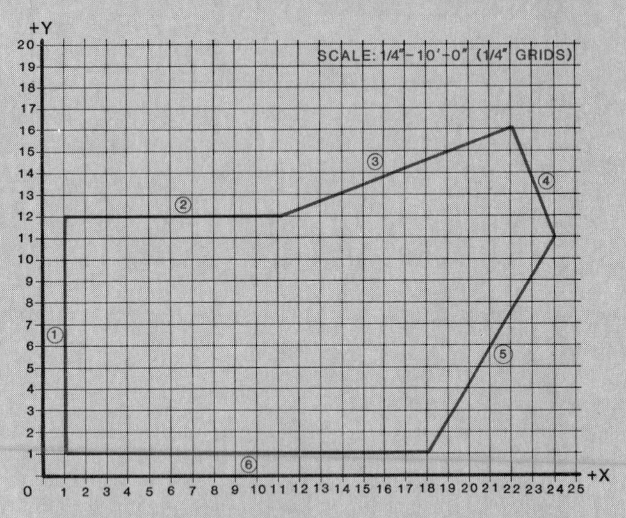

DATA TABLE					
LINE	X	Y	go to	X	Y
1	1	1	→	1	12
2	1	12	→		
3			→		
4			→		
5			→		
6			→		

Fig. 27-50 Complete the coordinate points for this plot plan.

DATA TABLE					
LINE	X	Y	go to	X	Y
1	3	2	→	3	7
2	2	6	→	8	13
3	8	13	→	13	6
4	12	7	→	12	2
5	12	2	→	3	2
6	6	2	→	6	6
7	6	6	→	9	6
8	9	6	→	9	2

Fig. 27-51 Layout coordinate points on a 1/4" grid system at a scale of 1/4" = 1'-0".

COORDINATE DATA TABLE					
LN	X	Y	go to	X	Y
1	22	7	→	9	7
2	9	7	→	7	5
3	7	5	→	7	2
4	7	2	→	2	2
5	2	2	→	2	9
6	2	9	→	9	22
			new start		
7	2	16	→	2	9
8	2	9	→	5	10
9	5	10	→	2	10
10	2	10	→	5	9
11	5	9	→	5	16

Fig. 27-52 Complete coordinate points on a 1/4" grid system (scale 1/4" = 1'-0").

UNIT 28

DRAFTING TIME SAVERS

Architectural drawings must frequently be prepared quickly because construction often begins immediately upon completion of the working drawings. Under these conditions, speed in the preparation of drawings is of utmost importance. For this reason, many timesaving devices are employed by architectural drafters. The purpose of these timesaving devices is to eliminate unnecessary time on the drawing board without sacrificing the quality of the drawing.

ARCHITECTURAL TEMPLATES

Templates are usually made of sheet plastic. But they are sometimes made of pieces of paper, cardboard, or metal. Openings in the template are shaped to represent the outline of various symbols and fixtures. A symbol or fixture is traced on the drawing by following the outline with a pencil or pen. This procedure eliminates the repetitious task of measuring and laying out the symbol each time it is to be used on the drawing. Remember the template scale must always be the same scale as the scale of the drawing.

Templates have openings that represent many different types of symbols and fixtures. A template is positioned to be used to outline a door symbol in Fig. 28-1. Four other general-purpose templates are shown in Fig. 28-2.

Many architectural templates are used to draw only one type of symbol. Special templates are available for doors, windows, landscape features, electrical symbols, plumbing symbols, furniture, structural steel, lettering, circle, and ellipse guides.

Each part of the symbol is drawn by using a different opening in the template. For example, in using the template shown in Fig. 28-3, the center part of the template is employed to outline the window. The horizontal lines are then added by using the openings at the right of the template as shown. The vertical lines are added by using the opening at the bottom of the template.

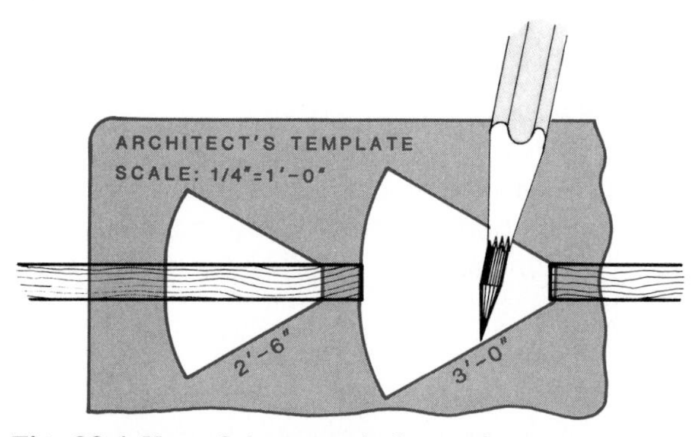

Fig. 28-1 Use of door-symbol template.

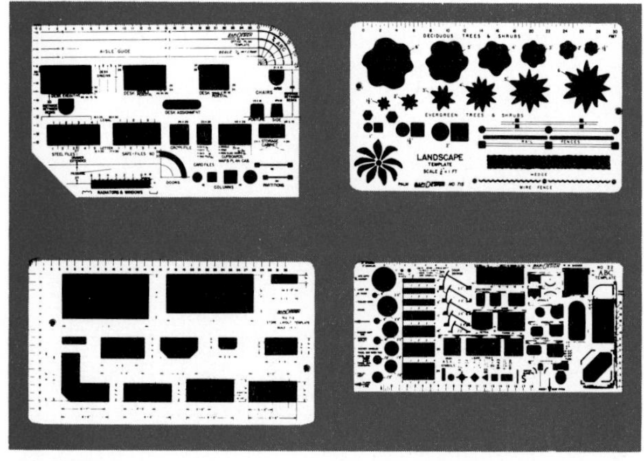

Fig. 28-2 General-purpose architectural templates. *(Rapidesign, Inc.)*

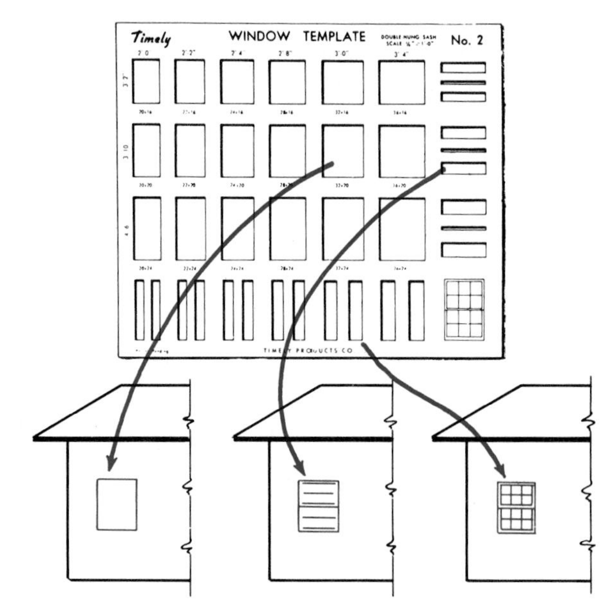

Fig. 28-3 Steps in using a window template. *(Timely Products Co.)*

When the major axis of a symbol is to be aligned with the lines of the drawing, it is necessary to use a T square, drafting machine, or parallel slide as a guide. This alignment is made by resting one true edge of the template against the blade of the T square, parallel slide, or drafting machine. This procedure is also necessary to ensure the alignment of symbols that are repeated in an aligned pattern.

OVERLAYS

An *overlay* is any sheet that is placed over the original drawing. The information placed on the overlay becomes part of the interpretation of the original drawing.

Most overlays are made by drawing on translucent material such as acetate, drafting film, or vellum. Overlays are used in the design process to add to or change features of the original drawing without marking the original drawing.

Overlays are also used to add features to a drawing that would normally complicate the original drawing. Lines that would become hidden and many other details can be made clear by preparing this information on an overlay. Figure 28-4 shows the use of an acetate overlay.

Overlays that adhere to the surface of the drawing save drawing-board time. Attaching a preprinted symbol or fixture symbol by this

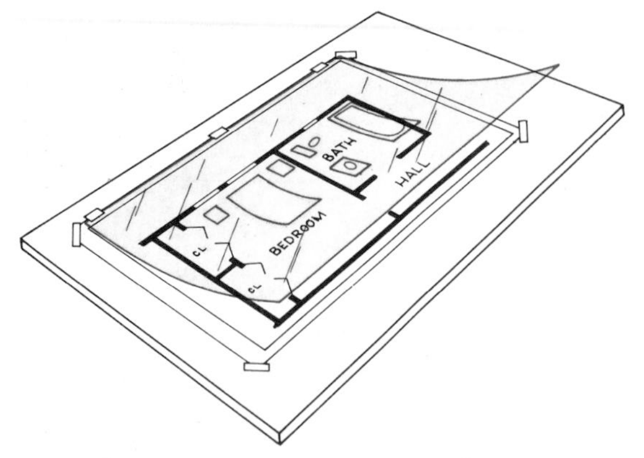

Fig. 28-4 The use of an acetate overlay.

method is considerably faster than drawing it.

In addition to fixture and symbol overlays, continuous material symbols are often used on architectural drawings. Figure 28-5 shows the use of a section-lining overlay. These overlays are self-adhering and can be cut to any desired size or shape.

UNDERLAYS

Underlays are drawings or parts of drawings that are placed under the original drawing and traced on the original.

Symbol Underlays

Many symbols and features of buildings are drawn more than once. The same style of door or window or the same type of tree or shrubbery may be drawn many times by the architectural drafter in the course of a day. It is a considerable waste of time to measure and lay out these features each time they are to be drawn. Therefore, many drafters prepare a series of underlays of the features repeated most often on their drawings. Figure 28-6 shows several underlays commonly used on architectural drawings. Underlays are commonly prepared for doors, windows, fireplaces, trees, walls, and stairs.

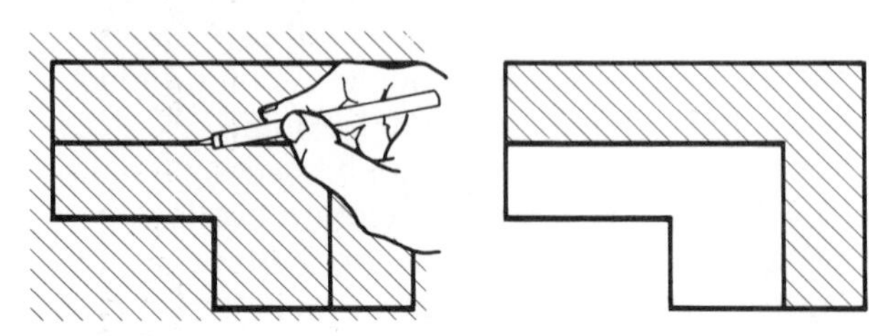

Fig. 28-5 A section-lining overlay.

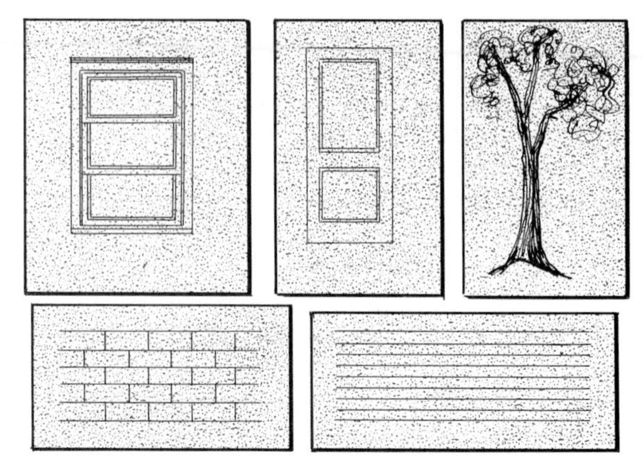

Fig. 28-6 Examples of common architectural underlays.

Lettering Underlays

Lettering guidelines are frequently prepared on underlays. When the guidelines are placed under the drawing, the drafter may trace the line from the original drawing, thus eliminating the measurement of each line. If the underlay remains under the drawing while it is being lettered, there is no need to draw guidelines on the drawing. The spacing of other lines, such as cross-hatching and brick-symbol lines, is also prepared on underlays.

Drawing paper preprinted with title blocks is a considerable time saver (Fig. 28-7). However, when printed title blocks are not available, the title-block underlay is often used to save valuable layout time and to ensure the correct spacing of lettering.

Use of Underlays

Underlays are master drawings. To be effective, they must be prepared to the correct scale and carefully aligned. The underlay is first positioned under the drawing and aligned (Fig. 28-8); then

Fig. 28-7 A title-block underlay.

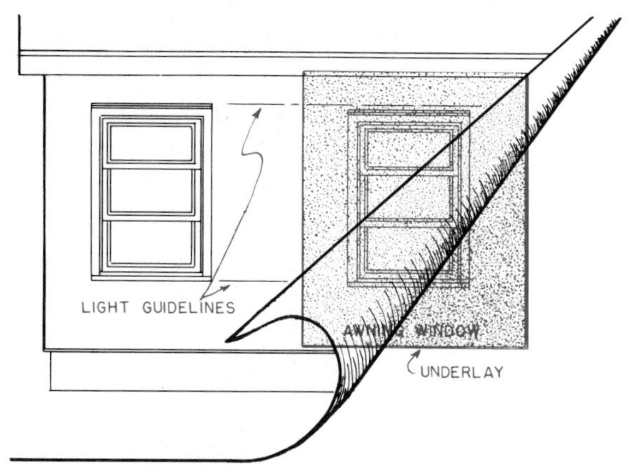

Fig. 28-8 The correct positioning of an underlay.

it is traced on the drawing. The underlay can now be removed or moved to a new location to trace the symbol or feature again if necessary. Architects use master underlays many times.

Underlays do not necessarily replace the use of instruments or scales in original design work. They are most effective when symbols are continually repeated. Figure 28-9 shows a comparison of the use of the scale, dividers, and underlay in laying out wall thicknesses. The use of the underlay in this case is only possible after the original wall dimensions have been established by the use of the scale.

GRIDS

Grid sheets are used under the tracing paper as underlays and are removed after the drawing is finished, or the drawing is prepared on nonreproducible grid paper. Nonreproducible grid paper does not reproduce when the original drawing is duplicated. Figure 28-10A and B shows an original drawing complete with nonreproducible grid lines and the print from this drawing without grid lines.

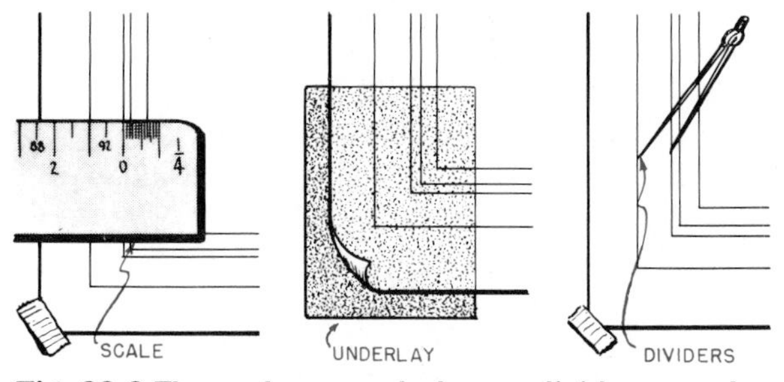

Fig. 28-9 The scale, an underlay, or dividers may be used to lay out wall thickness.

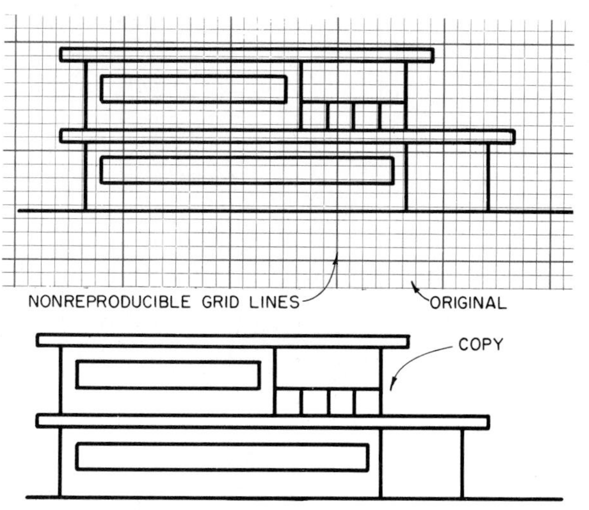

Fig. 28-10A An example of the use of nonreproducible grid paper.

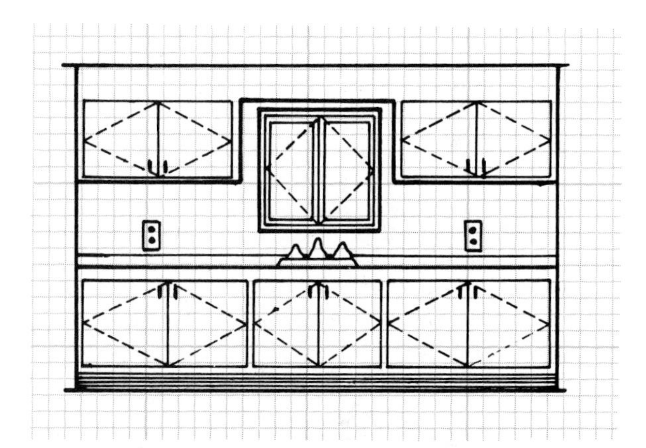

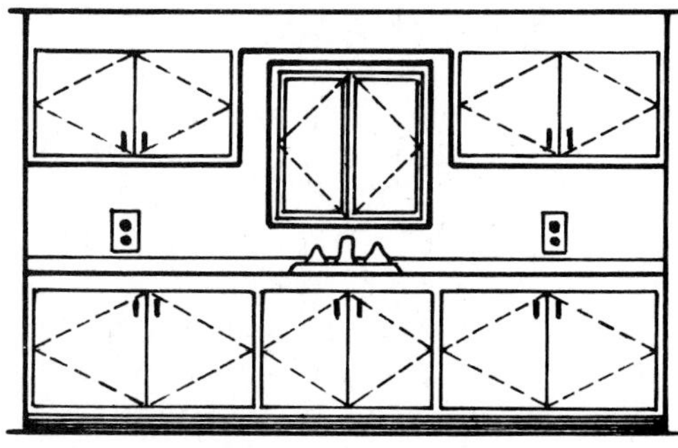

Fig. 28-10B Use of nonreproductible grid paper with interior detail.

Squared Paper

Squared (graph) paper is available in graduations of 4, 8, 16, and 32 squares per inch. Squared paper is also available in decimal-divided increments of 10, 20, and 30 or more squares per inch. Decimal-divided squared paper is used for the layout of survey and plot

Fig. 28-11 A drawing may be projected in many locations. *(Graphic Indicator Co.)*

plants. Metric graph paper is usually ruled in units of 2 or 5 mm.

Pictorial Grids

Grids prepared with isometric angles and preplotted to perspective vanishing points are used for pictorial illustrations. *Perspective* and *isometric* grid paper are available with many different angles of projection. An example of one perspective grid is shown in Fig. 28-11.

The use of the grid-pick function on a CAD system is an effective time-saver. When this mode is used, each input point or line is automatically snapped to the nearest intersection on the grid network.

TAPE

Many types of manufactured tape can be substituted for drawn lines and symbols on architectural drawings.

Pressure-Sensitive Tape

Tape with printed symbols and special lines is used to produce lines and symbols that otherwise would be difficult and time-consuming to construct. Figure 28-12 shows some of the various symbols and lines available in this kind of tape. A special roll-on applicator enables the drafter to draw lines by using tape, as shown in Fig. 28-13. This method is used extensively on

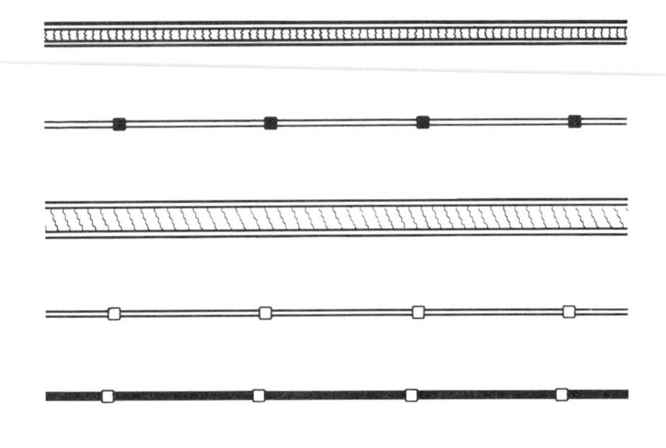

Fig. 28-12 Architectural symbol tape. *(Chart-Pak, Inc.)*

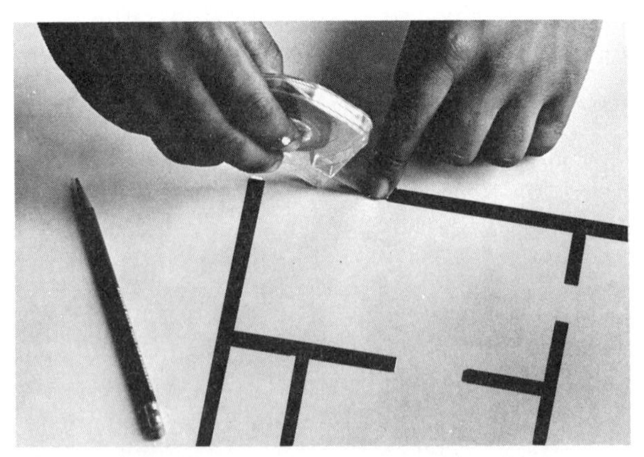

Fig. 28-13 Rolling on pressure-sensitive tape.

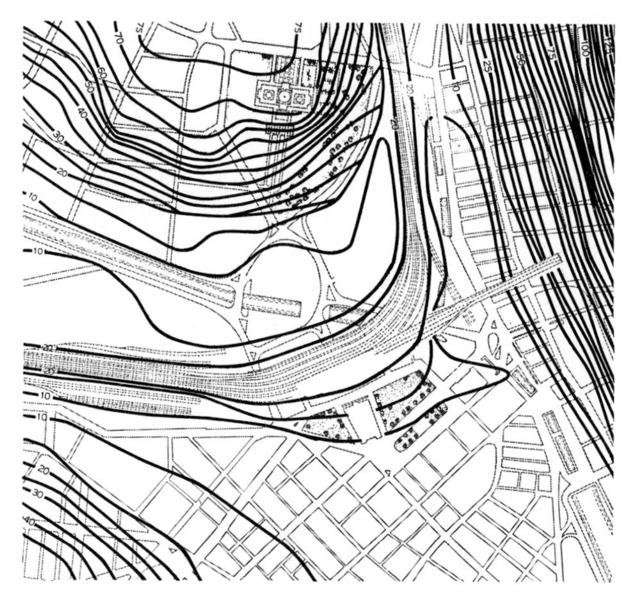

Fig. 28-14 Pressure-sensitive tape used on a map overlay. *(Chart-Pak, Inc.)*

overlays (Fig. 28-14). Figure 28-15 shows another application of rolled on tape.

Matte-Surface Tape

Temporary changes can be added to a drawing by drawing the symbol, note, or change on translucent matte-surface tape. If the drawing is changed, the tape can be removed and a new symbol added, or the symbol can be made permanent. The proposed closet wall in Fig. 28-16 was prepared on transparent tape. If the arrangement is unsatisfactory, the tape can be removed without destroying the drawing.

Drafting Tape

Drafting tape has other timesaving uses besides its use to attach the drawing to the drawing board. Strips of drafting tape help ensure the equal length of lines when ruling many close

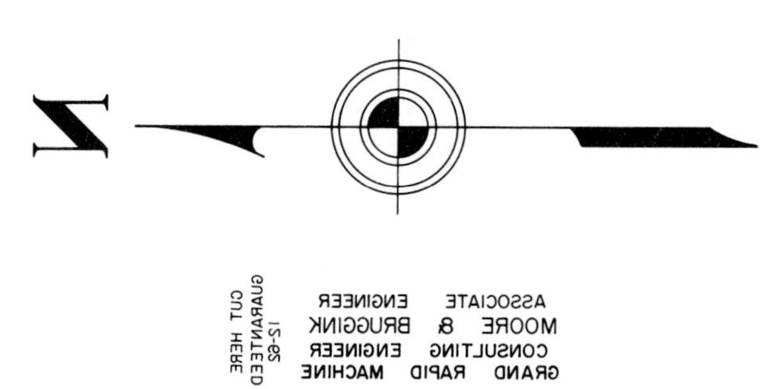

Fig. 28-15 Reverse uses of pressure-sensitive tape.

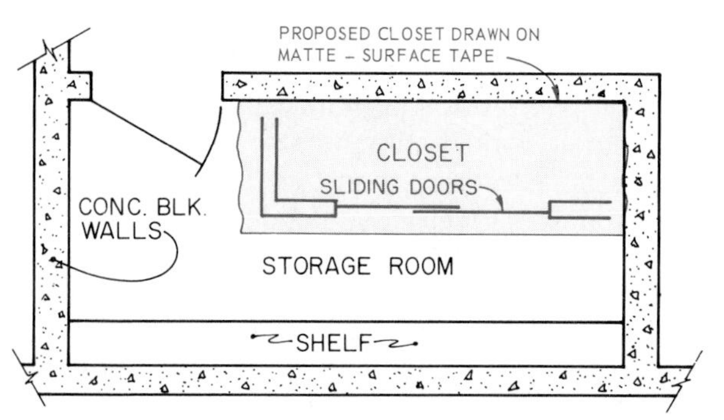

Fig. 28-16 A trial layout prepared on a matte-surface tape.

lines. Strips of tape are placed on the drawing to mask the areas not to be lined. The lines are then drawn on the paper and extended on the tape. When the tape is removed, the ends of the lines are even and sharp, as shown in Fig. 28-17. This procedure eliminates the careful starting and stopping of the pencil stroke with each line.

Since drafting tape will pull out some graphite, a piece of paper can be placed on drawings to perform the masking function for large areas. If a small area is to remain unlined, it is easier to line through the area and erase the small area, using an erasing shield as shown in Fig. 28-18.

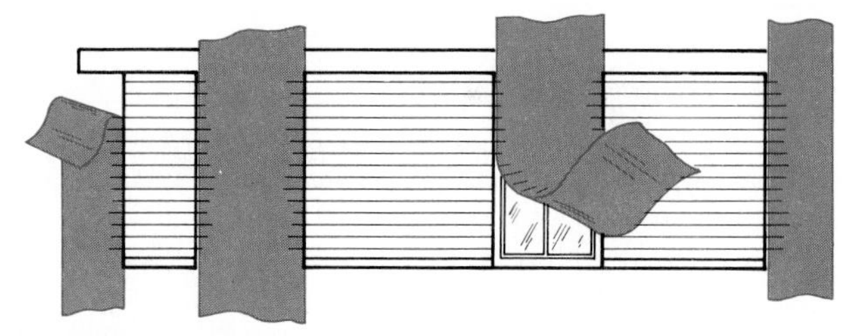

Fig. 28-17 The use of masking tape to save time drawing many similar lines.

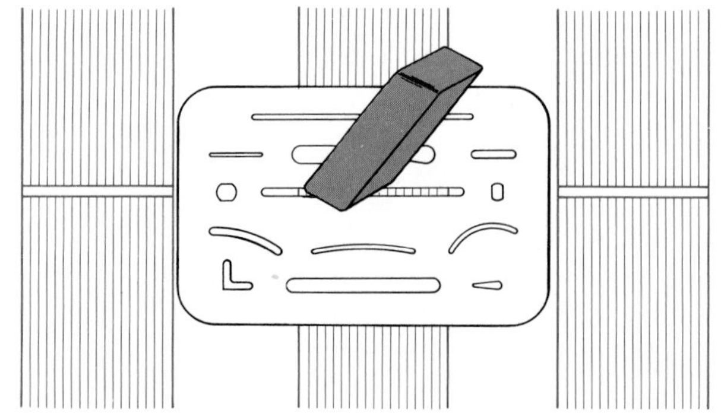

Fig. 28-18 Use of erasing shield to remove symbol lines.

ABBREVIATIONS

Stenographers use shorthand to speed and condense their work. Architects also use shorthand. Architects' shorthand consists of symbols and abbreviations. When a symbol does not describe an object completely, a word or phrase must be used. Words and phrases can occupy much space on a drawing. Abbreviations therefore should be used to minimize this space. Refer to Section 28 for a list of architectural abbreviations.

RUBBER STAMPS

For architectural symbols that are often repeated, the use of rubber stamps is effective and timesaving. Stamps can be used with any color ink, or stamps can be used in faint colors to provide an outline that can then be rendered with pencil or ink. Rubber stamps are used most often for symbols that do not require precise positioning on the drawing, such as landscape features, people, and cars. However, stamps may be used for furniture outlines and for labels. Figure 28-19 shows some common symbols used on rubber stamps. These images are also available as underlays or on pressure-sensitive material for attachment directly to a drawing.

Fig. 28-19 Examples of common rubber-stamp symbols.

OVERLAY DRAFTING

Overlay drafting is a method of aligning related drawings to ensure accuracy and eliminate much duplication of effort. First a base drawing is prepared. In architectural work the base drawing is usually a floor plan. Then related drawings are prepared directly over the base and aligned either with pins through holes or by index marks on each drawing. Overlay drafting is sometimes called *pin-bar drawing* or *layering*.

When large numbers of drawings are involved the overlay drawings are prepared on clear polyester film. Aligning the specialized plans in a set of drawings as shown in 28-20 insures all structural features such as walls, and columns align. It also eliminates the potential problems of overlapping mechanical, electrical and piping facilities on the same drawing base. When this aligning is done on a CAD system, the layering task is used. The use of 255 layers is possible on most complete CAD systems.

Overlay drawings also simplify the interpretation of drawings by subcontractors. For example if clear film is used for the plumbing plan then only the plumbing plan placed on the floor plan is duplicated. The plumbing contractor then receives a plumbing floor plan without the clutter of HVAC electrical or structural details. Although the contractor may be given a complete set for referral.

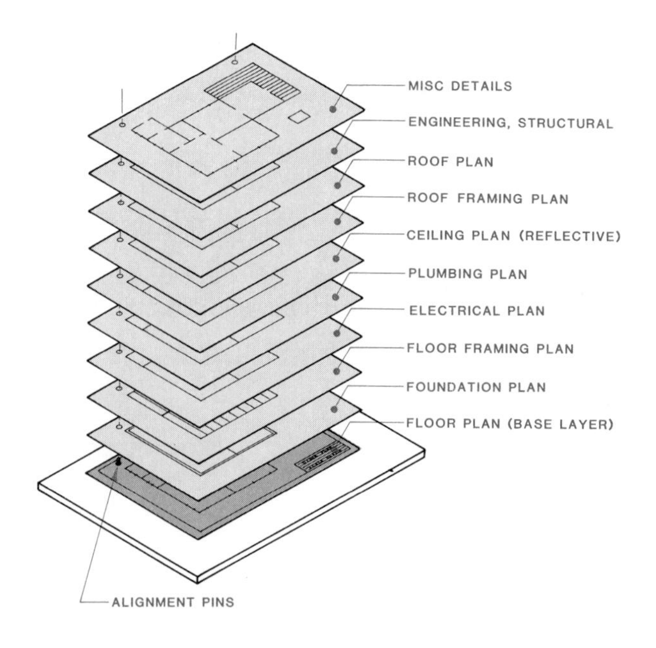

Fig. 28-20 A layered set of drawings.

BURNISHING PLATES

Burnishing plates are embossed sheets with raised areas representing an outline of a symbol or texture lines. The plates are placed under a drawing; then a soft pencil is rubbed over the surface of the drawing. This creates lines on the drawing over the raised portions of the plate. The use of burnishing plates allows the drafter to create consistent texture lines throughout a series of drawings with a minimum use of time.

REPRODUCTION

Often a section of a drawing needs to be changed, or a design element needs to be repeated on many drawings. The entire drawing need not be redrawn, nor must the design element be drawn repeatedly on each drawing. The section to be redrawn or repeated can be drawn once, attached to the drawing, and then the entire drawing can be reproduced through a variety of reproduction processes.

CAD ZOOM

The use of the CAD zoom task speeds drawing by allowing a plan to be repeatedly enlarged (zoom-in) to facilitate drawing details, then reduced (zoom-out) to enable the drafter to see the entire drawing.

Exercises

1. Prepare an underlay for a fireplace.
2. Prepare a lettering-guide underlay for $\frac{1}{8}$", $\frac{1}{4}$", and $\frac{3}{16}$" letters.
3. Add landscape features to the plan shown in Fig. 33-3, using a landscape-plan template.
4. Define the following terms: *template, pin graphics, overlay, underlay, graph paper, perspective grid, matte-surface tape, burnishing plates.*

UNIT 29

ARCHITECTURAL LETTERING

Architectural lettering differs greatly from lettering used on engineering drawings because most architectural drawings are shown to a client. Architectural drafters usually develop a stylized type of lettering that helps them work quickly yet produces an accurate and attractive drawing.

PURPOSE

Figure 29-1 shows a plan without any lettering. This plan does not communicate a complete description of the size and function of the various components. All labels, notes, dimensions, and descriptions must be legibly lettered on architectural drawings if they are to function as an effective means of graphic communication.

Legible, well-formed letters and numerals do more for a drawing than merely aid in communi-

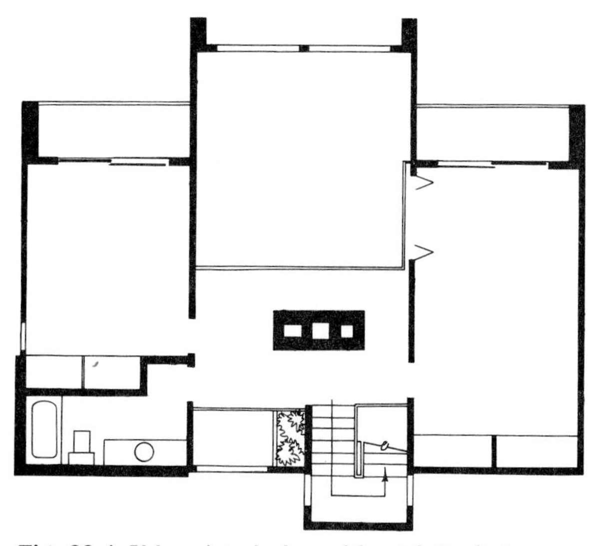

Fig. 29-1 Abbreviated plan without lettering.

cation. Effective lettering helps give the drawing a finished and professional look. Poor lettering is the mark of an amateur. The plan shown in Fig. 29-2 is more easily interpreted and appears more professional because lettering was used. In fact, Fig. 29-1 would be almost impossible to understand without the labeling of area functions and minimum dimensions.

STYLES

Because architectural designs are somewhat personalized, many lettering styles have been developed by various architects. Nevertheless, these personalized styles are all based on the *American National Standard Alphabet* shown in Fig. 29-3.

Although the American Standard Alphabet and numerals are recommended for architectural drawings, many deviations are used. No style should be used that is difficult to read or may be misinterpreted. Errors of this type can be very costly. This is especially true for numerals used for dimensioning.

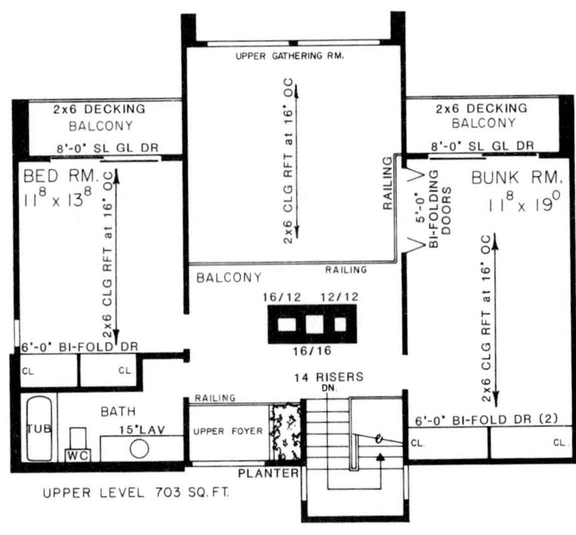

Fig. 29-2 The same plan as in Fig. 29-1 with lettering added.

ABCDEFGHIJKLMNOP
QRSTUVWXYZ &
1234567890

VERTICAL LETTERS

Fig. 29-3 The American National Standard Alphabet.

RULES FOR ARCHITECTURAL LETTERING

Much practice is necessary to develop the skills necessary to letter effectively. Although architectural lettering styles may be very different, all professional drafters follow certain basic rules of lettering. If you follow these rules, you will develop accuracy, consistency, and speed in lettering your drawings.

1. Always use guidelines in lettering. Notice what a difference guidelines make in the lettering shown in Fig. 29-4.

2. Choose one style of lettering, and practice the formation of the letters of that style until you master it. Figure 29-5 compares the effect of using a consistent style with that of using an inconsistent style. Each letter in the inconsistent style may be correct, but the effect is undesirable.

3. Make letters bold and distinctive. Avoid a delicate, fine touch.

4. Make each line quickly from the beginning to the end of the stroke. See the difference between letters drawn quickly and those drawn slowly in Fig. 29-6.

5. Practice with larger letters (about $\frac{1}{4}$", or 6 mm), and gradually reduce the size until you can letter effectively at $\frac{1}{16}$", or 2 mm.

6. Practice spacing by lettering words and sentences, not alphabets. Figure 29-7 shows the effect of uniform and even spacing of letters.

7. Form the habit of lettering whenever possible—as you take notes, address envelopes, or write your name.

USE GUIDELINES FOR GREATER
ACCURACY IN LETTERING.

LETTERING WITHOUT GUIDELINES
LOOKS LIKE THIS.

Fig. 29-4 Always use guidelines when lettering.

CONSISTENT STYLE
CONSISTENT STYLE
INCONSISTENT STYLE

Fig. 29-5 Always use a consistent lettering style.

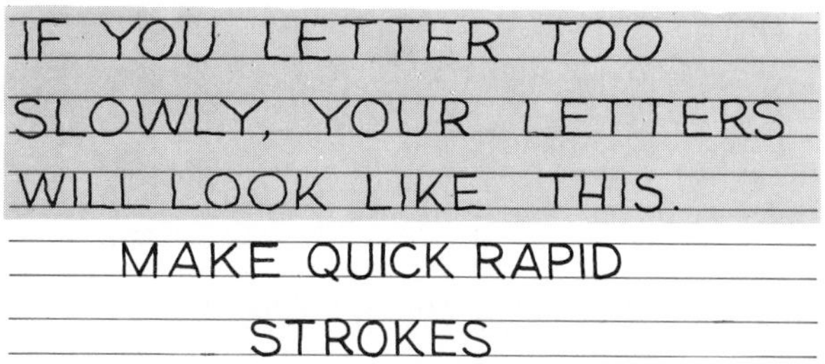

IF YOU LETTER TOO
SLOWLY, YOUR LETTERS
WILL LOOK LIKE THIS.
MAKE QUICK RAPID
STROKES

Fig. 29-6 Make each letter stroke quickly.

UNIFORM SPACING
UNEVEN SPACING

Fig. 29-7 Uniform spacing of letters is important.

8. Practice only the capital alphabet. Lowercase letters are rarely used in architectural work.

9. Do not try to develop speed at first. Make each stroke quickly, but take your time between letters and between strokes until you have mastered each letter. Then gradually increase your speed. You will soon be able to letter almost as fast as you can write script.

10. If your lettering has a tendency to slant in one direction or the other, practice making a series of vertical and horizontal lines, as shown in Fig. 29-8.

11. If slant lettering is desired, practice slanting the horizontal strokes at approximately 68°. The problem with most slant lettering is that it is difficult to maintain the same degree of slant continually. The tendency is for more and more variations of slant to creep into the style.

12. Letter the drawing last to avoid smudges and overlapping with other areas of the drawing. This procedure will enable you to space out your lettering and to avoid lettering through important drawing details.

13. Use soft pencil, preferably an HB or F. A soft pencil will glide and is more easily controlled than a hard pencil.

14. Numerals used in architectural drawing should be adapted to the style, just as the alphabet is adapted. Fractions also should be made consistent with the style. Fractions are 1⅔ times the height of the whole number. The numerator and the denominator of a fraction are each ⅔ of the height of the whole number as shown in Fig. 29-9A. Notice also that in the expanded style, the fraction is slashed to conserve vertical space. The fraction takes the same amount of space as the whole number (Fig. 29-9B).

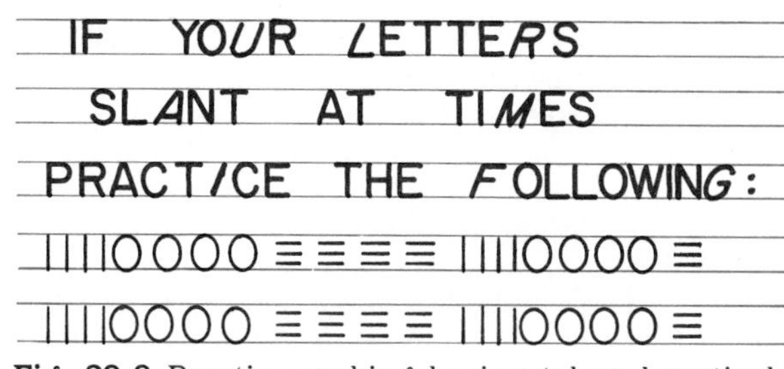

Fig. 29-8 Practice making horizontal and vertical lines.

15. The size of the lettering should be related to the importance of the labeling (Fig. 29-10).

16. Specialized lettering templates can also be used.

TYPESET LETTERING

Typeset letters, although more consistent in size and style than hand lettering, are much slower to apply. Some typeset letters are applied one letter at a time by the *pressure sensitive method.* This type of application is used primarily for major labels on architectural drawings. Words, sentences, and dimensions can be set in type on opaque paper or transparent tape and positioned in place on the drawing with rubber cement or wax-backed paper. In architectural work, transparent tape labels can be positioned without covering the lines that pass close to or through the label.

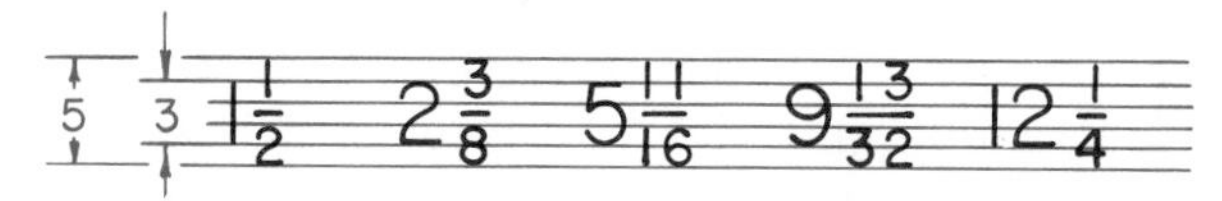

Fig. 29-9A Proper fraction proportions.

6 3/4" 7 9/16" 5'-6 1/2" 3 5/8"

Fig. 29-9B Alternate fraction style used to conserve vertical space.

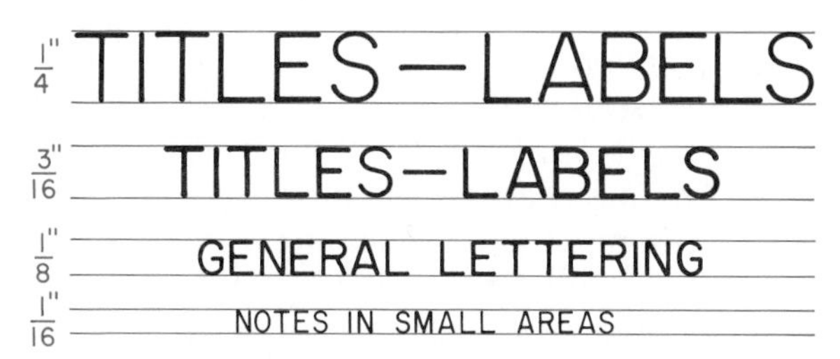

Fig. 29-10 Lettering height should relate to the importance of the label.

A drawing can be lettered on a CAD system using the typewriter and text tasks. Menu items include the option of selecting type font, slope angle, line weight (pen size), width, height, and spacing of characters.

Exercises

1. **Letter your name and your complete address, using both expanded and condensed lettering styles.**
2. **Letter the preceding rules for lettering, using any style you choose.**
3. **Define the following terms: *American National Standard Alphabet, condensed style, expanded style, consistent style, vertical strokes, horizontal strokes, lettering pencil, slant lettering.***

UNIT 30

ARCHITECTURAL DRAWING TECHNIQUES

In addition to the precise technical line work used on floor plans and elevations, other line techniques are used to create realism in architectural drawings. Some of the line techniques used are variations in the distance between lines or dots, variations in the width of lines, blending of lines, and use of gray tones or solid black areas (Fig. 30-1). These techniques are used to show materials, texture, contrast between areas, or light and shadow patterns. Some common combinations of line patterns are shown in Fig. 30-2. The use of these techniques to show texture on a perspective drawing is shown in Fig. 30-3.

Varying the interval between lines drawn with pencil or pen can indicate texture, light, and density pattern, as shown in Fig. 30-4. When less precise line identification is desired, the use of *wash-drawing* (water-color) techniques is effective. The drawing shown in Fig. 30-5 is a wash drawing. The drawing shown in Fig. 30-6 and 30-7 is rendered using a combination of wash techniques over a line drawing.

Floor plans and elevation drawings are prepared primarily for the builder. These drawings must be accurately scaled and dimensioned. However, some floor plans and elevation drawings are rendered to provide the prospective customer with a better idea of the final appearance of the building. These plans have no dimensions but include items such as plantings, floor surfaces, and material textures not usually found on floor plans or elevations used for construction purposes. Screens, lines, and patterns (Fig. 30-8) can also be applied by the use of appliqués.

Sketching is a communications medium used constantly by designers. In fact, most designers begin with a rough sketch. Sketches are used to record dimensions and the placement of existing objects and features prior to beginning a final

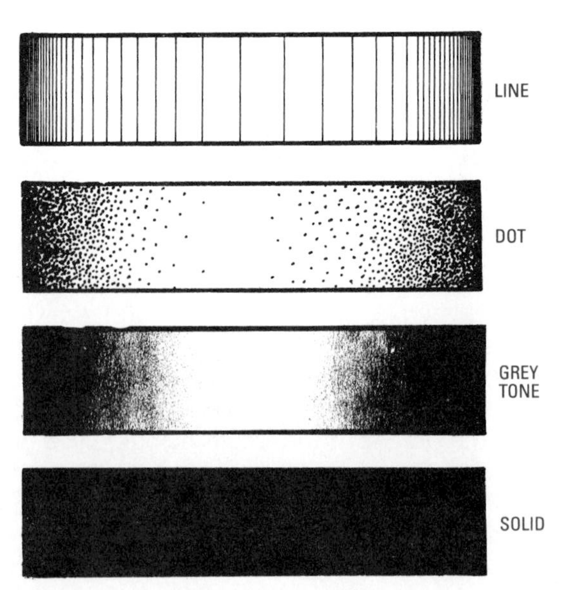

Fig. 30-1 Some types of line techniques.

Fig. 30-2 Examples of line-pattern combinations.

Fig. 30-3 Methods of illustrating textures.

Fig. 30-4 Line-rendering techniques.

Fig. 30-5 Wash-drawing technique used for landscape rendering.

Fig. 30-6 Combination of line and wash techniques. *(Home Planners, Inc.)*

Fig. 30-7 A presentation drawing.

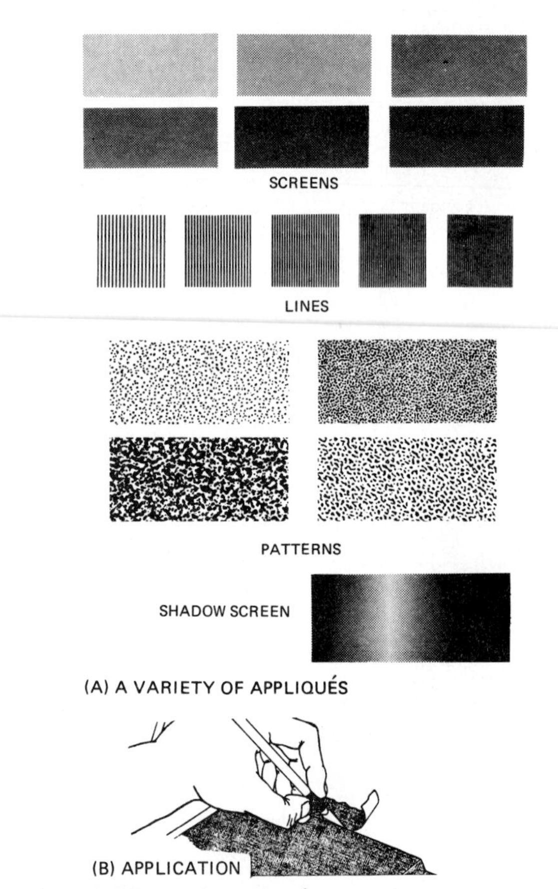

Fig. 30-8 Use of appliqués.

drawing. Sketches are used to show alternative possibilities concerning the approach to the design problem.

Sketching on graph paper as shown in Fig. 30-9 helps increase speed and accuracy. Sketches also help record ideas off the job and help the designer remember unique features about a structure or site so that the actual design activity can take place in a different location.

Fig. 30-9 Method of holding pencil for sketching.

Fig. 30-10B Sketching diagonal lines.

In sketching, use a soft pencil. Hold the pencil comfortably. Draw the pencil, do not push it. Position the paper so your hand can move freely. Sketch in short, rapid strokes. Long, continuous lines tend to bend on the arc line from elbow to fingers. Figure 30-10A shows how to sketch horizontal lines, and figure 30-10B shows how to sketch diagonal lines. To be effective, a sketch must be readable by another person without additional explanations.

The accuracy, effectiveness, and appearance of the finished drawing depend largely on the selection of the correct pencil and the pointing of that pencil. Figure 30-11 shows thickness and density of various types of lines used on architectural drawings. Figure 23-2 shows the degrees of hardness of drawing pencils ranging from 9H, extremely hard, to 7B, extremely soft. Pencils in the hard range are used for layout work. Pencils in the soft range are used for sketching and rendering. Basic architectural drawings of the type prepared by designers are usually drawn with pencils in a medium range. When the pencil is too soft, it will produce a line that smudges. If the pencil is too hard, a groove will be left in the paper and the line will be very difficult to erase. Since the density of the line also depends on the hardness of the lead, it is necessary to select the correct degree of hardness for pencil tracings. Different degrees of hardness react differently, depending on the type of drawing medium used.

Sharpen the drawing pencil by exposing approximately ½″ of lead. Then form the lead into a sharp conical point on a sanding pad. Slowly rotate the pencil while you rub it over the sanding pad. A mechanical lead pointer or smooth single-cut file may also be used. Mechanical pencils with leads of 0.3, 0.5, 0.7 and 0.9 mm are very thin and do not need to be sharpened.

Fig. 30-10A Sketching horizontal lines.

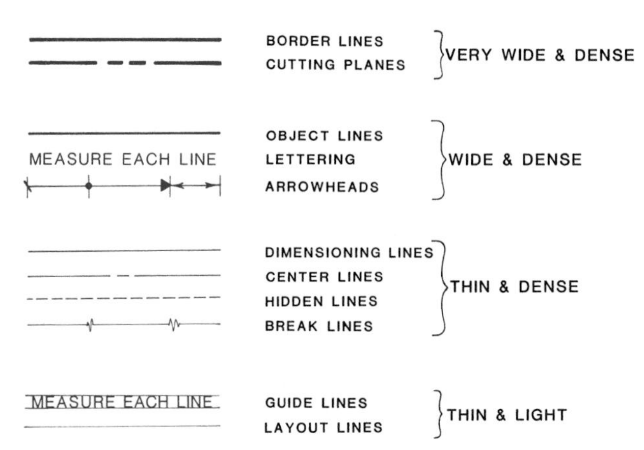

Fig. 30-11 Architectural line widths and densities.

Most designers and illustrators develop their own distinct drawing style and technique in preparing interpretive architectural drawings as shown in Figs. 30-3 through 30-7. However, working drawings are prepared for builders' use. Therefore drawings such as floor plans, elevations, details and sections must be prepared with more precision and control to assure a greater degree of accuracy and consistency of line weight and densities. To provide the geometric control needed for precise drawings, the practice shown in Fig. 30-12 should be followed in using a pencil with any straight edge such as a T square, triangle, drafting machine blade, or parallel slide.

Absorbent powder (pounce) dispersed from a porus bag or shaker can is also used to avoid smudging. The fine powder absorbs extra grains of smudge-causing graphite which can be swept away with a drafting brush. Instruments ride on the top of the powder, thus reducing smudging. Brush away the powder before darkening the lines or the pencil will skip, leaving spaces in the line.

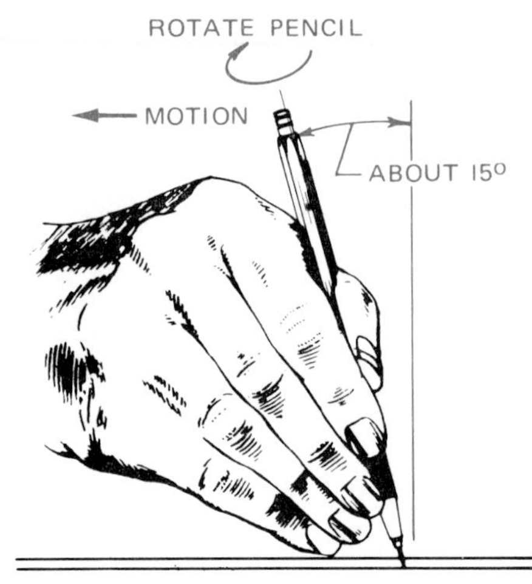

Fig. 30-12 Drawing pencil lines with a straight edge.

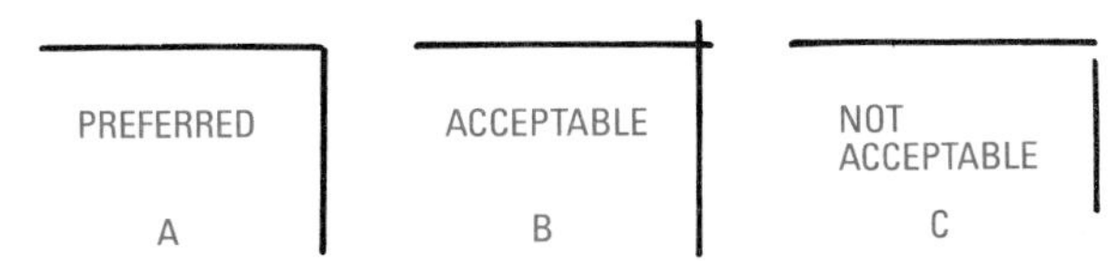

Fig. 30-13 Architectural line intersections.

In Fig. 30-13, drawing A shows a standard corner intersection used in both mechanical and architectural drawings. Drawing B shows an intersection standard that is acceptable only on some architectural drawings. Care must be taken in using this type of corner on detail drawings since the overlapping lines may intersect another material part or dimension and create confusion. Drawing C is not acceptable since no corner exists for interpretation or measurement.

Exercises

1. **Sketch the pictorial drawing shown in Fig. 32-8C.**
2. **Define these terms: *line, tone, wash drawing, presentation drawing.***
3. **Sketch a pictorial drawing of a building in your neighborhood using line rendering techniques.**
4. **Sketch and add surface textures to the drawing shown in Fig. 30-14.**

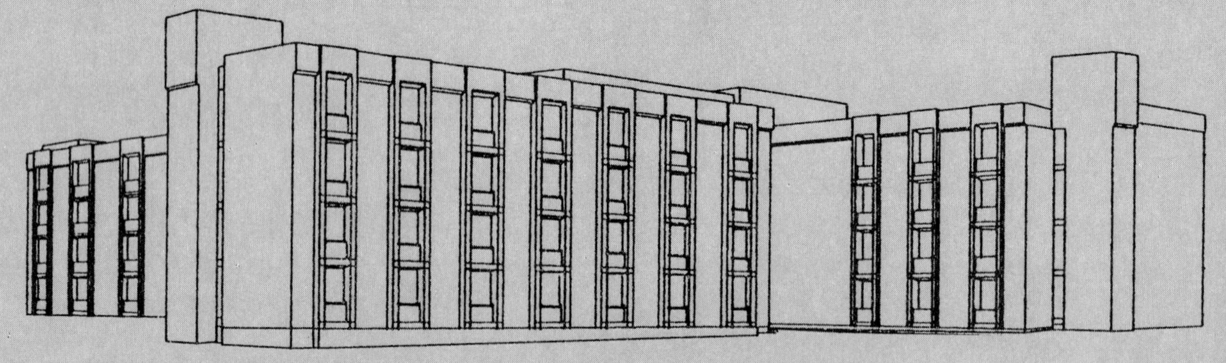

Fig. 30-14 Add surface treatment to this drawing.

SECTION 7

Drawing Floor Plans

The most commonly used architectural drawing is the floor plan. The floor plan is a drawing of the outline and partitions of a building as you would see them if the building were cut horizontally about 4′ (1.2 m) above the floor line. The floor plan provides more specific information about the design of the building than any other plan. To design a floor plan that will be accurate and functional, the designer must determine what facilities will be included in the various areas of the building. These various areas must be combined into an integrated plan. Only then can a final floor plan be prepared that includes a description and sizes of all the materials and areas contained in the design. The floor plan is used as a base for the projection of other drawings.

UNIT 31

FUNCTIONAL ROOM PLANNING

Not long ago, the outside of most homes was designed before the inside. A basic square, rectangle, or series of rectangles was established to a convenient overall size and then rooms were fitted into these forms. Today, the inside of most homes is designed before the outside, and the outside design is determined by the size and relationship of the inside areas. This is known as *designing from the inside out.* In designing from the inside out, the architect evolves the plan from basic room requirements. By learning about the living habits and tastes of the occupants, the architect determines what facilities are required for each room. In this way, the furniture, fixtures, and amount of space that will be appropriate for the activities are provided for.

Designing in this manner may seem to conflict with the bubble diagram approach described in Unit 2. But this is not the case. The bubble diagram approach (Fig. 31-1A) is an effective

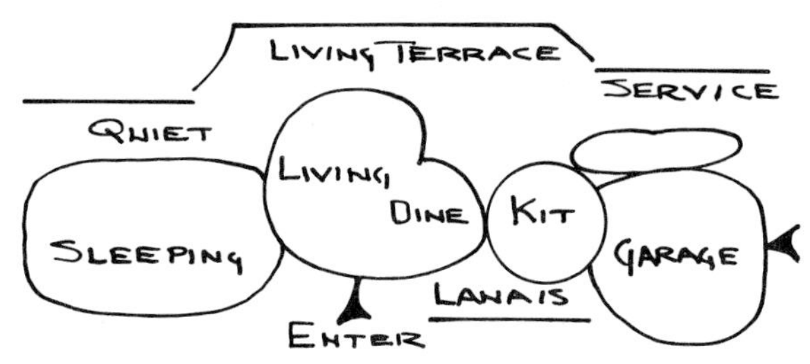

Fig. 31-1A Schematic floor plan design.

method of establishing the overall design concept. Floor plan designing from the inside out involves the finalization of the plan from the general concept. But during the entire process the wants and needs of the occupant must be kept clearly in mind. Providing *all* of the listed needs and as many of the listed wants as possible is the key to functional floor plan design. Figure 31-1B shows a list of vacation home wants and needs. Figure 31-1C shows a plan for providing all eight of the needs and four of the wants.

SEQUENCE OF DESIGN

When designing from the inside out, the home planner first determines what furniture and fixtures are needed. Next, the amount and size of furniture must be determined. The style selected will greatly affect the dimensions of the furniture. After the furniture dimensions are estab-

Needs	Wants
1. Kitchen with breakfast area	9. Fireplace
2. Full bath	10. Study
3. Master bedroom bath	11. Entry area
4. Half bath main level	12. Family room
5. 3 bedrooms	13. Fourth bedroom
6. Living area with adjacent dining room	14. Two car garage
7. One car garage	15. Hot tub
8. Two story house	16. Pool

Fig. 31-1B Needs and wants list.

lished, furniture templates can be made and arranged in functional patterns. Room sizes can then be established by drawing a perimeter around the furniture placements.

When the room sizes are determined, rooms can be combined into areas, and areas into the total floor plan. Finally, the outside is designed by projecting the elevations from the floor plan.

FURNITURE

Furniture styles vary greatly in size and proportion. Sizes of furniture therefore cannot be decided on until the style is chosen. The furniture style should be consistent with the style of architecture. Furniture styles should also be consistent within an area. Although some subtle mixing can occur, great care must be taken to achieve a harmonious result.

Selection

Furniture should be selected according to the needs of the occupants (Fig. 31-2). A piano should be provided for someone interested in music. A large amount of bookcase space must be provided for the avid reader. The artist, drafter, or engineer may require drafting furniture in the den or study, A good starting point in room planning is to list the uses to be made of each room. Then, make a list of furniture needed for each of these activities. From these requirements, a rather comprehensive list of needed

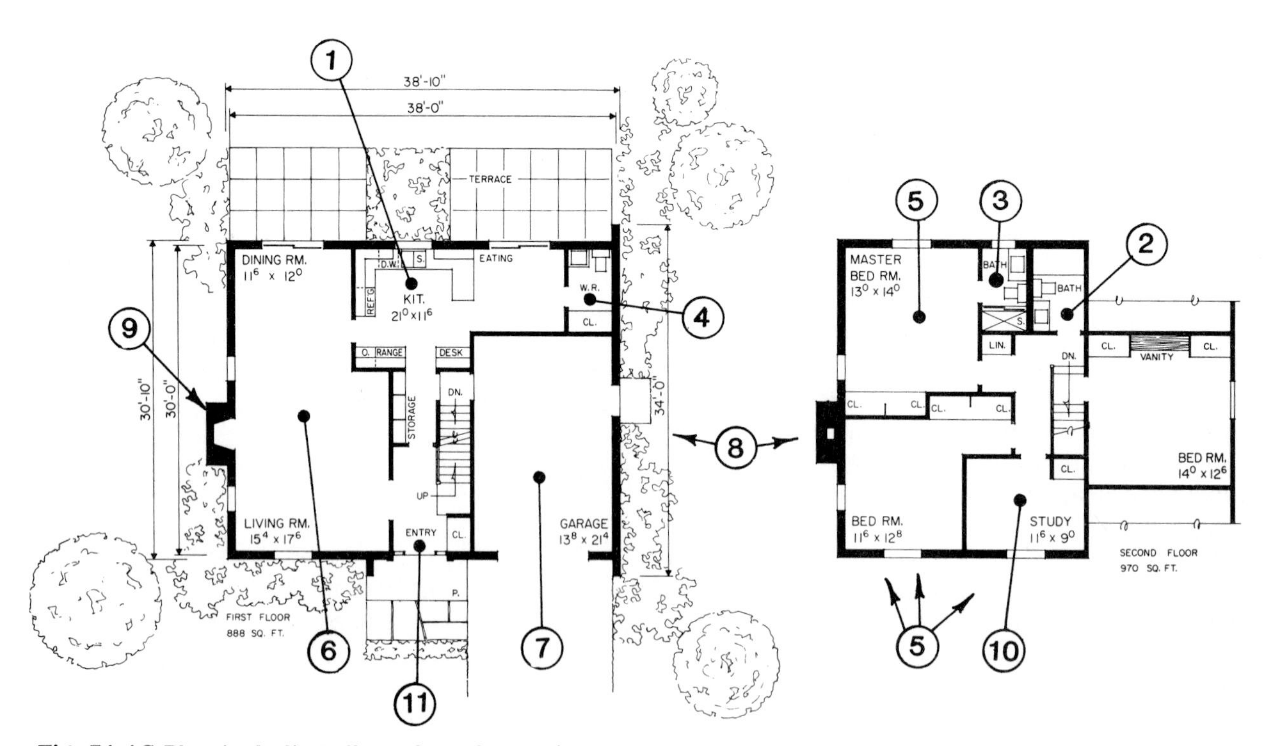

Fig. 31-1C Plan including all needs and most important wants.

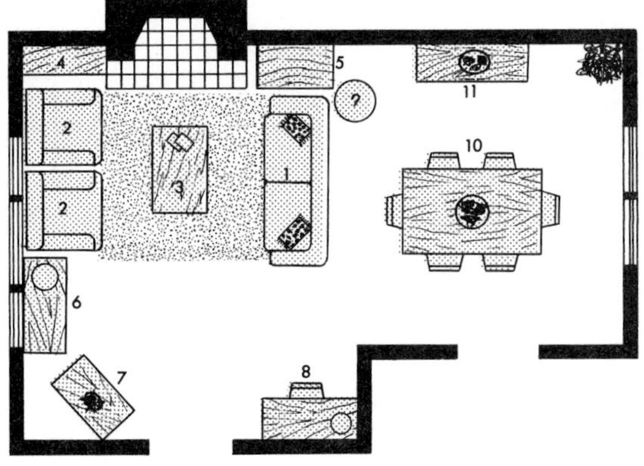

Fig. 31-2 Living needs determine the amount of furniture.

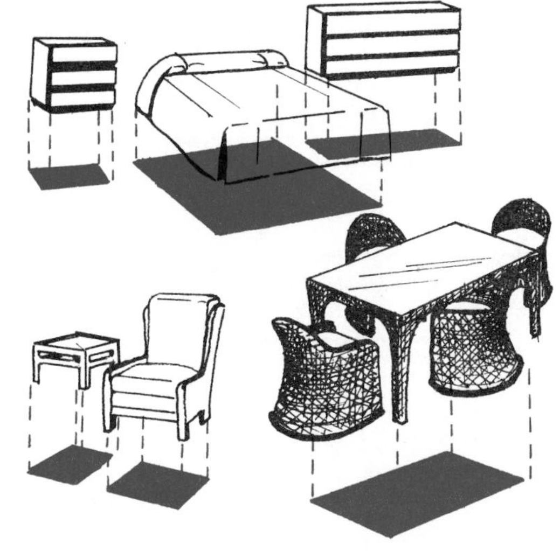

Fig. 31-3 Templates represent the width and length of each piece of furniture.

furniture can be compiled. When the exact style is determined, the width and length of each piece of furniture can also be added to the list, as shown below for a living room.

1 couch 34″ x 100″ (864 x 2540 mm)
2 armchairs 30″ x 36″ (762 x 914 mm)
1 chaise 28″ x 60″ (711 x 1524 mm)
1 TV 26″ x 24″ (660 x 610 mm)
1 stereo system 24″ x 56″ (610 x 1422 mm)
1 bookcase 15″ x 48″ (381 x 1219 mm)
1 floor lamp 6″ x 14″ (152 x 356 mm)
1 coffee table 18″ x 52″ (457 x 1321 mm)
2 end tables 14″ x 30″ (356 x 762 mm)
1 baby grand piano 60″ x 80″ (1524 x 2032 mm)

Similar lists should be prepared for the dining room, kitchen, bedrooms, nursery, bath, and all other rooms where furniture is required.

Furniture Templates

Arranging and rearranging furniture in a room is heavy work. It is simpler to arrange furniture by the use of templates (Fig. 31-3). *Furniture templates* are thin pieces of paper, cardboard, plastic, or metal that represent the width and length of pieces of furniture. They are used to determine exactly how much floor space each piece of furniture will occupy. One template is made for each piece of furniture on the furniture list.

Templates are always prepared to the scale that will be used in the final drawing of the house. The scale most frequently used on floor plans is 1/4″ = 1′-0″. Scales of 3/16″ = 1′-0″ and 1/8″ = 1′-0″ are sometimes used.

Wall-hung furniture, or any projection from furniture, even though it does not touch the floor, should be included as a template because the floor space under this furniture is not usable for any other purpose. Templates show only the width and length of furniture and the floor space covered (Fig. 31-4A and B). Figure 31-4C shows common furniture sizes that can be used as a guide in constructing furniture templates.

ROOM ARRANGEMENTS

Furniture templates are placed in the arrangement that will best fit the living pattern anticipated for the room. Space must be allowed for free flow of traffic and for opening and closing doors, drawers, and windows.

Determining Room Dimensions

After a suitable furniture arrangement has been established, the room dimensions can be determined by drawing an outline around the furniture, as shown in Fig. 31-5. *Room templates* are made by cutting around the outline of the room. Figure 31-6 shows some typical room templates constructed by cutting around furniture template arrangements.

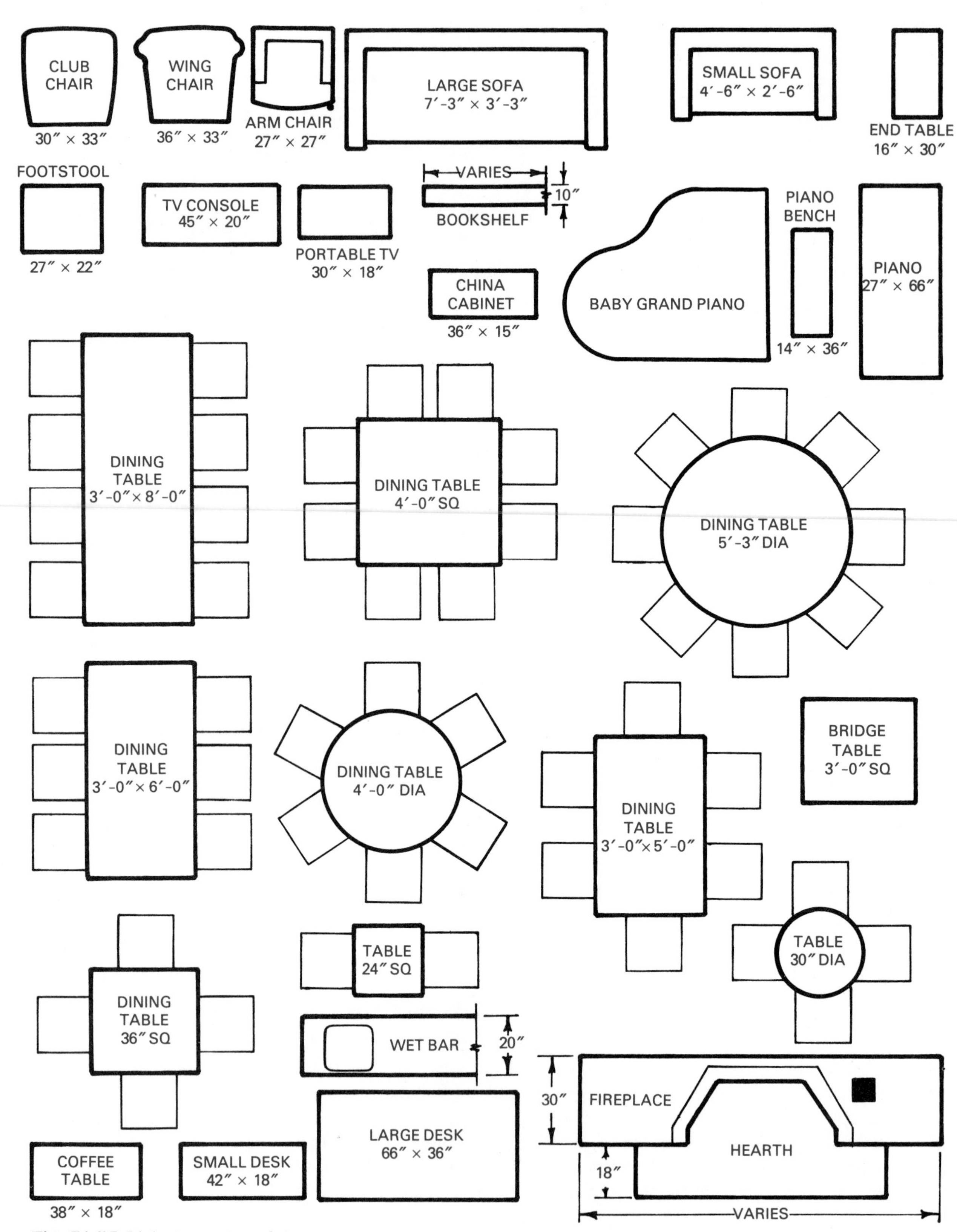

Fig. 31-4A Living-area templates.

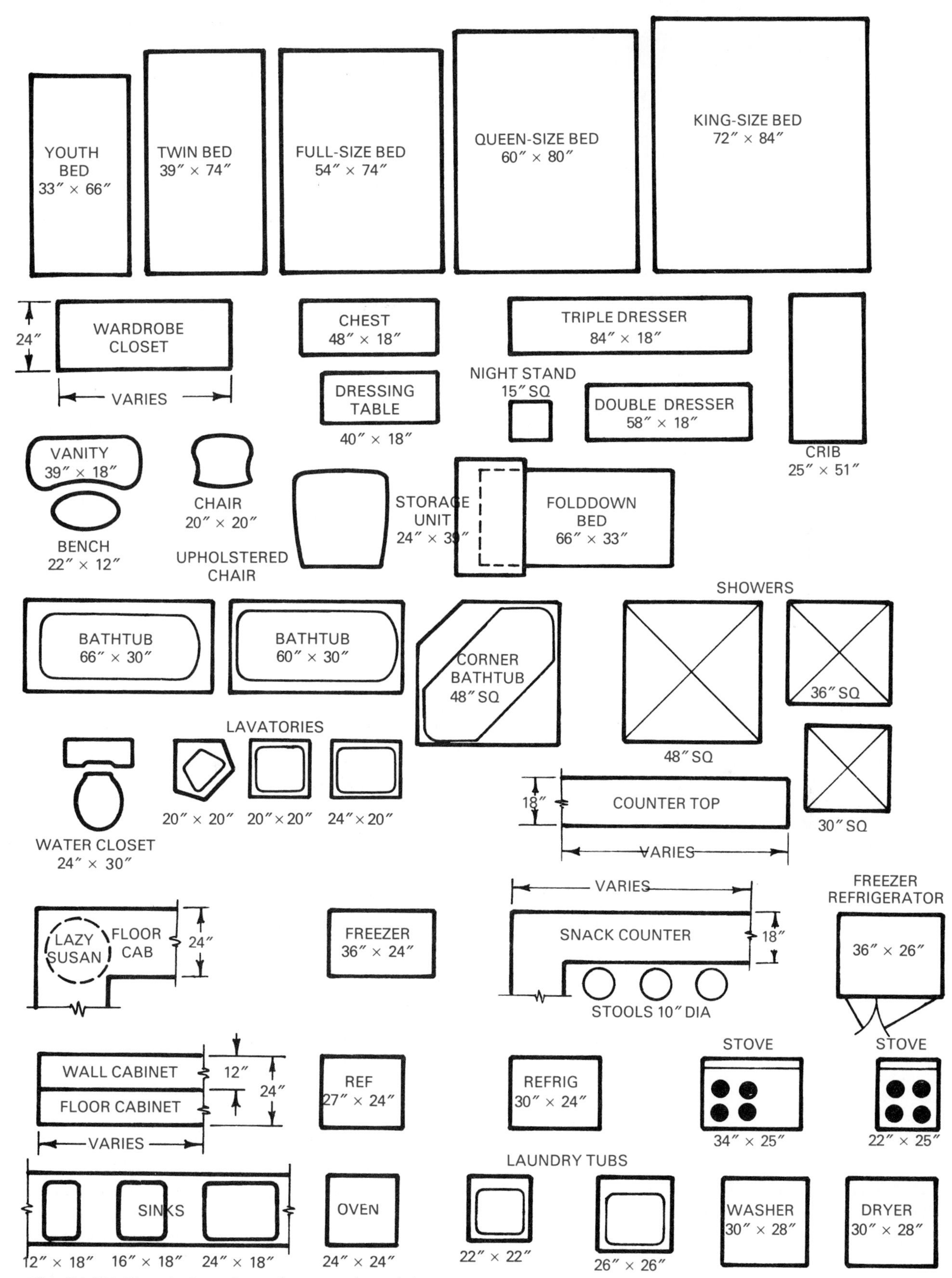

Fig. 31-4B Sleeping- and service-area templates.

Item	Length, in (mm)	Width, in (mm)	Height, in (mm)
COUCH	72(1829)	30(762)	30(762)
	84(2134)	30(762)	30(762)
	96(2438)	30(762)	30(762)
LOUNGE	28(711)	32(813)	29(737)
	34(864)	36(914)	37(940)
COFFEE TABLE	36(914)	20(508)	17(432)
	48(1219)	20(508)	17(432)
	54(1372)	20(508)	17(432)
DESK	50(1270)	21(533)	29(737)
	60(1524)	30(762)	29(737)
	72(1829)	36(914)	29(737)
STEREO CONSOLE	36(914)	16(406)	26(660)
	48(1219)	17(432)	26(660)
	62(1575)	17(432)	27(660)
END TABLE	22(559)	28(711)	21(533)
	26(660)	20(508)	21(533)
	28(711)	28(711)	20(508)
TV CONSOLE	38(965)	17(432)	29(737)
	40(1016)	18(457)	30(762)
	48(1219)	19(483)	30(762)
SHELF MODULES	18(457)	10(254)	60(1524)
	24(610)	10(254)	60(1524)
	36(914)	10(254)	60(1524)
	48(1219)	10(254)	60(1524)
DINING TABLE	48(1219)	30(762)	29(737)
	60(1524)	36(914)	29(737)
	72(1829)	42(1067)	28(711)
BUFFET	36(914)	16(406)	31(787)
	48(1219)	16(406)	31(787)
	52(1321)	18(457)	31(787)
DINING CHAIRS	20(508)	17(432)	36(914)
	22(559)	19(483)	29(737)
	24(610)	21(533)	31(787)

Item	Diameter, in (mm)	Height, in (mm)
DINING TABLE (ROUND)	36(914)	28(711)
	42(1067)	28(711)
	48(1219)	28(711)

Fig. 31-4C Common furniture sizes.

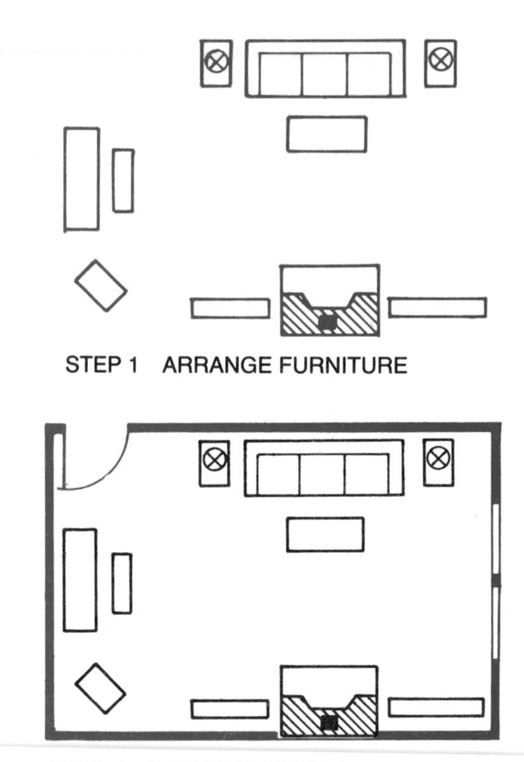

Fig. 31-5 The preferred method of determining room dimensions.

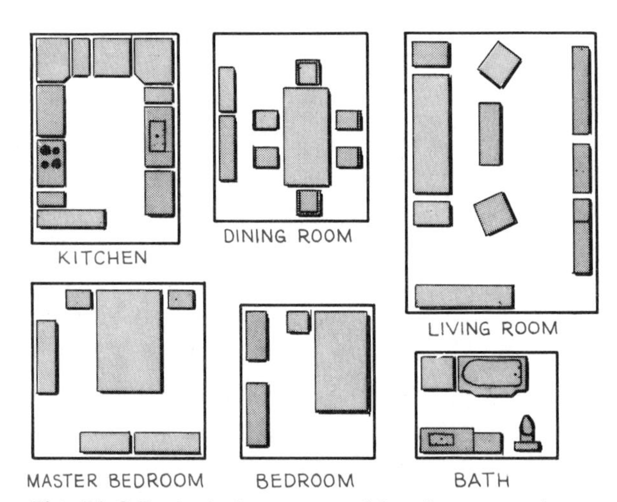

Fig. 31-6 Typical placement of furniture templates.

Common Room Sizes

Determining what room sizes are desirable is only one aspect of room planning. Since the cost of the home is largely determined by the size and number of rooms, room sizes must be adjusted to conform to the acceptable price range. Figure 31-7 shows sizes for each room in large, medium, and small dwellings. These dimensions represent only average widths and lengths. Even where no financial restriction exists, room sizes are limited by the functional requirements of the room. A room can become too large to be functional for the purpose intended.

	Living room, ft²/m²	Dining room, ft²/m²	Kitchen, ft²/m²	Bedrooms, ft²/m²	Bath, ft²/m²
SMALL HOME	200/18.6	155/14.4	110/10.2	140/13.0	40/3.7
AVERAGE HOME	250/23.2	175/16.2	135/12.5	170/15.8	70/6.5
LARGE HOME	300/27.8	195/18.1	165/15.3	190/17.7	100/9.3

Fig. 31-7 Small, average, and large room sizes.

Checking Methods

It is sometimes difficult to visualize the exact amount of real space that will be occupied by furniture or that should be allowed for traffic through a given room. One device used to give a point of reference is a template of a human figure, as shown in Fig. 31-8. With this template you can imagine yourself moving through the room to check the appropriateness of furniture placement and the adequacy of traffic allowances.

The experienced architect and home planner does not always go through the procedure of cutting out furniture templates and arranging them into patterns to arrive at room sizes. But the architect uses templates frequently to recheck designs. Until you are completely familiar with furniture dimensions and the sizes of building materials, the use of the procedures outlined in this unit is recommended.

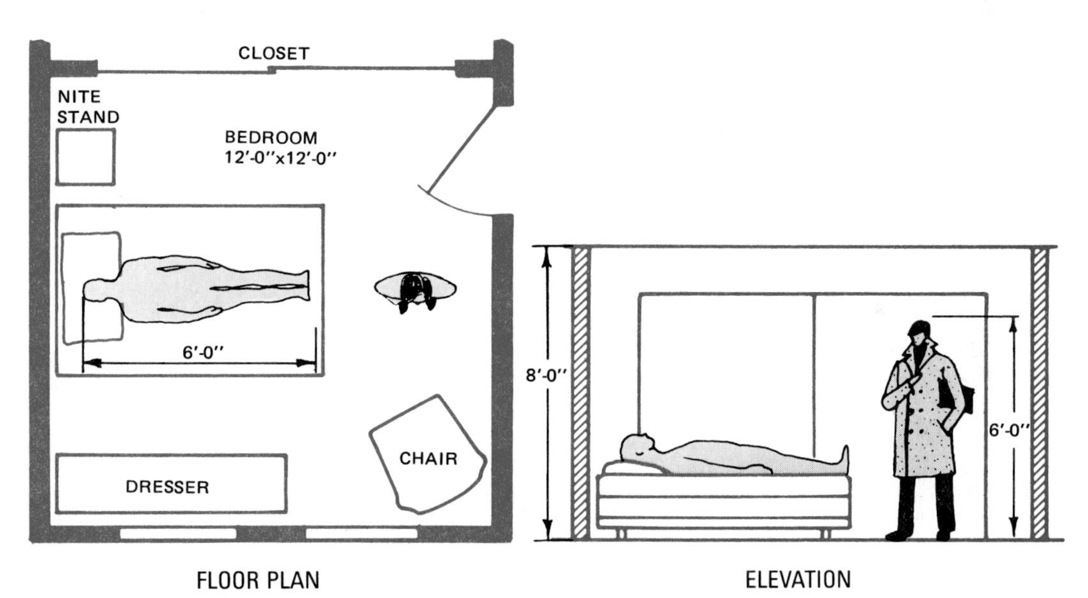

Fig. 31-8 Comparative size of a room with the size of its inhabitants.

Exercises

1. Rearrange the following steps in their proper order in room planning:

a. Make list of furniture needed.
b. Choose furniture style.
c. Choose home style.
d. Determine living habits.
e. Make furniture templates.
f. Determine room dimensions.
g. Arrange furniture templates.
h. Determine sizes of furniture.

2. **Define your living needs and activities. List each piece of furniture needed for each room in order to fulfill these needs.**
3. **Make a list of furniture you would need for a home you might design. The list should include the number of pieces and size (width and length) of each piece of furniture.**
4. **Make a furniture template (¼″ = 1′-0″) for each piece of furniture you would include in a home of your design.**
5. **Define the following terms: *furniture template, furniture dimensions, room dimensions, furniture style.***

UNIT 32

FLOOR-PLAN DESIGN

The architect or designer develops and records ideas through preliminary sketches that are later transformed into final working drawings. These ideas are directly translated into sketches that may approximate the final design. The architect knows through experience how large each room or area must be made to perform its particular function. He or she can mentally manipulate the relationships of areas and record design ideas through the use of sketches. This skill is attained after much experience.

LEARNING DESIGN PROCEDURES

Until you have gained experience in designing floor plans, you may have difficulty in creating floor plans by the same methods professionals use. In the beginning, you should rely on more tangible methods of designing.

The information in this unit outlines the procedures you should use in developing floor-plan designs. These procedures represent real activities that relate to the mental activities of a professional designer. Figure 32-1A shows the use of a bubble diagram as used in Unit 2. In this application only the interior of the structure is involved. Remember the inside and the outside

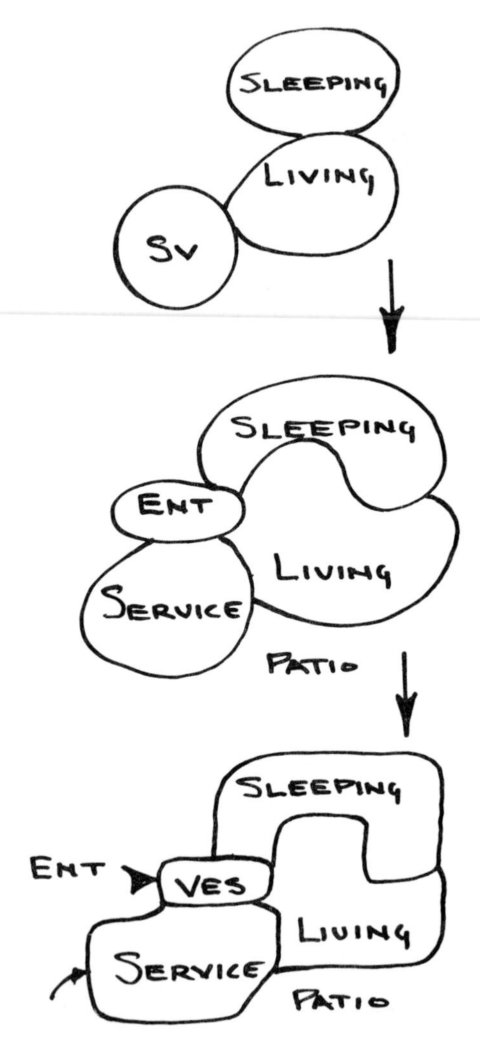

Fig. 32-1A Schematic (bubble) diagram of plan shown in Fig. 32-11.

designs must be correlated to function together. Once the general arrangement and relationship of areas of rooms is established with these diagrams, the design sequences shown in Fig.

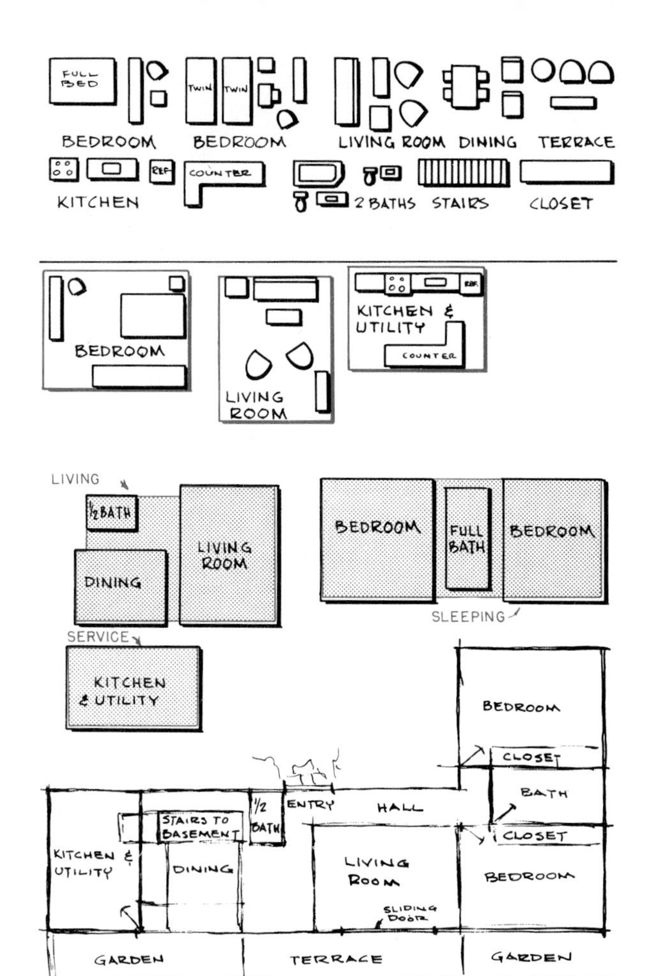

Fig. 32-1B Preliminary rough sketches of floor plans are prepared from template layouts of the living, service, and sleeping areas.

32-1B can be used. You will notice that the sequence includes the development of room templates through the use of furniture templates, as described in Unit 31. These room templates become the building blocks used in floor-plan designing. The room templates are arranged, rearranged, and moved into various patterns until the most desirable plan is achieved. Think of these room templates as pieces of a jigsaw puzzle that are manipulated to produce the final picture. Figure 32-2 shows a typical set of room templates prepared for use in floor-plan design.

PLANNING WITH ROOM TEMPLATES

A residence or any other building is not a series of separate rooms, but a combination of many activities areas. In the design of floor plans, room templates are first divided into area classifications, as shown in Fig. 32-3. The templates are then arranged in the most desirable plan for each area. Next, the area arrangements are combined in one plan (Fig. 32-4). The position of each room is then sketched and revised to produce a functional plan.

As you combine areas, you will need to rearrange and readjust the position of individual rooms. At this time, you should consider the traffic pattern, compass direction, street location, relationship to landscape features and all other orientation features that affect the design. Space must be allowed for stairways and halls. Figure 32-5 shows some common allowances for

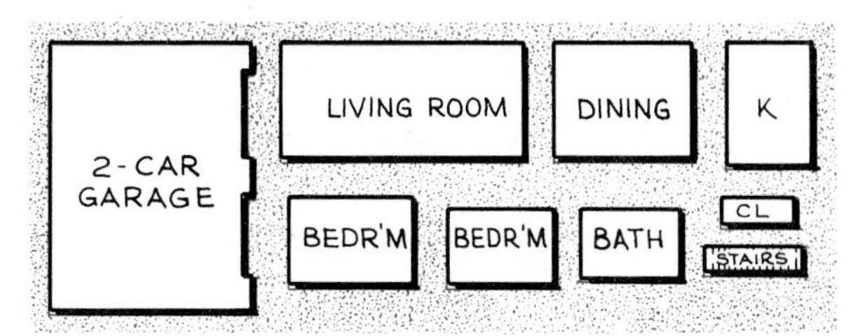

Fig. 32-2 Room templates prepared for floor-plan design use.

Fig. 32-3 Room templates divided into areas.

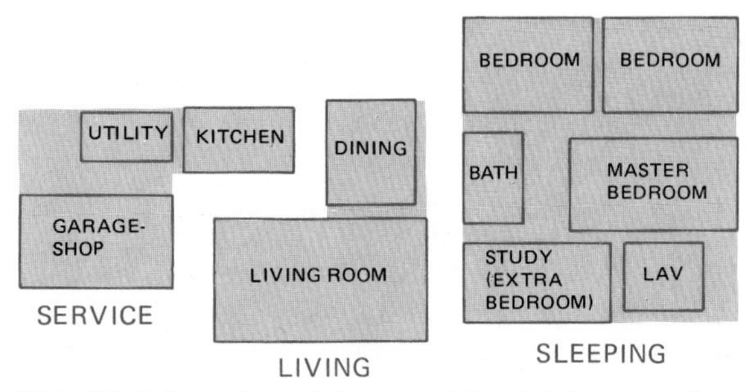

Fig. 32-4 Area templates combined into one plan.

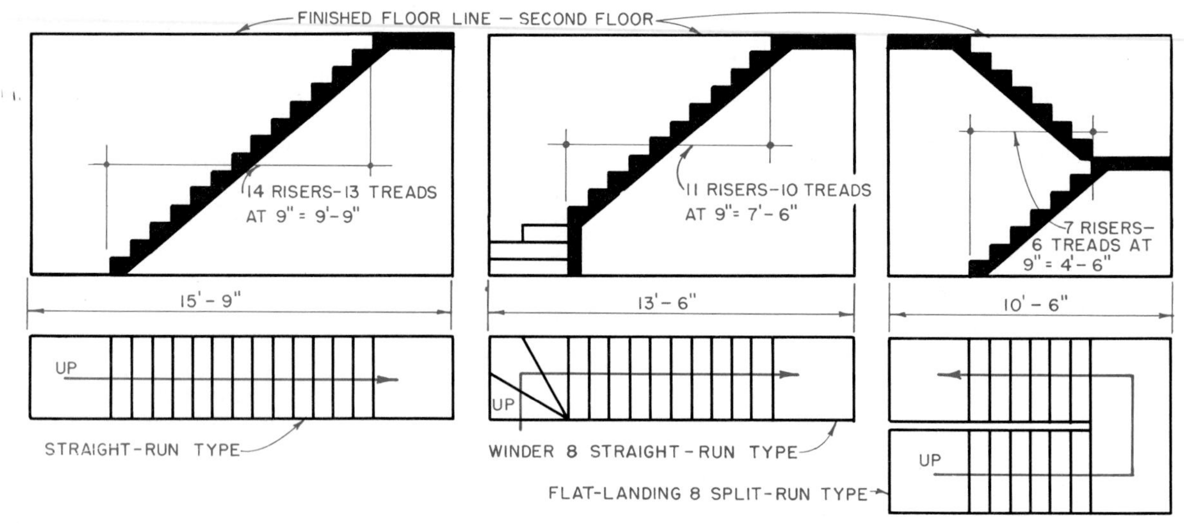

Fig. 32-5 Space must be allowed for halls and stairways.

stairwells on the floor plan. Unless closets have been incorporated in the room template, adequate space must also be provided for storage spaces.

Rooms and facilities such as the recreation room, laundry, workshop, and heating equipment must also be considered. The designer must be sure that the floor plan provides sufficient space for these facilities.

Templates can be created on a CAD system using the line as arc functions. Templates can then be arranged in many positions using the move-copy-delete tasks. Clusters of templates can be combined into groups and moved as one unit until the desired position is found.

FLOOR-PLAN SKETCHING

Preparing a layout for a room template is only a preliminary step in designing the floor plan. The template layout shows only the desirable size, proportion, and relationship of each room to the entire plan.

Preliminary Sketching

Template layouts such as the one shown in Fig. 32-4 usually contain many irregularities and awkward corners because room dimensions were established before the overall plan was completed. These offsets and indentations can be smoothed out by increasing the dimensions of some rooms and changing slightly the arrangement of others. These alterations are usually made by sketching the template layout. At this time, features such as fireplaces, closets, and divider walls can be added where appropriate.

Think of this first floor-plan sketch as only the beginning. Many sketches are usually necessary before the designer achieves an acceptable floor plan. In successive sketches the design should be refined further. Costly and unattractive offsets and indentations should be eliminated. Modular sizes can be established that will facilitate the maximum use of standard building materials and furnishings. The exact positions and sizes of doors, windows, closets, and halls can be determined.

Refinement of the design is done by resketching until a satisfactory sketch is reached. Except for very minor changes, it is always better to make a series of sketches than to erase and change the original sketch. Many designers use tracing paper to trace the acceptable parts of the design and then add design improvements on the new sheet. This procedure also provides the designer with a record of the total design process. Early sketches sometimes contain solutions to problems that develop later in the final design. Final sketches can still be improved as shown in Fig. 32-6.

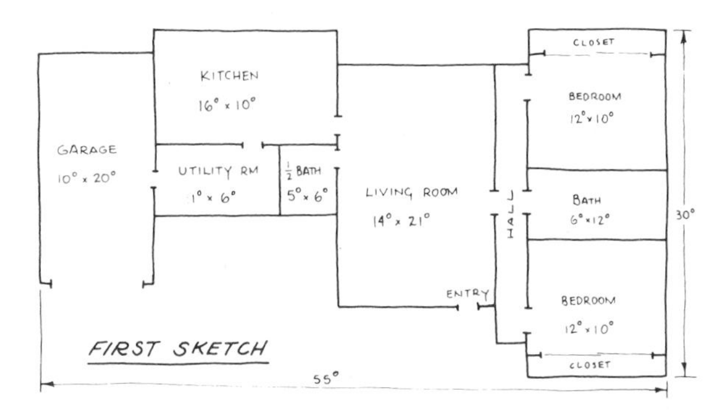

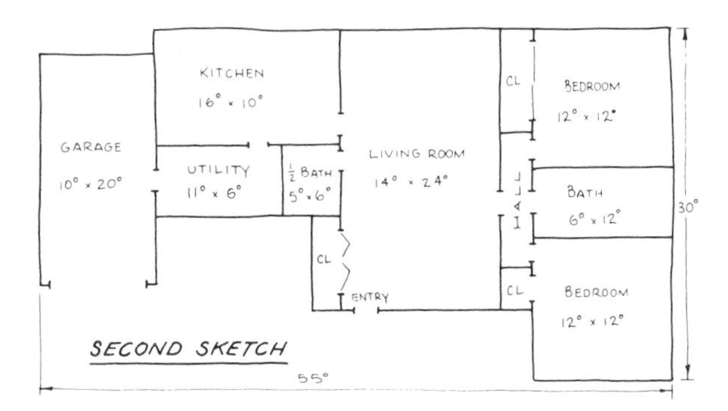

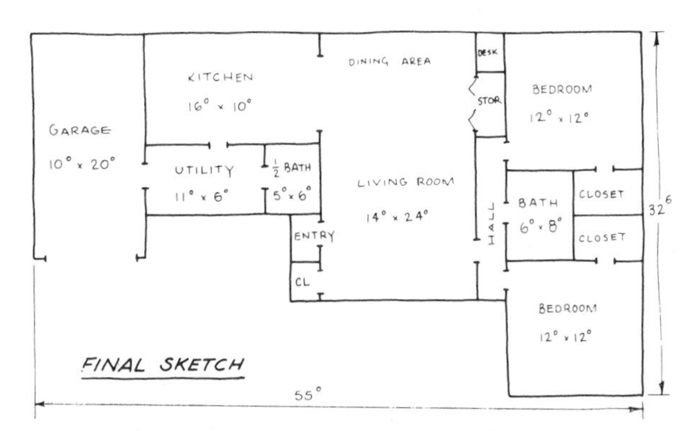

Fig. 32-6 Revisions may be made to final sketches.

Plan Variations

Many different room arrangements are possible within the same amount of space. Figure 32-7 shows two methods of rearranging and redistributing space within the same area.

Final Sketching

Single-line sketches are satisfactory for basic planning purposes. However, they are not adequate for establishing final sizes. A final sketch should be prepared on grid paper to provide a better detailed sketch. This sketch should include the exact positions of doors, windows, and partitions. It should also include the locations of shrubbery, trees, patios, walks, driveways, courts, pools, and gardens. Figure 32-8A shows the many plan variations possible by combining basic geometric shapes on different levels. Notice how the areas are arranged to achieve the most convenient plan. Figure 32-8B shows how a loft can be added without adding to the volume of the design. Vertical refinements such as this can be made in the final design stage, but it is better to anticipate this use of space as early in the design process as possible. Figure 32-8C shows a floor plan with a separate loft plan.

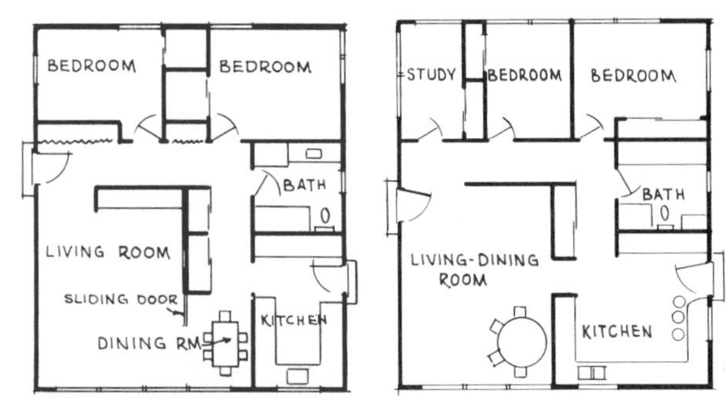

Fig. 32-7 Many different room arrangements are possible within the same amount of space.

OPEN PLANNING

When the rooms of a plan are divided by solid partitions, doors, or arches, the plan is known as a *closed plan.* A closed plan is shown in Fig. 32-9. This plan is a refinement of the conceptual plan shown in Fig. 2-26. If the partitions between the rooms of an area are eliminated, such as in Fig. 32-10, then the plan is known as an *open plan.*

The open plan is used mostly and to best advantage in the living area. Here the walls that separate the entrance foyer, living room, dining room, activities room, and recreation room can be removed or partially eliminated. These open areas are created to provide a sense of spaciousness, to aid lighting efficiency, and to increase the circulation of air through the areas.

Obviously, not all the areas of a residence lend themselves well to open-planning techniques. For example, a closed plan is usually used in the sleeping area. The floor plan for an open plan is developed in the same manner as for a closed plan. However, in the open plan the

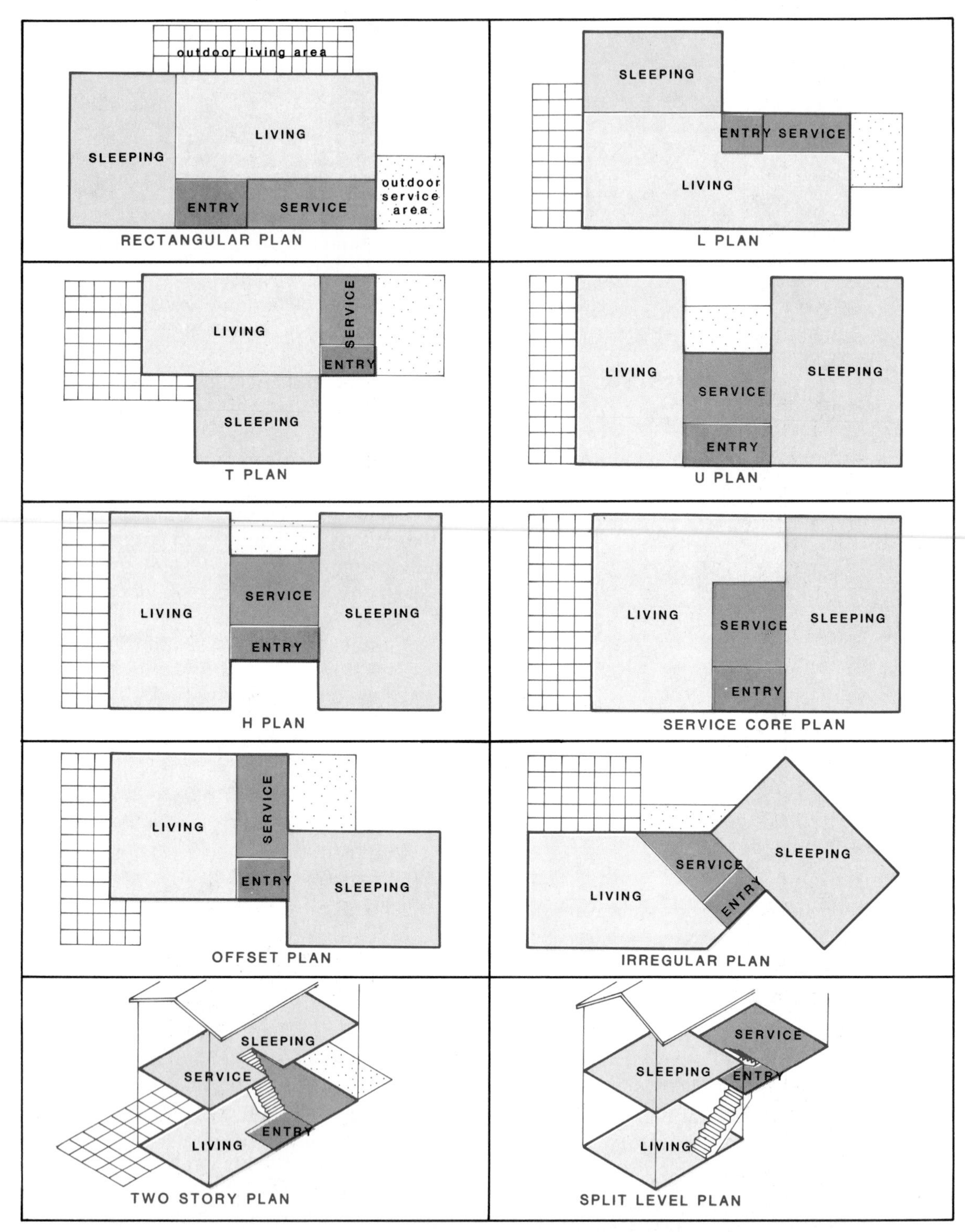

Fig. 32-8A Many different plans are possible by altering geometric space.

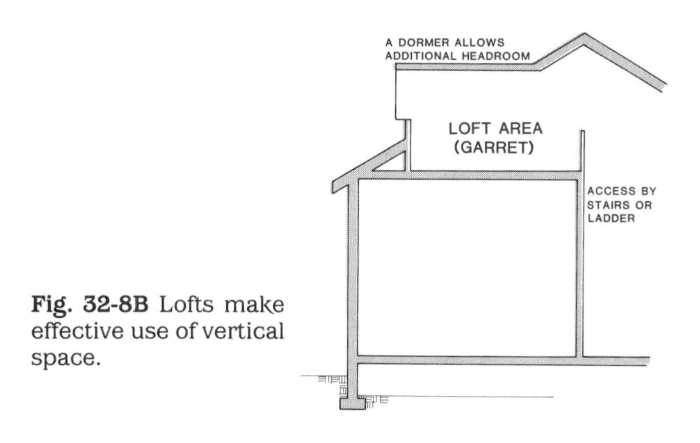

Fig. 32-8B Lofts make effective use of vertical space.

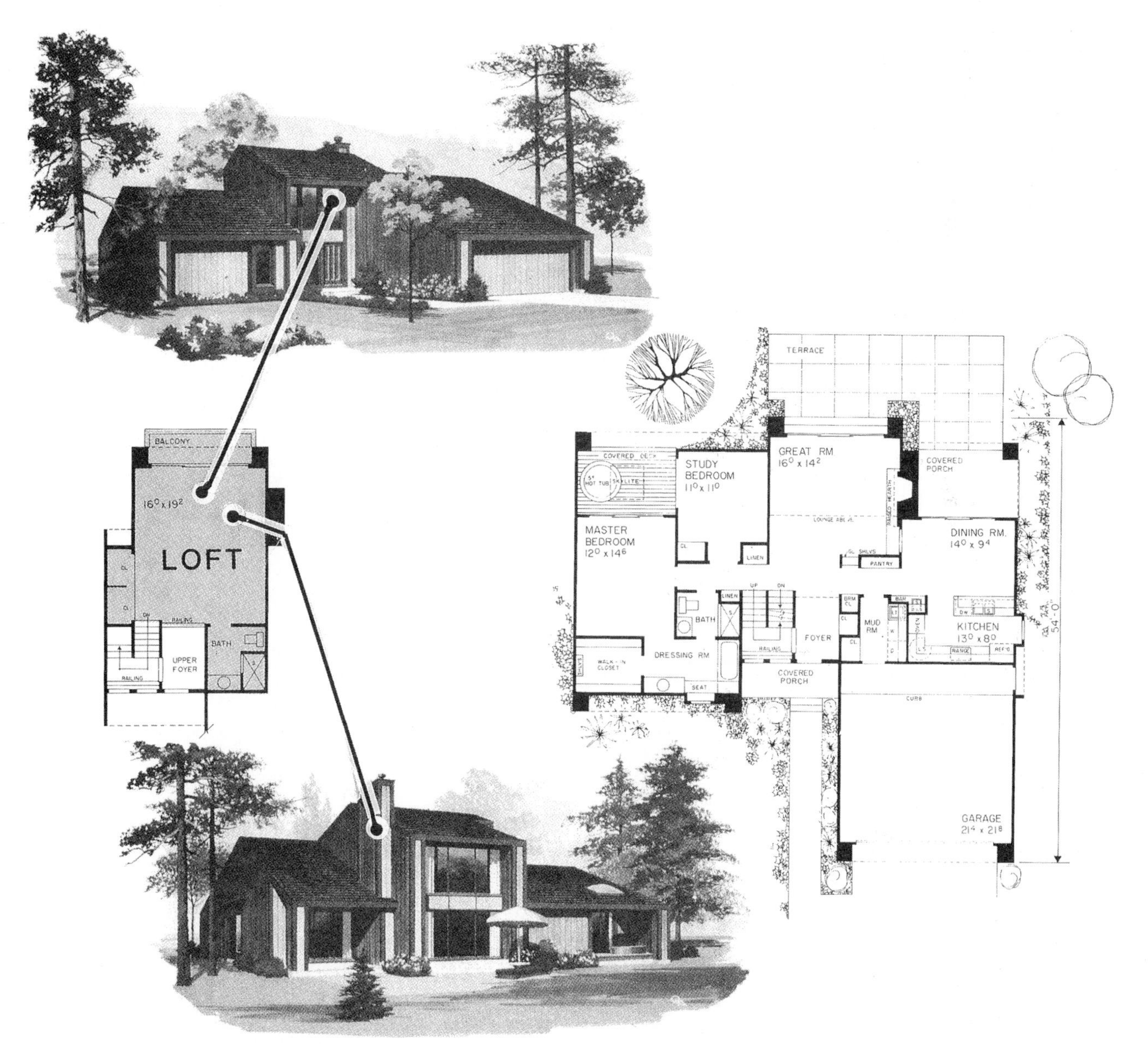

Fig. 32-8C Loft plan related to exterior.

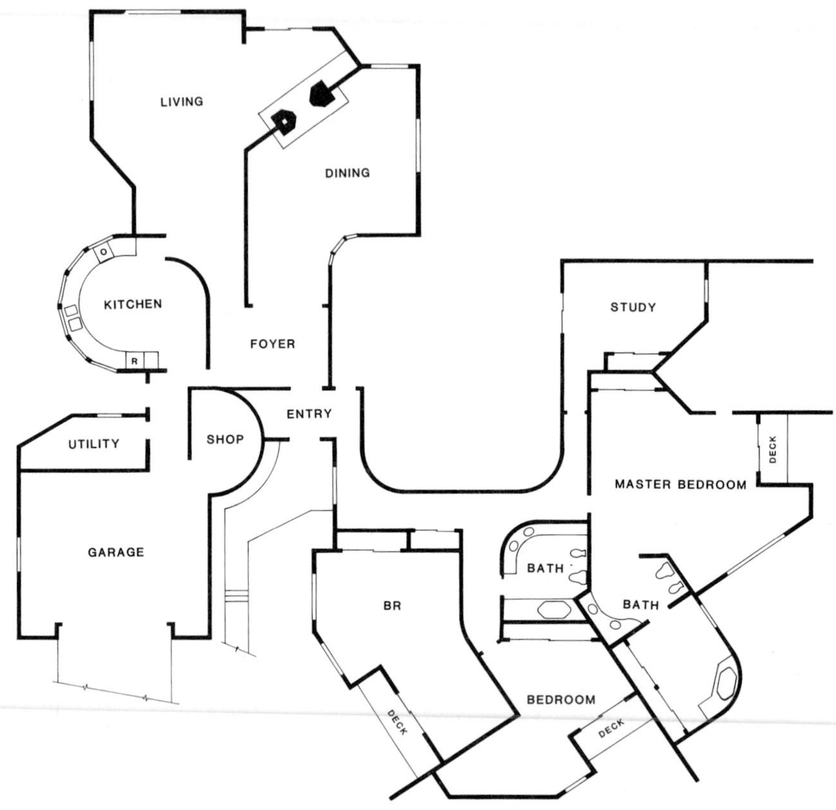

Fig. 32-9 This closed plan is the final floor plan design developed from the conceptual design sketch shown in Fig. 2-27.

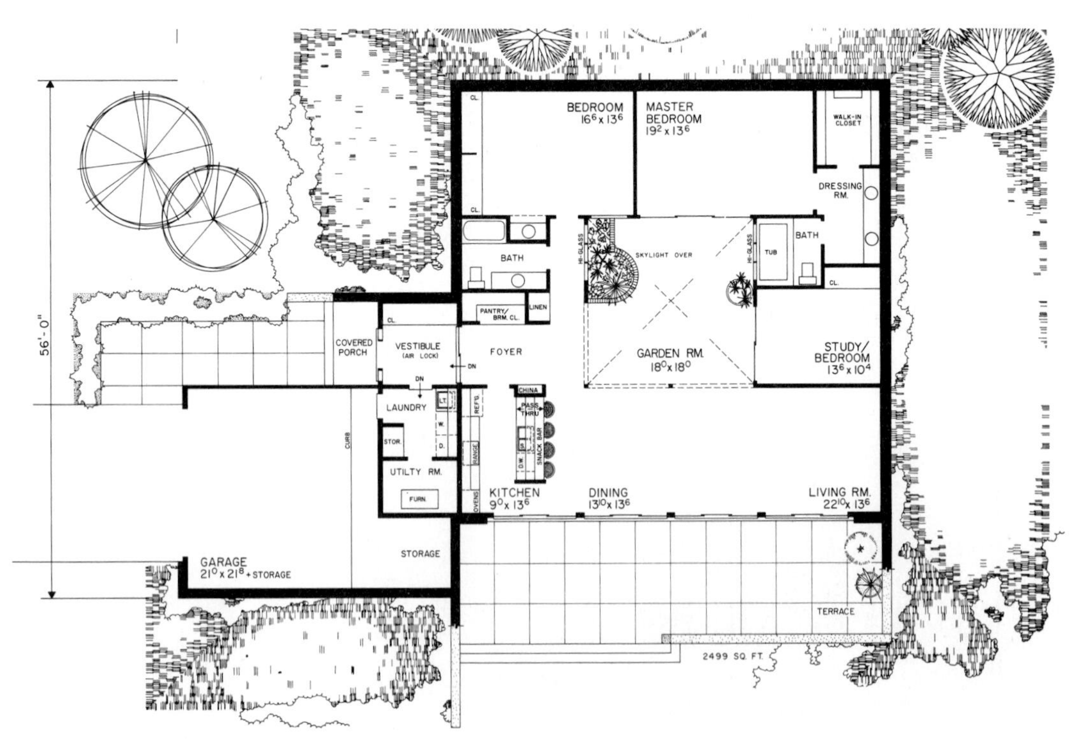

Fig. 32-10 An open plan. *(Home Planners, Inc.)*

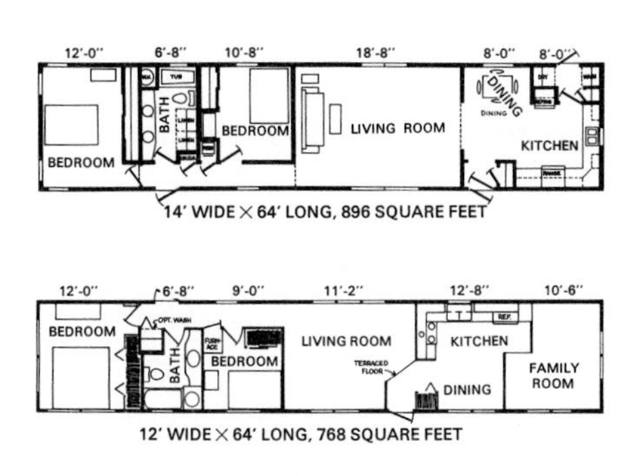

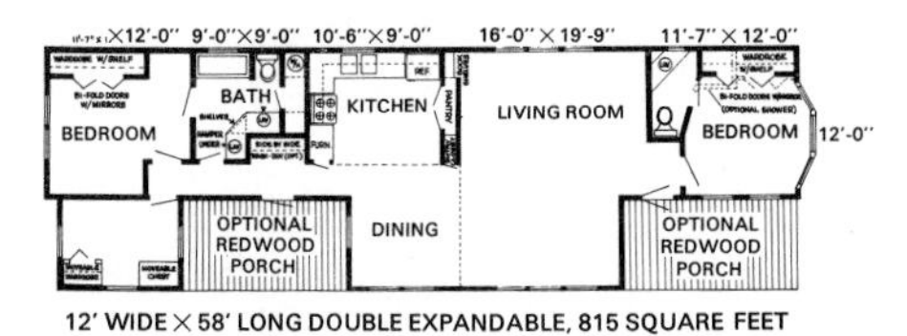

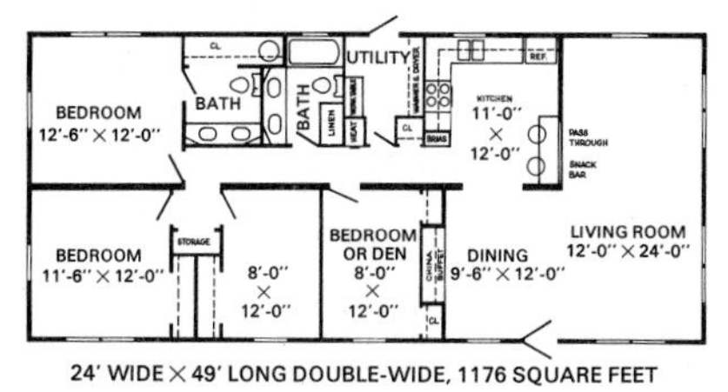

Fig. 32-11 A mobile-home design must fit within the limits of the overall size.

partitions are often replaced by dividers or by variations in level to set apart the various living functions.

When designing floor plans for modular units such as apartments or mobile homes (Fig. 32-11), the designer must develop a plan within predetermined dimensional limits.

EXPANDABLE PLANS

Because of limitations of time or money, it may be desirable to construct a house over a period of time. The house can be built in several steps. The basic part of the house can be constructed first. Then additional rooms (usually bedrooms) can be added in future years as the need develops.

When future expansion of the plan is anticipated, the complete floor plan should be drawn before the initial construction begins, even though the entire plan may not be complete at that time. If only part of the building is planned and built, and a later addition is made, the addition will invariably look tacked on. This appearance can be avoided by designing the floor plan for expansion, as shown in Fig. 32-12.

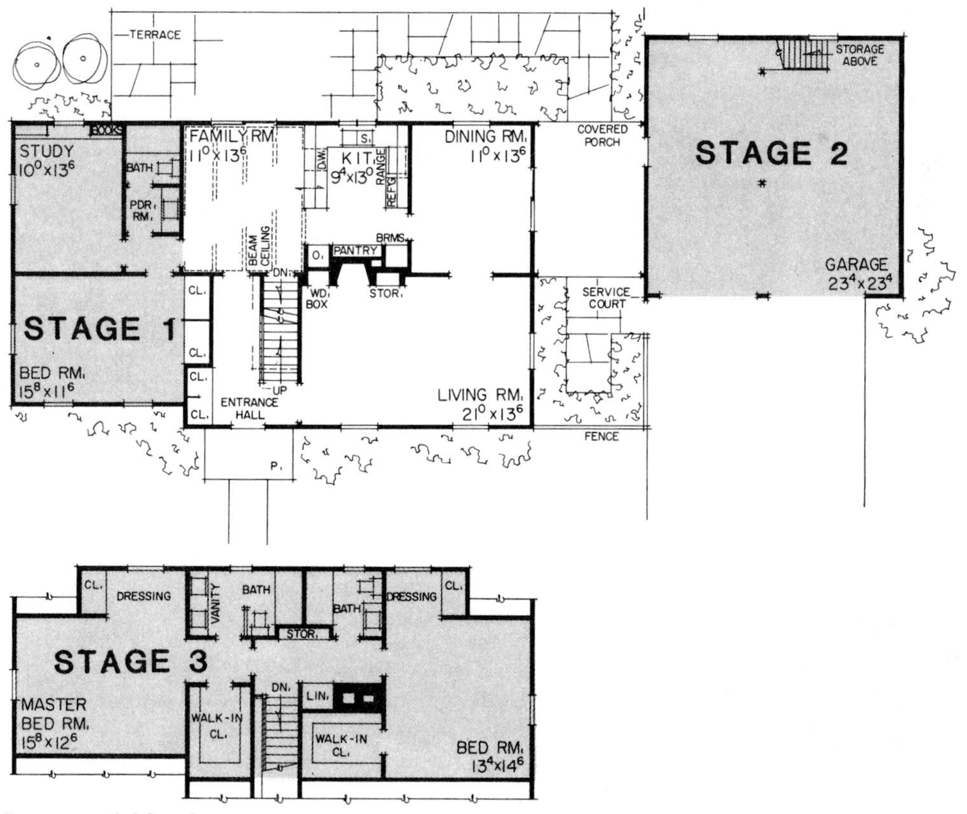

Fig. 32-12 An expandable plan.

Exercises

1. **Prepare room templates and use them to make a functional arrangement for the living area, service area, and sleeping area of a house.**

2. **Arrange templates for a sleeping area, service area, and living area of your own design in a total composite plan.**
3. **Make a floor-plan sketch of the arrangement you completed in Problem 2. Have your instructor criticize this sketch. Then revise it according to the recommendations.**
4. **After you have completed a preliminary line sketch, prepare a final sketch complete with wall thicknesses, overall dimensions, driveways, walks, and shrubbery.**
5. **Arrange the templates found in Fig. 32-13 in a floor-plan arrangement. Make a sketch of this arrangement, and revise the sketch until you arrive at a suitable plan.**
6. **Make room templates of each room in your own home. Rearrange these templates according to a remodeling plan, and make a sketch.**
7. **Define the following terms: *room template, floor-plan sketching, sketch, template layout, final sketching, open planning, closed plan, expandable plan.***

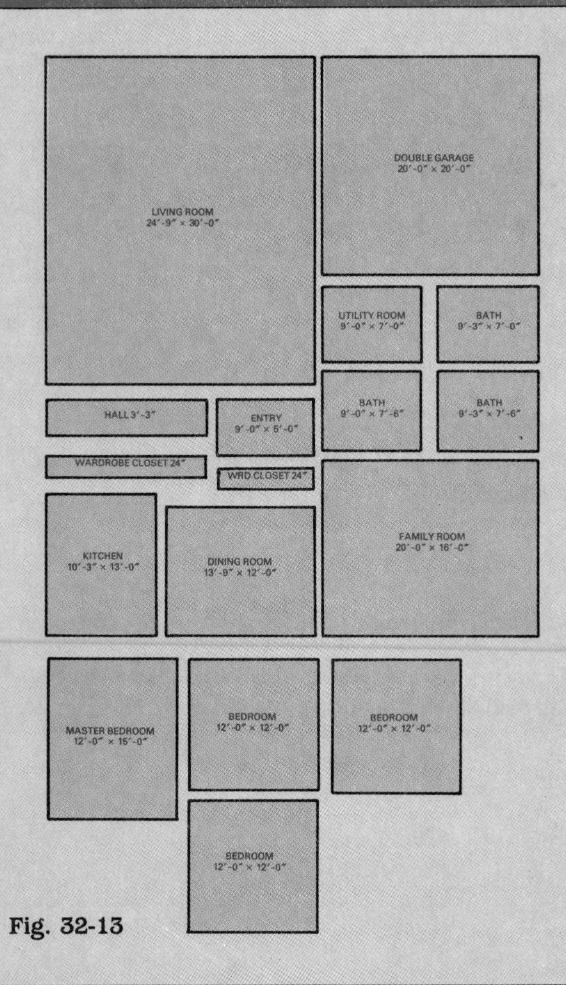

Fig. 32-13

UNIT 33

FLOOR-PLAN DRAWING

A *floor plan* is a drawing of the outline and partitions of a building as you would see them if the building were cut (sectioned) horizontally about 4′(1.2 m) above the floor line, as shown in Fig. 33-1. There are many types of floor plans, ranging from very simple sketches to completely dimensioned and detailed floor-plan working drawings. (See the special floor-plan symbols

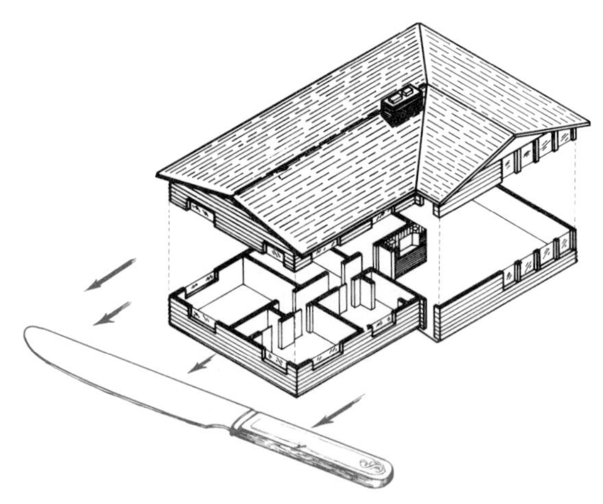

Fig. 33-1 A floor plan is a section view cut through the building 4′ above the floor line.

which are shown in these sections: 8, elevation symbols; 10, site plan symbols; 14, electrical symbols; 15, HVAC symbols; and 16, plumbing symbols.)

TYPES OF FLOOR PLANS

Some of the various types of floor plans commonly prepared for interpretation by the layman are the *single line drawing* (Fig. 33-2), the *abbreviated plan* (Fig. 33-3), and the *pictorial floor plan* (Fig. 33-4). Bird's-eye views such as the one shown in Fig. 33-5 are often prepared to convey a sense of depth to the viewer. Pictorial plans such as that shown in Fig. 33-6 are often prepared to show the relationship among various areas of the site. These plans are satisfactory for general use. However, a completely dimensioned floor plan is necessary to show the amount of detail necessary for construction purposes.

Floor-plan sketches such as those shown in Figs. 33-2 through 33-6 are sufficient for rough layout and preliminary design purposes but are not accurate or complete enough to be used as working drawings. An accurate floor plan, complete with dimensions and material symbols, must be prepared. When a plan of this type is developed, the contractor can interpret the desires of the designer without consultation. The prime function of a working-drawing floor plan is to communicate information to the contractor. A complete floor plan eliminates misunderstandings between designer and the builder. The builder's judgment must be used to fill in the omitted details if an incomplete floor plan is pre-

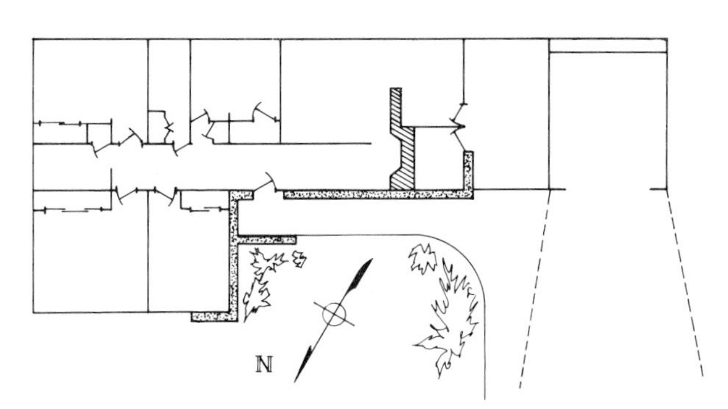

Fig. 33-2 A single-line floor plan.

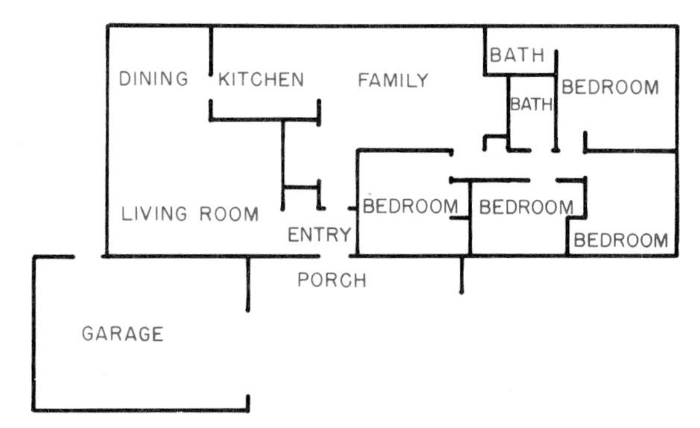

Fig. 33-3 An abbreviated floor plan.

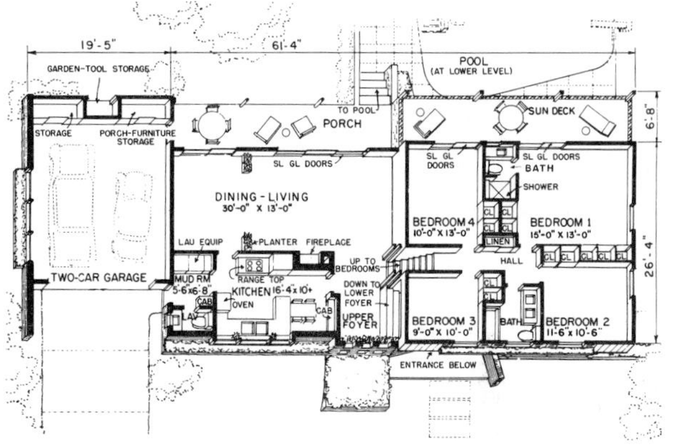

Fig. 33-4 A pictorial floor plan. *(Master Plan Service, Inc.)*

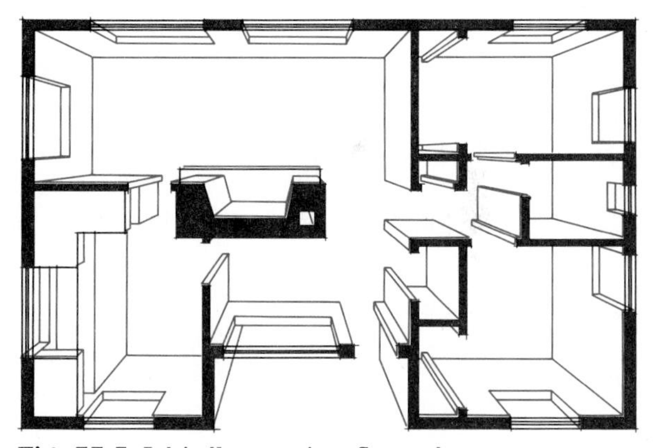

Fig. 33-5 A bird's-eye view floor plan.

Fig. 33-6 A flat-pictorial floor plan.

pared. The function of the designer is transferred to the builder when that happens.

FLOOR-PLAN SYMBOLS

Drafters substitute symbols for materials and fixtures, just as stenographers substitute shorthand for words. It is obviously more convenient and timesaving to draw a symbol of a material than to repeat a description every time that material is used. It would be impossible to describe all construction materials used on floor plans, such as fixtures, doors, windows, stairs and partitions, without the use of symbols.

Figures 33-7 through 33-11 show common symbols used on floor plans together with the related elevation symbol. These include symbols for doors, windows, appliances, fixtures and sanitation facilities, and building materials. Floor-plan symbols for plumbing, heating, air-conditioning, and electrical components are covered later in those specialized units. Figure 33-12 shows the application of some of these symbols to a floor plan, and Fig. 33-13A & B shows the relationship of floor-plan symbols to construction features.

Although architectural symbols are standardized, some variations of symbols are used in different parts of the country. Figure 33-14 shows several methods drafters use for drawing construction details and the outside walls of frame buildings. When walls are shaded in as shown in the two left walls in Fig. 33-14, the shading is often done on the underside of the vellum sheet. Wall openings indicated on floor plans show the relationship between the wall and ceiling, as shown in Fig. 33-15A. When using a CAD system, symbols can be stored in libraries, then called-up and inserted into the floor plan with a touch of a stylus.

Since it is impossible to show a complete plan at the normal scale of ¼″ = 1′-0″ on this page, Fig. 33-15B shows a portion of the plan found in Fig. 69-1 as it would be drawn at ¼″ = 1′-0″. Learning and remembering floor-plan symbols will be easier if you associate each symbol with the actual material or facility it represents. For

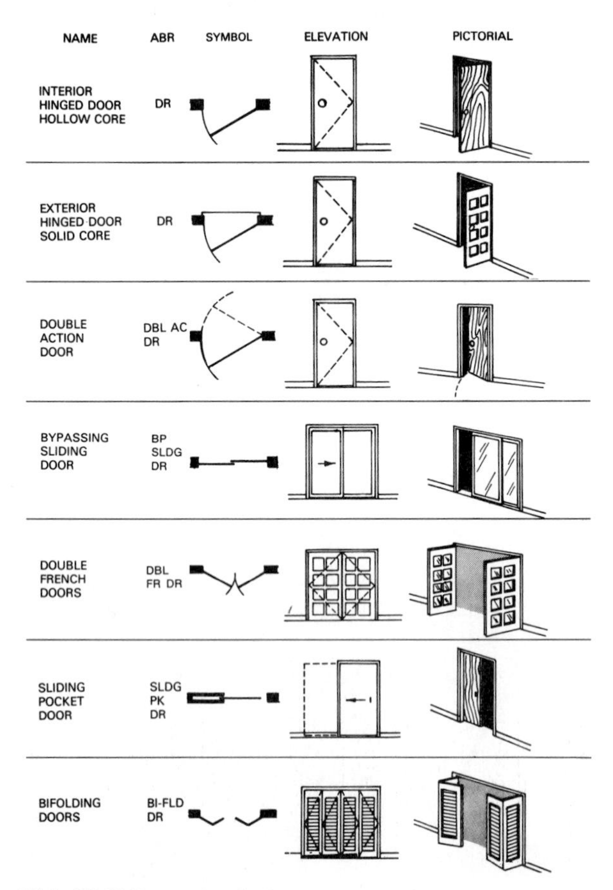

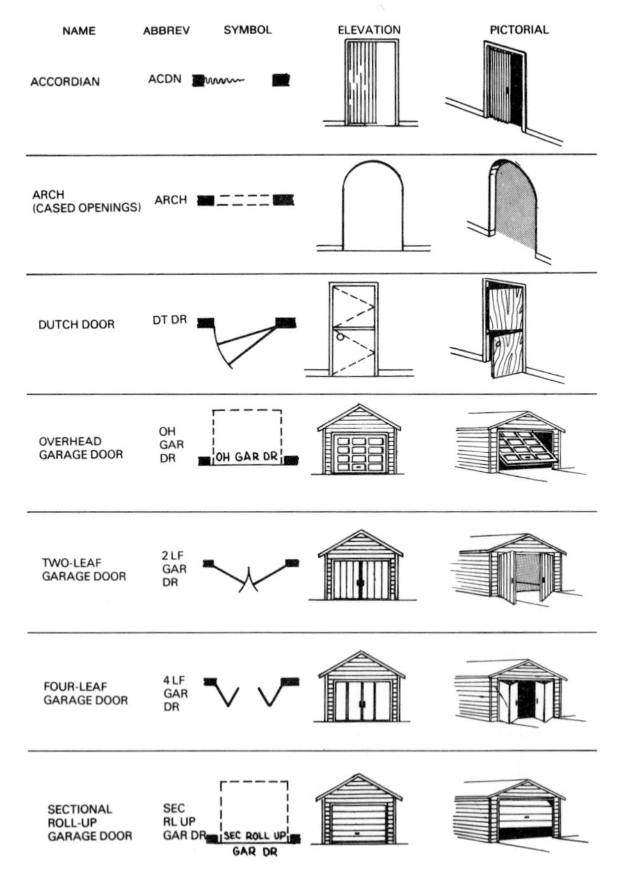

Fig. 33-7 Door symbols.

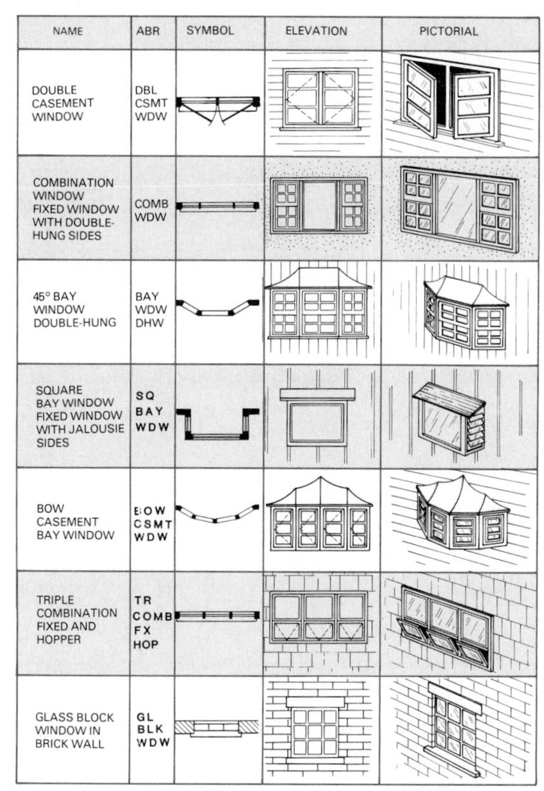

NAME	ABR	SYMBOL	ELEVATION	PICTORIAL
DOUBLE CASEMENT WINDOW	DBL CSMT WDW			
COMBINATION WINDOW FIXED WINDOW WITH DOUBLE-HUNG SIDES	COMB WDW			
45° BAY WINDOW DOUBLE-HUNG	BAY WDW DHW			
SQUARE BAY WINDOW FIXED WINDOW WITH JALOUSIE SIDES	SQ BAY WDW			
BOW CASEMENT BAY WINDOW	BOW CSMT WDW			
TRIPLE COMBINATION FIXED AND HOPPER	TR COMB FX HOP			
GLASS BLOCK WINDOW IN BRICK WALL	GL BLK WDW			

Fig. 33-8A Window symbols.

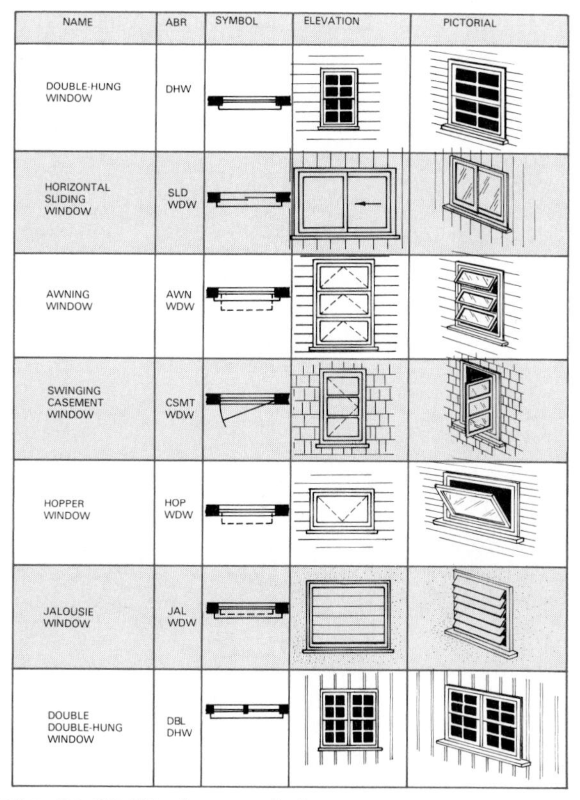

NAME	ABR	SYMBOL	ELEVATION	PICTORIAL
DOUBLE-HUNG WINDOW	DHW			
HORIZONTAL SLIDING WINDOW	SLD WDW			
AWNING WINDOW	AWN WDW			
SWINGING CASEMENT WINDOW	CSMT WDW			
HOPPER WINDOW	HOP WDW			
JALOUSIE WINDOW	JAL WDW			
DOUBLE DOUBLE-HUNG WINDOW	DBL DHW			

Fig. 33-8B Window symbols.

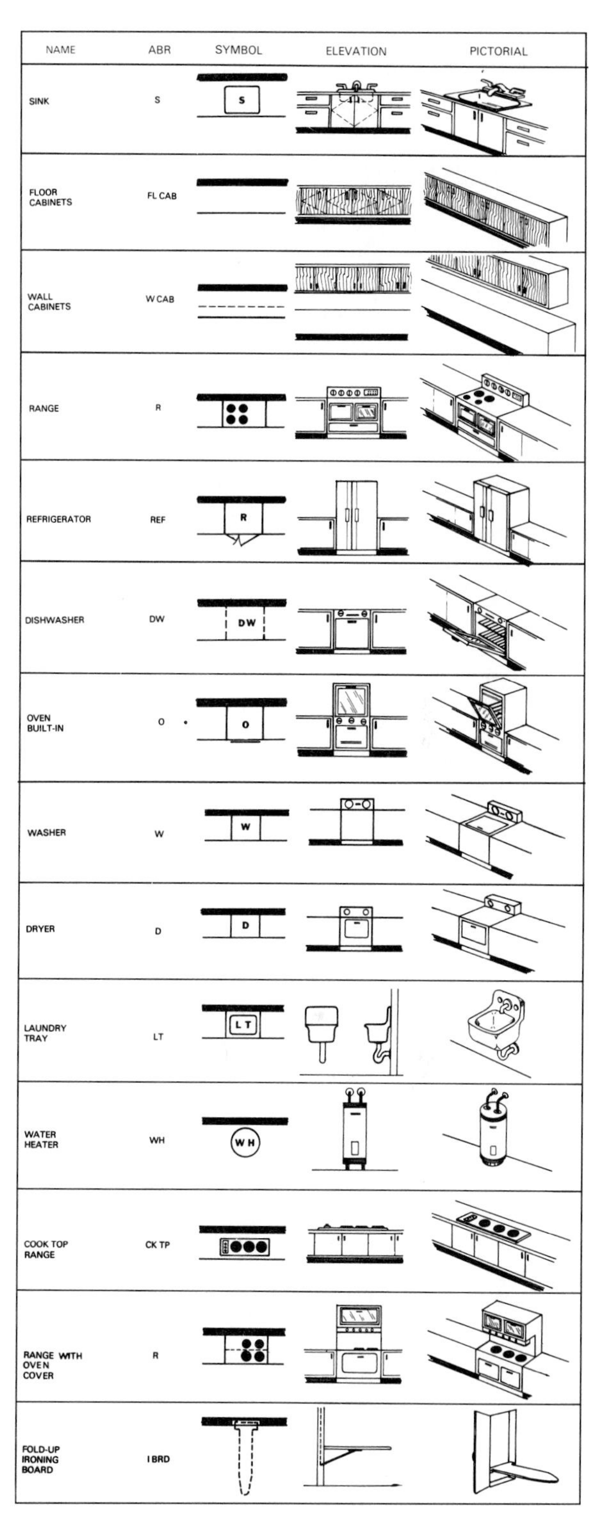

NAME	ABR	SYMBOL	ELEVATION	PICTORIAL
SINK	S	S		
FLOOR CABINETS	FL CAB			
WALL CABINETS	W CAB			
RANGE	R			
REFRIGERATOR	REF	R		
DISHWASHER	DW	DW		
OVEN BUILT-IN	O	O		
WASHER	W	W		
DRYER	D	D		
LAUNDRY TRAY	LT	LT		
WATER HEATER	WH	WH		
COOK TOP RANGE	CK TP			
RANGE WITH OVEN COVER	R			
FOLD-UP IRONING BOARD	I BRD			

Fig. 33-9 Appliance and fixture symbols.

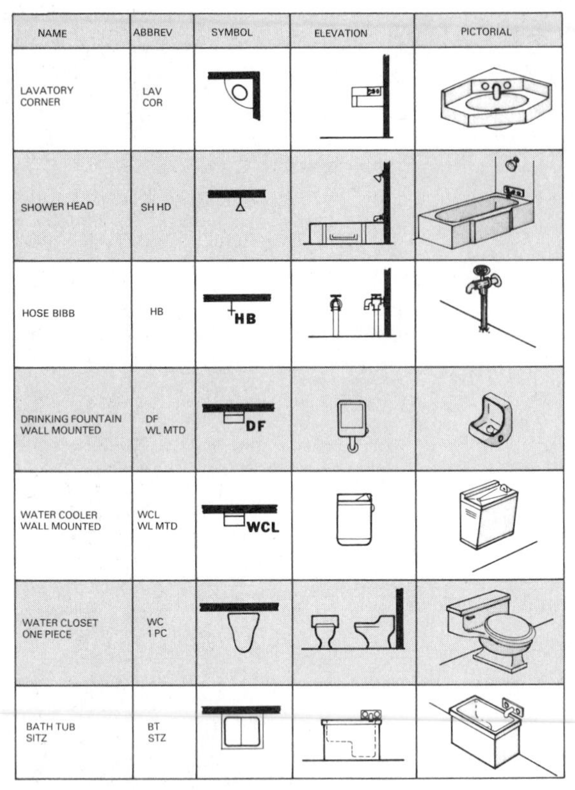

NAME	ABBREV	SYMBOL	ELEVATION	PICTORIAL
LAVATORY CORNER	LAV COR			
SHOWER HEAD	SH HD			
HOSE BIBB	HB	HB		
DRINKING FOUNTAIN WALL MOUNTED	DF WL MTD	DF		
WATER COOLER WALL MOUNTED	WCL WL MTD	WCL		
WATER CLOSET ONE PIECE	WC 1 PC			
BATH TUB SITZ	BT STZ			

Fig. 33-10A Sanitation facility symbols.

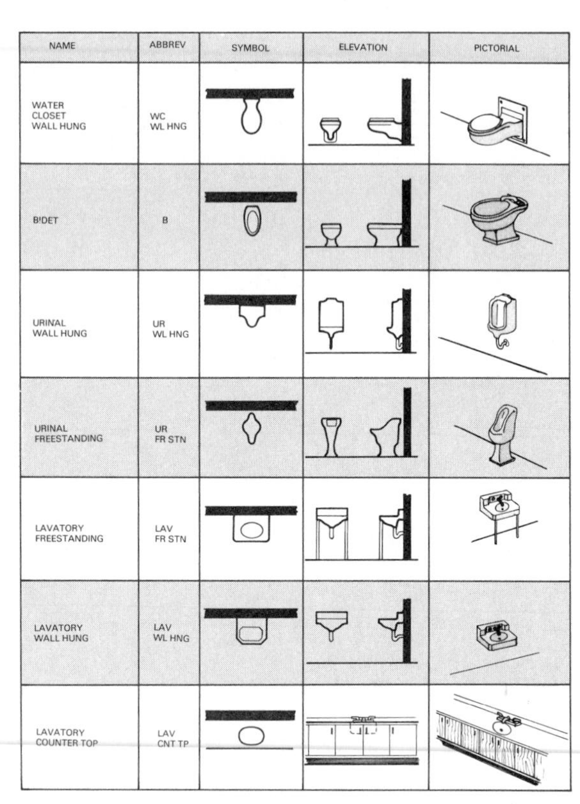

NAME	ABBREV	SYMBOL	ELEVATION	PICTORIAL
WATER CLOSET WALL HUNG	WC WL HNG			
BIDET	B			
URINAL WALL HUNG	UR WL HNG			
URINAL FREESTANDING	UR FR STN			
LAVATORY FREESTANDING	LAV FR STN			
LAVATORY WALL HUNG	LAV WL HNG			
LAVATORY COUNTER TOP	LAV CNT TP			

Fig. 33-11 Lavoratory symbols.

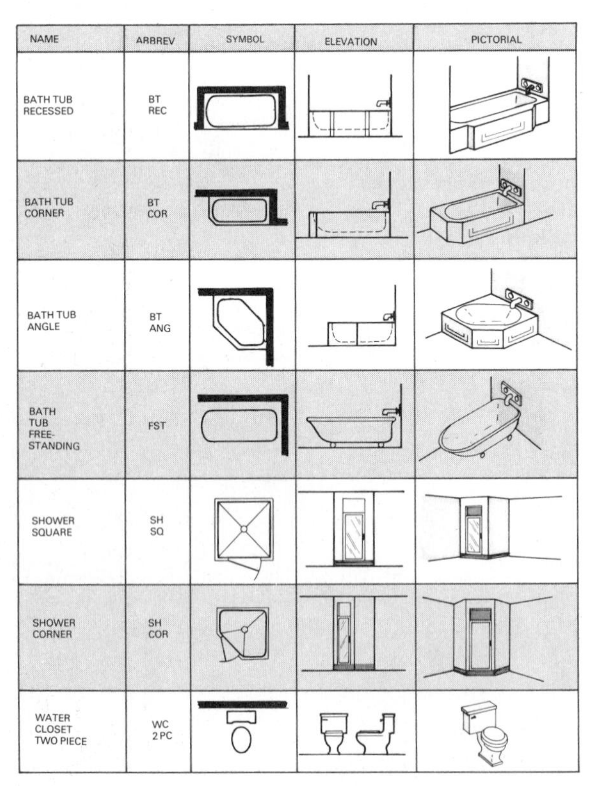

NAME	ABBREV	SYMBOL	ELEVATION	PICTORIAL
BATH TUB RECESSED	BT REC			
BATH TUB CORNER	BT COR			
BATH TUB ANGLE	BT ANG			
BATH TUB FREE-STANDING	FST			
SHOWER SQUARE	SH SQ			
SHOWER CORNER	SH COR			
WATER CLOSET TWO PIECE	WC 2 PC			

Fig. 33-10B Sanitation facility symbols.

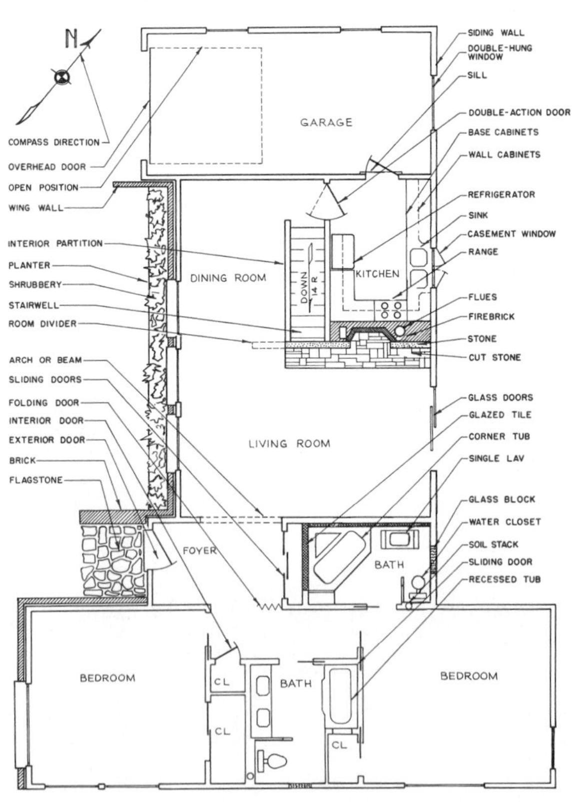

Fig. 33-12 The application of floor-plan symbols.

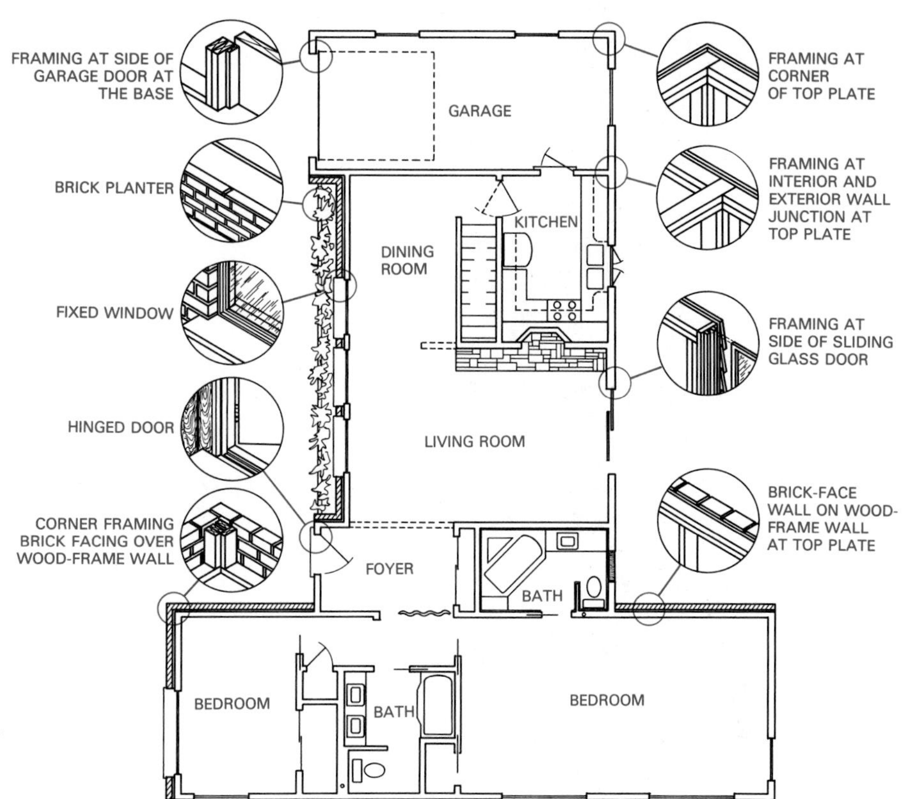

Fig. 33-13A Methods of showing construction details on a floor plan.

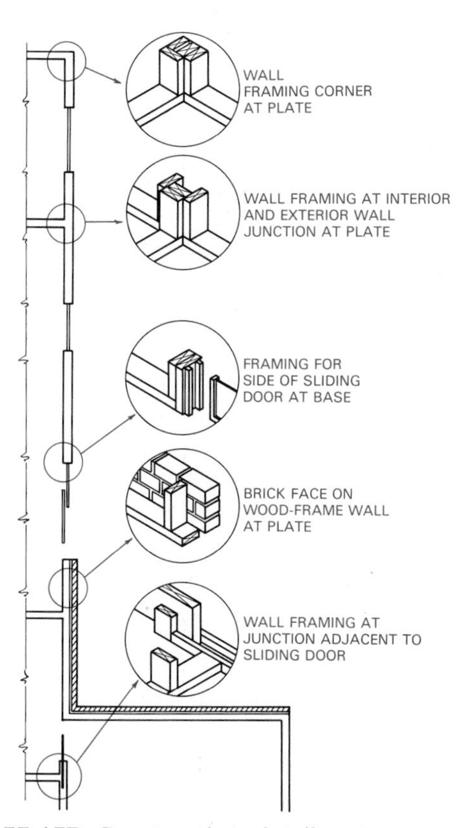

Fig. 33-13B Construction details represented by floor-plan symbols.

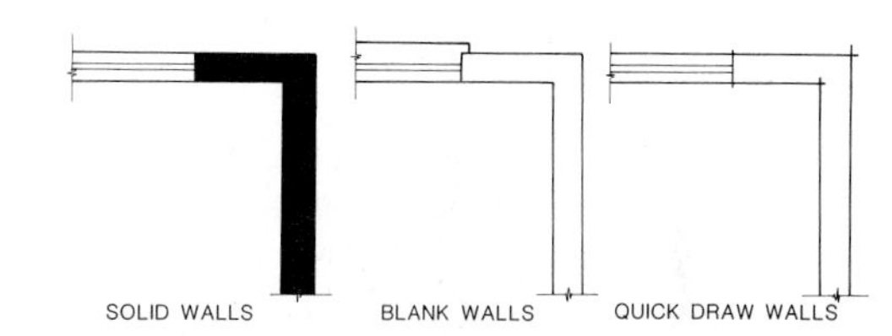

Fig. 33-14 Methods of showing wall construction on a floor plan.

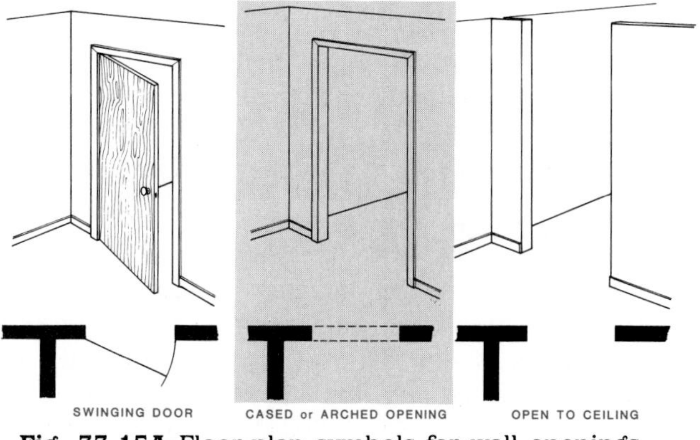

Fig. 33-15A Floor-plan symbols for wall openings.

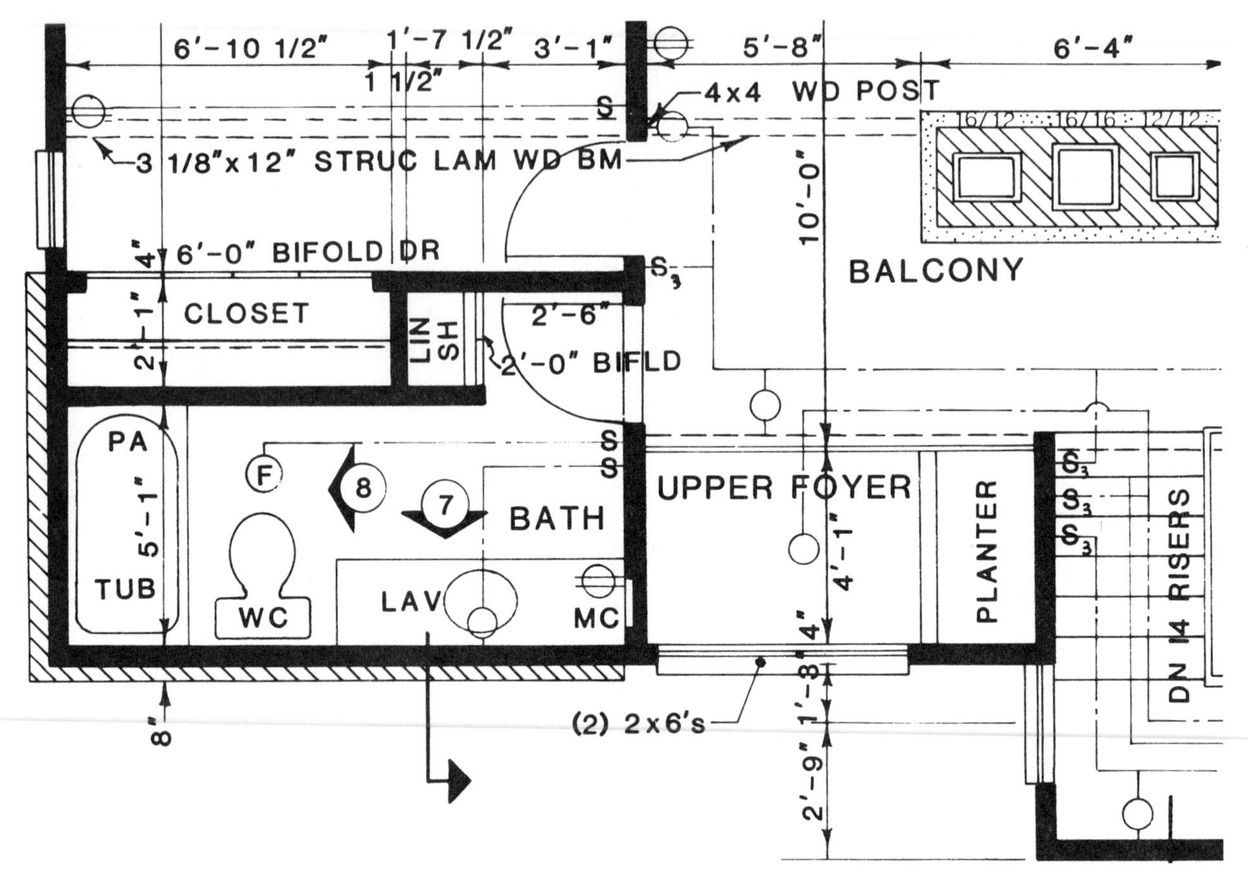

Fig. 33-15B Part of a full-size drawing prepared at ¼" = 1'-0".

example, as you learn the telephone-jack symbol, you should associate this symbol with the actual appearance of the telephone jack.

Floor-plan symbols often represent the exact appearance of the floor-plan section as viewed from above, but sometimes this representation is not possible. Many floor-plan symbols are too intricate to be drawn to the scale ¼" = 1'-0" or ⅛" = 1'-0". Therefore, many details are eliminated or simplified on the floor-plan symbols.

STEPS IN DRAWING FLOOR PLANS

For maximum speed, accuracy, and clarity, the following steps, as illustrated in Fig. 33-16A should be observed in laying out and drawing floor plans:

1. Block in the overall dimensions of the house and add the thickness of the outside walls with a hard pencil (4H).

2. Lay out the position of interior partitions with 4H pencil.

3. Locate the position of doors and windows by centerline and by their widths (4H). Double check all dimensions.

4. Darken the object lines with an F pencil.

5. Add door and window symbols with a 2H pencil. Be sure to draw the door swing away from perpendicular walls to provide the most convenient access (Fig. 33-16B).

6. Add symbols for stairwells.

7. Erase extraneous layout lines if they are too heavy. If they are extremely light, they can remain.

8. Draw the outlines of kitchen and bathroom fixtures.

9. Add the symbols and sections for any masonry work, such as fireplaces and planters.

10. Dimension the drawing (see Unit 34).

11. When establishing window dimensions be sure the window square footage is at least 10 percent of the room square footage.

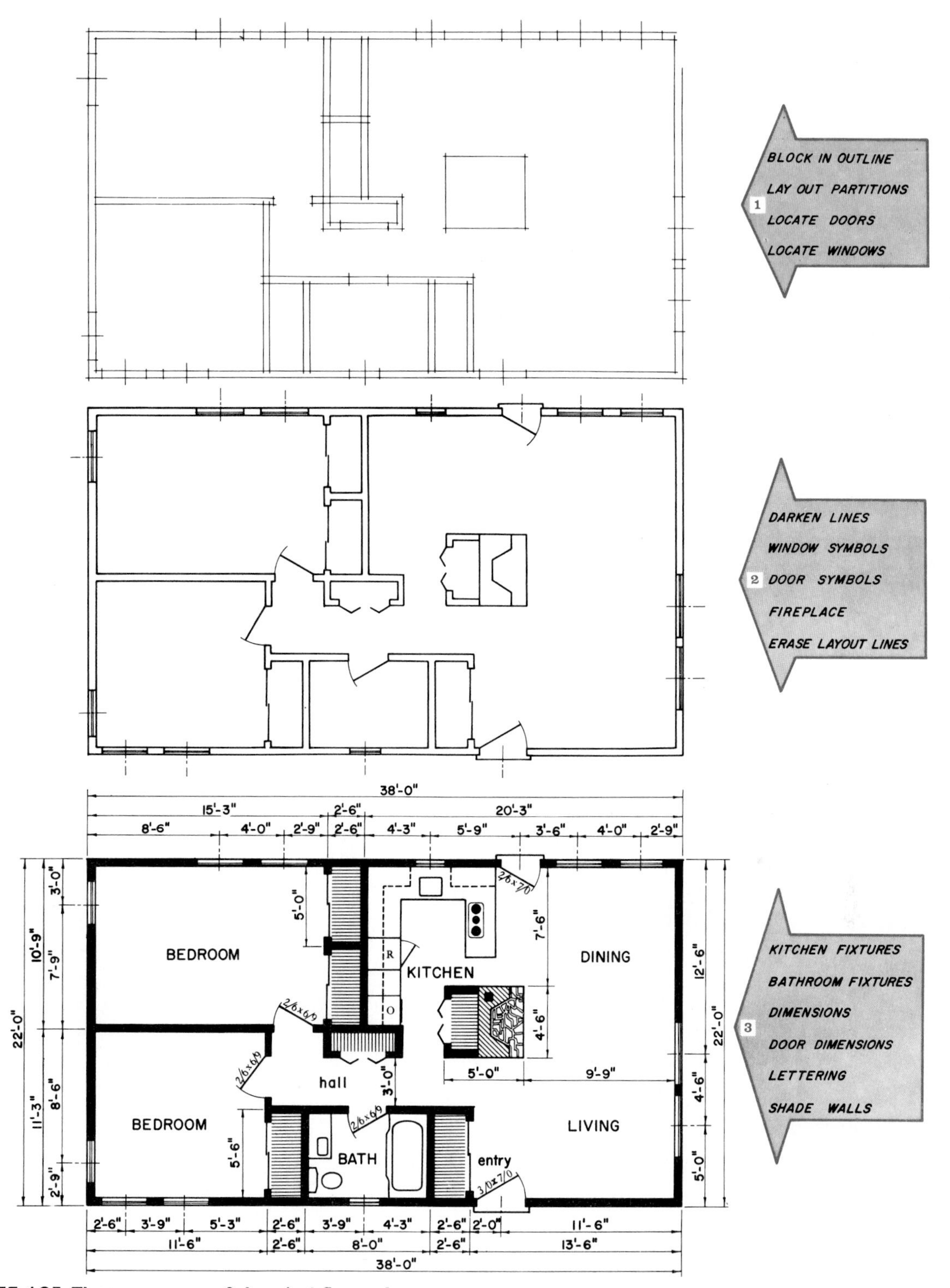

Fig. 33-16A The sequence of drawing floor plans.

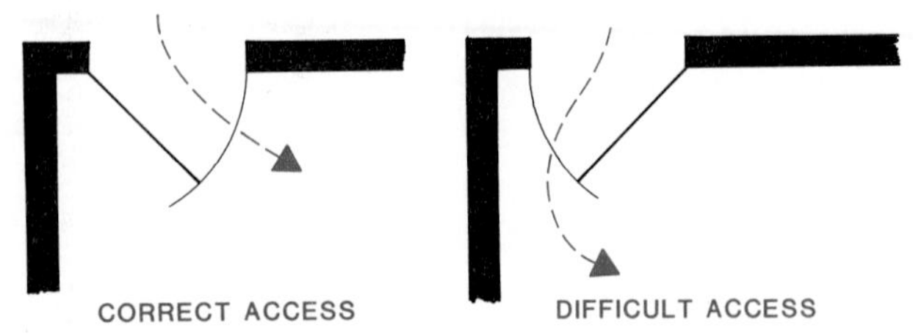

Fig. 33-16B Doors should open toward perpendicular wall.

When floor plans are prepared on a CAD system, follow the steps outlined in Figure 27-39.

SECOND-FLOOR PLAN

Bilevel, two-story, one-and-one-half-story, and split-level homes require a separate floor plan for each additional level, as shown in Fig. 33-17. This floor plan is prepared on tracing paper placed directly over the first-floor plan to ensure alignment of walls and bearing partitions. When the major outline has been traced, the first-floor plan is removed. Figure 33-18 shows a second-floor plan projected from the first-floor plan. Alignment of features such as stairwell openings (Fig. 33-19), outside walls, plumbing walls, and chimneys is critical in preparing the second-floor plan. Figure 33-20 shows a typical second-floor plan of a one-and-one-half-story house with the roof line broken. This drawing, in addition to revealing the second-floor plan, shows the outline

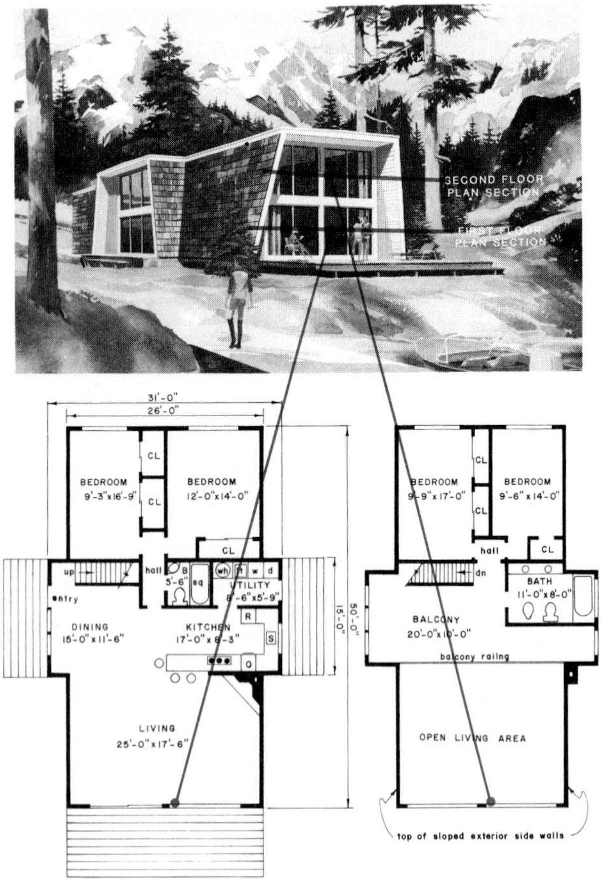

Fig. 33-17 Relationship of floor-plan locations to pictorial rendering.

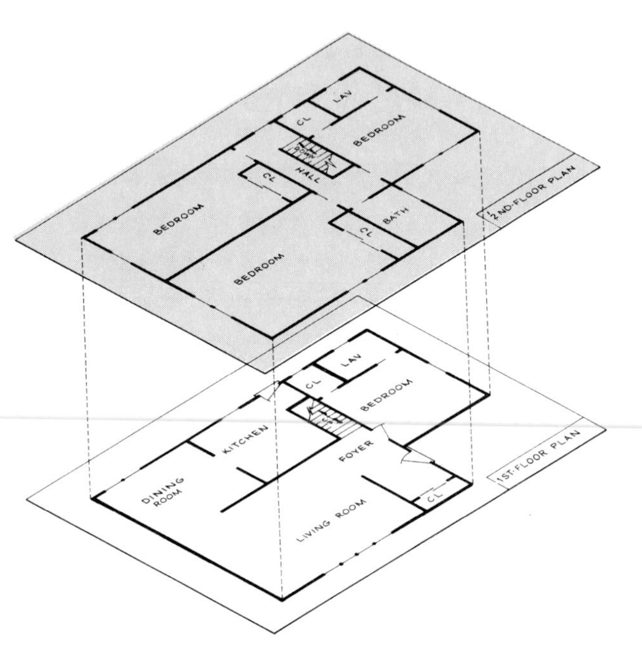

Fig. 33-18 Projection of the second-floor plan.

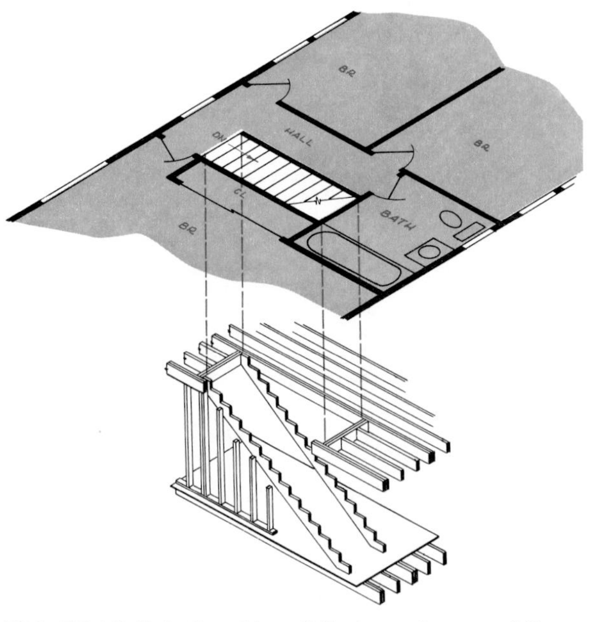

Fig. 33-19 Relationship of first- and second-floor openings.

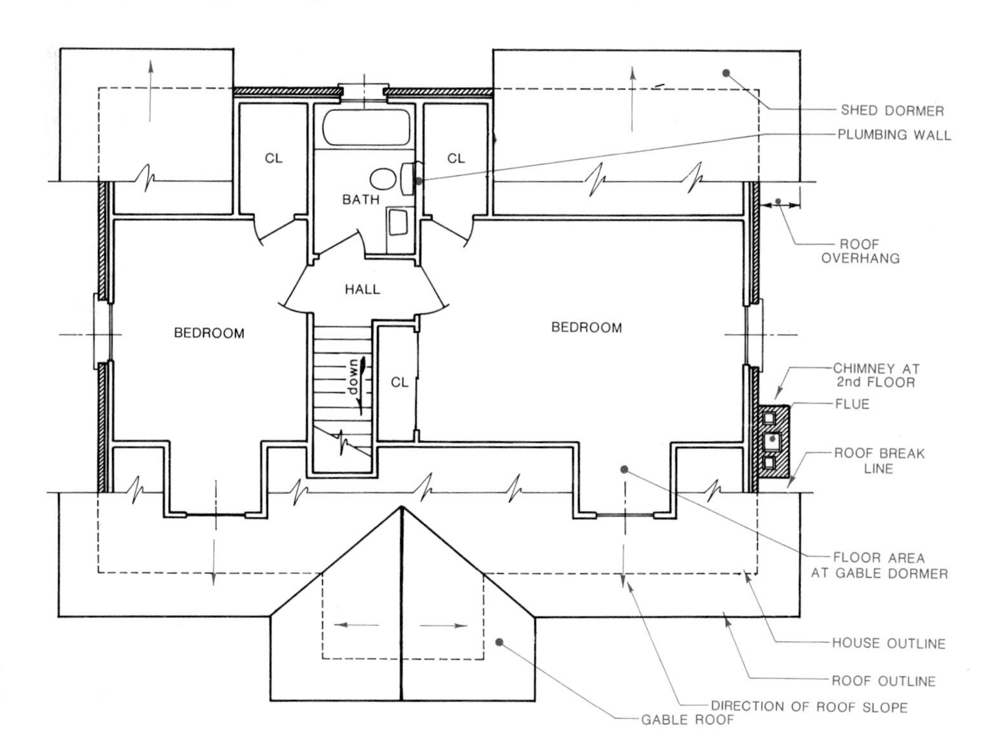

Fig. 33-20 A one-and-one-half-story second-floor plan.

of the roof. In this plan, dotted lines are used to show the outline of the building under the roof.

Figure 33-21 shows a second-floor plan of a two-story house. In this plan, there is no break, since the first-floor plan is the same size as the second-floor plan. This plan is prepared by tracing over the first-floor outline. In drawing multiple floor plans, use the layering task to avoid redrawing identical features.

ALTERNATIVE PLANS

A technique frequently used to alter or adapt the appearance of the house and the location of various rooms is the practice of reversing a floor plan. Figure 33-22A shows a floor plan with its reversed counterpart. This reversal is accomplished by turning the floor plan upside down and tracing the mirror image of the floor plan to provide either a right-hand or a left-hand plan.

For buildings with complex ceiling designs involving multiple lighting fixtures or levels, a reflected ceiling plan is often prepared. A reflected ceiling plan is prepared to the same dimensions and shape as the floor plan but drawn as the ceiling would be viewed if the floor was a mirror, as shown in Fig. 33-22B. A floor plan can be reversed on a CAD system by using the mirror function.

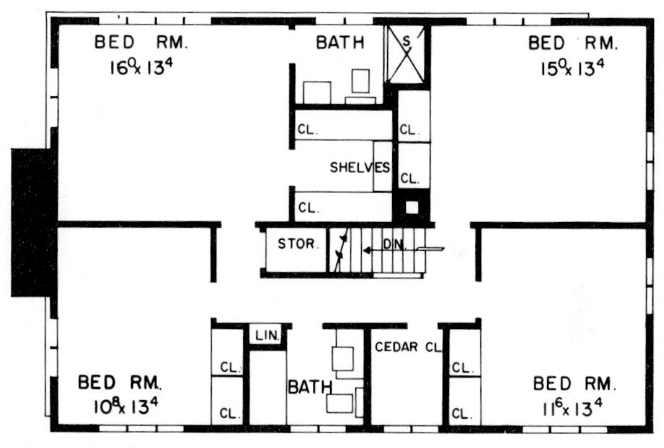

Fig. 33-21 A full second-floor plan. *(Home Planners, Inc.)*

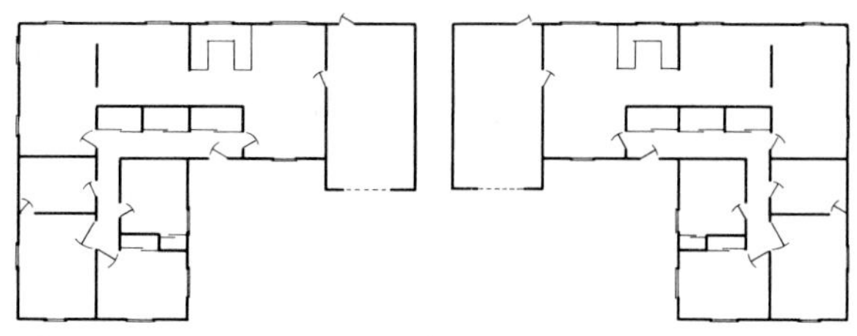

Fig. 33-22A A reversed plan.

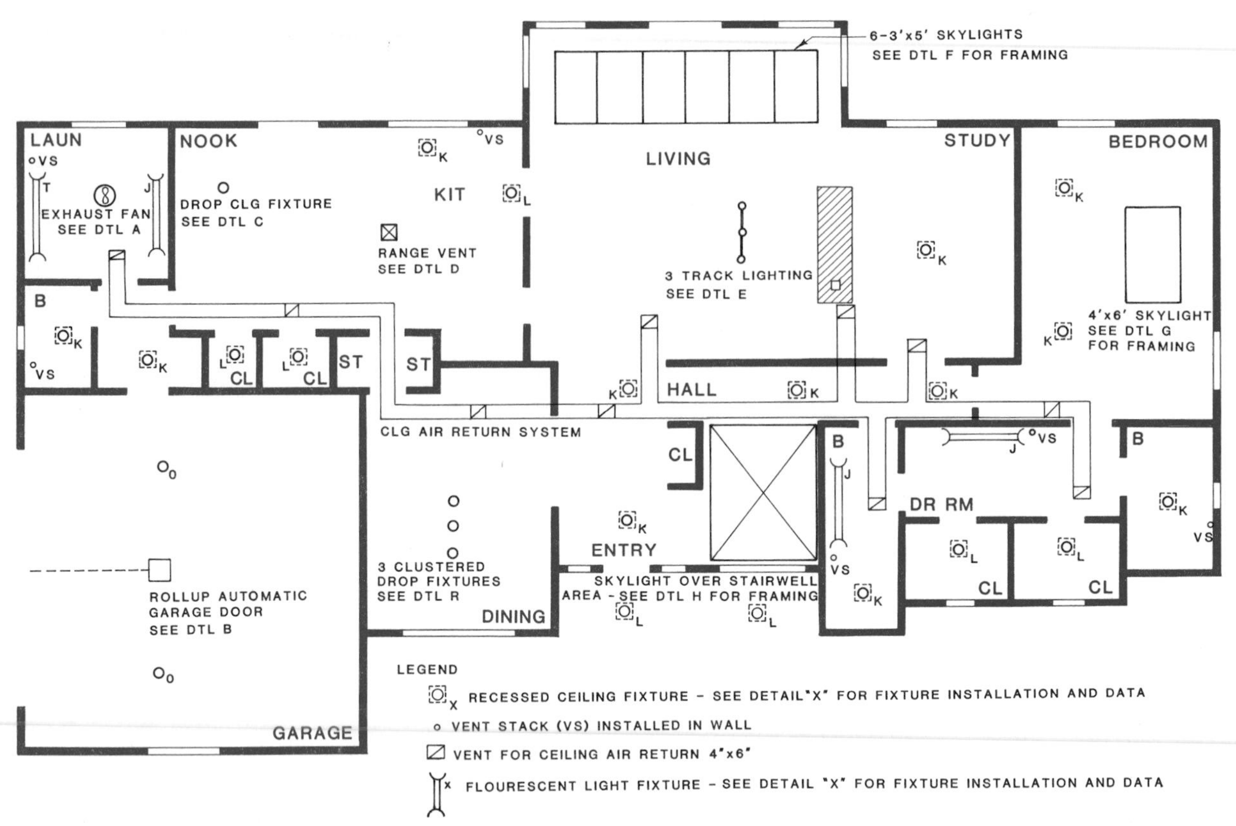

Fig. 33-22B Reflected ceiling plan.

Exercises

1. Draw a complete floor plan, using a sketch of your own design as a guide, and using the scale 1/4" = 1'-0".
2. Draw a complete floor plan from the sketch shown in Fig. 33-23, using the scale 1/4" = 1'-0".
3. Identify the symbols in Fig. 33-24.
4. Draw a complete floor plan, using the abbreviated drawing shown in Fig. 33-25 as a

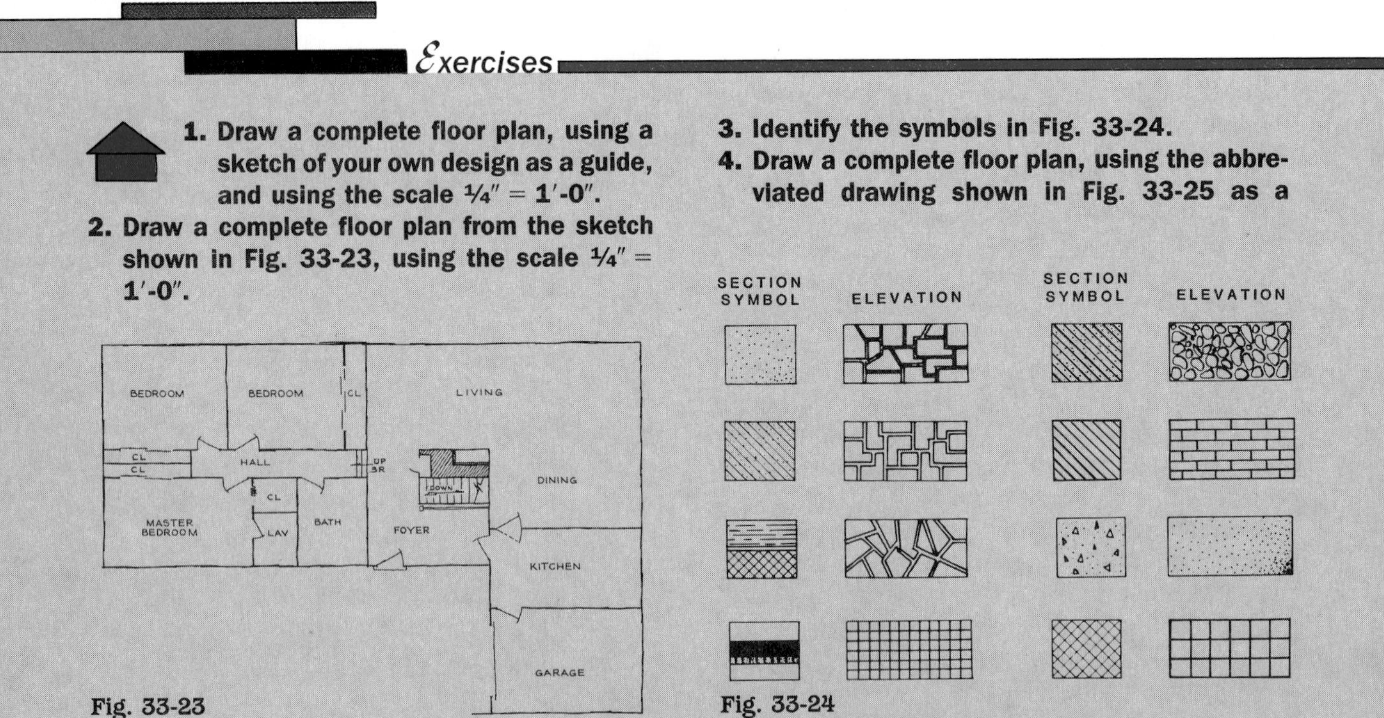

Fig. 33-23

Fig. 33-24

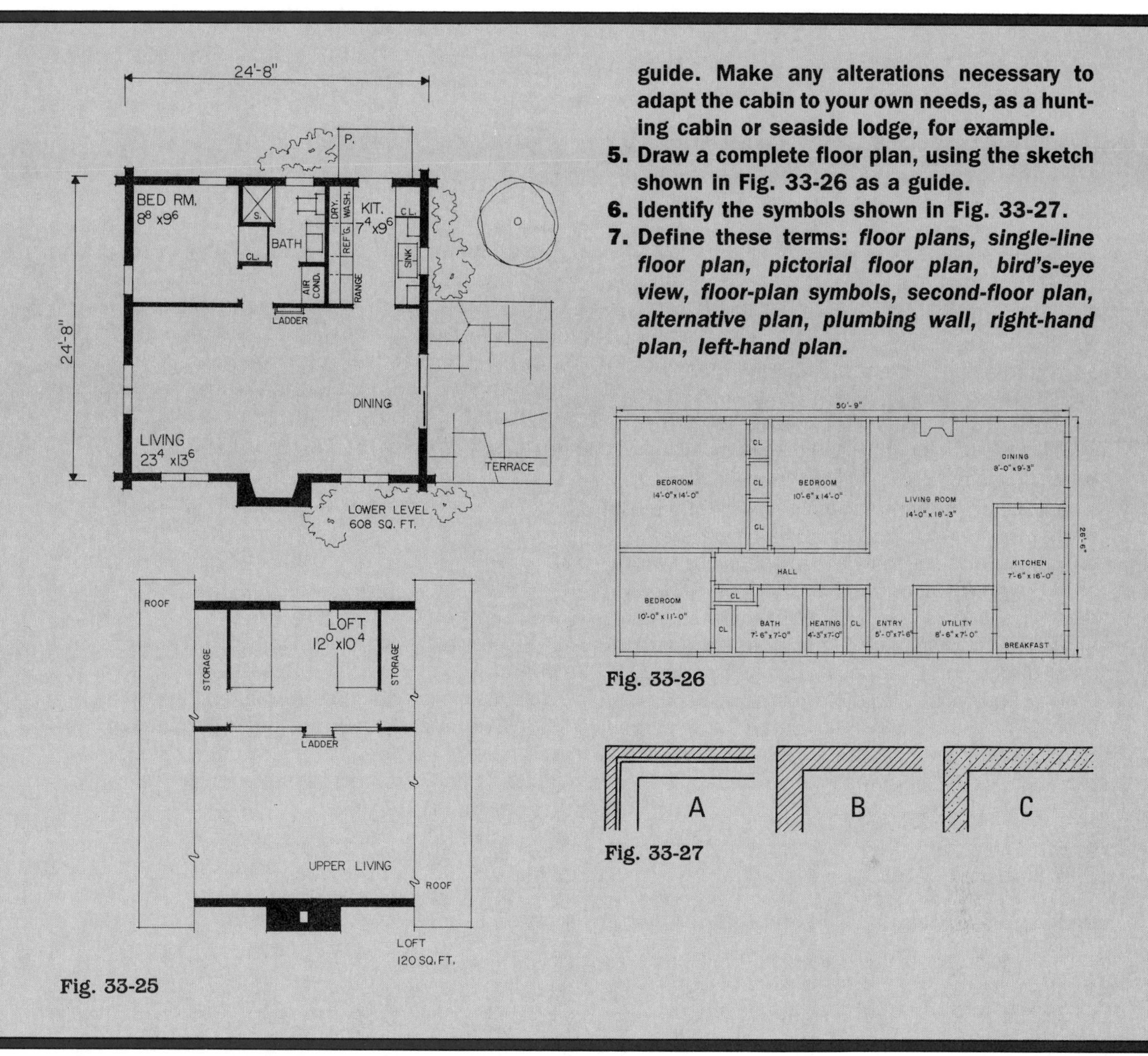

Fig. 33-25

guide. Make any alterations necessary to adapt the cabin to your own needs, as a hunting cabin or seaside lodge, for example.

5. **Draw a complete floor plan, using the sketch shown in Fig. 33-26 as a guide.**
6. **Identify the symbols shown in Fig. 33-27.**
7. **Define these terms: *floor plans, single-line floor plan, pictorial floor plan, bird's-eye view, floor-plan symbols, second-floor plan, alternative plan, plumbing wall, right-hand plan, left-hand plan.***

Fig. 33-26

Fig. 33-27

UNIT 34

FLOOR-PLAN DIMENSIONING

In colonial times, simple cabins could be built without architectural plans and without established dimensions. The outline of the house and the position of each room could be determined experimentally by pacing off approximate distances. The owner could then erect the dwelling, using existing materials and adjusting sizes and dimensions as necessary. The owner acted in the capacities of architect, designer, contractor, carpenter, and materials manufacturer.

Today, building materials are so varied, construction methods so complex, and design requirements so demanding that a completely dimensioned drawing is necessary to complete any building exactly as designed.

SIZE DESCRIPTION

Dimensions on the floor plan show the builder the width and length of the building. They show the location of doors, windows, stairs, fireplaces, planters, and so forth. Just as symbols and notes show exactly what materials are to be used in the building, dimensions show the sizes of materials and exactly where they are to be located.

Architectural dimensioning drawing style differs from that for mechanical drawing dimensioning. Dimensioning practices will also vary among designers. Several common methods of dimensioning floor plans are shown in Fig. 34-1. Because a large building must be drawn on a relatively small sheet, a small scale (¼″ = 1′-0″ or ⅛″ = 1′-0″) must be used. The use of such a small scale means that many dimensions must be crowded into a very small area. Therefore, only major dimensions such as the overall width and length of the building and of separate rooms, closets, halls, and wall thicknesses are shown on the floor plan. Dimensions too small to show directly on the floor plan are described either by a note on the floor plan or by separate, enlarged details. *Enlarged details* are sometimes merely enlargements of some portion of the floor plan. They may also be an allied section indexed to the floor plan. Separate details are usually necessary to interpret adequately the dimensioning of fireplaces, planters, built-in cabinets, door and window details, stair-framing details, or any unusual construction methods.

Complete Dimensions

The number of dimensions included on a floor plan depends largely on how much freedom of interpretation the architect wants to give to the builder. If complete dimensions are shown on the plan, a builder cannot deviate from the original design. However, if only a few dimensions are shown, then the builder must determine many of the sizes of areas, fixtures, and details. When you rely on a builder to provide dimensions, you place the builder in the position of a designer. A good builder is not expected to be a good designer. Supplying adequate dimensions will eliminate the need for guesswork.

Limited Dimensions

A floor plan with only limited dimensions is shown in Fig. 34-2. This type of dimensioning, which shows only the overall building dimensions and the width and length of each room, is sufficient to summarize the relative sizes of the building and its rooms for the prospective owner.

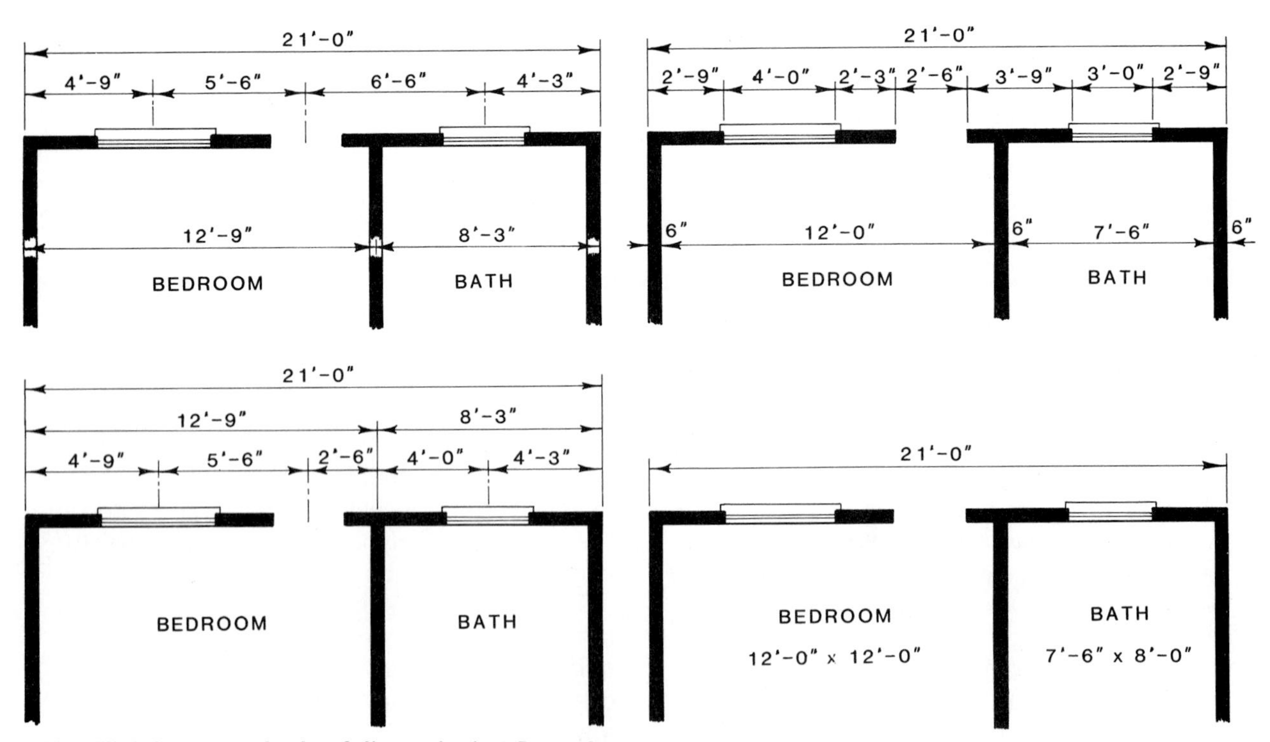

Fig. 34-1 Some methods of dimensioning floor plans.

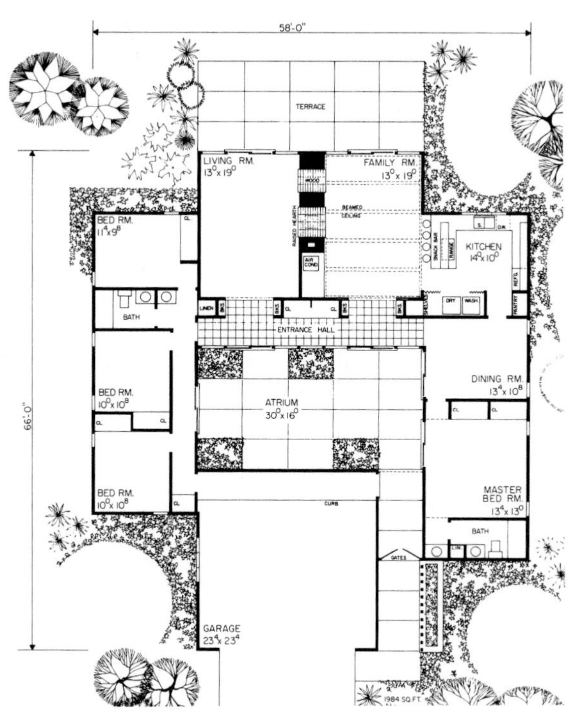

Fig. 34-2 A floor plan with minimum amount of dimensions.

These dimensions are not sufficient for building purposes.

A floor plan must be completely dimensioned (Fig. 34-3) to ensure that the house will be constructed precisely as designed. These dimensions convey the exact wishes of the architect and owner to the builder, and little tolerance is allowed the contractor in interpreting the size and position of the various features of this plan. The exact size of each room, closet, partition, door, or window is given.

RULES FOR DIMENSIONING

Many construction mistakes result from errors in architectural drawings. Most errors in architectural drawing result from mistakes in dimensioning. Dimensioning errors are therefore costly in time, efficiency, and money. Familiarization with the following rules for dimensioning floor plans will eliminate much confusion and error. These rules are illustrated by the numbered arrows in Fig. 34-3.

1. Architectural *dimension lines* are unbroken lines with dimensions placed above the line. Arrowheads styles are optional (Fig. 34-4).

2. Foot and/or inch marks are used on all architectural dimensions.

3. Dimensions over 1′ are expressed in feet and inches.

4. Dimensions less than 1′ are shown in inches.

5. A slash is often used with fractional dimensions to conserve vertical space.

6. Dimensions should be placed to read from the right or from the bottom of the drawing.

7. Overall building dimensions are placed outside the other dimensions.

8. Line and arrowhead weights for architectural dimensioning are the same as those used in dimensioning mechanical drawings.

9. Room sizes may be shown by stating width and length.

10. When the area to be dimensioned is too small for the numerals, they are placed outside the extension lines.

11. Rooms are sometimes dimensioned from center lines of partitions; however, rule 13 is preferred.

12. Window and door sizes may be shown directly on the door or window symbol or may be indexed to a door or window schedule.

13. Rooms are dimensioned from wall to wall, exclusive of wall thickness.

14. Curved leaders are sometimes used to eliminate confusion with other dimension lines.

15. When areas are too small for arrowheads, dots may be used to indicate dimension limits.

16. The dimensions of brick or stone veneer must be added to the framing dimension (Fig. 34-5).

17. When the space is small, arrowheads may be placed outside the extension lines.

18. A dot with a leader refers to the large area noted.

19. Dimensions that cannot be seen on the floor plan or those too small to place on the object are placed on leaders for easier reading.

20. In dimensioning stairs, the number of risers is placed on a line with an arrow indicating the direction (down or up).

21. Windows, doors, pilasters, beams, construction members, and areaways are dimensioned to their centerlines.

22. Use notes when symbols do not show clearly what is intended.

23. Subdimensions must add up to overall dimensions. For example: 14′-9″ + 11′-9″ = 26′-6″.

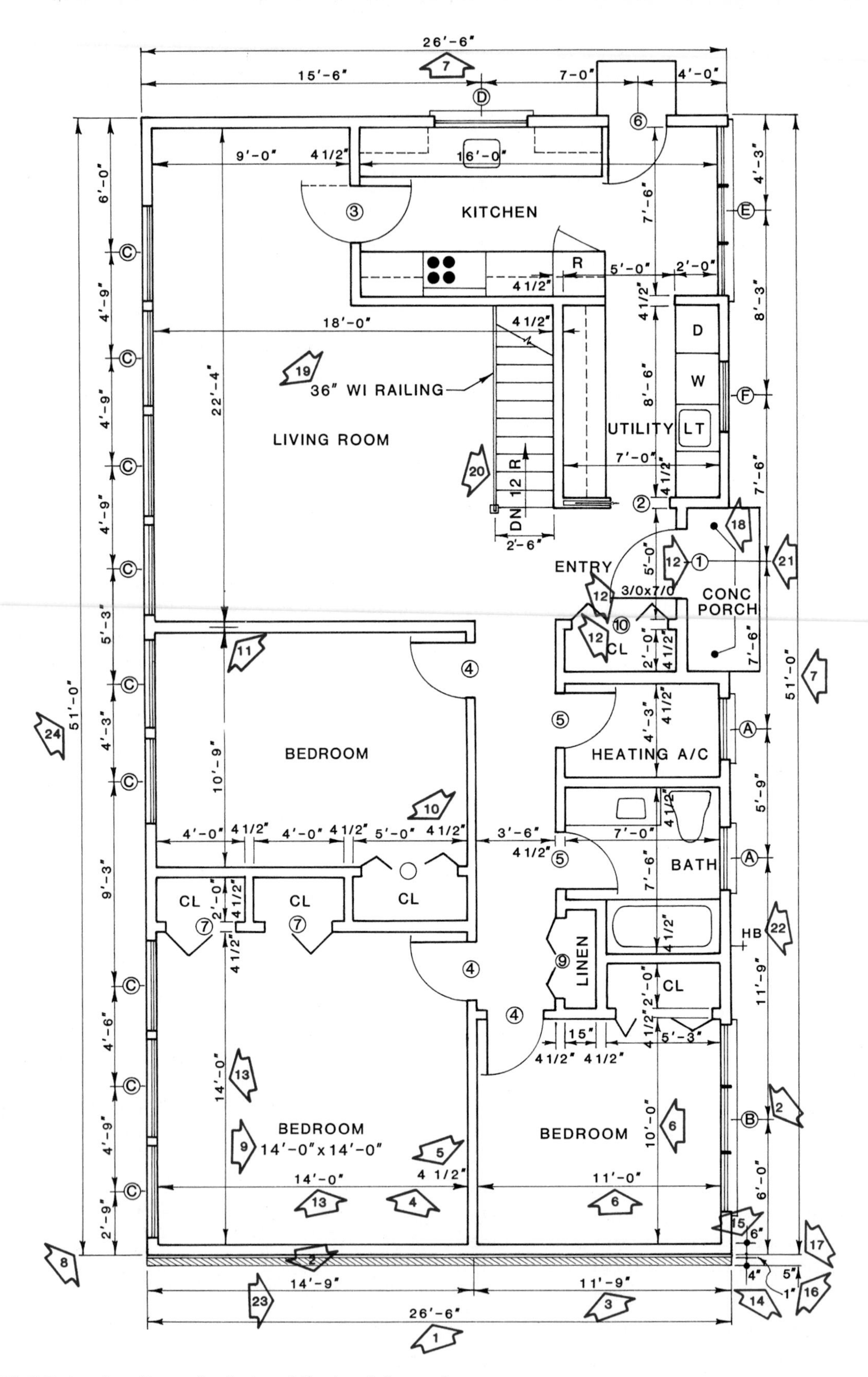

Fig. 34-3 Rules for dimensioning architectural floor plans.

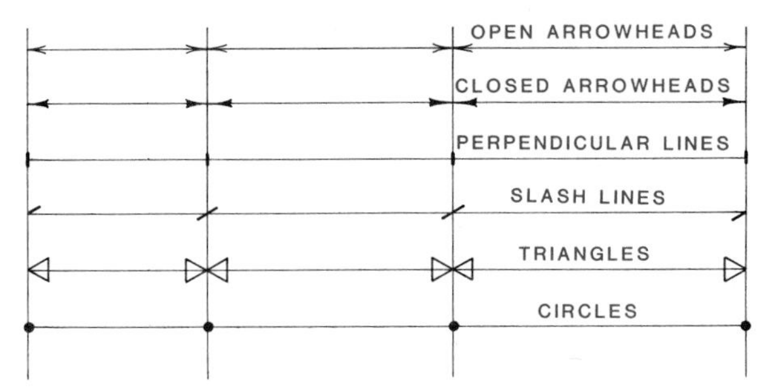

Fig. 34-4 Different styles of arrowheads used on dimension lines.

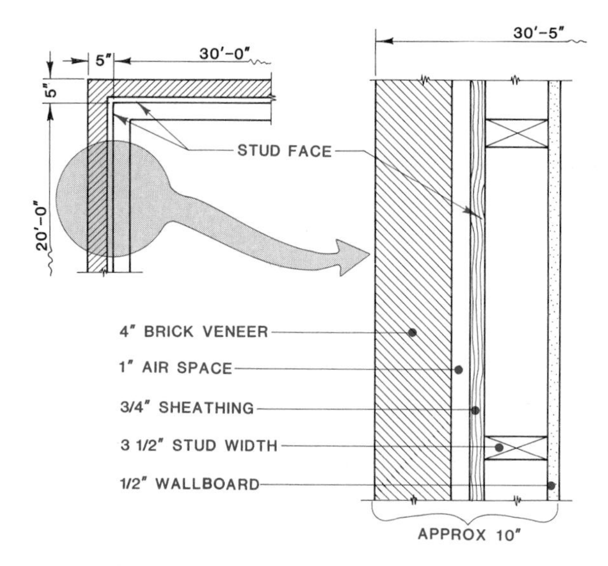

Fig. 34-5 Masonry dimensions must be added to the framing dimensions.

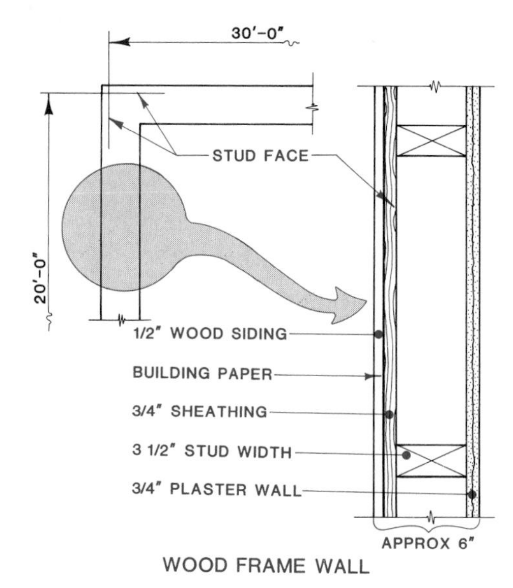

Fig. 34-6 When framing dimensions alone are desired, the dimensions should read to the face of the stud.

24. Architectural dimensions always refer to the actual size of the building regardless of the scale of the drawing. The building in Fig. 34-3 is 26′-6″ wide.

25. When framing dimensions are desirable, rooms are dimensioned by distances to the outside face of the studs in the partitions. Refer to Fig. 34-6.

26. Since building materials vary somewhat in size, first establish the thickness of each component of the wall and partition, such as furring thickness, panel thickness, plaster thickness, stud thickness, brick and tile thicknesses. Add these thicknesses together to establish the total wall thickness. Common thicknesses of wall and partition materials are shown in Figs. 34-5 and 34-6.

A floor plan can be dimensioned on a CAD system by placing the stylus on the end of each distance to be dimensioned. The computer will then measure and calculate this distance. When the stylus is placed on the location desired, the system will draw the dimension, lines, and arrowheads. The drafter selects the dimensioning mode (customary, metric, decimal, fraction), arrowhead style, text size, and distance from object to dimension location.

The floor plan dimensions in this text adhere to the American National Standards Institute (ANSI) drafting manual.

MODULAR CONSTRUCTION

Buildings to be erected with modular components must be designed within modular limits. In dimensioning by the modular system, building dimensions are expressed in standard sizes. This procedure ensures the proper fitting of the various components. Planning rooms to accommodate standard materials also saves considerable labor, time, and material. The modular system of coordinated drawings is based on a standard grid placed on the width, length, and height of a building, as shown in Fig. 34-7.

In modular designing, an effort is made to establish all building dimensions (width, length, and height) to fall on a 4″ module. Building materials should be selected to conform to this module. As new building materials are developed, their sizes are established to conform to

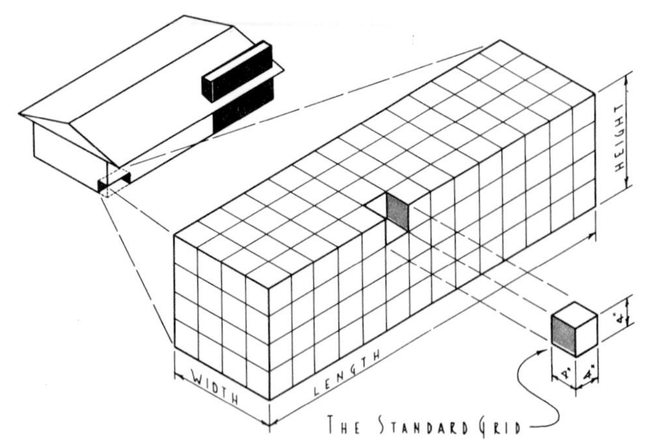

Fig. 34-7 A modular grid.

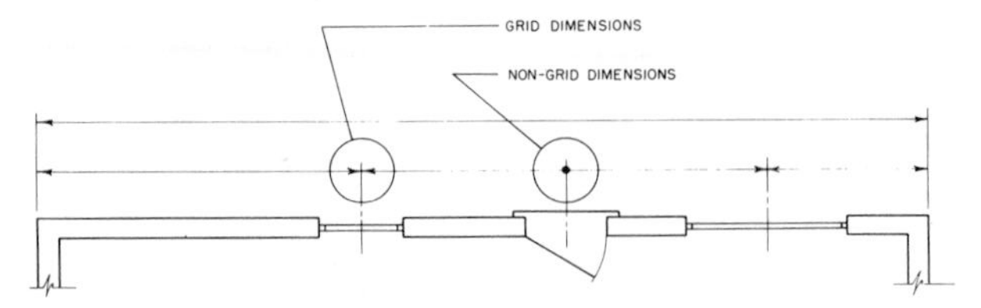

Fig. 34-8 Modular dimensioning methods.

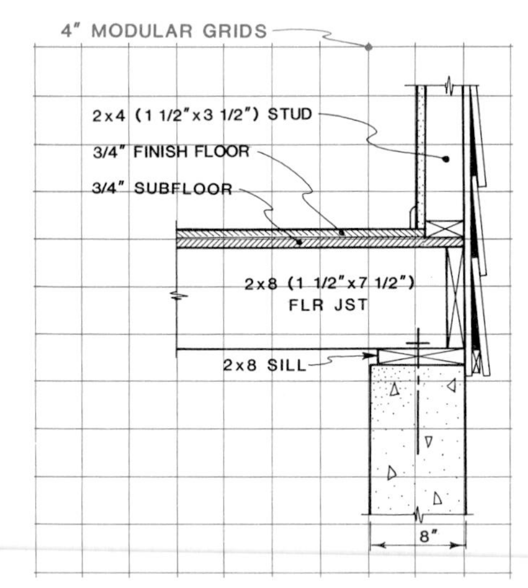

Fig. 34-9 Modular dimensions, as applied to detail drawings.

modular sizes. However, many building materials do not conform to the 4″ grid, and therefore the dimensioning procedure must be adjusted accordingly. Dimensions that align with the 4″ module are known as *grid dimensions.* Dimensions that do not align with the 4″ module are known as *nongrid dimensions.* Figure 34-8 shows the two methods of indicating grid dimensions and nongrid dimensions. Grid dimensions are shown by conventional arrowheads, and nongrid dimensions are shown by dots instead of arrowheads on the dimension lines.

When using a CAD system, grid dimensions are aligned on the grid-pick network. Non-grid dimensions are drawn in the free pick mode.

In many detail drawings, it is possible to eliminate the placement of some dimensions by placing the grid lines directly on the drawing, as shown in the detail given in Fig. 34-9. When the 4″ grid lines coincide exactly with the material lines, no dimensions are needed, since each line represents 4″ and any building material that is an increment of 4″ is reflected by placement on this grid. Other dimensions that do not coincide, such as the 3½″ stud dimension shown in Fig. 34-9, must be dimensioned by conventional methods.

Exercises

1. Dimension an original floor plan that you have completed for a previous assignment.

2. Find the dimensioning errors in Fig. 34-10. List the dimensioning rule violated in each case. (Example: L violates rule 7.)

3. Draw and dimension the floor plan of your own home.

4. Define the following terms: *dimension line, fractional dimensions, overall dimensions, centerlines, extension lines, leader lines, subdimensions, framing dimensions, modular dimensioning, grid dimension, ANSI.*

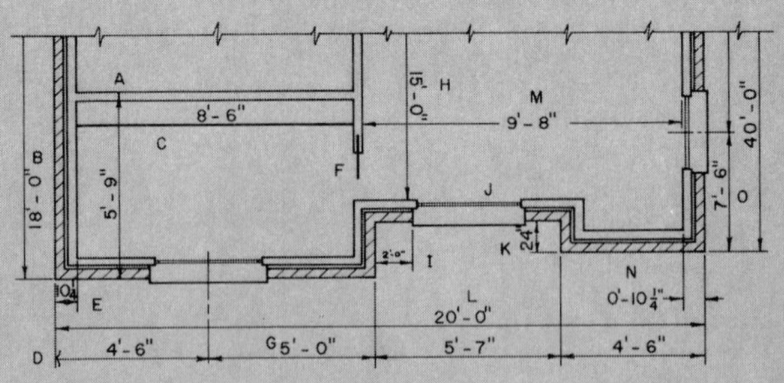

Fig. 34-10

SECTION 8

Elevation Drawings

The main features of the interior of a building are shown on the floor plan. The main features of the outside of a building are shown on elevation drawings. Elevation drawings are orthographic drawings of the exterior of a building. They are prepared to show the design, materials, dimensions, and final appearance of the exterior of a building.

UNIT 35

ELEVATION DESIGN

Designing the elevation of a structure is only one part of the total design process. However, this is the part of the building that people observe, and it is the part they use to judge the entire structure.

RELATIONSHIP WITH THE FLOOR PLAN

Since a structure is designed from the inside out, the design of the floor plan normally precedes the design of the elevation. The complete design process requires a continual relationship between the elevation and the floor plan through the entire process.

Flexibility is possible in the design of elevations, even in those designed from the same floor plan. Figure 35-1 shows the development of two different elevations from the same basic floor plan. When the location of doors, windows, and chimneys has been established on the floor plan, the development of an attractive and functional elevation for the structure still depends on the factors of roof style, overhang, grade-line position, and relationship of windows, doors, and chimneys to the building line. Figure 35-1 clearly shows that choosing a desirable elevation design is not an automatic process that follows the floor-plan design, but a development which calls on the imagination.

The designer should keep in mind that only horizontal distances can be established on the floor plan and that the vertical heights, such as heights of windows, doors, and roofs must be shown on the elevation. As these vertical heights are established, the appearance of the outside and the functioning of the heights as they affect the internal functioning of the building must be considered, as shown in Fig. 35-2.

FUNDAMENTAL SHAPES

The basic architectural style of a building is more closely identified with the design of the elevation than with any other factor. Consequently the selection of the basic type of structure must be compatible with the architectural style of the elevation. The elevation design can be changed to create the appearance of different architectural styles if the basic building type is consistent with that style. Within basic styles of architecture there is considerable flexibility in the type of structure. The basic types of structures include the *one-story* (Fig. 35-3), the *one-and-one-half-story* (Fig. 35-4), the *two-story* (Fig. 35-5), the *split-level* (Fig. 35-6), and the *bilevel* (Fig. 35-7).

Roof Styles

Nothing affects the silhouette of a house more than the roof style. The most common roofs are the gable roof, hip roof, flat roof, and shed roof.

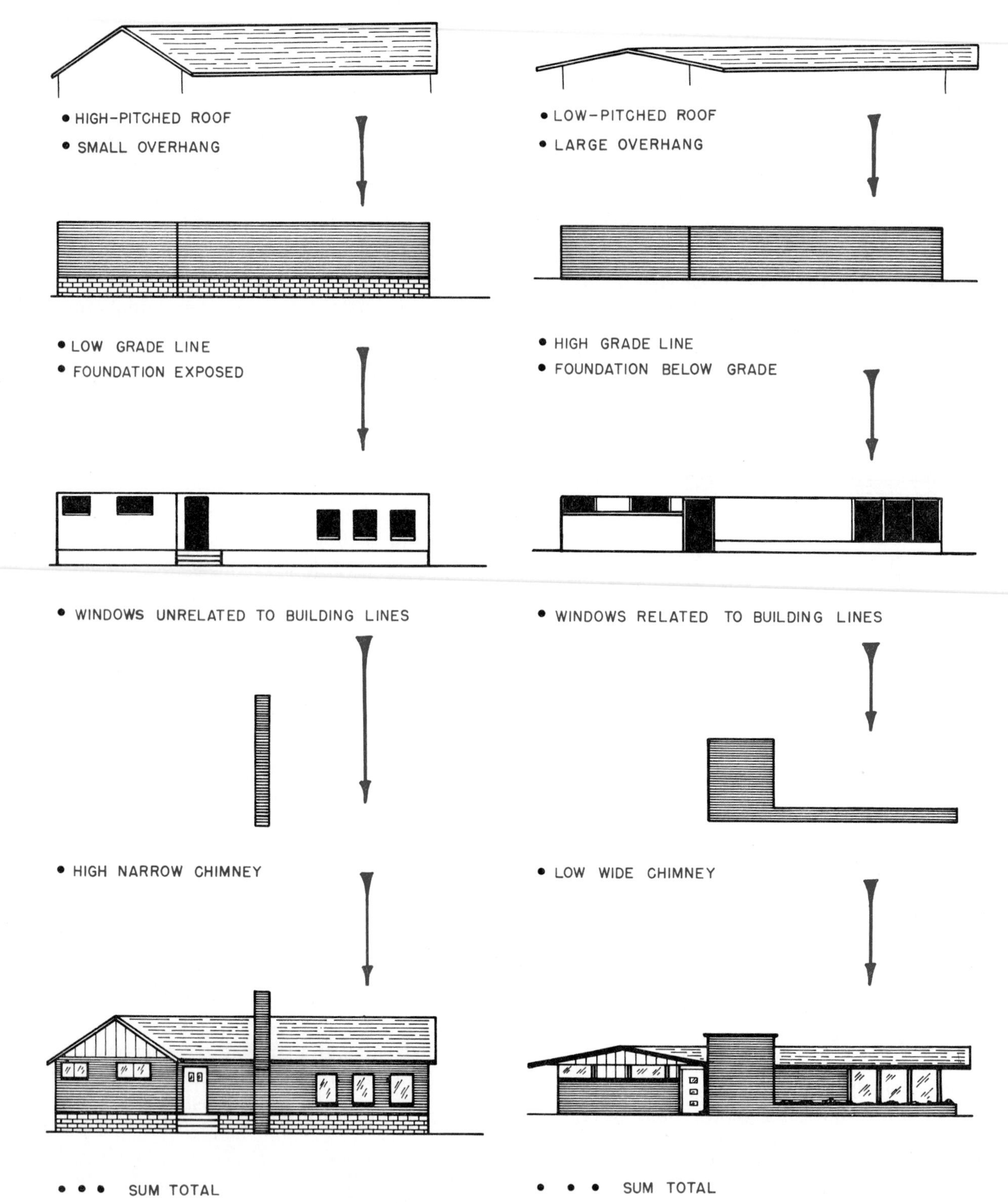

Fig. 35-1 Many factors affect the total appearance of the elevation.

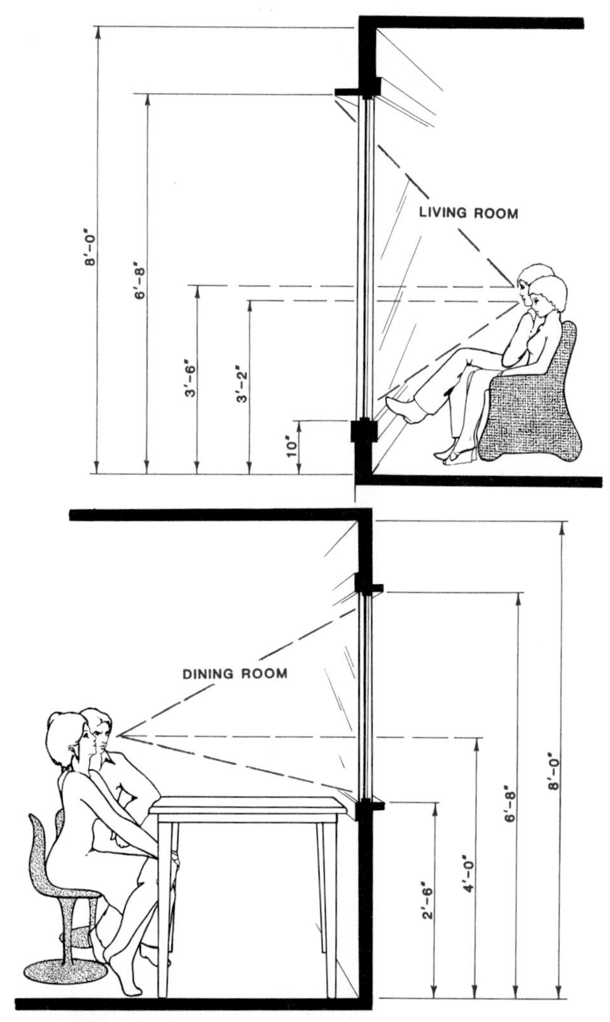

Fig. 35-2 Vertical distances, such as the heights of windows and doors, can be shown only on an elevation drawing. *(Small Homes Council)*

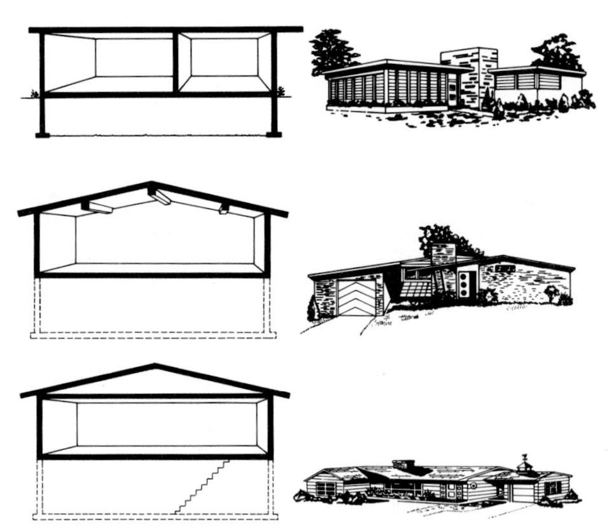

Fig. 35-3 Types of one-story structures. *(National Lumber Manufacturers Assoc.)*

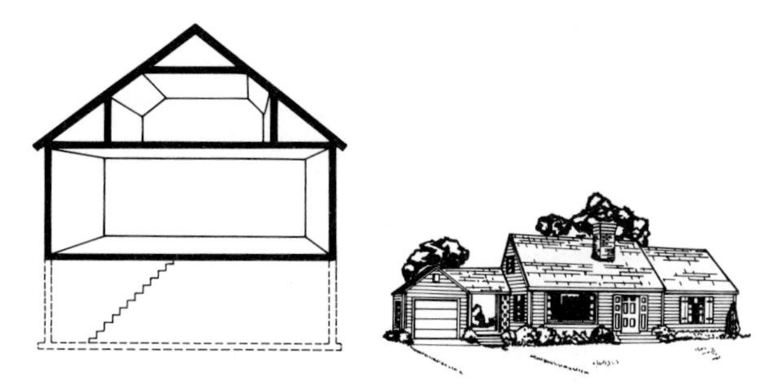

Fig. 35-4 A one-and-one-half-story home. *National Lumber Manufacturers Assoc.)*

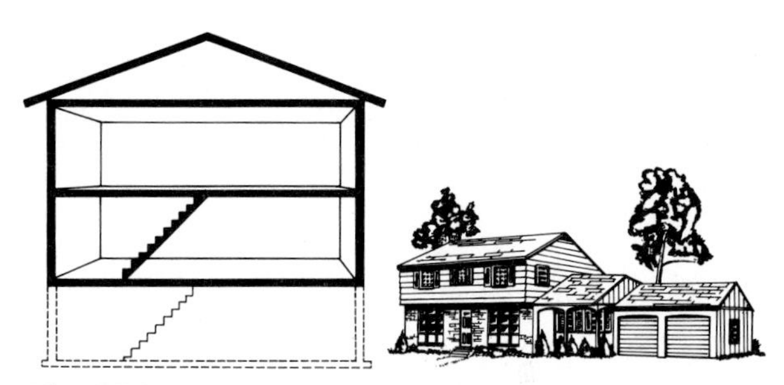

Fig. 35-5 A two-story house. *(National Lumber Manufacturers Assoc.)*

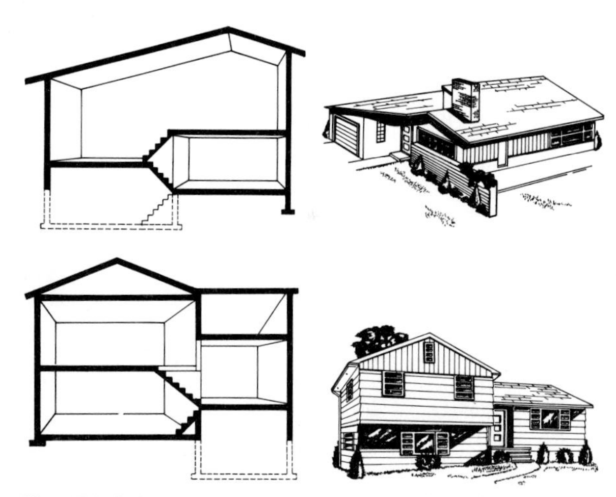

Fig. 35-6 Split-level homes. *(National Lumber Manufacturers Assoc.)*

Figure 35-8 shows examples of different types of roof styles and how they appear in plan and elevation. A change in the roof style of a structure greatly affects the appearance of the elevation even when all other factors remain constant, as shown in Fig. 35-9.

Gable roofs, as shown in Fig. 35-10, are used extensively on Cape Cod and ranch homes. The *pitch* (angle) of a gable roof varies from the high-pitch roofs found on chalet and A-frame style

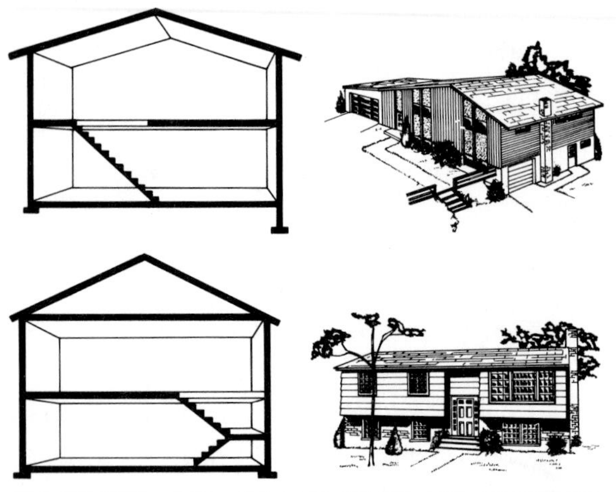

Fig. 35-7 The bilevel house. *(National Lumber Manufacturers Assoc.)*

buildings (Fig. 35-11A) to the low-pitch roofs found on most ranch homes (Fig. 35-11B). Figure 35-12 shows the relationship between the actual appearance of a gable roof and the appearance of a gable roof on an elevation drawing.

Hip roofs are used when eave-line protection is desired around the entire perimeter of the building. Notice how the hip-roof overhang shades the windows of the house in Fig. 35-13. For this reason, hip roofs are very popular in warm climates. Figure 35-14 shows the way a hip roof appears on an elevation drawing. Hip roofs are commonly used on Regency and French provincial homes.

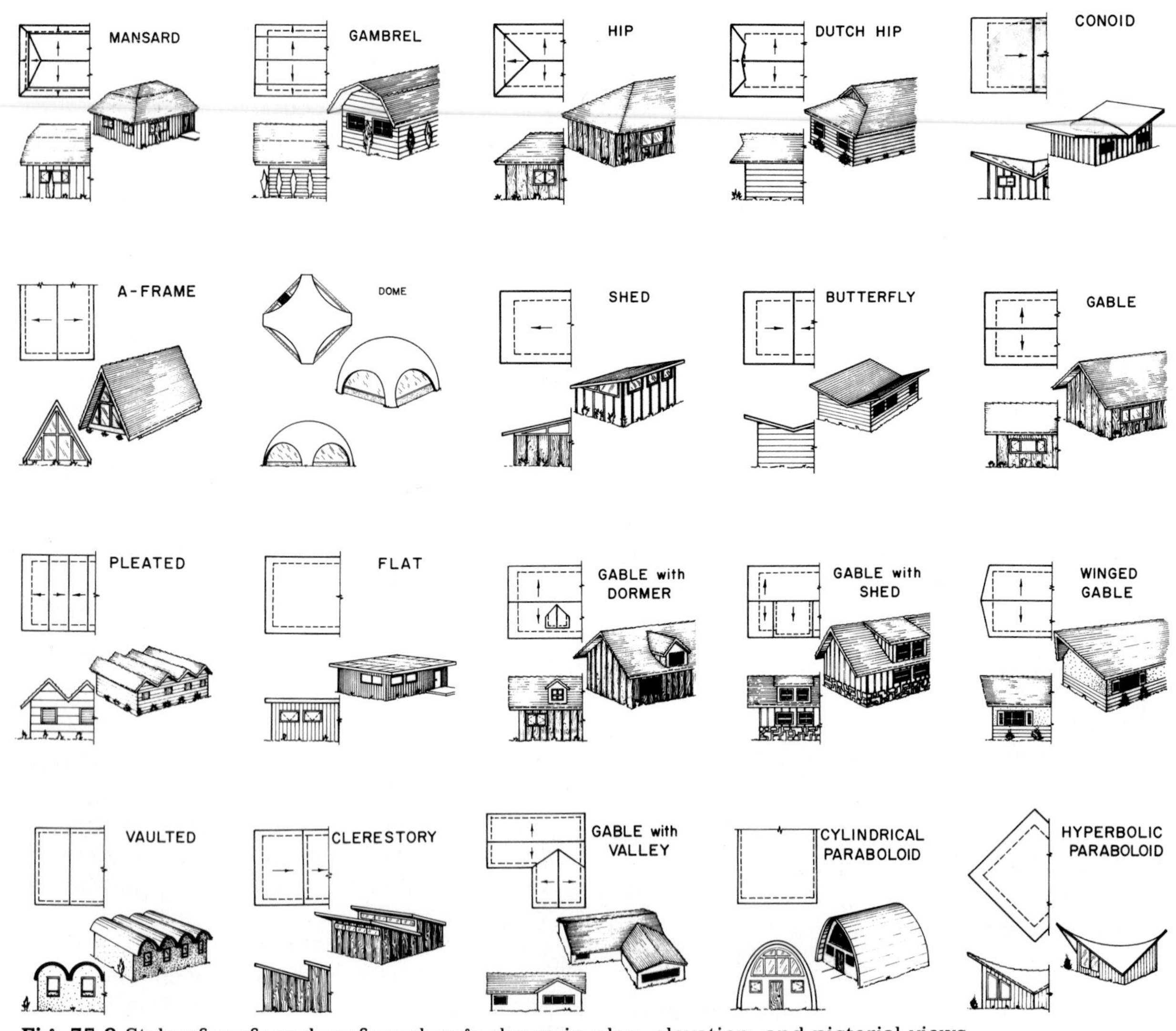

Fig. 35-8 Style of roofs and roof overhangs shown in plan, elevation, and pictorial views.

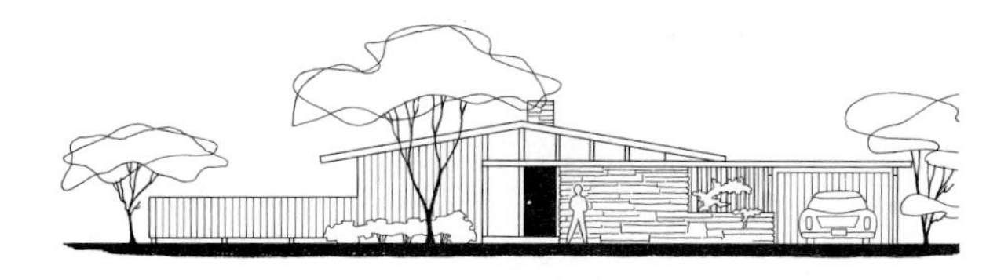

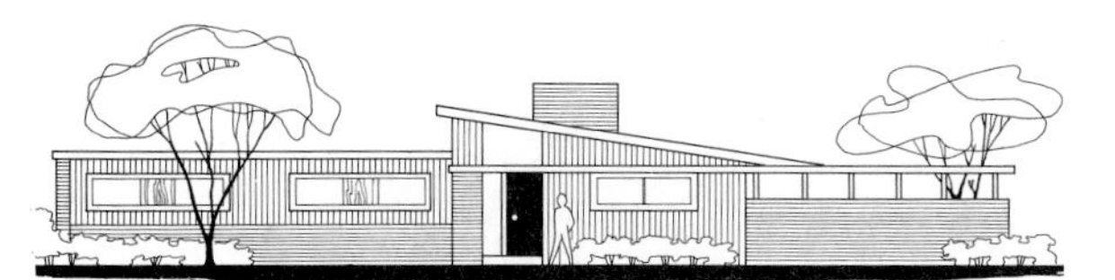

Fig. 35-9 The type of roof greatly affects the appearance of the elevation.

Fig. 35-10 A house with many gables. *(Home Planners, Inc.)*

Fig. 35-11A High-pitch gable roof. *(Master Plan Services, Inc.)*

Fig. 35-11B Low-pitch gable roof.

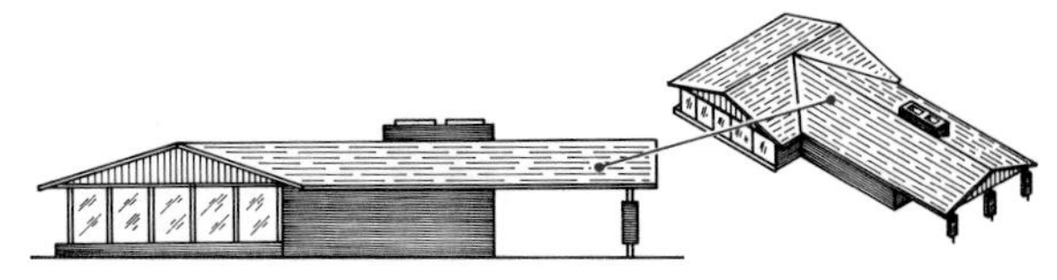

Fig. 35-12 The appearance of a gable roof on an elevation drawing.

Fig. 35-13 A house with a hip roof. *(Home Planners, Inc.)*

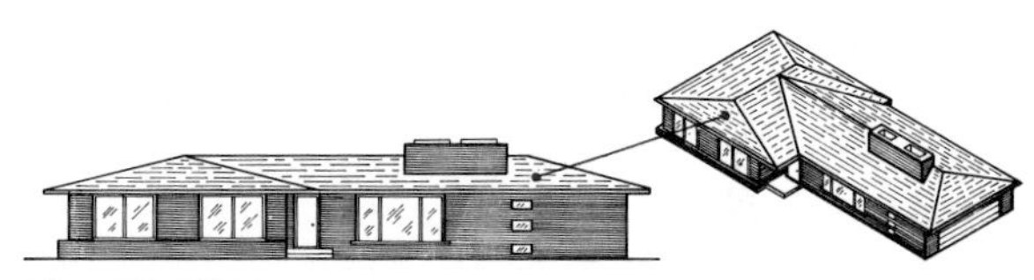

Fig. 35-14 The appearance of a hip roof on an elevation drawing.

Flat roofs are used to create a low silhouette on many modern homes (Fig. 35-15). Since no support is achieved by the leaning together of rafters, slightly heavier rafters are needed for flat roofs. Built-up asphalt construction is often used on flat roofs. Water may be used as an insulator and solar heater on flat roofs. Figure 35-16 illustrates a flat roof on an elevation drawing.

Shed roofs are flat roofs that are higher at one end than at the other. They may be used effectively when two levels exist and where additional

Fig. 35-15 A home with a flat roof. *(Home Planners, Inc.)*

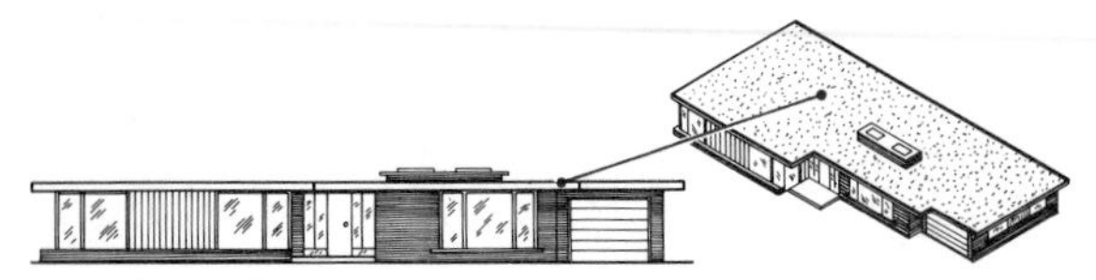

Fig. 35-16 The appearance of a flat roof on an elevation drawing.

light is needed. The use of *clerestory windows* between the two sheds (Fig. 35-17) provides skylight illumination. The double shed is really very advantageous on hillside split-level structures. Figure 35-18 shows a shed roof on an elevation. The *geodesic dome* may constitute the roof (Fig. 35-19) or may extend completely to the ground and be part of the side wall of a structure as well.

Overhang

Sufficient roof overhang should be provided to afford protection from the sun, rain, and snow. The length and angle of the overhang will greatly affect its appearance and its functioning in providing protection. Figure 35-20 shows that when the pitch is low, a larger overhang is needed to provide protection. However, with a high-pitch roof, the overhang may block the view from the inside, if extended to equal the protection of the low-pitch overhang. To provide protection and at the same time allow sufficient light to enter the windows, slatted overhangs may be used. Figure

Fig. 35-17 A shed roof with clerestory windows.

Fig. 35-19 A geodesic dome. *(Kaiser Aluminum and Chemical Corp.)*

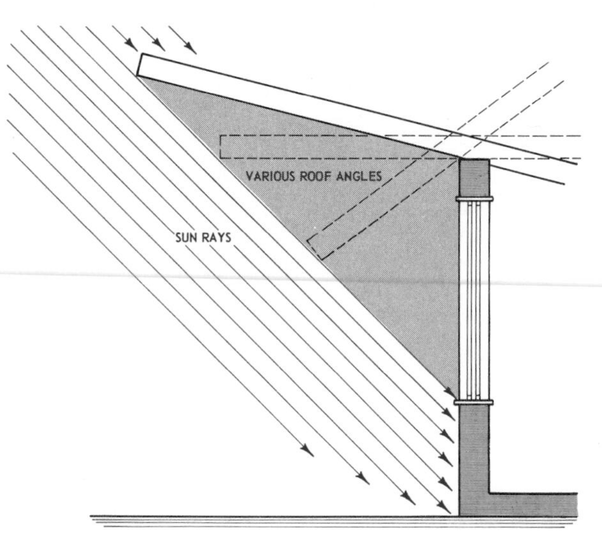

Fig. 35-20 The angle of the overhand determines its length.

35-21 shows the effect of a small overhang. Figure 35-22 shows the protection afforded by a large overhang. The fascia edge of the overhang does not always need to be parallel to the sides of the house. The amount of overhang is also determined by architectural style.

FORM AND SPACE

The total appearance of the elevation depends upon the relationship among the areas of the

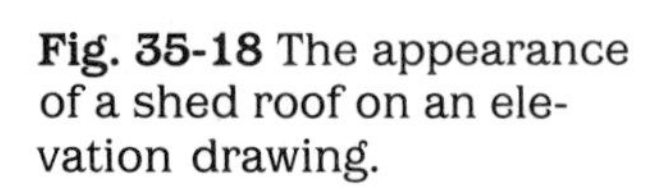

Fig. 35-18 The appearance of a shed roof on an elevation drawing.

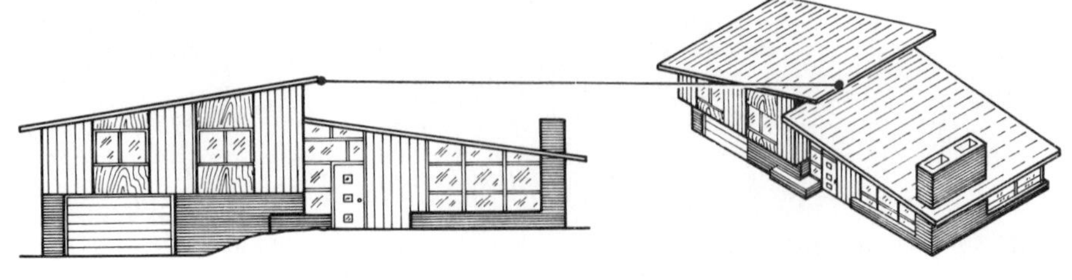

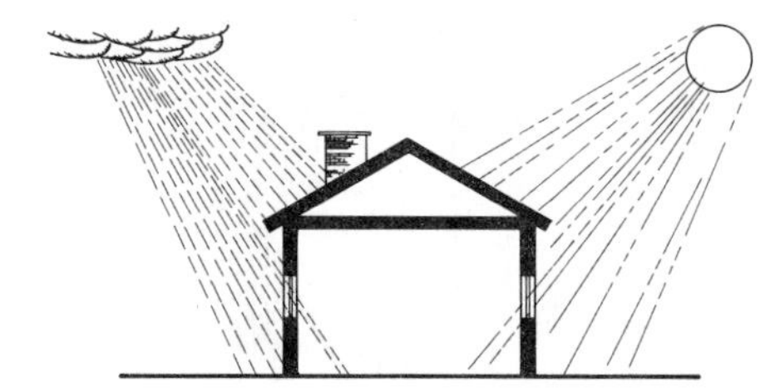

Fig. 35-21 The effect of a small overhang.

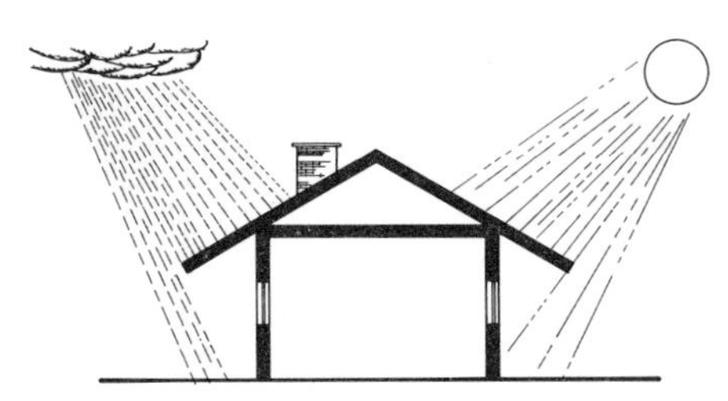

Fig. 35-22 The advantage of a large overhang.

elevation such as surfaces, doors, windows, and chimneys. The balance of these areas, the emphasis placed on various components of the elevation, the texture, the light, the color, and the shadow patterns all affect greatly the general appearance of the elevation.

RELATED AREAS

The elevation should appear as one integral and functional facade rather than as a surface in which holes have been cut for windows and doors and to which structural components, such as chimneys, have been added without reference to the other areas of the elevation. Doors, windows, and chimney lines should constitute part of the general pattern of the elevation and should not exist in isolation. Figure 35-23 shows an elevation in which the windows and doors are related to the major lines of the elevation. Figure 35-24 shows the same elevation with unrelated doors and windows.

Windows

When the vertical lines of the windows are extended to the eave line from the ground line or planter line or some division line in the separation of materials, the vertical lines become related to the building. Horizontal lines of a window extending from one post to another, or from one vertical separation in the elevation to another, as between a post and a chimney, also help to relate the window to the major lines of

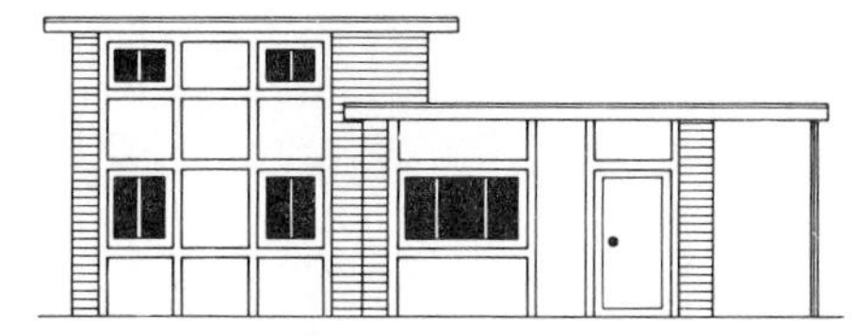

Fig. 35-23 An elevation drawing, showing related lines.

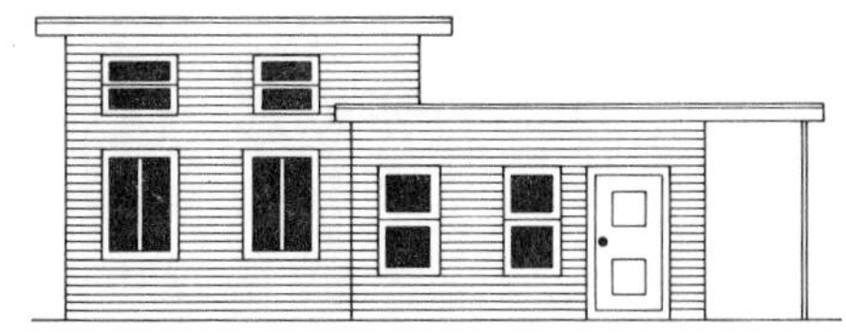

Fig. 35-24 Unrelated elevation lines.

the elevation. For example, the windows in Fig. 35-25A were related by being extended to fill the area between posts and between the door and eave line. Figure 35-25B shows the interior view of different window styles.

Doors

Doors can easily be related to the major lines of the house by extending the area above the door to the *roof line,* or divider, and making the side

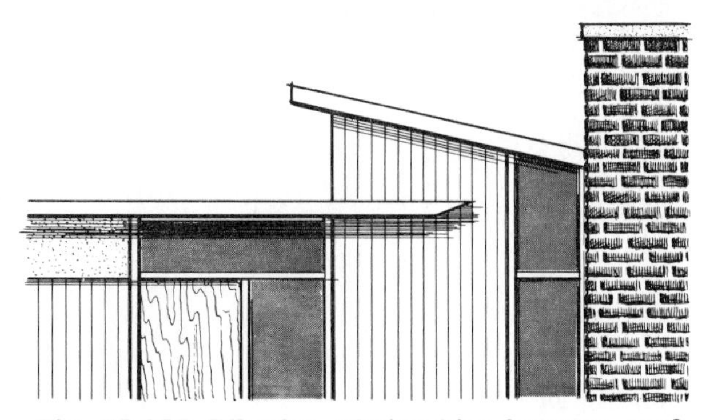

Fig. 35-25A Effective relationships between roof, windows, siding, and chimney. *(Home Planners, Inc.)*

Fig. 35-25B Interior view of window types.

panels consistent with the door size, as shown in Fig. 35-26. A comparison between Fig. 35-27 and Fig. 35-28 shows the effect of extending the door lines vertically and horizontally to integrate the door with the window line and the overall surface of the elevation.

SHAPE

Although it is important to relate the lines of the elevation to each other, nevertheless the overall shape of the elevation should reflect the basic shape of the building. Do not attempt to camouflage the shape of the elevation. An example of such camouflage is the old western building (Fig. 35-29), whose builder tried to disguise the actual elevation by constructing a false front. Patio walls, fences, or other structures may block the view of the elevation wall. If such blocking occurs, it is advisable to draw the elevation with the wall in position, to show how the elevation will appear when viewed from a distance. It is advisable also to draw the elevation as it will exist inside the wall.

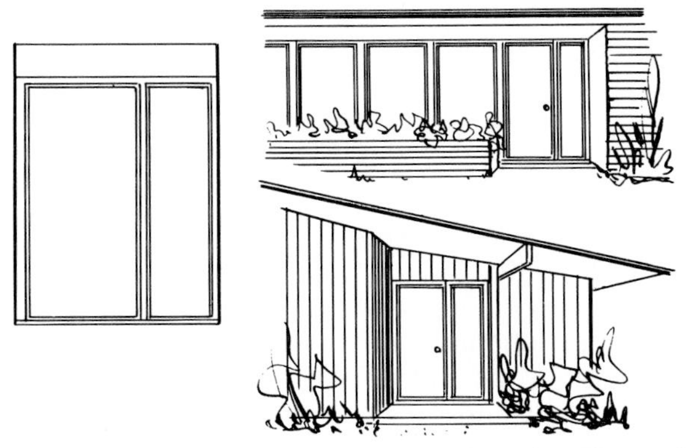

Fig. 35-26 A door unit related to the other lines of the elevation.

Fig. 35-29 Avoid elevation camouflage.

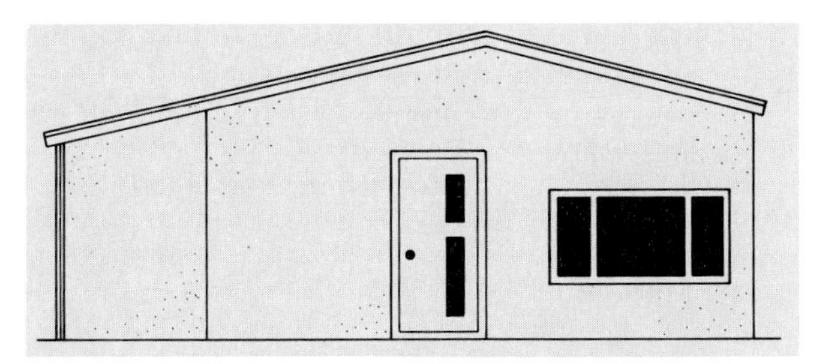

Fig. 35-27 Unextended door and window lines.

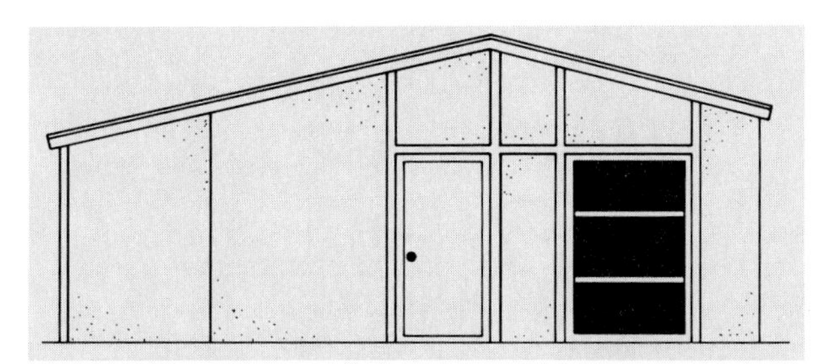

Fig. 35-28 Door and window lines extended horizontally and vertically.

Balance

The term *balance* refers to the symmetry of the elevation. An elevation is either *formally* or *informally* balanced (Fig. 35-30). Formal balance is used extensively in colonial and period styles of architecture. Informal balance is more widely used in modern residential architecture. Informally balanced elevations are frequently used in modern designs such as ranch and split-level styles.

Emphasis

Elevation emphasis, or *accent*, can be achieved by several different devices. An area may be accented by mass, by color, or by material. Every elevation should have some point of emphasis.

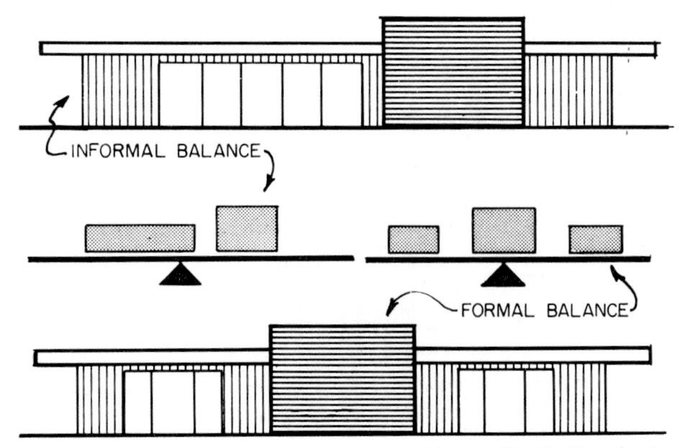

Fig. 35-30 A formally and an informally balanced elevation.

Compare the two elevations in Fig. 35-31. Note that in the second example the chimney and the roof have been accented to provide a focal point.

Light and Color

An elevation that is composed of all light areas or all dark areas tends to be uninteresting and neutral. Some balancing of light, shade, and color is desirable in most elevations. This can be achieved through developing shadow patterns by depressing areas, by using overhang, by texturing, and by color variation.

Texture

An elevation contains many kinds of materials such as glass, wood, masonry, and ceramics. These must be carefully and tastefully balanced to be effective. An elevation composed of too few materials is ineffective and neutral. Likewise, an elevation that uses too many materials, especially masonry, is equally objectionable. In choosing the materials for the elevation, the designer should not mix horizontal and vertical siding or different types of masonry. If brick is the primary masonry used, brick should be used throughout. It should not be mixed with stone. Similarly, it is not desirable to mix several types of brick or several types of stone. Figure 35-32 shows the effect of mixing too many materials in an elevation.

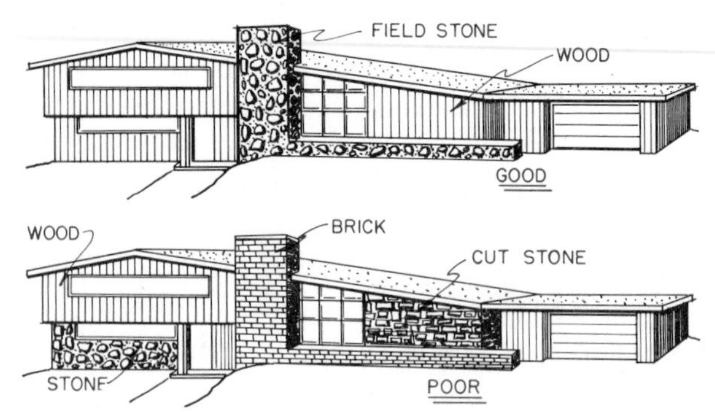

Fig. 35-32 The effect of combining too many materials.

LINES

The major horizontal lines of an elevation are the *ground line*, *eave line*, and *ridge line*. One of these lines should be emphasized. The lines of an elevation can help to create horizontal or vertical emphasis. If the ground line, ridge line, and eave line are accented, the emphasis will be placed on the horizontal. If the emphasis is placed on vertical lines such as corner posts and columns, the emphasis will be vertical. Figure 35-33 shows a comparison between placing the emphasis on horizontal lines and placing the emphasis on vertical lines. In general, low buildings will usually appear longer and lower if the emphasis is placed on horizontal lines.

Lines should be consistent. The lines of an elevation should appear to flow together as one integrated line pattern. It is usually better to continue a line through an elevation for a long distance than to break the line and start it again. Figure 35-34 shows the difference between a

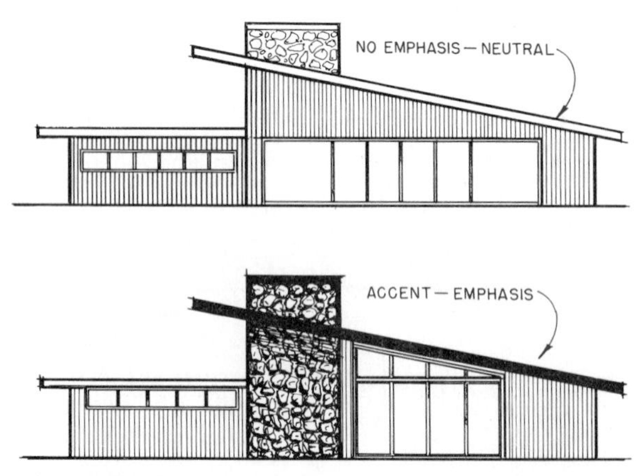

Fig. 35-31 Every elevation should have some point of emphasis.

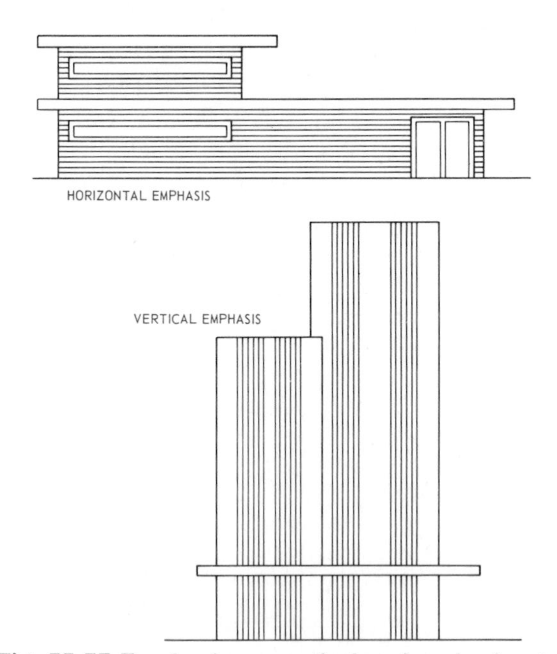

Fig. 35-33 Emphasis on vertical and on horizontal lines.

building with consistent lines and a building with inconsistent lines. Rhythm can be developed by use of lines, and lines can be repeated in various patterns. When a line is repeated, the basic consistency of the elevation design is considerably strengthened.

When additions are made to an existing design, care must be taken to ensure that the lines of the addition are consistent with the established lines of the structure.

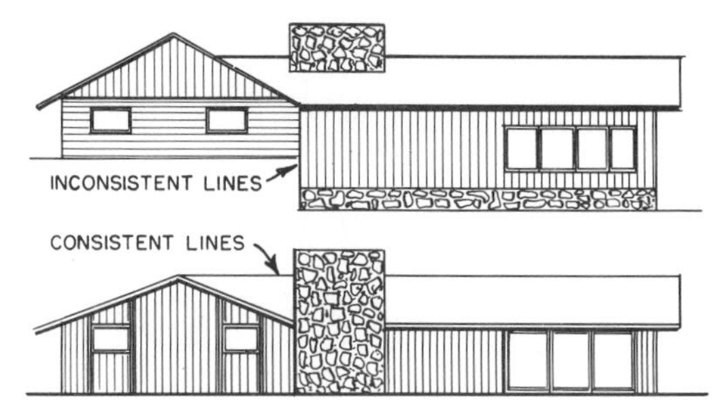

Fig. 35-34 Consistent lines are essential for good elevation design.

Exercises

1. **Sketch an elevation of your own design. Trace the elevation, adding a flat roof, gable roof, shed roof, and butterfly roof. Choose the one you like best and the one that is most functional for your design.**
2. **Sketch the front elevation of your home. Vary the roof style, making it consistent with the major lines of the elevation. Redesign the elevation, relating the door and window lines to the major lines of the building.**
3. **Too many materials are used in the building shown in Fig. 35-35. Sketch an elevation of this house and change the building materials to be consistent with the design.**
4. **Resketch the formally balanced elevation shown in Fig. 35-30. Convert this elevation to an informally balanced elevation.**
5. **Redesign the elevation in Fig. 35-36 with a different roof.**
6. **Define the following terms: *one-story house, one-and-one-half-story house, bilevel, split-level, hip roof, flat roof, gable roof, shed roof, high pitch, low pitch, celestial windows, gambrel roof, mansard roof, butterfly roof, overhang, related lines, unrelated lines, ridge lines, eave lines, ground lines, formal balance, informal balance, texture emphasis.***
7. **Redesign the elevation shown in Fig. 35-37.**

Fig. 35-35

Fig. 35-36

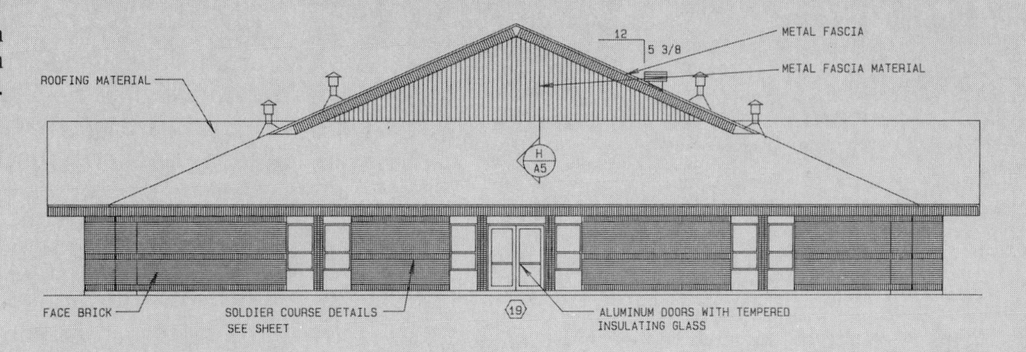

Fig. 35-37 Redesign this elevation with different materials.

UNIT 36

ELEVATION PROJECTION

Elevation drawings are projected from the floor plan of an architectural drawing just as the side views are projected from the front view of an orthographic drawing. To visualize and understand multiview (orthographic) projection, imagine a building surrounded by a transparent box, as shown in Fig. 36-1. If you draw the outline of the structure on the transparent planes that make up the box, you create the orthographic views you need. These are the front view on the front plane, the side view on the side plane, and the top view on the top plane. When the planes of the top, bottom, and sides are hinged (swung) out from the front plane, as shown in Fig. 36-2, the six views of the house are shown exactly as they are positioned on an orthographic drawing. Study the position of each view as it relates to the front view. The right side is to the right of the front view. The left side is to the left, the top(roof) view is on the top, the bottom view is on the bottom. The rear view is to the left of the left-side view, since, when this view hinges around to the back, it would fall into this position.

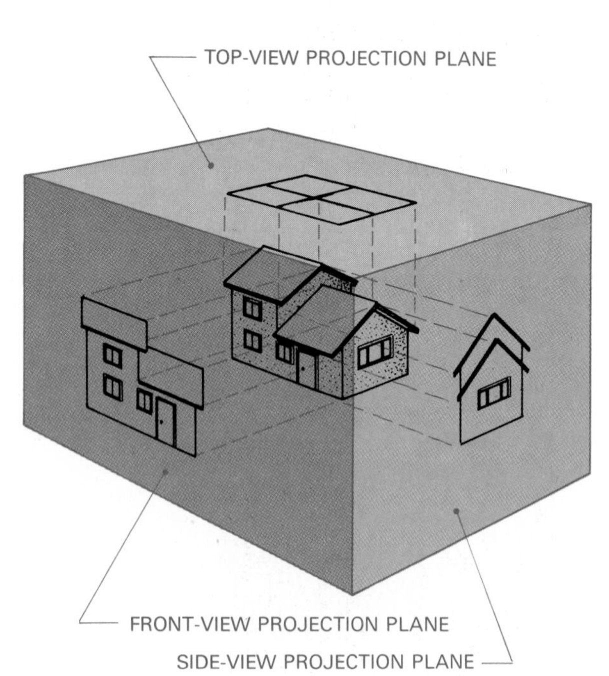

Fig. 36-1 Projection box shows three planes of a building.

Notice that the length of the front view, top (roof) view, and bottom view are exactly the same as the length of the rear view. Notice also that the heights and alignments of the front view, right side, left side, and rear view are the same. Memorize the position of these views and remember that the lengths of the front, bottom, and top views are *always* the same. Similarly, the heights of the rear, left, front, and right side are *always* the same.

All six views are rarely used to depict architectural structures. Instead, only four elevations (sides) are usually shown, and the top view is usually replaced with a section through the structure called a floor plan. The roof plan is also developed from the top view. The bottom view is never developed in a construction drawing. Figure 36-3 shows how elevations are projected from the floor plan. The positions of the chimney, doors, windows, overhang, and building corners are projected directly from the floor plan outward to the elevation plane.

ELEVATION PLANES

You may think of the elevation as a drawing placed on a vertical plane. Figure 36-4 shows how the vertical planes are related to the floor-plan projection.

Functional Orientation

Four elevations are normally projected from the floor plan. When these elevations are classified according to their function, they are called the front elevation, the rear elevation, the right elevation, and the left elevation. When these elevations are all *projected* on the same drawing sheet, the rear elevation appears to be upside down and the right and left elevations appear to rest on their sides. Because of the large size of most elevation drawings, and because of the need to show elevations as we normally see them, each elevation is usually drawn with the ground line on the bottom (Fig. 36-5).

Compass Orientation

The north, east, south, and west compass points are often used by architects to describe and label elevation drawings. This method is pre-

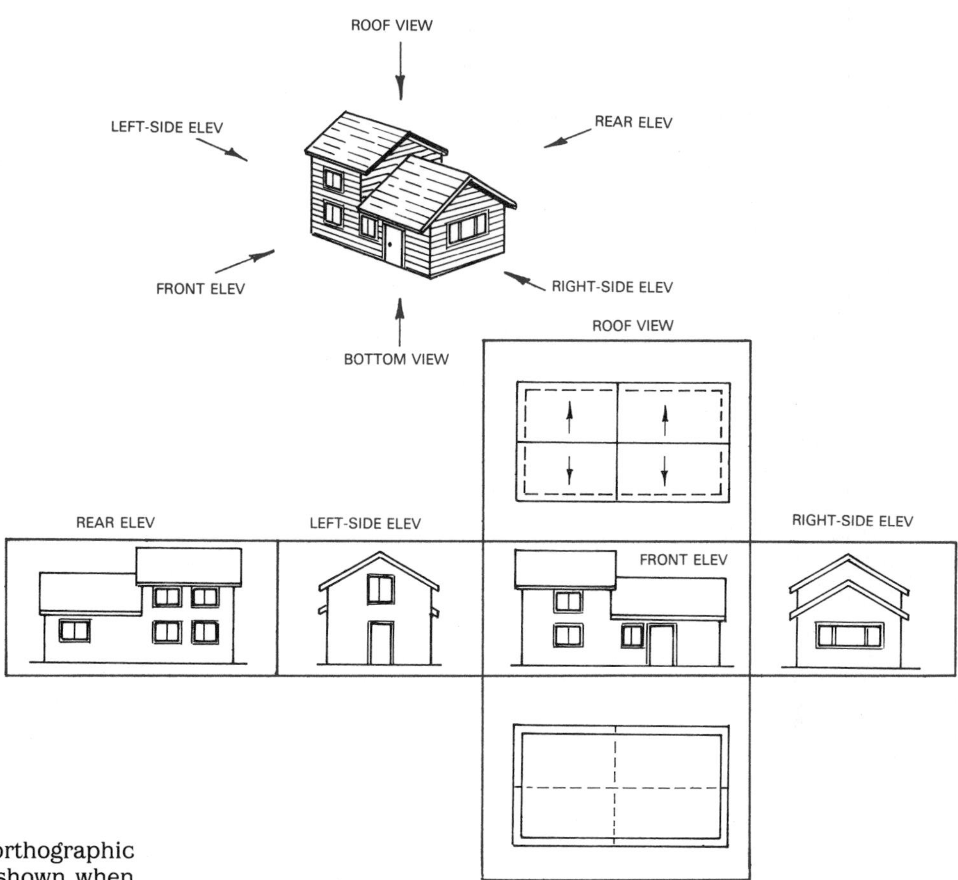

Fig. 36-2 The six orthographic sides of a house are shown when the box is opened and laid flat.

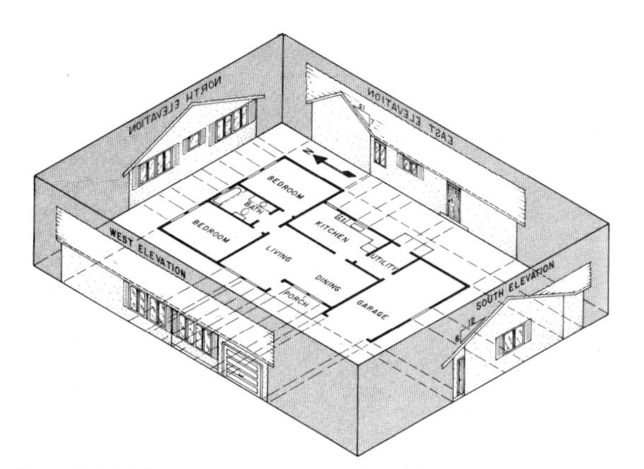

Fig. 36-3 Elevation planes of projection.

ferred because it reduces the chance of elevation callout error. When this method is used, the north arrow on the floor plan is the key to the designation of the elevation title. For example, in Fig. 36-5 the rear elevation is facing north. Therefore, the rear elevation is also the north elevation. The front elevation is the south elevation, the left elevation is the west elevation, and the right elevation is the east elevation.

Auxiliary Elevations

When a floor plan has sides that deviate from the normal 90° projection, an *auxiliary elevation* view is often necessary. To project an auxiliary elevation, follow the same rules for projecting orthographic auxiliaries. Project the auxiliary elevation perpendicular to the wall of the floor plan from which you are projecting (Fig. 36-6). When an auxiliary elevation is drawn, it is usually prepared in addition to the standard elevations and does not replace them. It merely clarifies the foreshortened lines of the major elevations caused by the receding angles.

STEPS IN PROJECTING ELEVATIONS

The major lines of an elevation are derived by projecting vertical lines from the floor plan and measuring the position of horizontal lines from the ground line.

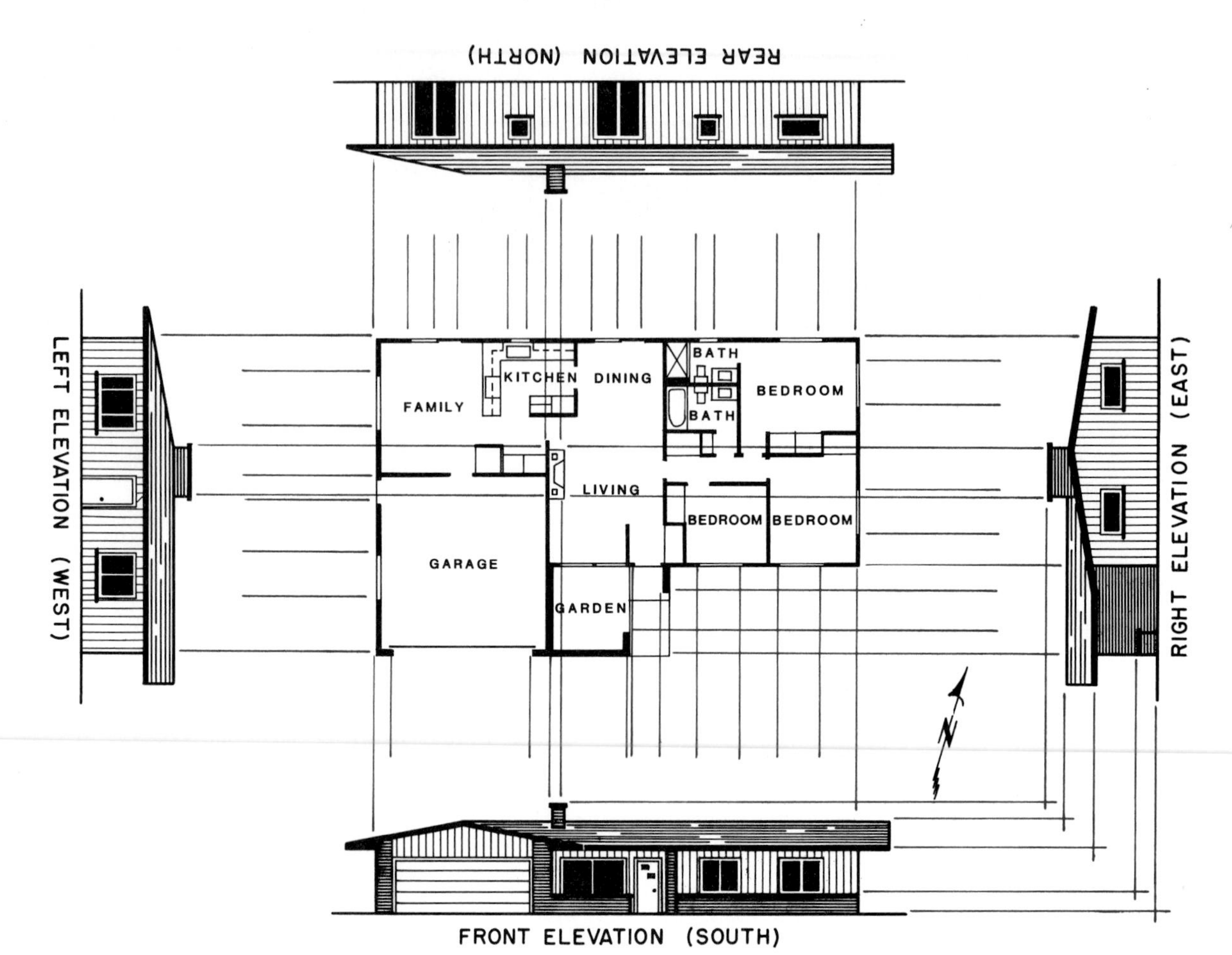

Fig. 36-4 The relationship of side and front elevation views is with the ground line on the bottom of the sheet.

Vertical-Line Projection

Vertical lines representing the main lines of the building should first be projected as shown in Fig. 36-7. These lines show the overall length and width of the building. They also show the length of the major parts or offsets of the building. When projecting an elevation on a CAD system, use the grid pick function to project the major lines from the floor plan to the elevation planes. The floor plan can be rotated 90° to enable the drafter to always see each elevation with the ground line on the bottom during the drawing process.

Horizontal-Line Projection

Horizontal lines that represent the height of the eave line, ridge line, and chimney line above the ground line are projected to intersect with the vertical lines drawn from the floor plan, as shown in Fig. 36-7. The intersection of these lines provides the overall outline of the elevation.

Roof-Line Projection

The ridge line and eave line cannot be accurately located until the *roof pitch* (angle) is established. On a high-pitch roof, there is a greater distance between the ridge line and the eave line than on a low-pitch roof. Figure 36-8 shows a high-pitch roof and a low-pitch roof. *Pitch* is the angle of the roof and is described in terms of the ratio of the *rise over the run* (rise/run). *Run* is the horizontal distance covered by a roof. *Rise* is the vertical distance. The run is always expressed in units of 12. Therefore, the pitch is the ratio of the rise to 12.

After the pitch is established a slope diagram must be drawn on the elevation as shown in Fig-

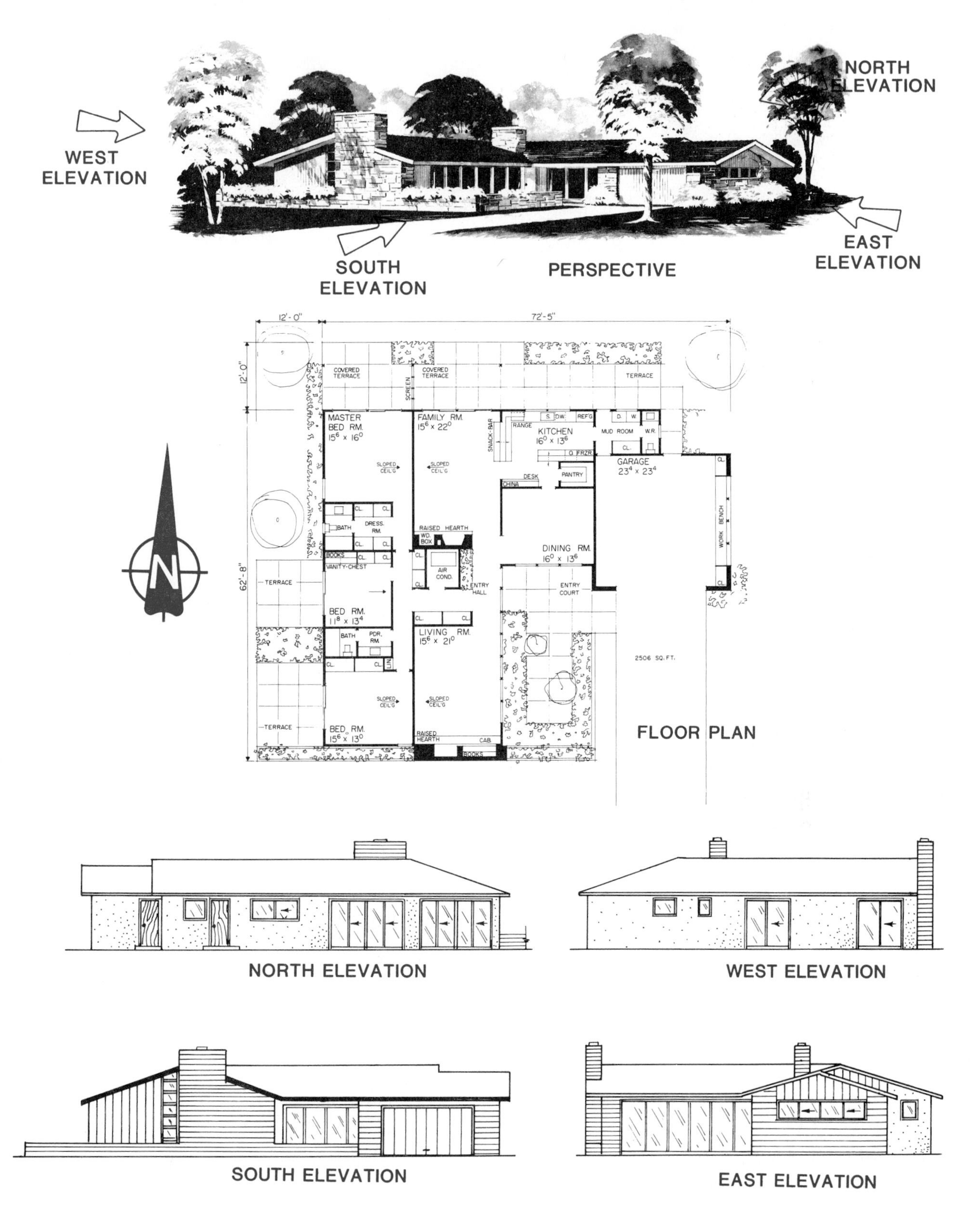

Fig. 36-5 Compass direction used to identify elevation. *(Home Planners, Inc.)*

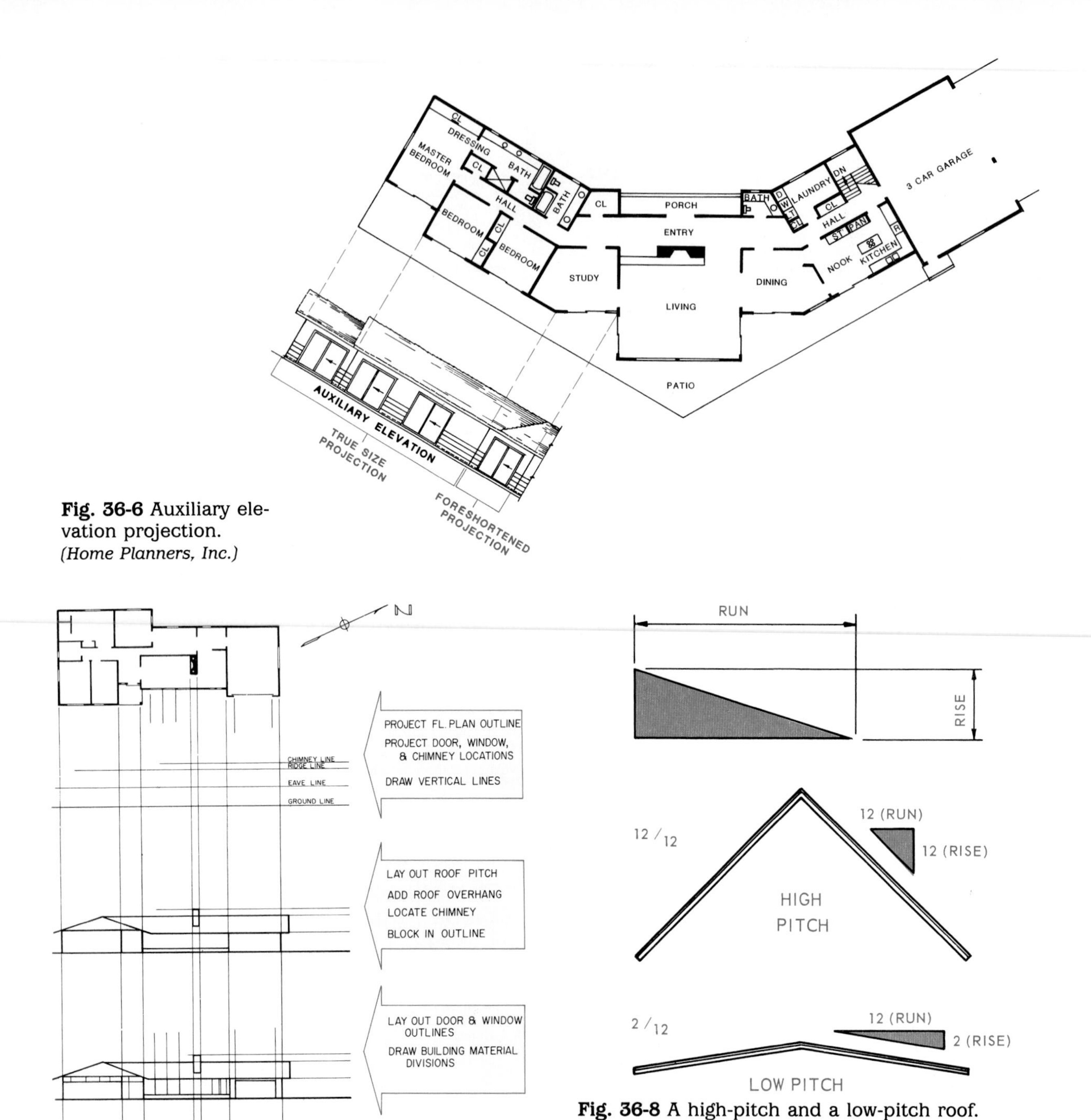

Fig. 36-6 Auxiliary elevation projection. *(Home Planners, Inc.)*

Fig. 36-7 The sequence of projecting elevations.

Fig. 36-8 A high-pitch and a low-pitch roof.

ure 36-9. The *slope diagram* is developed on the working drawings by the drafter. The carpenter must work with the *pitch fraction* to determine the angle of the rafters from a pitch angle table, so the ends of the rafters can be correctly cut.

Double the run and it will become the span. This is a constant of 24. Place the rise over the span (24) and reduce if necessary. This fraction is used by the carpenter to determine the rafter angle in degrees.

Blocking-in the Outline

After the roof outline has been established, the major lines of the house are drawn. This drawing is made by following the outline developed from the intersection of horizontal and vertical lines. The outlines of materials, doors, windows, chimney, and roof are drawn in their final line weight.

Adding Elevation Symbols

Details of an elevation can be drawn in detail separately or on the elevation drawing. Another option is to show the outline of the detail with a reference callout to a schedule with a full description of the outlined detail (Fig. 36-10). A description of the different types of schedules is shown in Section 19.

FLEXIBILITY

It is possible to project many different elevation styles from one floor plan. The pitch, size of overhang, position of the grade line, window position and style, chimney size and style, and the door style and position all can be manipulated to create different effects. Figure 36-11 shows two different elevations projected from the same floor plan. Here the change was accomplished primarily by varying the major lines of the elevation and changing the siding material and roof line.

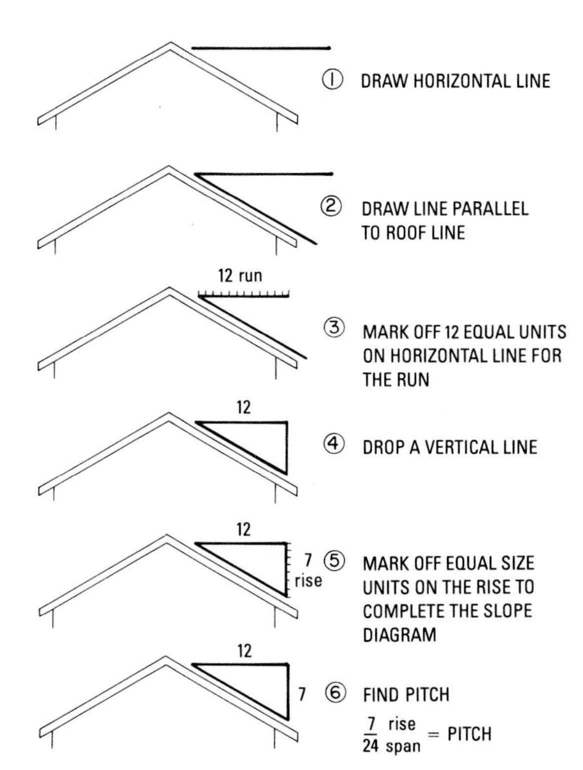

Fig. 36-9 Steps in drawing a slope diagram.

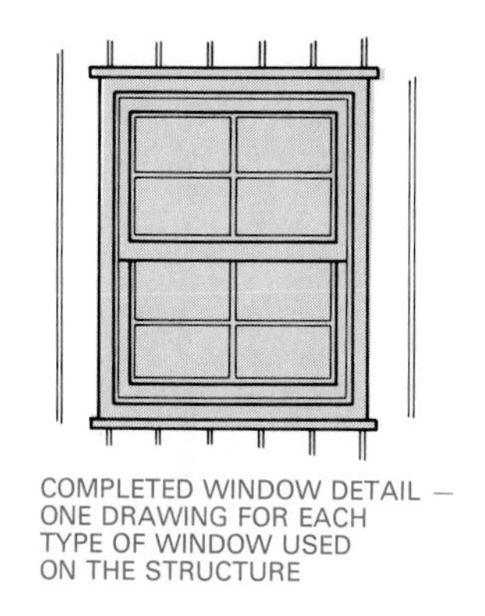

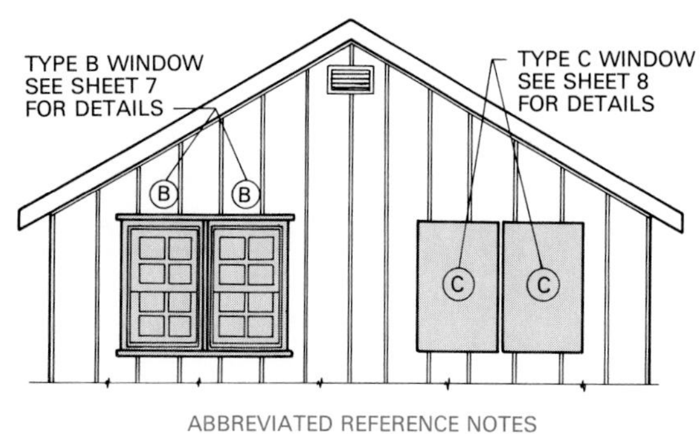

Fig. 36-10 Alternate methods of drawing elevation details.

INTERIOR ELEVATIONS

Floor plans show horizontal arrangements of partitions, fixtures, and appliances, but do not show the design of interior walls. *Interior elevations* are necessary to show the design of interior vertical planes. Because of the need to show cabinet height and counter arrangement detail, interior wall elevations are most often prepared for kitchen and bathroom walls as shown in Fig. 36-12. An interior wall elevation shows the appearance of the wall as viewed from the center of the room.

A coding system is used in place of labels to identify the walls on the floor plans for which interior elevations have been prepared when many are drawn. The code symbol shows the direction of the view, the elevation detail number, and the page in the set where found. If only a few interior elevations are prepared, then the title of the room and the compass location of the wall is the only identification needed. The compass system is shown in Fig. 36-12 and the coding system in Fig. 36-13. The following steps in drawing an interior elevation are outlined in Fig. 36-14.

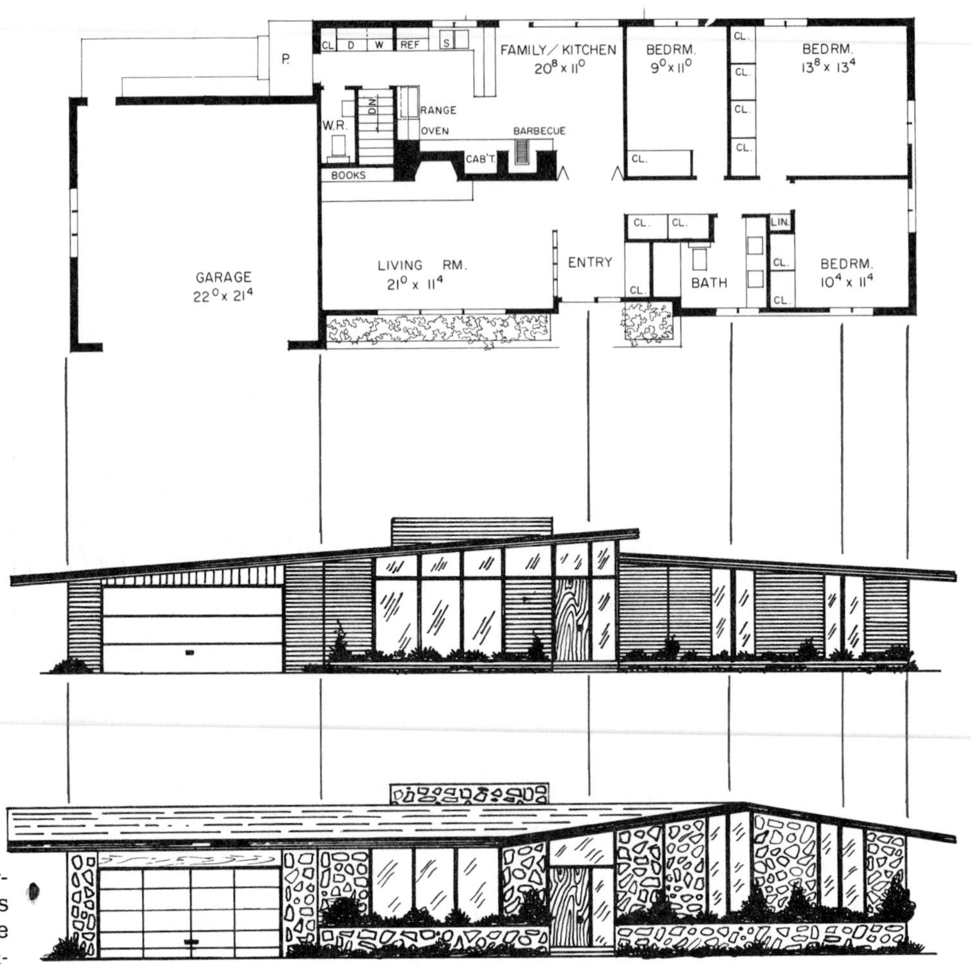

Fig. 36-11 Two different elevation styles projected from one floor plan. *(Home Planners, Inc.)*

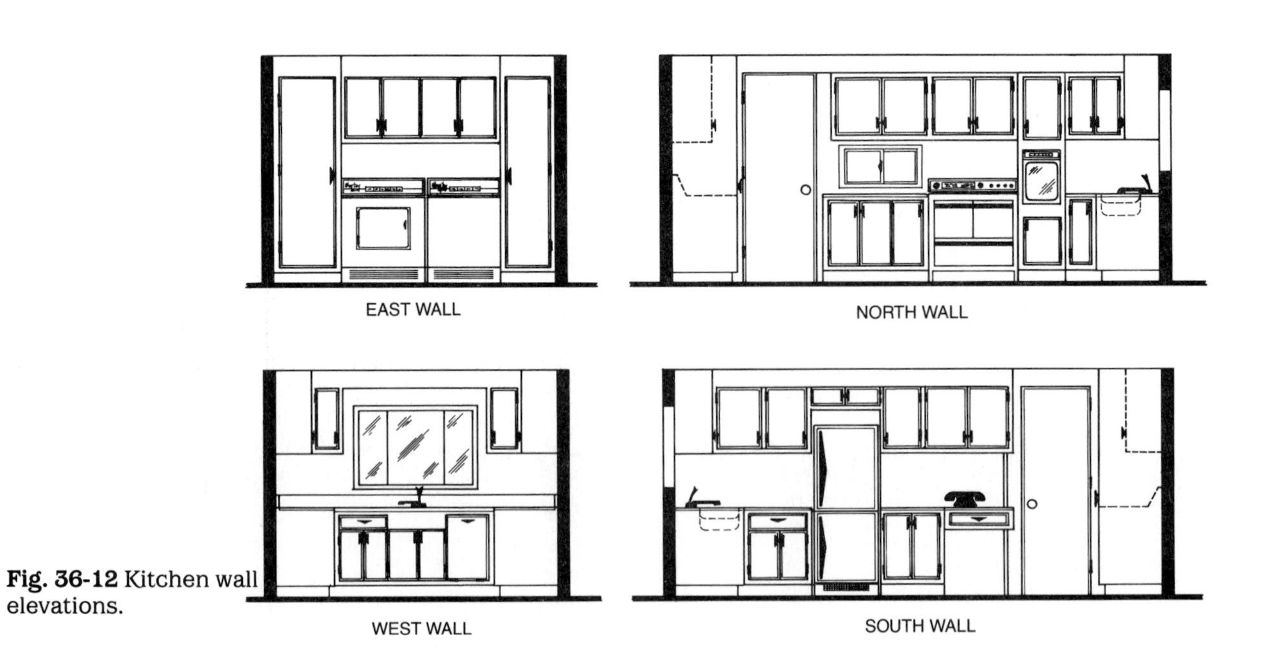

Fig. 36-12 Kitchen wall elevations.

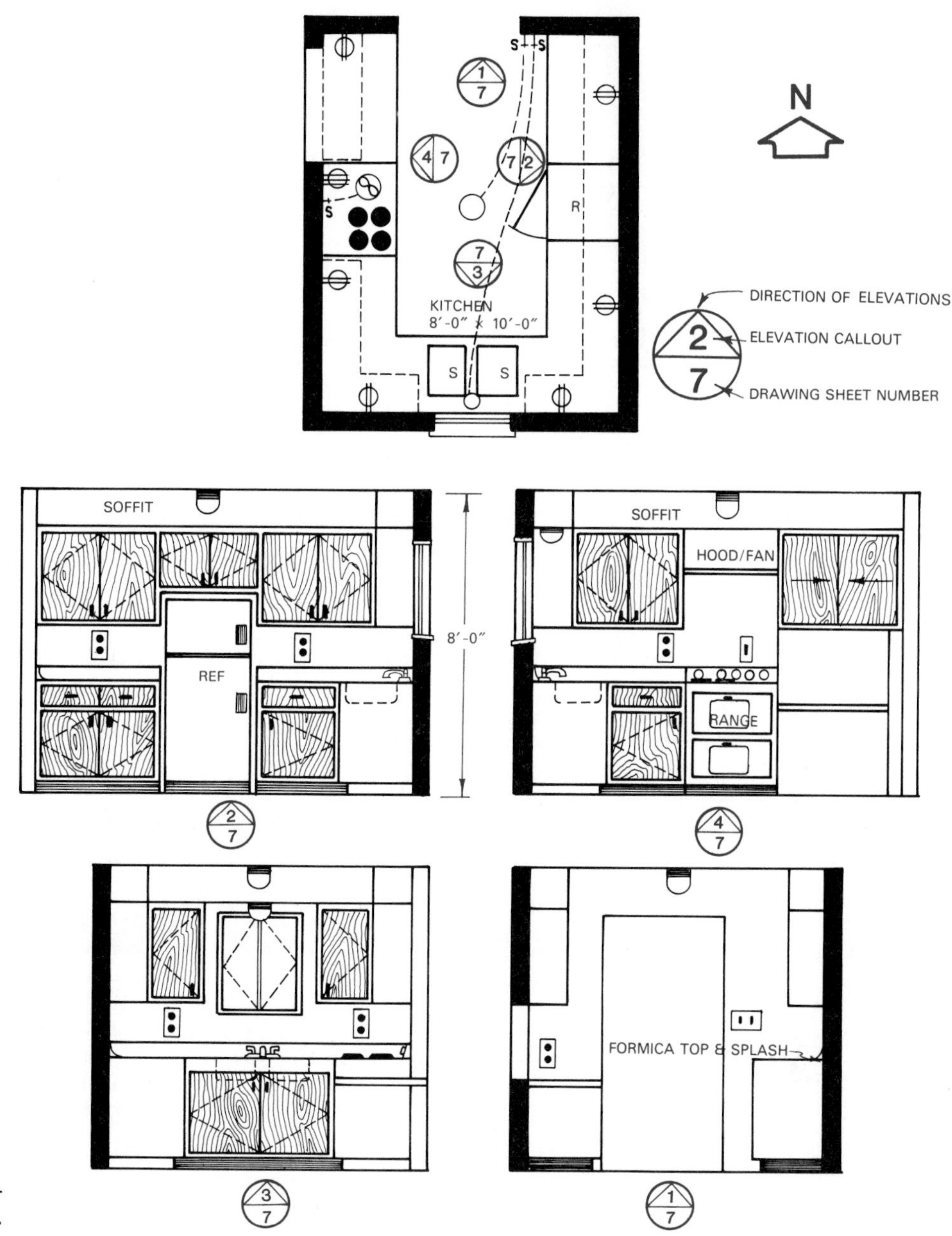

Fig. 36-13 Interior elevation coding system.

1. Outline the floor plan.

2. Projections are made from the floor-plan outline perpendicular from each corner.

3. Then ceiling lines are added to give each wall its specified height.

4. Details are then projected directly from the floor plan to each elevation drawing.

Projecting the interior elevation in this manner results in an elevation drawn on its side or upside down. Interior elevation drawings, like exterior elevations, are not prepared in the original position as they are projected from the floor plans. Interior elevations are positioned with the floor line on the bottom as normally viewed. Once the features of the wall are added to the drawing projected from the floor plan, dimensions, instructional notes, and additional features can be added to the drawing as shown in Fig. 36-15.

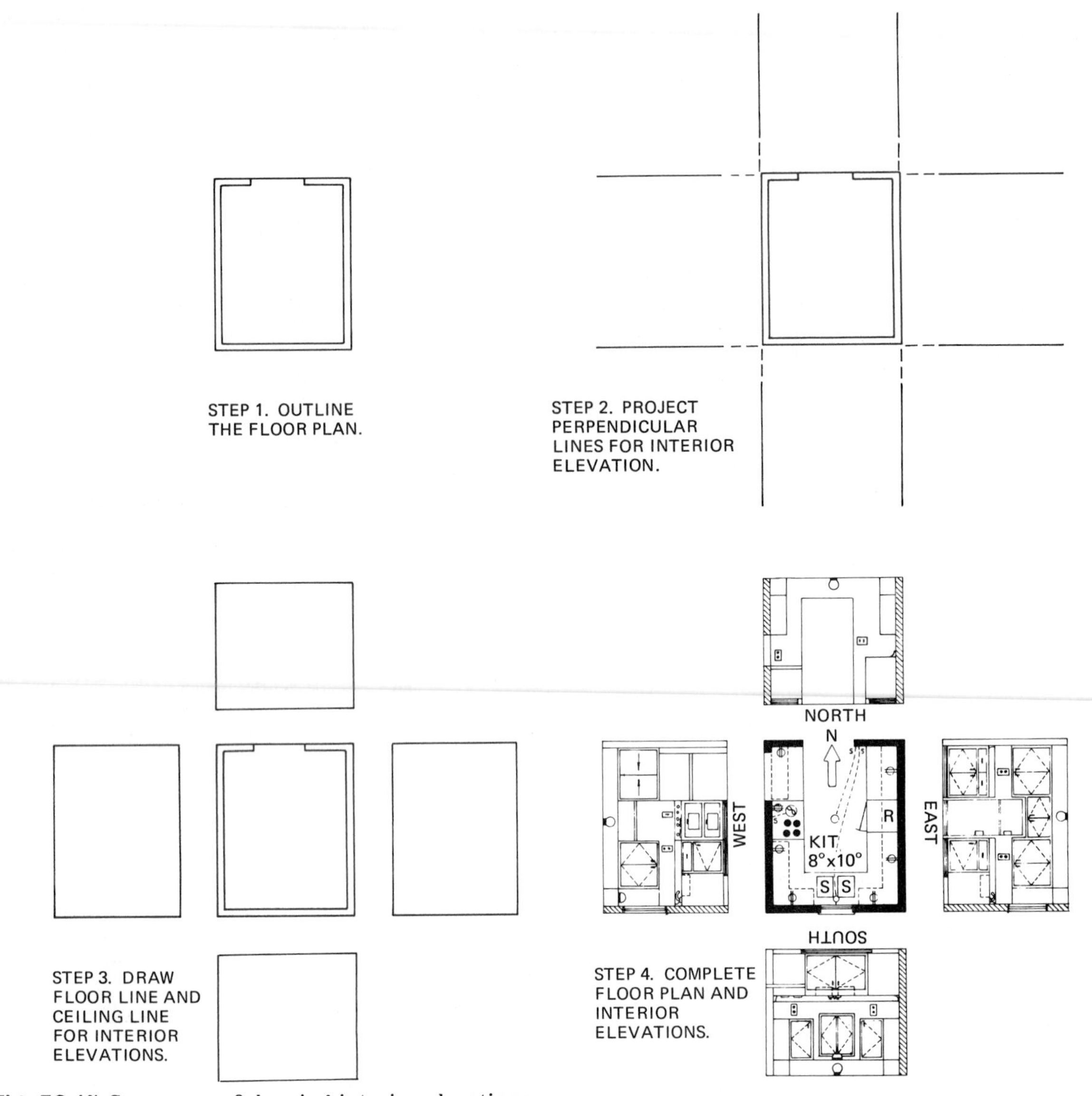

Fig. 36-14 Sequence of drawing interior elevations.

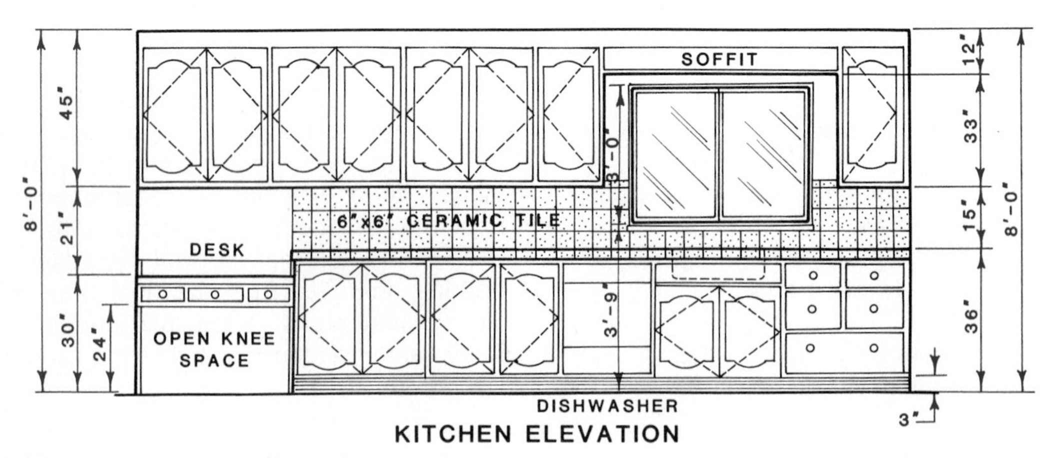

Fig. 36-15 Interior elevation dimensions.

Exercises

1. **Project the front, rear, right, and left elevations of a floor plan of your own design.**
2. **Sketch the front elevation of your home.**
3. **Project and sketch or draw the front elevation suggested in the pictorial drawing and floor plan in Fig. 36-16.**
4. **Define the following terms: *front elevation, rear elevation, right elevation, left elevation, north elevation, east elevation, south elevation, west elevation, auxiliary elevations, pitch, rise, run.***

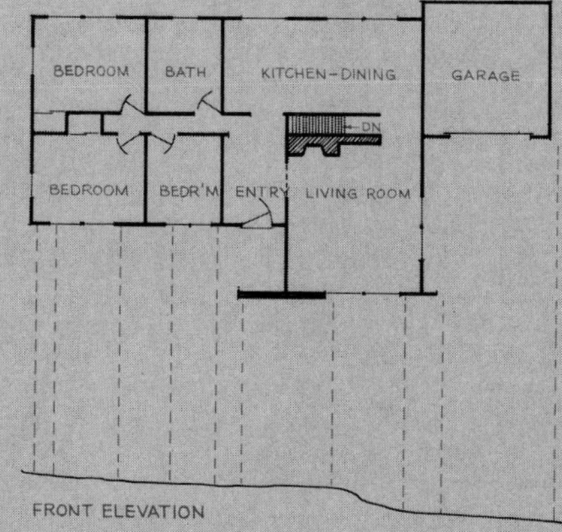

Fig. 36-16

UNIT 37

ELEVATION SYMBOLS

Symbols are needed to clarify and simplify elevation drawings. Symbols help to describe the basic features of the elevation. They show what building materials are used, and they describe the style and position of doors and windows. Symbols also help to make the elevation drawing look realistic. Some of the most common elevation symbols are shown on the elevation drawing in Figs. 37-1 and 37-2.

MATERIAL SYMBOLS

Figure 37-3 shows the relationship between material symbols used on an elevation and the actual material as it is used in construction. Most architectural symbols look very similar to the material they represent. However, in many cases the symbol does not show the exact appearance of the material. For example, the symbol for brick, as shown in Fig. 37-3, does not include all the lines shown in the pictorial drawing. Representing brick on the elevation drawing exactly as it appears is a long, laborious, and unnecessary process. Therefore, like the symbol for brick, many elevation symbols are simplifications of the actual appearance of the material. The symbol often resembles the appearance of the material at a distance. Figure 37-4 shows a full-size portion, at $\frac{1}{4}'' = 1'\text{-}0''$ scale, of the plan shown in Fig. 69-5. For a more detailed list of elevation and related plan symbols, refer back to Figs. 33-7 through 33-11. When using a CAD system, elevation symbols such as doors and windows can be stored in symbols libraries. However, the crosshatch function can often be used to add siding materials symbols on elevation surfaces.

WINDOW SYMBOLS

The position and style of windows greatly affect the appearance of the elevation. Windows are, therefore, drawn on the elevation with as much detail as the scale of the drawing permits. Parts of windows that should be shown on all elevation drawings include the sill, sash, mullions, and muntins (Fig. 37-5). Figure 37-6 shows the method of illustrating casement, awning, and sliding windows. Figure 37-7 shows the parts of a double-hung window in more detail.

In addition to showing the parts of a window, it is also necessary to show the direction of the hinge for casement and awning windows. Figure 37-8 shows the method of indicating the direc-

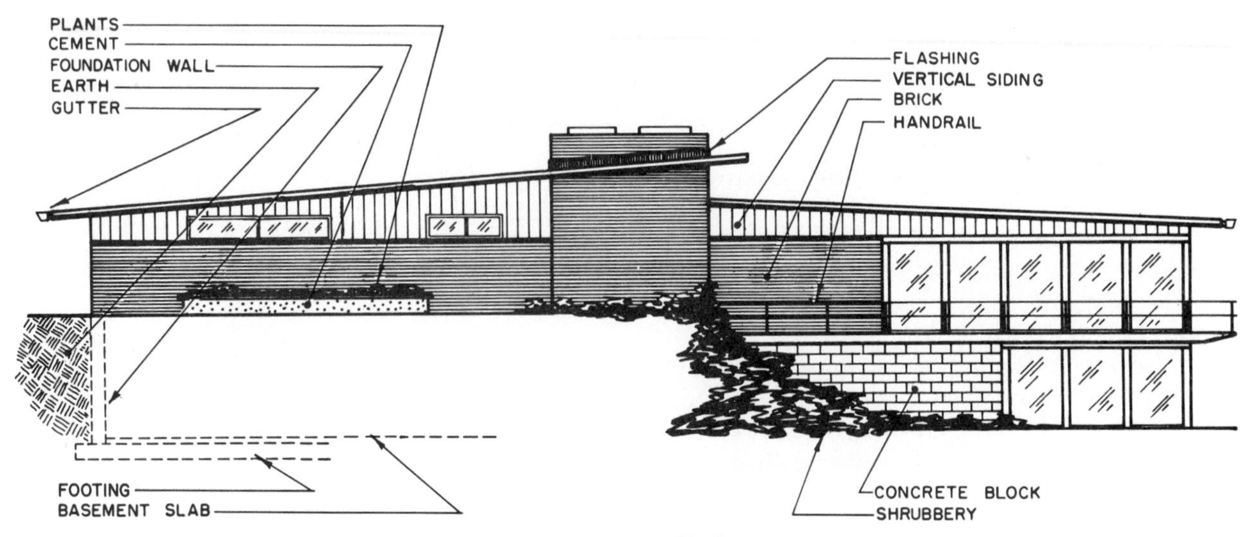

Fig. 37-1 Symbols help make the elevation look more realistic.

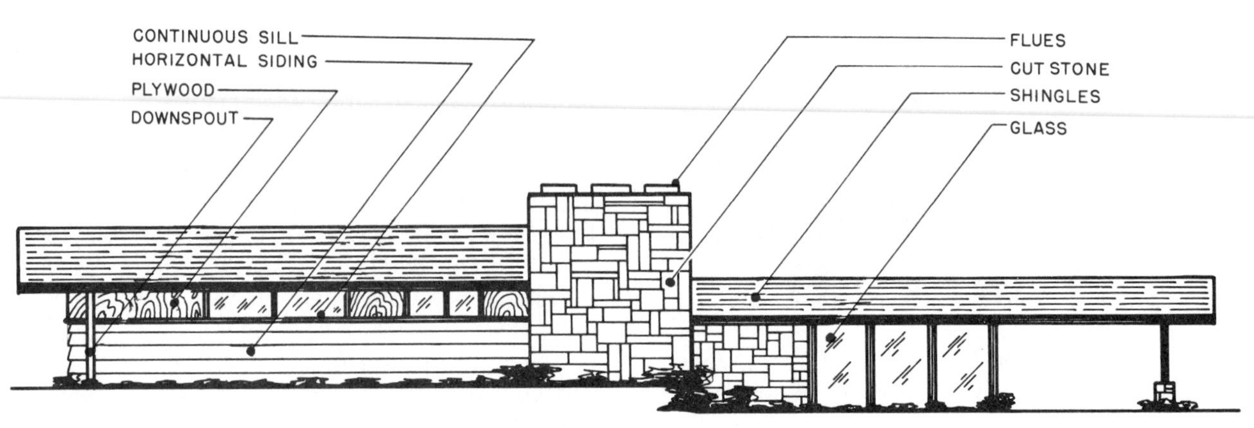

Fig. 37-2 Some common elevation symbols in use.

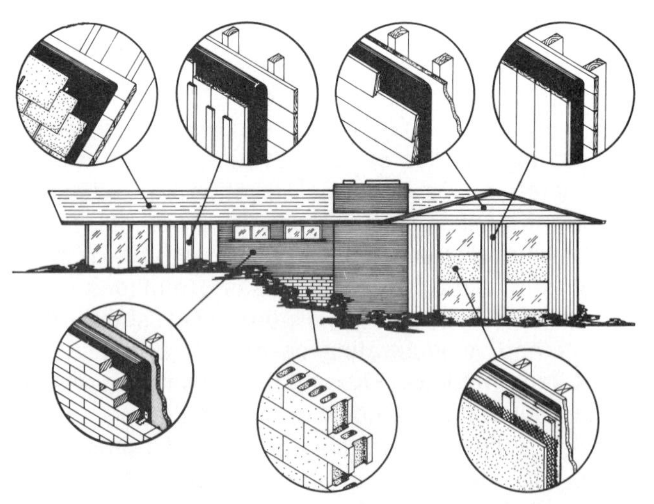

Fig. 37-3 The relationship between material symbols and materials used in construction.

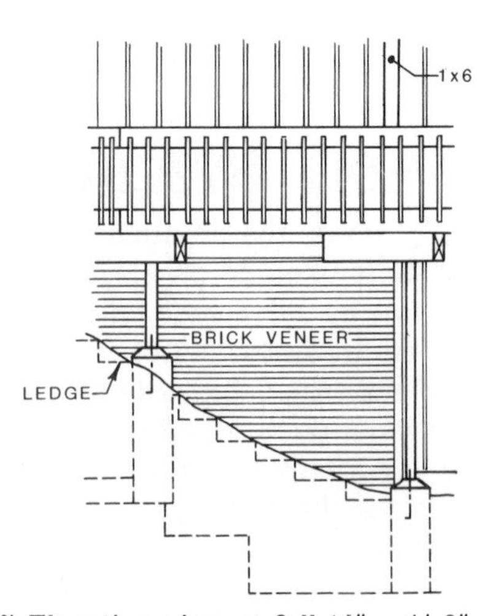

Fig. 37-4 Elevation view at full ⅛″ = 1′-0″.

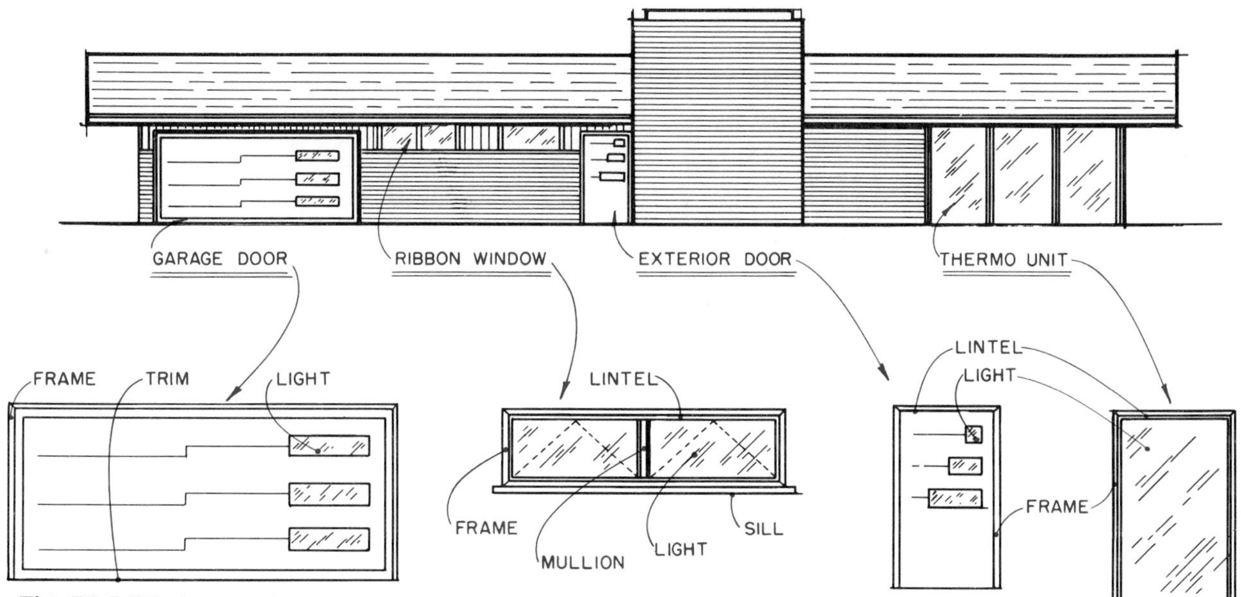

Fig. 37-5 Window symbols as they appear on elevation drawings.

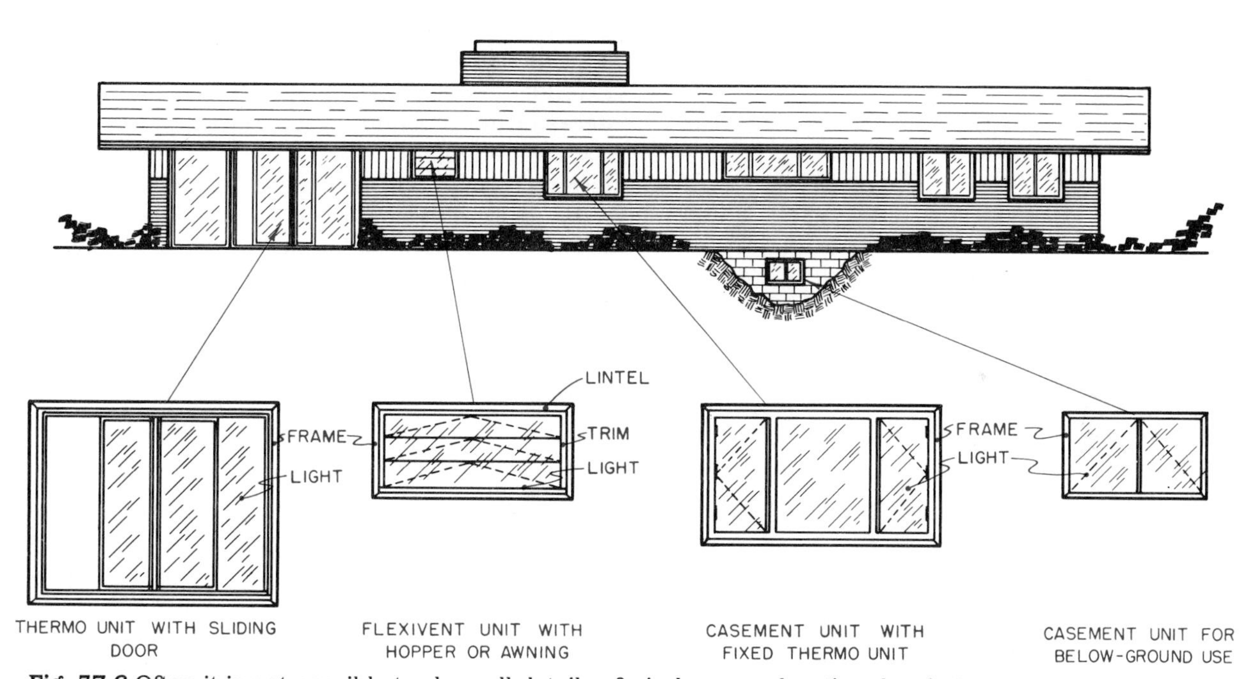

Fig. 37-6 Often it is not possible to show all details of windows on elevation drawings.

tion of the hinge on elevation drawings. The direction of the hinge is shown by dotted lines. The point of the dotted line shows the part of the window to which the hinge is attached.

Many different styles of windows are available, as shown in Fig. 33-8. These illustrations also show the normal amount of detail used in drawing windows on elevations. Architects often use an alternative method of showing window styles on elevation drawings. In this alternative method, the drafter prepares a window-detail drawing, as shown in Fig. 36-10, to a larger scale. The designer then prepares a separate window drawing in detail for each different style of window to be used. When the elevation is drawn, only the position of the window is shown. The style of window to be included in this opening is then shown by a letter or number indexed to the letter or number used for the large detail drawing. Sometimes, the window symbol is ab-

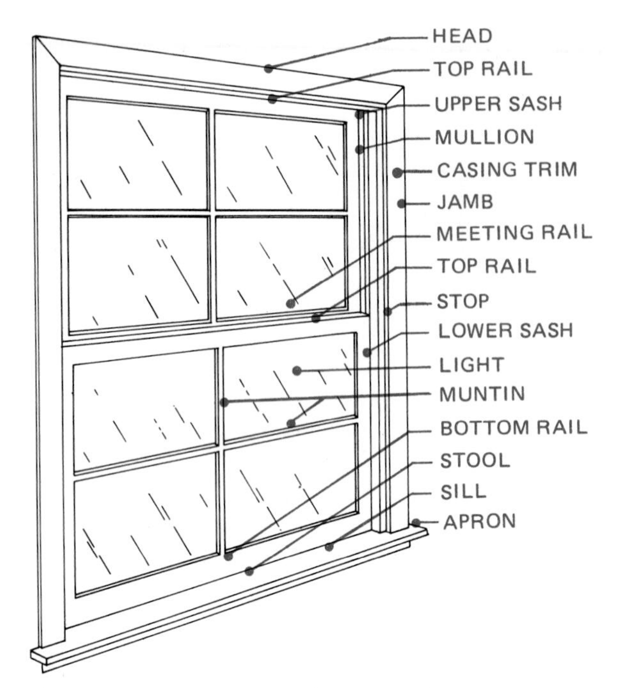

Fig. 37-7 Double-hung window parts.

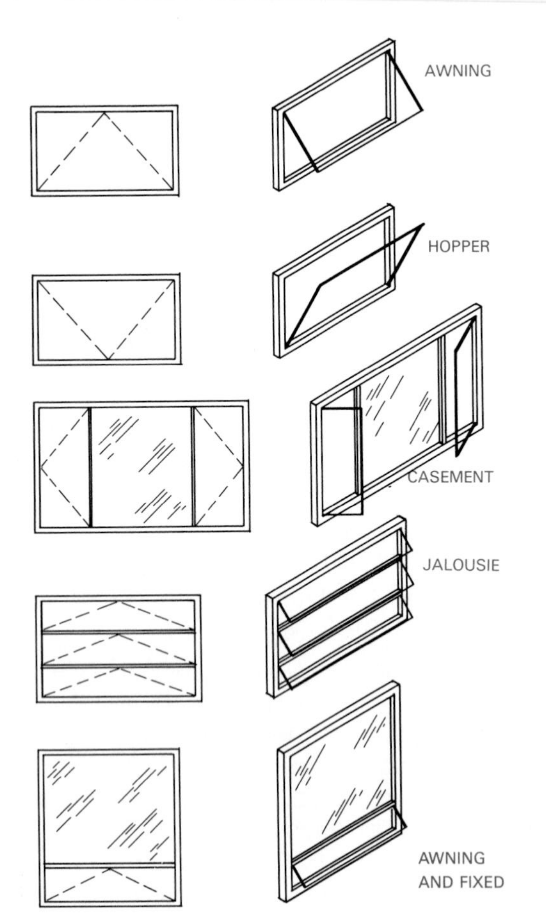

Fig. 37-8 The placement of the hinge is shown by the dotted line.

breviated and indexed in the same way to a more complete detail. Unit 70 contains further treatment of door and window schedules.

DOOR SYMBOLS

Doors are shown on elevation drawings by methods similar to those used for illustrating window style and position. They are either drawn completely, if the scale permits, or shown in abbreviated form. Sometimes the outline is indexed to a door schedule. The complete drawing of the door, whether shown on an elevation or on a separate detail, should show the division of panels and lights, sill, jamb, and head-trim details. Separate door details are sometimes indexed to the elevation outline of the door location as shown in Fig. 37-9.

Many exterior door styles are available (Fig. 37-10). The total relationship of the door and trim to the entire elevation cannot be seen unless the door trim is also shown (Fig. 37-11). Exterior doors are normally larger than interior doors. Exterior doors must provide access for larger amounts of traffic and be sufficiently large to permit the movement of furniture. They must also be thick enough to provide adequate safety, insulation, and sound barriers. Common exterior door sizes include widths of 2′-8″, 3′-0″, and 3′-6″ (0.8, 0.9 and 1.1 m). Common exterior door heights range from 6′-8″ to 7′-6″ (2.0 to 2.3 m).

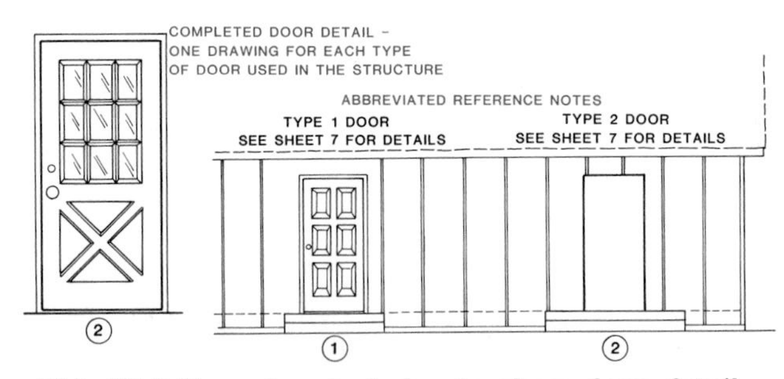

Fig. 37-9 Use of code index to show door detail.

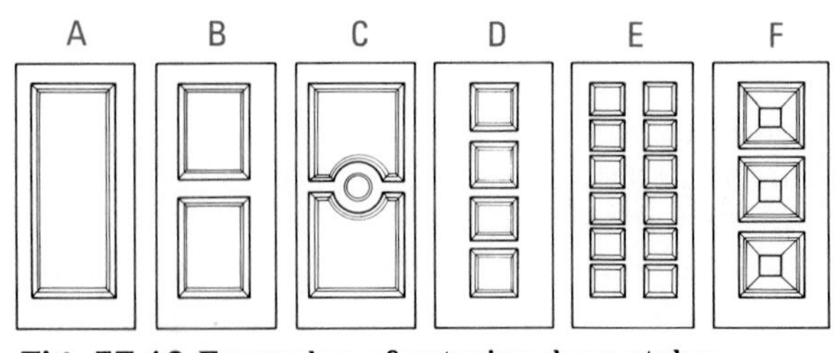

Fig. 37-10 Examples of exterior-door styles.

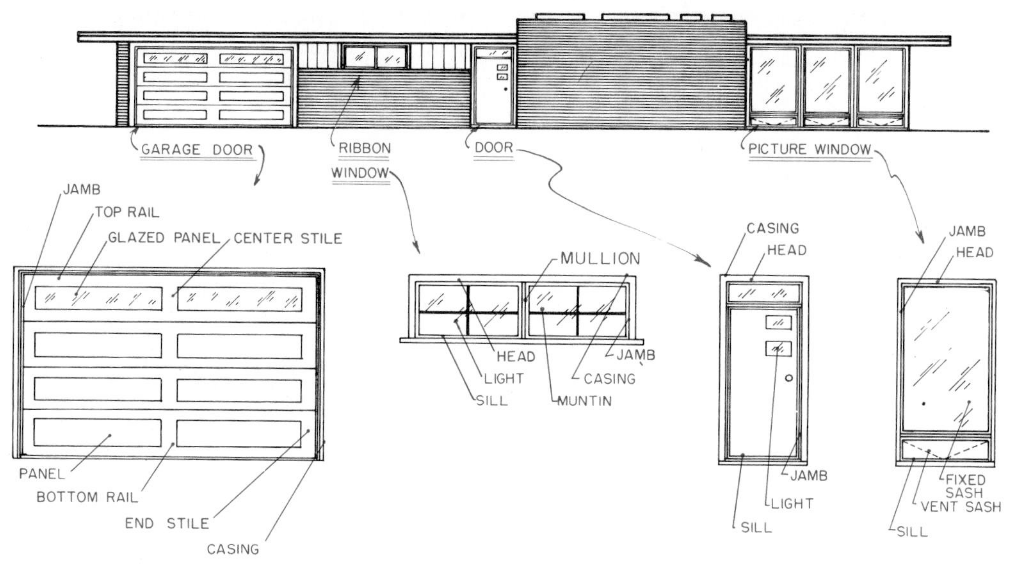

Fig. 37-11 Methods of drawing door and window trim.

Exercises

1. **Draw and add symbols to the elevation outline shown in Fig. 37-12.**
2. **Add elevation symbols to an elevation of your own design.**
3. **Identify the elevation symbols shown in Fig. 37-13.**
4. **Draw the front elevation of the house shown in Fig. 37-14. Show elevation symbols.**

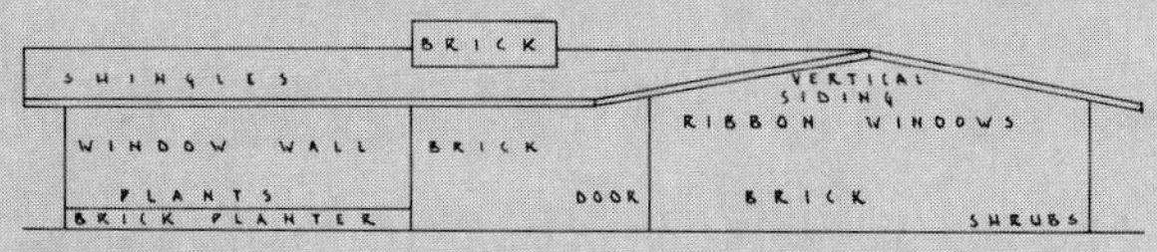

Fig. 37-12

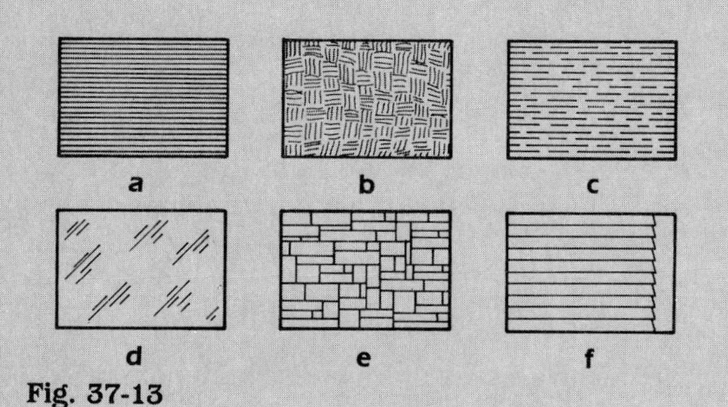

Fig. 37-13

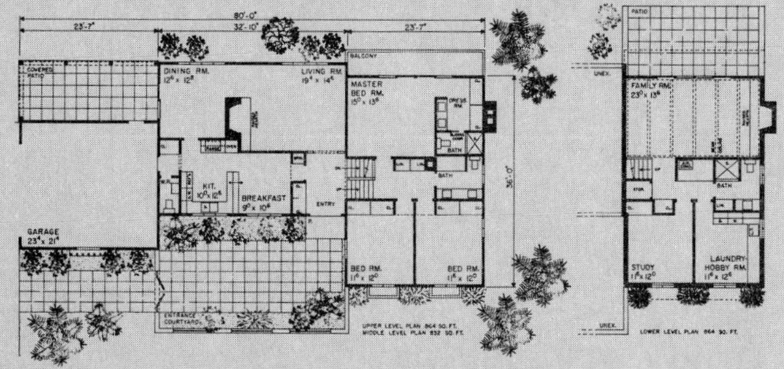

Fig. 37-14 Draw the front elevation of this house, complete with symbols. *(Home Planners, Inc.)*

5. **Redesign and finish the front elevation of the house shown in Fig. 37-15, using brick as the basic siding material.**
6. **Draw the front and left-side elevations of the house shown in Fig. 37-16. Use symbols for the siding materials shown.**

Fig. 37-15 Redesign and finish this elevation. *(Home Planners, Inc.)*

Fig. 37-16 Draw a front and left-side elevation of this house. *(Boise Cascade)*

UNIT 38

ELEVATION DIMENSIONING

Horizontal (width and length) dimensions are placed on floor plans. Vertical (height) dimensions are placed on elevation drawings.

Many dimensions on elevation drawings show the vertical distance from a datum line. The *datum line* is a reference that remains constant. Sea level is commonly used as the datum for many drawings, although any distance above sea level can be conveniently used for a vertical reference.

Dimensions on elevation drawings show the height above the datum of the ground line. They also show the distance from the floor line to the ceiling and ridge and eave lines, and to the tops of chimneys, doors, and windows. Distances below the ground line are shown by dotted lines.

RULES FOR ELEVATION DIMENSIONING

Elevation dimensions must conform to basic standards to ensure consistency of interpretation. The arrows on the elevation drawing in Fig. 38-1 show the application of the following rules for elevation dimensioning:

1. Vertical elevation dimensions should be read from the right of the drawing.

2. Levels to be dimensioned should be labeled with a note, term, or abbreviation.

3. Room heights are shown by dimensioning from the floor line to the ceiling line.

4. The depth of footings ("*footer*") is dimensioned from the ground line.

5. Heights of windows and doors are dimensioned from the floor line to the top of the windows or doors.

6. Elevation dimensions show only vertical distances (height). Horizontal distances (length and width) are shown on the floor plan.

7. Windows and doors may be indexed to a door or window schedule, or the style of the windows and doors may be shown on the elevation drawing.

8. The roof pitch is shown by indicating the rise over the run.

9. Dimensions for small, complex, or obscure areas should be indexed to a separate detail.

10. Ground-line elevations are expressed as heights above a datum point.

11. Heights of chimneys above the ridge line are dimensioned.

12. Floor and ceiling lines are shown with hidden lines.

13. Heights of planters and walls are dimensioned from the ground line.

14. Thicknesses of slabs are dimensioned.

15. Overall height dimensions are placed on the outside of subdimensions.

16. Thicknesses of footings are dimensioned.

17. Where space is limited, the alternative method in Fig. 38-2 can be used to show feet and inches.

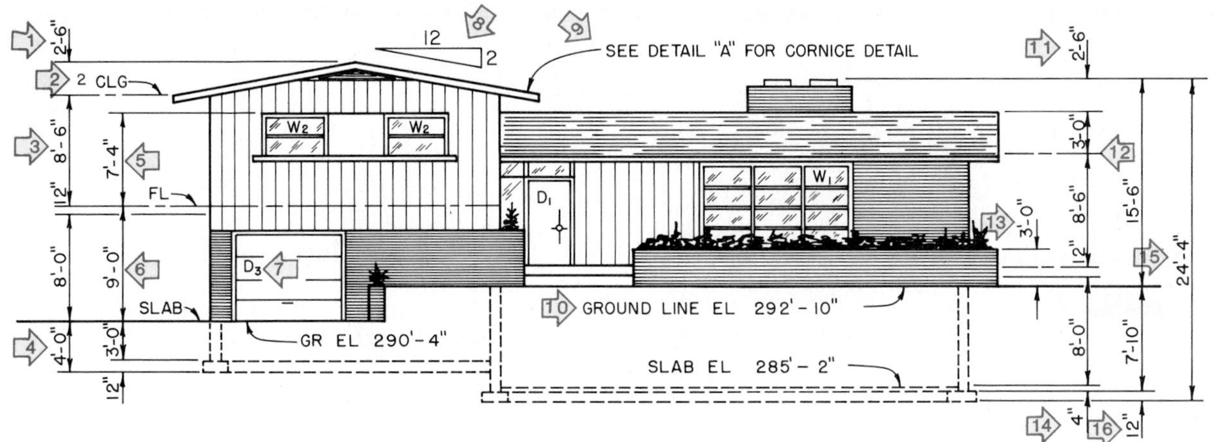

Fig. 38-1 Rules for elevation dimensioning.

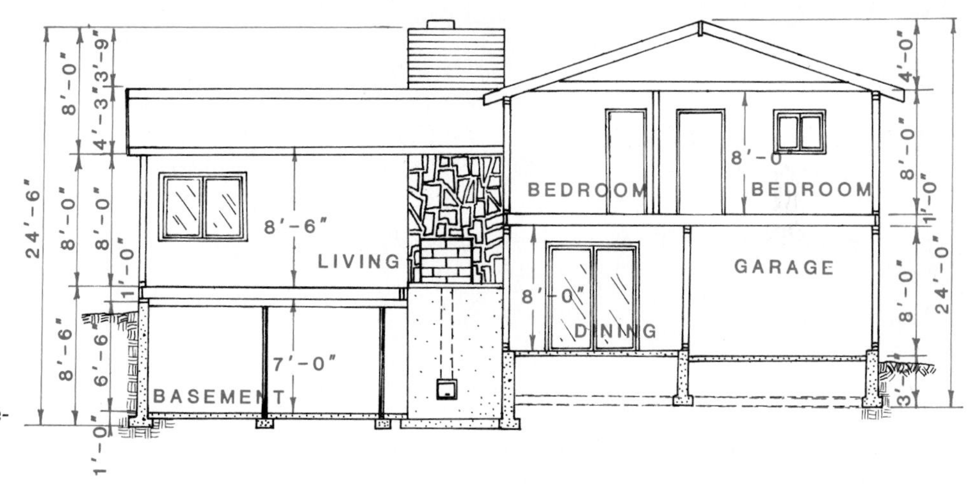

Fig. 38-2 Alternate elevation-dimensioning method.

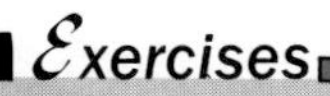

Exercises

1. **Add the elevation dimensions to an elevation drawing of your own design. Estimate dimensions.**
2. **Add dimensions to the elevation drawing shown in Fig. 38-3.**
3. **Dimension an elevation drawing of your home.**
4. **Draw an elevation of the home shown in Fig. 38-4. Completely dimension this elevation, following the rules for dimensioning outlined in this unit.**
5. **Record the missing dimensions on the elevation shown in Fig. 38-5. Estimate dimensions.**
6. **Finish and dimension the elevation shown in Fig. 38-6. Estimate dimensions.**
7. **Define these terms: *datum line, sea level, vertical dimensions, ground line, ceiling line, ridge line, eave line, chimney line, room height, door schedule, window schedule, slab thickness, footing thickness, overall dimensions, subdimensions.***

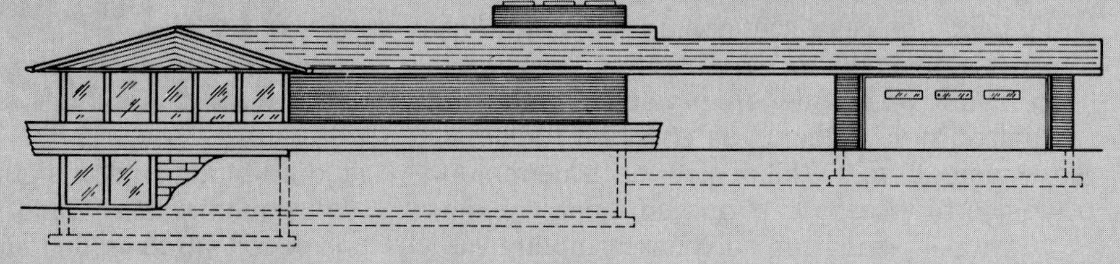

Fig. 38-3

Fig. 38-4 Draw and dimension one elevation of this home. *(Home Planners, Inc.)*

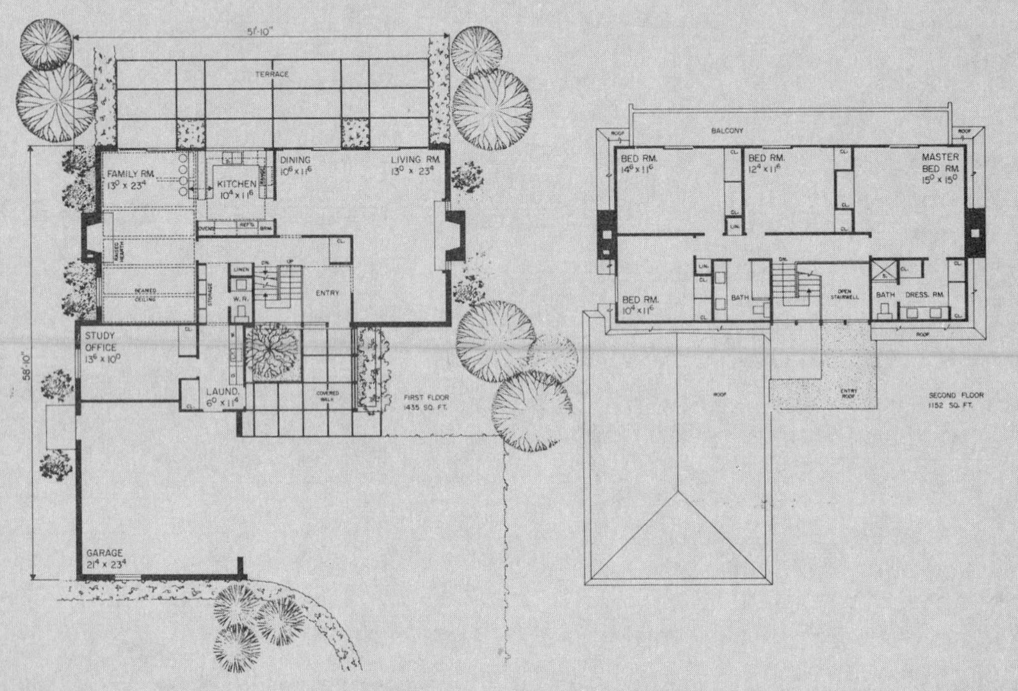

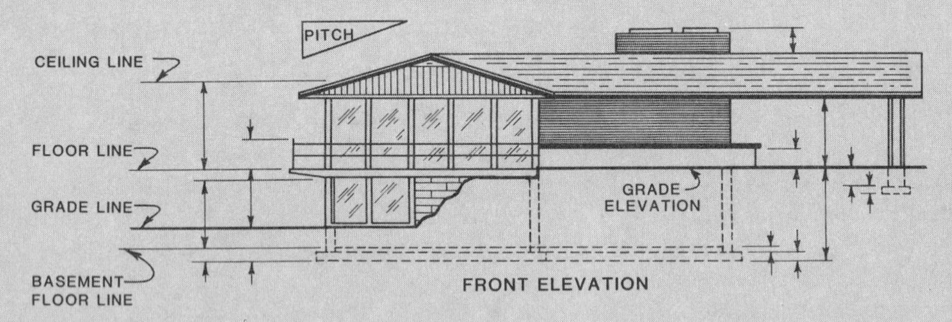

Fig. 38-5

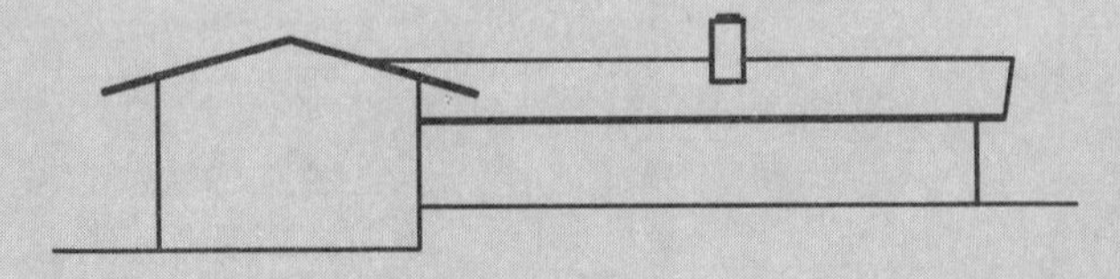

Fig. 38-6

LANDSCAPE RENDERING

Elevation drawings, although accurate in every detail, do not show exactly how the building will appear when it is complete and landscaped. The reason is that elevation drawings do not show the position of trees, shrubbery, and other landscape features that would be part of the total elevation design. Adding these landscape features to the elevation drawing creates a more realistic drawing of the house.

INTERPRETIVE DRAWING

Figure 39-1 shows some of the advantages of adding landscape features to an elevation drawing. The elevation shown in Fig. 39-1 at A, when dimensioned, would be adequate for construction purposes. However, the illustration shown in Fig. 39-1 at B more closely resembles the final appearance of the house.

Dimensions and hidden lines are omitted when landscape features are added to elevation presentation drawings. Drawings of this kind are prepared only to interpret the final appearance of the building (Fig. 39-2). They are not used for construction purposes.

Sequence

An elevation drawing is converted into a landscape elevation drawing in several basic steps, as shown in Fig. 39-3. After material symbols are added to the elevation, the positions of trees and shrubs are added. The elevation lines within the outlines of the trees and shrubs are erased, and details are added. Finally, shade lines are added to trees, windows, roof overhangs, chimneys, and other major projections of the house. The addition of landscape features should not hide the basic lines of the house. If many trees or shrubs are placed in front of the house, it is best to draw them in their winter state.

Figure 39-4A shows several methods of drawing trees and shrubs using single-line techniques on elevation drawings. Figure 39-4B shows the trees and shrubs rendered with watercolor techniques. The drafter should use the medium that best suits the elevation drawing to be rendered.

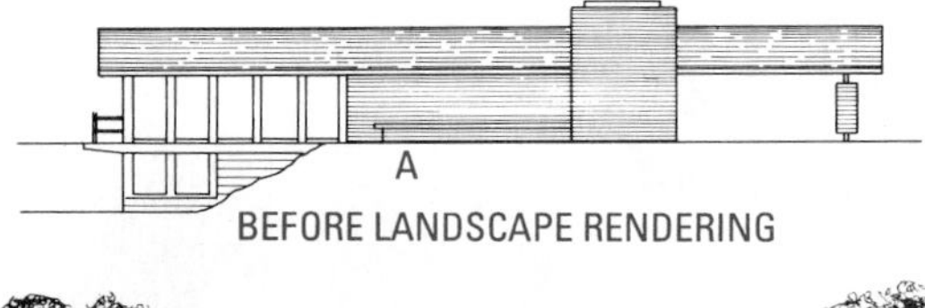

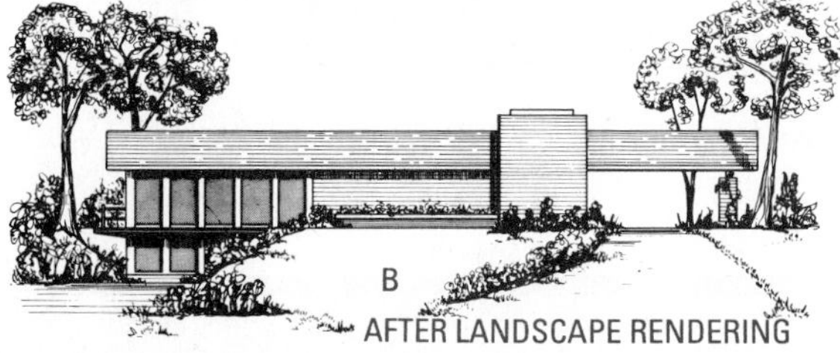

Fig. 39-1 An elevation drawing before (A) and after (B) landscape rendering.

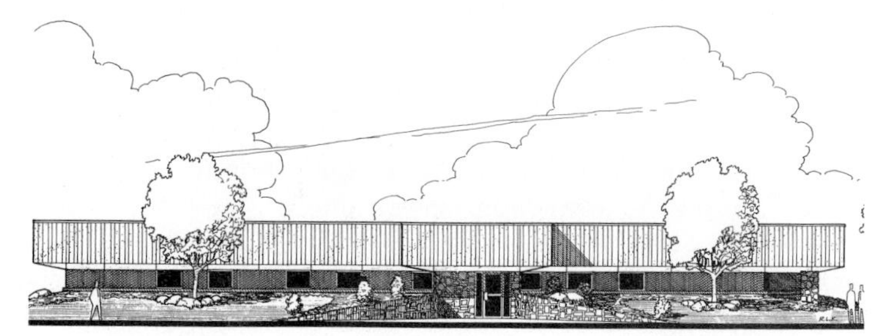

Fig. 39-2 Rendered elevations make the building appear complete and desirable. *(Killian and Associates)*

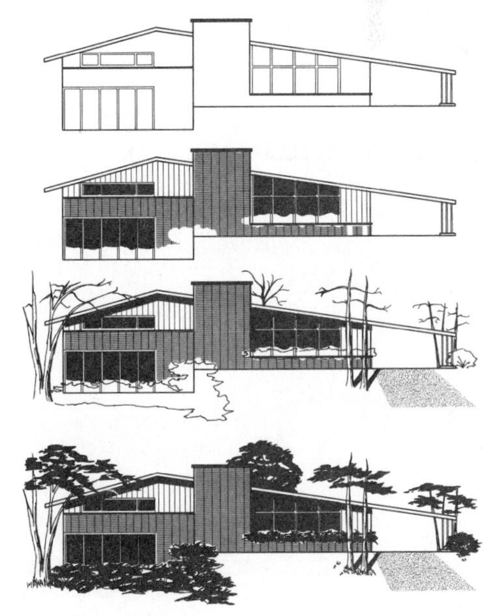

Fig. 39-3 The sequence of adding landscape features to an elevation.

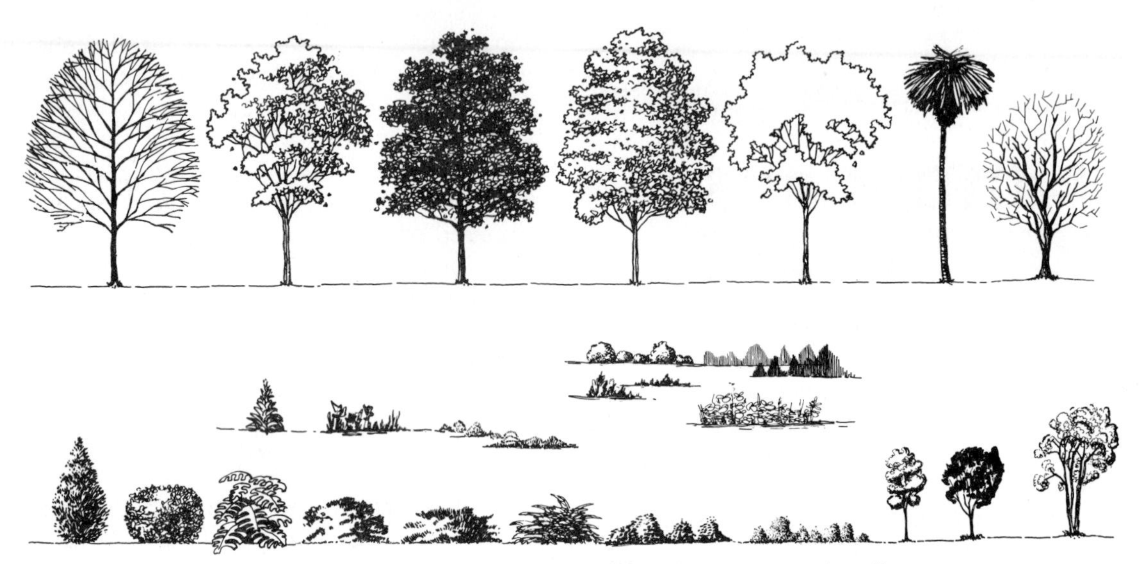

Fig. 39-4A Examples of sketches for drawing trees and shrubbery on an elevation.

Fig. 39-4B The same trees and shrubs rendered with watercolor techniques.

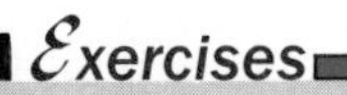

Exercises

1. Add landscape features to the elevation shown in Fig. 39-5.

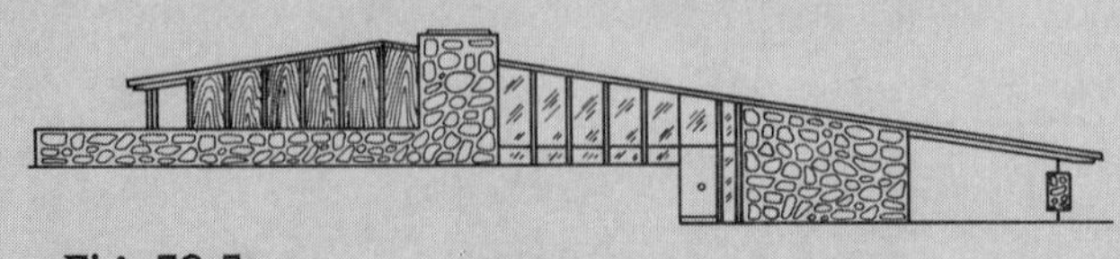

Fig. 39-5

2. Add trees, shrubs, plants, and shadows to an elevation of your own design.

3. Add landscape features to an elevation drawing of your home. Improve the present landscape treatment.

4. Define these terms: *landscape rendering, interpretive drawings, shade lines.*

SECTION 9

Pictorial Drawings

Pictorial drawings, both isometric and perspective, are picturelike drawings. They show several sides of an object in one drawing. Isometric drawings are used extensively in mechanical engineering drawings. The perspective drawing is more popular as an architectural pictorial drawing. Since the subject of most architectural pictorial drawings is much larger than that of most engineering drawings, perspective techniques are necessary to eliminate distortion of the drawing.

UNIT 40

EXTERIOR PICTORIALS

ISOMETRIC DRAWINGS

The receding lines in *isometric drawings* have constant angles (30°) from the horizon. Thus all receding lines are parallel. Figure 40-1 shows a comparison of isometric, oblique, and perspective drawings. Because of the size, especially length, of most architectural structures, isometric drawings result in great visual distortion of receding areas. Notice the more realistic appearance of the perspective drawing compared with the isometric and oblique drawings in Fig. 40-1. Figure 40-2 also shows the different distortions by superimposing an isometric outline of a building over a perspective drawing of the same building.

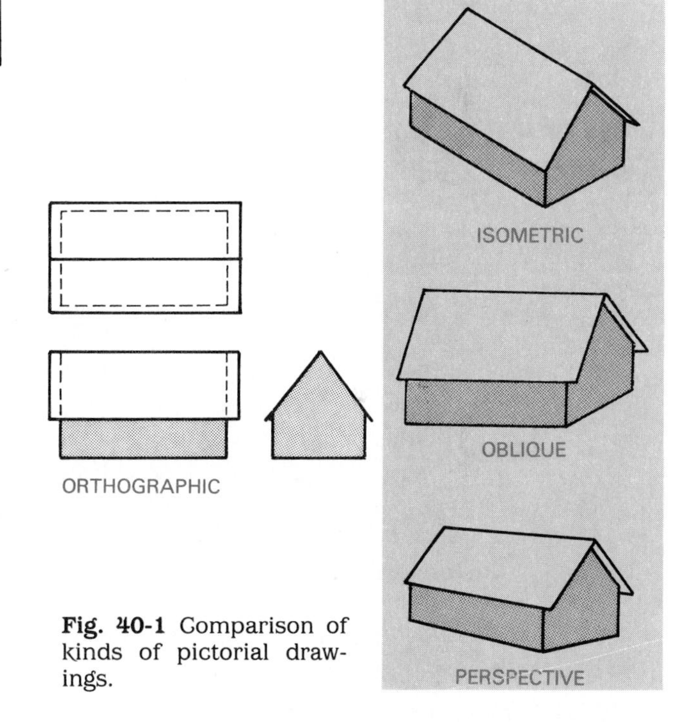

Fig. 40-1 Comparison of kinds of pictorial drawings.

PERSPECTIVE DRAWINGS

In *perspective drawings*, receding lines of a building will appear to meet. They are not drawn parallel. To give an example of a perspective view, as you look down a railroad track, the tracks appear to come together and vanish at a point on the distant horizon. Similarly, the horizontal lines of the building shown in Fig. 40-3 appear to be coming together. A perspective drawing, more than any other kind of drawing, resembles a photograph.

On a perspective drawing, the receding lines of a building are purposely drawn closer together on one or several sides of the building to create the illusion of depth. The point at which these lines intersect is known as the *vanishing point*. Just as railroad tracks would appear to come together on the horizon, the vanishing points in

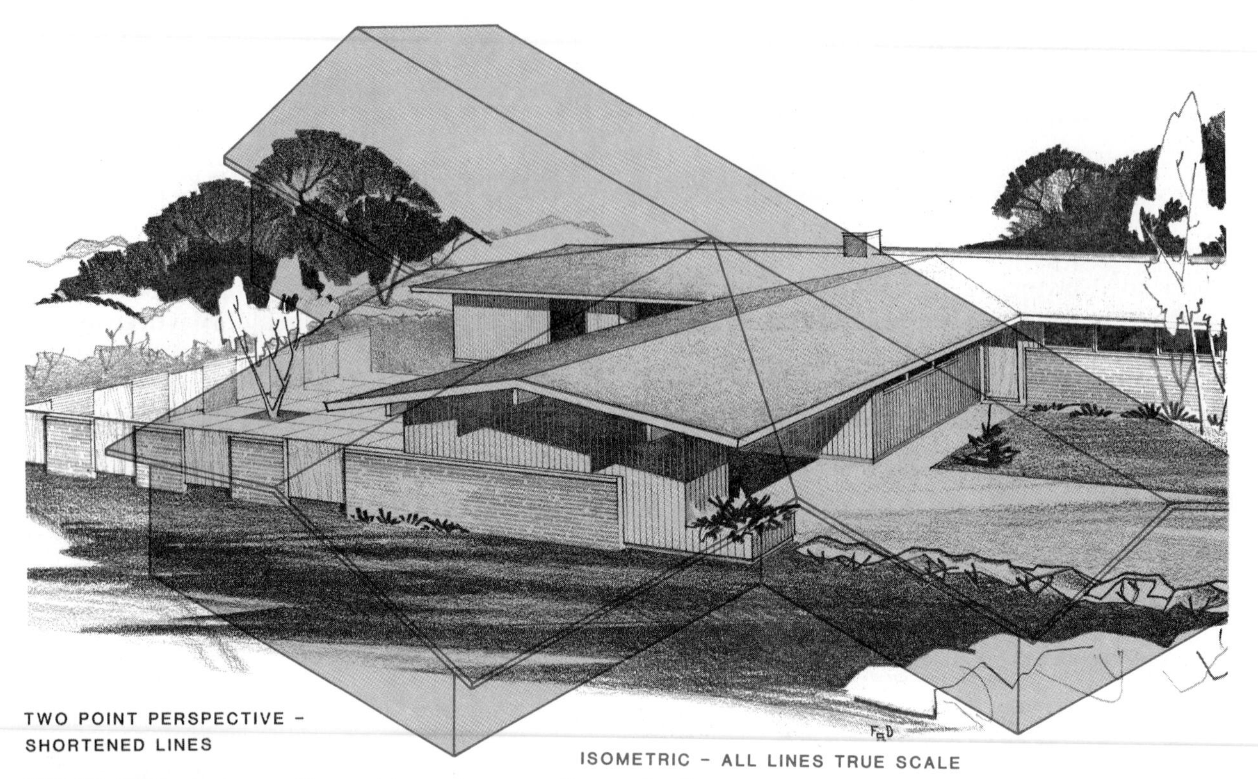

Fig. 40-2 Comparison of isometric and perspective projections.

Fig. 40-3 Long horizontal lines appear to meet. *(Home Planners, Inc.)*

a perspective drawing are always placed on a horizon line.

In preparing perspective drawings, the *horizon line* is the same as your line of sight. If the horizon line is placed through the building, the building will appear at your eye level. If the horizon line is placed below the building, the building will appear to be above your eye level. If the horizon line is placed above the building, it will appear to be below your line of sight. Figure 40-4 shows the effect of horizon-line placement on, above, and below the building. Applications of horizon-line placement are shown in Figs. 40-5 (through the building), 40-6 (below the building), and 40-7 (above the building).

Because perspective drawings do not reveal the true size and shape of the building, perspective drawings are never used as working drawings. To make the drawing appear more realistic, the actual length of the receding sides of the drawing are shortened. Figure 40-8 shows a perspective drawing with shortened sides and an isometric drawing that is prepared to the true dimension of the building. The two sides of the isometric drawing appear distorted because we are accustomed to seeing areas decrease in depth from our point of vision.

ONE-POINT PERSPECTIVE

A *one-point perspective* is a drawing in which the front view is drawn to its true scale and all receding sides are projected to a single vanishing point located on the horizon. If the vanishing point is placed directly behind the object, as can

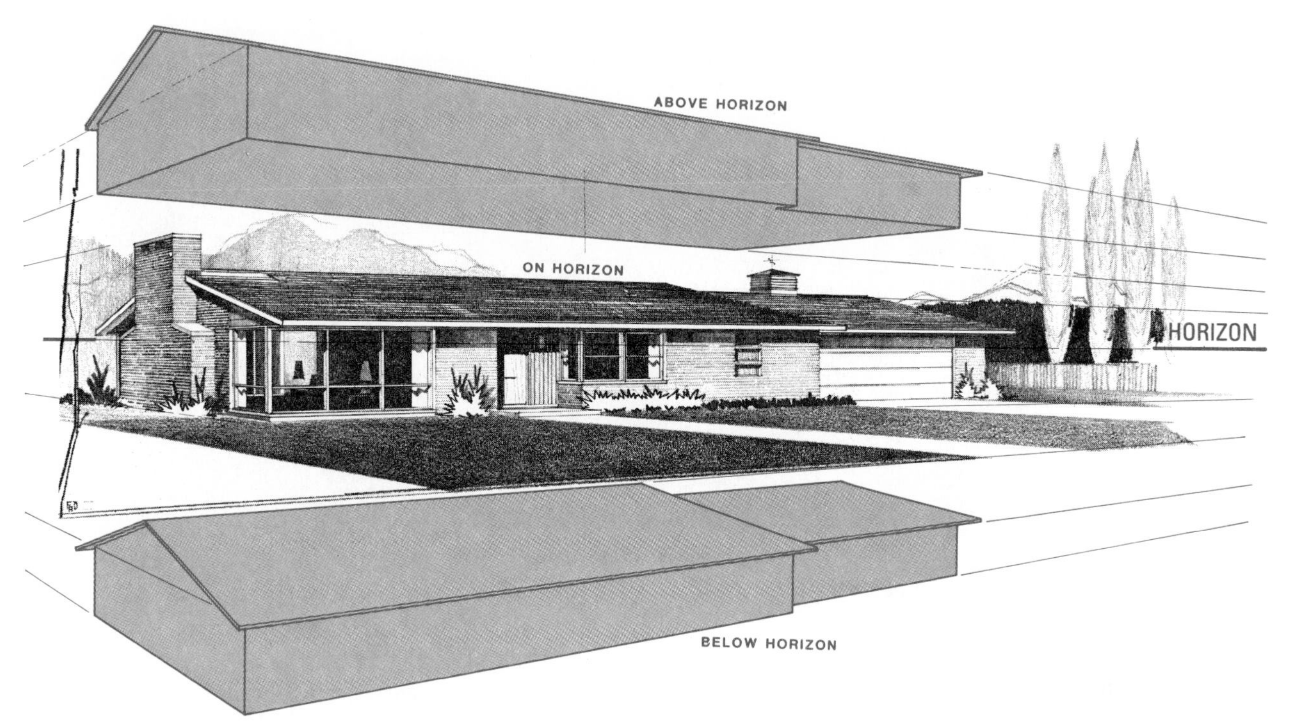

Fig. 40-4 The effect of horizon-line placement. *(Home Planners, Inc.)*

Fig. 40-5 The horizon extends through this building. *(Home Planners, Inc.)*

Fig. 40-6 The horizon is below this building. *(Boise Cascade)*

be seen in the center of Fig. 40-9, no sides would show unless they were drawn with hidden lines. If the vanishing point is placed directly to the right or to the left of the object, with the horizon passing through the object, only one side (left or right) will show. If the object is placed above the horizon line and vanishing point, the bottom of the object will show. If the object is placed below the horizon line and vanishing point, the top of the object will show. The one-point perspective is relatively simple to draw. The front view is drawn to the exact scale of the building. The corners of the front view are then projected to one vanishing point. Follow

HORIZON

Fig. 40-7 The horizon line is placed above this structure. *(Home Planners, Inc.)*

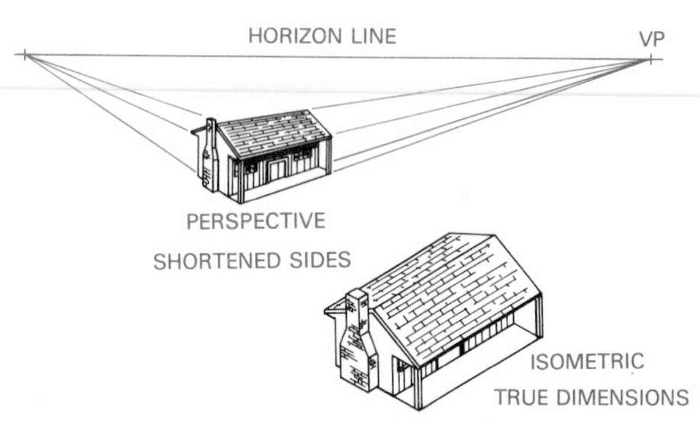
HORIZON LINE
VP
PERSPECTIVE
SHORTENED SIDES
ISOMETRIC
TRUE DIMENSIONS

Fig. 40-8 The length of receding lines should be shortened on perspective drawings.

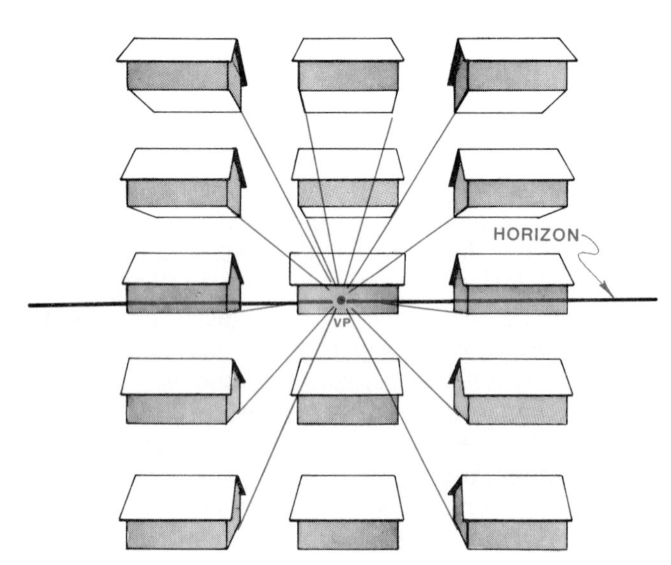
HORIZON
VP

Fig. 40-9 A one-point perspective can show any three sides.

these steps in drawing or sketching a one-point perspective as shown in Fig. 40-10.

1. Draw the horizon line and mark the position of the vanishing point. With the vanishing point to the left, you see the left side of the building, to the right you see the right side of the building, and to the rear you see only the front of the

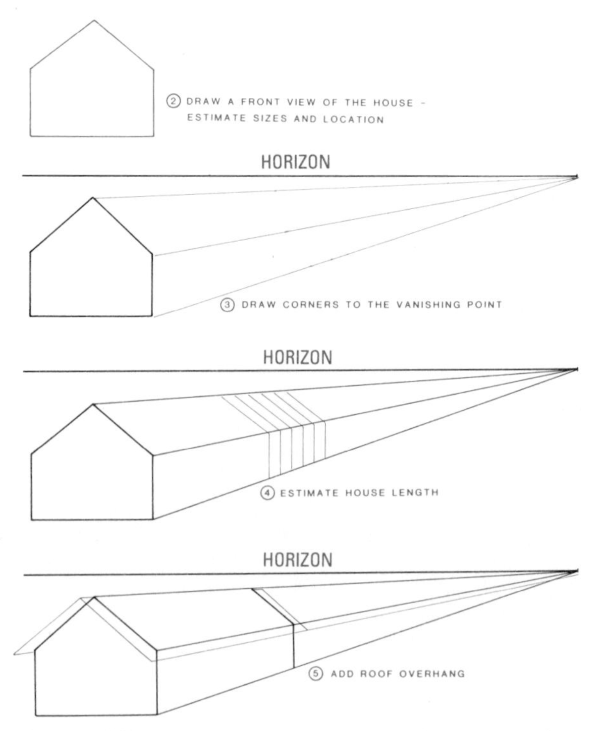
② DRAW A FRONT VIEW OF THE HOUSE - ESTIMATE SIZES AND LOCATION
HORIZON
③ DRAW CORNERS TO THE VANISHING POINT
HORIZON
④ ESTIMATE HOUSE LENGTH
HORIZON
⑤ ADD ROOF OVERHANG

Fig. 40-10 Sequence of one-point pictorial projection.

building if the horizon line extends through the building.

2. Draw the front view of the building to a convenient scale.

3. Project all visible corners of the front view to the vanishing point.

4. Estimate the length of the house. Draw lines parallel with the vertical lines of the front view to indicate the back of the building.

5. Make all object lines heavy, such as roof overhang. Erase the projection lines leading to the vanishing point.

Remember that vanishing points need not always fall outside the building outline. However, when they are located within, only the frontal plane will show, as in Fig. 40-11. Figure 40-12 shows different views of the area shown in Figs. 4-3A and B with the views placed below and above the horizon.

TWO-POINT PERSPECTIVE

A *two-point perspective* drawing is one in which the receding sides are projected to two vanishing points, one on each end of the horizon line (Fig. 40-13). In a two-point perspective, no sides are drawn exactly to scale. All sides recede to vanishing points. Therefore, the only true-length line on a two-point perspective is the vertical corner of the building from which the sides are projected. When the vanishing points are placed close together on the horizon line, considerable distortion results because of the acute receding angles (Fig. 40-14). When the vanishing points are placed farther apart, the drawing looks more realistic. One vanishing point is often placed farther from the building than the other vanishing point. This placement allows one side of the building to recede at a sharp angle while the other recedes less sharply. The vanishing points for the perspective drawing shown in Fig. 40-15 are placed closer to the left of the building than to the right. Consequently vanishing points placed nearer the sides of the drawing will result in smaller receding angles and therefore the sides will appear shorter. Moving the vanishing points further away from the drawing will result in the side of the structure appearing longer.

VERTICAL PLACEMENT

The distance an object is placed above or below the horizon line also affects the amount of distortion in the drawing. Moving an object a greater distance vertically from the horizon line has the same effect as moving the vanishing points closer together. Objects placed close to the horizon line, either on it, just above it, or just below it, are less distorted than objects placed a great distance from the horizon (Fig. 40-16).

Fig. 40-11 Vanishing point within the building outline.

Fig. 40-12A View on the horizon.

Fig. 40-12B View below horizon.

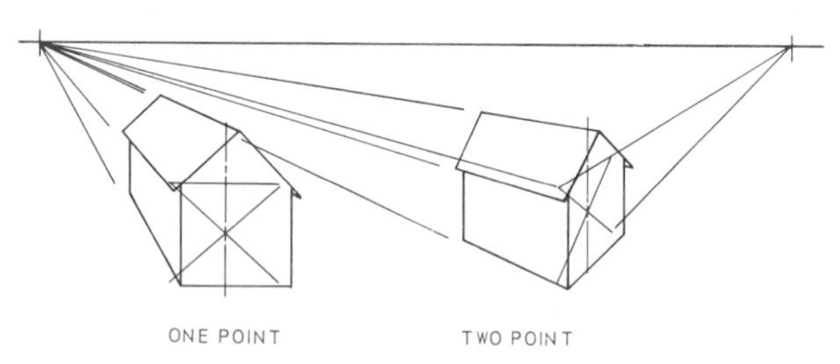

Fig. 40-13 The use of two vanishing points.

Fig. 40-15 The effect of placing the left vanishing point closer to the building than the right vanishing point. *(New Homes Guide)*

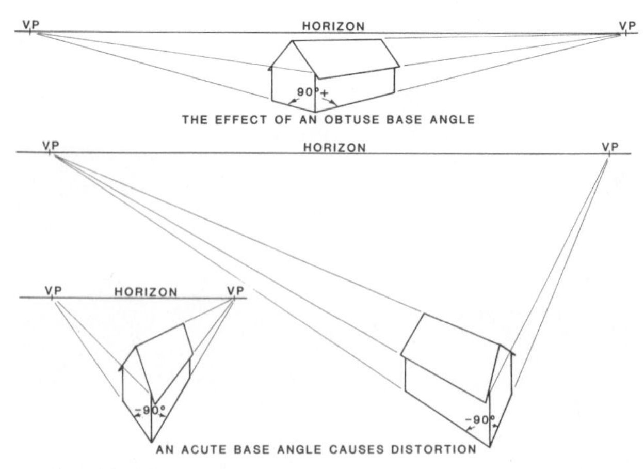

Fig. 40-14 The distance between vanishing points affects the angles of the object.

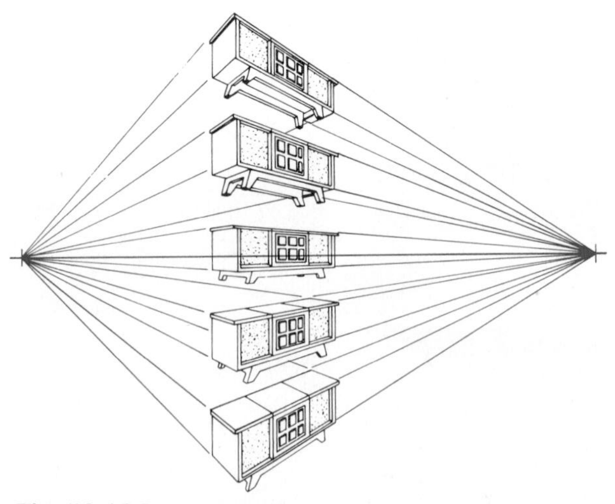

Fig. 40-16 Less distortion occurs close to the horizon.

SEQUENCE

In drawing or sketching a simple two-point perspective, the steps outlined in Fig. 40-17 can be followed. However, in projecting a two-point perspective from an established floor plan and to develop a more accurate perspective, the steps in Fig. 40-18A through D should be followed:

1. Draw a horizontal picture-plane line (Fig. 40-18A).

2. Position the long side of the floor plan at an angle with the picture plane.

3. Locate the station point down from the picture plane about twice the width of the floor plan.

4. Project lines from the station point to the picture plane parallel to the front and end of the floor plan.

5. Draw the ground line (Fig. 40-18B).

6. Draw a horizon line about 6′ or 2 m above the ground, depending on your viewing position.

7. Project lines down from the picture-plane intersection found in step 4 that intersect the horizon line to establish the vanishing points.

8. Position the elevation drawing on the ground line (Fig. 40-18C).

9. Project key elevation lines horizontally.

10. Project key floor-plan lines toward the station point.

11. Where the floor-plan lines of step 10 intersect the picture plane, project lines down vertically.

12. Establish the corners of the building, the doors, and the windows at the points of intersection of the vertical and horizontal lines.

13. Connect intersections to the vanishing points to establish perspective outline.

14. Add the building extensions (Fig. 40-18D).

15. Locate and draw the chimney.

THREE-POINT PERSPECTIVE

Three-point-perspective drawings are used to overcome the height distortion of tall buildings. In a one-or two-story building, the vertical lines recede so slightly that, for practical purposes, they are drawn vertically. However, the top or bottom of extremely tall buildings appears smaller than the area nearest the viewer. A third vanishing point, as shown in Fig. 40-19, may be used to provide the desired recession. The

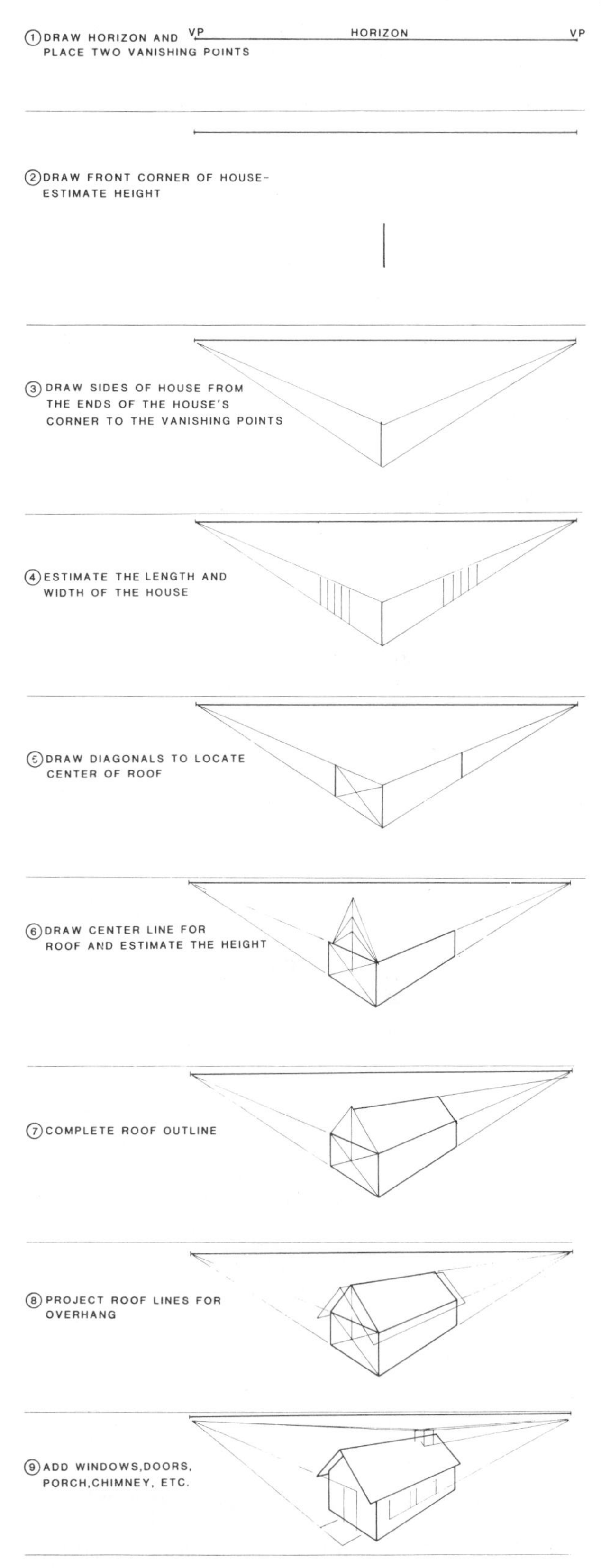

Fig. 40-17 Steps in preparing a two-point perspective.

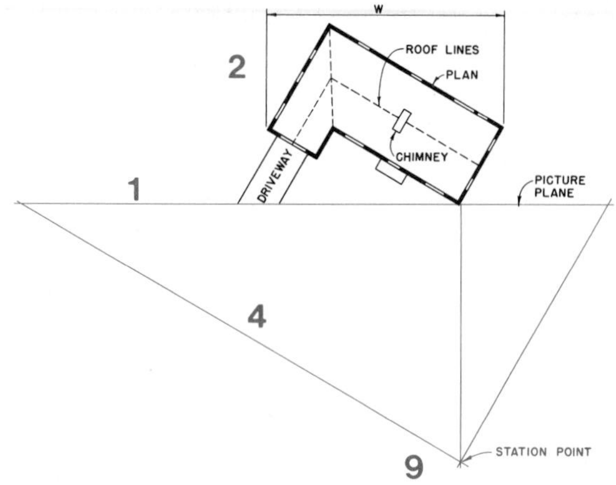

Fig. 40-18A Establish the picture plane and station point.

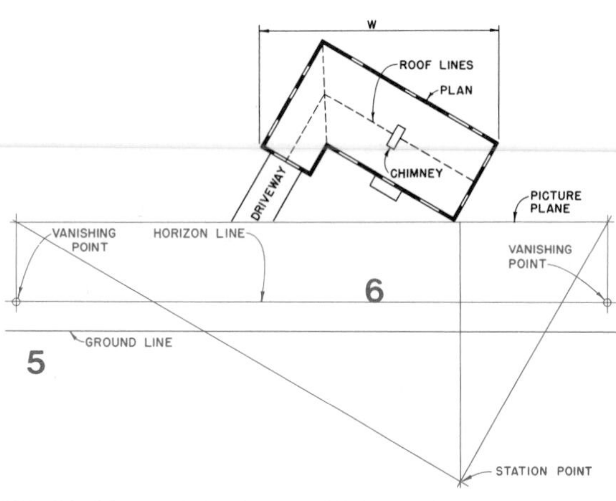

Fig. 40-18B Locate the vanishing points and ground line.

greater the vertical distance between the horizon and the lower vanishing point, the more closely the vertical lines approach a parallel state and the less is the distortion. The farther the third vanishing point is placed away from the object, the less the distortion. If the lower vanishing point is placed so far below or above the horizon that the angles are hardly distinguishable, then the advantage of a three-point perspective is lost because the vertical lines are almost parallel.

DETAIL PICTORIALS

Pictorial drawings are normally used for presentation purposes. However, pictorial drawings may be used to show construction details. Figure

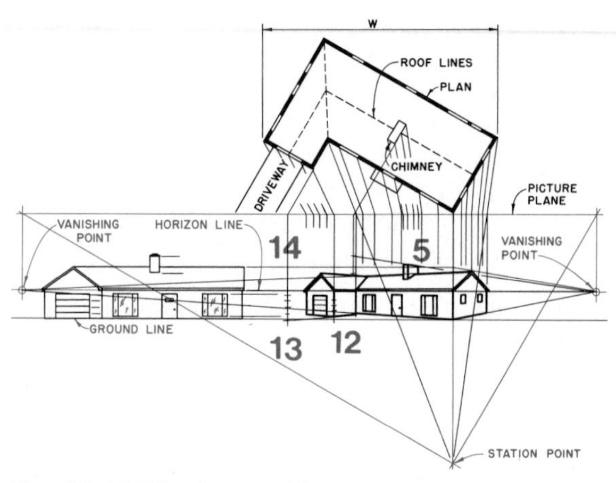

Fig. 40-18C Project and intersect similar floor-plan and elevation lines.

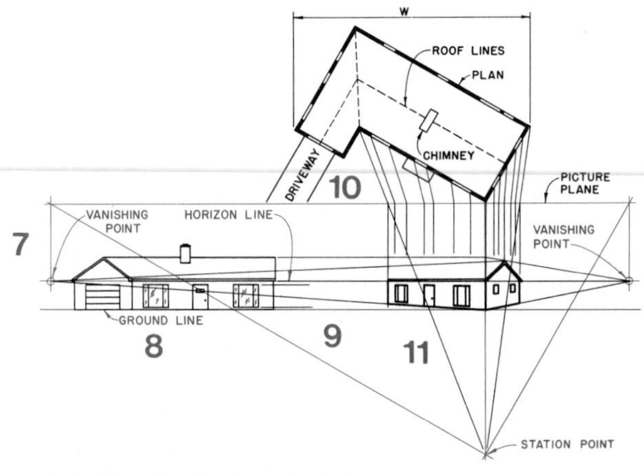

Fig. 40-18D Project building extensions from the floor plan and elevation.

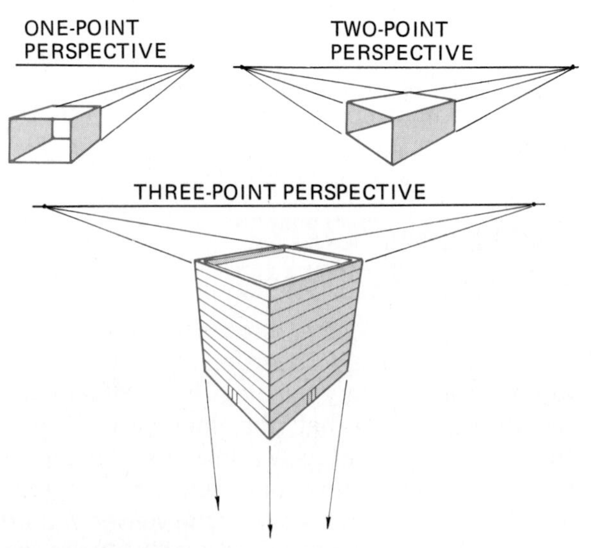

Fig. 40-19 Comparison of one-, two-, and three-point-perspective drawings.

67-6 illustrates how an exploded view can be used to show the assembly and relationship of building components.

PROJECTION DEVICES

Instead of a straight edge, underlay perspective grids are often used to connect the lines of a perspective drawing with the vanishing points, as shown in Fig. 28-14. Perspective grids are preprinted, therefore the positions of the horizon lines and angle of the vanishing point lines are fixed. The specific grid sheet must be selected to produce the drawing view desired.

Mechanical devices such as the perspective-aided drawing (PAD) system can be used to prepare various types of pictorial drawings. In the system shown in Fig. 40-20, a wide range of alternatives in vanishing points, angles of view, station points, and scaling are possible.

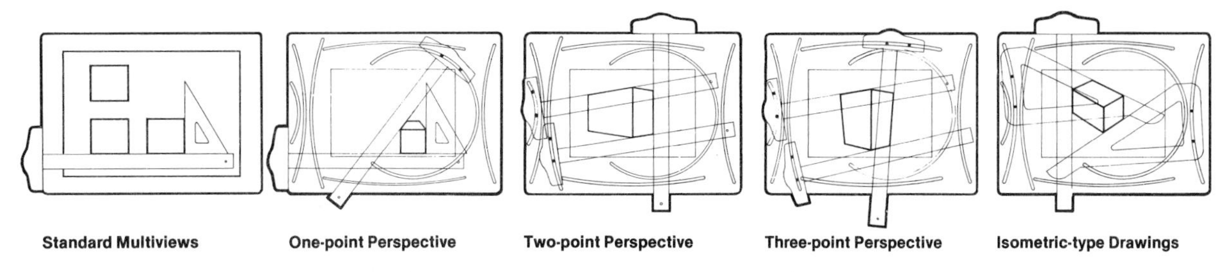

Fig. 40-20 Perspective-aided drawing system. *(PAD International Corp.)*

Exercises

1. **Draw a two-point perspective of a building of your own design.**
2. **Project a two-point-perspective drawing from the floor plan and elevation shown in Fig. 36-3.**
3. **Trace the building shown in Fig. 40-21. Find the position of the vanishing points and the horizon.**
4. **Project a one- and two-point-perspective drawing of the house shown in Fig. 36-1.**
5. **Use the layout and floor plan shown in Fig. 40-22 to project a two-point perspective.**

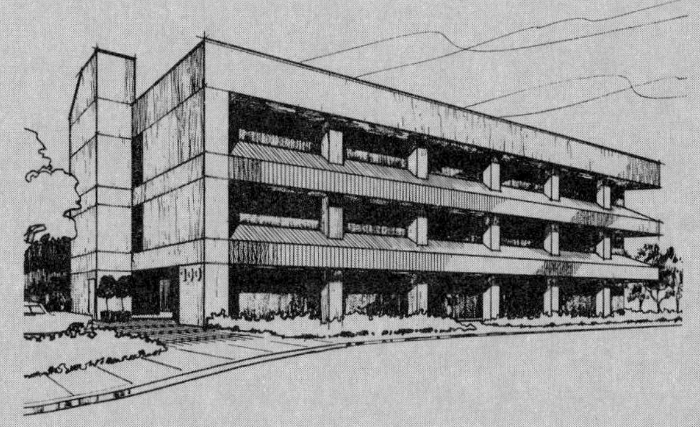

Fig. 40-21

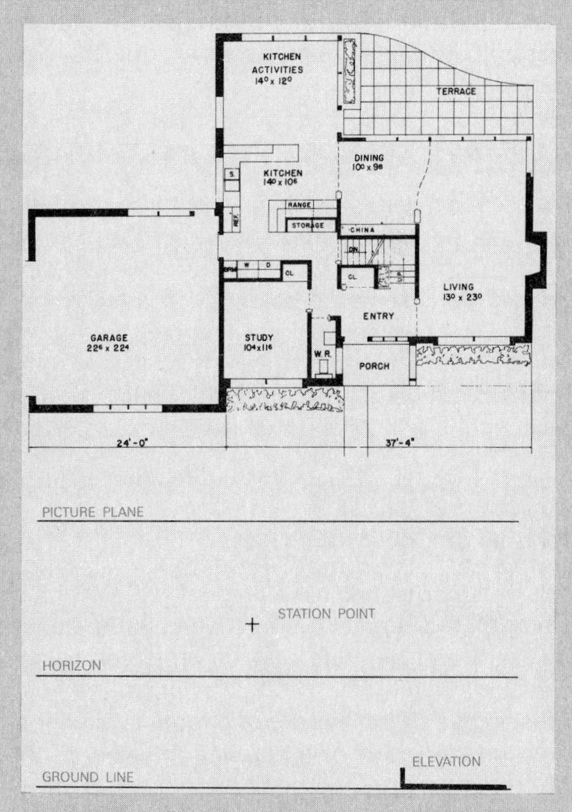

Fig. 40-22

Sketch a design of the elevation prior to projection.
6. Draw a one-point and a two-point perspective of your own home.
7. Sketch a three-point perspective of the tallest building in your community.
8. Define these terms: *pictorial, one-point perspective, two-point perspective, three-point perspective, isometric, vanishing point, horizontal, vertical, parallel, horizon, station point, base line.*

UNIT 41

INTERIOR PICTORIALS

A pictorial drawing of the interior of a building may be an isometric drawing, a one-point-perspective drawing, or a two-point-perspective drawing. Pictorial drawings may be prepared for the entire floor plan. More commonly, however, pictorial drawings are prepared for a partial view of a single room.

ISOMETRIC DRAWINGS

Isometric drawings using constant angles of 30° from the horizontal are satisfactory for pictorial floor plans (Fig. 41-1). There are no receding lines on an isometric drawing. Isometric lines are always parallel and may be prepared to an exact scale. Isometric drawings of single room interiors are usually not desirable, because they lack receding lines and therefore look distorted.

ONE-POINT PERSPECTIVE

A *one-point perspective* of a room is a drawing in which all the intersections between walls, floors, ceilings, and furniture may be projected to one vanishing point (Fig. 41-2). Drawing a one-point perspective of the interior of a room is similar to drawing the inside of a box with the front of the box removed. In a one-point interior perspective, walls perpendicular to the plane of projection, such as the back wall, are drawn to scale. The vanishing point on the horizon line is then placed somewhere on this wall (actually behind this wall). The points of intersection where this wall intersects the ceiling and floor are then projected from the vanishing point to form the intersection between the side walls and the ceiling and the side walls and the floor.

Vertical Placement

If the vanishing point is placed high, very little of the ceiling will show in the projection, but much of the floor area will be revealed (Fig. 41-3A). If the vanishing point is placed near the center of the back wall, an equal amount of ceiling and floor will show (Fig. 41-3B). If the vanishing point is placed low on the wall, much of the ceil-

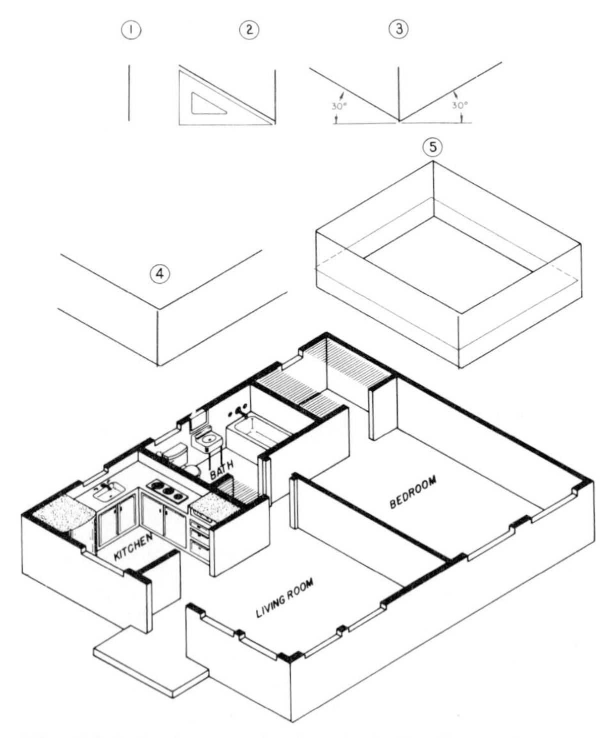

Fig. 41-1 An isometric drawing of a floor plan.

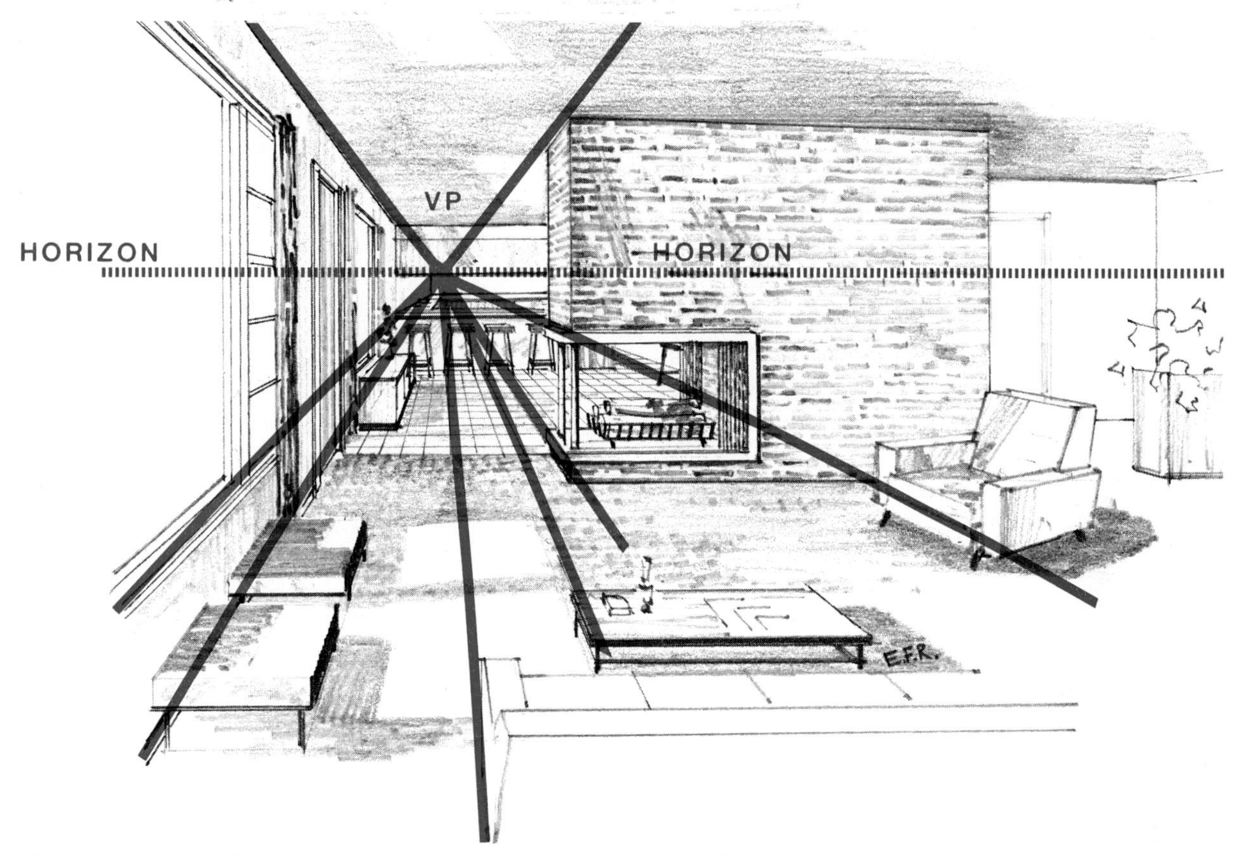

Fig. 41-2 A one-point interior perspective. *(Home Planners, Inc.)*

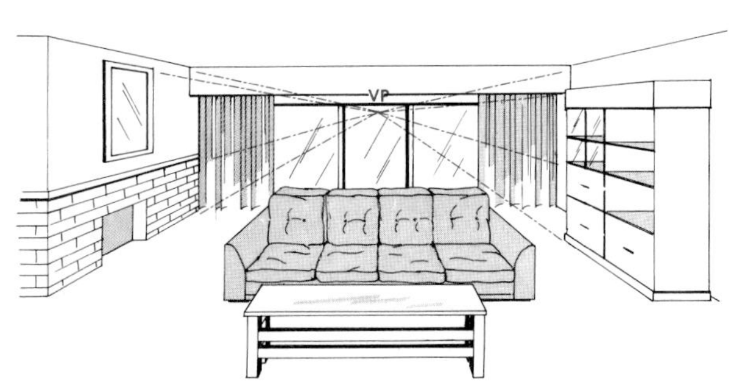

Fig. 41-3A The effect of a high central vanishing point.

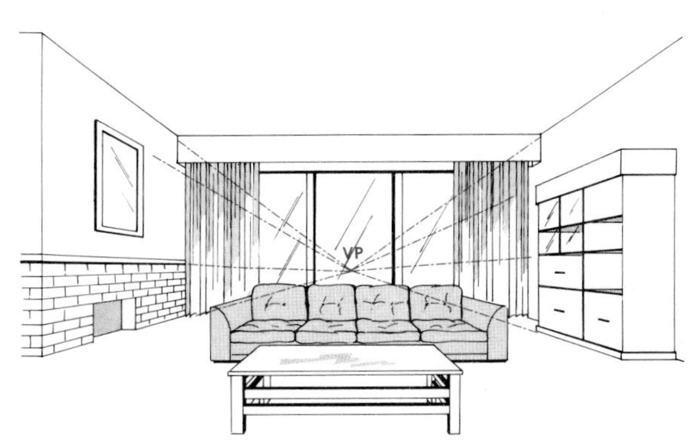

Fig. 41-3B The effect of a centrally located vanishing point.

ing but very little of the floor will be shown (Fig. 41-3C). Since the horizon line and the vanishing point are at your eye level, you can see that the position of the vanishing point affects the angle from which you view the object.

Horizontal Placement

Moving the vanishing point from right to left on the back wall has an effect on the view of the side walls. If the vanishing point is placed to-

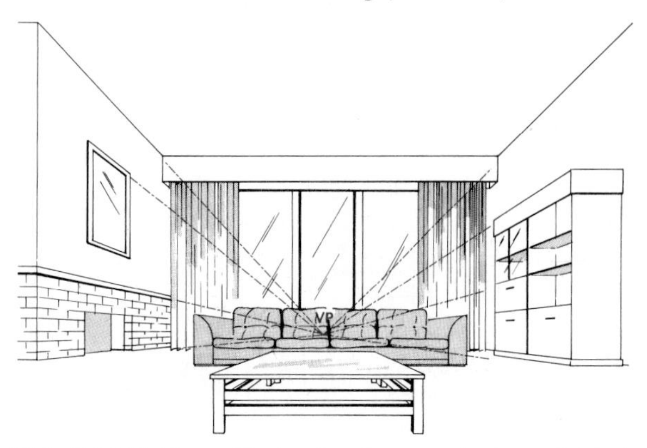

Fig. 41-3C The effect of a low central vanishing point.

ward the left, more of the right wall will be revealed (Fig. 41-4A). Conversely, if the vanishing point is placed near the right side, more of the left wall will be revealed in the projection (Fig. 41-4B). If the vanishing point is placed in the center, an equal amount of right wall and left wall will be shown. When one wall should dominate, place the vanishing point on the extreme end of the opposite wall. When projecting wall offsets and furniture, always block in the overall size of the item to form a perspective view, as shown in Fig. 41-5A, B, and C. The details of furniture or closets or even of persons can then be completed within this blocked-in cube or series of cubes.

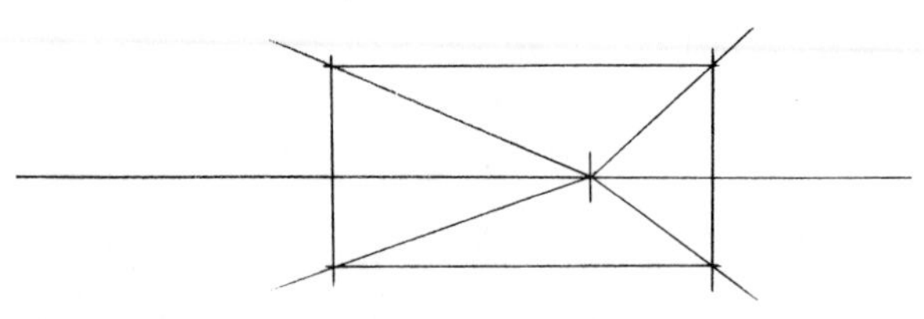

Fig. 41-5A First step in developing one-point perspective.

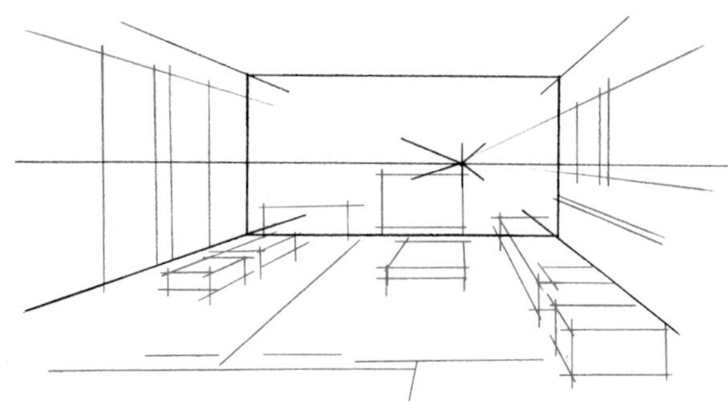

Fig. 41-5B Second phase in developing one-point perspective.

Fig. 41-4A The effect of the left-wall placement of the vanishing point.

Fig. 41-5C Final steps in blocking in one-point perspective.

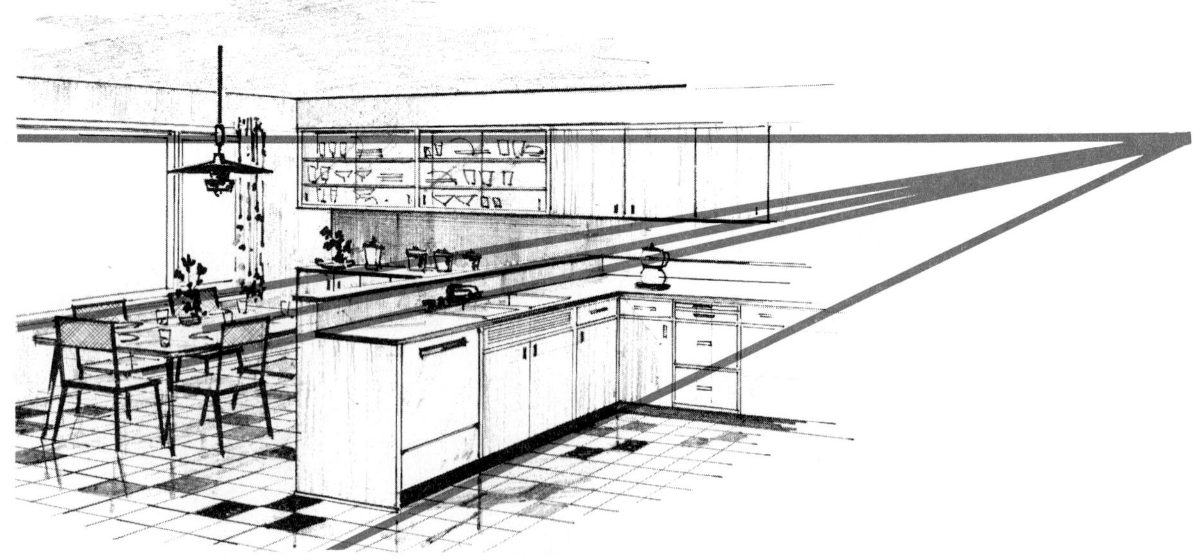

Fig. 41-4B Left-wall emphasis is obtained by placing the vanishing point to the extreme right. *(Home Planners, Inc.)*

TWO-POINT PERSPECTIVE

Two-point perspectives are normally prepared to show the final design and decor of two walls of a room. The base line on an interior two-point perspective is similar to the base line on an exterior two-point perspective. The base line in the drawing shown in Fig. 41-6 is the corner of the cabinet. Once the walls are projected to the vanishing points in the two-point perspective, each object in the room can also be projected to the vanishing point as in external two-point perspectives. Projecting the fireplace shown in Fig. 41-7 to the vanishing point is the same as drawing an exterior perspective.

The sequence of steps in drawing two-point interior perspectives is shown in Fig. 41-8. Figure 41-9 shows the relationships between floor plan drawings and perspective drawings. Pictorial grids are recommended to eliminate the projection of horizon and vanishing points. However, the proper grid must be selected to provide the horizon and vanishing-point position necessary to reveal the room at the angle desired. Figure 41-10 shows the application of a perspective grid to an interior two-point perspective drawing.

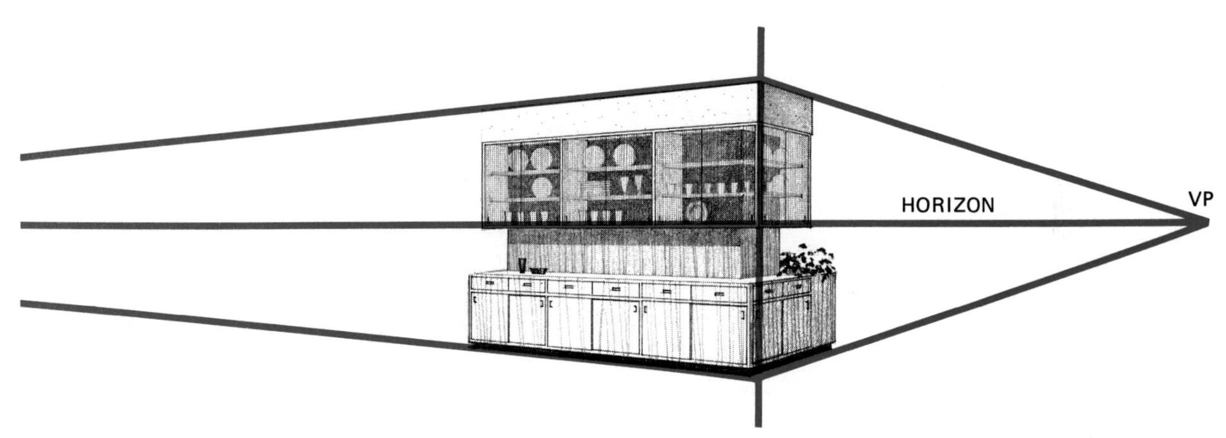

Fig. 41-6 A two-point interior perspective.

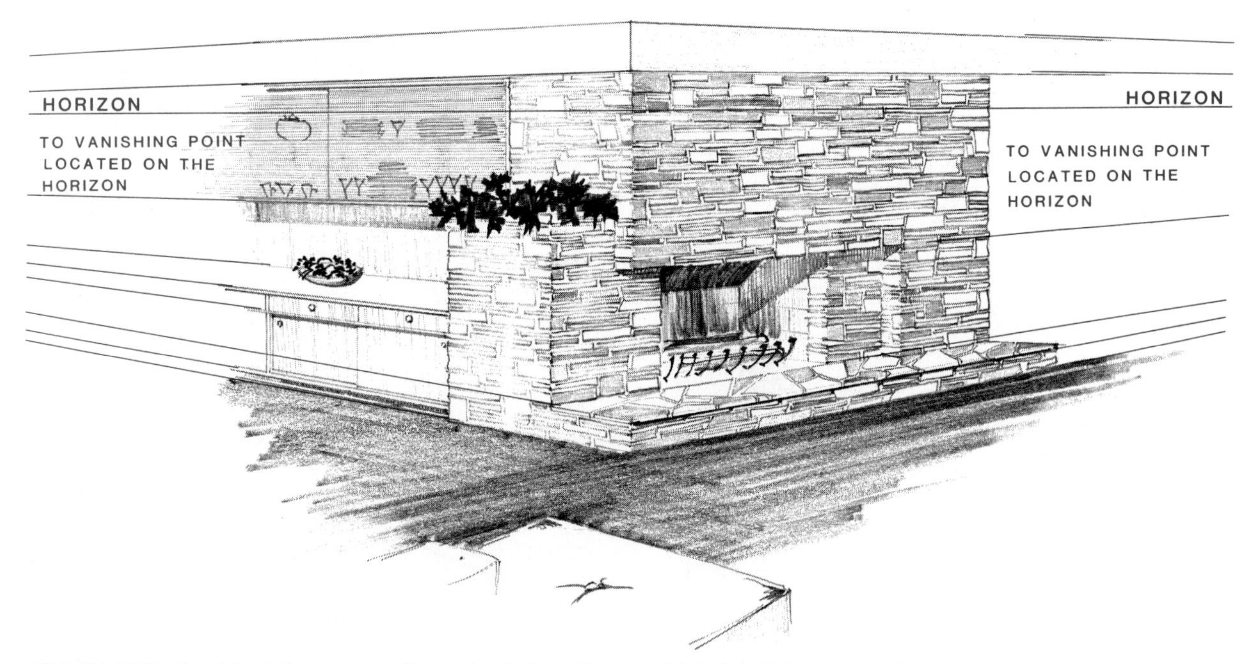

Fig. 41-7 Each object in a room is projected to the vanishing point.

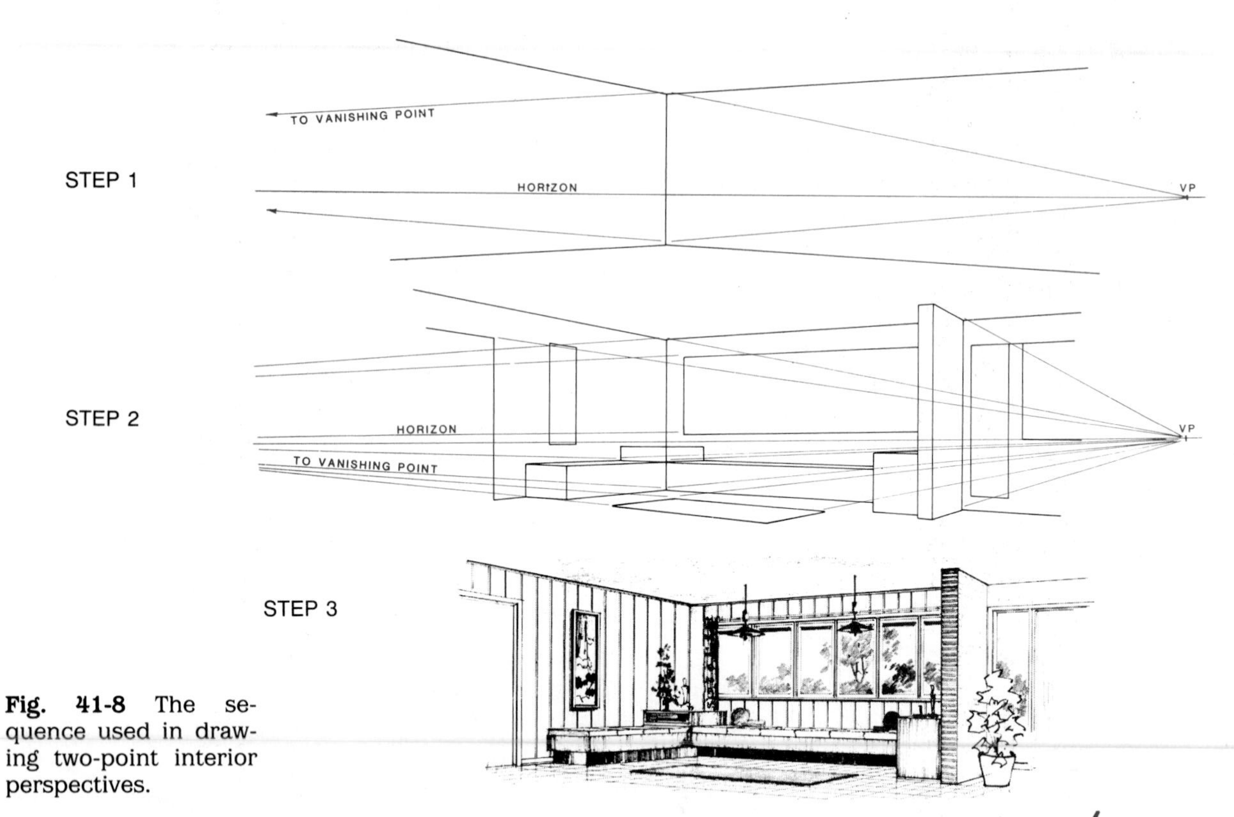

Fig. 41-8 The sequence used in drawing two-point interior perspectives.

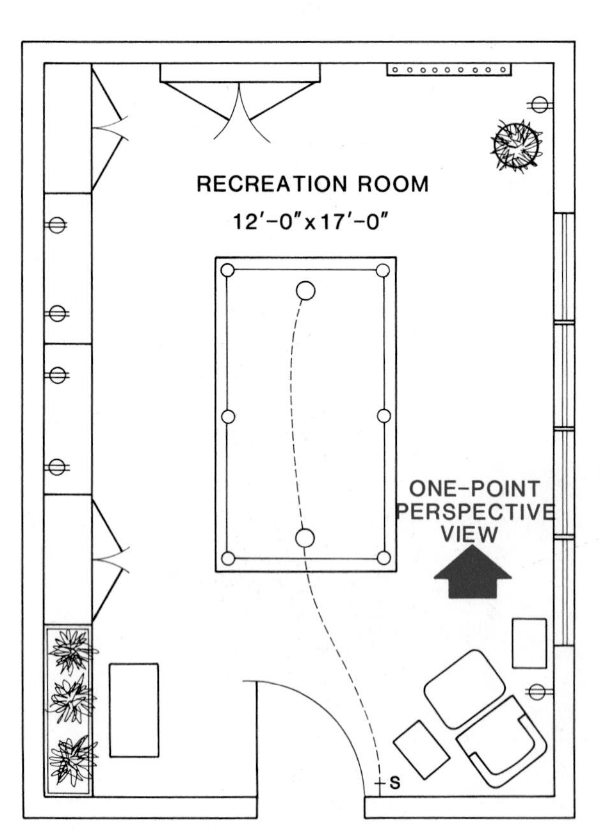

Fig. 41-9A Floor plan of perspective shown in Fig. 41-9B.

Fig. 41-9B Perspective floor plan of Fig. 41-9A.

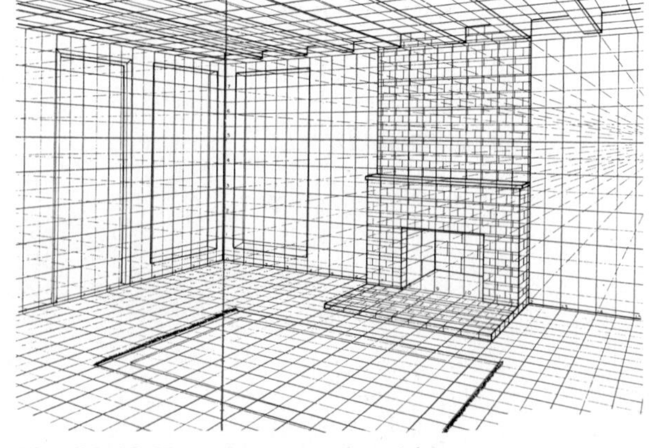

Fig. 41-10 Use of perspective grid.

Exercises

1. Trace the drawing shown in Fig. 41-11. With a colored pencil, project the ceiling and floor lines to find the vanishing points and the horizon line.
2. Trace the perspective shown in Fig. 41-12. Find the position of the vanishing points. Extend the drawing to include the right wall of the living area.
3. Prepare a one-point interior perspective of your own room.
4. Prepare a one-point perspective of a room of your own design.
5. Draw a one-point interior perspective of a classroom. Prepare one drawing to show more of the ceiling and left wall. Prepare another drawing to show more of the floor and right wall.
6. Trace the lines in Fig. 41-13 and sketch in furniture.
7. Define these terms: *isometric, interior perspective.*

Fig. 41-11

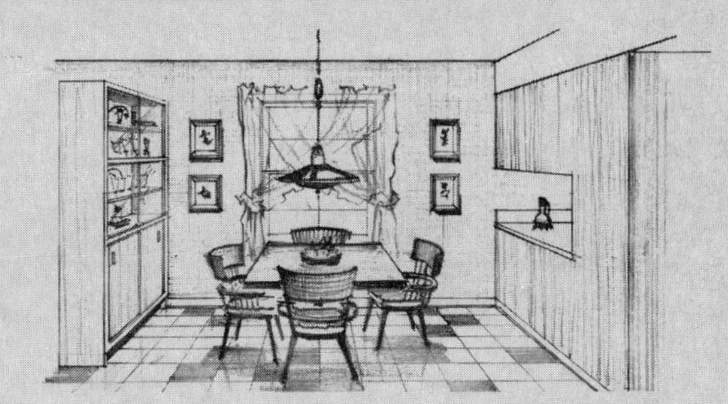

Fig. 41-12

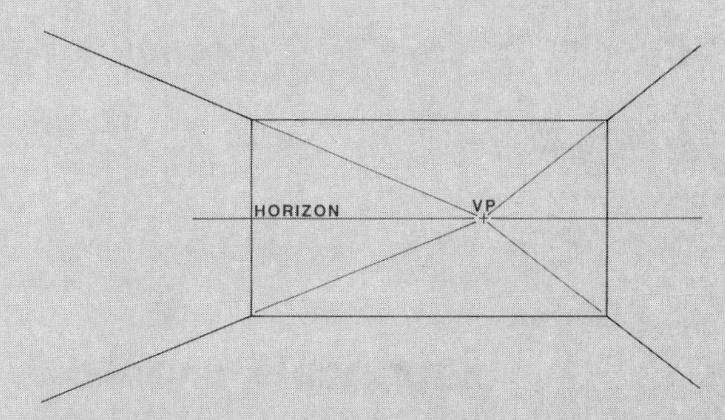

Fig. 41-13

UNIT 42

ARCHITECTURAL RENDERING

To *render* a pictorial drawing is to make the drawing appear more realistic. This may be done through the media of pencil, pen and ink, watercolors, pastels, or airbrush. Drawings are rendered by adding realistic texture to the materials and establishing shade and shadow patterns.

MEDIA

Soft pencils are one of the most effective media for rendering architectural drawings because tones can be greatly varied by the weight of the line used. Smudge blending to add tone can be accomplished by rubbing a finger over penciled areas. Figure 42-1A shows pencil techniques used to create an architectural rendering.

Watercolor techniques are also popular in architectural rendering. Fig. 42-1B shows wash techniques added to an exterior line drawing to produce realistic effects. Pen-and-ink renderings of architectural drawings vary greatly. Strokes must be placed farther apart to create

Fig. 42-1A Pencil rendering.

Fig. 42-1B Watercolor wash rendering. *(Home Planners, Inc.)*

Fig. 42-1C Pen-and-ink rendering. *The Port Authority of New York and New Jersey)*

Fig. 42-1D Combination of line and wash techniques. *(Home Planners, Inc.)*

light effects and closer together to produce darker effects (Fig. 42-1C). Figure 42-1D shows a wash rendering of an interior perspective drawing. Figure 42-1E shows the same building rendered in pencil and in pen and ink.

SHADE

When you *shade* an object, you lighten the part of the object exposed to the sun or other light sources and darken the part of the object not exposed to the sun or light source. Notice how the lower part of Fig. 42-2 is shaded darker, and the top shaded lighter, to show the different exposure to the sun. Likewise, the right portion of Fig. 42-3 is shaded darker to denote distance and lack of direct sunlight. When objects with sharp corners are exposed to strong sunlight, one area may be extremely light and the other side of the object extremely dark. However, when objects and buildings have areas that are round (cylindrical), so that their parts move gradually from dark to light areas, a gradual shading from extremely dark to extremely light must be made.

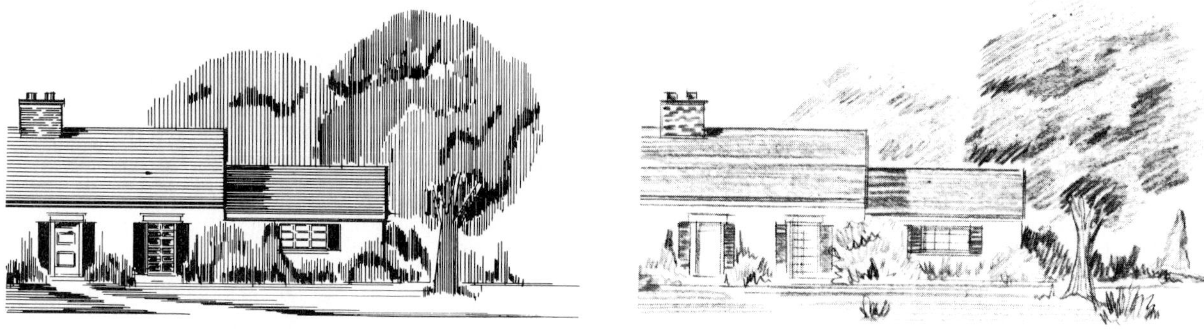

Fig. 42-1E Comparison of pencil and pen-and-ink techniques.

Fig. 42-2 The use of shading to show depth and light angles. *(The Port Authority of New York and New Jersey)*

Fig. 42-3 Methods of shading to show distance. *(The Port Authority of New York and New Jersey)*

SHADOW

In order to determine what areas of the building will be drawn darker to indicate *shadowing,* the angle of the sun in the illustration must be established. When the angle of the sun is established, all shading should be consistent with the direction and angle of the shadow. Figure 42-4 shows the shadowing effect of a light source directed on objects from the middle left. Figure 42-5 shows the shadow patterns resulting from a light source shining from the top left. In drawing shadows on buildings always keep the light source location in mind. On buildings that are drawn considerably below the horizon line, shadow patterns will often reveal more than the actual outline can reveal. Notice how the shadow patterns of the overhangs, walls, and tree positions are revealed and reinforced by the use of shadows in Fig. 42-6A. Figure 42-6B shows how shadowing is used to reveal the length of the roof overhang not otherwise apparent on this drawing.

TEXTURE

Giving *texture* to an architectural drawing means making building materials appear as rough or as smooth as they actually are. Smooth surfaces are no problem since they are very reflective and hence are very light. Only a few reflection lines are usually necessary to illustrate smoothness of surfaces such as aluminum, glass, and

Fig. 42-4 The use of shading to show sunlight on high areas.

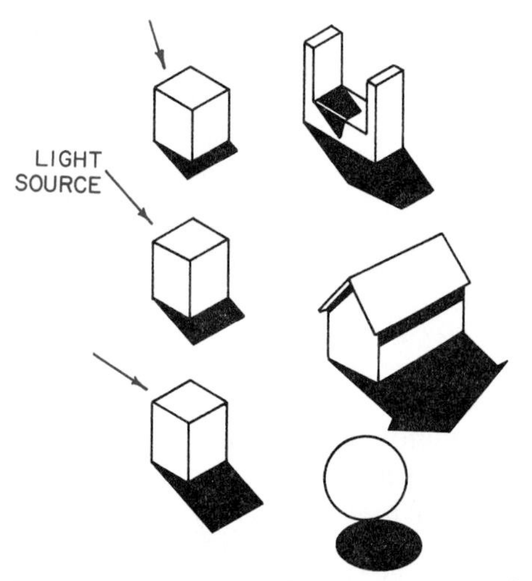

Fig. 42-5 Shadow patterns in the area opposite the light source.

painted surfaces. On rough surfaces, the thickness or roughness of the material can often be shown by shading. Figure 42-7 shows texturing of an external surface.

SEQUENCE

In preparing pictorial renderings, proceed in the following sequence, as shown in Fig. 42-8A through D.

1. Block-in with single lines the projection of the perspective (Fig. 42-8A).

2. Sketch the outline of building materials in preparation for rendering (Fig. 42-8B). This work can be done with a soft pencil, a ruling pen, or a crow-quill pen. Establish the light source and sketch shadows and shading. Darken windows, door areas, and underroof overhangs.

3. Add texture to the building materials. For example, show the position of each brick with a chisel-point pencil. Leave the mortar space white and lighten the pressure for the areas that are in direct sunlight (Fig. 42-8C).

4. Complete the rendering by emphasizing light and dark areas and establishing more visible contrasts of light and dark shadow patterns (Fig. 42-8D). Figure 42-9A, B, and C show a similar sequence in the development of an interior wash rendering.

TECHNIQUES

Figures 42-10 through 42-17 show the application of various rendering techniques to architectural drawings. Figure 42-10A shows some of the techniques used to draw shrubs on pictorial drawings. Figure 42-10B shows methods of ren-

Fig. 42-6A Shadows reveal hidden outlines.

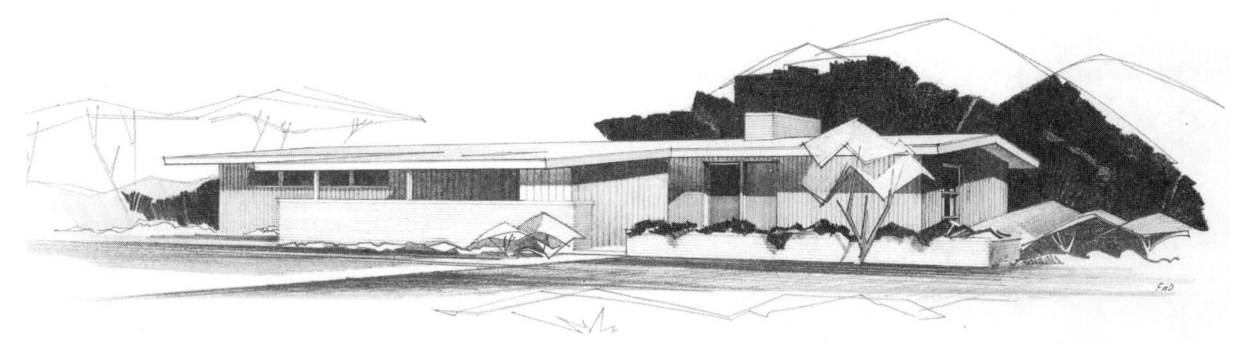

Fig. 42-6B Shadows reveal length of overhang.

Fig. 42-7 Exterior texture rendering.

Fig. 42-8A Block in the basic outline.

Fig. 42-8B Prepare the sketch for rendering.

Fig. 42-8C Texture is added to building materials.

Fig. 42-8D The completed rendering. *(Home Planners, Inc.)*

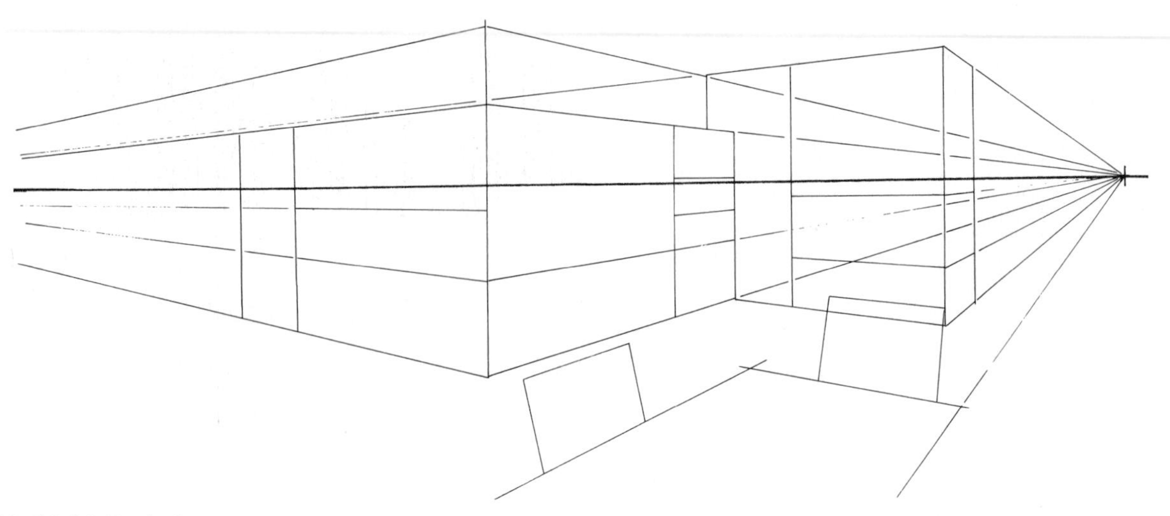

Fig. 42-9A Basic layout.

Fig. 42-9B Added detail.

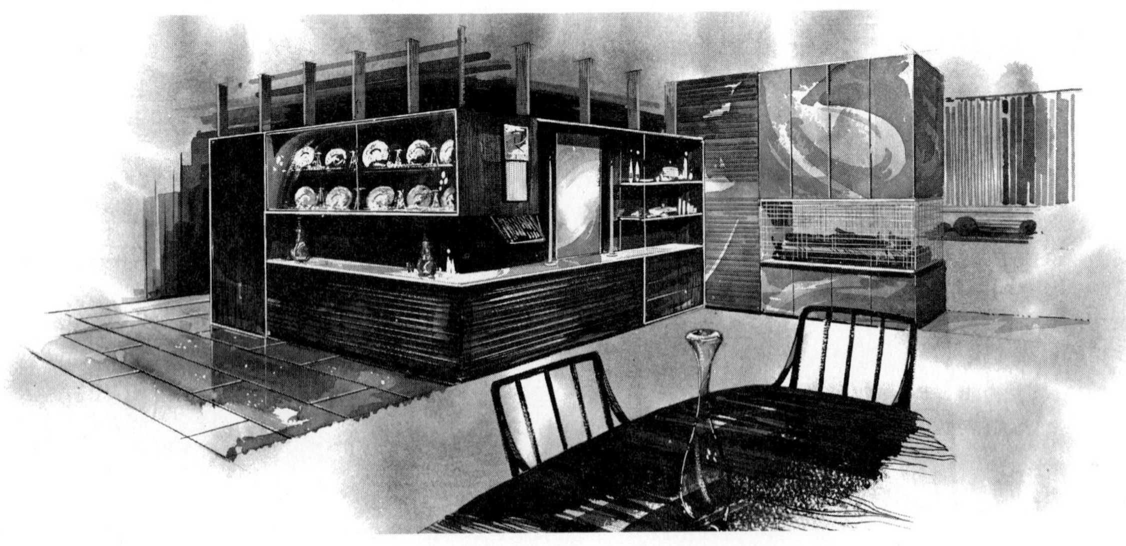

Fig. 42-9C Finished rendering.

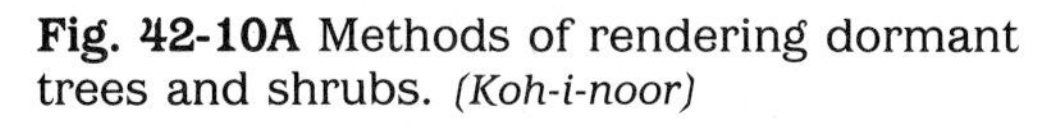

Fig. 42-10A Methods of rendering dormant trees and shrubs. *(Koh-i-noor)*

Fig. 42-10B Methods of rendering ground cover.

dering ground cover. Notice how the darker trees are placed behind the structure. Lighter, more open trees are placed in front to allow the details of the building to be observed without restriction. These methods are used to show tree placement without blocking out the view of the buildings as shown in Fig. 42-10C.

Figure 42-11A shows depth and shadows in rendering windows. Notice that some windows are rendered to show reflected light, and others are drawn to reveal the room behind, as though the window were open. Keep in mind most windows look dark except at night. This is because it is darker on the inside than on the outside. So to produce a realistic window, dark or even black surfaces are often used as shown in Fig. 42-11B. However, at times outlines of interior features are drawn as shown in Fig. 42-11C. Rendering fences and walls (Fig. 42-12) and rendering chimneys (Fig. 42-13) require a combination of shade, shadow and texture techniques.

Sketches of people are often necessary to

Fig. 42-10C Tree rendering techniques.

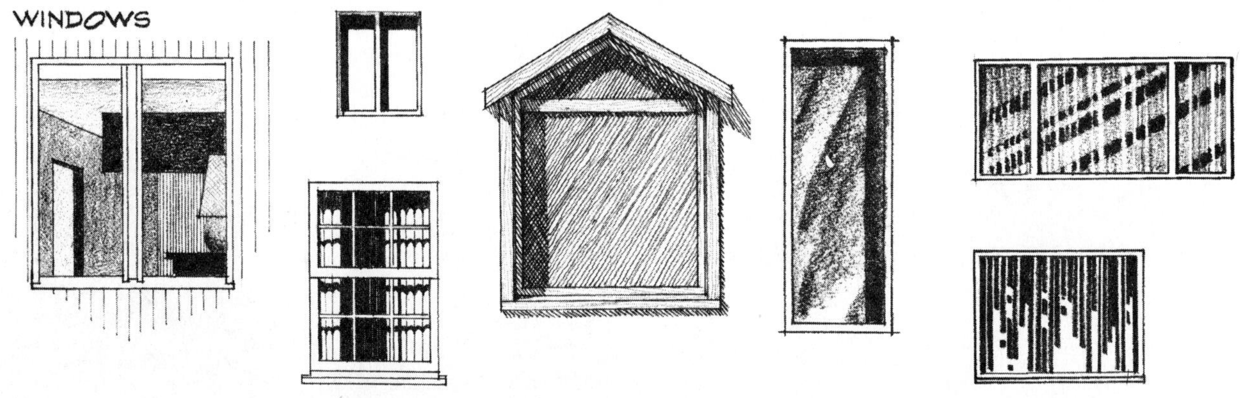

Fig. 42-11A Use of shadows in rendering windows.

Fig. 42-11B Dark areas used in window rendering.

Fig. 42-11C Use of faint interior outlines.

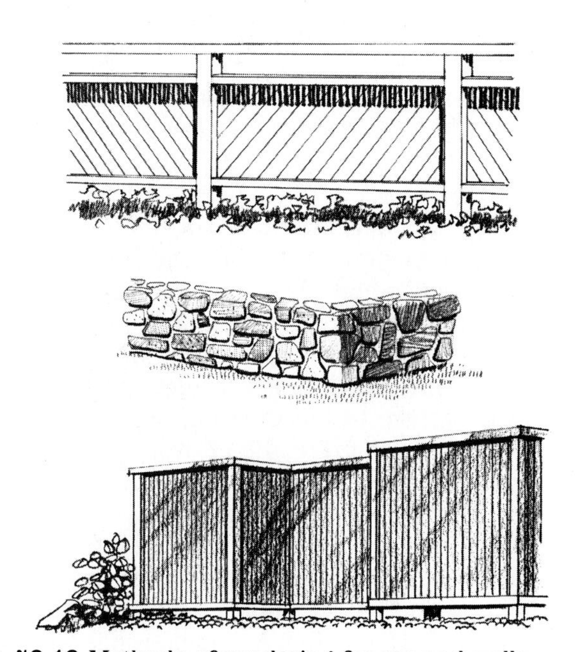

Fig. 42-12 Methods of rendering fences and walls.

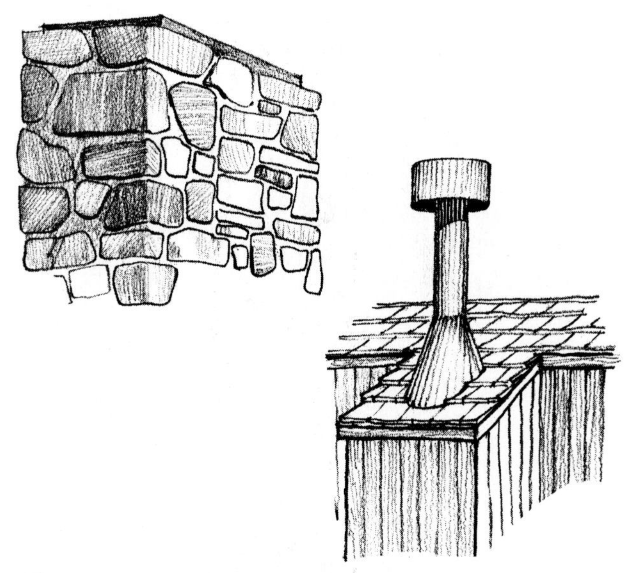

Fig. 42-13 Pencil renderings of chimneys.

show the relative size of a building and to put the total drawing in proper perspective. Since people should not interfere with the view of the building, architects frequently draw people in outline or in extremely simple form (Fig. 42-14A). People are also used to show traffic patterns (Fig. 42-14B) and size differences and to provide a feeling of perspective and depth. Automobiles are added to architectural renderings (Fig. 42-15) to provide proper perspective and to give a greater feeling for external traffic patterns.

It is also important to utilize proper rendering techniques to show texture and shadows on flat

Fig. 42-14A Architectural sketches of people.

Fig. 42-15 Automobile sketches for use on architectural drawings. *(McGraw-Hill:* Entourage *by Ernest Burden)*

Fig. 42-14B The use of people to show traffic patterns. *(The Port Authority of New York and New Jersey)*

surfaces, such as siding (Fig. 42-16A). Figure 42-16B shows examples of rendering techniques for exterior materials used on siding and roofs. Notice how the shading of the texture varies depending on the light source. Also note how the use of shadows from trees and overhangs are incorporated into the texture values and add more realism to the drawing. The rendering of doors is shown in Fig. 42-17. Notice that most of this is done by light shading and also by variation of the pencil-stroke widths.

One popular technique that can be used to convert a perspective line drawing into a rendering is to apply screen tones. The drawings shown in Fig. 42-18 are simple line drawings with various shades of screens added to denote texture and shadow. The use of more realistic pencil technique to show texture and shadows is illustrated in Fig. 42-19. Though rare, construction detail drawings are sometimes rendered to promote clarity when prepared for laymen or workers unfamiliar with architectural detail drawings. A rendered isometric detail drawing is shown in Fig. 42-20.

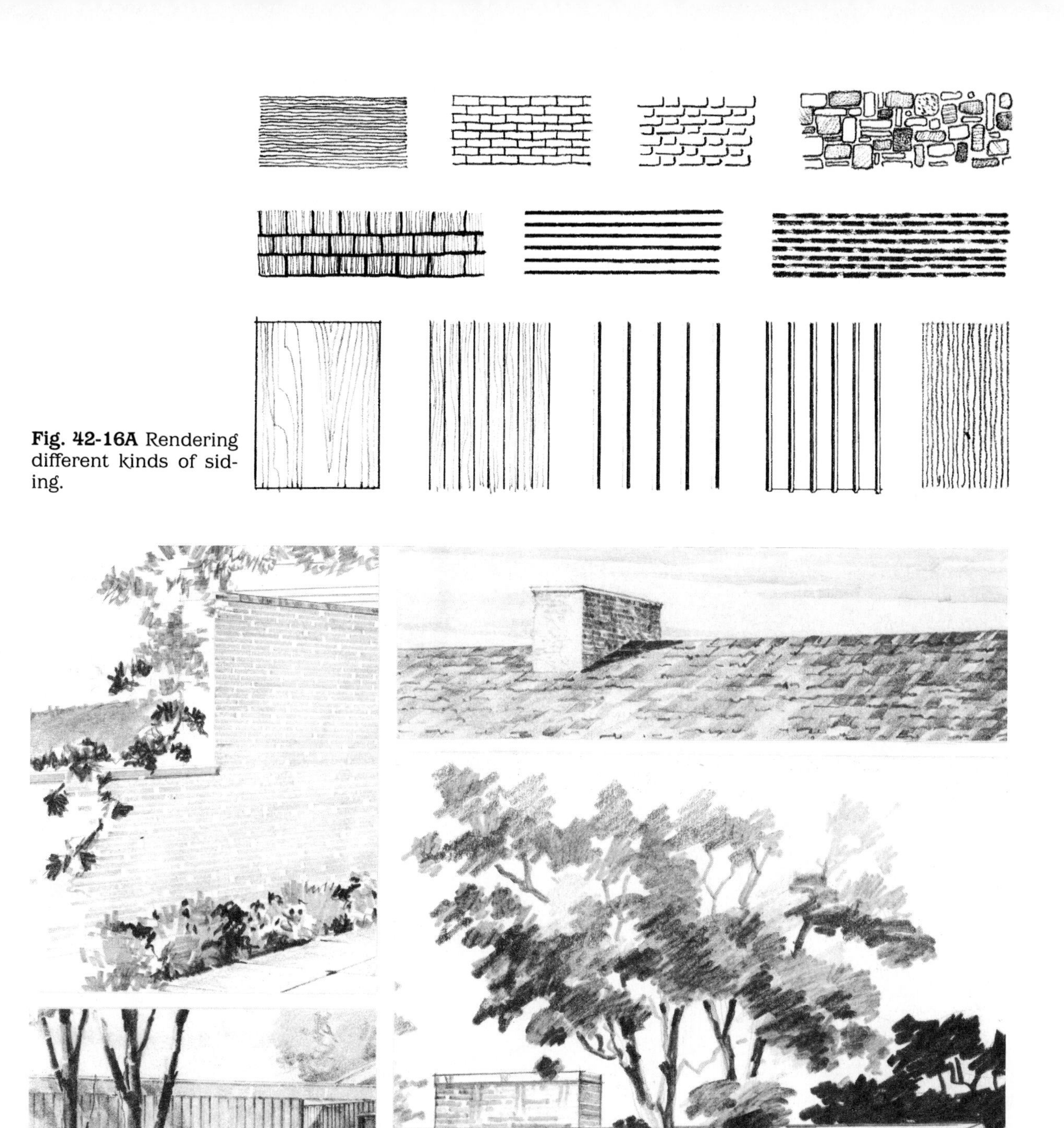

Fig. 42-16A Rendering different kinds of siding.

Fig. 42-16B Methods of rendering exterior siding materials.

Fig. 42-17 Methods of rendering doors.

Fig. 42-18 The area-tone method of rendering.

Black pencil or ink lines on white or light-colored stock are used most often for architectural rendering because of the ease of reproduction with the remainder of the set of drawings. However, for presentation purposes, renderings on colored stock with white and black lines are sometimes prepared. When colored illustration board is used, all lines, shades, or tints are added with a variety of white water color or acrylic paint. Notice how the combination of black, white, and gray tones is used in Fig. 42-21 to emphasize textures, shading, and landscape features.

When full-color architectural renderings are prepared, wash drawing (water color) techniques are usually used. However, acrylics are often used as the primary medium.

If drawing on a CAD system screen, tones can be added by using crosshatch procedures to fill areas with different textured surface treatments.

Fig. 42-19 A combination of pencil techniques showing shade, shadow, and texture.

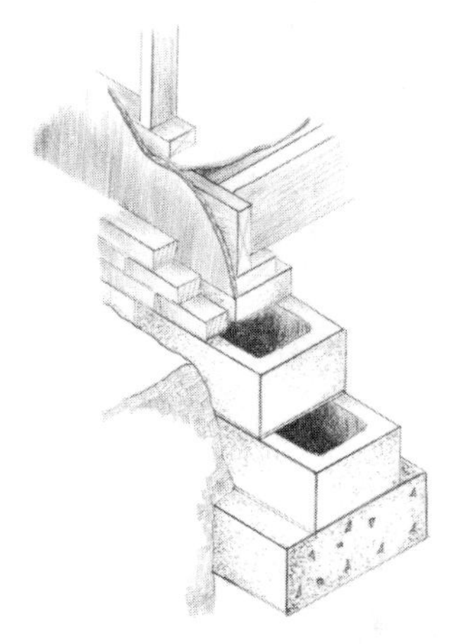

Fig. 42-20 Rendered detail drawing.

ABBREVIATED RENDERING

Often it is necessary or desirable to render only one part of a building. In such cases, the other attached parts may be only outlined and the rendering gradually diminished, instead of abruptly stopped. Figure 42-22 was prepared to show only the porch of the house. However, the relationship to the remainder of the house is important and, therefore, the house is shown in outline.

Fig. 42-21 Rendering on colored stock.

Fig. 42-22 A partial rendering. *(Home Planners, Inc.)*

Exercises

1. **Render a perspective drawing of your own house.**
2. **Render a perspective drawing of a house of your own design.**
3. **Render a perspective sketch of your school. Choose your own medium: pencil, pen and ink, watercolors, pastels, or airbrush.**
4. **Complete the perspective shown in Fig. 42-23 and completely render the drawing in pencil.**
5. **Define these terms: *render, texture, shade, shadow, chisel-point, crow-quill.***

Fig. 42-23 Complete this rendering. *(Home Planners, Inc.)*

PART FOUR

Specialized Architectural Plans

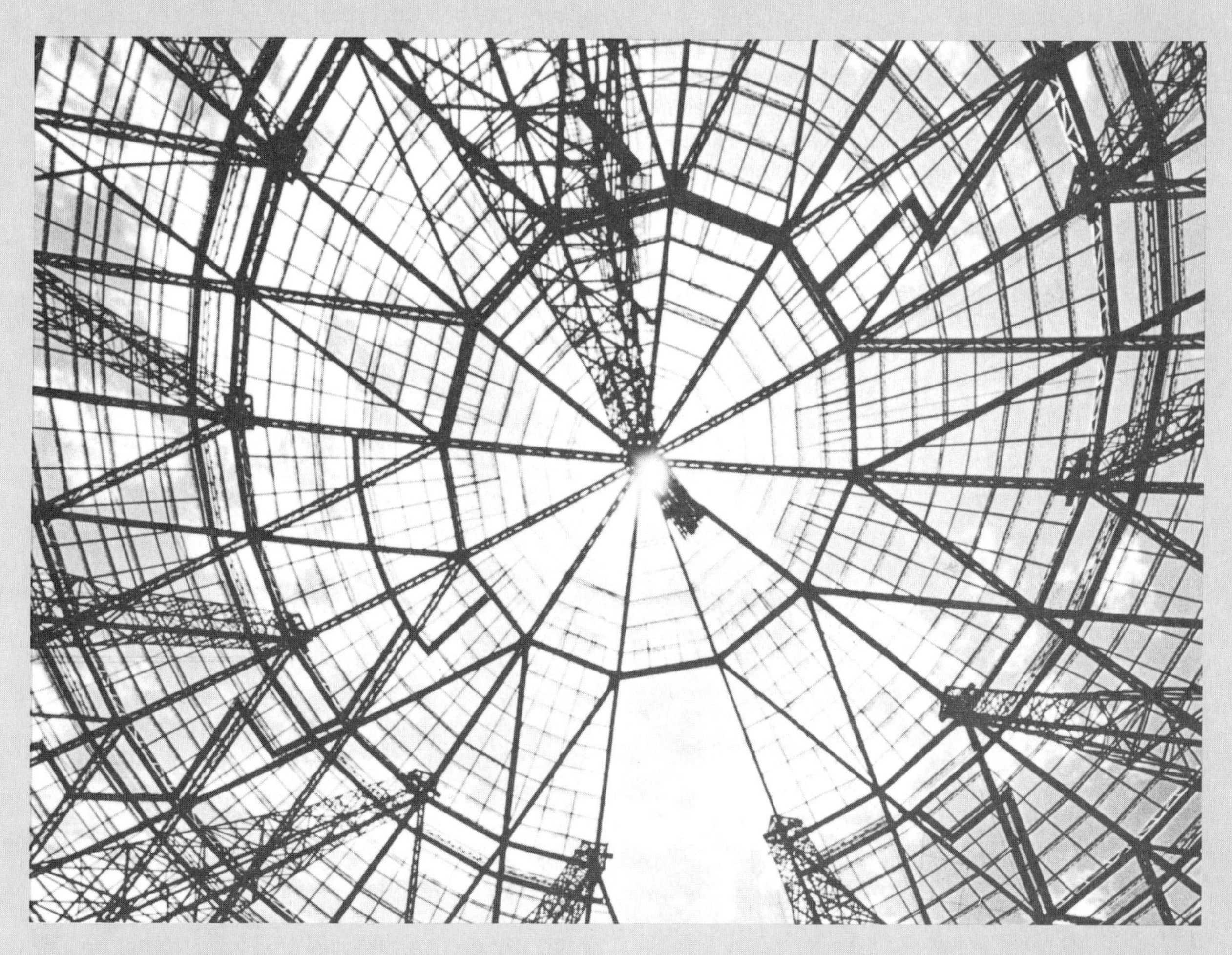

Basic architectural plans, such as floor plans, elevations, and pictorial drawings, are adequate for describing the general design of a structure. However, to ensure that the building will be completed as specified, a more complete and detailed description of the construction features of the design must be prepared. Technical architectural plans are prepared for this purpose. These include site development plans, sectional drawings, foundation plans, framing plans, electrical plans, climate-control plans, plumbing diagrams, and modular-construction plans. In Part Four you will learn the basic practices and procedures in preparing technical architectural plans, and the relationship of a set of architectural plans.

SECTION 10

Site Development Plans

Site plans are necessary to show the relationship of all structures to the site. The features shown on site plans include the outline and dimensions of all buildings and their exact position on the site. Also included on site plans is the final contour of the landform and the position and size of all other constructed features such as patios, walks, driveways, pools, easements and horticultural elements. Site plans are divided into three types: survey plans, plot plans, and landscape plans. However, the features of these three types of plans are often incorporated into one composite site plan. On large projects the development of the site plan is under the direction of a licensed landscape architect. The development of the site and the integration of all structures with the landform is an integral part of the organic design process as described in Unit 2.

UNIT 43

PLOT PLANS

Plot plans are used to show the location and size of all buildings on the lot. Overall building dimensions and lot dimensions are shown on plot plans. The position and size of walks, drives, patios, and courts are also shown. Compass orientation of the lot is given, and contour lines are sometimes shown. Figure 43-1 shows the key figures and the symbols commonly used on plot plans. Plot plans may also include details showing site construction features. These include details and typical sections for walks, driveways, patios, culverts, decking, and pool construction.

When a separate survey and landscape plan are prepared for a project, contour lines, utility lines and planting details may be omitted from the plot plan. But often only the most dominant of these features are included on a plot plan.

GUIDES FOR DRAWING PLOT PLANS

When plot plans are prepared, the numbered features shown in Fig. 43-2 and Fig. 43-3A should be drawn according to these matching guides for drawing plot plans:

1. Draw only the outline of the main structure on the lot. Cross-hatching is optional.

2. Draw the outlines of other buildings on the lot.

3. Show overall building dimensions.

4. Locate each building by dimensioning perpendicularly from the property line to the closest point on a building. On curved property lines, dimension to points of tangency as shown in Fig. 43-3B. The *property line* shows the legal limits of the lot.

5. Show the position and size of driveways.

6. Show the location and size of walks.

7. Indicate grade elevation of key surfaces such as patios, driveways, and courts.

8. Outline and show the symbol for surface material used on patios and terraces.

9. Label streets adjacent to the outline.

10. Place overall lot dimensions either on extension lines outside the property line or directly by the property line.

11. Show the size and location of courts.

12. Show the size and location of pools, ponds, or other bodies of water.

13. Indicate the compass orientation of the lot by the use of a north arrow.

14. Use a decimal scale such as 1″ = 10′, or 1″ = 20′, for preparing the plot plan.

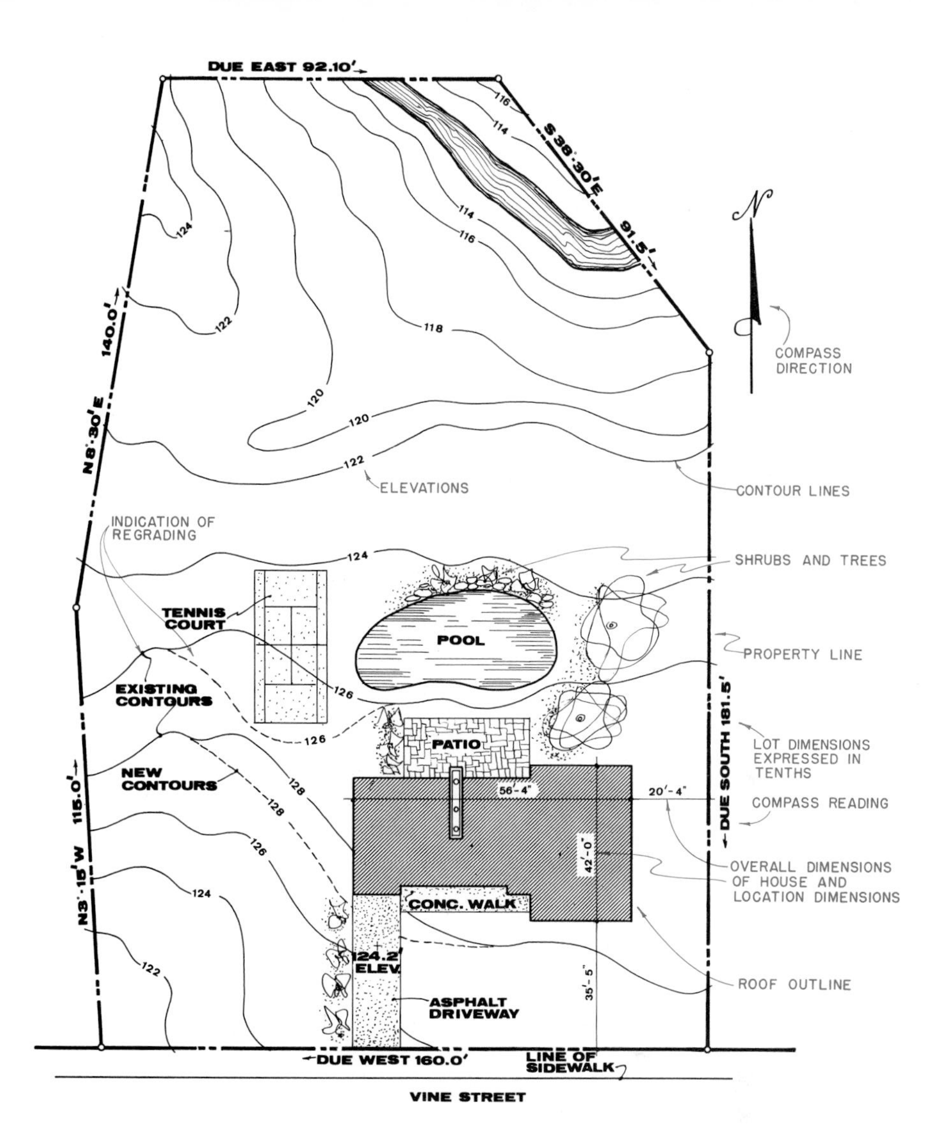

Fig. 43-1 Plot-plan symbols.

15. Show the position of utility lines on a plot plan or a survey plan (Fig. 43-4A).

16. The perimeter dimensions include the compass direction for each property line (Fig. 43-4B).

17. Show trunk base location and coverage of all major trees.

18. Label and dimension all landscape construction features.

19. Draw and dimension the location and minimum distance allowed from septic system components to nearest building as shown in Fig. 45-10.

ALTERNATIVES AND VARIATIONS

Although plot plans should be prepared according to the standards shown in Fig. 43-2, many optional features also may be included in plot plans. For example, sometimes the interior partitions of the residence are given to show a correspondence between the outside living areas and those inside. Some designers prefer to include only the outline of the building on the plot plan, while others favor crosshatching or shading the buildings.

The positions of entrances to buildings are sometimes noted on plot plans, as shown in Fig. 43-5. This device provides an interpretation of the access to the house from the outside, without requiring a detailed plan of the inside. Contour lines are another optional feature on plot plans. On lots that slope greatly in contour, contour lines are necessary. The plot plan is also often used to show the outline of the roof, as in Fig. 43-6. There are many ways to place buildings on a lot. Sometimes alternative plot plans

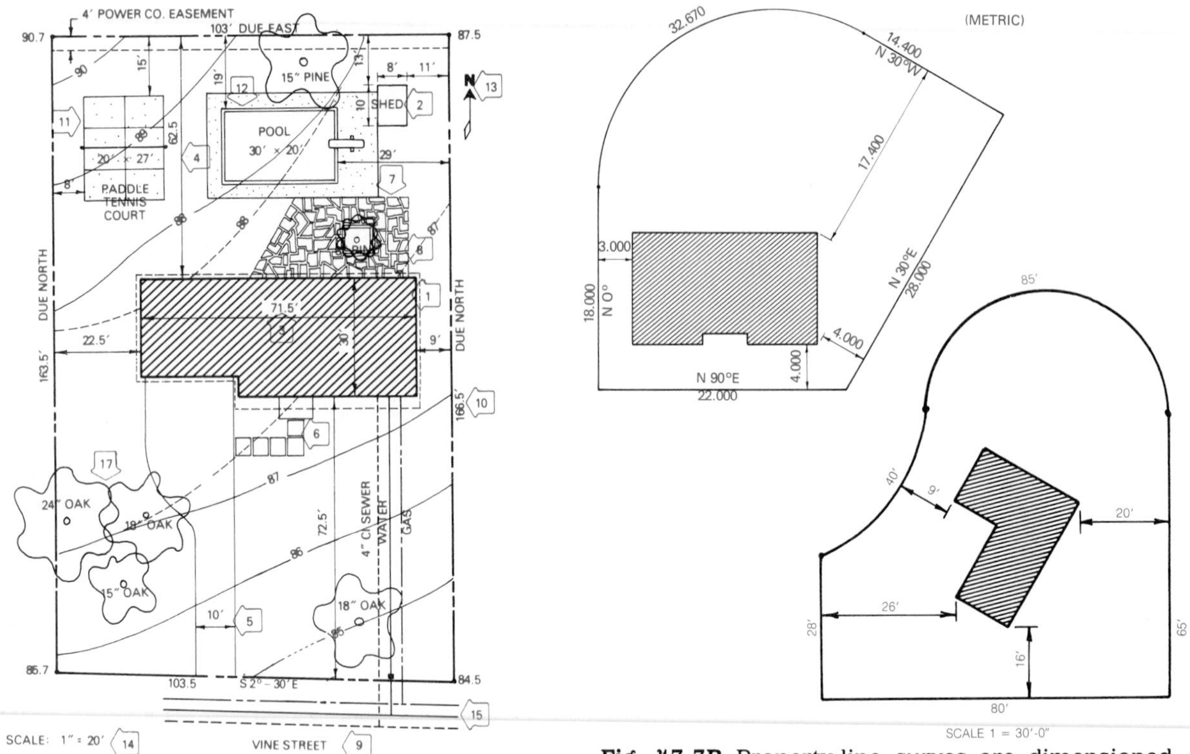

Fig. 43-2 Guides for drawing plot plans.

Fig. 43-3B Property-line curves are dimensioned from point of tangency.

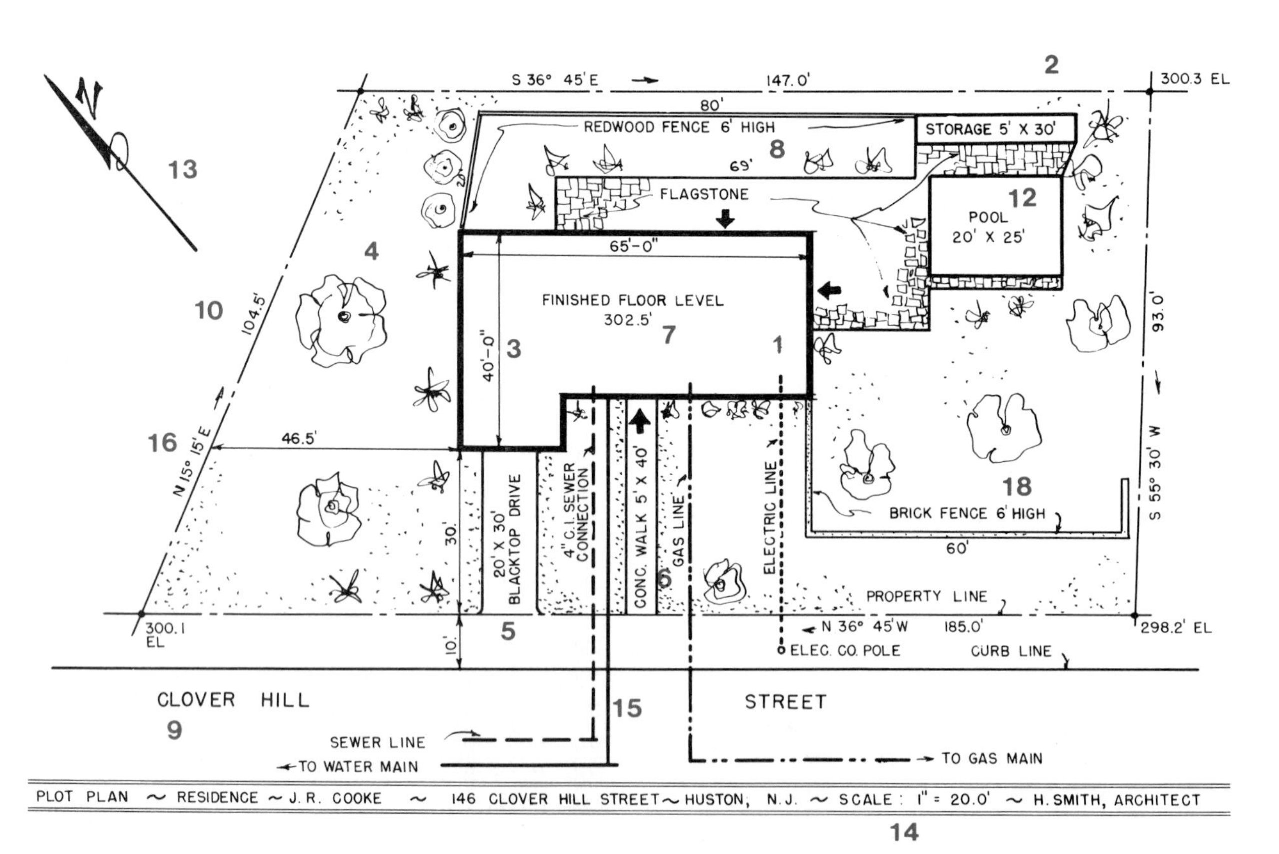

Fig. 43-3A Plot-plan dimensions.

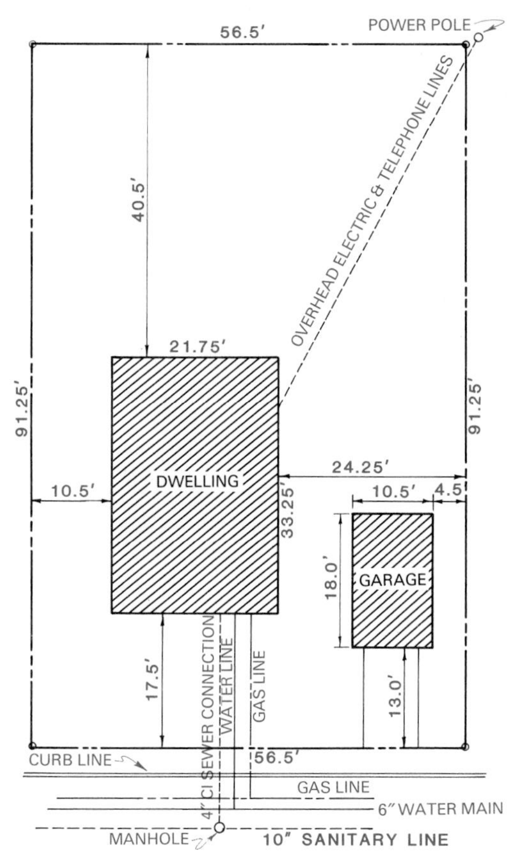

Fig. 43-4A Utility line shown on plot plan.

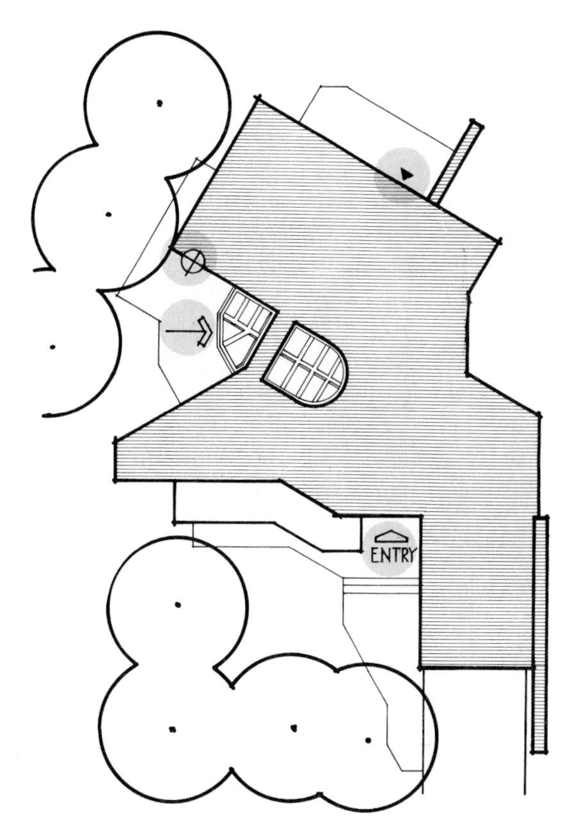

Fig. 43-5 Entrance symbols.

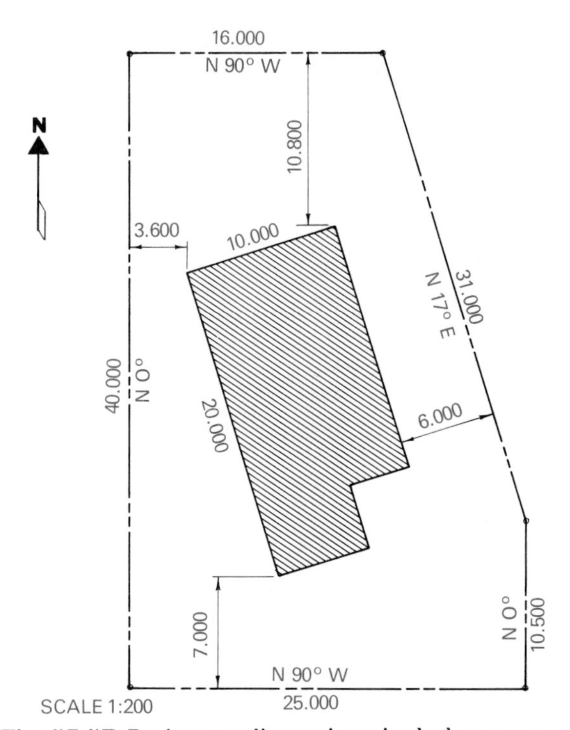

Fig. 43-4B Perimeter dimensions include compass direction.

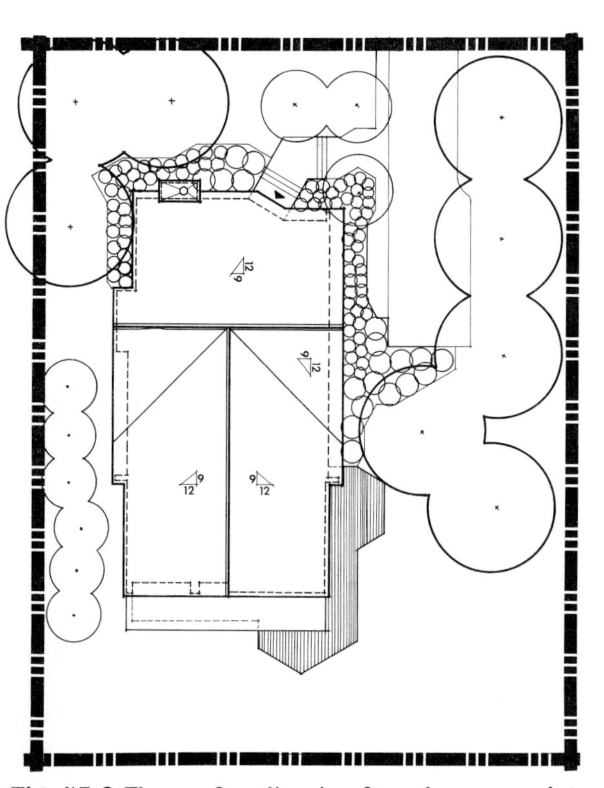
Fig. 43-6 The roof outline is often shown on plot plans.

are also possible in developing almost any detail of a plot plan.

In Fig 43-7 notice how open space is used as an important design feature and how site vegetation was preserved. Also note how the geometric forms of structures are related and acute angles are avoided in design.

In site designing, the sequences outlined in Unit 2 should be followed to ensure the proper orientation and utilization of the natural features of the site. In drawing plot plans, these procedures should be reviewed and followed to ensure that the final design still works as conceived. Figure 43-8 shows the steps in drawing plot plans which include this review.

SETBACK REQUIREMENTS

Property lines show the legal limits of a lot on all sides. However, most building codes require that buildings be located (set back) specified minimum distances from property lines. Figures 43-9A, B, and C show typical building-line setback requirements for different shapes and types of lots. Setback requirements for different-size dwellings vary among building codes. However, Fig. 43-10 shows typical code requirements. The designer must be aware of these restrictions before beginning to develop a site design.

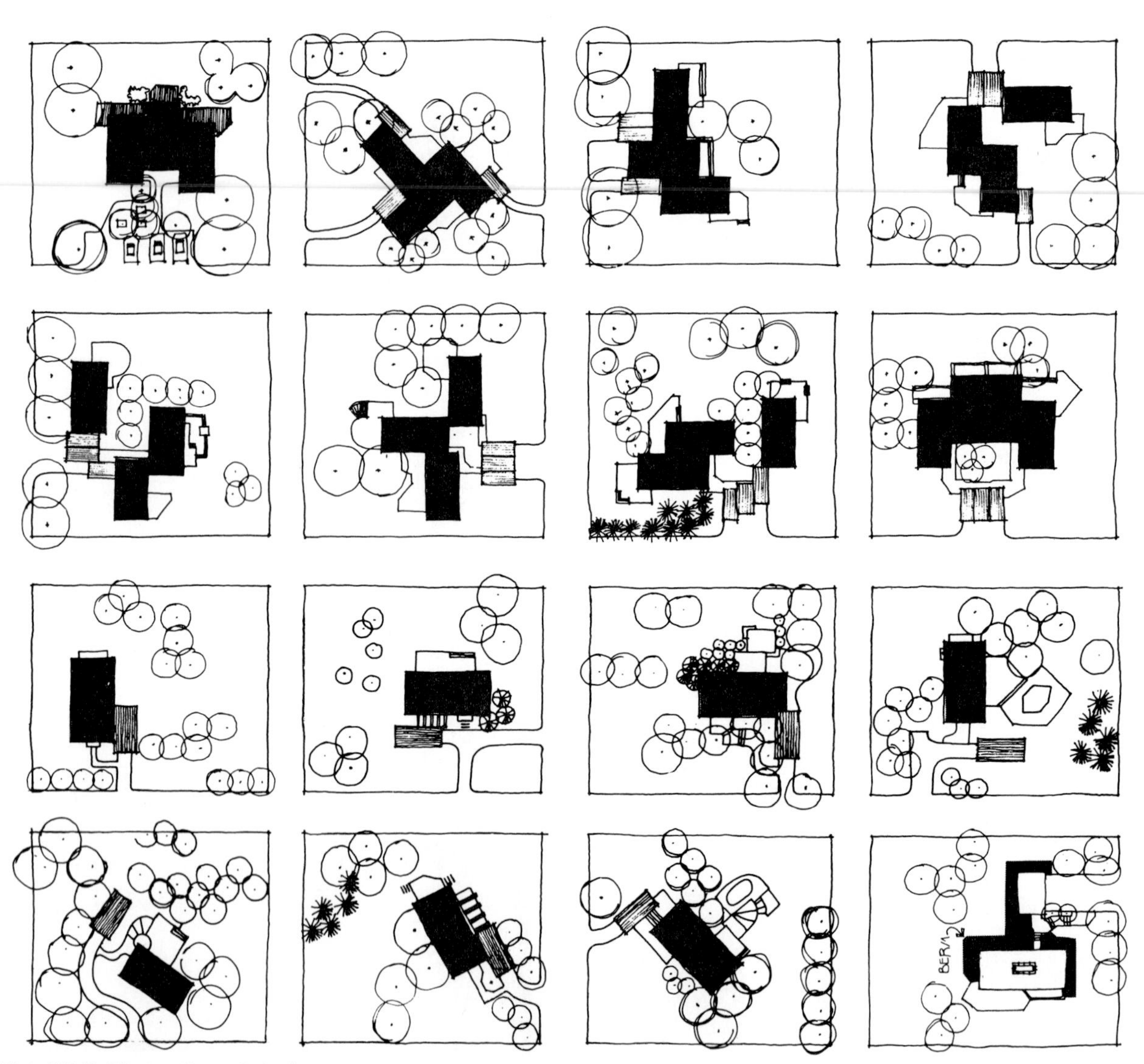

Fig. 43-7 Alternative plot plans.

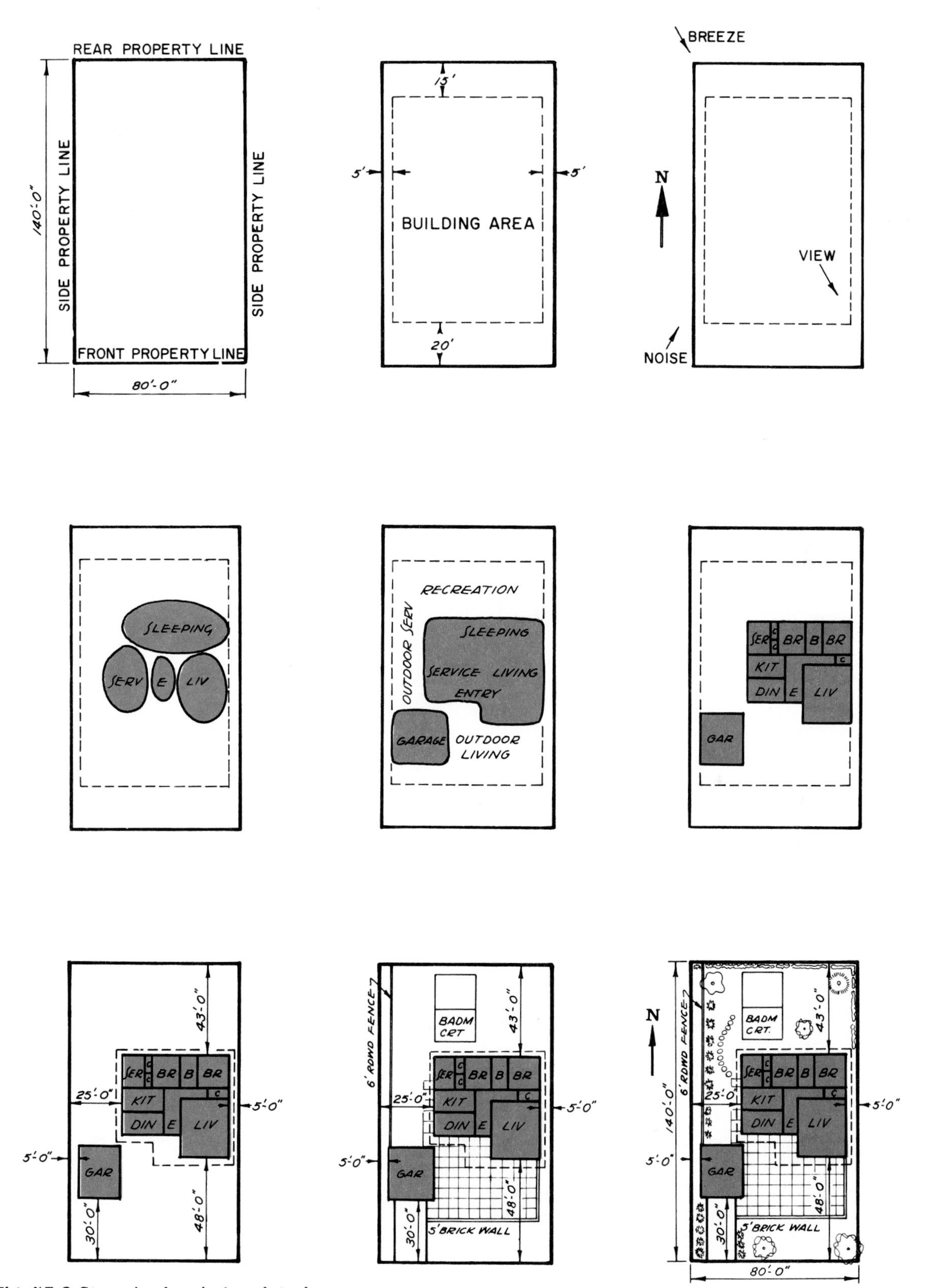

Fig. 43-8 Steps in drawing a plot plan.

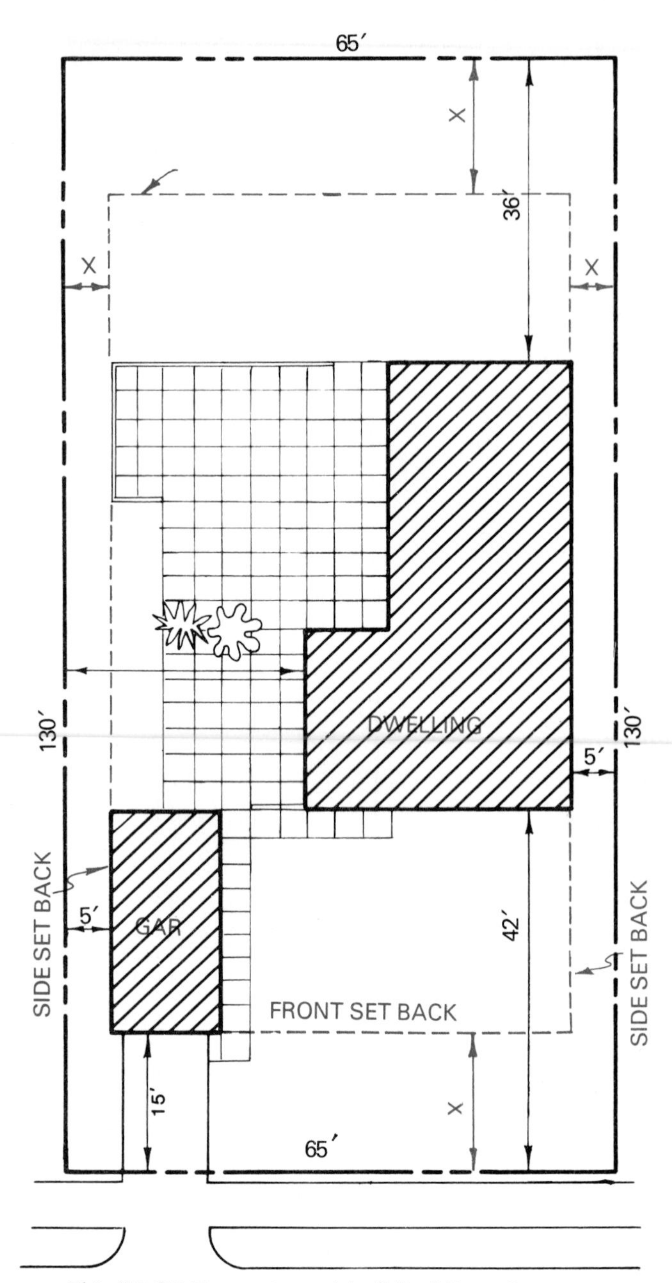

Fig. 43-9A Property and building lines for a rectangular, interior lot.

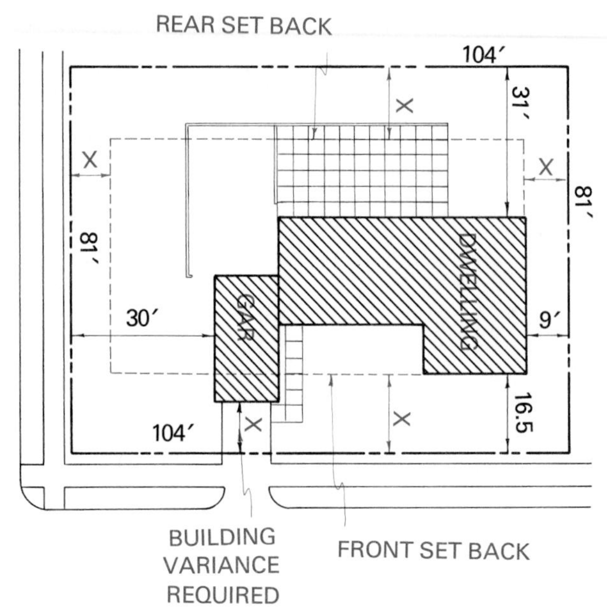

Fig. 43-9B Property and building lines for a corner lot.

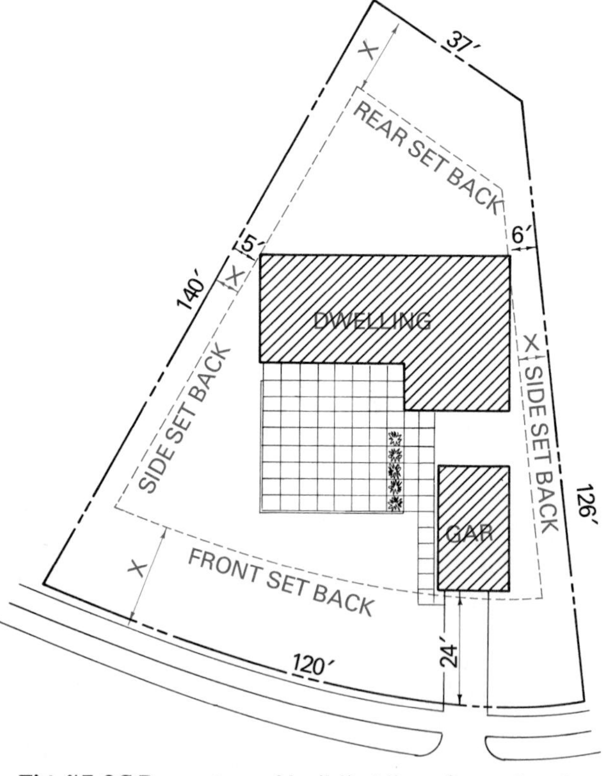

Fig. 43-9C Property and building lines for an irregular lot.

SITE DETAILS

To ensure that the construction of site features meets the needs of the designer, details are usually prepared. These details should be indexed to the plot survey and/or landscape plan.

General site description	Zone	Building site: Minimum area (sq ft)	Building site: Minimum width (ft)	Building site: Minimum depth (ft)	Building site: Maximum coverage (percent)	Required yards: Front (ft)	Required yards: Rear (ft)	Required yards: Side: Interior	Required yards: Side: Corner	Lot area per family unit	Building height
Very large home site	I-6	20,000	110	130	30	60	20	10 (Total 30)	15 (Total 30)	20,000	
Large home site	I-5	15,000	100	100	30	20	20	10 (Total 25)	15 (Total 25)	15,000	35 Ft
Average home site	I-4	10,000	80	100	35	20	20	10	12	10,000	
Small home site	I-3	7,000	65	100	35	20	20	5	12	7,000	
Site for duplex units	I-2	7,000	65	100	40	20	20	5	10	3,500	

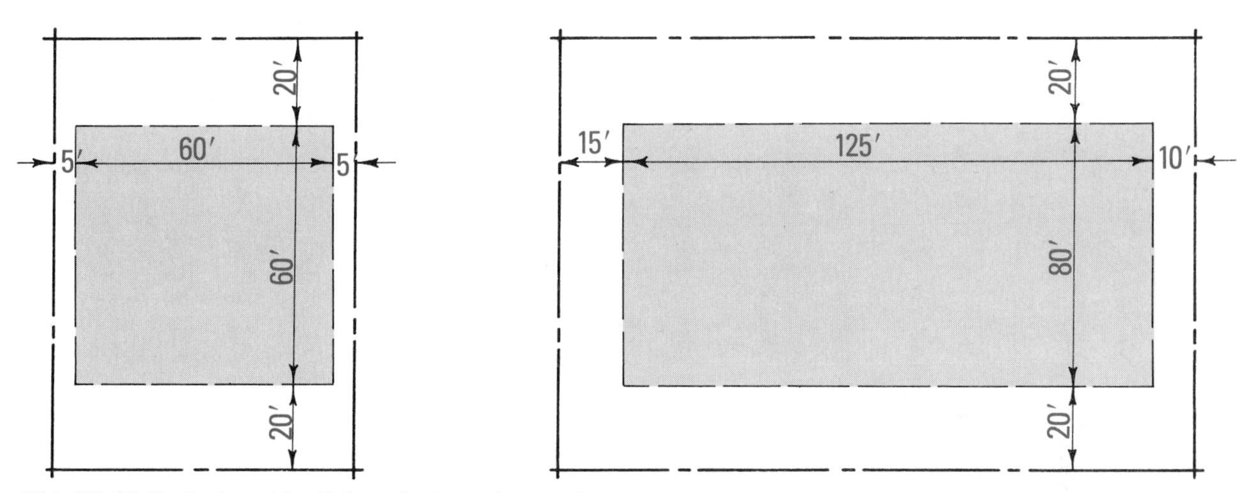

Fig. 43-10 Typical residential zoning requirements.

Exercises

1. **Draw a plot plan of your own home.**
2. **From a survey plan you have developed, complete a plot plan showing the position of the residence you are designing.**
3. **Place the outline of a residence on the plot plan shown in Fig. 43-11. Include a two-car garage, swimming pool, and tennis court on this plan. Remember to take full advantage of existing environmental factors. Use a scale of 1″ = 100′.**
4. **Sketch one of the properties shown in Fig. 4-6A, B, C, or D. Prepare a template of the house to the same scale and place it on the property in the most desirable location.**
5. **Sketch the lot shown in Fig. 43-13. Add a pool, a cabana, patios, walks, and a driveway.**

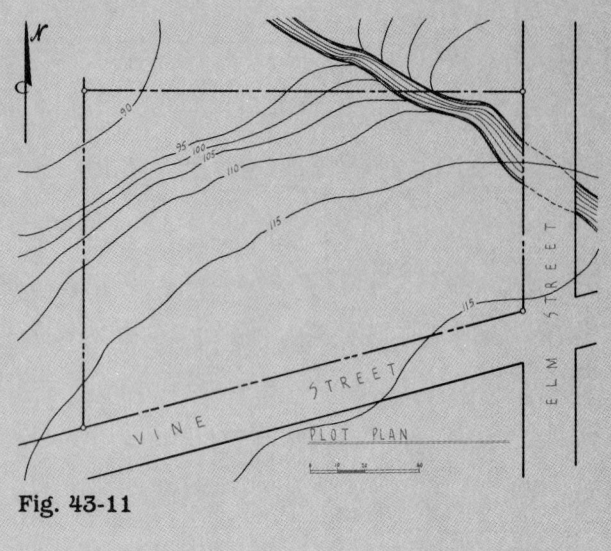

Fig. 43-11

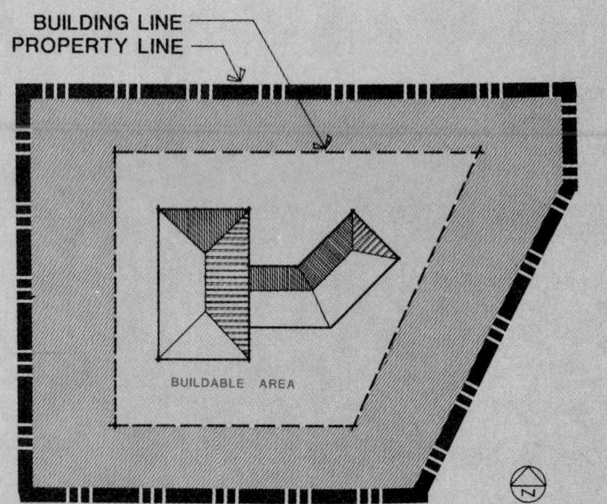

Fig. 43-12

6. **Sketch the lot layout shown in Fig. 43-13. Place templates of the house and garage on this lot in the most desirable position. Also sketch the position of driveways, walks, and other landscape features you would add to this design. Use a scale of 1″ = 40′.**
7. **Define these terms: *plot plan, lot, compass orientation, contour lines, property line, grade elevation.***

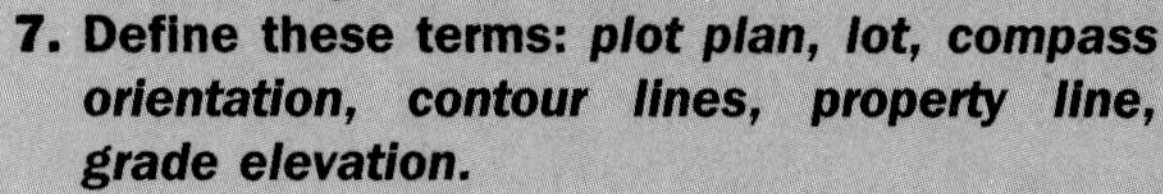

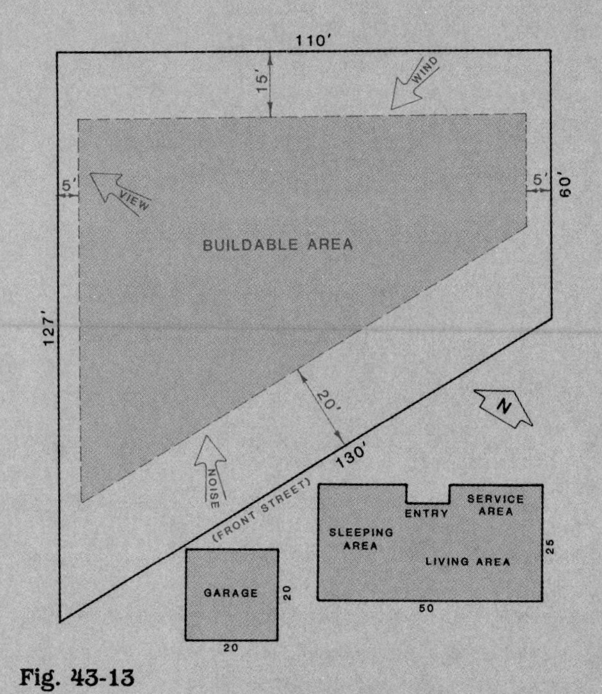

Fig. 43-13

UNIT 44

LANDSCAPE PLANS

The primary function of the *landscape plan* is to show the types and location of vegetation for the lot. It may also show the contour of the land and the position of buildings. Such features are often necessary to make the placement of the vegetation meaningful. Symbols are used on landscape plans to show the position of trees, shrubbery, flowers, vegetable gardens, hedgerows, and lawns. Figure 44-1 shows some common symbols used on a landscape plan. A landscape architect or gardening contractor designs and prepares the landscape plan in cooperation with the designing architect. The landscape architect specifies the type and location of all trees, shrubs, flowers, hedges, and ground cover. He or she often proposes changes in the existing contour of the land to enhance the site appearance and function.

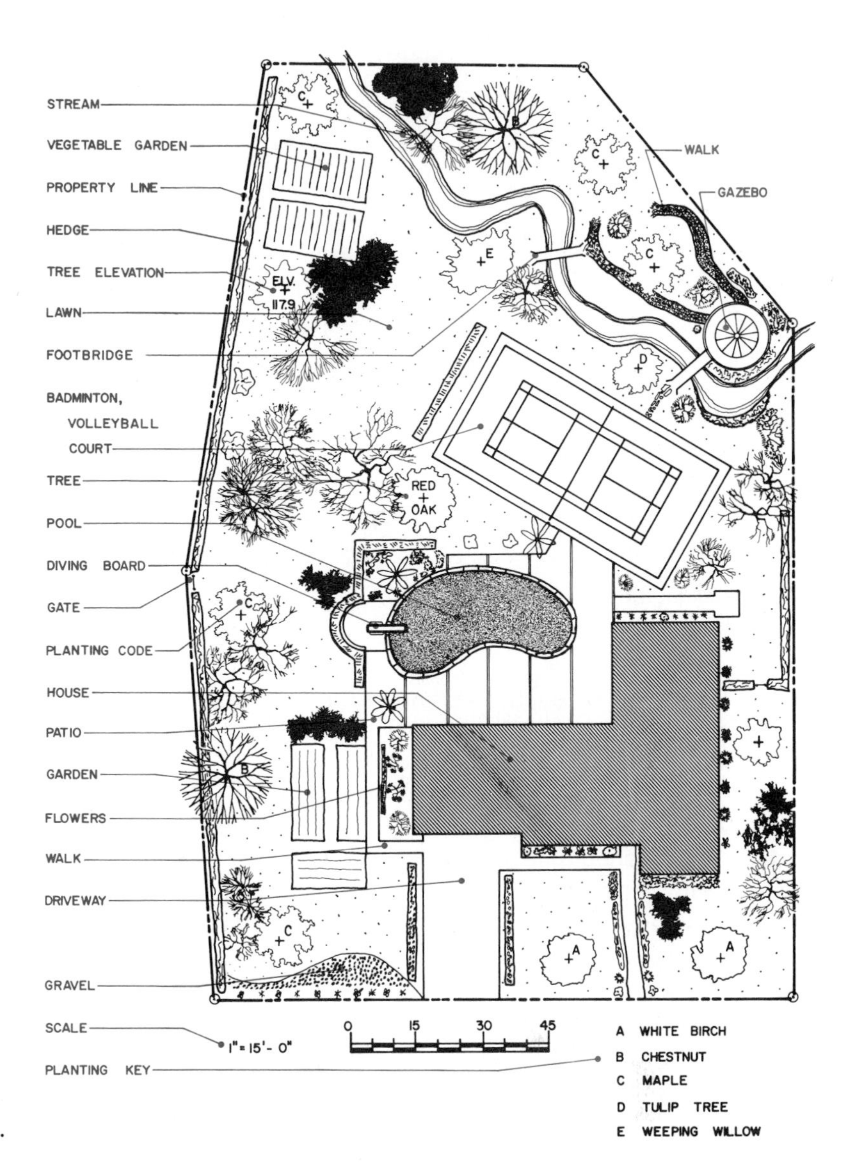

Fig. 44-1 Landscape-plan symbols.

GUIDES FOR PREPARING LANDSCAPE PLANS

The following guides in preparing landscape plans are illustrated by the numbered arrows shown in Fig. 44-2A:

1. The elevation of all major trees is noted to show the datum level.

2. Vegetable gardens are shown by outlining the planting furrows.

3. Orchards are shown by outlining each tree in the pattern.

4. The property line is shown to define the limits of the lot.

5. Trees are located to provide shade and windbreaks and to balance the decor of the site.

6. Shrubbery is used to provide privacy, define boundaries, outline walks, conceal foundation walls, and balance irregular contours.

7. The outlines of courts are shown.

8. Flower gardens are shown by the outline of their shapes.

9. Lawns are shown by small, sparsely placed dots.

10. The outlines of all walks and planned paths are shown.

11. Conventional map symbols are used for small bridges.

12. The outline and surface covering of all patios and terraces are indicated.

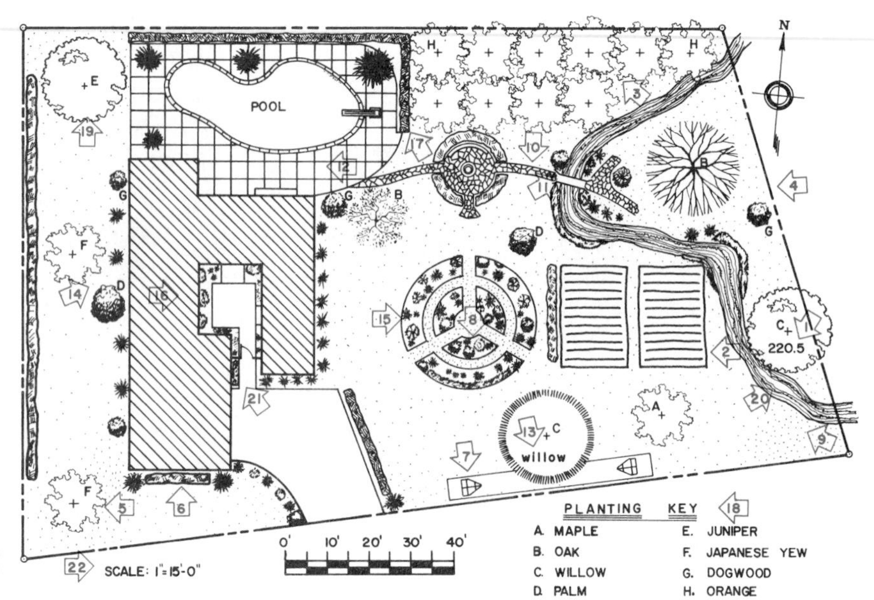

Fig. 44-2A Guide for preparing landscape plans.

13. The name of each tree and shrub is labeled by the symbol or indexed to a planting schedule.

14. All landscaping should enhance the function and appearance of the site.

15. Flowers should be located to provide maximum beauty and ease of maintenance.

16. Buildings are outlined, crosshatched, or shaded. In some cases, the outline of the floor plan is shown in abbreviated form. This helps to show the relationship of the outside to the inside living areas.

17. Hedges are used as screening devices to provide privacy, to divide areas, to control traffic, or to serve as windbreaks.

18. Each tree or shrub is indexed to a planting schedule, if there are too many to be labeled on the drawing, as suggested in 13.

19. A tree is shown by drawing an outline of the area covered by its branches. This symbol varies from a perfect circle to irregular lines representing the appearance of branches (Fig. 44-2B). A plus sign (+) indicates the location of the trunk.

20. Water is indicated by irregular parallel lines.

21. Shrubbery in front of the house should be low in order not to interfere with traffic or with window location.

22. An engineer's scale is used to prepare landscape plans. This is the measure surveyors use. The engineer's scale divides the foot into decimal parts.

Fig. 44-2B Methods of drawing trees on landscape plans.

23. Contour lines may be shown on landscape plans.

24. Separate planting plan schedules may be prepared for complex designs.

25. Planting details may be needed to ensure planting quality.

PHASING

The complete landscaping of a lot may be prolonged through several years. This procedure is sometimes followed because of a lack or time to accomplish all the planting necessary, or for financial reasons. Figure 44-3 shows a landscape plan divided into three different phases for completion. When a landscape plan is *phased,* the total plan is drawn, and then different shades or colors are used to designate the items that will be planted in the first year, in the second year, and in the third year. A plan can be phased over many years or several months, depending on the schedule for completion of the landscaping.

PRESENTATION PLANS

Many landscape plans, such as the one shown in Fig. 44-4, are strictly interpretive and contain few or no dimensions. The plan shown in Fig.

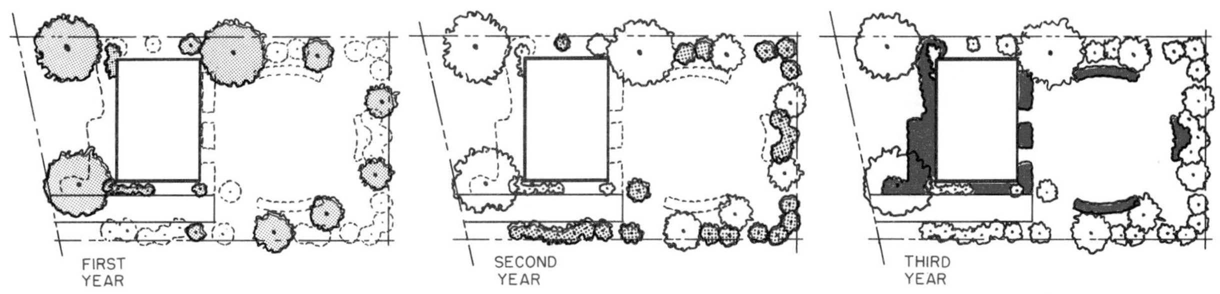

Fig. 44-3 A phased landscape plan.

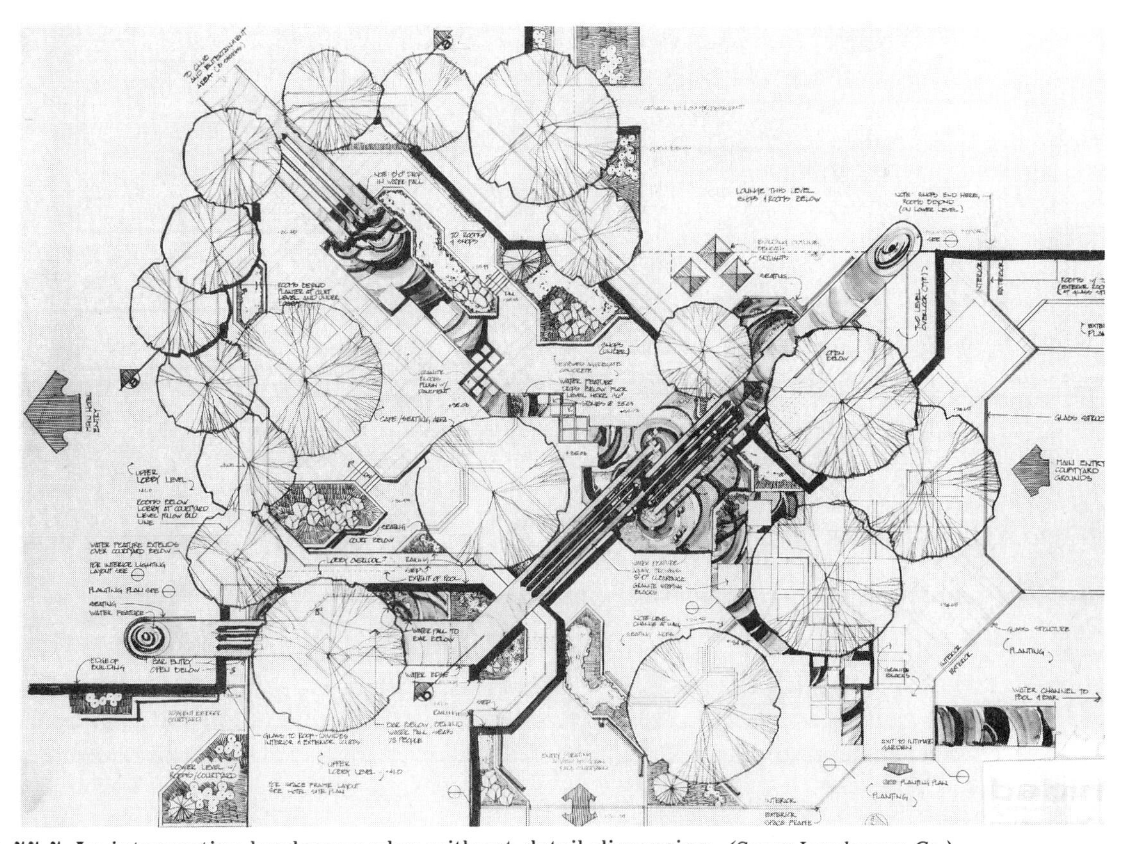

Fig. 44-4 An interpretive landscape plan without detail dimension. *(Secor Landscape Co.)*

44-5A also contains no dimensions. The perspective position drawn in Fig. 44-5B is shown by the arrow in Fig. 44-5A.

COMBINATION PLANS

Often, the lot or estate is too large to be shown accurately on a standard landscape plan. A scale such as 1″ = 20′ may not show the entire estate on the drawing; or a scale must be used that is so small that the features cannot be readily identified, labeled, and dimensioned. One solution to such a problem is to prepare a total plan of the large estate to a large scale such as 1″ = 200′. This is indexed to a drawing of the immediate area around the building with a scale such as 1″ = 20′. Figure 44-6 shows the total estate plan as an insert in the drawing showing the lower right-hand corner of the estate developed in more detail. It is sometimes desirable or necessary to combine all the features of the survey, plot plan, and landscape plan in one plan. In such a combination, all the symbol dimensions are incorporated in one location plan, as shown in Fig. 44-6. This includes contour lines and the exact position of vegetation and buildings. Figure 44-7 shows a composite site plan.

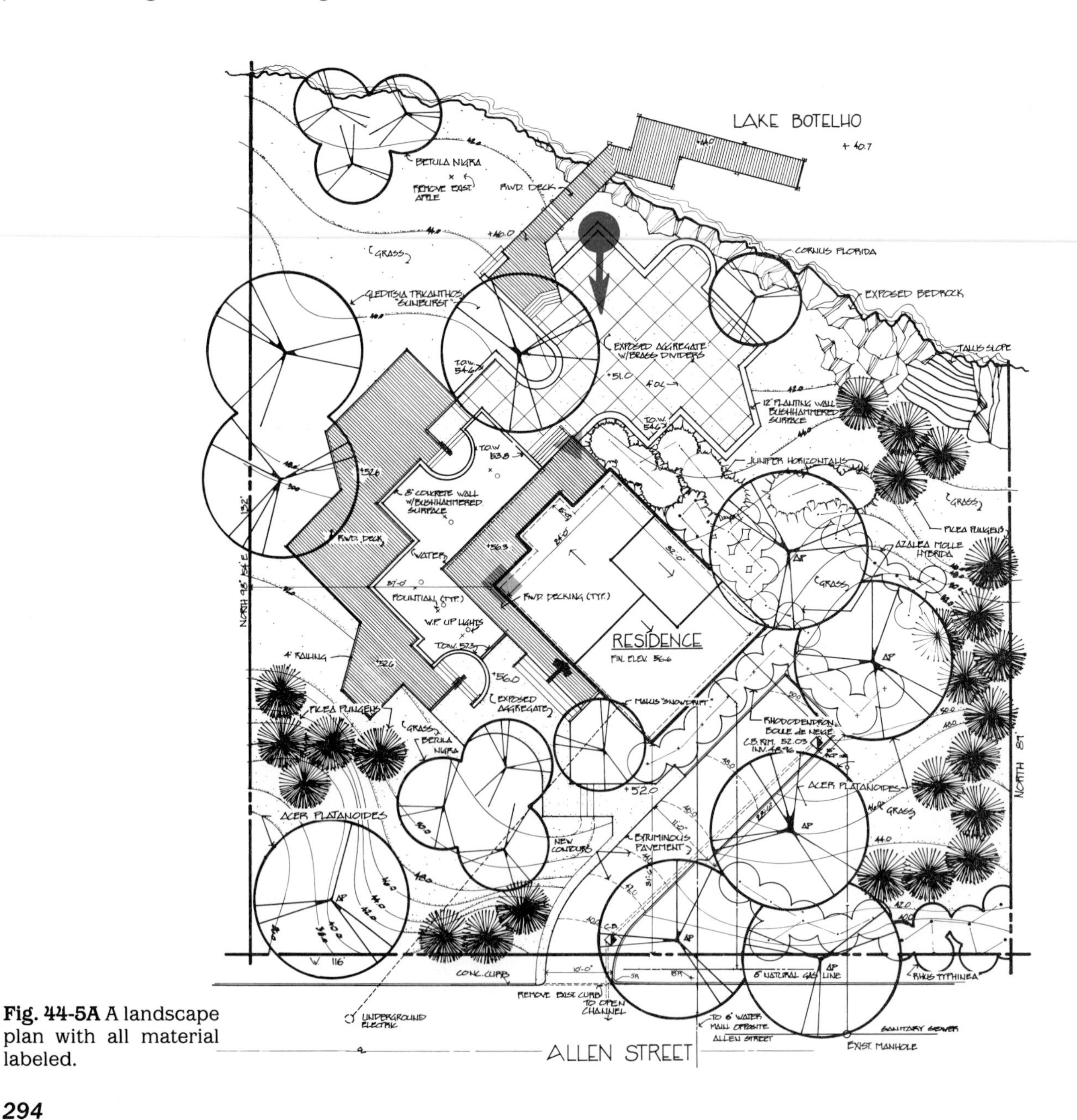

Fig. 44-5A A landscape plan with all material labeled.

Fig. 44-5B A perspective drawing of the area shown in Fig. 44-5A.

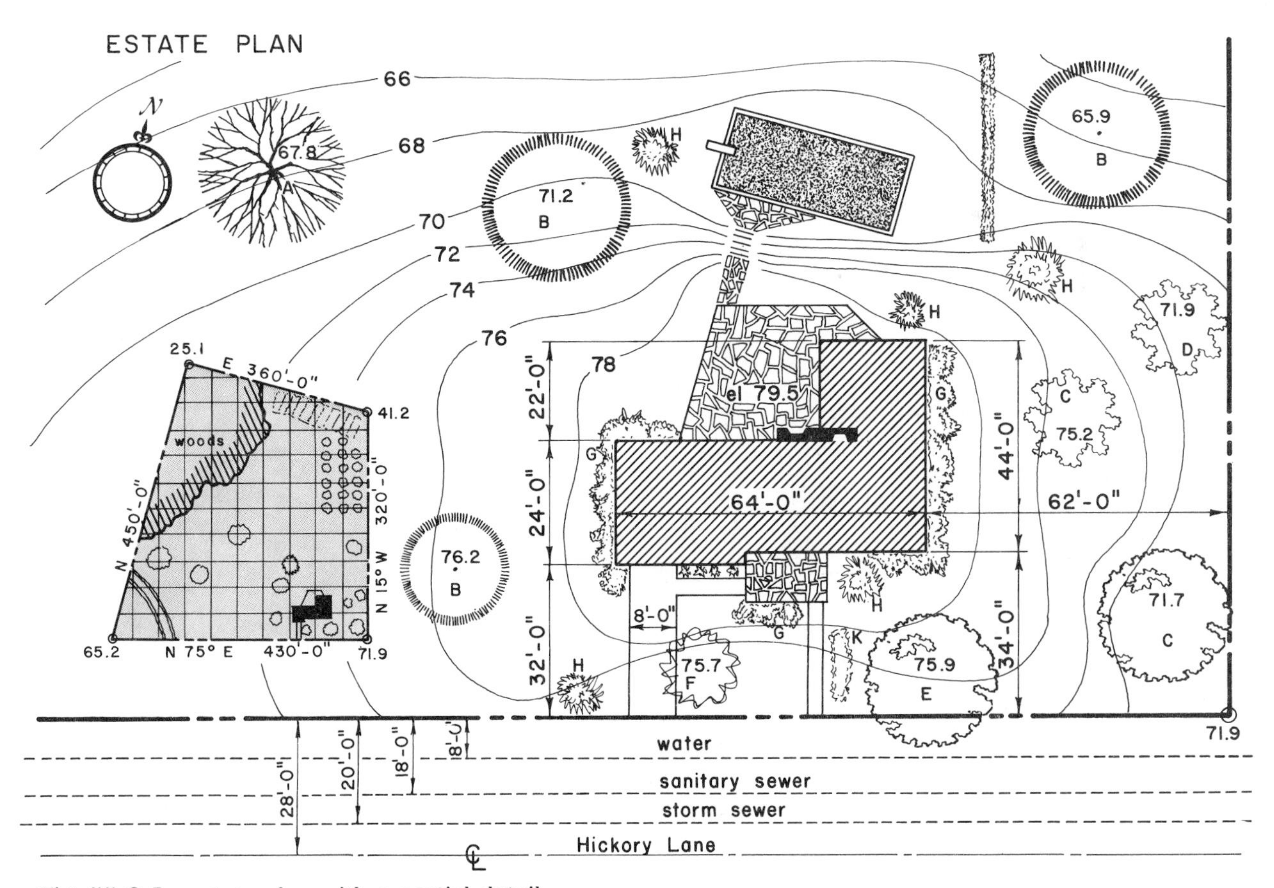

Fig. 44-6 An estate plan with a partial detail.

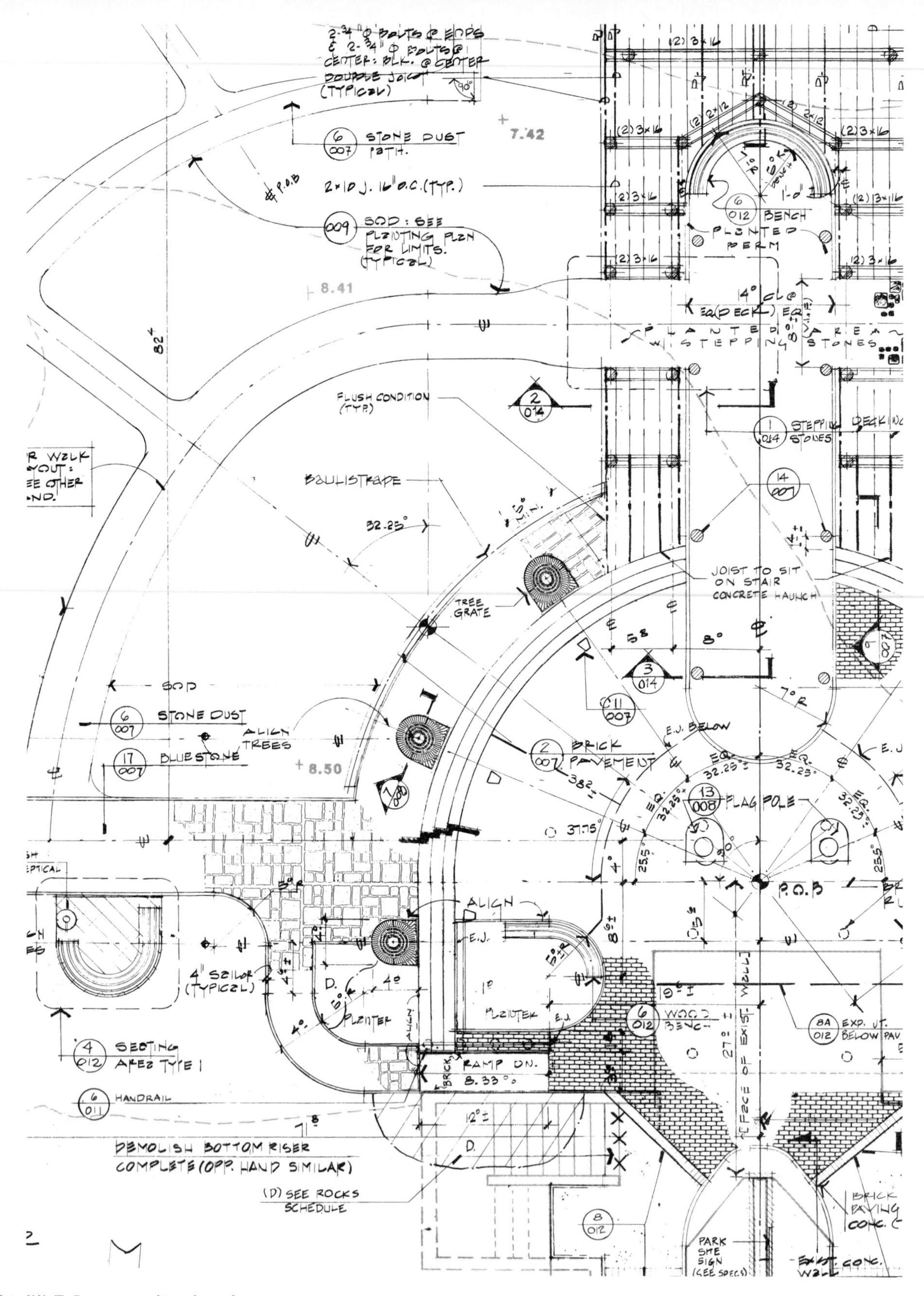

Fig. 44-7 A composite site plan.

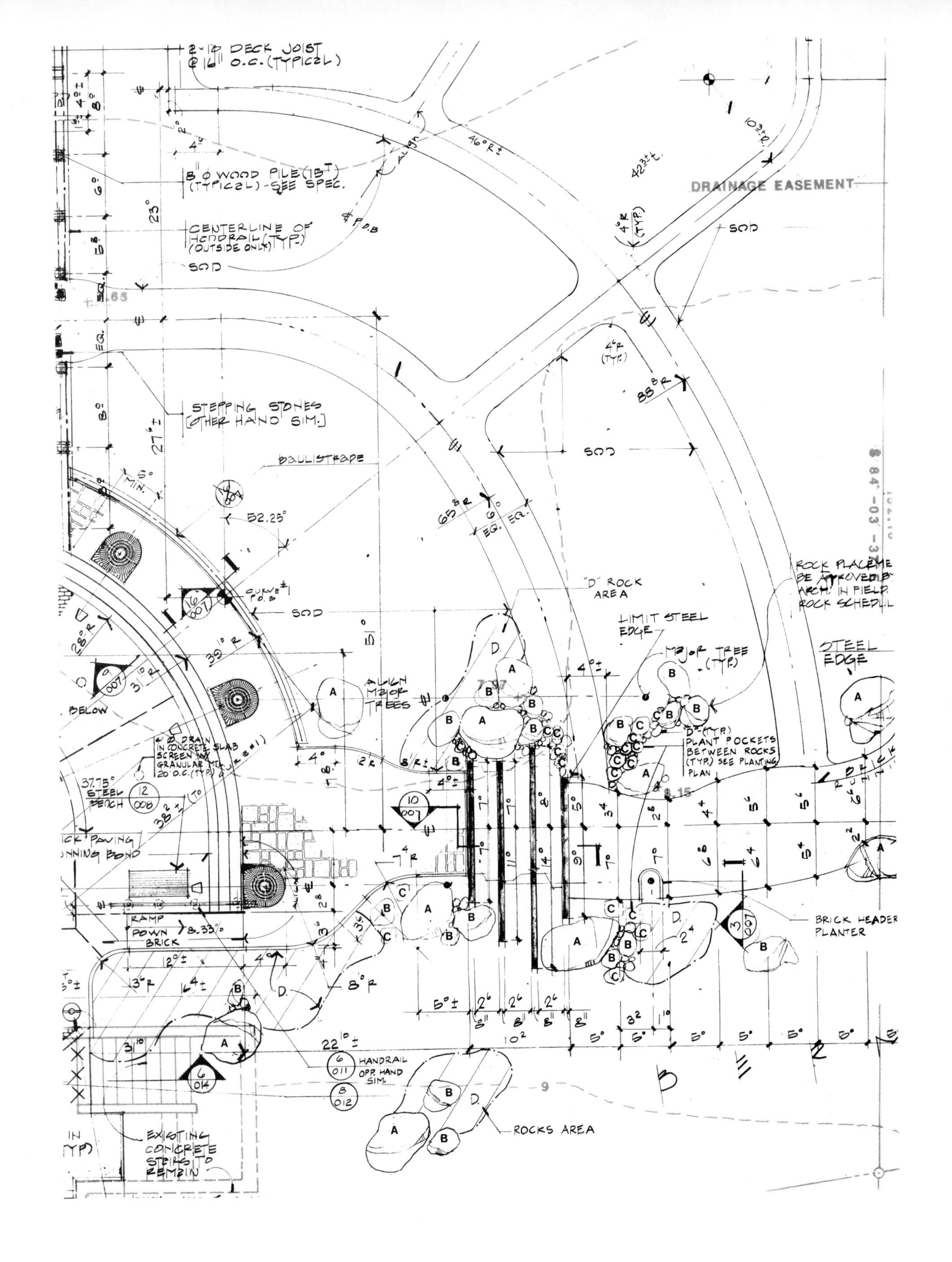
2-10 DECK JOIST @ 16" O.C. (TYPICAL)
8" ø WOOD PILE (18T) (TYPICAL) - SEE SPEC.
CENTERLINE OF HANDRAIL (TYP.) (OUTSIDE ONLY)
SOD
DRAINAGE EASEMENT
STEPPING STONES [OTHER HAND SIM.]
"D" ROCK AREA
LIMIT STEEL EDGE
MAJOR TREE (TYP.)
STEEL EDGE
ALIGN MAJOR TREES
D-(TYP.) PLANT POCKETS BETWEEN ROCKS (TYP.) SEE PLANTING PLAN
BRICK HEADER PLANTER
RAMP DOWN 8.33% BRICK
STEEL BENCH
HANDRAIL OPP. HAND SIM.
ROCKS AREA
EXISTING CONCRETE STAIRS TO REMAIN

1. Add landscape features to your own plot plan.
2. Sketch the property in Fig. 43-13 and add the landscape symbols.
3. Place the house shown in Fig. 44-8 on a 100′ × 200′ lot. Prepare a landscape drawing of the lot according to your own taste.
4. Add landscape symbols as indicated by the labels in Fig. 44-9.
5. Define these terms: *landscape plan, landscape symbol, landscape architect, gardening contractor, datum level, orchards, tree-location symbol, map symbols.*

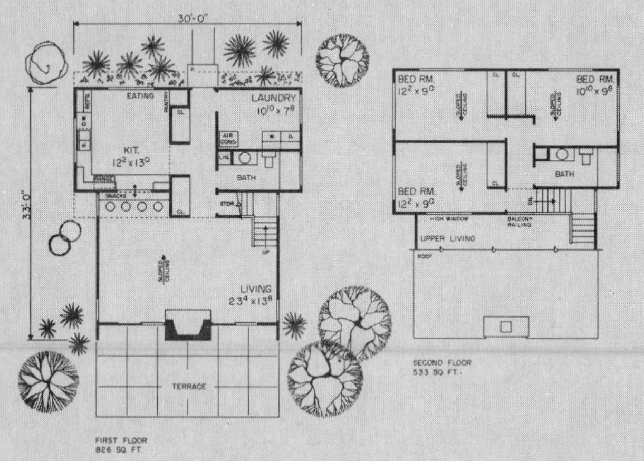

Fig. 44-8 Place this house on a 100′ × 200′ lot. *(Home Planners, Inc.)*

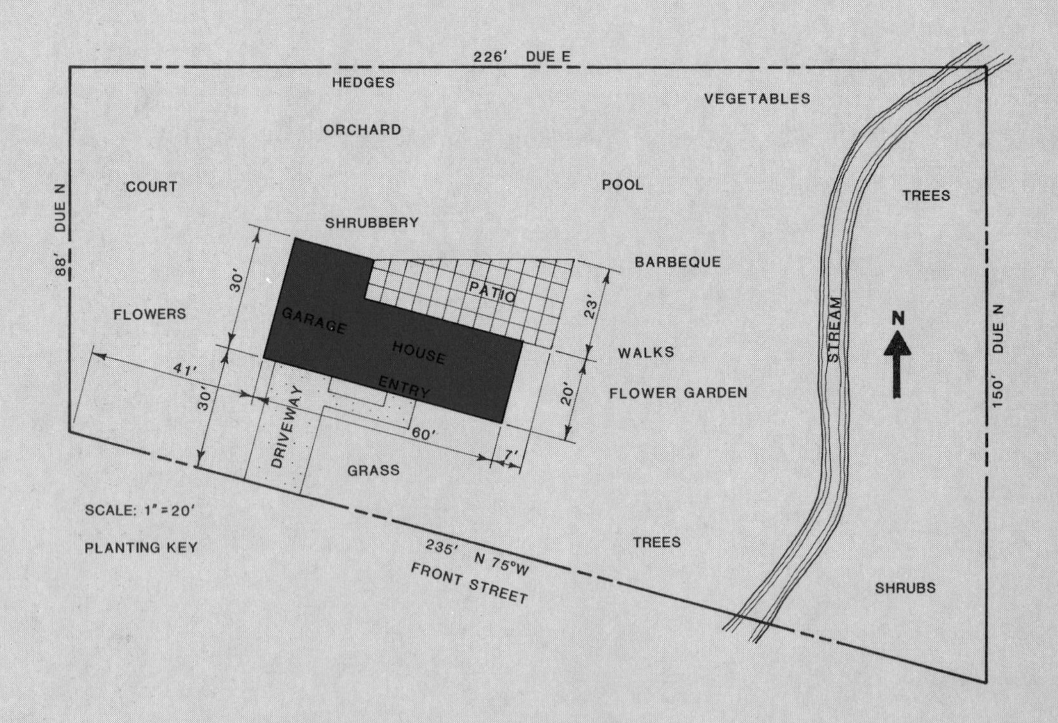

Fig. 44-9

UNIT 45

SURVEY PLANS

A *survey* is a drawing showing the exact size, shape, and levels of a lot. When prepared by a licensed surveyor, the survey can be used as a legal document. It is filed with the deed to the property. The lot survey includes the length of each boundary, tree locations, corner elevations, contour of the land, and position of streams, rivers, roads or streets, and utility lines. It also lists the name of the owner of the lot and the owner or title of adjacent lots.

A survey drawing must be accurate and must communicate a complete description of the features of the lot. Symbols are used extensively to describe the features of the terrain. Figure 45-1 shows the survey symbols most frequently used. Some symbols depict the appearance of a feature. Most survey symbols are *schematic* representations of some feature. Figure 45-2 shows topographic symbols used on survey, plat, and geographical survey plans.

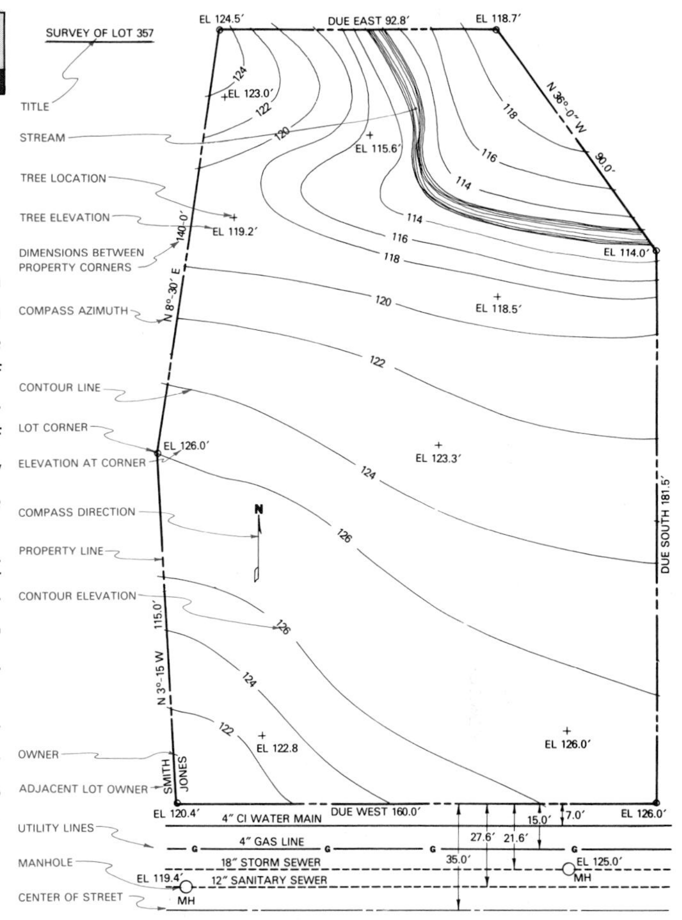

Fig. 45-1 Survey-plan symbols.

GUIDES FOR DRAWING SURVEYS

The numbered arrows in Fig. 45-3 correspond to the following guides for preparing survey drawings:

1. Record the elevation above the datum of the lot at each corner.

2. Represent the size and location of streams and rivers by wavy lines (blue lines on geographical surveys).

3. Use a cross to show the position of existing trees. The elevation at the base of the trunk is shown.

4. Indicate the compass direction of each property line by degrees, minutes, and seconds. Figure 45-4 shows the relationship of each property-line compass direction to the compass orientation of the lot.

5. Use a north arrow to show compass direction.

6. Break contour lines to insert the height of contour above the datum.

7. Show lot corners by small circles.

8. Draw the property line symbol by using a heavy line with two dashes repeated throughout.

9. Show elevations above the datum or sea level by contour lines (brown lines on geographical survey maps—see Fig. 45-18).

10. Show any proposed change in grade line by contour lines. Figure 45-5 shows how a lot is planned for recontour and how the recontoured lines are established to show new heights above the datum. Dotted contour lines show original grade. Solid contour lines show the new proposed grading levels.

11. Show lot dimensions directly on the property line. The dimension on each line indicates the distance between corners.

12. Give the names of owners of adjacent lots outside the property line. The name of the owner of the property is shown inside the property line.

13. Dimension the distance from the property line to all utility lines.

NAME	ABBREV	SYMBOL
TREES	TR	
GROUND COVER	GRD CV	
BUSHES SHRUBS	BSH SH	
OPEN WOODLAND	OP WDL	
ORCHARD	OR	
MARSH	MRS	
SUBMERGED MARSH	SUB MRS	
DENSE FOREST	DN FR	
SPACED TREES	SP TR	
TALL GRASS	TL GRS	
LARGE STONES	LRG ST	
SAND	SND	
DRY CRACKED CLAY	DRY CRK CLY	

NAME	ABBREV	SYMBOL
GRAVEL	GRV	
CULTIVATED AREA	CULT	
WATER	WT	
WELL	W	
NORTH-MERIDIAN ARROWS	N MER ARR	N NORTH
PROPERTY LINE	PR LN	
SURVEYED CONTOUR LINE	SURV CON LN	75
ESTIMATED CONTOUR	EST CON	
FENCE	FN	x—x—x—x
RAILROAD TRACKS	RR TRK	
PAVED ROAD	PV RD	
UNPAVED ROAD	UNPV RD	
POWER LINE	POW LN	

NAME	ABBREV	SYMBOL
POWER TRANSMISSION LINE	PW TR LN	
GENERAL LINE LABEL TYPE	GN LN	oil line
WELL LABEL TYPE	WL	oil
TANK LABEL TYPE	TK	water
MINING AREA	MIN AR	
SHAFT	SHF	
TUNNEL ENTRANCE	TUN ENT	
BOUNDRY, STATE	BND ST	
BOUNDRY, COUNTY	BND CNTY	
BOUNDRY, TOWN	BND TWN	
BOUNDRY, CITY INCORPORATED	BND CTY	
BOUNDRY, NATIONAL OR STATE RESERVATION	BND NAT OR ST RES	
BOUNDRY, SMALL AREAS: PARKS, AIRPORTS, ETC.	BND	

NAME	ABBREV	SYMBOL
BOUNDRY, LAND GRANT	BND LD GR	
BOUNDRY, U.S. LAND SURVEY TOWNSHIP	BND US LD SUR TWN	
BOUNDRY, TOWNSHIP APPROXIMATED	BND TWN	
BOUNDRY, SECTION LINE U.S. LAND SURVEY	BND SEC LN US LD SUR	
BOUNDRY, SECTION LINE APPROXIMATED	BND SEC LN	
BOUNDRY, TOWNSHIP NOT U.S. LAND SURVEY	BND TWN	
INDICATION CORNER SECTION	COR SEC	
BOUNDRY MONUMENT	BND MON	
U.S. MINERAL OR LOCATION MONUMENT	U.S. MIN MON	
DEPRESSION CONTOURS	DEP CONT	
FILL	FL	
CUT	CT	
LEVEE	LEV	

NAME	ABBREV	SYMBOL
WATER LINE	WT LN	
GAS LINE	G LN	
SANITARY SEWER	SAN SW	
SEWER TILE	SW TL	
SEPTIC-FIELD LEACH LINE	SP FLD LCH LN	
PROPERTY CORNER WITH ELEVATION	PROP CR EL	EL 70.5
SPOT ELEVATION	SP EL	+ 78.8
WATER ELEVATION	WT EL	80
BENCH MARKS WITH ELEVATIONS	BM/EL	BM X 84.2 BM △ 84.2
HARD-SURFACE HEAVY DUTY ROAD — FOUR OR MORE LANES	HRD SUR HY DTY RD	
HARD-SURFACE HEAVY DUTY ROAD — 2 OR 3 LANES	HRD SUR HY DTY RD	
HARD-SURFACE MEDIUM DUTY ROAD — FOUR OR MORE LANES	HRD SUR MED DTY RD	
HARD-SURFACE MEDIUM DUTY ROAD — 2 OR 3 LANES	HRD SUR MED DTY RD	

NAME	ABBREV	SYMBOL
IMPROVED LIGHT DUTY ROAD	IMP LT DTY RD	
TRAIL UNIMPROVED DIRT ROAD	TRL UNIM DRT RD	
ROAD UNDER CONSTRUCTION	RD CONST	
BRIDGE OVER ROAD	BRG OV RD	
RAILROAD TUNNEL	RR TUN	
ROAD OVERPASS	RD OVP	
ROAD UNDERPASS	RD UNP	
SMALL DAM	SM DA	
LARGE DAM WITH LOCK	LRG DM LK	
BUILDINGS	BLDGS	
SCHOOL	SCH	
CHURCH	CH	
CEMETARY	CEM	† CEM

NAME	ABBREV	SYMBOL
LEVEE WITH ROAD	LV RD	
MINE DUMP	MN DP	
RIVER	RV	
STREAM PERENNIAL	ST PRE	
STREAM INTERMITTENT	ST INT	
AQUEDUCT ELEVATED	AQ EL	
AQUEDUCT TUNNEL	AQ TUN	
STREAM DISAPPEARING	ST DIS	
SMALL RAPIDS	SM RP	
SMALL WATER FALL	SM WT FL	
LARGE RAPIDS	LRG RP	
WASH	WSH	
LARGE WATER FALL	LRG WT FL	

NAME	ABBREV	SYMBOL
LAKE, INTERMITTEN	LK INT	
LAKE, DRY	LK DRY	
REEF	RF	
SOUNDING WATER DEPTH CURVE	SND WT DPT CUR	30
EXPOSED WRECK	EX WRK	
SUNKEN WRECK	SUN WRK	
EXPOSED ROCK IN WATER	EX RK WK	
EXPOSED ROCK NAVIGATION DANGER	EX RK NAV DAN	
SPRING	SP	
PILINGS	PLG	
CANAL WITH LOCK	CN LK	
SWAMP	SWP	
SHORELINE	SH LN	

Fig. 45-2 Topographic symbols.

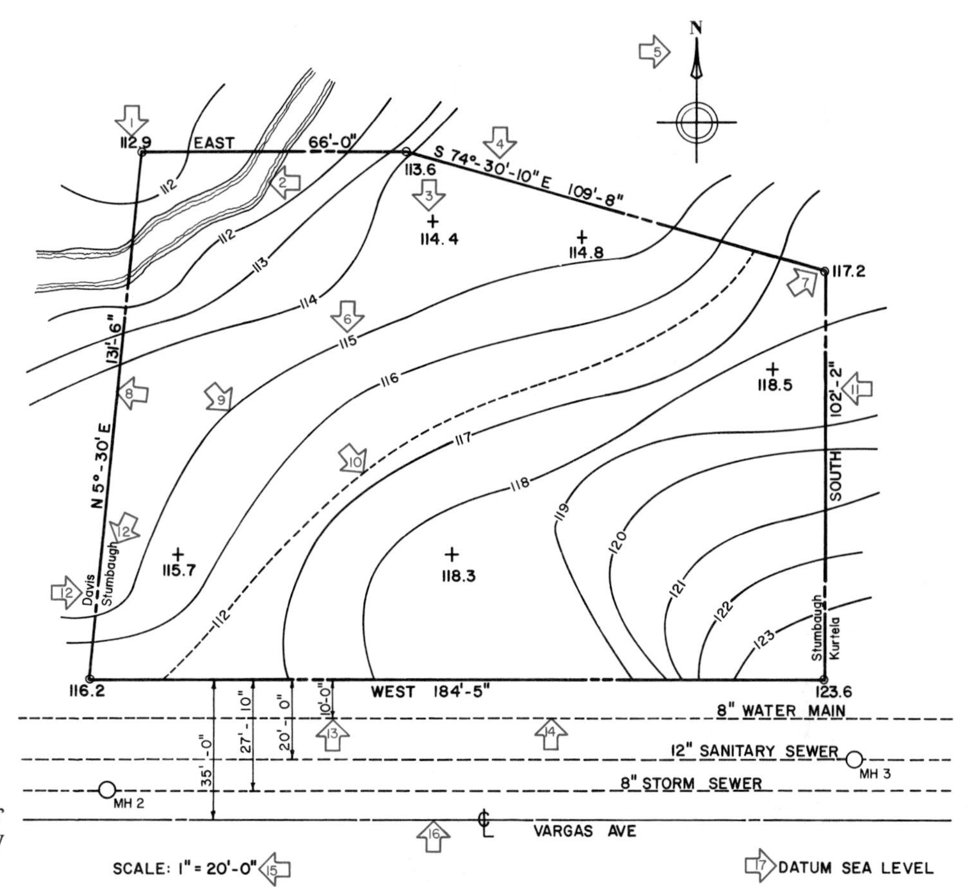

Fig. 45-3 Guides for preparation of survey plans.

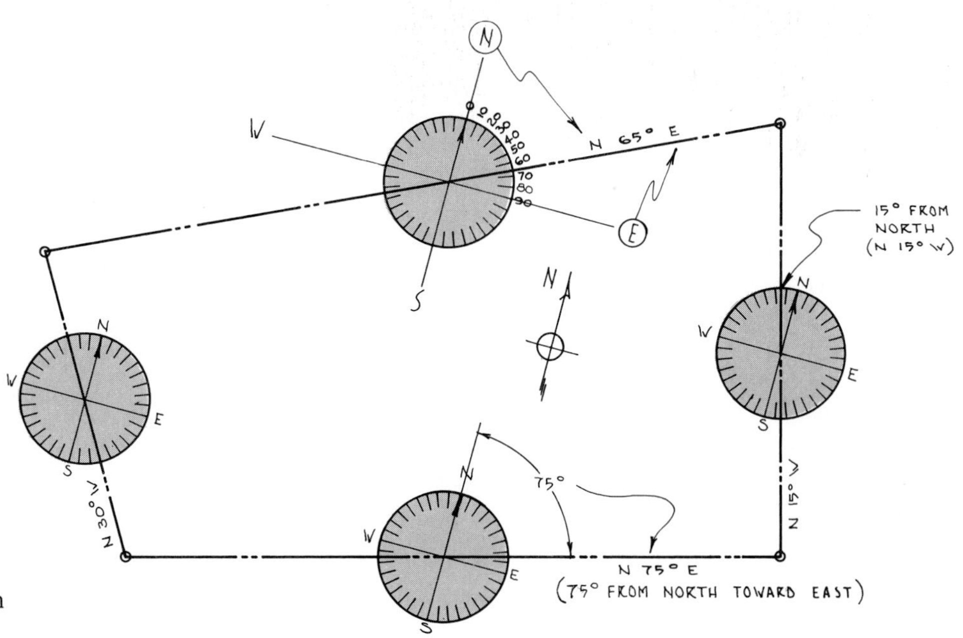

Fig. 45-4 An azimuth projection.

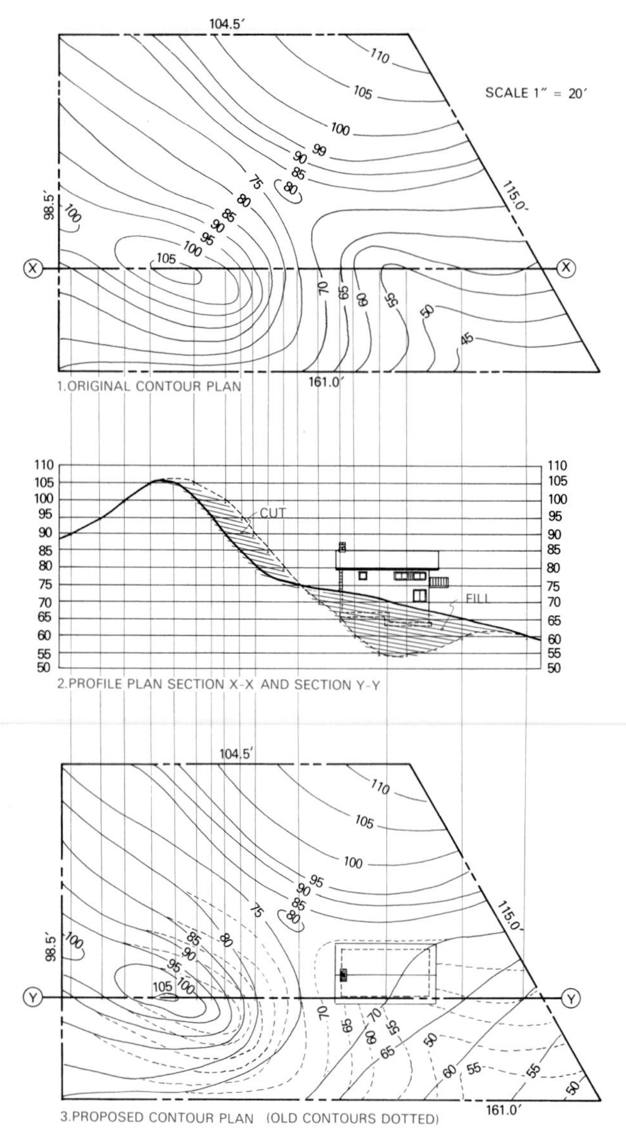

Fig. 45-5 Dotted lines show contour before grading.

14. Show the position of utility lines by dotted lines. Utility lines are labeled according to their function.

15. Draw surveys with an engineer's scale. Common scales for surveys are 1″ = 10′ and 1″ = 20′.

16. Show existing streets and roads either by center lines or by curb or surface outlines.

17. Indicate the datum level used as reference for the survey.

LOT LAYOUT

The size and shape of lots can be determined by several different methods. However, the methods of dimensioning lots are the same.

Lot Dimensions

The exact shape of the lot is shown by the property line. The property line is dimensioned by its length and angle. The angle of each property line from north is known as an *azimuth*. Figure 45-4 shows how the azimuth of each line is determined with a compass or protractor. In Fig. 45-4, on the upper property line, **N** indicates that the bearing of that property line reads from north; 65° means that the property is 65° from north; **E** means that the line is between north and east. Hence, **N** 65° **E** means 65° from north heading east.

The angle of the property line is established by intersecting the property line with the center of the compass when the compass needle is aligned with north. The degree of the angle of this property line is then read on the circumference of the compass, as shown in Fig. 45-6.

Transit Method

Surveyors use a transit to establish the angle (azimuth) of each property line. The *transit* is a telescope that can be set at any desired angle. Once an angle is set, the line may be projected to a rod at any visible distance, as shown in Fig. 45-7. A second line can then be projected by rotating the transit to the desired angle between the property lines. The rotation and projection are shown in position 2, Fig. 45-7. In measuring the length of each property line, surveyors use a steel tape or chain. Figure 45-8 shows the sequence of using the transit and chain to establish the angle and the length of each property line. Step 1 shows the projection of a line 50′ long. Step 2 shows the rotation of an angle 45° from this line. Step 3 shows a line projected 85° from the other end of the first line at a distance of 25′. Step 4 shows the projection of 95° to intersect the line established in step 2. This is called *closing*.

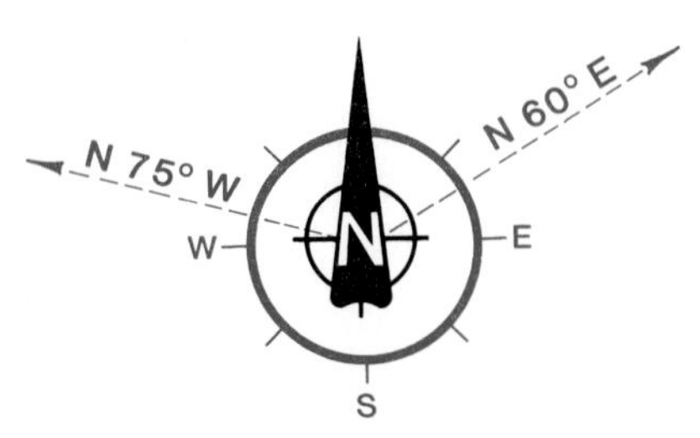

Fig. 45-6 Property-line angles are read on the circumference of the compass.

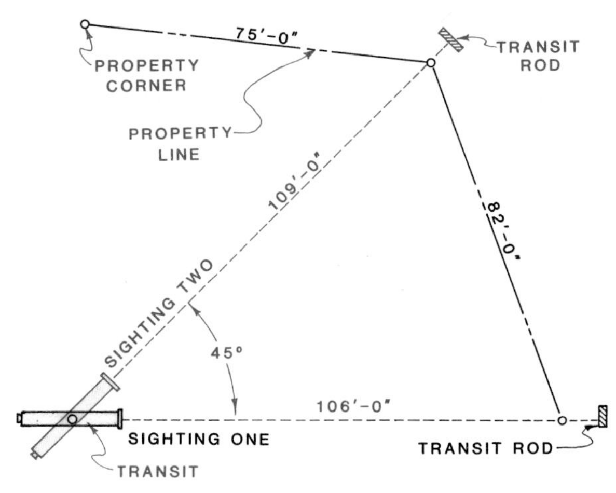

Fig. 45-7 A line can be projected to a rod at any visible distance.

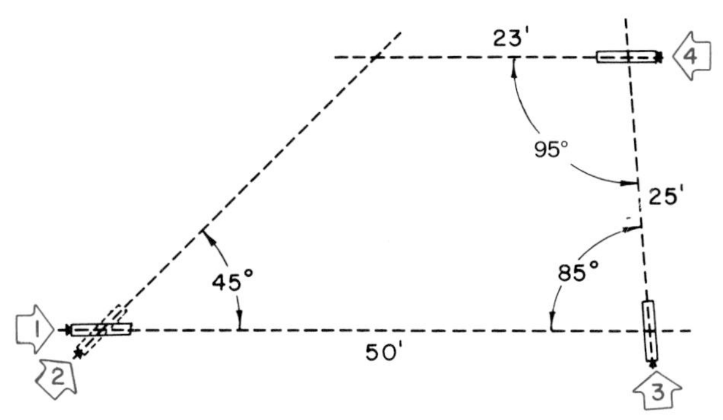

Fig. 45-8 The sequence of establishing the angle and length of each property line.

Heights of various parts of the line are established by sighting through a level from a known to an unknown distance. Figure 45-9A and B shows how a level is used to measure a distance from the known to an unknown distance. In Fig. 45-9A, position A is 0.5′ lower than position B. Figure 45-9B shows how several elevations can be projected from one known point. For example, if position A is a known value above the datum, a level line can be projected to position C. By measuring any distance up on point C, a level line can be established between C and D, and likewise between D and B. To establish levels with the transit, the surveyor sets up the instrument so that all points can be seen through the telescope. The reading from the rod on the cross hairs is recorded. The rod is then moved to the second position to be established. The rod is raised or lowered until the original reading is located. The bottom of the rod is then on the same level with the original point.

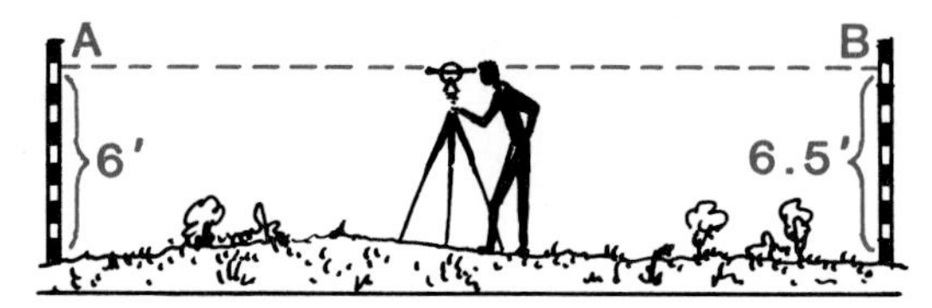

Fig. 45-9A Establishing height with a level.

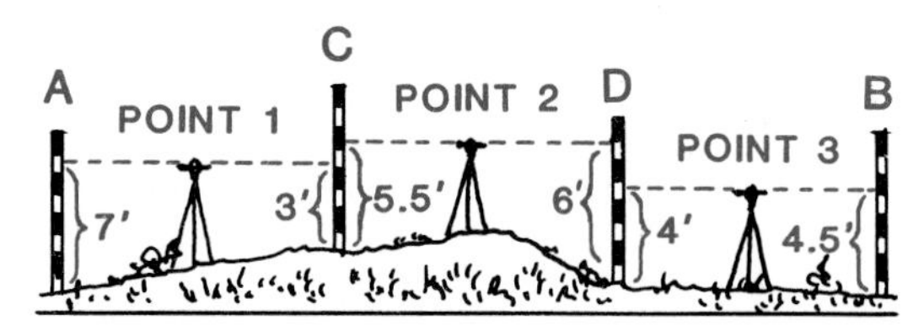

Fig. 45-9B Projecting several elevations from one point.

To find the difference in elevation between two points, such as points A and B in Fig. 45-9a, sight on a rod held over point A. Note the reading where the horizontal cross hairs of the telescope cut the graduation on the rod. Then with the rod held at point B, rotate the telescope in a horizontal plane. Again sight on the rod. Note where the horizontal cross hairs cut the graduation on the rod. The difference between reading A (6′) and reading B (6.5′) will give the difference in elevation between the two points. The ground at point B is 0.5′ lower than at point A. When you cannot find the difference in elevation of two points by keeping the transit in one place, several different points must be used. This is shown in Fig. 45-9B.

PLANE-TABLE METHOD

The plane-table method is an alternative method of plot layout. This method is less accurate than the transit method since it relies on the unaided eye and not on a graduated telescope. Furthermore, this method is generally used to draw a lot that has been established rather than to layout a lot.

The *plane table* is a drawing board mounted on a tripod. The plane table is placed on a starting point, as shown in step 1, Fig. 45-10. The first line is established by sighting from corner A to corner B. The distance from corner A to corner B is then measured and drawn to scale on the plane table. Next, the line AE is drawn by sighting from the starting point A to corner E. In step 2, the plane table is moved to a point over corner B. Line AB is kept in the same position. Line BC

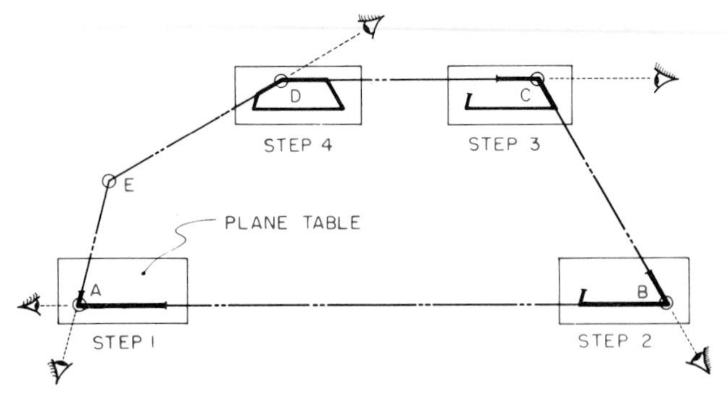

Fig. 45-10 The plane-table method of property-line layout.

is then established by sighting from corner B to corner C and measuring this distance. Step 3 establishes line CD by moving the plane table to corner C and sighting to corner D. Step 4 completes the layout by sighting from corner D back to corner E.

CONTOUR LINES

Contour lines show the various heights of the lot above an established plane known as the *datum.* Sea level is the universal datum line, although many municipalities have established datum points to aid surveyors. The datum is always zero.

Contour lines result from imaginary cuts made horizontally through the terrain at regular intervals. Figure 45-11 shows how this cut forms contour lines. The *contour interval,* or the vertical distance between contour lines, can be any convenient distance. It is usually an increment of 5′. Contour intervals of 5′, 10′, 15′, and 20′ are common on large surveys. The use of smaller contour intervals gives a more accurate description of the slope and shape of the terrain than does the use of larger intervals. Contour intervals as small as 1′-0″ may be used on small, relatively level sites.

Land on any part of a contour line has the same altitude, or height, above the datum. Contour lines are therefore always continuous. The area intersected by contour lines may be so vast that the lines may go off the drawing. However, if a larger geographical area were drawn, the contour lines would ultimately meet and close. Figure 45-12A shows how contours are projected from a profile through a hill. Contour lines that are very close together indicate a very steep slope. Contour lines that are far apart indicate a more gradual slope (Fig. 45-12B). Figure 45-12C shows the contour of a hill may appear the same as a valley until the elevations are added to the contour interval lines.

Surveyors often note elevation heights on a grid, as shown in Fig. 45-13A. This can be done easily in the field. A drafter can then use these grid heights to prepare a contour map, as shown in Fig. 45-13B. For example, the 4200′ contour line (see Fig. 45-13B) represents the location on the drawing that is at a level 4200′ above the datum. Follow the 4200′ intersection points shown in Fig. 45-13A and you will see how the contour map was developed. A profile of part of the contour map is shown in Fig. 45-13C. An alternative method of showing contours without developing contour lines is shown in Fig. 45-13D. In this drawing the height of each point is recorded and the arrow is used to show the downward direction of the slope.

GEOGRAPHICAL SURVEYS

Geographical survey maps are similar to surveys except that they cover extremely large areas. The entire world is divided into geographical survey regions. However, not all these regions have been surveyed. When large areas are to be covered, a small scale is used. When smaller areas are to be covered, a larger scale—such as 1 to 25,000—can be used. Figure 45-14A shows a typical portion of a geographical survey map. Geographical survey maps show the general

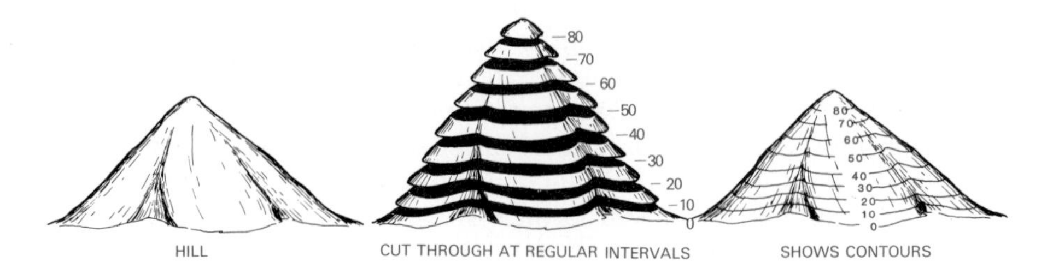

Fig. 45-11 Contour lines result from imaginary cuts made through the terrain.

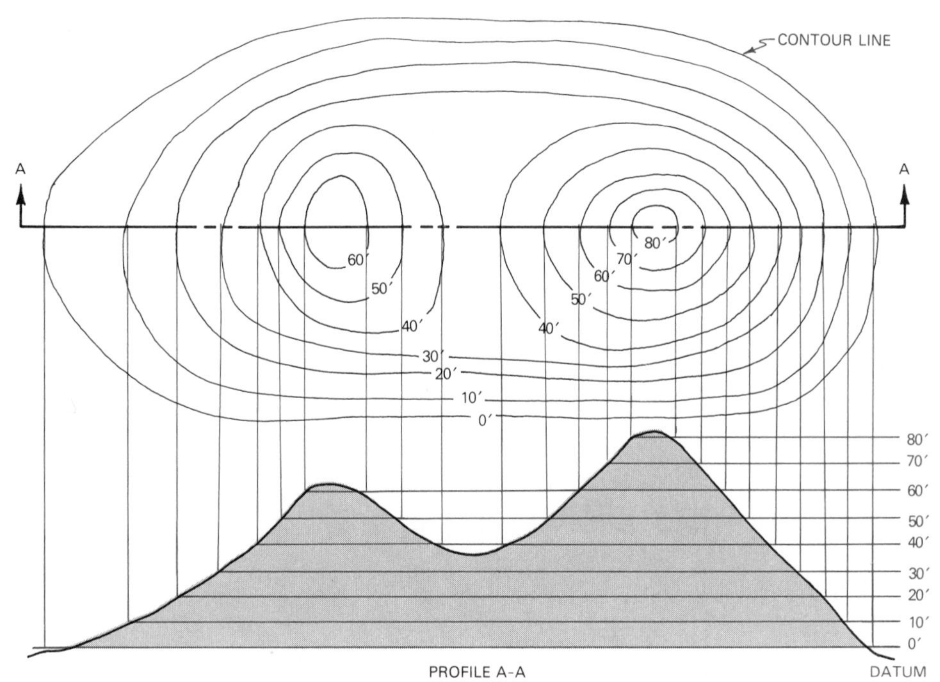

Fig. 45-12A The projection of a profile from contour lines.

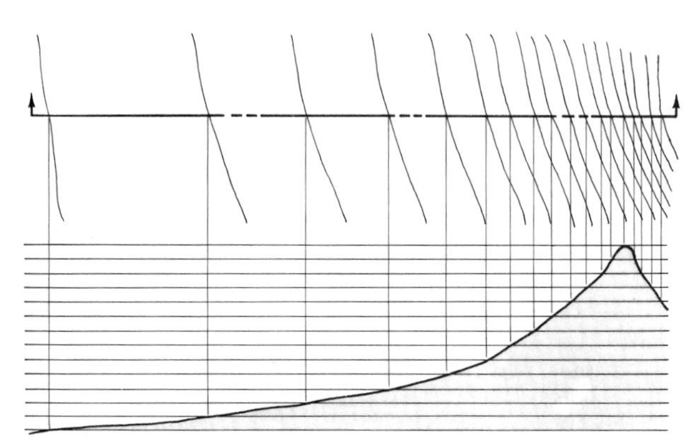

Fig. 45-12B Closer contour spacing denotes steeper slopes, wider spacing denotes flatter areas.

contour of the area, natural features of the terrain, and constructed features. Figure 45-14B shows an aerial photo of the area covered by the map in Fig. 45-14A.

Computer programs are now available which can be used to create pictorial contour maps as shown in Fig. 45-14C. This is done using X-Y coordinate input data and the corresponding datum level (Z) for each coordinate. Land-form profiles can also be generated on a computer using a wide variety of geographical input data as shown in Fig. 14-14D. This profile is used in the same way the composite analysis drawing data (Fig. 2-22) is used in the design process.

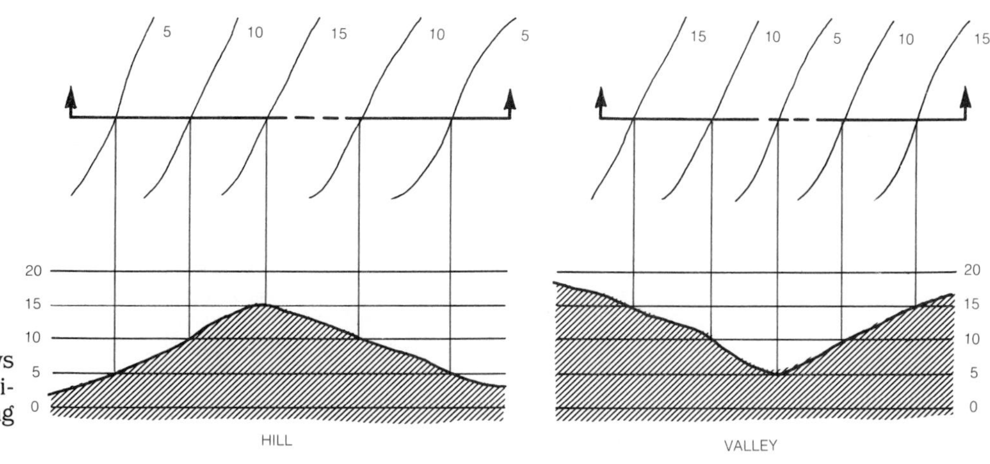

Fig. 45-12C Valleys and hills appear similar before adding contour height.

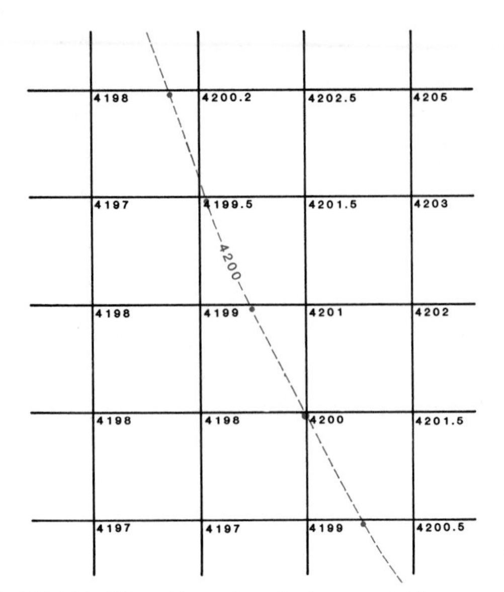

Fig. 45-13A Elevations located on a grid.

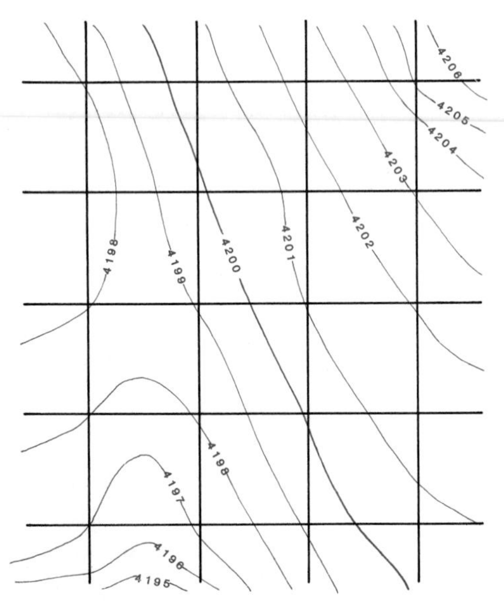

Fig. 45-13B Grid elevations converted to contour lines.

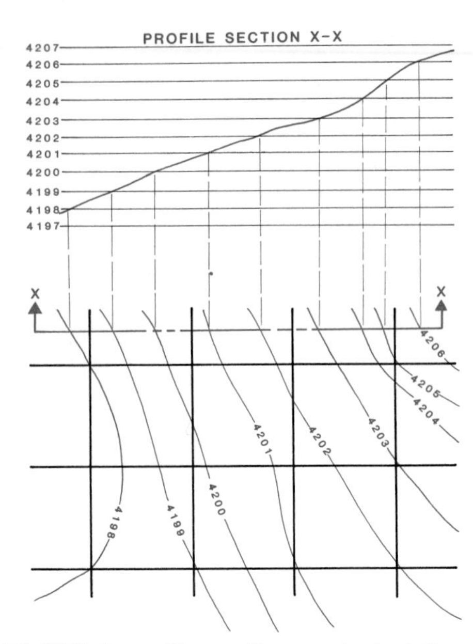

Fig. 45-13C A profile section projected from contour lines.

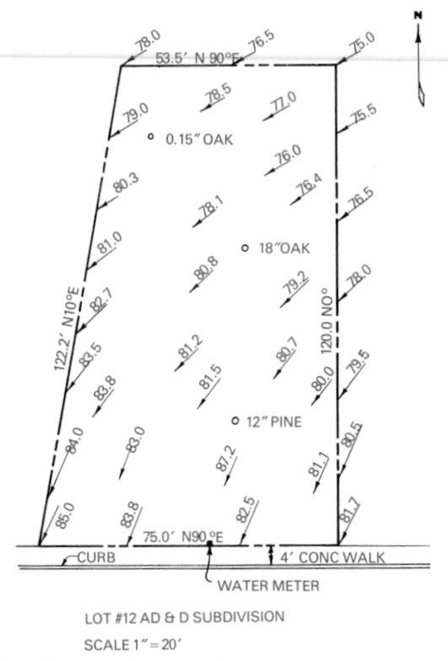

Fig. 45-13D Elevation of terrain shown by height dimension and downward slope direction.

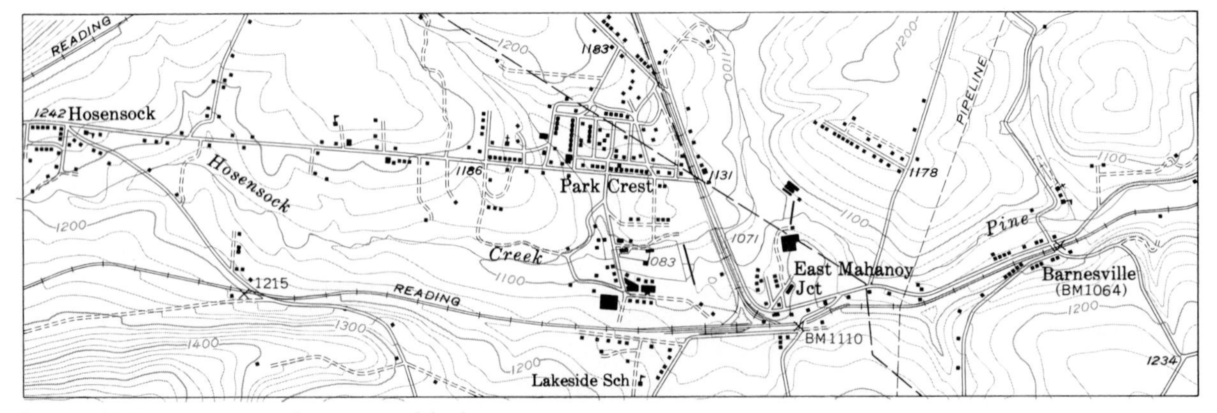

Fig. 45-14A A segment of a geographical survey map.

Fig. 45-14B Aerial photo of area shown in Fig. 45-14A.

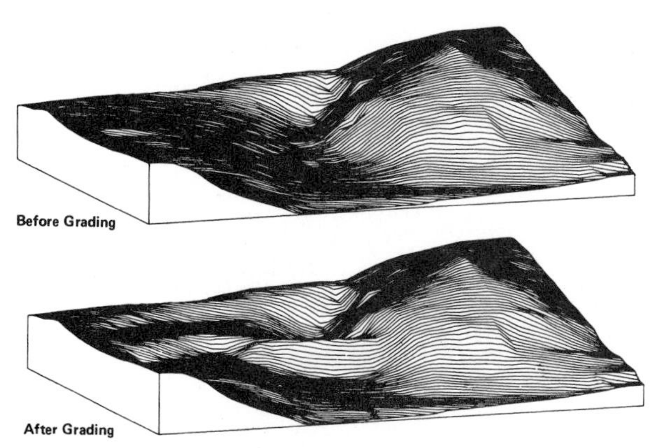

Fig. 45-14C Computer-generated pictorial contour maps. *(Comart Design Systems)*

PLAT PLANS

Plat plans are outlines of land subdivisions. They may be of home developments, industrial parks, or urban neighborhoods. They show the size and shape of each parcel of property in the area. Geographical survey regions are further subdivided into township grids. A township grid is 24 square miles. Each township grid is divided into 16 townships, each 6 mi^2; and each township is divided into 36 sections, each 1 mi^2. Figure 45-15 shows the township grid subdivided into townships and sections.

Plat plans are further subdivisions of township sections and are identified by section. Plat plans are identified in the following order:

1. Name of plat
2. Section
3. Township
4. County
5. State

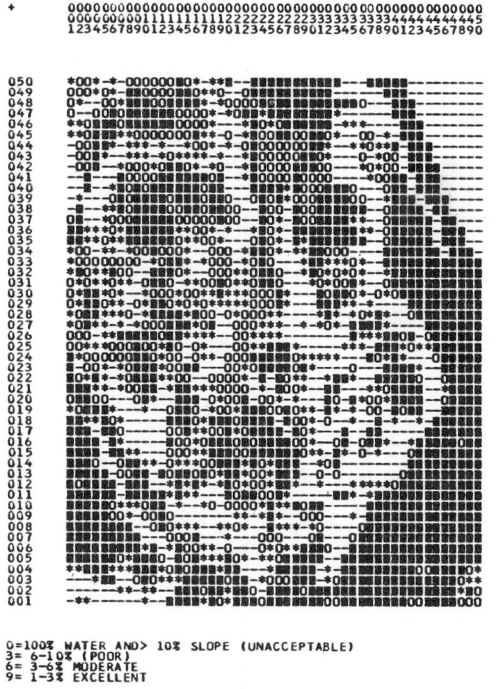

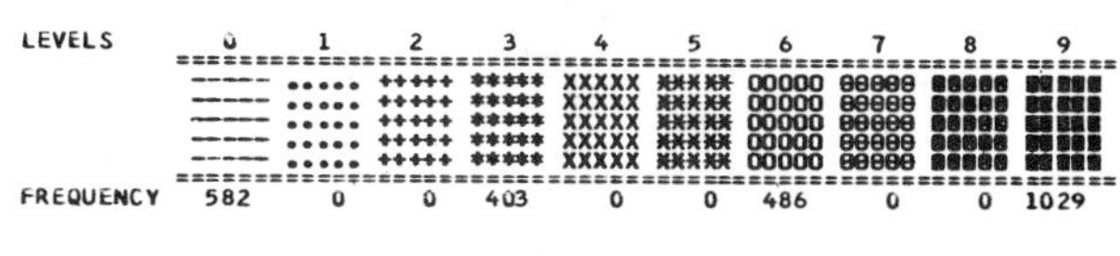

Fig. 45-14D Computer-generated land form profile. *(Department of Landscape Architecture, Harvard. Department of Landscape Architecture, Ohio State University)*

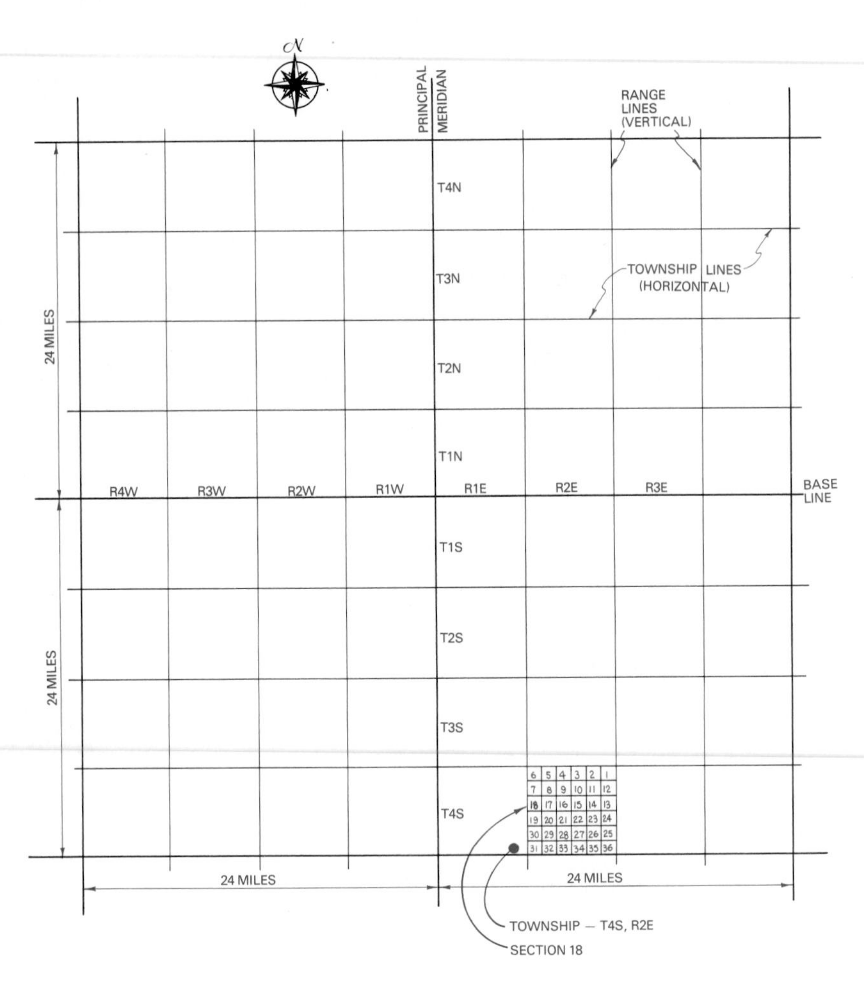

Fig. 45-15 Division of township grids.

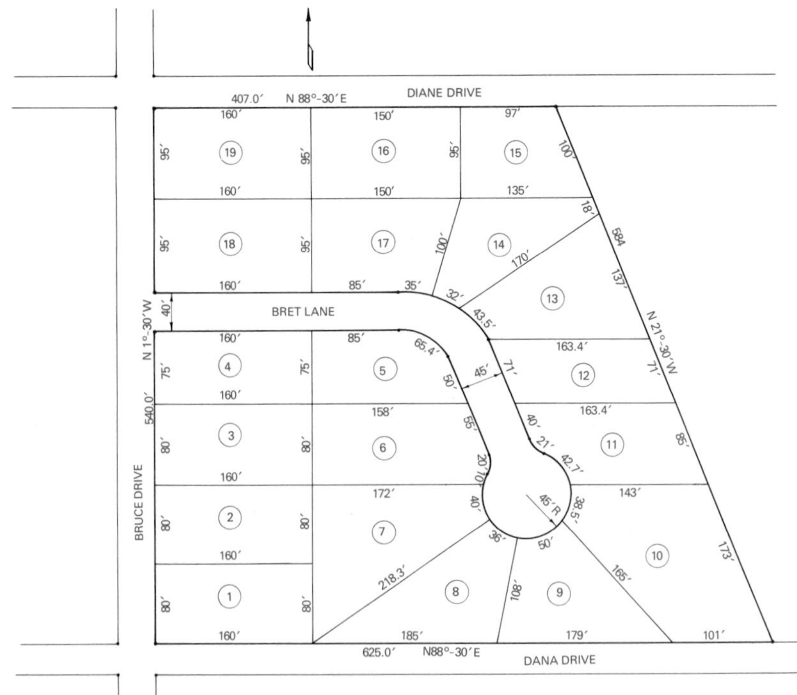

Fig. 45-16A Dimensions of grid properties.

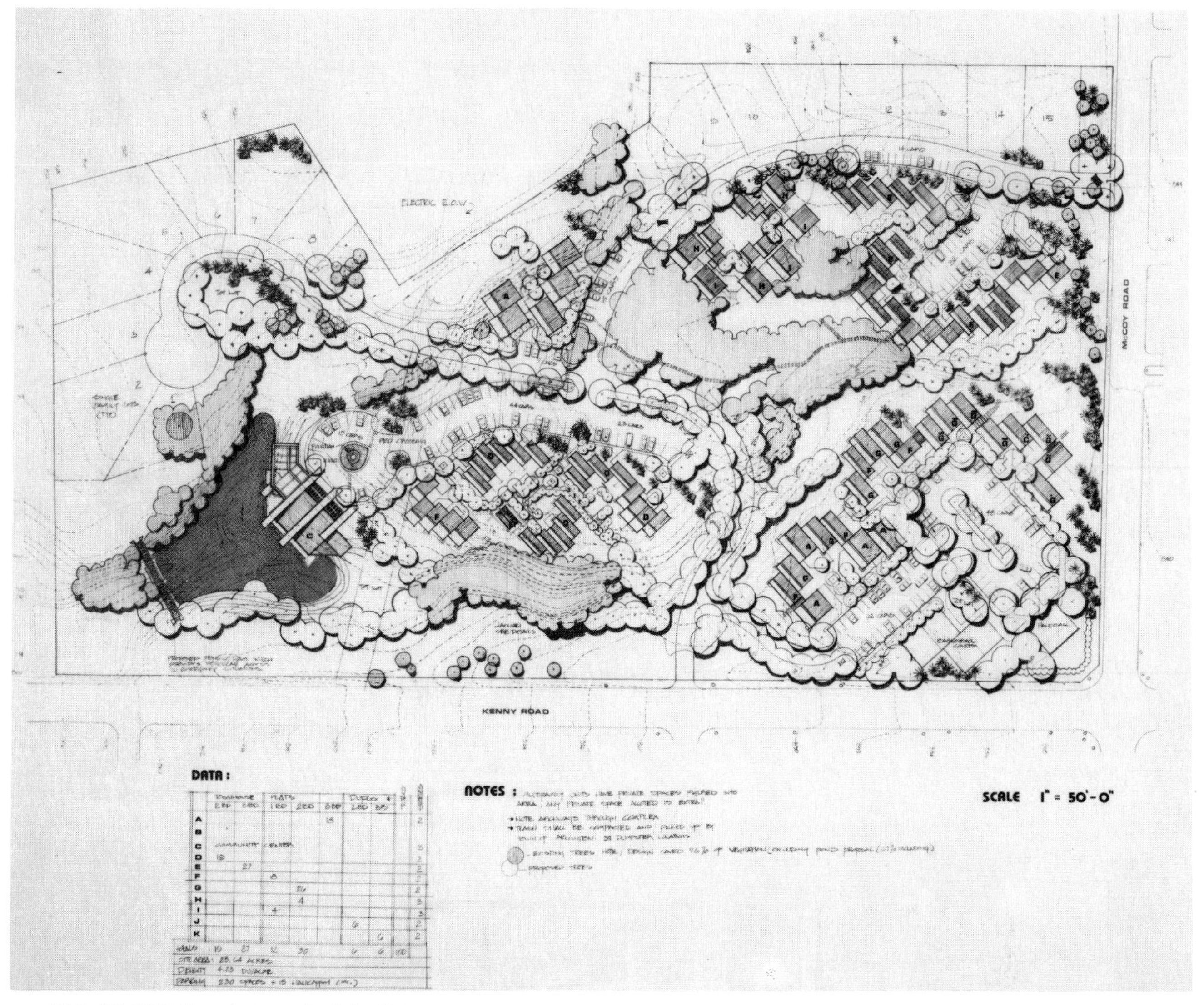

Fig. 45-16B Development plat plan.

Plat plans are located from a specific corner of a particular section in order to identify the plat in relation to the entire township grid. The dimensions of grid properties within the plat are described in the identical manner in which survey plans describe individual lots (Fig. 45-16A).

Figure 45-16B shows a plat plan for a development including the location of all structures and major landscaping features. This type of drawing is often used for sales presentation and conceptual design purposes.

1. Make a survey of a lot in your neighborhood, using the plane-table method.
2. Determine the azimuth of each property line shown in Fig. 45-17. Use a protractor or compass.
3. What is the contour interval used in Fig. 45-17?
4. Draw a profile for the area shown by the cutting plane line in Fig. 45-18.
5. Draw a survey of the ideal property on which you would build the home of your design.
6. Identify the following terms: *survey, contour lines, lot cornice, dotted contour lines, utility lines, azimuth bearing, transit, angle, surveyor, plane table, contour interval, geographical surveys, plot, parcel, township, subdivision, parcel grid property, section.*

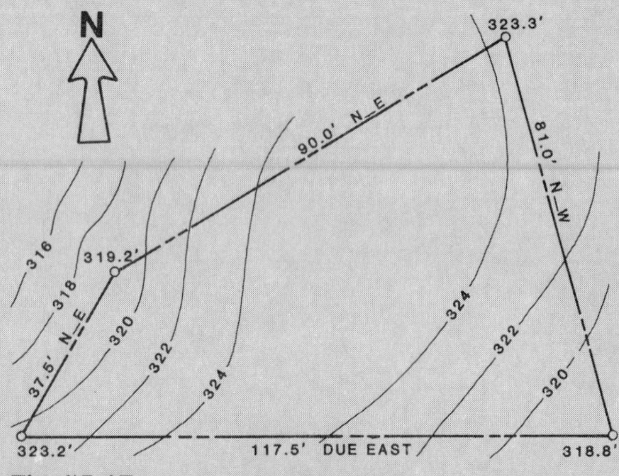

Fig. 45-17

Fig. 45-18

7. Develop a profile for sections X-X and Y-Y in Fig. 45-19 using contour intervals of 1′-0″.

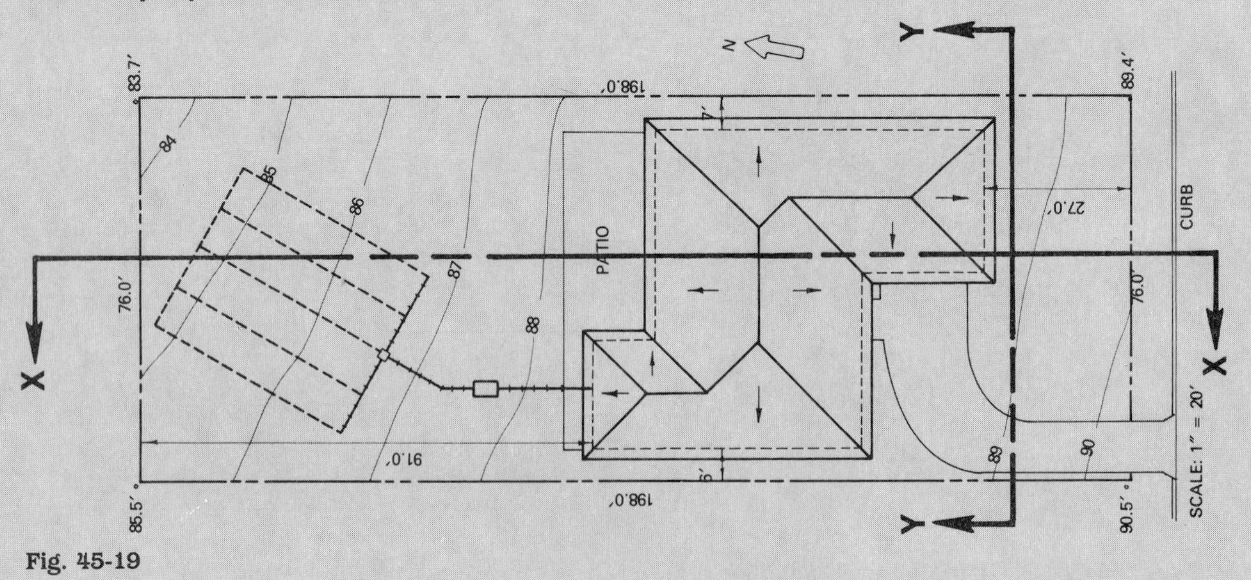

Fig. 45-19

SECTION 11

Sectional Drawings

Sectional drawings reveal the internal construction of an object. Architectural sectional drawings are prepared for the entire structure (full sections), or are prepared for specific parts of the building (detail sections). The size and complexity of the part usually determines the type of section. This section introduces sectional drawings which are used extensively throughout the remainder of the text.

UNIT 46

FULL SECTIONS

Architects frequently prepare drawings that show a building cut in half. Their purpose is to show how the building is constructed. These drawings are known as *longitudinal* or *transverse sections*. *Longitudinal* means lengthwise. A longitudinal section is one showing a lengthwise cut through the house. *Transverse* means across. A transverse section is one showing a cut across the building.

Transverse (Fig. 46-1) and longitudinal (Fig. 46-2) sections have the same outlines as the elevation drawings of the building. Figure 46-1 is a section cut parallel to the short axis of the building. Figure 46-2 is a section cut parallel to the major axis of the building.

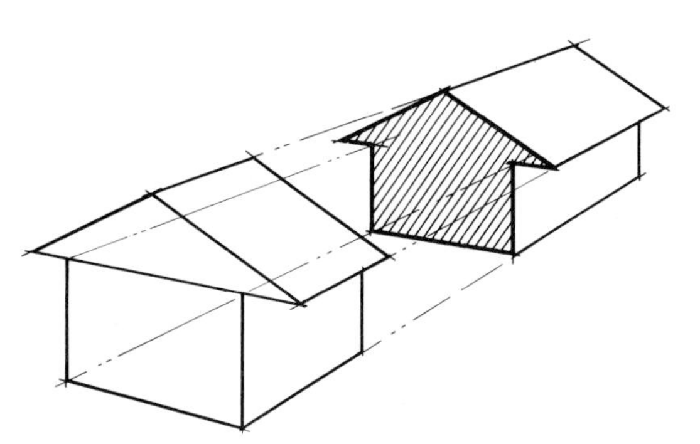

Fig. 46-1 A transverse section through the minor axis of the building.

THE CUTTING PLANE

The *cutting plane* is an imaginary plane that passes through the building. The position of the cutting plane is shown by the cutting-plane line. The cutting-plane line is made up of a long heavy line followed by two dashes. Figure 46-3 shows a cutting-plane line and the cutting plane it represents. The cutting-plane line is placed on the

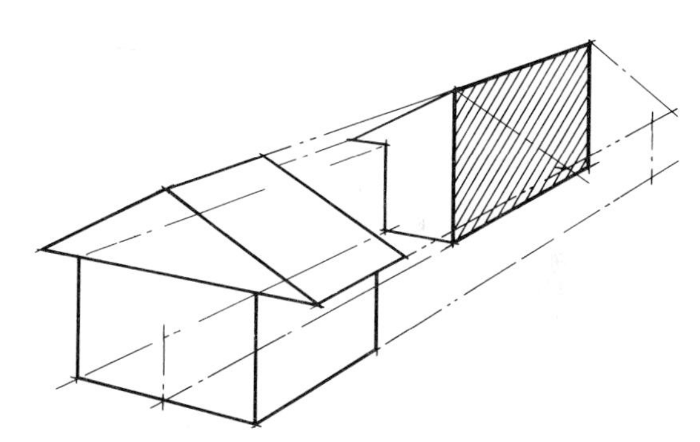

Fig. 46-2 A longitudinal section through the major axis of the building.

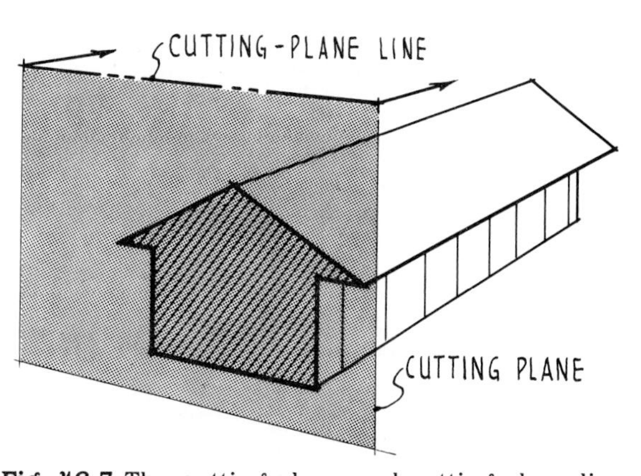

Fig. 46-3 The cutting-plane and cutting-plane line.

part to be sectioned, and the arrows at its ends show the direction from which the section is to be viewed. For example, in Fig. 46-4, section BB is a west section; section AA is an east section, and section CC is a south elevation.

The cutting-plane line often interferes with dimensions, notes, and details. An alternative method of drawing cutting-plane lines is used to overcome this interference. The alternative method is shown in Fig. 46-5. Notice that only the extremes of the cutting-plane line are used. The cutting-plane line is then *assumed to be a straight line* between these extremes.

When a cutting-plane line must be offset to show a different area, the offsetting corners are drawn as shown in Fig. 46-6. The *offset cutting plane* is often used to show different wall sections on one sectional drawing.

SYMBOLS

Section-lining symbols sometimes represent the way building materials look when they are cut through. Many, however, are purely symbolic in order to conserve time on the drawing board. A floor-plan drawing is actually a horizontal section. Many of the symbols used in floor plans also apply to section symbols. Symbols for many building materials are shown in Fig. 46-7A and B.

Section lining is added to a drawing using a CAD crosshatch task by touching the perimeter line of each enclosed area to be crosshatched. Different patterns can be generated by varying the width and offset value of the crosshatch lines.

A building material is only sectioned when the cutting-plane line passes through it. The outline

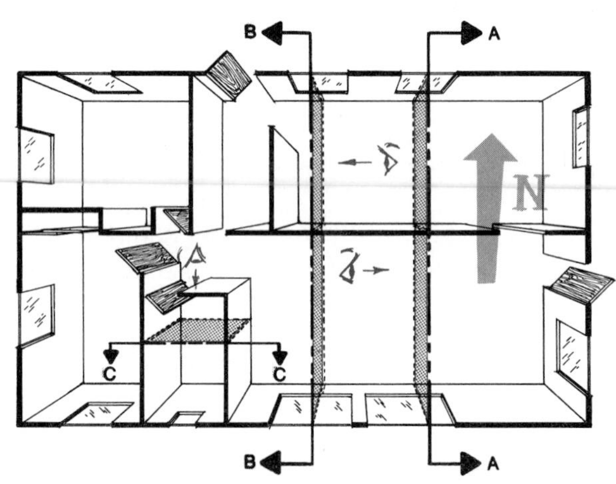

Fig. 46-4 Arrows on the cutting-plane line determine the position from which these sections are viewed.

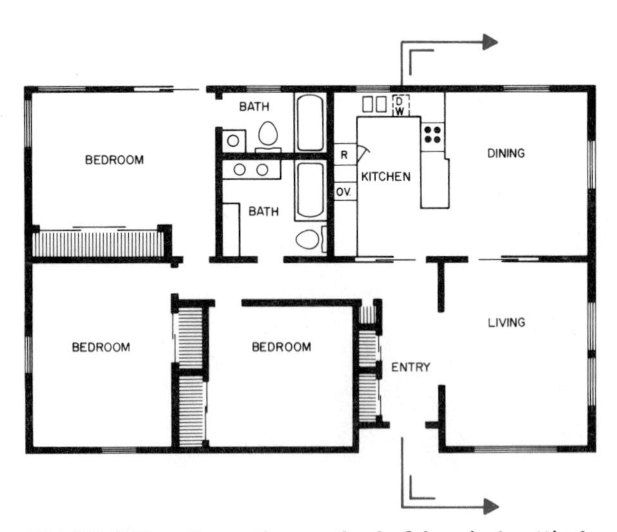

Fig. 46-5 The alternative method of drawing cutting-plane lines.

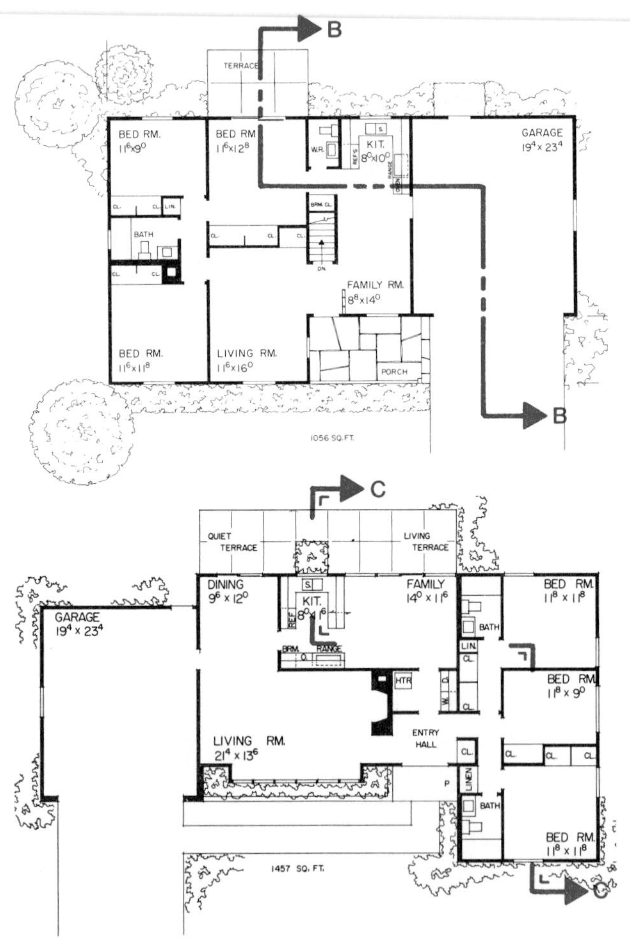

Fig. 46-6 The use of an offset cutting-plane line.

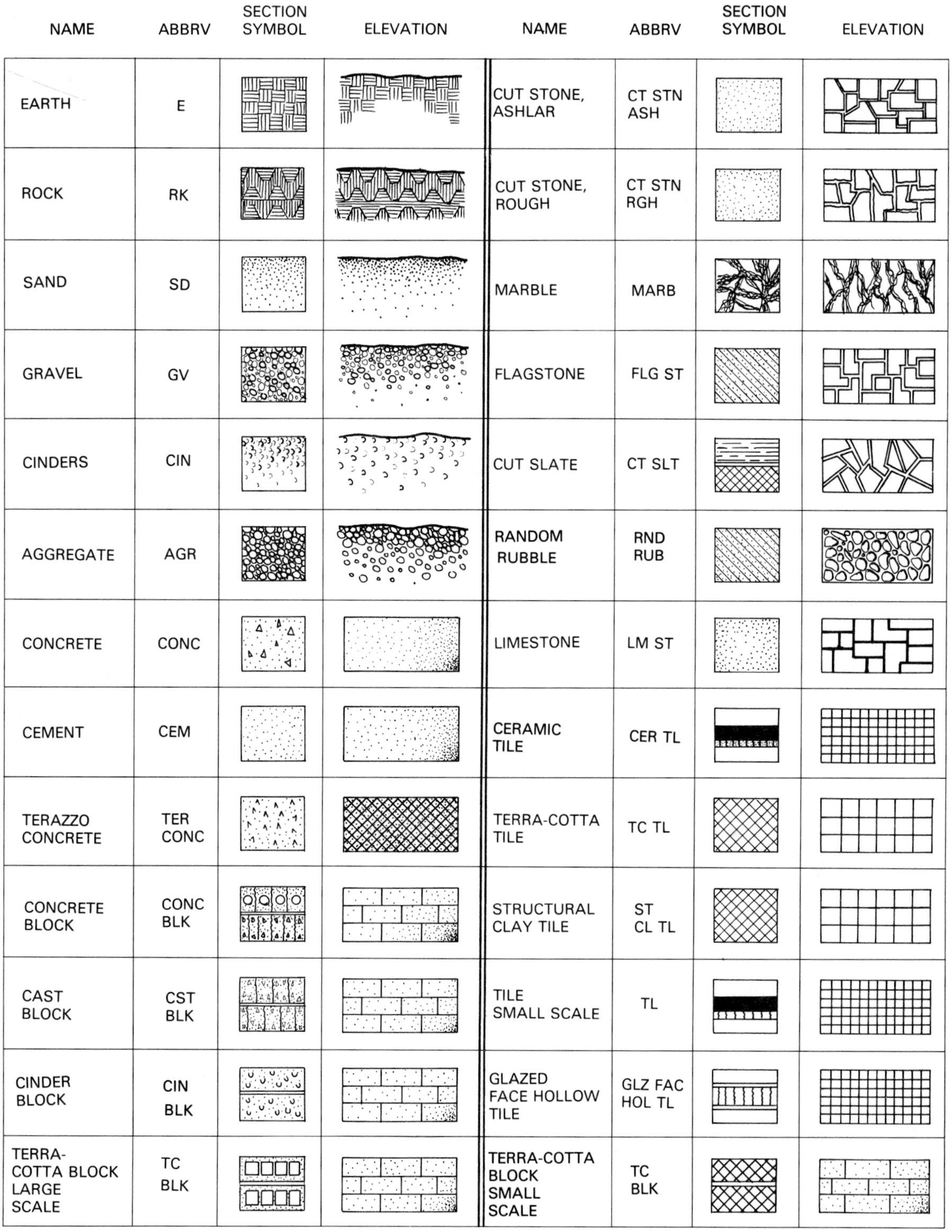

NAME	ABBRV	SECTION SYMBOL	ELEVATION	NAME	ABBRV	SECTION SYMBOL	ELEVATION
EARTH	E			CUT STONE, ASHLAR	CT STN ASH		
ROCK	RK			CUT STONE, ROUGH	CT STN RGH		
SAND	SD			MARBLE	MARB		
GRAVEL	GV			FLAGSTONE	FLG ST		
CINDERS	CIN			CUT SLATE	CT SLT		
AGGREGATE	AGR			RANDOM RUBBLE	RND RUB		
CONCRETE	CONC			LIMESTONE	LM ST		
CEMENT	CEM			CERAMIC TILE	CER TL		
TERAZZO CONCRETE	TER CONC			TERRA-COTTA TILE	TC TL		
CONCRETE BLOCK	CONC BLK			STRUCTURAL CLAY TILE	ST CL TL		
CAST BLOCK	CST BLK			TILE SMALL SCALE	TL		
CINDER BLOCK	CIN BLK			GLAZED FACE HOLLOW TILE	GLZ FAC HOL TL		
TERRA-COTTA BLOCK LARGE SCALE	TC BLK			TERRA-COTTA BLOCK SMALL SCALE	TC BLK		

Fig. 46-7A Common sectional symbols.

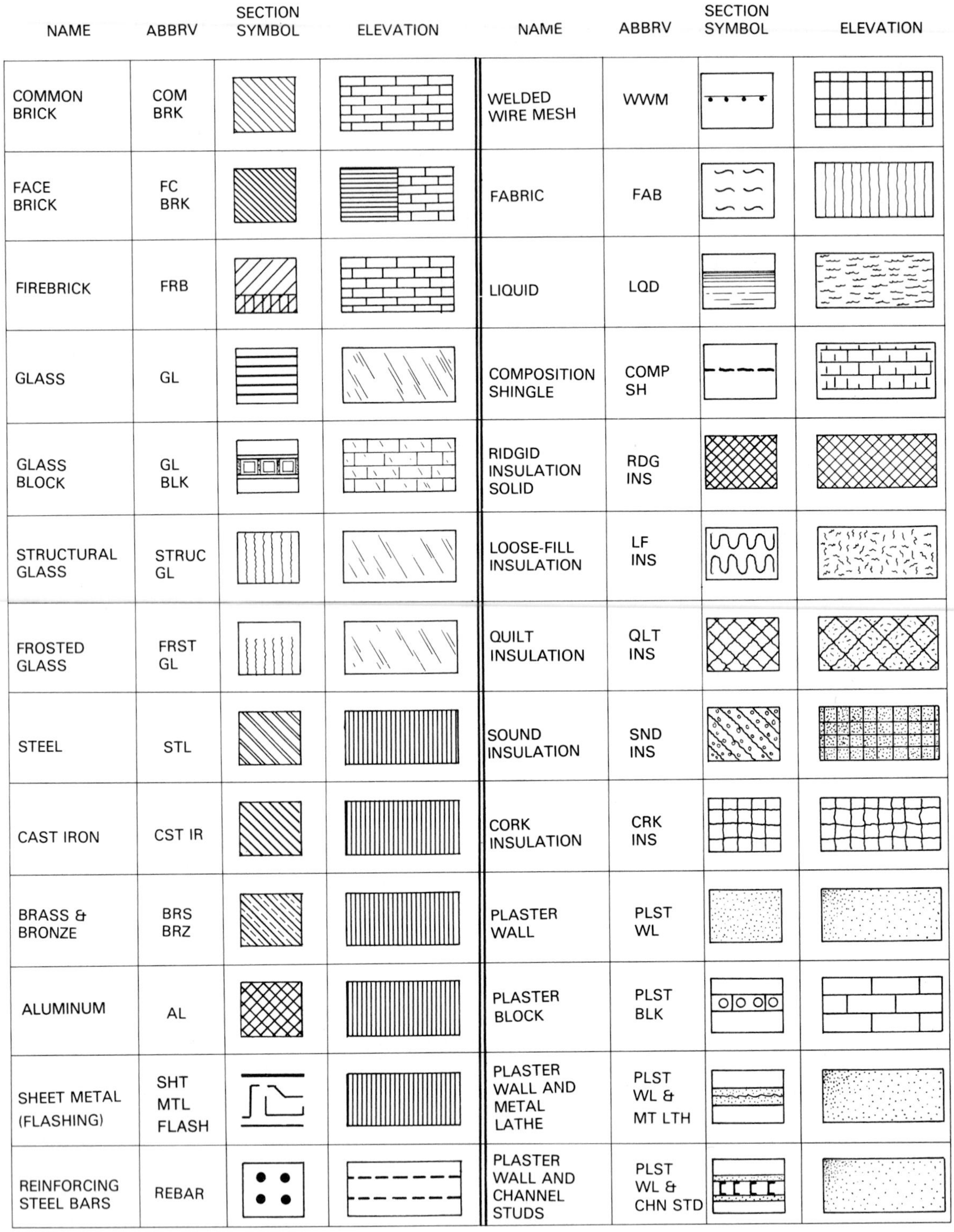

NAME	ABBRV	SECTION SYMBOL	ELEVATION	NAME	ABBRV	SECTION SYMBOL	ELEVATION
COMMON BRICK	COM BRK			WELDED WIRE MESH	WWM		
FACE BRICK	FC BRK			FABRIC	FAB		
FIREBRICK	FRB			LIQUID	LQD		
GLASS	GL			COMPOSITION SHINGLE	COMP SH		
GLASS BLOCK	GL BLK			RIDGID INSULATION SOLID	RDG INS		
STRUCTURAL GLASS	STRUC GL			LOOSE-FILL INSULATION	LF INS		
FROSTED GLASS	FRST GL			QUILT INSULATION	QLT INS		
STEEL	STL			SOUND INSULATION	SND INS		
CAST IRON	CST IR			CORK INSULATION	CRK INS		
BRASS & BRONZE	BRS BRZ			PLASTER WALL	PLST WL		
ALUMINUM	AL			PLASTER BLOCK	PLST BLK		
SHEET METAL (FLASHING)	SHT MTL FLASH			PLASTER WALL AND METAL LATHE	PLST WL & MT LTH		
REINFORCING STEEL BARS	REBAR			PLASTER WALL AND CHANNEL STUDS	PLST WL & CHN STD		

Fig. 46-7B Common elevation symbols.

of all other materials visible behind the plane of projection must also be drawn in the proper position and scale. Open space is never sectioned. Figure 46-8 shows various building materials as they appear in a transverse section across the gable end of a residence. Figure 46-9 shows similar building materials as they appear on a longitudinal section. Because full sections show the construction method used in the entire building, they must be drawn to a relatively small scale. The use of this small scale often makes the drawing and interpretation of minute details extremely difficult. Removed detail sections are often used to clarify small details as shown in Fig. 46-10.

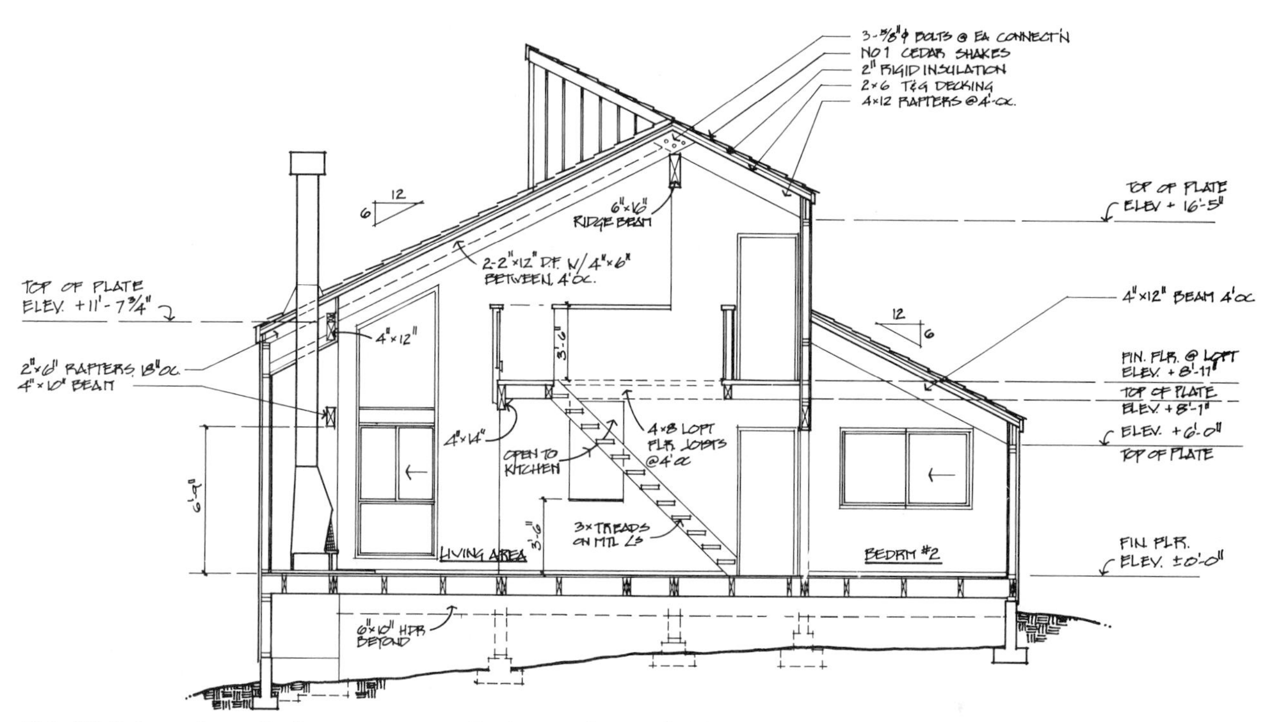

Fig. 46-8 A section of a house, perpendicular to the roof ridge.

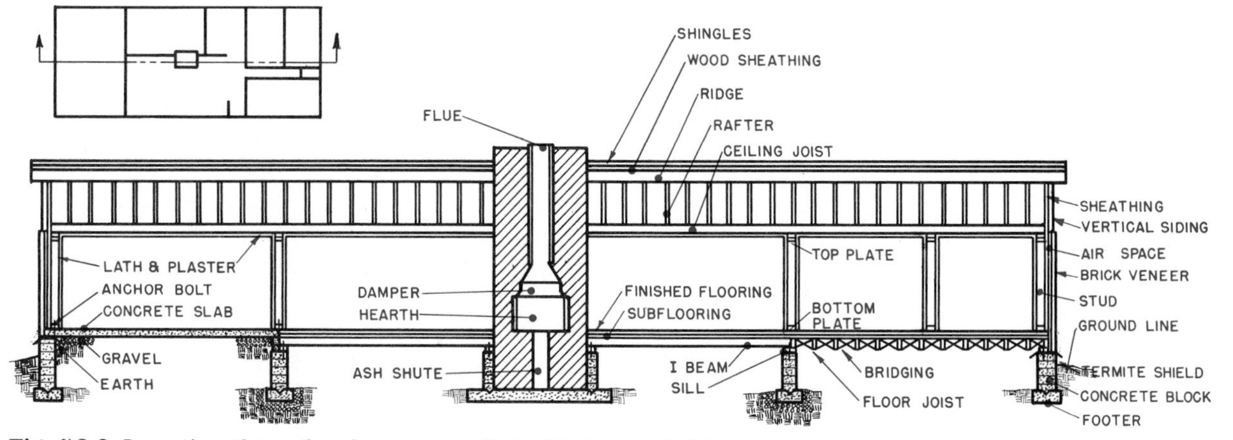

Fig. 46-9 A section through a house, parallel with the roof ridge.

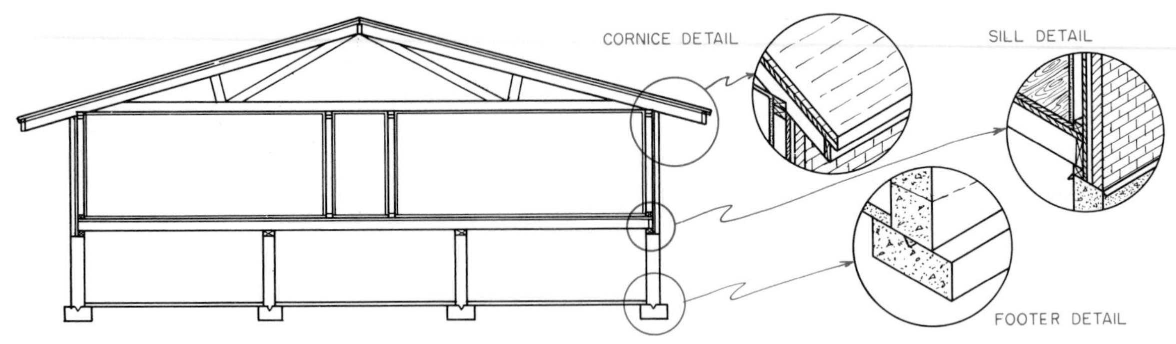

Fig. 46-10 Three common sectional details are the cornice, sill, and footing.

STEPS IN DRAWING FULL SECTIONS

In drawing full sections, the architect actually constructs the framework of a house on paper. Figure 46-11 shows the progressive steps in the layout and drawing of a gable-end section:

1. Lightly draw the floor line approximately at the middle of the drawing sheet.

2. Measure the thickness of the subfloor and of the joist and draw lines representing these under the floor line.

3. From the floor line, measure up and draw the ceiling line.

4. Measure down from the floor line to establish the top of the basement slab and footing line, and draw in the thickness of the footing.

5. Draw two vertical lines representing the thickness of the foundation and the footing.

6. Construct the sill detail and show the alignment of the stud and top plate.

7. Measure the overhang from the stud line and draw the roof pitch by projecting from the top plate on the angle that represents the rise over the run.

8. Establish the ridge point by measuring the distance from the outside wall horizontally to the center of the structure.

9. Add details and symbols representing siding and interior finish.

SECTIONAL DIMENSIONING

Since full sections expose the size and shape of building materials and components not revealed on floor plans and elevations, these sections are an excellent place on which to locate many detail dimensions. Full-section dimensions primarily show specific elevations, distances, and the exact size of building materials.

Figure 46-12 shows some of the more important dimensions that can be placed on full sections. The rules for dimensioning elevation drawings apply also to full elevation sections (see Unit 38).

MULTILEVEL SECTIONS

Full sections are especially effective and necessary for showing the various methods of con-

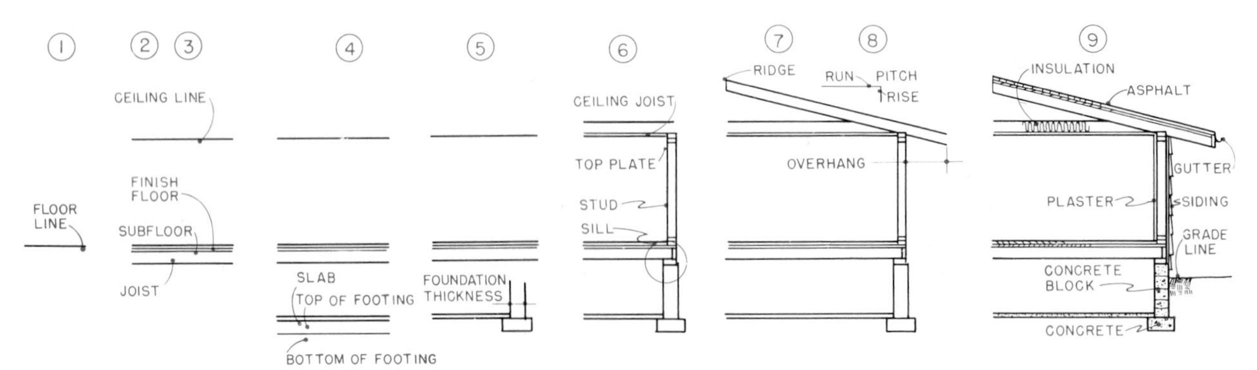

Fig. 46-11 The sequence of projecting an elevation section.

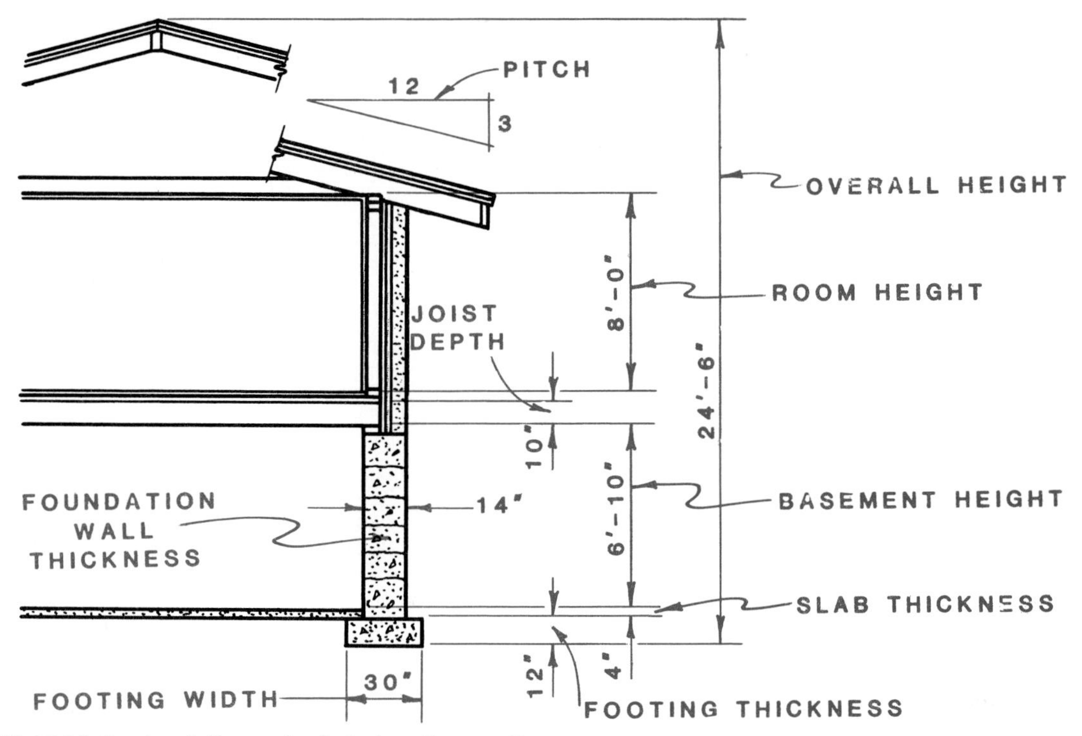

Fig. 46-12 Methods of dimensioning elevation sections.

structing multilevel buildings, since footings, grade lines, slabs, and floor lines vary greatly. Figure 46-13 shows a section of a split-level home. It is difficult to show the relationship of the various levels and the construction of each without using this kind of section.

Figure 46-14 shows the use of sections to accurately describe the shape of the Statue of Liberty at nine different levels. Computer-aided drafting and design methods (see Unit 27) were used in the planning stages of the statue restoration. Light beams were bounced off the interior of the shell and computer-generated configurations of the skeleton were generated. The sections shown were drawn from this data.

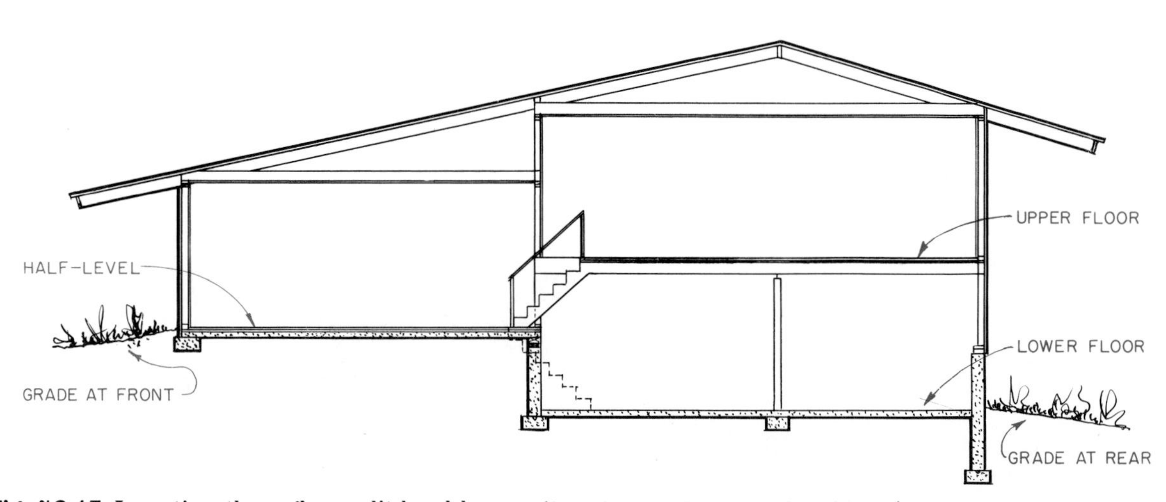

Fig. 46-13 A section through a split-level home. *(Swanke Hayden Connell Architects)*

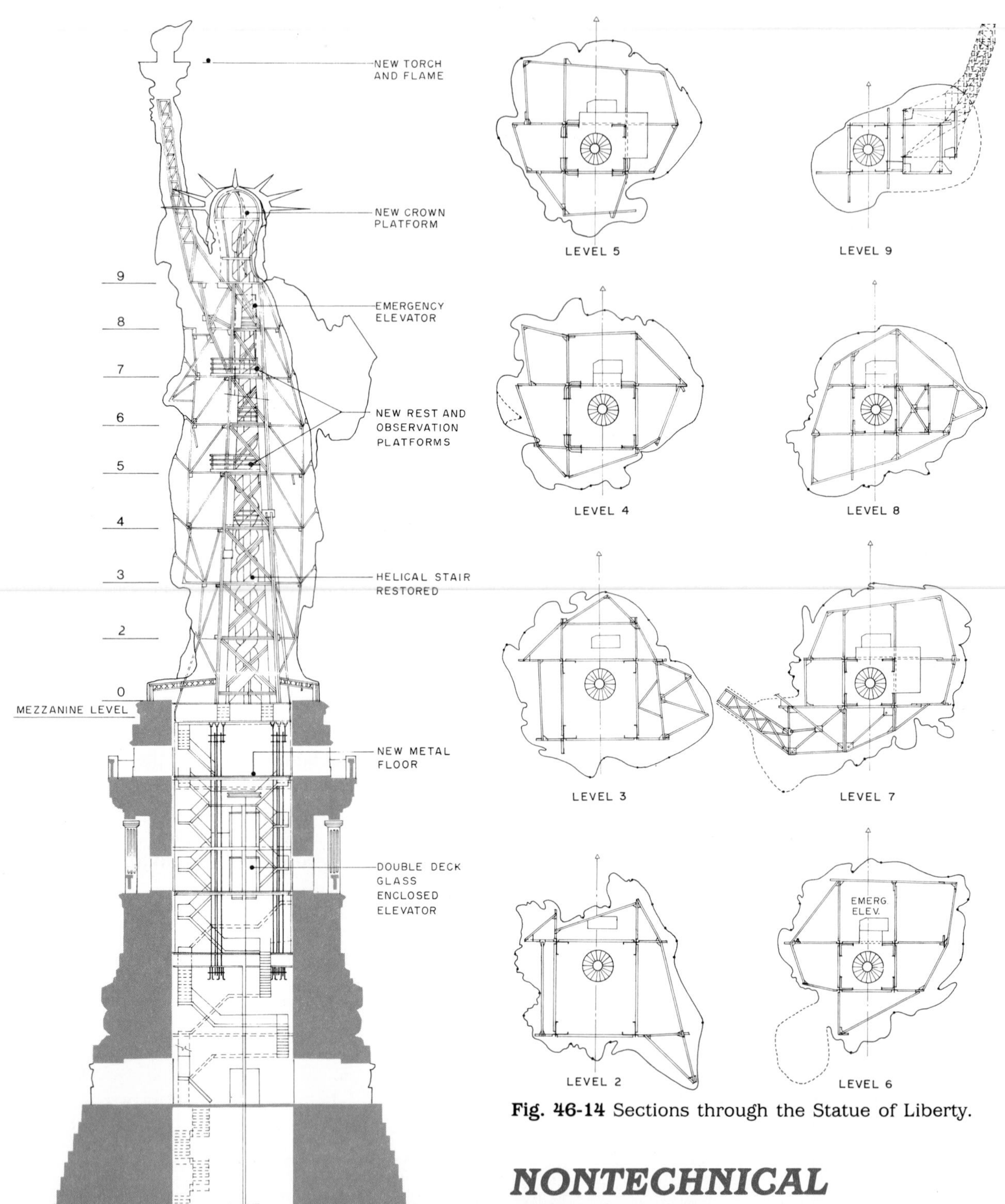

Fig. 46-14 Sections through the Statue of Liberty.

NONTECHNICAL SECTIONS

Full sections without structural details or dimensions, as shown in Fig. 46-15, are used to show size and proportional relationships. Notice how cars and people are used to relate the size of the structure to the various design features.

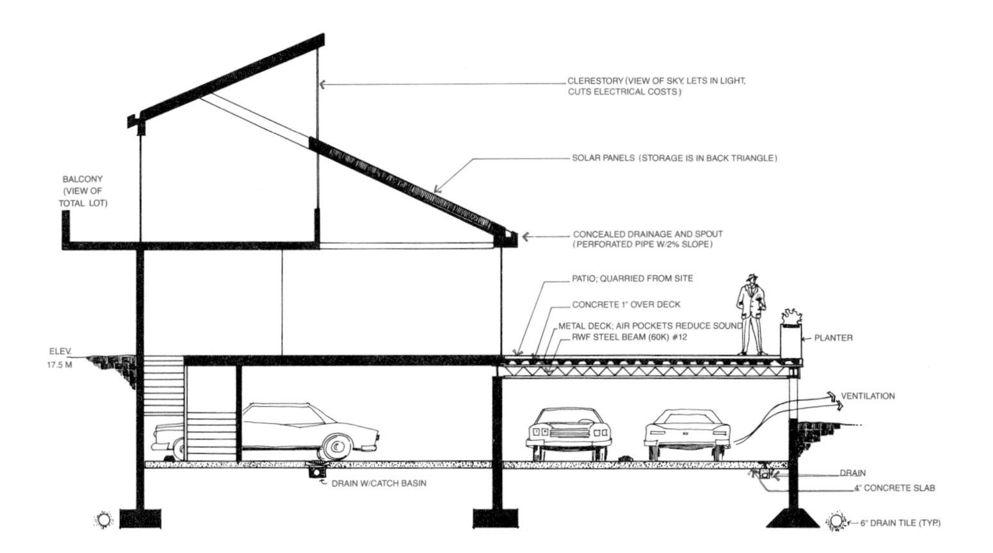

Fig. 46-15 Full section without structural details.

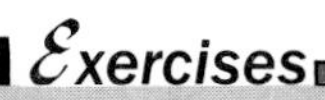

Exercises

1. Draw a full section of a house you have designed.

2. Draw and dimension a longitudinal section AA of the plan shown in Fig. 46-16.

3. Draw section BB, Fig. 46-16.

4. Identify the symbols found in Fig. 46-17.

5. Define these terms: *longitudinal section, transverse section, cutting plane, offset cutting plane, section lining, full section.*

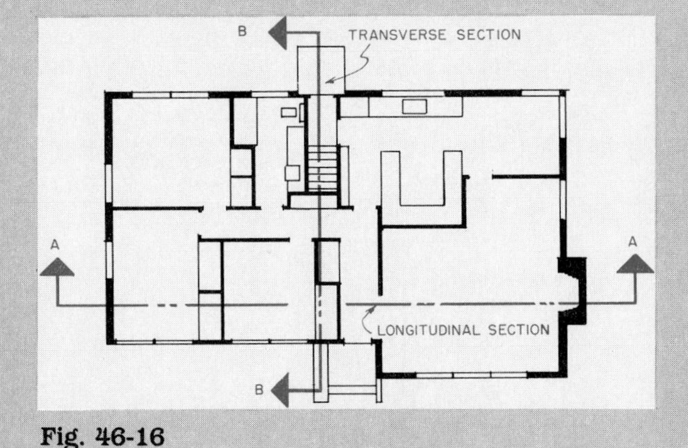

Fig. 46-16

1 2 3 4 5

6 7 8 9 10

11 12 13 14 15

16 17 18 19 20

Fig. 46-17

UNIT 47

DETAIL SECTIONS

Because full sections of a floor plan are usually drawn to the small scale of 1/4" = 1'-0", many parts are difficult to interpret and dimension. In order to reveal the exact position and size of many small members, the drafter needs an enlarged section.

VERTICAL WALL SECTIONS

One method of showing sections larger than is possible in the full section is through the use of *break lines.* Break lines are used to reduce vertical distances on exterior walls. Using break lines allows the drafter to draw the area larger than is possible when the entire distance is included in the drawing. Break lines are placed where the material does not change over a long distance. Figure 47-1 shows the difference between a brick-veneer wall drawn completely to a small scale and the same wall enlarged by the use of break lines. Figure 47-2 shows the use of break lines to enlarge a frame-wall section.

Some sections need to be drawn for interpretation or dimensioning. Sometimes it is impossible to draw an entire wall section to a large enough scale, even when using break lines. When that is the case, you draw a removed section. A *removed section* is one drawn away from the original location on the same sheet or on another sheet. Removed sections are frequently

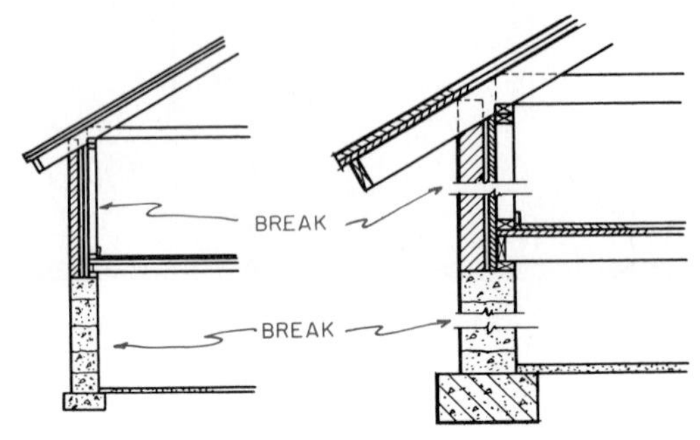

Fig. 47-1 Full wall section.

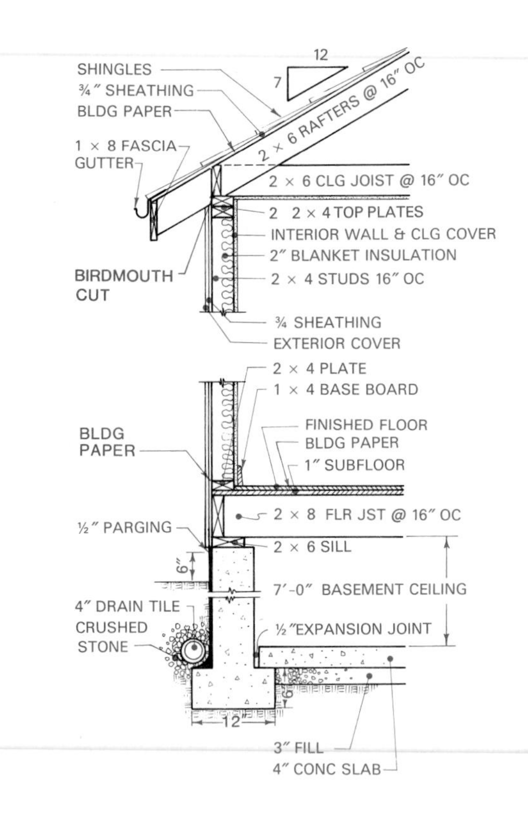

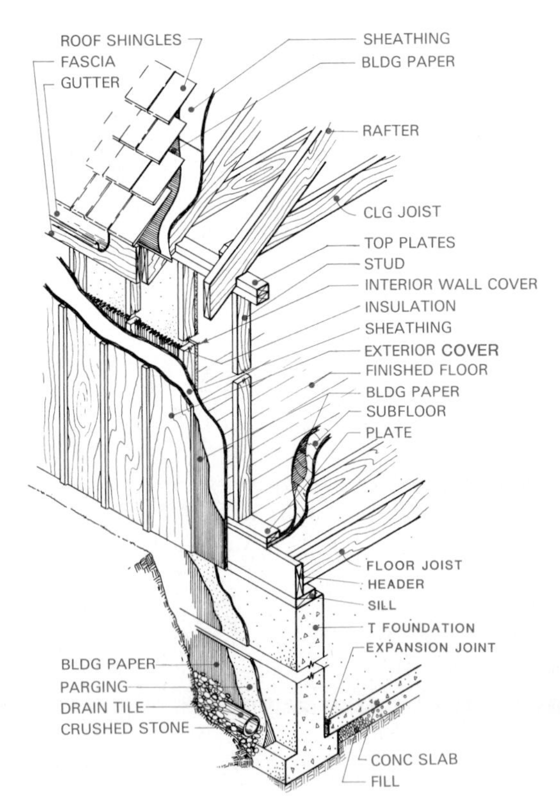

Fig. 47-2 The use of break lines on a frame wall in brick veneer section.

drawn for the ridge, cornice, sill, footing, and beam areas, as shown in Fig. 47-3.

Cornice Sections

Figure 47-4 shows a typical cornice section prepared on a CAD system. *Cornice sections* are used to show the relationship between the outside wall, top plate, and rafter construction. Some cornice sections show gutter details.

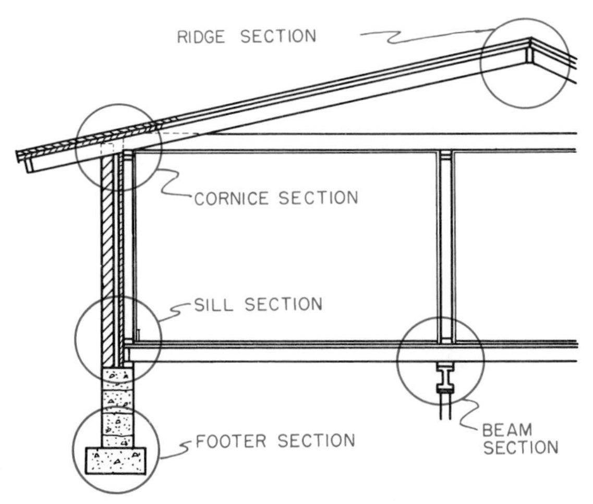

Fig. 47-3 Removed sections are often drawn of the ridge, cornice, sill, footing, and beam areas.

Sill Sections

Sill sections, as shown in Fig. 47-5, show how the foundation supports and intersects with the floor system and the outside wall.

Footing Sections

A *footing section* is needed to show the width and length of the footing, the type of material used, and the position of the foundation wall on the footing. Figure 47-6 shows several footing details and the pictorial interpretation of each type.

Beam Details

Beam details are necessary to show how the joists are supported by beams and how the columns or foundation walls support the beams. As in all other sections, the position of the cutting-plane line is extremely important. Figure 47-7 shows two possible positions of the cutting plane. If the cutting-plane line is placed parallel to the beam, you see a cross section of the joist, as shown at A. if the cutting-plane line is placed perpendicular to the beam, you see a cross section of the beam, as shown at B. Figure 47-8

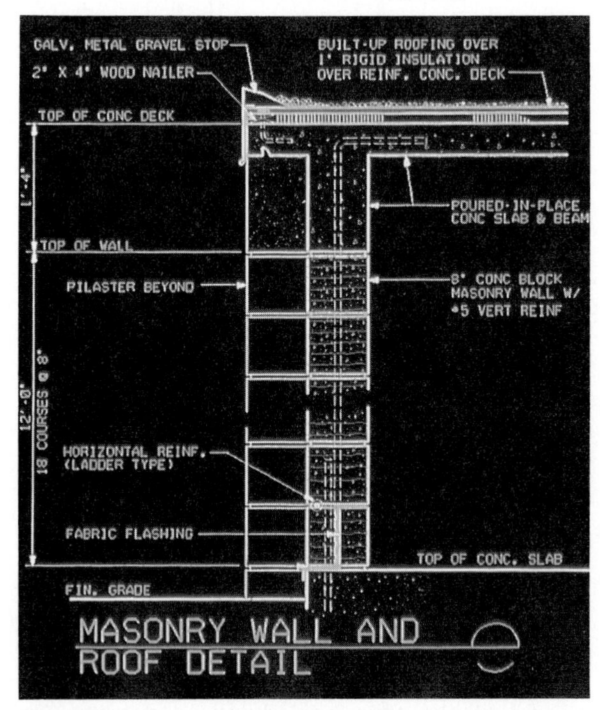

Fig. 47-4 CAD generated cornice section. *(Intergraph Corp.)*

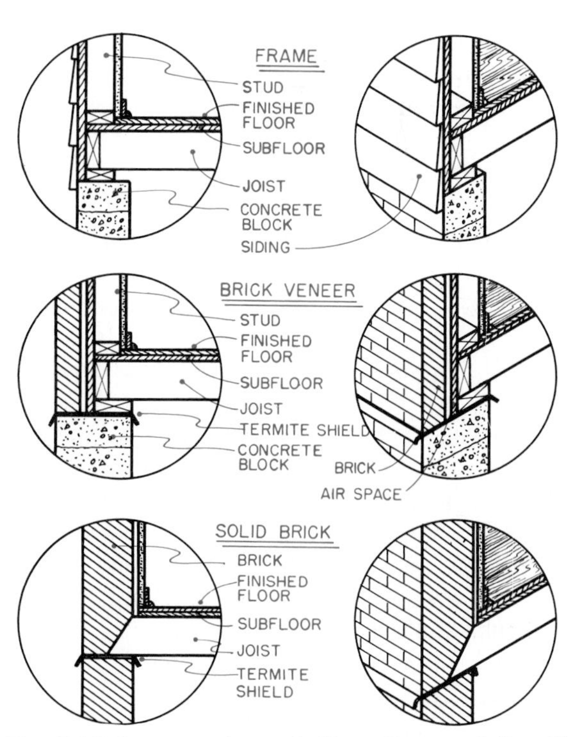

Fig. 47-5 A comparison of sill sections and the sill working drawings they represent.

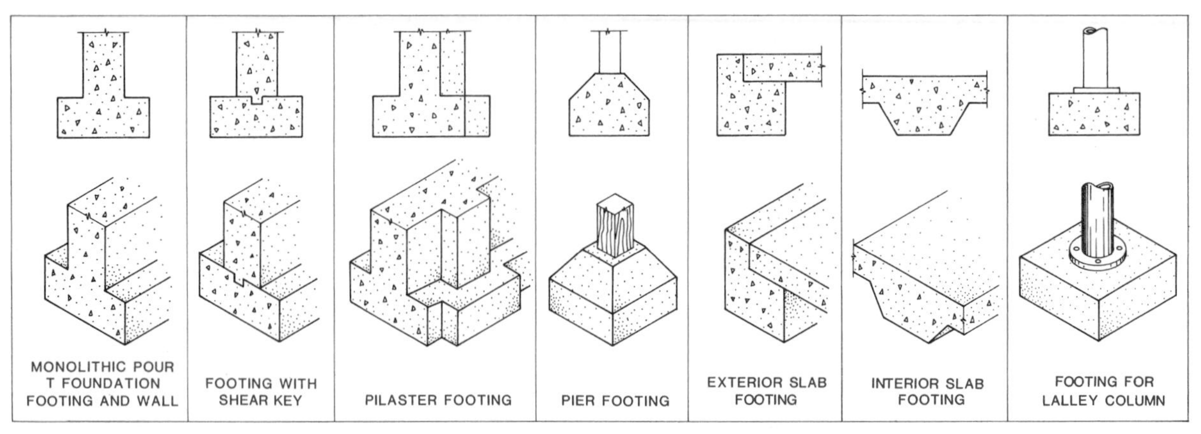

Fig. 47-6 A comparison of pictorial and orthographic footing sections.

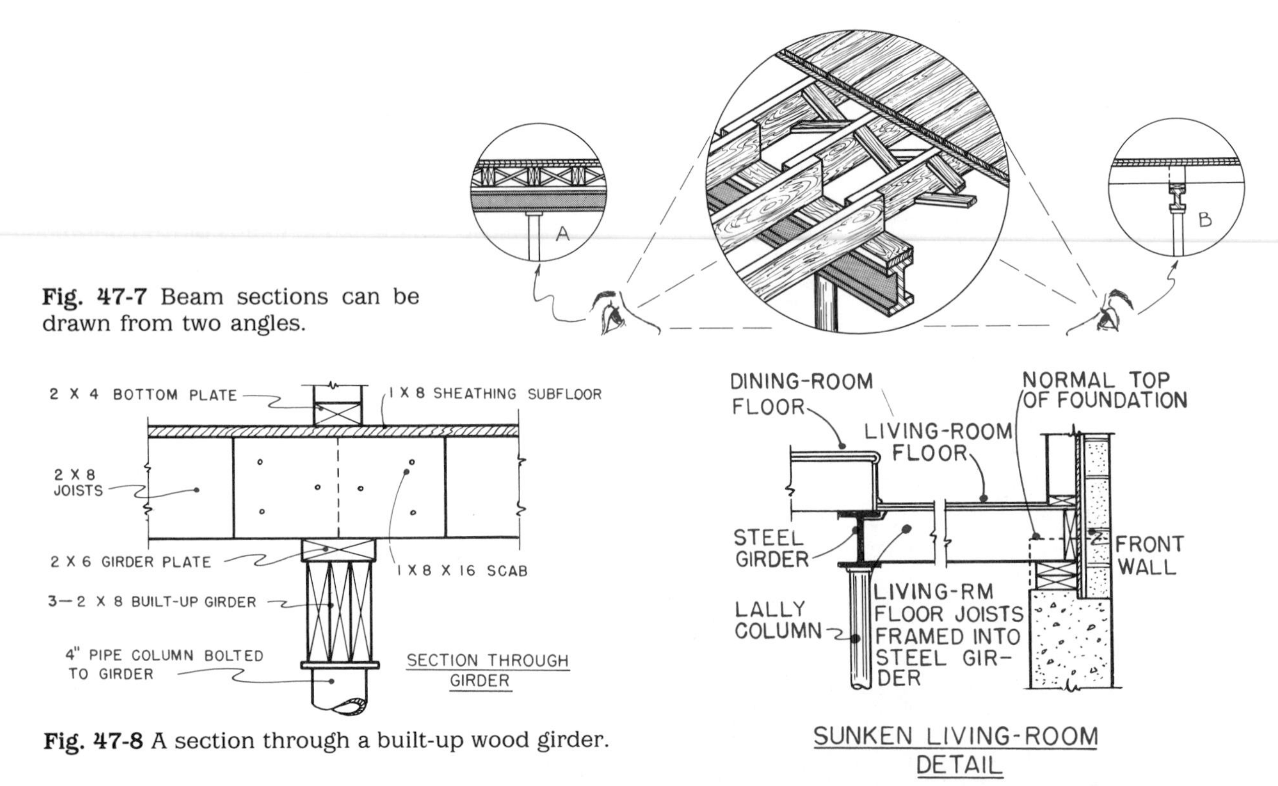

Fig. 47-7 Beam sections can be drawn from two angles.

Fig. 47-8 A section through a built-up wood girder.

Fig. 47-9 The beam section and sill section can be shown on the same drawing by use of break lines.

shows a similar section through a built-up wood girder.

Sometimes it is necessary to show the relationship between the beam detail and a detail of the sill area. This relationship is especially needed when a room is sunken or elevated. In Fig. 47-9, the beam and sill areas on one sectional drawing are shown by breaking the area between the sill and the beam.

Interior-Wall Sections

To illustrate the methods of constructing inside partitions, sections are often drawn of interior walls at the base and at the ceiling. *Base sections* (Fig. 47-10A) show how the wall-finishing materials are attached to the studs and how the intersection between the floor and wall is constructed. The section at the ceiling, as shown in Fig. 47-10B, is drawn to show the intersection between the ceiling and the wall and to show how the finishing materials of the wall and ceiling are related. Vertical wall sections are also prepared for stair and fireplace details (Fig. 47-11).

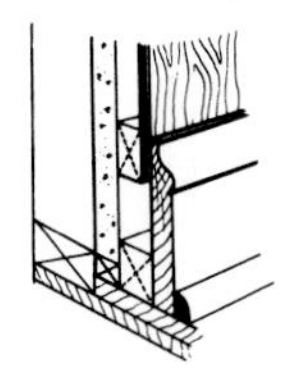
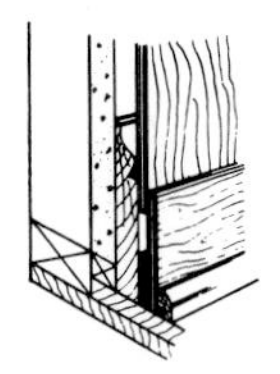
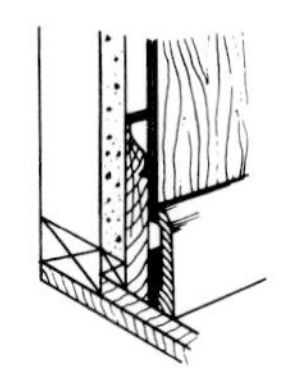

Fig. 47-10A Pictorial sections through the base of an interior partition.

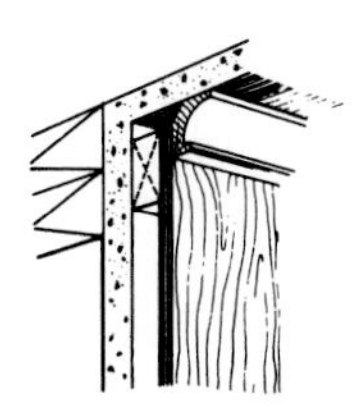
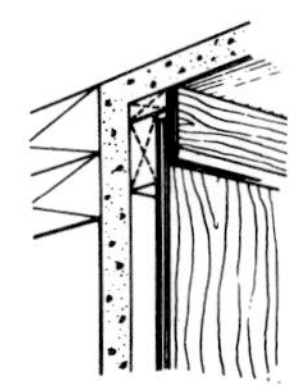
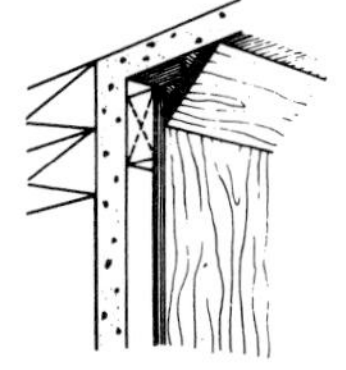

Fig. 47-10B Pictorial sections at the intersection of interior wall and ceiling. *(United States Plywood Corp.)*

HORIZONTAL-WALL SECTIONS

Horizontal-wall sections of interior and exterior walls can be drawn to clarify wall-framing construction.

Exterior Walls

A floor plan is a horizontal section. However, many details are omitted from the floor plan because of the small scale used. Very few construction details are necessary to interpret adequately the floor plan. If the floor plan is drawn exactly as a true horizontal section, it will appear similar to the sections shown in Fig. 47-12. When more information is needed to describe the exact construction of the outside corners and the intersections between interior partitions and outside walls, horizontal sections of the type shown in Fig. 47-12 are prepared. These are known as stud layout details.

Interior-Wall Sections

Typical sections are often drawn of interior-wall intersections. Unusual wall-construction methods are *always* sectioned. For example, a horizontal section is needed to show the inside corner construction of a paneled wall (Fig. 47-13). An outside corner section of paneling construction is shown in Fig. 47-14. Horizontal sections are also used extensively to show how paneled joints and other building joints are constructed (Fig. 47-15). Horizontal sections are also very

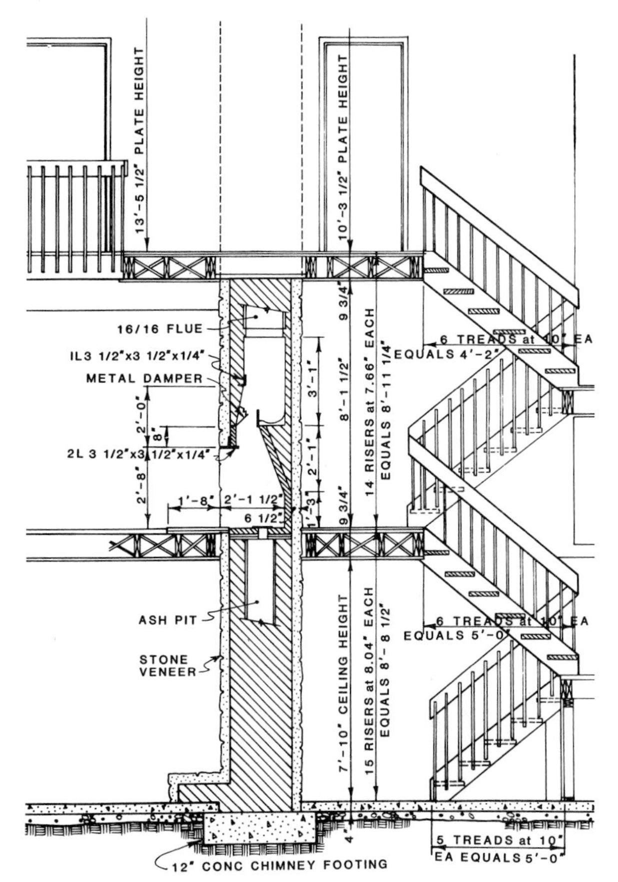

Fig. 47-11 Vertical wall section.

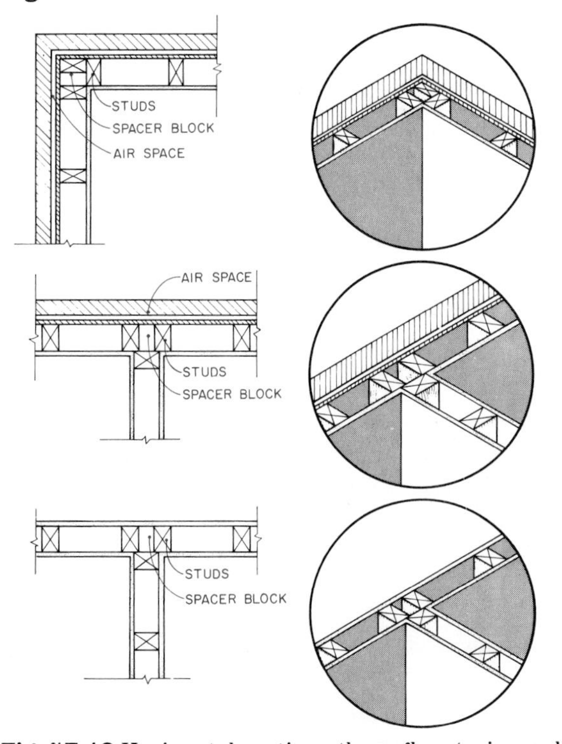

Fig. 47-12 Horizontal sections through exterior-wall intersections.

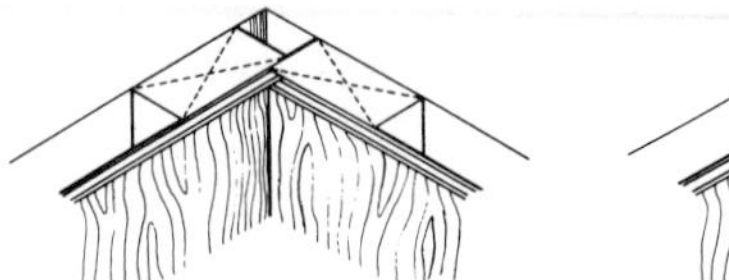
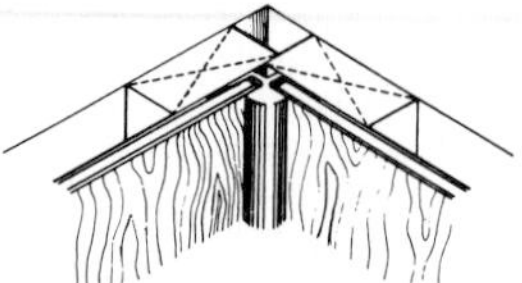

Fig. 47-13 Two sections of an inside paneled-wall corner. *(United States Plywood Corp.)*

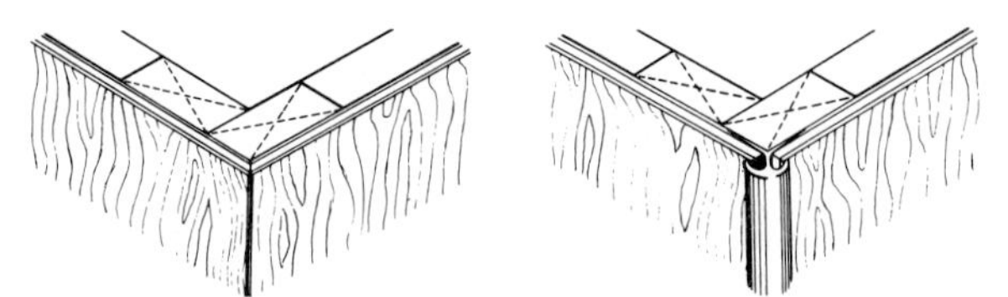

Fig. 47-14 Two sections of an outside paneled-wall corner. *(United States Plywood Corp.)*

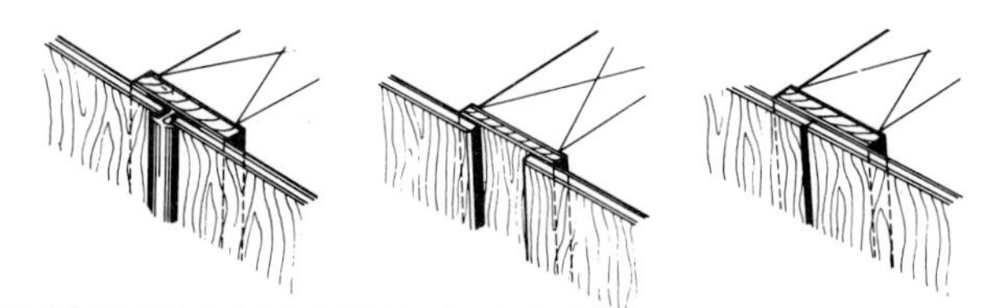

Fig. 47-15 Panel-joint details.

effective in illustrating the various methods of attaching building materials together. For example, the sections shown in Fig. 47-16 illustrate the various methods for attaching furring and paneling to interior walls.

WINDOW SECTIONS

Because much of the actual construction of most windows is hidden, a section is necessary for the correct interpretation of window-construction methods. Figure 47-17 shows the window areas commonly sectioned. These include the head, jamb, and sill construction.

Vertical Sections

Sill and head sections are vertical sections. Preparing sill and head sections on the same drawing is possible only when a small scale is used.

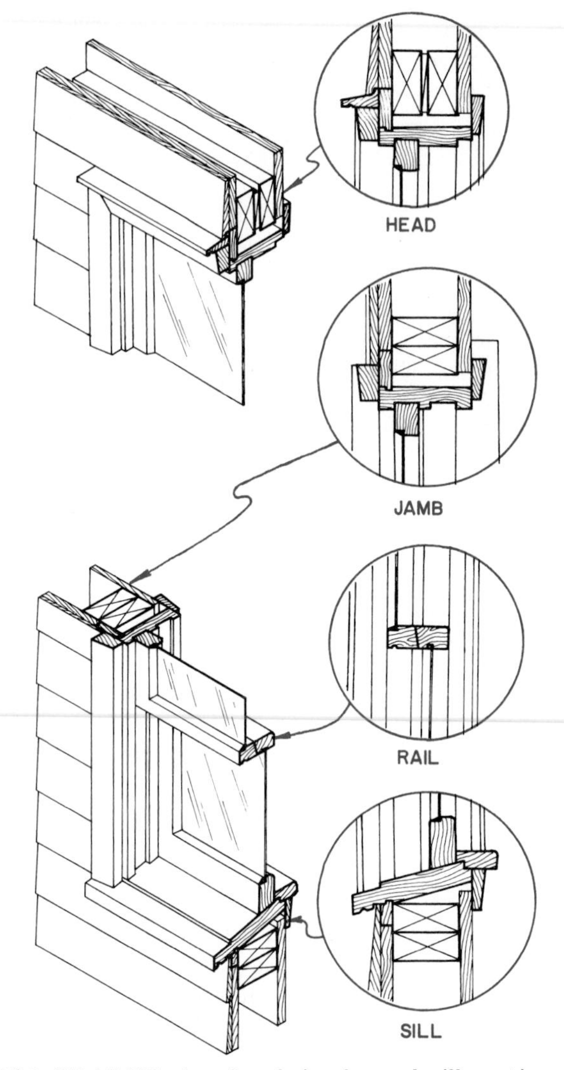

Fig. 47-17 Window head, jamb, and sill sections.

If a larger scale is needed, the sill and head must be drawn independently, or a break line must be used. Figure 47-18 shows the relationship between the cutting-plane line and the sill and head sections. The circled areas in Fig. 47-19 show the areas that are removed when a separate head and sill section is prepared.

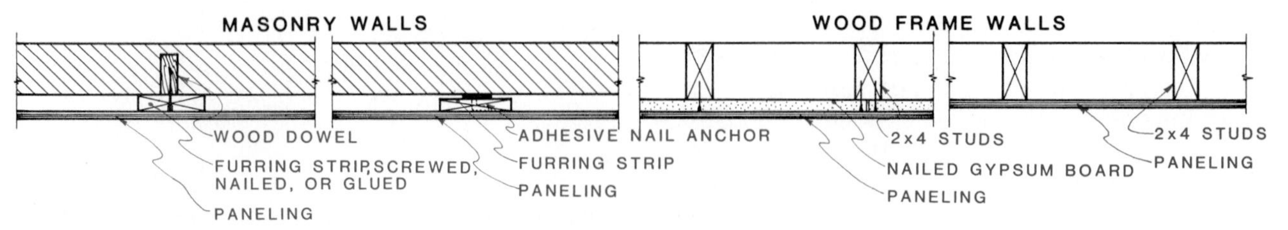

Fig. 47-16 Methods of attaching paneling shown in section.

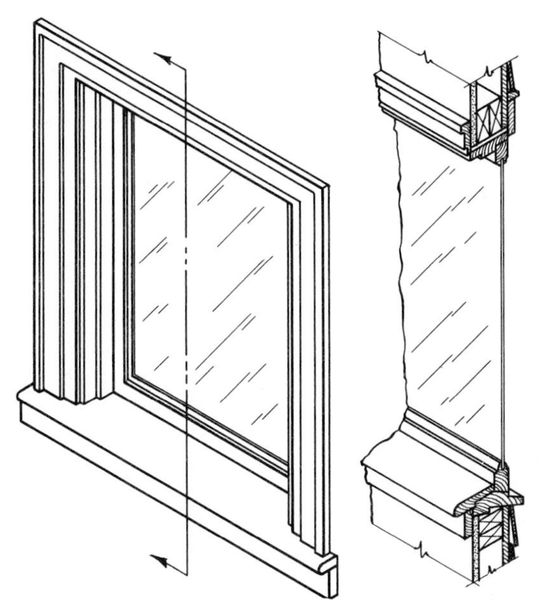

Fig. 47-18 Head and sill sections are in the same plane.

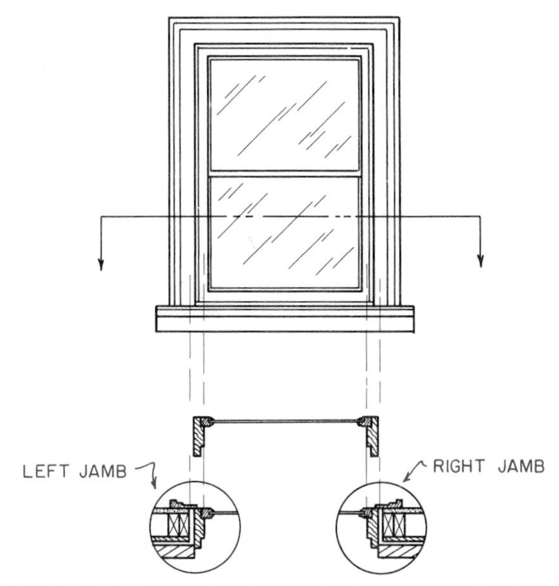

Fig. 47-20 The projection of the left- and right-jamb section.

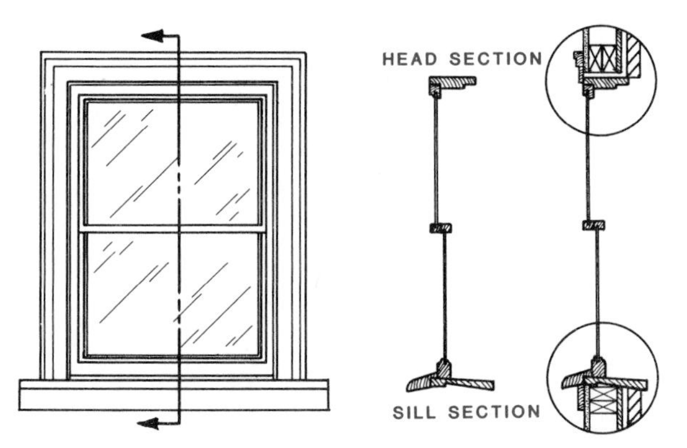

Fig. 47-19 The projection of the head and sill section.

Horizontal Sections

When a cutting-plane line is extended horizontally across the entire window, the resulting sections are known as *jamb sections*. Figure 47-20 shows the method of projecting the jamb details from the window-elevation drawing. Since the construction of both jambs is usually the same, the right jamb drawing is the reverse of the left. Only one jamb detail is normally drawn. The builder interprets the one jamb as the reverse of the other.

Commercial Details

Many window manufacturers use pictorial sectioning techniques to show the correct installation of windows. The relationship between a manufactured window and the framing methods necessary for correct fitting is shown in Fig. 47-21.

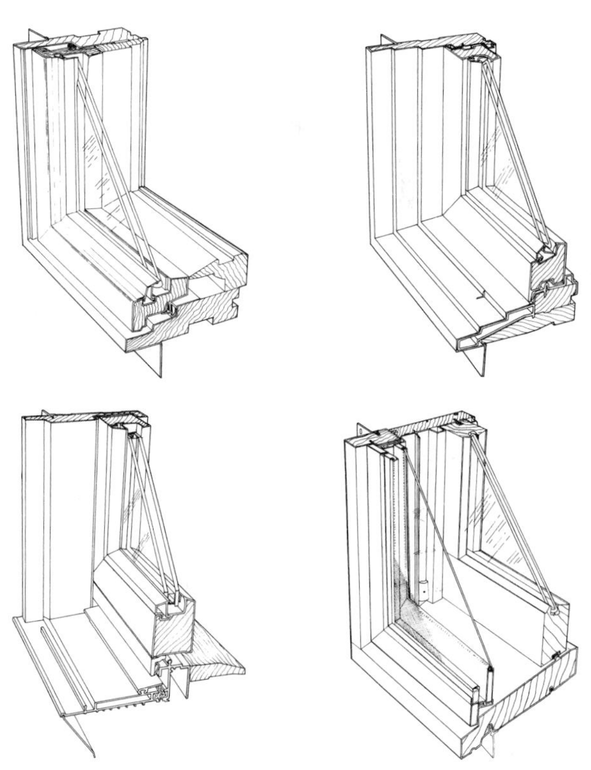

Fig. 47-21 A pictorial section showing sill, jamb, and head details. *(Perma-shield)*

DOOR SECTIONS

A horizontal section of all doors is shown on a floor plan. However, this section is symbolic and lacks sufficient detail for installation. An enlarged jamb, head, and sill section, as shown in Fig. 47-22, is necessary to show door construction. When a cutting-plane line is extended vertically through the sill and head, a section similar to the one shown in Fig. 47-23 is revealed. However, these sections are often too small to show the desired degree of detail necessary for construction. A removed section, as shown in Fig. 47-24, is drawn to show the enlarged head and sill sections.

Since doors are normally not as wide as they are high, an adequate jamb detail can be projected, as shown in Fig. 47-25, without the use of break lines or removed sections. Figure 47-26 shows the method of projecting the left and right jamb sections from the door elevation drawing. Occasionally, architectural drafters prepare sectional drawings of the rough framing details of the door head, sill, and jamb, exclusive of the door and door frame assembly. In such drawings, the drafter draws the framing section with

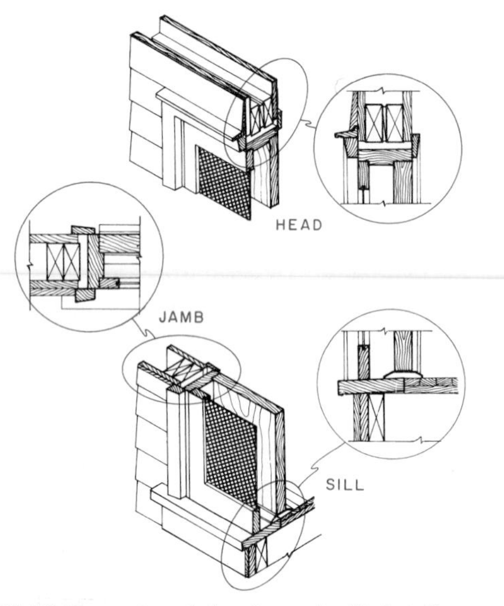

Fig. 47-22 Door, head, jamb, and sill details can be removed or drawn pictorially.

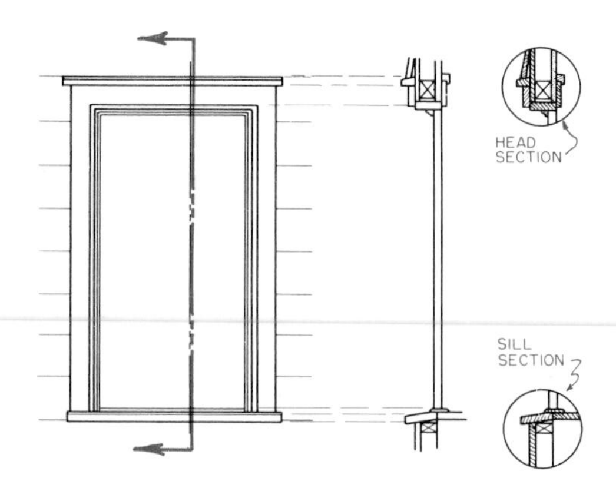

Fig. 47-24 The projection of the head and sill sections of a door frame.

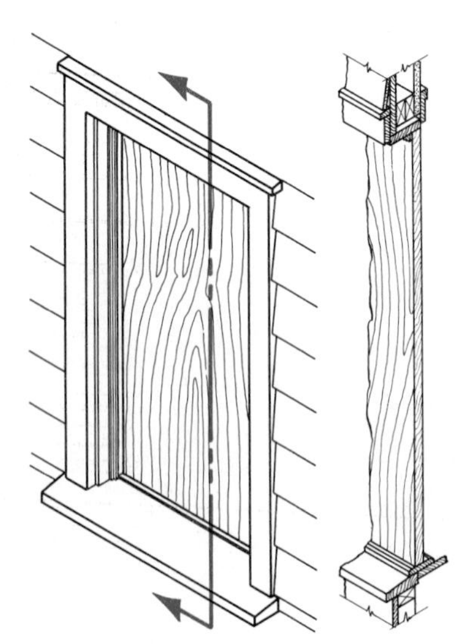

Fig. 47-23 The head and sill sections of a door, in the same plane.

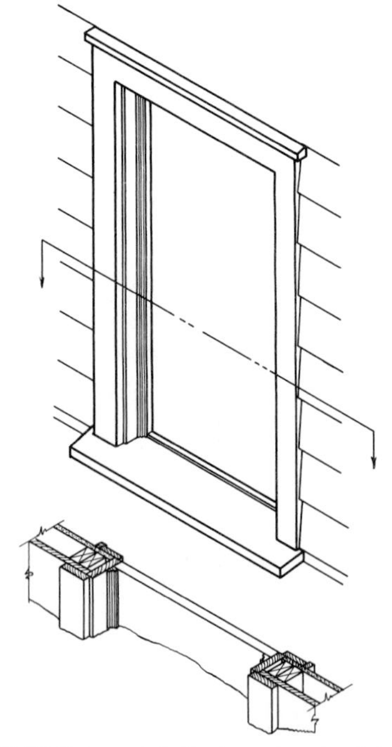

Fig. 47-25 Right and left door-jamb details are in the same plane.

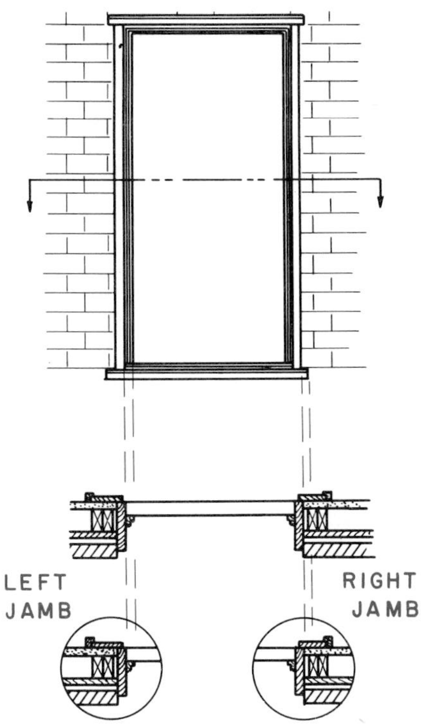

Fig. 47-26 The projection of the left and right door-jamb section.

the door frame and door removed. Usually, however, door sections are prepared with the framing trim and door in their proper locations.

Drawings of garage doors and industrial-size doors are usually prepared with sections of the brackets and apparatus necessary to house the door assembly, as shown in Fig. 47-27. This is done even when stock doors are used. Rarely is the architectural drafter called upon to prepare sectional drawings of internal door-construction details. Most doors are purchased from manufacturers' stock. Occasionally, doors are supplied by the manufacturer specifically for a building. Only when a special door is to be manufactured is a sectional drawing prepared of the internal detail of the door construction (Fig. 47-28).

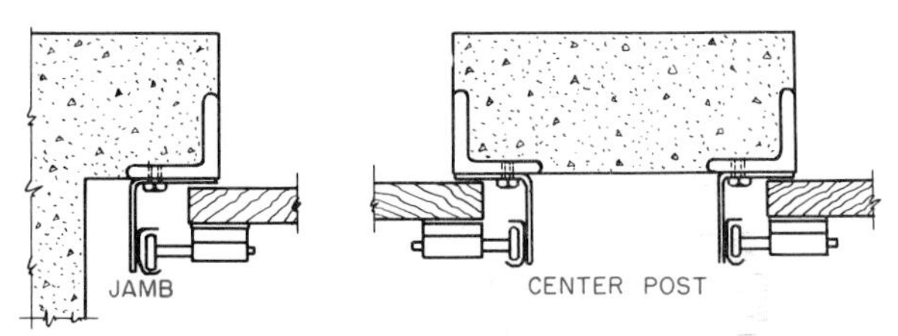

Fig. 47-27 Special brackets and devices often require detailed sectional drawing.

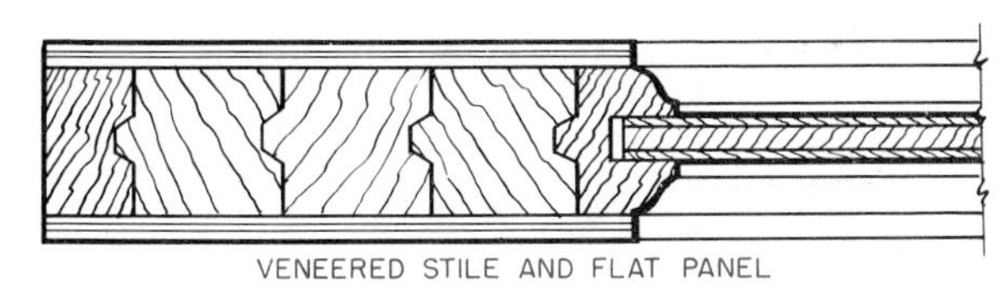

Fig. 47-28 Interior door construction is shown by a sectional drawing.

Exercises

1. **Draw a large cornice, sill, and footing section from the circled sections shown in Fig. 47-3.**
2. **Draw a section through the girder, as shown in Fig. 47-8, revolving the cutting plane line 90°.**
3. **Draw a head, jamb, and sill section of a typical window and a typical door of the house you have designed.**

4. **Draw a sill, cornice, and footing section of the house you have designed.**
5. **Draw a detail section of the intersection of the inside foundation-support wall, I beam, and interior partition, as shown in Fig. 47-9.**
6. **Define these terms: *break line, removed section, interior-wall sections, vertical-wall sections, horizontal-wall sections, jamb sections, head sections, sill sections.***

SECTION 12

Foundation Plans

The methods and materials used in constructing foundations vary greatly in different parts of the country and are continually evolving. The basic principles of foundation construction are the same, regardless of the application.

Every structure needs a foundation. The function of a foundation is to provide a level and uniformly distributed support for the structure. The foundation must be strong enough to support and distribute the load of the structure, and remain level to prevent the walls from cracking and the doors and windows from sticking. The foundation also helps to prevent cold air and dampness from entering the house. The foundation waterproofs the basement and forms the supporting walls of the basement.

UNIT 48

FOUNDATION MEMBERS

The structural members of the foundation vary according to the design and size of the foundation.

FOOTING

The *footing*, or *footer* (Fig. 48-1), distributes the weight of a building over a large area. Concrete is commonly used for footings because it can be poured to maintain a firm contact with the supporting soil. Concrete is also effective because it can withstand heavy weights and is a relatively decay-proof material. Steel reinforcement is sometimes added to the concrete footing to keep the concrete from cracking and to provide additional support. The footing must be laid on solid ground to support the weight of the building effectively and evenly. In cold climates, the footing must be placed below the frost line. Always consult the local building code for frost line foundation depth requirements before establishing footing depths.

FOUNDATION WALLS

The function of the *foundation wall* is to support the load of the building above the ground line and to transmit the weight of the house to the footing. Foundation walls are normally made of concrete, stone, brick, or concrete block (Fig. 48-2). When a complete excavation is made for

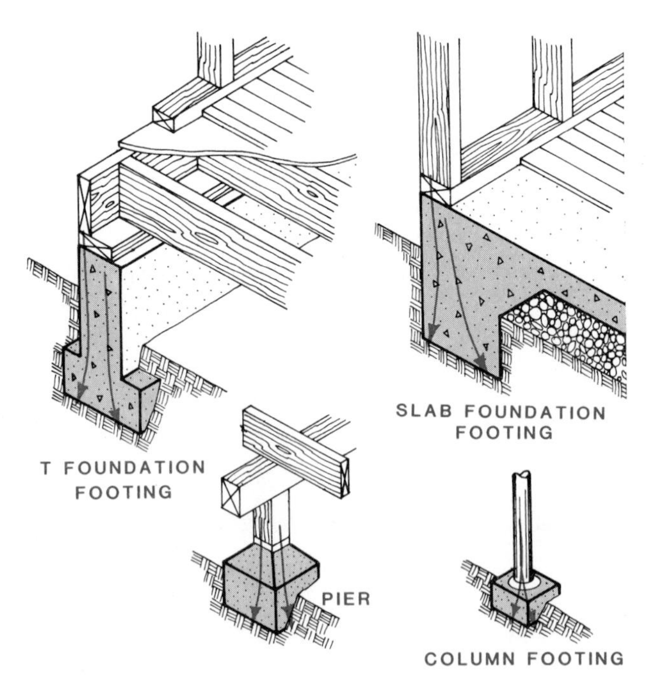

Fig. 48-1 Footings distribute the weight of the building over a wide area.

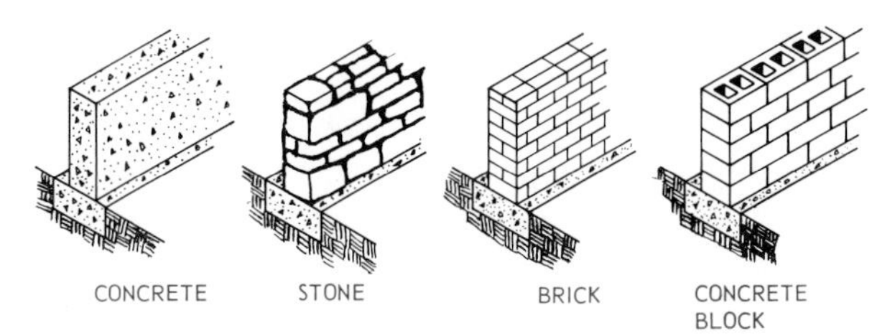

Fig. 48-2 Foundation walls are constructed of concrete, stone, brick, or concrete block.

a basement, foundation walls also provide the walls of the basement (Fig. 48-3).

PIERS AND COLUMNS

Piers and *columns* are vertical members, usually made of concrete, brick, steel, or wood. They are used to support the floor systems (Fig. 48-4). Piers or columns may be used as the sole support of the structure. They also may be used in conjunction with the foundation wall and provide only the intermediate support between girders or beams, as shown in Fig. 48-5.

ANCHOR BOLTS

Anchor bolts are embedded in the top of the foundation walls or piers (Fig. 48-6A and B). The exposed part of the bolt is threaded so that the first wood member (the sill) can be bolted onto the top of the foundation wall. Anchor bolts for

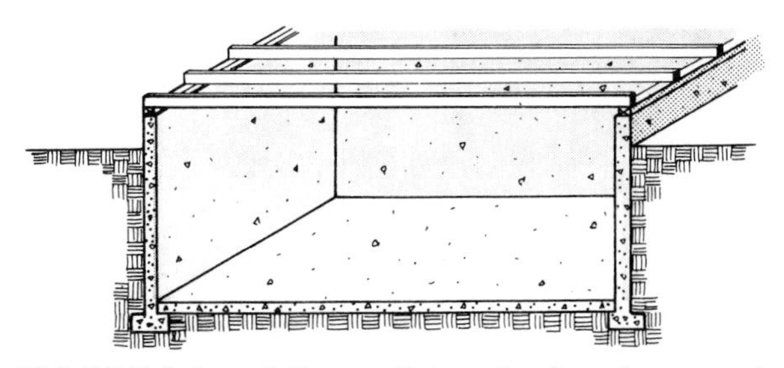

Fig. 48-3 A foundation wall can also be a basement wall.

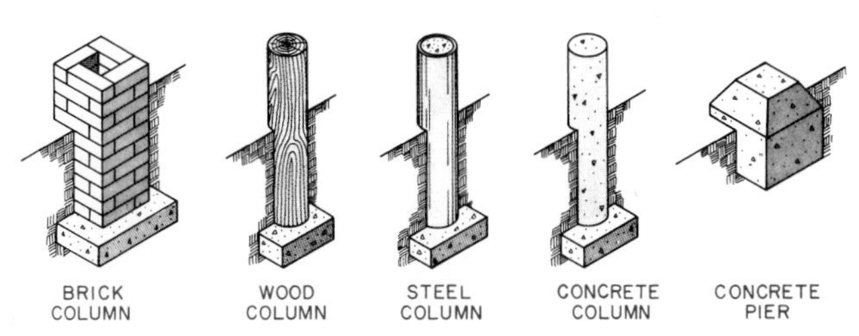

Fig. 48-4 Piers and columns are made of concrete, brick, steel, or wood.

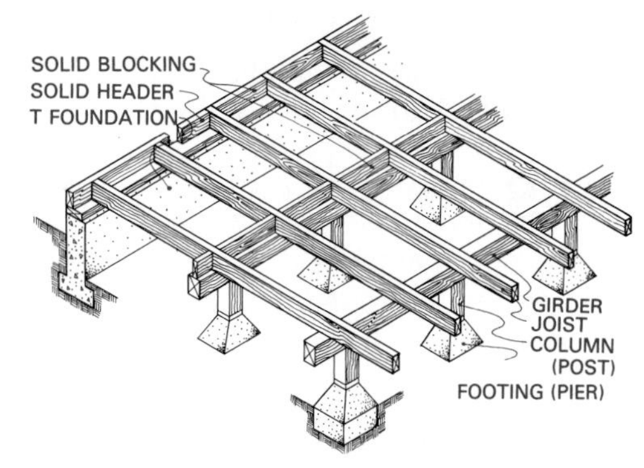

Fig. 48-5 Piers used as intermediate support.

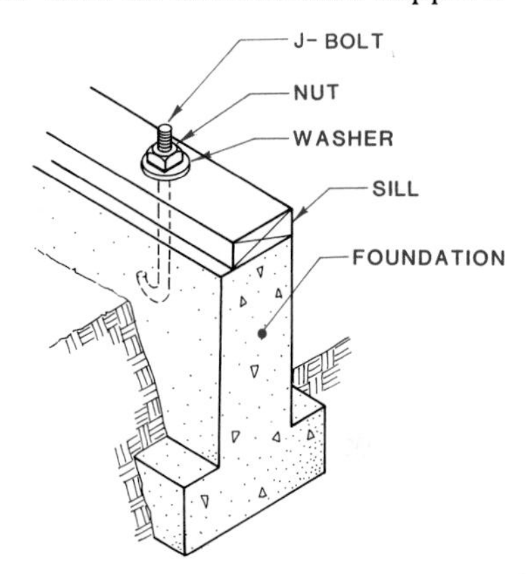

Fig. 48-6A Anchor bolts hold the sill to the foundation.

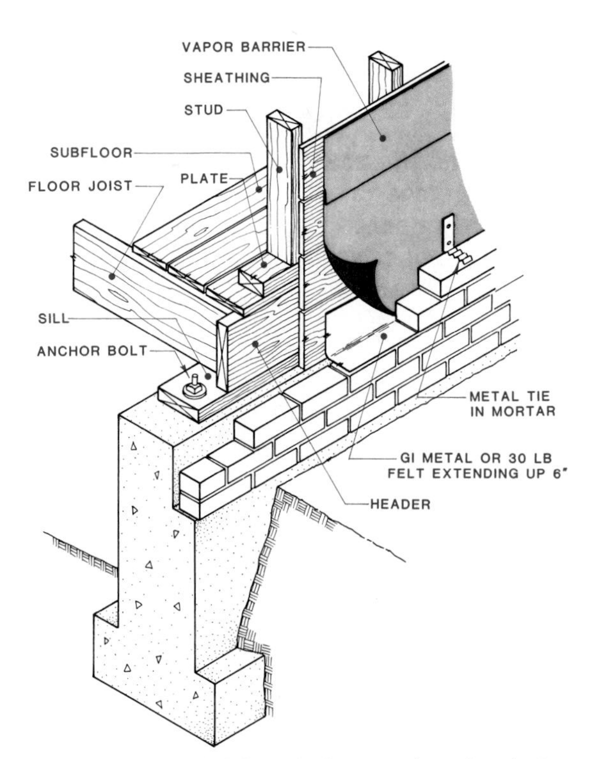

Fig. 48-6B Pictorial foundation section showing an anchor bolt.

residential use are ½″ or ⅝″ in diameter and 10″ long. They are spaced approximately 6′ or 8′ apart, starting 1′ from each corner.

SILLS

Sills are wood members that are fastened with anchor bolts to the foundation walls (Fig. 48-7). Sills provide the base for attaching the floor system to the foundation. A galvanized iron sheet is often placed under the sill to check termites (Fig. 48-8), since the sill is normally the lowest wood member used in the construction. If galvanized iron is not used then the sill must be treated to prevent dry rot or termite damage. This also applies to all wood in contact with masonry, concrete, or soil. Building laws specify the distance required from the bottom of the sill to the grade line inside and outside the foundation.

POSTS

Posts are vertical wood members that support the weight of girders or beams and transmit the weight to the footings (Fig. 48-9). The terms, post or column, are often used interchangeably.

CRIPPLES

Cripples are used to raise the floor level without the use of a higher foundation wall (Fig. 48-10). Since the load of the structure must be transmitted through the cripples, these are usually heavy members, often four-by-four (4 x 4′s) spaced at close intervals or two-by-fours (2 x 4′s) spaced closer than the normal 16″ OC.

GIRDERS

Girders are major horizontal support members upon which the floor system is supported. They

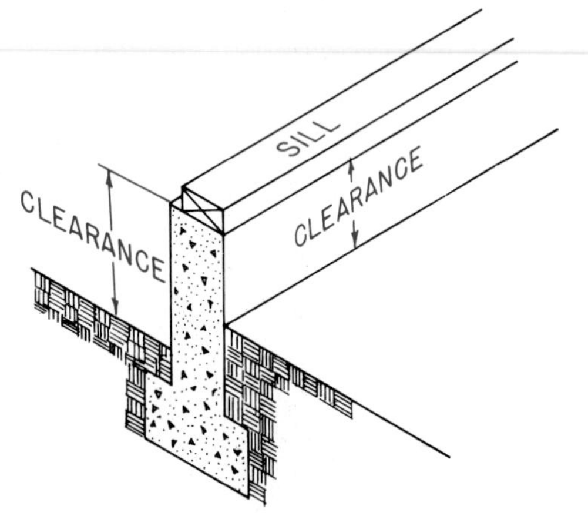

Fig. 48-7 The sill is the point of contact between the foundation and the framework of the building.

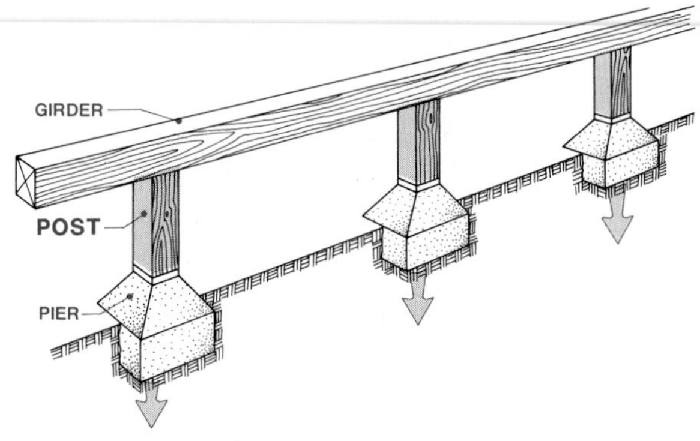

Fig. 48-9 Posts transmit the weight of girders and beams to the footings.

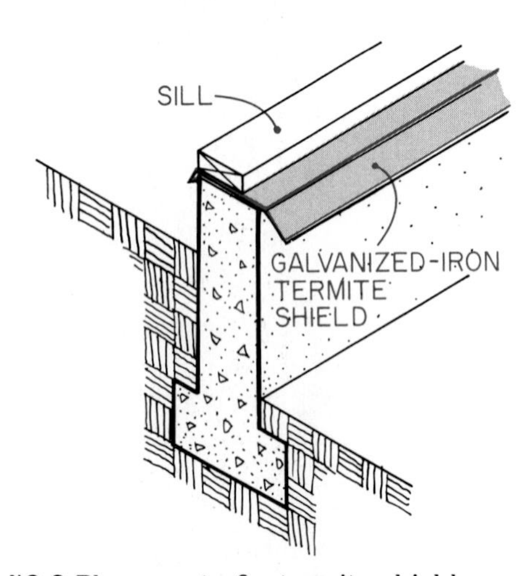

Fig. 48-8 Placement of a termite shield.

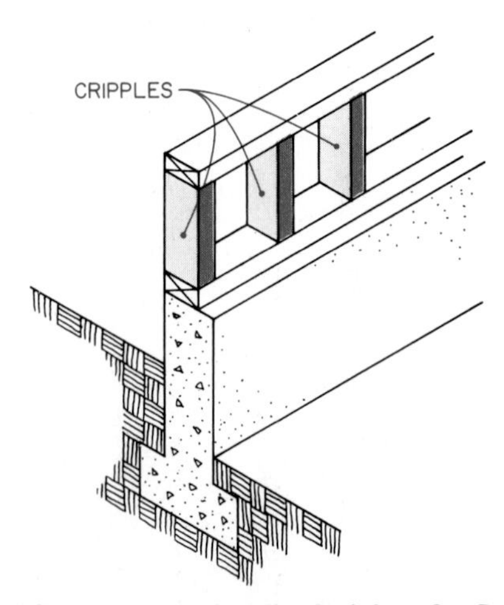

Fig. 48-10 Cripples raise the height of a floor without raising the foundation height.

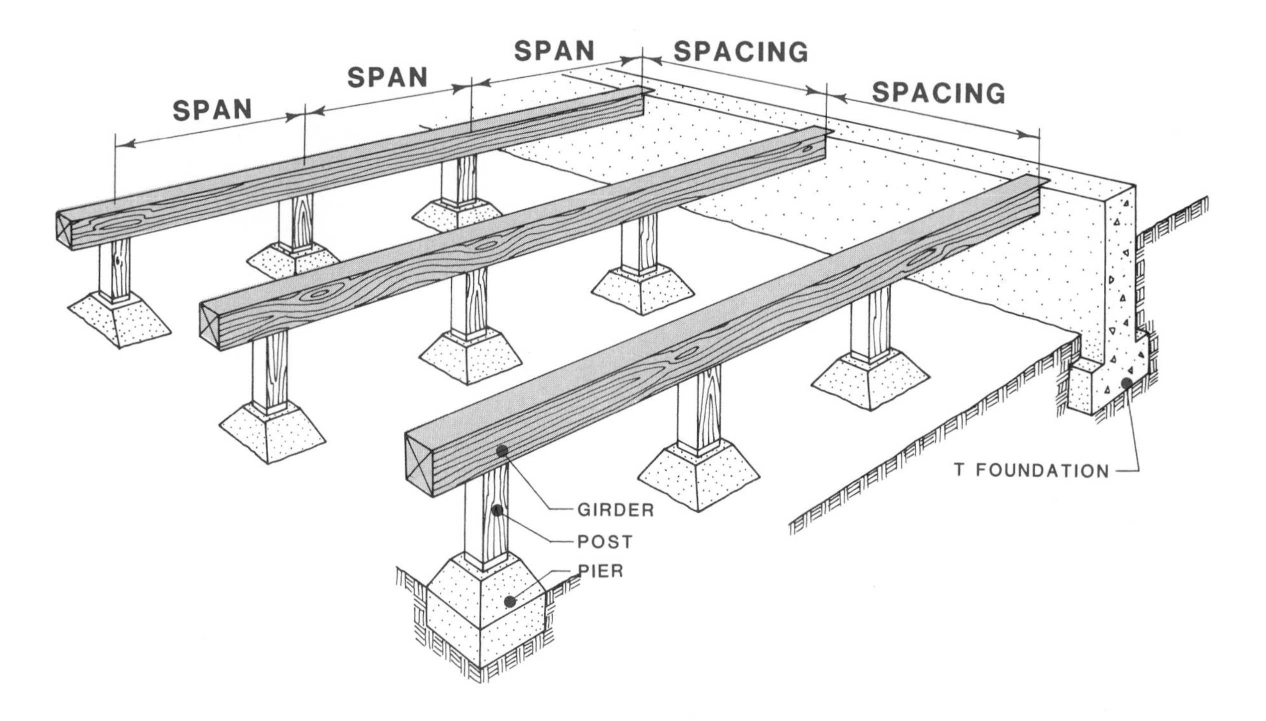

TYPICAL SPANS FOR WOOD GIRDERS

SIZE	I STORY	2 STORY
4" x 6"	5'-0"	4'-0"
6" x 6"	6'-0"	5'-0"
4" x 8"	6'-6"	5'-6"
6" x 8"	8'-0"	7'-0"
4" x10"	8'-0"	7'-0"
6" x10"	9'-0"	7'-0"

Fig. 48-11 Girders are major horizontal support members.

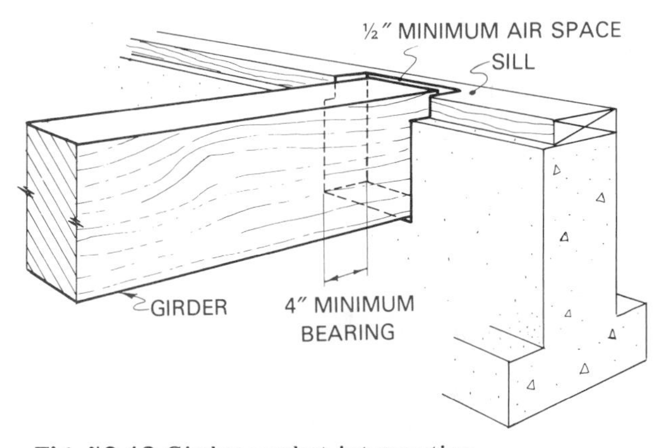

Fig. 48-12 Girder pocket intersection.

in turn are supported by posts and piers and are secured to the foundation wall as shown in Fig. 48-11. Girder sizes are closely regulated by building codes. The allowable span of the girder depends on the size of the girder. A decrease in the size of a girder means that the span must be decreased by adding additional column supports under the girder. Built-up wood girders for residential construction are normally made from 2 x 6's, 2 x 8's or 2 x 10's spiked together. Figure 48-12 shows a girder intersecting a foundation wall in a girder pocket.

STEEL BEAMS

Steel beams perform the same function as wood girders. But, steel beams can span larger areas than can wood girders of an equivalent size.

JOISTS

Joists are the parts of the floor system that are placed perpendicular to the girders as shown in Fig. 48-13. Joists span either from girder to girder or from girder to the foundation wall. The ends of the joists butt against a header or extend to the end of the sill, with blocking usually placed between them.

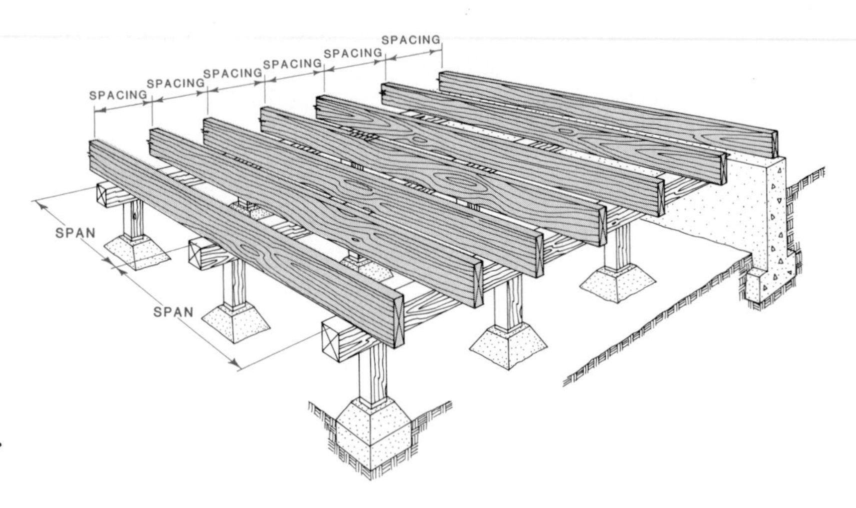

Fig. 48-13 Joists support the floor and rest on girders.

PILASTERS

Pilasters are reinforcements in a wall designed to provide more rigidity without increasing the width of the wall along the entire length. Pilasters are also used for girder end support instead of making a girder pocket in the foundation. Figure 48-14 shows a typical pilaster in a concrete foundation wall. Building codes normally specify the size and spacing of pilasters, depending on the wall width, height, and material.

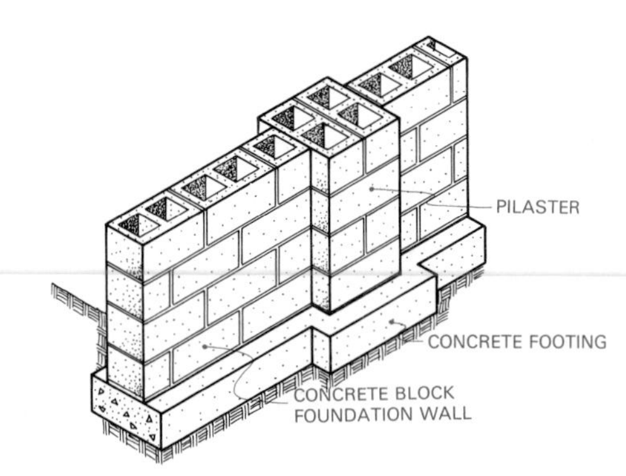

Fig. 48-14 Concrete block foundation pilaster.

Exercises

1. **Draw a slab-foundation plan to the scale ¼″ = 1′-0″ for the floor plan shown in Fig. 6-2.**
2. **Draw a T-foundation plan for Fig. 48-15, using the scale ¼″ = 1′-0″.**
3. **Draw a slab foundation for Fig. 48-15, using the scale ¼″ = 1′-0″.**
4. **Draw the foundation plan for the house you are designing.**
5. **Know these architectural terms: *foundation, structural members, footing, concrete, foundation wall, pier, column, anchor, sill, post, cripple, girder, span, spacing, joists, pilaster, beam, spiked.***

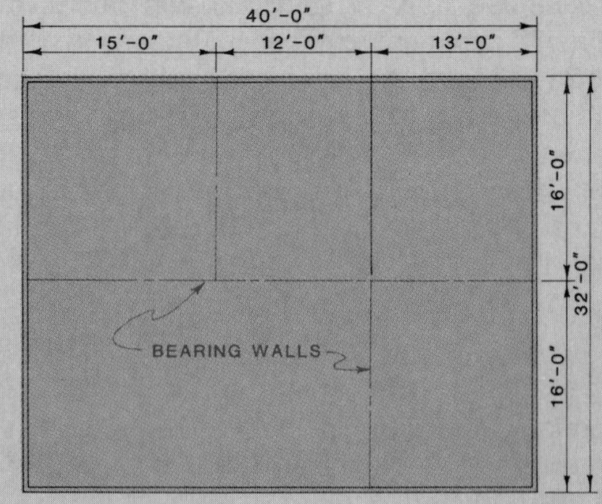

Fig. 48-15

UNIT 49

FOUNDATION TYPES

The type of foundation the architect selects for a structure depends on the nature of the soil, the size and weight of the structure, the climate, building laws, and the relationship of the floor to the grade line (Fig. 49-1). Foundations are divided into three basic types: the T foundation, the slab foundation, and the pier-and-column foundation (Fig. 49-2).

T FOUNDATIONS

The *T foundation* consists of a footing upon which is placed a concrete wall or a concrete-block wall. The combination of the footing and the wall forms an inverted T. The T foundation is popular in structures with basements or when the bottom of the first floor must be accessible. Figure 49-3 shows T-foundation intermediate supports used to provide basement headroom.

The details of construction relating to the T foundation are shown in Fig. 49-4A. The methods of representing this construction on a foundation plan are shown in Fig. 49-4B.

SLAB FOUNDATIONS

A *slab foundation* is a poured solid slab of concrete. A slab is a monolithic object. A slab is poured directly on the ground, with footings placed where extra support is needed. The depth and width of footings will vary with soil conditions, size of structure and building codes.

A slab is poured above grade for multilevel structures. Slab foundations are ideally suited

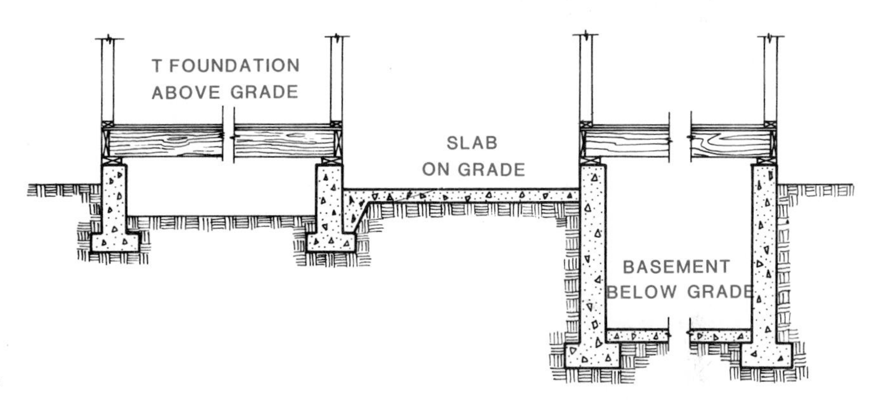

Fig. 49-1 Foundation positions.

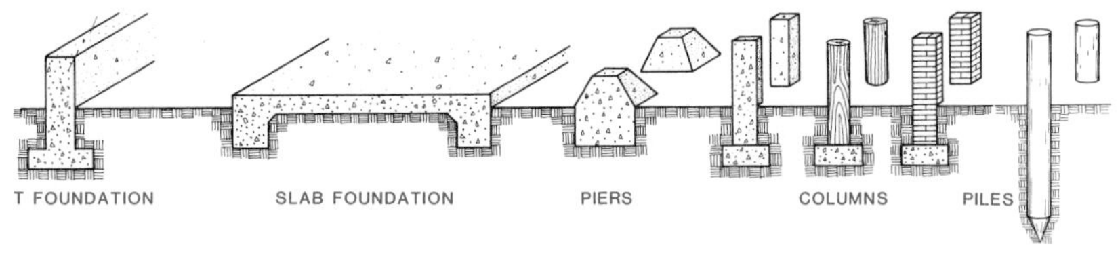

Fig. 49-2 Types of foundations.

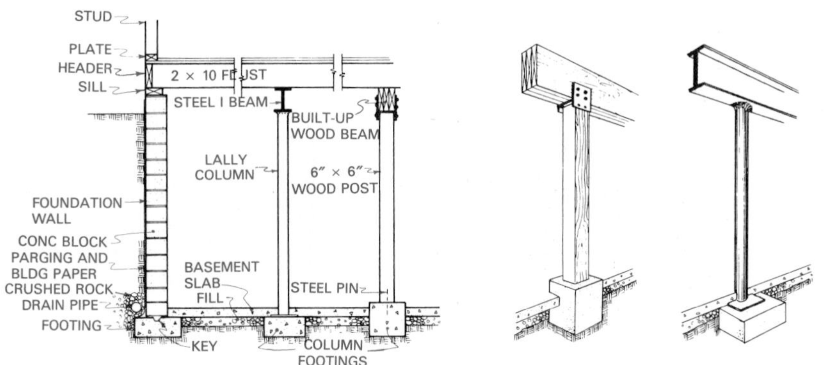

Fig. 49-3 T foundation.

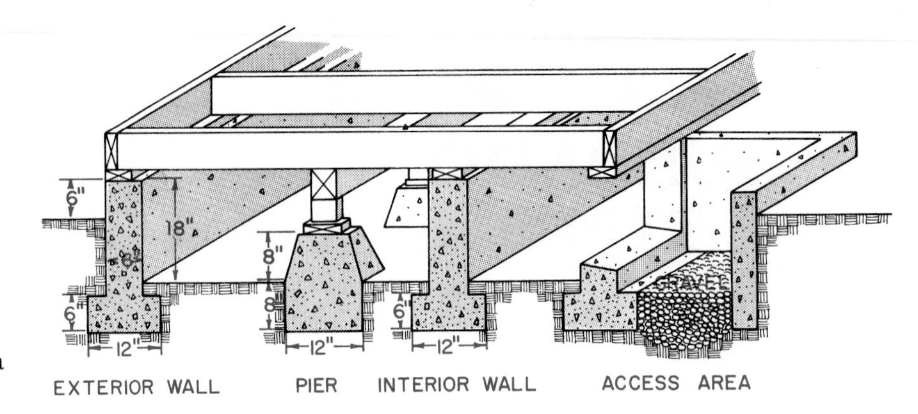

Fig. 49-4A Elements of a T-foundation.

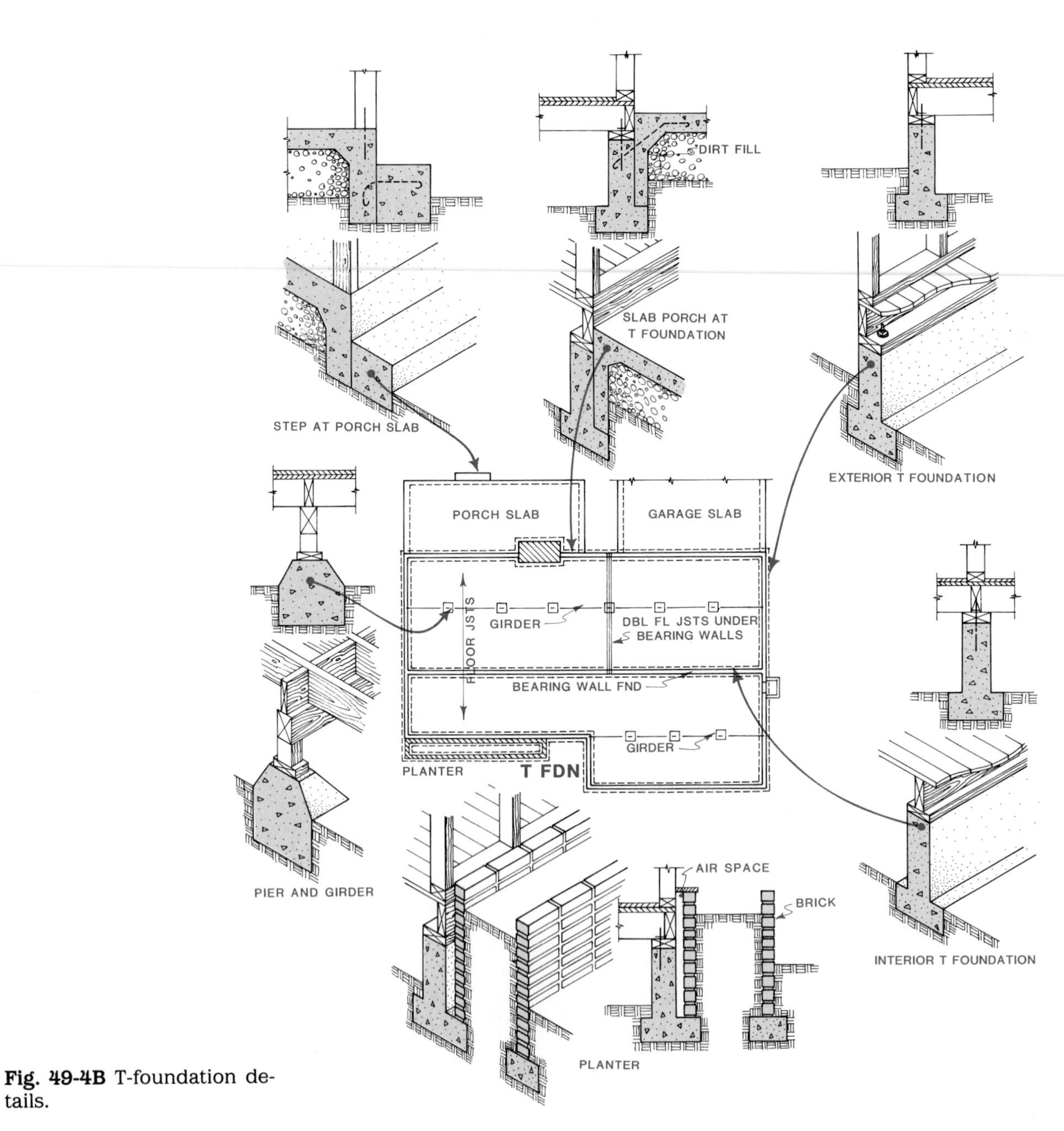

Fig. 49-4B T-foundation details.

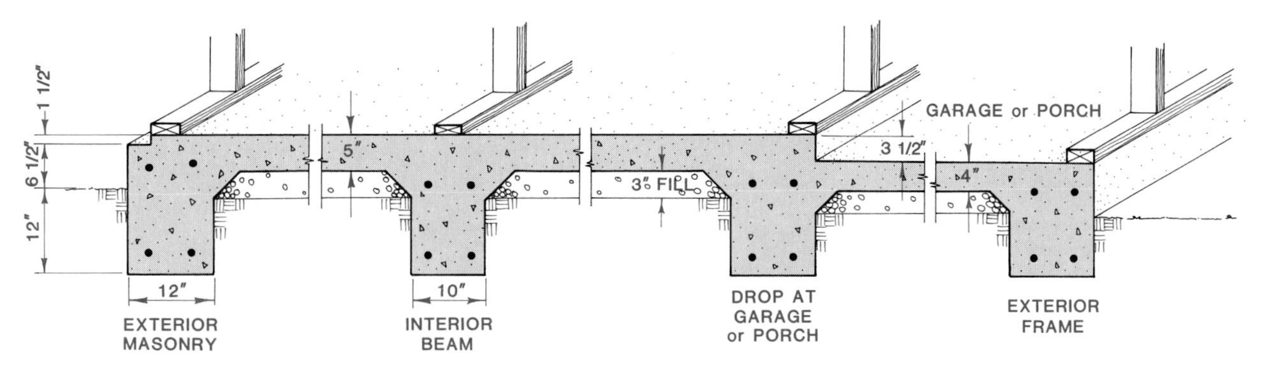

Fig. 49-5A Relationship of T-foundation to slab.

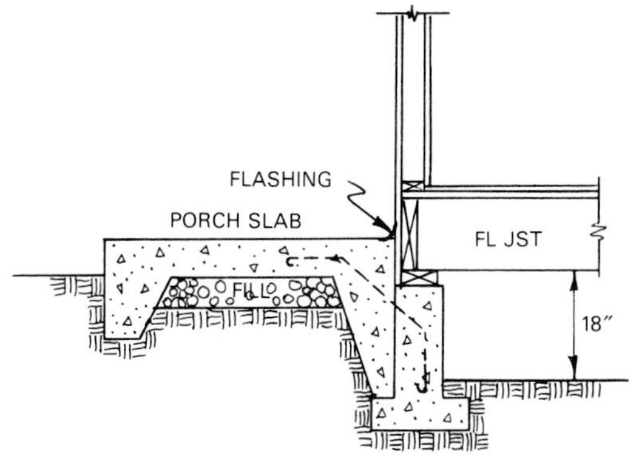

Fig. 49-5B Slab-foundation support methods.

where soil and climate make other types of foundations less practical and less economical. The slab is poured directly on the ground, with footings placed where extra support is needed (Fig. 49-5A). Figure 49-5B shows how exterior slabs tie-in with foundation walls.

A slab foundation requires considerably less labor to construct than do most other foundation types. Details of the slab foundation and methods of drawing slab foundations are shown in Fig. 49-6. Slab foundations lose heat around their perimeter (Fig. 49-7). Therefore special

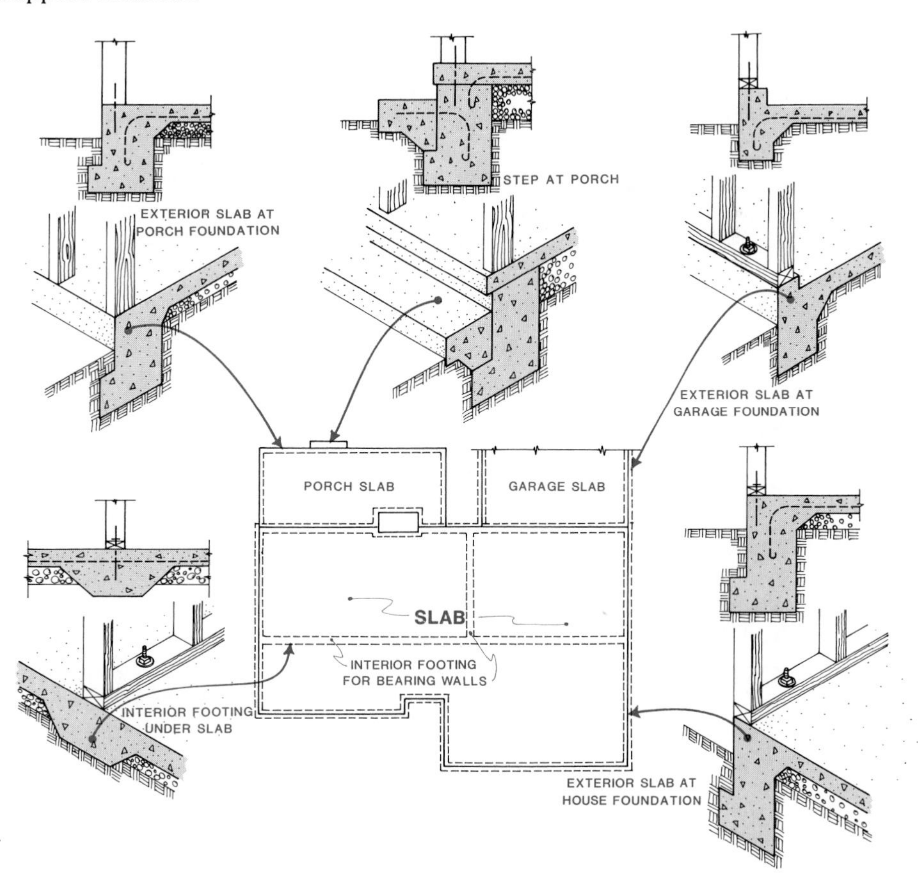

Fig. 49-6 Slab-foundation details.

attention should be paid to design adequate insulation around the perimeter.

PIER-AND-COLUMN FOUNDATIONS

The *pier-and-column foundation* consists of individual footings upon which columns are placed.

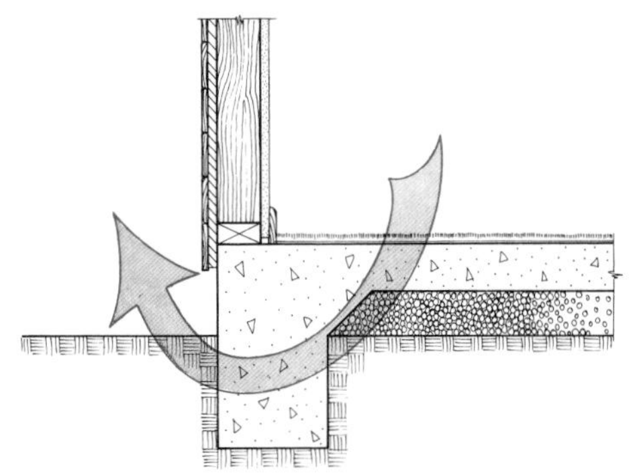

Fig. 49-7 Slabs lose heat around the perimeter.

Fewer materials and less labor are needed for the pier-and-column foundation (Fig. 49-8). The main objection to using pier-and-column foundations for most residence work is that a basement is not possible and the perimeter around the crawl space is open.

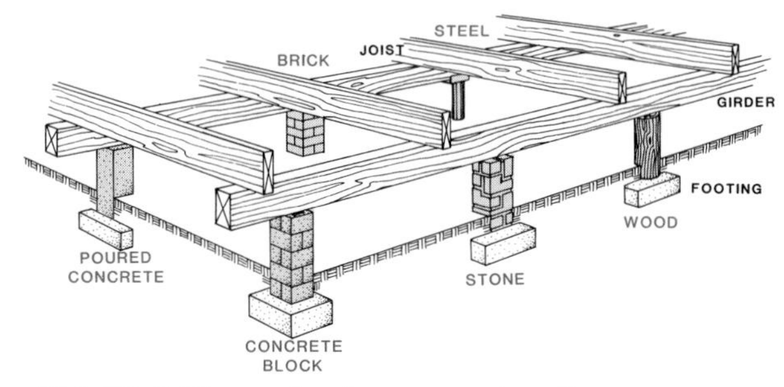

Fig. 49-8 Pier-and-column construction.

Exercises

1. **Sketch Fig. 49-9, using the ¼″ = 1′-0″ for a T foundation.**
2. **Sketch Fig. 49-9, using the ¼″ = 1′-0″ for a slab foundation.**
3. **Sketch Fig. 49-9, using the ¼″ = 1′-0″ for a pier foundation.**
4. **Draw a foundation plan for a building of your design.**
5. **Know these architectural terms: *contractor, grade, slab, T foundation, column, frost line.***

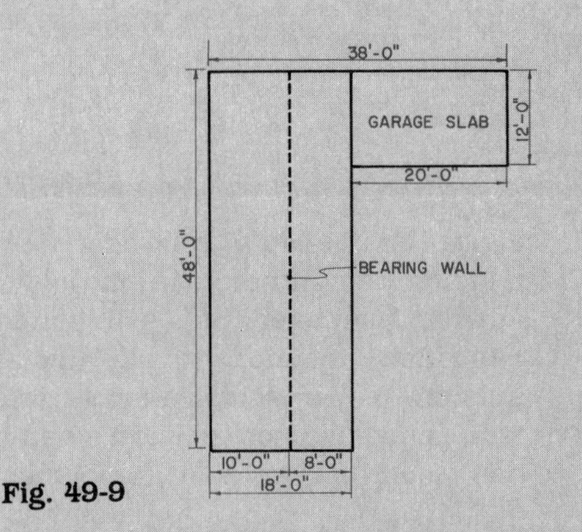

Fig. 49-9

UNIT 50

FOUNDATION DRAWINGS

The designer must be familiar with all methods of foundation construction in order to design the most practical and economical foundation. The designer must choose the most appropriate foundation for the type of soil, climate, and structure to be supported (Fig. 50-1). He or she must prepare working drawings that will facilitate the layout, excavation, and construction of the foundation.

EXCAVATIONS

Foundation plans should clearly show what parts of the foundation are to be completely excavated for a basement, partly excavated for crawl space, or unexcavated. The depth of the excavation should be shown also on the elevation draw-

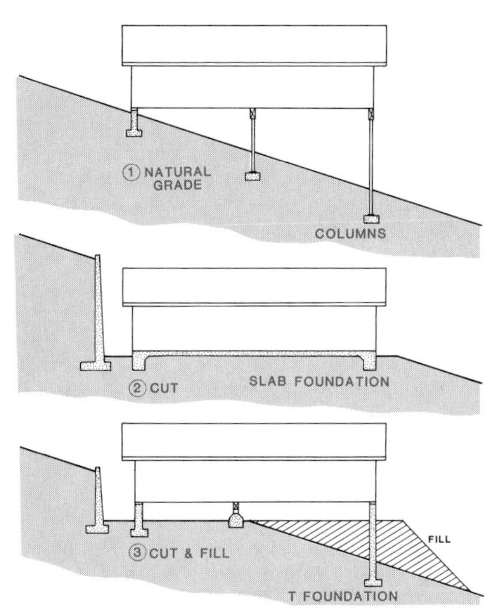

Fig. 50-1 Conditions of the terrain determine the type of foundation used.

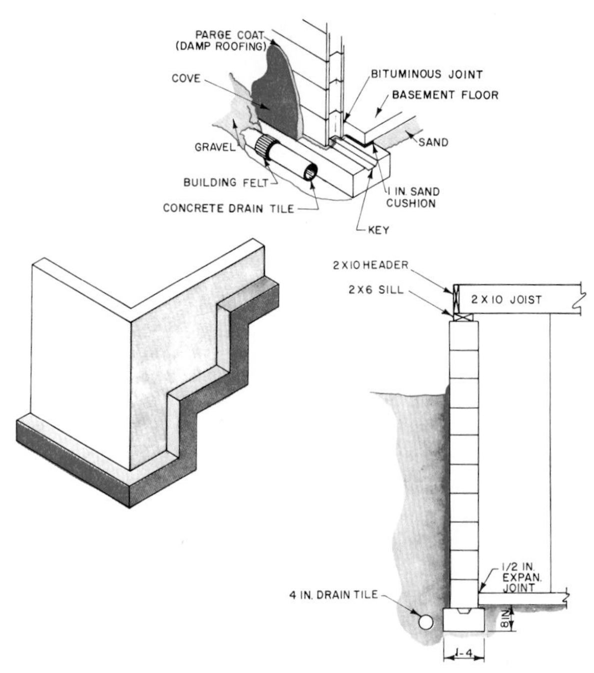

Fig. 50-2 Extending the length of a T foundation forms basement walls.

ings. If a basement is planned, the entire excavation for the basement is dug before the footings are poured. If there is to be no basement, a trench excavation is made.

T FOUNDATIONS

The T foundation is prepared by pouring the footing in an excavated trench, leveling the top of the footing, and erecting a concrete block or masonry wall on top of the footing. If concrete foundation walls are to be used, building forms are erected on top of the footing. Concrete is poured into these forms. After the concrete dries, the wood is removed and may be reused for other forms. When a poured foundation wall is to be used, the concrete mix is sand, gravel, water, and cement. After the forms are filled, the concrete is leveled with a strike board so that it has a rough, nonslip surface. By continuing the depth of the T foundation, a basement area is formed (Fig. 50-2).

SLAB FOUNDATIONS

The excavation for a slab foundation is made for the footings only. Most codes require the top of the slab to be at least 6″ to 8″ above grade. Figure 50-3A shows a slab foundation plan with footing detail sections indexed to Fig. 50-3B. Specifications which must also be included with slab plans include the thickness of the slab, size and spacing of reinforcing bars, and concrete pounds per square inch (lb/in^2) minimums.

Slab foundations and basement floors in T foundations should be waterproofed. This can be done by putting a waterproof membrane between the slab and the ground. Slabs are often reinforced with steel-wire mesh placed inside the slab before pouring. Types and sizes and mixtures of materials for foundations are rigidly controlled by most building codes. The designer must check the building code for the area.

SILL DETAILS

A drawing showing the intersection between exterior foundation walls and the floor system is known as a *sill detail.* The *sill* is the link between the framework of the structure and the foundation. Sill details are sections through the foundation extending through the floor line. Figure 50-4 shows sill details for a brick-veneer structure, and Fig. 50-5 shows a sill detail for solid masonry and cavity masonry walls. Notice that in solid masonry or cavity masonry walls, a *fire cut* is included on all wood members that extend into the masonry. This fire cut is made to eliminate destruction of the masonry wall if the beam collapses, as shown in Fig. 50-6.

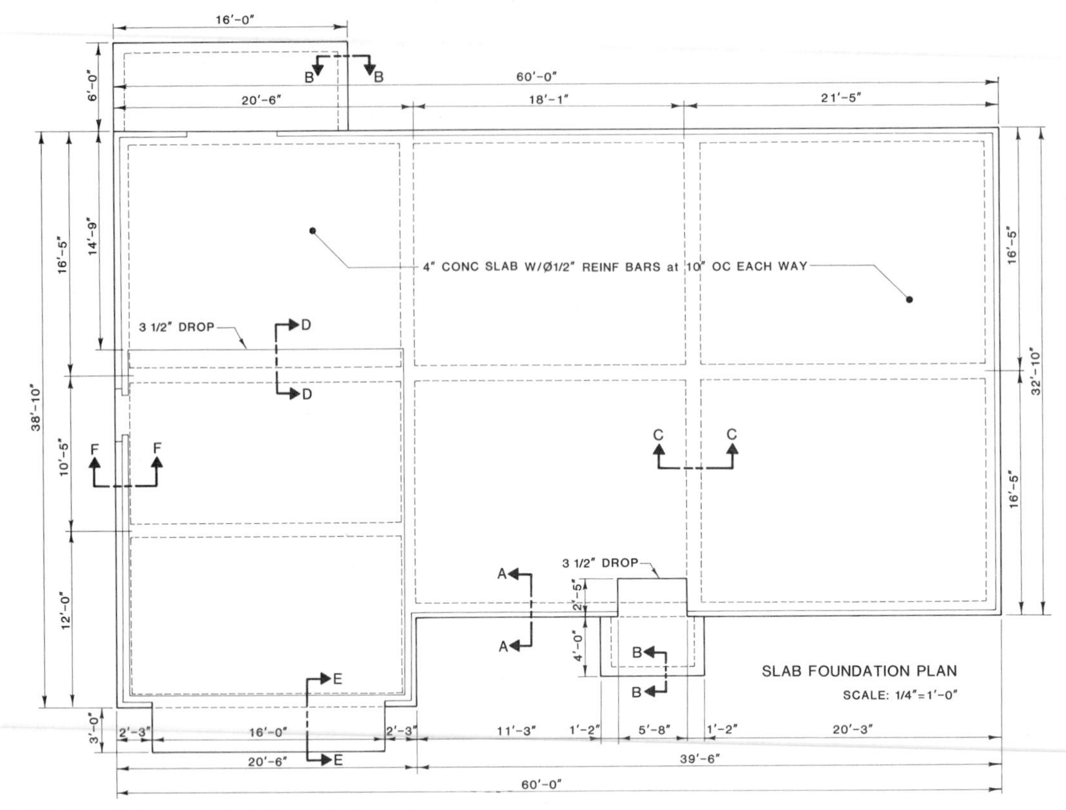

Fig. 50-3A A reinforced slab foundation plan.

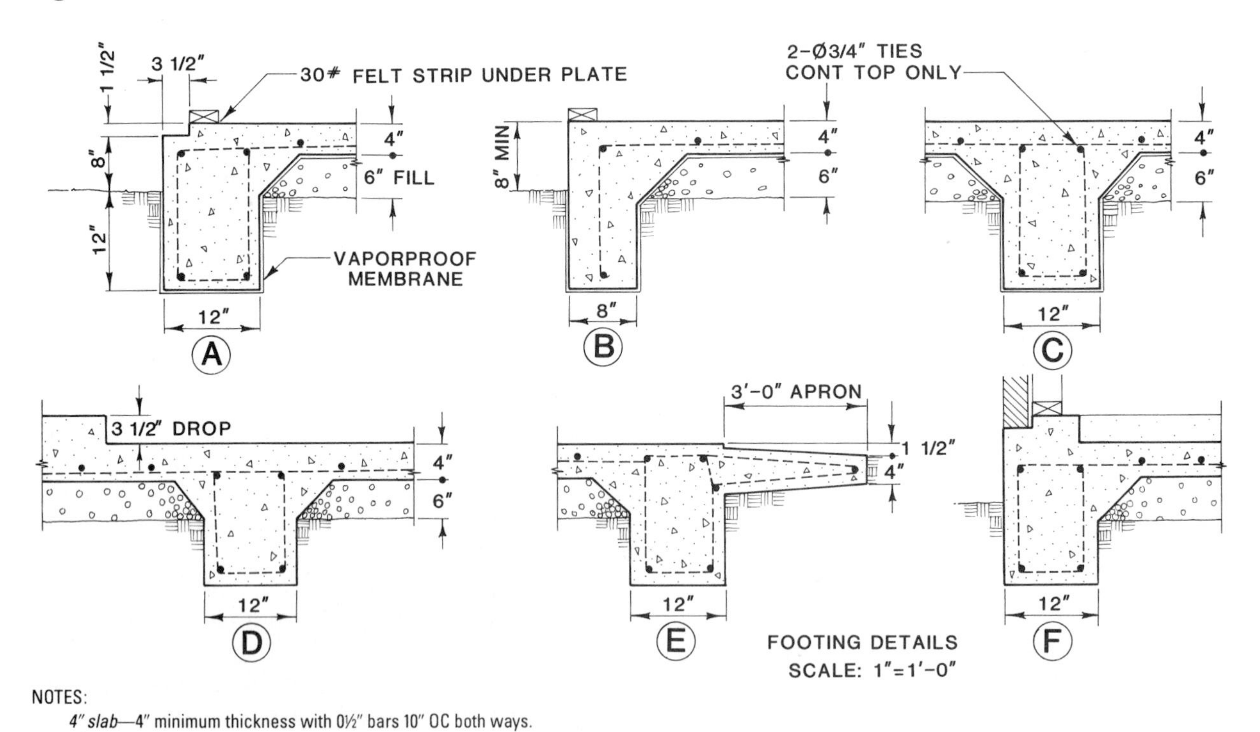

NOTES:

4" slab—4" minimum thickness with 0½" bars 10" OC both ways.
Maximum clear panel between beams is 18' -0".
Concrete—25000 PSI minimum.
Corner bars—provide #6 bars in all corners of the perimeter or exterior beams one each at top and bottom each side.

Fig. 50-3B Slab foundation footing details. *(Vardy Vincent)*

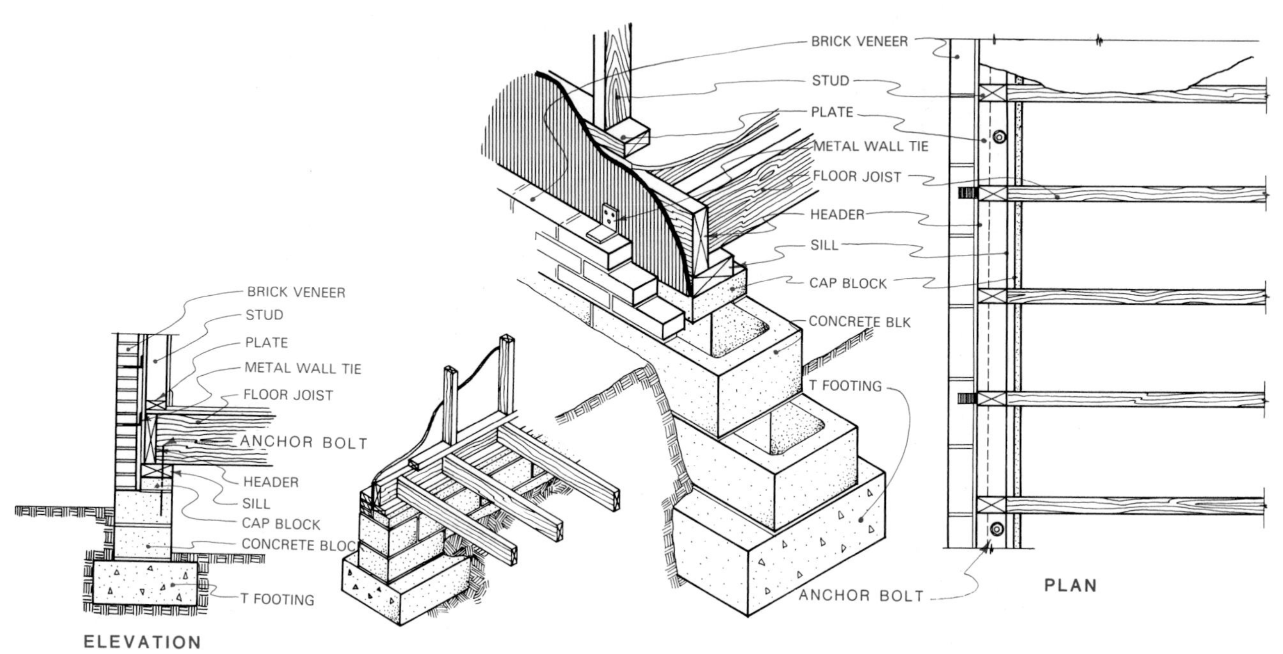

Fig. 50-4 Sill details for brick veneer.

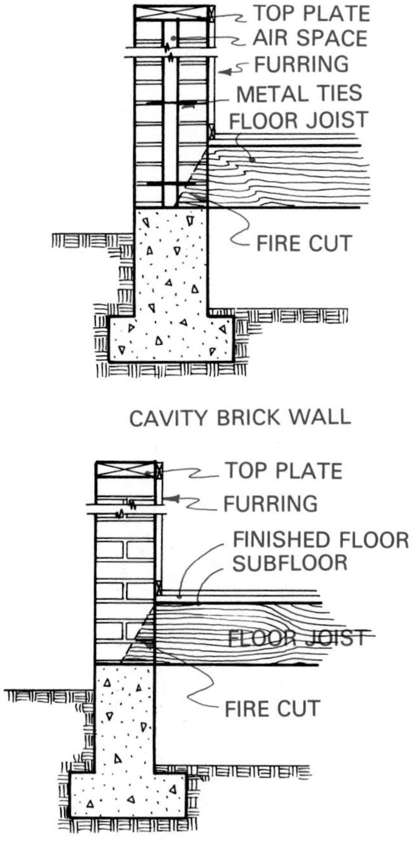

TOP PLATE
AIR SPACE
FURRING
METAL TIES
FLOOR JOIST
FIRE CUT

CAVITY BRICK WALL

TOP PLATE
FURRING
FINISHED FLOOR
SUBFLOOR
FLOOR JOIST
FIRE CUT

Fig. 50-5 Sill detail for solid masonry.

FOUNDATION PLANS

Foundation plan views are similar to other floor plans, except that they represent a section through the foundation just below the top of the foundation, as shown in Fig. 50-7. Other sections (Fig. 50-8) are keyed to the foundation plan through the use of isometric section details. Figure 50-9 shows a portion of the plan found in Fig. 69-3 at full 1/4″ = 1′-0″ scale.

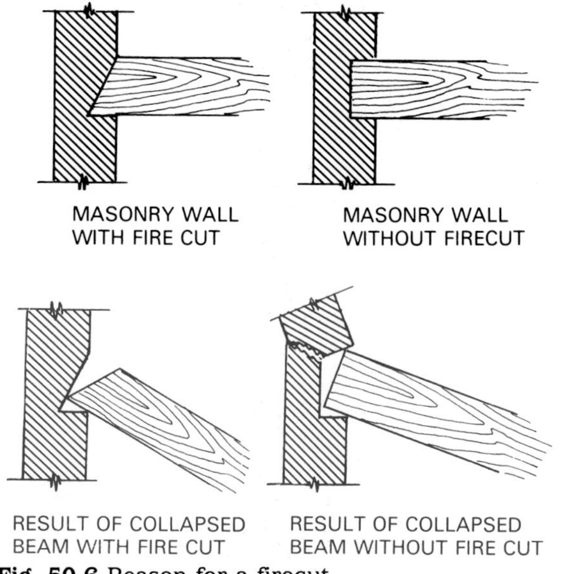

Fig. 50-6 Reason for a firecut.

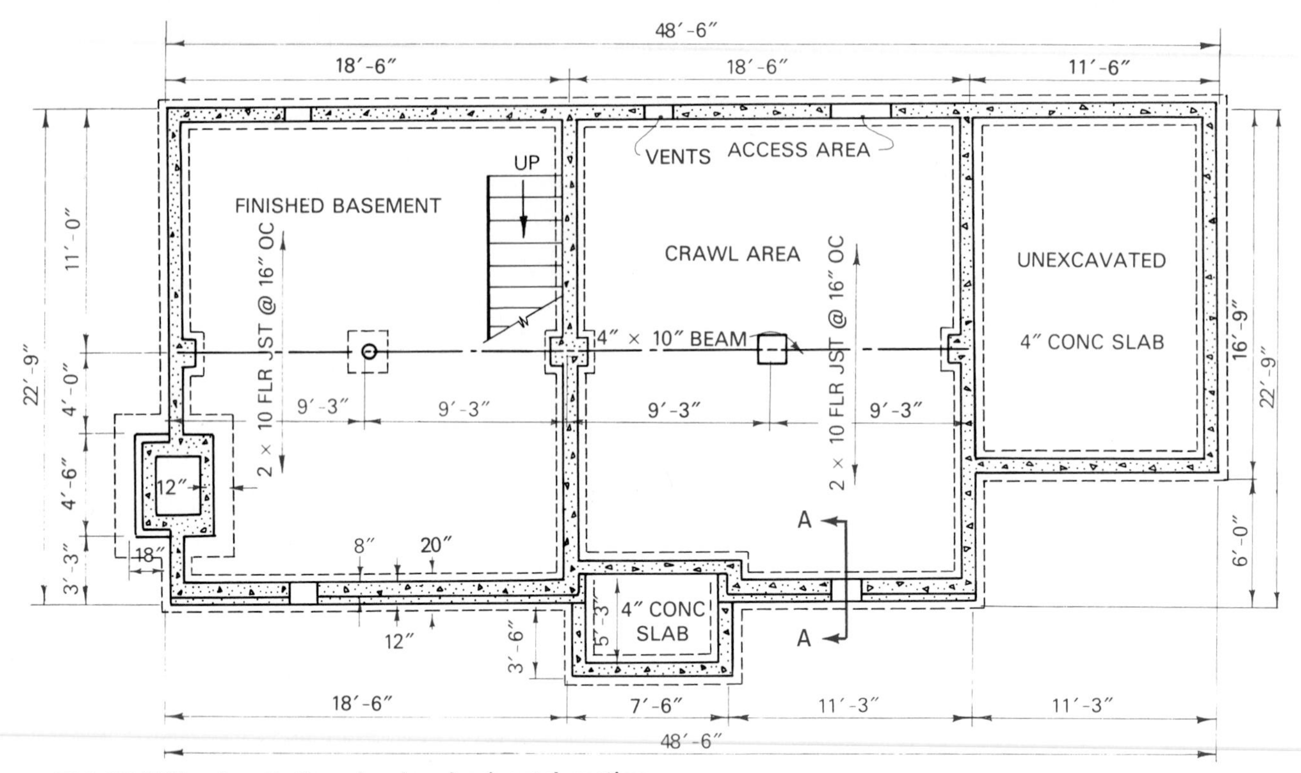

Fig. 50-7 The foundation plan is a horizontal section.

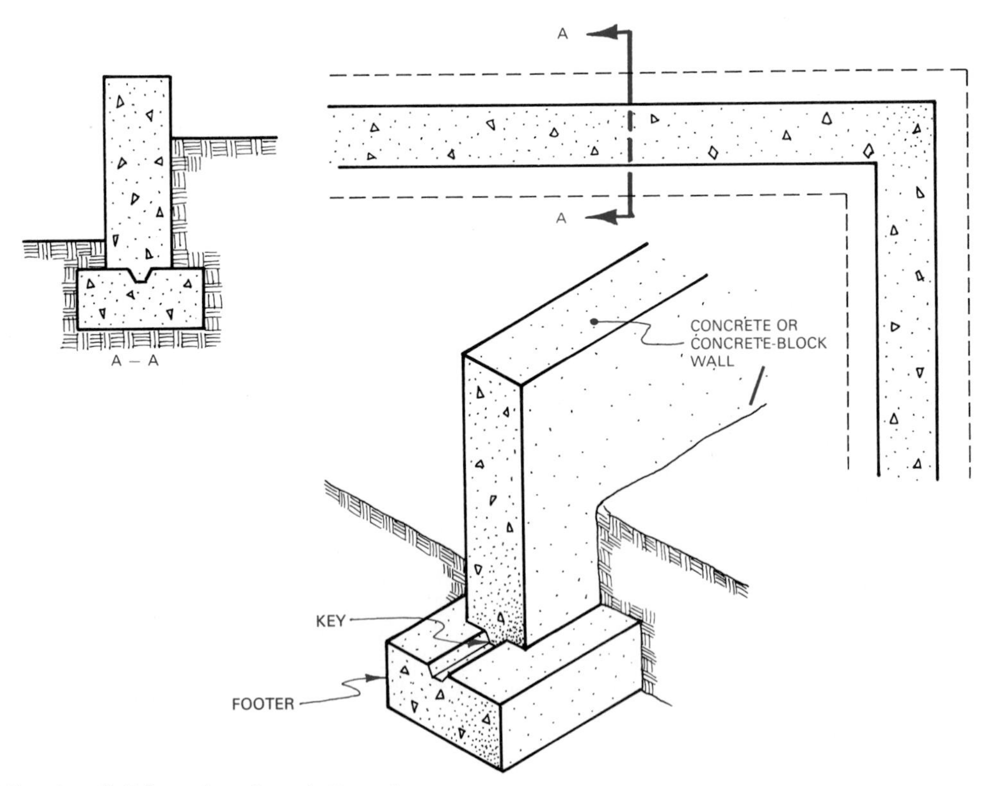

Fig. 50-8 Section A-A keyed to foundation plan.

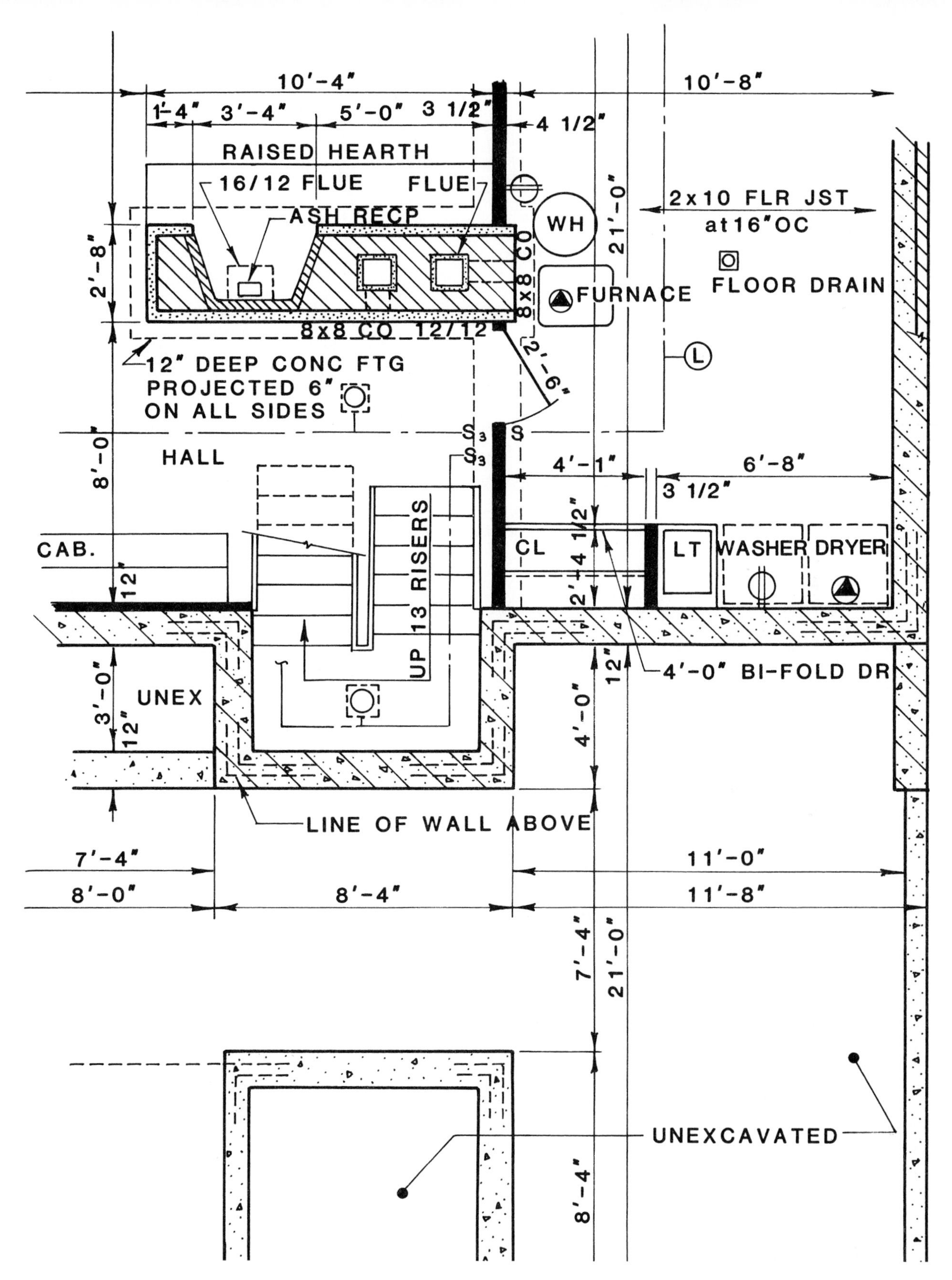

Fig. 50-9 Full-size (¼" = 1'0") portion of plan shown in Fig. 69-2.

Exercises

1. **In Fig. 50-10, how many cubic feet of dirt must be excavated? How many cubic yards? How many cubic feet of concrete must be poured for the footing? How many cubic yards for the slab?**
2. **Draw sections A, B, and C of the foundation plan shown in Fig. 50-11.**
3. **Calculate the material needed for a foundation of your design.**
4. **Define these architectural terms: *excavations, foundation forms, steel-wire mesh, strike board, waterproof membrane.***

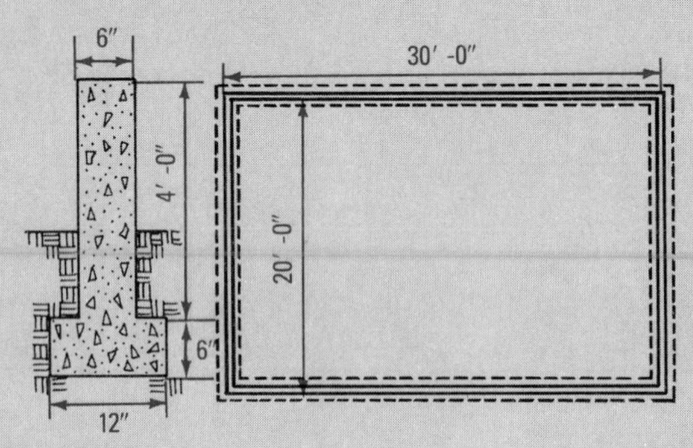

Fig. 50-10

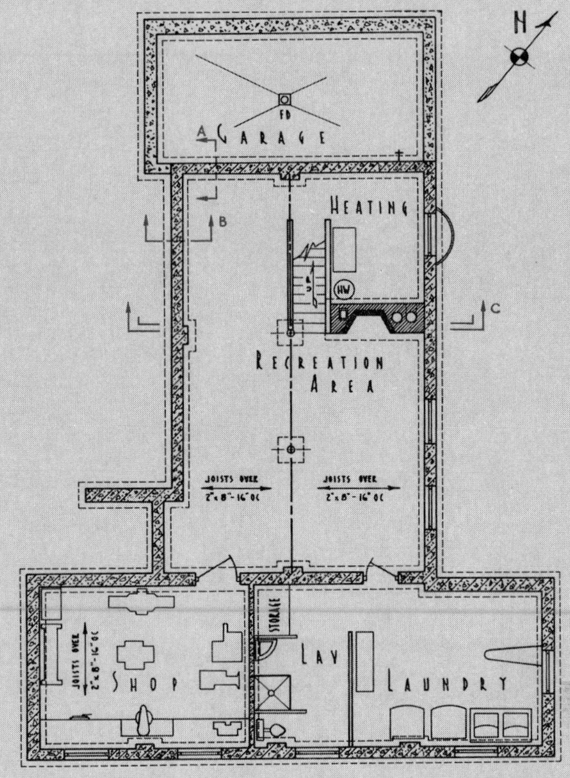

Fig. 50-11

UNIT 51

FIREPLACES

In design of fireplaces, the style and type, support, framing, size and material of components, size and ratio of the opening and firebox, and height of the chimney must be considered.

FIREPLACE DESIGN TYPES

Fireplace types are divided into five basic types depending on the opening: flush, two-sided, three-sided, see-through, and open on all sides. Figure 51-1A shows a flush-mounted fireplace, sometimes called a single-faced fireplace. Flush fireplaces vary in the design of a mantle and height of the hearth. Two-sided fireplaces like the one shown in Fig. 51-1B are appropriate for L-shaped living-dining areas. Three-sided fireplaces (Fig. 51-1C) are often used as a separation between rooms; usually between the dining and living areas. A variation of the flush fireplace is the double-flush, or see-through, fireplace. This design provides the benefits of the

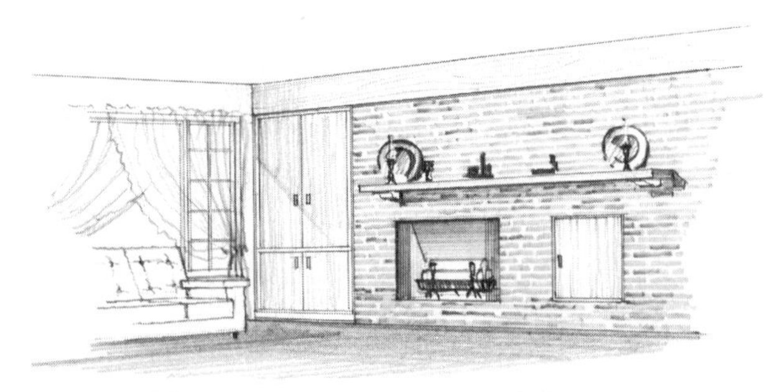

Fig. 51-1A Flush-type, single-faced fireplace.

Fig. 51-1B Corner, double-faced fireplace.

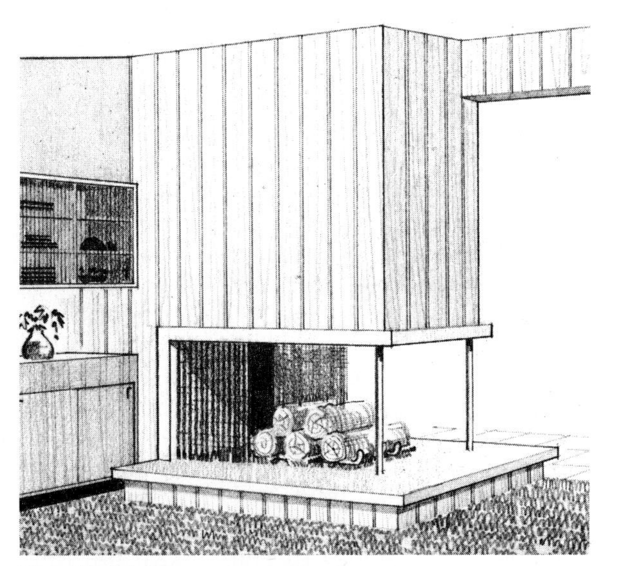

Fig. 51-1C Three-sided fireplace.

Fig. 51-1D Double-faced see-through fireplace.

fire to two rooms simultaneously (Fig. 51-1D). Open-pit fireplaces (Fig. 51-1E) must be designed with adequate hood space to provide sufficient draft to prevent smoke from entering the room.

FUNCTIONING

Oxygen is the vital ingredient needed for effective fireplace functioning. Since warm air rises it is drawn into the fireplace (fireplace draw), feeding the fire with needed oxygen. However, since cold air continually replaces rising warm air, much of the heat produced by many fireplaces goes up the chimney. To reduce this heat loss and redirect some of the heat back inside, warm-air outlets, balanced by cold-air outlets, can be

Fig. 51-1E Open-pit fireplace. *(Home Planners, Inc.)*

used. Using outlets of this type allows heat to reenter the room while smoke, debris, and toxic fumes are directed outside through the chimney. In well insulated buildings, often a fresh supply of oxygen is scarce. In this case a thermosiphoning fireplace can be used.

SIZING

Designing for adequate warm-air rise (draw) is critical for proper fireplace functioning. Inadequate draw either from an undersized flue or from improper chimney placement can result in smoke leaking into the inside rather than being drawn up the chimney. It is important to choose the appropriate fireplace size and the correct ratio of flue to firebox size.

Since the size of a fireplace should match its capacity, care must be taken to design the most efficient ratio of flue size to firebox size. Figure 51-2A shows the key dimensional areas used to compute the proportions of an efficiently operating fireplace. Figure 51-2B shows the actual dimensions which can be substituted for the letters in Fig. 51-2A to arrive at the correct proportions.

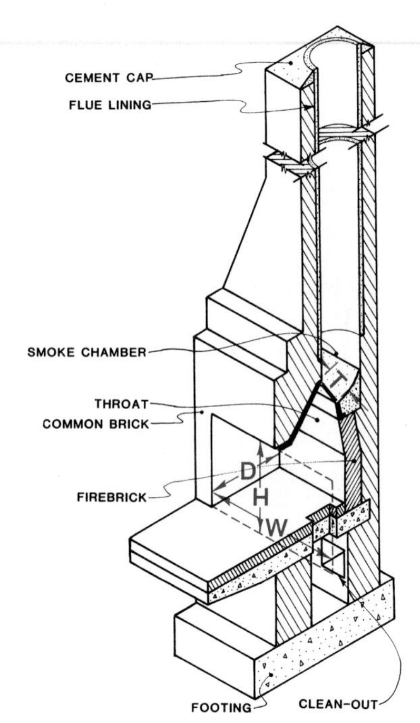

Fig. 51-2A Fireplace key dimensions.

FIREPLACE

The main part of the fireplace is the *firebox*. The firebox reflects heat and draws smoke up the chimney. Included in the firebox are the sides, back, smoke chamber, flue, throat, and damper (Fig. 51-3). Most fireboxes are constructed in a factory. The mason places the firebox in the proper location in the chimney construction and lines it with firebrick. Common brick is a popular fireplace material, though concrete block or natural stone may also be used. Firebrick or other refractory material must be used to line the firebox, and firebrick or flue tile should also make up the inner course of the flue. Firebrick or flue tile should be laid in fireclay, with outer walls laid in regular masonry mortar. Any noncombustible material may be used to surround the firebox opening and as a surface on the exterior hearth. Any combustible material may be used as trim as long as safe clearance distances of at least 15″ from the firebox opening are maintained. The hearth also should be constructed of fire-resistant material such as brick, tile, marble, or stone. The best method of drawing construction details of a fireplace is to prepare a sectional drawing. This gives the position of the firebox and the size of materials used in the footing, hearth, face, flue, and cap of the chimney. It also shows the relationship of the chimney to the floor and ceiling lines of the structure.

CHIMNEY

A chimney extends from the footing through the roof of the house. The footing must be of sufficient size to support the entire weight of the chimney. The chimney extends above the roof line to provide a better draft for drawing the smoke and to eliminate the possibility of sparks igniting the roof. The height of the chimney above the roof line varies somewhat, according to local building codes. In most areas the minimum distance is 2′ (610 mm) if the chimney is closer than 15′ to the nearest ridge (Fig. 51-4A). Often caps or uneven flue projections (Fig. 51-4B) helps prevent downdrafts.

The chimney is secured to ceiling and floor joists by iron straps embedded in the brickwork. Floor joists and ceiling joists around the chimney and fireplace and hearth should have sufficient clearance to protect them from the heat. Most building codes specify this distance.

The designer must also indicate the type and size of flues to be inserted in the chimney. One flue is necessary for each fireplace or furnace

Fireplace dimensions *(in inches)*		Recommended flue sizes *(in inches)*					
		Fireplace width W	**Rectangular flues**			**Equivalent round**	
			Nominal or Outside Dimension	Inside Dimension	Effective Area	Inside Diameter	Effective Area
W	24 to 84						
H	⅔ to ¾ W	24	8½ × 8½	7¼ × 7¼	41□″	8	50.3□″
D	½ to ⅔ H{16 to 24 (Rec) for Coal; 18 to 24 (Rec) for Wood}	30 to 34	8½ × 13	7 × 11½	70□″	10	78.54□″
FLUE *(effective area)*	⅛ WH for unlined flue; 1/10 WH for rectangular lining; 1/12 WH for circular lining	36 to 44	13 × 13	11¼ × 11¼	99□″	12	113.0□″
		46 to 56	13 × 18	11¼ × 6¼	156□″	15	176.7□″
T *(area)*	¾ to ½ Flue area	58 to 68	18 × 18	15¾ × 5¾	195□″	18	254.4□″
T *(width)*	3″ minimum to 4½″ minimum	70 to 84	20 × 24	17 × 21	278□″	22	380.13□″

Fig. 51-2B Dimensional fireplace proportions.

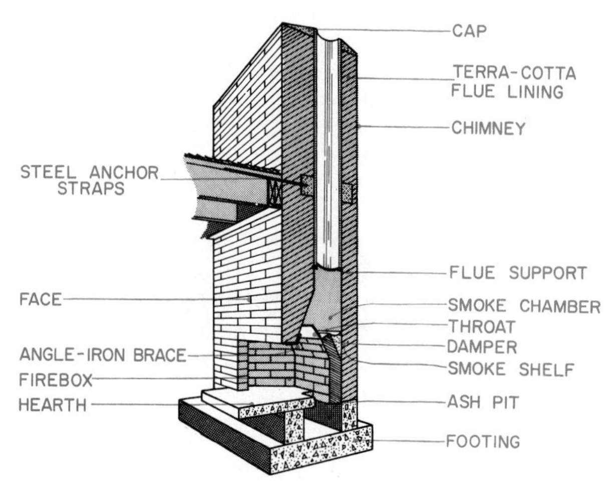

Fig. 51-3 The major components of fireplace and chimney structure.

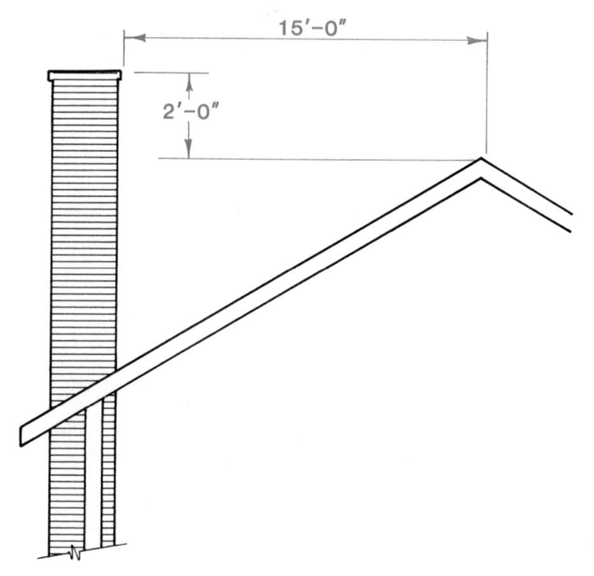

Fig. 51-4A Chimney height code requirements.

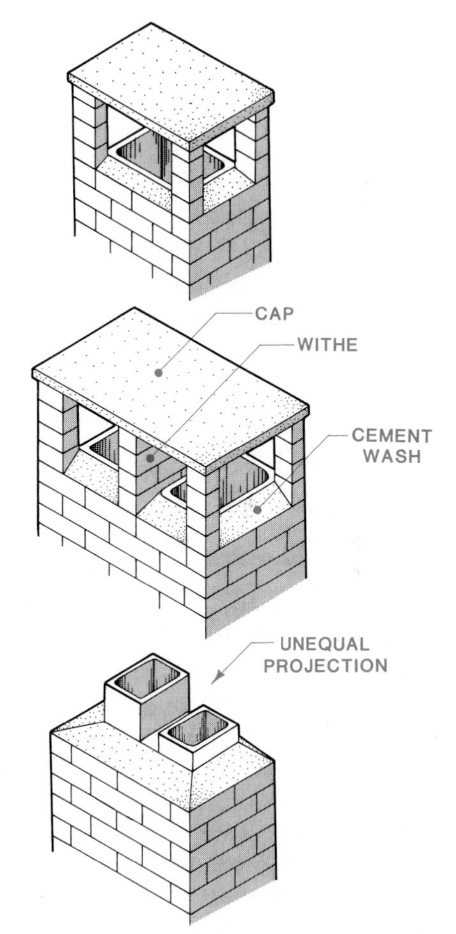

Fig. 51-4B Use of caps or offset flues for downdraft protection.

leading into the chimney. Note the three flues serving the three fireplaces in Fig. 51-4C. The flue from the fireplace or from the furnace in the basement extends directly to the top of the chimney, completely bypassing the first-floor fireplaces. The first-floor fireplace flues completely bypass the second-floor fireplace flues, and so forth.

Figure 51-5 shows the method of detailing a floor plan of a fireplace. This plan can be included on the floor plan drawing or as a separate detail if the scale is too small to include all dimensions easily.

STRUCTURAL SUPPORT

In the structural design process provisions must be made for the support of the heavy load of a fireplace system. Fireplace and chimney assemblies must rest on a solid footing and cannot be supported by the normal building footings. A solid reinforced footing, as shown in Fig. 51-6, is most often used for residential construction. To calculate the proper footing size the same procedure should be used as in designing a footing for a pier or column as covered in Unit 83. As with all footings, the conditions of the soil and slope of the terrain must be considered.

PREFABRICATED FIREPLACES

Freestanding metal fireplaces constructed of heavy-gauge steel are available in a variety of shapes. They are relatively light wood-burning stoves and therefore need no concrete foundation for support. A stovepipe leading into the chimney provides the exhaust flue. Since metal units reflect more heat than masonry, the metal fireplace is much more efficient, especially if centrally located. These fireplaces can be mounted on the walls or on legs, or they can be built into the chimney. They are complete, ready-to-install fireplaces. Nevertheless, a fire-resistant material such as concrete, brick, stone, or tile must be used beneath and around these fireplaces.

Many fireplaces are built by using some manufactured components and constructing the remainder on the site. The design of the fireplace usually determines the amount of manufactured components that can be used. Unconventional fireplaces usually use few standard components.

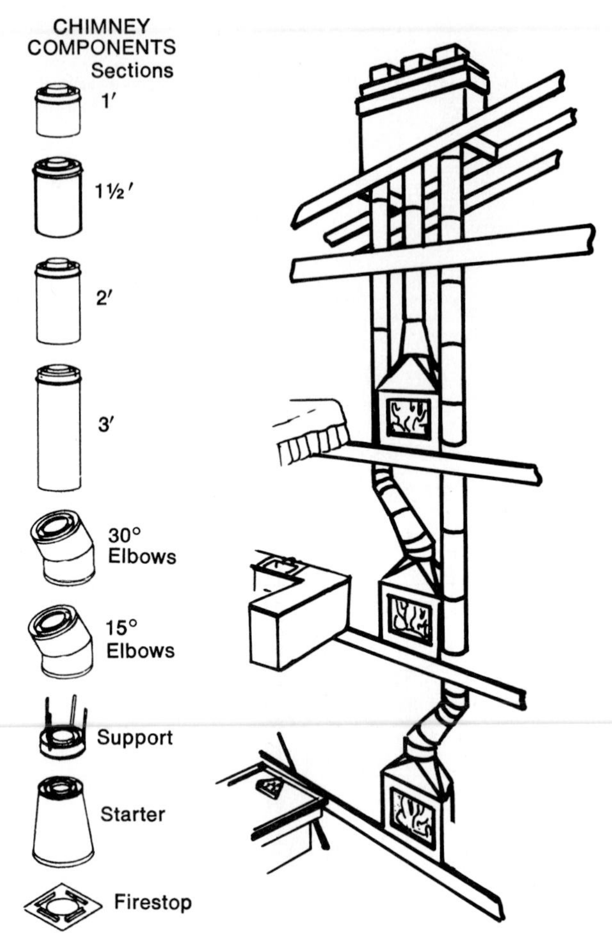

Fig. 51-4C Multiple-flu design.

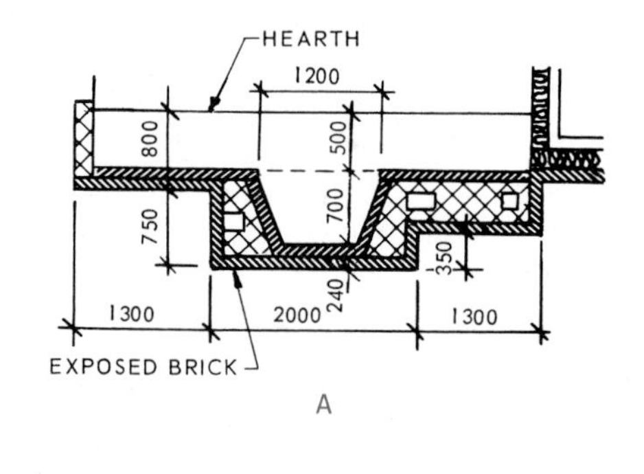

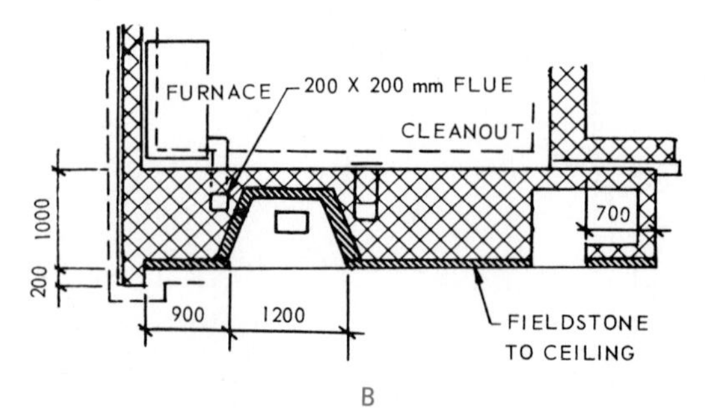

Fig. 51-5 Floor plan fireplace details in metric sizes.

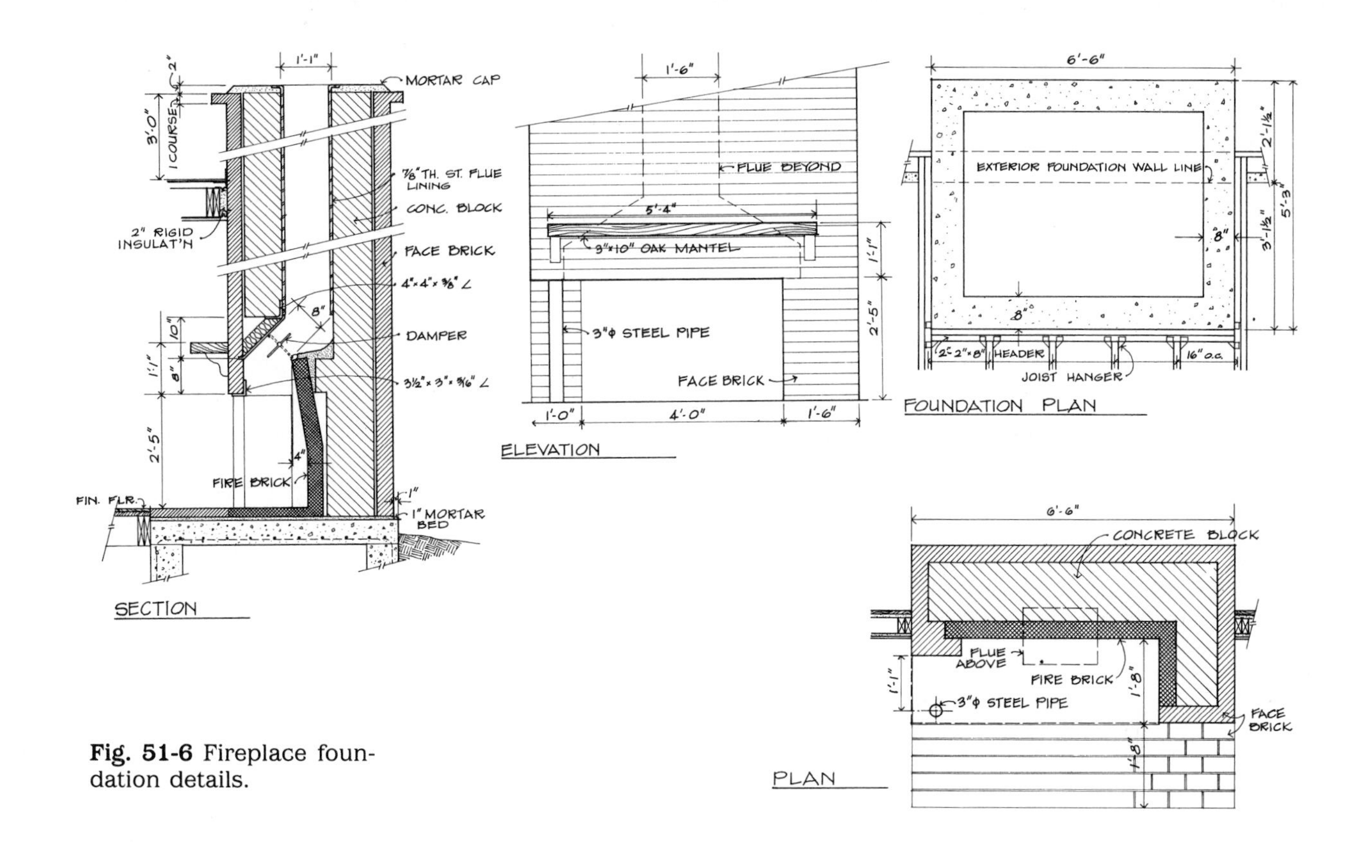

Fig. 51-6 Fireplace foundation details.

Exercises

1. **Redraw the fireplace in Fig. 51-3, using the scale ½″ = 1′-0″. See Fig. 51-6 for typical dimensions.**
2. **Design a fireplace for the house you are designing.**
3. **Identify the parts of the fireplace shown in Fig. 51-7.**
4. **Define these architectural terms: *firebox, chimney, smoke chamber, flue, throat, damper, draft, draw, ceiling joist, firebrick, prefabricated fireplace, chimney sections.***

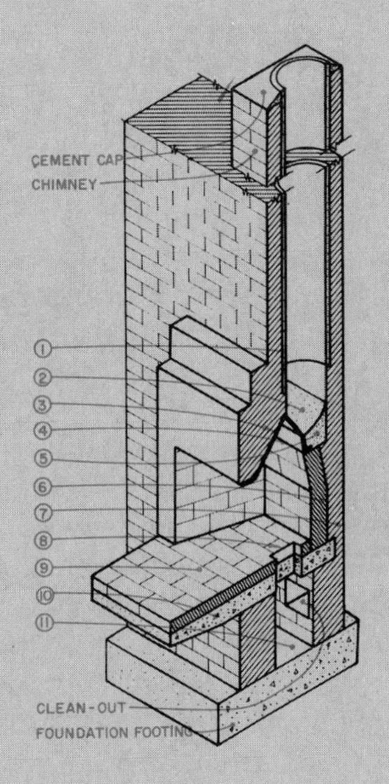

Fig. 51-7

SECTION 13

Framing Plans

Most of the basic engineering principles upon which modern framing methods are based have been known for centuries. However, it has not been until recent years that the development of materials and construction methods has allowed the full use of these principles. Today's designers can choose among many basic materials in the design of the basic structural framework of a building. New and improved methods of erecting structural steel, new developments in laminating and processing preformed wood structural members, developments and refinements in the use of concrete and masonry products such as prestressed concrete slabs, and continual progress in standardization in the design of structural components all provide the designer with the flexibility to design the most appropriate structural system for a building at the lowest possible cost and with a minimum waste of materials and time.

UNIT 52

CONSTRUCTION SYSTEMS

New construction materials and new methods of using conventional materials now provide designers with great flexibility in construction design. Stronger buildings can now be erected with lighter and fewer materials.

Regardless of the materials and methods used, the physical principles of structural design remain constant. In most structures, the roof is supported by wall framework, interior partitions, or columns. Each exterior wall and bearing partition is supported by the foundation, which in turn is supported by footings. Footings distribute building loads over a wide area of load-bearing soil and thus tie the entire structural system to the ground. Figure 52-1 shows these major lines of force.

Although the basic principles involved in preparing construction drawings are the same for all types of construction, the use of symbols, conventions, and terms changes drastically from system to system. Construction systems are broadly divided into four material groups: skeleton wood frame, heavy timber, structural steel, and masonry and concrete. Although these are logical categories for study, most contemporary designs include a combination of these systems and materials. Coverage of these groups is introduced in this unit. Specific drawings relating to these groups are covered in Units 53 through 58.

PRINCIPLES OF CONSTRUCTION

A knowledge of the relationship between loads, forces, and strength of building materials is vital to every architectural designer. Structurally, buildings are divided into two types: bearing-

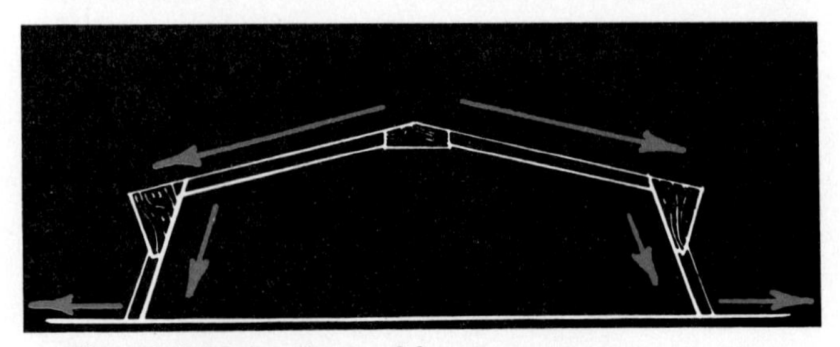

Fig. 52-1 Major lines of force.

wall structures and skeleton-frame structures. Bearing-wall structures have solid walls that support the weight of the walls, floors, and roof. Skeleton-frame structures are constructed with an open, self-supporting framework covered by an outer, nonbearing surface. Most contemporary buildings are of the skeleton-frame type. In both types of construction, structural stability is based on the strength and placement of structural members. The selection and use of materials depends on the loads and forces acting on these materials.

Loads

Loads which are supported by all buildings include *live loads* and *dead loads*. The total weight of all live and dead loads is known as the *building load.*

Live loads include the weight of all movable objects, such as people and furniture, as shown in Fig. 52-2. Live loads also include the weight of snow and the force of wind (Fig. 52-3), which varies greatly among regions.

Most loads follow lines of gravity; however, lateral (horizontal) wind, earth, and earthquake loads also act on walls and roofs. These vary according to roof type and building materials. To counteract the force of these loads, some structural stability is achieved by the attachment of the roof to walls as shown in Fig. 52-4. Diagonal bracing (Fig. 52-5), knee bracing, or wall panels are also used to provide maximum rigidity. Reinforced concrete and pilasters are often used to withstand lateral earth loads which act against foundation walls, as shown in Fig. 52-6.

Dead loads are those loads created by the weight of building materials and permanently installed components. They include the weight of all roof materials (Fig. 52-7), which are supported by walls, which in turn are supported by

A STRUCTURE IS NOT COMPLETE
WITHOUT A ROOF.

WALLS CANNOT RESIST FORCES—
FORCE FROM OUTSIDE. .

FORCES FROM INSIDE...

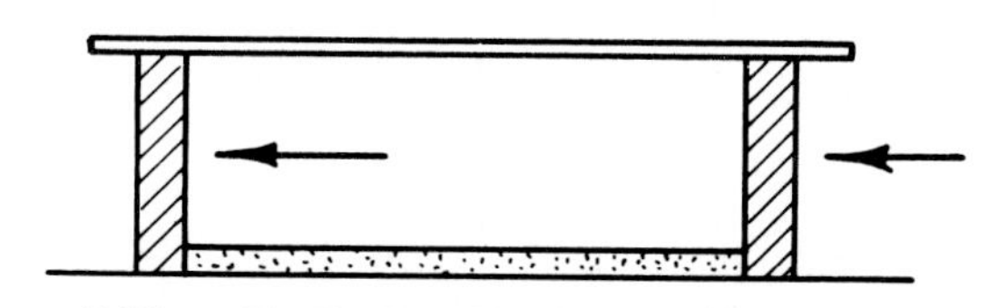
UNTIL THE ROOF IS ADDED

Fig. 52-4 Roofs add stability.

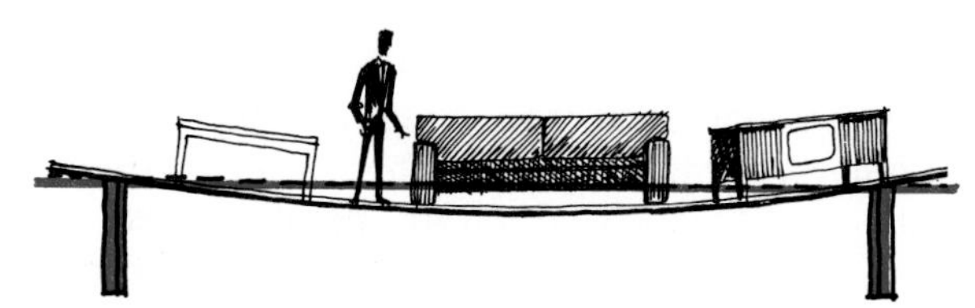

Fig. 52-2 Live floor loads.

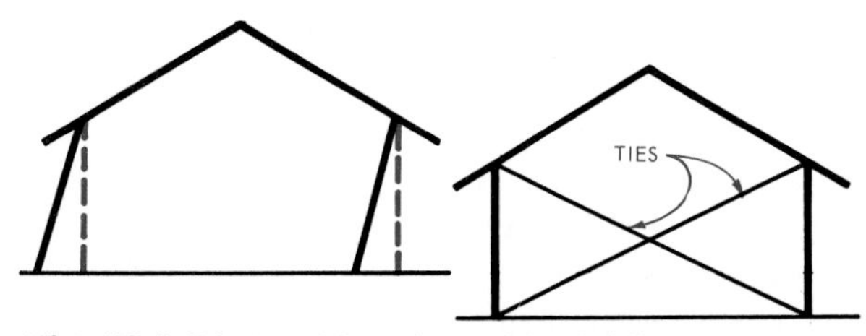
TIES

Fig. 52-5 Diagonal bracing adds rigidity.

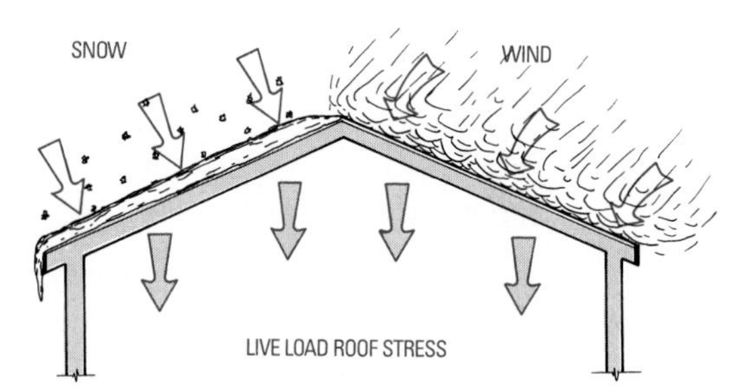
SNOW
WIND
LIVE LOAD ROOF STRESS

Fig. 52-3 Live roof loads.

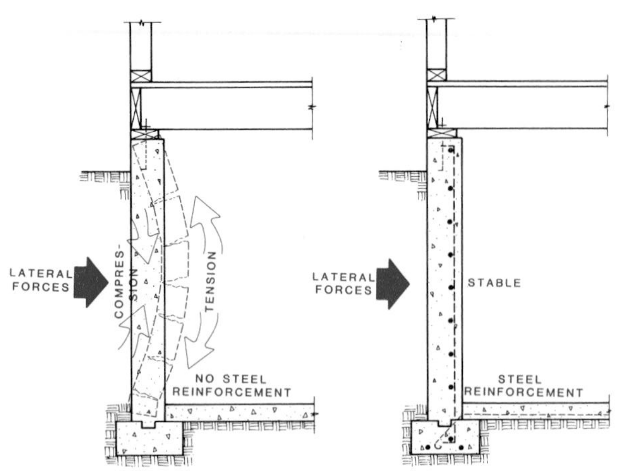

Fig. 52-6 Heavy lateral earth loads require additional reinforcement.

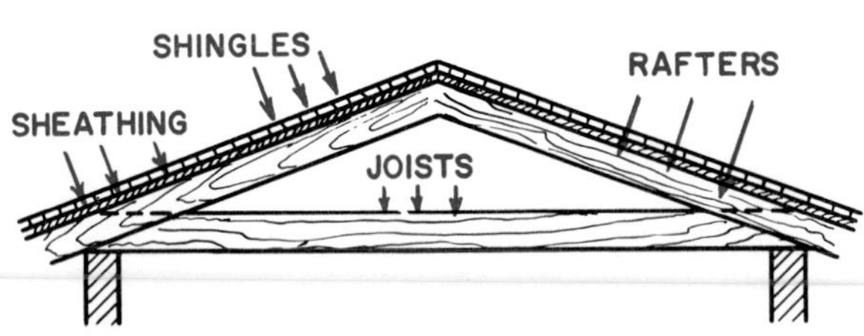

Fig. 52-7 All roof materials are part of the total dead load.

footings. Every piece of lumber, brick, glass, and nail adds to the dead load of a structure. Since buildings must support their own weight, dead loads are calculated for each building or specified in precalculated building codes. Material types and the size and spacing of members are also calculated or specified by code to prevent the building of unstable structures due to excess loading.

Structural Forces

Live and dead loads create four types of forces that exert stress on building materials: compression, tension, shear, and torsion, as illustrated in Fig. 52-8. *Compression* force is caused by loads pushing on an object. Compression forces tend to flatten materials as a roller flattens a roadbed. *Tension* force is created by loads pulling on an object. Tension forces stretch materials. A supporting chain on a hanging light fixture is in tension. *Shear* force results from loads which cause a member to break with a slicing action. Shear loads may cause material fractures by abrupt action such as a large member falling on a lighter object. *Torsion* force is the result of loads twisting an object. *Torsion* force can twist a member out of shape or fracture it completely by overloading an end or by movement of a connecting member.

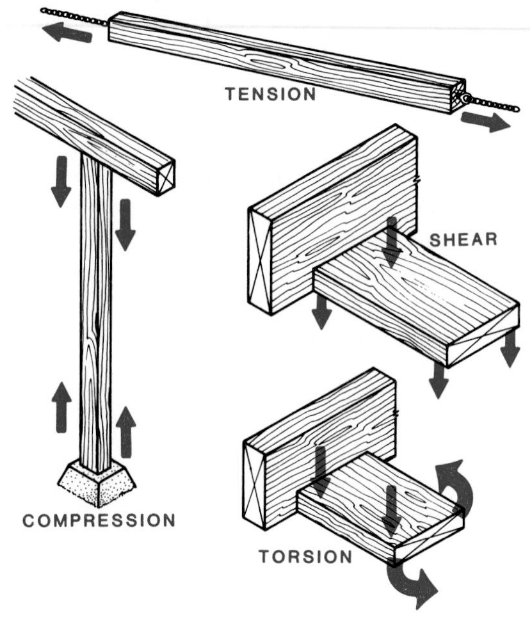

Fig. 52-8 Types of forces.

Structural Members

Compression, tension, shear, and torsion forces create enormous stress on building materials. If a material is sufficiently strong, this stress will create little or no damage. However, if the material is weaker than the forces applied, the resulting stress can compress, stretch, slice, twist, or completely fracture a member. Therefore the resistance of each structural member must always be equal to or greater than the force applied (Fig. 52-9). To ensure that every structural

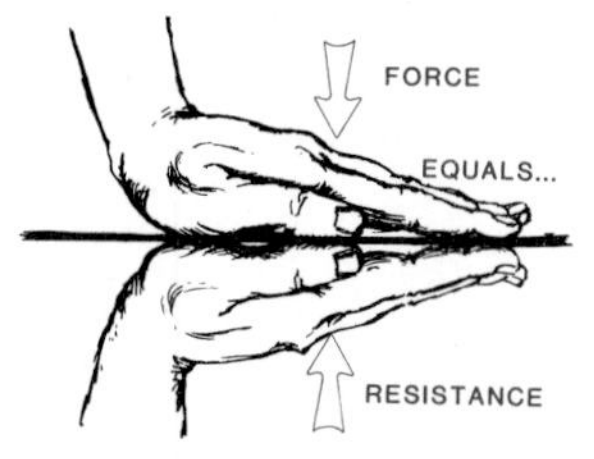

Fig. 52-9 The resistance moment must be equal to or greater than the bending moment.

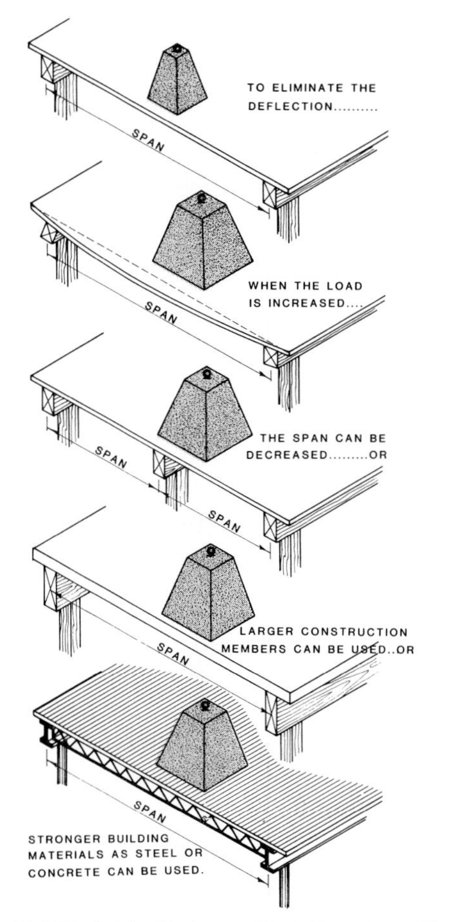

Fig. 52-10A Variable factors affect the strength of a bulding.

member can resist the forces and stresses created by building loads, the material, size, shape, placement, and spacing must be carefully planned. If any of these factors change during the building process, another design element may need to be changed to maintain the structural integrity of the building, as shown in Fig. 52-10A. However, design changes must always be approved by a licensed architect or engineer and adhere to local building code requirements.

Strength of Materials

The strength of a construction material is the material's capacity to support loads by resisting compressive, tension, shear, and torsion stresses. The structural strength of a member depends on the type, size, and shape of the material. Different structural materials have varying capacities to resist stress and support building loads. For example, a steel member can support more weight than a wood member of the same size. The load-bearing capacity will also vary among species of wood because of different fiber stress levels. Figure 52-10B shows the relationship of member size to load.

Figure 52-11 shows the range of construction materials and their relative ability to resist bending stress. *Bending stress* results from both compressive and tension forces acting on a member at the same time. Figure 52-12 shows

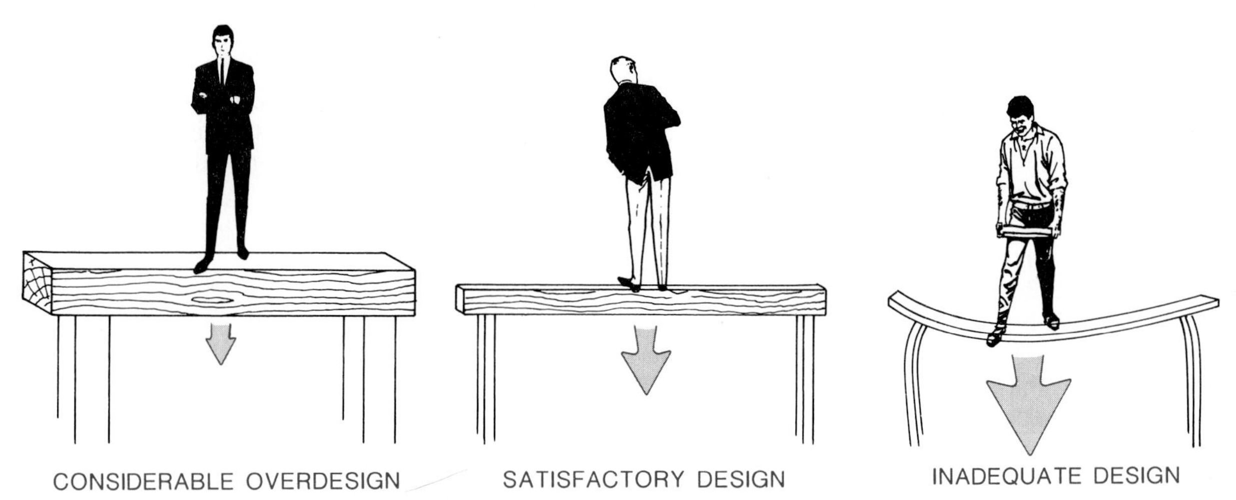

Fig. 52-10B Relationship of member size to load.

STRUCTURAL MEMBER 10'-0" SPAN	APPROX DEFLECTION	STRUCTURAL MEMBER 10'-0" SPAN	APPROX DEFLECTION
500 LB FORCE; 10'-0"; 2"x6" LAID FLAT	10"	500 LB FORCE; 4"x8" LAMINATED WOOD BEAM	.07"
500 LB FORCE; 2"x6" ON EDGE	.75"	500 LB FORCE; 4"x6" REINFORCED CONCRETE BEAM	.05"
500 LB FORCE; 2"x12" ON EDGE	.10"	500 LB FORCE; STEEL S BEAM	.02"
500 LB FORCE; 6"x8" ON EDGE	.10"	500 LB FORCE; 24' FLAT ROOF TRUSS	.02"
500 LB FORCE; 8" DIAM LOG	.08"	500 LB FORCE; 48' HOWE TRUSS	.01"

Fig. 52-11 Deflection of structural materials.

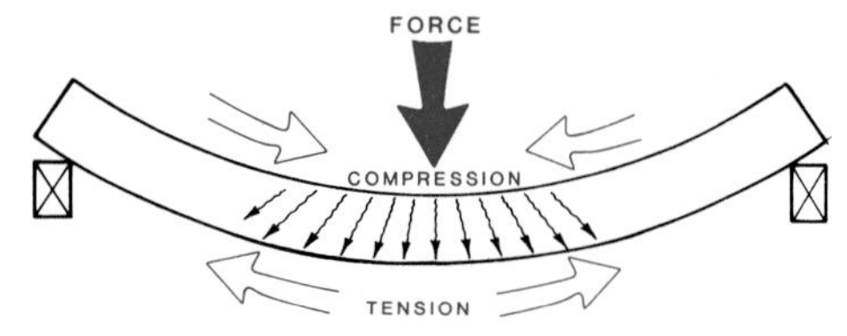

Fig. 52-12 Loads can create both compression and tension stress.

how a load creates compressive stress on the top part of a member while creating tension stress on the bottom. Reinforcing the bottom half of a member balances this stress. Larger members can obviously support greater loads than smaller members of the same material. For this reason builders must avoid excess notching of structural members since notching reduces the structural size of the member, as shown in Fig. 52-13.

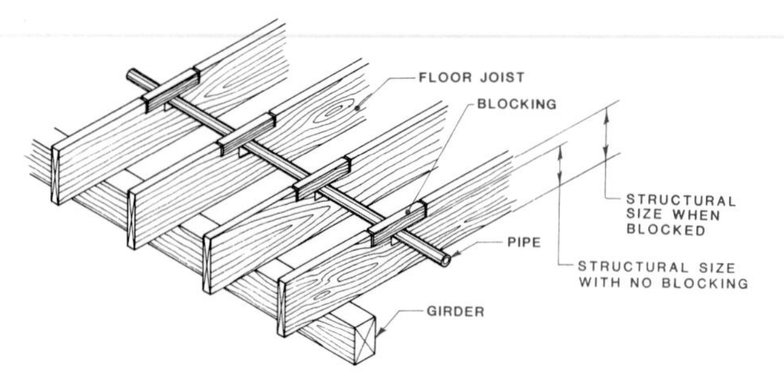

Fig. 52-13 Notching reduces the structural size of a member.

Member Shape

Different shapes of the same material can support greater loads. For example, hold a sheet of thin paper between your fingers as shown in Fig. 52-14. The paper will not support itself and will drop. Now fold the paper in half. It will now support a light object. Fold it again several times and this same piece of paper will support a much heavier object. Figure 52-15 shows how this principle is used to increase both the lateral rigidity and the load-bearing capacity of a structural member.

Member Placement

The strength of a building material is significant only when it becomes an integral part of a structure. Most materials are somewhat flexible until tied in to a structure. Member orientation is also an important factor; for example, Fig. 52-16 shows that turning a member on its side will increase the vertical deflection (bending) but the horizontal deflection will be unaffected. Likewise, a structural member placed on its edge will increase the horizontal deflection but the vertical deflection will be unaffected. Therefore members with their widest dimension positioned parallel to the load direction will resist greater loads than members placed with their smallest

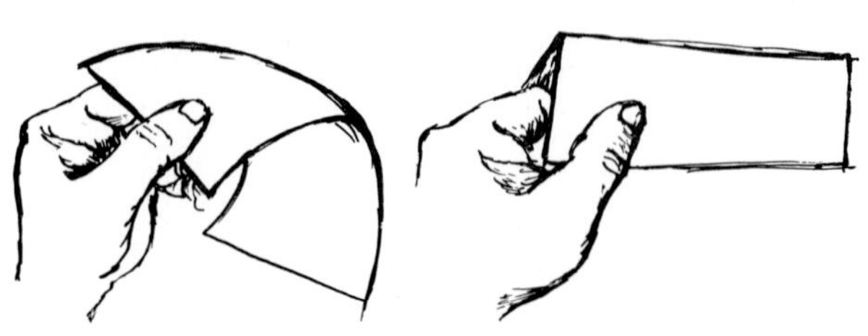

Fig. 52-14 Material strength is related to its shape.

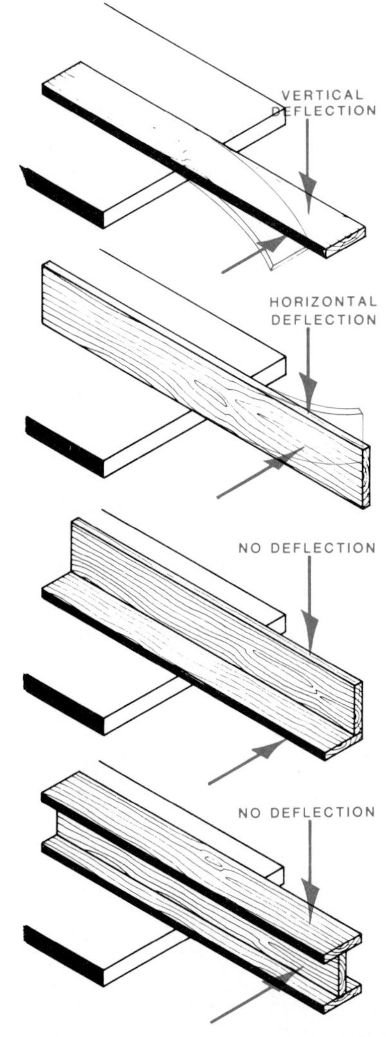

Fig. 52-15 Material shape related to deflection.

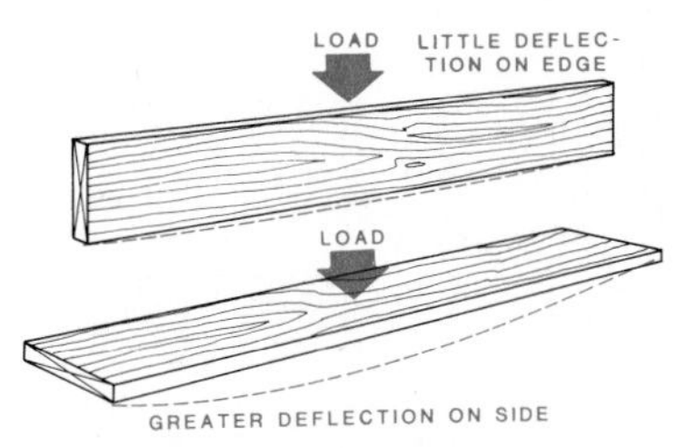

Fig. 52-16 Stability is related to member position.

dimension aligned in the load direction. For this reason, combining the horizontal and vertical components of the member to make a channel or I beam reduces both the vertical and the horizontal deflection.

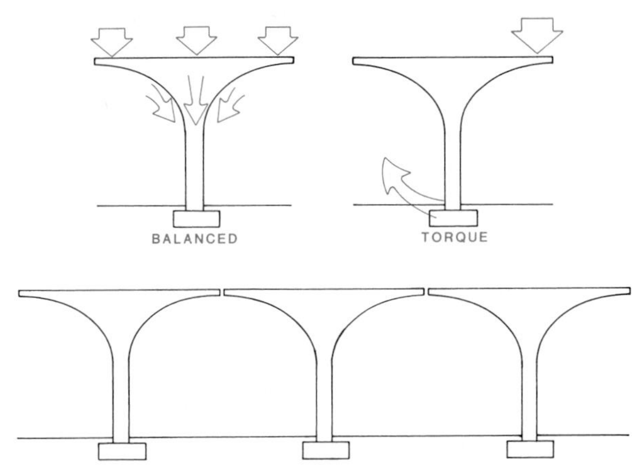

Fig. 52-17 Center-supported cantilevered decks.

Spans and Spacing

Spacing is the distance between parallel members. A *span* is the distance a member extends between supports. The maximum allowable span (length) of a support member is directly related to loads applied and the strength of the material. Obviously, stronger materials can support greater loads at greater distances with less deflection. Decreasing the span while using the same material can increase the load-bearing capacity of a member.

When only one end of a horizontal structural member is supported, the condition is known as *cantilevering*. Cantilevered members can be center-supported or eccentric. Figure 52-17 shows a center-supported concrete slab construction. Center-supported members are in equilibrium since the center supports equal dead loads on all sites. Eccentric cantilevered members are supported on one side opposite an unsupported end. Eccentric cantilevering requires stronger materials and stronger anchorage on the supported side, as shown in Fig. 52-18. This is because deflection increases as the distance and/or loads from the support increase.

SKELETON WOOD FRAME

With the passing of the log cabin (Fig. 52-19), which is a bearing-wall design, the *skeleton frame* became the most popular type of light

Fig. 52-18 Relationship of loads, support, and length of cantilevered members.

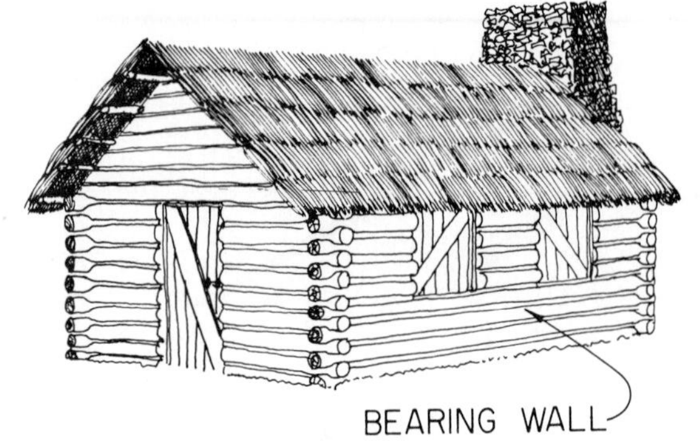

Fig. 52-19 A log cabin is the simplest type of bearing-wall construction.

construction. In skeleton-frame construction all walls, floors, and roofs are composed of small structural members which combine to make a skeleton frame, as shown in Fig. 52-20. A tee-pee (Fig. 52-21) is the simplest form of skeleton

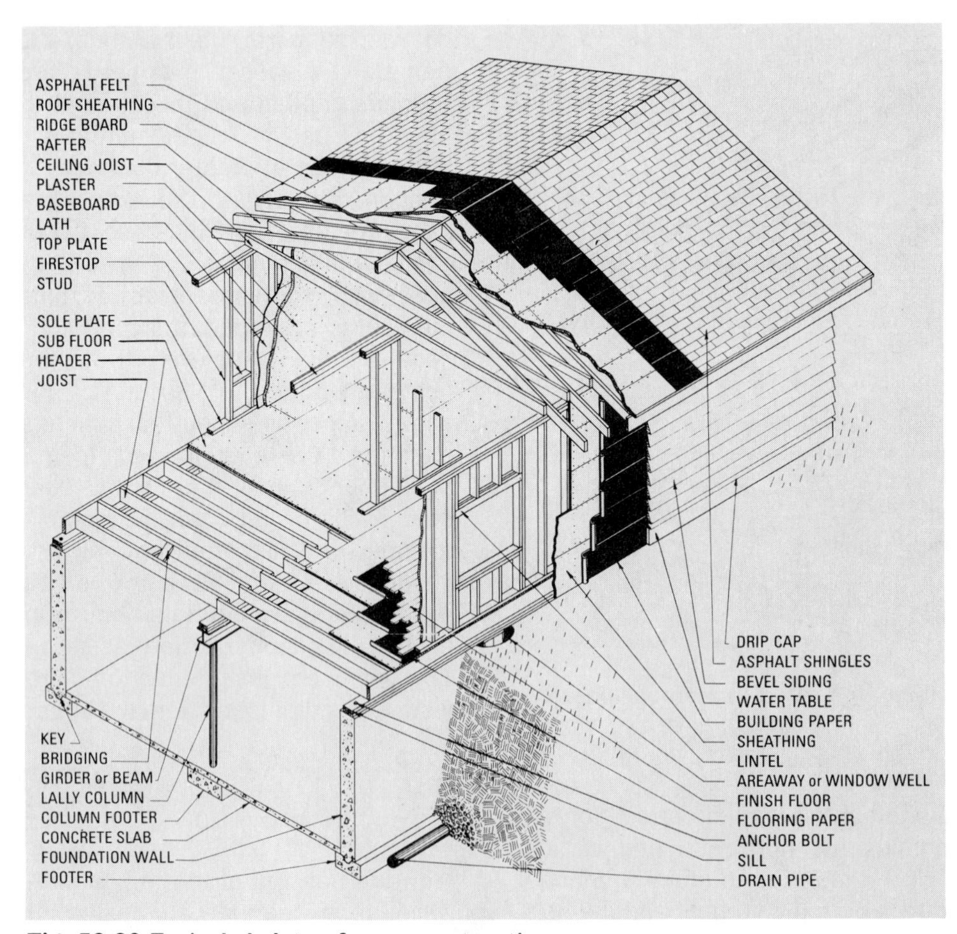

Fig. 52-20 Typical skeleton-frame construction.

Fig. 52-21 A teepee is the simplest type of skeleton-frame construction.

WOOD GRADE	QUALITY LEVEL
FAS FIRSTS	HIGH QUALITY
FAS SECONDS	
SELECT	
COMMON #1	
COMMON #2	
COMMON #3A	
COMMON #3B	POOR QUALITY

Fig. 52-23 Hardwood lumber grades.

frame. Because of the limited size of materials that can be handled on most light construction sites, skeleton-frame buildings are erected using an open structural framework covered by layers of sheathing, insulation and weatherproofing material. In this respect, the skeleton frame resembles the structural makeup of most vertebrates, as shown in Fig. 52-22.

Materials

Light construction materials include lumber, plywood, fasteners, and multimaterial components such as doors, windows, cabinets, plumbing, ductwork, and electrical fixtures.

Lumber Construction lumber is graded by intended use, strength, and appearance. Lumber grades are determined by the number and location of knots, checks, and splits, and by the degree of warp or wind. Lumber is also classified by species, grade, size, and special treatment. For grading purposes, lumber is divided into two broad categories, hardwoods and softwoods.

Hardwoods are used for surfaces with great wear potential, such as flooring and railings. Hardwoods are also used to make items that require a fine natural finish, such as cabinets and furniture. Hardwood species most commonly used in construction include oak, walnut, birch, cherry, mahogany, and maple. Hardwood is graded from highest quality to lowest quality depending on the amount of usable material in each piece, as shown in Fig. 52-23. Hardwood

Fig. 52-22 Skeleton frame in building and vertebrate construction.

Grade	Use
Selects and finish	Graded from the best side. Used for interior and exterior trim, molding, and woodwork where appearance is important.
B & BTR	Used where appearance is the major factor. Many pieces clear, but minor appearance defects allowed which do not detract from appearance.
C Select	Used for all types of interior woodwork. Appearance and usability slightly less than B & BTR.
D Select	Used where finishing requirements are less demanding. Many pieces have finish appearance on one side with larger defects on back.
Boards	Lumber with defects that detract from appearance but suitable for general construction.
No. 1 common (WWPA) Select merchantable (WCLIB)	All sound tight knots, with use determined by size and placement of knots. Used for exposed interior and exterior locations where knots are not objectionable.
No. 2 common (WWPA) Construction (WCLIB)	All sound tight knots with some defects, such as stains, streaks, and patches of pitch, checks, and splits. Used as paneling and shelving, subfloors, and sheathing.
No. 3 common (WWPA) Standard (WCLIB)	Some unsound knots and other defects. Used for rough sheathing, shelving, fences, boxes, and crating.
No. 4 common (WWPA) Utility (WCLIB)	Loose knots and knotholes, up to 4″ wide. Used for general construction purposes, such as sheathing, bracing, low-cost fencing, and crating.
No. 5 common (WWPA) Economy (WCLIB)	Large knots or holes, unsound wood, massed pitch, splits, and other defects. Used for low-grade sheathing, bracing, and temporary construction. Pieces of higher-grade wood may be obtained by crosscutting or ripping boards without defects.

Fig. 52-24 Range of yard lumber grades.

Grade	Use
LF (Light Framing)	Used in thicknesses from 2″ to 4″, and widths from 3″ and 4″, for studs, joists, and rafters in light framing.
JP (Joints and Planks)	Used in thicknesses from 2″, to 4″, and widths over 2″, for joists and rafters to be loaded on either side, or for planking when laid flat.
B&S (Beams and Stringers)	Used in thicknesses from 2″ to 4″. Widths over 2″ but must be loaded on narrow edge.
P&T (Posts and Timbers)	Used for posts or columns 5″ x 5″ and larger or where bending resistance is not critical.

Fig. 52-25 Structural lumber grades.

lumber is available in lengths of 4′ to 16′, widths up to 12″, and thicknesses up to 2″.

Softwoods are used for structural members such as joists, rafters, studs, sheathing, and formwork. Most skeleton-frame lumber is softwood. Softwood lumber is divided into three grading classes: *yard*, *structural*, and *factory* (shop) lumber. *Yard lumber* is used for most light framing members, such as sheathing, bracing, subfloors, and casings. The range of yard lumber grades is shown in Fig. 52-24. *Structural lumber*, as the name implies, is used for load-bearing members. Structural lumber is classified according to use (Fig. 52-25) and grades according to stress-resistance characteristics. *Factory*, or *shop lumber*, consists of light members which are finished at a mill and used for trim, molding, and door and window sashes.

All lumber is graded at a lumber mill by a regional authority which subscribes to the *American Lumber Standards*. Lumber is *grade marked* with a variety of information, including the mill identification number, certification association, grade number, moisture content, and species classification, as shown in Fig. 52-26.

The size of structural lumber is defined as either *rough* or *finished*. *Rough* sizes represent the width and thickness of a piece of lumber as cut from a log before surfacing. Rough lumber is also called *nominal* size lumber. *Finished* lumber sizes represent the actual dimensions of a

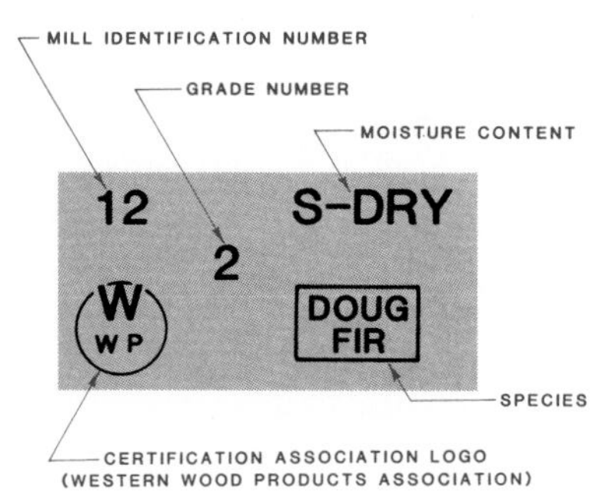

Fig. 52-26 Typical lumber grade marks.

member after final surfacing. Finished lumber is also known as *surfaced, dressed, dimensional,* or *actual size* lumber.

When a drawing callout reads: 2 × 4, builders know that the rough size is 2″ × 4″ but the actual size of the member is 1 ½″ × 3 ½″. Figure 52-27A shows the U.S. Customary range of rough and finished standard lumber sizes. Figure 52-27B shows the range of metric lumber sizes. Since lumber is not always surfaced on all sides, symbol designations, shown in Fig. 52-28, are used to indicate the number of sides to be surfaced.

Plywood Solid lumber is limited in width and has a tendency to warp, wind, split, and check. Conversely, plywood has large width capacity and is very structurally stable. Plywood is manufactured from thin sheets (0.10″ to 1.25″) of wood laminated together, with an adhesive, under extremely high pressure. The number of layers *(plys)* varies from three to seven. The grain of each ply is laid perpendicular to the grain of each adjacent layer. The grain of both outside sheets always faces the same direction. This greatly reduces the tendency of plywood to warp, twist, check, split, splinter, and shrink compared to solid wood. Plywood is available in individual 4′ × 8′ sheets or in continuous panels up to 50′ in length, and in thicknesses of ¼″, 5/16″, ½″, ⅝″, ¾″, 1″, 1⅛″, and 1¼″.

All plywood is divided into two broad categories: *exterior* (waterproof) or *interior.* Plywood surfaces are made from softwood for structural use and hardwood for cabinets and furniture. Because of the different uses, construction-grade plywood (softwood) and veneer-grade plywood (hardwood) are graded with different quality-rating systems.

Construction-grade plywood is unsanded and identified by grade levels based on structural

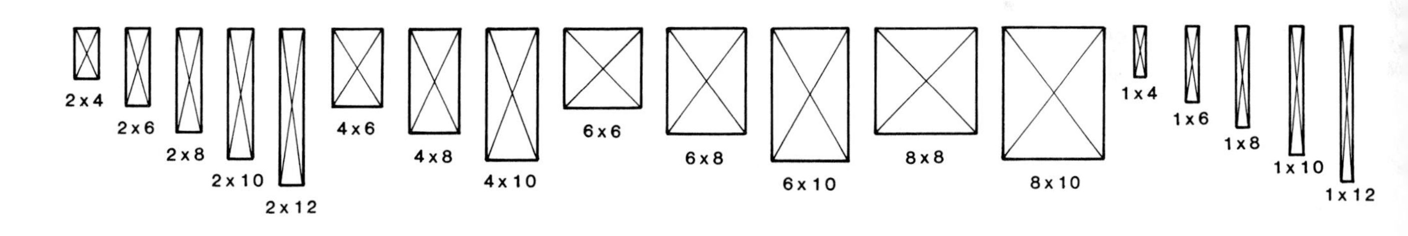

Lumber Sizes in Inches

Nominal Size	2x4	2x6	2x8	2x10	2x12	4x6	4x8	4x10	6x6	6x8	6x10	8x8	8x10
Dressed Size	1½x3½	1½x5½	1½x7½	1½x9½	1½x11½	3 9/16x5½	3 9/16x7½	3 9/16x9½	5½x5½	5½x7½	5½x9½	7½x7½	7½x9½

Board Sizes in Inches

Nominal Size	1x4	1x6	1x8	1x10	1x12
Actual Size—Common	¾x3 9/16	¾x5 9/16	¾x7½	¾x9½	¾x11½
Actual Size—Shiplap	¾x3	¾x4 15/16	¾x6⅞	¾x8⅞	¾x10⅞
Actual Size—T&G	¾x3¼	¾x5 3/16	¾x7⅛	¾x9⅛	¾x11⅛

Fig. 52-27A U.S. Customary lumber sizes.

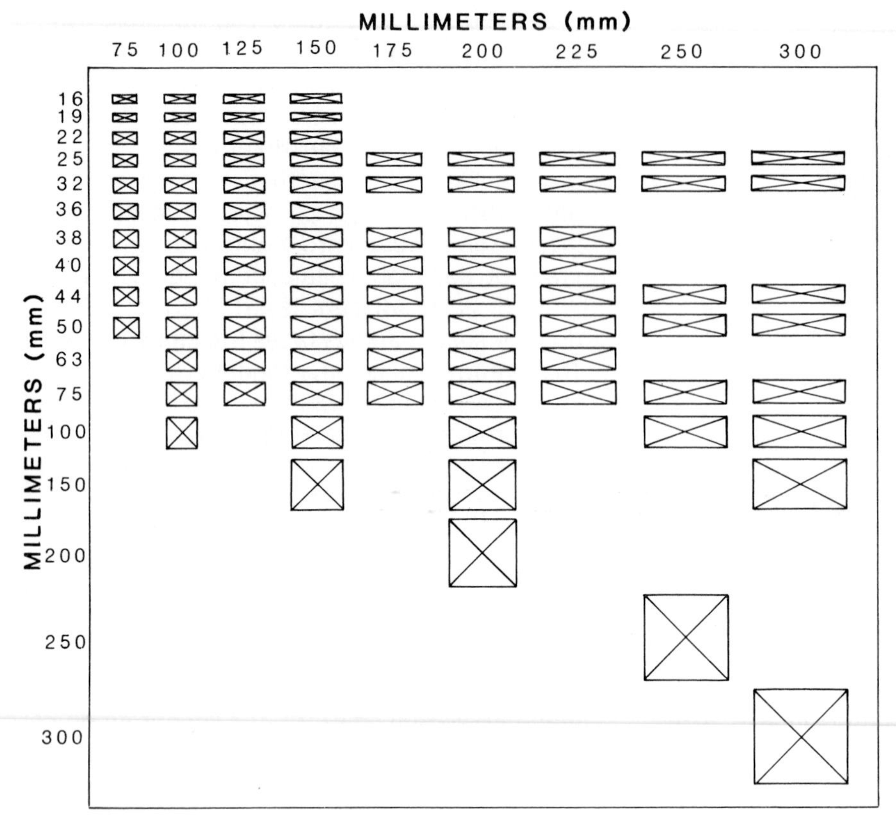

Fig. 52-27B Metric lumber sizes.

strength, as shown in Fig. 52-29A. Because of live-load differences, plywood panels used for structural purposes are marked with two index numbers indicating the structural rating (Fig. 52-29B). The first number represents the maximum span (in inches) possible between supporting roof members. The second number represents the maximum allowable span when used for flooring.

Designation	*Description*
S1S	Surfaced on one side
S2S	Surfaced on two sides
S1E	Surfaced on one edge
S2E	Surfaced on two edges
S1S1E	Surfaced on one side, one edge
S2S1E	Surfaced on two sides, one edge
S4S	Surfaced on four sides

Fig. 52-28 Lumber surfacing codes.

Grade	*Description*
Standard	For use as subflooring, roof sheathing, wall sheathing, and structural interior applications.
Structural Class I and II	For uses requiring resistance to tension, compression, and shear stress including box beams, stressed skin panels, and engineered diaphragms. High nail-holding quality and controlled grade and glue bonds.
CC Exterior	Meets all exterior plywood requirements.
BB Concrete-Form Panels, Class I and II	Edges sealed and oiled at the mill and used for concrete form panels.

Fig. 52-29A Plywood construction grades.

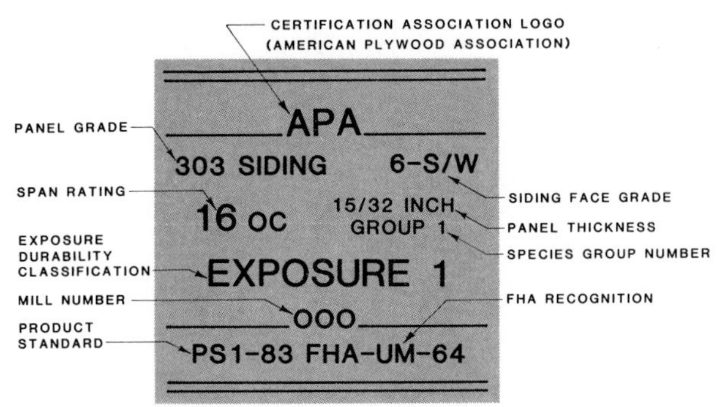

Fig. 52-29B Construction plywood grade marks.

Since hardwood plywood is used for cabinet and furniture making, veneer plywood grades are classified by a letter indicating the number of knots, checks, stains, and open sections in each panel, as shown in Fig. 52-30. These letters are used to show only the grade of the front and back plys. For example, a B-D grade means the front veneer is B grade and the back veneer is D grade. In addition, panel grading, span rating, research report number, exposure class, thickness, mill number, species number (Fig. 52-31), and product standard grade marks are stamped on each sheet, as shown in Fig. 52-32. Unlike the procedure followed in selecting softwood plywood for use in rough construction, grain patterns, color and texture consistency, specific species, smoothness, and finishability must be matched in selecting hardwood plywood for cabinets, paneling, and furniture.

Grade	*Description*
Grade A	Paintable and smooth with no more than 18 neat boat, sled, or router type repairs made parallel with grain, and will accept natural finish.
Grade B	Solid surface with shims, circular repair plugs, or tight knots less than 1″ wide permitted.
Grade C	Tight knots of less than 1½″, knotholes less than 1″ wide, synthetic or wood repairs, limited splits, slitching, discoloration, and sanding defects that do not impair strength permitted.
Grade C Plugged	Some broken grain, synthetic repairs, splits up to 1″ wide, knotholes and bareholds up to ¼″ × ½″ permitted.
Grade D	Knots and knotholes up to 2½″ wide across grain or 3″ wide if within limits permitted but restricted to interior use.

Fig. 52-30 Veneer plywood grades.

Framing Types

When multiple-level buildings are constructed using a wood skeleton frame, either the *platform* (western) framing method (Fig. 52-33) or the *balloon* (eastern) framing method (Fig. 52-34) is used. In platform framing the second floor rests directly on first-floor exterior walls. In balloon-

Group 1	*Group 2*	*Group 3*	*Group 4*	*Group 5*
Beech	Cedar, Port	Alder, Red	Aspen	Basswood
Birch	Cypress	Birch, Paper	Cedar	Poplar
Sweet	Douglas Fir 2 (b)	Cedar, Alaska	Incense	Balsam
Yellow	Fir	Fir	Western Red	
Douglas Fir 1 (a)	Balsam	Subalpine	Cottonwood	
Maple, Sugar	California Red	Hemlock	Pine	
Pine	White	Maple	Eastern	
Caribbean	Hemlock	Bigleaf	White	
Ocote	Lauan	Pine	Sugar	
Pine, South	Maple, Black	Jack		
Loblolly	Pine	Ponderosa		
Longleaf	Red	Spruce		
Shortleaf	Western White	Redwood		
Slash	Spruce	Spruce		
	Yellow Poplar			

Fig. 52-31 Wood species groups.

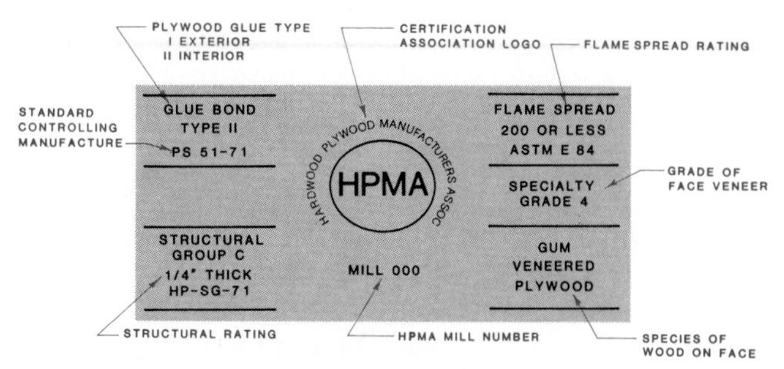

Fig. 52-32 Typical hardwood plywood grade marks.

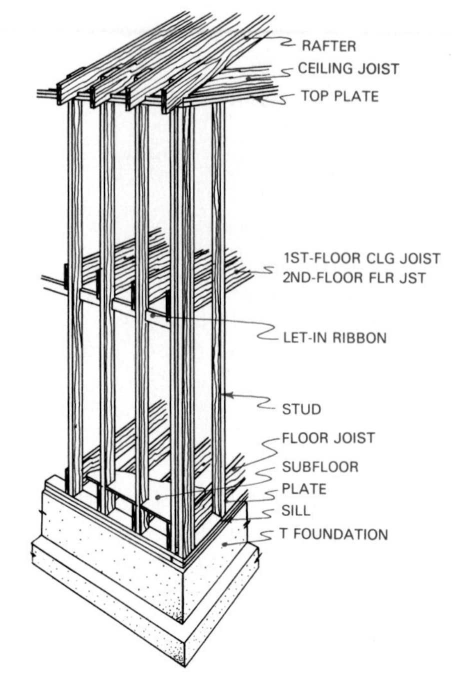

Fig. 52-34 Balloon framing.

framed buildings, the first-floor joists rest directly on a sill plate, and the second floor joists bear on ribbon (leger) strips set into the studs. The studs are continuous for the full height of the building, and floor joists butt against the sides of the studs.

HEAVY TIMBER CONSTRUCTION

Large timbers have been used in construction for centuries, usually for floor and roof systems in

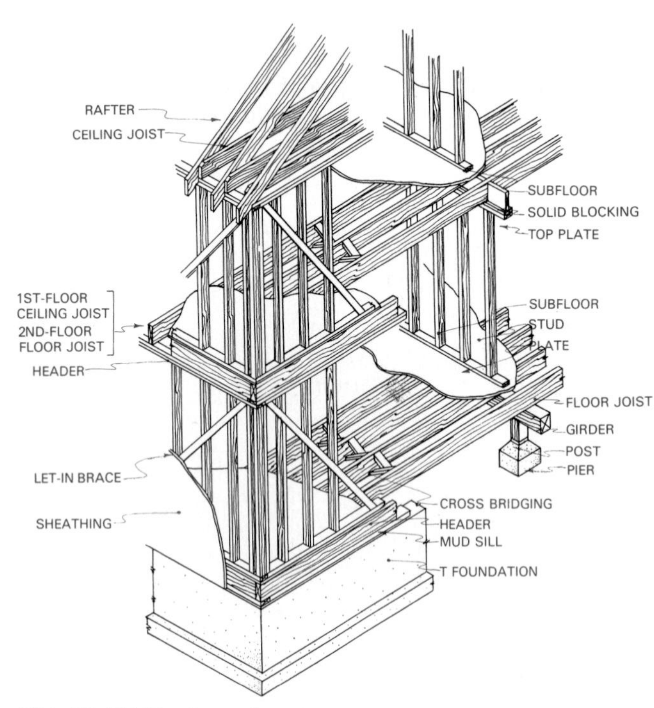

Fig. 52-33 Platform framing.

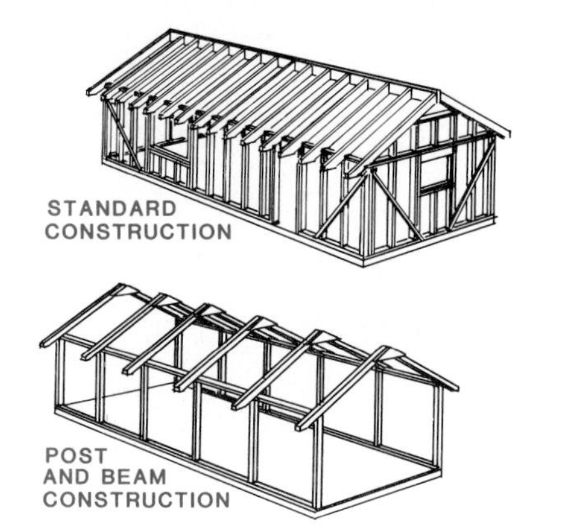

Fig. 52-35 Heavy timber and skeleton frame construction.

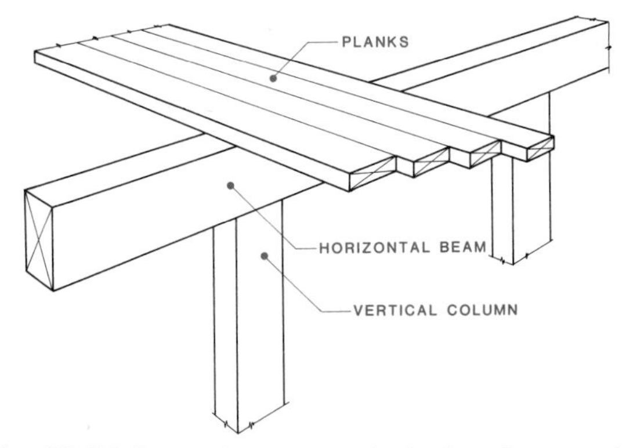

Fig. 52-36 Posts, beams, and planks characterize heavy timber construction.

buildings of bearing-wall design. The development of large glass sheets, sheathing materials, and improvements in the manufacture and transportation of large wood members have given rise to the use of heavy timbers for walls as well as for long-span floors and roof systems. When heavy timbers are used in this manner, the building type is known as *post-and-beam* construction. Figure 52-35 shows a comparison of post-and-beam and skeleton-frame construction. In post-and-beam framing, larger members are spaced farther apart than in skeleton-frame construction.

Although post-and-beam members are larger than those used in skeleton-frame construction, wider modular spacing actually results in the use of less building materials. There is also considerable savings in labor since fewer members are handled and fewer intersections connected. Since many post-and-beam members are exposed, a better grade of lumber is therefore specified to create a more pleasing visual effect. Rigid insulation must be used above, not under, the roof planks to expose the natural plank ceiling. And plumbing and electrical lines must be passed through cavities in columns and/or beams. Bearing partitions and other dead loads, such as bathtubs, must be located over beams, or additional support framing must be used.

Post-and-beam construction is based on the relationship of three basic components: *columns* (posts), *beams*, and *planks*. Vertical *columns* support *beams*, which support *planks* placed perpendicular to the beams, as shown in Fig. 52-36. Floor and roof systems are supported by beams, which are supported by posts (columns), which transfer loads to footings. Member sizes and spaces vary depending on load requirements. Figures 52-37 and 52-38

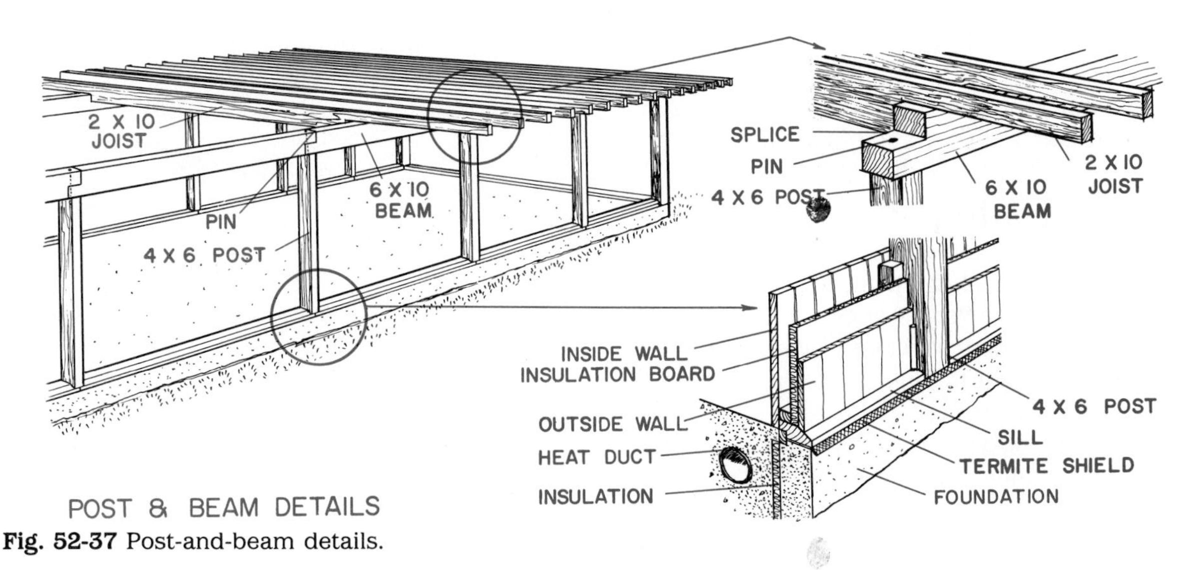

Fig. 52-37 Post-and-beam details.

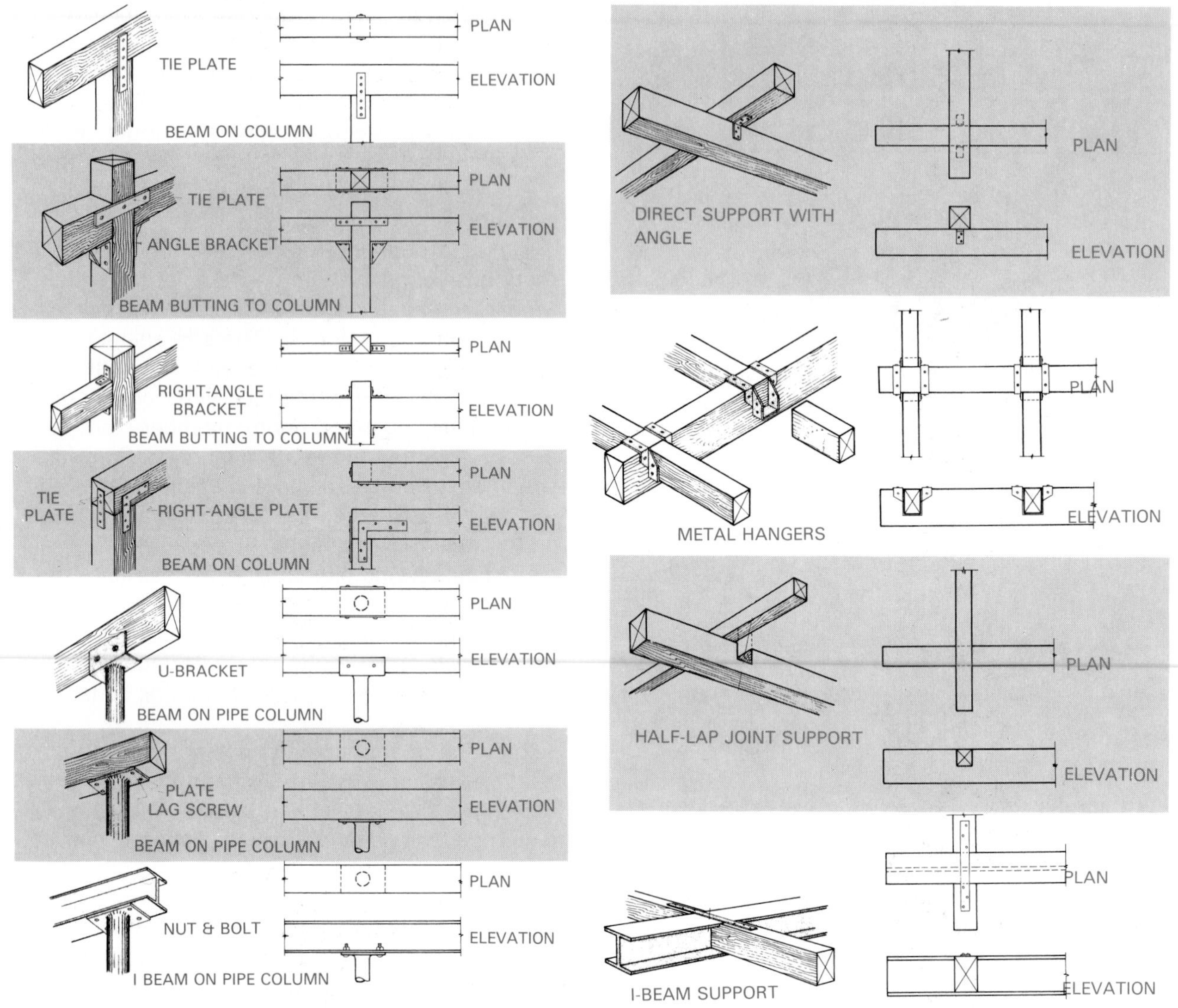

Fig. 52-38 Post-and-beam joints.

shows some of the common details, members, and joints used in post-and-beam construction.

Floor Framing

Timber floor systems involve the use of heavy wood planks (or thick plywood) placed over widely spaced beams, as shown in Fig. 52-39. This type of system is called *plank-and-beam construction.* Each modularly spaced beam replaces several intermediate joists in a skeleton-framed floor system. For example, a 24′ distance may require 19 conventional floor joists spaced at 16″ OC. But only 7 beams placed 4′ OC may be needed to support the same loads.

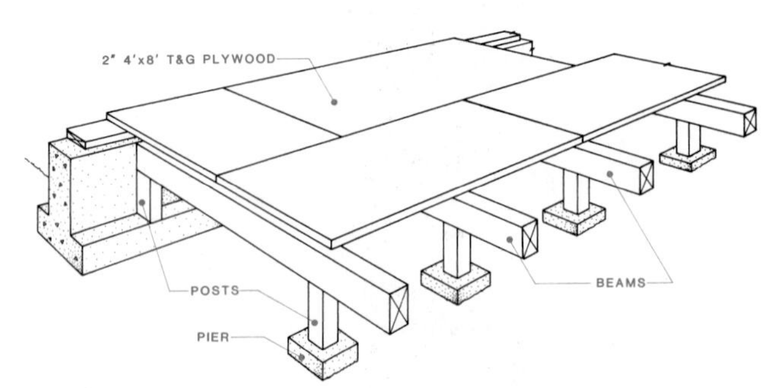

Fig. 52-39 Timber-plank and beam floor system.

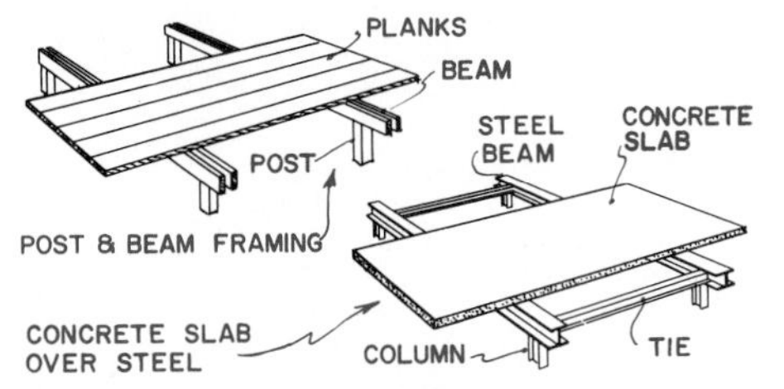

Fig. 52-40 Wood planking functions are similar to concrete slab decking.

However, in this system floor planks must also be sufficiently strong to avoid deflection at midspan. Wood planking, like reinforced concrete slabs, reinforces and ties beams together, as shown in Fig. 52-40.

Wall Framing

Just as beams replace conventional joists in plank-and-beam floor systems, posts replace conventional studs in post-and-beam wall construction. This enables the creation of large open spans (Fig. 52-41). The spanned space between the wall posts can be filled with any nonbearing material or components such as windows, doors, or insulating material. For this reason nonstructural elements in a post-and-beam outside wall are known as *curtain walls*. Figure 52-42 shows a wall-elevation framing drawing with wall column intersections detailed.

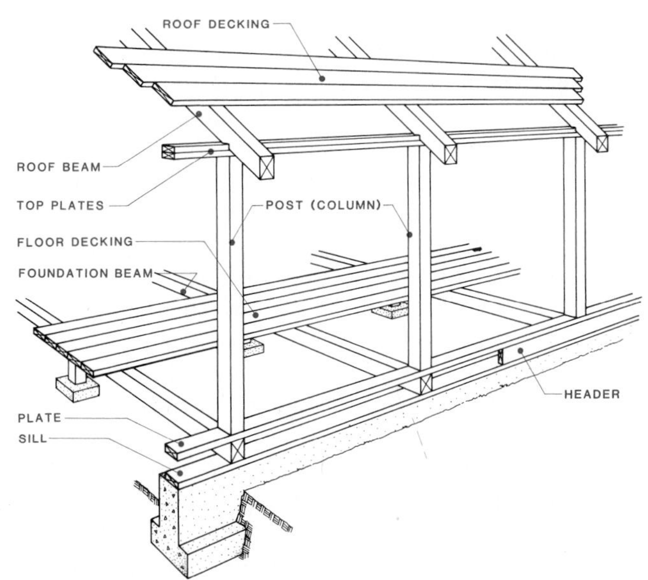

Fig. 52-41 Post-and-beam framing creates open wall areas.

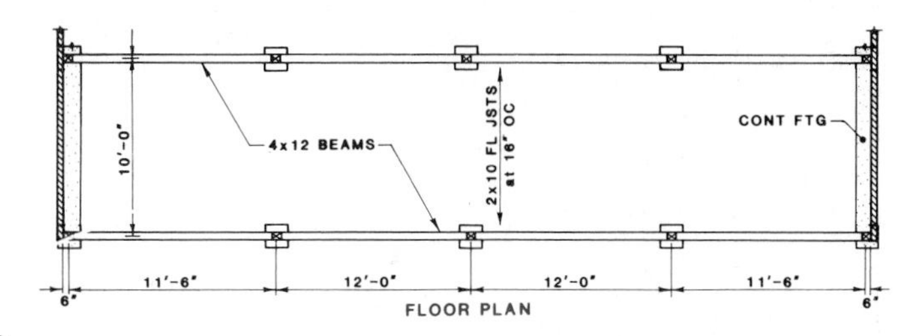

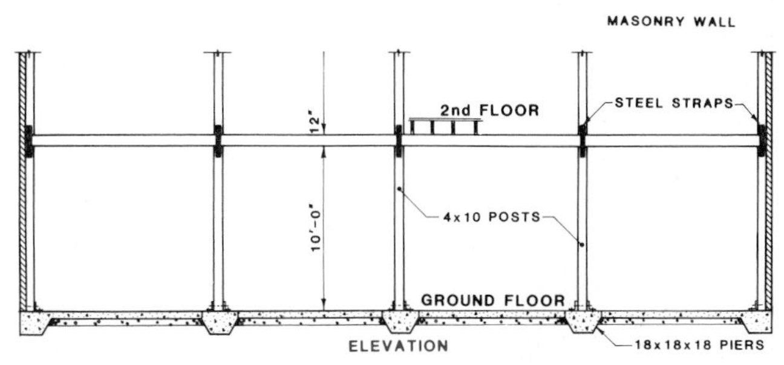

Fig. 52-42 Post-and-beam wall framing.

Roof Framing

Plank-and-beam roofs are similar to plank-and-beam floors since planks are used instead of conventional roof joists and sheathing. There are two types of plank-and-beam roof systems: *longitudinal* and *transverse*.

In longitudinal systems roof beams are aligned parallel with the long axis of the building (Fig. 52-43). In transverse systems the beams are laid across the short width of the building. As illustrated in Fig. 52-44A, transverse beams either intersect a ridge beam on pitched roofs or lie flat across the span on flat roofs. One end of

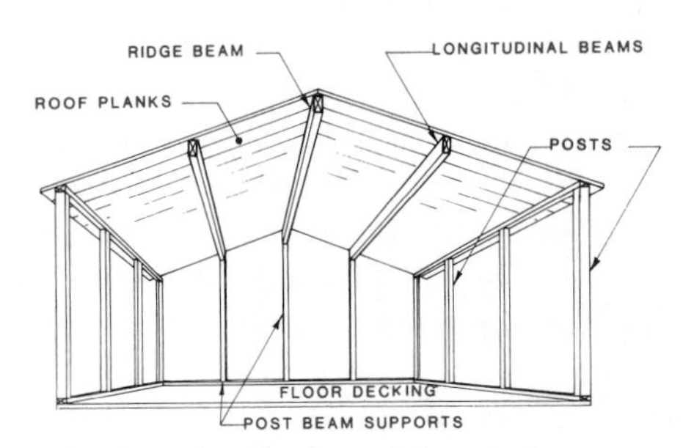

Fig. 52-43 Longitudinal roof framing.

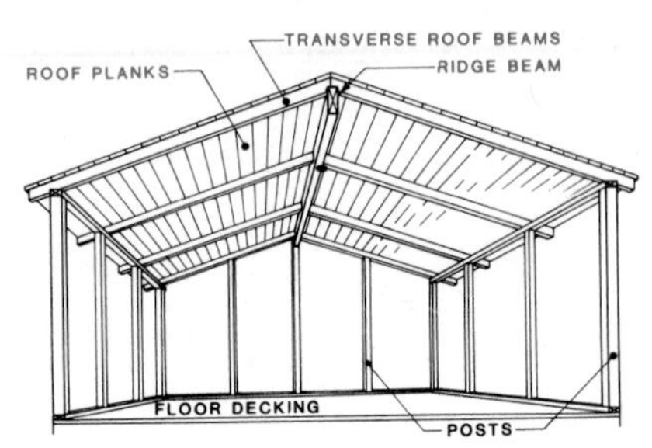

Fig. 52-44A Transverse roof framing.

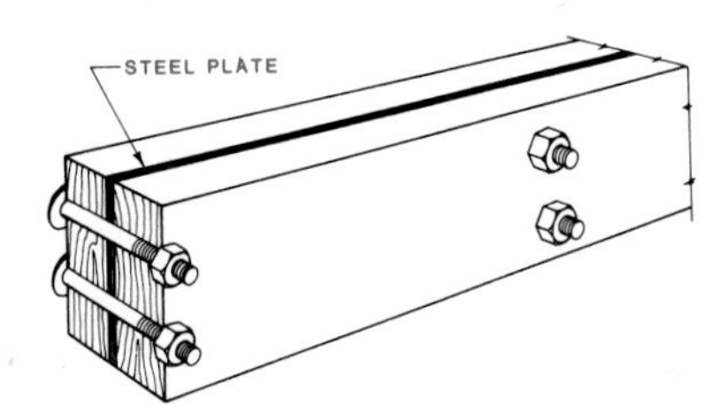

Fig. 52-44B Flitch plate beam.

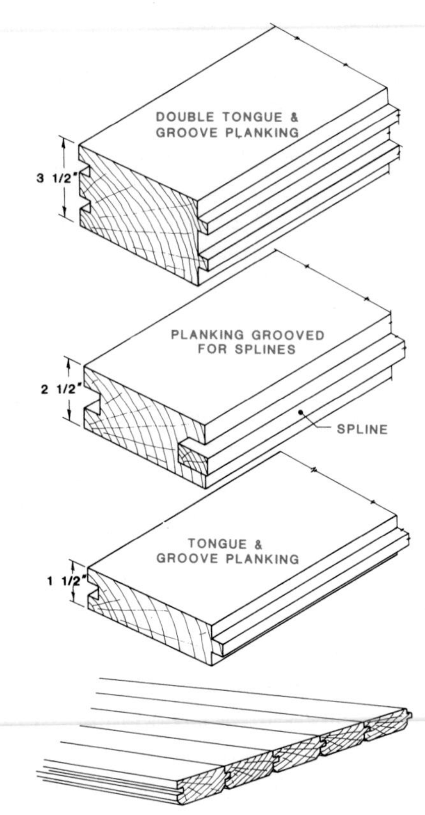

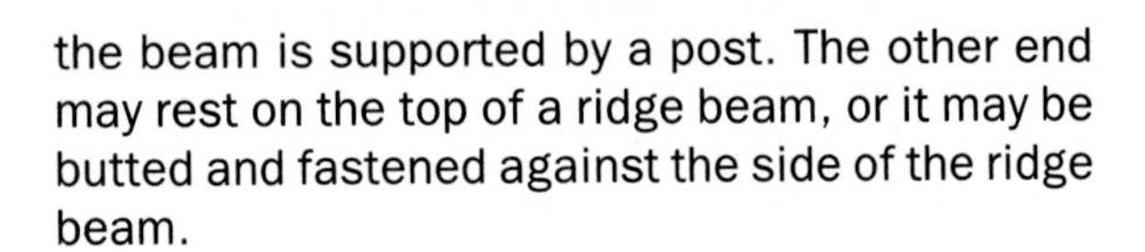
Fig. 52-45 Tongue-and-groove planking.

the beam is supported by a post. The other end may rest on the top of a ridge beam, or it may be butted and fastened against the side of the ridge beam.

Structural Timber Members

There are three types of structural members used in contemporary post-and-beam construction: solid, laminated, and fabricated components.

Solid Members Solid wood timbers are available in thicknesses which range from 3″ to 12″; however, the use of sizes over 8″ is hampered by the tendency of large solid wood timbers to warp and wind. One method of stabilizing larger solid wood members is the addition of steel plates known as *flitch plates* (Fig. 52-44B) to help hold the timber in a straight position.

Because planking is used in smaller widths (2″ to 4″), solid flooring is commonly used, although laminated flooring is available. Because of the impact of live load thrusts, solid planking is usually specified as tongue and groove (T & G). The T & G joint (Fig. 52-45) reduces deflection by tying the flooring planks together into one monolithic unit.

Laminated Members When larger timbers are needed to support heavier weights or greater spans, laminated timbers are often used. Laminated timbers are made from thin layers (less than 2″) of wood, glued together under pressure. Members are laminated either vertically or horizontally (Fig. 52-46). Laminated timbers are stronger than solid timbers because the grain direction is reversed in alternate layers as in plywood. The coefficient of expansion is lower in laminated members due to a more consistent moisture content than solid wood.

In addition to the use of straight laminated members for post and beams, laminated decking is also available in 2″ to 6″ thicknesses and in 6″ to 12″ widths. Laminated decking is specified in nominal sizes on construction drawings. When decking material is laminated, the layers are offset to create a variety of T & G patterns, as shown in Fig. 52-47. Another method of adding strength to wood decking is the lamination of members on their side, as shown in Fig. 52-48. This method aligns the wood grain vertically with more depth to create greater resistance to loads.

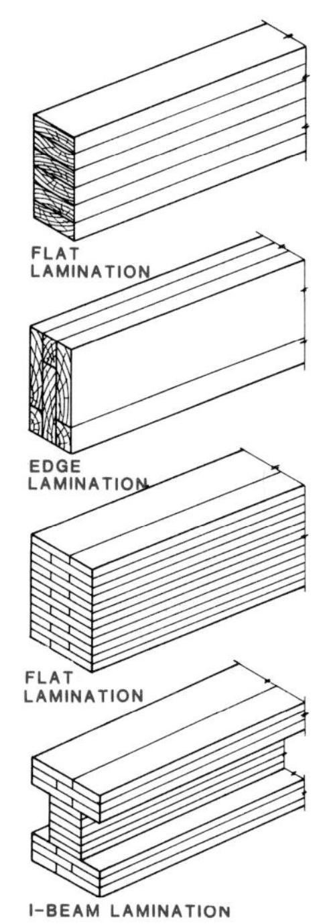

Fig. 52-46 Laminated beams.

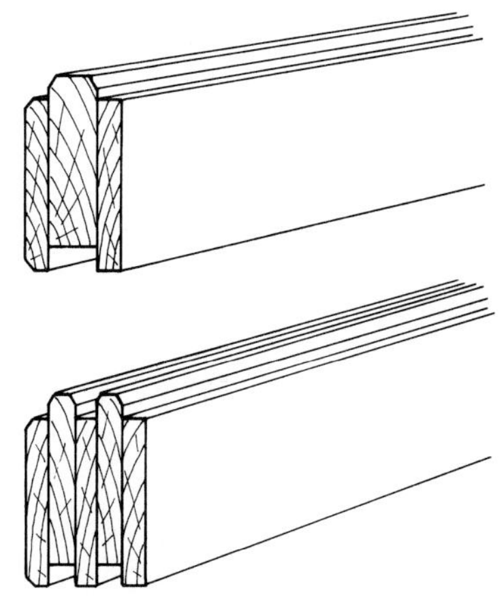

Fig. 52-47 Offset laminated members.

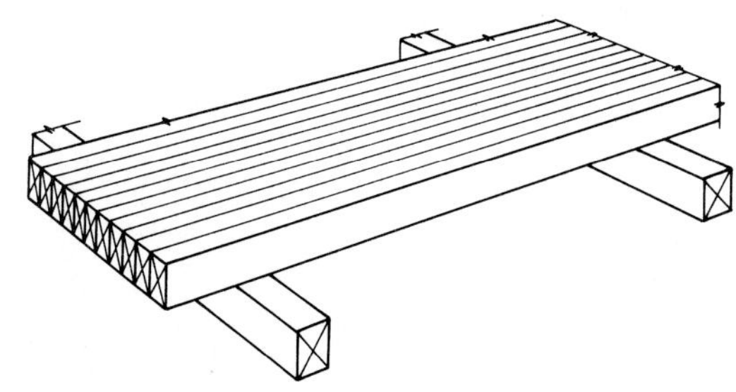

Fig. 52-48 Side laminated decking.

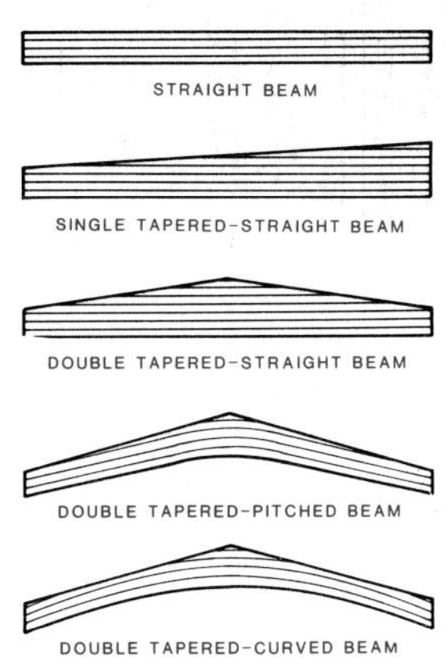

Fig. 52-49 Laminated beam forms.

Although lamination can create stronger, larger, and more structurally stable members, its most popular feature is the ability to be bent into a wide variety of structurally sound and aesthetically pleasing shapes. A variety of beam forms (Fig. 52-49), including arches (Fig. 52-50), are created by first bending thin, parallel layers of wood to a desired shape. Then the layers are glued and clamped together under extreme pressure. When the glue dries, the member retains the new, bent form as shown in Fig. 52-51.

In drawing arches, the base location of each arch is shown in the ground floor plan. The profile shape, including height and width dimensions, is drawn on elevation drawings and/or on elevation sectional drawings (Fig. 52-52).

Fabricated Members When heavy load bearing is not critical, lightweight stressed-skin (sandwich) panels are often used in place of solid or laminated members. These prebuilt or site-built panels are made by gluing and/or nailing plywood sheets to structural member frames. Since panels can easily be constructed using standard plywood sizes, their use for floor, wall, and roof panels is popular. Stressed skin panels can also be used to make box beams in spans up to 120′ depending on load factors. Folded plate roofs and curved panels can also be either prefabricated on the site or factory-built.

Timber Connectors Because of heavy timber sizes, concentrated loads, and lateral thrust from winds and earthquakes, special joints and fasteners are required to attach post-and-beam members to the foundation and to

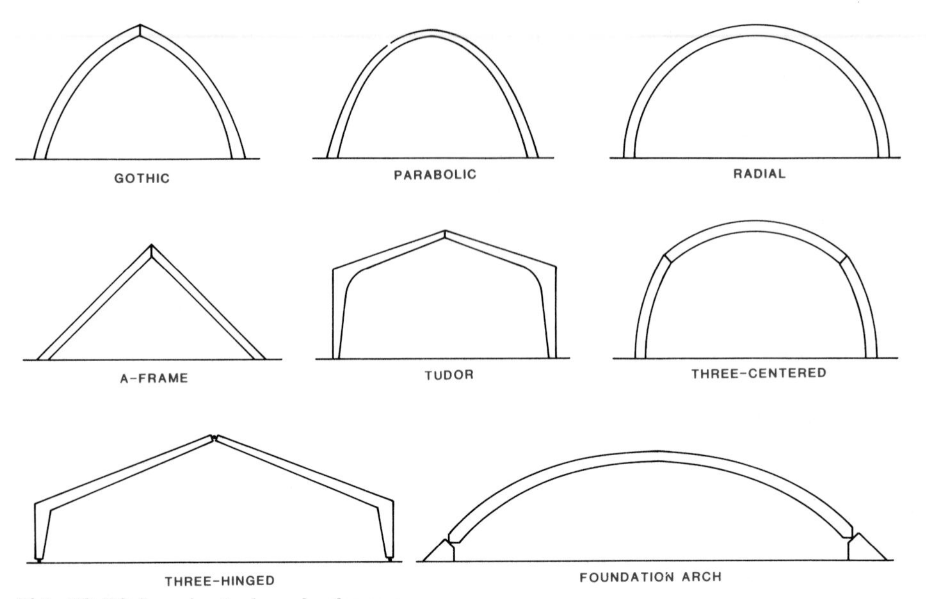

Fig. 52-50 Laminated arch shapes.

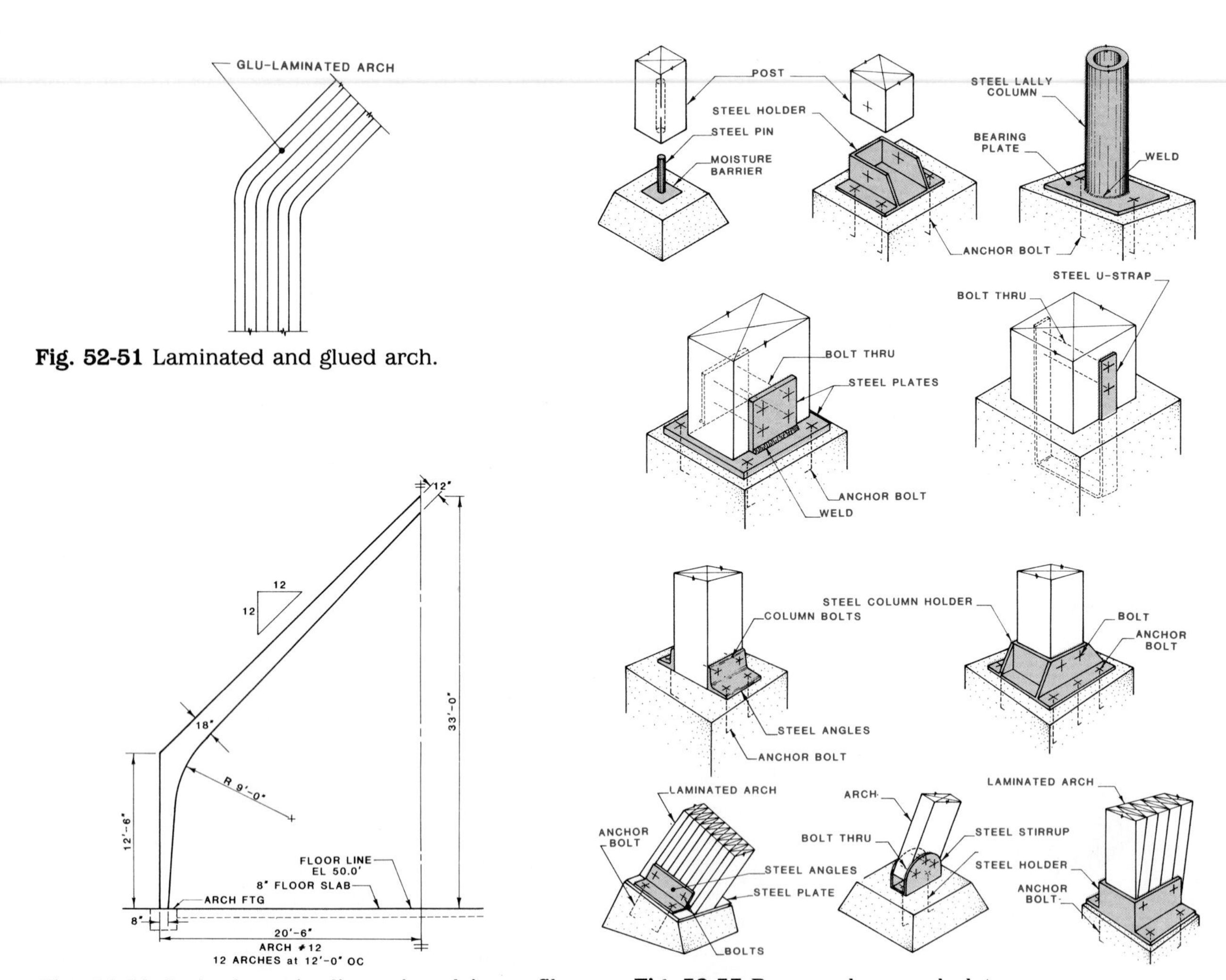

Fig. 52-51 Laminated and glued arch.

Fig. 52-52 Arch shape is dimensioned in profile.

Fig. 52-53 Base anchors and plates.

other members. Nails are inadequate, except as a temporary holding device, and lag screws can only be used in areas of limited stress.

Base anchors and *plates* (Fig. 52-53) are used to attach the base of heavy timber posts to the foundation and to prevent wood deterioration. *Metal strap ties* and *gusset plates* (Fig. 52-54) are used to attach posts to beams and to keep traverse roof beams aligned with ridge beams. To fasten perpendicular intersections, *angle brackets* (Figs. 52-55 and 52-56) are commonly used to prevent lateral movement between members. Keep in mind that post-and-beam construction has good resistance to dead loads, which exert pressure directly downward. However, because of the large unsupported wall areas, lateral live loads can be a problem, as shown in Fig. 52-57. *Diagonal ties* or *sheathing*

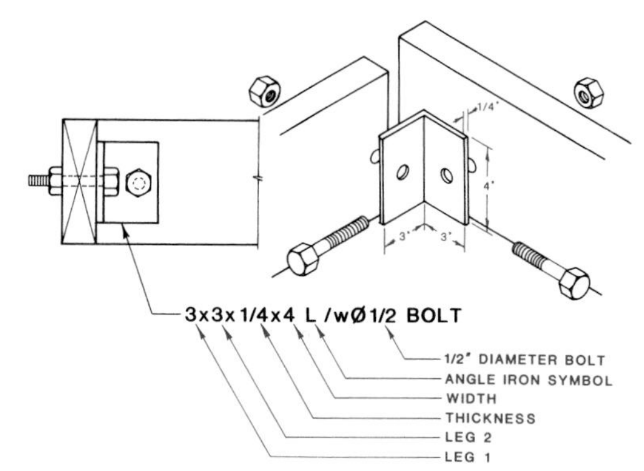

Fig. 52-55 Angle brackets.

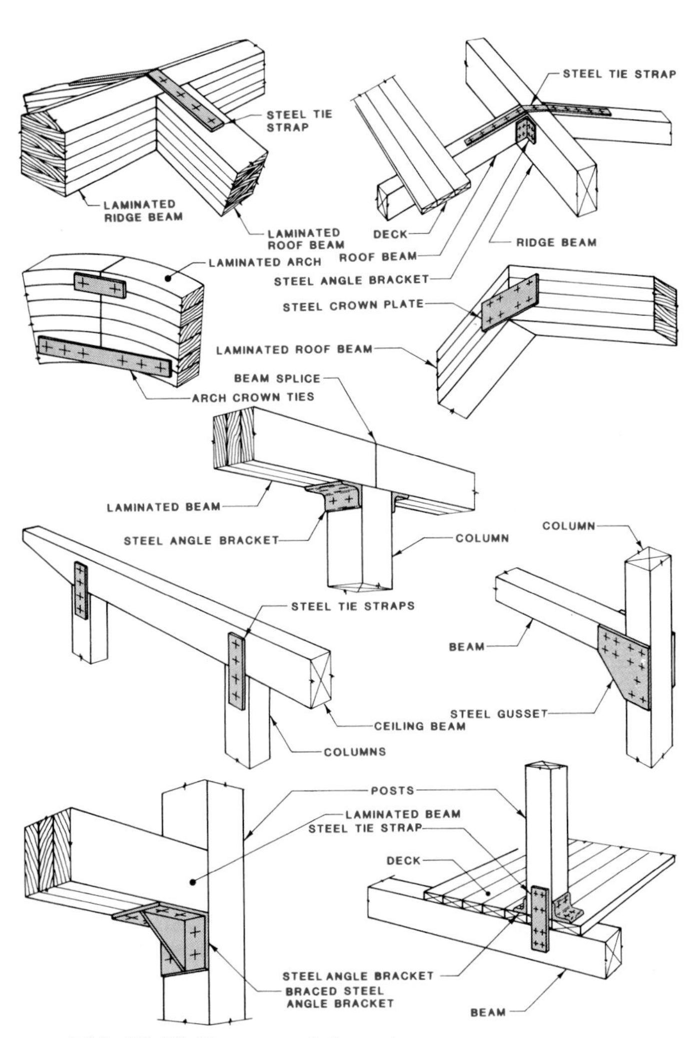

Fig. 52-54 Straps and gussets.

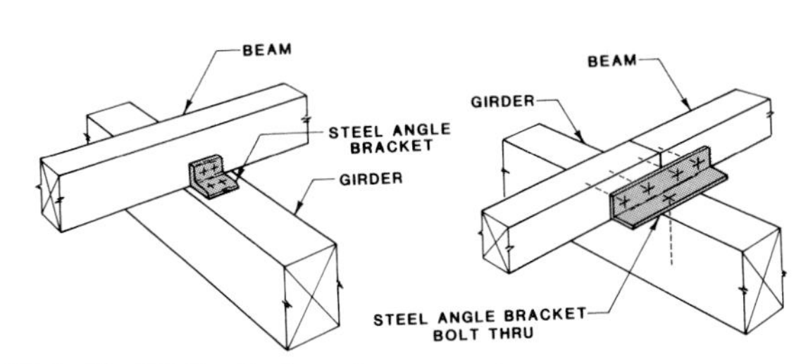

Fig. 52-56 Angle-bracket notations.

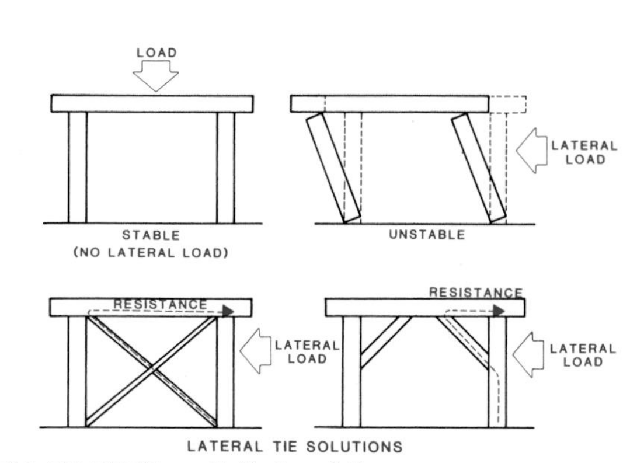

Fig. 52-57 Use of diagonal ties.

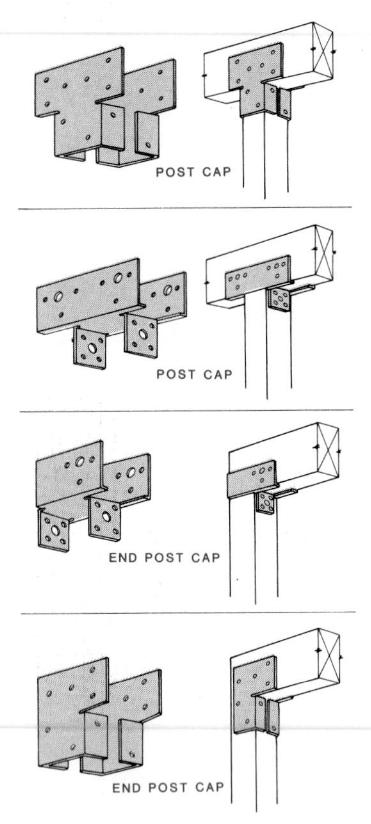

Fig. 52-58 Post caps.

helps control the lateral thrust, but angle brackets also play a big role in providing rigidity to the joints.

Post caps are used extensively when posts intersect beams at a beam joint, as shown in Fig. 52-58. However, when the end of a member intersects the side of another member, without resting on it, metal hangers (Fig. 52-59) are usually specified. In some cases special clips (Fig. 52-60) may be used to prevent movement where two members intersect at angles other than 90°.

Because interior members in post-and-beam construction are often exposed, the use of hidden joints and fasteners is often needed. Dowels, rods, and half-lap joints, as shown in Fig.

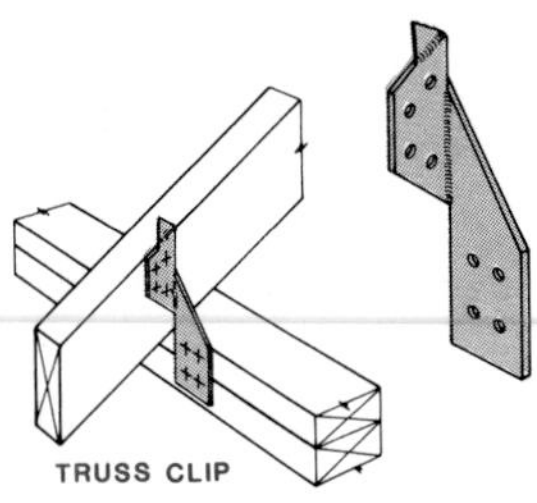

Fig. 52-60 Truss clips.

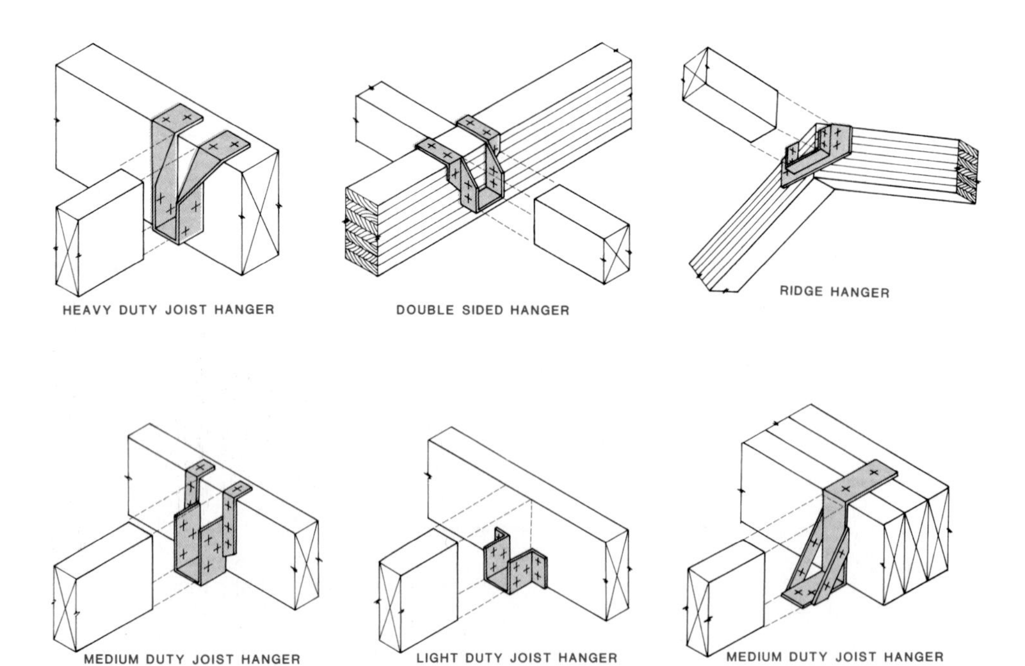

Fig. 52-59 Joint-and-beam hangers.

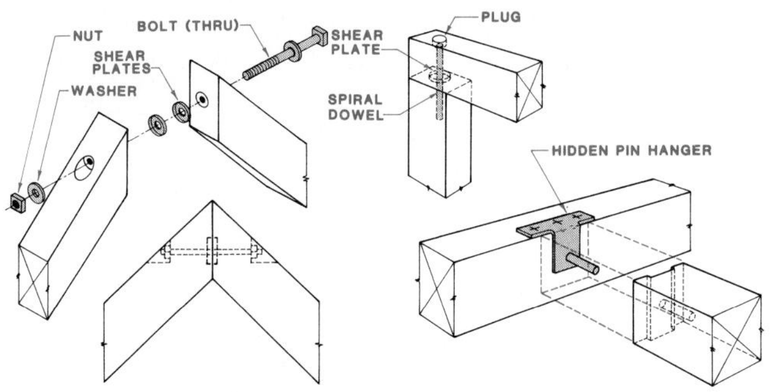

Fig. 52-61 Hidden connectors.

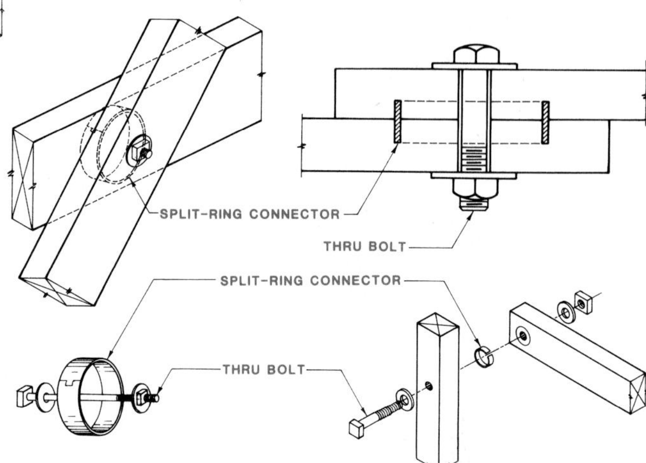

Fig. 52-62 Ring connectors.

52-61, are used for this purpose. However, these methods are often costly to construct and install. Split-ring connectors, as shown in Fig. 52-62, are extremely strong. They can be assembled with relative ease and the ring can be concealed with the use of dowels or bolts. When bolts are used the ends can be counterbored and plugged once the final alignment is achieved.

STRUCTURAL STEEL SYSTEMS

Steel can span greater distances and support greater loads than any other conventional building material. However, steel is extremely heavy and creates additionally heavy dead loads. Structural steel is available in hot-rolled plates, bars, pipes, tubing, and a variety of rolled shapes which are used for girders, beams, columns, decking, and trusses. Preparing structural steel drawings requires a working knowledge and understanding of steel symbols, notations, identifications, drawing conventions, measurements, and fastening and intersection methods.

Structural steel types are specified by metallurgical characteristics and minimum stress yields as designated by the American Society for Testing Materials (ASTM). Figure 52-63 shows the descriptions of ASTM structural steel types.

Structural Steel Members

Steel used for structural purposes is manufactured in plates, bars, tubing, and rolled shapes, such as angles, channels, S-beams, wide flange beams, and tee sections.

Plates Structural plates are flat sheets of rolled steel and range in thickness from ⅛″ to 3″ and in widths from 8″ to 60″. Plates are specified by thickness, width, and length in that order; for example a 1″ × 9″ × 1′-6″ plate is specified as shown in Fig. 52-64. Plates are used as webs in built-up girders and columns (Fig. 52-65) and to reinforce webs or flanges of structural steel shapes. Bearing plates also provide bearing surfaces between columns and concrete footings, as shown in Fig. 52-66.

Bars Steel bars used for structural purposes are available in round, square, hexagonal, and flat (rectangular) cross-section shapes as shown in Fig. 52-67. Square, hexagonal, and round bars are rolled in ¹⁄₁₆″ increments from ¹⁄₁₆″ to 12″. Flat bars are rolled in ¼″-width increments up to 8″. Round bars are specified by diameter. Square bars are specified by width or gauge number, and flat bars are specified by width and

ASTM Type*	*Min Yield Stress Point*	*Manufactured Forms*	*Description*
A36	36,000 PSI	Sheets Plates Bars Shapes Rivets Nuts Bolts	A medium carbon steel that is the most commonly used structural steel. Suitable for buildings and general structures, and capable of welding and bolting.
A440	42,000 PSI	Plates Bars Shapes	A high strength, low alloy steel suitable for bolting and riveting, but not welding. Used for lightweight structures—high resistance to corrosion.
A441	40,000 PSI	Plates Bars Shapes	A high strength, low alloy steel modified to improve welding capabilities in lightweight buildings and bridges.
A572	41,000 PSI	Limited Types of Shapes Bars & Plates	A high strength, low alloy economical steel suitable for boltings, riveting, and welding with lightweight high toughness for buildings and bridges.
A242	42,000 PSI	Plates Bars Shapes	A durable, corrosion resistant, high strength, low alloy steel which is lightweight and used for buildings and bridges exposed to weather. Can be welded with special electrodes.
A588	42,000 PSI	Plates Bars Shapes	A lightweight corrosion, high strength, low alloy steel with high durability in high thicknesses used for exposed steel.
A514	90,000 PSI	Limited shapes & Plates	A quenched and tempered alloy steel with varying strength, width, thickness, and type.
A570	25,000 PSI	Plates Light Shapes	A light gauge steel used primarily for decking, siding, and light structural members.
A606	45,000 PSI	Plates	A high strength, low alloy sheet and strip steel with high atmospheric corrosion resistance.

*American Society for Testing Materials

Fig. 52-63 Structural steel types.

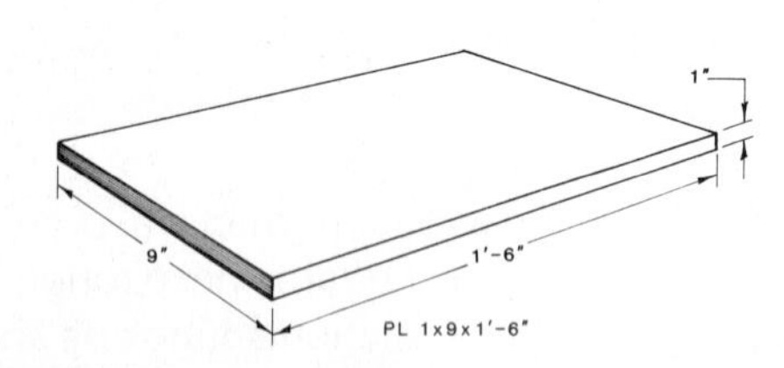

Fig. 52-64 Plate specifications.

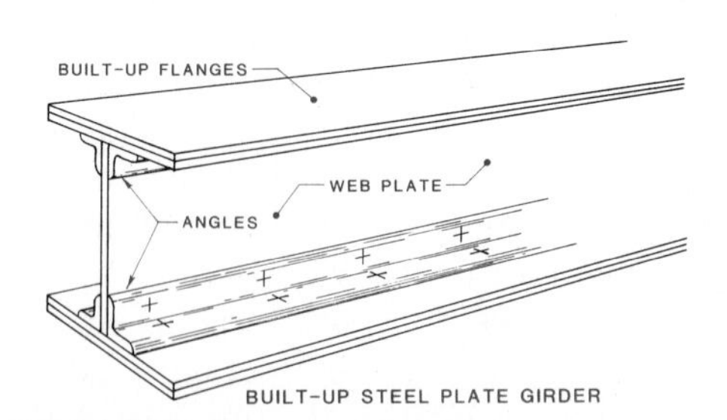

Fig. 52-65 Built-up plate girder.

thickness. Hexagonal bars are specified by the distance across the flats (AF). Steel bars are used primarily for bracing other structural components and for concrete reinforcement.

Steel Pipe and Structural Tubing

Steel pipe and tubing are used extensively in exposed areas because of their clean aesthetic lines. Structural pipe and tubing is available in

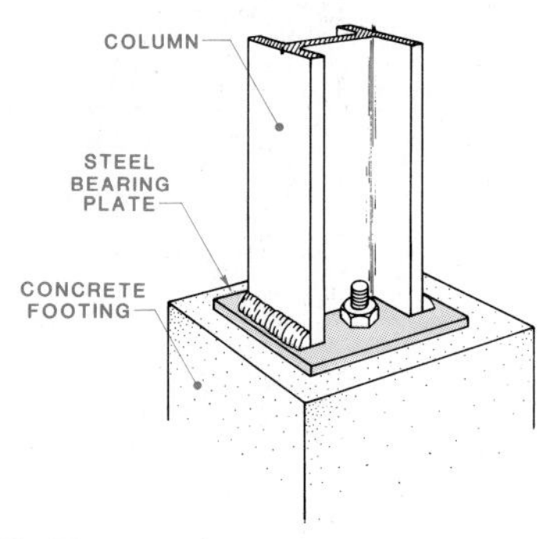

Fig. 52-66 Bearing plate.

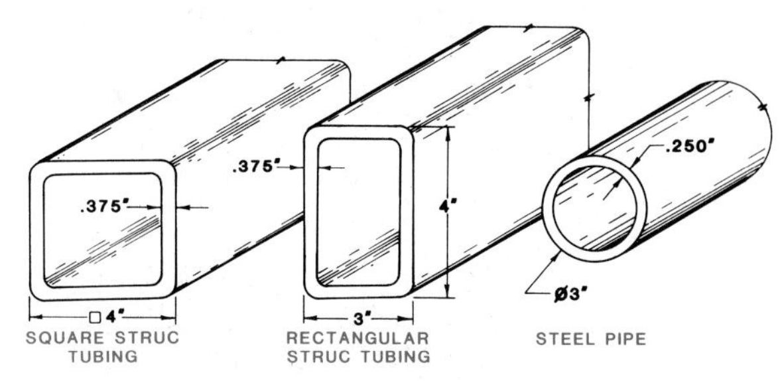

Fig. 52-68 Steel pipe and tubing.

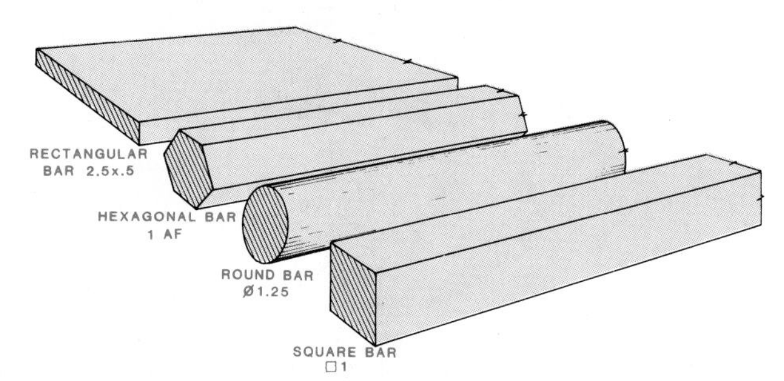

Fig. 52-67 Types of steel bars.

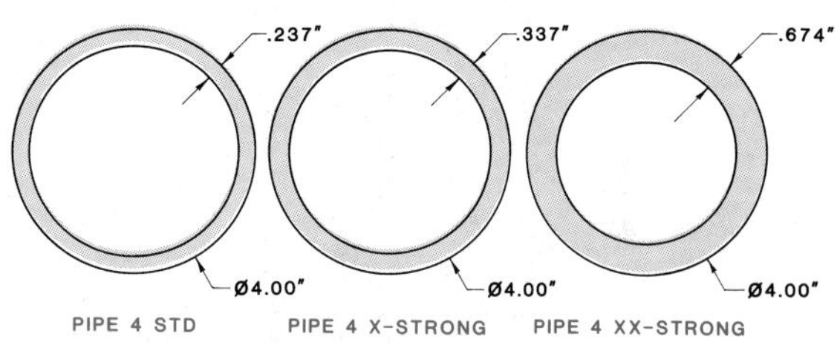

Fig. 52-69 Steel pipe thickness.

round, square, and rectangular cross-section shapes, as shown in Fig. 52-68. Hollow steel pipe is manufactured in sizes from ½″ to 12″ (inside diameter) and in three strength classes. Strength classes relate to wall thicknesses and are either standard weight (STO), extra strong (x-strong) or double extra strong (xx-strong). On structural drawings, pipe is specified by diameter and strength. Thus, a 4″ extra-strong steel pipe is labeled: pipe 4 xx-strong (Fig. 52-69).

Square structural tubing is specified by cross-section width and thickness and is available in sizes from 2″ × 2″ to 10″ × 10″ (outside dimension). Rectangular tubing sizes range from 3″ × 2″ to 12″ × 8″ (OD). Structural tubing is specified by the symbol TS followed by the width, thickness, and wall thickness. A rectangular structural tube 4″ wide, 3″ thick, with a wall thickness of ¼″, is therefore labeled: TS 4 × 3 × .25. Round structural tubing is specified by outside diameter and wall thickness. For example, a 4″-diameter tube with a ¼″ wall thickness is labeled: 4 OD × .25.

Structural Steel Shapes Steel is rolled into angles, channels, S-beams, wide flange beams, and tee sections. These shapes, with accompanying notations, are shown in Fig. 52-70. Steel shapes are designated on construction drawings by shape symbol, depth in inches, and weight in foot pounds of length. Structural steel is used for girders, beams, columns, and truss components. Figure 52-71 shows the most standard structural steel shapes and the related drawing symbol.

L-shape structural steel members (angles) are rolled in the (cross-section) shape of the letter L with legs of equal or unequal length. Equal L-shapes are available with leg lengths of 1″ to 8″. Unequal leg lengths range from 1¾″ to 9″. Whether equal or unequal, the thickness of each leg is always the same. L-shapes (angles) are specified on drawings by the symbol L followed by the length of each leg, followed by the wall thickness. For example, an L-shape member with one 2″ and one 3″ leg and a wall thickness of ½″ is specified: L 2 × 3 × .5. L-shape members are used as components in built-up beam,

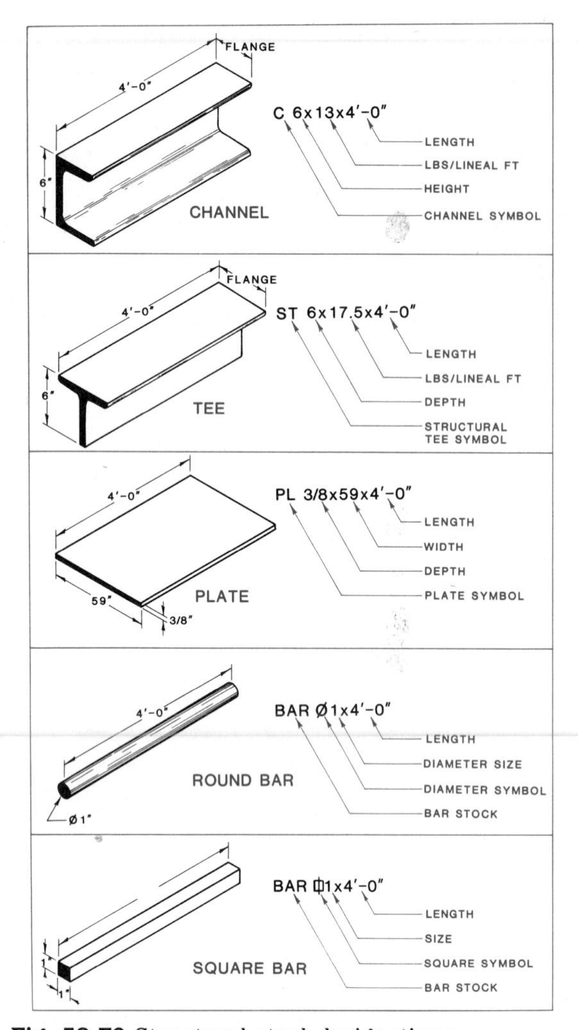

Fig. 52-70 Structural steel designations.

NAME	SECTIONAL FORM	SYMBOL	PICTORIAL
WIDE FLANGE		W	
AMERICAN STANDARD BEAM		S	
TEE		T	
ANGLE		L	
ZEE		Z	
AMERICAN STANDARD CHANNEL		C	
BULB ANGLE		BL	
LALLY COLUMN		◎	
SQUARE BAR		⌷	
ROUND BAR		Φ	
PLATE		℄	

Fig. 52-71 Structural steel shapes.

columns, and trusses. They are also used for connectors and as lintels in light- or short-span construction.

Channels are rolled into a cross-section shape resembling the letter U, with the inner faces of flanges shaped with a $2/12$ pitch. Channels are classified by depth from 3″ to 15″. There are two types of channels specified for structural use: American Standard channels (C) and Miscellaneous channels (MC). Channels are specified by symbol (C or MC), followed by the depth times the weight per foot. An 8″-deep Standard channel that weighs 11.2 lb/ft is labeled: S 8 × 11.2. Channels are used for roof purlins, lintels, truss chords, and to frame-in floor and roof openings.

S-shapes (formerly I beams) are rolled in the shape of a capital letter I. American Standard shapes have narrow flanges with a $2/12$ inside pitch. S-shapes are classified by the depth of the web and the weight per foot. Web depths range from 3″ to 24″. S-shapes are designated by their symbol (S) followed by the web depth and the weight per lineal foot, as shown in Fig. 52-72. For example, an S-shape member with a 14″-deep web that weighs 56 lb/ft is labeled: S 14 × 56. On some drawings the length may be added to the designation rather than as a dimension on the drawing. S-shapes are used extensively as columns because of their symmetry. Their narrow flanges are applicable to many designs where size restrictions are a problem.

W-shapes (formerly wide-flange or H beams) are similar to S-shapes but with wider flanges and comparatively thinner webs; their capacity to resist bending is greater than S-shapes. W-shapes are designated in the same manner as S-shapes, as shown in Fig. 52-73. For example,

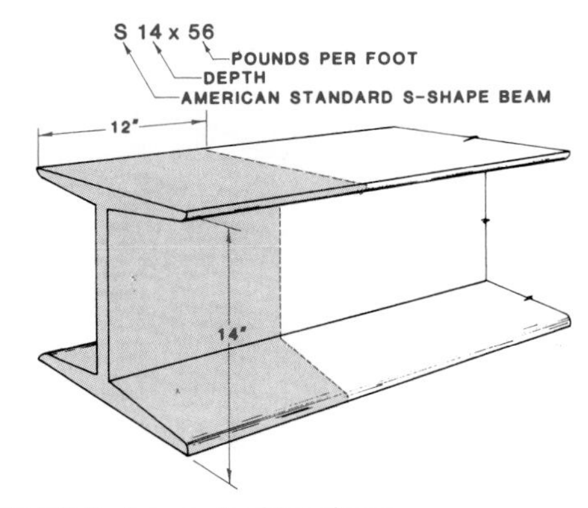

Fig. 52-72 S-shape designations.

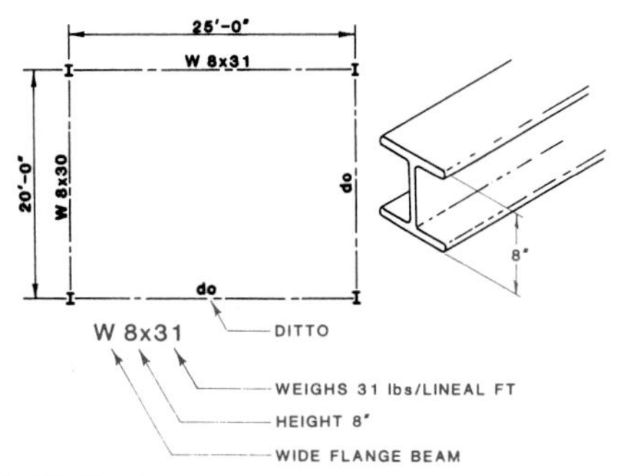

Fig. 52-73 W-shape designations.

W 18 × 62 describes a W-shape member with an 18″-deep web weighing 62 lb/ft foot. W-shapes are available in depths from 4″ to 36″. Lighter-weight versions of W-shape members are known as M-shapes.

Structural tees are made by cutting through the web of an S, W, or M-shape, although some tees are rolled to order. If the web is cut exactly through the center, two identical tees result. The symbol for a tee is the capital letter T. On structural drawings the tee symbol includes the shape from which the tee was cut (S, W, or M) followed by the letter T, the depth of cut (from web to flange), and the weight per foot. Therefore a tee-shape member cut in half from a W 12 × 50 would be specified WT 6 × 25. (6 is half the depth and 25 is half the weight per foot.) Tees are most commonly used for truss cords and to support concrete reinforcement rods.

All inch marks are omitted on shape notations used on structural drawings since all sizes are assumed to be in inches unless otherwise specified.

Steel Construction Systems

In steel construction, plates, bars, tubing, and rolled shapes are used for columns, girders, beams, and bases in a variety of construction systems. There are three general types of steel construction systems: steel cage, large-span, and cable-supported.

When steel members are used in a manner similar to skeleton-frame wood members, the system is known as *steel cage construction*. Therefore, the terms *steel skeleton-frame* and *steel cage* construction are often used interchangeably. The major components used in steel cage construction, as shown in Fig. 52-74, consist of columns, girders, and beams. *Columns* are vertical members which rest on footings or piers. *Girders* are horizontal members which extend between columns. They are sometimes called *sprandrel beams* when connecting perimeter columns (Fig. 52-75). Beams are horizontal members placed on or between girders.

Beams are supported by girders, which in turn are rigidly attached to columns, through which all loads are transmitted through bearing plates to footings. Since all live and dead loads are transmitted through the columns there is no need for additional exterior or interior bearing walls in steel cage construction. This enables buildings to be built extremely high with a minimum of interior obstruction. It also allows the use of large glass exterior walls (curtain walls) that have no structural value.

Even steel cage construction cannot provide the amount of unobstructed space required for structures such as aircraft hangars, sports stadiums, and convention centers. For these structures, large trusses or arches are necessary to span long distances. For even larger areas, cable-supported construction may be necessary.

Structural Steel Drawing Conventions

Structural steel drawings are of several types: design drawings, erection drawings, and working (shop) drawings. *Design (schematic) drawings* are very symbolic and show only the position of each structural member with a single line, as shown in Fig. 52-76. Notations describing each

1. Anchors or hangers for open-web steel joists
2. Anchors for structural steel
3. Bases of steel and iron for steel or iron columns
4. Beams, purlins, girts
5. Bearing plates for structural steel
6. Bracing for steel members or frames
7. Brackets attached to the steel frame
8. Columns, concrete-filled pipe, and struts
9. Conveyor structural steel frame work
10. Steel joists, open-web steel joists, bracing, and accessories supplied with joists
11. Separators, angles, tees, clips, and other detail fittings
12. Floor and roof plates (raised pattern or plain (connected to steel frame)
13. Girders
14. Rivets and bolts
15. Headers or trimmers for support of open-web steel joists where such headers or trimmers frame into structural steel members
16. Light-gage cold-formed steel used to support floor and roofs
17. Lintels shown on the framing plans or otherwise scheduled

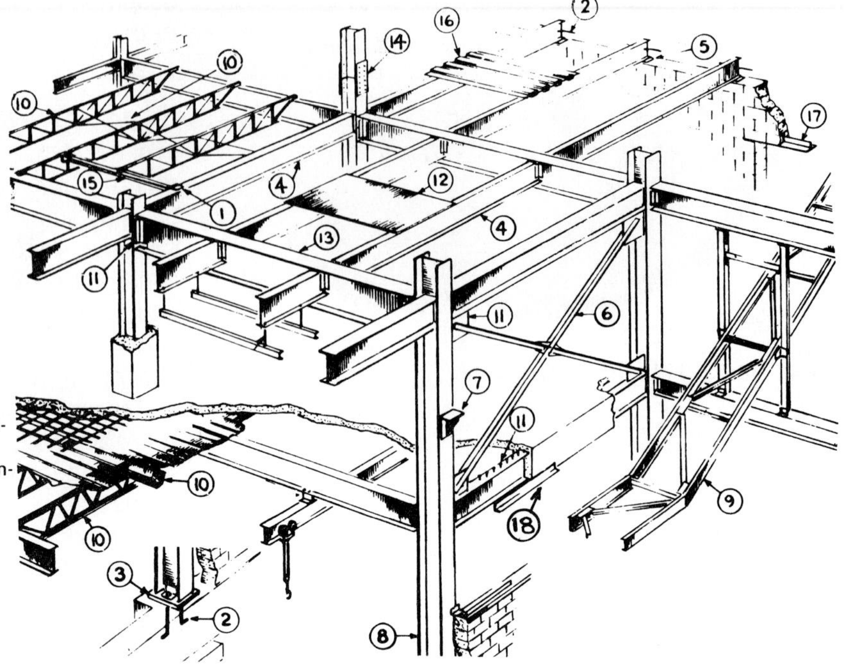

Fig. 52-74 Steel cage construction.

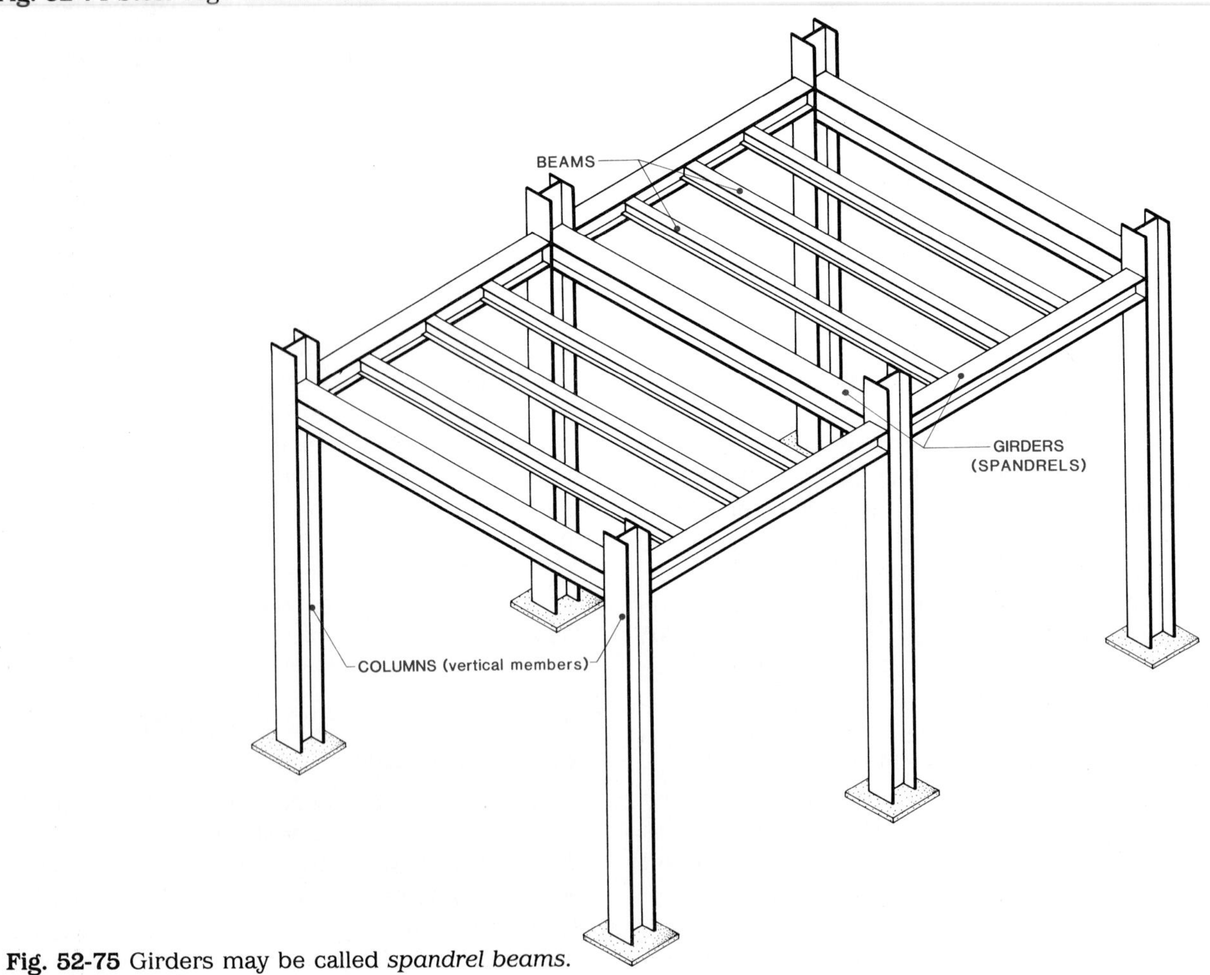

Fig. 52-75 Girders may be called *spandrel beams.*

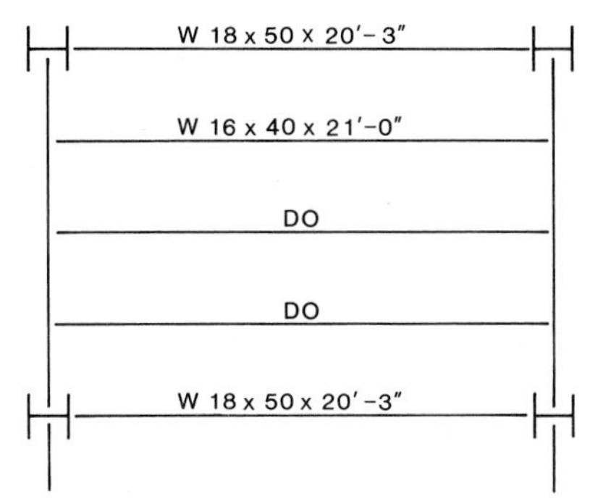

Fig. 52-76 Symbolic structural steel plan.

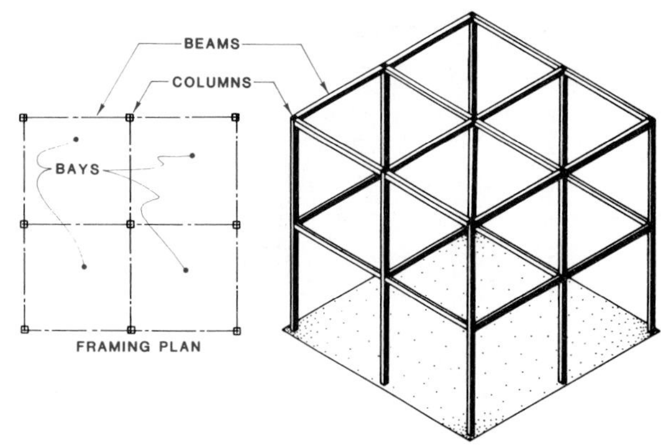

Fig. 52-78 Construction bays.

member's shape, size, and weight are included on each line. When a series of members with identical characteristics are aligned, the successive lines are labeled with a ditto symbol (DO) indicating that the shape, size, and weight of the member are identical to the previous member's. *Working (shop) drawings* are complete orthographic engineering drawings showing the exact size and shape of each member, including every cut, hole, and method of fastening. *Erection drawings* show the method and order of assembling each member, which is coded for easy field identification.

Floor Plans Structural steel floor plan drawings are dimensioned by showing the horizontal distances between columns. To make reading and referencing of dimensions easier, a grid system is often used to identify the positions of columns. Grid systems are sequenced numerically or alphanumerically as shown in Fig. 52-77. Regularly repeated spaces between beams, girders, and their supports are called *bays* (Fig. 52-78). The dimensioning of the framing for a bay module is shown in Fig. 52-79.

Elevations Structural elevation height dimensions are placed on an elevation drawing, as in Fig. 52-80, or as an elevation note indicating the height above a datum point. This is known as *datum point referencing.* Figure 52-81 shows a datum point reference on a plan view. Figure 52-82 shows the same reference on an elevation view.

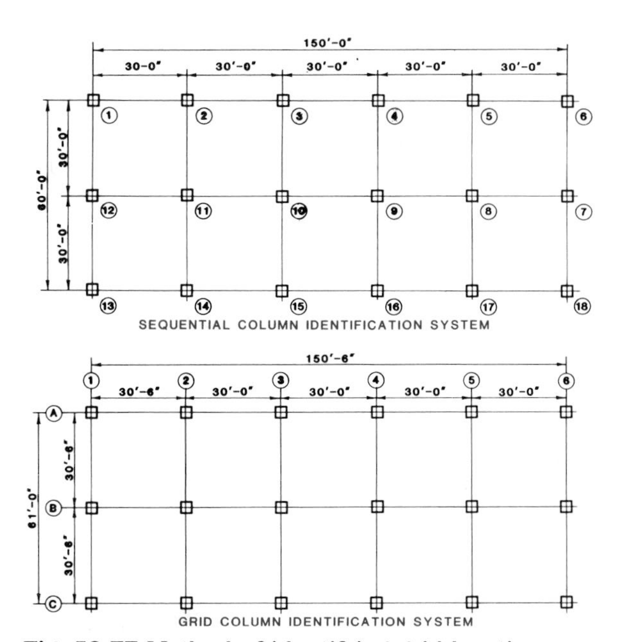

Fig. 52-77 Method of identifying grid locations.

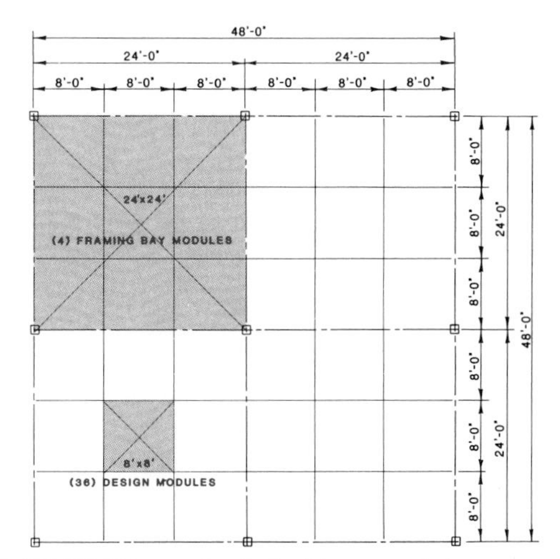

Fig. 52-79 Dimensioning of bay modules.

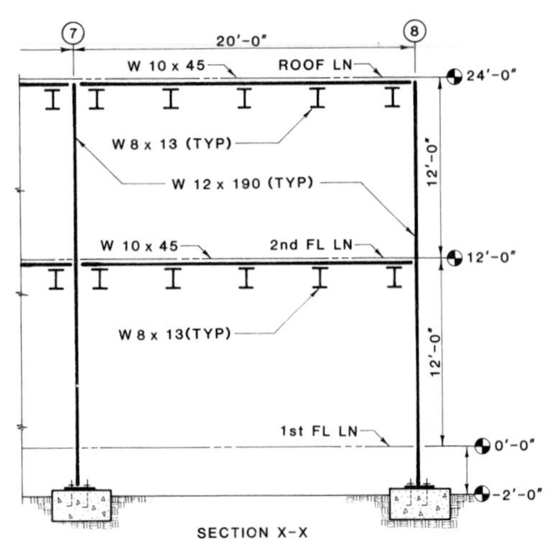

Fig. 52-80 Height dimensions on a structural steel drawing.

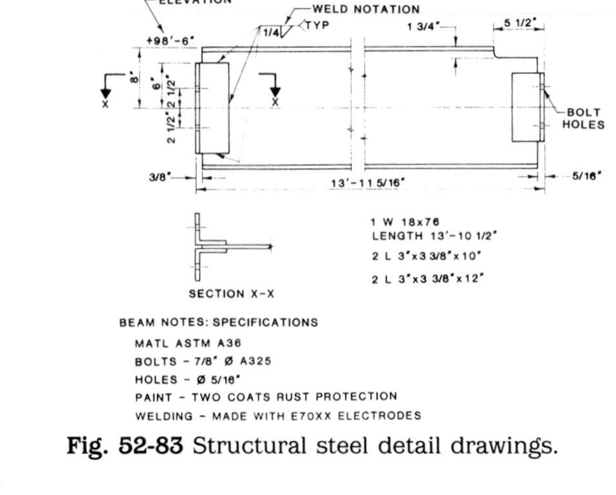

Fig. 52-83 Structural steel detail drawings.

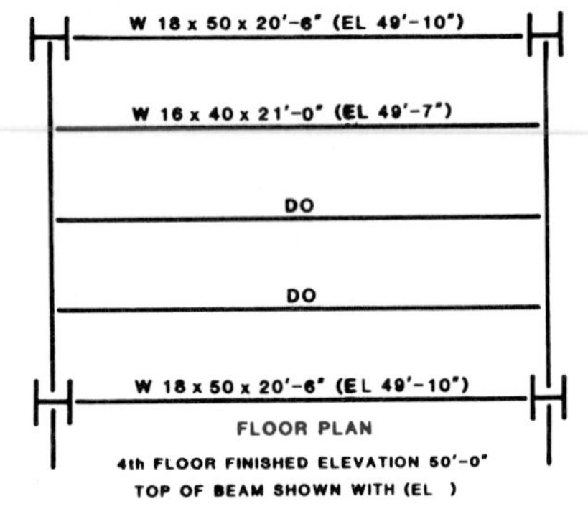

Fig. 52-81 Plan view datum point referencing.

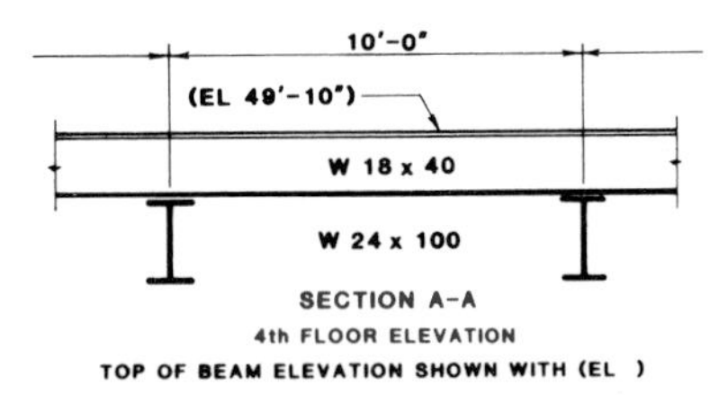

Fig. 52-82 Elevation datum point referencing.

Dimensioning Three methods of dimensioning are used on structural drawings. The first method involves a description of each member, which includes the length placed directly on each schematic line. The second method uses notations to show only the shop size (width) and weight, with dimension lines used to show the position and length of each member. The third method uses a coding system which relates each member to a schedule which contains all pertinent information. In a coding system the first number identifies the floor or level.

Detail Drawings Structural detail drawings are used to clarify details relating to the exact shape, size, and relationship of structural members and other building materials, as shown in Fig. 52-83. Longitudinal or transverse sections are drawn to show both the length (longitudinal) of the structure and the width (transverse) of the structure in sections.

Foundations

Foundation plans for structural steel buildings are essentially the same as for other types of construction except for the intersections of columns, base plates, and footings. Foundation plans, as shown in Fig. 52-84, show the position of each column and accompanying footing or pier. These are usually indexed with a grid system in which footings and columns share the same number.

Structural Steel Floor Framing Plans

Structural steel floor framing systems are comprised of girders, beams, joists, and decking materials. A floor framing plan is prepared for each floor of a multilevel building. Although the

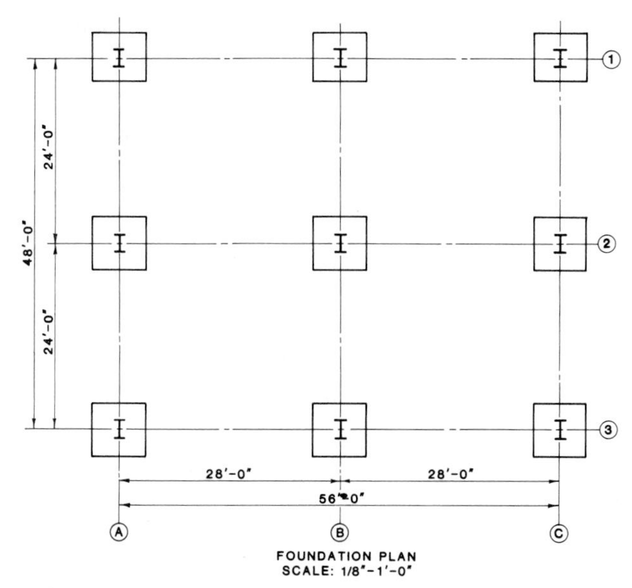

Fig. 52-84 Foundation plan showing position of columns, footings, and girders.

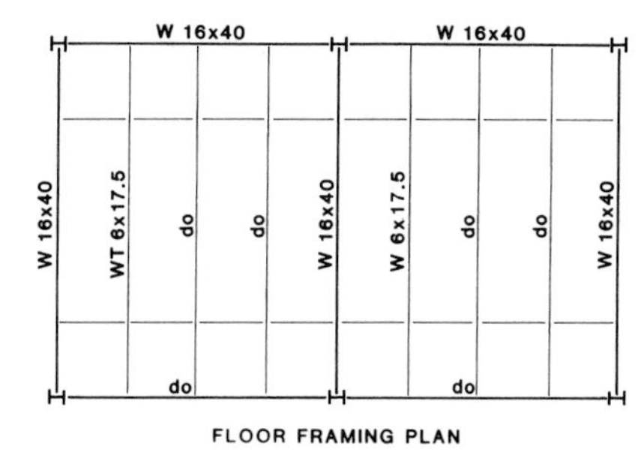

Fig. 52-85 Continuous and intersecting member designations.

framing for many floors may be nearly identical, this cannot be assumed unless specified. Usually there are slight differences on each floor plan. For this reason CAD layering is ideal for preparing high-rise structural steel floor framing plans. Layering, or manual pin graphics, allows the drafter to draw a base floor plan and make specific floor changes without redrawing each floor separately. The same spacing must be used between grid lines to ensure alignment of columns and other vertically oriented features, such as stairwells, plumbing lines, HVAC ducts, and electrical conduits. Each floor framing plan shows the position of each column that passes through the floor. Major members, such as girders and beams, are shown with a solid heavy line with the identifying notation placed directly on or under the line. The length of each line represents the length of each member.

If a continuous beam passes over a girder, as in Fig. 52-85, a solid unbroken line is drawn through the girder line. However, if the beam stops and is connected to the girder, the beam line is broken as shown in Fig. 52-86. Remember that solid lines represent continuous members. Broken lines indicate that the member intersects or is under a continuous member.

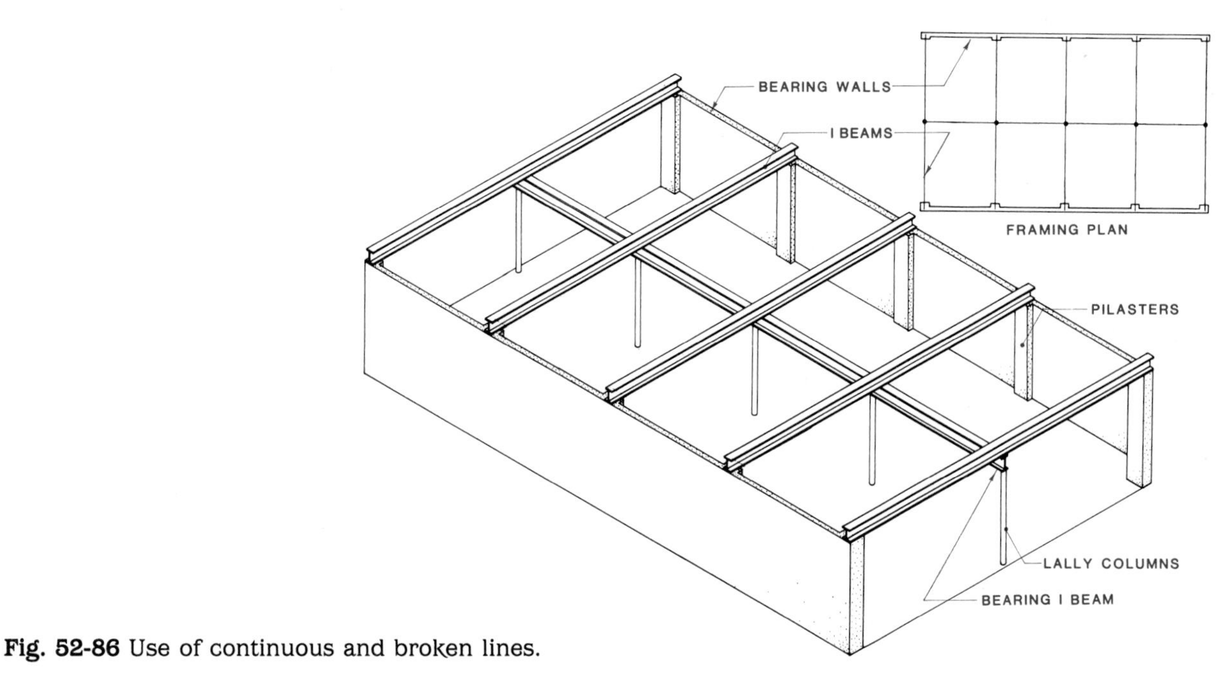

Fig. 52-86 Use of continuous and broken lines.

Girders and Beams Steel *girders* and *beams* are the largest horizontal structural members in a steel floor system. Often their large width creates a problem with headroom or window opening size. Built-up girders and columns (Fig. 52-87) may be shop- or field-assembled. In either case dimensioned drawings, including methods of assembly, must be prepared.

Joists Steel joists, as shown in Fig. 52-88, are small horizontal structural members supported by beams or girders. Joists directly support the roof or floor system decking and transmit this weight to the beams. There are three types of steel joists: short-span, long-span, and deep long-span. Open-web steel joists are actually small welded trusses made from steel bars and shapes.

Since joists are always spaced closely, only one note is needed to give the number and spacing of each classification of joists. Only the first few joists in a series are usually noted. For example, if there are 8 short-span joists spaced at 3′ intervals over a 24′ distance, the note should read: 8 SP @ 3′-0″ = 24′-0″. In addition, a notation is placed on a line representing the joists' direction and includes the length class of the joist and load table range.

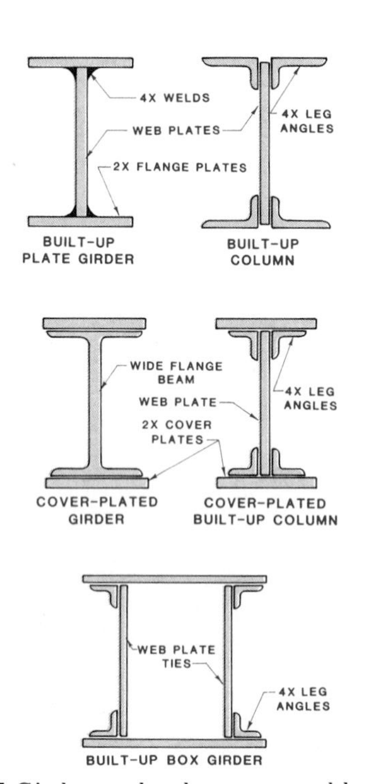

Fig. 52-87 Girder and column assembly methods.

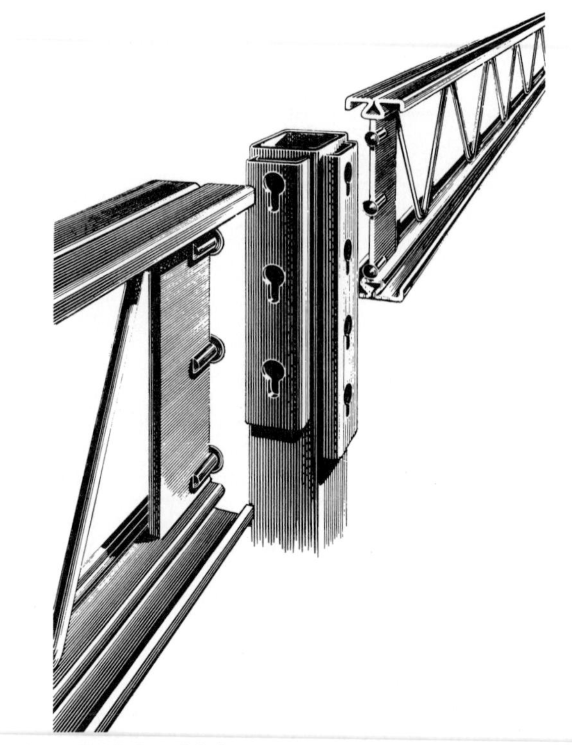

Fig. 52-88 Steel joists.

Decks. *Steel decks* for floors and roofs use corrugated sheets, interlocking galvanized steel panels, or cellular units over steel joists. Steel deck details or sectional drawings, as shown in Fig. 52-89, are usually prepared to show the relationship of the decking to the structural support members.

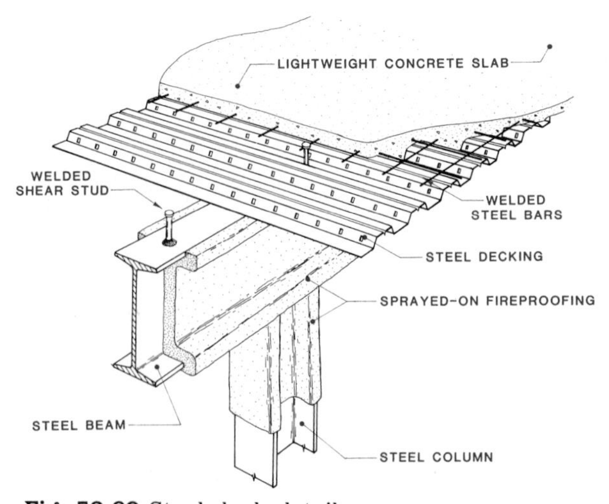

Fig. 52-89 Steel deck details.

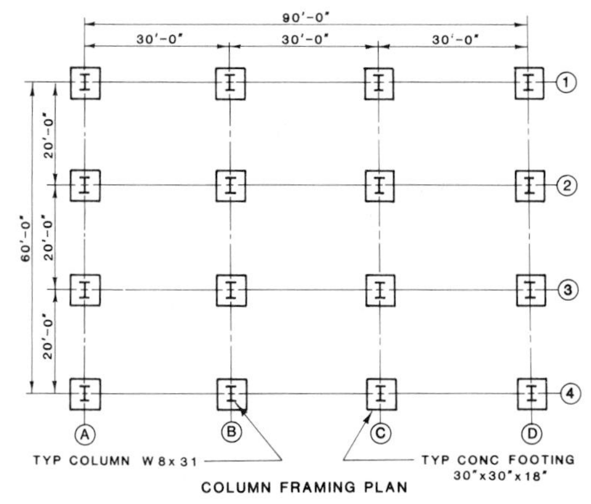

Fig. 52-90 Column and footing outlines shown on a plan view.

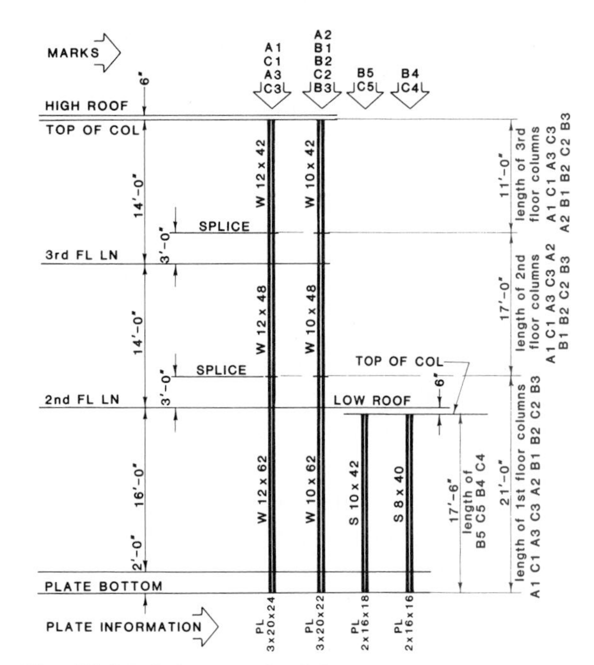

Fig. 52-92 Column schedule.

Columns. *Steel column* positions are shown on floor plans or column framing plans with the outline of the columns and footings, as shown in Fig. 52-90. Columns are usually located on grid lines but may be dimensioned the same as other floor plan features if their position does not coincide with a grid intersection. The style, size, and weight may be noted on each column on a floor plan, or this information may be shown on a column schedule. *Column schedules* are schematic elevation drawings showing the entire height of a building and the elevation of each floor, base plate, and column splice. The type, depth, weight, and length of columns with common characteristics are shown under the column mark for each column. Figure 52-91 shows a floor plan with column locations related to the column schedule shown in Fig. 52-92. There are 13 columns in this plan. Columns with common specifications are grouped together at the top of the schedule. Under each grouping a heavy vertical line represents the height of each column, with the type, size, and weight noted on each. For example, columns with marks A1, C1, A3, and C3 are all 12″-wide flange shapes and extend vertically 46′-0″ from base to top (2′ + 16′ + 14′ + 14′). Three individual columns comprise each of these. The bottom length is

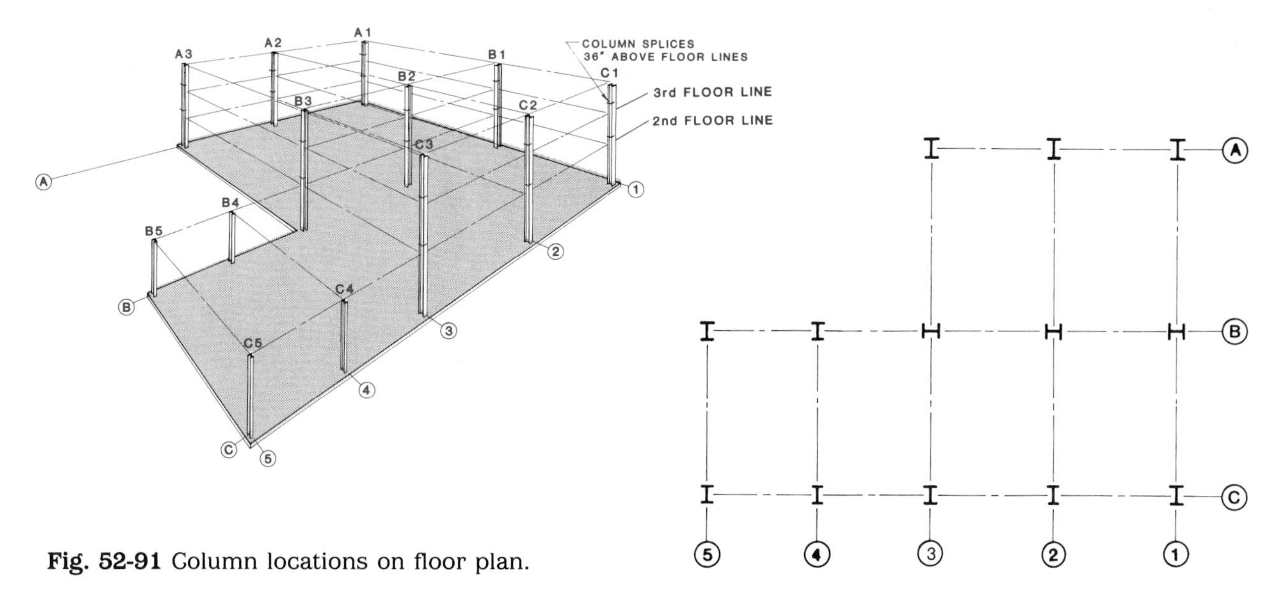

Fig. 52-91 Column locations on floor plan.

21′-0″ from plate to first splice (3′ above the floor line), 17′-0″ from the first splice to the second splice, and 11′-0″ from the second splice to the top. The second row of column marks (A2, B1, B2, C2, and B3) have the same lengths as the first row but are 10″-wide flange shapes. The third-row and fourth-row columns are continuous 17′-6″, without splices; row three columns (B5, C5) are 10″ S-shapes and row four columns (B4, C4) are 8″ S-shapes.

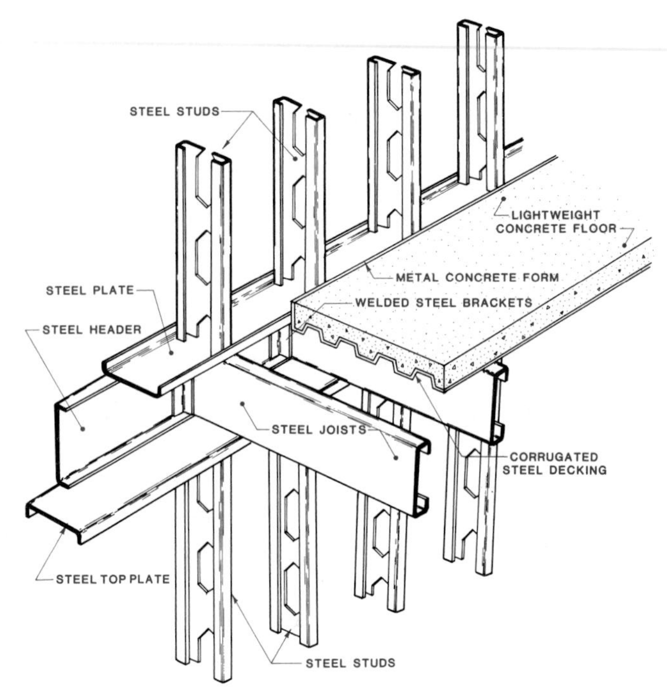

Fig. 52-94 Lightweight steel wall framing.

Curtain Walls

One of the greatest advantages of steel cage construction is the unobstructed space provided by curtain walls. Since all building loads are transmitted through columns, the remaining open wall space can be filled with any type of nonbearing (curtain) panels, as shown in Fig. 52-93. Units of this type are usually prefabricated in modular units; therefore, wall framing plans show only the position of the module and not the construction details. Details such as light metal wall framing (Fig. 52-94) are usually shown on an elevation and/or plan sectional view. Nonbearing interior partitions of this type can be moved at any time without sacrificing the structural integrity of the building.

Structural Steel Roof Systems

Structural steel roof systems are very similar to floor systems if girders, beams, joists, and flat decking are used. However, if larger spans are required and the size limit of the largest girder (36″) is reached, trusses, domes, space frames, arches, folded plates, or cable-supported systems may be necessary to span the distance.

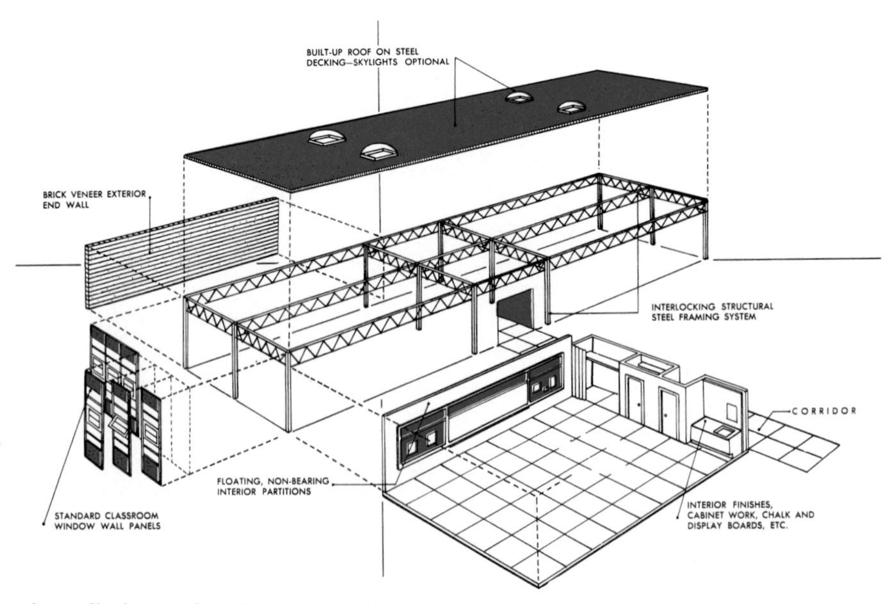

Fig. 52-93 Curtain walls in steel cage construction.

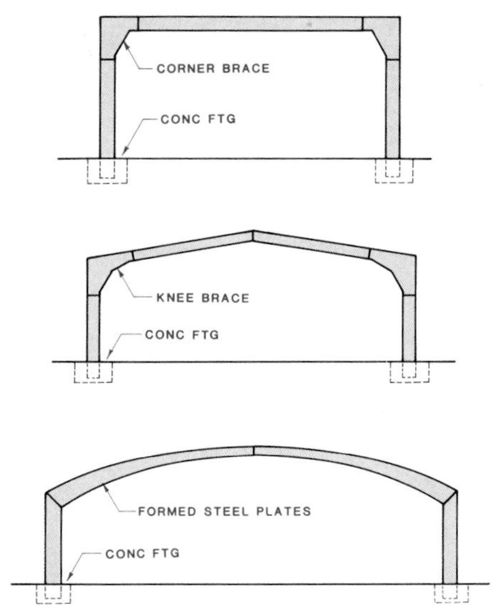

Fig. 52-95 Rigid steel bents.

Girders, beams, joists, and decking are used in roof systems in the same manner as in floor systems. The only difference is the application of roof covering materials in lieu of flooring materials.

Steel roof trusses can span much longer distances (up to 80′) than comparable solid members. Trusses consist of a horizontal bottom and top chord, connected with vertical and/or diagonal webs or struts. An additional horizontal member known as a *purlin* connects the top chords of adjacent trusses, which stabilizes and locks the entire roof system together.

Rigid steel bent frames are also used to span long distances. Bents are either straight, single-span, shaped, or multiple-span, as shown in Fig. 52-95. Whenever longer spans must be covered, *arches* (or *vaults*) may be used. Arches are bent trusses which are either hingeless, two-hinged, or three-hinged, as shown in Fig. 52-96. Arch details are shown on structural drawings the same way truss details are shown.

Space frames are three-dimensional trusses formed by connecting series of triangular polyhedrons, as shown in Fig. 52-97. Space frames, because of their light weight and ability to resist bending, can span extremely large distances.

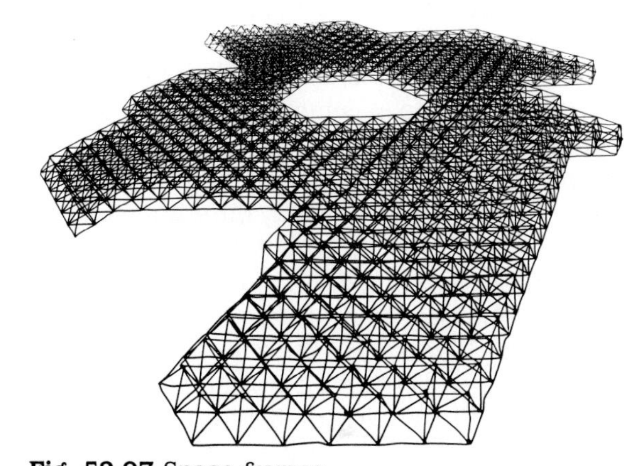

Fig. 52-97 Space frames.

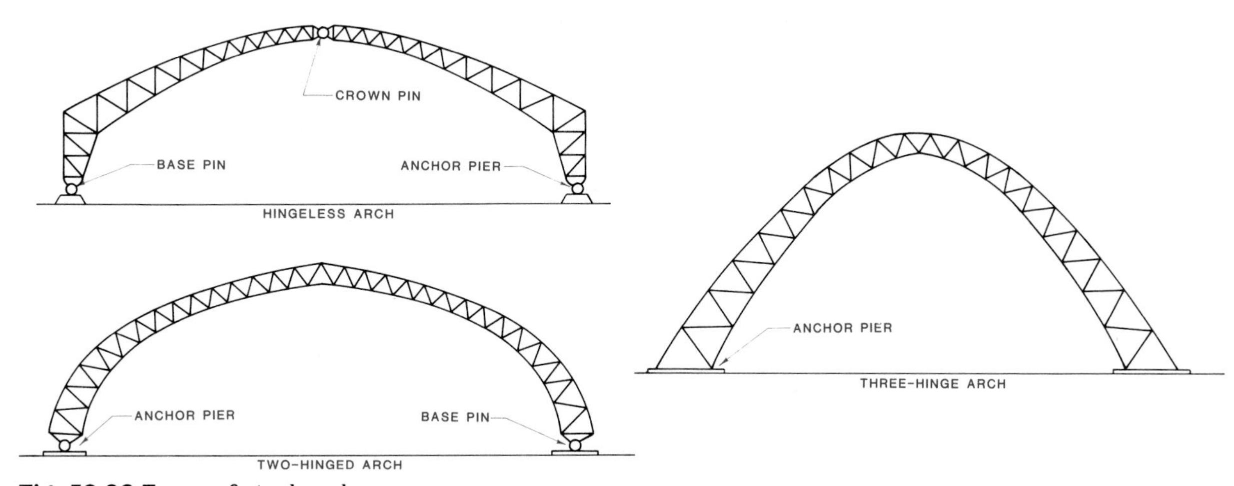

Fig. 52-96 Types of steel arches.

However, their load-bearing capacity is limited so they are not often used for floor systems. Domes or cable-supported roof construction are the most economical for the largest spans, such as sports arenas and convention centers.

Fasteners and Intersections

Major structural steel members depend on a wide variety of joining methods and devices to function as a structurally stable frame. This includes the use of brackets, rivets, bolts, and welds to attach members together and to foundation piers and footings. Figure 52-98 shows these three major methods of assembling structural steel components. Steel components that can be assembled before shipping to a site are usually assembled and welded at a fabrication shop. All other members are assembled and permanently fastened at the building site. For example, brackets for bolting or riveting girders to columns are welded to the girder at the shop, then bolted to the column in the field during erection.

Brackets Most structural steel members intersect at right angles. Many different types of brackets are used to provide a perpendicular surface for bolting, riveting, or welding. *Angles*, *L-shapes*, and *bent* or *welded plates* are used for this purpose. Figure 52-99 shows the use of angles, nuts, and bolts to join a girder to the top of a column. Bracket information on a structural drawing shows the size of the bracket legs followed by the thickness, width, shape, symbol, and fastening device information. If brackets are to be welded to a member at a fabrication shop, a detail drawing is not provided in the field. Only the assembled intersection of the joint is drawn. Only shop fabricators are provided with a complete set of details.

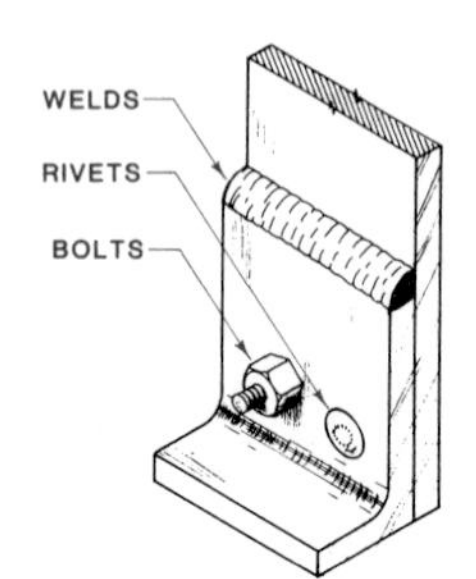

Fig. 52-98 Methods of joining steel members.

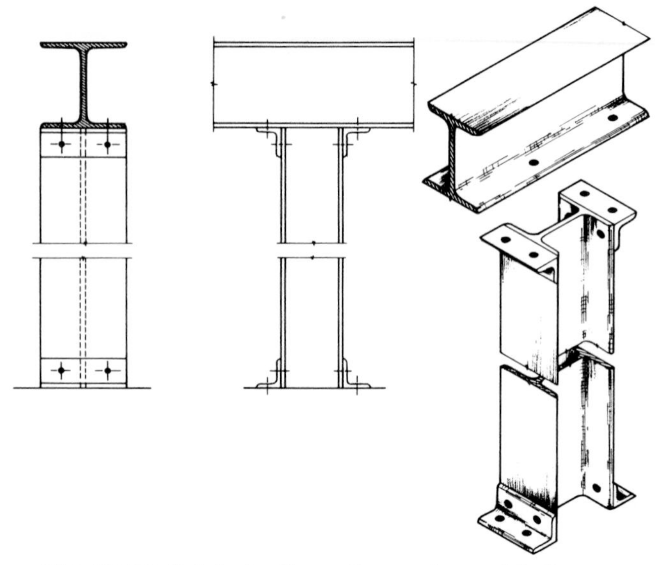

Fig. 52-99 Joining with angles, nuts, and bolts.

Rivets Rivets are used to connect steel members, as shown in Fig. 52-100. Four common types of bucked rivets are shown in Fig. 52-101. The type of rivet specified is shown at the end of the drawing notation. Rivets are made of soft steel and when cooled tend to shrink. Consequently, bolts are now used more extensively than rivets in the erection of structural steel.

Bolts High-strength bolts and nuts (Fig. 52-102) can carry loads equal to rivets of the same size but can be turned tighter because of their high tensile strength. Bolts used in steel construction are either high-strength or unfinished bolts. High-strength bolts are used to connect extremely heavy load-bearing members such as girders, beams, and columns. They are also used to attach members where shear loads are transmitted through the bolts. Unfinished bolts are used for lighter connections where loads are transmitted directly from member to member. For example, when a beam rests directly on a girder there is no vertical shear load on the bolts holding the two members together. Unfinished bolts are also used to anchor column base plates to footings, as shown in Fig. 52-103, since there is also no shear stress at this location.

Welds Welding is a popular method of connecting structural steel members and has some advantage over bolting. For example, fabrication

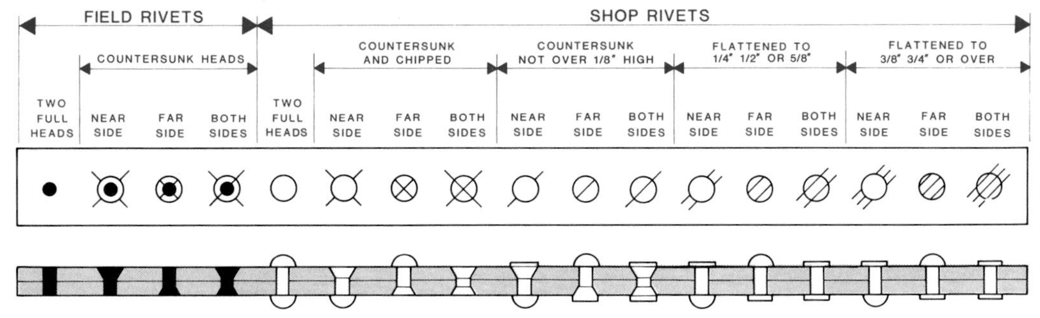

Fig. 52-100 Types of rivets.

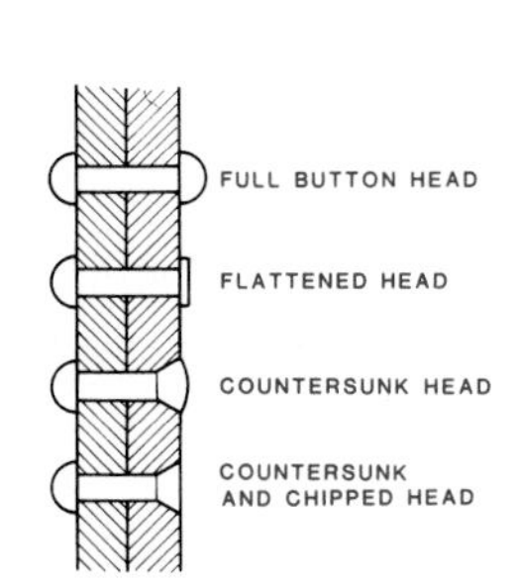

Fig. 52-101 Bucked rivets.

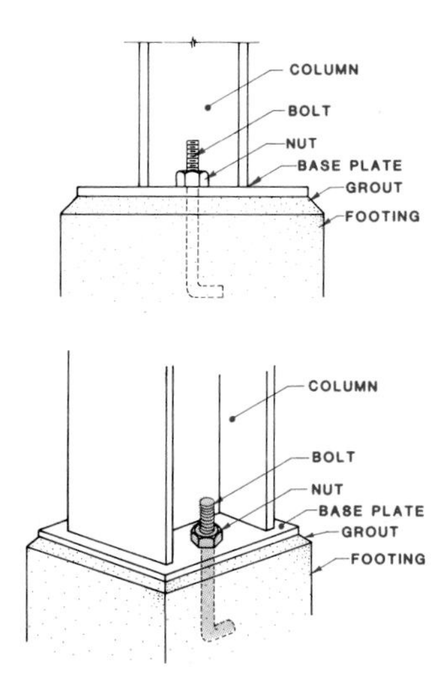

Fig. 52-103 Bolts used on column bases.

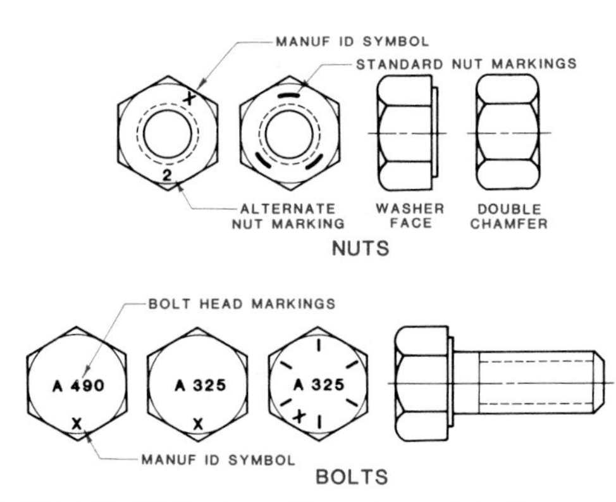

Fig. 52-102 High-strength nuts and bolts.

is simplified by reducing the number of individual parts to be cut, punched with holes, handled, and installed. The major types of welds are illustrated in Fig. 52-104.

The convention used to locate welding information on drawings is a horizontal reference line with a sloping arrow directed to the joint (Fig.

WELD SYM	WELD NAME
◺	FILLET
▭	PLUG/SLOT
○	SPOT/PROJECTION
⊖	SEAM
◡	BACK/BACKING
◡◡	SURFACING
//	SCARF (BRAZING)
JL	FLANGE - EDGE
|L	FLANGE - CORNER
||	GROOVE - SQUARE
V	GROOVE - V
|/	GROOVE - BEVEL
Y	GROOVE - U
ꟾ	GROOVE - J
)(	GROOVE - FLARE V
|(	GROOVE - FLARE BEVEL

Fig. 52-104 Types of walls.

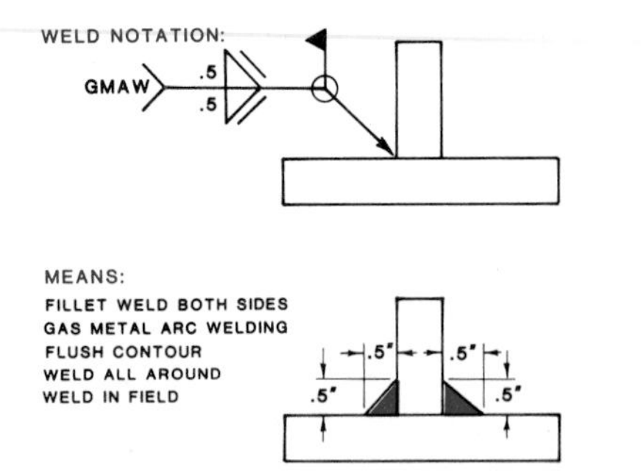

Fig. 52-105 A welding symbol showing the information contained in a detail drawing.

52-105). The arrow may be directed right or left, upward or downward, but always at an angle to the reference line. If no extra marking is shown, a shop weld is assumed. A triangular flag indicates a field weld. An open circle means weld all around the member. If the open circle is shown around the base of the flag, it means weld all around in the field.

The basic weld symbols (Fig. 52-106) or supplementary weld symbols (Fig. 52-107) are located midway on the horizontal reference line. The symbol is located below the line if the weld is to be placed on the near side where the arrow points. The symbol is placed above the line if the weld is to be placed on the far side, and above and below if both sides are to be welded. The size of the weld (or its depth) is indicated to the left of the basic symbol. The length of the weld is shown to the right of the symbol. When long joints are used, intermittent welds are often specified. These are indicated by the length of weld followed by the center-to-center spacing (pitch). Such welds are usually staggered on either side of a joint.

The tail of the reference line may contain information about the kind of material or process required. This feature is not often used on structural steel details. When no information is required the tail is omitted. Figure 52-108 shows the position of information on a welding symbol.

Special Fastener In addition to the conventional fastening methods of riveting, bolting, and welding, specially designed systems are often used in light steel construction, as shown in Fig. 52-109. In these systems male and female brackets are aligned to enable easy erection by alignment of parts. Bridging is another auxiliary device used to make steel construction

LEGEND OF WELDING SYMBOLS

F	– Finish symbol
⌒	– Contour symbol
A	– Groove angle: included angle of countersink for plug welds
R	– Root opening: depth of filling for plug and slot welds
S	– Depth of preparation
	– Size or strength for specific welds
	– Height of weld reinforcement
	– Radii of flare-bevel grooves
	– Radii of flare-V grooves
	– Angle of joint (brazed welds)
(E)	– Effective throat
T	– Specific process or reference
L	– Length of weld
	– Length of overlap (brazed joints)
P	– Pitch of welds (center-to-center spacing)
1	– Weld located on opposite side of arrow
2	– Weld located on same side of arrow
(N)	– Number of spot or projection welds
◄	– Weld made in field
o	– Weld all around

Fig. 52-106 Weld symbols.

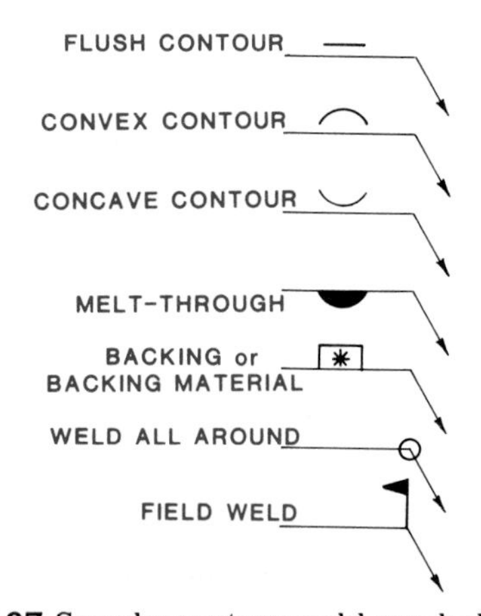

Fig. 52-107 Supplementary weld symbols.

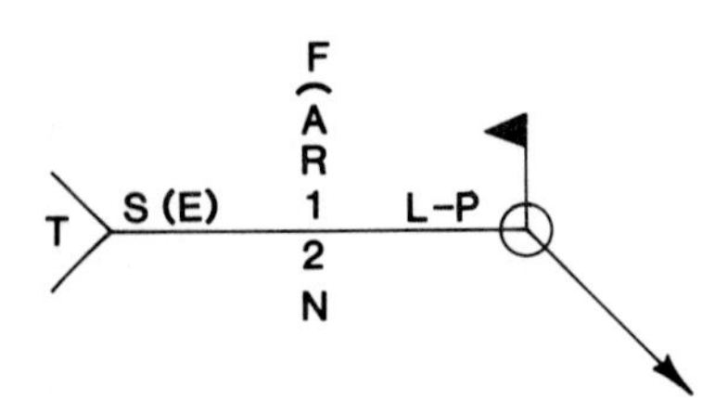

Fig. 52-108 Position of information on a weld symbol.

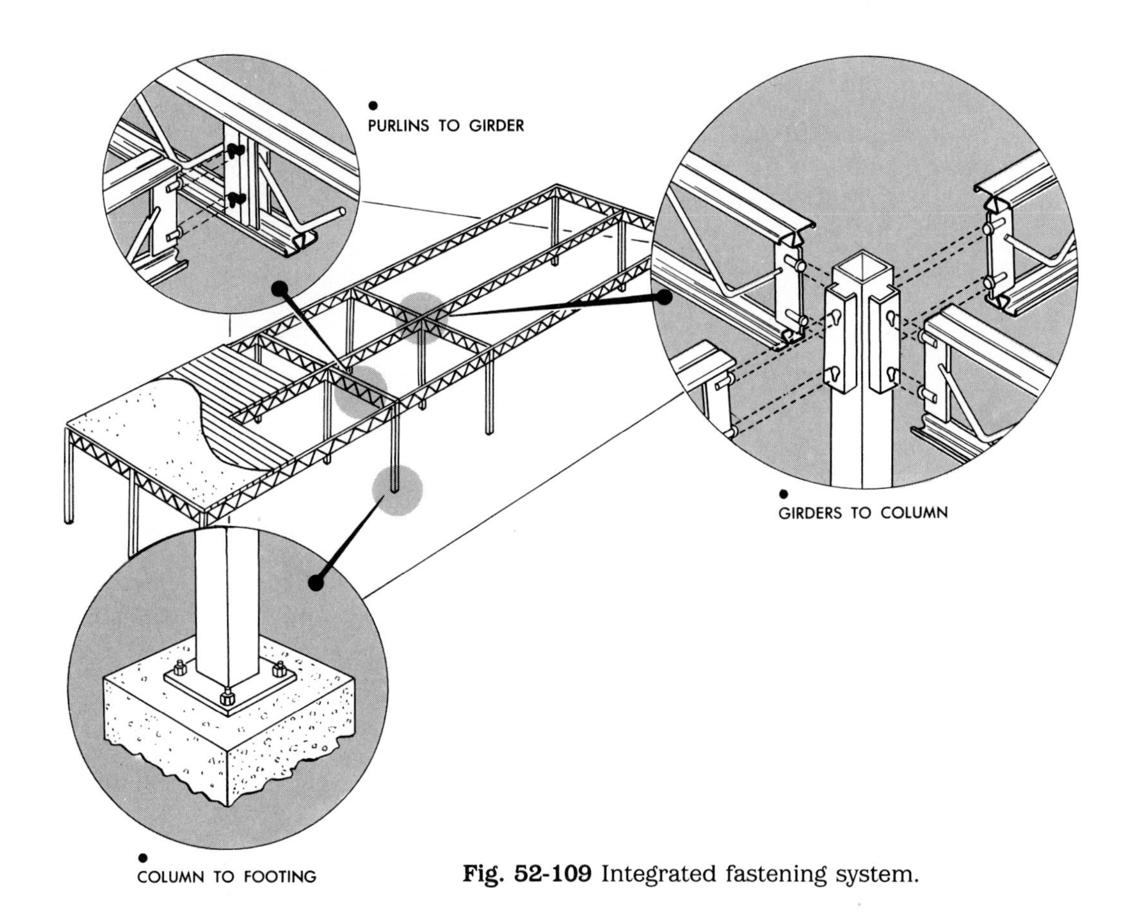

Fig. 52-109 Integrated fastening system.

systems structurally sound. Steel bridging between joists is used in the same manner as wood bridging in wood skeleton-frame systems. On structural drawings bridging is shown with a dotted line. Although steel construction is extremely rigid, special cross-bracing is often required to counteract lateral wind loads, as shown in Fig. 52-110.

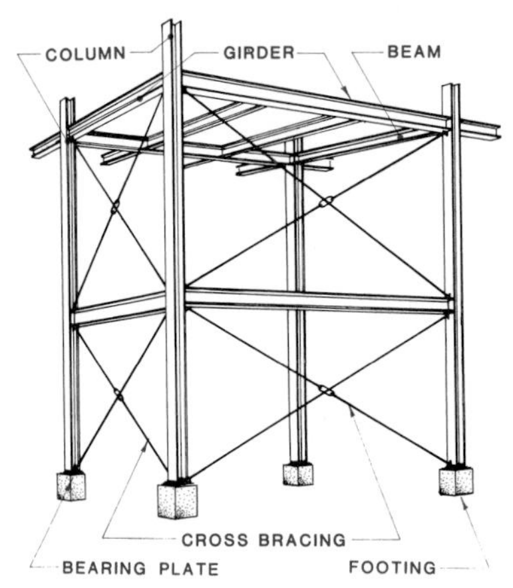

Fig. 52-110 Structural cross-bracing.

MASONRY CONSTRUCTION SYSTEMS

Construction systems which use masonry as a primary material are usually combined with other systems such as structural steel or skeleton-frame construction. Buildings are not constructed with masonry materials alone since wood, steel, or reinforced concrete is needed to span floor and roof areas.

Masonry Materials

Many different types, sizes, shapes, and grades of brick, concrete block, stone, and structural clay products are used in contemporary construction.

Brick Masonry Bricks are divided into two general categories by grade: *common brick* and *face brick*. The color, texture, and dimensional tolerance of common brick are less consistent and critical than for face bricks. Common

Grade	*Use*
SW	Used for *maximum* exposure to heavy snow, rain, and/or continuous freezing conditions.
MW	Used for average exposure to rain, snow, and moderate freezing conditions.
NW	Used for exposure to *minimum* rain, snow, and freezing conditions.

Fig. 52-111 Grades of common brick.

Type	*Use*
FBX	Used where minimum size and color variations, and high mechanical standards are required.
FBS	Used where wide color variations and size variations are permissible or desired.
FBA	Used where wide variations in color, size, and texture are required or permissible.

Fig. 52-112 Grades of face brick.

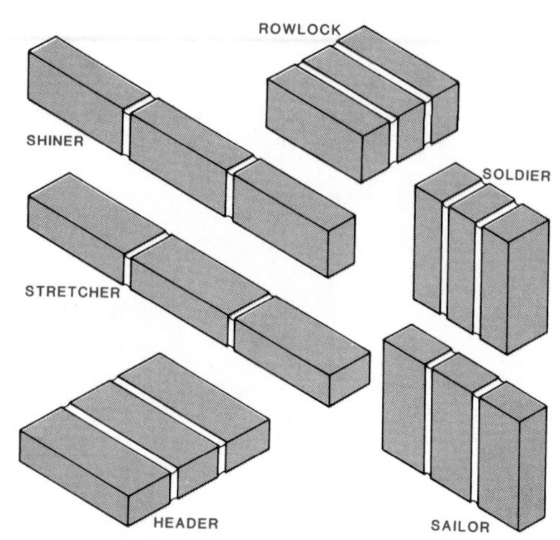

Fig. 52-113 Brick classified by laid position.

brick is therefore less expensive and is generally used in unexposed construction areas. Common brick is graded according to structural characteristics, as shown in Fig. 52-111.

Face brick is used in exposed areas that require dimensional accuracy, absorption control, color and texture consistency. Face brick is therefore graded according to these characteristics, as shown in Fig. 52-112. Many special types of face brick are available for specific construction needs, for example, glazed brick, firebrick, cored brick, and paving brick.

Most bricks are rectangular. Special shapes for sills, corners, and thresholds are also available or can be made to order. The position of the brick as laid in construction is also used to classify bricks, as shown in Fig. 52-113. Most bricks have holes to reduce weight (Fig. 52-114) and increase bonding. Solid bricks are available in standardized sizes, and sizes differ among brick types. Face bricks are divided into standard, Norman, and Roman types (Fig. 52-115). Common bricks are divided into standard, oversized, and modular types, as shown in Fig. 52-116. Modular brick is sized to align on 4″ grids when mortar joint dimensions are added. Therefore increments of 4″, 8″, 12″, and so forth fit any modular space.

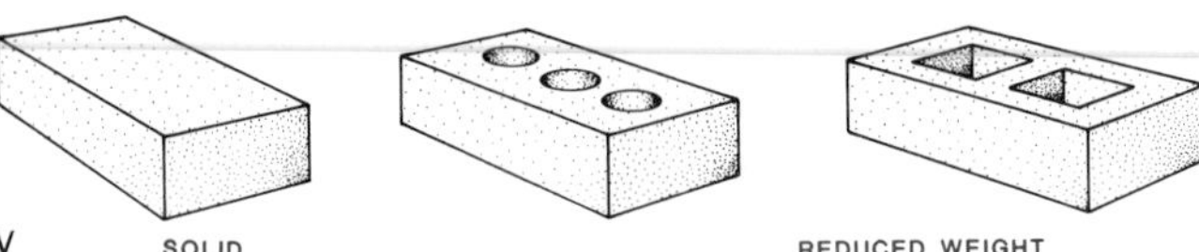

Fig. 52-114 Some bricks have holes to reduce weight.

Type	*Size*	
Standard	2½″ × 3½″ × 11½″	
Norman	2³⁄₁₆″ × 3½″ × 11½″	2¼″ × 3″ × 11¹¹⁄₁₆″
Roman	1½″ × 3½″ × 11½″	

Fig. 52-115 Face brick sizes.

Type	*Size*
Standard	2½″ × 3⅞″ × 8¼″
	2¼″ × 3¾″ × 8″
Oversized	3¼″ × 3¼″ × 10″
Modular	2½″ × 3¾″ × 7¾″
(¼″ Joints)	2⁵⁄₁₆″ × 3¾″ × 7¾″
	2½″ × 3¾″ × 11¾″
Modular	2¼″ × 3½″ × 7½″
(½″ Joints)	2¼″ × 3½″ × 11½″
	2¹⁄₁₆″ × 3½″ × 7½″

Fig. 52-116 Common brick sizes.

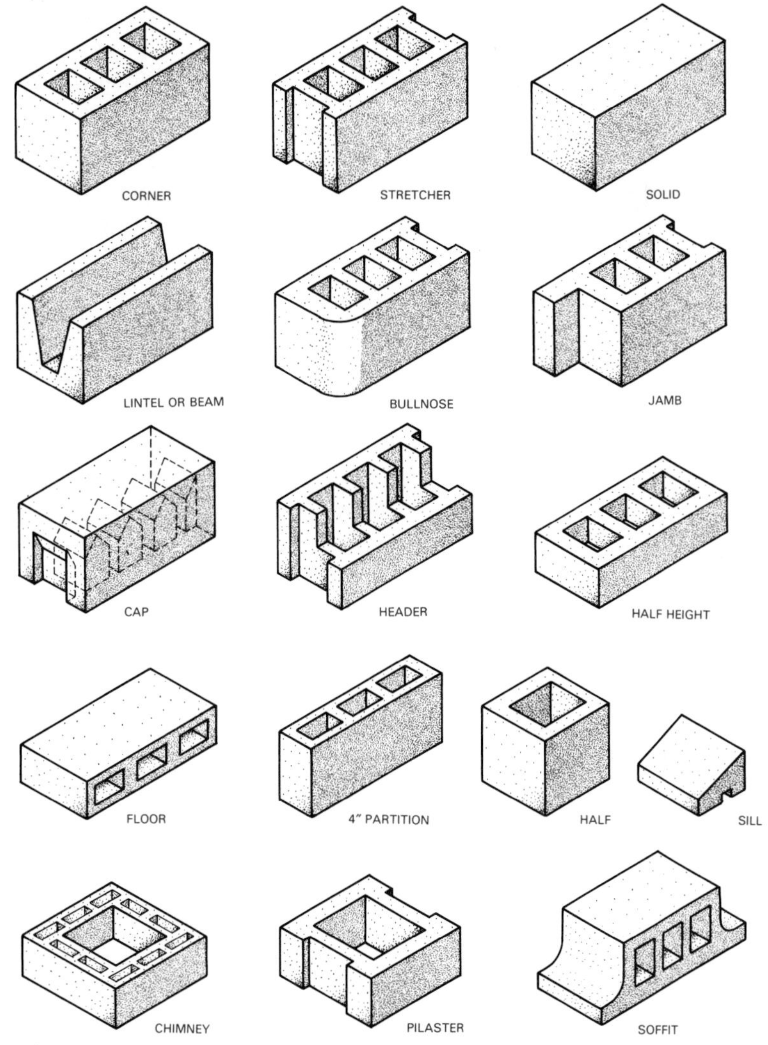

Fig. 52-117 Common concrete block shapes.

Concrete Masonry Concrete is precast in many different shapes for a wide variety of construction purposes, as shown in Fig. 52-117. Concrete blocks are either solid, hollow-core, or split-face for exposed surfaces. Concrete block aggregate is a combination of sand and crushed rocks, slate, slag, or shale. The weight, texture and color of each block are determined by the types of aggregate used.

Concrete block is manufactured in modular sizes. The actual size of each block is ⅜″ smaller than the space to be filled to allow for the thickness of the mortar joint. For example, the dimensions of an 8″ × 12″ × 16″ concrete block are actually 7⅝″ × 11⅝″ × 15⅝″. Modular dimensions also allow blocks to be used in conjunction with modular brick, as shown in Fig. 52-118. One example of a standard concrete block and brick modular dimensioning is shown in Fig. 52-119.

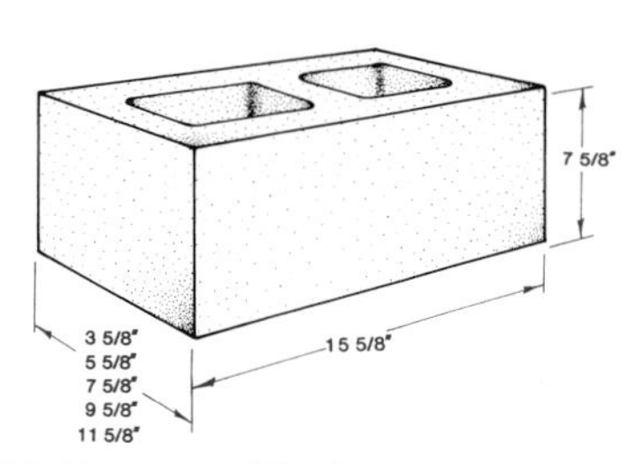

Fig. 52-118 Concrete block sizes.

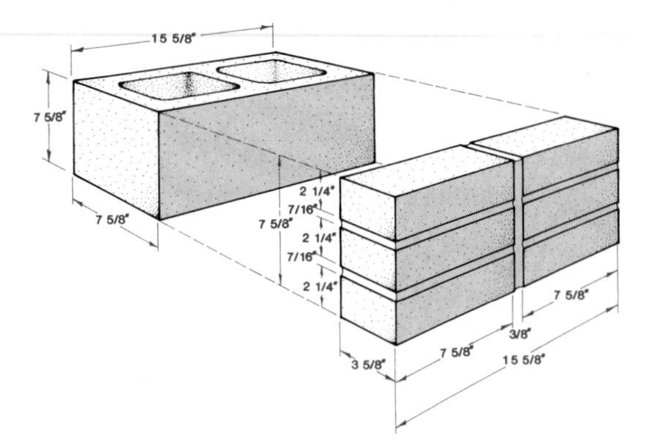

Fig. 52-119 Relationship of modular brick and concrete block dimensions.

Stone Masonry For centuries natural stones were used as a major structural material. Today stone is primarily used decoratively except for landscape construction. Stone masonry is classified by the type of material, shape of the cut, finish, and laying pattern. The most common types of stone material used in construction are sandstone, limestone, granite, slate, and marble. These stone materials can be cut and arranged into a variety of patterns.

Structural Clay Tile Hollow-core structural tile units are larger than bricks and are either load-bearing or non-load-bearing. Structural tiles are used for partitions, fireproofing, surfacing, or furring. Figure 52-120 shows the wide variety of structural tile sizes and shapes. Load-bearing tile is graded according to structural characteristics: LBX for tile exposed to weathering and LB for tile not exposed to weathering or frost. Non-load-bearing facing tile is graded by clearability, stain resistance, color consistency, and dimensional accuracy: FTX for high quality, FTS for low quality.

Masonry Construction Types

There are four basic types of masonry construction: solid, cavity, facing, and veneer.

Solid Masonry Most masonry bearing-wall construction is solid. Solid masonry construction can utilize any masonry material if cut flat to support loads. However, the material used is restricted by the bearing capacity of the material for the loads involved. Concrete block is most commonly used for solid load-bearing walls. For heavy loads and/or high walls, steel reinforcing rods (Fig. 52-121) are specified to add structural stability. When solid masonry walls are constructed with combinations of materials such as concrete block and brick, steel reinforcement, as shown in Fig. 52-122, is mandatory. Reinforcement between courses is also necessary for walls subject to earthquakes, heavy storms, wind, and lateral earth loads. Figure 52-123 shows steel reinforcement between brick and concrete block courses.

Masonry-Cavity Walls To reduce live loads and materials costs and to improve temperature and humidity insulation, cavity walls are often used in preference to solid masonry walls. In cavity wall construction, two separate and parallel walls are designed several inches apart with a structural tie, usually metal, which bonds the walls together, as shown in Fig. 52-124.

Masonry-Faced Walls Walls are often faced with different masonry materials to meet internal and external exposure requirements.

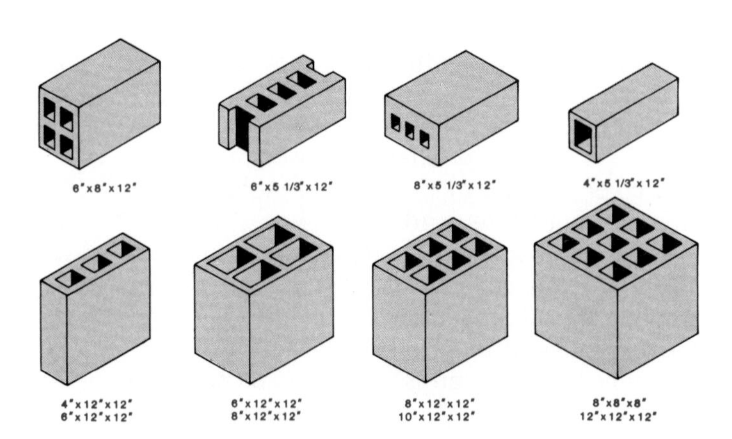

Fig. 52-120 Types of structural clay tile.

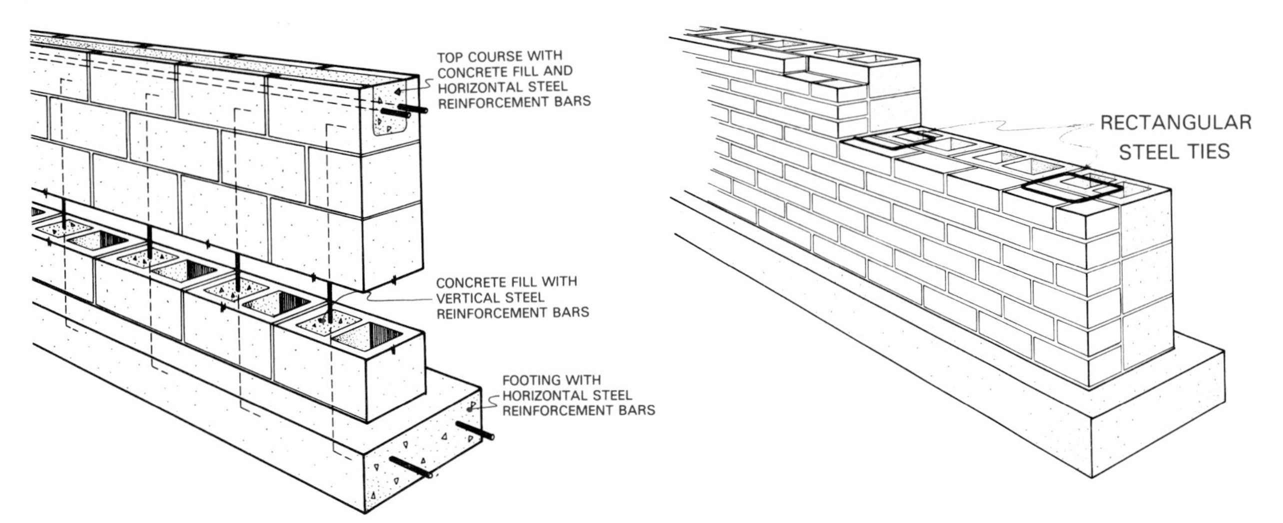

Fig. 52-121 Use of rebars in solid masonry and concrete block walls.

Fig. 52-123 Steel reinforcement between masonry materials.

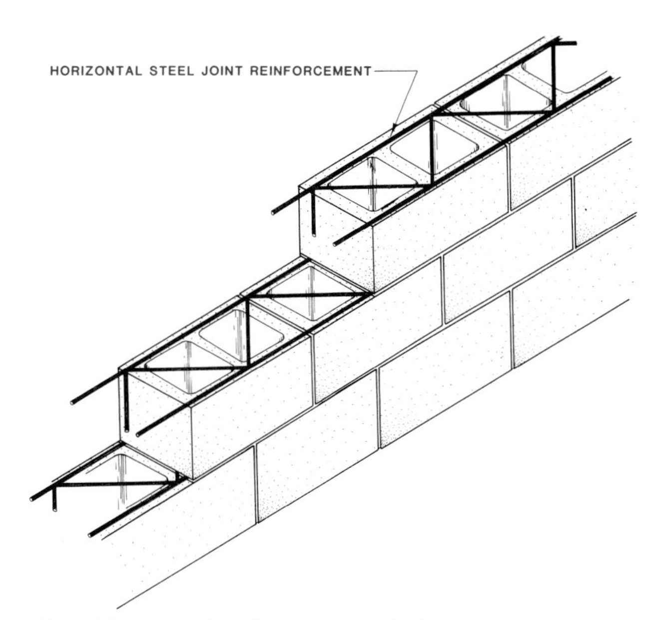

Fig. 52-122 Reinforcement between masonry courses.

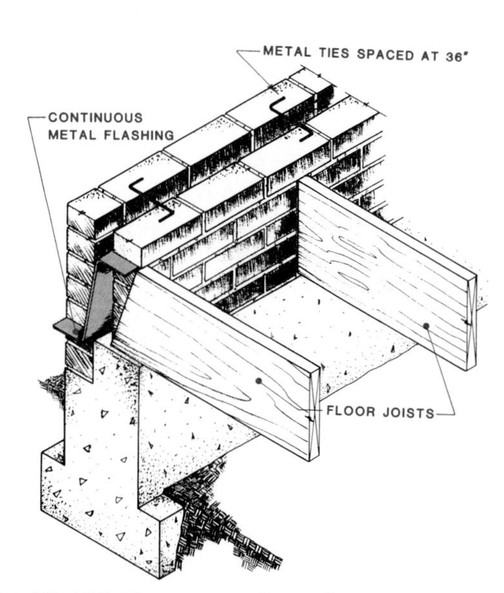

Fig. 52-124 Masonry-cavity wall.

This is primarily because the specifications of the wall and facing materials are usually different. For example, a faced wall may consist of common bricks faced with structural tile, or concrete block faced with brick. Regardless of the material, the two walls are always bonded so they become one wall structurally. The bonding material can be metal ties, steel reinforcing rods, or masonry units laid on end to intersect the opposite wall. Always remember that walls with different coefficients of expansion are never faced together because of differing rates of expansion and contraction under extreme temperature-change conditions.

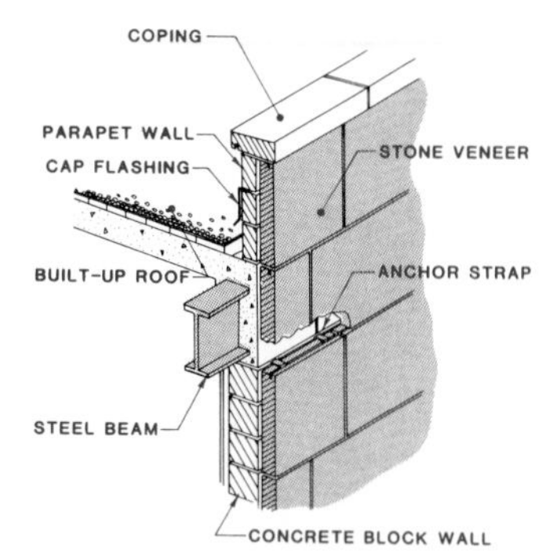

Fig. 52-125 Masonry-veneer wall.

Masonry Veneers Veneer walls, like faced masonry walls, include two separate walls constructed side by side. Unlike faced walls, the veneer wall is not tied to the other wall to form a single structural unit. The veneer wall is simply a non-load-bearing decorative facade, although the two walls are connected with masonry ties. A veneer wall may include two different masonry materials, (Fig. 52-125) or often includes a skeleton-frame wall veneered with a masonry material. The space between the wood and masonry walls (usually 1″) may remain empty or may be filled with insulation depending on climatic conditions. A wall detail or sectional drawing is usually prepared to show this information.

Masonry Bonds A masonry bond is the pattern of arranging and attaching masonry units in courses (rows). Masonry can be laid in a variety of bond patterns, which can make the use of the same size, shape, and material appear completely different, as shown in Fig. 52-126. Figure 52-127 shows the various types of mortar joints specified on construction drawings.

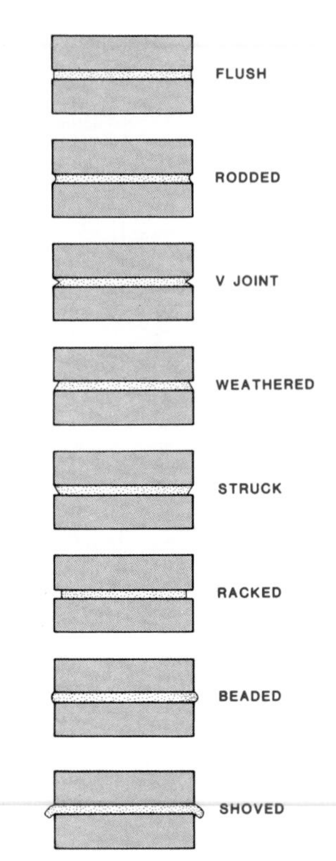

Fig. 52-127 Types of mortar joints.

Drawing Notations

Most specific information relating to masonry construction is placed either on sectional detail drawings and/or in the set of specifications. Some sets of drawings may call out and dimension masonry construction directly on floor plans and/or elevations. Figure 52-128 shows masonry construction details references from the floor plan found on drawing sheet IB.

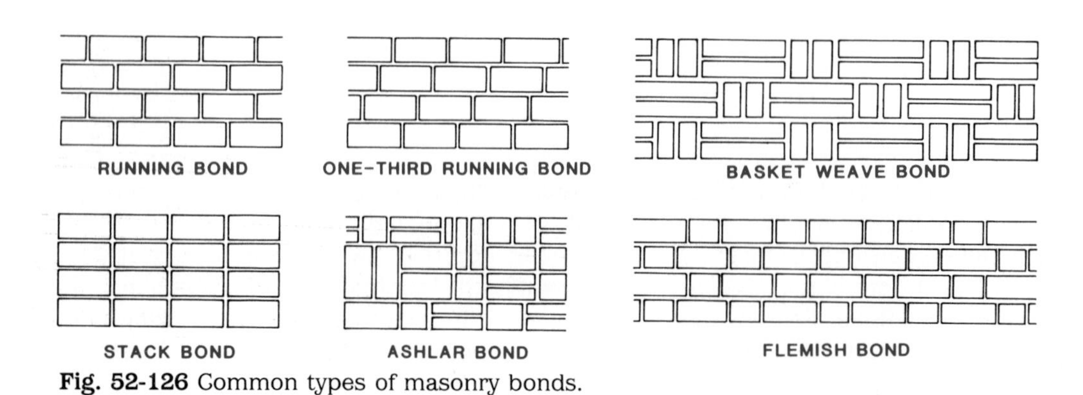

Fig. 52-126 Common types of masonry bonds.

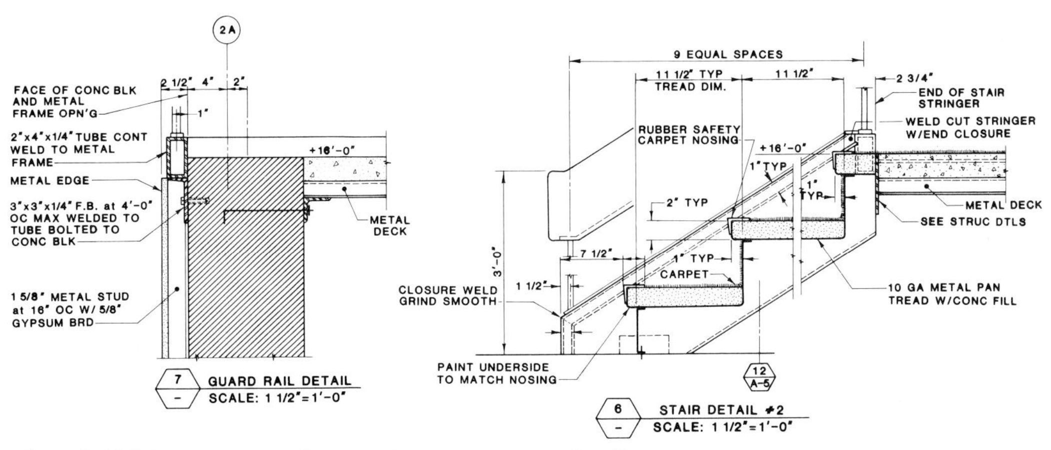

Fig. 52-128 Floor plan referenced to a masonry detail.

CONCRETE CONSTRUCTION SYSTEMS

Early Romans crushed and processed natural rocks to create the cement used to bond their structures. Today cement is manufactured primarily from clay and limestone. *Concrete* used in contemporary construction is a combination of cement, water, aggregate, and chemical setting retarders.

Concrete Structural Members

Concrete made from portland cement has been used for over 100 years for the pouring of foundations, walls, and ground-supported slabs. However, not until low-tensile-strength concrete was reinforced with high-tensile-strength steel could concrete be used for structural components such as columns, beams, girders, and suspended-slab floor and roof systems.

Rebars Steel bars used to reinforce concrete slabs, beams, and columns are known as reinforcing bars or *rebars*. Figure 52-129 shows the application of rebars in reinforced concrete construction. Steel rebars are either smooth or deformed (grooved or embossed). Deformed bars create a stronger bond between bar and concrete because concrete is held in place by the grooves or depressions.

Rebars are sized by numbers (1 through 18) representing ⅛″ increments, up to 2½″, as shown in Fig. 52-130. The bar size number, mill number, steel type symbol, and grade are marked on each rebar. Beam rebars are located horizontally near the bottom side of beams to provide maximum tension resistance. Slab rebars are placed horizontally and in parallel rows also close to the bottom of the slab. To prevent cracking due to temperature and moisture changes, bars are also placed perpendicular to the load-supporting bars. These bars are known as *temperature bars*. Figure 52-131 shows the positioning of temperature bars.

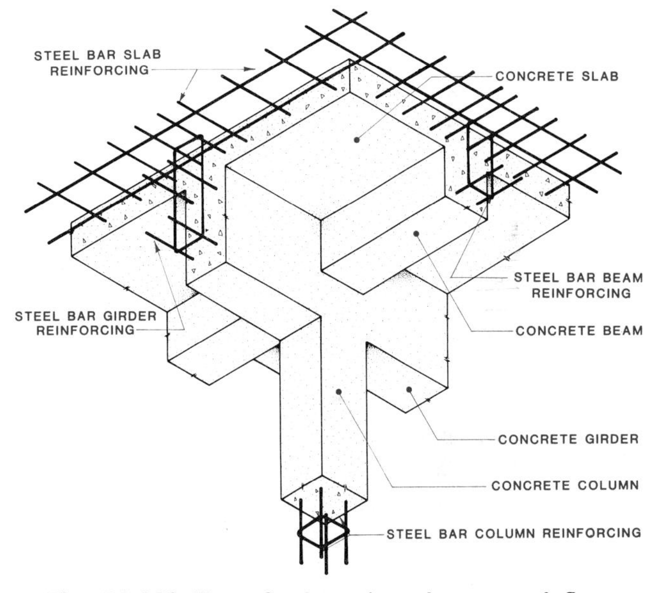

Fig. 52-129 Use of rebars in columns and floor slabs.

Bar Size	Area-Sq. In.	Weight Lbs. Per Ft.	Diameter-Inches
3	.11	.376	7/16
4	.20	.668	9/16
5	.31	1.043	11/16
6	.44	1.502	7/8
7	.60	2.044	1
8	.79	2.670	1 1/8
9	1.00	3.400	1 1/4
10	1.27	4.303	1 7/16
11	1.56	5.313	1 5/8
14	2.25	7.650	1 7/8
18	4.00	13.600	2 1/2

Fig. 52-130 Rebar sizes.

Since minimum thickness (1″ to 3″) must be maintained between rebars and the concrete surface, fixtures known as *bolsters* (saddles) and *chairs* are used to hold the bars in place during slab pouring. U-shaped rods, known as *stirrups*, are used for this purpose in beam pouring.

The exact position of rebars in a slab is shown on a floor plan. The location of rebars in a wall is shown on a plan detail and/or elevation section.

Wire Mesh Steel-welded wire mesh (wire fabric) is frequently used in slabs in place of rebars. Square, rectangular, and triangular patterns are available. Square patterns are the most commonly used. Wire mesh is specified on construction drawings by the spacing (in inches) between wire strands and by the gauge of the wire. The two intersecting wires in a pattern are known as longitudinal (long way) and transverse (short way) wires. A rectangular pattern with transverse wires spaced 4″ apart and longitudinal wires spaced 12″ apart is labeled 4 × 12. This is followed by the gauge of the transverse and longitudinal wires separated with a slash. For example, a rectangular pattern wire mesh with #6 transverse wire spaced 4″ apart and #10 longitudinal wires spaced 12″ apart is noted on the drawing as: 4 × 12-6/10. Common wire mesh sizes are shown in Fig. 52-132.

Columns Concrete columns are vertical members which support weights transferred from beams and girders. Concrete columns are made structurally sound by the addition of rebars. Figure 52-133 shows a typical reinforced concrete column. When the exact position of each column is not dimensioned on a floor plan, column schedules are prepared. To prepare a concrete column schedule, a grid coordinate system is used to identify and relate each column to the column schedule. Information in a concrete column schedule includes the size and number of vertical and horizontal reinforcement bars and connector data. The schedule coding is indexed to a column plan; however, sectional drawings are usually used to show the exact relationship of column, beam, and reinforcement material, as shown in Fig. 52-134.

Beams and Girders Concrete girders are major horizontal members which rest on columns. Beams are horizontal members supported by girders or columns. Concrete beams and girders are reinforced with steel rebars to increase tensile strength. Some reinforced concrete

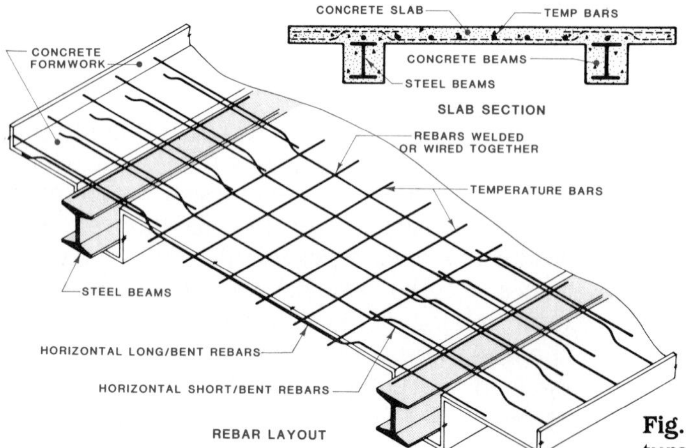

Fig. 52-131 Placement of horizontal and temperature rebars.

Square	Rectangular
6×6-10/10	6×12-4/4
6×6-8/8	6×12-2/2
6×6-6/6	6×12-1/1
6×6-4/4	
	4×12-8/12
4×4-10/10	4×12-6/10
4×4-8/8	
4×4-6/6	4×16-8/12
4×4-4/4	4×16-6/10

Fig. 52-132 Wire mesh sizes.

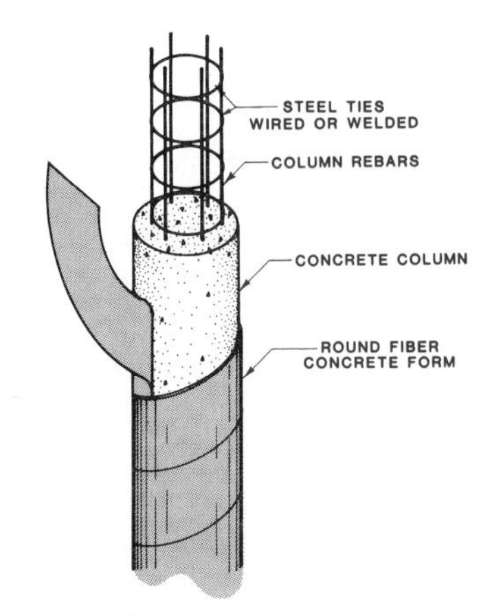

Fig. 52-133 Reinforced concrete columns.

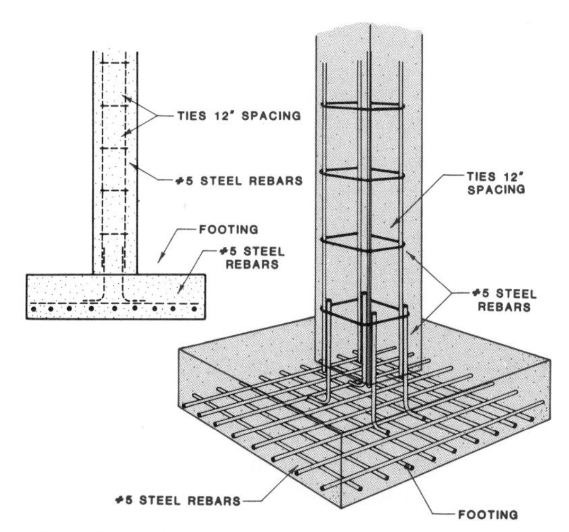

Fig. 52-134 Columns and footing rebars.

beams are rectangular, but most are wider at the top. The position of girders and beams is shown on floor plans with dotted lines or indexed to a beam schedule similar to a column schedule.

Concrete Slab Members Once slabs could only be poured in place at ground level. However, with the advent of steel reinforcement methods, structural slab members can now be manufactured off-site and positioned during the building process in the same manner as columns, beams, and girders.

Concrete Construction Types

Concrete can either be cast in place on-site or precast off-site and shipped to the site as finished girders, beams, slabs, columns, or components. Either type may be reinforced, prestressed, posttensioned, or poured plain depending on the construction application and/or site conditions.

Reinforced Concrete Most concrete used in contemporary construction is reinforced with steel rebars or wire mesh. Developments in reinforcing concrete are mainly responsible for the increased use of concrete in all types of building.

Because concrete is weak in tensile strength but has a strong resistance to compression stress, it was previously used only for nontension applications such as ground-level slabs, walks, or roadways. However, when rebars or wire mesh with high tensile strength are added to the concrete, the tensile strength of concrete is greatly increased and the compression strength is doubled.

Prestressed Concrete When loads are added to concrete members, some deflection (sag) occurs in the center of the member. This happens to all materials under load; however, since concrete has very low tensile strength, excessive deflection can result in tension cracking or complete member failure. This is caused by compression of the upper side and tensioning (stretching) of the lower side, as illustrated in Fig. 52-135. To counteract these unstable tension and compression stresses, concrete is often *prestressed.*

Prestressing is a method of compressing concrete so that both upper and lower sides of a member remain in compression during loading. Prestressing can be accomplished either by *pretensioning* or *posttensioning.*

Pretensioning When concrete is pretensioned, deformed steel bars, called *tendons,* are stretched (tensioned) between the anchors and concrete is poured around the bars.

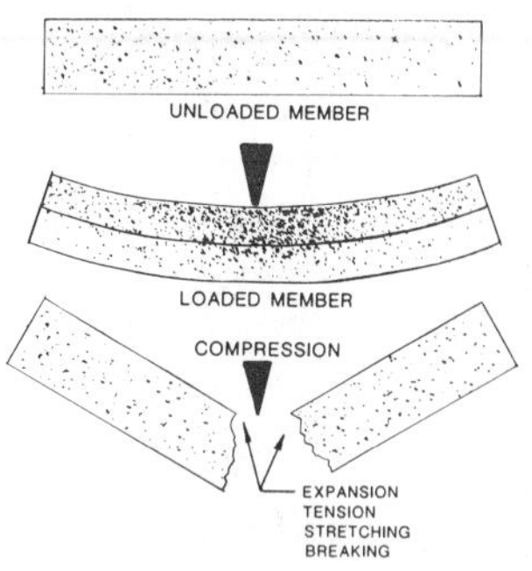

Fig. 52-135 Effect of loading on concrete.

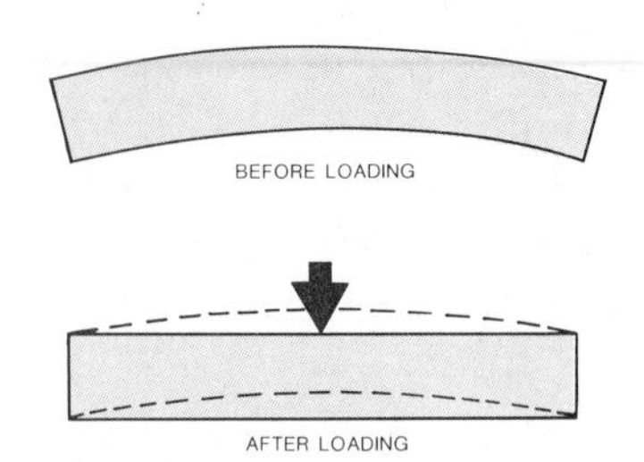

Fig. 52-137 Prestressing with draped tendons.

Once the concrete has cured, the tension is released and the bars attempt to return to their original, shorter length. But the concrete hardens around the bar grooves and holds the deformed bars at near their stretched length. This creates a continual state of compressive stress that can be compared to holding a row of blocks as shown in Fig. 52-136.

As a further aid to prevent bending, concrete members are prestressed by draping tendons near the bottom of the member, as shown in Fig. 52-137. This bottom tension buckles the member upward so that when the anticipated loads are added, the beam straightens to a level position.

Posttensioning In posttensioning, tendons are either placed inside tubes embedded in the concrete or greased to allow slippage. The tendons are then stretched with hydraulic jacks and the ends anchored. This creates compressive stress by the ends of the tendons pulling toward the center. Posttensioning can be done at a factory, or on-site to reduce shipping weight, especially for very large members.

Cast-in-Place Concrete Forms for pouring concrete for footings, foundations, slabs, and walls have been used for a long time. However, new developments in reinforced and prestressed concrete construction techniques enable builders to erect structures with extremely complex contours. The building shown in Fig. 52-138 is an example of concrete shell construction. Concrete shells are a type of concrete system which uses poured reinforced concrete. A light steel structure is erected. Then concrete is poured or sprayed and the casting of concrete holds the steel in place after hardening.

Poured concrete is no longer restricted to ground-level work because of the need for support and elaborate systems of formwork. Drawings for cast-in-place concrete systems include the outline and dimensional size of the finished job, including the position of rebars and joints.

Although it is the responsibility of the contractor to build forms that will produce the desired result, formwork drawings are sometimes included in a complete set of architectural drawings. Some sets of specifications also include minimum sizes for formwork members and reinforcement but do not include drawings of construction details.

Precast Concrete Systems Precast concrete is the opposite of cast-in-place concrete. Precasting of concrete is simply the pouring of concrete into wood, metal, or plastic molds. Once set, the molded concrete is then placed in position in the hardened form.

Although concrete block is the most commonly used precast concrete material, it is considered a masonry building block like brick. Precast concrete structural members include wall panels, girders, and a variety of slabs, as shown in Fig. 52-139. Precast concrete is usually reinforced.

Precast slabs are available in solid, hollow-core, and single and double tee shapes, and are

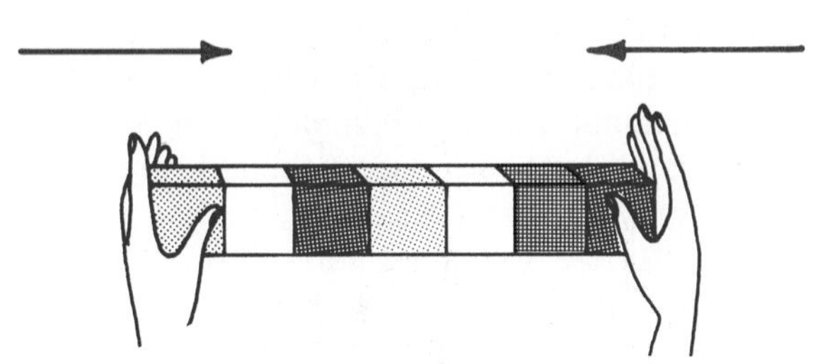
Fig. 52-136 Principle of pretensioning concrete.

Fig. 52-138 Application of cast-in-place concrete construction.

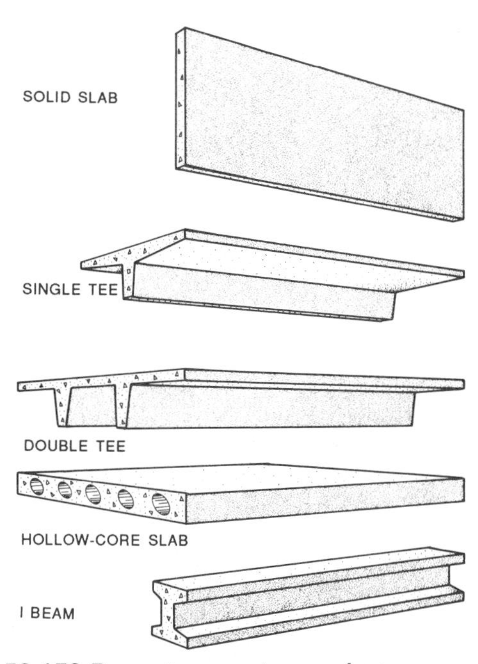

Fig. 52-139 Precast concrete members.

used for walls, floors, and roof decks. Wall panels are solid precast units used either for bearing or nonbearing walls, depending on the amount of reinforcement. Since the exterior sides of concrete wall panels are usually exposed, special textured finishes are often applied during the casting process. These panels may be combined with layers of insulation to form a complete monolithic wall unit.

Slab Component Systems Precast slab components or cast-in-place slabs are divided into two types: one-way systems and two-way systems.

In *one-way systems* the rebars are all parallel. One-way system girders, which rest on columns, are also parallel to the rebar alignment. One-way solid slabs are extremely heavy and are therefore impractical for most spans over 12′. To lighten the dead load, ribbed one-way slabs are often used. The ribbed slab is a thin slab 2″ to 3″, supported by cast ribs. These units are constructed of precast slab tees or cast in place.

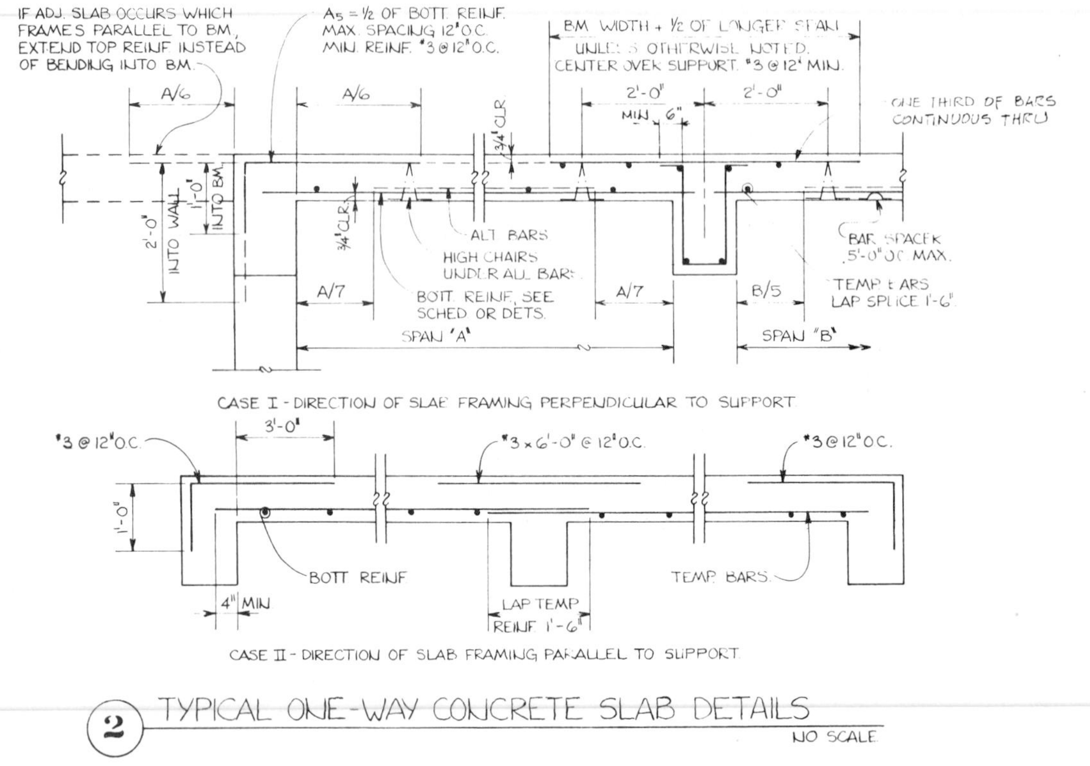

Fig. 52-140 Rebar position shown with dotted lines.

When ribbed slabs are to be poured in place, a ribbed slab plan shows the horizontal rib positions with dotted lines. Dotted lines are also used on detail drawings to show the position of rebars in the slab and ribs, as shown in Fig. 52-140.

In *two-way slab* systems (Fig. 52-141) the rebars, girders, and beams are aligned in perpendicular directions. When ribs extend in both directions the system is known as a *waffle slab*, as shown in Fig. 52-142. *Pan and waffle slabs* are now used to cast suspended floor and roof systems for spans up to 60 ft. Temporary fiberglass or metal pans (domes) are placed, open side down, 4″ to 7″ apart on a temporary floor. But no pans are placed around columns. Rebars are added and concrete is then poured to a depth of several inches over the pans. When the concrete is cured the pans and temporary flooring are removed, and a suspended waffled floor (or roof) results. This type of cast-in-place system is lightweight (for concrete), sound-resistant, fireproof, and relatively economical.

A *flat slab* is a two-way slab unit which rests

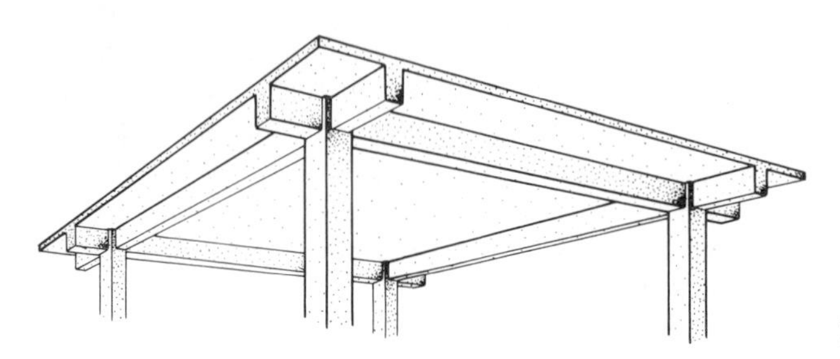

Fig. 52-141 Two-way slab system.

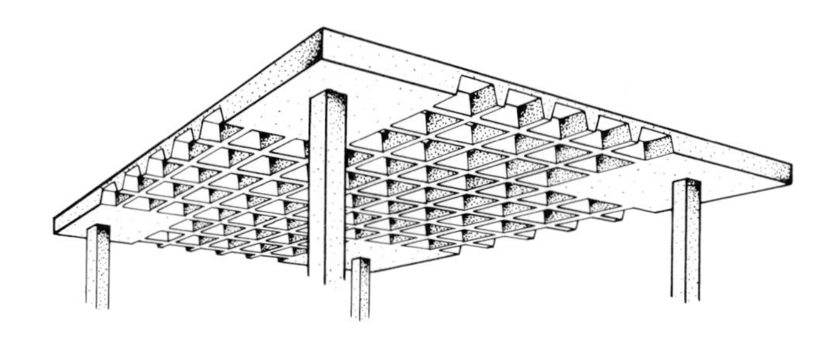

Fig. 52-142 Waffle slab system.

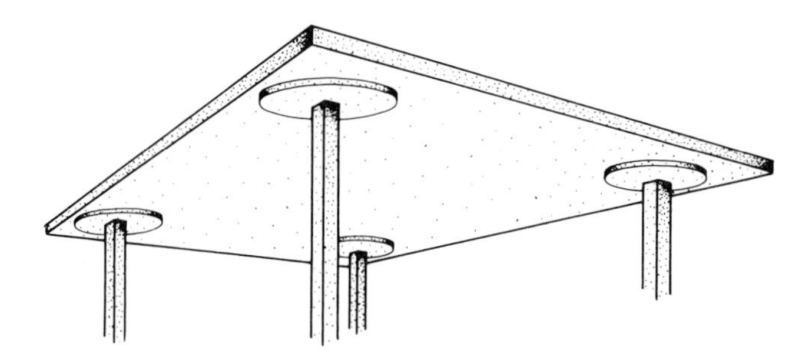

Fig. 52-143 Flat slab construction.

directly on columns without a girder or beam support. A flat slab floor (or roof) system is actually a series of individual slabs, with the center of each slab resting independently on a column. All of the slabs' weight is directed through the columns to footings. When the slabs are joined together, a unified floor is created, as shown in Fig. 52-143. In flat slab systems the supporting columns are strengthened by the addition of a thicker slab area (*drop panel*) around columns. A column capitol or flared head also helps spread the slab loads in this type of construction.

Concrete Joints Since concrete expands and contracts with changes in moisture levels and temperature, relief joints are required to allow for these fluctuations. Some construction drawings and/or specifications indicate minimum dimensions for placement of expansion joints. And some drawings show specifically where joints are required or must be avoided.

PREFABRICATION

Throughout the early twentieth century, from 1900 to about 1940, prefabrication became popular for houses. Its use was a modified do-it-site yourself approach to home building. A few companies ventured into prefabrication of a complete house package including ceiling, wall, and floor panels, complete with plumbing and electrical work installed in the walls. At the same time, conventional builders were accepting prefabrication for some parts of a house. They recognized, for example, that a better and less expensive window sash could be produced in a plant than could be handmade on the job site. As builders became more aware of the time, labor, and materials that could be saved by prefabrication, they began to use preassembled cabinets, prefitted doors, prefinished sink tops, prefinished floors, and other prefabricated parts.

Today, the most successful companies producing factory-made homes rely on some use of conventional framing methods. They simply apply the techniques of mass production to their production methods. The goal is to minimize custom-job work without sacrificing the quality of the construction.

All structures are factory-built to some extent; that is, not all the materials or components are manufactured or put together on the site. Some structural components are simply precut. This means that all the materials are cut to specification at the factory, and then assembled on the site by conventional methods. With the most

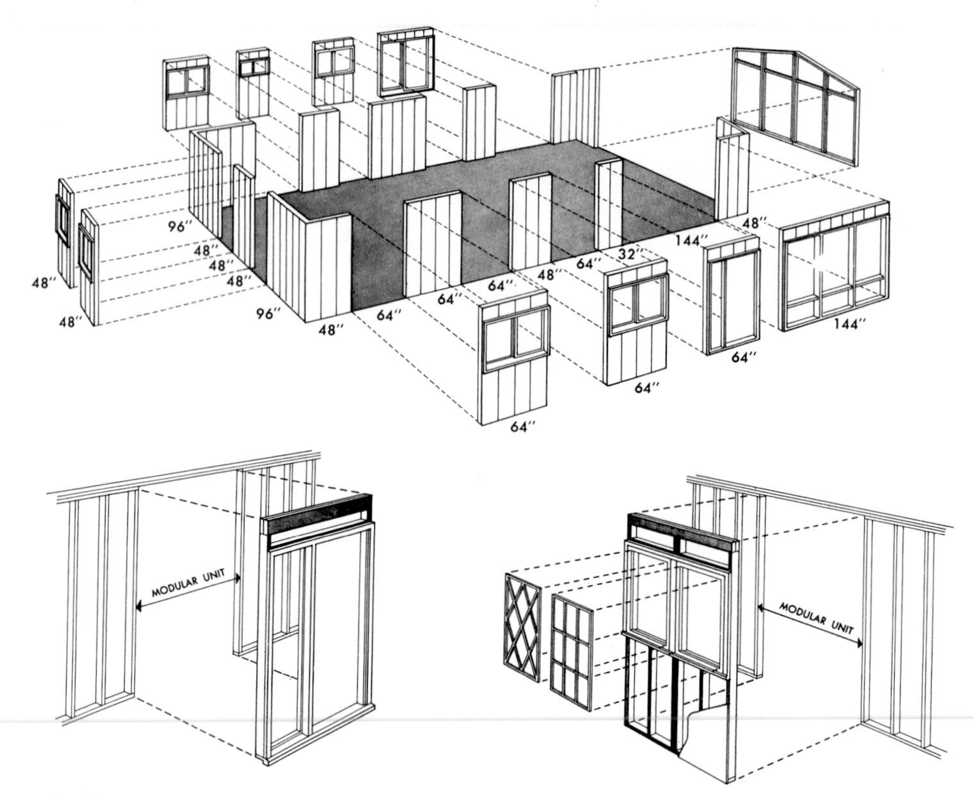

Fig. 52-144 Use of factory-built components.

common type of prefabricated homes, the major components, such as the walls, trusses, decks, and partitions, are assembled at the factory. The utility work, such as installation of electrical, plumbing, and heating systems, is completed on site. The final finishing work, such as installation of prehung floors, prefinished roof coverings, and prefinished walls, is also done on site.

There are some factory-built homes, however, that are constructed in complete modules at the factory and require only final electrical-outlet, roof-overhang, and assembly-fastening work on site to complete the job.

With the exception of mobile homes, most factory-built homes require some on-site preparation. Designers have been working for years to develop residential designs that would eliminate or greatly curtail the amount of on-site preparation.

MANUFACTURED BUILDINGS

The first completely factory-built homes were trailers and mobile homes. The mobile-home buyer, unlike the buyer of a conventional factory-build home, has little option to adjust or customize the design of the residence. He or she does have opportunities to select from many different sizes, models, and interior components. The mobility of our population, the increasing cost of real estate, and rising real-estate taxes have contributed to the growth of the mobile-home industry. One study estimates that approximately 25 percent of all one-family houses in the United States are mobile homes. Thus, the mobile-home designer must plan homes that can be mass-produced but also provide a variety of options for the prospective buyer. A variety of sizes must also be designed to span the price range of various consumers. The major limiting factor in mobile-home design is the width restriction for highway transportation.

HURRICANE AND EARTHQUAKE CONSTRUCTION

Extreme climate or environmental conditions require special construction methods. For example, in hurricane- or earthquake-prone areas, additional tie-down brackets (Fig. 52-145) can help anchor the structural frame to the foundation. The use of additional blocking (Fig. 52-146) on the sill and between vertical members can also help anchorage and provide additional rigidity to the structural frame. The wall-and-cornice-area design shown in Fig. 52-147 is designed keep the wall stable and hold the roof to the wall and the sill to the foundation during extreme lateral force conditions.

To minimize structural damage during earthquake shocks, foundation supports should extend to bedrock. Where buildings must be erected on landfill or less stable soil areas, the footings or slab should cover the largest horizontal area possible. Architectural engineers are also exploring the use of spring shock absorbers between floors to help absorb tremor-shock stress before structural failure occurs. Structural systems using suspended floor systems which could collapse as one unit on the floors below should be avoided near known fault lines.

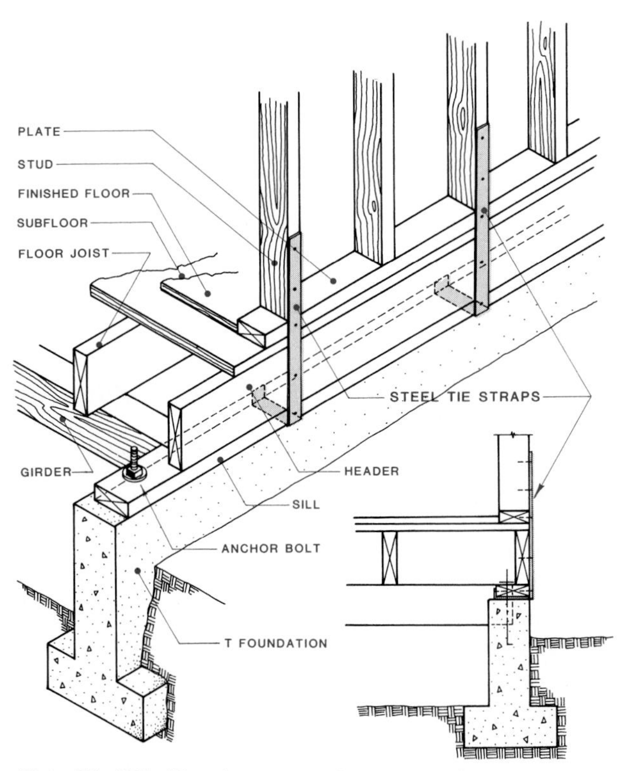

Fig. 52-145 Hurricane- and earthquake-resistant sill tie downs.

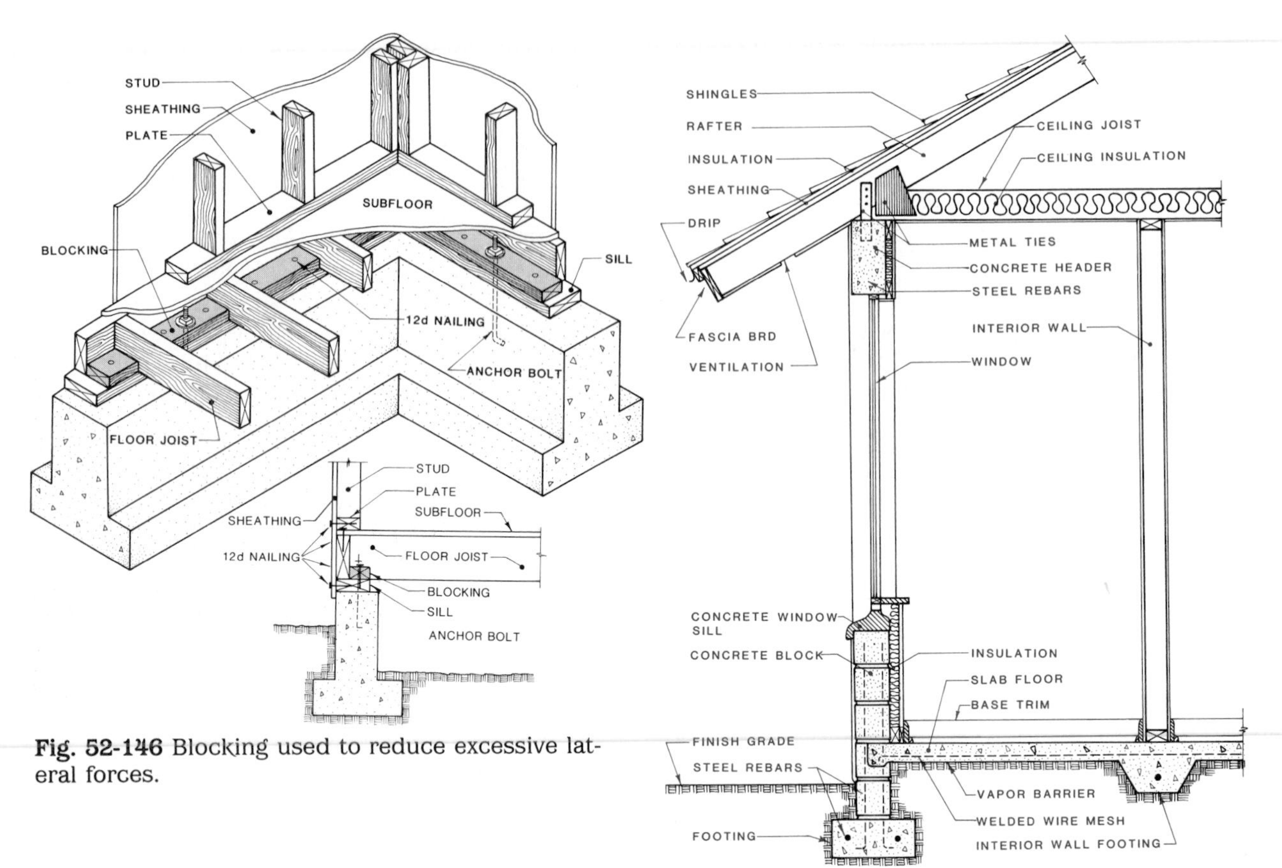

Fig. 52-146 Blocking used to reduce excessive lateral forces.

Fig. 52-147 Hurricane- and earthquake-resistant wall designs.

Exercises

1. **Identify the structural members shown in Fig. 52-148.**
2. **Identify the wall and floor members shown in Fig. 52-149.**
3. **Identify the truss types and roof members shown in Fig. 52-150.**
4. **Identify the members and components shown in Fig. 52-151.**
5. **Identify the roof members and components shown in Fig. 52-152.**
6. **Identify the cornice area materials and members shown in Fig. 52-153.**
7. **Identify the stair members shown in Fig. 52-154.**

8. **Choose the detail drawings from this unit you will use in a building of your own design. Sketch any alteration needed to apply specifically to your design.**
9. **Define the following architectural terms: *span, framework, skeleton frame, structural tie, sheathing, live load, dead load, tension, compression, shear, torsion, deflection, equilibrium, conventional wood framing, post-and-beam framing, column, beam, post, plank, cantilever, bearing partition, nonbearing partition.***

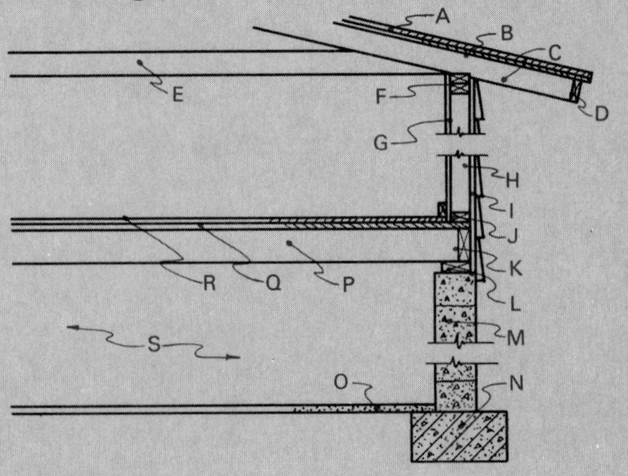

Fig. 52-148

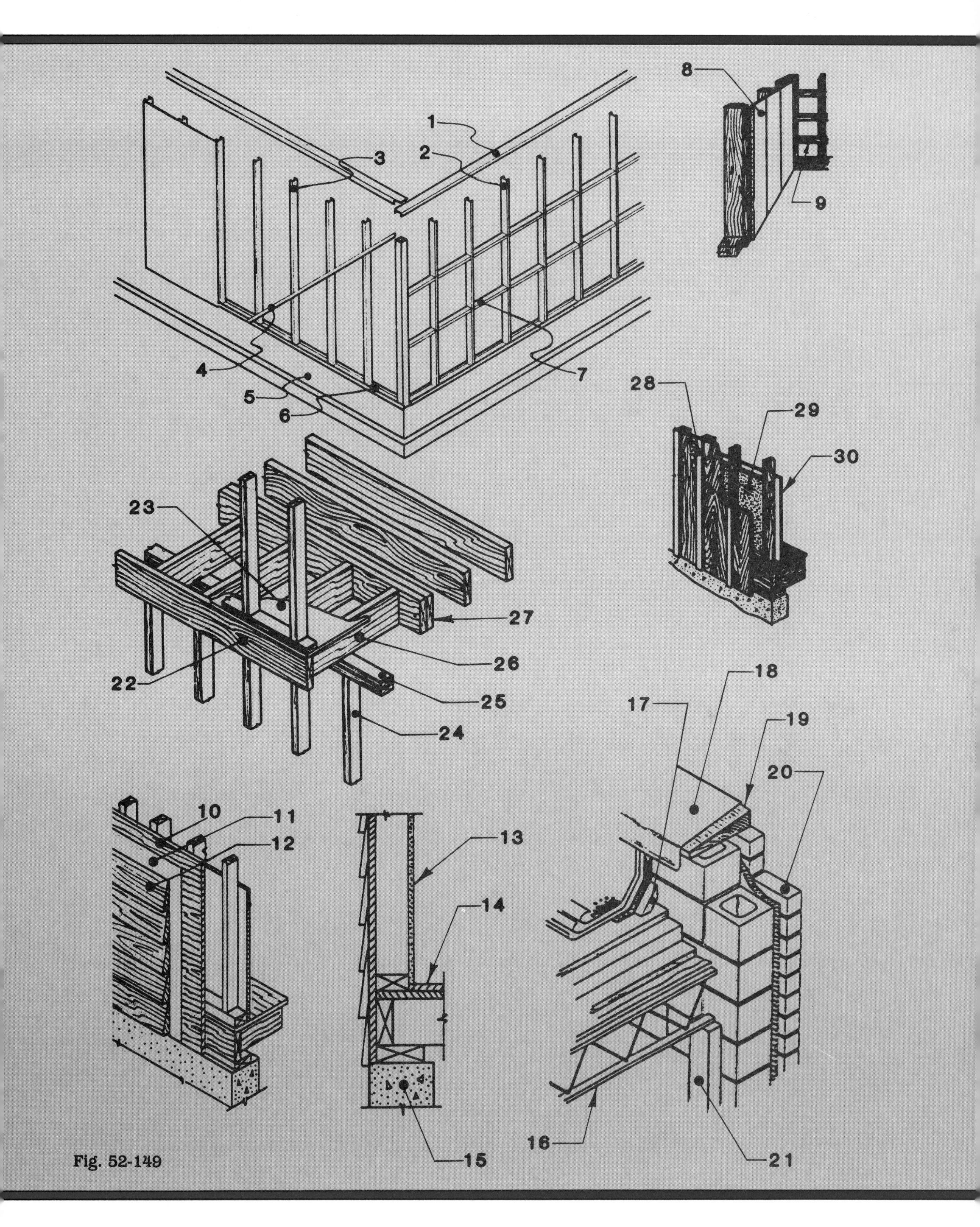
1
2
3
4
5
6
7
8
9
10
11
12
13
14
15
16
17
18
19
20
21
22
23
24
25
26
27
28
29
30
Fig. 52-149

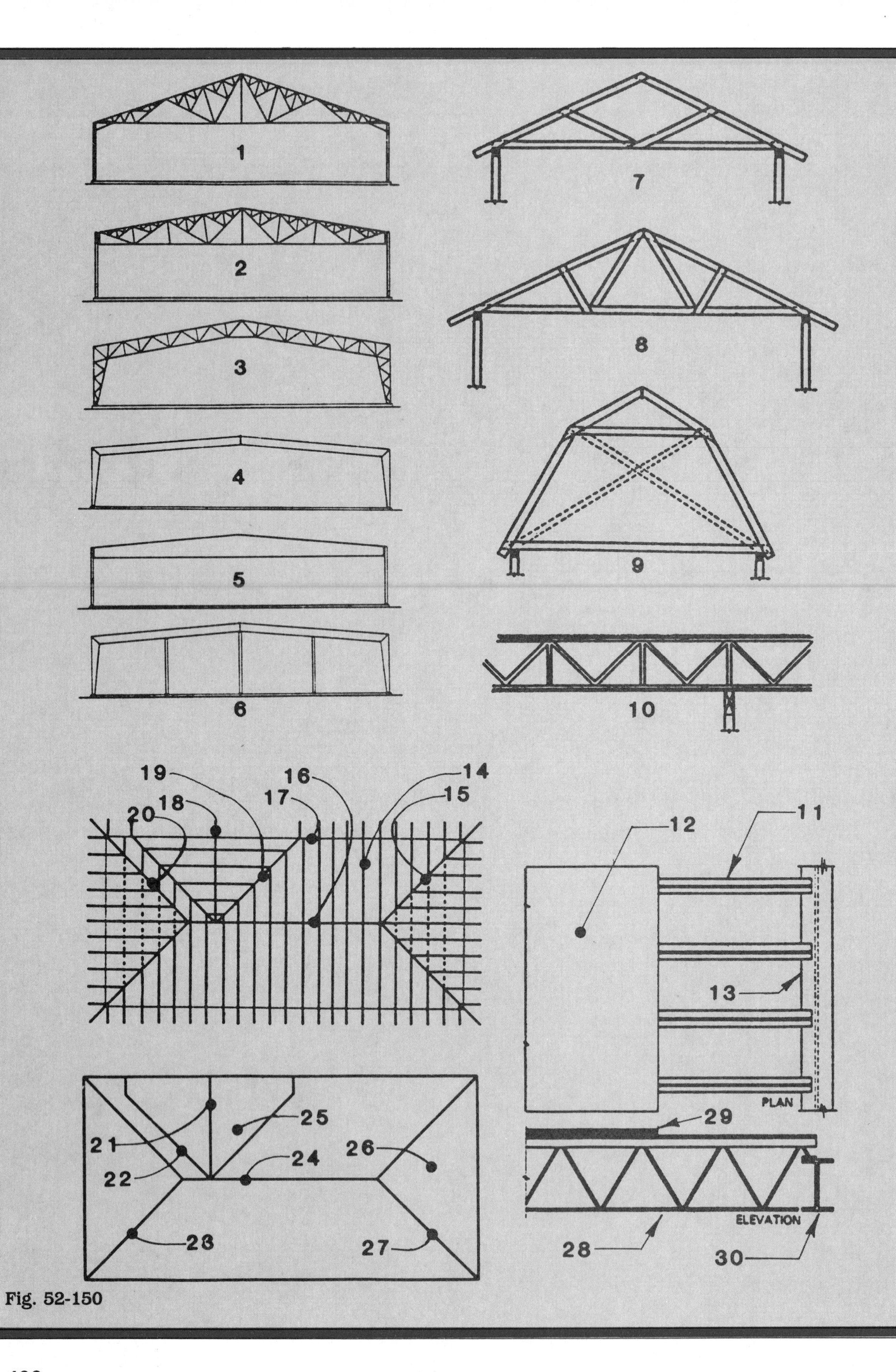
1
2
3
4
5
6
7
8
9
10
19
16
14
17
15
20
18
11
12
13
PLAN
29
21
25
24
26
22
ELEVATION
28
27
28
30
Fig. 52-150

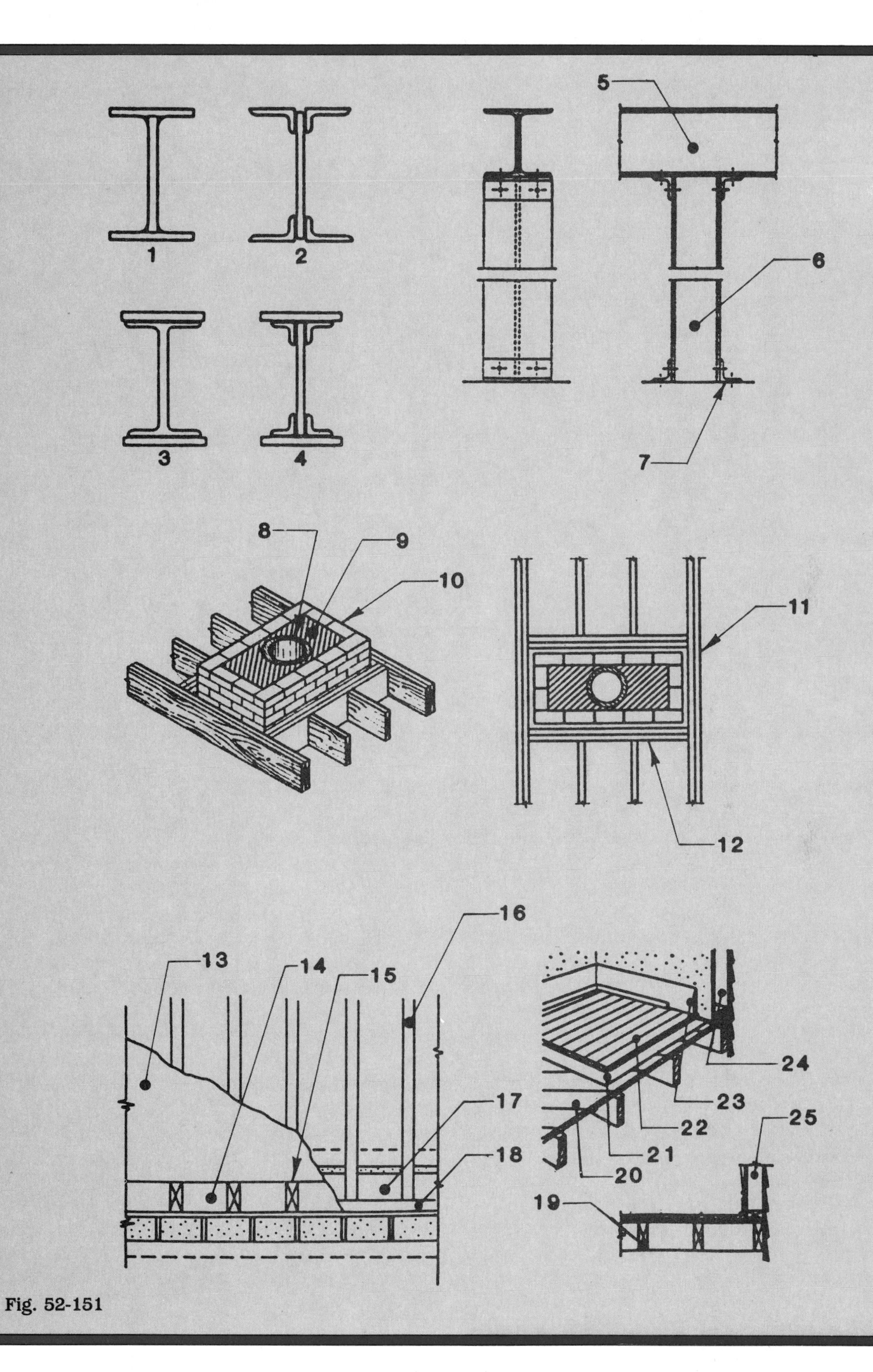

Fig. 52-151

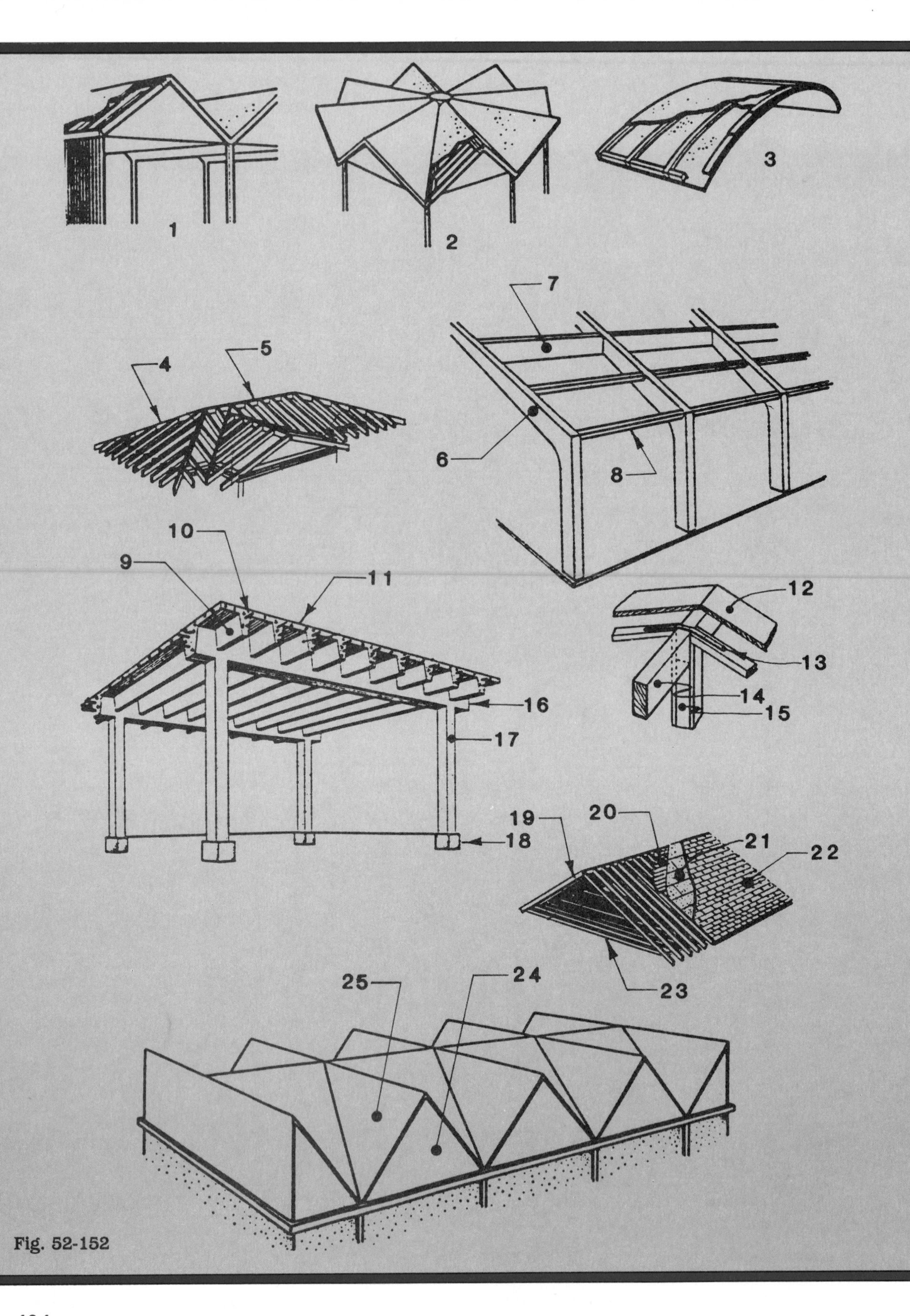

Fig. 52-152

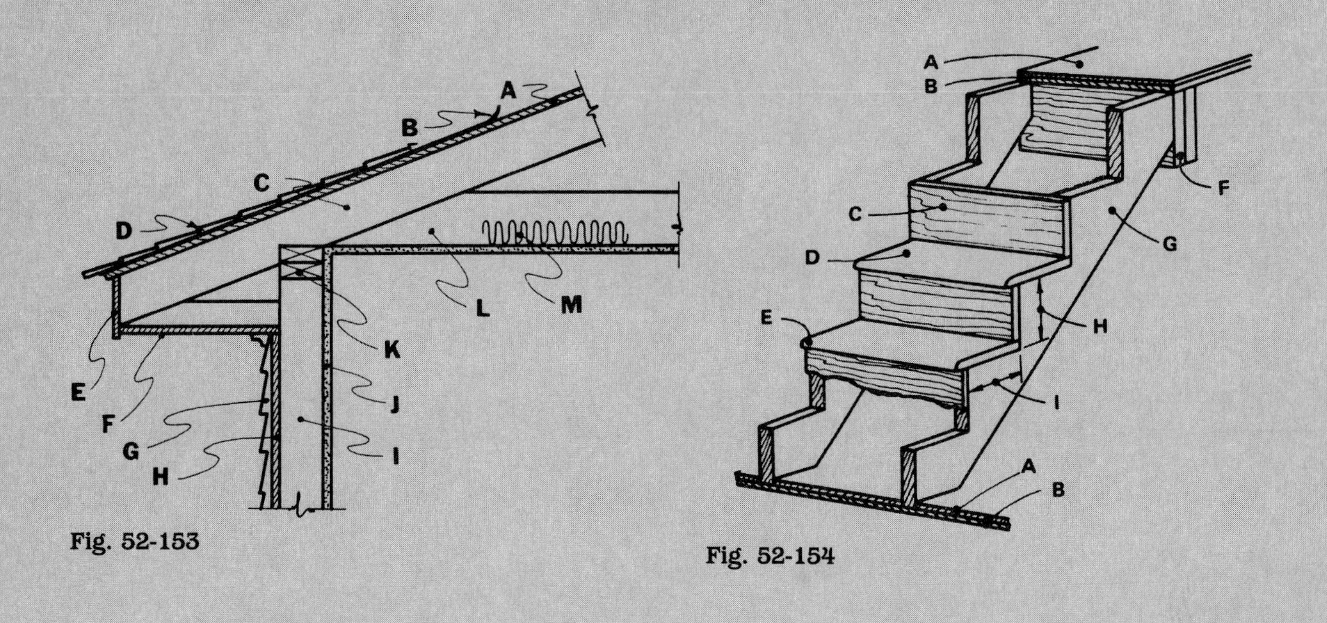

Fig. 52-153

Fig. 52-154

UNIT 53

FLOOR FRAMING PLANS

Floor framing plans range from those plans that show the structural support for the floor platform, to drawings that show the construction details of the intersections of the floor system with foundation walls, fireplaces, stairwells, and so forth. *Platform-floor systems* are those systems that are suspended from foundation walls and/or beams.

TYPES OF PLATFORM-FLOOR SYSTEMS

Platform-floor systems are divided into three types: conventional, plank-and-beam, and panelized floor systems. These types are shown in Fig. 53-1.

Conventional Systems

The conventionally framed platform system provides a flexible method of floor framing for a wide variety of design conditions. Floor joists are usually spaced at 16″ (406 mm) intervals and are supported by the side walls of the foundation and/or by beams.

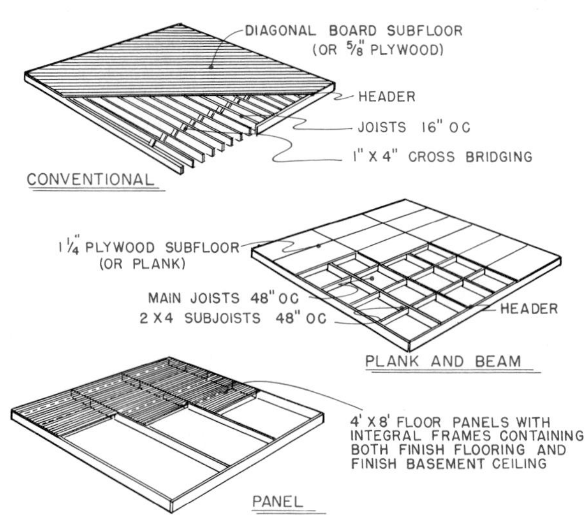

Fig. 53-1 Types of platform-floor systems.

Post-and-Beam Systems

The *plank-and-beam* (post-and-beam) method of floor framing uses fewer members that are larger than conventional framing members. Because of the size and rigidity of the members, the need for bridging for stability between joists is eliminated.

Panelized Systems

Panelized floor systems are composed of preassembled sandwich panels of a variety of skin and core materials. Core-panel systems are used for long clear spans over basement construction and for shorter spans in nonbasement houses. Other experimental methods of core-component design and construction are continually being developed and refined to reduce the on-site construction costs. Research in engineering and wood technology is continually extending the use of components for support systems.

DESIGN

The design of the floor system depends on load, type of material, size of the members, spacing of the support members, and distance between the major support for members (*span*). Figure 53-2 shows that as the load is increased, the span must be decreased to compensate for the increase, or the member must be made larger or of a stronger material. The design of floor systems, therefore, demands very careful calculation in determining the live and dead loads acting on the floor. The most appropriate material for posts, beams or girders, and blocking must be selected. The design also requires determination of the exact size of the posts, beams, deck materials, and joists and establishment of the exact spacing between posts, girders, and joists. The parts of a floor system that must be selected on the basis of the loads, material, size, and spacing include the deck, joists, girders or beams, and posts or columns.

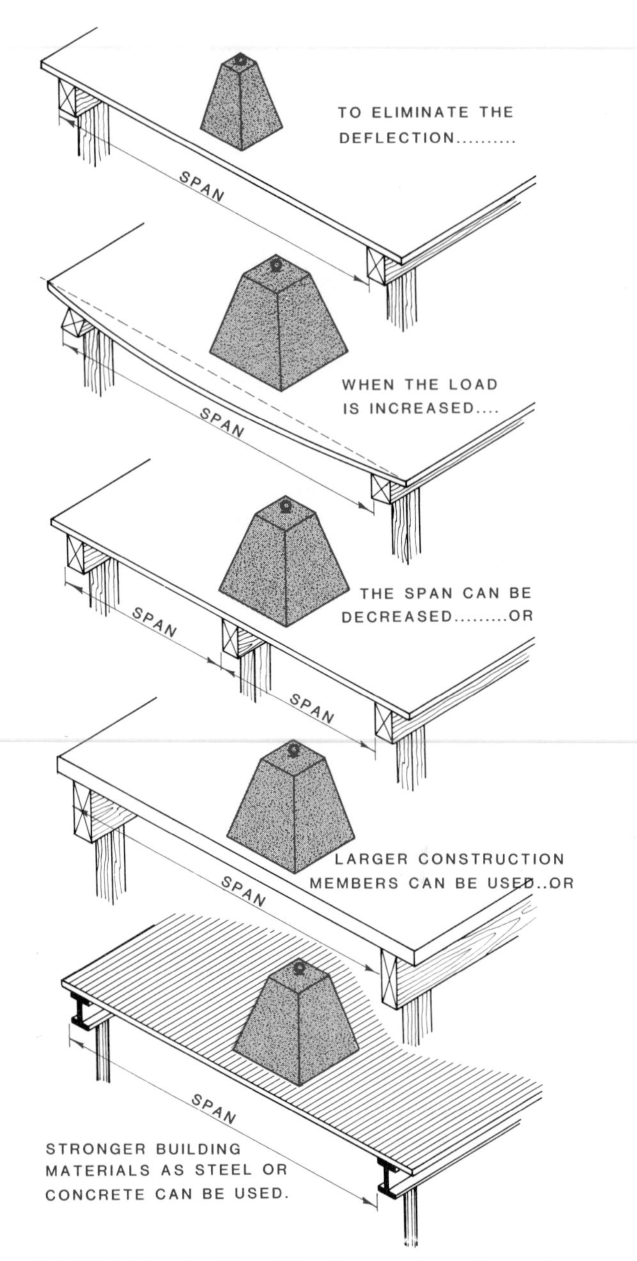

Fig. 53-2 The design of the floor system depends on many factors.

FLOOR DECKING

Decking is the top surface of a floor system. The floor deck, in addition to bridging, provides lateral support for the joists. Decking usually consists of a subfloor and a finished floor, although in some systems they are combined. Subfloor decking materials consist of diagonal wood members, plywood sheets, prefabricated panels, plank boards, concrete slabs (Fig. 53-3A), or corrugated steel sheets. When wood planks (Fig. 53-3B) or plywood sheets (Fig.53-3C) are used as subflooring, they are laid directly over the joists. Exterior and interior sole plates are laid directly on the subflooring. The functions of the subfloor are as follows:

1. It increases the strength of the floor and provides a surface for the laying of a finished floor.

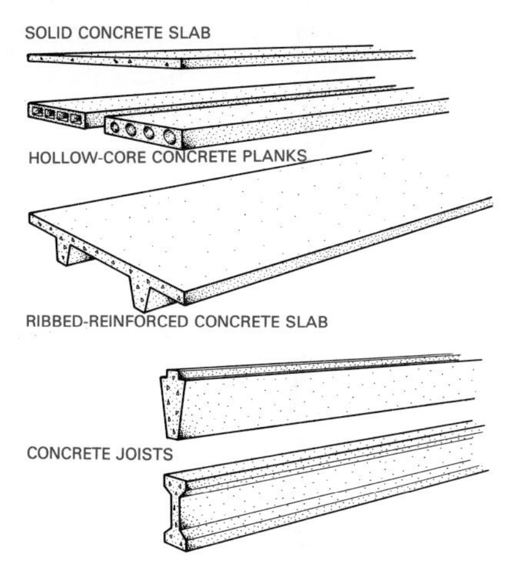

Fig. 53-3A Precast concrete floor system components.

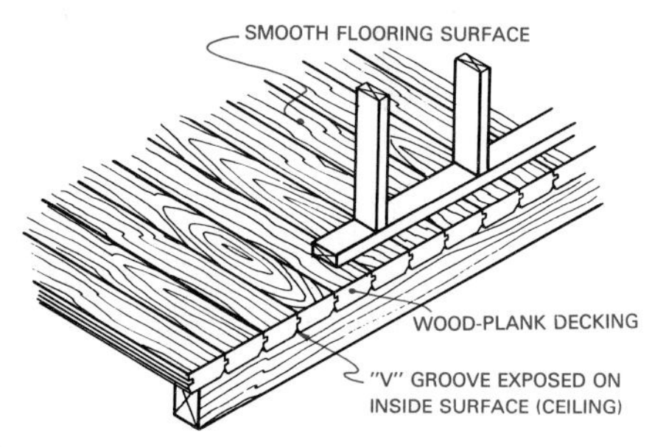

Fig. 53-3B Wood-plank subfloor.

2. It helps to stiffen the position of the floor joist.

3. It serves as a working surface during construction.

4. It helps to deaden sound.

5. It prevents dust from rising through the floor.

6. It helps to insulate.

7. It acts as a buffer to soften and reduce the hard impact of slab floor construction, as shown in Fig. 53-3D.

Finished flooring is installed over the subfloor and butted against the partition framing members. The finished floor provides a wearing surface over the subfloor, or over the joist if there is no subfloor. In this type of construction, the finished floor must be tongue-and-groove boards 1½″ to 2″ thick. Hardwood, such as oak, maple, beech, and birch, is used for finished floors. Tile is often used as a finished floor.

When steel subfloor decks are used, they are usually constructed of corrugated sheet steel. These subfloors act as platform surfaces during construction and also provide the necessary subfloor surface for a concrete slab floor. When precast-concrete floor systems are used, concrete slab members function as subflooring.

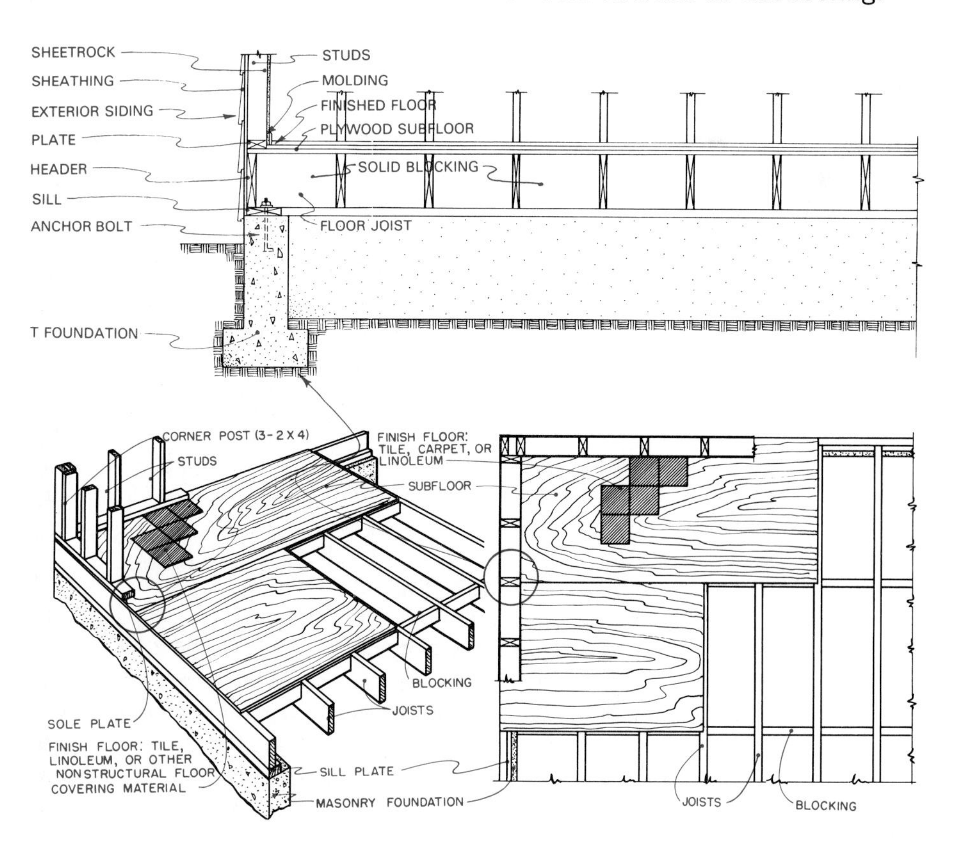

Fig. 53-3C Plywood subfloor system.

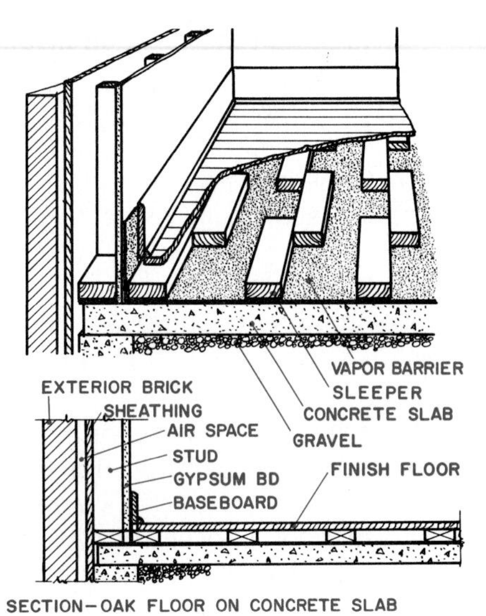

Fig. 53-3D Slab subfloor.

Girder Size	Supporting Walls	No Wall Support
4 × 4	3′-6″ 3′-0″	4′-0″ 3′-6″
4 × 6	5′-6″ 4′-6″	6′-6″ 5′-6″
4 × 8	7′-0″ 6′-0″	8′-6″ 7′-6″

Fig. 53-3E Allowable spans for girders.

JOISTS

Floor joists are horizontal members which rest on a wall and/or beam and on which the floor decking is laid. Floor joists must support the maximum live loads of the floor at its midspan.

To determine the proper joists to use, you must consider the load, spacing, and strength of the joist material.

Loads

Only live loads bear directly on the decking and joists. Therefore, the total live load for the room having the heaviest furniture and the heaviest traffic should be used to compute the total load for the entire floor. To find the live load in pounds per square foot, divide the total room load in pounds by the number of square feet supporting the load. To find the live load for a floor in kilograms per square meter, divide the total room load in kilograms by the number of square meters supporting the load. The dead load is all the permanent entities built into the structure. The total load is the sum of live load and dead load. Unit 83 presents construction loads in detail.

Figure 53-3E is an engineered table that is used to select joist sizes and spacing. First select the wood group, 1,2,3 or 4. Next select the smallest span (points of support) the joist must cross. The joist size and the joist spacing (distance between members) is shown on the left. For example in selecting a joist for group 2 (wood quality) to span 14′-0″, look for 14′-0″ or the next larger span (you can overbuild, but you cannot underbuild). The next largest span is 14′-6″. Following the table to the left, the entry shows 2″ × 10″ joists spaced at 16″ OC (on center). Figure 53-3F shows the normal girder span for standard grade wood, with and without supporting walls. Figure 53-4 shows methods of drawing both plan and elevation views of joists with solid blocking or wood cross-bridging.

You can see, therefore, that as the size, spacing, and load vary, the spans must vary accordingly. Or if the span is changed, the dimensions of the spacing of the joist must change accordingly. The illustration in Figure 53-3C indicates the method of drawing part of the floor framing plan that shows the size and position of joists and the blocking between joists. In this particular detail, the relative position of the subfloor, finished floor, sill, and exterior walls is also shown. Figure 53-5 shows the relationship between the joists of the floor and the finished floor and sill.

In some systems, the floor system has no joists, as shown in Fig. 53-6. However, in this system the girder and blocking perform the function of the joist. The girders rest directly on posts, and the subflooring rests directly on the girders. In this construction, these girders are spaced more closely than most girders that support joists. The subflooring can be nailed or glued with adhesives to bond structural members together.

HEADERS

Whenever it is necessary to cut regular joists to provide an opening for a stairwell or a hearth, it is necessary to provide auxiliary joists called *headers*. Headers are placed at right angles to the regular joists, to carry the ends of joists that

Size of floor joists	Spacing of floor joists	Maximum span (feet and inches) Group 1	Group 2	Group 3	Group 4
2″ × 6″	12″	10′-6″	9′-0″	7′-6″	5′-6″
	16″	9′-6″	8′-0″	6′-6″	5′-0″
	24″	7′-6″	6′-6″	5′-6″	4′-0″
2″ × 8″	12″	14′-0″	12′-6″	10′-6″	8′-0″
	16″	12′-6″	11′-0″	9′-0″	7′-0″
	24″	10′-0″	9′-0″	7′-6″	6′-0″
2″ × 10″	12″	17′-6″	16′-6″	13′-6″	10′-6″
	16″	15′-6″	14′-6″	12′-0″	9′-6″
	24″	13′-0″	12′-0″	10′-0″	7′-6″
2″ × 12″	12″	21′-0′	21′-0″	17′-6″	13′-6″
	16″	18′-0″	18′-0″	15′-6″	12′-0″
	24″	15′-0″	15′-0″	12′-6″	10′-0″

Fig. 53-3F Allowable spans for spacing of floor joists.

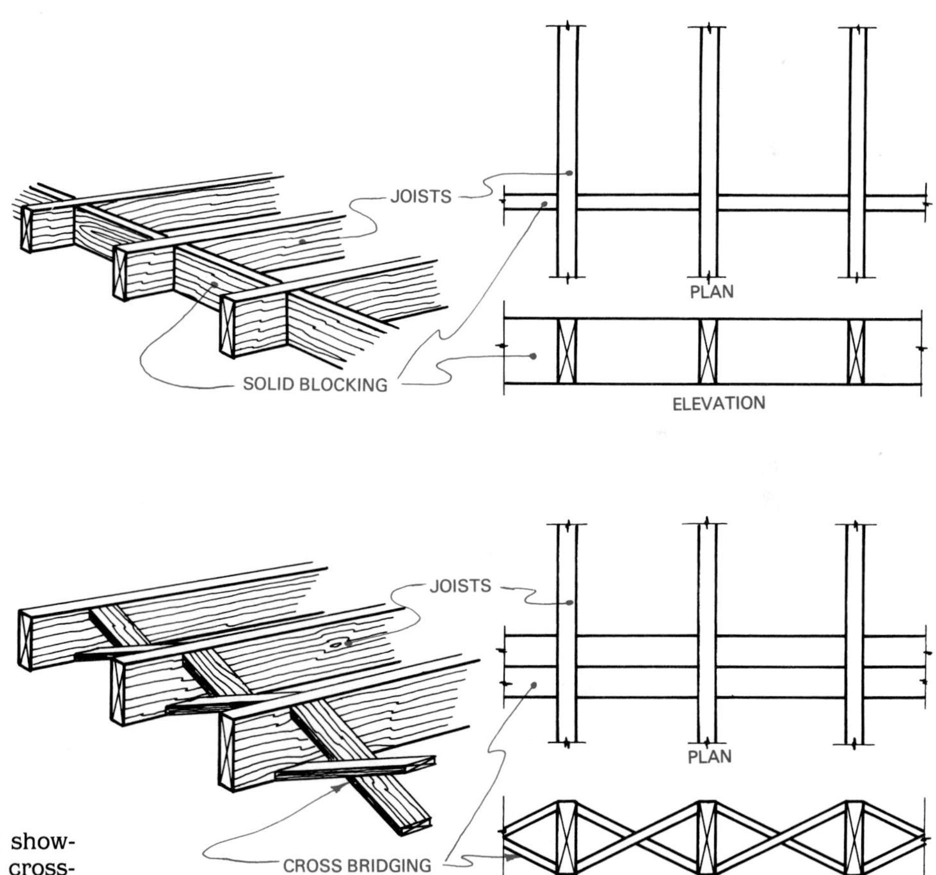

Fig. 53-4 Floor-framing showing solid blocking and cross-bridging.

are cut. A header cannot be of greater depth than any other joist; therefore, headers are usually *doubled* (placed side by side) to compensate for the additional load. Figure 53-7 shows the use of the double header as compensation for the joists that are cut to provide space for the fireplace. Figure 53-8 shows similar use of headers around chimney openings. Additional support is also needed under bearing partitions. Figure 53-9 shows the use of double joists under

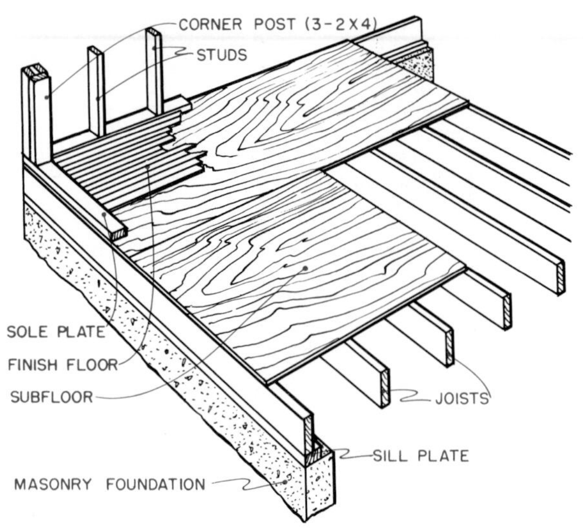

Fig. 53-5 A method of drawing floor-framing systems to show joists and deck relationship.

partitions, with and without spacers. If space for pipes or wires is needed, spacers are used.

GIRDERS AND BEAMS

Wood girders or steel beams are horizontal supports that are perpendicular to the floor joist. They are primarily used to shorten the span of the floor joists. All the floor system dead load weights, all of the live load weights acting on the floor, plus the additional live and dead weights of any floors above and the weight of the roof are transmitted through bearing partitions or columns to the foundations. Bearing partitions may be exterior and/or interior walls although not all interior walls are bearing partitions.

Loads are further transmitted to the foundation through intermediate supports (Fig. 53-10) such as columns, intermediate foundation walls,

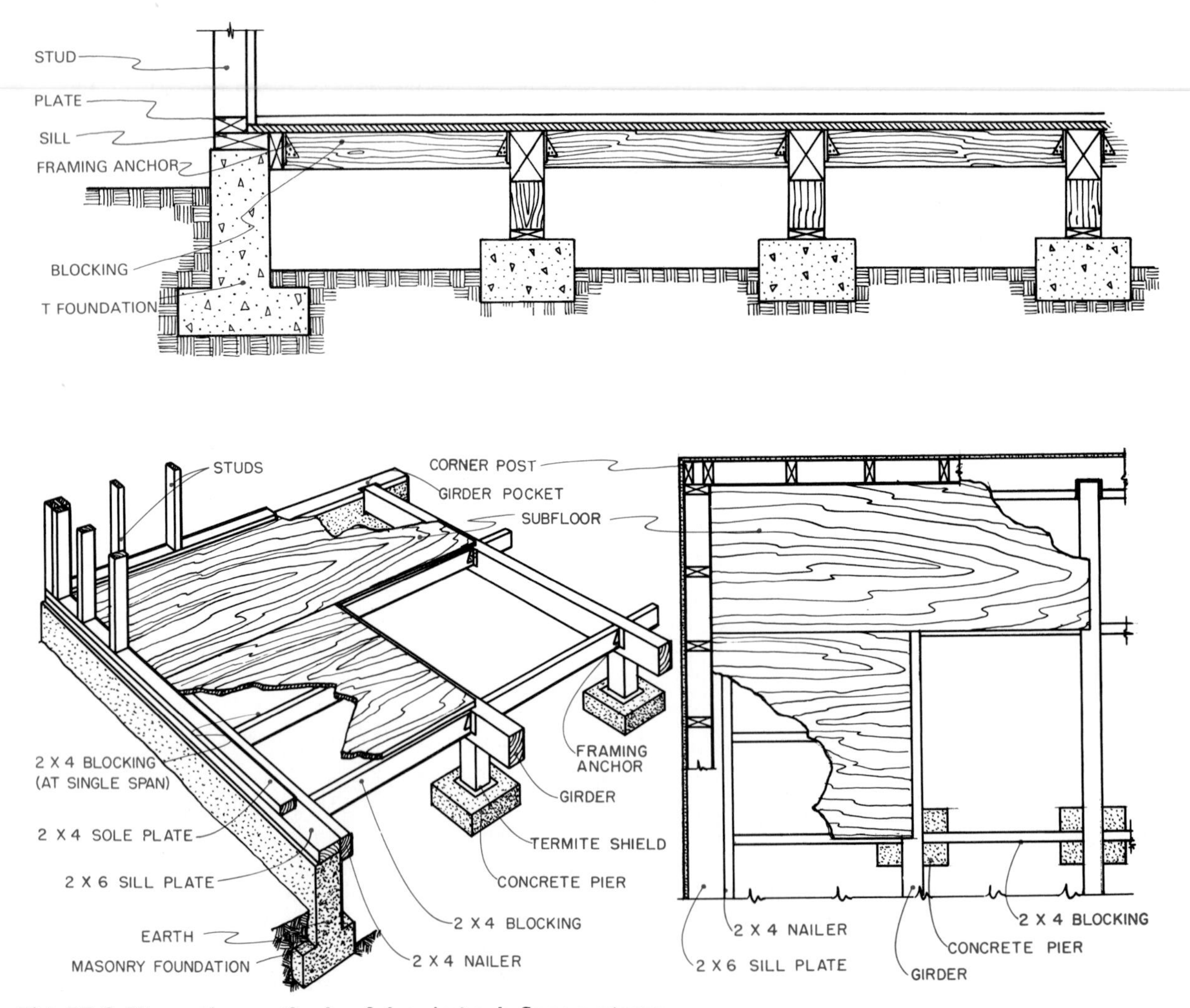

Fig. 53-6 Alternative methods of drawing sub-floor systems.

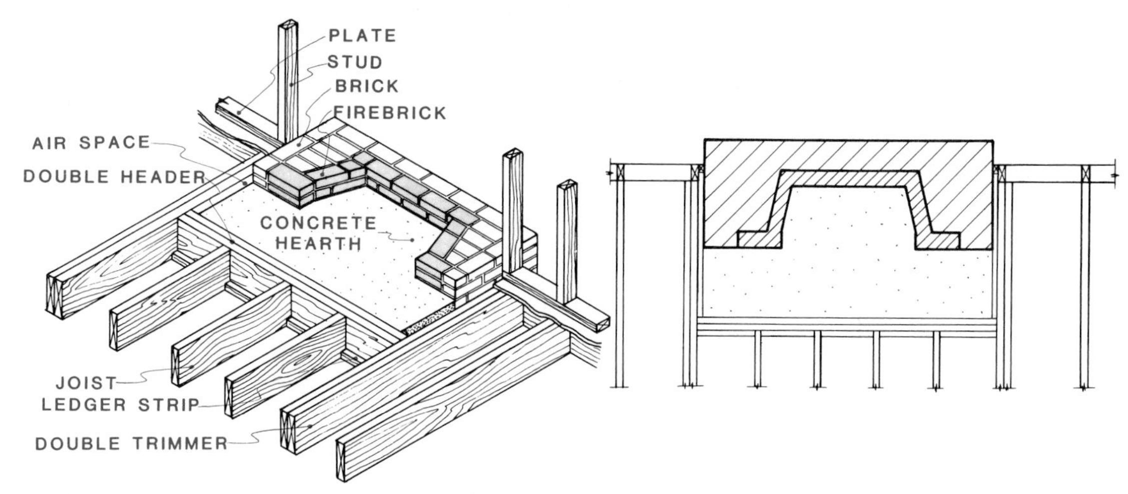

Fig. 53-7 Double headers are used around fireplaces.

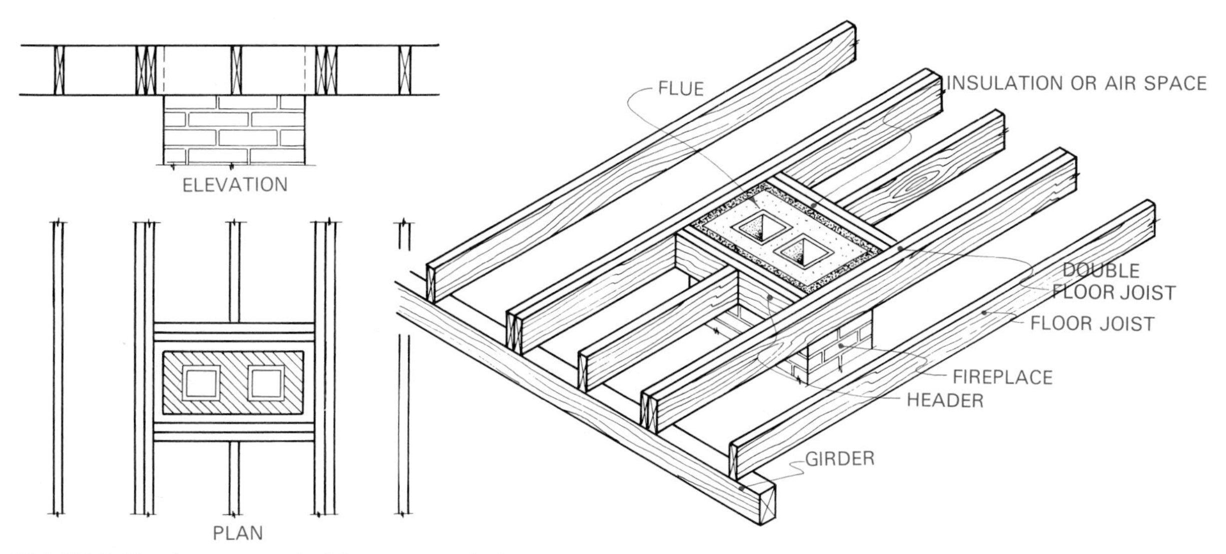

Fig. 53-8 Headers around chimney openings.

and posts or horizontal intermediate supports (beams or girders). Figure 53-11 shows loads transmitted through a built-up wood girder. Figure 53-12 shows a steel beam used as an intermediate support for floor joists.

Figure 53-13 shows the application of these support methods to a typical section through an interior wall. Steel beams, as shown in Fig. 53-14, are also used in the same manner with lightweight steel joists to provide intermediate floor system support.

COLUMNS

When girders or beams do not completely span the distance between foundation walls, then wood posts, steel-pipe columns, masonry columns, or steel-beam columns must be used for intervening support. In selecting and locating posts or columns, keep in mind that the height of the post is related to the load it can support. The 4″ × 4″ post on the right side of Fig. 53-15 may be adequate to support a load which the 4″ × 4″ post on the left cannot support because of the increased height. To support the load on the right, a larger post, stronger material or shorter height, is required.

PLANS

The more complete the architectural plan, the better the chances are that the building will be

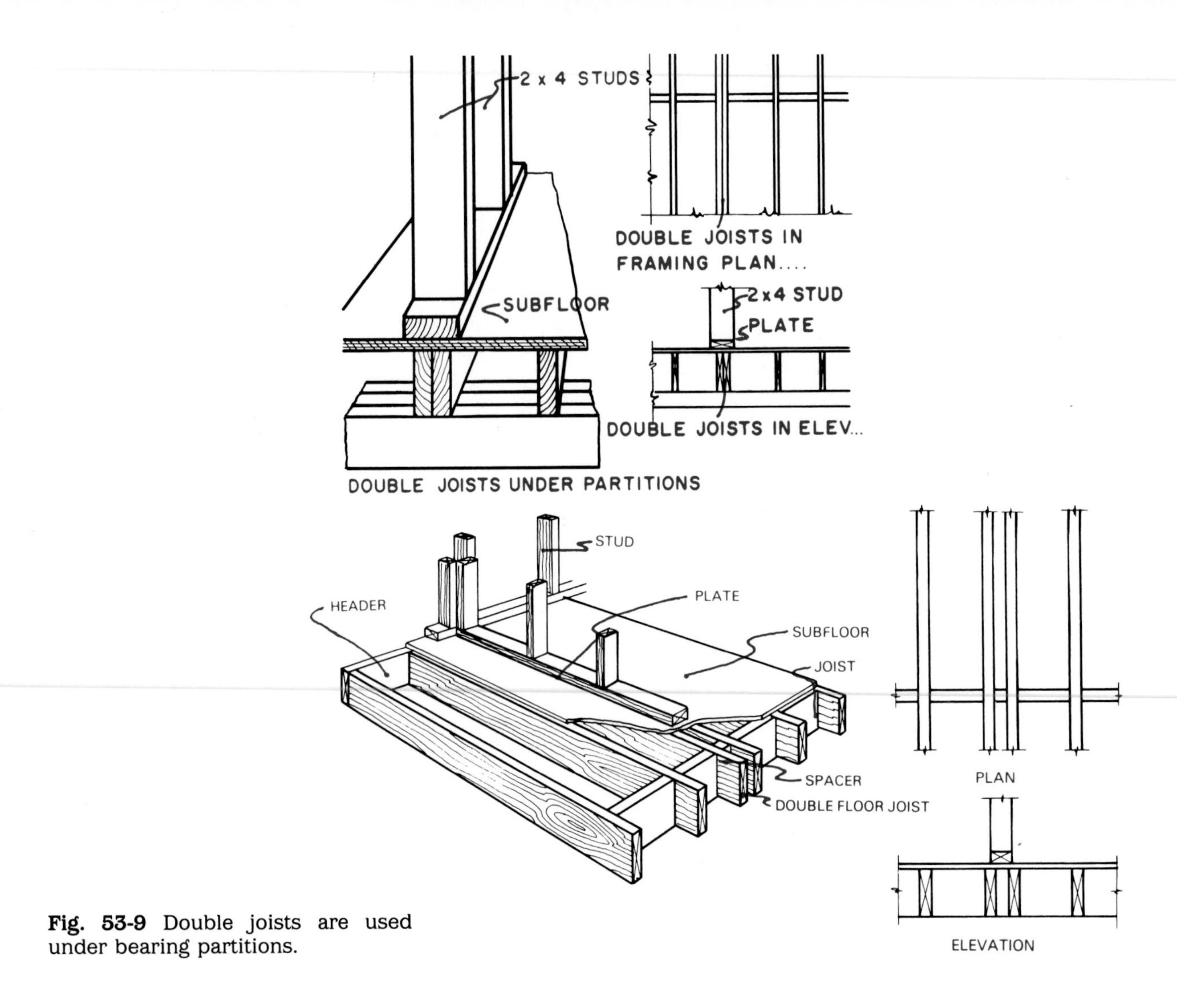

Fig. 53-9 Double joists are used under bearing partitions.

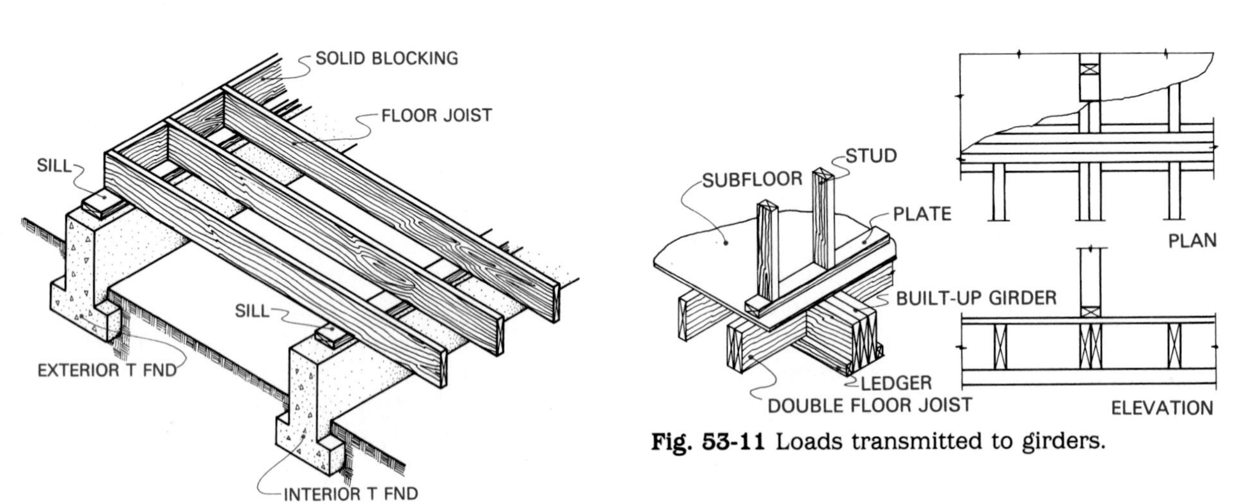

Fig. 53-10 Loads transmitted to intermediate supports.

Fig. 53-11 Loads transmitted to girders.

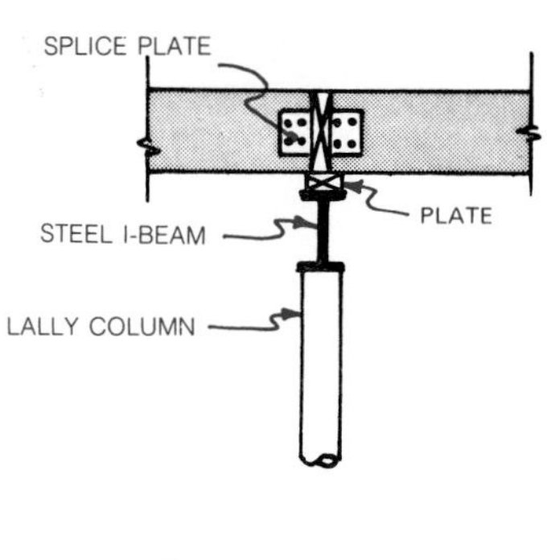

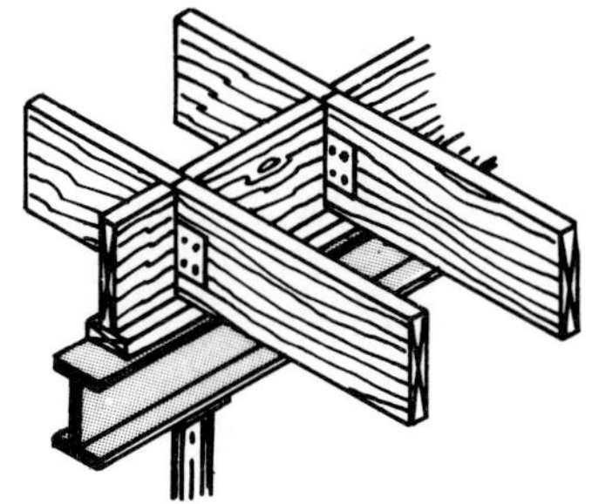

Fig. 53-12 Steel beam supporting wood joists.

constructed exactly as designed. If a floor framing plan is not prepared to accompany the basic architectural plans, then the framing of the floor system is left entirely to the desires of the builder. Some architectural plans do not include a floor framing plan. Only the direction of joists and the possible location of beams or girders are shown on the floor plan. Figure 53-16A shows a plan of this type on a floor plan. Figure 53-16B and C shows other methods of drawing floor framing plans. All are related to the basic floor plan shown in Fig. 53-16A. The most complete and most detailed method of drawing floor framing plans is shown in Fig. 53-16B. Each structural member is represented by a double line that shows its exact thickness.

The more abbreviated plan shown in Fig. 53-16C is a short-cut method of drawing floor framing plans. A single line is used to designate each member. Chimney and stair openings are shown by diagonals. Only the outline of the foundation

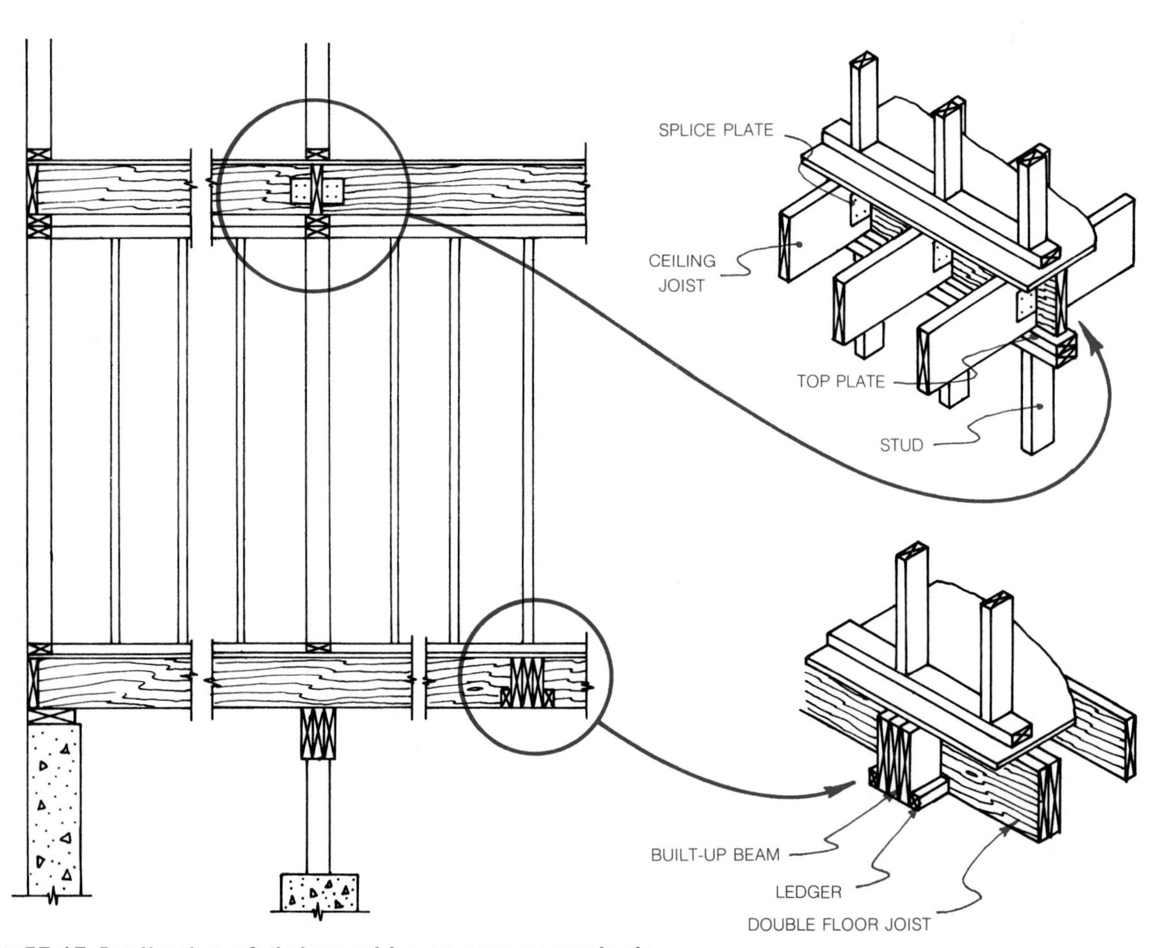

Fig. 53-13 Application of girder and beam support methods.

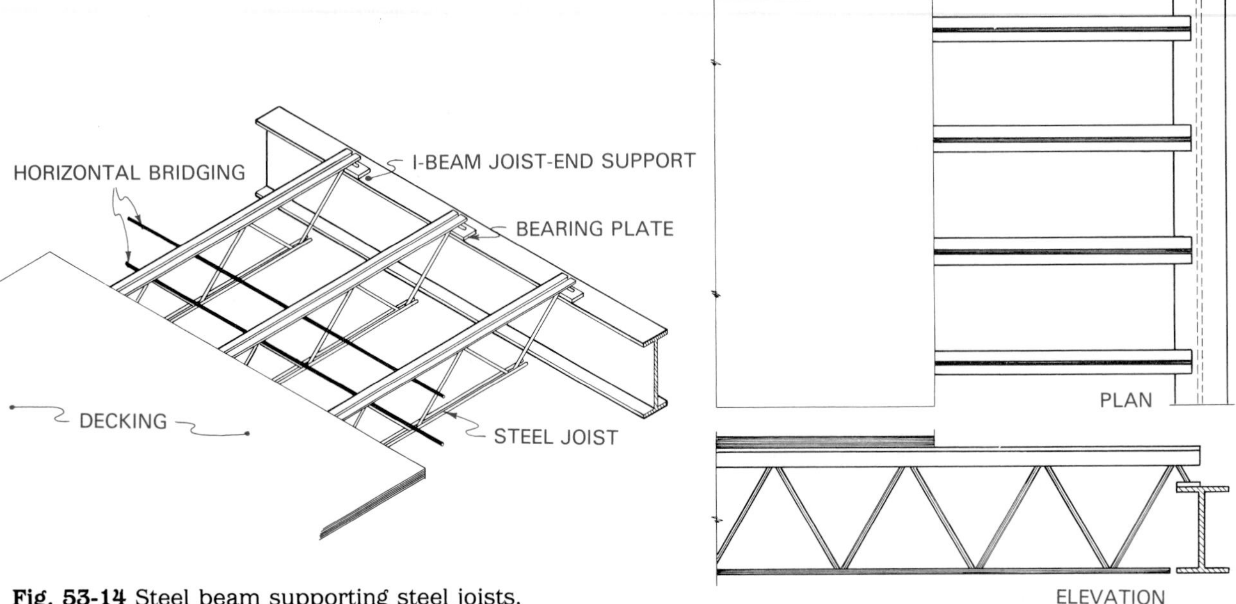

Fig. 53-14 Steel beam supporting steel joists.

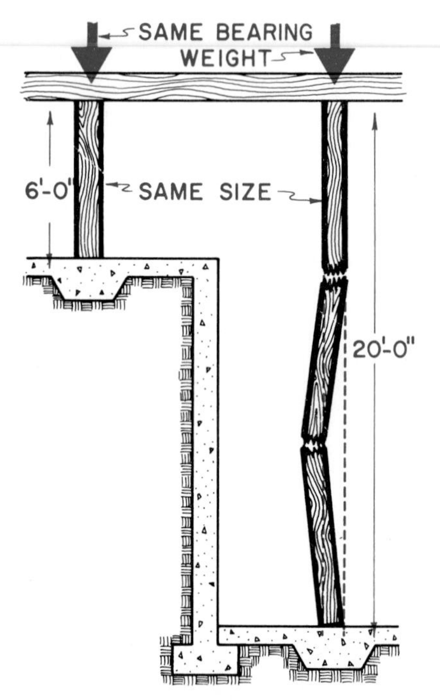

Fig. 53-15 A heavier post is needed to support the same load when the height is increased.

and post locations is shown. The abbreviated floor framing plan given in Fig. 53-16D uses a technique similar to the one used in floor plans to show the entire area where uniformly distributed joists are placed. The direction of joists is shown by an arrow. The size and spacing of joists are shown by notes placed on the arrow. This type of framing plan is usually accompanied by numerous detail drawings such as the ones shown in Figs. 53-3 and 53-5. The method of cutting and fitting subfloor and finished floor panels is usually determined by the builder. However, where off-site or mass-produced floor systems are built, a plan similar to the one shown in Fig. 53-17 is often prepared to ensure a maximum utilization of materials with a minimal amount of waste.

DETAILS

Although many floor framing plans are easily interpreted by the experienced builder, others may require that the detail of some segment of the plan be prepared separately to explain more clearly the construction methods recommended. The detail is drawn to eliminate the possibility of error in interpretation or to explain more thoroughly some unique condition of the plan. Details may be merely enlargements of what is already on the floor framing plan. They may be prepared for dimensioning purposes, or they may show a view from a different angle to reveal the underside or elevation view for better interpretation.

Figure 53-18A and B show a floor framing plan and several details that have been removed for clarity. Detail 1 shows the position of cross-bridging. Detail 2 shows the relationship of the

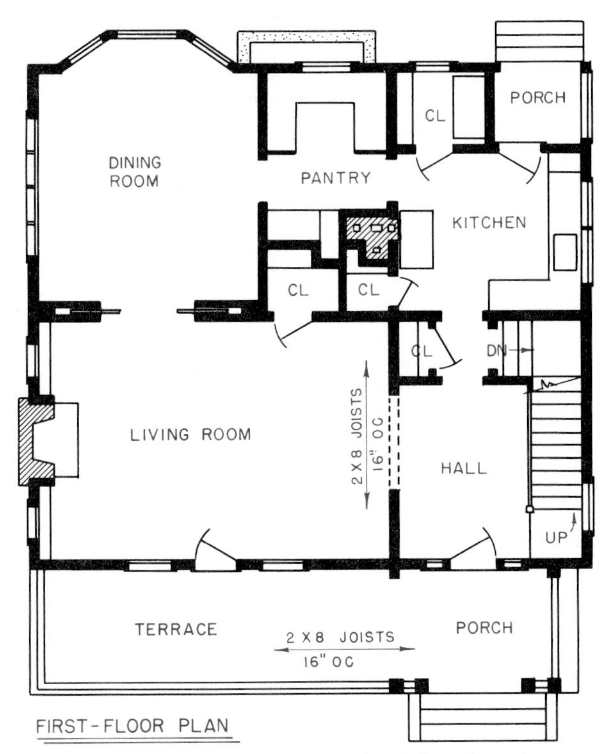

Fig. 53-16A A method of showing joist direction on a floor plan.

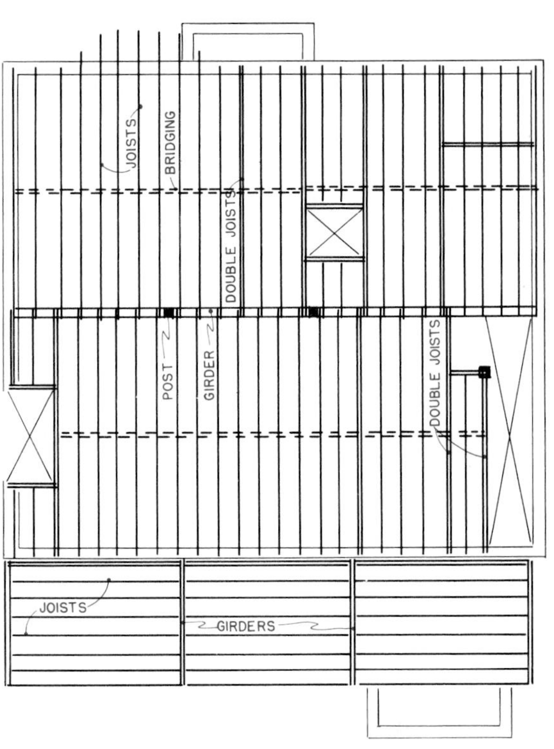

Fig. 53-16C A simplified method of drawing floor-framing plans.

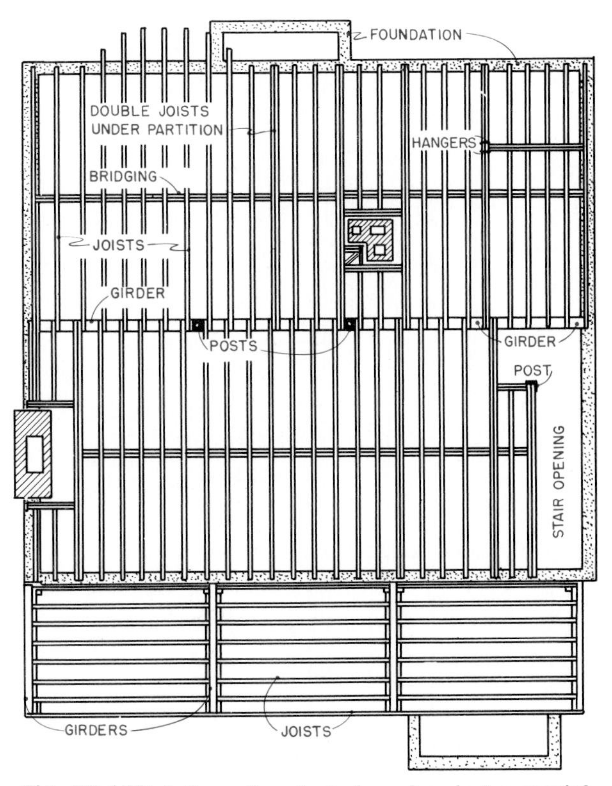

Fig. 53-16B A floor-framing plan showing material thickness.

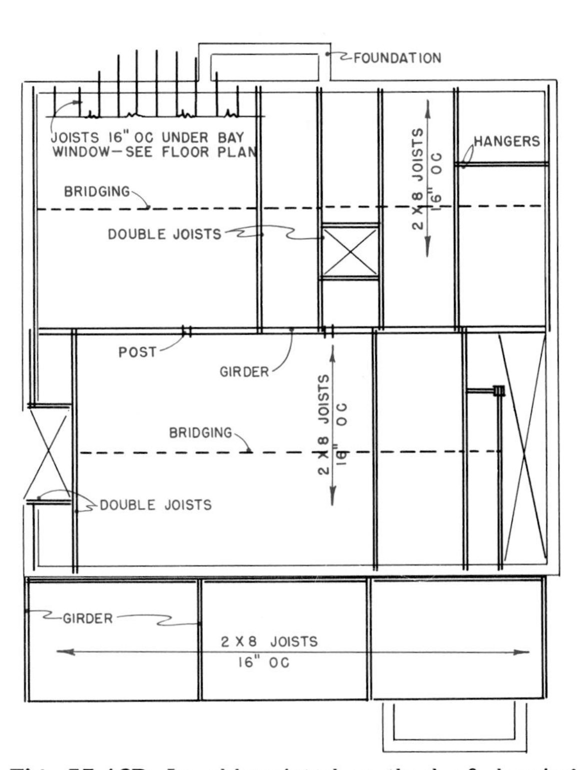

Fig. 53-16D An abbreviated method of drawing floor-framing plans.

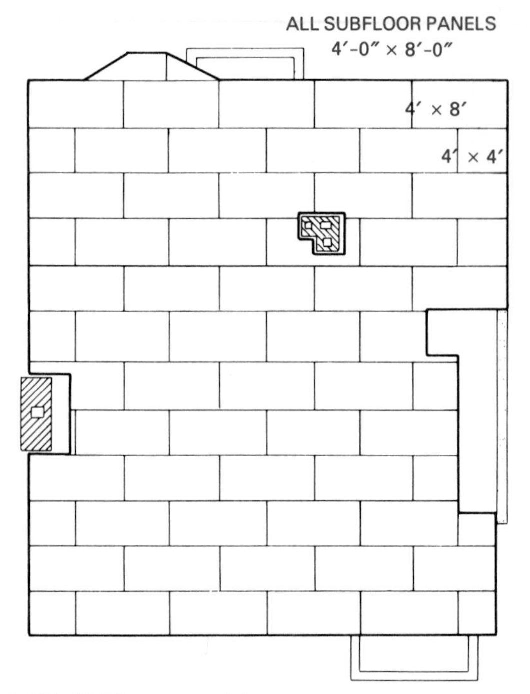

Fig. 53-17 Floor panel layout.

built-up beam, the double joist under the partition, and the solid bridging. Detail 3 shows the sill construction in relation to the floor joist and rough flooring, and to the foundation. Detail 4 shows the method of supporting the built-up beam by the *lally* (steel) *column* and the joist position on the beam. Detail 5 shows several alternative methods of supporting the joist over a built-up beam or an I beam; thus the builder is given an option. Detail 6 shows the attachment of the typical box sill to the masonry foundation. Detail 7 shows the method of supporting the built-up beam with a pilaster, and the tie-in with the box sill and joist.

Sill Support Details

Detail drawings showing sill construction details reveal not only the construction of the sill but also the method of attaching the sill to the foundation. Since the *sill* is the transition between the foundation and the exterior walls of a structure, a sill detail is usually included in most sets of architectural plans.

Some sill details are shown in pictorial form, as in Figs. 53-19 and 53-20. However, if pictorial drawings are used, two drawings must be used to show the exterior and interior views. Figure 53-19 shows an exterior pictorial view of a sill corner, and Fig. 53-20 shows an interior pictorial view of a sill corner. Pictorial drawings are easy to interpret but are more difficult and time-consuming to draw and dimension; therefore, most sill details are prepared in sectional form as shown in Fig. 53-21.

The floor area in a sill detail sectional drawing usually shows at least one joist. This is done to

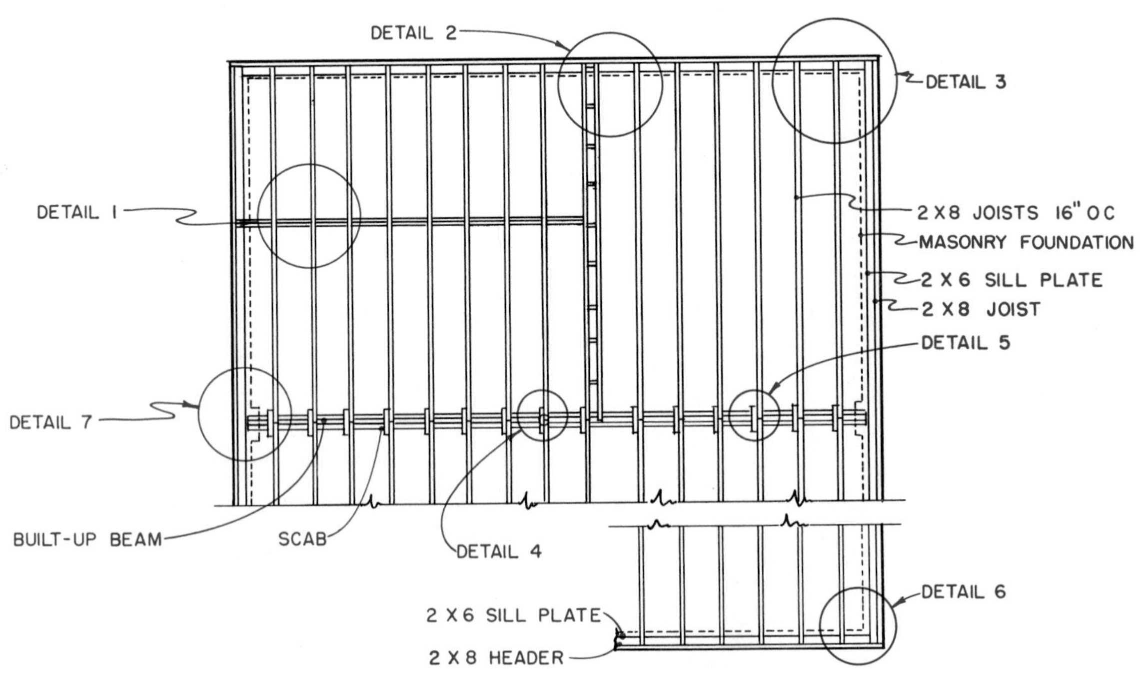

Fig. 53-18A Floor framing plan with details noted.

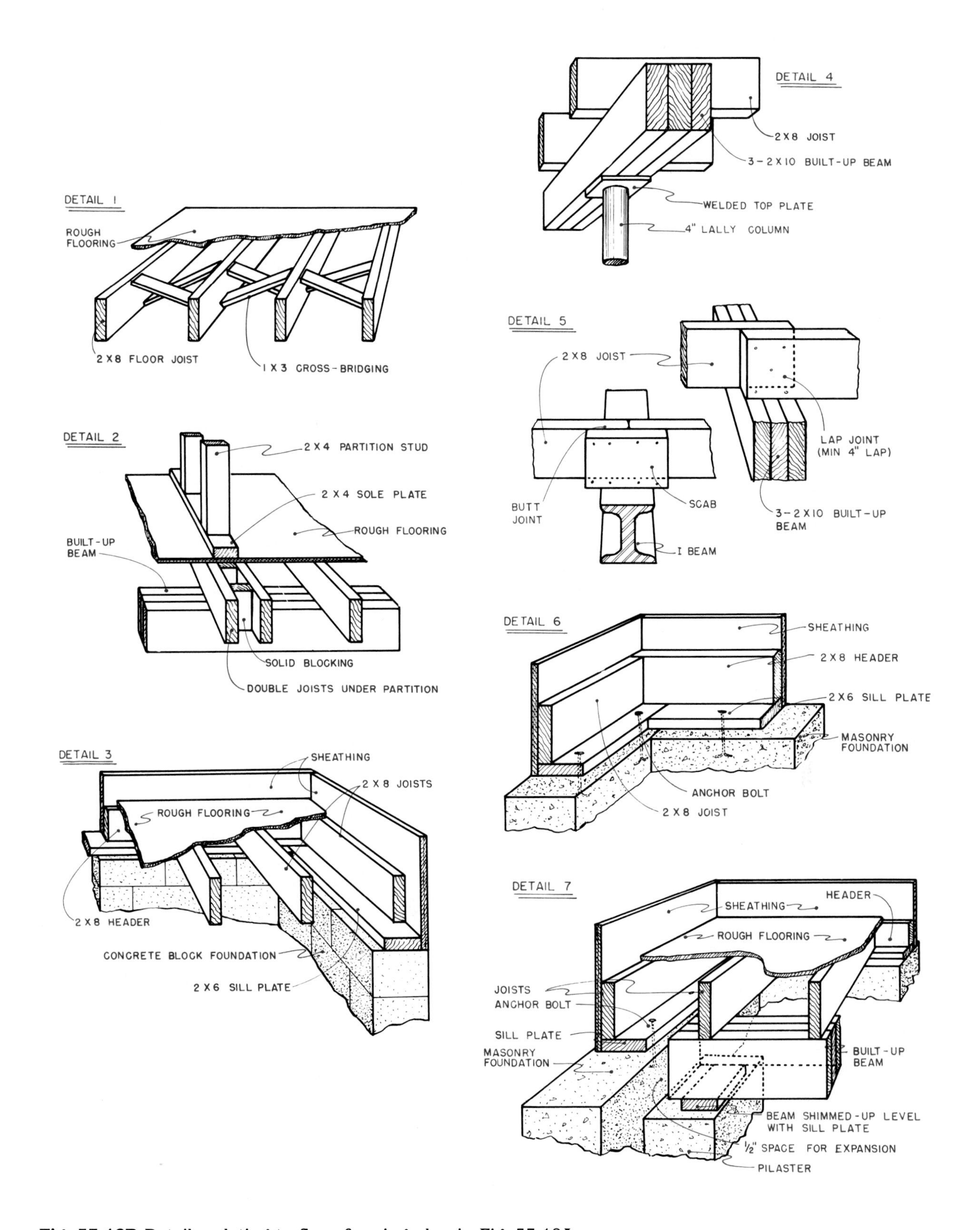

Fig. 53-18B Details relating to floor framing plan in Fig. 53-18A.

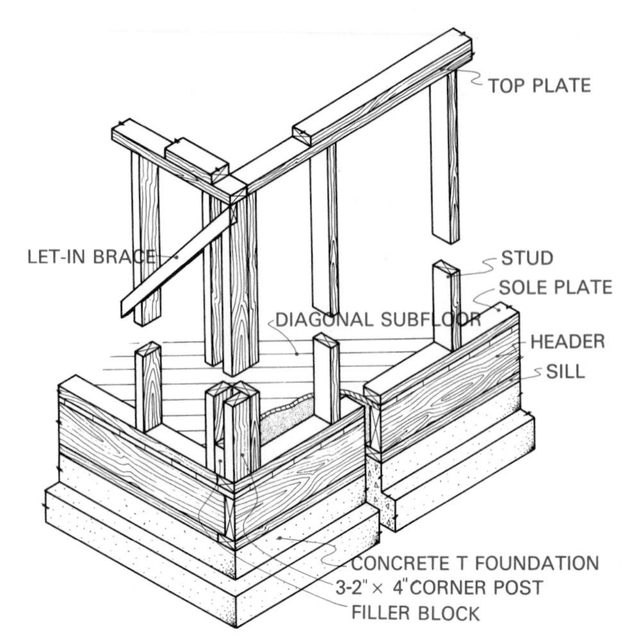

Fig. 53-19 Exterior view of sill details.

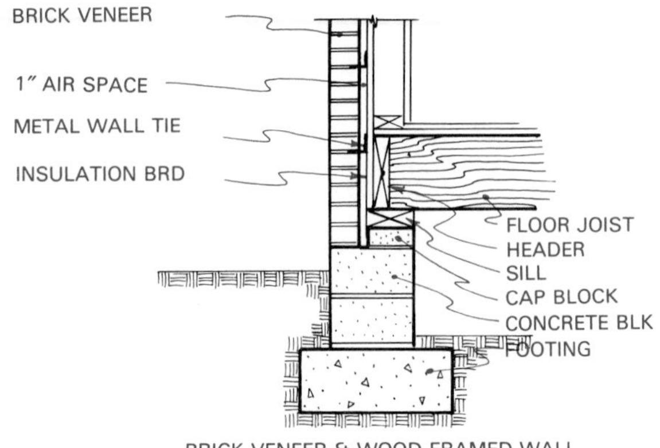

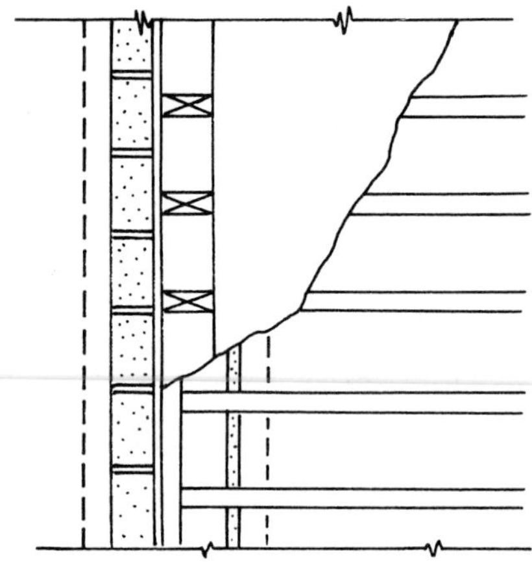

Fig. 53-21 Sill section detail.

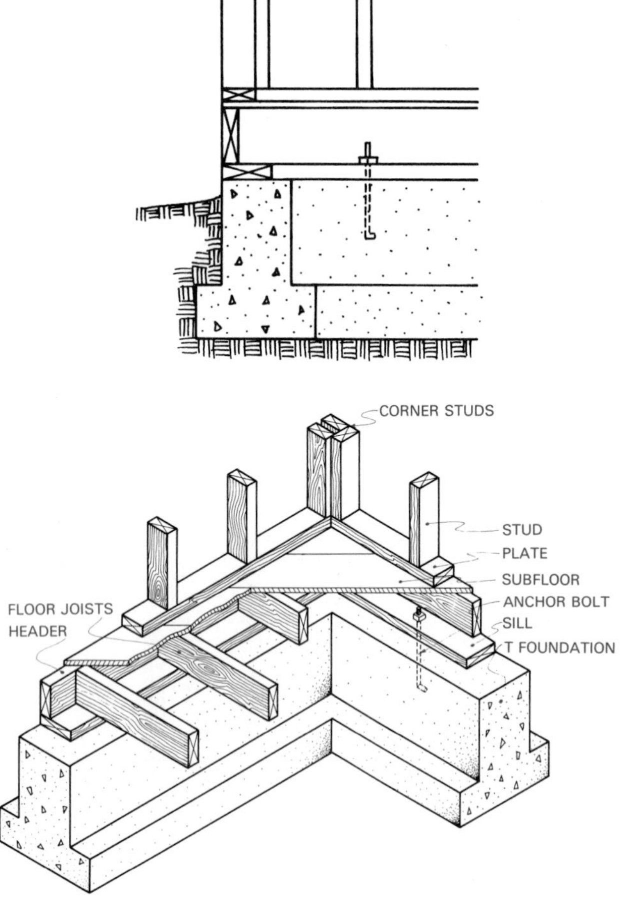

Fig. 53-20 Interior view of sill details.

show the direction of the joist and its size and placement in relationship to the placement of the subfloor and finished floor. This information is also sometimes shown on a floor framing plan. However, the floor framing plan shows neither the attachment of the floor system and sill to the foundation, nor the intersection of the exterior walls with the floor and sill. Figure 53-22 shows pictorial sill details and floor framing sections related to a balloon framing sill. The floor framing plan in this illustration shows the spacing of girders, blocking, and piers more clearly. However, the elevation section shows the intersections between the floor-system foundation sill and exterior wall more clearly. For this reason, sometimes both drawings are used to describe fully the type of construction required.

Sill details of this type are also required to show the relationship and joining of materials, such as masonry, wood, precast concrete, and structural steel. Figure 53-23 shows a typical masonry sill detail. This section is necessary to show the fire-cut portion of the joist on the foun-

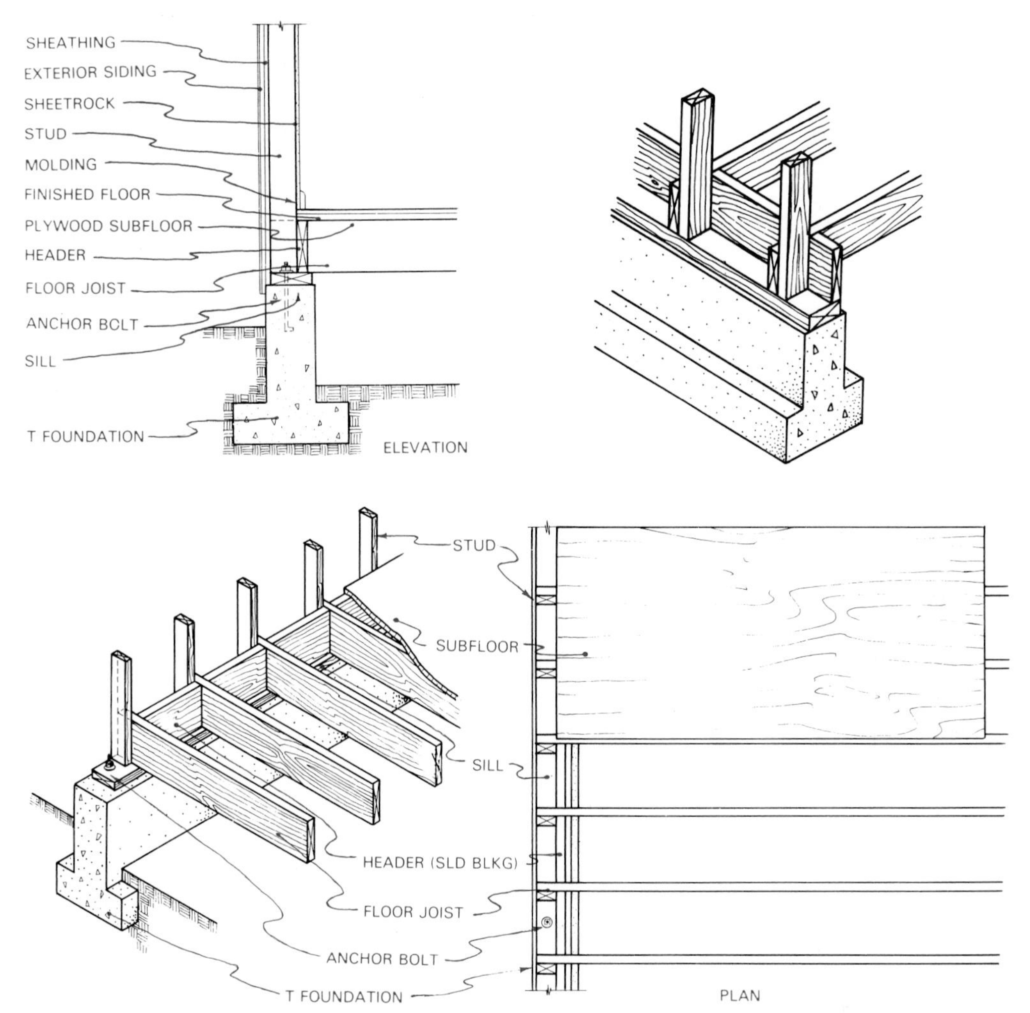

Fig. 53-22 Sill detail of balloon-framed building.

dation because the fire-cut detail does not show on the floor framing plan.

Precast-concrete and structural-steel construction also require sill and floor details to show the size and spacing of structural members. These details are included in sets of drawings to show the size and spacing of members and the fastening methods and devices used to anchor concrete and masonry to wood, structural concrete, or steel members.

Intermediate Support Details

These show the position and method of attachment of *intermediate support members*, such as girders and beams. These details are often shown by either a pictorial drawing, a floor framing plan detail, or an elevation section, as shown in Fig. 53-24A, B, and C. In these three illustrations, notice how the elevation section through the sill and floor-plan detail is used to show the difference between construction methods used with a standard girder. Figure 53-24B shows a girder supported with a box sill (pilaster). Figure 53-24C shows a fire-cut girder and Fig. 53-24A shows a girder supported in a pocket in a masonry foundation wall.

If a beam or girder cannot span the distance between foundation walls, an intermediate vertical support, such as a wood or steel column, is used. Figure 53-25 also shows methods of attaching the column to a girder or beam.

Figure 53-26 shows several methods of connecting intersecting horizontal members. Other intersections between perpendicular beams and floor joists are shown in Fig. 53-27. The joist in

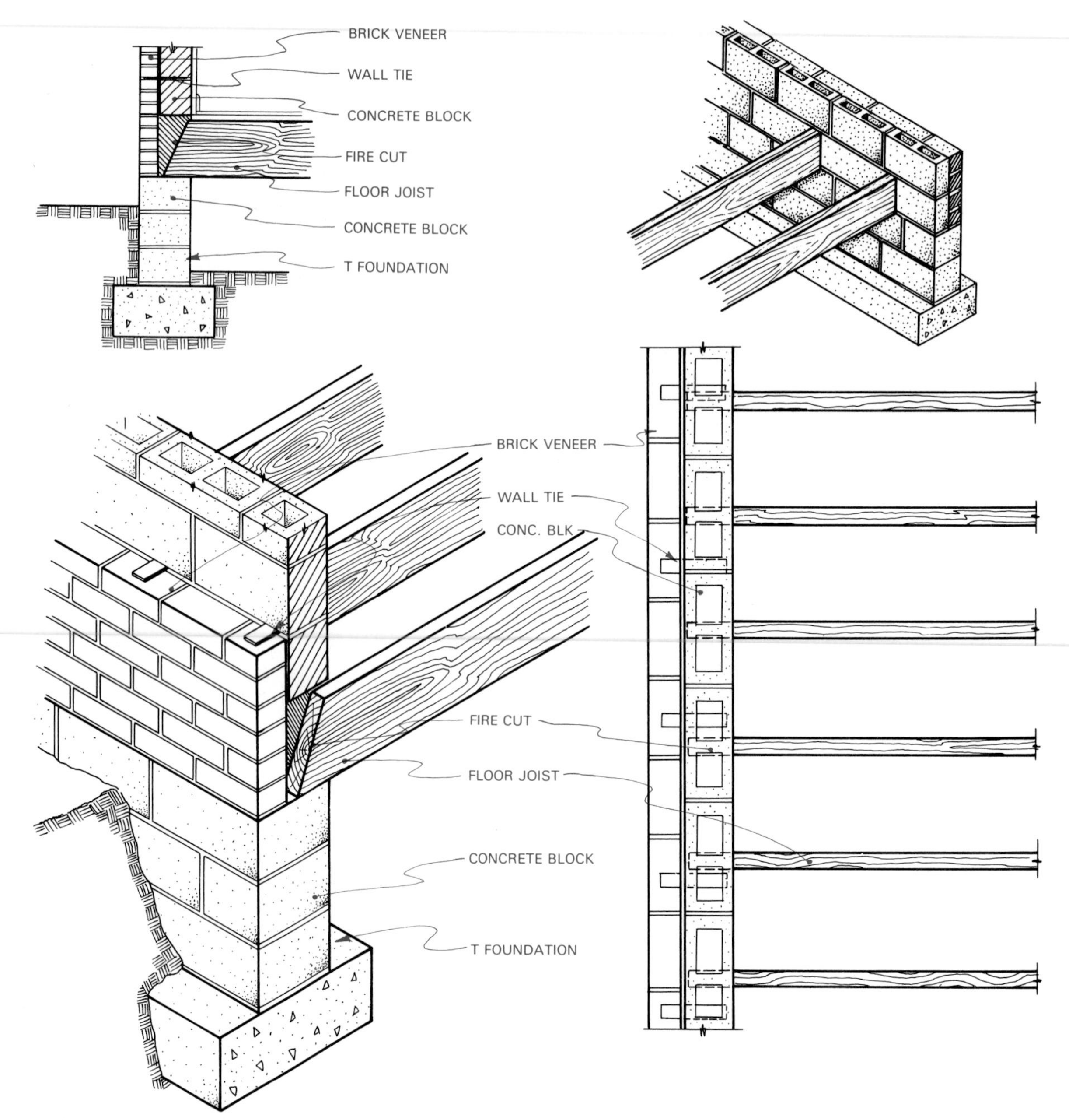

Fig. 53-23 Masonry sill detail.

the left column rests directly on the beam or girder, while the joist in the right column intersects the girder, thus allowing the top of the girder and the top of the joist to be at or near the same level.

Recommended methods of splicing lumber when necessary should also be detailed. Spliced members should be as strong as single members to eliminate building failures. The splices shown in Fig. 53-28 will resist compression, tension, and bending.

STAIRWELL FRAMING

The stairwell opening as drawn on the floor framing plan shows the relative position of the double joists and headers. Frequently, more information is needed concerning the relationship of the other parts of the stair assembly to the stairwell opening shown in Fig. 53-29. Figure 53-29A shows the stairwell opening for straight-run stairs and Fig. 53-29B shows the opening for L-shaped stairs. Figure 53-29C shows a plan

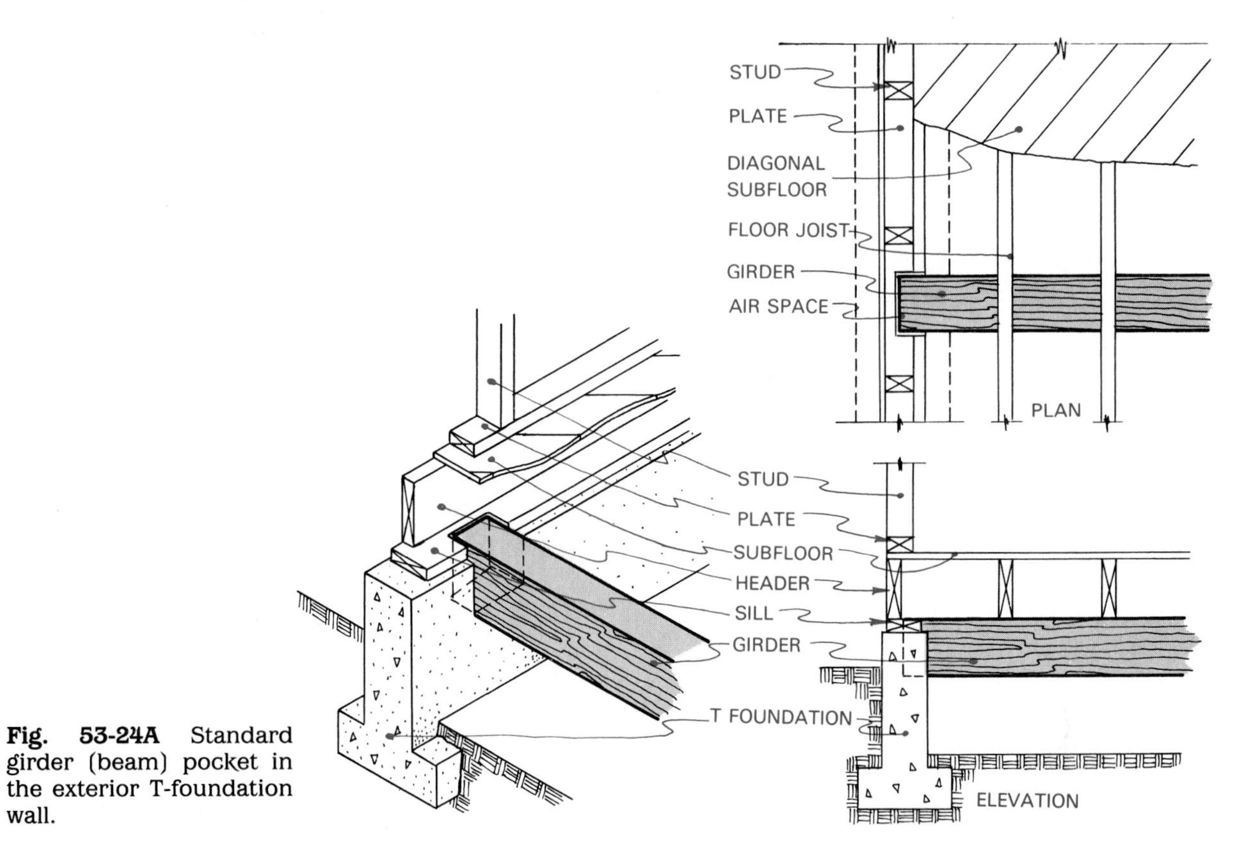

Fig. 53-24A Standard girder (beam) pocket in the exterior T-foundation wall.

PLAN

GIRDER

BOX SILL
(PILASTER)

ELEVATION

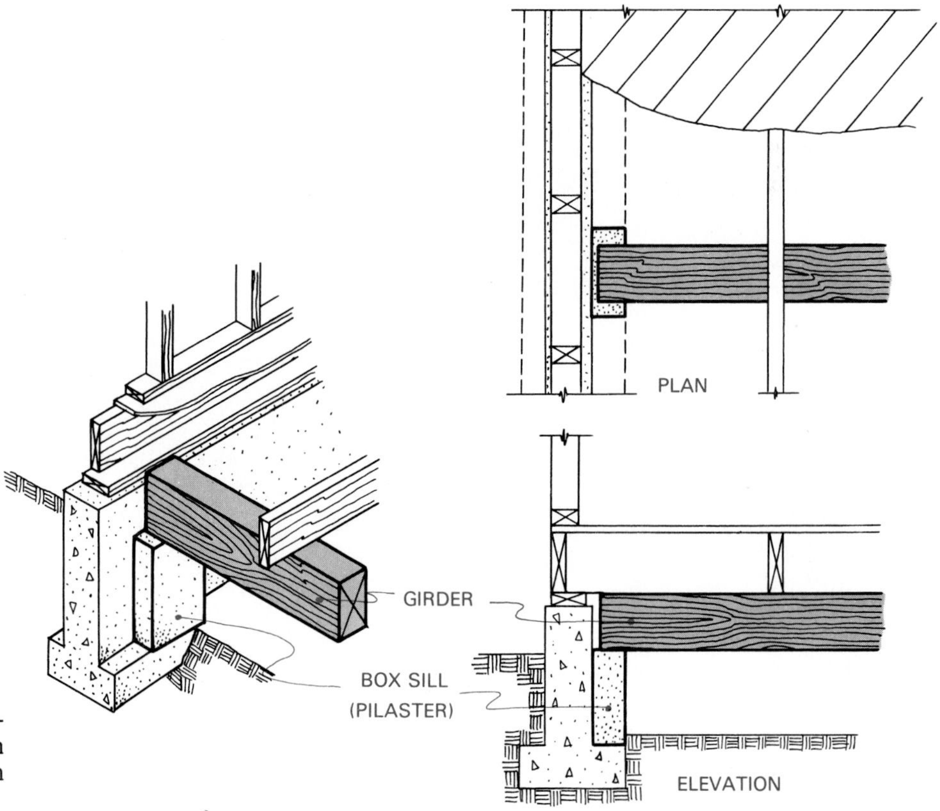

Fig. 53-24B Girder supported with a box sill on an exterior T-foundation wall.

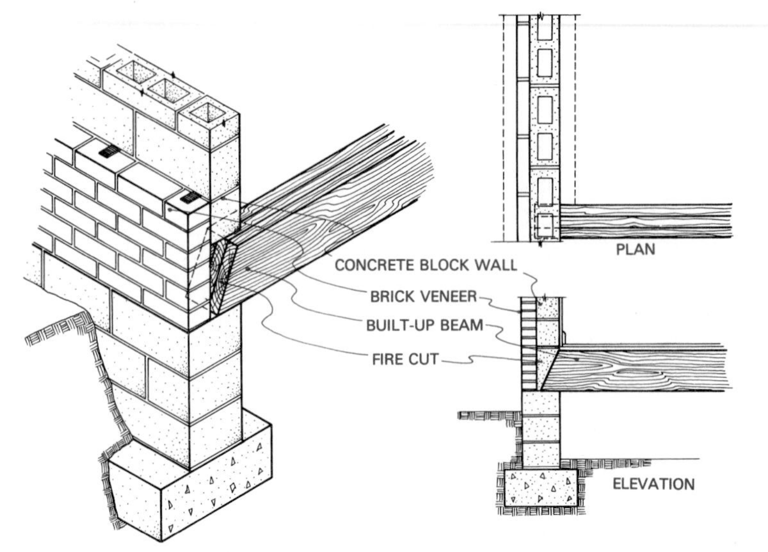

Fig. 53-24C Fire-cut girder supported in a pocket.

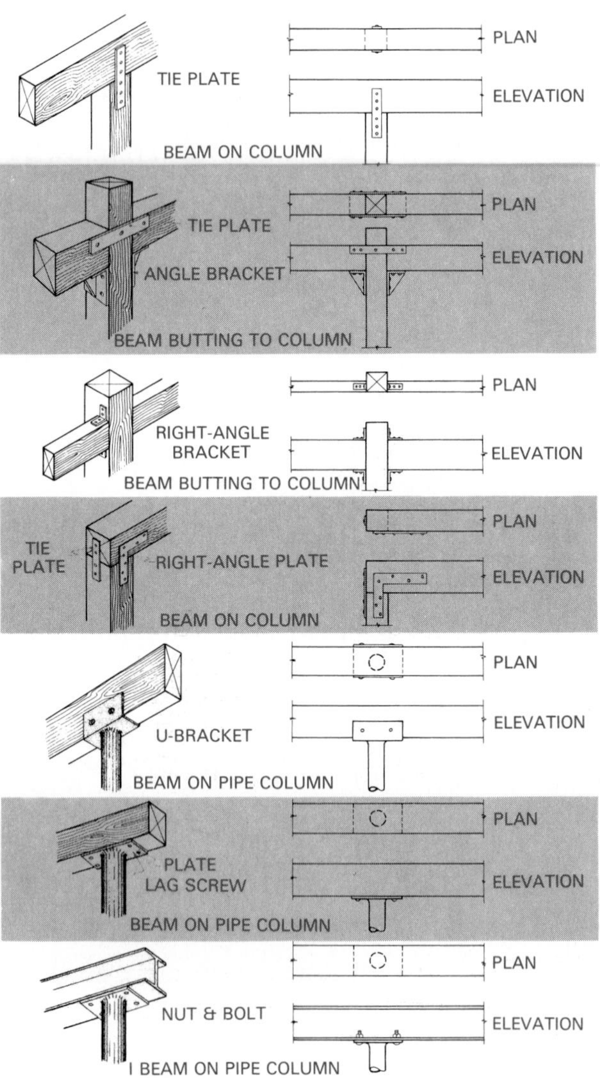

Fig. 53-25 Methods of connecting girders or beams to columns.

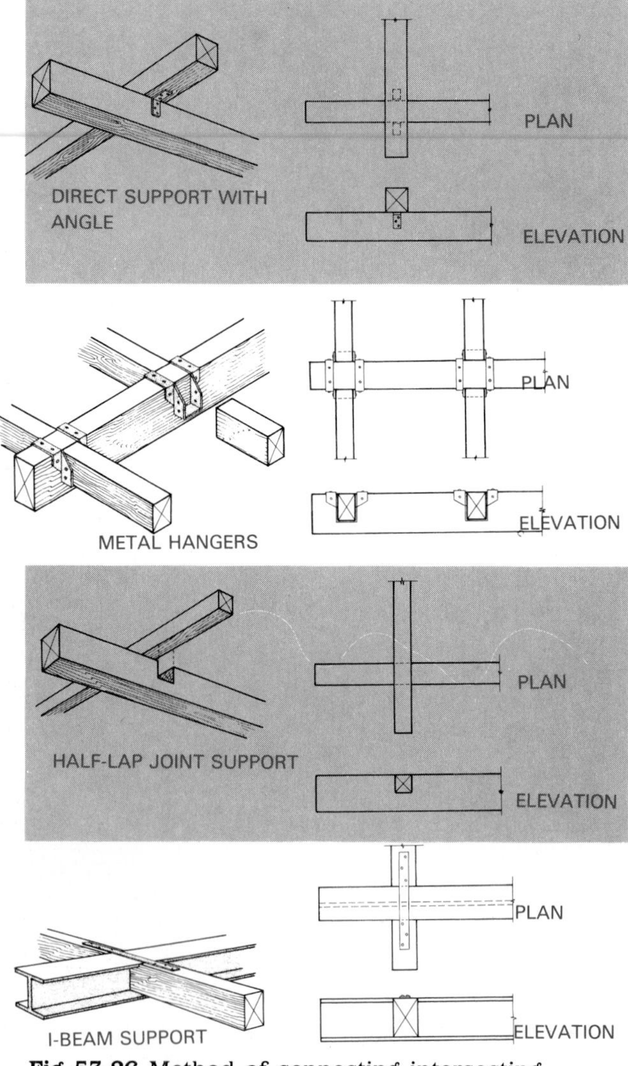

Fig 53-26 Method of connecting intersecting girders and beams to a vertical post.

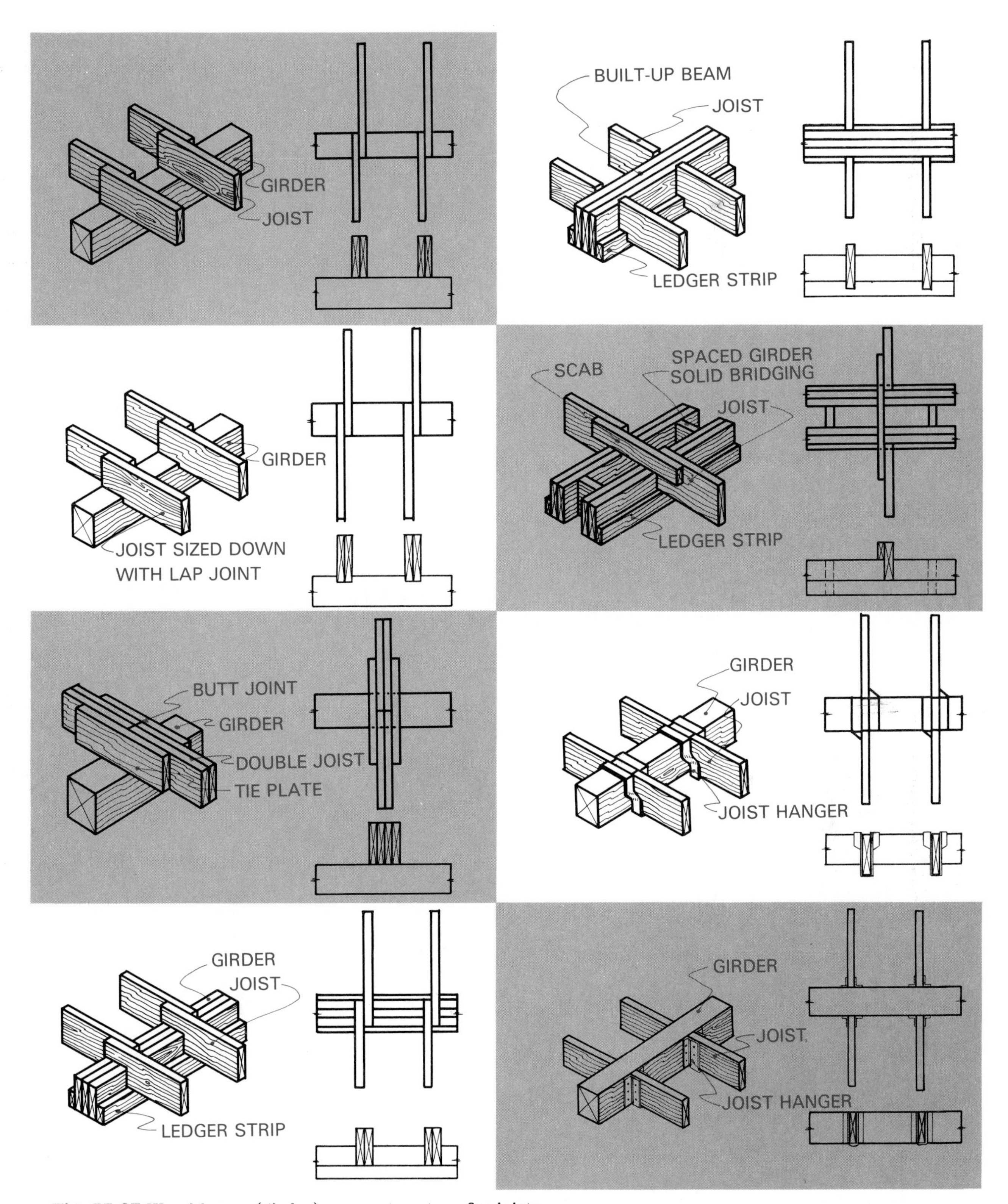

Fig. 53-27 Wood beam (girder) support system for joists.

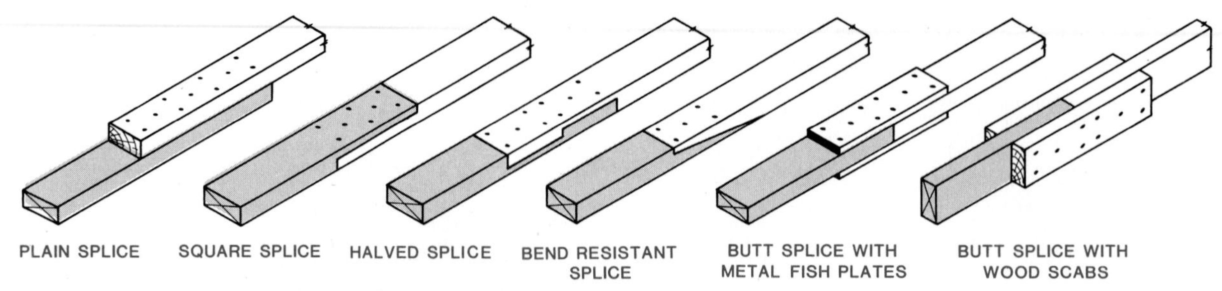

Fig. 53-28 Splices that resist compression, tension, and bending.

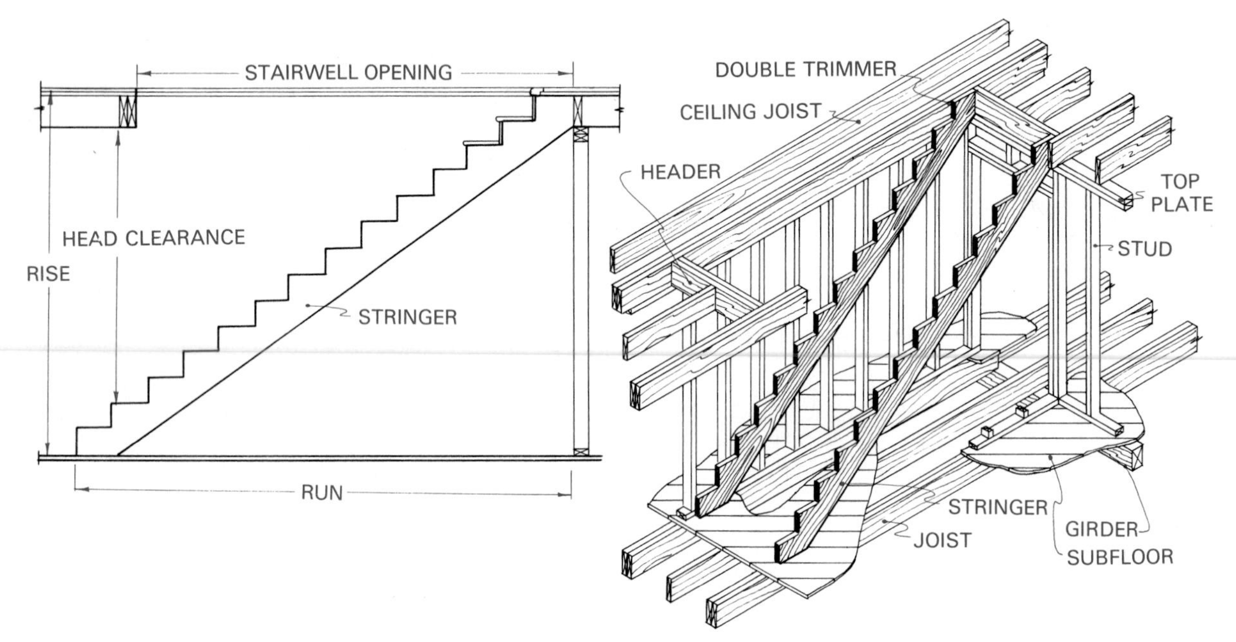

Fig. 53-29A Straight run stairwell.

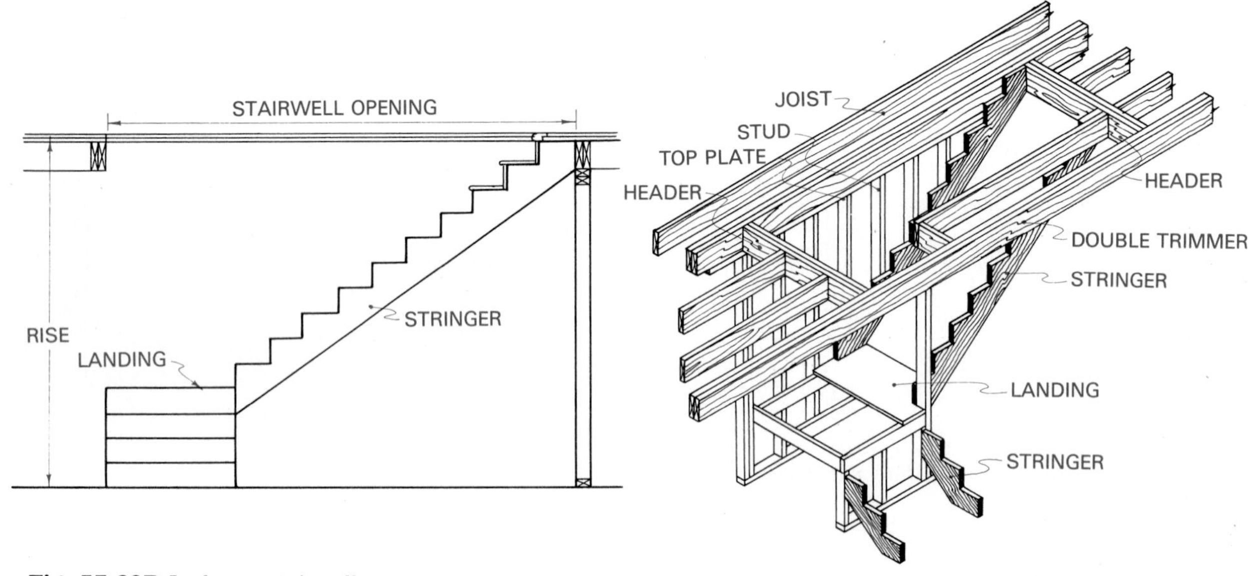

Fig. 53-29B L-shape stairwell.

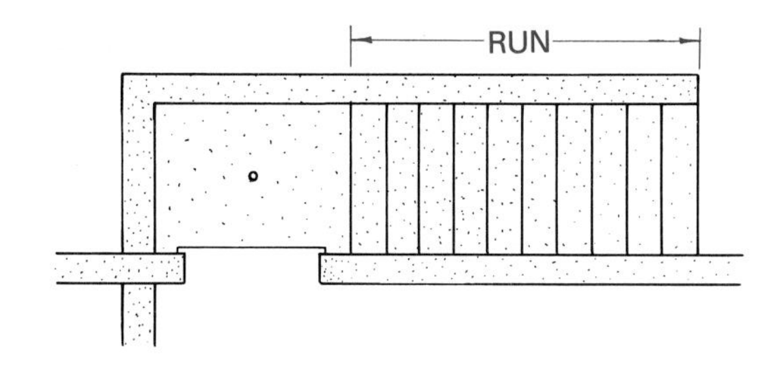

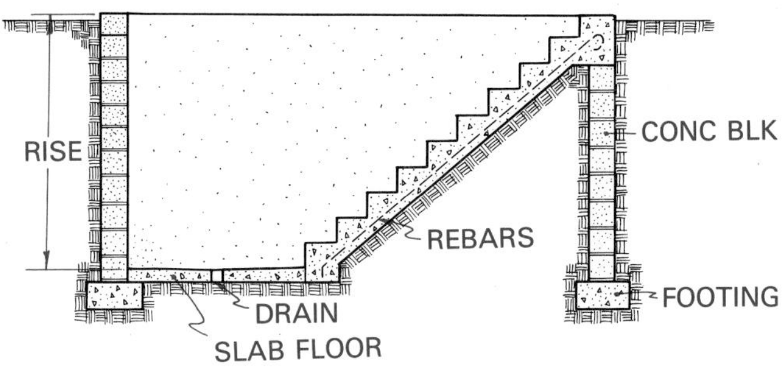

Fig. 53-29C Concrete construction stairway.

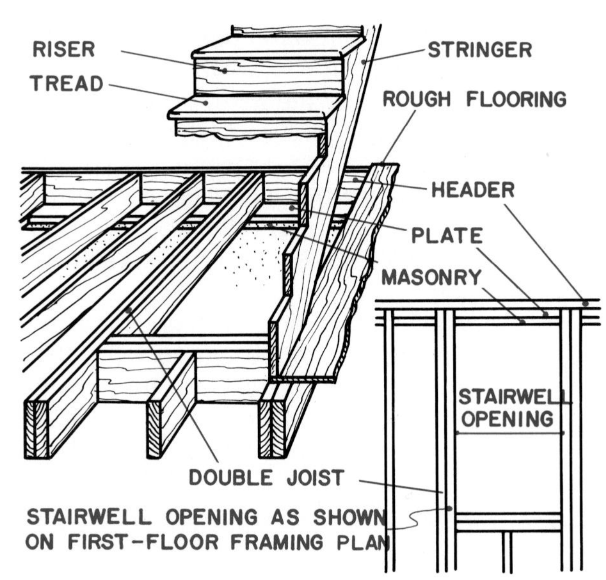

Fig. 53-30 Parts of the stair assembly.

and elevation of a concrete stair system. Information concerning the size and position of the various parts of the stair assembly is shown in Fig. 53-30. Such information is often shown in a separate detail.

Since the stairwell opening must be precisely shown on the floor framing plan, a complete design of the stair system should precede the preparation of the floor framing plan. The steps outlined in Figs. 53-31A through G show the sequences necessary for determining the exact dimensions of the entire stair structure.

1. Lay out the distance from the first-floor level to the second-floor level exactly to scale (Fig. 53-31A). Convert this distance to inches and add the position of the ceiling line.

2. Determine the most desirable riser heights (7½″, or 190 mm, is normal). Divide the number of inches (millimeters) between floor levels by the desired riser height to find the number of risers needed (Fig. 53-31B). Divide the area between the floors into spaces equaling the number of risers needed. This work can be done by inclining the scale.

3. Extend the riser-division lines lightly for about an inch, or 25 mm.

4. Determine the total length of the run (Fig. 53-31C). Lay out this distance from a starting point near the top riser line and measure the

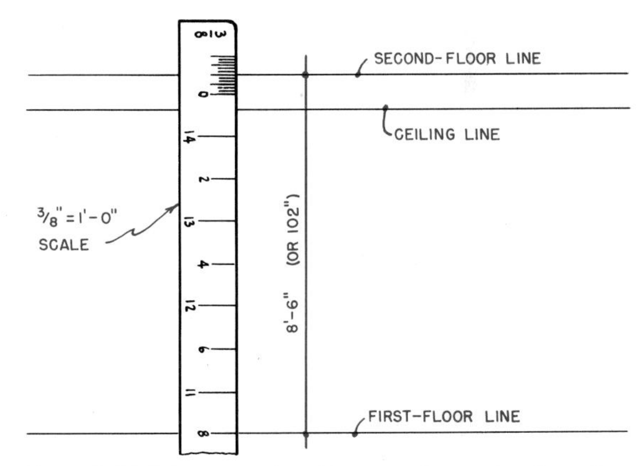

Fig. 53-31A Lay out the distance from the first-floor level to the second-floor level.

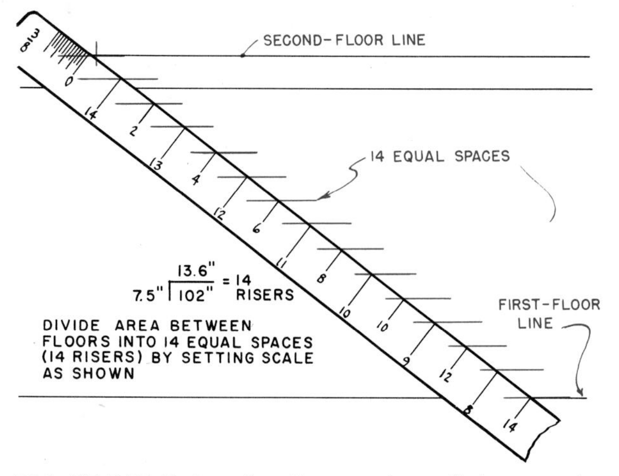

Fig. 53-31B Determine the number of risers and extend the riser lines.

total run horizontally. Extend this line vertically to the first-floor line. The total run is the number of treads multiplied by the width of each tread. There is always one less tread than riser.

5. Locate the top and bottom nosing points (Fig. 53-31D). Mark the intersection between the starting point of the total run and the intersection between the end of the total run and the first riser line.

6. Draw the nosing line by connecting the bottom nosing point with the top nosing point.

7. Draw riser lines intersecting the nosing points and the light riser lines (Fig. 53-31E).

8. Make the tread lines and the riser lines heavy.

9. Draw the soffit line the same as the thickness of the stringer. Establish headroom clearances. Draw a parallel line 6′-6″ (1.98 m) above the nosing line (Fig. 53-31F). Establish the stairwell opening by cutting the joists where the headroom clearance line intersects the bottom of the joist.

10. Show the outline of the carriage or stringer assembly.

11. Erase all layout lines and make all object lines heavier (Fig. 53-31G).

12. Add dimensions to describe the length of the stairwell opening, the size of the tread widths, the riser height, the minimum headroom, the total rise, and the total run.

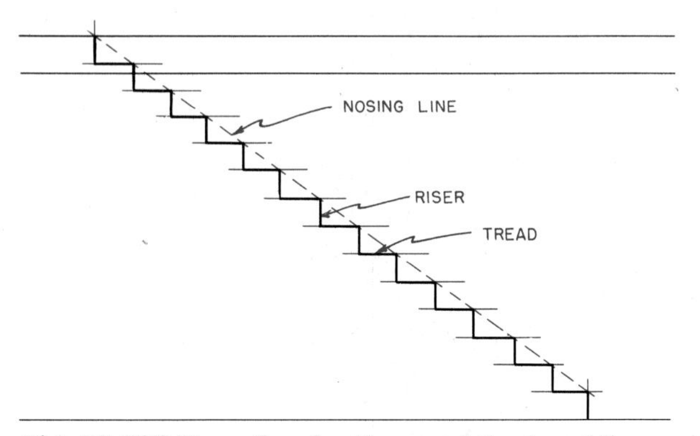

Fig. 53-31E Draw the rise lines and the tread lines.

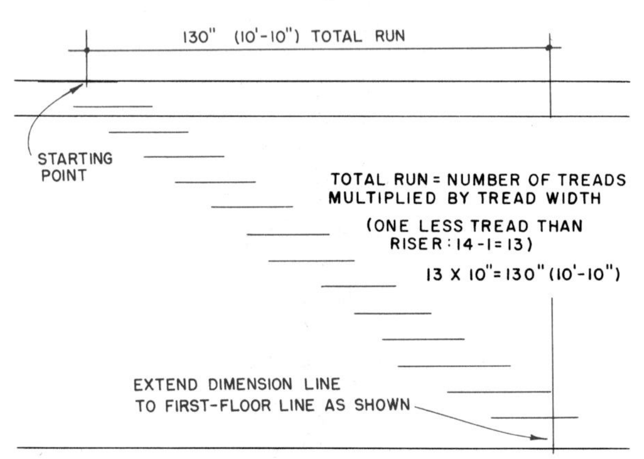

Fig. 53-31C Lay out the total run.

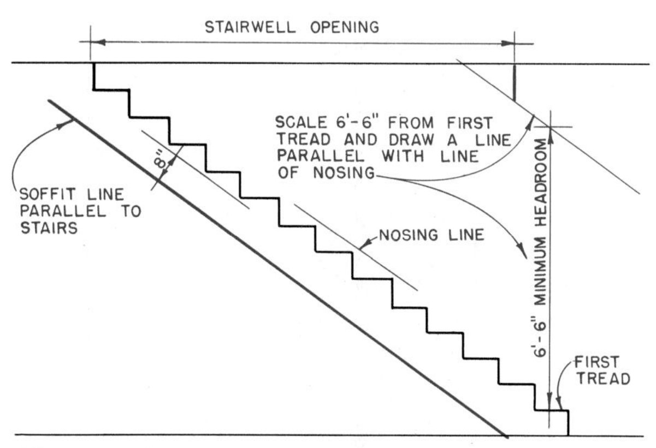

Fig. 53-31F Establish the headroom clearance, stairwell opening, and soffit line.

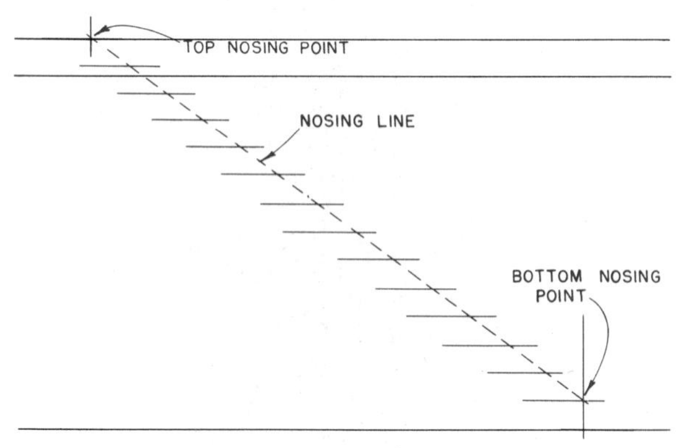

Fig. 53-31D Locate the nosing points and draw the nosing line.

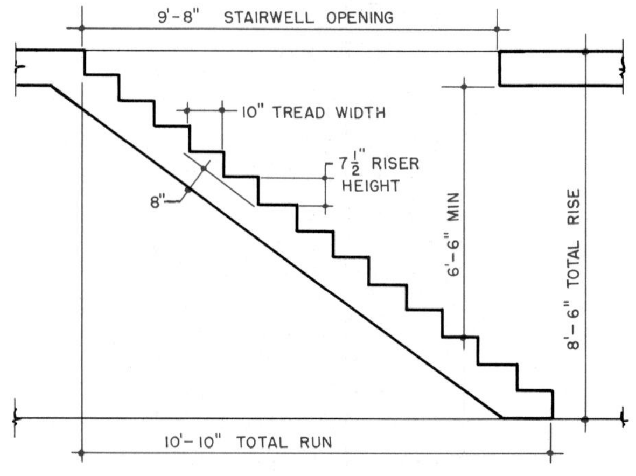

Fig. 53-31G Erase guidelines and add dimensions.

When the basic information pertaining to the overall dimensions and relationships of the stair assembly is established, a complete sectional drawing showing thicknesses and floor framing tie-ins can be prepared, as shown in Fig. 53-32. In this sectional drawing, the headers and the position of the bearing walls are shown.

STEEL

Floor framing plans for steel construction (Fig. 53-33) are prepared like other floor framing plans. The exact positions of columns, beams, and *purlins* (horizontal members) are dimensioned and the classification of each member indicated on the plan. Details should accompany steel-framing drawings to indicate the method of attaching steel members to each other. Lightweight concrete can be poured over steel or wood subfloors.

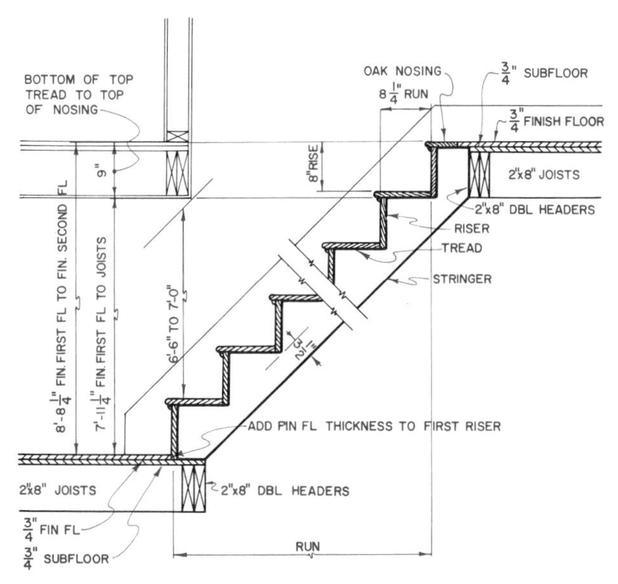

Fig. 53-32 Sectional drawing of a stair assembly.

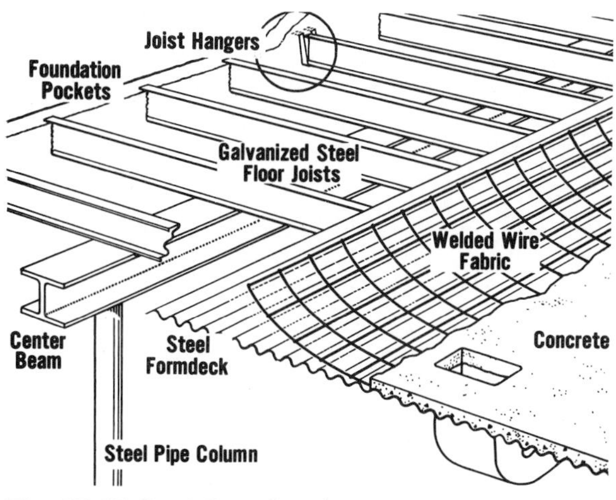

Fig. 53-33 Steel-floor framing.

SECOND-FLOOR PLANS

Second-floor framing details are usually shown with a full section through the exterior wall. Figure 53-34 shows the intersection of the second-floor joists in *balloon (eastern) framing.* In this style of framing, the studs are continuous from the foundation to the eave. The second-story joists are supported by a ribbon board that is recessed and nailed directly to the studs. Figure 53-35 shows second-floor construction for *western (platform) framing.* In this style of framing, the second-floor joists rest directly on a top plate that rests directly on first-floor studs. When a combination of exterior covering materials is used, the relationship between the floor system and the exterior wall is shown on the elevation. In Fig. 53-36, a brick veneer covers the

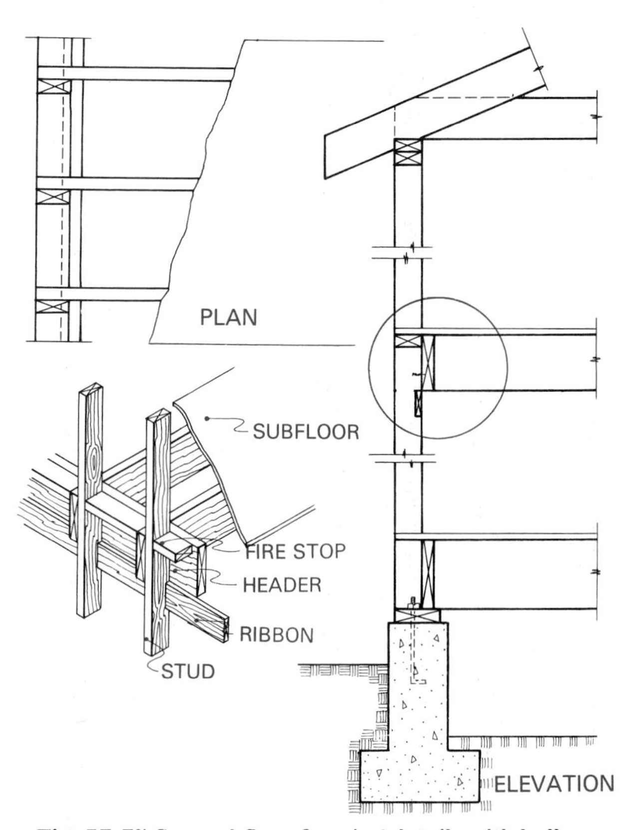

Fig. 53-34 Second-floor framing details with balloon (eastern) framing.

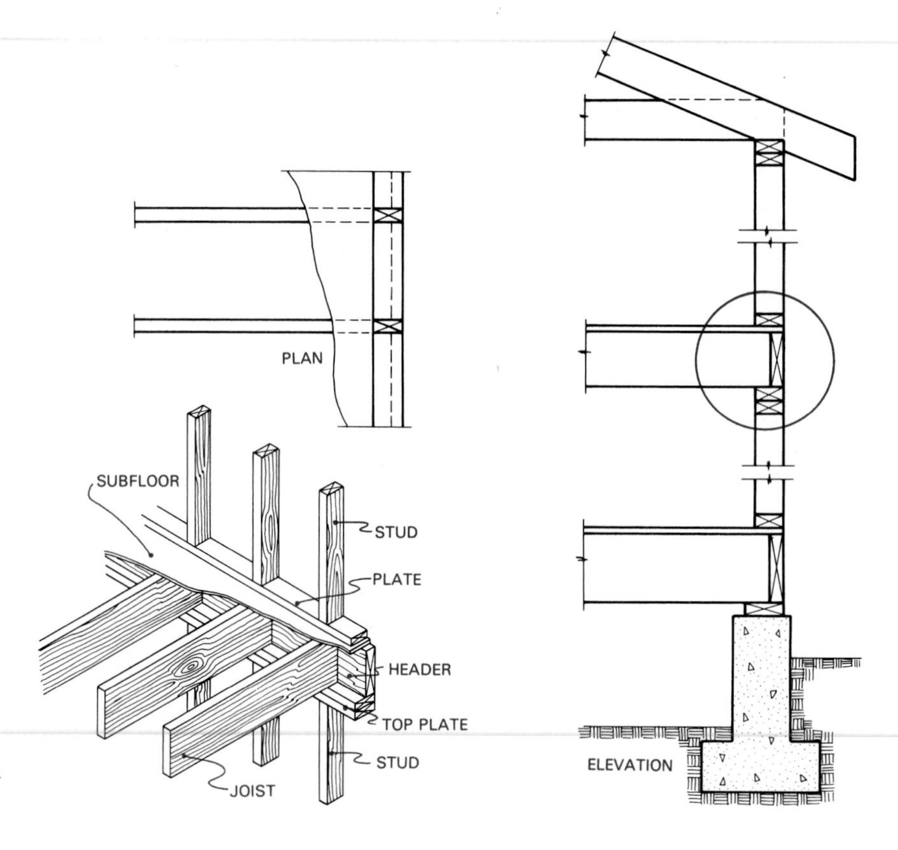

Fig. 53-35 Second floor with platform (western) framing.

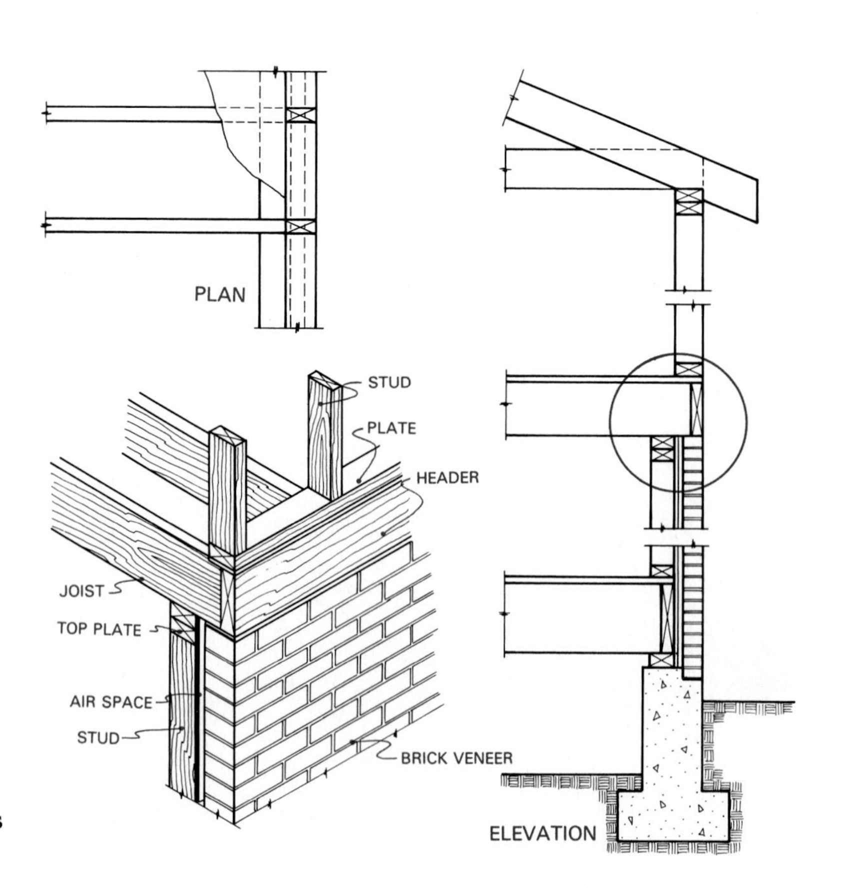

Fig. 53-36 Combination of materials shown for first and second levels.

first story. The second-floor header and joists rest directly on top of the brick veneer.

If the upper story is cantilevered over the first floor, the second-floor joists will run either parallel or perpendicular to the first-floor top plate that supports the second floor. When the joists are perpendicular to the wall, the construction is simple, as shown in Fig. 53-37. When the joists run parallel to the wall, a short lookout joist must be used to support the second floor, as shown in Fig. 53-38.

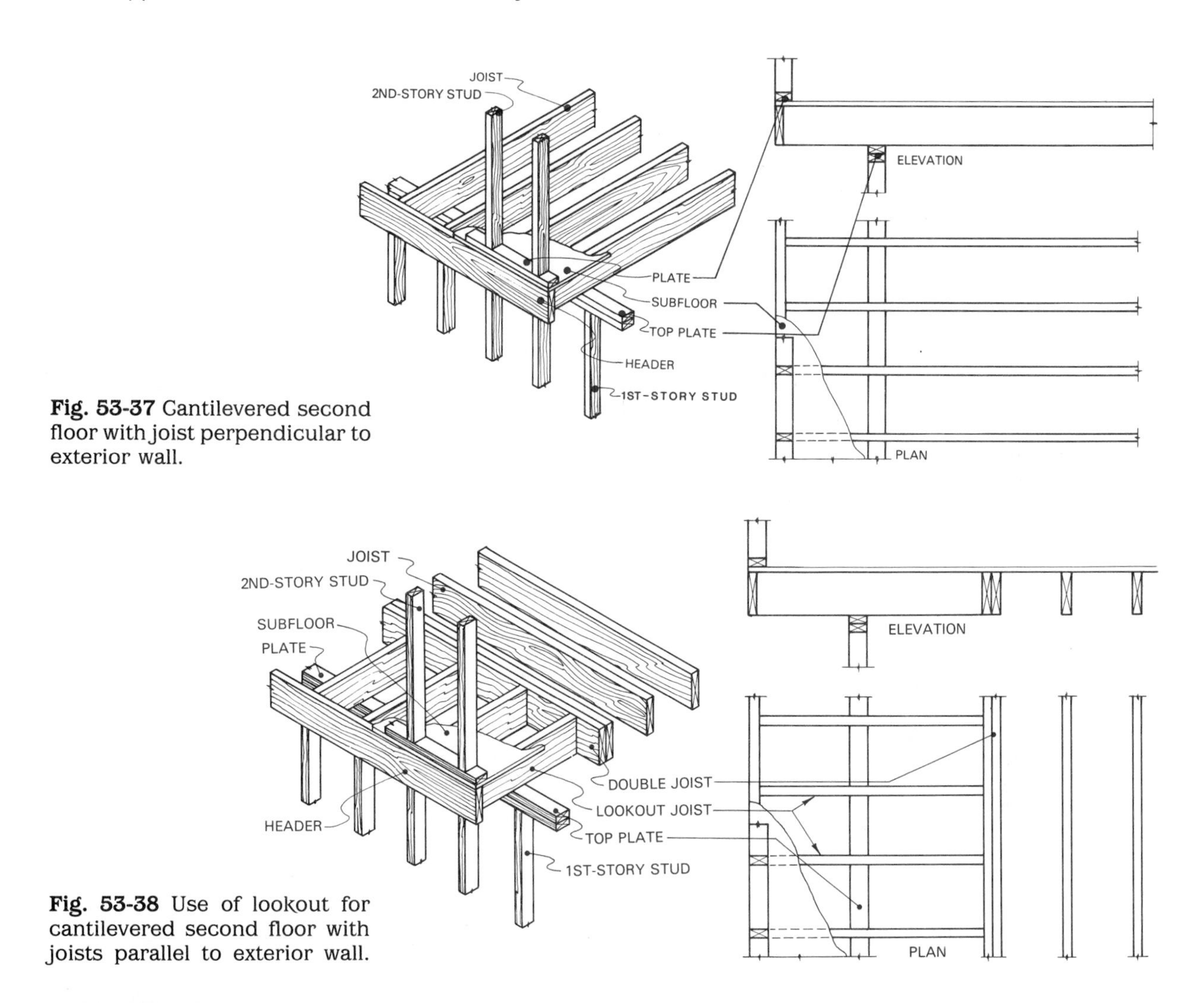

Fig. 53-37 Cantilevered second floor with joist perpendicular to exterior wall.

Fig. 53-38 Use of lookout for cantilevered second floor with joists parallel to exterior wall.

Exercises

1. Identify the floor framing terms illustrated in Fig. 53-39.

2. Develop a floor framing plan for as many floors as you are designing in your house plans.

3. Redesign the floor framing plan shown in Fig. 53-40 with the joists aligned in the opposite direction.

4. Add a 12′ x 14′ room to the upper left corner of the 20′ wall in Fig. 53-40. Draw the foundation wall and footing outlines and show the position of joists and girders.

5. **Design and draw a plan view of the floor system shown in Fig. 53-41.**
6. **Define the following architectural terms:** ***panel framing deflection, load, spacing, modulus of elasticity, fiber stress, maximum span, blocking, bridging, girder pocket, girder, built-up girder, pier, post, column, beam, header, double header, double joists, girder load area, post-load area, lally column, I beam, channel, joist hanger, scab, ledger strip, butt joint, square splice, butt splice, halved splice, bent splice, stairwell opening, tread, riser, nosing, unit run, unit rise, stringer, top nosing point, bottom nosing point, nosing line.***

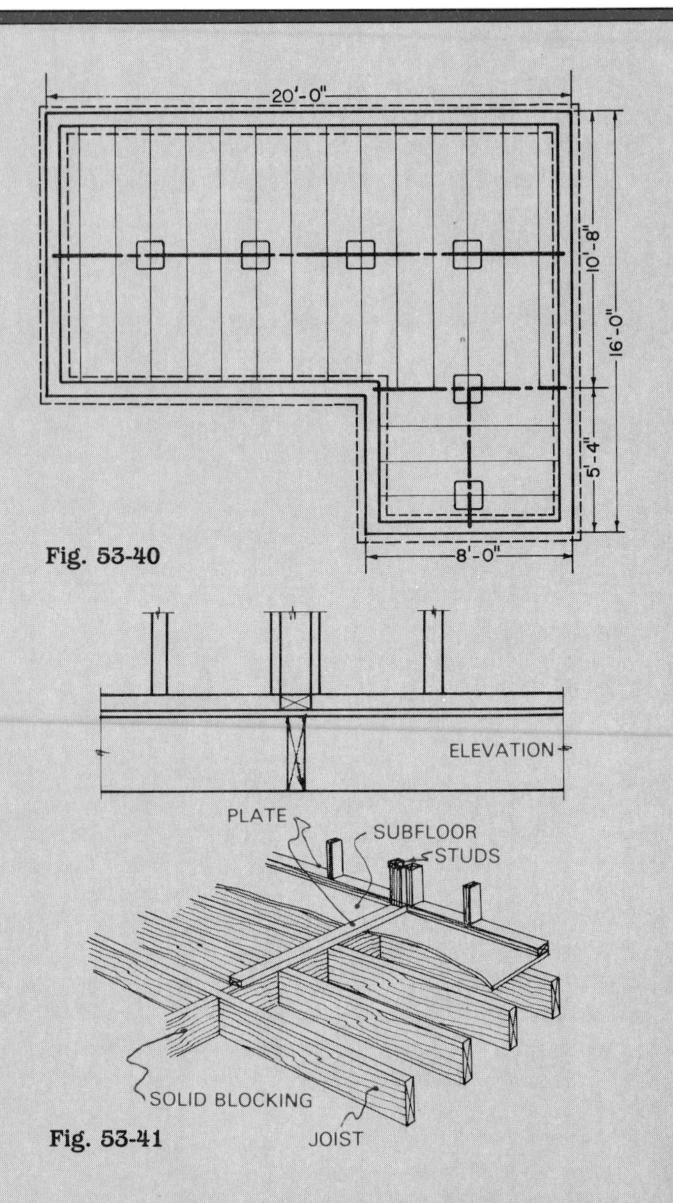

Fig. 53-40

Fig. 53-41

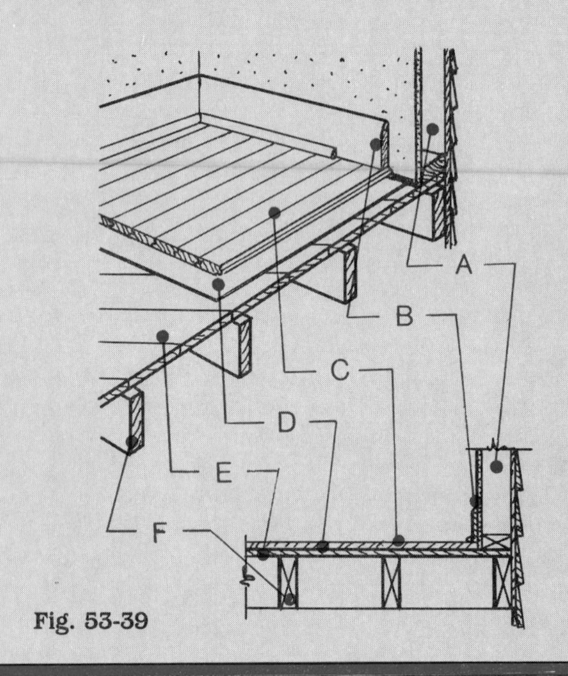

Fig. 53-39

UNIT 54

EXTERIOR-WALL FRAMING PLANS

Exterior walls for most residential buildings are of either conventional skeleton or post-and-beam construction. The typical method of erecting walls for most conventional buildings follows the braced-frame system. Prefabrication methods have led to variations in the erection of exterior walls, ranging from the panelization of just a basic frame to the complete panelized exterior wall, including plumbing, electrical work, doors, and windows, as shown in Fig. 54-1.

Large commercial structures usually use the curtain wall for their exterior wall. In this type of construction, nonstructural wall panels cover a steel framework. Figure 54-2 shows types of steel-framework structures and their relationship to height. Regardless of the method of construction or fabrication, the preparation of exte-

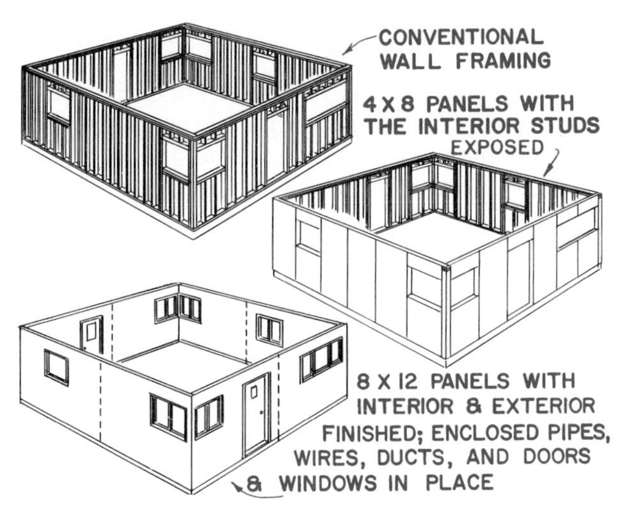

Fig. 54-1 Methods of wall paneling.

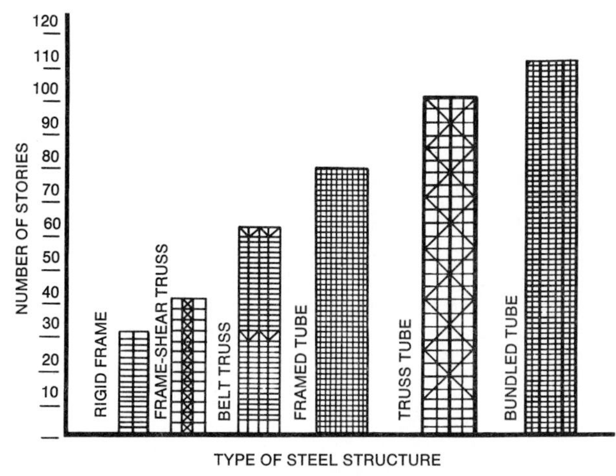

Fig. 54-2 Types of steel framework compared to height. *(American Iron and Steel Institute)*

rior-panel drawings is relatively the same, whether they are prepared for factory use or for field use.

FRAMING ELEVATIONS

Exterior-wall framing panels are best constructed by using a framing elevation drawing as a guide. The wall-framing elevation drawing is the same as the north, south, east, or west elevation of the building, with all the building materials removed except the basic framing. Figure 54-3 shows a wall-framing elevation compared with a pictorial drawing of the same wall. Notice that the framing elevation is an orthographic projection and does not reveal a second dimension or angle of projection. Figure 54-4 shows

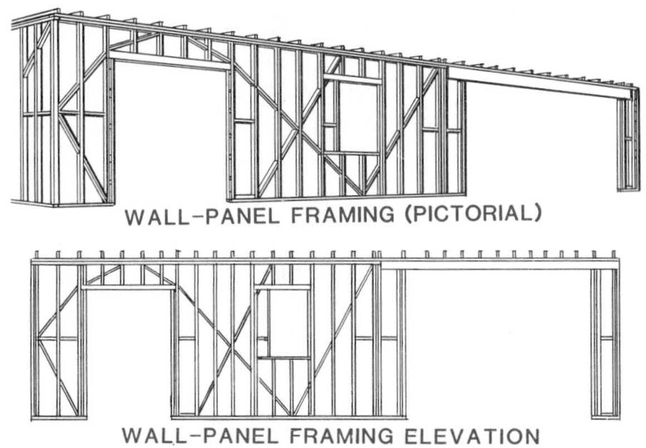

Fig. 54-3 Wall-framing elevation.

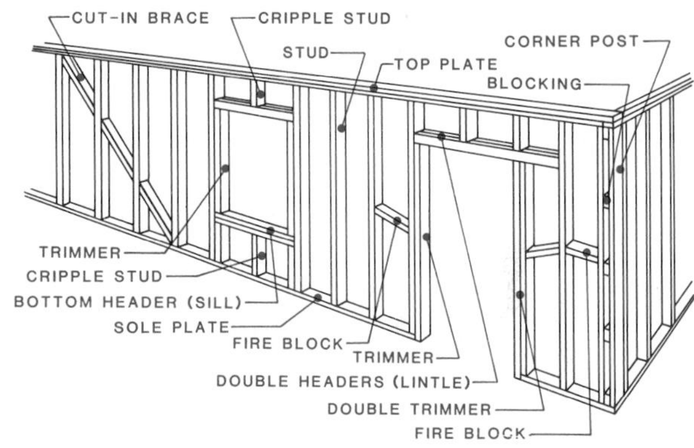

Fig. 54-4 Basic framing members shown in framing elevations.

some of the basic framing members included in framing elevations.

The framing elevation is projected from the floor plan and elevation, as shown in Fig. 54-5. Since floor-plan wall thicknesses normally include the thickness of siding materials, care should be taken to project the outside of the framing line to the framing drawing and not the outside of the siding line. When drawing door and window framing openings, the sizes as given on the manufacturing specifications or on the door or window schedules should be rechecked for rough frame openings and finished window sizes. Then project the openings from the floor plan and elevation. When aligned correctly, the elevation will supply all the projection points for the horizontal framing members, and the floor plan will provide all the points of projection for the location of vertical members.

COMPLETE SECTIONS

Another method of illustrating the framing methods used in wall construction is shown in Fig.

54-6. In this drawing, the elevation-framing information is incorporated in a complete sectional drawing of the entire structure. The advantage of this drawing is that it shows the relationship of the elevation-panel framing to the foundation-floor system and roof construction. Since this is a sectional drawing, blocking, joists, or any other member that is intersected by the cutting-plane line is shown by crossed diagonals.

BRACING

One of the problems in preparing and interpreting framing-elevation drawings is to determine whether bracing is placed on the inside of the wall, on the outside of the wall, or between the studs. Figure 54-7 at A shows the method of illustrating *let-in braces* that are notched on the outside of the wall so that the outer face of the brace is flush with the stud. Figure 54-7 at B

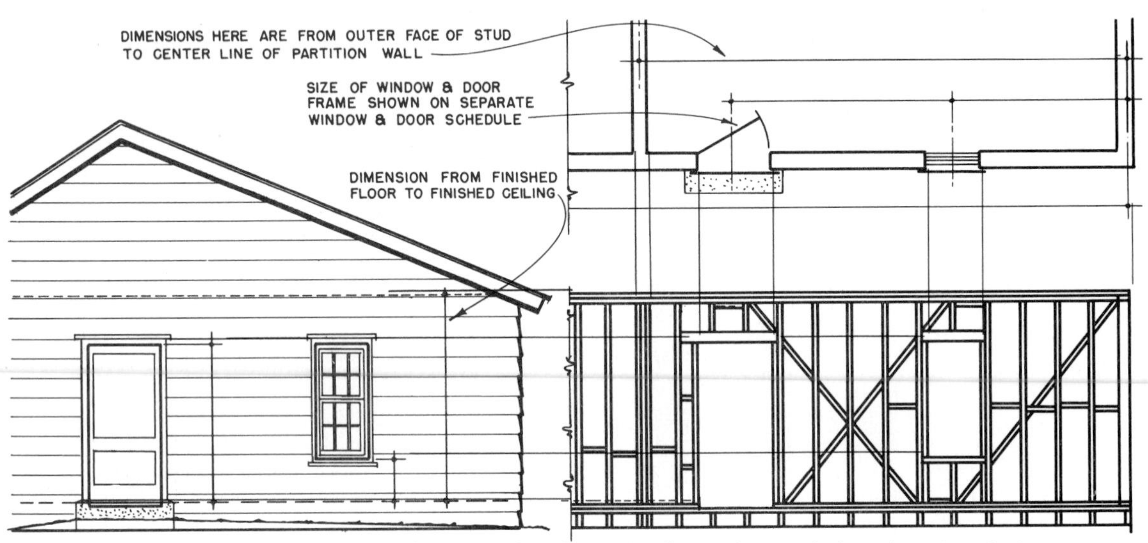

Fig. 54-5 Projection of an exterior-framing elevation from the floor plan and elevation drawing.

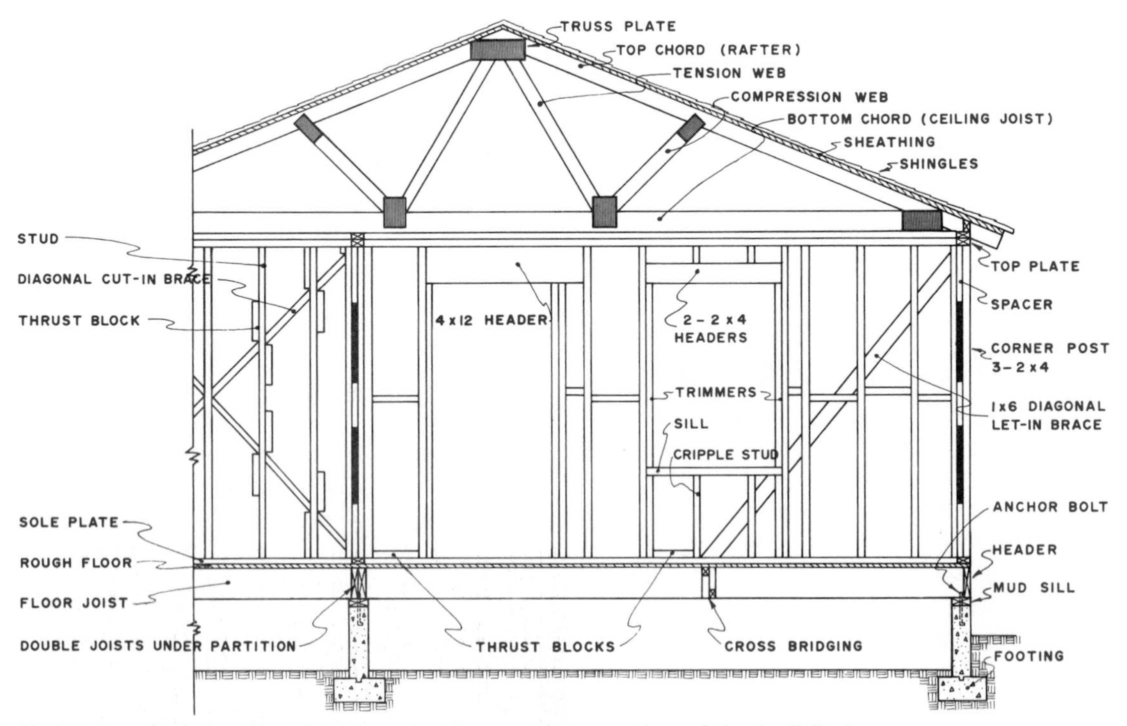

Fig. 54-6 A framing elevation incorporated in complete section of the building.

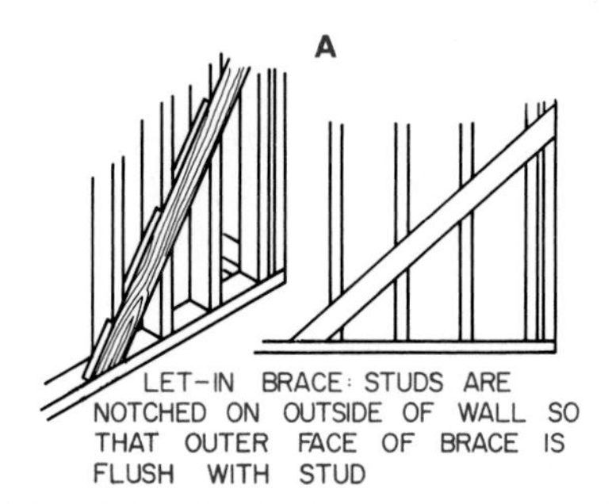

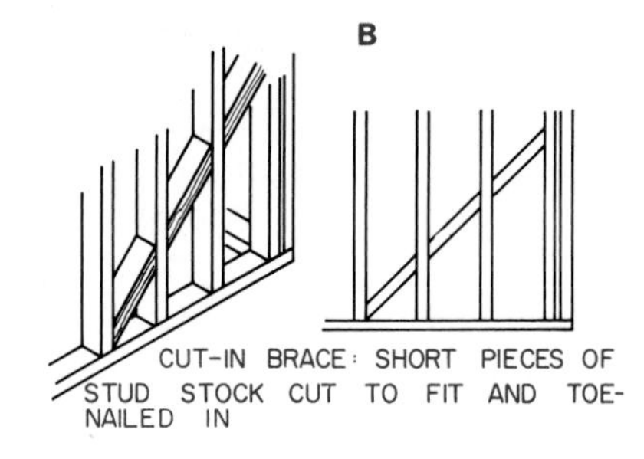

B
CUT-IN BRACE: SHORT PIECES OF STUD STOCK CUT TO FIT AND TOE-NAILED IN

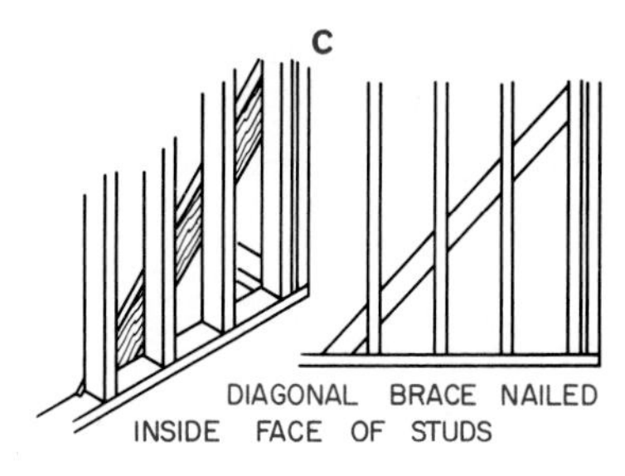

Fig. 54-7 Methods of illustrating braces.

shows the method of illustrating *cut-in braces* that are nailed between the studs; Fig. 54-7 at C shows the method of illustrating *diagonal braces* that are nailed on the inside faces of the studs that form the wall. Similar difficulties often occur in interpreting the true position of headers, cripple studs, plates, and trimmers. Figure 54-8 shows the method of illustrating the position of these members on the framing-elevation drawing to eliminate confusion and to simplify the proper interpretation.

PANEL ELEVATIONS

Panel elevations show the attachment of sheathing to the framing. It is often necessary to show the relationship between the *panel layout* and the *framing layout* of an elevation when the panel drawing and the framing drawing are combined in one drawing. The diagonals that indicate the position of the panels are drawn with dotted lines, as shown in Fig. 54-9. When only the panel layouts are shown, the outline of the

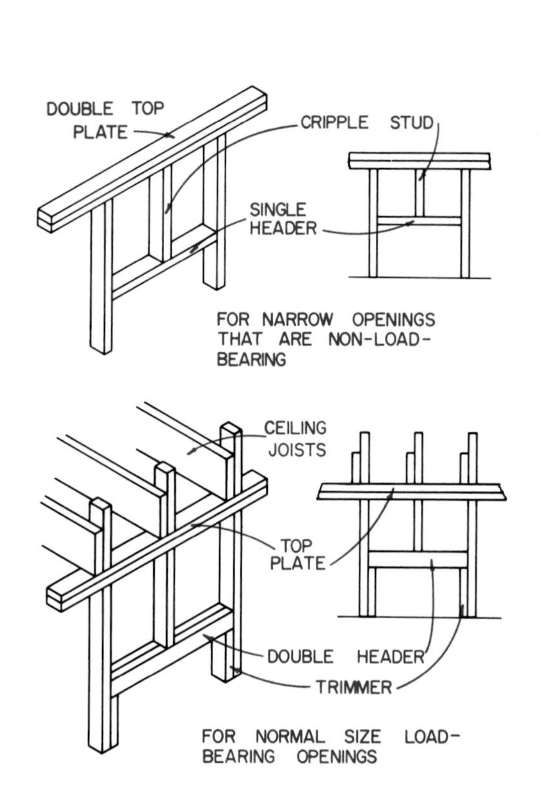

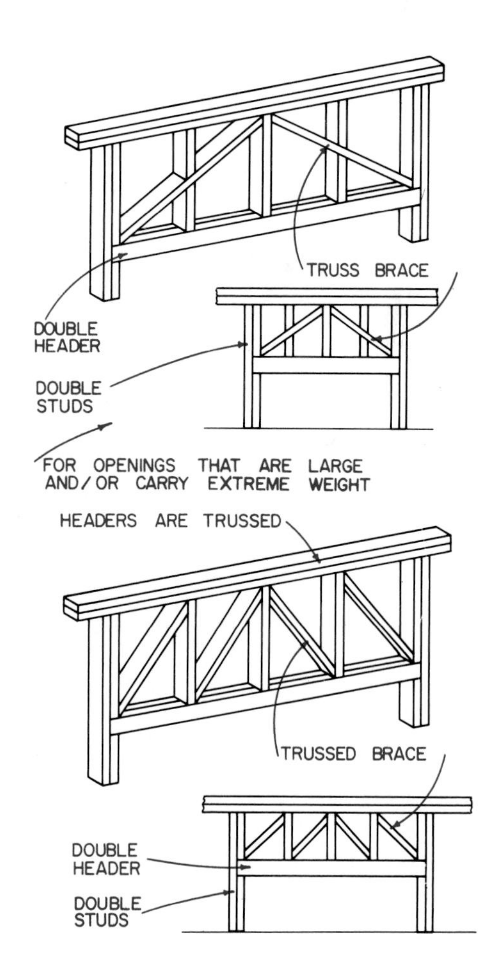

Fig. 54-8 Methods of illustrating the positions of headers and cripples.

panels and diagonals are drawn solid, as shown in Fig. 54-9. In this case, a separate framing plan must also be prepared and correlated with the panel elevation.

DIMENSIONS

The method of dimensioning panel and framing-elevation drawings is shown in Fig. 54-10. Overall widths, heights, and spacing of studs should be given. Control dimensions for the openings of horizontal members and *rough openings* (framing openings) for windows should also be included (Fig. 54-11A). If the spacing of studs does not automatically provide the rough opening necessary for the window, the rough-opening width of the window should also be dimensioned. Figure 54-11B shows standard framing dimensions to produce a standard ceiling height using standard 93″ studs.

DETAILS

Not all the information needed to frame an exterior wall can be shown on the elevation drawing. One of the most effective means of showing information at right angles to the elevation drawing is by *removed sections.*

Removed Sections

Removed sections may be indexed to the floor plan or elevation, as indicated in Section 11. They may be removed sections from a pictorial drawing, as shown in Fig. 54-12. In this example, section A describes the framing method employed on the wall and roof intersections, using break lines to expose the framing. Section B shows a wall section at the sill, revealing the intersection between the foundation-floor system and the exterior wall. These sections also

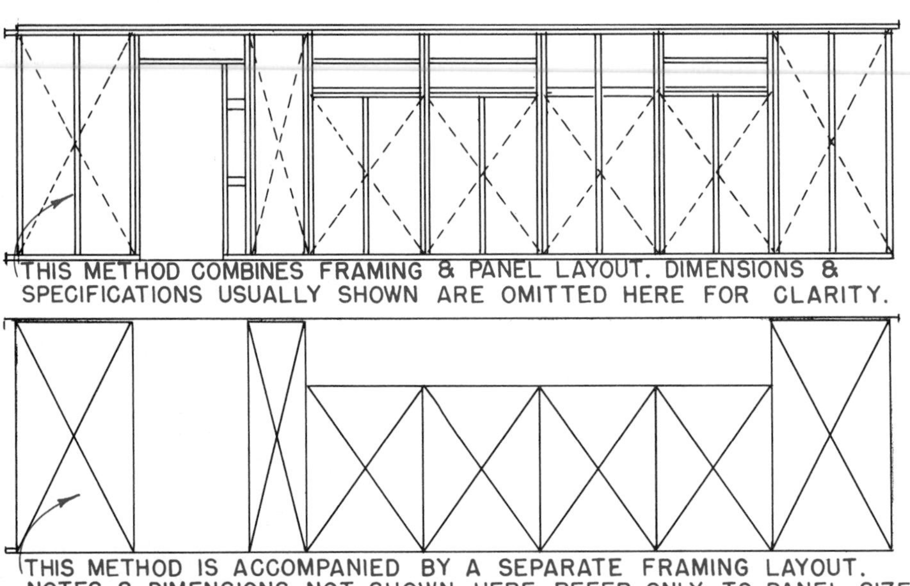
THIS METHOD COMBINES FRAMING & PANEL LAYOUT. DIMENSIONS & SPECIFICATIONS USUALLY SHOWN ARE OMITTED HERE FOR CLARITY.
THIS METHOD IS ACCOMPANIED BY A SEPARATE FRAMING LAYOUT. NOTES & DIMENSIONS NOT SHOWN HERE REFER ONLY TO PANEL SIZES & SPECIFICATIONS.

Fig. 54-9 Diagonal lines are used to show the positions of panels.

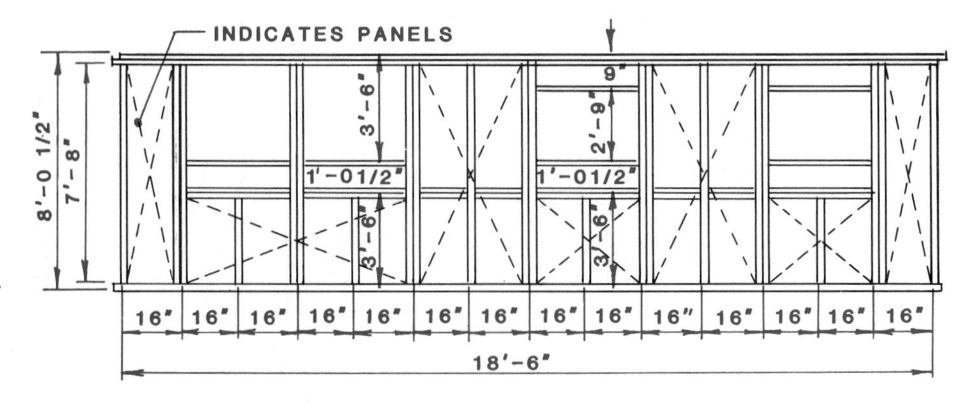
INDICATES PANELS
8'-0 1/2"
7'-8"
3'-6"
9"
2'-9"
1'-01/2"
1'-01/2"
3'-6"
3'-6"
16" 16" 16" 16" 16" 16" 16" 16" 16" 16" 16" 16" 16" 16"
18'-6"

Fig. 54-10 Methods of dimensioning panel and framing elevations.

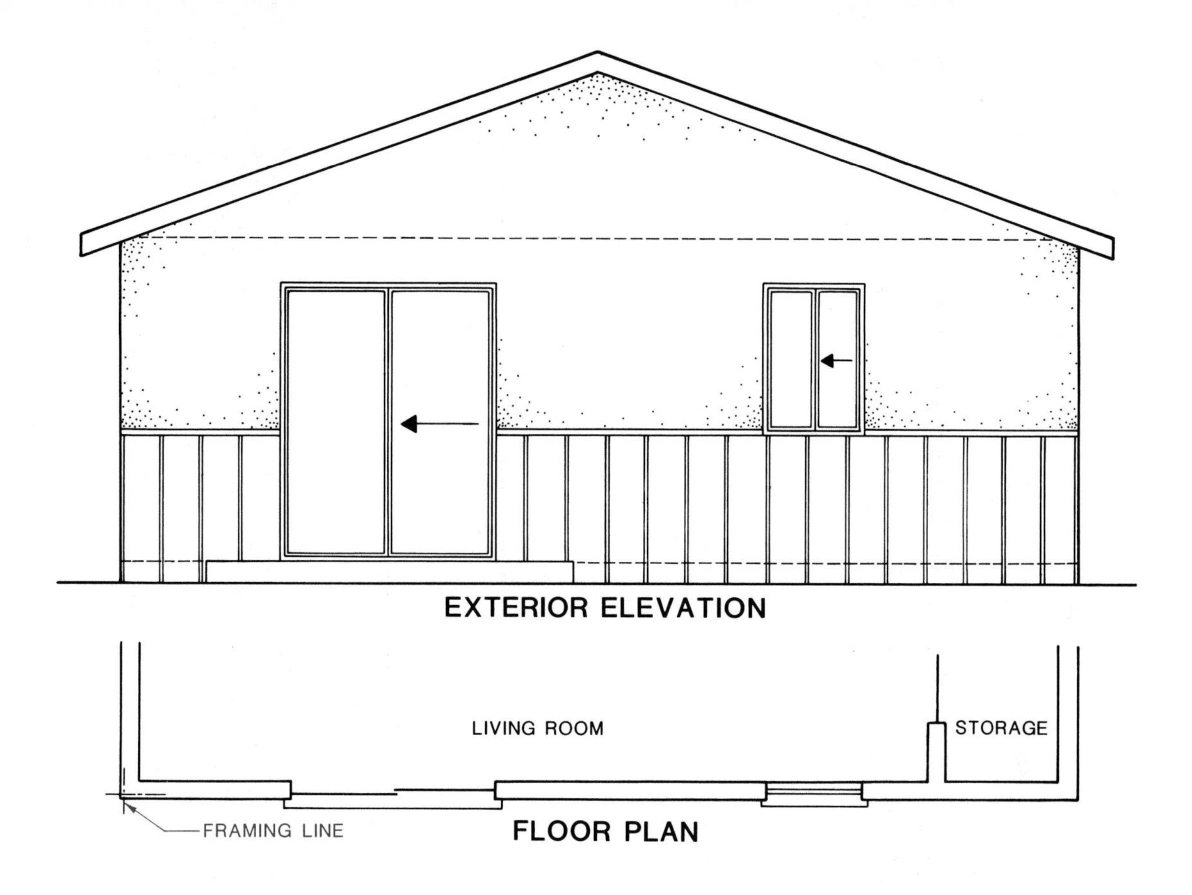

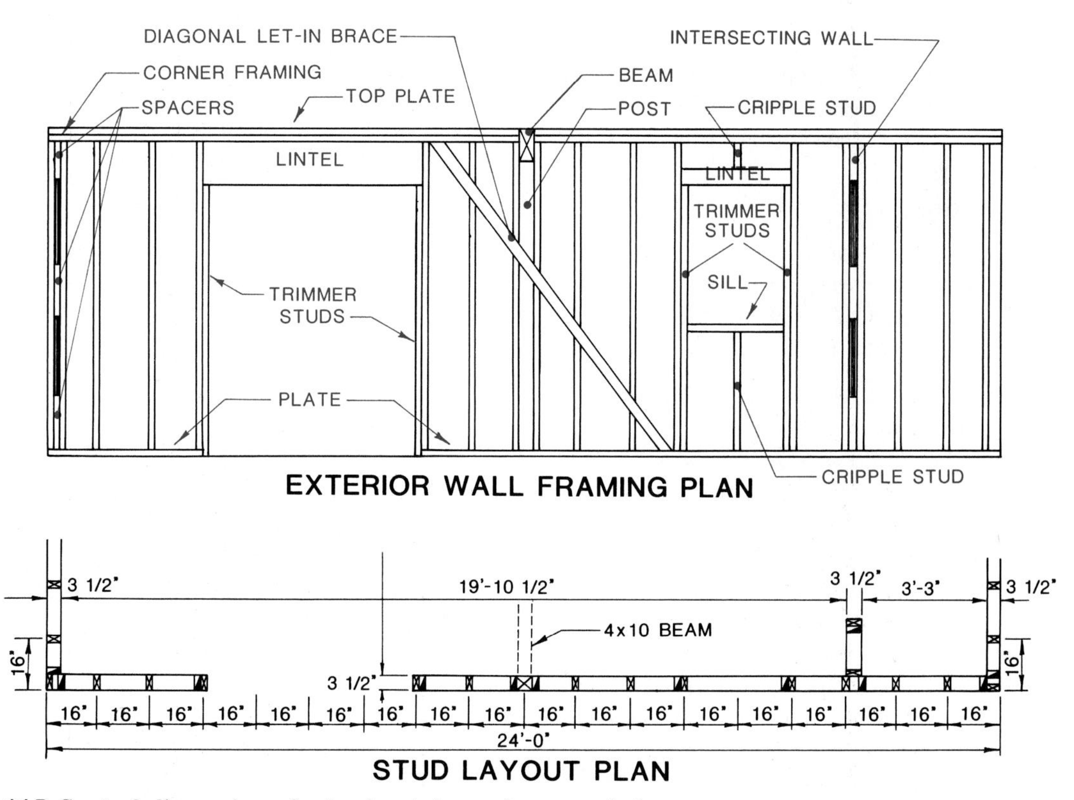

Fig. 54-11A Control dimensions for horizontal member openings.

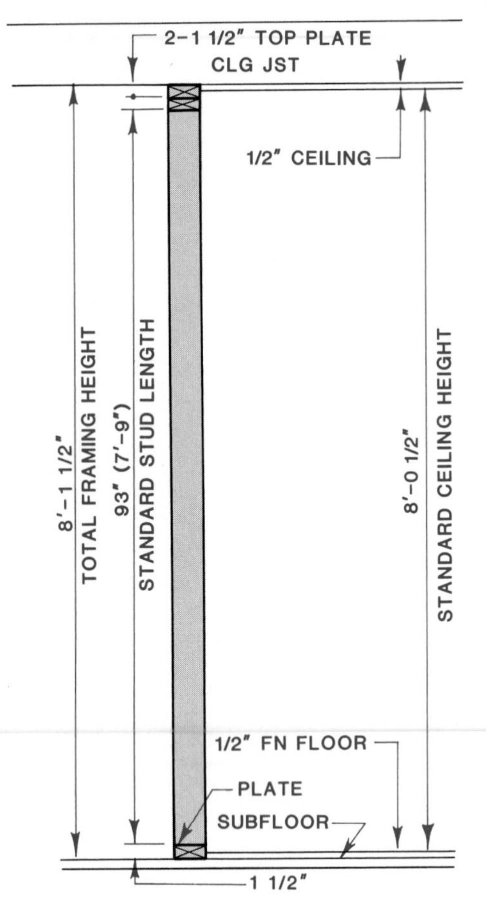

Fig. 54-11B Standard framing dimensions.

Rendering by George A. Parenti for Masonite Corporation

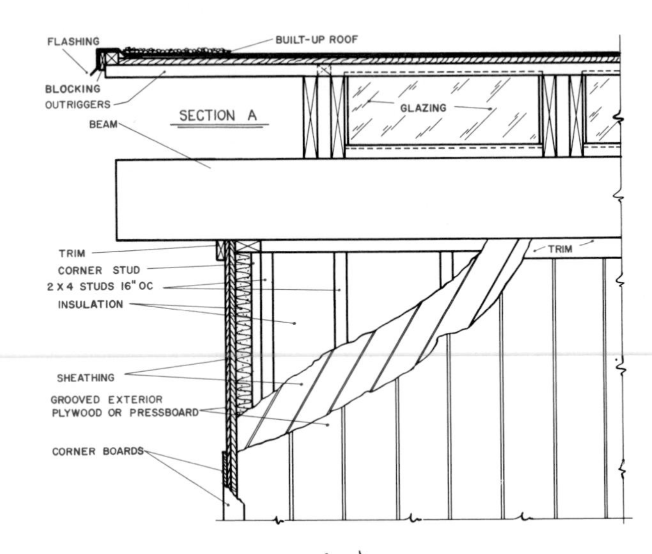

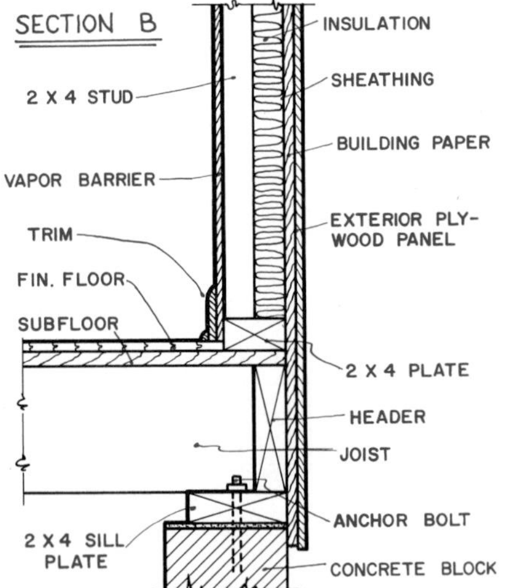

Fig. 54-12 Removed sections may be indexed to a floor plan and to elevation or pictorial drawings.

show the inside wall treatment, insulation, sheathing, and exterior siding. Removed sections are effective in showing enlarged details.

Sectional Breaks

A larger scale is used on wall sections if break lines are employed. Figure 54-13 shows the sequence of steps used to lay out and draw a typical external wall section.

1. Determine the width of the walls, foundation, and footing.

2. Lay out the angle of the roof and point of intersection of the roof and top plate. Lay out the width of the joist and sill.

3. Block in the position of roof rafters, top plates, sole plate, and roof floor lines.

4. Draw vertical lines to indicate the width of stud, insulation, air space, and brick.

5. Add details of the outlines of roof boards and shingles. Show outline of cornice construction.

6. Draw horizontal lines representing break lines. Add section-lining symbols.

Sectional breaks can also be used to show a rotated section of the building material in the break area, as shown in Fig. 54-14.

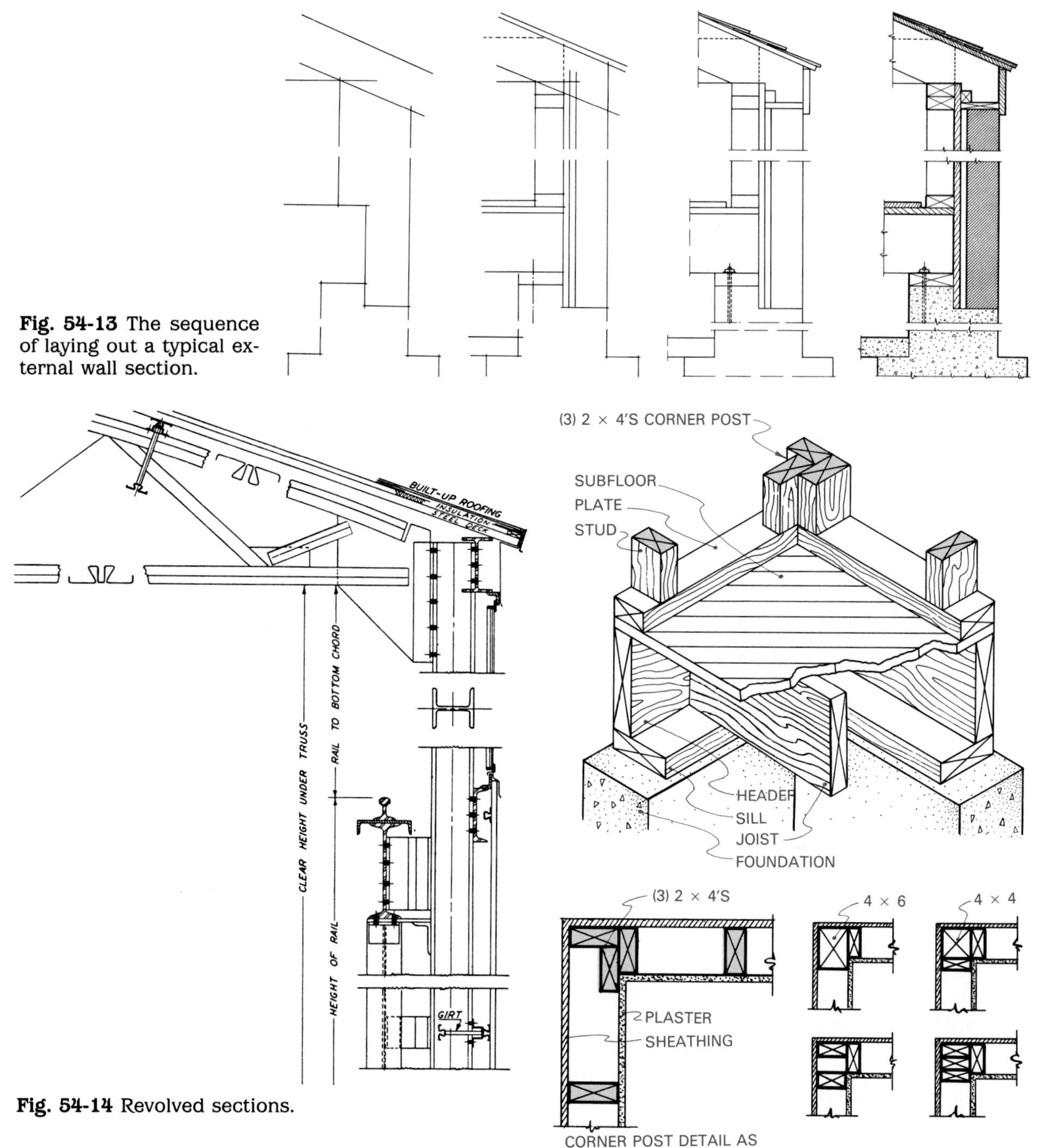

Fig. 54-13 The sequence of laying out a typical external wall section.

Fig. 54-14 Revolved sections.

Fig. 54-15 Pictorial and horizontal sections of corner-post construction.

Pictorial Details

A *pictorial detail* or horizontal section of a wall is often used to clarify the relationship of framing members. This method is especially helpful in describing the layout of corner posts, as shown in Fig. 54-15. The horizontal section is more accurate in showing exact size and position of studs, but the pictorial drawing is more effective in showing the total relationship between sole plate, corner-post studs, and box-sill construction. See Fig. 53-34 for the use of a pictorial detail to clarify the section of a balloon framing elevation; Fig. 53-35 shows the same detail relating to a platform-system drawing. Pictorial drawings are also often used to expand a detail for clarification, as shown in Fig. 54-16.

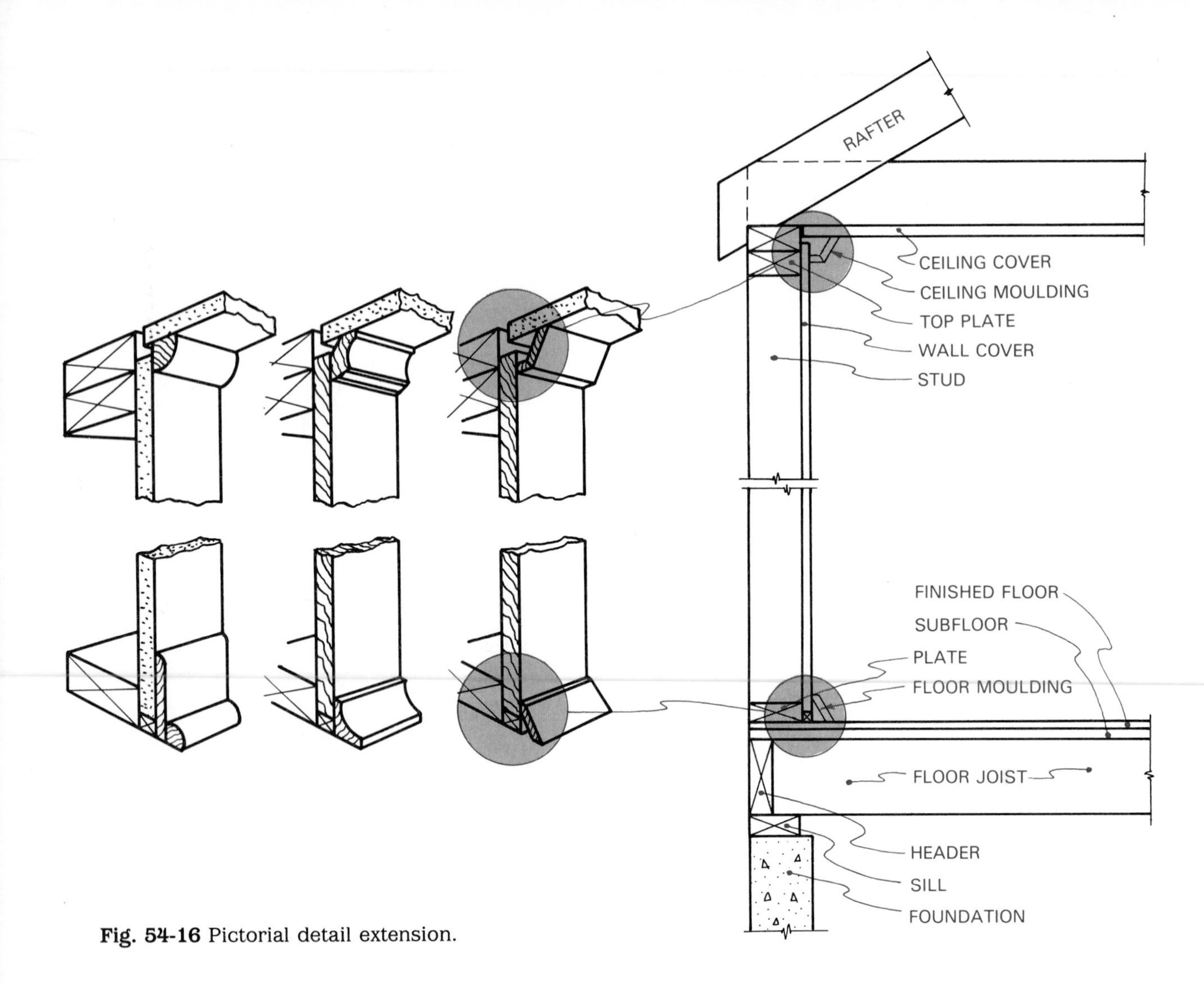

Fig. 54-16 Pictorial detail extension.

Exploded Views

Exploded views are most effective in showing internal construction that is hidden when the total assembly is drawn in its completed form. Figure 54-17 is an exploded view of a corner post that shows its construction on the sole plate and the position of the top plate. This method of detailing is also extensively used in cabinet work. Figure 54-18 shows steel-wall framing methods in an exploded view.

Siding Details

New siding materials are constantly being developed, and new applications found for existing materials. Aluminum is a good example. Today, the residential use of aluminum is most common for door and window frames, gutters and downspouts, and siding.

One method of showing the relationship between the basic framing and siding materials is the *breakaway* pictorial drawing, as shown in Fig. 54-19. This kind of drawing can be most effectively interpreted by the layman. However, it is most difficult to dimension for construction purposes. A more effective means of showing the exact position of siding materials is the vertical or horizontal section. Figure 54-20A through I shows the sectional method of representing typical external walls. Figure 54-20A shows brick veneer on frame backing; Fig. 54-20B, solid brick wall; Fig. 54-20C, stucco; Fig. 54-20D, horizontal wood siding; and Fig. 54-20E, board-and-batten construction. Figure 54-20F shows plan and elevation sections for plywood lap siding. Figures 54-20G and H show the same type drawing for masonry wall details. Sometimes full wall sections are prepared from foundation to

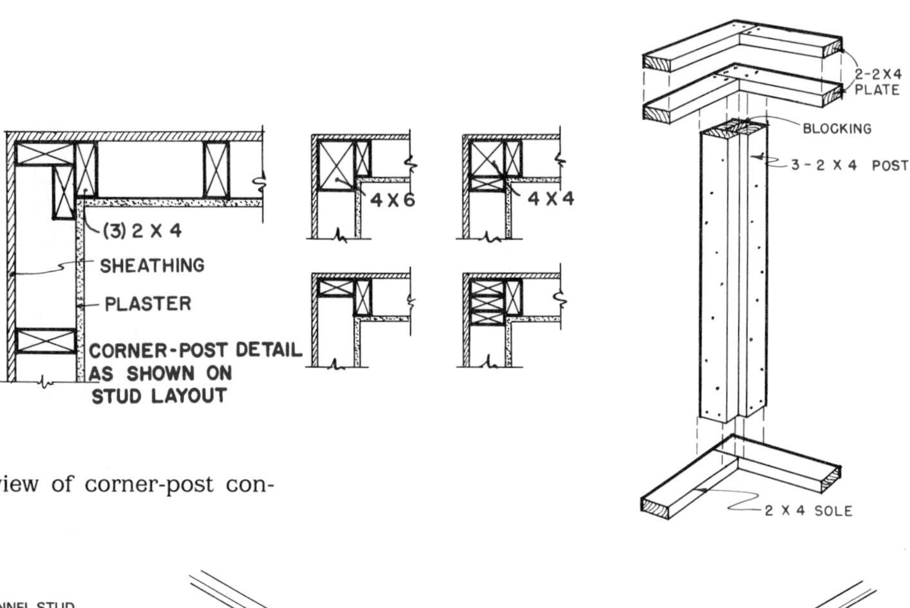

Fig. 54-17 An exploded view of corner-post construction.

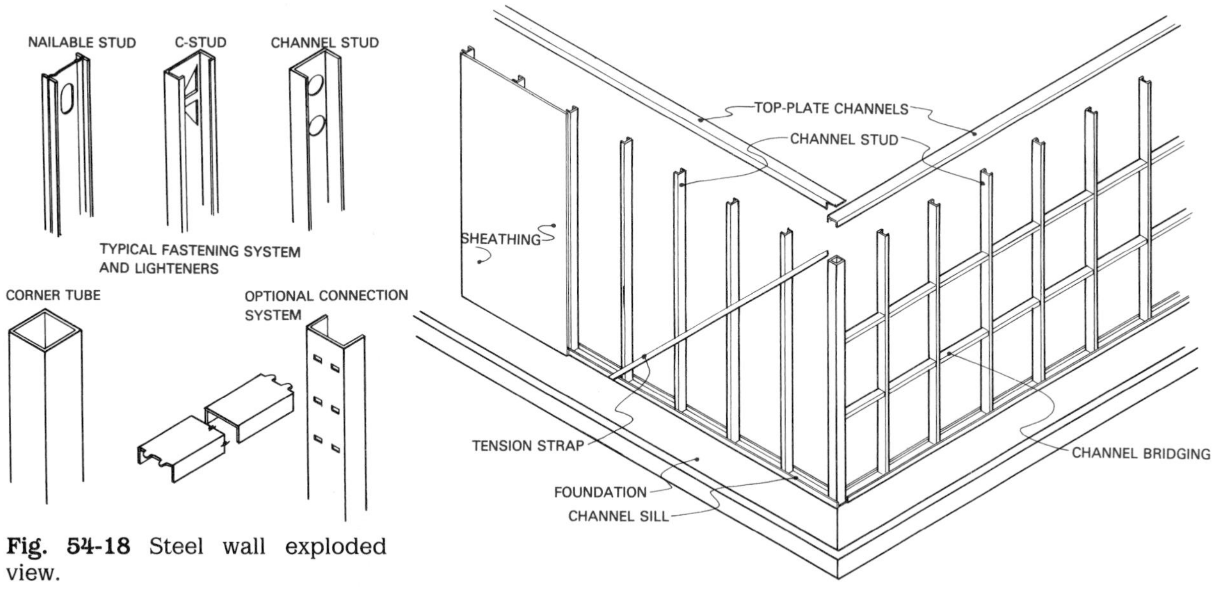

Fig. 54-18 Steel wall exploded view.

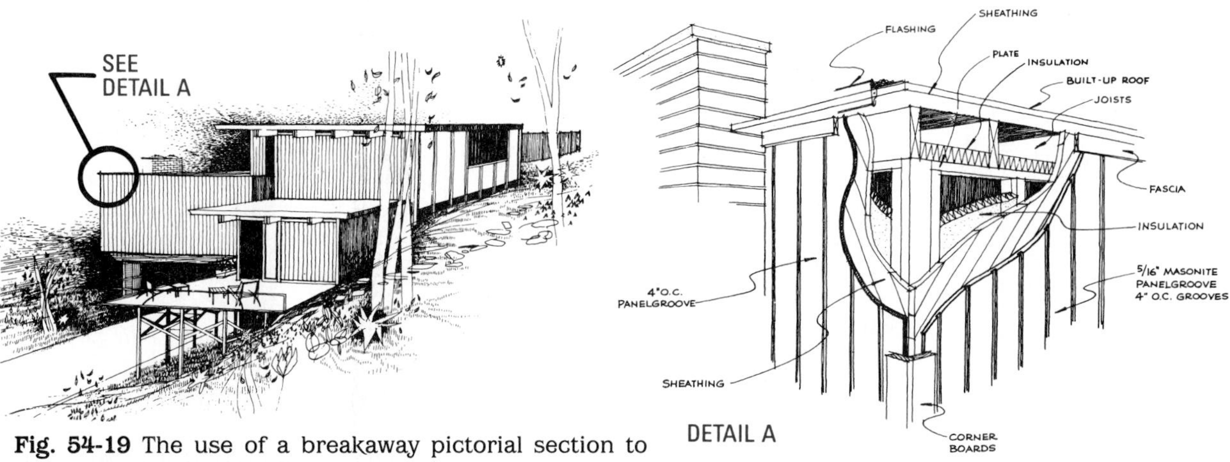

Fig. 54-19 The use of a breakaway pictorial section to show construction details. *(Rendering by George A. Parenti for the Masonite Corporation)*

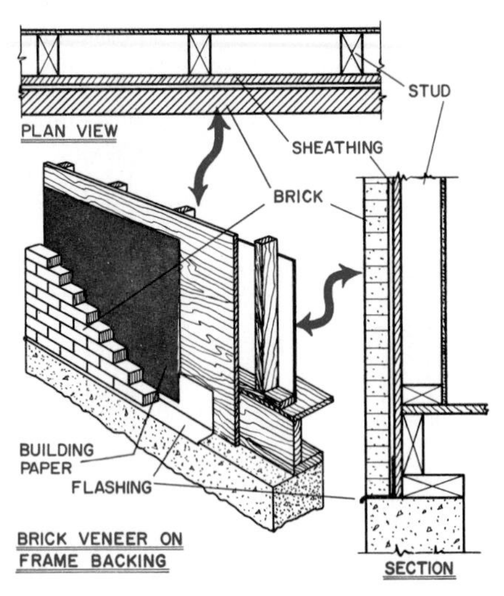

Fig. 54-20A Construction details can be shown on a plan section or on an elevation section.

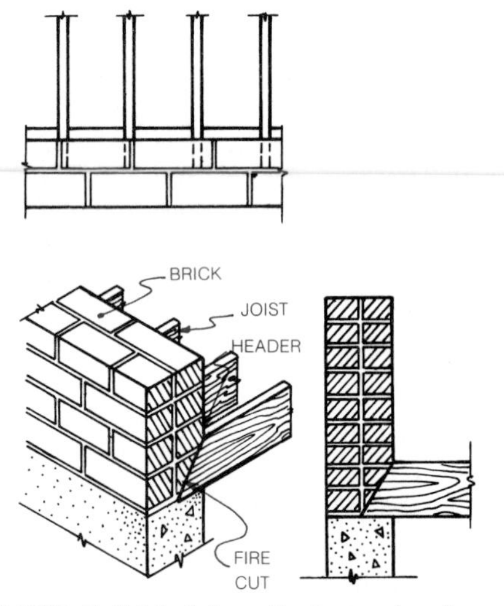

Fig. 54-20B Solid brick wall shown in plan, elevation, and pictorial views.

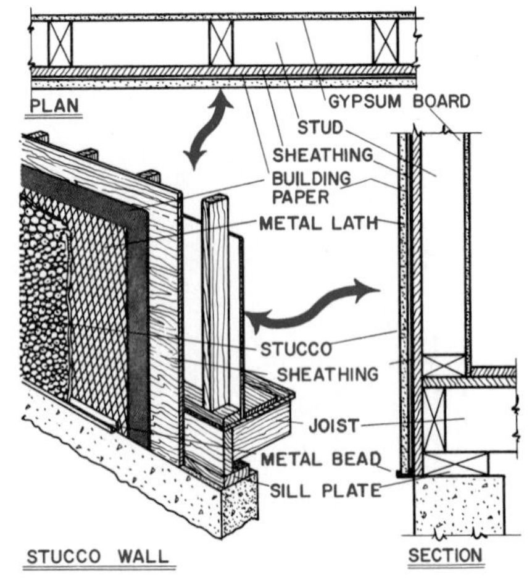

Fig. 54-20C The relationship between a plan and an elevation section of a stucco wall.

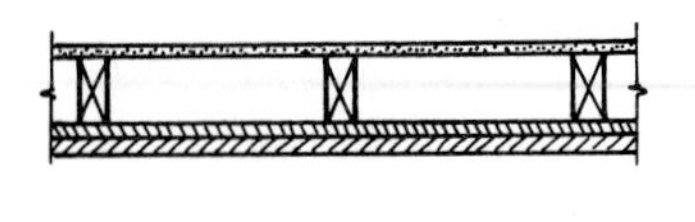

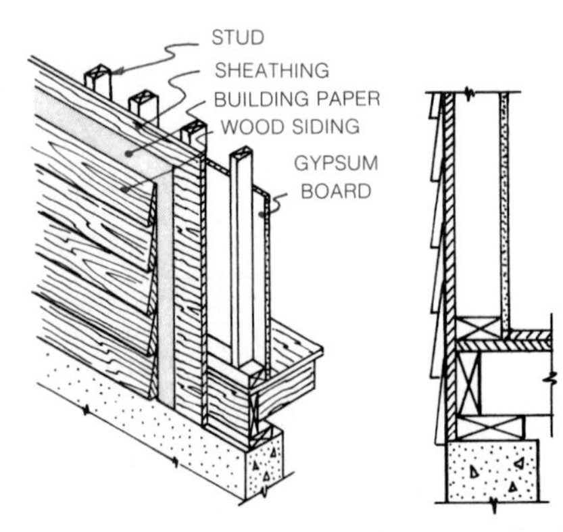

Fig. 54-20D Horizontal wood siding wall shown in plan, elevation, and pictorial views.

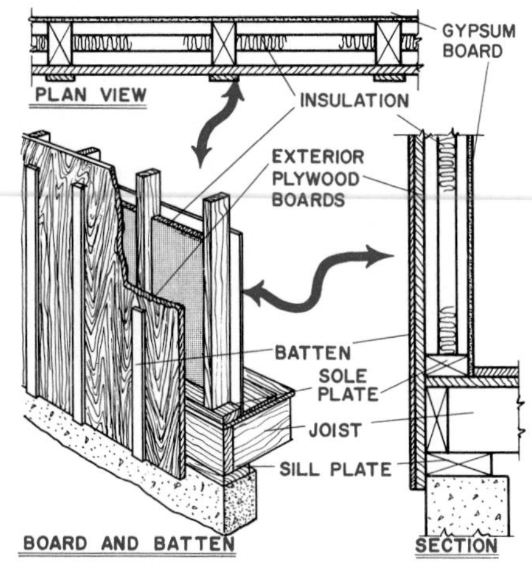

Fig. 54-20E The relationship between a plan and an elevation section of a board-and-batten wall.

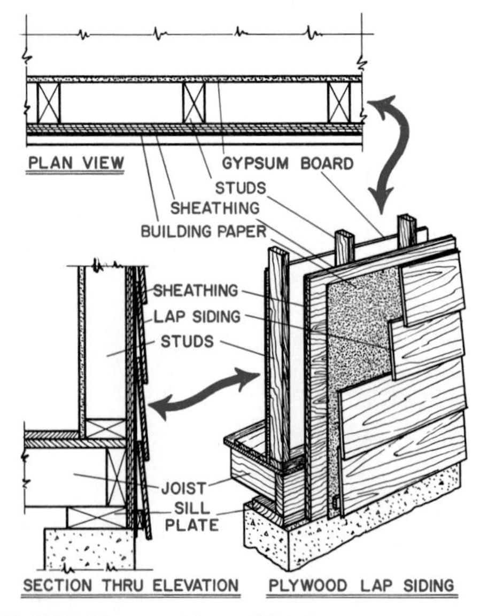

Fig. 54-20F Plywood lap siding.

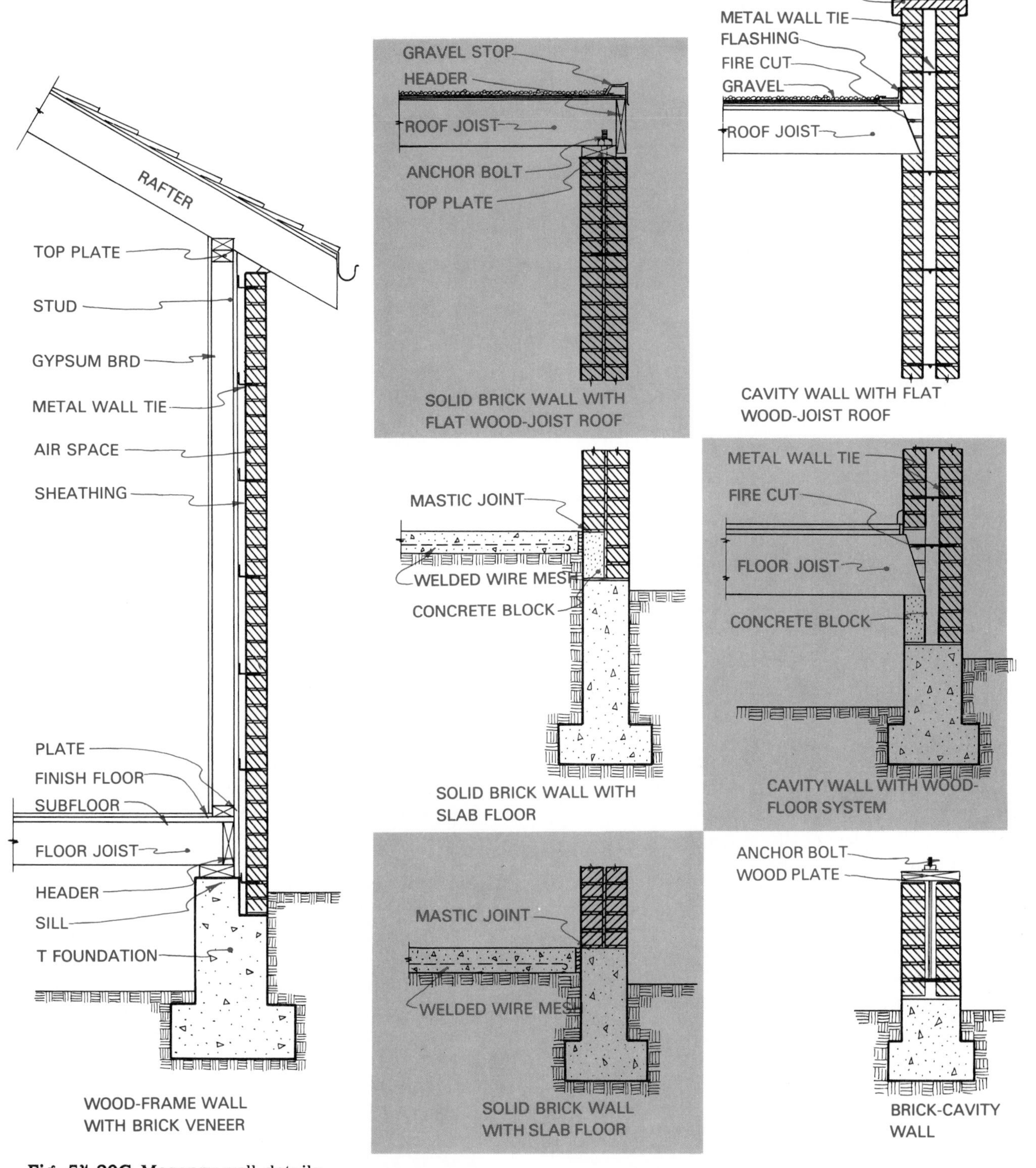

Fig. 54-20G Masonry wall details.

roof as shown in Fig. 54-20I. Compare the plan section and the elevation section with the related pictorial drawing. Follow the relationship of each material as it exists in each drawing. You should be able to visualize the pictorial drawing by studying the sectional drawings of the plan and elevation for these and other similar siding materials.

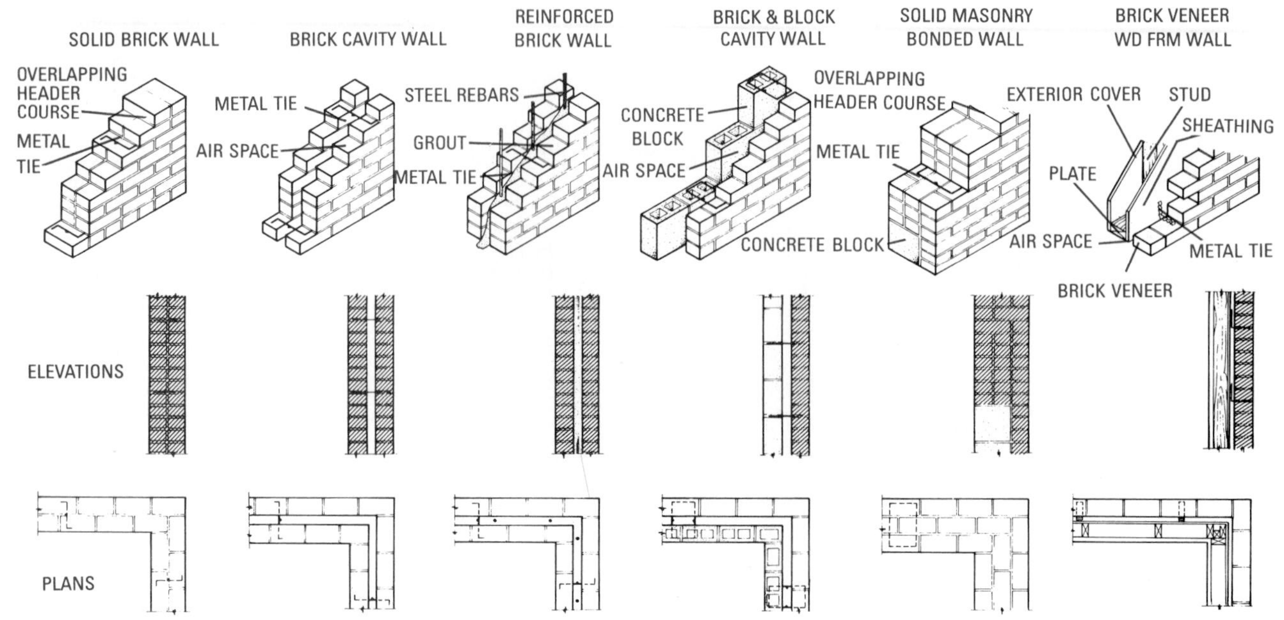

Fig. 54-20H Brick wall details.

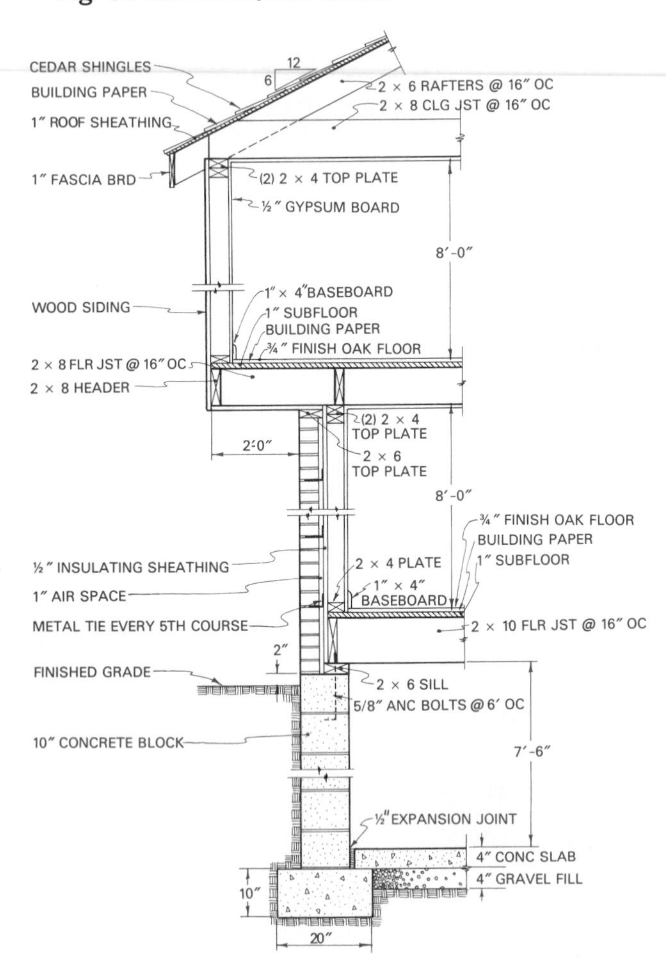

Fig. 54-20I Full wall section.

WINDOW-FRAMING DRAWINGS

One of the most effective methods of showing window-framing details is through head, jamb, and sill sections, as is described in Section 11 and summarized in Fig. 54-21. Most windows are factory-made components ready for installation. Therefore, the most critical framing dimensions are those that describe the exact size of the framing opening. Window-framing drawings should include the sash-opening dimensions in addition to the dimensions of the framing opening (Fig. 54-22). Figure 54-23 shows rough frame openings and sash openings for some of the more common sizes of windows.

When fixed windows or unusual window treatments are constructed in the field or even at the factory for a specific building, complete framing details must be drawn similar to the detailed drawing shown in Fig. 54-24. The more unusual the use of nonstandard sizes and components, the more complete must be the detail framing drawings that accompany the design.

DOOR-FRAMING PLANS

The upper part of Fig. 54-25 shows the door and wall relationships for conventional door framing

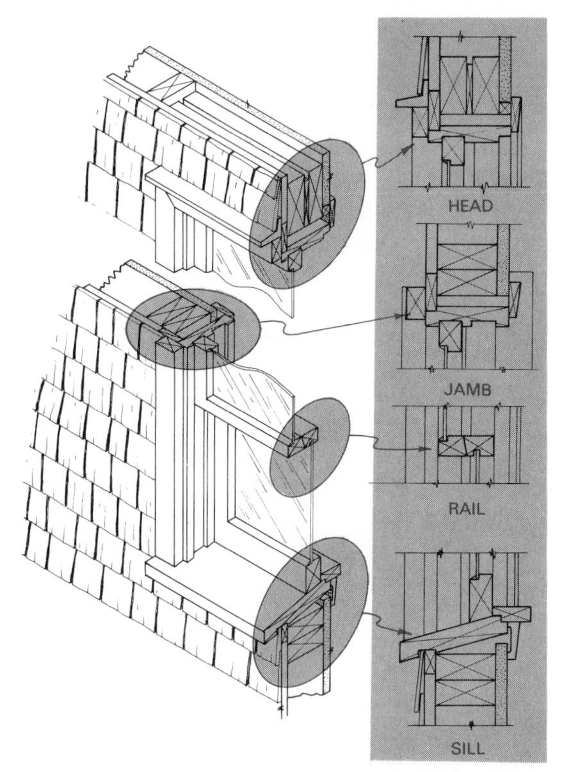

Fig. 54-21 Head, jamb, and sill sections are most commonly used to show window-framing details.

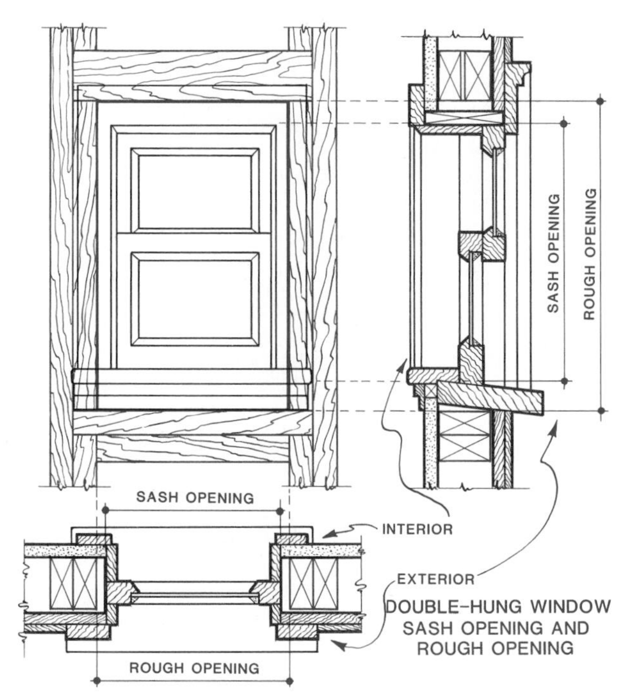

Fig. 54-22 Details are often needed to show the rough openings for doors and windows.

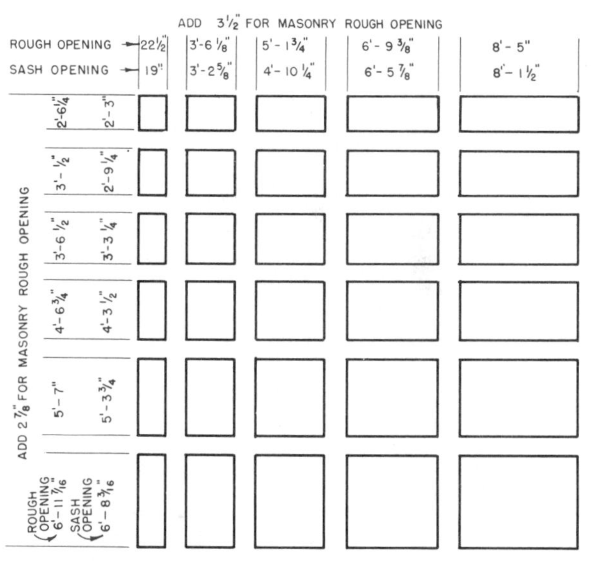

Fig. 54-23 Rough opening dimensions for common size of windows.

which are described in wall-framing details. The use of modular-component door units, as shown in the lower part of Fig. 54-25, is increasing throughout the home-building industry. Maintaining accurate rough-opening dimensions for these units is most critical to their installation. Whether the door framing is conventional or of a component design, the exact position of the opening and the dimensions of the rough opening must be clearly illustrated and labeled on the framing drawing. Figure 54-26A shows some rough-opening dimensions for standard-sized doors that are used in various locations through a residence. Figure 54-26B shows rough opening dimensions compared to actual door sizes. Note 3½″ is added to the width and height to provide for the rough opening size of the door.

Head, sill, and jamb sections, as shown in Section 11, are as effective in describing the door-framing construction as they are in showing window-framing details. Since the door extends to the floor, the relationship of the floor-framing system to the position of the door is critical. The method of intersecting the door and hinge with the wall framing, as shown in Fig. 54-27, is also important. Section A (head) shows the intersection of the door and the header framing. Section B (sill) shows how the door relates to the floor framing. Section C (jamb) shows how the hinged side is constructed.

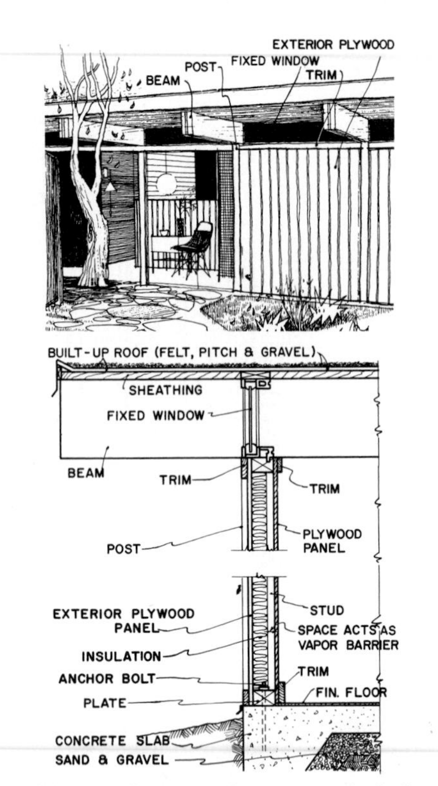

Fig. 54-24 Details are always needed for fixed-window construction.

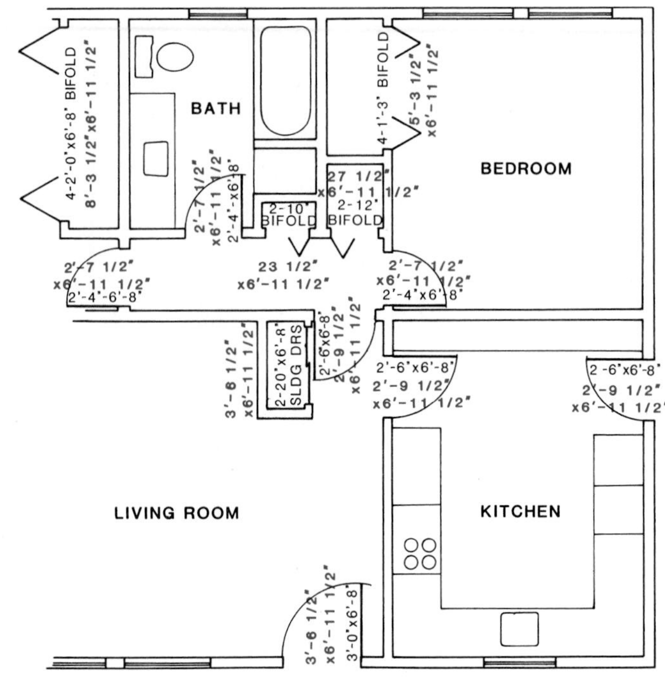

Fig. 54-26A Rough-opening dimensions for standard-sized doors.

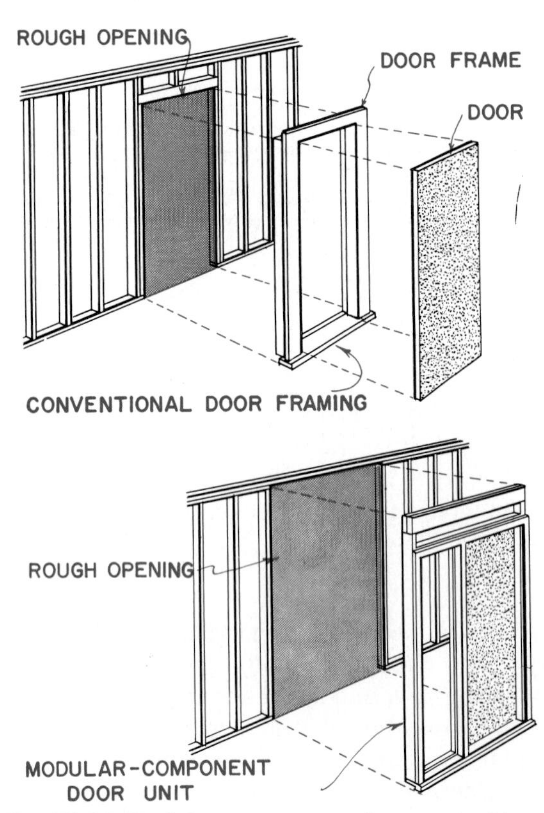

Fig. 54-25 Modular-component door assembly.

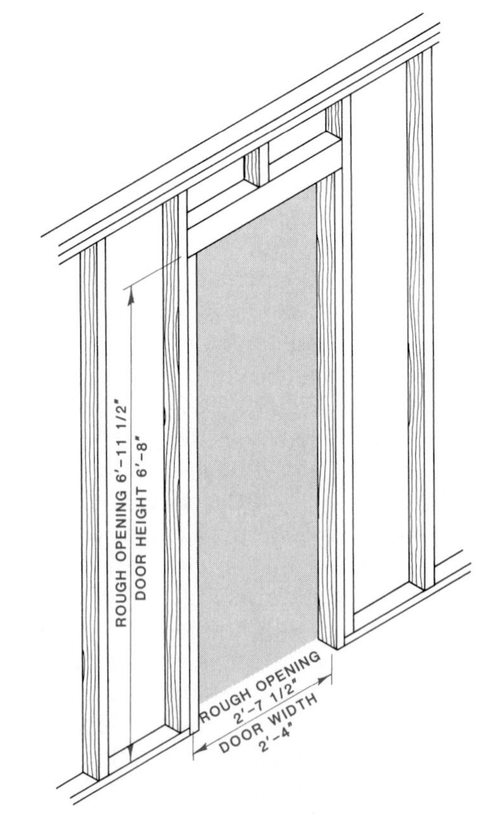

Fig. 54-26B Add 3½" to width and height for rough opening.

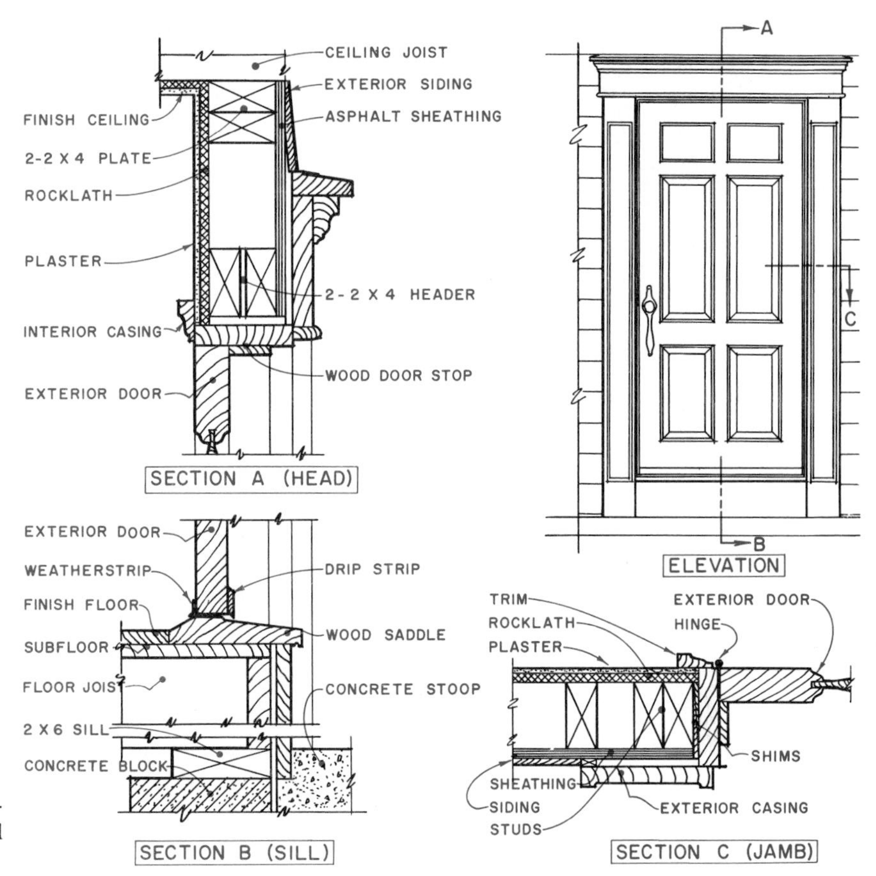

Fig. 54-27 The relationship between the door assembly and the wall-framing method.

Exercises

1. **Draw a plan section view of the wall shown in Fig. 54-24.**
2. **Identify the framing members shown in Fig. 54-28.**
3. **Draw a wall section, using Fig. 54-28 as a guide. Plan to use stone veneer.**
4. **Prepare an exterior panel-framing plan for a home of your own design.**
5. **Prepare an exterior panel-framing plan for your own home.**
6. **Define the following terms: *curtain wall, prefabrication, panel, stud, cripples, header, trimmer, plate, framing, weight, corner post, siding, top plate, rough opening, head section, sill section, jamb section, curtain wall.***

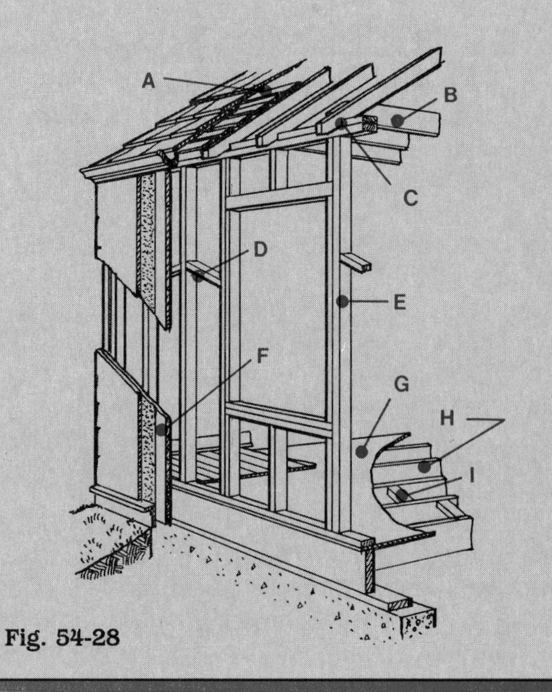

Fig. 54-28

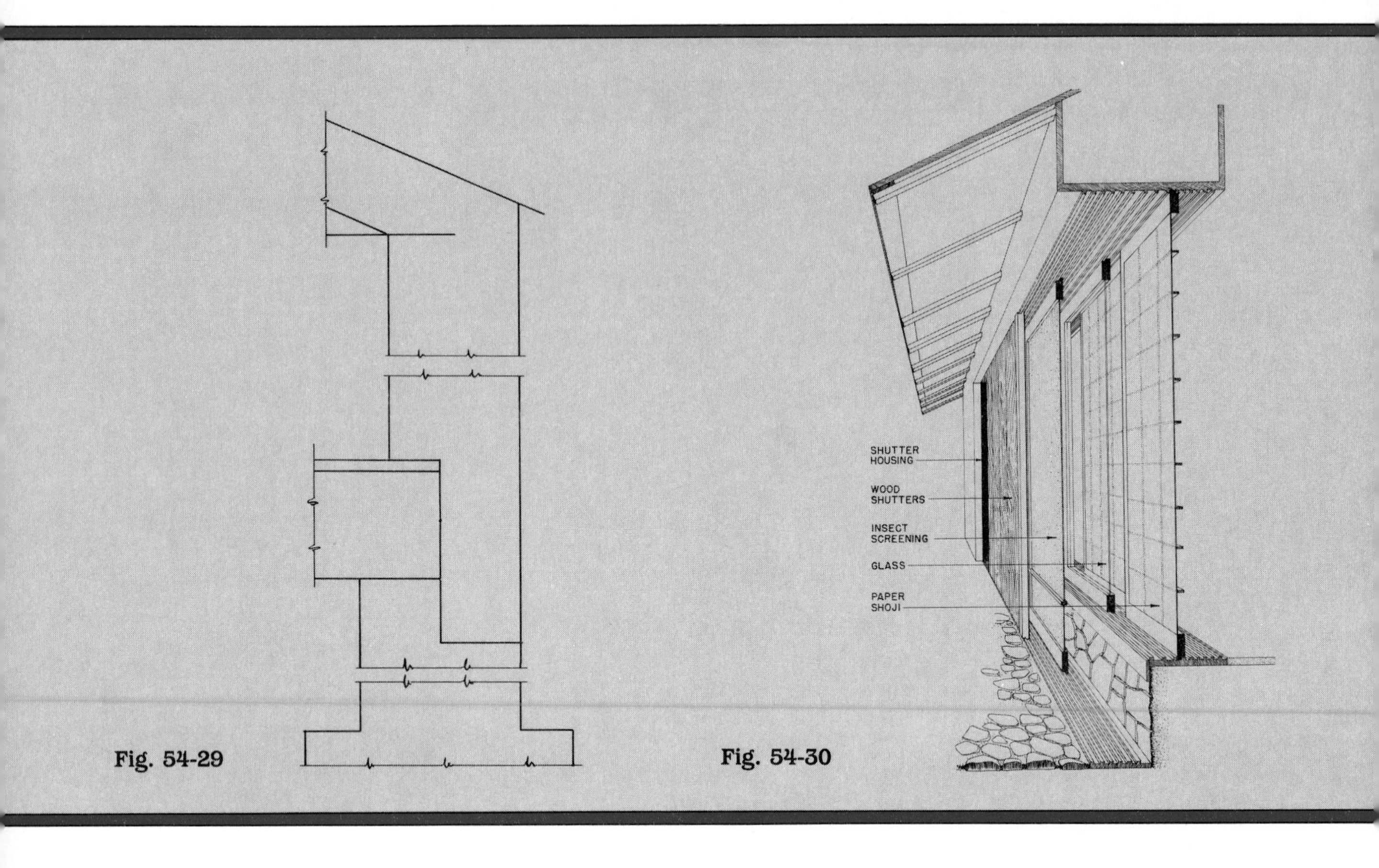

Fig. 54-29

Fig. 54-30

UNIT 55

INTERIOR-WALL FRAMING PLANS

Interior-framing drawings include plan, elevation, and pictorial drawings of partitions and wall coverings. Detail drawings of interior partitions may show intersections between walls and ceilings, floors, windows, and doors, or they may only show rough framing as in Fig. 55-1A.

PARTITION-FRAMING PLANS

Interior partition-framing elevations are most effective in showing the construction of interior partitions. Interior partitions are projected from the partition on the floor plan in a manner similar to the projection of exterior partitions. To ensure the correct interpretation of the partition elevation, each interior-elevation drawing should include a label indicating the room and compass direction of the wall. For example, the elevation shown in Fig. 55-1B should be labeled *North Wall Living Room*. If either the room name or the compass direction is omitted, the elevation may be misinterpreted and confused with a similar wall in another room. The elevation drawing is always projected from the room it represents.

A complete study of the floor plan, elevation, plumbing diagrams, and electrical plans should be made prior to the preparation of the interior wall-framing drawings. Provision must be made in the framing drawings for soil stacks and other large plumbing facilities and for special electrical equipment (Fig. 55-2). When a stud must be broken to accommodate various items, the framing drawing must show the recommended construction (Figs. 55-3 and 55-4). A structural stud should never have more than half its thickness removed.

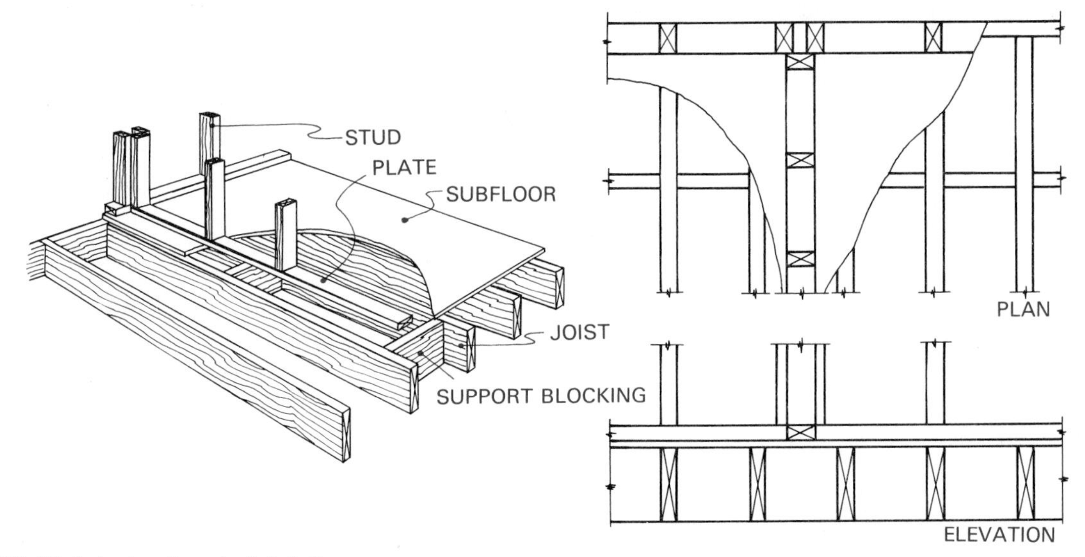

Fig. 55-1A Interior framing detail.

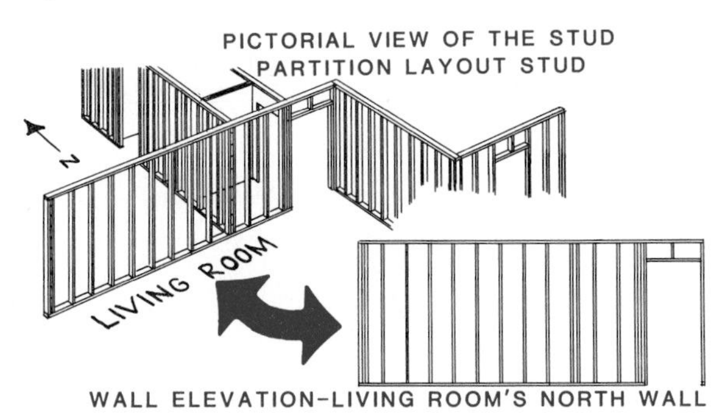

Fig. 55-1B Panel elevation of an interior wall.

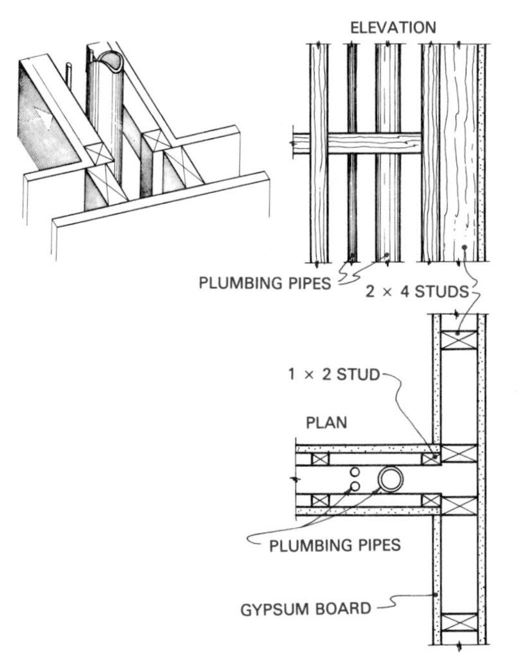

Fig. 55-2 Space must be allowed for plumbing and electrical equipment.

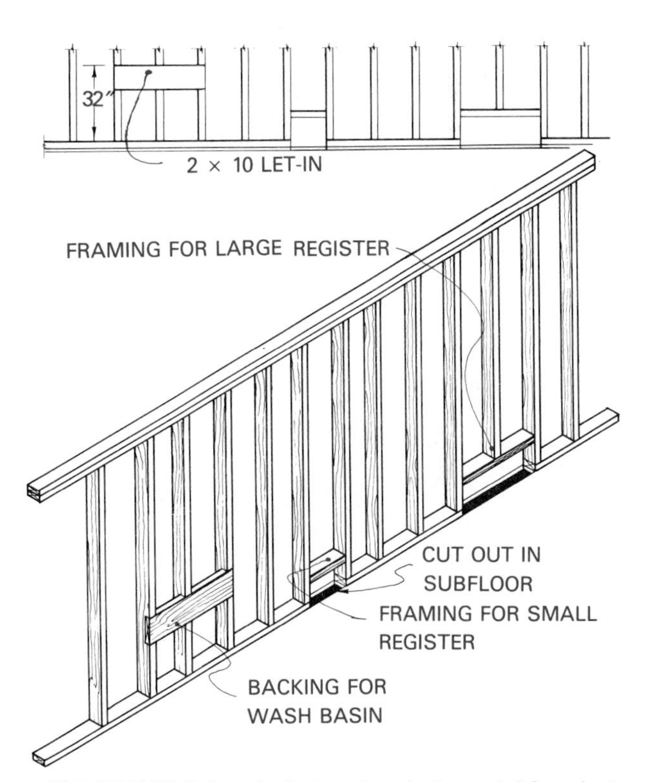

Fig. 55-3 Wall-framing plan showing special framing needs.

WALL-COVERING DETAILS

Basic types of wall-covering materials used for finished interior walls include plaster, drywall construction, paneling, tile, and masonry.

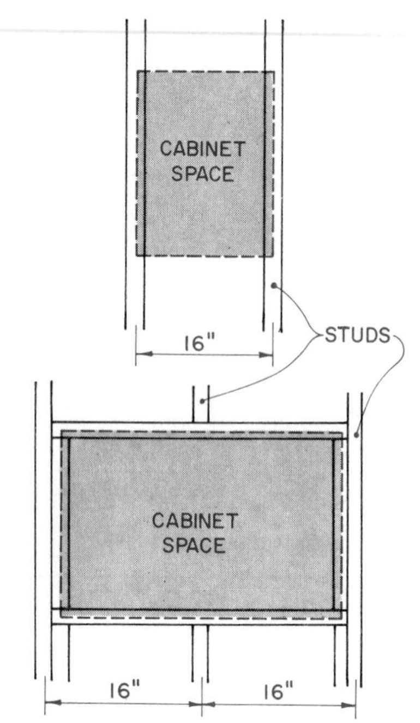

Fig. 55-4 Framing-elevation details show provisions for built-in items.

Plaster

Plaster is applied to interior walls by using wire lath or gypsum sheet lath, as shown in Fig. 55-5. Plaster walls are very strong and sound-absorbing. Plaster is also decay-proof and termite-proof. However, plaster walls crack easily and take months to dry. Also, installation costs are high.

Drywall Construction

Materials applicable to drywall construction include fiber boards, gypsum wallboards, sheet-rock, and plywood. Dry-wall wallboards are normally nailed directly to the studs, as shown in Fig. 55-6. When this construction is used, furring strips may be placed over the joints and nail holes. However, the more common practice is to camouflage the joints by sanding a depression in the wallboard and applying a perforated tape covered with Swedish putty and sanded smooth, as shown in Fig. 55-7.

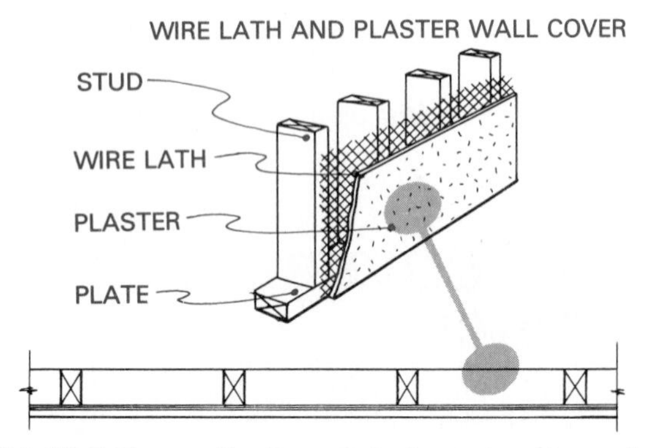

Fig. 55-5 The application of plaster to interior walls.

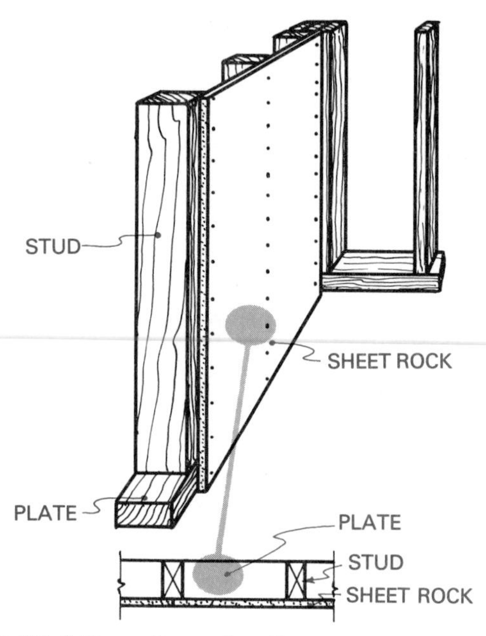

Fig. 55-6 Drywall construction.

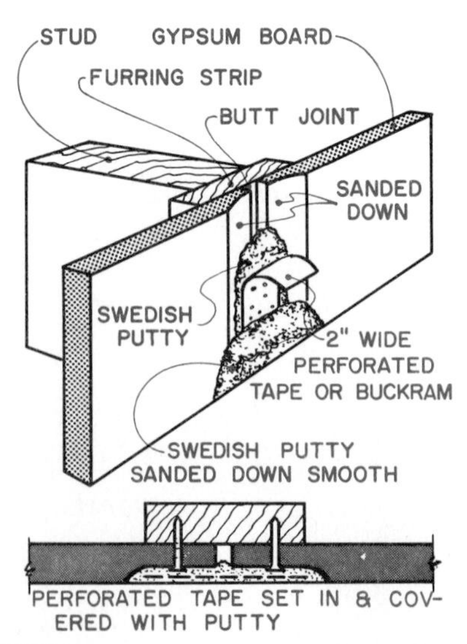

Fig. 55-7 The most common method of concealing drywall joints.

Paneling

When paneling is used as an interior finish, horizontal furring strips should be placed on the studs to provide a nailing or gluing surface for the paneling (Fig. 55-8). Determining the type of joint that should be used between panels is a design problem that should be solved through the use of a separate detail, as shown in Fig. 55-9. The joint may be exposed by use of the butt joint or cross-lap joint. A series of furring strips can be used between the joints or on the outside of them

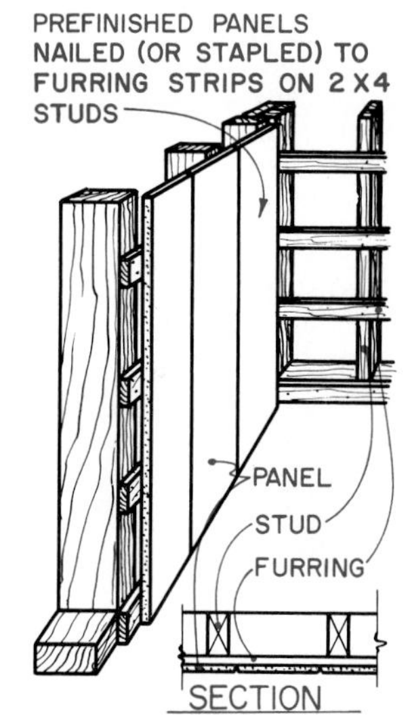

Fig. 55-8 Furring strips provide a horizontal surface for attaching paneling.

The method of intersecting the outside corners of paneling must also be detailed. Outside corners can be intersected by mitering or overlapping and exposing the paneling. Corner boards, metal strips, or molding may be used on the intersection (Fig. 55-10). Inside-corner intersections can be constructed as shown in Fig. 55-11.

BASE INTERSECTIONS

The method of intersecting the finished wall materials and the floor should be detailed. The details may be a section or a pictorial drawing, as shown in Fig. 55-12. The position of the sole plate, wallboard or lath and plaster, baseboard, and molding should be shown.

CEILING INTERSECTIONS

Details should also be prepared to show the intersection between the ceiling and the wall. Details should show the position of the top plate, wallboard-ceiling finish, and position of molding used at the intersection (Fig. 55-13). Care should be taken when designing the base treatment and ceiling-intersection treatment to ensure that the intersections are consistent in style, as shown in Fig. 55-14.

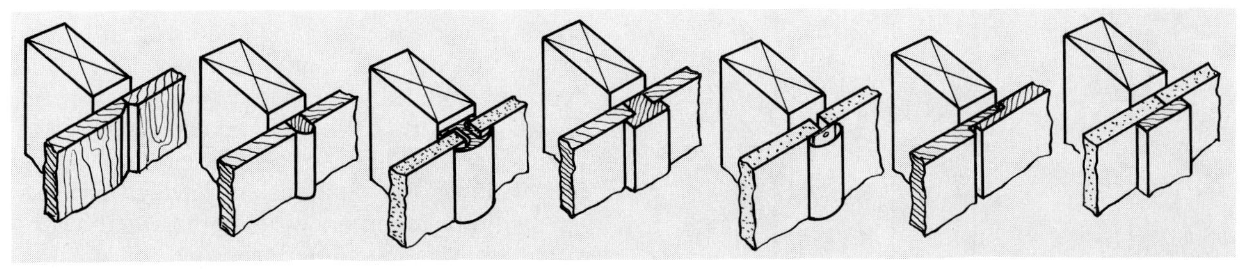

Fig. 55-9 Panel-joint details.

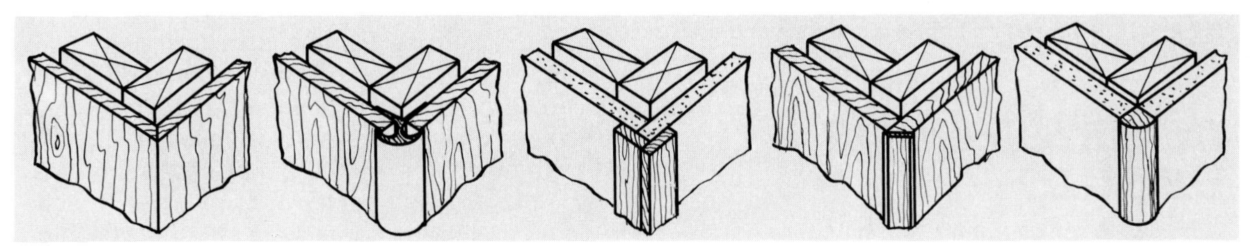

Fig. 55-10 Outside-corner panel joints.

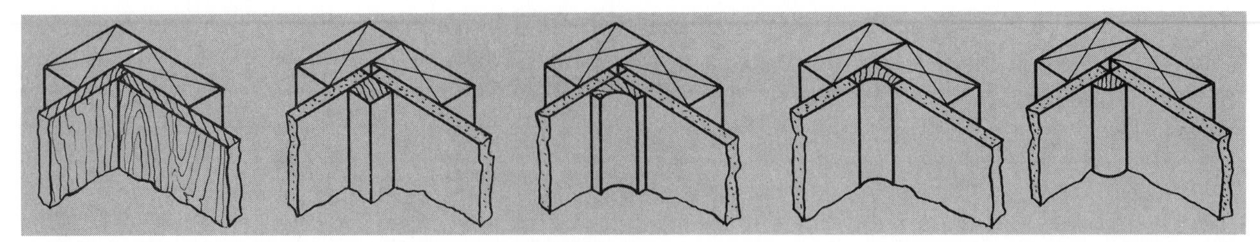

Fig. 55-11 Inside-corner panel joints.

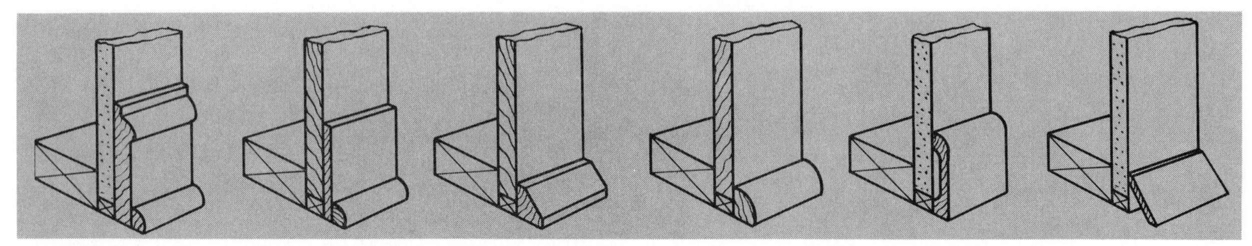

Fig. 55-12 A method of intersecting the panel wall with the floor.

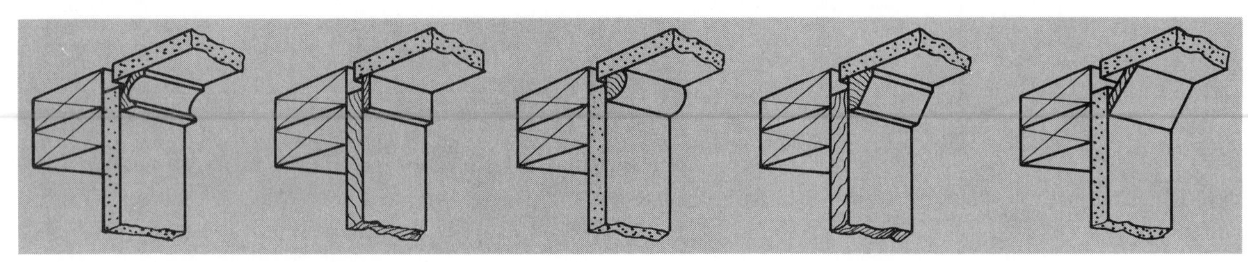

Fig. 55-13 A method of intersecting the panel wall with the ceiling.

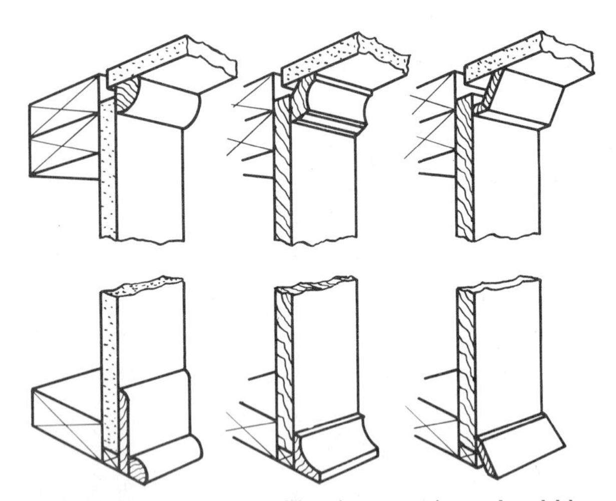

Fig. 55-14 Base and ceiling intersections should be consistent.

INTERIOR-DOOR DETAILS

Pictorial or orthographic jamb, sill, and head sections should be prepared to illustrate methods of framing used around interior doors. Figure 55-15 shows the methods of framing split-jamb, surface-mounted, bifolding, sliding-pocket, sliding-bypass, and folding doors. A detailed drawing need not be prepared for each door but should be prepared for each type of door used in the house and should be keyed to the door schedule for identification. Figure 55-16 shows the relationship between a pictorial section of an interior-door jamb and the variations of this section necessary for plaster, gypsum-board, or paneled wall coverings. Details are not usually necessary for the actual construction of a door, for this is an item that is normally outlined in the specifications. However, it is important to select a proper door from manufacturers' specifications or to prepare a detailed drawing to ensure compliance with minimum standards. Figure 55-17 shows the cutaway drawing and section of two types of solid doors. Hollow-core doors are generally used on interior partitions.

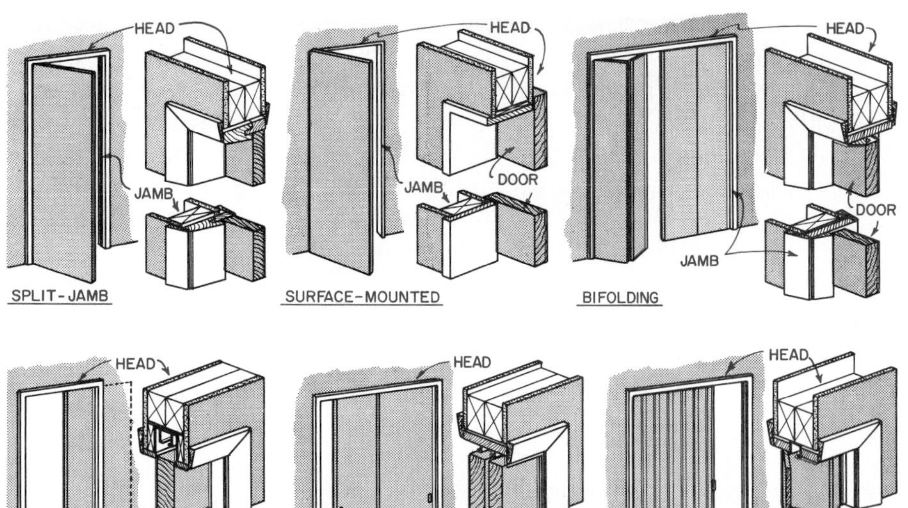

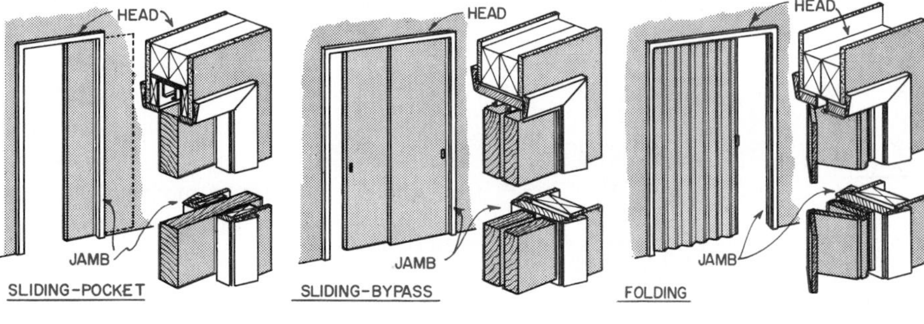

Fig. 55-15 The framing methods used for different types of interior doors.

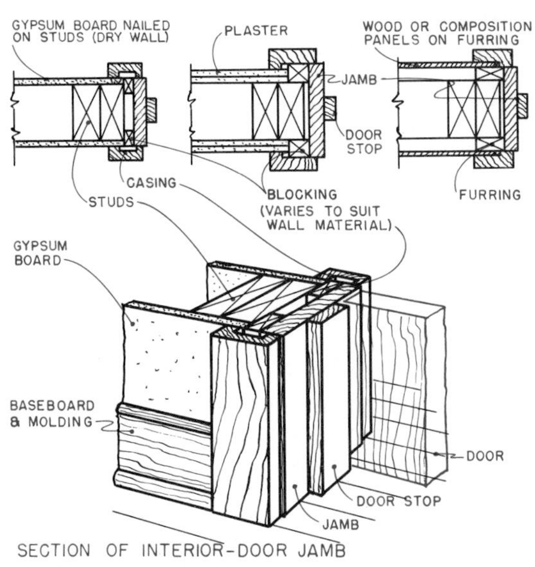

Fig. 55-16 The door-framing methods needed for different types of construction.

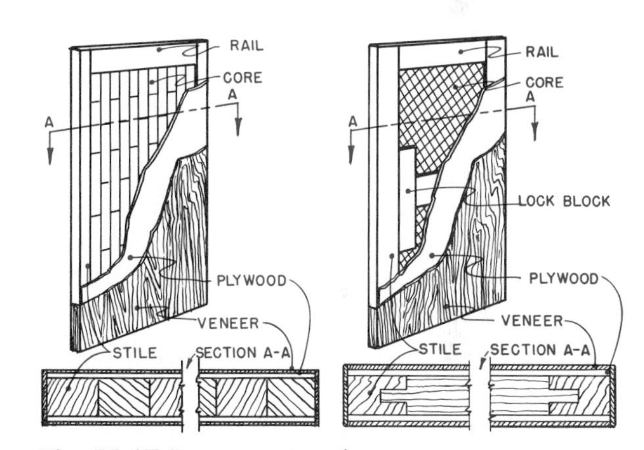

Fig. 55-17 Door construction.

Exercises

1. **Project a panel-framing elevation drawing of the plumbing wall of the master bedroom bathroom, as shown in Fig. 37-14.**
2. **Draw a vertical section of the wall shown in Fig. 55-8.**
3. **Draw plan and elevation sections of one of the joints shown in Fig. 55-9.**
4. **Draw plan sections for one of the intersections shown in Fig. 55-10 and Fig. 55-11.**
5. **Draw a typical interior-wall framing plan for a kitchen, bath, and living-area wall of the house you are designing. Use as many detail drawings**

as you think necessary to support the framing plans.

6. Draw a complete horizontal wall section of the rear dining room wall shown in Fig. 7-4. As part of the fireplace treatment, plan to make the entire wall brick from the fireplace to the door opening.
7. Prepare a plan section similar to the plan shown in Fig. 55-16. Show the position of brick, blocking, casing, and jamb.
8. Define the following terms: *drywall, plaster, paneling, base detail, interior partition, furring strip, wall intersection.*

UNIT 56

STUD LAYOUTS

A *stud layout* is a plan similar to a floor plan, showing the position of each wall-framing member. The stud layout is a section through each interior elevation drawing, as shown in Fig. 56-1A. Figure 56-1B shows an exploded view of a wall intersection and the exposed stud sections. The cutting-plane line for purposes of projecting the stud layout is placed approximately at the midpoint of the panel elevation. Figure 56-2 shows a stud layout that represents the framing plan of the panel.

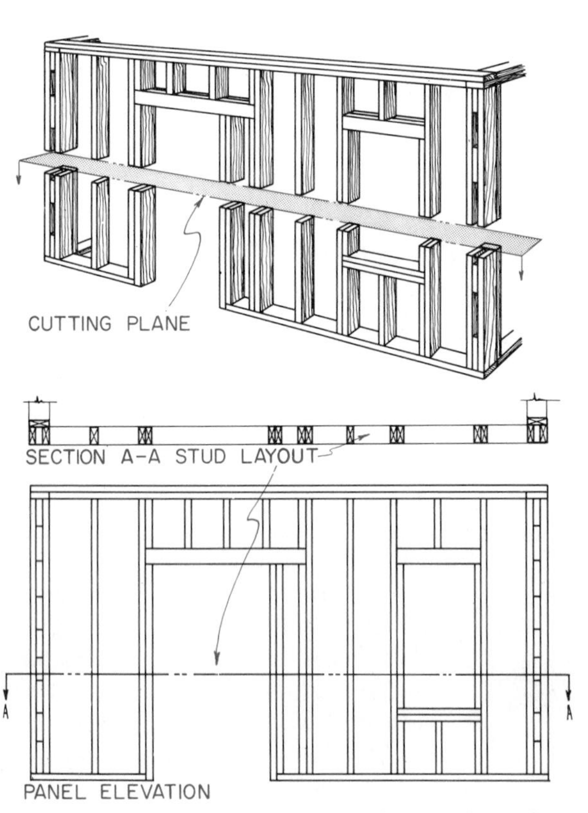

Fig. 56-1A The stud layout is a plan section taken through the panel elevation.

STUD DETAILS

Stud layouts are of two types: the *complete plan,* which shows the position of all framing members on the floor plan, and the *stud detail,* which

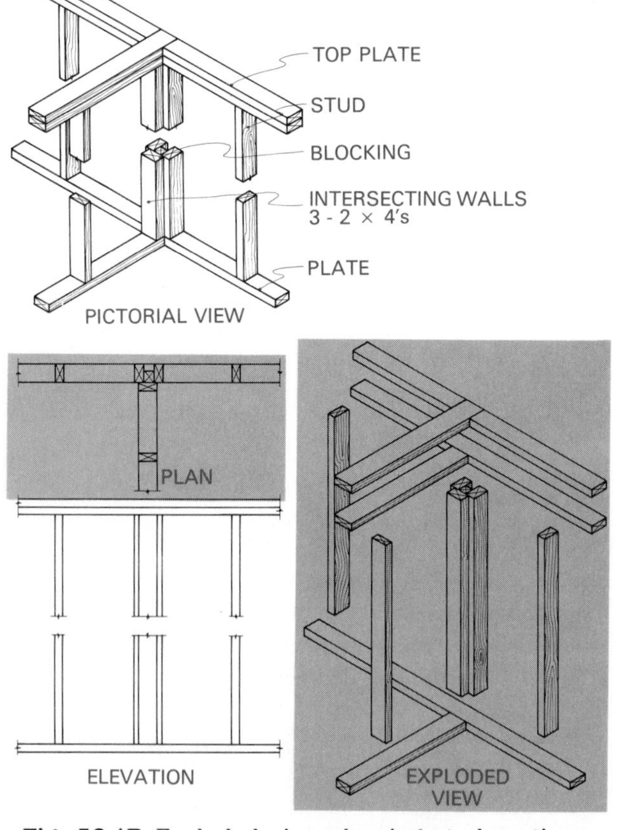

Fig. 56-1B Exploded view showing stud sections.

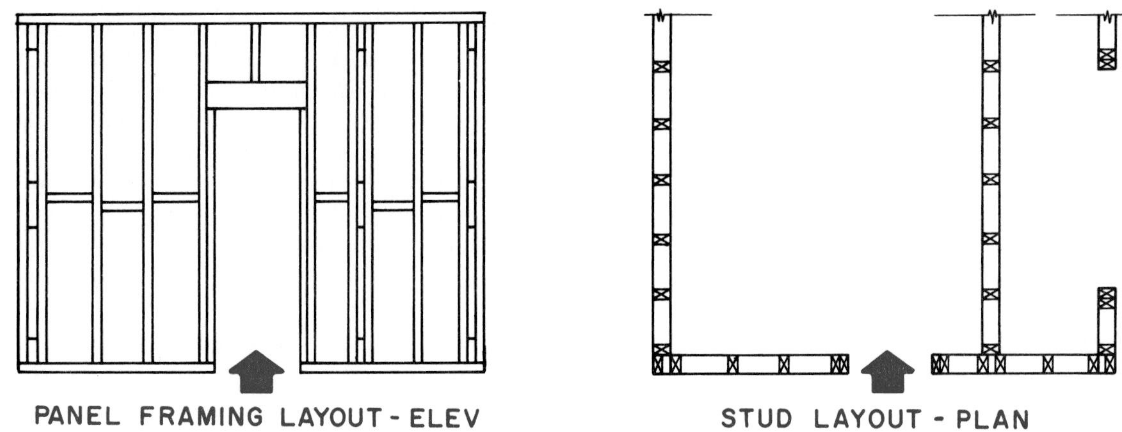

Fig. 56-2 The relationship of a stud layout to a panel elevation.

shows only the position and relationship of several studs or framing intersections.

CORNER POST

The position of each stud in a corner-post layout is frequently shown in a plan view, as illustrated in Fig. 56-3. Occasionally, siding and inside-wall covering materials are shown on this plan. Preparing this type of detail without covering materials is the quickest way to show corner-post construction.

PARTITION INTERSECTIONS

Details of the exact position of each stud and blocking in an intersection are shown by a plan section (Fig. 56-4). If wall coverings are shown on the detail, the complete wall thickness can be drawn (Fig. 56-5A). In preparing the plan section, care should be taken to show the exact position of blocking or short pieces of stud stock that may not pass through the cutting-plane line. Blocking should be labeled to prevent the possibility of identifying the blocking as a full-length stud. When laying out the position of all studs, remember that the finished dimensions of a 2 × 4 stud are actually 1½″ × 3½″. The exact dressed sizes of other rough stock are as follows:

Rough	Dressed
2 × 3	1½ × 2½
2 × 4	1½ × 3½
2 × 6	1½ × 5½
2 × 8	1½ × 7¼
2 × 10	1½ × 9¼
2 × 12	1½ × 11¼
1 × 6	¾ × 5½
1 × 8	¾ × 7¼

COMPLETE PLAN

To conserve space where full partition width is not important, as between closets, studs are sometimes turned so they are flat. This rotation should be reflected in the stud layout (Fig. 56-5B).

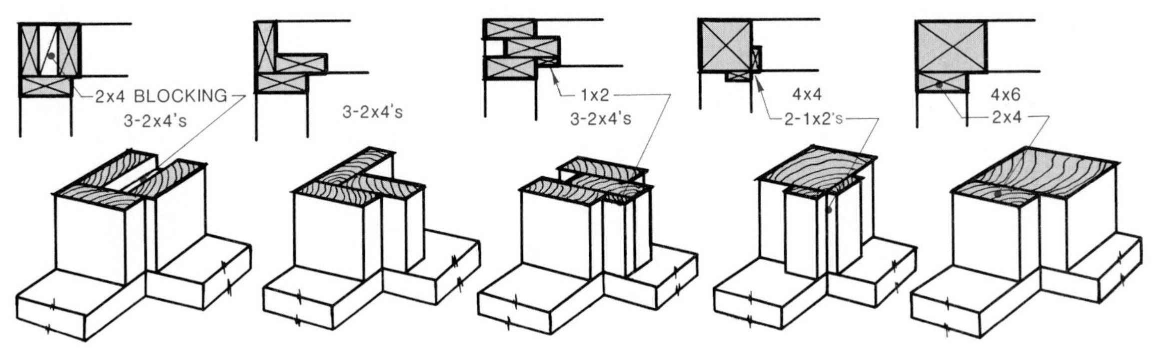

Fig. 56-3 Stud details of several corner-post layouts.

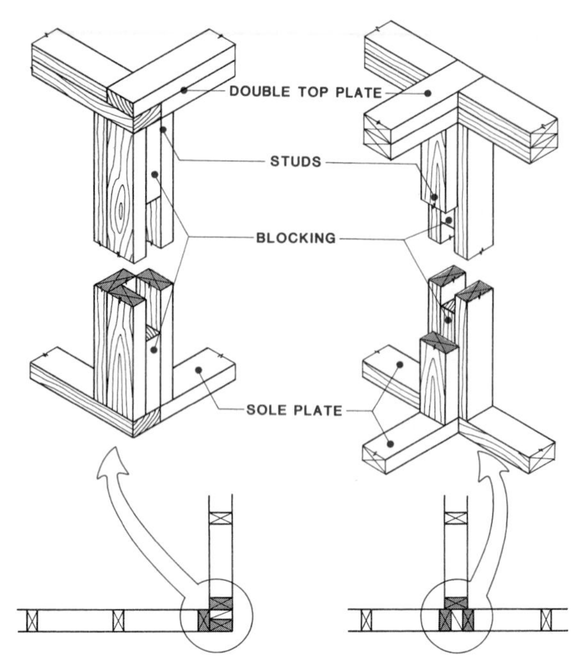

Fig. 56-4 Position of studs in a partition intersection.

The main purpose of a stud layout is to show how interior partitions fit together and how studs are spaced on the plan. Figure 56-6 shows part of a stud layout. The outline of the plate and the exact position of each stud that falls on an established center (16″, 32″, or 48″) are normally identified by diagonal lines. Studs other than those that are on regular centers are shown by different symbols. Studs that are short, blocking, or different in size are identified by a different key (Fig. 56-7). Using a coding system of this type eliminates the need for dimensioning the position of each stud if it is part of the regular partitioned pattern. The practice of coding studs and other members shown on the stud plan also eliminates the need for showing detailed dimensions of each stud. Detailed dimensions are normally shown on the key or on a separate enlarged detail. Distances that are dimensioned on a stud layout include the following:

1. Inside framing dimensions of each room
2. Framing width of the halls
3. Rough openings for doors and arches
4. Length of each partition
5. Width of partition where dimension lines pass through from room to room. (This provides a double check to ensure that the room dimensions plus the partitioned dimensions add up to the overall dimension.) When a stud layout is available, it is used on the job to establish partition positions. Figure 56-8 shows the application of these dimensional practices to a typical stud layout plan.

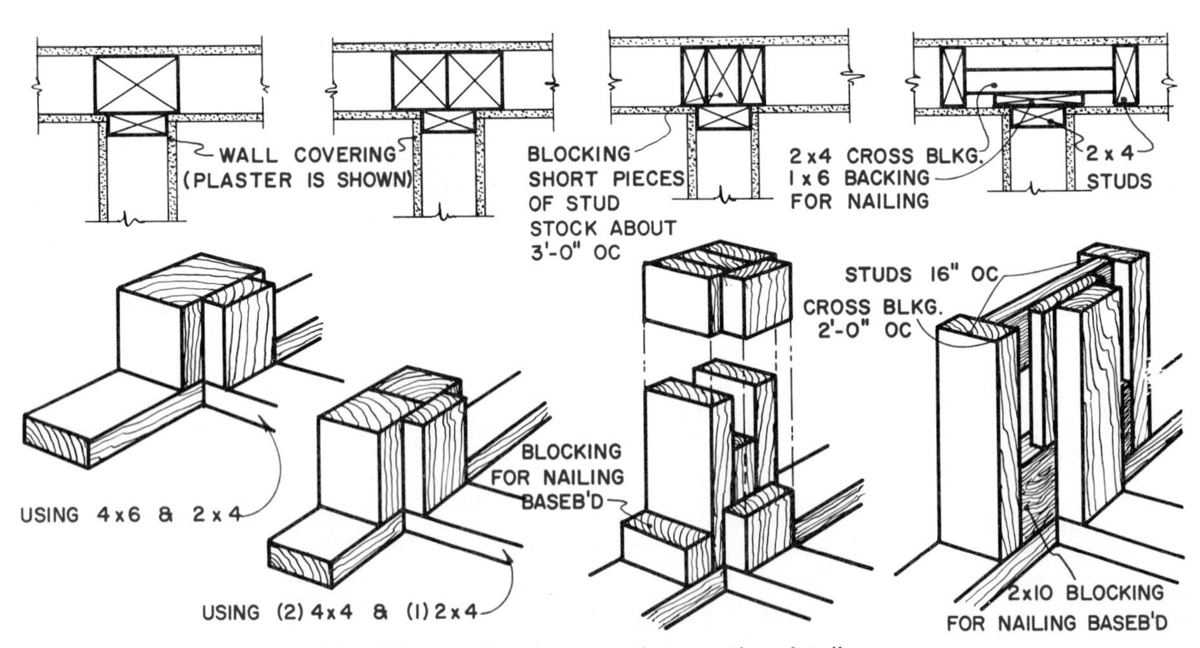

Fig. 56-5A Wall covering and blocking can be shown on intersection details.

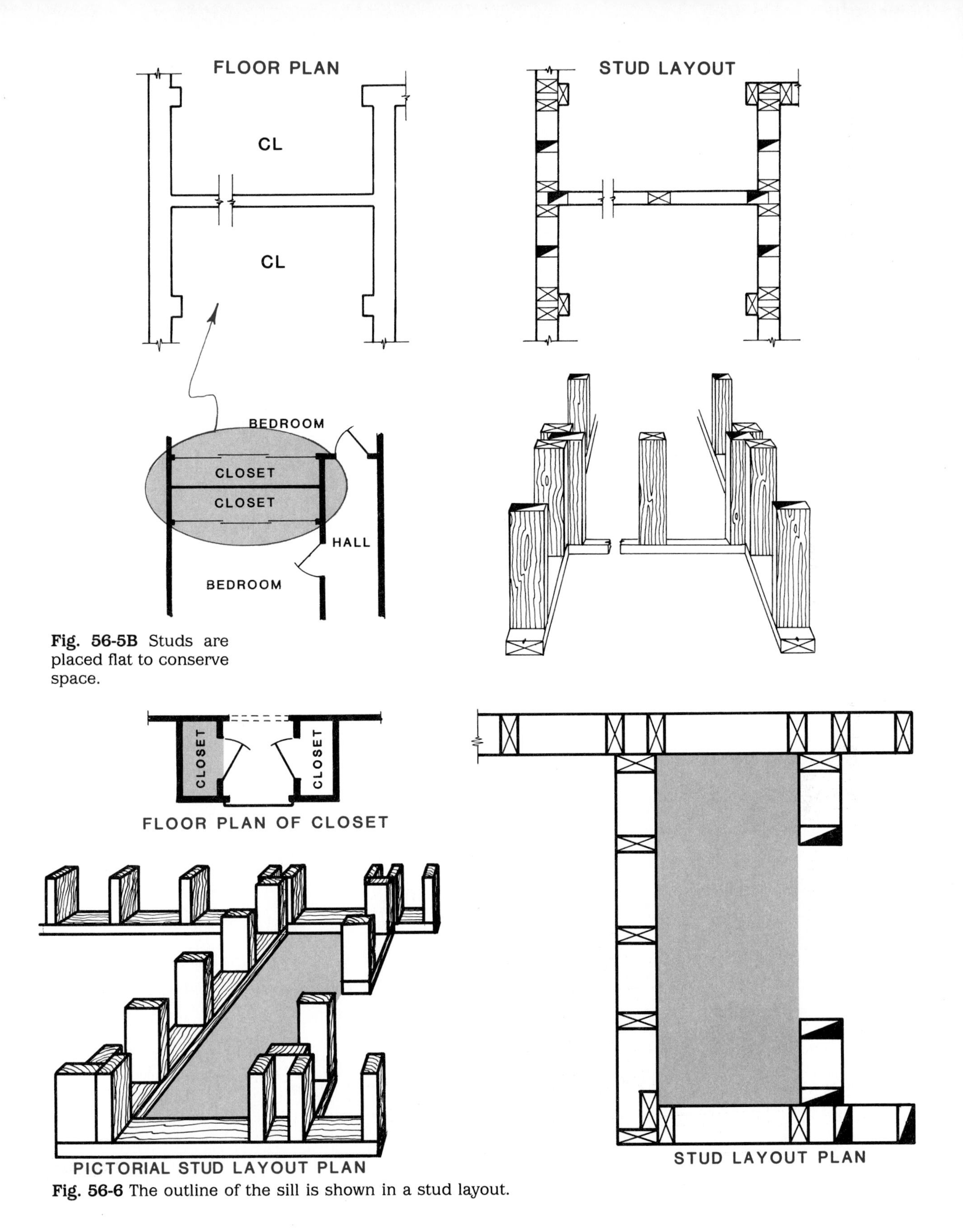

Fig. 56-5B Studs are placed flat to conserve space.

Fig. 56-6 The outline of the sill is shown in a stud layout.

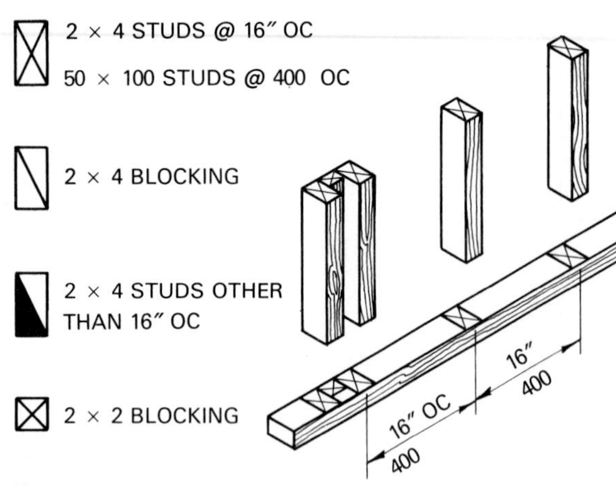

Fig. 56-7 Stud-layout symbols.

When stud layouts are prepared for modular plans modular grids as shown in Fig. 56-9 are used. Figure 56-10 shows some of the complexities of modular design of intersections. Partitions must fit between the basic exterior modular increments with allowances for exterior wall thicknesses.

Studs placed on 16″ centers will only align with every other 24″ modular grid since increments of 24″ and 48″ are used in most modular planning. In detailing a modular stud layout, space must be provided for such items as backup members, door and window placement, plumbing runs, medicine cabinets, closets, fireplace, and so forth.

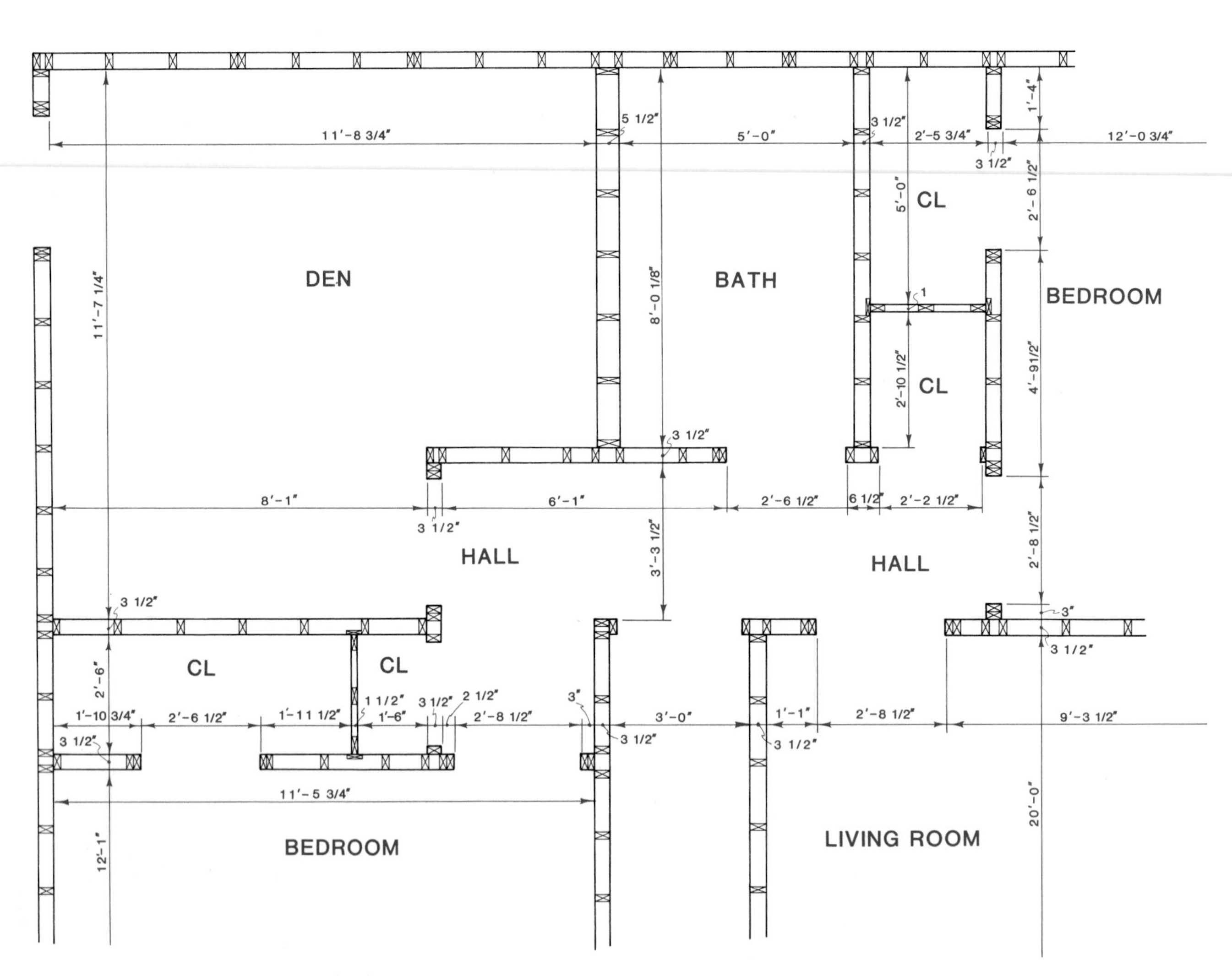

Fig. 56-8 Stud-layout dimensioning.

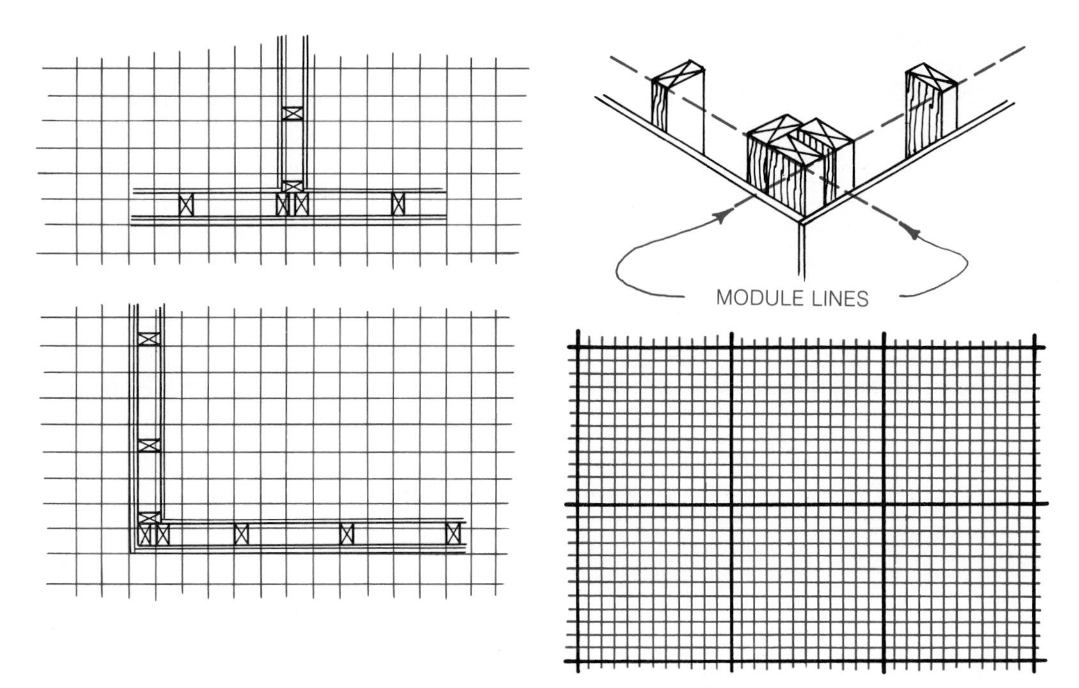

Fig. 56-9 Stud layout on modular grids.

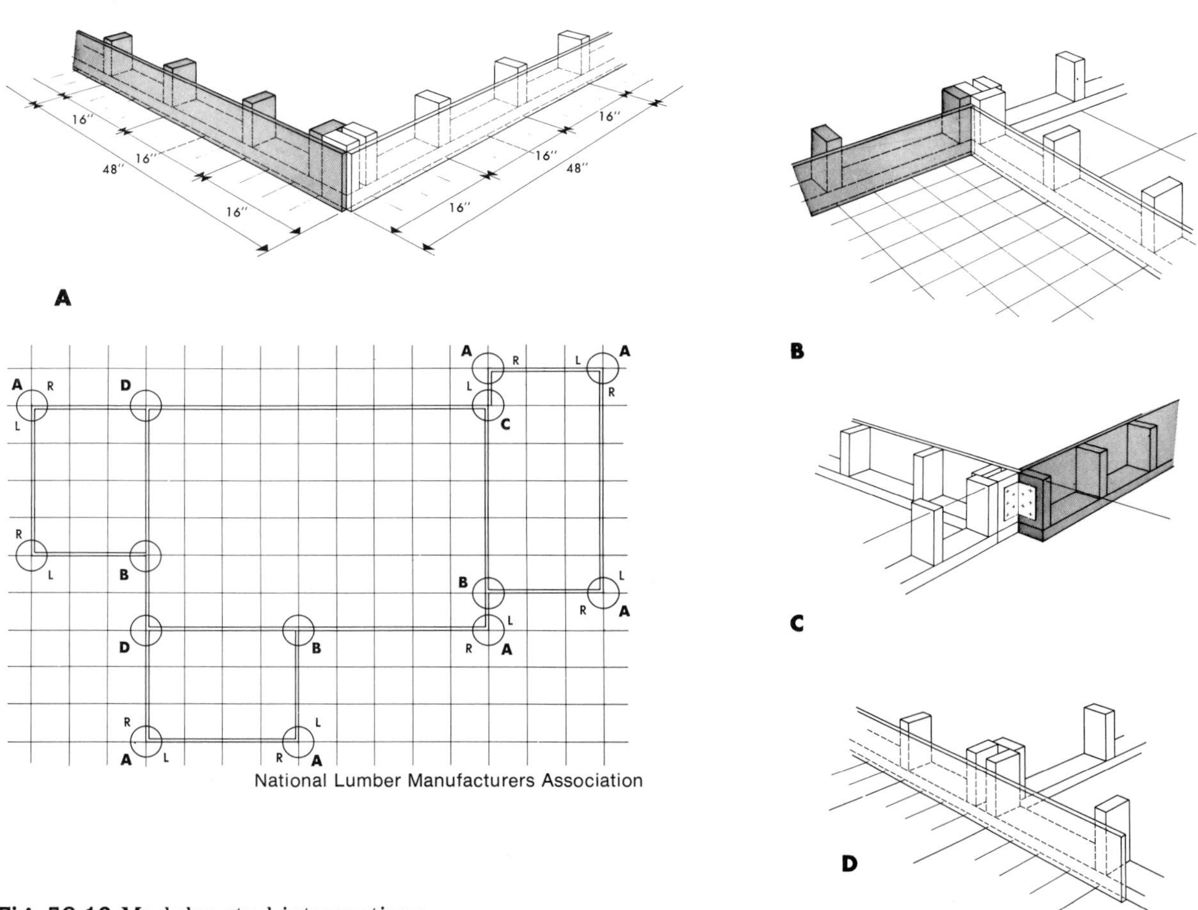

Fig. 56-10 Modular stud intersections.

Exercises

1. **Prepare a stud layout of the floor plan shown in Fig. 33-12.**
2. **Prepare a corner-post detail for the corner posts shown in Fig. 34-6.**
3. **Prepare a stud layout (1/4" = 1'-0") for a home of your own design.**
4. **Draw a stud layout of the east living-room partition shown in Fig. 34-3.**
5. **Define the following terms: *stud layout, stud detail, corner post, rough lumber, dressed lumber, blocking, inside framing dimensions.***

UNIT 57

ROOF FRAMING PLANS

The first structure made for shelter was probably a lean-to roof supported by posts. As structures became larger and more complex, the composition and shape of the roof also changed.

The main function of a roof is to provide protection from rain, snow, sun, and very hot or cold temperatures. The roof of a northern building is designed to withstand heavy snow loads. The thatched roof of a tropical native hut provides protection only from sun and rain. As roof styles developed through the centuries, *pitches* (angles) were changed, gutters and downspouts were added for better drainage, and overhangs were extended to provide more protection from the sun. As the size of roofs changed, the size and types of material changed accordingly. In any building, the roof is an integral part of the total design.

STRUCTURAL DESIGN

The walls of the structure are given stability by their attachment to the ground and to the roof. Most buildings are not structurally sound without roofs. Walls cannot resist outside or inside forces unless some horizontal support (roof) is given to the upper part of the wall. This principle is shown in Fig. 57-1.

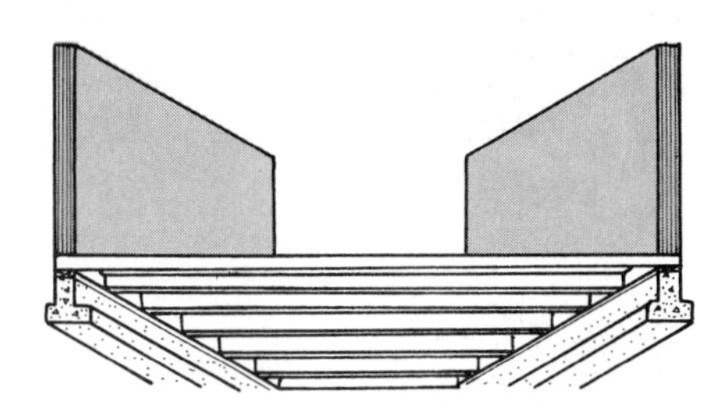

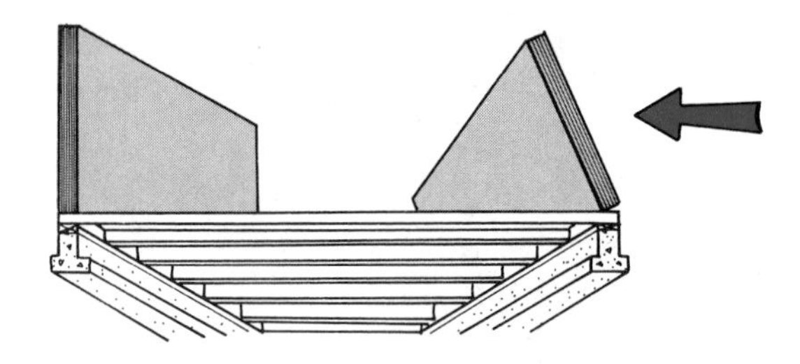

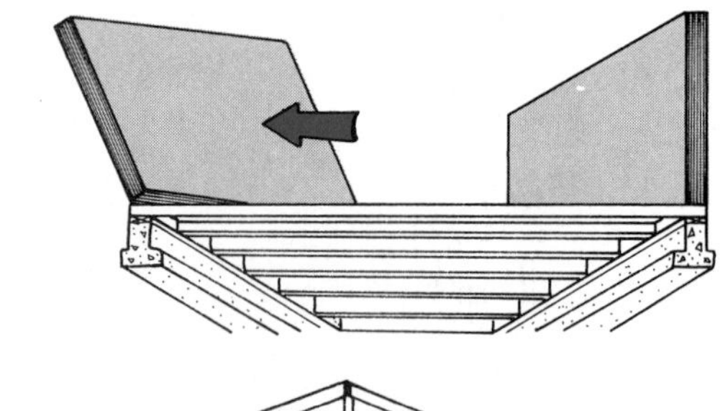

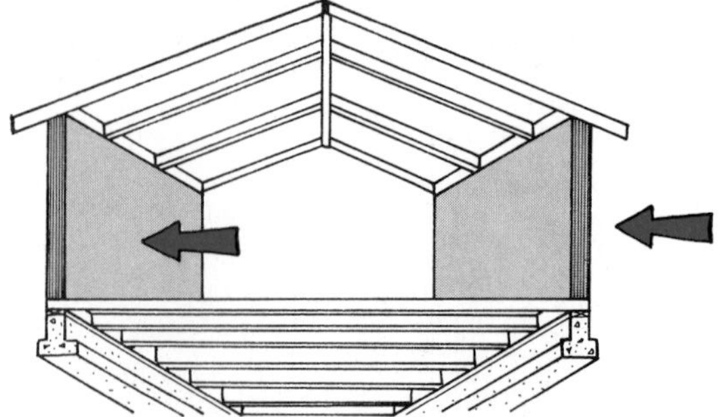

Fig. 57-1 A roof adds stability to a structure.

SUPPORT

The weight of the roof is normally supported by the exterior walls of the structure (Fig. 57-2). Roofs may be supported by a combination of the exterior walls plus the load-bearing partitions or beams (Fig. 57-3). In a frame or continuous-arch design, the roof is supported by direct connection with the foundation (Fig. 57-4).

LOADS

The structural members of a roof must be sufficiently strong to withstand the loads that bear upon it.

Dead Loads

Dead loads that bear upon most roofs include the weight of shingles, sheathing, and rafters.

All loads are computed on the basis of pounds per square foot, or kilograms per square meter if metric measurements are used. The typical asphalt-shingle roof weighs approximately 10 lb/ft^2 (Fig. 57-5), and a typical asbestos roof weighs approximately 12 lb/ft^2. A Spanish tile roof weighs 17 lb/ft^2. Thus a 40′ x 20′ (800 ft^2) asphalt-shingle roof would be designed to carry an 8000-lb load (800 ft^2 x 10 lb/ft^2).

Live Loads

Live loads that act on the roof include wind loads and snow loads, which vary greatly from one geographical area to another. For example, the combined wind and snow loads in the South Pacific are approximately 20 lb/ft^2. In the central and western parts of the United States, these loads are 30 lb/ft^2, and in the northeastern and northwestern parts of the United States, they are 40 lb/ft^2 (Fig. 57-6).

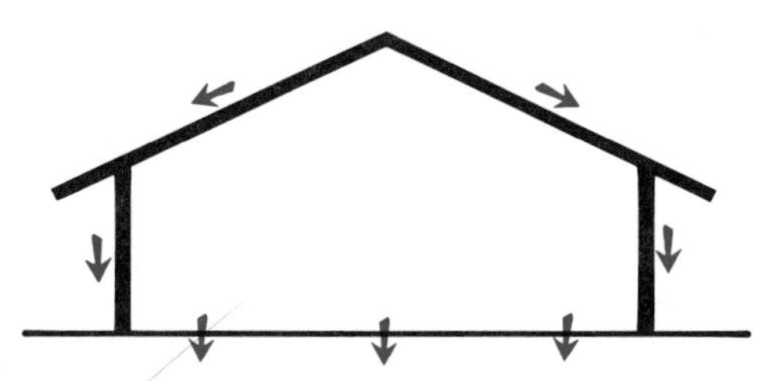

Fig. 57-2 The weight of the roof is transmitted to the outside walls.

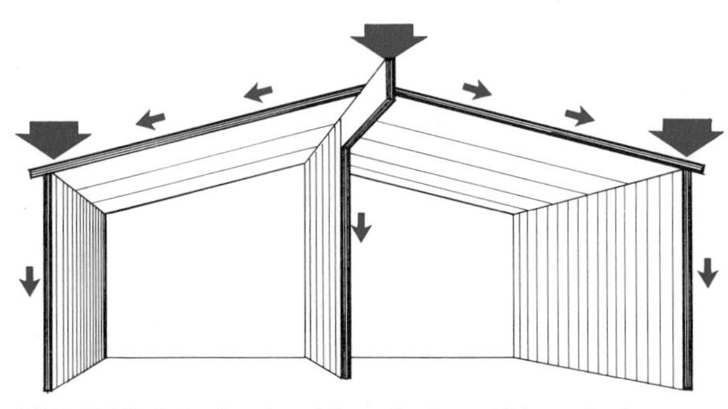

Fig. 57-3 Interior load-bearing partitions help support roofs.

Fig. 57-4 A roof can be connected directly to the foundation.

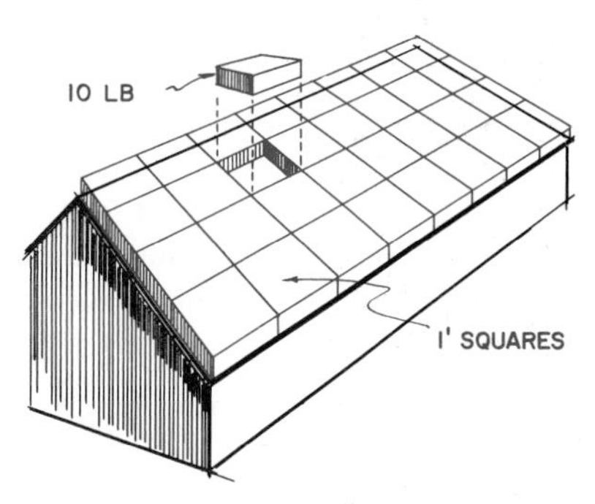

Fig. 57-5 Roof loads are measured in pounds per square foot.

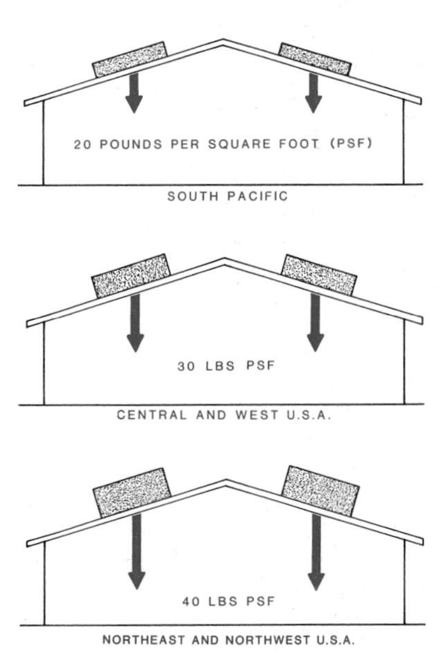

Fig. 57-6 Live roof loads vary from region to region.

Snow and wind loads vary greatly as the pitch of the roof is changed. Snow loads are exerted in a vertical direction; wind loads are exerted in a horizontal direction. A high-pitch roof will withstand snow loads better than a low-pitch roof (Fig. 57-7). The reverse is true of wind loads. There is virtually no wind load exerted on a flat roof, and moderate wind loads (15 lb/ft^2) are exerted on a low-pitch roof. Approximately 35 lb/ft^2 is exerted on a high-pitch roof. An excessively resistant wind load is exerted on a completely vertical wall, approximately 40 lb/ft^2, which is equivalent to hurricane force, as shown in Fig. 57-8. For all practical purposes, snow and wind loads are combined in one total live load. Live loads and dead loads are then combined in the total load acting on the roof.

ROOF FRAMING TYPES

The conventional method of roof framing consists of roof rafters or trusses spaced at small intervals such as 16″ (406 mm) on center. These roof rafters run perpendicular to the ridge board and align with the partition studs placed on the same centers (Fig. 57-9). The second method is the post-and-beam or plank-and-beam (Fig. 57-10). The plank-and-beam roof consists of posts supporting beams that run parallel with the ridge

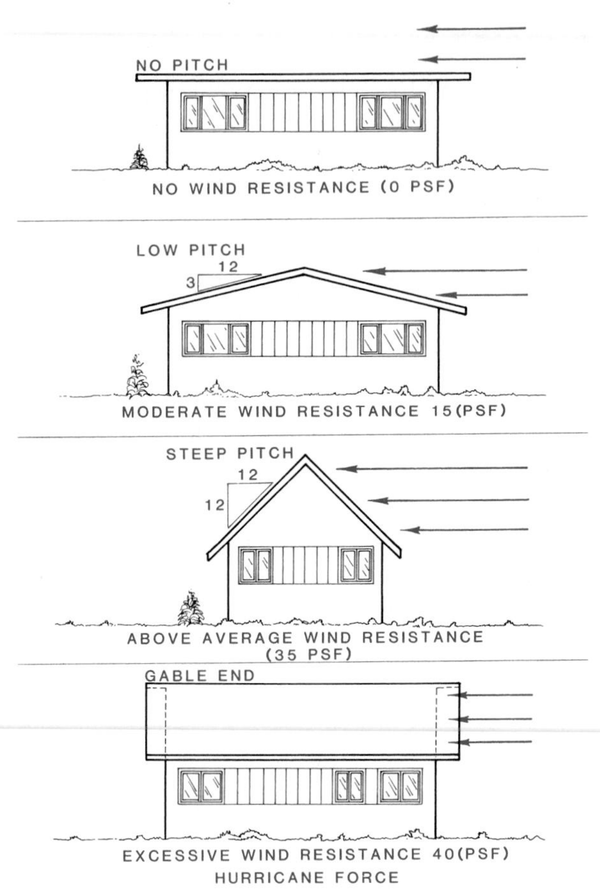

Fig. 57-8 High-pitch roofs contribute to high wind loads.

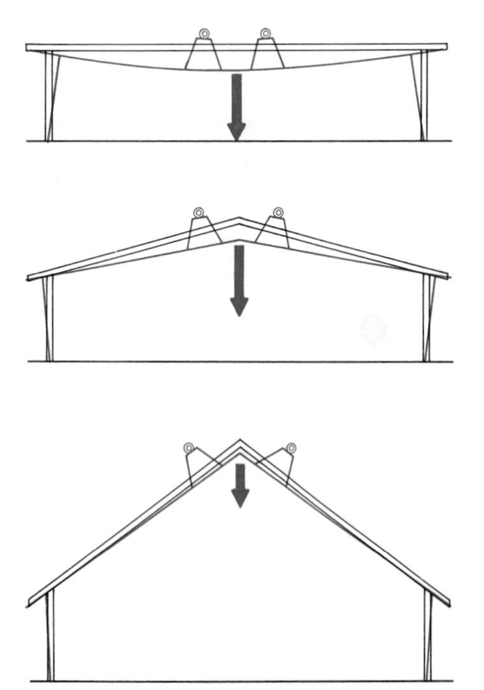

Fig. 57-7 Low-pitch roofs need heavier support or shorter spans to withstand snow and wind loads.

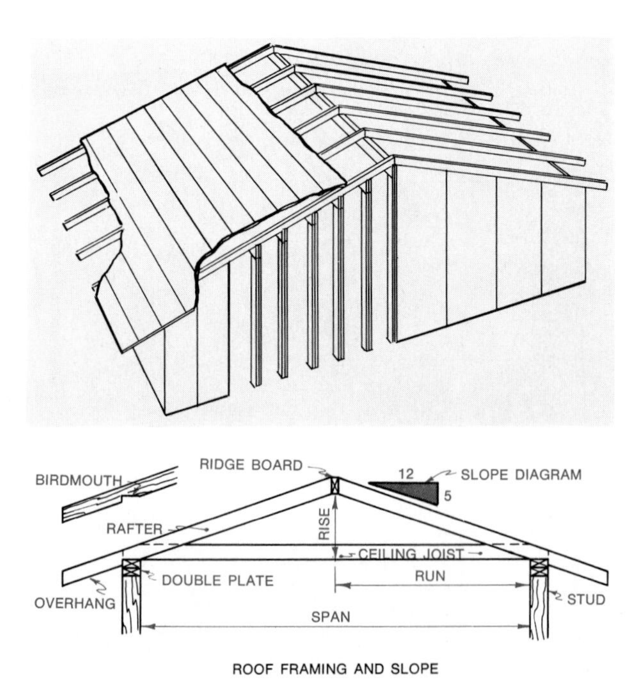

Fig. 57-9 Conventional roof framing.

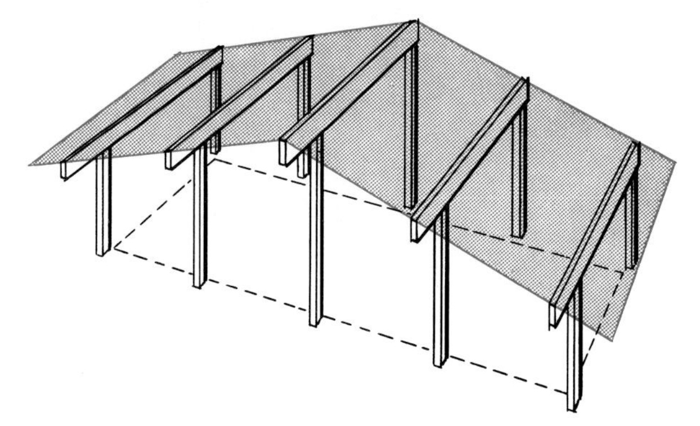

Fig. 57-10 Plank-and-beam framing.

Size of ceiling joists	Spacing of ceiling joists	Maximum span			
		Wood Group 1	Wood Group 2	Wood Group 3	Wood Group 4
2″ × 4″	12″	11′-6″	11′-0″	9′-6″	5′-6″
	16″	10′-6″	10′-0″	8′-6″	5′-0″
2″ × 6″	12″	18′-0″	16′-6″	15′-6″	12′-6″
	16″	16′-0″	15′-0″	14′-6″	11′-0″
2″ × 8″	12″	24′-0″	22′-6″	21′-0″	19′-0″
	16″	21′-6″	20′-6″	19′-0″	16′-6″

Fig. 57-11 Maximum span for wood ceiling joists.

board, or beam. A *ridge beam* is the center of a gable roof. Planks are then placed across the beams. Roof planking can then be used as a ceiling and a base for roofing that will shed water. Exposed plank-and-beam ceilings achieve a distinctive and pleasing architectural effect. When planks are selected for appearance, the only ceiling treatment needed is a desired finish on the planks and beams. Longitudinal beam sizes vary with the span and spacing of the beams. Design variations of end walls are achieved by an extensive use of glass and protecting roof overhangs. The use of larger members in post-and-beam construction allows the designer to plan larger open areas unobstructed by bearing partitions.

GABLE ROOF

Gable roofs are constructed with conventional rafters and ceiling joists. Figure 57-11 shows typical rafter sizes, spans, and spacing, and Fig. 57-12 shows the same information for ceiling joists. These are usually spaced 16″ on center and covered with sheathing, felt, and shingles (Fig. 57-13). An adaptation of this conventional method of constructing roofs is the use of roof trusses to replace the conventional rafters and ceiling joists (Fig. 57-14). Trusses provide a much more rigid roof but eliminate the use of a space between the joists and rafters for an attic or crawl-space storage. An increasingly popular gable construction is a prefabricated gable end.

Size rafter	Spacing rafter	Maximum span			
		Group 1	Group 2	Group 3	Group 4
2″ × 4″	12″	10′-0″	9′-0″	7′-0″	4′-0″
	16″	9′-0″	7′-6″	6′-0″	3′-6″
	24″	7′-6″	6′-6″	5′-0″	3′-0″
	32″	6′-6″	5′-6″	4′-6″	2′-6″
2″ × 6″	12″	17′-6″	15′-0″	12′-6″	9′-0″
	16″	15′-6″	13′-0″	11′-0″	8′-0″
	24″	12′-6″	11′-0″	9′-0″	6′-6″
	32″	11′-0″	9′-6″	8′-0″	5′-6″
2″ × 8″	12″	23′-0″	20′-0″	17′-0″	13′-0″
	16″	20′-0″	18′-0″	15′-0″	11′-6″
	24″	17′-0″	15′-0″	12′-6″	9′-6″
	32″	14′-6″	13′-0″	11′-0″	8′-6″
2″ × 10″	12″	28′-6″	26′-6″	22′-0″	17′-6″
	16″	25′-6″	23′-6″	19′-6″	15′-6″
	24″	21′-0″	19′-6″	16′-0″	12′-6″
	32″	18′-6″	17′-0″	14′-0″	11′-0″

Fig. 57-12 Maximum rafter spans for slopes of 4/12 or more.

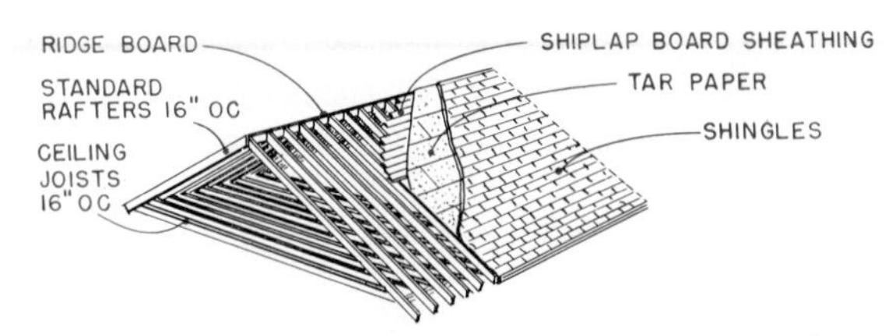

Fig. 57-13 A conventionally framed gable roof.

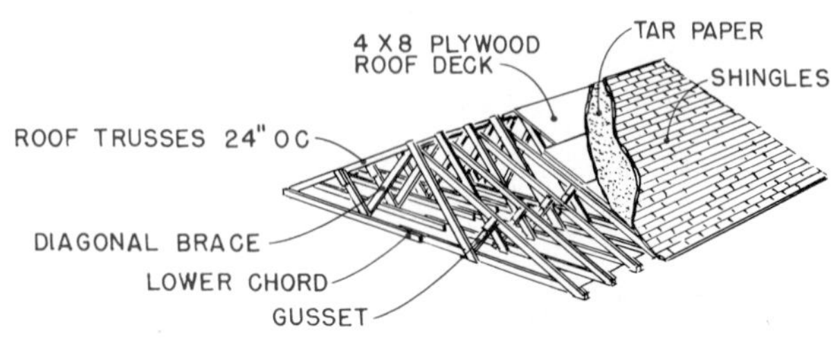

Fig. 57-14 A trussed gable roof.

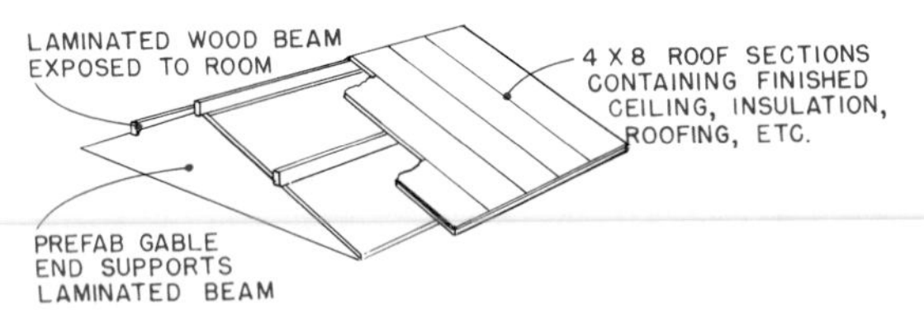

Fig. 57-15 A prefabricated post-and-beam gable roof.

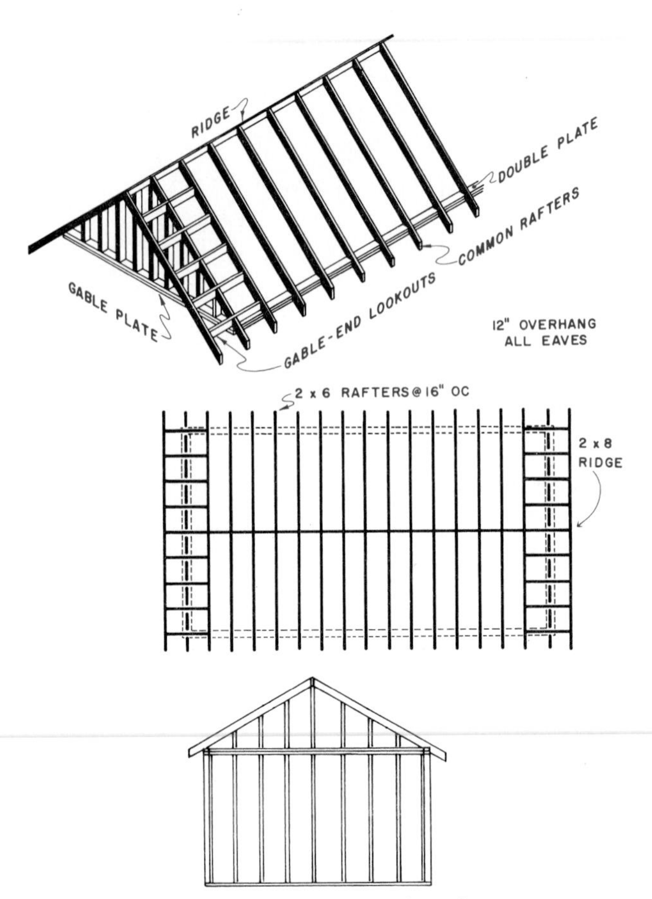

Fig. 57-16 Gable roof plan and elevation.

Beams and roof sections with insulation and finishing are attached to the gable ends (Fig. 57-15). This is one variation of the post-and-beam method of roof construction. Figure 57-16 shows an example of a roof plan and elevation drawing of a typical gable roof framing plan.

Pitch

A gable roof has pitch on two sides but no pitch on the gable ends. The *pitch* is the angle between the top plate and the ridge board. *Rise* is the vertical distance from the top plate to the ridge. *Run* is the horizontal distance from the top plate to the ridge. The *pitch* is referred to as the *rise over the run.* In a gable roof with the ridge board in the exact center of the building, the run is one-half the span (Fig. 57-17). The run is always expressed in units of 12. Therefore, the rise is the number of inches the roof rises as it moves 12″ horizontally. A $\frac{6}{12}$ pitch means that the roof rises 6″ for every 12″ of horizontal distance (run). Pitch can also be expressed as a fraction indicating the total vertical distance to the total horizontal distance. To easily compute this fraction, use the rise over the span. The span is the run doubled. These numbers are constants. The run is 12 and the span is 24. For example, in Fig. 57-17, the $\frac{6}{12}$ pitch is the fraction $\frac{6}{24}$, or $\frac{1}{4}$.

Gable-roof pitches vary greatly. The $\frac{1\frac{1}{2}}{12}$ pitch ($\frac{1}{16}$) shown in Fig. 57-18 is almost a flat roof. The $\frac{8}{12}$ pitch ($\frac{1}{3}$), however, is a moderately steep roof. The angle of a roof with a $\frac{12}{12}$ pitch ($\frac{1}{2}$) is 45°. The carpenter's square is used on the construction site to lay out the roof pitch on the rafters, as shown in Fig. 57-19.

Ridge Beams

The ridge board or ridge beam as shown in Fig. 57-20 is the top member in the roof assembly. Rafters are fixed in their exact position by being secured to the ridge board. The ridge beam in a cathedral ceiling of post-and-beam construction may be seen from the inside. Figure 57-21 shows various methods of intersecting rafters with the ridge beam, and Fig. 57-22 shows ridge beam assembly details.

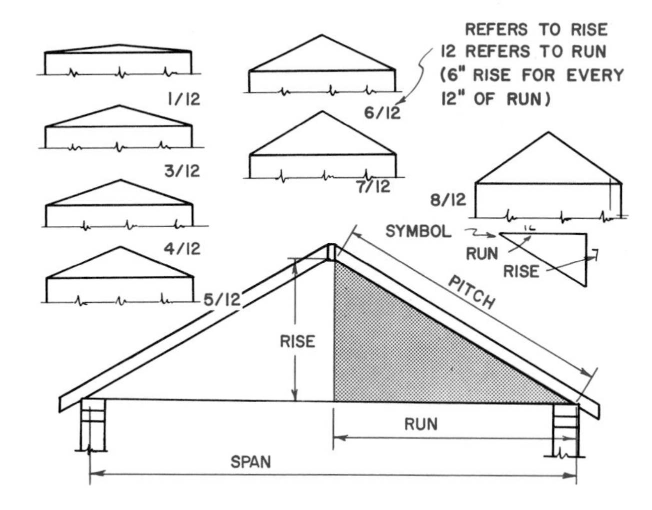

Fig. 57-17 Roof pitch can be expressed as rise over run or as the fraction rise over span.

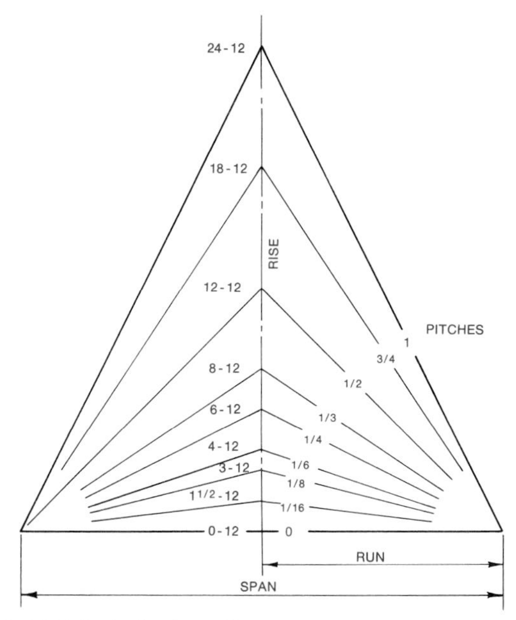

Fig. 57-18 Typical roof spans.

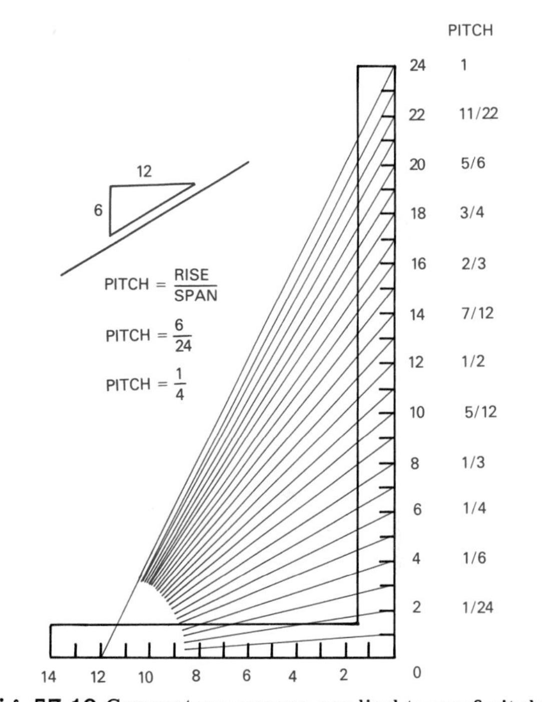

Fig. 57-19 Carpenters square applied to roof pitch.

Gable End

The gable end is the side of the house that rises to meet the ridge (Fig. 57-23). In some cases, especially on low-pitch roofs, the entire gable-end wall from the floor to the ridge can be panelized with varying lengths of studs. However, it is more common to prepare a rectangular wall panel and erect separate studs that project from the top plate of the panel to the rafter. Sheathing and siding are then added to the entire gable end of the house. With post-and-beam construction, clerestory windows in the gable end can be used. This is because studs are unnecessary on the gable end. An increased use of windows in gable ends has necessitated the use of larger

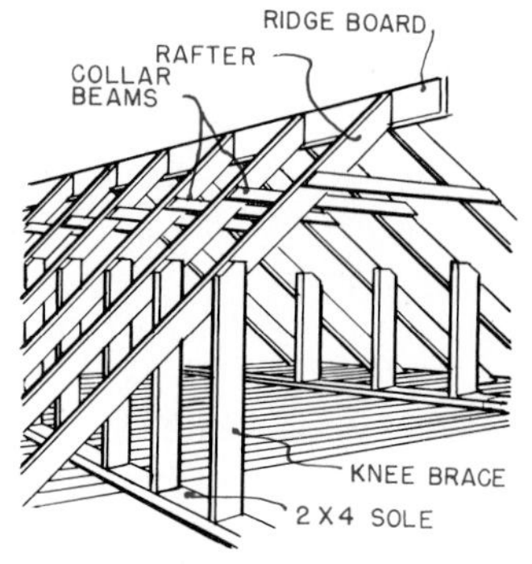

Fig. 57-20 The ridge board is the top member in the roof assembly.

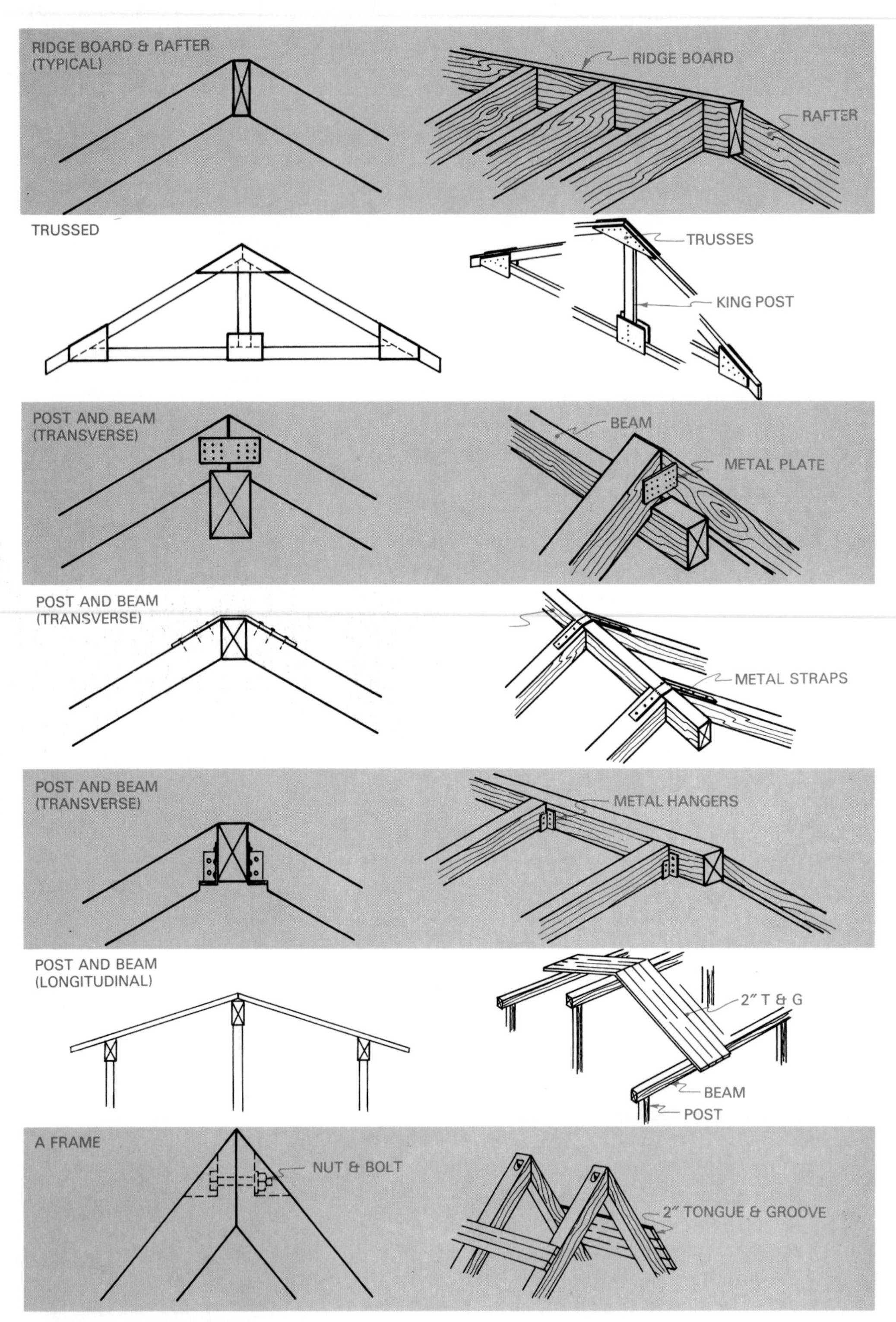

Fig. 57-21 Ridge beam assemblies.

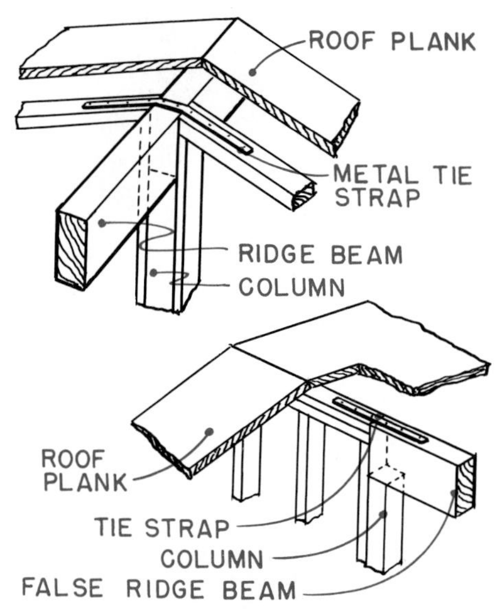

Fig. 57-22 Ridge beam assembly detail.

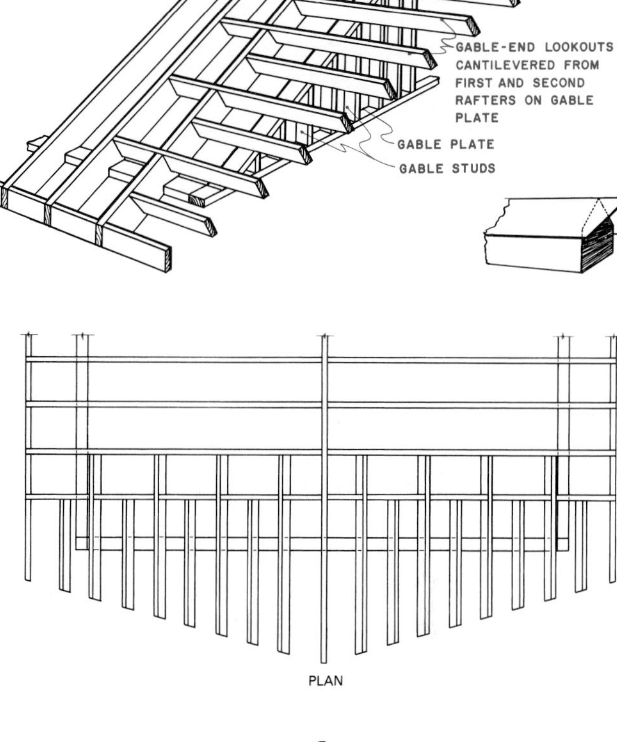

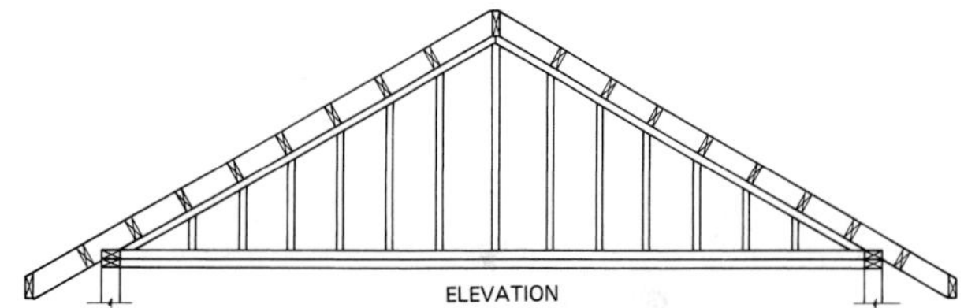

Fig. 57-24 Winged-gable lookout construction.

overhangs on the gable end. Gable-end lookouts can be framed from the first or second rafters on the gable plate, as shown in Fig. 57-24.

Overhangs

Several variations of cuts are used to finish the rafter end. A comparisn of the plumb, level, combination, and square cuts is given in Fig. 57-25. Notice also that the rafters are notched with a *plate* (seat or birdmouth) *cut*, to provide a level surface for the intersection of the rafters and top plates. The area bearing on the plate should not be less than 3″ (75 mm) (Fig. 57-26).

Overhangs are normally larger on the side walls than on gable ends. Overhangs should be designed to provide maximum protection from sun and rain without restricting the light and view. Figure 57-27 shows some of the difficulties in designing large overhangs. On a low-pitch roof, the problem is not acute since the rise is relatively small compared with the run. But on a step-pitch roof, where the rise is large, the light

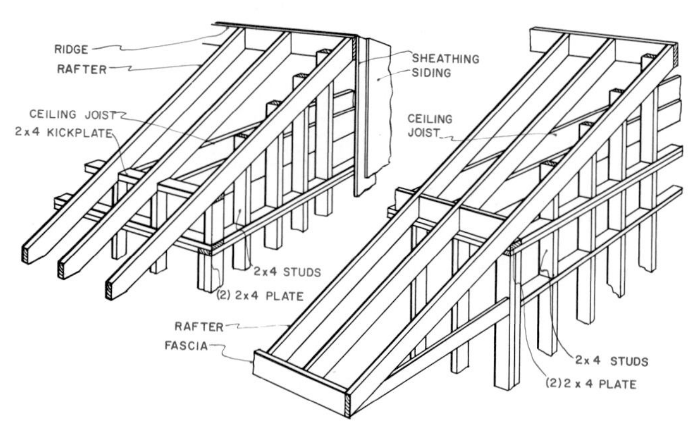

Fig. 57-23 Gable-end construction.

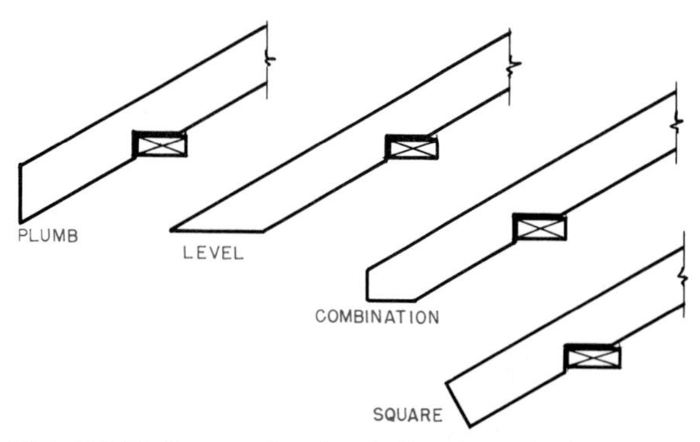

Fig. 57-25 Types of rafter tail cuts and plate cuts.

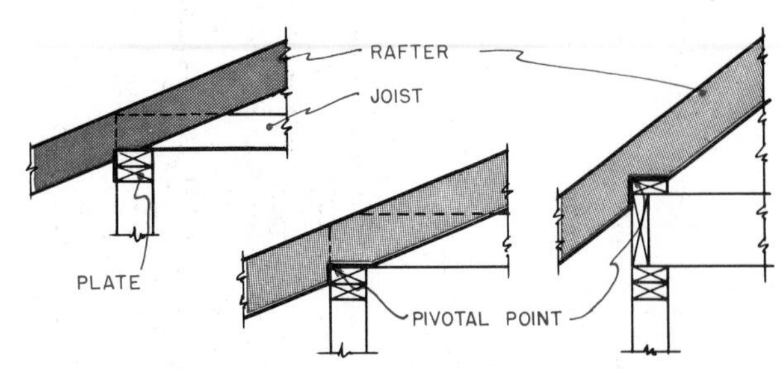

Fig. 57-26 Method of intersecting rafters and top plates.

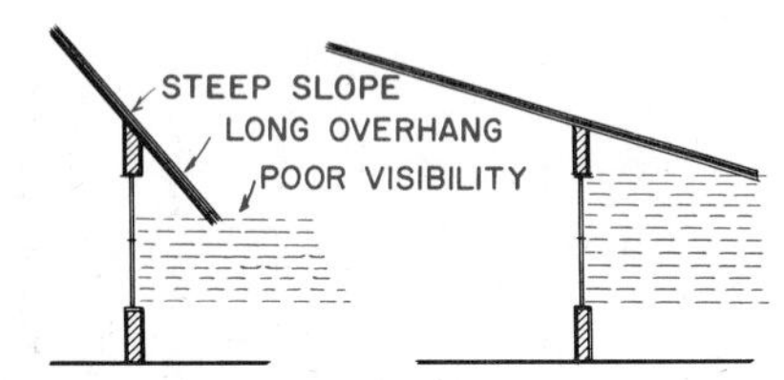

Fig. 57-27 Large overhangs are desirable only if designed correctly.

might be completely shut off if the same amount of overhang were used. Furthermore, when the end of the overhang extends below the level of the window, there is no possibility of using a flat soffit. Figure 57-28 shows several alternative soffits.

Collar Beams

Collar beams provide a tie between rafters. They may be placed on every rafter or only on every other rafter. Collar beams are used to reduce the rafter stress that occurs between the top plate and the rafter cut. They may also act as ceiling joists for finished attics (Fig. 57-29).

Knee Walls

Knee walls are vertical studs that project from an attic floor to the roof rafters, as shown in Fig. 57-30. Knee walls add rigidity to the rafters and may also provide wall framing for finished attics.

TRUSSES

Lightweight wood trusses have become increasingly popular for small buildings. Roof trusses allow complete flexibility for interior spacing. They save approximately 30 to 35 percent on materials, compared with conventional framing methods. Trusses can be fabricated and erected in one-third of the time required for rafter and ceiling-joist construction. Truss construction

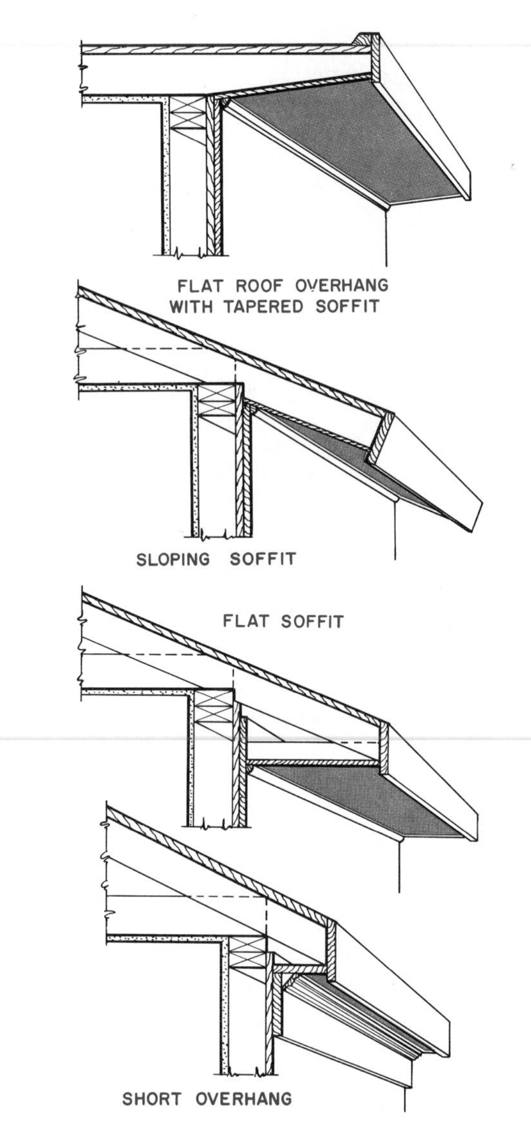

Fig. 57-28 Types of soffit design. *(National Lumber Manufacturers Assoc.)*

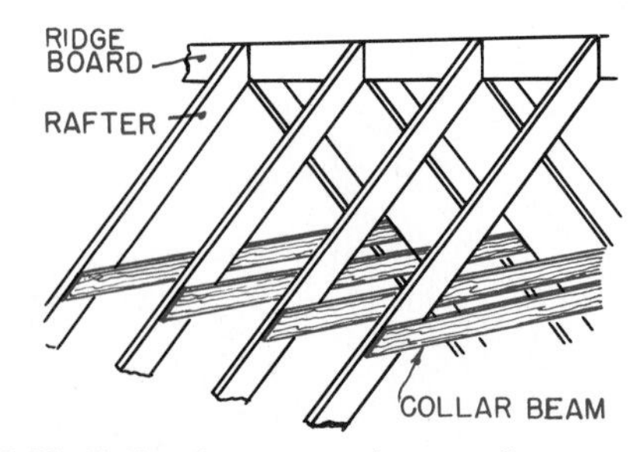

Fig. 57-29 Collar beams reduce rafter stress.

helps to put the building under cover very quickly. The use of trusses saves material and erection time and eliminates normal interior load-bearing partitions. Standard types of

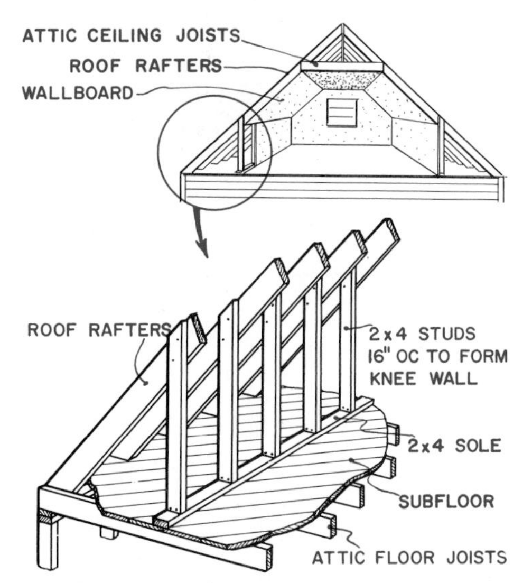

Fig. 57-30 Knee walls add rigidity to the rafters.

trusses are shown in Fig. 57-31. Truss construction methods are as applicable to large steel-framed buildings (Fig. 57-32) as to small wood-framed buildings (Fig. 57-33). Figure 57-34 provides engineering design data for common dry-wall truss construction based on normal residential construction loading.

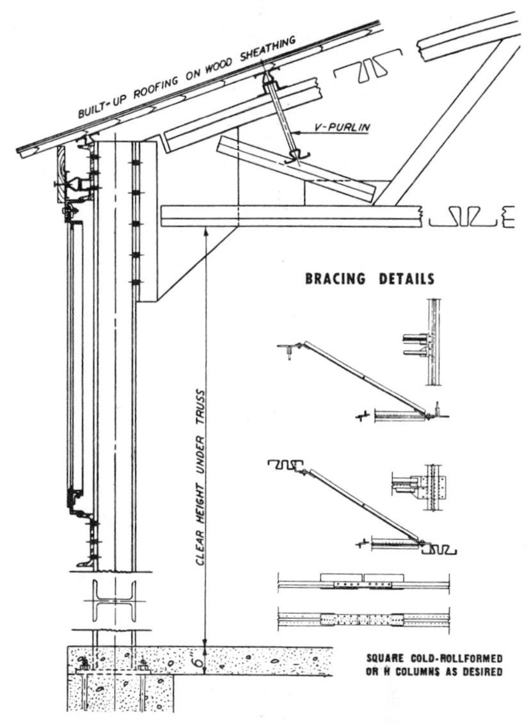

Fig. 57-32 A structural-steel truss. *(Macomber, Inc.)*

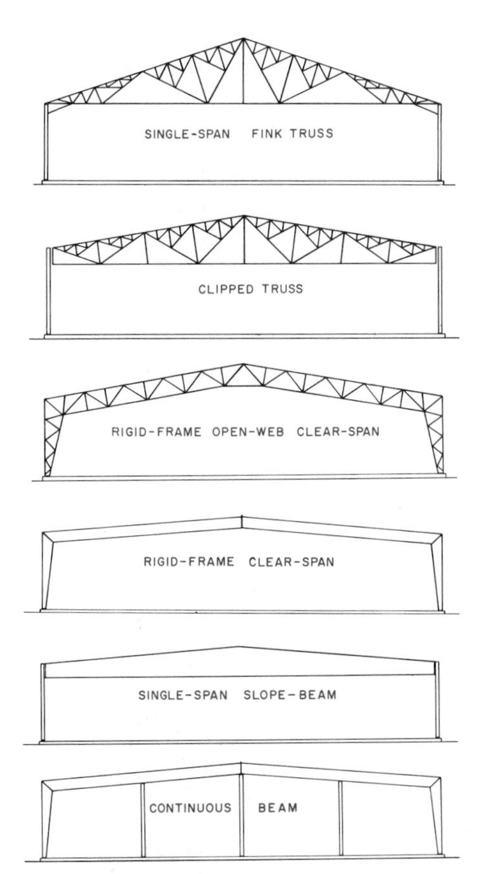

Fig. 57-31 Standard types of trusses.

HIP ROOF

Hip-roof framing is similar to gable-roof framing except that the roof slopes in two directions instead of intersecting a gable-end wall. Where two adjacent slopes meet, a *hip* is formed on the external angle. A *hip rafter* extends from the ridge board over the top plate to the edge of the overhang. The hip rafter does the same job as the ridge board.

The internal angle formed by the intersection of two slopes of the roof is known as the *valley*. A *valley rafter* is used on the internal angle as a hip rafter is used on the external angle. Hip rafters and valley rafters are normally 2″ (50 mm) deeper or 1″ (25 mm) wider than the regular rafters, for spans up to 12′ (3.7 m). For spans of over 12′, the hip and valley rafters should be doubled in width. *Jack rafters* are rafters that extend from the wall plate to the hip or valley rafter. They are always shorter than *common rafters*. Figure 57-35 illustrates the use of these various framing members in hip-roof construction.

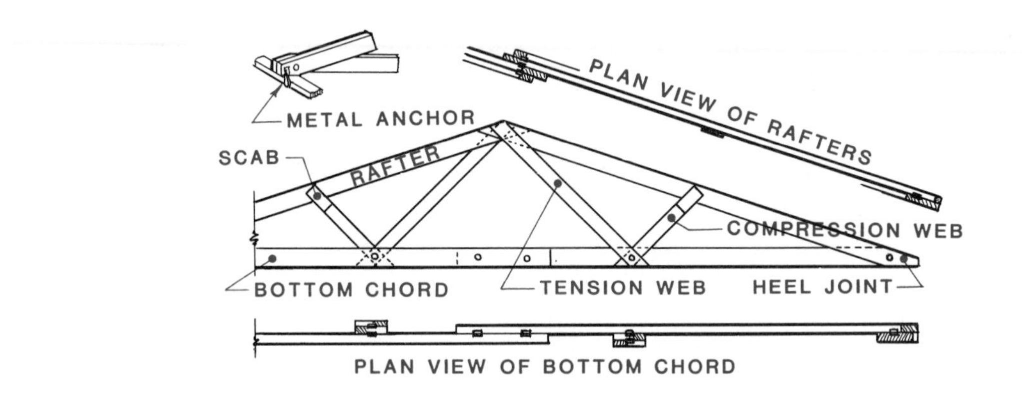

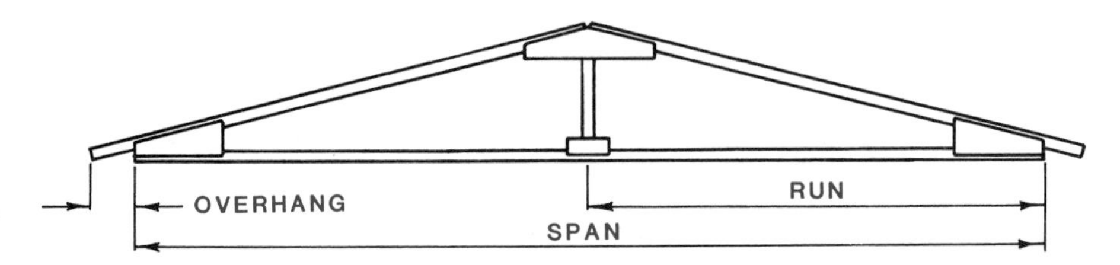

Fig. 57-33 Wood-frame trusses.

SPAN			26′			28′			30′			32′		
			0″	4″	8″	0″	4″	8″	0″	4″	8″	0″	4″	8″
TOP CHORDS	OVERHANG A	2/12	33½″	31½″	29½″	45½″	43½″	41½″	33½″	31½″	29½″	45½″	43½″	41½″
		3/12	30¼″	28¼″	26¼″	42¼″	40¼″	38¼″	30¼″	28¼″	26¼″	42¼″	40¼″	38¼″
		4/12	26⅛″	24⅛″	22⅛″	36⅞″	34⅞″	32⅞″	24⅞″	22⅞″	20⅞″	34¾″	32¾″	30¾″
BOTTOM CHORD	B		13′-0″	12′-8″	14′-0″	14′-0″	14′-4″	14′-8″	14′-0″	14′-4″	14′-8″	16′-0″	16′-4″	16′-8″

Fig. 57-34 Common truss specifications.

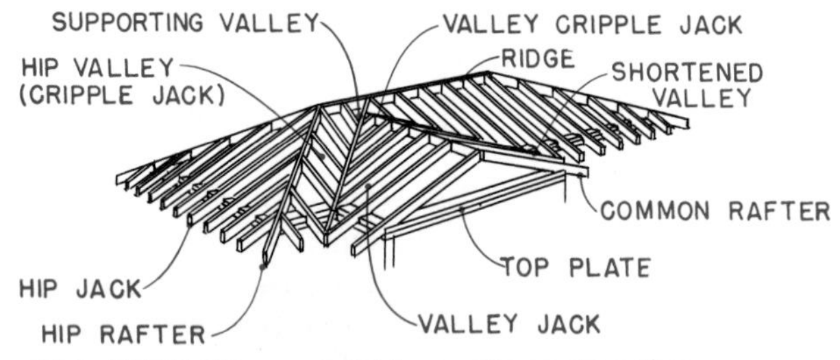

Fig. 57-35 Hip and gable roof construction.

SHED ROOF

A *shed roof* is a roof that slants in only one direction (a gable roof is actually two shed roofs, sloping in opposite directions). Shed-roof rafter design is the same as rafter design for gable roofs, except that the run of the rafter is the same as the span. The shed rafter differs from the common rafter in the gable roof in that the shed rafter has two plate cuts, a tail cut, and a top-end cut (Fig. 57-36).

FLAT ROOF

A *flat roof* has no slope. Therefore, the roof rafters must span directly from wall to wall or from an exterior wall to an interior bearing partition. When rafters also serve as ceiling joists, the size of the rafter must be computed on the basis of both the roof load and the ceiling load.

Overhang

Large overhangs are possible on flat roofs. The roof joists can be extended past the top plate far enough to provide sun protection and, yet, not block the view. This extension is possible be-

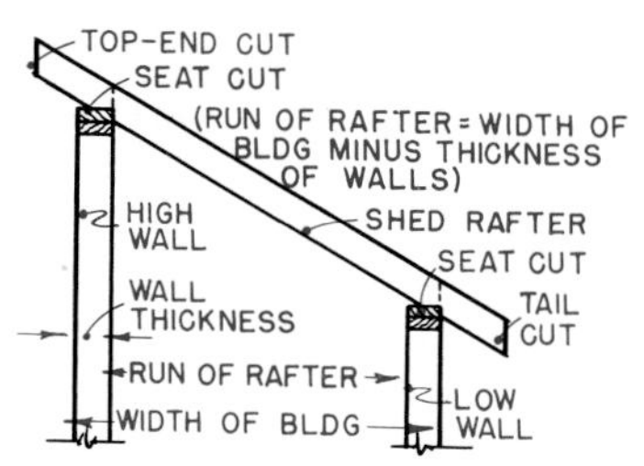

Fig. 57-36 Shed-roof framing.

cause there is no slope. When overhangs are desired on all sides, lookout rafters, as shown in Fig. 57-37, are used to extend the overhang on the side of the building perpendicular to the rafter direction. Figure 57-38 shows a flat roof cornice, or eave, section.

Drainage

Since there is no slope to a flat roof, drainage must be provided by downspouts extending through the overhang. Flat roofs must be designed for a maximum snow load, since snow will not slide off the roof but must completely melt and drain away. A built-up roof consisting of sheathing, roofing paper, and crushed gravel can be used (Fig. 57-39). If a flat roof is so designed, a gravel stop and cant strip can be made high enough to hold water at a specific level. Water could then lie on the roof at all times and provide additional insulation. Additional waterproofing and larger structural members must be used to support the additional weight.

Steel

Steel-construction methods are especially applicable to flat roofs. The simplicity of erecting a steel roof results from the great strength of steel joists. The cross-bracing between widths of a steel *purlin* (horizontal member), as shown in Fig. 57-40, provides the purlin with a strength comparable to that of a truss.

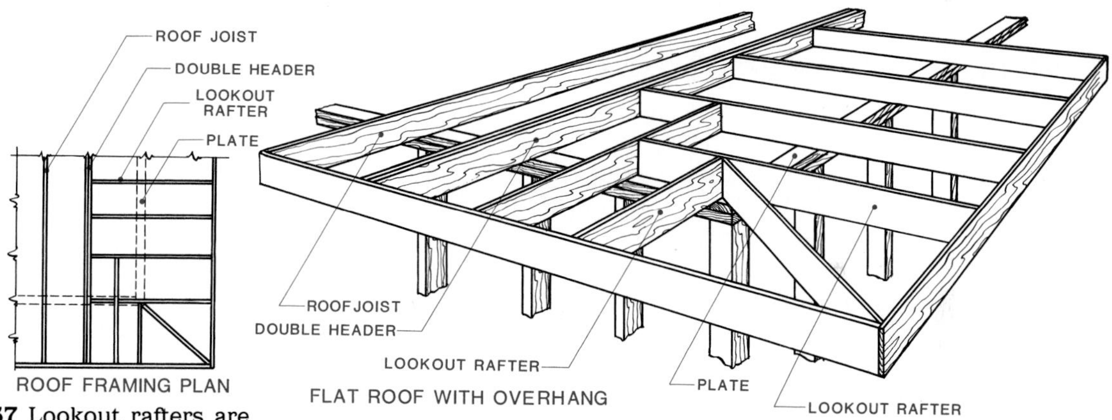

Fig. 57-37 Lookout rafters are used to extend the overhang perpendicular to the common rafters.

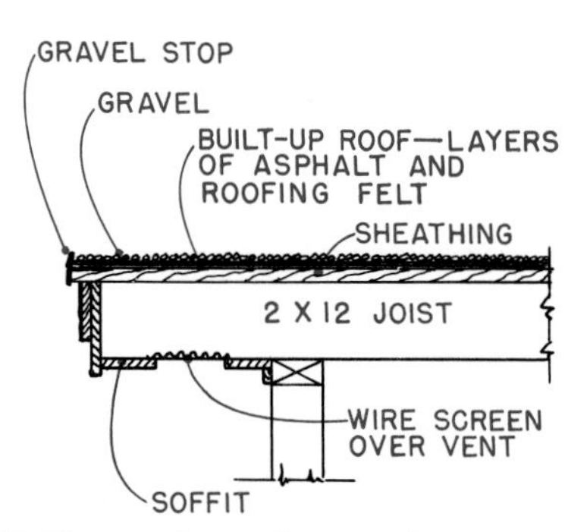

Fig. 57-38 Flat-roof cornice section.

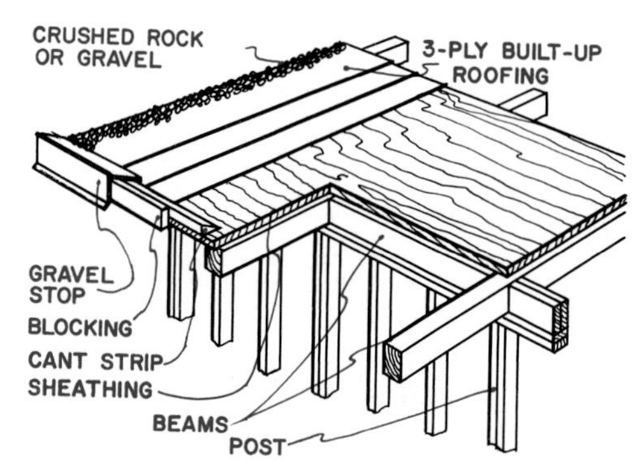

Fig. 57-39 Built-up roof construction.

Fig. 57-40 Typical steel roof construction. *(Mitchell Construction Co.)*

PLAN DEVELOPMENT

A *roof plan* (Fig. 57-41) is one showing the outline of the roof and the major object lines indicating ridges, valleys, hips, and openings. The roof plan is not a framing plan, but a plan view of the roof. To develop a roof framing plan, a roof must be stripped of its covering to expose the position of each structural member and each header (Fig. 57-42). The roof plan can be used as the basic outline for the roof framing plan. The roof framing plan must show the exact position and spacing of each member. Figure 57-43 shows a comparison of a roof plan and a roof framing plan of the same roof.

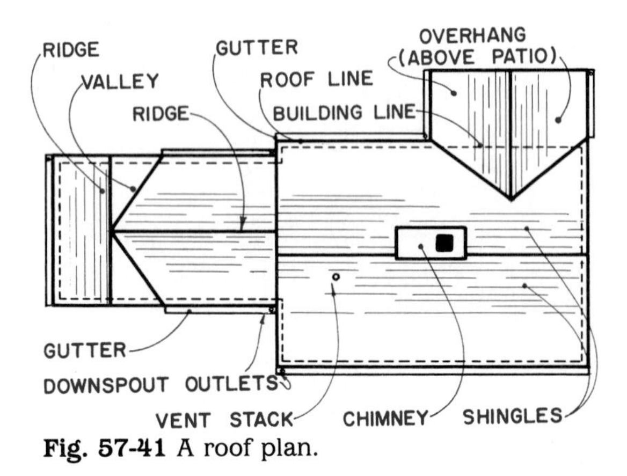

Fig. 57-41 A roof plan.

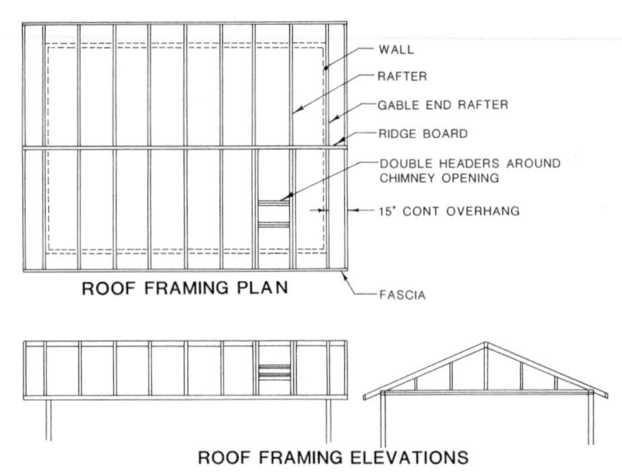

Fig. 57-42 A roof framing plan.

Single-Line Plans

In the roof framing plan shown in Fig. 57-43, each member is represented by a single line. This method of preparing roof framing plans is acceptable when only the general relationship and center spacing are desired. Figure 57-44 shows a single-line hip-roof framing plan.

Steel Construction

Roof framing plans for steel construction are prepared like those for wood construction. However, in plans for structural steel, single-line drawings are almost universally used. A com-

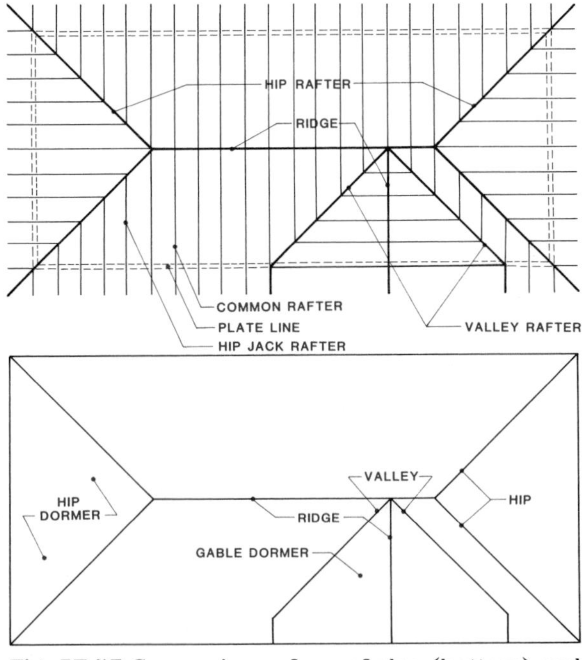

Fig. 57-43 Comparison of a roof plan (bottom) and a roof framing plan(top).

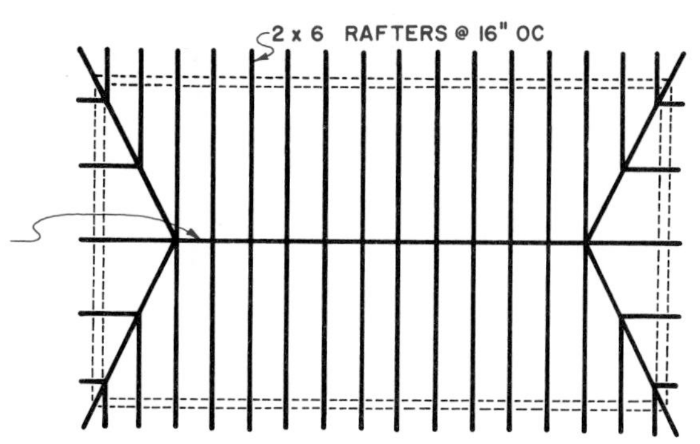

Fig. 57-44 Single-line hip-roof framing plan.

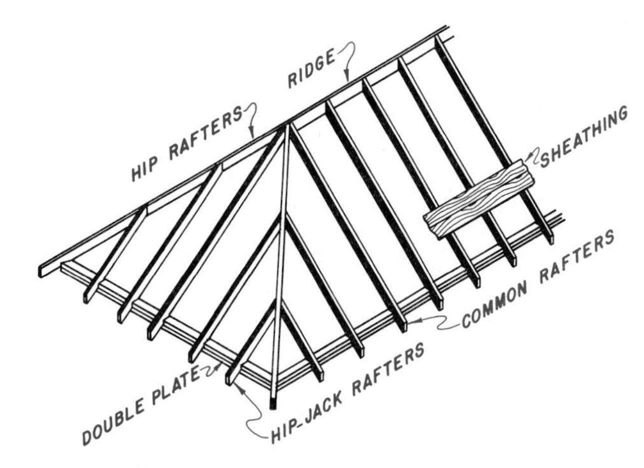

plete classification of each steel member is shown on the drawing. This includes the size, type, and weight of the beams and columns. Each different type of member is shown by a different line weight, which relates to the size of the member. Structural-steel roof framing plans frequently show *bay areas* (areas between columns) indicating the number of spaces and the spacing of each purlin between columns. Bays are frequently shown in circles: numerically in one direction and alphabetically in the other direction (Fig. 57-45).

Complete Plan

When more details concerning the exact construction of intersections and joints are needed, a plan showing the thickness of each member, as shown in Fig. 57-46, should be prepared. This type of plan is necessary to show the relative height of one member compared with another; that is, to determine whether one member passes over or under another. In this plan, the width of ridge boards, rafters, headers, and plates should be shown to the exact scale.

When a complete roof framing plan of this type cannot fully describe construction framing details, then additional removed pictorial or elevation drawings should be prepared, as shown in Fig. 57-47. A similar technique of removing details can also be used to increase the size of a particular area in detail for dimensioning purposes.

On roof framing plans, only the outline of the top of the rafters is shown. All areas underneath, including the gable-end plate, are shown by dotted lines. When a gable-end lookout slopes, a true orthographic projection of the plan would indicate three lines—two lines for the top of the rafter and one line for the bottom. However, roof framing plans are normally simplified to show only the outline of the top of each member (Fig. 57-48). The angle or vertical position of any roof framing member should be shown on a roof framing elevation.

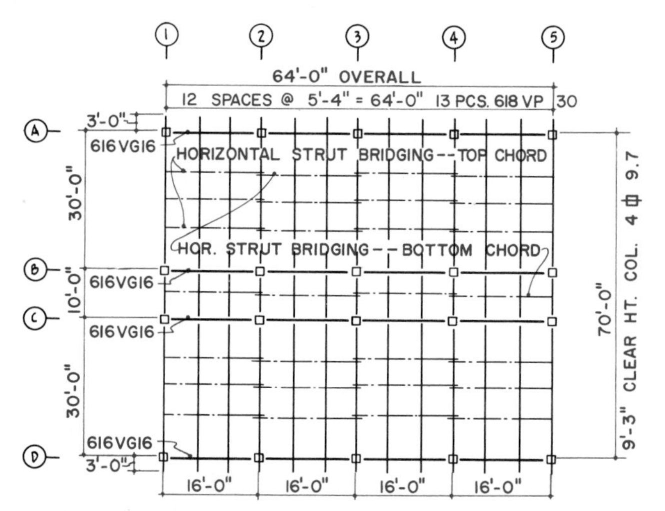

Fig. 57-45 Steel roof framing plan areas are often identified by numbers and letters. *(Macomber, Inc.)*

ROOF FRAMING ELEVATIONS

Roof framing plans show horizontal relationships of members such as thickness, length, and horizontal spacing. In a top view (plan), you cannot show vertical dimensions such as structural heights and pitches. Transverse sections supply this information through one part of the structure. However, if a comparison of different heights and pitches is desired, a composite

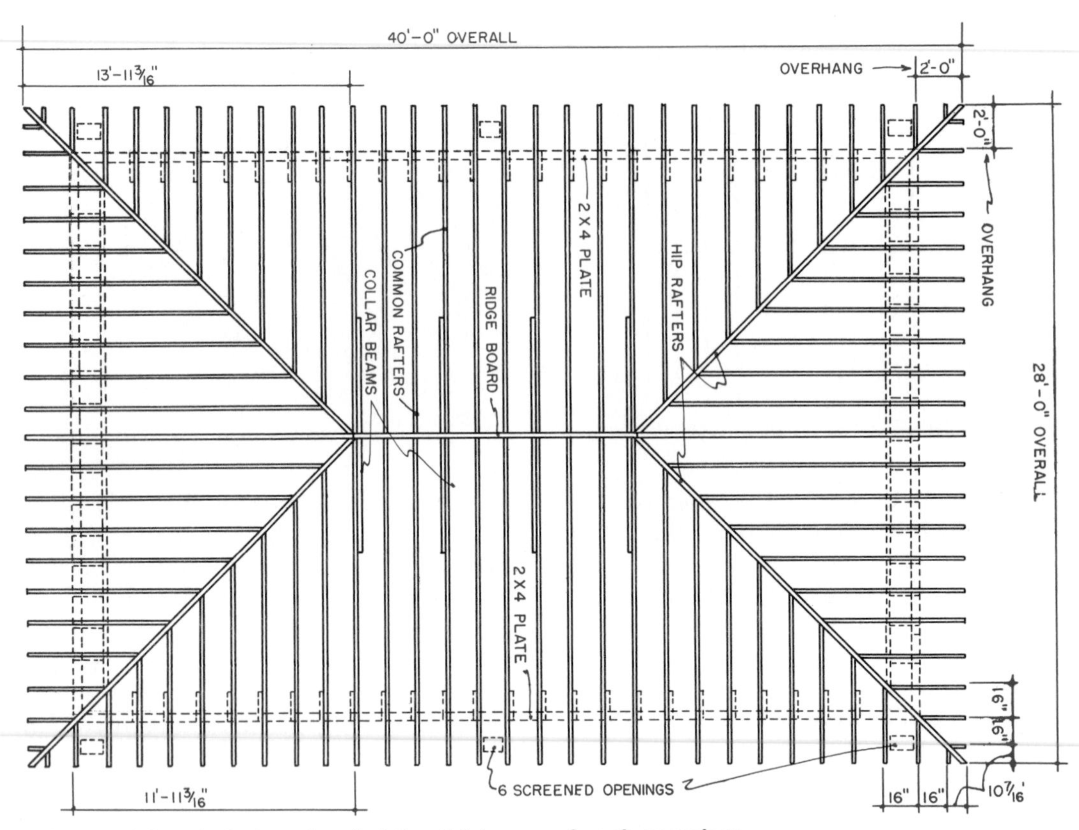

Fig. 57-46 A roof framing plan showing the thickness of each member.

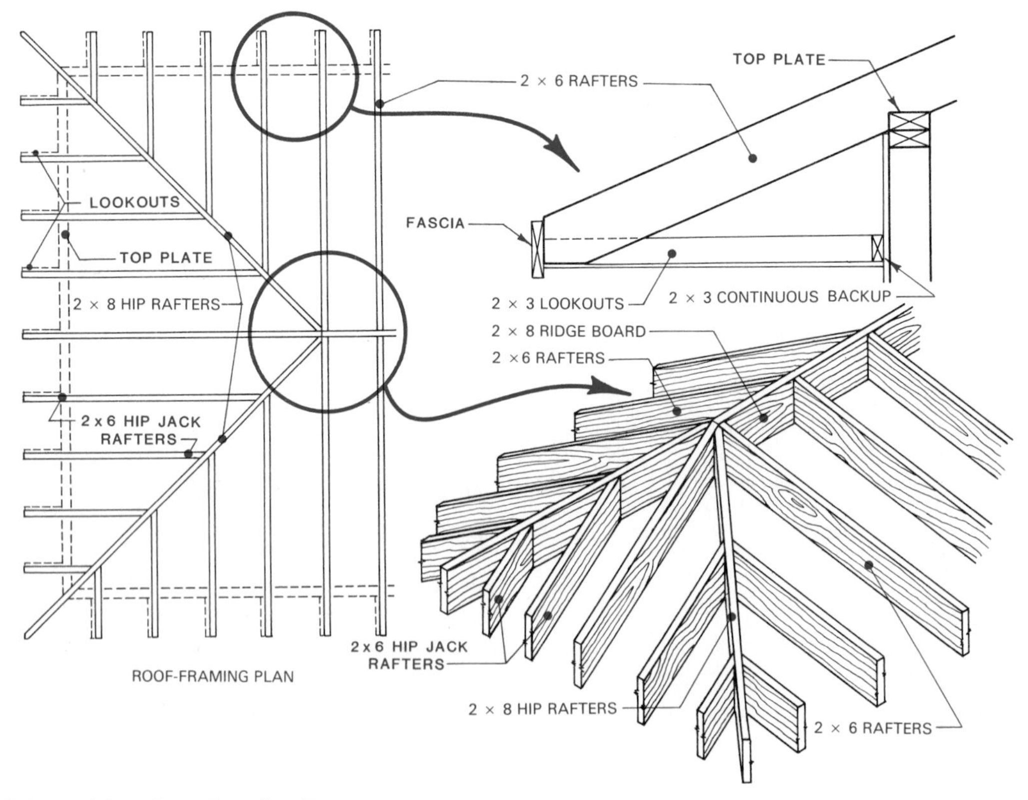

Fig. 57-47 Roof framing plan details.

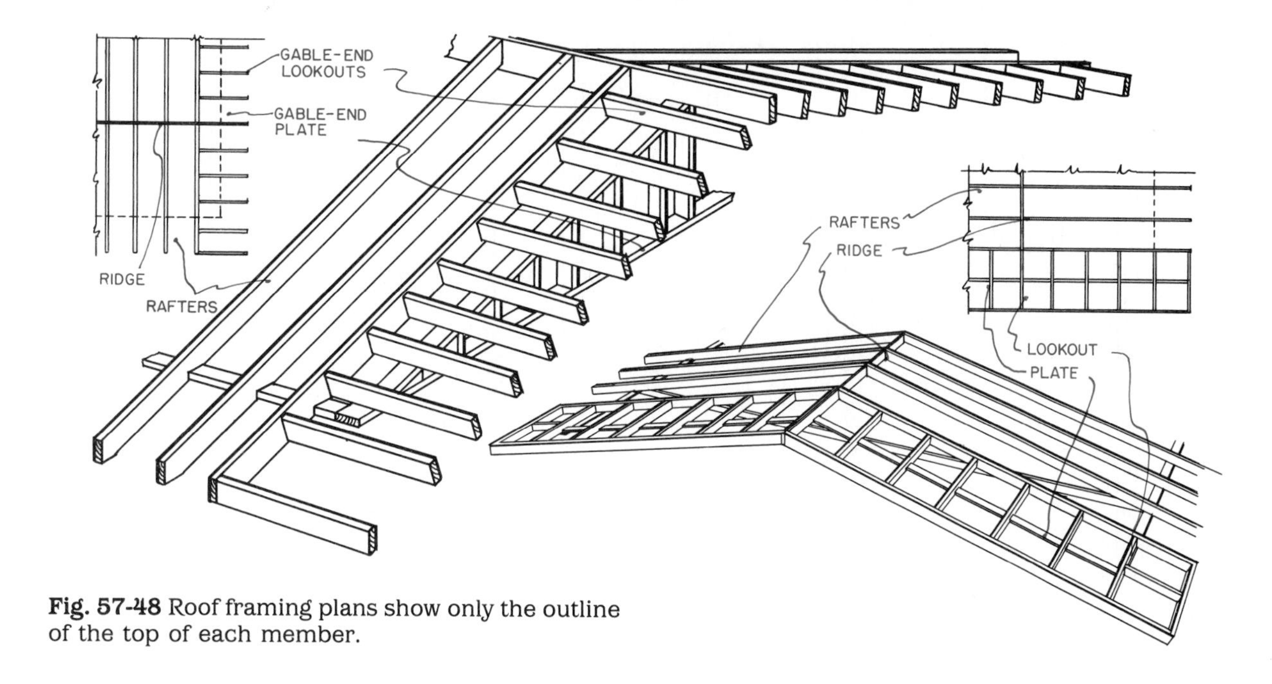

Fig. 57-48 Roof framing plans show only the outline of the top of each member.

framing-elevation drawing should be prepared. This elevation can be projected from the floor framing plan and corresponding lines on the elevation drawings, as shown in Fig. 57-49.

DORMERS

Parts of the roof framing plan that extend above the normal plane of projection, such as dormer rafters, are drawn with dotted lines. This device indicates that the parts do not directly intersect with the other framing members shown on the plan. It is also quite common to show the position of dormer rafters or ridge as illustrated in Fig. 57-50. The details of intersecting the dormer-roof framing with dormer walls may be shown on dormer-roof framing plans or other details (Fig. 57-51).

Dormer rafters and walls do not lie in the same plane as the remainder of the roof rafters. A framing-elevation drawing is needed to show the exact position of the dormer members and their tie-in with the common roof rafters. Figure 57-52 illustrates a side framing elevation of an individual gable dormer and shows how it is structurally related to other roof framing members. Figure 57-53 shows the sidewall framing of a shed dormer, in which the position of the dormer studs is revealed by the elevation drawing.

Plates in roof framing elevations are actually parts of sections. They are used to show the

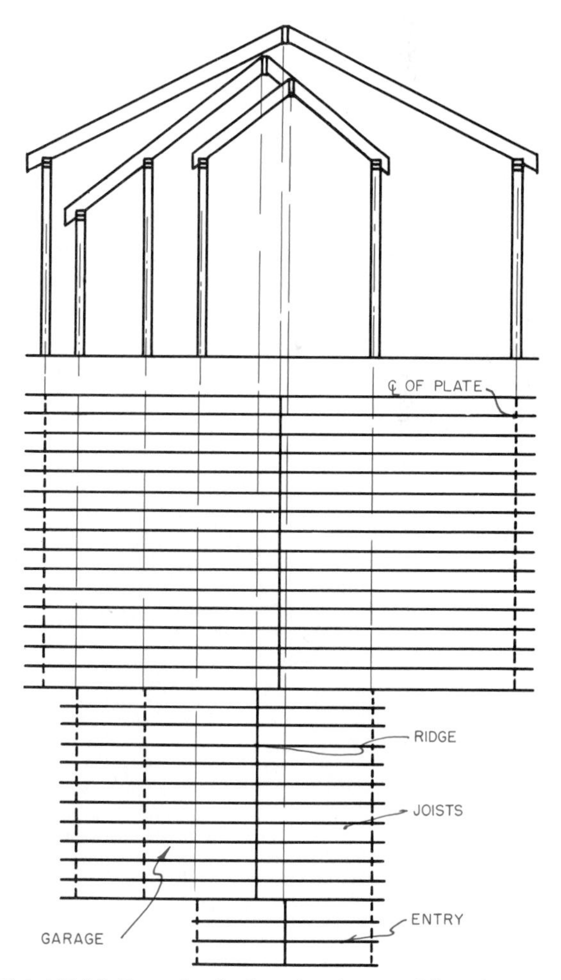

Fig. 57-49 A method of projecting roof framing elevations.

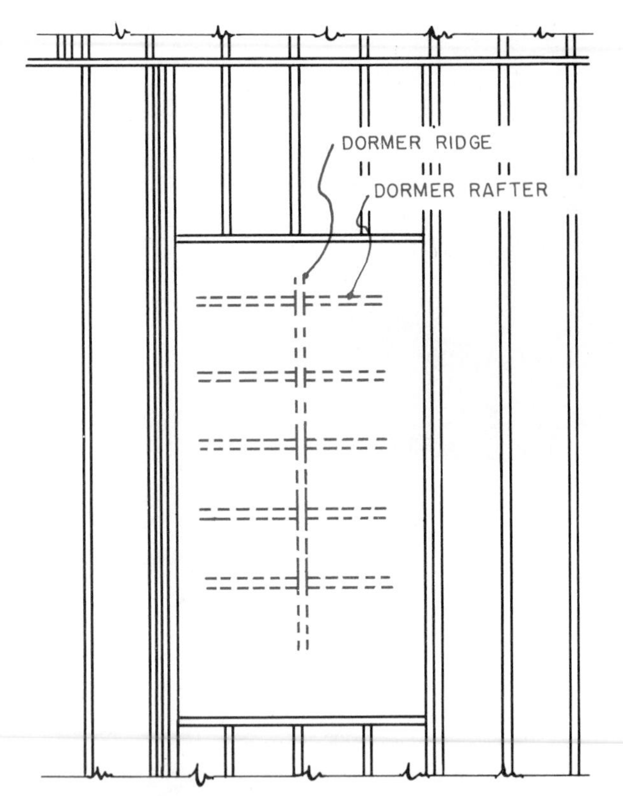

Fig. 57-50 A method of drawing dormer rafters.

basic relationship between major framing members and the roof covering and trim details. Figure 57-54 shows the steps in laying out one of the most widely used drawings of this type, a cornice detail. Other cornice details that show the relationship between roof framing members and trim materials are illustrated in Fig. 57-55.

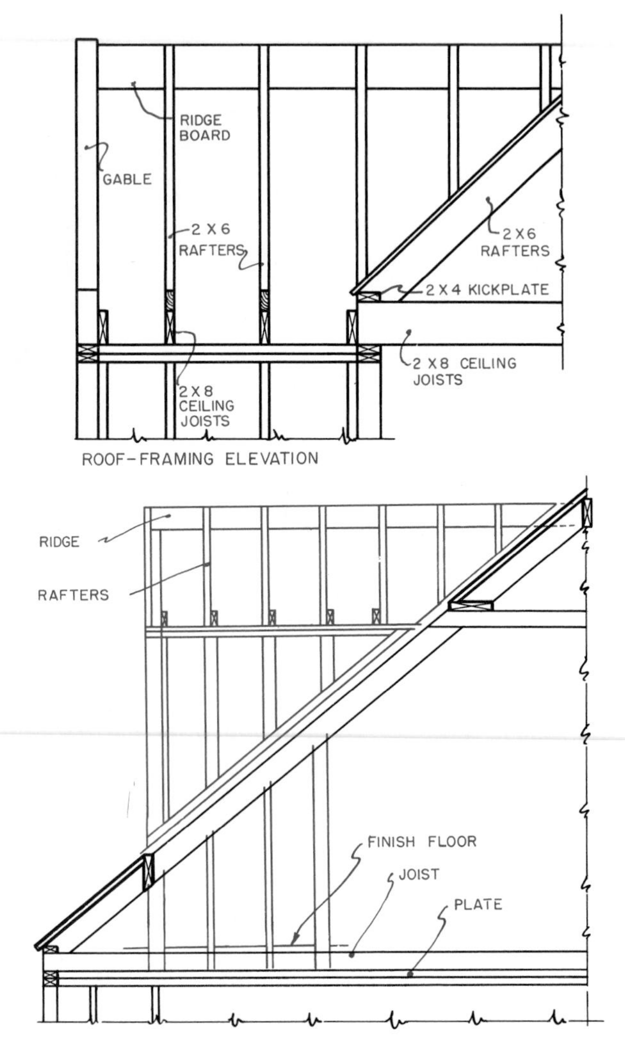

Fig. 57-52 A frame elevation of an individual dormer.

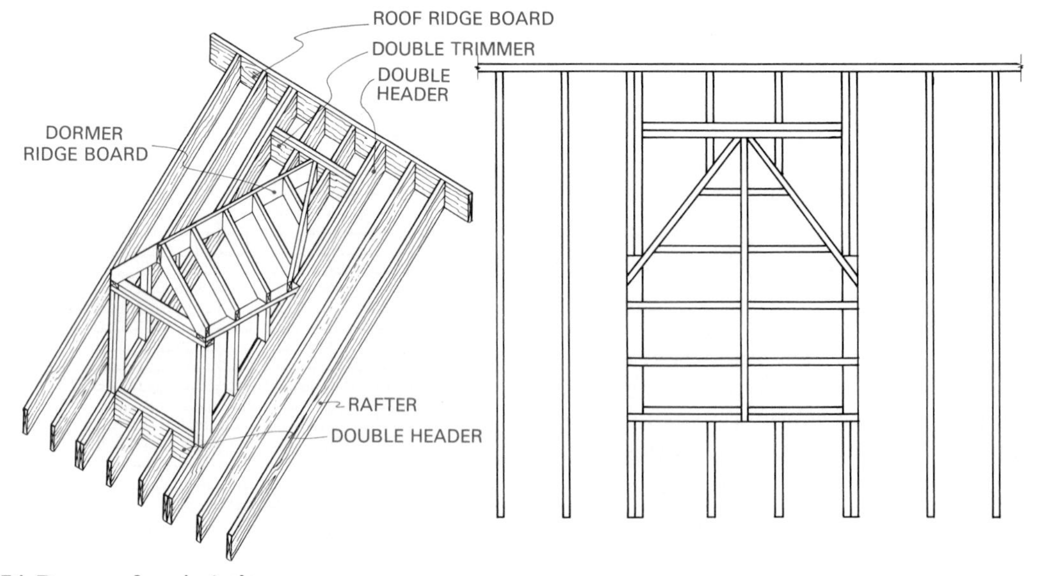

Fig. 57-51 Dormer framing plan.

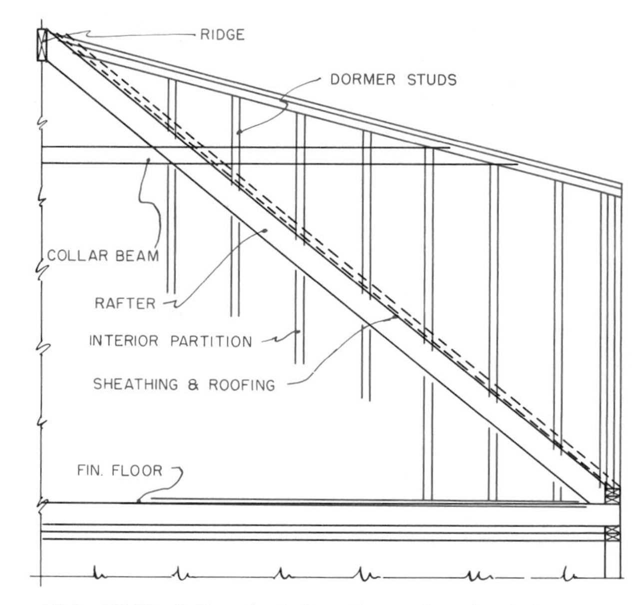

Fig. 57-53 A framing elevation of a shed dormer.

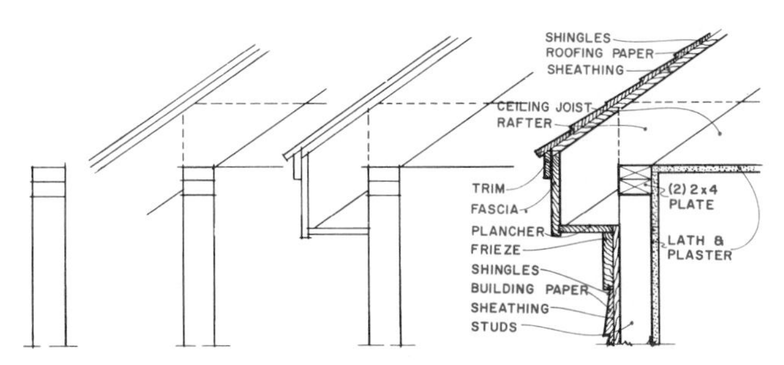

Fig. 57-54 Steps in laying out a cornice detail.

CHIMNEY DETAILS

Chimney roof framing construction details are normally shown on detail drawings. Figure 57-56 shows ceiling joists framed around a chimney. Figure 57-57 shows the framing necessary for the intersection of the chimney and the roof rafters. Figure 57-58 shows a plan view and pictorial view of a saddle construction that is used to divert water away from a chimney.

BEAMS

The *beams* are the major support of the roof. Beams have been developed that are lighter in weight and stronger than a solid beam (Fig. 57-59). The *laminated beam*, or the *built-up beam*, has great strength and can be formed into graceful arches or other shapes (Fig. 57-60).

The *box beam* offers lightness, low cost, and good supporting strength. The strongest of the built-up beams is the *steel-reinforced beam* (Fig. 57-61). It can support heavy weights and span long distances because of the steel plate bolted between the wood members.

ROOF PANELS

Many forms of lightweight, prefabricated roof units have been developed. Some of the more commonly used units are described.

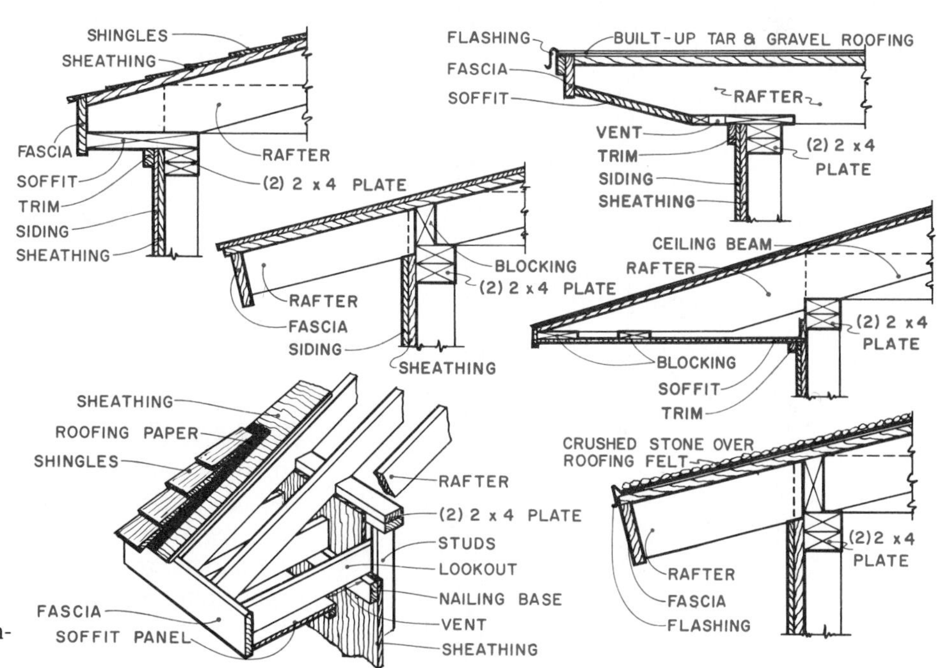

Fig. 57-55 Cornice framing details.

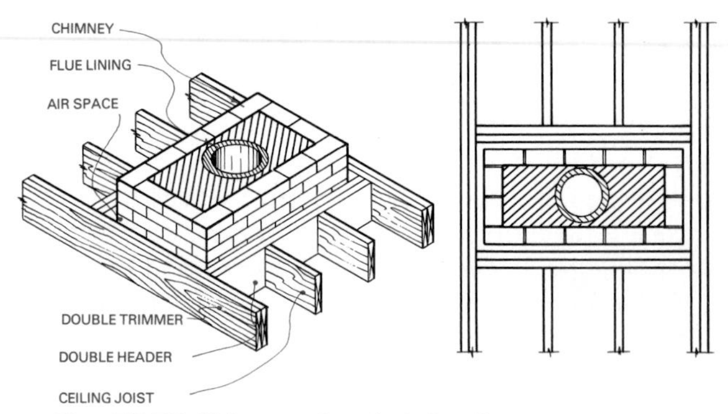

Fig. 57-56 Chimney framing details.

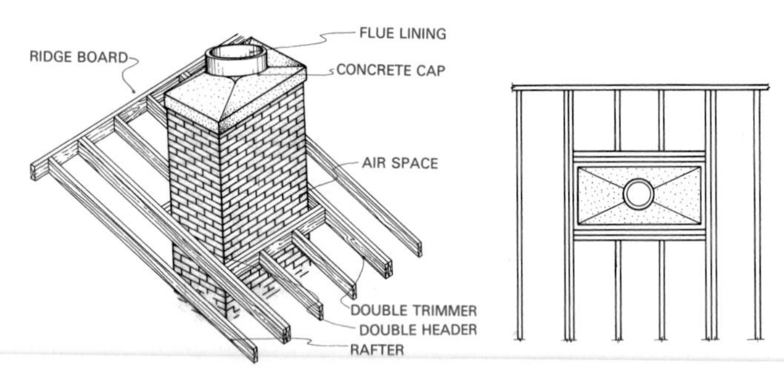

Fig. 57-57 Chimney and roof rafter intersection details.

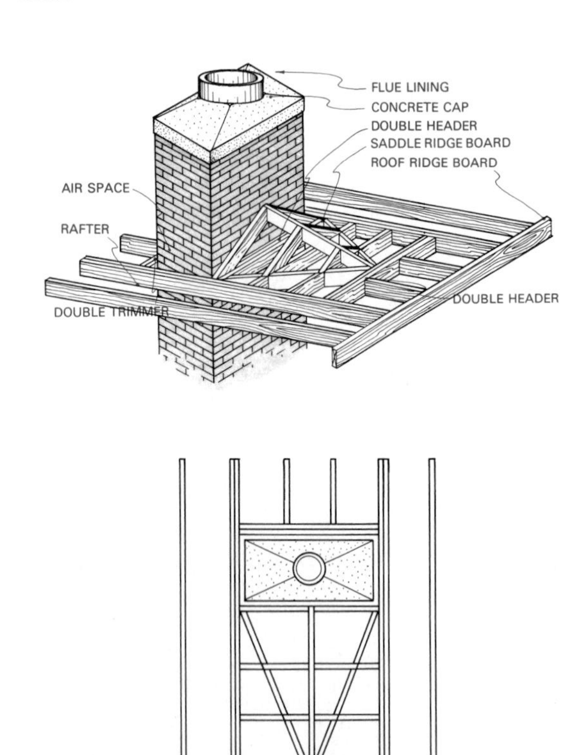

Fig. 57-58 Saddle framing diverts water from chimney.

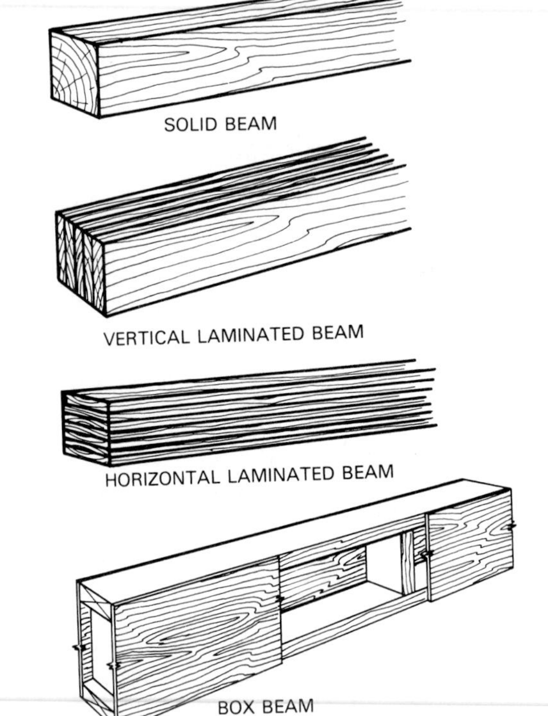

Fig. 57-59 Kinds of wood beams.

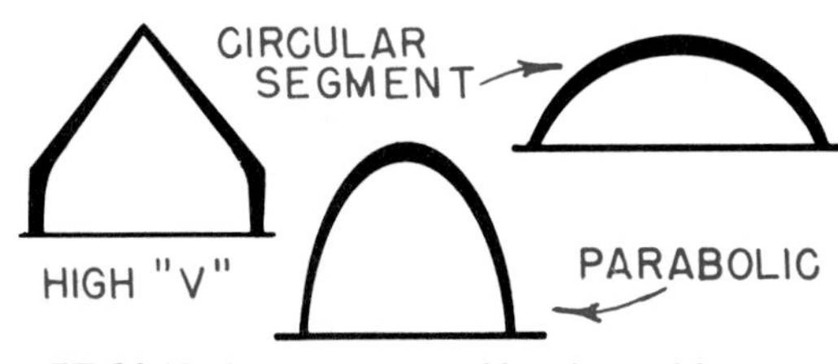

Fig. 57-60 Various shapes of laminated beams and arches.

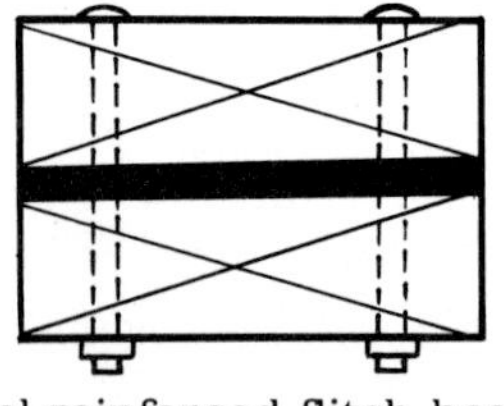
Fig. 57-61 Steel-reinforced flitch beam.

Stressed Skin Panels

Stressed skin panels are constructed of plywood and seasoned lumber. The simple framing and the plywood skin act as a unit to resist loads. Glued joints transmit the shear stresses, making it possible for the structure to act as one piece. Stressed skin panels (Fig. 57-62) are used for floors and walls, as well as in roofs.

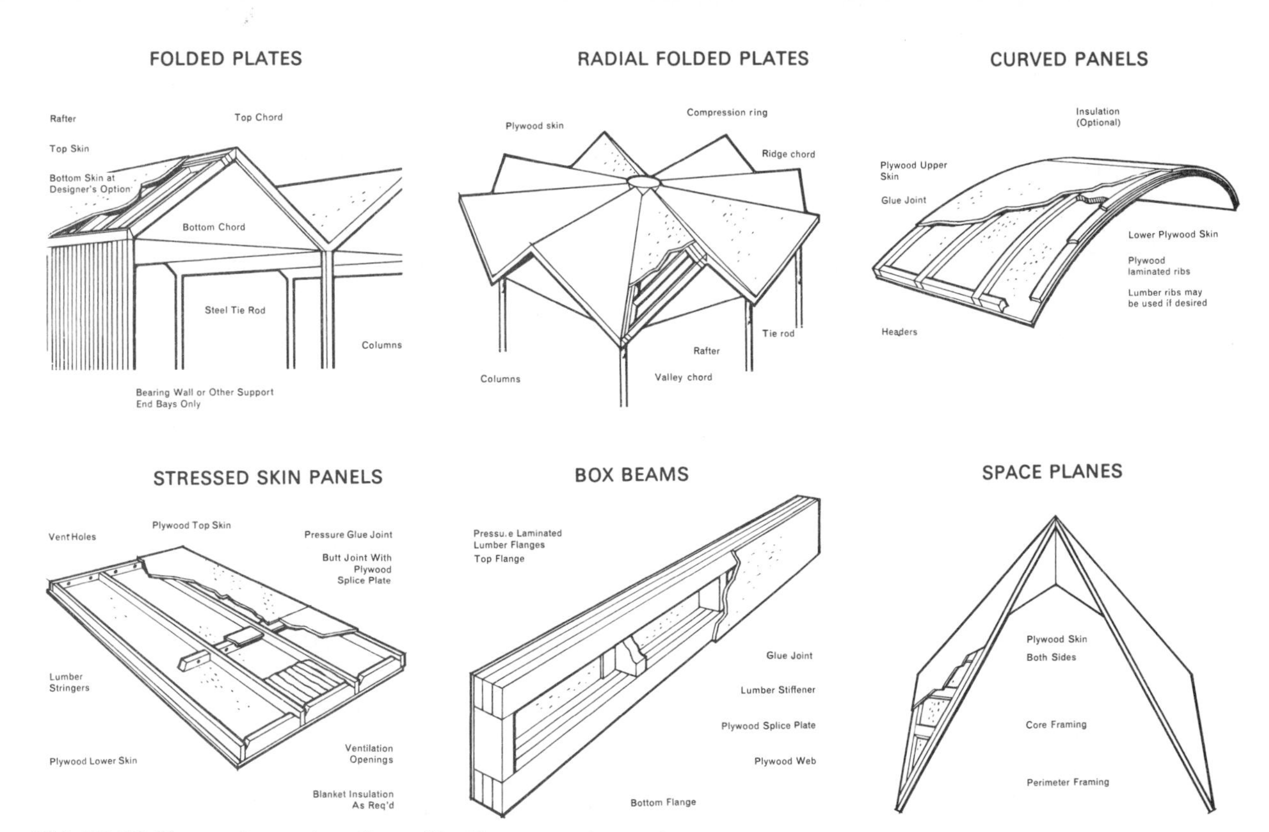

Fig. 57-62 Use and construction of built-up wood members.

Curved Panels

Curved panels are constructed in three types: the *sandwich,* or *honeycomb paper-core,* panel; the *hollow-stressed* end panel; and the *solid-core* panel. The arching action of these panels (Fig. 57-62) permits spanning great distances with a relatively thin cross section.

Folded Plate Roofs

Folded plate roofs are thin skins of plywood reinforced by purlins to form shell structures that can ulilize the strength of plywood. The use of folded plate roofs eliminates trusses and other roof members. The tilted plates lean against one another, acting as giant V-shaped beams supported by walls or columns (Fig. 57-62).

CONCRETE

Pouring concrete into forms is certainly not new to the building industry. But the preparation of concrete building components away from the site is relatively new. The use of reinforced and prestressed concrete for floors, roofs, and walls, and the fabrication of concrete into shells account for an increase in the use of concrete as a building medium.

Prestressed Concrete

When a beam carries a load, it bends, and its center sags lower than the ends. The bottom fibers are stretched and the top fibers are compressed. Concrete can be *prestressed* to eliminate sag. A prestressed beam is made by first stretching wires longer than the beam between two anchors. Concrete is poured around these wires and allowed to cure, and then the stretched wires are released. The wires want to return to their original shape, and so create a force on the concrete. The wires inside are held under stress.

Prestressing can be compared to holding a row of blocks or books between your hands. As long as sufficient pressure can be exerted, as shown in Fig. 57-63A, no sag can occur. The use of prestressed concrete prevents cracks because this concrete is always in compression. Prestressing permits less depth of beams as related to the span. Prestressed concrete has remarkable elastic properties and develops con-

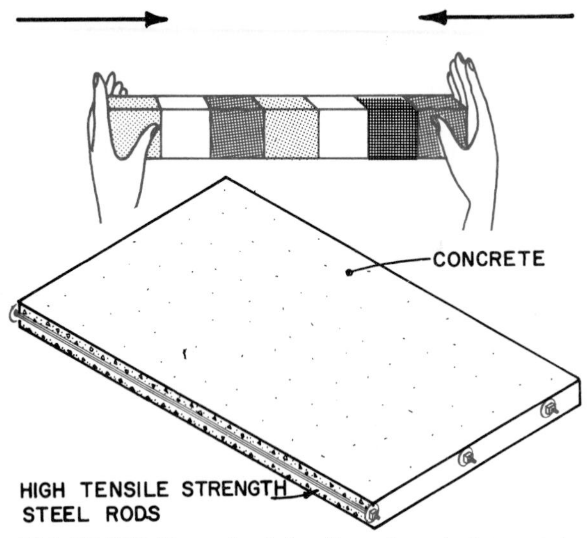

Fig. 57-63A The principle of prestressing concrete.

siderable resistance to shear stresses. Figure 57-63B shows an application of prestressed concrete to a structural frame.

Precast Concrete

Precast concrete is concrete that has been poured into molds prior to its use in construction. When high stresses will not be incurred, precasting without prestressing will suffice for most construction.

Reinforced Concrete

Reinforced concrete is precast or poured-on-site concrete with steel reinforcing rods inserted for stability and rigidity. The rods are not placed under stress as are the wires in prestressed concrete. Reinforced concrete slabs are used extensively for floor systems where short spans make prestressing unnecessary.

Concrete Shells

Concrete shells are curved, thin sheets of concrete usually poured or sprayed on some material that provides a temporary form until the concrete hardens. One method is the use of reinforcing mats that are first laid flat on the ground and then lifted into the desired position. The mats are then sprayed with a coating of concrete that holds the steel rods in place after the concrete solidifies. Concrete-shell construction is becoming increasingly popular in the design of air terminals, auditoriums, and gymnasiums.

ROOF FRAMING DIMENSIONS

In dimensioning roof framing plans, the size and spacing of framing members and major distances between framing components must be shown. Figure 57-64 shows a typical framing plan dimensioned with overall dimensions, subdimensions, and sizes of all framing materials labeled. Regular interval spacings of structural members, such as roof rafters, floor joists, and wall studs, are not dimensioned if they fall on modular increments. Notes are used on framing drawings to show all spacing of members. On detail drawings, overall dimensions are not given; only key distances between structural lev-

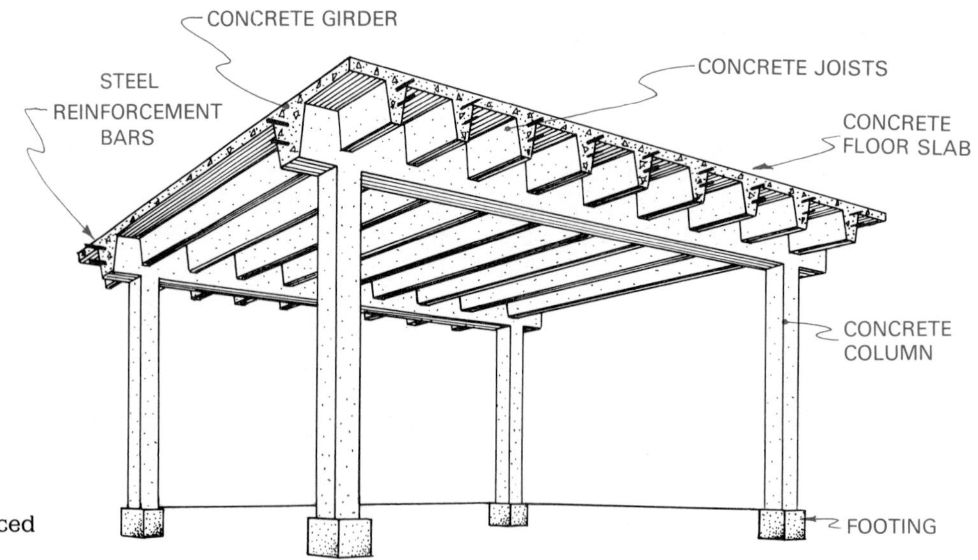

Fig. 57-63B Reinforced concrete frame.

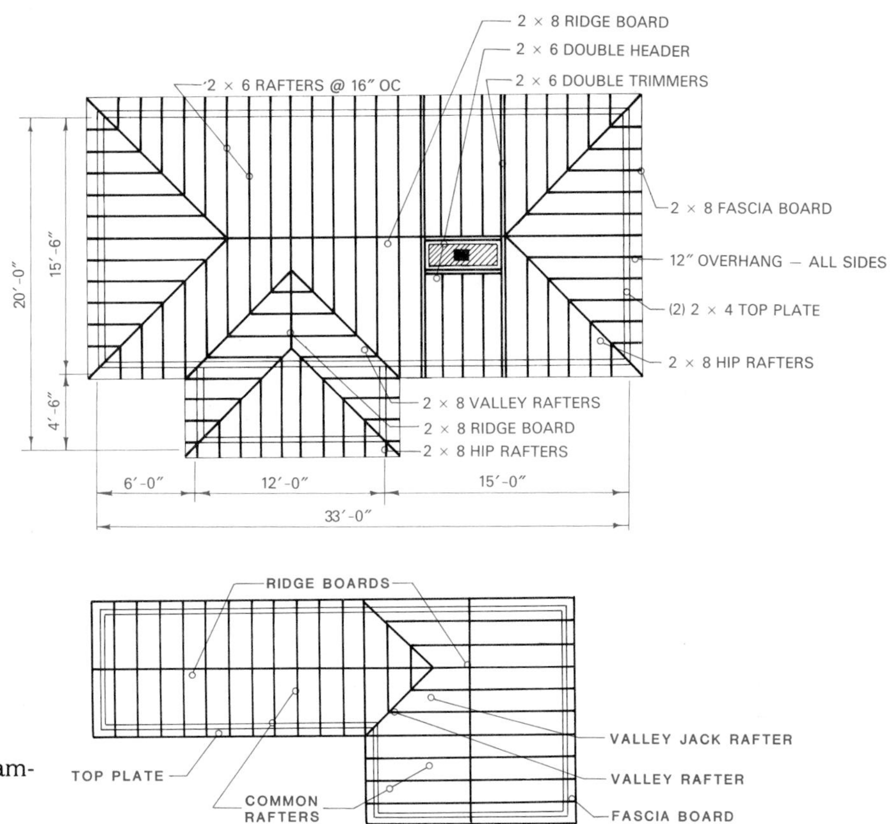

Fig. 57-64 Dimensioned framing plan.

els and horizontal distances are shown. If material sizes are not given on the framing drawing, refer to the specifications list for materials description, which includes sizes.

On modular framing drawings, dimensions are sometimes omitted and a modular grid is used instead. Framing members that fall on the centerlines of the grid are modular and can be determined by counting the number of blocks from one member to another. However, framing members that do not fall on the modular grid are dimensioned separately.

Exercises

1. **Identify the roof framing members shown in Fig. 57-65.**
2. **Draw a roof framing plan for the dormer shown in the roof framing elevation in Fig. 57-66.**
3. **Identify the roof framing members shown in Fig. 57-67.**
4. **Identify the roof framing members shown in Fig. 57-68.**
5. **Match the members to the letters in Fig. 57-69.**

6. **Prepare a roof framing plan for a house of your own design.**

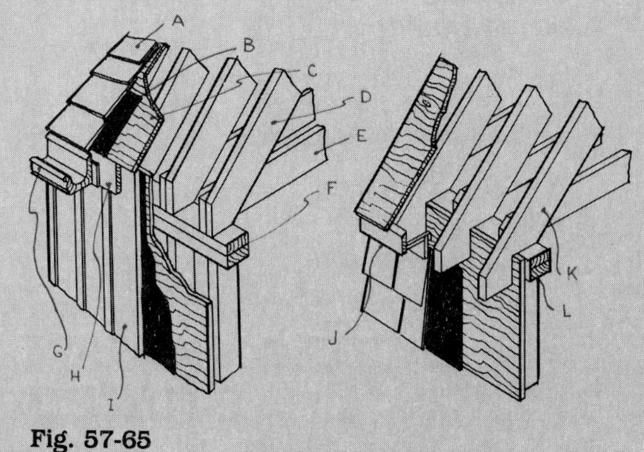

Fig. 57-65

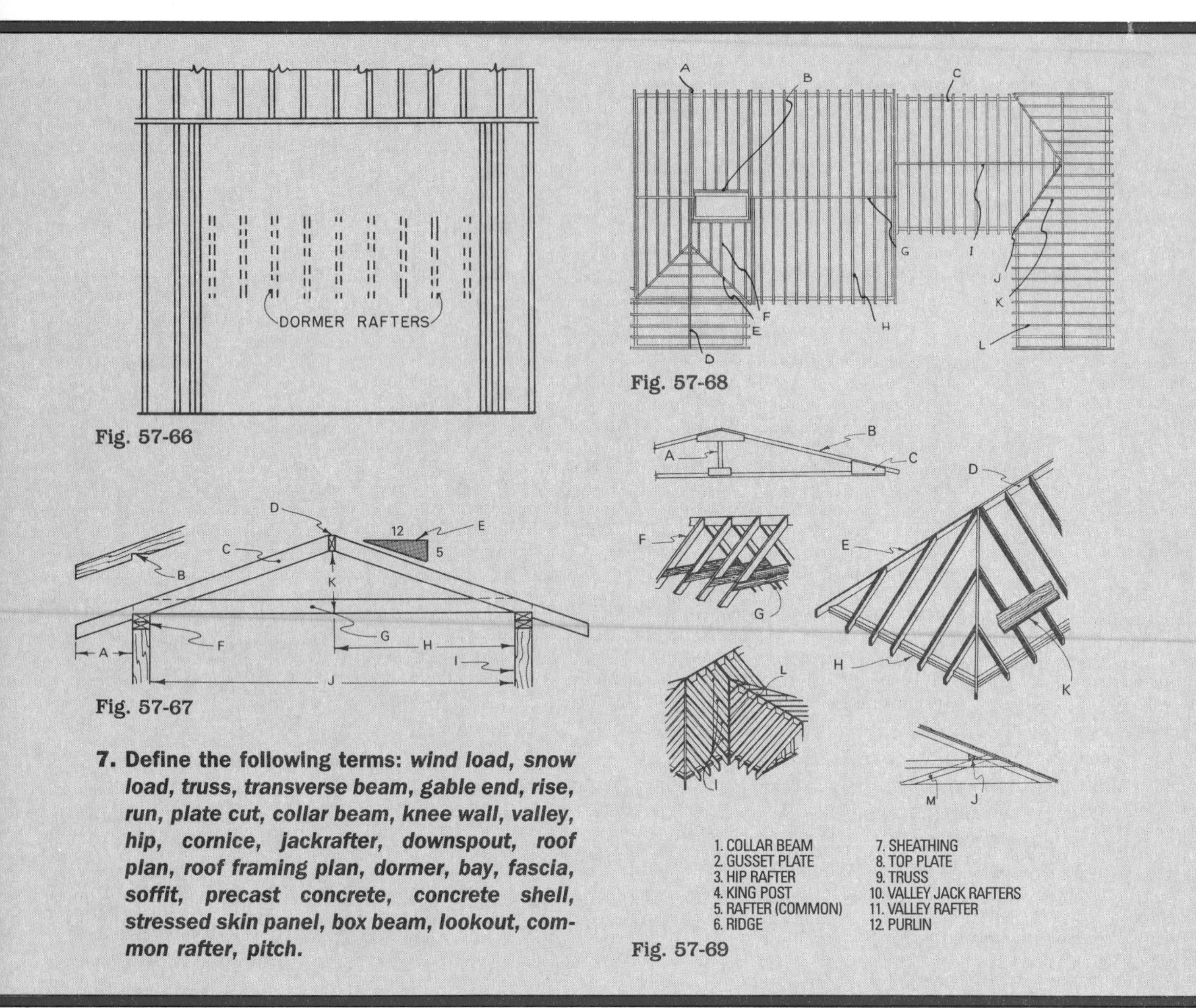

Fig. 57-66

Fig. 57-67

Fig. 57-68

Fig. 57-69

7. Define the following terms: *wind load, snow load, truss, transverse beam, gable end, rise, run, plate cut, collar beam, knee wall, valley, hip, cornice, jackrafter, downspout, roof plan, roof framing plan, dormer, bay, fascia, soffit, precast concrete, concrete shell, stressed skin panel, box beam, lookout, common rafter, pitch.*

UNIT 58

ROOF COVERING MATERIALS

Roof covering protects the building from rain, snow, wind, heat, and cold. Materials used to cover pitched roofs include wood shingles, asphalt shingles, and asbestos shingles. On heavier roofs, tile or slate may also be used. Roll roofing or other sheet material, such as galvanized iron, aluminum, copper, and tin, may also be used for flat or low-pitched roofs. A built-up roof consisting of layers of roofing felt covered with gravel topping may also be used on low-pitched or flat roofs. If a built-up roof is used on high-pitched roofs, the gravel will weather off.

SHEATHING

Roof *sheathing* consists of 1 x 6 lumber or plywood sheets nailed directly to the roof rafters. Sheathing adds rigidity to the roof and provides a surface for the attachment of waterproofing materials. In humid parts of the country, sheathing boards are sometimes spaced slightly apart to provide ventilation and to prevent shingle rot.

ROLL ROOFING

Roll roofing may be used as an *underlayment* for shingles or as a finished roofing material (Fig. 58-1). Roll roofing used as an underlayment includes asphalt and saturated felt. The underlayment serves as a barrier against moisture and wind. Mineral surface and selvage roll-roofing can be used as the final roofing surface.

WEIGHT

The weight of roofing materials is important in computing dead loads. A heavier roofing surface makes the roof more permanent than does a lighter surface. Generally, heavier roofing materials last longer than lighter materials. Therefore, heavy roofing, such as strip or individual shingles, is superior. Roof covering materials are classified by their weight per 100 ft^2 (100 ft^2 equals 1 *square*). Thus 30-lb roofing felt weighs 30 lb per 100 ft^2.

SHINGLES

Shingles are commonly made from asphalt, wood, tile, and slate and are available in a variety of patterns and shapes. Shingles and underlayment are overlapped when applied, as shown in Fig. 58-2. Shingles are best for pitches steeper than 4/12. For pitches less than 4/12, roll roofing is satisfactory.

BUILT-UP ROOFS

Built-up roof coverings are used on flat or extremely low-pitched roofs. Because rain or snow may not be immediately expelled from these roofs, complete waterproofing is essential.

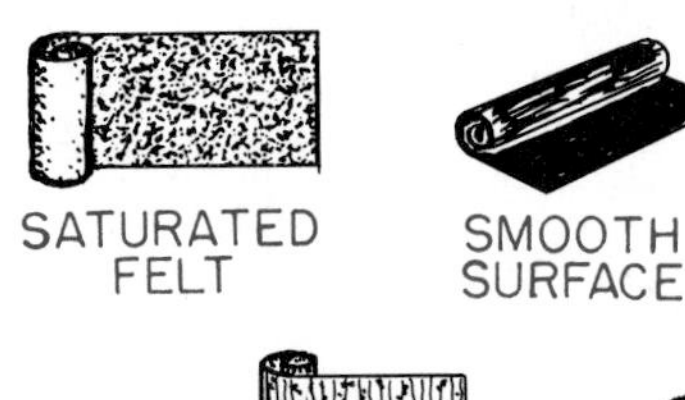

Fig. 58-1 Kinds of roll roofing.

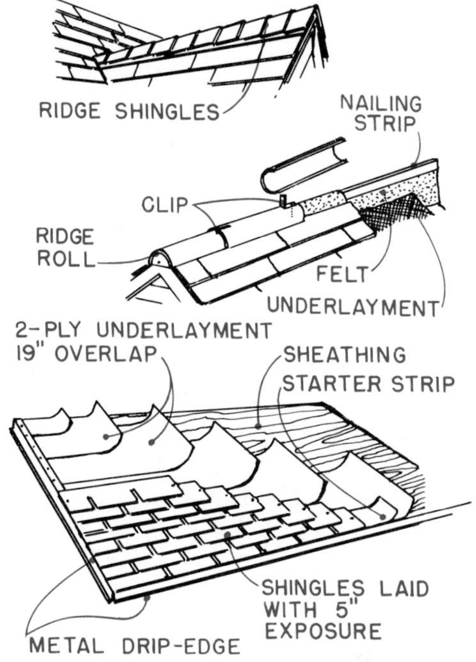

Fig. 58-2 Methods of shingle application.

Built-up roofs may have three, four, or five layers of roofing felt, sealed with hot-mopped tar or asphalt, between each two coatings. The final layer of tar or asphalt is then covered with roofing gravel or a top sheet of roll roofing (Fig. 58-3).

FLASHING

Joints where roof covering materials intersect at the ridge, hip, valley, chimney, and parapet joints must be flashed. *Flashing* is additional covering used on a joint to provide complete waterproofing. Roll-roofing, shingles, or sheet metal is used as the flashing material. For flashing hip and valley joints, shingle flashing is best.

When sheet-metal flashing is used, watertight sheet-metal joints should be used, as shown in Fig. 58-4. Chimney flashing is frequently bonded into the mortar joint with the other end caulked under shingles to provide a waterproof joint.

GUTTERS

Gutters are troughs designed to carry water to the downspouts, where it can be emptied into the sewer system. The two materials used most commonly for gutters are sheet metal and wood such as red cedar and redwood. Plastic gutters and downspouts are growing in use. Gutters may be built into the roof structure, as shown in the

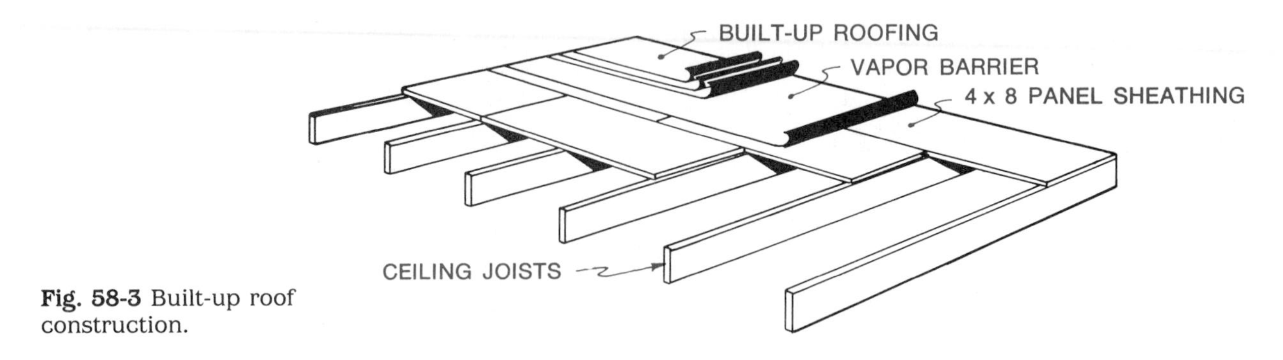

Fig. 58-3 Built-up roof construction.

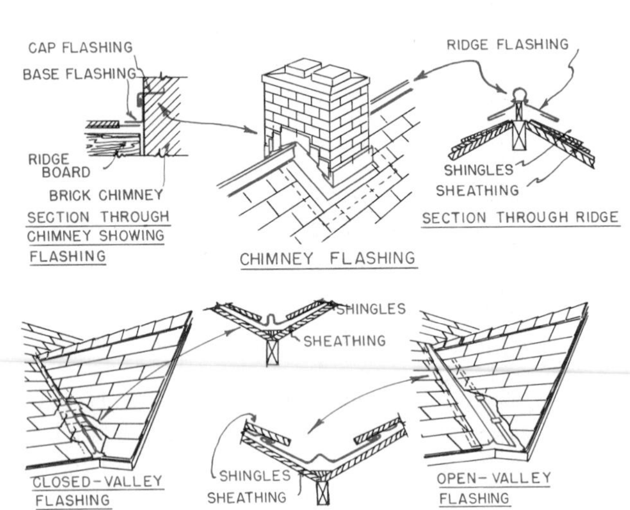

Fig. 58-4 Application of sheet-metal flashing.

fascia-board gutter and the pole gutter in Fig. 58-5. Gutters may be made of additional sheet metal or wood attached or hung from the fascia board. All gutters should be pitched sufficiently to provide for drainage to the downspout. In selecting gutters and downspouts, care must be taken to ensure that their size is adequate for the local rainfall.

SHEATHING PLAN

In large developments often a roof sheathing plan is prepared to plan the best possible arrangement of 4′ x 8′ sheets with the minimum amount of waste. Roof sheathing plans as shown in Fig. 58-6 are prepared to ensure that the sheets are laid as planned. Figure 58-7 shows additional roof covering details.

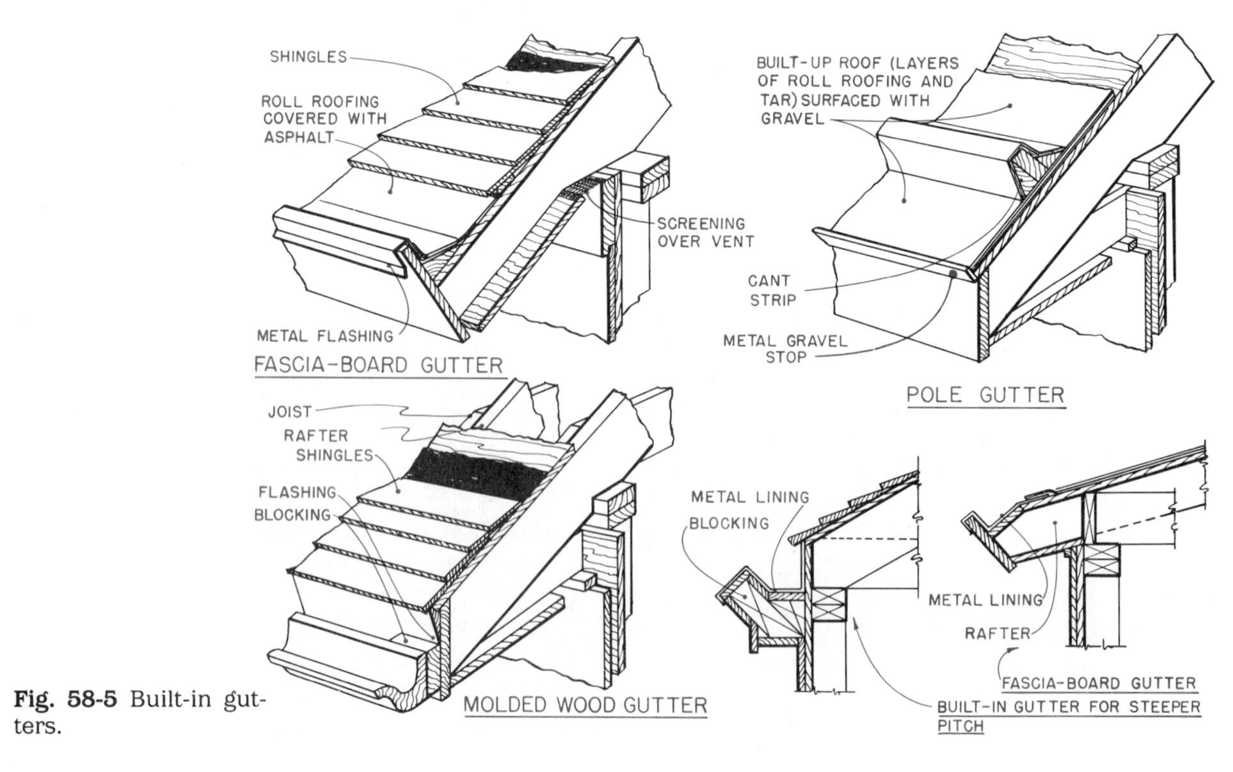

Fig. 58-5 Built-in gutters.

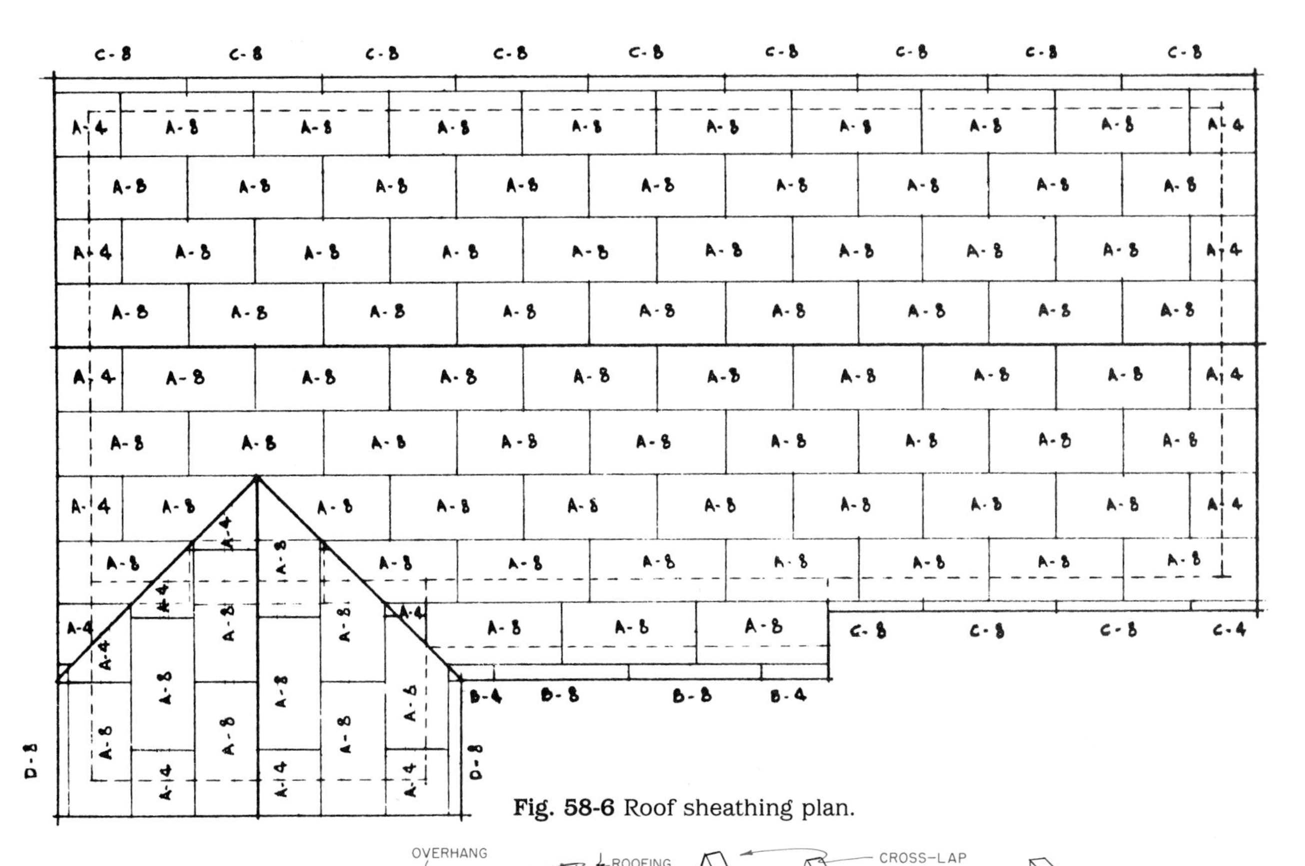

Fig. 58-6 Roof sheathing plan.

Fig. 58-7 Roof construction details.

Exercises

1. Name two types of roll roofing suitable for finished roof covering.
2. Name two types of roll roofing suitable for underlayment.
3. Specify the type of roofing to be used in the house of your design.

4. Draw a cornice section of the house of your design, showing the gutters you have chosen.
5. Define these terms: *roof sheathing, roll roofing, slate, shingles, underlayment, square, roofing felt, flashing, pole gutter, fascia board, louvers.*

SECTION 14

Electrical Planning

Electricity is the major source of energy for most dwellings. Electricity helps us cook, wash, clean, heat, air-condition, light, preserve, and entertain. But the best-designed building is not practical if the wiring is not adequate to bring sufficient power to appliances, lighting fixtures, equipment, and machinery.

UNIT 59

LIGHTING DESIGN

Planning for adequate lighting involves the eyes, the object, and the light source. Planning to light any area involves three questions: How much light is needed? What is the best quality of light? How should this light be distributed? Whether planning lighting for a large commercial building or for a residence, the same design factors must be considered.

TYPES OF LIGHT

Candles and oil and natural-gas lamps were once the major sources of light. Today's major source of light in the home comes from the incandescent lamp and the fluorescent lamp. *Incandescent lamps* use a filament inside the bulb to provide a small, concentrated glow of light when an electric current heats the filament to the glowing point. *Fluorescent lamps* give a uniform glareless light that is ideal for large working areas. Fluorescent lamps give more light per watt, last as much as seven times longer, and generate less heat than incandescent lamps. In the fluorescent lamp, current flows through mercury vapor and activates the light-giving properties of the coating inside the tube.

LIGHT MEASUREMENTS

You can read in bright sunlight or in a dimly lit room because your eyes are adaptable to varying intensities of light (Fig. 59-1). However, you must be given enough time to adjust slowly to different light levels. Sudden extreme changes of light may cause great discomfort.

Light is measured in customary units called *footcandles* (candelas). A footcandle is equal to the amount of light a candle throws on an object 1′ away (Fig. 59-2). Ten footcandles (10 fc) equals the amount of light that 10 candles throw on a surface 1′ away. A 75-watt (75-W) bulb provides 30 fc of light at a distance of 3′. It provides 20 fc at a distance of 6′.

In the metric system the standard unit of illumination is the *lux* (lx). One lux is equal to 0.093 fc. To convert footcandles to lux, multiply by 10.764.

On a clear summer day, the sun delivers 10,000 fc (107,640 lx) of light to the Earth (Fig. 59-3). This is found at the beaches and in open

Fig. 59-1 Eyes will adjust to extreme intensities of light.

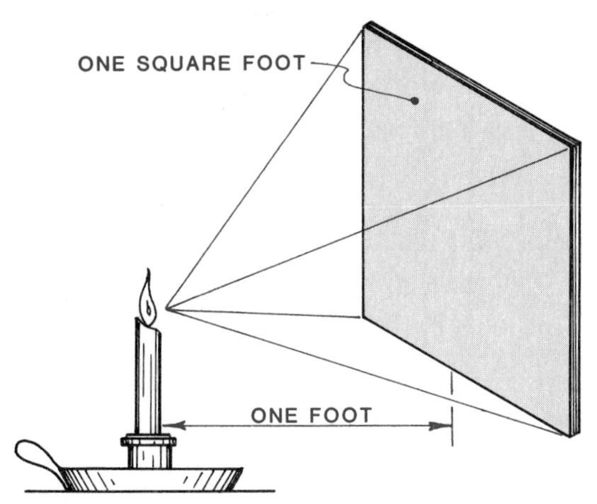

Fig. 59-2 One footcandle of light (candela) is the amount of light striking a 1 ft^2 surface, 1′ away.

fields. In the shade of a tree, there will be 1000 fc (10,764 lx). In the shade on an open porch, there will be 500 fc (5382 lx). Inside the house, a few feet from the window, there will be 200 fc, (2153 lx); and in the center of the house, 10 fc (108 lx).

Accepted light levels for various living activities are as follows: 10 to 20 fc (108 to 215 lx): casual visual tasks, card playing, conversation, television, listening to music; 20 to 30 fc (215 to 320 lx): easy reading, sewing, knitting, house cleaning; 30 to 50 fc (320 to 540 lx): reading newspapers, doing kitchen and laundry work, typing; 50 to 70 fc (540 to 750 lx): prolonged reading, machine sewing, hobbies, homework; 70 to 200 fc (750 to 2150 lx): prolonged detailed tasks such as fine sewing, reading fine print, drafting.

DISPERSAL OF LIGHT

After the necessary amount of light is known, the method of spreading, or *dispersing,* the light through the rooms must be determined.

Types Of Lighting

There are five types of lighting dispersement (Fig. 59-4): direct, indirect, semidirect, semi-indirect, and diffused. *Direct* light shines directly on an object from the light source. *Indirect* light is reflected from large surfaces. *Semidirect* light shines mainly down as direct light, but a small portion of it is directed upward as indirect light. *Semi-indirect* light is mostly reflected but some shines directly. *Diffused* light is spread evenly in all directions.

Reflectance

All objects absorb and reflect light. Some white surfaces reflect 94 percent of the light that strikes them. Some black surfaces reflect only 2 percent. The rest of the light is absorbed. The proper amount of reflectance is determined by the color and type of finish. The amounts of reflectance that are recommended for a room are

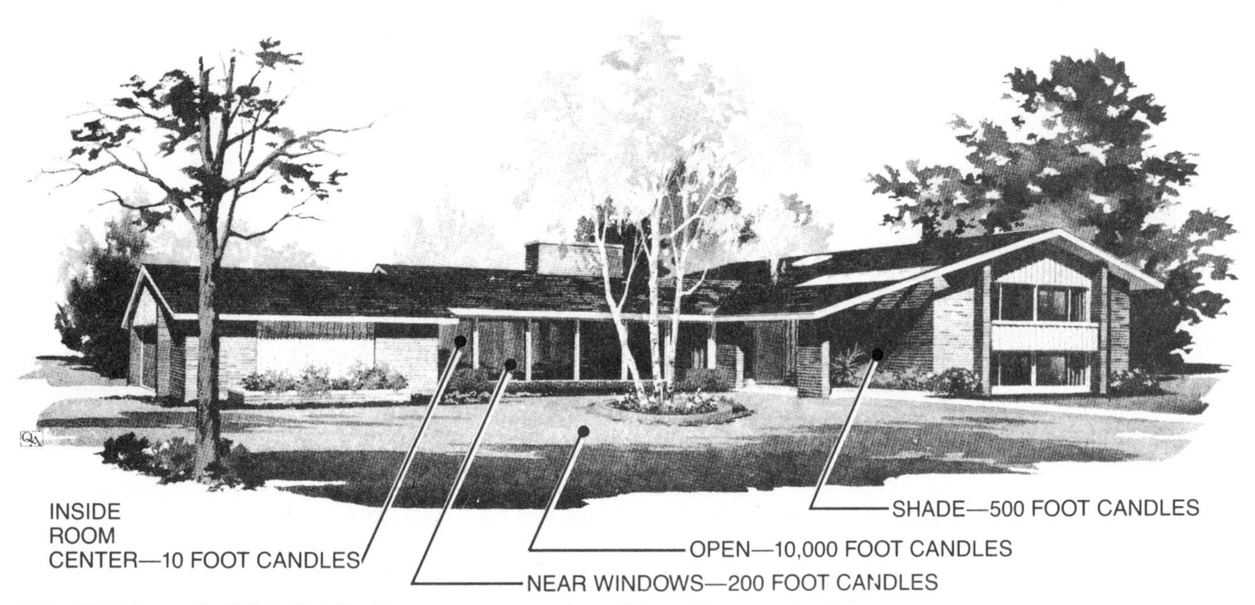

Fig. 59-3 Average light distribution on a sunny day. *(Home Planners, Inc.)*

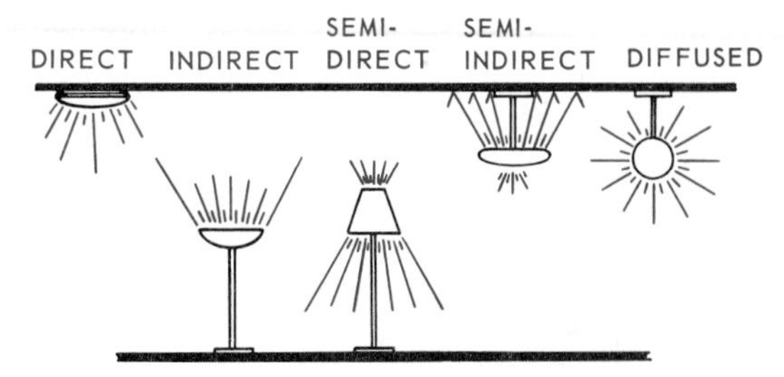

Fig. 59-4 Methods of light dispersement.

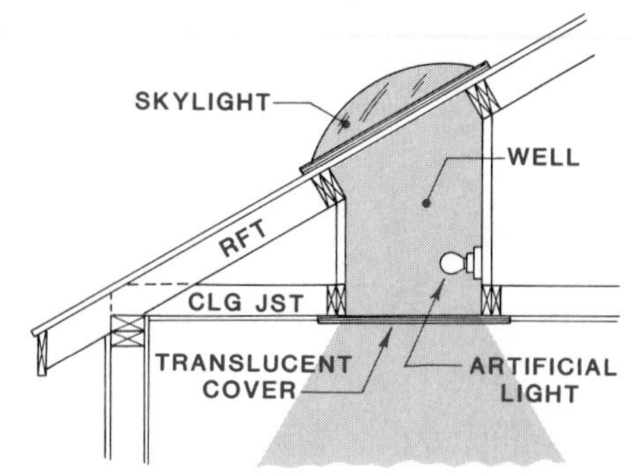

Fig. 59-5 Skylight with artificial light option.

from 60 percent to 90 percent for the ceiling, from 35 to 60 percent for the walls, and from 15 to 35 percent for the floor.

All surfaces in a room will act as a secondary source of light when the light is reflected. Glare can be eliminated from this secondary source of light by having a dull, or matte, finish on surfaces and by avoiding strong beams of light and strong contrasts of light. Eliminating excessive glare is essential in designing adequate lighting.

LIGHTING METHODS

Good lighting in a home depends upon three methods. *General lighting* spreads an even, low-level light throughout a room. *Specific (local) lighting* directs light to an area used for a specific visual task. *Decorative lighting* makes use of lights to develop different moods and to accent objects for interest.

General Lighting

General lighting is achieved by direct or indirect methods of light dispersement. General light can also be produced by portable lamps, ceiling fixtures, or lengths of light on the walls. In the living and sleeping areas, the intensity of general lighting should be between 5 and 10 fc (54 to 108 lx). A higher level of general lighting should be used in the service area and bathrooms. In addition to artificial general light sources, windows and skylights can be used to admit light during the day. If the skylight is covered with translucent panels, it can contain an artificial light source for nighttime use as shown in Fig. 59-5.

Specific Lighting

Specific (local) lighting (Fig. 59-6) for a particular visual task is directed into the area in which the task will be done. The specific light in a room will also add to the general lighting level.

Decorative Lighting

Decorative lighting (Fig. 59-7) is used for atmosphere and interest when activities do not require much light. Bright lights are stimulating; low levels of lighting are quieting. Decorative lighting strives for unusual effects. Some of these can be obtained with candlelight, lights behind draperies, lights under planters, lights in the bottoms of ponds, lights controlled with a dimmer switch, and different types of cover materials over floor lights and spotlights.

ELECTRICAL FIXTURES

The average two-bedroom home should have between 10 and 15 light fixtures. It should also have from 5 to 10 floor, table, or wall lamps. Light fixtures fall into three groups: ceiling fixtures (Fig. 59-8), wall fixtures (Fig. 59-9A), and portable plug-ins. Types of building lighting fixtures include valances, wall brackets, and cornices. A *valance* (Fig. 59-9B) is a covering over a long source of light over a window. Its light illuminates the wall and draperies for the spacious effect that daylight gives a room. A *wall bracket* balances the light of a valance. It gives an upward and downward wash of light difficult to obtain on an inner wall. A *cornice* (Fig. 59-9C) is attached to the ceiling and can be used with or without drapes. All light from this fixture is directed downward, to give an impression of height to the room. A variation of cornice lighting is soffit construction as shown in Fig. 59-9D. The cornice can be framed with a translucent or

Fig. 59-6 Specific lighting.

Fig. 59-7 Decorative lighting.

webbed opening to direct the light and control the intensity.

ILLUMINATION PLANNING

Following are general rules to observe when planning the lighting of each room.

The kitchen requires a high level of general lighting from ceiling fixtures. Specific lighting for all work areas—range, sink, tables, and counters—is also recommended.

The bathroom requires a high level of general lighting from ceiling fixtures. The shower and water closet, if compartmented, should have a recessed, vaporproof light. The mirror should have lights on two sides.

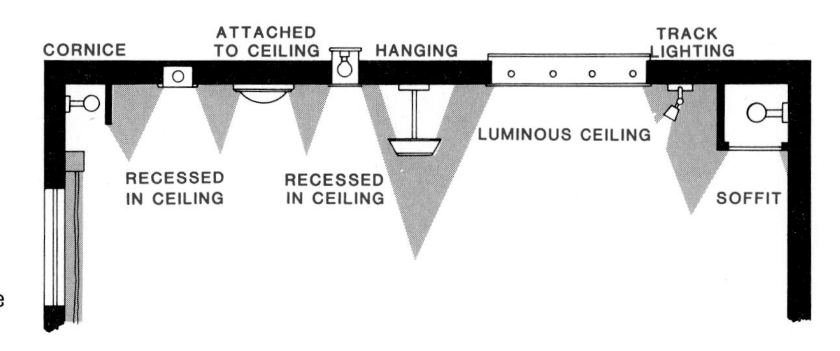

Fig. 59-8 Examples of ceiling fixture types.

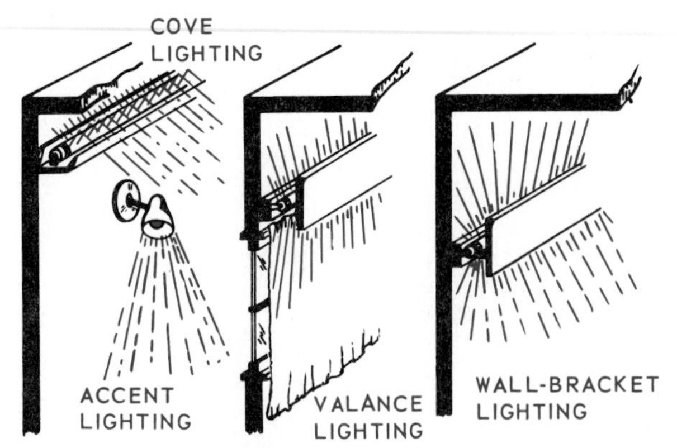

Fig. 59-9A Examples of wall fixture types.

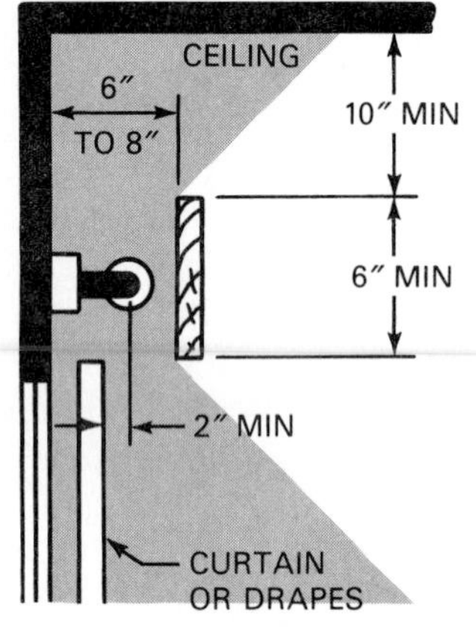

Fig. 59-9B Valance lighting.

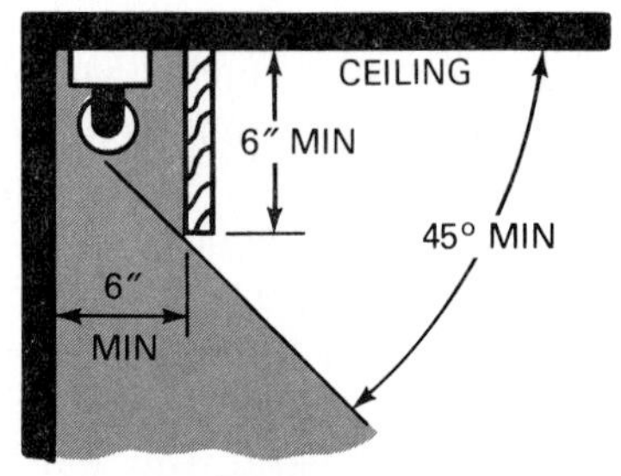

Fig. 59-9C Cornice lighting.

The living room requires a low level of general lighting but should have specific lighting for areas for reading and other visual tasks. Decorative lighting should be used.

The bedroom requires a low level of general lighting but should have specific lighting for reading in bed and on both sides of the dressing-table mirror.

The dressing area requires a high level of general lighting.

Children's bedrooms require a high level of general lighting. Closets should have a fixture placed high at the front.

The dining area requires a low level of general lighting, with local lighting over the dining table.

The entrance and foyer require a high level of general and decorative lighting.

Traffic areas require a high level of general lighting for safety.

Reading and desk areas require a high level of general light and specific light that is diffused and glareless. There should be no shadows.

Television viewing requires a very low level of general lighting. Television should not be viewed in the dark because the strong contrast of dark room and bright screen is tiring to the eyes.

Outdoor lighting is accomplished by waterproof floodlights and spotlights. Extensive outdoor lighting will provide convenience, beauty, and safety. Areas that could be illuminated are the landscaping, game areas, barbecue area, patio, garden, front of picture window, pools, and driveways. Outdoor lights should not shine directly on windows. Lights near the windows should be placed above the windows to eliminate glare. Ground lights should be shielded by bushes to keep them from shining into windows.

		CAVITY DIMENSIONS		
LOCATION	**USE**	**DEPTH**	**WIDTH**	**LENGTH**
KITCHEN	Over sink or work center	8″ to 12″	12″	38″ Min.
BATH OR DRESSING ROOM	Over large mirror	8″	18″ to 24″	Length of mirror
LIVING AREA	Over piano, desk, sofa, or other seeing area	10″	Fit space available 12″ Min.	Fit space available 50″ Min.

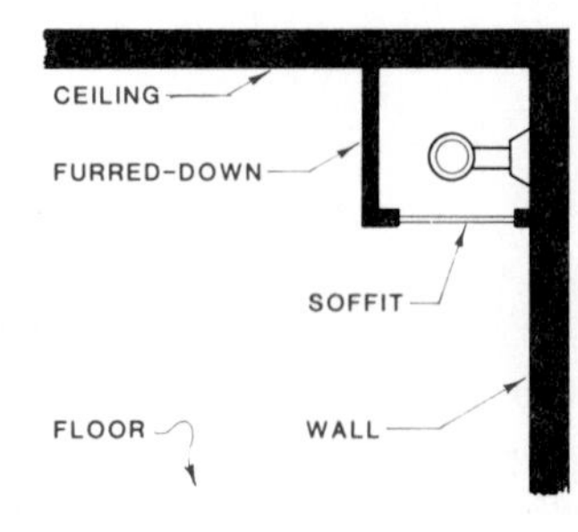

Fig. 59-9D Soffit lighting construction data.

Exercises

1. **List several sources of general, specific, and decorative lighting.**
2. **Plan the lighting needs of the house you are designing.**
3. **Plan the lighting for the floor plan in Fig. 70-1.**
4. **Define these terms: *incandescent bulb, fluorescent tube, filament, direct light, indirect light, semidirect light, diffused light, reflectance, valance lighting, wall-bracket lighting, cornice lighting.***

UNIT 60

ELECTRICAL PRINCIPLES

In this age of technological innovation and vanishing supplies of inexpensive oil, electrical systems must be designed to serve today's needs and also be adaptable to future requirements. Sources of electrical energy, methods of distribution, and devices that consume energy are all important concerns in the design of modern building.

POWER DISTRIBUTION

One of Thomas Edison's greatest inventions was the system of distributing electrical energy. Electrical utility companies and agencies that have grown from his scheme now serve most of the industrial world with vast, interconnected electrical power grids, or networks.

Electrical power is generated from a few basic sources of energy: wind, water, nuclear, fossil fuel, and geothermal. In practically every case, the energy source is harnessed to produce a rotary mechanical motion that will drive a generator. Figure 60-1 shows how the various sources generate electricity. Wind causes windmills (wind turbine) to drive generators. Hydroelectric power uses water falling from a high level to turn a turbine which turns a generator. Nuclear energy heats liquid to make steam to turn a turbine which turns a generator. Fossil fuels, wood agricultural wastes, or garbage burn and heat liquid

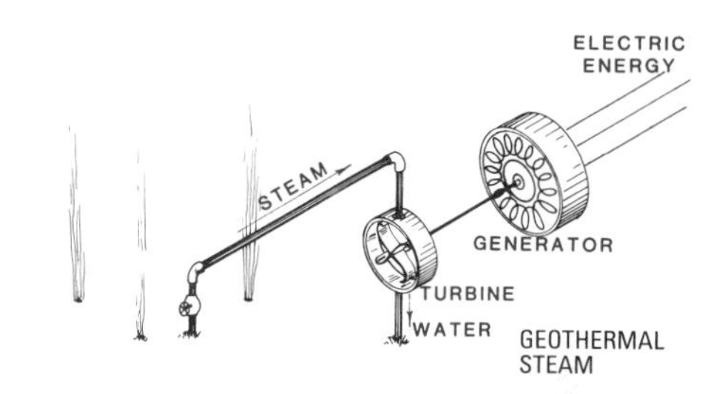

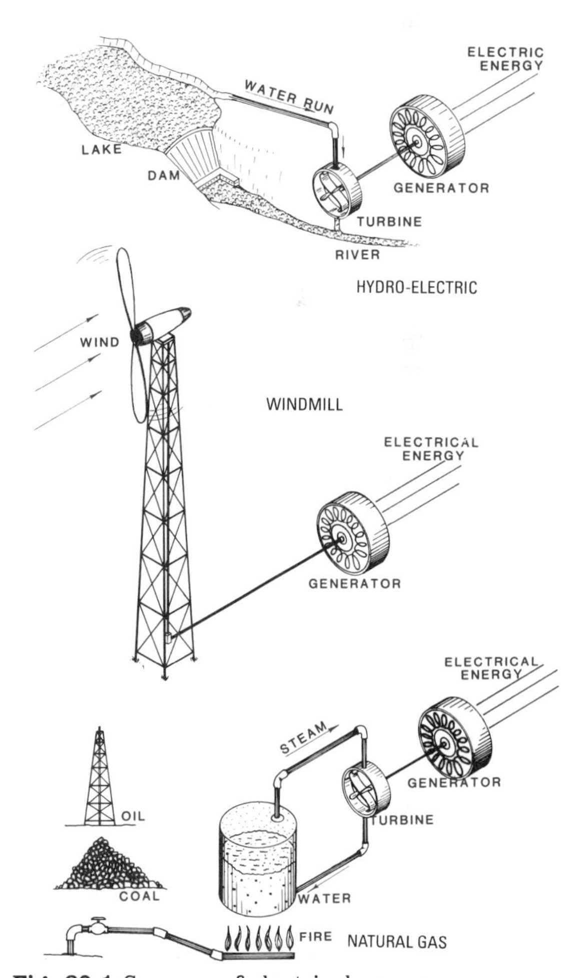

Fig. 60-1 Sources of electrical power.

to steam which drives a turbine. Geothermal steam from natural geysers drives a turbine.

Electrical power is stepped up to very high voltages (hundreds of thousands of volts) for transmission by wires over long distances. High-voltage transmission greatly reduces the loss of power caused by electrical heating of the wires. Wherever the transmission lines enter an industrial or residential community for local power distribution, large transformers are used to step down the voltage to a few thousand volts. Smaller transformers are then used on poles or in underground vaults for final distribution to small groups of houses or individual factories. Usually 110 V is used for most residential circuits and 220 V is used for more heavy duty electrical needs, for example, heat-generating appliances such as air-conditioning units, dryers, and ranges.

WIRING TERMS

Terms used in describing electrical systems include the units of measure and names of electrical equipment. Fortunately, there is only one system of units used for common electrical measurements. Both metric and customary systems use volts, amperes (amps), and watts as the basic units of measurement.

The *volt* is the unit of electrical pressure or potential. It is this pressure that makes electricity flow through a wire. For a particular electrical load, the higher the voltage, the greater will be the amount of electricity that will flow. The technical term for flow of electricity is *current*. The *ampere* is the unit used to measure the magnitude of an electric current. An ampere is defined as the specific quantity of electrons passing a point in 1 second. The amount of current in amperes in a circuit is used to determine wire sizes and the current rating of circuit breakers and fuses. The amount of power required to light lamps, heat water, turn motors, and do all types of work is measured in *watts*. Power depends on both potential (volts) and current (amperes). Current in amperes multiplied by potential in volts equals power in watts (amperes x volts = watts).

The actual work done, or energy used, is the basis for the cost of electricity. The unit used to measure the consumption of electrical energy is the *kilowatt-hour*. A kilowatt is 1000 watts. Time is usually measured in hours. Thus, a 1000-W electric hand iron operating for 1 hour will consume 1 kilowatt-hour (1 kWh). The device used to measure what is consumed is the *watt-hour meter*.

WIRING

Electrical systems in buildings are usually divided into two parts: service and branch circuits. The *service* part consists of all wiring and apparatus needed to bring electricity into the building. *Branch* circuits distribute the electricity throughout the structure.

For safety, convenience, and legal purposes, individual bare wires are not placed within the walls of a building. Wires are covered with insulating materials. Groups of wires may be further covered with plastic or metal jackets. Such groups of wires are called *cables*. Individual wires may be run in pipes or conduits made of steel, aluminum, or plastic. Rigid steel conduit is usually used underground. The steel is heavy-gauge and threaded at both ends. A thinner-gauge conduit called *electric metallic tubing* is often used when the conduit is left exposed—on concrete block walls, for example. Electric metallic tubing is too thin to accept threads, so that force fittings must be used on its ends. Flexible steel or aluminum conduit is commonly used for branch circuits located between walls, in ceilings, or under floors (but not in concrete).

Plastic-jacketed multiwire cable is extremely popular for branch circuits because of its low material and installation costs. Its major disadvantage is that it cannot be changed or expanded easily once it is in place in the walls of a structure. However, its lower cost and ease of initial installation usually offset that disadvantage.

SERVICE

Power is supplied to a building via the *service drop*. Three heavy wires, together called the *drop*, extend from a utility pole or an underground source to the structure. These wires are often twisted into an unobtrusive cable. At the building intersection, the overhead wires are fastened to the structure and spliced to service-entrance wires that enter a conduit through a service weather head. If the service is supplied underground, three wires are run in a rigid conduit. An underground service conduit is either brought directly to the meter socket or to a pulling gutter, depending upon local codes and utility-company requirements. The pulling gutter is used so that the heavy service wires can be grasped easily and pulled with sufficient force, without scraping them on the meter-socket housing as shown in Fig 60-2A. In preparing service drop paths on plot plans, minimum height requirements for connector lines must be carefully planned. Figure 60-2B shows the normal minimum line clearance required by electrical code. If these distances cannot be maintained, rigid conduit, electrical metallic tubing, or busways (channels, ducts) must be specified.

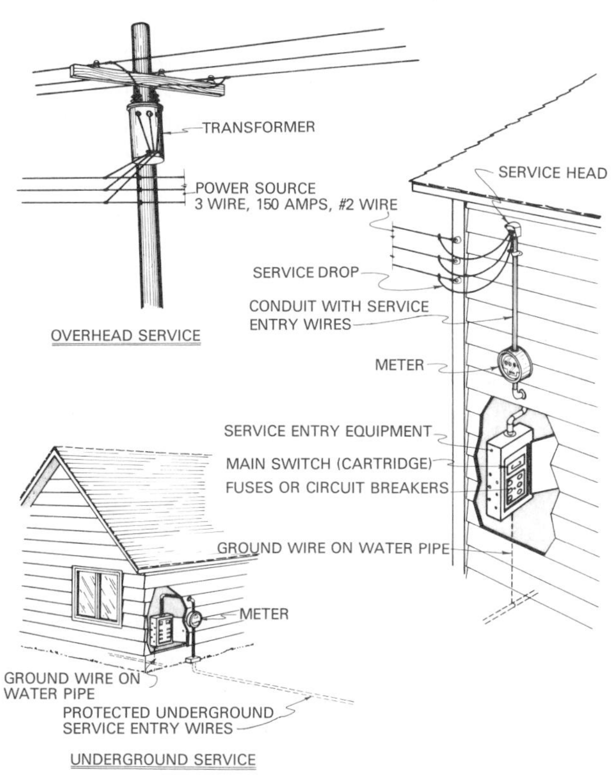

Fig. 60-2A Service entrance equipment.

Modern service-entrance equipment is usually built so that the pulling gutter (if required), the meter socket, the main breaker (or switch and fuses), and a bank of branch circuit breakers are mounted in one steel enclosure.

Branch circuits are distributed throughout the home from a distribution panel (Figure 60-3). Each circuit is protected with its own circuit breaker. Some circuits require a two-pole circuit breaker for protection. *Circuit breakers* are used to protect wiring against overheating and possible fire due to overloading. If the current (amperes) in a branch circuit exceeds the rating of its circuit breaker, the breaker will disconnect (*trip*). When the breaker trips, power to the branch circuit is disconnected. If the sum of the current drawn by the individual branch circuits exceeds the rating of the main circuit breaker,

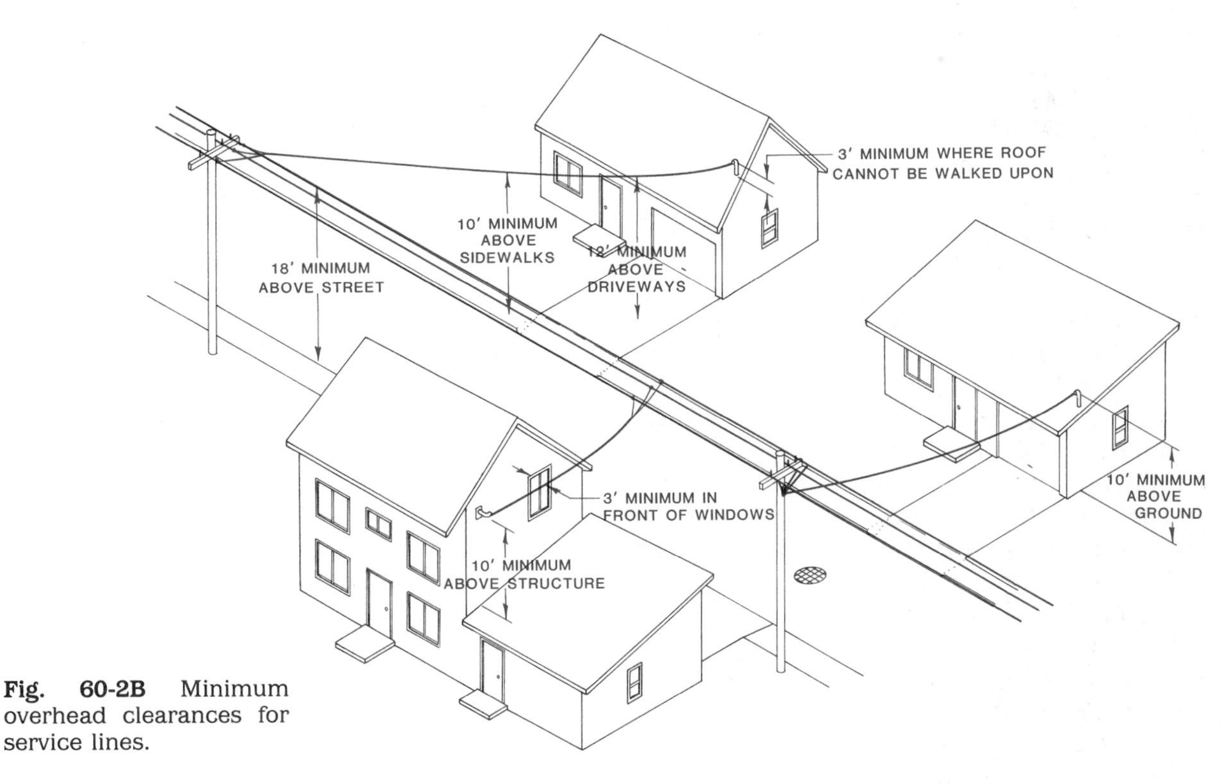

Fig. 60-2B Minimum overhead clearances for service lines.

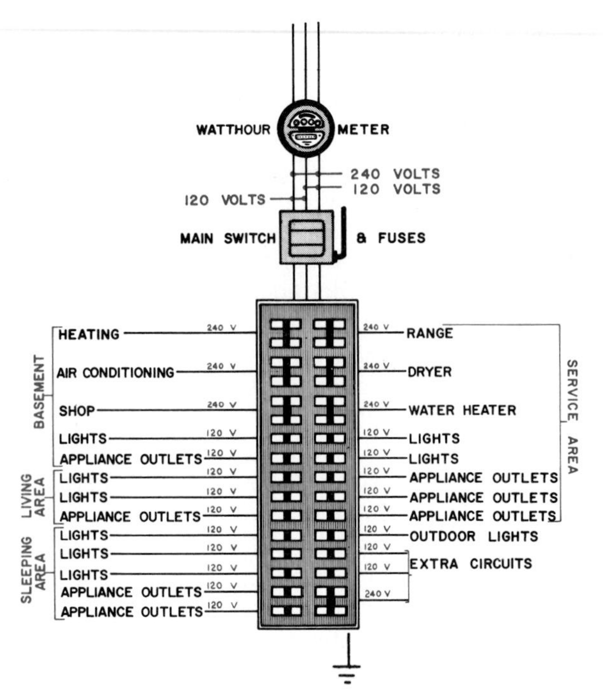

Fig. 60-3 Distribution panel.

the main breaker will trip. This protects the service-entrance wires and equipment from overheating.

SAFETY

In addition to protecting wiring and equipment from overloading, it is important to protect people from shock and electrocution. All electrical codes require some type of safety device to ensure an automatic cutoff of power if the current exceeds the capacity of the circuit. Several devices are used to provide protection against dangerous circuit faults. These include circuit breakers, ground-fault circuit interrupters, fuses and interlocks. *Circuit breakers* are spring-loaded devices in which a strip of metal bends when heated and trips a switch which disconnects power to the circuit. *Ground-fault circuit interrupters* are monitoring devices which turn off the power to the circuit when a preset amount of current is reached in any conductor in a circuit. *Fuses* are simple devices in which a piece of metal melts when the circuit is heated and interrupts the current. Interlocks are switches which automatically disconnect the power when a circuit cabinet is opened.

Designers must ensure that all electrical circuits are properly grounded. This involves not only the use of grounding wires in all circuits but also the grounding of the entire system and all electrical equipment. A system is grounded by connecting one of the wires to a cold-water pipe or a copper rod driven into the ground. Equipment grounding is accomplished by connecting a grounding wire from the non-current-carrying metal part of a piece of equipment to some ground point.

Figure 60-4 shows a simplified residential electrical system with a main circuit breaker and eight branch circuits. A receptacle is shown connected to circuit #7. The black wire connected to the circuit breaker is not grounded. It is called a *hot lead* since there is a voltage between it and the ground or neutral. The white wire is the branch-circuit neutral wire. The branch-circuit grounding system (green or bare wires or the conduit in an all-metallic system) is also connected to the neutral bus. A bus is a copper bar used to connect the neutral and ground wires from all circuits. The neutral bus in turn is connected to a ground stake (or stakes) driven into the ground outside the building, usually near the service entrance. The unmetered sides of incoming water lines were commonly used as the earth ground; however, they are no longer considered adequate grounds.

The black and white leads connected to the same receptacle carry the same current.

The refrigerator freezer shown in Fig. 60-4 is connected to the branch circuit by a three-wire cord with a grounding plug. The center prong of the plug is connected directly to the metal frame or cabinet of the appliance. If an uninsulated black (hot) wire should touch the cabinet, excess current would flow in the black wire, and the circuit breaker would trip. If the appliance cabinet was not grounded (that is, if there was no green wire), no excess current would flow, and the appliance would continue to operate. With the hot wire touching the cabinet, any person touching ground (a water pipe, a faucet, a radiator, for example) and the refrigerator would receive a painful shock, possible injury, or electrocution.

Large appliances, including refrigerators, freezers, washers, dryers, and the like, are required to have grounding cords and plugs. Some small appliances and power tools also use the three-wire grounding plug or double insulation for shock protection.

In any properly functioning branch circuit, the current in the black (hot) lead is the same as the

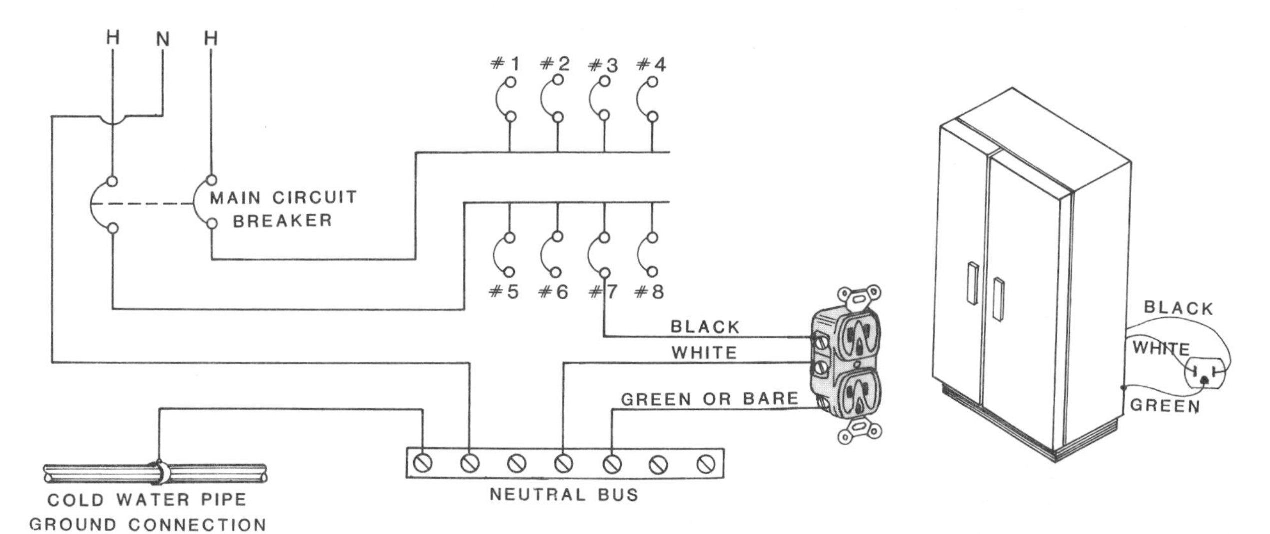

Fig. 60-4 Residential electrical system.

current in the white (neutral) lead. In a device such as an electric shaver, there is no grounding (green) wire. If the black wire touches an exposed metal part of the shaver, a path to ground could exist. The current path would be from the black wire to the metal part on the shaver, to the skin of the person touching the metal part, and through the water faucet touched by the person to the ground. This potentially dangerous situation can be avoided by using a ground-fault circuit interrupter (GFCI).

The purpose of a GFCI receptacle, or a ground-fault detector, is to cut off the current at the convenience outlet. When the GFCI receptacle senses any change of current it will immediately trip a switch stopping the current. It operates faster and is safer than the circuit breaker switch or fuse at the power entry panel.

The GFCI operates by sensing the current in the black and white leads. These currents must be equal in magnitude and opposite in direction. If they are not—suppose that the black wire accidentally touches an alternate return path—the GFCI opens the circuit. What makes the GFCI important is that it can detect very low currents that would not trip the branch circuit breaker. These low currents (sometimes called *leakage current*) do not need solid metal paths but can travel along damp surfaces. Though low, they can cause injury and even death.

GFCIs are available in two basic forms. Both can be used to protect more than one receptacle. One form of GFCI incorporates a circuit breaker and mounts in the distribution panel. All outlets on the branch circuit connected to this GFCI circuit breaker are protected. The other form of GFCI is part of a receptacle occupying the same space as a duplex receptacle. It has no circuit breaker. It can be used alone or at the beginning of a branch circuit to protect all the outlets that follow. All GFCIs contain a built-in self-testing feature. Regularly, usually monthly, the test button should be pressed to see if the GFCI is functioning properly. GFCIs are available for 15- and 20-A circuits.

SERVICE AND BRANCH REQUIREMENTS

The National Electric Code (NEC) provides the basis for calculating the requirements for a minimum electrical service. For this example, consider a typical 1500-ft^2 single-family house. This house has two bedrooms, two bathrooms, a kitchen, a laundry room/service porch, a dining room, and a living/family room. The house has an unimproved attic and an attached garage. It has no basement. Natural gas is used for cooking and heating. Water is heated by solar energy augmented by a natural-gas hot-water heater. No air conditioning is provided. The calculations start with the individual branch circuits.

General Lighting Loads

In all dwellings other than hotels, the NEC requires a minimum general lighting load of 3 W/ft^2 of floor space. However, the amount of wattage demanded at one time (demand factor) is calculated at 100 percent for only the first 3000 W; 35 percent is used for the second 117,000 W and 25 percent is used for any demand over 120,000 W. Thus the general lighting load planned for a 1500-ft^2 house would be 3525 W, not the full 4500 W:

$$\begin{array}{rrcl} & 1500\text{ft}^2 \times 3\text{ W} & = & 4500\text{ W} \\ \text{First} & 3000\text{ W} \times 100\% & = & 3000\text{ W} \\ \text{Next} & \underline{1500\text{ W}} \times 35\% & = & \underline{\ \ 525\text{ W}} \\ & 4500\text{ W} & & 3525\text{ W} \end{array}$$

Since each branch circuit can supply 2400 W (120 V × 20 A = 2400 W), a 1500-ft^2 house should have two 1860 W general lighting circuits as shown in Fig. 60-5. Note that the general lighting load includes general-purpose receptacles. The total load is distributed between the two hot legs to equalize the load and to prevent a total blackout in any one room, where practical. Since two circuits feed each bedroom and the living room, a failure on one circuit will not darken all lights in each room. A three-wire arrangement with a common neutral is employed.

Small-Appliance Circuits

A minimum of two branch circuits for small appliances is required by the NEC. These circuits feed only the kitchen, the dining room, and the family room. The two circuits required in the kitchen may be wired alternately to adjacent receptacles or by using split-link receptacles; that is, each half of a duplex receptacle may be on a different circuit. GFCI outlets or circuit breakers offer life-saving protection at kitchen counter receptacles, especially those near sinks.

Laundry Circuit

A separate 20-A circuit is required in a laundry area to provide power for the washing machine and the motor and controls of the gas dryer. Again, GFCI is not required by many electrical codes, but because of the danger of leakage currents, one is recommended.

Required GFCI Circuits

A GFCI receptacle must be located wherever there is a possibility for people to ground themselves and be shocked with the electrical current flowing through the body and into the ground. Therefore, in new construction GFCI receptacles must be located with each convenience outlet near water sources and/or pipes in the bathroom, kitchen, garage, and outdoors (Fig. 60-6). Since outside receptacles are used for such devices as electric garden tools, auto polishers, barbecue starters, and rotisseries, they should be located for maximum convenience. A receptacle located not less than 10 or more than 15′ from the inside wall of a permanently installed swimming pool must also be wired through a GFCI. GFCIs are also recommended for receptacles along kitchen counters and in laundry

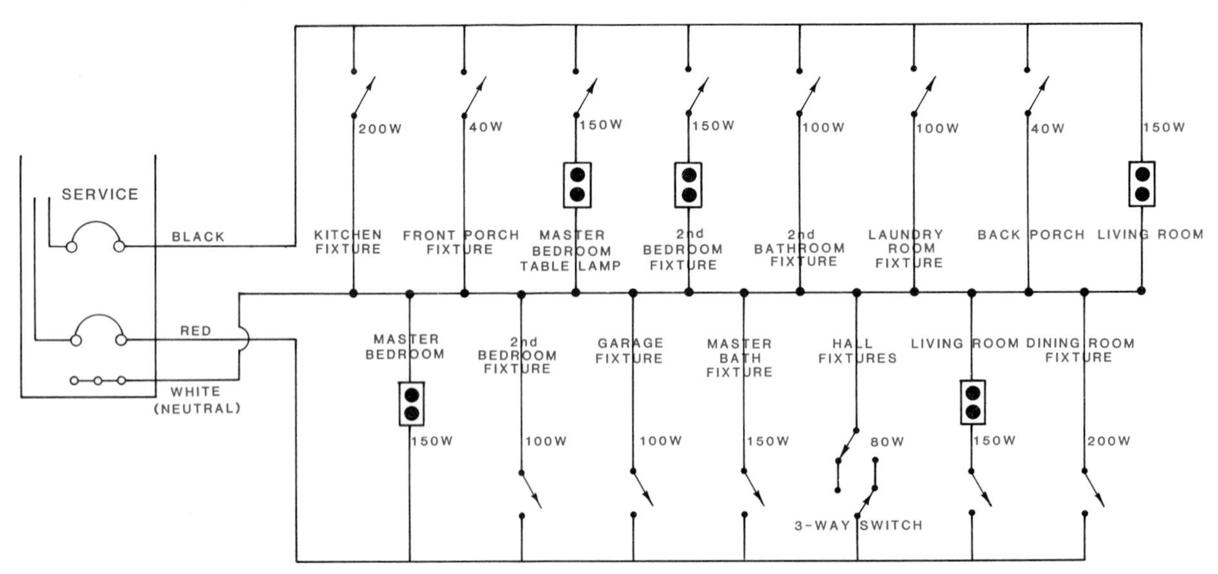

Fig. 60-5 Typical lighting load arrangement.

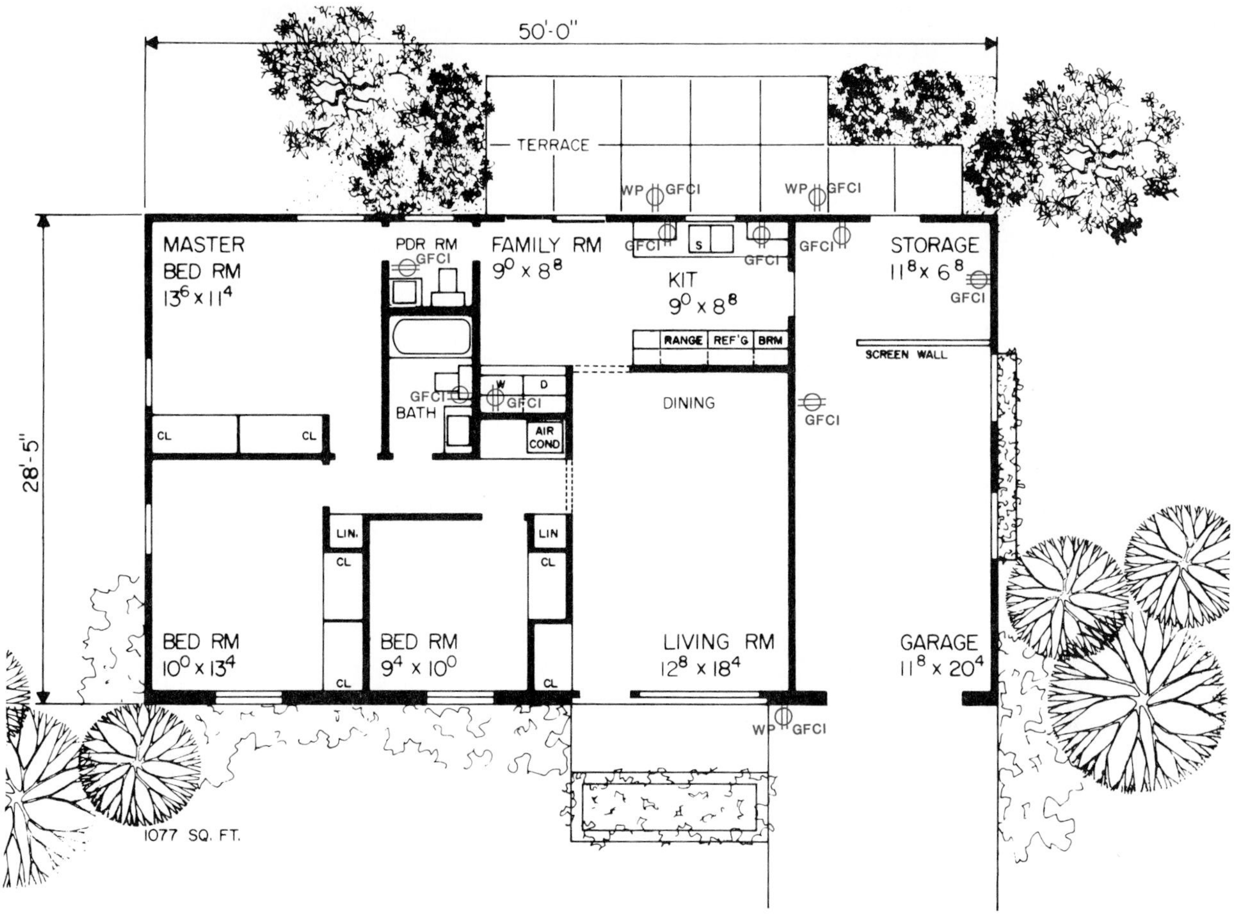

Fig. 60-6 Recommended location for ground-fault circuit interrupters. *(GFCIs)*

rooms. A minimum of four more receptacles is required by the NEC. They may be on one circuit protected by a single 20-A GFCI circuit breaker, or they may use individual GFCI receptacles. A GFCI valve will trip in 1/40 second when a ground fault of 0.005 A is reached.

Service Size Requirements

Since all lights and all appliances are probably not going to be used at the same time, it is not economical to provide a service capable of supplying the full load. The NEC permits each small-appliance circuit and each laundry circuit to be computed as a 1500-W load.

The minimum service includes:

3 W/ft^2 for general lighting load	4500 W
1500 W for each small-appliance circuit	3000 W
1500 W for laundry circuit	1500 W
	9000 W

If this building has six or more two-wire branch circuits, the NEC requires a minimum of 100-A conductors. That is, the wire size must be capable of carrying 100 A in a three-wire service (two hot wires and a neutral) leg. Complete procedures for loads of various types are outlined in the National Electrical Code.

Large Appliances and Window Air Conditioners

Branch circuits for large appliances must be matched to the appliance. The common 120-V, 15-A circuit is often inadequate for certain motor loads, for example. Large appliances sometimes require individual circuits and operate on 240 V. The 240-V potential is obtained by connecting the two hot legs to the load through circuit breakers in each leg. Figure 60-7 shows acceptable electrical loads and circuits for residential wiring systems.

	Typical connected watts	Volts	Wires	Circuit breaker or fuse	Outlets on circuit	Outlet type	Notes
KITCHEN							
RANGE	12,500	120/240	3 #6 + GND	50A	1	14-50R	
OVEN (BUILT-IN)	4,500	120/240	3 #10 + GND	30A	1	14-30R	#1
RANGE TOP	6,000	120/240	3 #10 + GND	30A	1	14-30R	#1
DISHWASHER	1,500	120	2 #12 + GND	20A	1	5-15R	#2
WASTE DISPOSER	800	120	2 #12 + GND	20A	1	5-15R	#2
TRASH COMPACTOR	1,200	120	2 #12 + GND	20A	1	5-15R	#2
MICROWAVE OVEN	1,450	120	2 #12 + GND	20A	1 or more	5-15R	
BROILER	1,500	120	2 #12 + GND	20A	1 or more	5-15R	#3
FRYER	1,300	120	2 #12 + GND	20A	1 or more	5-15R	#3
COFFEEMAKER	1,000	120	2 #12 + GND	20A	1 or more	5-15R	#3
REFRIGERATOR/ FREEZER 16-25 cubic feet	800	120	2 #12 + GND	20A	1 or more	5-15R	#4
FREEZER chest or upright 14-25 cubic feet	600	120	2 #12 + GND	20A	1 or more	5-15R	#4
LAUNDRY							
WASHING MACHINE	1,200	120	2 #12 + GND	20A	1 or more	5-15R	#5
DRYER all-electric	5,200	120/240	3 #10 + GND	30A	1	14-30R	#1
DRYER gas/electric	500	120	2 #12 + GND	20A	1 or more	5-15R	#5
IRONER	1,650	120	2 #12 + GND	20A	1 or more	5-15R	
HAND IRON	1,000	120	2 #12 + GND	20A	1 or more	5-15R	
WATER HEATER	3,000-6,000					DIRECT	#6

Fig. 60-7 Load requirements of electrical appliances.

	Typical connected watts	Volts	Wires	Circuit breaker or fuse	Outlets on circuit	Outlet type	Notes
LIVING AREAS							
WORKSHOP	1,500	120	2 #12 + GND	20A	1 or more	5-15R	#7
PORTABLE HEATER	1,300	120	2 #12 + GND	20A	1	5-15R	#3
TELEVISION	300	120	2 #12 + GND	20A	1 or more	5-15R	#8
PORTABLE LIGHTING	1,200	120	2 #12 + GND	20A	1 or more	5-15R	#9
FIXED UTILITIES							
FIXED LIGHTING	1,200	120	2 #12	20A			#10
WINDOW AIR CONDITIONER							
14 000 BTU	1,400	120	2 #12 + GND	20A	1	5-15R	#11
25 000	3,600	240	2 #12 + GND	20A	1	6-20R	#11
29 000	4,300	240	2 #10 + GND	30A	1	6-30R	#11
CENTRAL AIR CONDITIONER							
23 000 BTU	2,200	240					#6
57 000 BTU	5,800	240					#6
HEAT PUMP	14,000	240					#6
SUMP PUMP	300	120	2 #12	20A	1 or more	5-15R	#1
HEATING PLANT oil or gas	600	120	2 #12	20A			#6
FIXED BATHROOM HEATER	1,500	120	2 #12	20A			#6
ATTIC FAN	300	120	2 #12	20A	1	5-15R	

NOTES

#1 May be direct-connected.

#2 May be direct-connected on a single circuit; otherwise, grounded receptacles required.

#3 Heavy-duty appliances regularly used at one location should have a separated circuit. Only one such unit should be attached to a single circuit at a time.

#4 Separate circuit serving only refrigerator and freezer is recommended.

#5 Grounding-type receptacle required. Separate circuit is recommended.

#6 Consult manufacturer for recommended connections.

#7 Separate circuit recommended.

#8 Should not be connected to appliance circuits.

#9 Provide one circuit for each 500 sq. ft (46 m^2). Divided receptacle may be switched.

#10 Provide at least one circuit for each 1200 watts of fixed lighting.

#11 Consider 20-amp, 3-wire circuits to all window-type air conditioners. Outlets may then be adapted to individual 120- or 240-volt units. This scheme will work for all but the very largest units.

Exercises

1. **Calculate the number of general lighting circuits needed for the plan shown in Fig. 60-8. Also indicate the location of GFCIs.**
2. **Draw Fig. 67-1 and complete the electrical plan. Show all the circuits and label each type. Label the voltage, wattage, wire size, and fuse size for each circuit.**

3. **List the rooms on each circuit for the house you are designing. Draw each circuit. Locate the position of the service drop and distribution panel.**
4. **Define these terms: *wiring system, generator, transformer, voltage, ampere, watt, circuit, electric current, overloading, conductor, insulator, kilowatt, kilowatt-hour, cables, conduit, electric metallic tubing, service drop, service entrance, distribution panel, service circuit, branch circuits, fuse, circuit breaker, outlet, lighting circuit, small-appliance circuit, grounding receptacles, ground-fault circuit interrupter.***

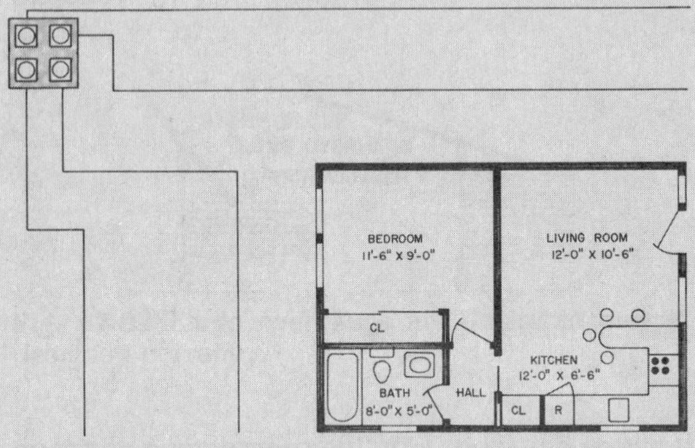

Fig. 60-8

UNIT 61

DRAWING ELECTRICAL PLANS

Wiring methods are controlled by building codes. Wiring is performed or approved by licensed electricians. However, wiring plans are prepared by the designers. For large structures, a consulting electrical contractor may prepare the final detailed electrical plans. Electrical plans include data on the type and location of all switches, fixtures, and controls.

PLANNING RULES

Basic rules to follow when planning electrical systems are as follows:

1. The main source of light in a room should be controlled by a wall switch located on the latch side of the room's entrance. It should not be necessary to walk into a dark room to find the light switch (Fig. 61-1).

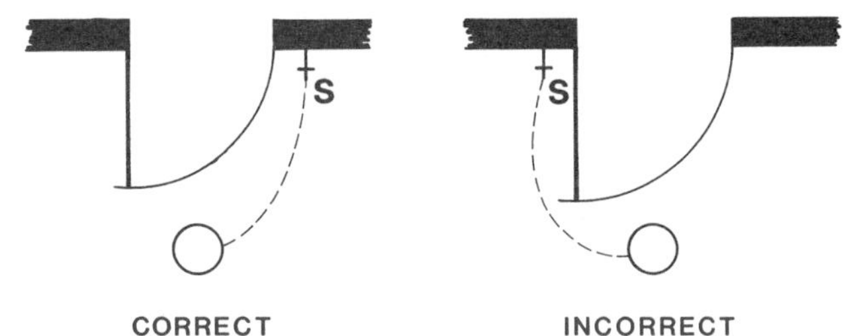

Fig. 61-1 The light switch should be located on the latch side of the door.

2. Outlets (except in kitchens) should average one for every 6′ (1.8 m) of wall space.

3. Electrical outlets in kitchens should average one for every 4′ (1.2 m) of wall space.

4. Walls between doors should have an outlet, regardless of the size of the wall space (Fig. 61-2).

5. Each room should have a major source of light for the entire room, controlled from one switch located near the entry (Fig. 61-3).

6. Each room should have adequate lighting for all visual tasks.

7. Each room should have at least one easy-to-reach outlet for the vacuum cleaner or other appliances that are often used.

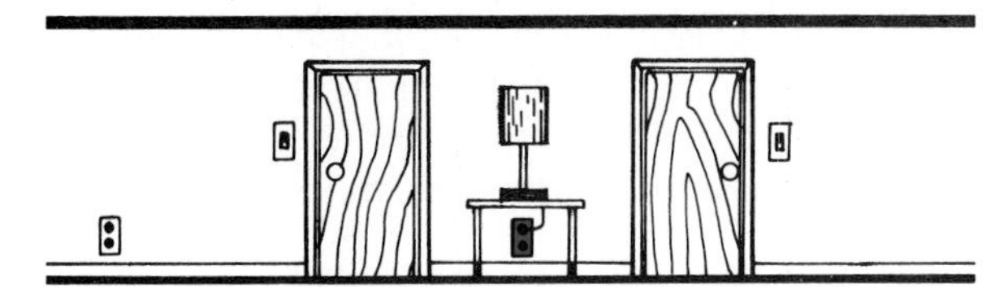

Fig. 61-2 Wall spaces between doors should have a convenience outlet.

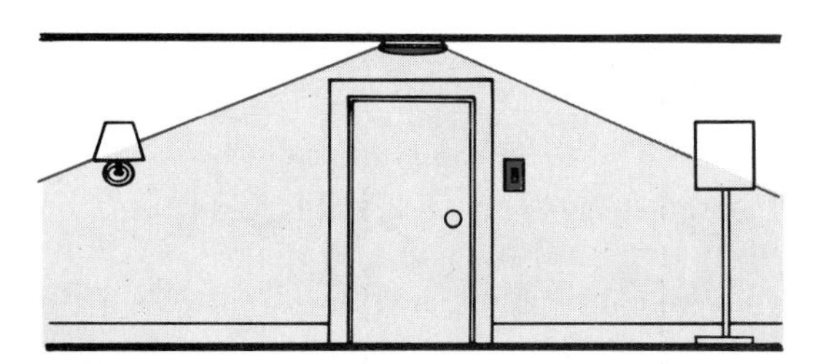

Fig. 61-3 A switch by a door should control the main source of light.

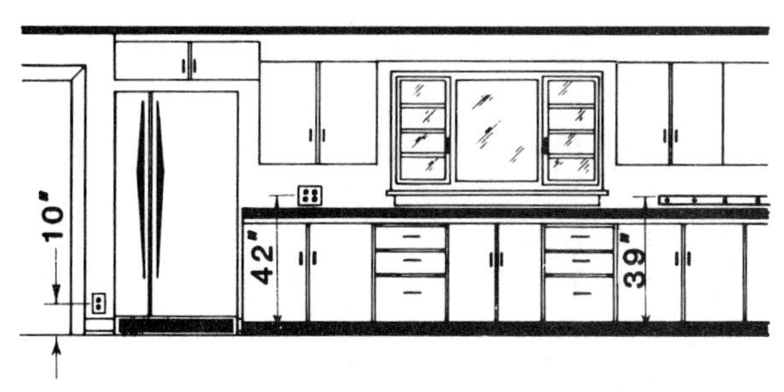

Fig. 61-4 The height of all outlets should be noted on wall elevations or in the specifications.

8. Not all the lights in one room should be on the same circuit.

9. The height of all outlets in the house should be listed on the plans (Fig. 61-4).

10. GFCI receptacles should be provided as outlined in Unit 60.

SWITCH LOCATION

Switches should be located according to the following guides:

1. Plan the switches needed for all lights and electrical equipment. Toggle switches are available in several different types: single-pole, three-way, and four-way (Fig. 61-5).

2. Show location and height of switches on drawings and specifications.

3. Select the type of switches, type of switch-plate cover, and type of finish.

4. If there are only lamps in a room, the entry switch should control the outlet into which at least one lamp is plugged.

5. Lights for stairways and halls must be controlled from both ends (Fig. 61-6) with three-way switches.

6. Bedroom lights should be controlled from bedside and entrance with a three-way switch.

7. Outside service area lights must be controlled with a three-way switch from the garage and from the exit of the house.

8. Basement lights should be controlled by three-way switches and a pilot light in the house at the head of the basement stairs (Fig. 61-7).

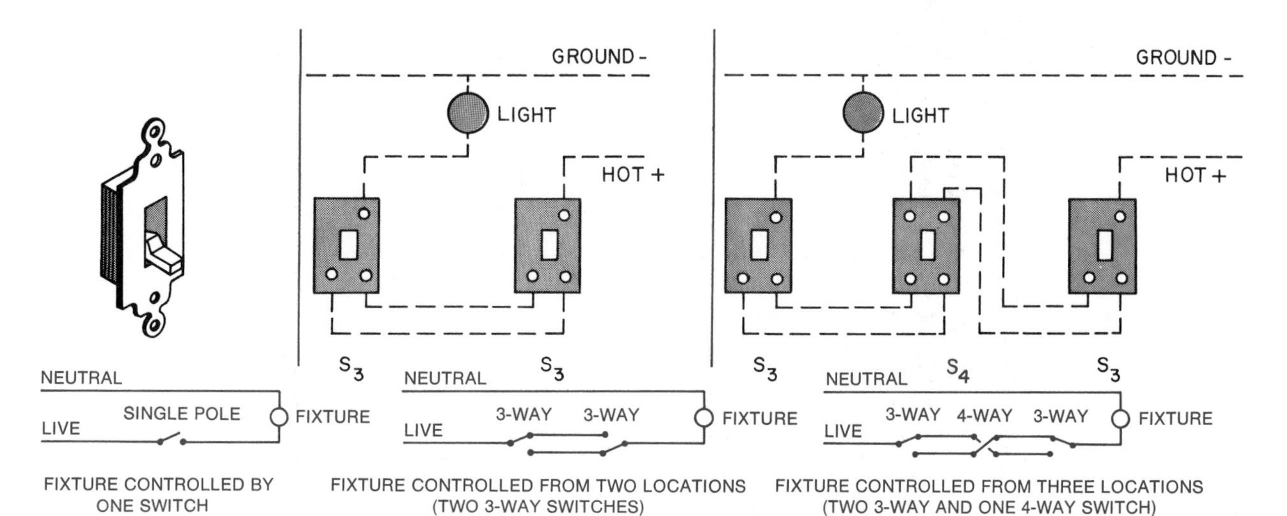

Fig. 61-5 Types of switching controls.

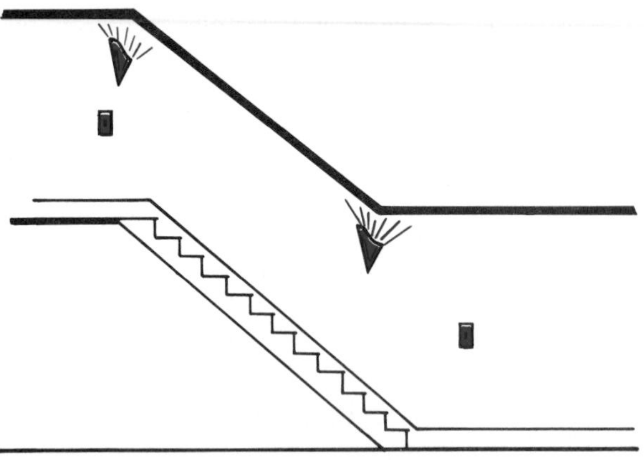

Fig. 61-6 Three-way switches should be used on stairway lights.

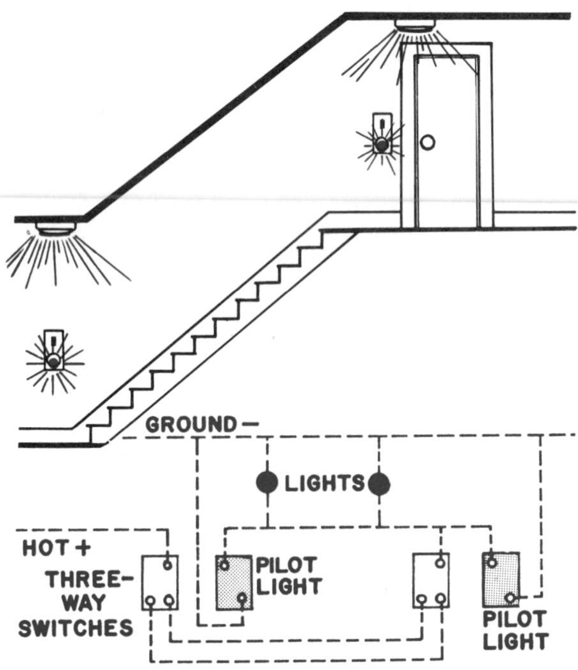

Fig. 61-7 Three-way switches with pilot lights should be used on basement stairs.

9. Install wall switches in preference to pull-string switches in closets.

10. Describe all special controls to be used.

SPECIAL CONTROLS

Special controls make appliances and lighting systems more efficient. Some special controls for electrical equipment include:

Mercury switches are silent, shockproof, long-lasting, and easy to install.

Automatic cycle controls, as on washers, can be installed on appliances to make them perform their functions on a time cycle.

Automatic controls adjust heating and cooling systems.

Clock thermostats adjust heating or cooling units for day and night.

Aquastats keep water heated to selected temperatures.

Dimmers control intensity of light.

Time switches control lights and/or watering systems.

Safety-alarm systems activate a bell when a circuit on a door or window is broken.

Master switches control switching throughout the home from one location.

Low-voltage switching systems (Fig. 61-8) provide economical long runs.

The low-voltage method of switching offers convenience and flexibility. A *relay* isolates all switches from the 120-V system. The voltage from the switch to the appliance is only 24 V. At the appliance, a *magnetic-controlled switch* opens the full 120 V to the appliance. The magnetic-controlled switch is more commonly called a *touch switch*. The low, 24-V system permits long runs of inexpensive wiring that is easy to install and safe to use. This makes it ideal for master-control switching from one location in the house.

ELECTRICAL OUTLETS

There are several types of electrical receptacles. The *convenience receptacle* is used for small appliance and lamp plugs. It is available in single, multiple, or strip outlets. Electricians use the terms outlet and receptacle interchangeably. However, the NEC defines an *outlet* as a point in a circuit where other devices can be connected. A *receptacle* is a device (at an outlet box) to which any plug-in extension line, appliance, or device can be attached.

Lighting receptacles are for the connection of lampholders, surface-mounted fixtures, flush or recessed fixtures, and all other types of lighting fixtures.

The *special-purpose* receptacle is the connection point of a circuit for one special piece of electrical equipment.

The wires that hook up the whole electric system are installed during the construction of the

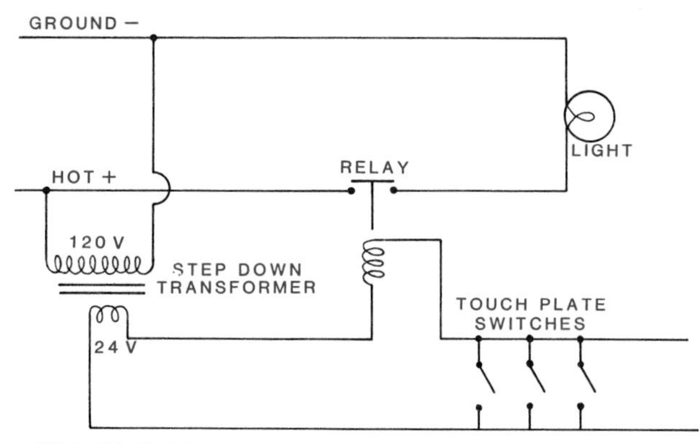

Fig. 61-8 A low-voltage control system.

building, in the walls, floors, and ceilings. In a finished house, the entire system is hidden. The conventional wiring system used for receptacles, lights, and small appliances consists of a black wire (the hot wire) and a white wire (the neutral or common wire). A third, green, wire is a grounding wire. For large appliances, the wiring consists of a black wire and a red wire both of which are hot wires, and a white wire. All three wires connect through a switch to the appliance.

If the wire is too long or too small, there will be a voltage drop because of the wire's resistance. Another cause of voltage drop is the drawing of too much current from the branch circuit. This will cause heating appliances such as toasters, irons, and electric heaters to work inefficiently. Motor-driven appliances will overwork and possibly burn out. Sufficient circuits with large enough wire must be provided for all appliances.

ELECTRICAL WORKING DRAWINGS

Complete electrical plans will ensure the installation of electrical equipment and wiring exactly as planned. If electrical plans are incomplete and sketchy, the completeness of the installation is largely dependent upon the judgment of the electrician. The designer should not rely upon the electrician to design the electrical system, but only to install it.

Preparing the Electrical Plan

After the basic floor plan is drawn, the designer should determine the exact position of all appliances and lighting fixtures on the plan, as shown in Fig. 61-9. The exact position of switches and outlets to accommodate appliances and fixtures should be determined. Next, the electrical symbols representing the switches, outlets, and electrical devices should be drawn on the floor plan. A line is then drawn from each switch to the connecting fixture. Figure 61-10A through I shows electrical symbols used for residential wiring plans. The exact position of each wire may be determined by the electrician unless specified on the plan. The designer usually indicates only the position of the fixture and the switch and the connecting line.

The positions of all outlets and controls are shown on the electrical plan by using electrical wiring symbols (Fig. 61-11); however, the entire circuit is not drawn on the electrical plan. A true wiring diagram shows the manner in which a light or fixture is wired to the switch and how the

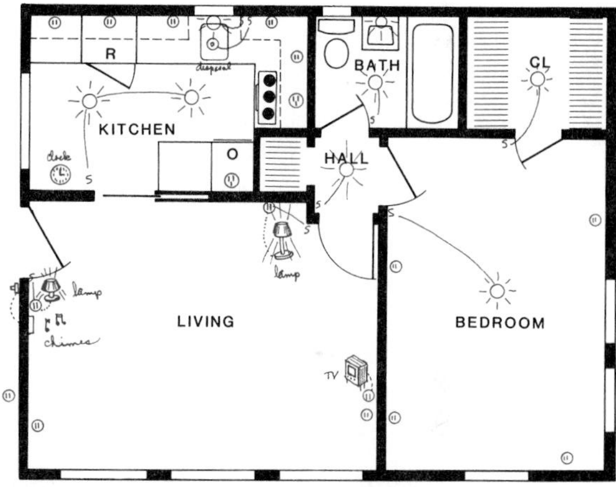

STEP 1. SKETCH IN ELECTRICAL REQUIREMENTS

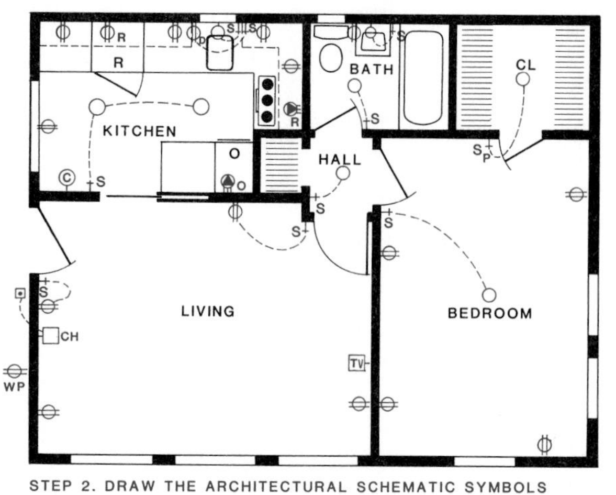

STEP 2. DRAW THE ARCHITECTURAL SCHEMATIC SYMBOLS

Fig. 61-9 Planning home wiring to anticipate needs.

NAME	ABBREV	SYMBOL	ELEVATION	PICTORIAL
SWITCH SINGLE-POLE	S	S OR $		
SWITCH DOUBLE-POLE	S_2	S_2		
SWITCH THREE-WAY	S_3	S_3		
SWITCH FOUR-WAY	S_4	S_4		
SWITCH WEATHERPROOF	S_{WP}	S_{WP}		
SWITCH AUTOMATIC DOOR	S_D	S_D		
SWITCH PILOT LIGHT	S_P	S_P		

Fig. 61-10A Electrical symbols.

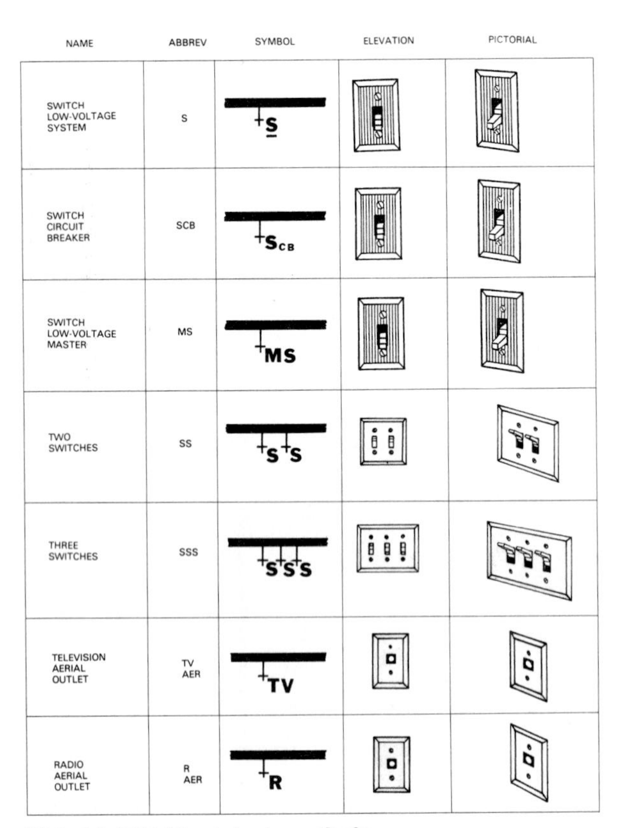

NAME	ABBREV	SYMBOL	ELEVATION	PICTORIAL
SWITCH LOW-VOLTAGE SYSTEM	S	S		
SWITCH CIRCUIT BREAKER	SCB	S_{CB}		
SWITCH LOW-VOLTAGE MASTER	MS	MS		
TWO SWITCHES	SS	S S		
THREE SWITCHES	SSS	S S S		
TELEVISION AERIAL OUTLET	TV AER	TV		
RADIO AERIAL OUTLET	R AER	R		

Fig. 61-10B Electrical symbols.

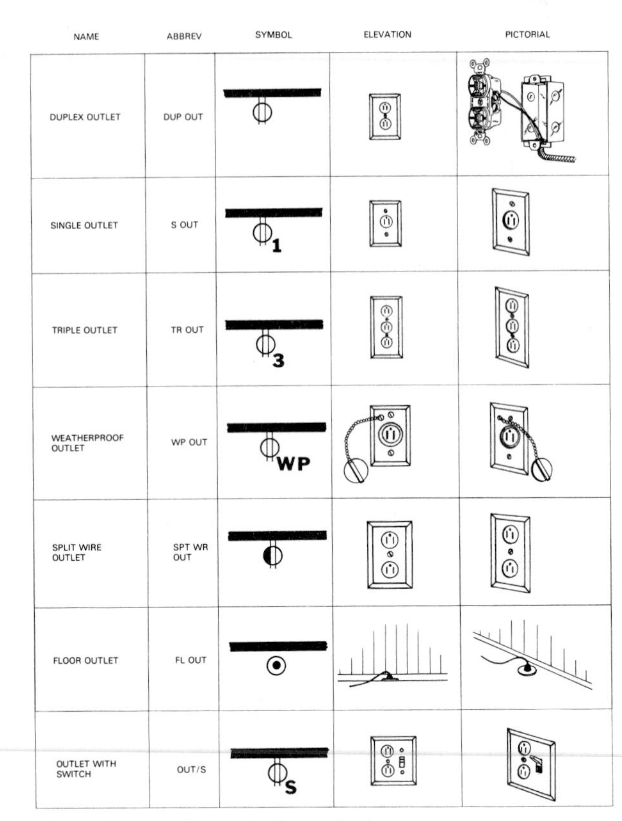

NAME	ABBREV	SYMBOL	ELEVATION	PICTORIAL
DUPLEX OUTLET	DUP OUT			
SINGLE OUTLET	S OUT	1		
TRIPLE OUTLET	TR OUT	3		
WEATHERPROOF OUTLET	WP OUT	WP		
SPLIT WIRE OUTLET	SPT WR OUT			
FLOOR OUTLET	FL OUT			
OUTLET WITH SWITCH	OUT/S	S		

Fig. 61-10C Electrical symbols.

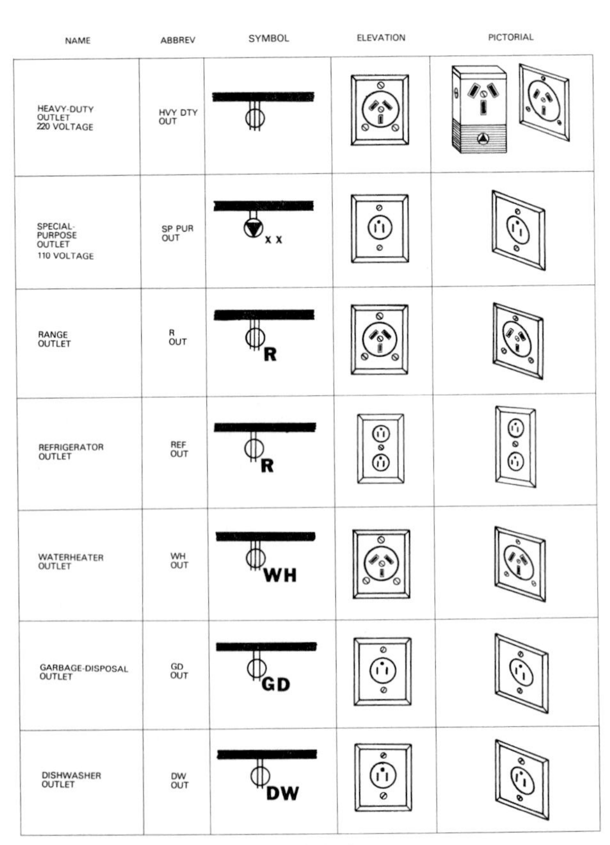

NAME	ABBREV	SYMBOL	ELEVATION	PICTORIAL
HEAVY-DUTY OUTLET 220 VOLTAGE	HVY DTY OUT			
SPECIAL-PURPOSE OUTLET 110 VOLTAGE	SP PUR OUT	X X		
RANGE OUTLET	R OUT	R		
REFRIGERATOR OUTLET	REF OUT	R		
WATERHEATER OUTLET	WH OUT	WH		
GARBAGE-DISPOSAL OUTLET	GD OUT	GD		
DISHWASHER OUTLET	DW OUT	DW		

Fig. 61-10D Electrical symbols.

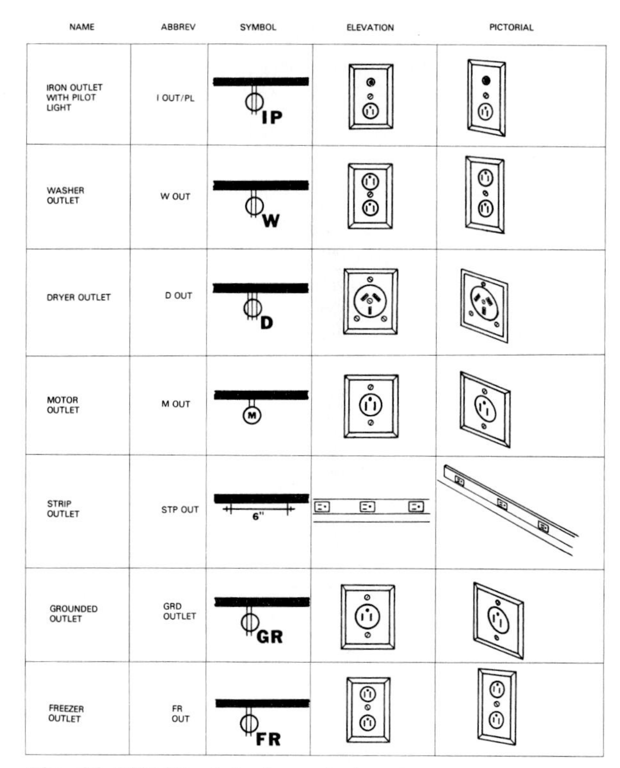

NAME	ABBREV	SYMBOL	ELEVATION	PICTORIAL
IRON OUTLET WITH PILOT LIGHT	I OUT/PL	IP		
WASHER OUTLET	W OUT	W		
DRYER OUTLET	D OUT	D		
MOTOR OUTLET	M OUT	M		
STRIP OUTLET	STP OUT	6"		
GROUNDED OUTLET	GRD OUTLET	GR		
FREEZER OUTLET	FR OUT	FR		

Fig. 61-10E Electrical symbols.

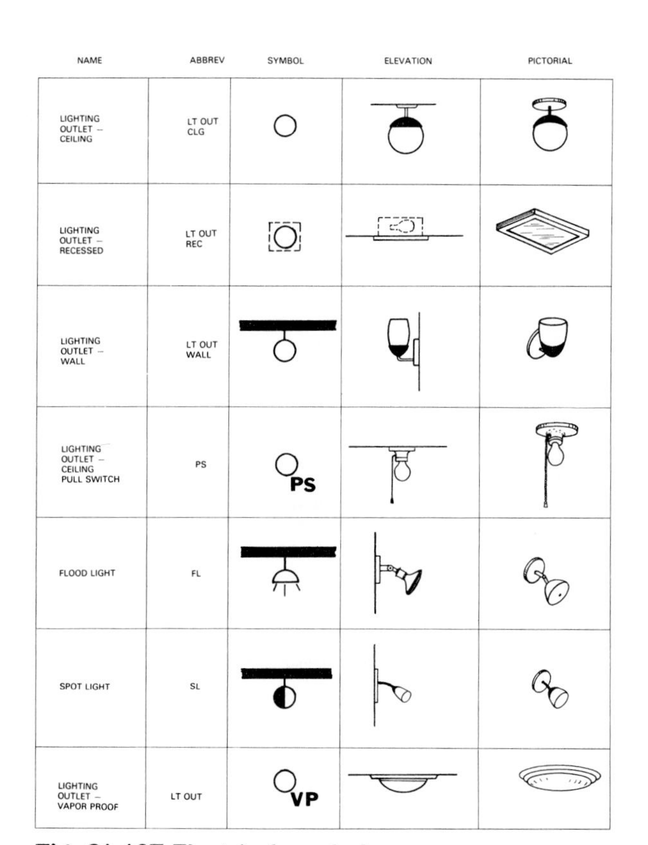

NAME	ABBREV	SYMBOL	ELEVATION	PICTORIAL
LIGHTING OUTLET – CEILING	LT OUT CLG			
LIGHTING OUTLET – RECESSED	LT OUT REC			
LIGHTING OUTLET – WALL	LT OUT WALL			
LIGHTING OUTLET – CEILING PULL SWITCH	PS	PS		
FLOOD LIGHT	FL			
SPOT LIGHT	SL			
LIGHTING OUTLET – VAPOR PROOF	LT OUT	VP		

Fig. 61-10F Electrical symbols.

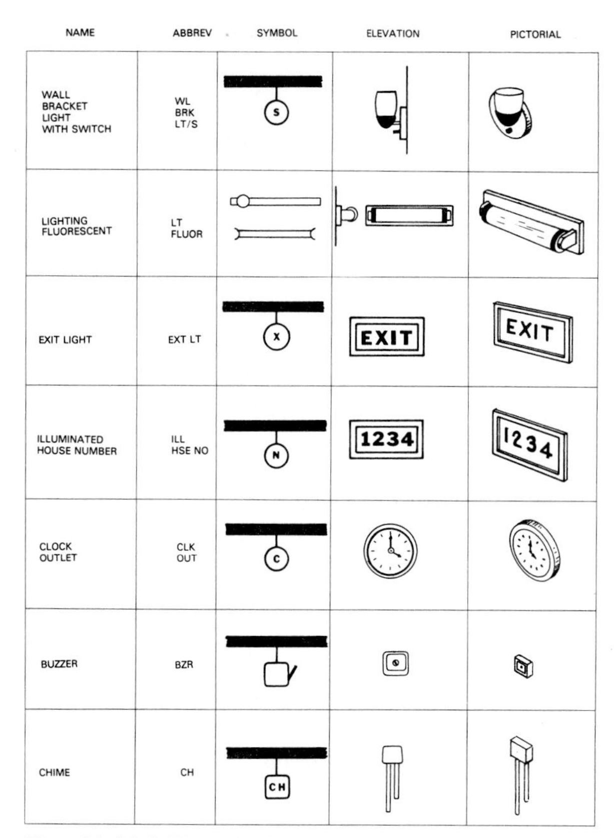

NAME	ABBREV	SYMBOL	ELEVATION	PICTORIAL
WALL BRACKET LIGHT WITH SWITCH	WL BRK LT/S	S		
LIGHTING FLUORESCENT	LT FLUOR			
EXIT LIGHT	EXT LT	X	EXIT	EXIT
ILLUMINATED HOUSE NUMBER	ILL HSE NO	N	1234	1234
CLOCK OUTLET	CLK OUT	C		
BUZZER	BZR			
CHIME	CH	CH		

Fig. 61-10G Electrical symbols.

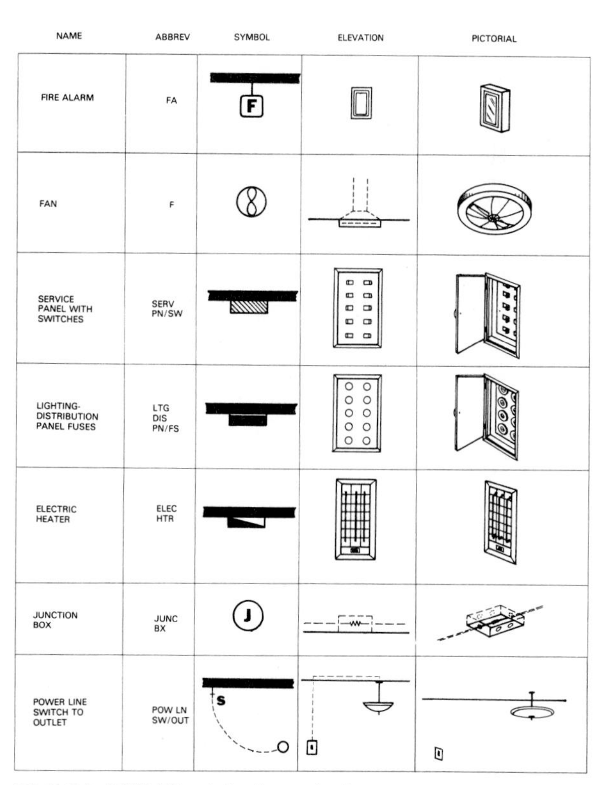

NAME	ABBREV	SYMBOL	ELEVATION	PICTORIAL
FIRE ALARM	FA	F		
FAN	F			
SERVICE PANEL WITH SWITCHES	SERV PN/SW			
LIGHTING-DISTRIBUTION PANEL FUSES	LTG DIS PN/FS			
ELECTRIC HEATER	ELEC HTR			
JUNCTION BOX	JUNC BX	J		
POWER LINE SWITCH TO OUTLET	POW LN SW/OUT	S		

Fig. 61-10H Electrical symbols.

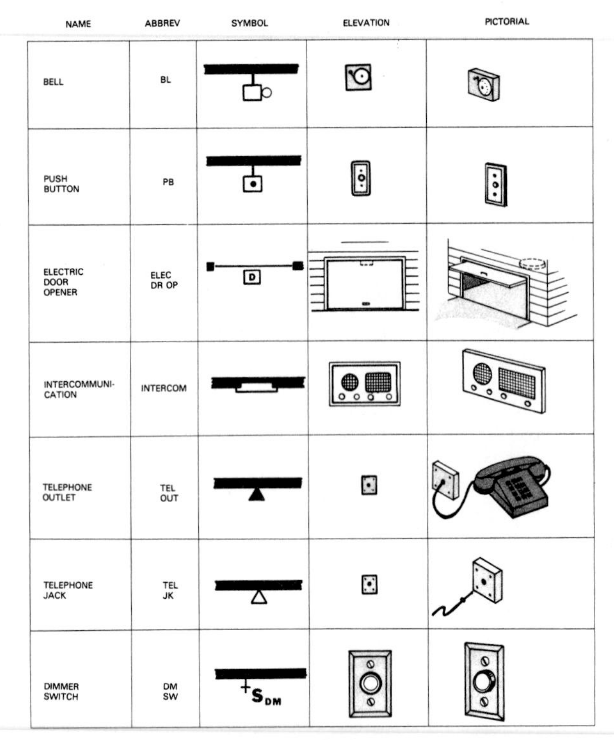

NAME	ABBREV	SYMBOL	ELEVATION	PICTORIAL
BELL	BL			
PUSH BUTTON	PB			
ELECTRIC DOOR OPENER	ELEC DR OP			
INTERCOMMUNICATION	INTERCOM			
TELEPHONE OUTLET	TEL OUT			
TELEPHONE JACK	TEL JK			
DIMMER SWITCH	DM SW	S_{DM}		

Fig. 61-10I Electrical symbols.

wire in the switch is actually connected. In the architectural abbreviated method, shown in Fig. 61-12, only the position of the fixture and the switch is shown on the drawing with a line connecting the outlet with the switch which controls it. The line does not represent the path of the actual wire. Figure 61-11, for example, is a complete wiring plan which shows the positions of all switches, outlets, and fixtures. This plan also shows the position of each lighting fixture with a dotted line connecting that fixture to the switch used to control it.

Room Wiring Diagrams

Figures 61-13 through 61-20 show some typical wiring diagrams of various rooms in the home. Refer to the symbols shown in Fig. 61-10 to identify the various symbols. You will notice you can trace the control of each fixture to a switch.

In the plan shown in Fig. 61-13, switches at both ends of the living room control selected receptacles. In the kitchen plan shown in Fig. 61-14 a wide range of receptacles are provided

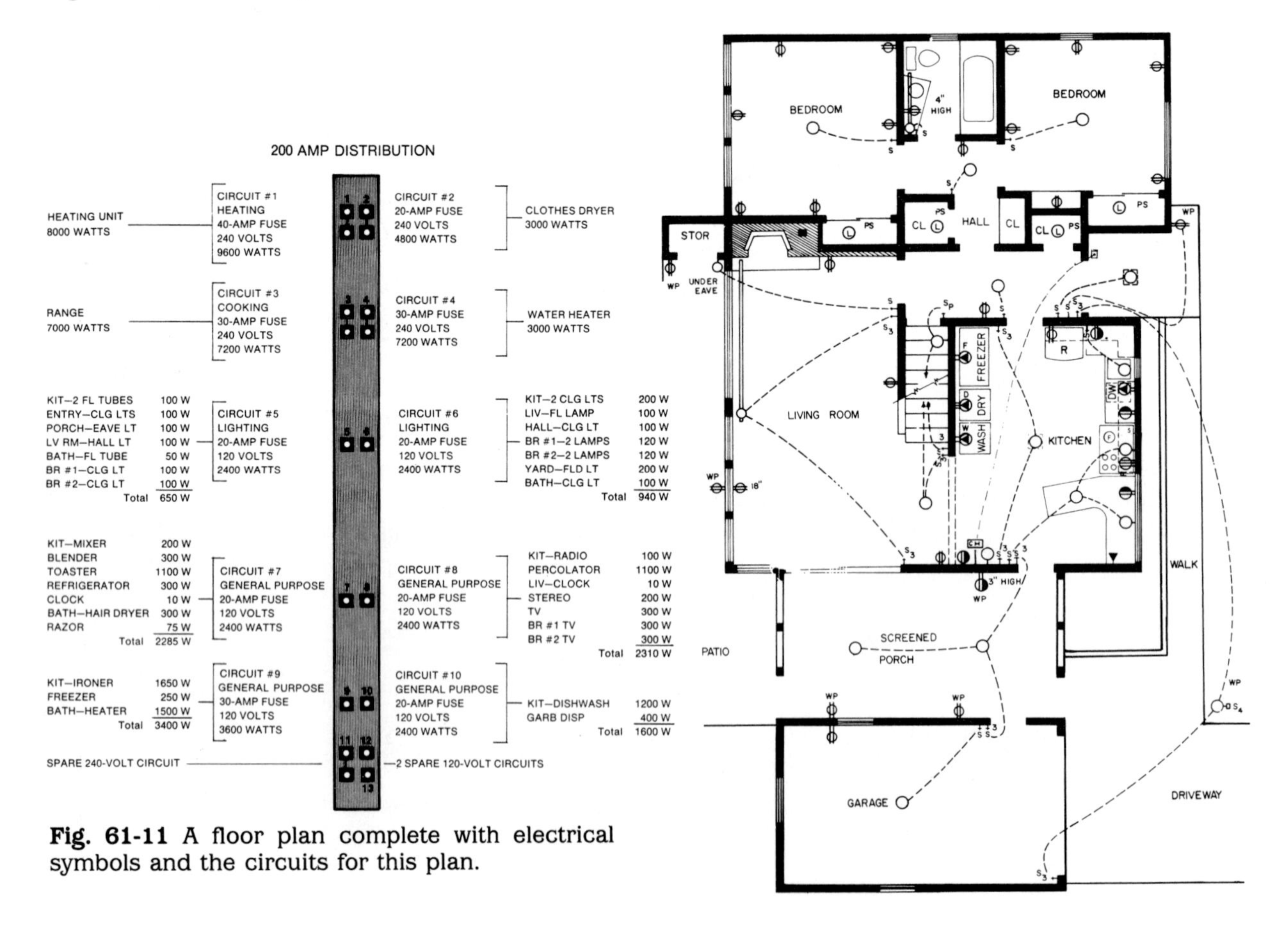

Fig. 61-11 A floor plan complete with electrical symbols and the circuits for this plan.

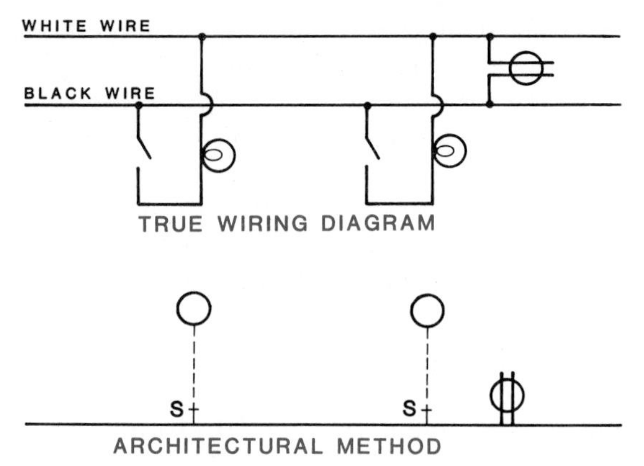

Fig. 61-12 Architectural wiring plans do not show the position of each separate wire.

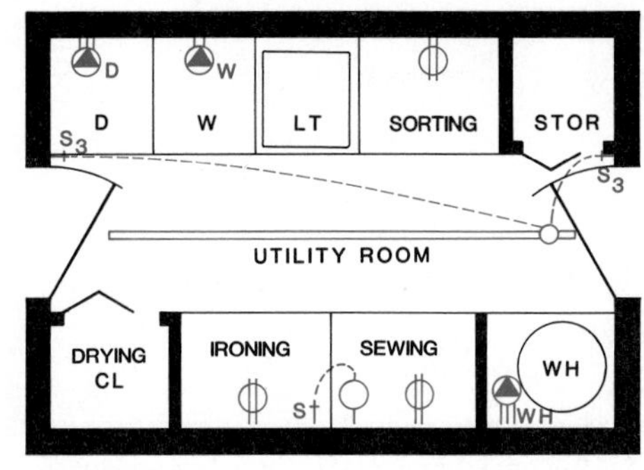

Fig. 61-15 The wiring plan for a utility room.

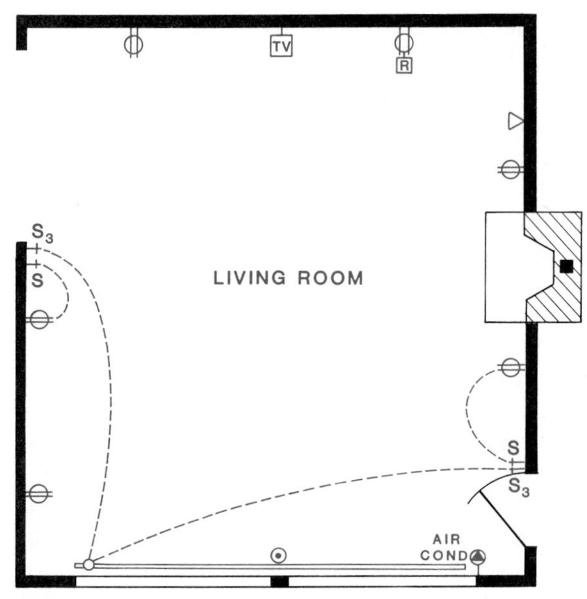

Fig. 61-13 The wiring plan for a living room.

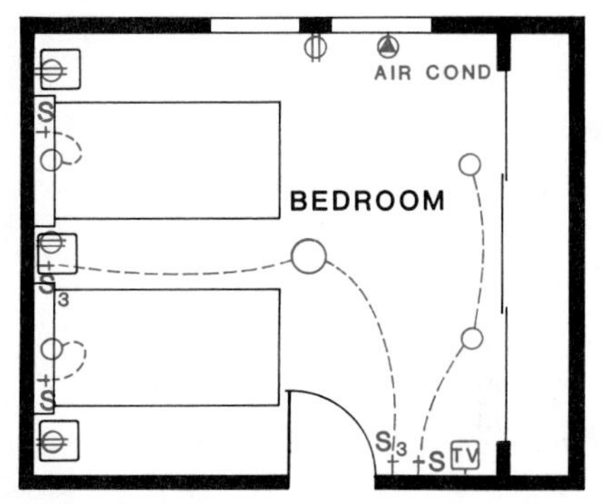

Fig. 61-16 The wiring plan for a bedroom.

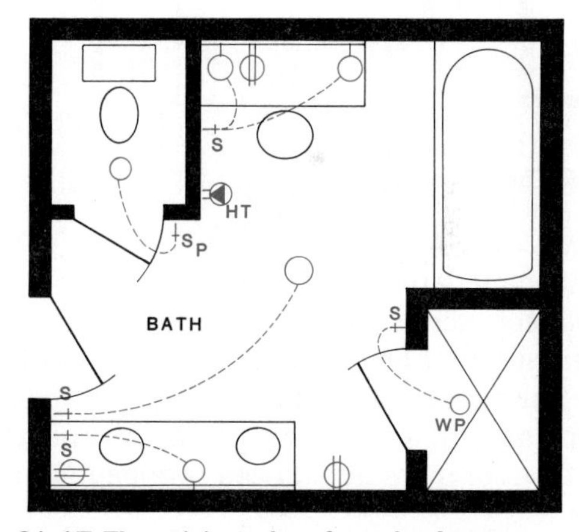

Fig. 61-17 The wiring plan for a bathroom.

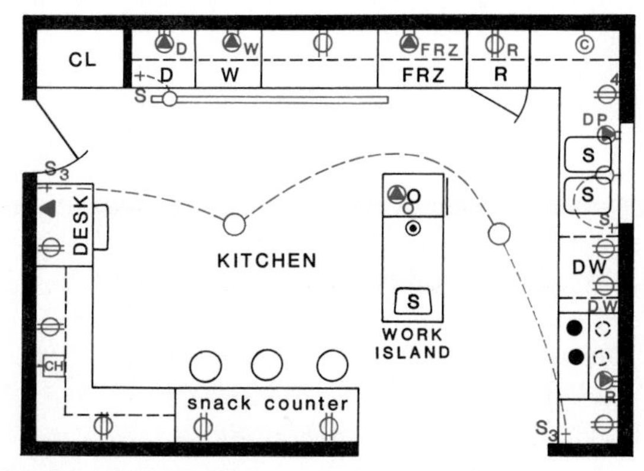

Fig. 61-14 The wiring plan for a kitchen.

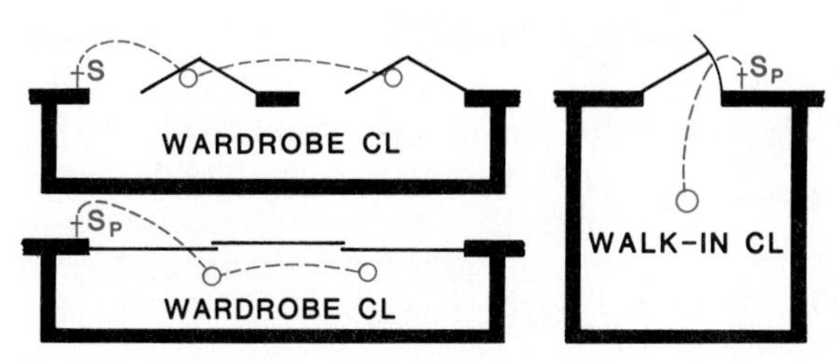

Fig. 61-18 A wiring plan for closets.

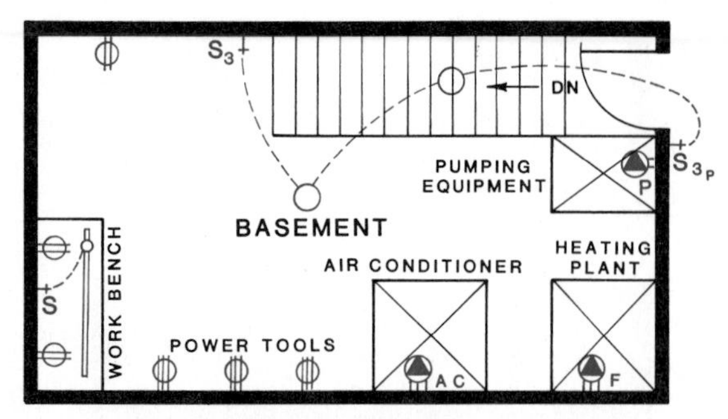

Fig. 61-19 The wiring plan for a basement.

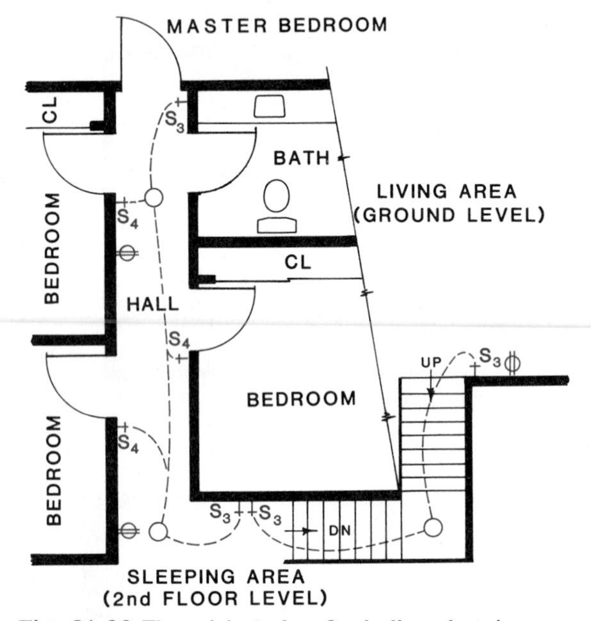

Fig. 61-20 The wiring plan for hall and stairs.

to accommodate the many appliances and electrical accessories. Because many kitchen appliances are heat-producing and therefore require high wattages, kitchen outlets should be divided among several circuits to eliminate the possibility of two or more appliances overloading a circuit when used at maximum load. Utility rooms also require heavy-duty outlets for motor-driven and heat-producing appliances, as shown in Fig. 61-15. Conversely, bedrooms (Fig. 61-16), baths (Fig. 61-17), and closets (Fig. 61-18) require comparatively low wattage levels. However, three-way switches which enable the bedroom lights to be controlled either at the entrance or at the bed are an added convenience. In the bath, adequate wattage must be provided if heat lamps are required. In basements or shop areas (Fig. 61-19), sufficient outlets and wattage to serve HVAC (heating, ventilating and air conditioning) and/or power tool equipment must be provided. Stairs and halls pose special problems in electrical planning, since the position of three-way switches must be carefully located to provide control at many locations to eliminate unnecessary backtracking. Figure 61-20 shows the use of multiple three-way switches to control a wide variety of options.

In planning the distribution of circuits among rooms, avoid placing all the outlets in one room on one circuit. Each room should be served by at least two circuits to eliminate a total room blackout if one circuit is disconnected. Also circuits should be balanced to avoid the overloading of any single circuit at one time. For example, outlets in the bedroom and kitchen are usually not used at the same time so some bedroom circuits can be placed on the same circuit as some of the kitchen outlets.

In drawing electrical plans on a CAD system, symbols can be called up from a symbol library. Then the move-copy task can be used to locate each symbol in its correct position. Electrical symbols are often added using the layering function. This enables the symbols to be plotted in a different color or omitted totally from the floor plan.

Exercises

1. Draw the complete electrical plan for the house you are designing. Show all circuits and label the capacity of each. Identify the circuits protected by a GFCI device.

2. Match the symbols to the terms listed in Fig. 61-21.

3. Define these terms: *switch, outlet, toggle switch, single-pole switch, three-way switch, four-way switch, switch plate, switch and*

pilot light, mercury switch, automatic-cycle control, photoelectric cells, clock thermostats, aquastats, dimmers, time switch, master-control switch, low-voltage switching system, hot wire, black wire, red wire, white wire, strip outlets, special-purpose outlets, convenience outlets, lighting outlets.

1 SINGLE-POLE SWITCH
2 DOUBLE-POLE SWITCH
3 THREE-WAY SWITCH
4 FOUR-WAY SWITCH
5 WEATHERPROOF SWITCH
6 AUTOMATIC DOOR SWITCH
7 SWITCH WITH PILOT LIGHT
8 LOW-VOLTAGE SYSTEM SWITCH
9 CIRCUIT BREAKER
10 CEILING OUTLET
11 WALL OUTLET
12 CEILING OUTLET-PULL SWITCH
13 RECESSED LIGHT
14 FLOOD LIGHT
15 SPOT LIGHT
16 VAPORPROOF CEILING LIGHT
17 FLUORESCENT LIGHT
18 FLUORESCENT LIGHT
19 TELEPHONE
20 TELEPHONE JACK
21 BUZZER
22 CHIME
23 PUSH BUTTON
24 BELL
25 DOUBLE OUTLET
26 SINGLE OUTLET
27 TRIPLE OUTLET
28 SPLIT-WIRE OUTLET
29 WEATHERPROOF OUTLET
30 FLOOR OUTLET
31 OUTLET WITH SWITCH
32 STRIP OUTLET
33 HEAVY-DUTY OUTLET
34 SPECIAL-PURPOSE OUTLET
35 RANGE OUTLET
36 REFRIGERATOR OUTLET
37 WATERHEATER OUTLET
38 GARBAGE-DISPOSAL OUTLET
39 DISHWASHER OUTLET
40 IRON OUTLET-PILOT LIGHT
41 WASHER OUTLET
42 DRYER OUTLET
43 MOTOR OUTLET
44 ELECTRIC DOOR OPENER
45 LIGHTING DISTRIBUTION PANEL
46 SERVICE PANEL
47 JUNCTION BOX
48 ELECTRIC HEATER
49 METER

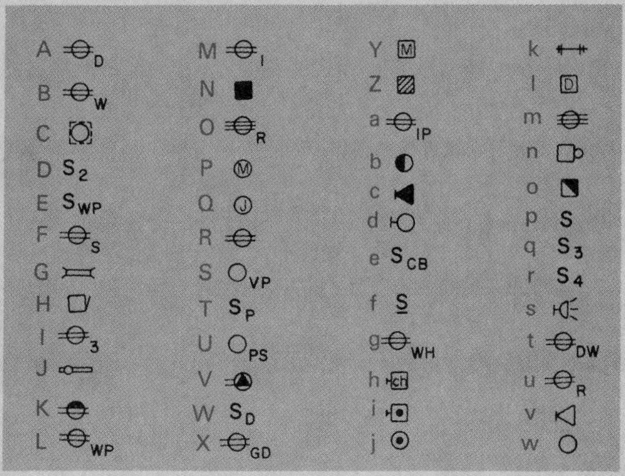

A ⊖D
B ⊖W
C
D S_2
E S_{WP}
F ⊖S
G
H
I ⊖3
J
K
L ⊖WP
M ⊖I
N
O ⊖R
P Ⓜ
Q Ⓙ
R ⊖
S O_{VP}
T S_P
U O_{PS}
V
W S_D
X ⊖GD
Y [M]
Z
a ⊖IP
b
c
d
e S_{CB}
f S
g ⊖WH
h CH
i
j
k
l [D]
m
n
o
p S
q S_3
r S_4
s
t ⊖DW
u ⊖R
v
w O

Fig. 61-21

SECTION 15

*H*eating, Ventilating, and Air Conditioning (HVAC)

Comfort requires more than just providing warmth in winter and coolness in summer. True comfort means a correct temperature, correct humidity, or amount of moisture in the air, and clean, fresh, odorless air. Air-conditioning provides this ideal comfort, and is achieved through the use of a heating system, a cooling system, air filters, and humidifiers. Climate-control plans show systems of maintaining specific degrees of temperature, amounts of moisture, and the exchange of odorless air.

UNIT 62

HVAC CONVENTIONS

Heating, ventilating, and air-conditioning (HVAC) equipment is drawn on floor plans using symbols (Fig. 62-1). They show the location and type of equipment, and also the movement of hot and cold air and water. The location of horizontal ducts on a heating and ventilating duct plan is shown by outlining the position of the ducts. Since vertical ducts pass through the plane of projection, diagonal lines are used to indicate the position of vertical ducts. The flow of air through the ducts can be easily traced because the direction of airflow is shown by an arrow (Fig. 62-2). Airflow coming from the heating-cooling unit is shown by an arrow pointing out from the diffusers. Return air is indicated by an arrow pointing into the duct.

CLIMATE-CONTROL METHODS

Many different systems can be used to heat and cool a building. The effective use of passive solar design and insulation, ventilation, roof overhang, caulking, weather stripping, and solar orientation helps to increase the efficiency of all climate-control systems.

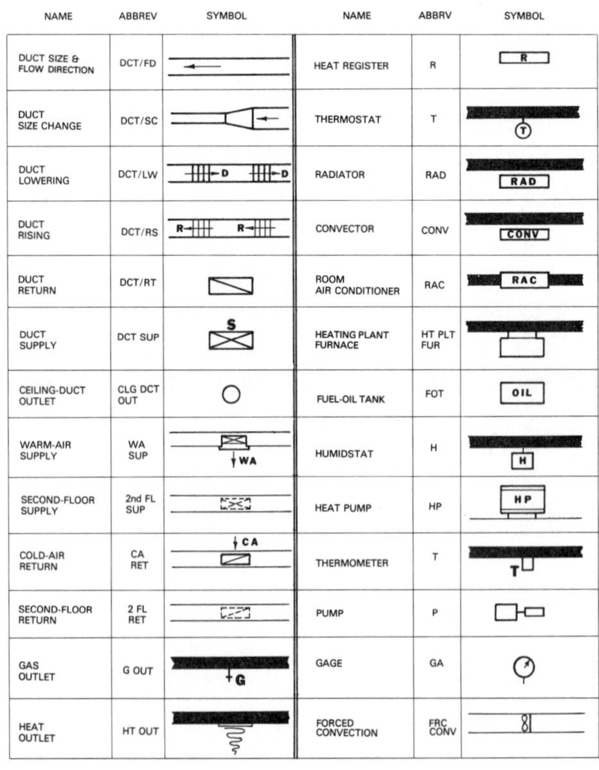

NAME	ABBREV	SYMBOL	NAME	ABBRV	SYMBOL
DUCT SIZE & FLOW DIRECTION	DCT/FD		HEAT REGISTER	R	R
DUCT SIZE CHANGE	DCT/SC		THERMOSTAT	T	T
DUCT LOWERING	DCT/LW	D D	RADIATOR	RAD	RAD
DUCT RISING	DCT/RS	R R	CONVECTOR	CONV	CONV
DUCT RETURN	DCT/RT		ROOM AIR CONDITIONER	RAC	RAC
DUCT SUPPLY	DCT SUP	S	HEATING PLANT FURNACE	HT PLT FUR	
CEILING-DUCT OUTLET	CLG DCT OUT		FUEL-OIL TANK	FOT	OIL
WARM-AIR SUPPLY	WA SUP	WA	HUMIDSTAT	H	H
SECOND-FLOOR SUPPLY	2nd FL SUP		HEAT PUMP	HP	HP
COLD-AIR RETURN	CA RET	CA	THERMOMETER	T	T
SECOND-FLOOR RETURN	2 FL RET		PUMP	P	
GAS OUTLET	G OUT	G	GAGE	GA	
HEAT OUTLET	HT OUT		FORCED CONVECTION	FRC CONV	

Fig. 62-1 Climate-control symbols.

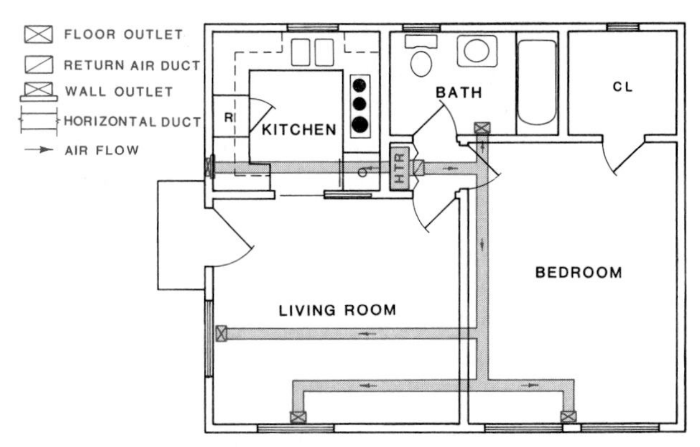

Fig. 62-2 Arrows indicate airflow direction.

HEAT TRANSFER

Heat is transferred from a warm to a cool surface by three processes: radiation, convection, and/or conduction (Fig. 62-3). In *radiation,* heat flows to a cooler surface through space in the same manner light travels. The air is not warmed but the cooler object it strikes becomes warm. The object, in turn, warms the air that surrounds it. In *convection,* a warm surface heats the air about it. The warmed air rises, and cool air moves in to take its place, causing a convection current. Figure 62-4 shows an application of the process of convection. In *conduction,* heat moves through a solid material. The denser the material, the better it will conduct heat. For example, iron conducts heat better than wood.

Heat loss or gain is the amount of heat that passes through the exterior surface of a building. All building materials will block some heat transfer from the inside to the outside and visa versa. Regardless of the material, some heat gain or loss will always occur. When this happens the temperature transfer is always from hot to cold. Factors influencing the amount of heat transfer in a building are the difference between indoor and outdoor temperatures, the type and thickness of building materials, the amount and type of insulation, and the amount of air leaking into or out of the structure. This latter process,

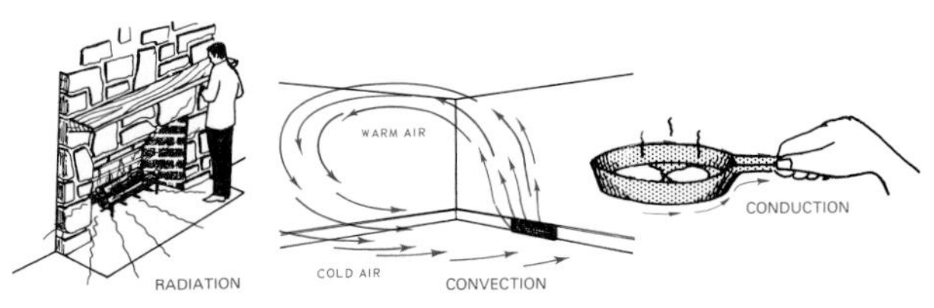

Fig. 62-3 Methods of heat transfer.

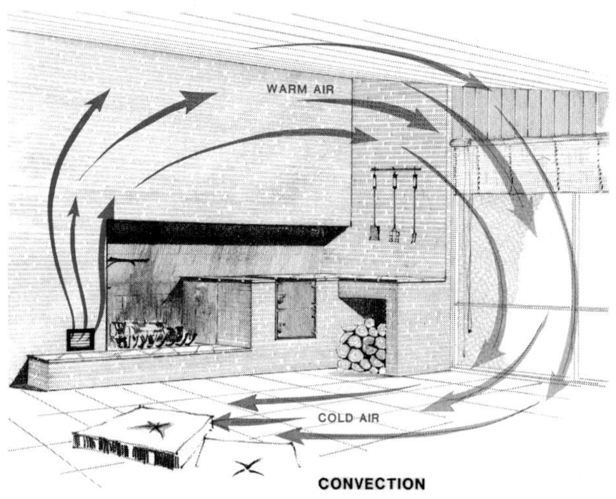

Fig. 62-4 Convection currents.

through which heat is lost through spaces around windows, doors, fireplaces, etc., is known as infiltration.

Thermal Conductivity

Thermal conductivity is the amount of heat that flows from one face of a material, through the material, to the opposite face. Thermal conductivity is defined as the amount of heat transferred through a 1-square-foot area, 1 inch thick, with a temperature difference of 1 degree Fahrenheit. The unit of measurement for heat is the British thermal unit (Btu). A Btu is the unit of heat needed to raise the temperature on 1 pound of water 1 degree Fahrenheit. Figure 62-5 shows a graphic interpretation of thermal conductivity.

Resistivity

Resistivity is the ability of materials to resist the transfer of heat. Materials with low resistance

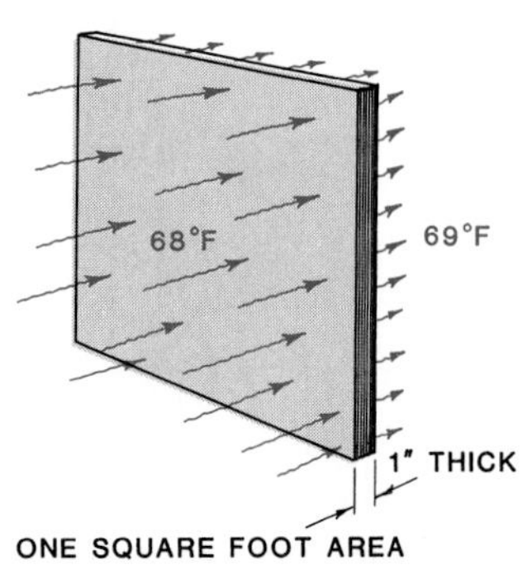

Fig. 62-5 Measurement of thermal conductivity.

qualities are known as conductors. Materials with high resistance qualities are known as insulators.

INSULATION

Insulation is a material used to stop the transfer of heat. It helps keep heat inside in the winter and outside in the summer. Without insulation, a heating or cooling system must work harder to overcome the loss of warm air or cool air through the walls, floors, and ceilings. Full insulation—6″ (150 mm) in the roof, 3″ (75 mm) in the walls, and 2″ (50 mm) in the floors—can save 40 percent of heating and cooling costs. Properly insulating walls and floors above can reduce 25 percent of the heat transfer. Because most roofs cannot be sheltered from the sun, 40 percent of all heat transfer is through the roof. Six-inch insulation, with an area for ventilation above the insulation, is most effective. The use of light-colored roofs also helps in reflecting the heat and preventing the absorption of excessive heat. Windows alone can allow 25 percent of the heat within a house to transfer to the outside (Fig. 62-6). Some deterrents to this transfer are the use of large roof overhangs, trees and shrubbery, drapes and window blinds, and double- or triple-paned thermal glass.

In an imperfectly constructed house, cracks around doors, windows, and fireplaces can combine to make the equivalent of a hole of sufficient size to lose all internal heat in less than an hour. Weather stripping and caulking can prevent this heat loss. To prevent this type of loss, a layer of insulation should be placed between the foundation and the earth below. This insulation should be outside the structure, thus placing the building in an insulation envelope. Such an arrangement not only conserves heat but prevents rapid changes in inside temperature in all seasons. Figure 62-7 shows a house totally enveloped in insulation.

Insulation is available in different forms (Fig. 62-8). Insulation is made from a wide variety of vegetable, mineral, plastic, and metal materials to stop heat transfer, block moisture (Fig. 62-9), stop sound, resist fire, and resist insects:

1. Flexible batt: paper-covered insulating materials that are attached between structural members. The batts are 2″ to 6″ (50 to 150 mm) thick.

2. Flexible blanket: paper-covered insulating materials that are attached between structural members. The blankets are long sheets, 1″ to 3″ (25 to 75 mm) thick.

3. Loose fill: materials poured or blown into walls or attic floors.

4. Reflective: multiple spaces of reflecting metal attached between construction members. Reflective material is often mounted on other types of insulation. It is excellent for reflecting heat and for retarding fire, decay, and insects.

5. Rigid board: thin insulating sheathing cover that is manufactured in varying sizes.

6. Additives: lightweight aggregates that are mixed with construction materials to increase their insulating properties.

7. Spray on: insulating materials mixed with an adhesive and sprayed on.

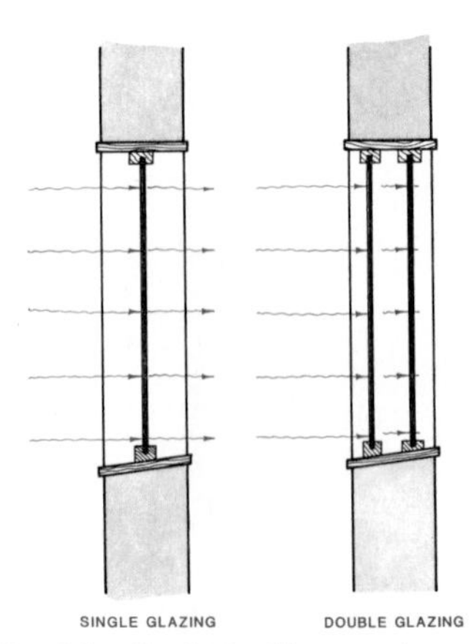

Fig. 62-6 Double glazing effect on heat loss.

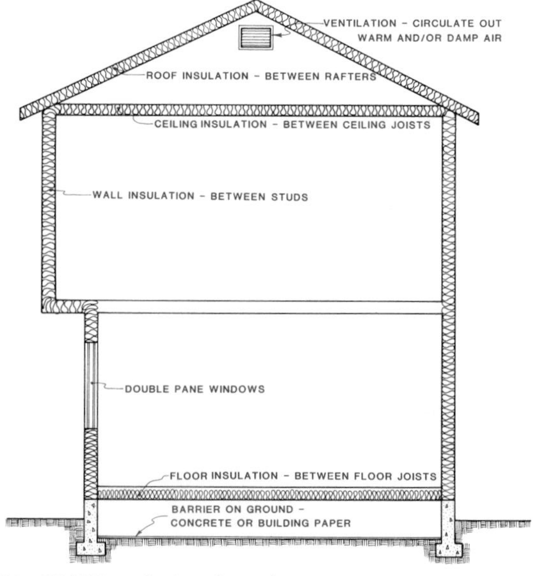

Fig. 62-7 Insulation locations.

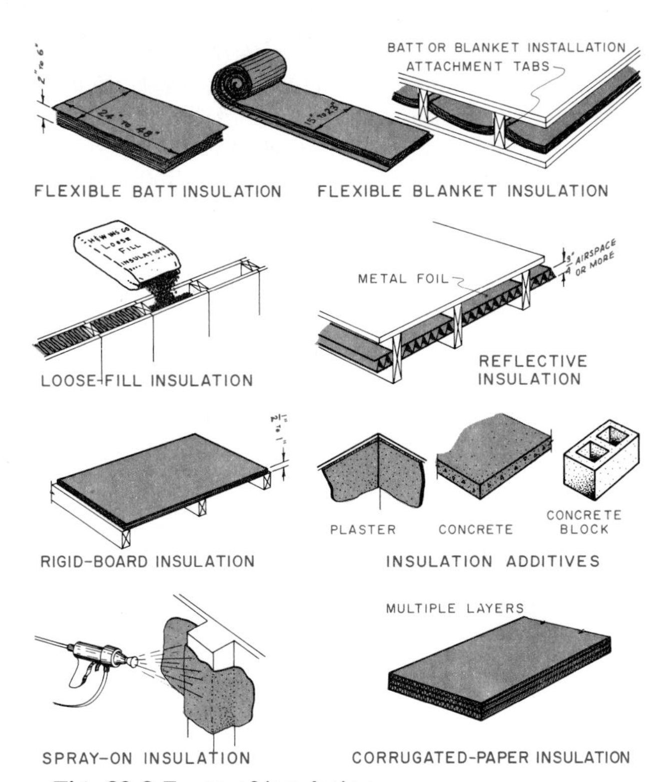

Fig. 62-8 Types of insulation.

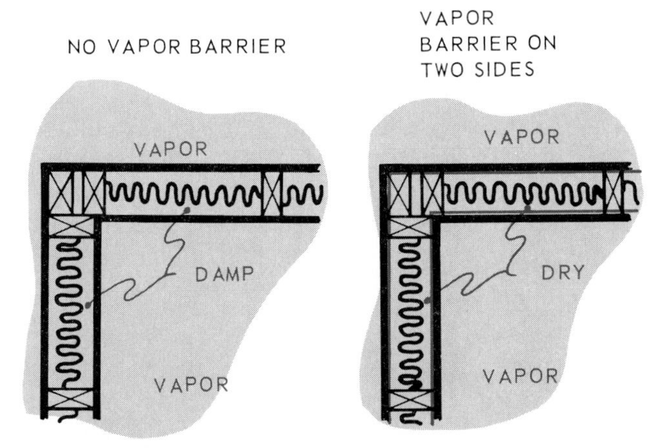

Fig. 62-9 Effect of vapor barrier.

8. Corrugated paper: multiple layers of corrugated paper that are easy to cut and install.

9. Surface air film: a film of air that clings to surfaces. The amount of air clinging varies with the amount of air movement and the type of surface. Since air is an excellent insulator, air film contributes an added insulation factor.

R Values

The effectiveness of an insulating material is measured by its R value. The R value is a uniform method of rating the resistance to heat flow through building materials. The higher the R number, the greater the resistance to the heat flow. Building materials have been tested and their thermal resistance (R value) listed on the product. For example the R value of 2.5″-thick glass fiber insulation is R-7. The R value of 6.5″-thick glass fiber insulation is R-22. When building materials are combined in layers, the sum of their R values is the total R value for the component. For example, the R value of the batt insulation used in the wall shown in Fig. 62-10 is R-8. But when the R values of the other materials used in the wall are added, the R value of the wall is R-12.5.

To more accurately determine the combined thermal conductivity of all materials in a structure, including air spaces, the U factor is used. The U factor is the reciprocal (1/*n*) of the R factor. Figure 62-11 shows the calculation of the U factor from a combination of R factors, ($1 \div 13 = 0.077$).

Heat-Loss Calculations

To calculate the heat loss, the following factors are used as shown in Fig. 62-10.

1. The area of the interior in square feet.

2. The type of construction and insulation (U factor).

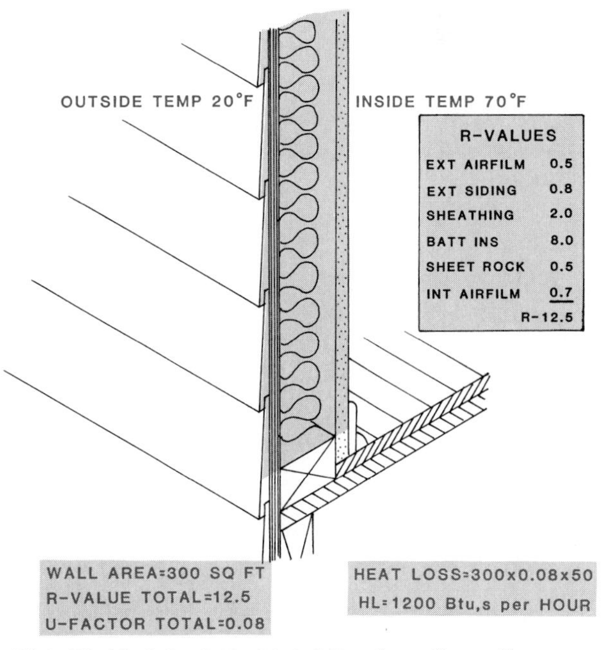

Fig. 62-10 Calculating total R value of a wall.

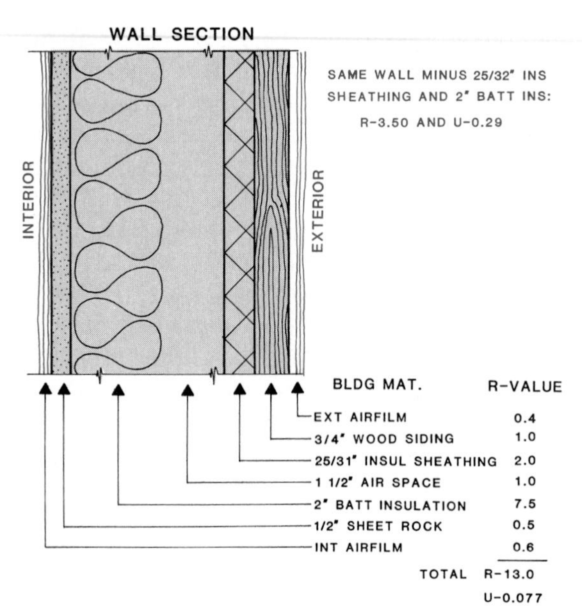

Fig. 62-11 Calculating the coefficient of heat transfer for a wood-frame wall.

3. The difference in temperature from the inside to the outside (degrees fahrenheit).

4. Amount of air transfer (infiltration).

The formula used to calculate the heat loss is: area × U factor × temperature difference (HL = A × U × T) The heat loss is calculated in Btu per hour. Added to this are the Btu lost through infiltration. The infiltration is calculated by multiplying the interior volume by the constant 0.018 to find the average amount of lost Btu per hour due to infiltration.

Figure 62-12 shows a sample of various thermal resistances (R values) which produce given indoor temperatures at various outdoor temperature levels. Figure 62-13 shows the R and U values of common building materials. Common building materials differ greatly in R value ratings, as do insulation materials. The thickness of material and type of insulation (batt, blown, board, or injected) greatly affects the R value of insulation as shown in Fig. 62-14. Even slight variations in airflow and reflective surface qualities of air spaces and surfacefilms causes wide differences in R and U values as shown in Fig. 62-15A and B. Windows and doors account for much heat loss in cold climates. Figure 62-16 shows the R and U values for different types of windows, and Fig. 62-17 shows the R and U values of common types of exterior doors. Naturally outdoor doors must be selected for their R and U value rating, which is not important for interior doors.

CONVENTIONAL HEATING SYSTEMS

The two most efficient types of heating systems are perimeter heating and radiant heating (Fig. 62-18). In *perimeter heating,* the heat outlets are located on the outside walls of the rooms. The heat rises and covers the coldest areas in the house. Warm air rises, passes across the ceiling, and returns while still warm. Base-

Outdoor temperature	Indoor surface temperature				
	Cool 60°F	Fair 64°F	Medium 66°F	Warm 68°F	Minimum for floor
+30°F	R-2.3	R-3.4	R-5.1	R-10.0	R-1.7
+20°F	R-2.8	R-4.2	R-6.4	R-12.5	R-2.2
+10°F	R-3.4	R-5.1	R-7.8	R-14.5	R-2.6
0°F	R-3.9	R-6.0	R-9.2	R-17.0	R-3.0
−10°F	R-4.4	R-6.8	R-10.1	R-20.0	R-3.4
−20°F	R-5.1	R-7.8	R-11.3	R-23.0	R-3.9
−30°F	R-5.7	R-8.4	R-12.8	R-25.0	R-4.4
−40°F	R-6.4	R-10.2	R-14.5	R-28.0	R-4.8

Fig. 62-12 Thermal resistances (R) required for given indoor temperatures.

Type of building material	Thickness	Conductance R-value (high value is more efficient)	Resistance U-factor (low value is more efficient)
Roof decking insulation	2″	5.56	0.18
Mineral wool fibrous insul	1″	3.12	0.32
Loose fill insulation	1″	3.00	0.33
Acoustical tile	1″	2.86	0.35
Carper and pad, fibrous	1″	2.08	0.48
Wood fiber sheathing	25/32″	2.06	0.49
Wood door	1¾″	1.96	0.51
Fiber board sheathing	½″	1.45	0.69
Soft woods (pine, fir, etc.)	1″	1.25	0.80
Wood subfloor	25/32″	0.98	1.02
Hardwoods	1″	0.91	1.10
Wood shingles, 16″ 7½″ exp	standard	0.87	1.15
Hardboard, wood fiber	1″	0.72	1.39
Plywood	½″	0.65	1.54
Asphalt shingles	standard	0.44	2.27
Sheet rock/plasterboard	½″	0.44	2.27
Built-up roofing	⅜″	0.33	3.03
Concrete/stone	4″	0.32	3.13
Asbestos cement board	⅜″	0.32	3.13
Gypsum plaster (light weight)	½″	0.32	3.13
Common brick	1″	0.20	5.00
Stucco	1″	0.20	5.00
Cement plaster	1″	0.20	5.00
Face brick	1″	0.11	9.10
Felt building paper (15 lb)	standard	0.06	16.67
Steel	1″	0.0032	312.50
Aluminum	1″	0.0007	1428.57

Fig. 62-13 R values and U factors of typical building materials.

Insulating materials	Material's thickness	R-value
Glass Fiber	2″ Batt	R-7
	4″ Batt	R-11
	6″ Batt	R-19
	6″ Blown	R-13
	8½″ Blown	R-19
	12″ Blown	R-26
	18″ Blown	R-38
	2-6″ Batts	R-38
Rock Wool	4″ Blown	R-11
	6½″ Blown	R-19
	13″ Blown	R-38
Cellulose Fiber	4″ Blown	R-11
	6½″ Blown	R-19
	13″ Blown	R-38
Polystrene	1″ Board	R-5
	1½″ Board	R-7.5
Polyurethane Foam	1½″ Board	R-9.3
	4″ Injected	R-25
Expanded MICA	Loose	R-2.5

Fig. 62-14 R values for insulation materials.

¾″ Air spaces	R-values	U-factor
Heat flow UP		
Non-reflective	R-0.87	1.15
Reflective, one surface	R-2.23	0.45
Heat flow DOWN		
Non-reflective	R-1.02	0.98
Reflective, one surface	R-3.55	0.28
Heat flow HORIZONTAL		
Non-reflective (also same for 4″ thickness)	R-1.01	0.99
Reflective, one surface	R-3.48	0.29

Fig. 62-15A Resistance values for air spaces.

boards, convectors, and radiators can be used to project the heat in the perimeter system.

Radiant heating functions by heating an area of the wall, ceiling, or floor. These warm surfaces in turn radiate heat to cooler objects. The heating surfaces may be lined with pipes con-

Air surface films	R-value	U-factor
INSIDE (still air)		
Heat flow UP (through horizontal surface) Non-reflective	R-0.61	1.64
Reflective	R-1.32	0.76
Heat flow DOWN (through horizontal surface) Non-reflective	R-0.92	1.09
Reflective	R-4.55	0.22
Heat flow HORIZONTAL (through vertical surface) Non-reflective	R-0.68	1.47

Fig. 62-15B Resistance values for air-surface films.

Material	U-factor		R-value	
	Cold climate (winter)	Warm climate (summer)	Cold climate (winter)	Warm climate (summer)
SINGLE GLASS	1.13	1.06	0.88	0.94
INSULATED GLASS				
¼" Air space	0.65	0.61	1.54	1.64
½" Air space	0.58	0.56	1.72	1.79
STORM WINDOWS				
1" to 4" Air space	0.56	0.54	1.79	1.85

Fig. 62-16 R values and U factors for windows.

Type of door	U-factor	R-value
Hollow core wood	1.00	R-1.0
Hollow core wood and storm door	0.67	R-1.5
Solid core wood	0.43	R-2.3
Solid core wood and storm door	0.28	R-3.5
Metal with urethane core	0.07	R-13.5

Fig. 62-17 R values for exterior doors.

taining hot water or hot air, or with electric resistance wires covered with plaster.

HEATING DEVICES

Devices that produce the heat used in the various heating systems include the following: warm-air units, hot-water units, steam units, electrical units, and solar systems.

Warm-Air Units

In a *warm-air unit*, the air is heated in a furnace. Air ducts distribute the heated air to outlets throughout the house (Fig. 62-19). The air supply can be taken from the outside, from the furnace room, or from return-air ducts in heated rooms. Warm-air units can operate either by gravity or by forced air, and they provide almost instant heat. Air filters and humidity control can be combined with the heating unit. The cooling system can use the same ducts as the heating system if the ducts are rustproof.

In *forced-air systems*, the air is blown through ducts by use of a fan in the furnace. *Gravity systems* rely on allowing the warm air to rise naturally to higher levels without the use of a fan. Therefore, the furnace in a gravity system must be located on a level lower than the area to be heated. Warm-air system plans include the location of each heating outlet and the location of all duct work from the furnace to the outlet, as shown in Fig. 62-20.

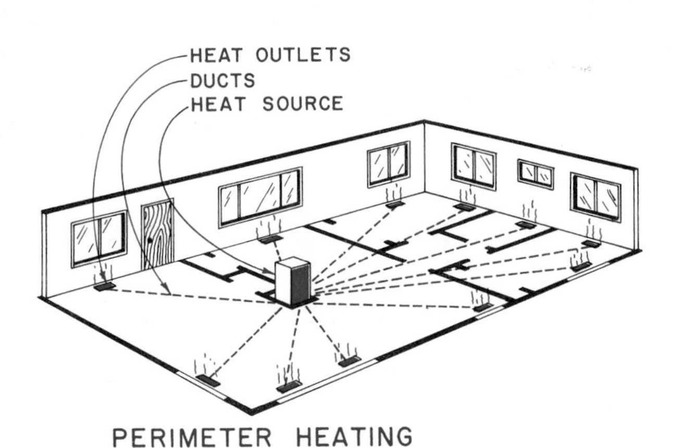

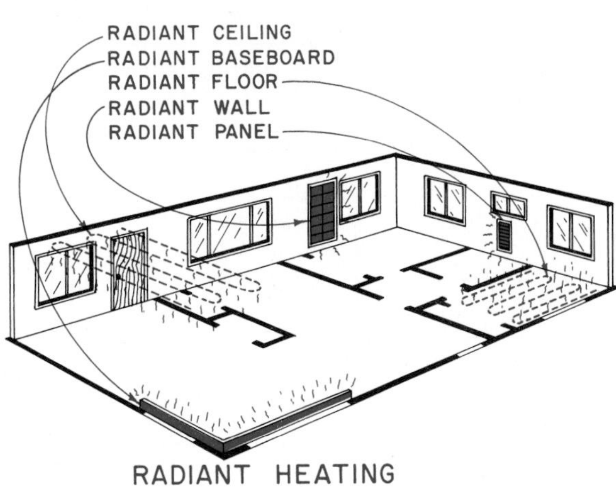

Fig. 62-18 Two types of heating systems.

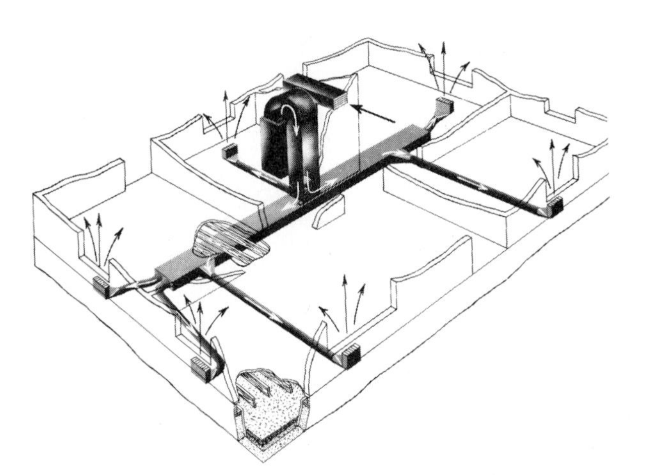

Fig. 62-19 Air ducts distribute heated air.

Warm-air systems fall into several categories with respect to duct work: *individual duct* systems, as shown in Fig. 62-20; *extended plenum* systems, as shown in Fig. 62-21, *perimeter loop* systems, such as the one shown in Fig. 62-22; and *perimeter radial* systems, as shown in Fig. 62-23.

When an abbreviated heating and air-conditioning plan is prepared, only the locations of warm-air outlets (Fig. 62-24) are shown on the plan. Figure 62-25 shows locations for heat distribution from ceiling walls and floor sources. Figure 62-26 is an abbreviated heating and air-conditioning plan showing only the positions of outlets. When this kind of plan is provided, the

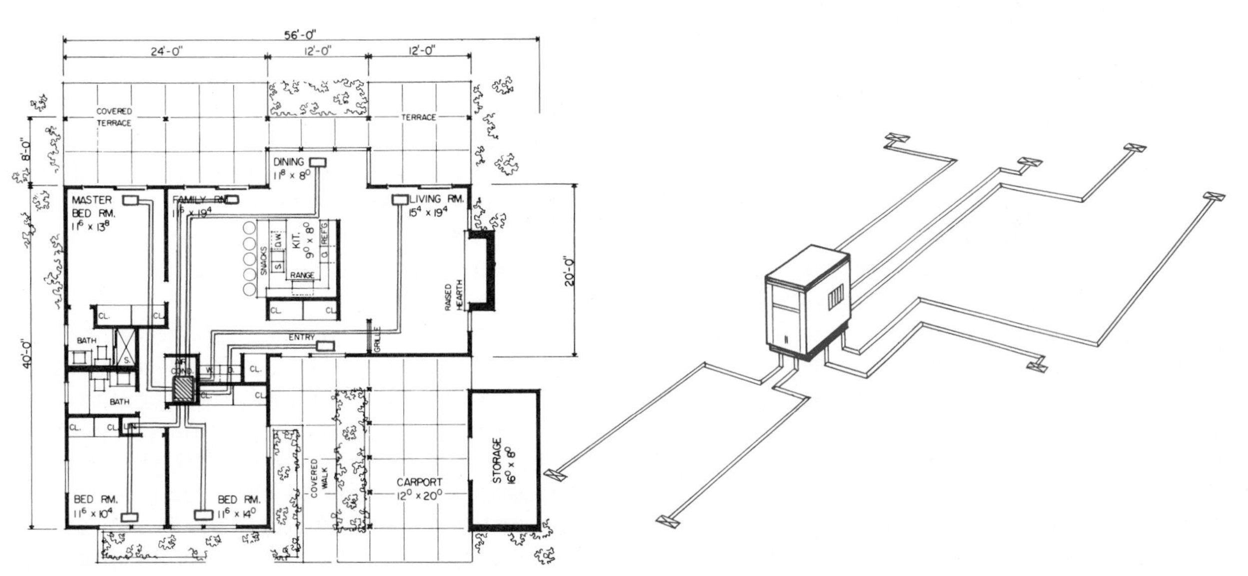

Fig. 62-20 Individual duct system.

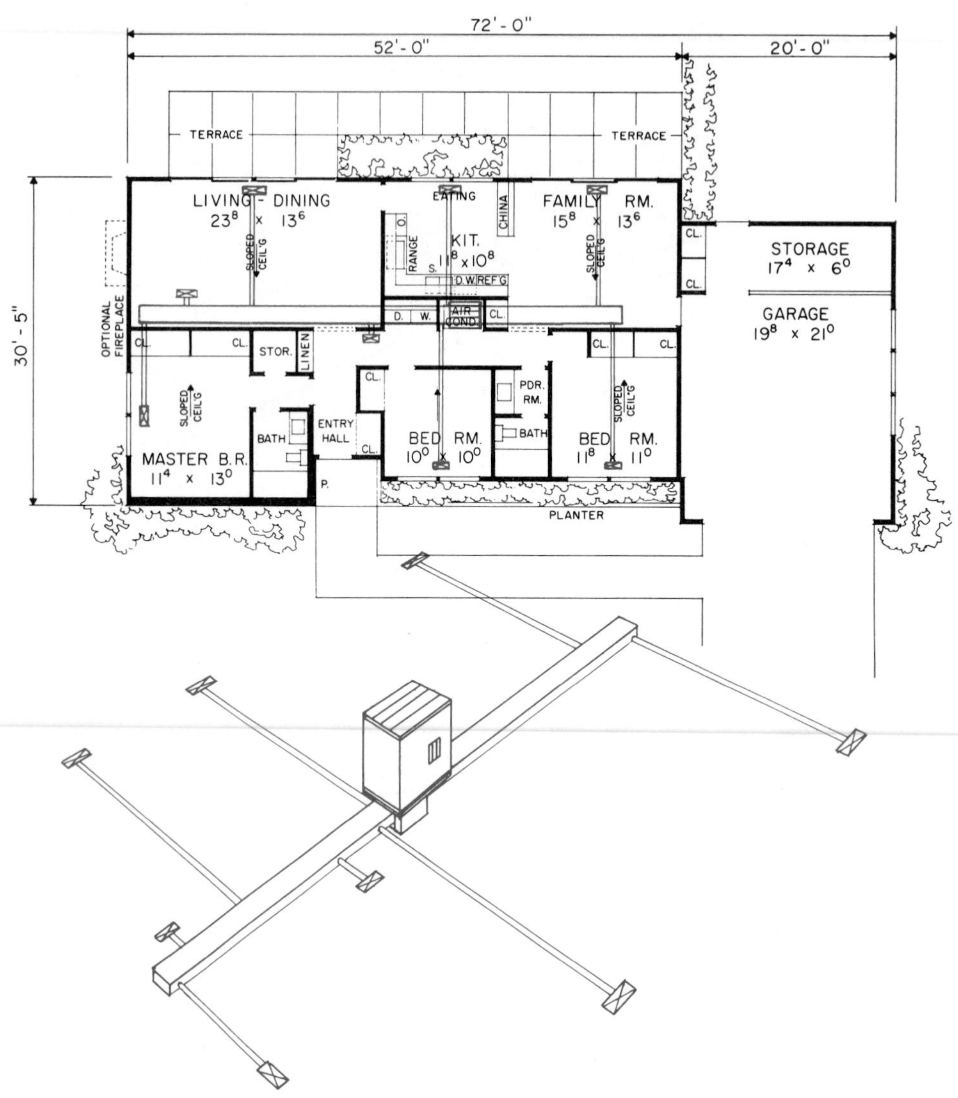

Fig. 62-21 Extended plenum system. *(Home Planners, Inc.)*

builder must determine the type, size, and location of all duct work connecting the furnace with the room outlets.

Hot-Water Units

A *hot-water unit* uses a boiler to heat water and a water pump to send the heated water to radiators, finned tubes, convectors, or baseboard outlets (Fig. 62-27). Forcing the water through the pipes with a pump is faster than allowing gravity to make it flow. Hot-water heating provides even heat and keeps heat in the outlets longer than warm-air units. The hot-water boiler is smaller than the warm-air furnace. The hot-water pipe is smaller and easier to install than warm-air ducts. Hot-water and cold-water lines in a hot-water unit is shown in Fig. 62-28. Hot-water units, however, are incompatible with air-conditioning systems, which require the installation of air ducts. There are several types of hot-water systems used to supply heat from the boiler to heating units: the series-loop system, the one-pipe system, the two-pipe system, and the radiant system.

The *series-loop* system, as shown in Fig. 62-29, is a continual loop of pipes containing hot water that passes through baseboard units. Hot water flows continually from the boiler through the baseboard units and back again to the boiler for reheating. The heat in a series-loop system cannot be controlled except at the source of the loop. Thus, it is effective for only small areas where radiators or convectors pro-

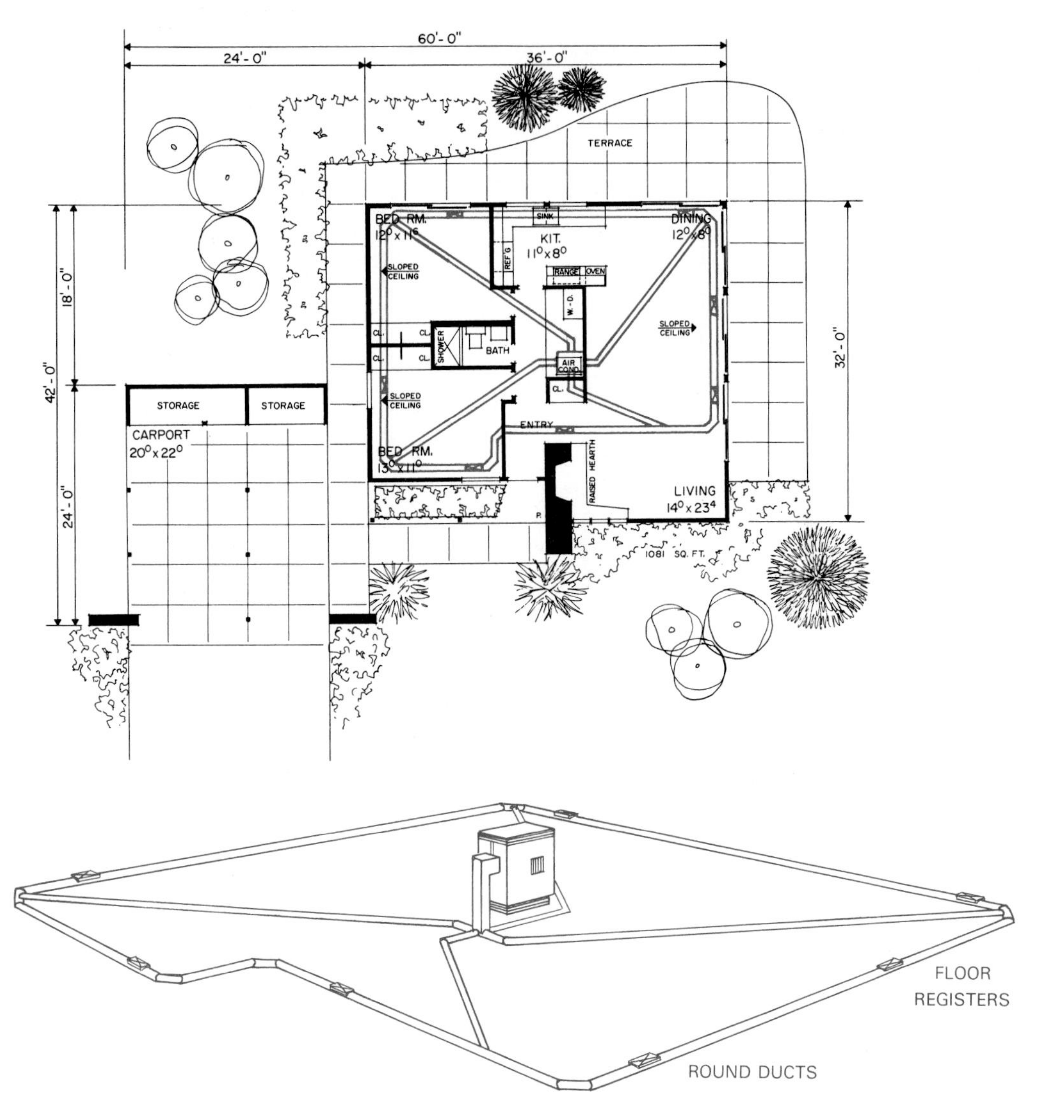

Fig. 62-22 Perimeter loop system.

duce heat at the same temperature throughout the system. The only way to vary temperature in this type of system is to increase the number of loops in a building. The temperature for each loop can then be varied at each individual loop boiler.

In *one-pipe* systems, heated water is circulated through pipes that are connected to radiators or convectors by means of bypass pipes. This allows each radiator to be individually controlled by valves. Water flows from one side of each radiator to the main line and returns to the boiler for reheating.

In a *two-pipe* system, as shown in Figs. 62-28 and 62-29, there are two parallel pipes; one for the supply of hot water from the boiler to each radiator, and the other for the return of cooled water from each radiator to the boiler. The heated water is directed from the boiler to each radiator but returns from each radiator through the second pipe to the boiler for reheating. In this system, all radiators receive water at nearly the same temperature.

Hot-water heating plans are sometimes shown without piping details. If only the position of the radiators is shown, then the plumbing contractor must determine the exact position of piping and also ensure that the radiators or convectors are located as determined on the piping plan.

The *radiant* hot-water heating system distributes hot water through a series of pipes in floors or ceilings. The warm surfaces of the floor or

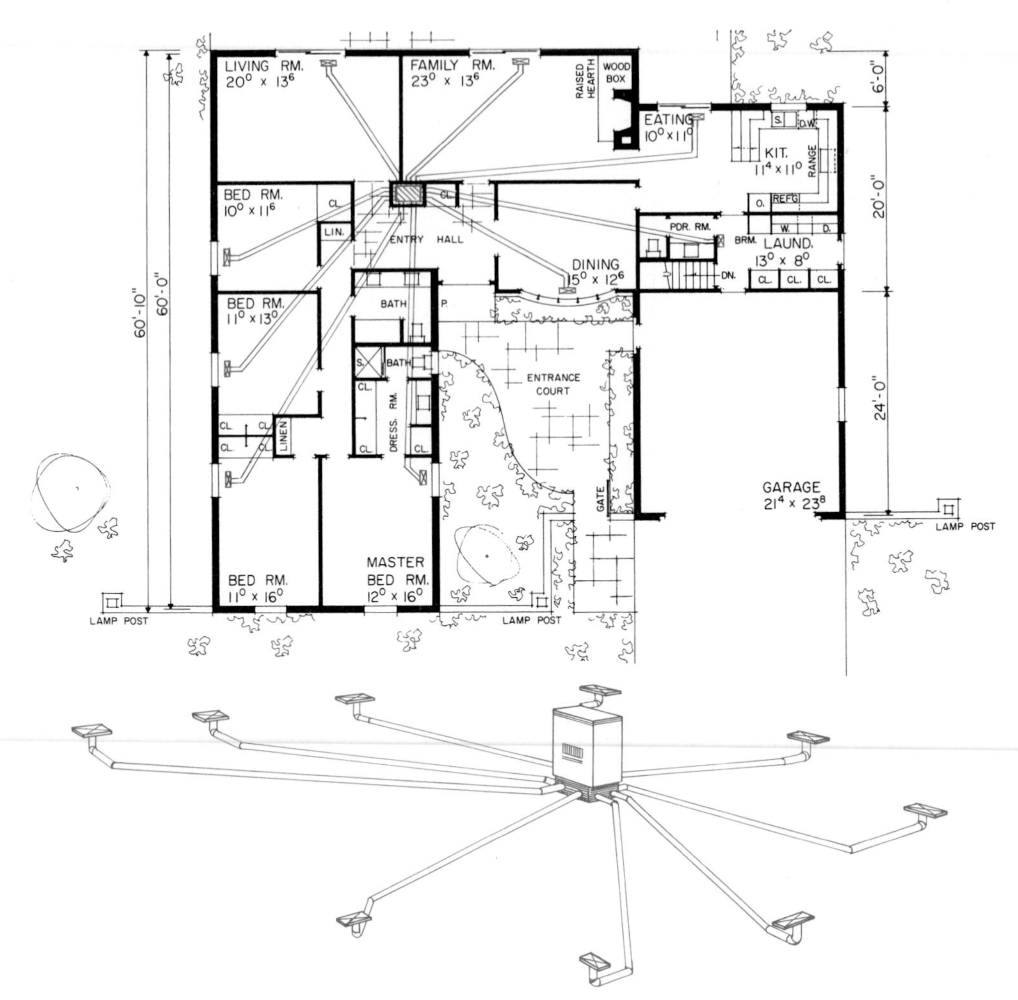

Fig. 62-23 Perimeter radial system.

ceiling radiate heat to cooler objects. Ceilings are often used for radiant hot-water heating since furniture and rugs restrict the distribution of heat in floors and walls. Figure 62-30 shows a radiant heating system.

Steam Units

The *steam-heating unit* operates by a boiler used to make steam. The steam is then transported by pipes to radiators or convectors and baseboards that give off the heat. The steam condenses to water, which returns to the boiler to be reheated to steam. The boiler must always be located below the level of the rooms being heated. Although steam-heating systems function on water vapor rather than hot water, drawings for steam systems are identical with those prepared for hot-water systems. Steam systems are easy to install and maintain, but they are not suitable for use with most convector radiators. They are most popular for large apartments, commercial buildings, and industrial complexes where separate steam generation facilities are provided. Steam heat is delivered through either perimeter or radial systems.

Electric Heat

Electric heat is produced when electricity passes through resistance wires. This heat is usually radiated although it could be fan-blown (convection). Resistance wires can be placed in panel heaters, wired into the wall or ceiling, placed in baseboards, or set in plaster to heat the walls, ceilings, or floors (Fig. 62-31). Electric heaters use very little space and require no air for combustion. Electric heat is very clean. It requires no storage or fuel and no duct work. Complete ventilation and humidity control must accompany electric heat, since it provides no air circulation and tends to be very dry.

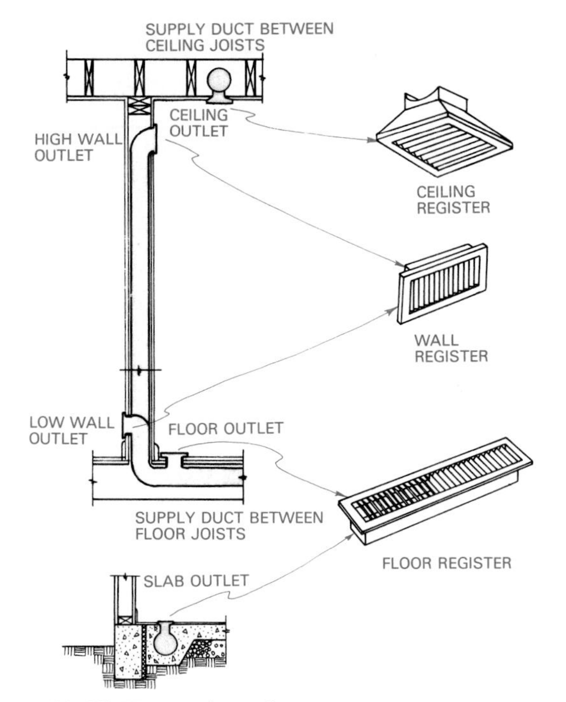

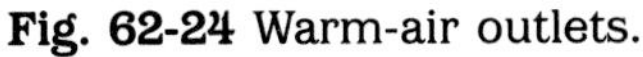

Fig. 62-24 Warm-air outlets.

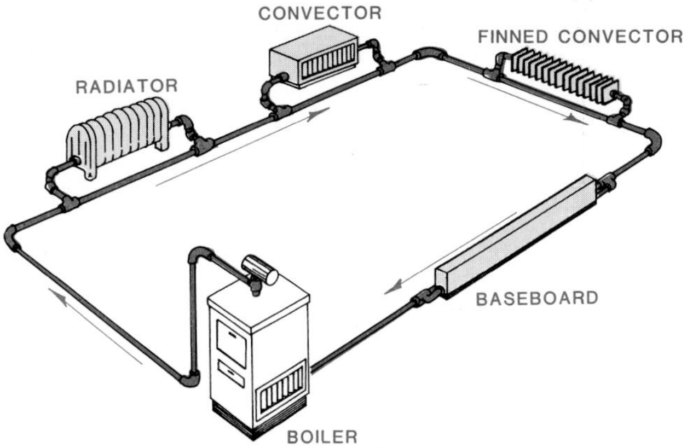

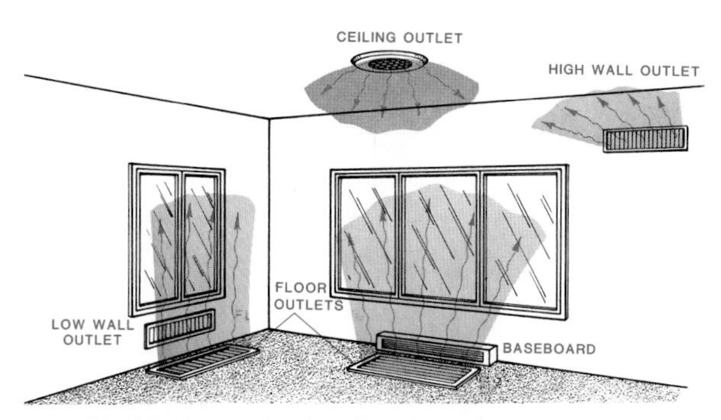

Fig. 62-25 Heat distribution locations.

Fig. 62-27 One-pipe hot-water system with different distribution methods.

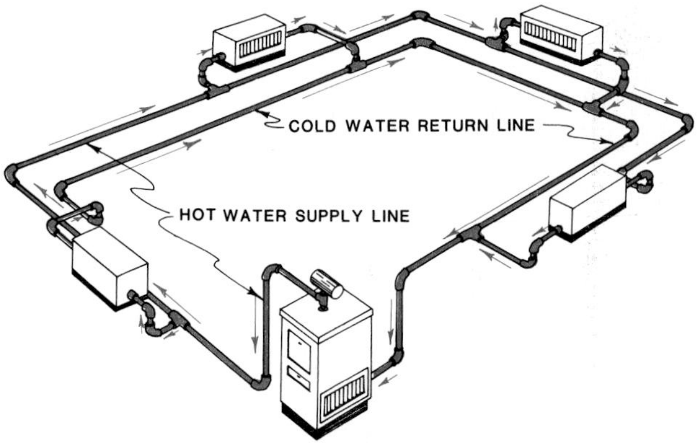

Fig. 62-28 Two-pipe hot-water series loop system.

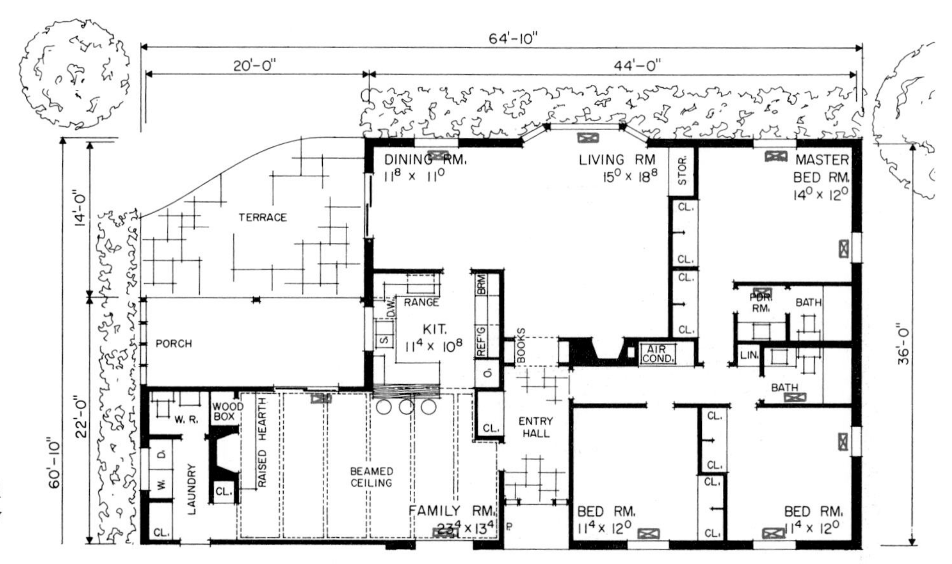

Fig. 62-26 Abbreviated method of showing only outlet locations.

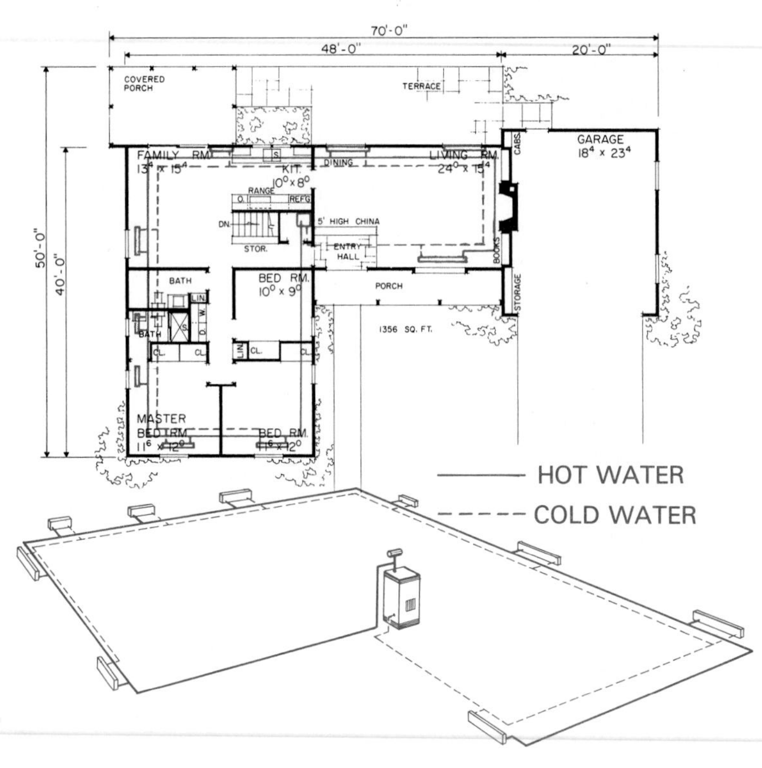

Fig. 62-29 Two-pipe hot-water system.

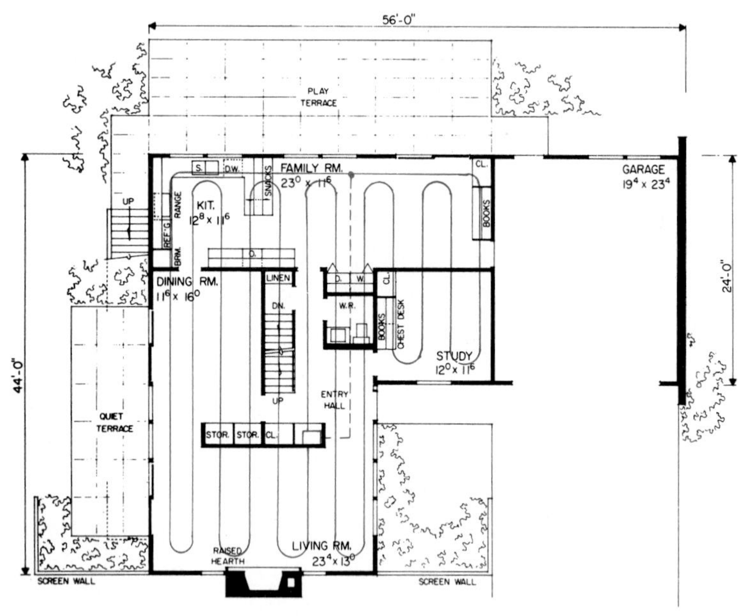

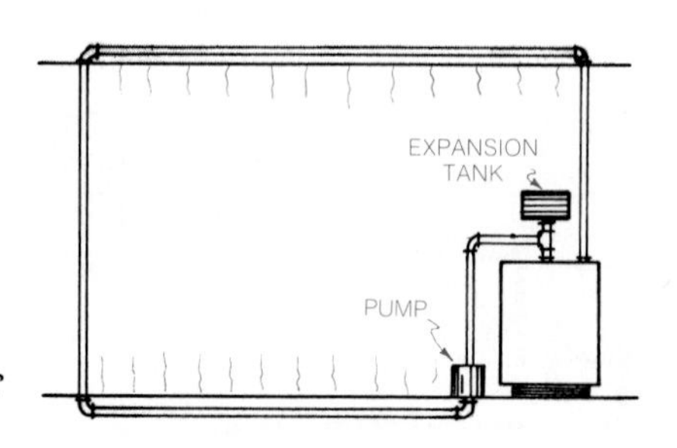

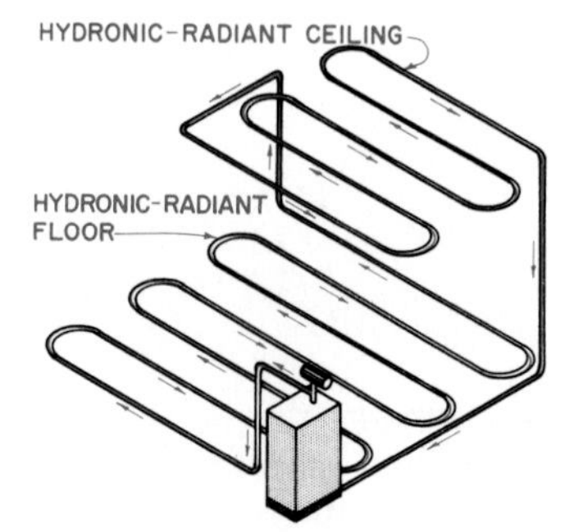

Fig. 62-30 Floor radiant hot-water system.

No plans are drawn specifically for electric heat, but notations are made on the floor plans concerning the location of either resistance wires or electric panels. On electrical plans, the location of facilities for power supply and thermostating is shown. Another use of electricity for air conditioning is to operate the *heat pump*. The heat pump is a year-round air-conditioner. In winter it takes heat from the outside air and pumps it into the house. There is always some heat in the air regardless of the temperature. In summer the pump is reversed and the heat in the house is pumped outside. Thus the pump works like a reversible refrigerator. Drawings for heat-pump duct-work layouts are identical with those used for forced warm-air systems.

The fifth heating device, employing *solar systems*, is discussed in detail in Unit 63.

CONVENTIONAL COOLING SYSTEMS

A building is air-conditioned by removing heat. Heat can be transferred in one direction only, from the warmer object to the cooler object. Therefore, to cool a building comfortably, a central air-conditioning system absorbs heat from the house and transfers it to a liquid refrigerant, usually freon. Warm air is carried away from rooms through ducts to the air-conditioning unit where a filter removes dust and other impurities. A cooling coil containing refrigerant then absorbs heat from the air passing through it. Then the blower that pulled the heat-laden air from the rooms pushes heat-free, or cool, air back to the rooms. There are four main methods of cooling structures: wastewater, cooling-tower, evaporation, and air-cool, as shown in Fig. 62-32.

The cooling unit can be part of the heating unit, using the same blower, vent, and perimeter ducts (Fig. 62-33). In this kind of system, the cool air rises against the warm walls in the summer and cools the house, as shown in Fig. 62-34. The cooling system can also be separate from the heating system.

When cooling systems are combined with heating systems, a combined heating-and-cooling-system plan is usually prepared, as shown in Fig. 62-35. This plan is read exactly the same as a warm-air heating-duct plan except for the addition of the cooling unit.

The size of air-conditioning equipment is usually rated in Btu. Although metric standards for air-conditioning units have yet to be established, British thermal units can be converted to joules (J). To convert Btu to joules, multiply the Btu value by 1055.

The average small house can be comfortably cooled with central air-conditioning units of

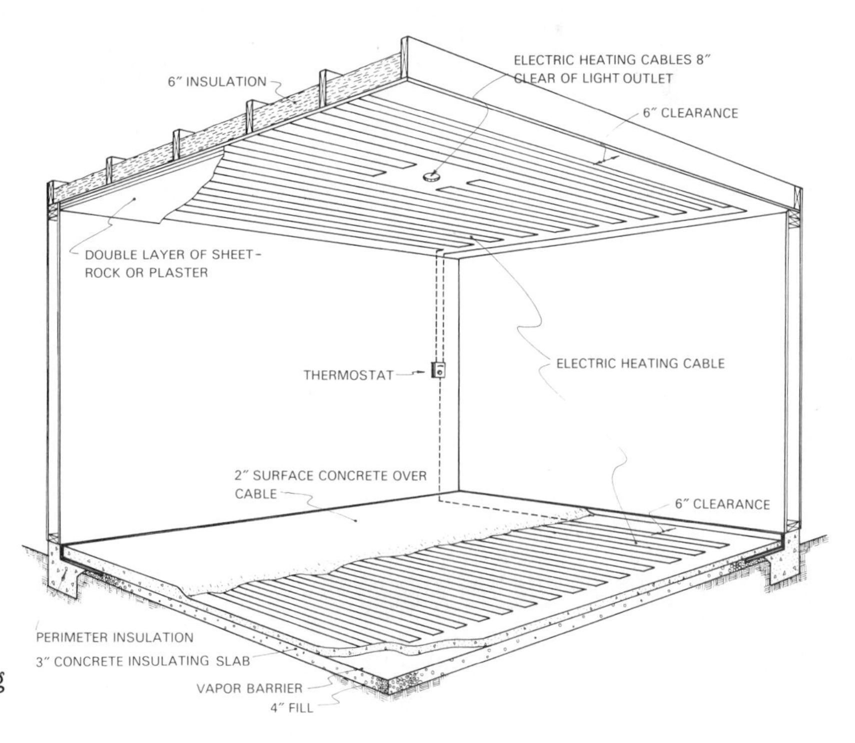

Fig. 62-31 Electric radiant heating floor and ceiling system.

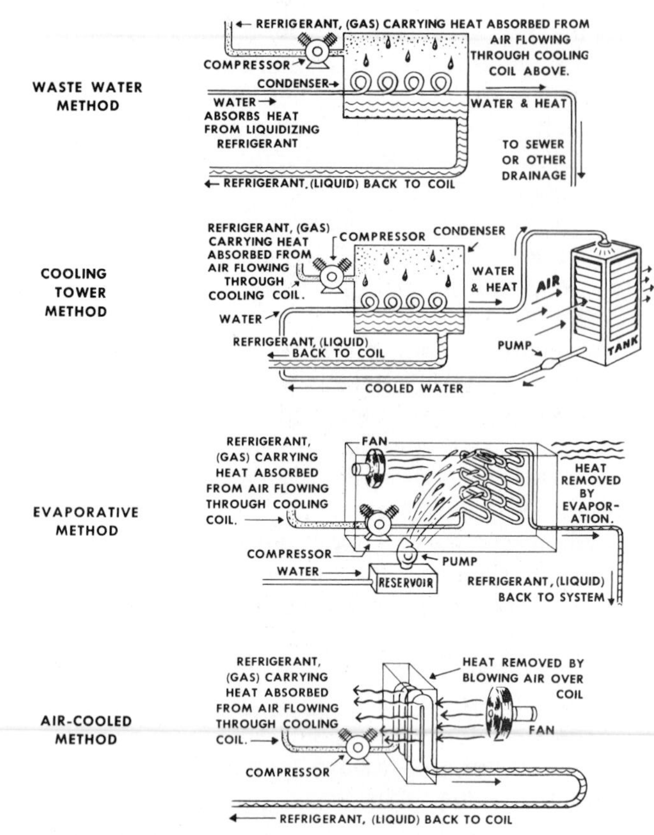

Fig. 62-32 Four methods of cooling.

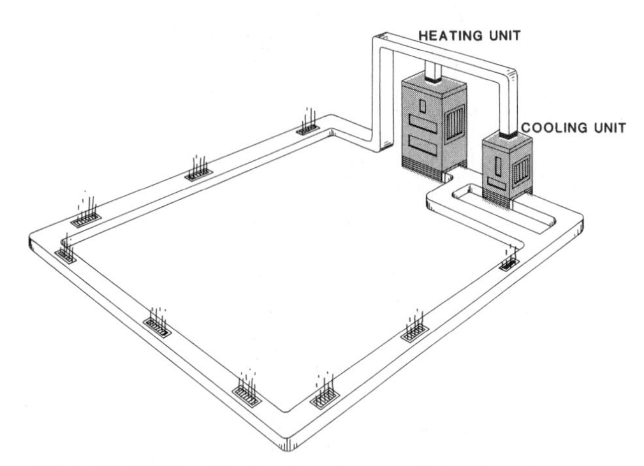

Fig. 62-33 Cooling unit as part of the heating system.

24,000 to 36,000 Btu. Larger homes may require a 60,000-Btu unit.

HUMIDITY CONTROL

The proper amount of moisture in the air is important for good air conditioning. Excessive moisture in the home comes from many sources,

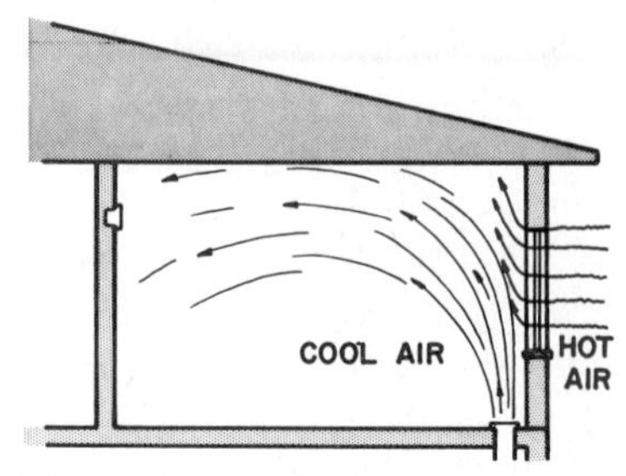

Fig. 62-34 Cool-air circulation.

such as cooking, cleaning, and washing, and from the outside air. To remove excessive moisture from the air, adequate ventilation and a humidification system are necessary. A *humidification system* takes the moisture from the damp air and passes it over cold coils. When the moisture-laden air passes over these coils, it deposits its excess moisture on the coils by condensation. Conversely, if the air is too dry, the humidification system adds moisture to the air. A device used only to remove the humidity from the air is known as a *dehumidifier*. A device used only to add humidity to the air is a *humidifier*.

Vapor barriers applied separately or to the face of insulation help repel the entrance of moist air into the structure. When applied to insulation the barrier material should be facing the inside of the building. Effective vapor barriers are polyethylene film sheets, building paper, and metal-foil-backed insulation.

VENTILATION

Ventilation is necessary to keep fresh air circulating. Effective ventilation also controls moisture and keeps air relatively dry. Constant air circulation is almost as important as constant temperature. These two work hand in hand to produce total comfort. All air in a building must be circulated gently and constantly, 24 hours a day. If air circulation stops, comfort stops. However, care must be taken to prevent the creation of drafts. A draft condition usually results from cold air entering through outside windows, cascading down window surfaces and sliding across the floor. This is eliminated by the use of perimeter heating systems.

Although structural tightness is a desirable method of controlling heat loss, some amount of air must be allowed to enter the structure to provide the oxygen needed for health, flame feeding, and moisture reduction. To provide ade-

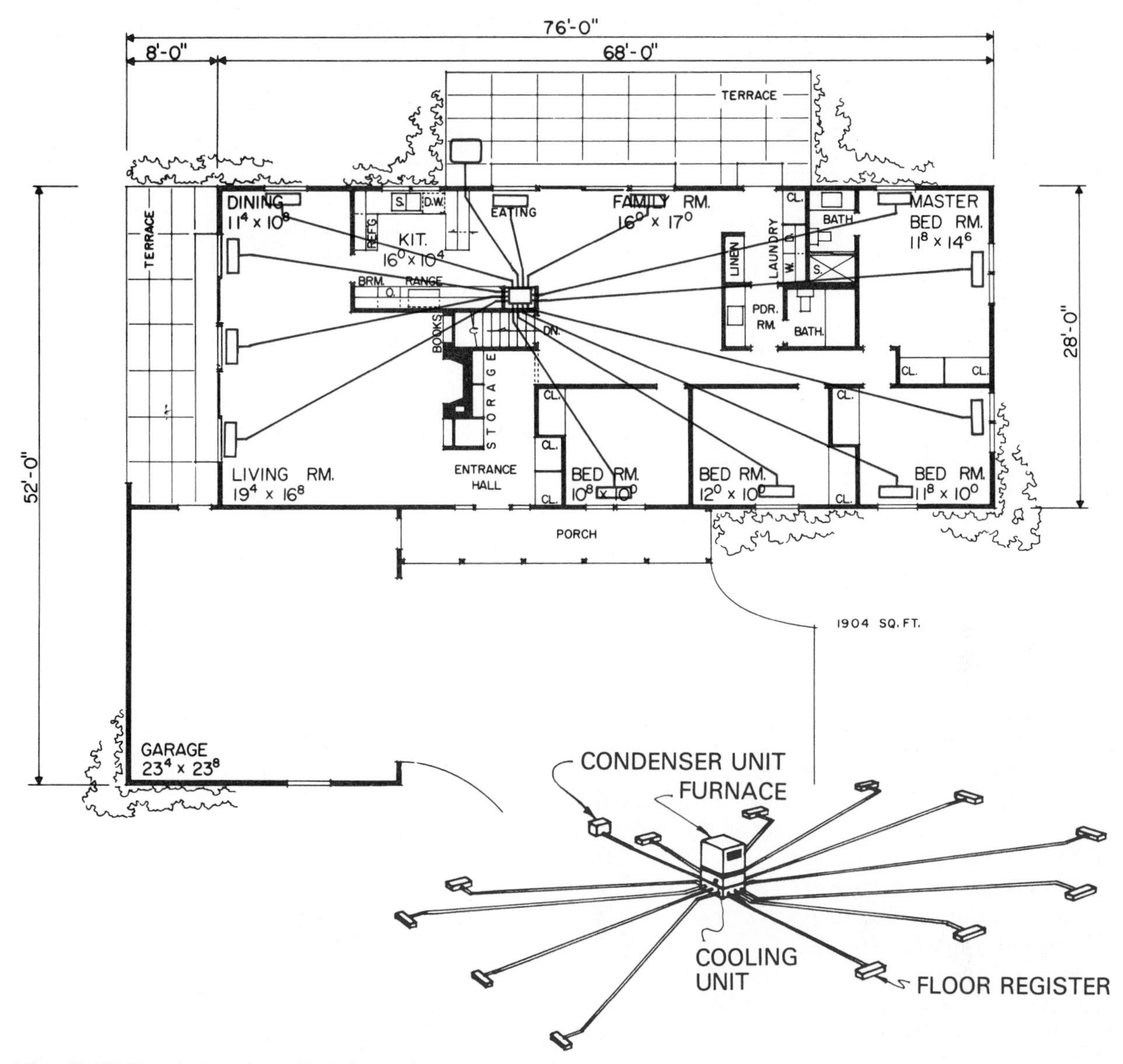

Fig. 62-35 Heating and cooling duct plan.

quate ventilation, all unheated areas, such as attics, garages, and crawl spaces, should be properly ventilated to provide air movement to prevent condensation. The vent openings should have screens to prevent birds, animals, or insects from entering.

Approximately 1 ft^2 of vent opening space should be provided for every 120 ft^2 of attic floor space. Fifty percent of this ventilating area should be placed at or near the top of the attic, either along the ridge or at both ends of the gables, and 25 percent should be placed in each opposing soffit. The key to effective air movement is to ensure a low continuous intake (from the soffits) up and out through the ridge exhaust (Fig. 62-36).

The simplest type of ventilation system is cross-ventilation through open windows. However, exhaust fans should be provided in the kitchen, bathroom, and attic to remove moisture, fumes, and warm air. Since large appliances tend to raise the heat of the house by 15 percent, vents for them should be provided.

CONTROL DEVICES

Thermostatic controls keep buildings at a constant temperature by turning climate-control

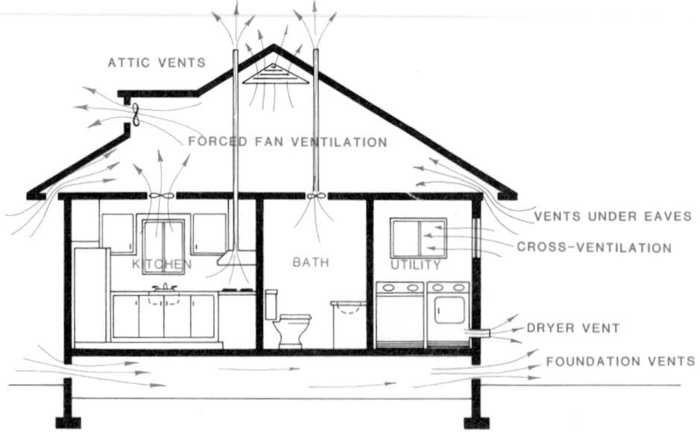

Fig. 62-36 Effective ventilation helps control moisture.

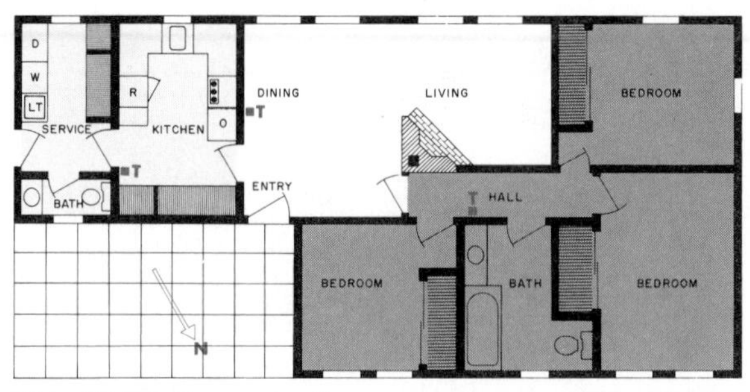

Fig. 62-37 Separate thermostats are desirable for large areas.

systems on or off when the temperature is beyond a certain setting. Thermostatic controls may be used with any heating or cooling system. The automatic thermostat control should be located on an interior wall away from any sources of heat or cold such as fireplaces or windows. Larger homes may need two or more separate heating or cooling zones that work on separate thermostats (Fig. 62-37). One advantage of electrical heating is that each room may be thermostatically controlled. This is especially important in regulating the temperature of children's rooms.

Exercises

1. **Sketch the floor plan shown in Fig. 69-2. Sketch a warm-air heating unit in the most appropriate location and locate the outlets for this system.**
2. **Sketch the house shown in Fig. 69-2. Locate the position of all duct work for a forced warm-air perimeter system.**

3. **Draw a plan of the climate-control system appropriate for the house of your design.**
4. **Answer the following questions related to Fig. 62-38.**
   - ***a.*** **What does the abbreviation F.A.U. represent?**
   - ***b.*** **How many supply registers are specified?**
   - ***c.*** **How many air-return registers are specified?**
   - ***d.*** **How is the bathroom heated?**
   - ***e.*** **What heating system places supply registers under windows?**
5. **Identify the symbols shown in Fig. 62-39.**
6. **Define these terms: *air-conditioning, humidity, air filter, humidifier, humidification, radiation, convection, conduction, insulation, perimeter, radiant heating, warm air, hot water, steam, electric heating, thermostat, solar heating.***

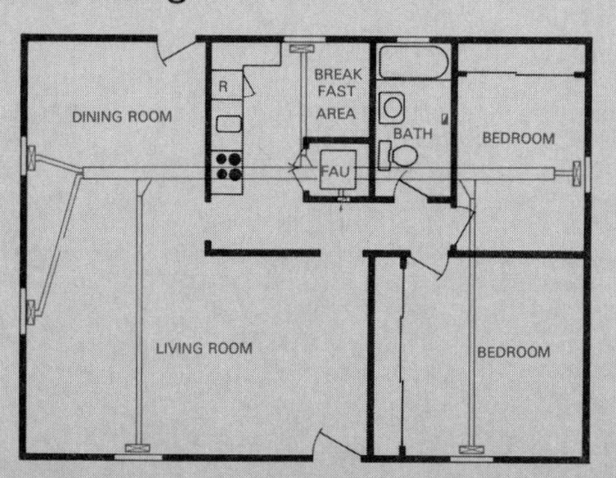

Fig. 62-38 Answer the questions in Exercise 4 relating to this plan.

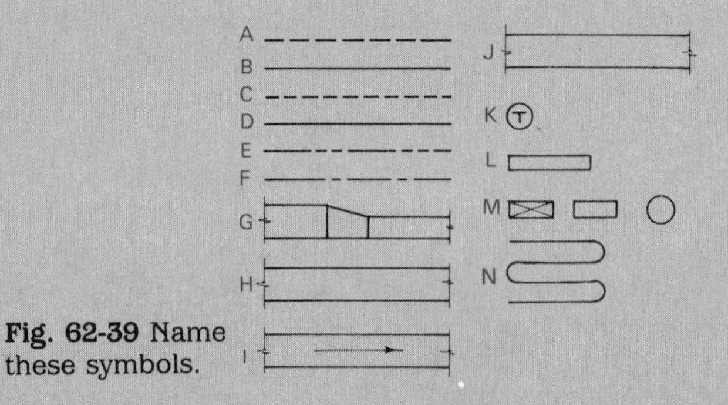

Fig. 62-39 Name these symbols.

UNIT 63

ACTIVE SOLAR SYSTEMS

Solar heating and/or cooling involves the sun to the fullest extent possible. There are two general classifications of solar systems, active and passive. Planning for *active (indirect)* solar systems requires knowledge of both mechanical systems and thermal principles. Active solar systems use mechanical devices to drive the components needed for solar collection, storage, distribution, and control. *Passive* systems, sometimes called *direct* systems, were covered in Unit 3 and include the total design of the structure and its surroundings. Passive systems are integrated with, and relatively indistinguishable from, the basic design of the structure. Passive systems operate without the use of special mechanical or electronic devices to heat or cool a structure.

Active solar systems use south-facing solar collectors set at an angle perpendicular to the sun. However, these collectors should have full access to the sun from at least 10 a.m. until 2 p.m., since most solar heat is emitted during these middle hours of the day. The solar energy is collected and stored. Then, either hot air can be blown or hot water piped from the storage tanks to other parts of the building as heat is needed. In active solar systems some thermostatic control is required to automate the distribution of heat to the various rooms as required. All active solar systems also require a passive structure to be truly effective, as shown in Fig. 63-1. For example, the ecologically planned dwelling shown in Fig. 63-2 uses active solar energy and recycles waste matter and gray water for additional efficiency.

There are three basic types of active solar systems: the solar furnace, electrical conversion, and individual air or liquid collector systems.

SOLAR FURNACE

The *solar furnace* is a collection of mirrors that focuses the sun's heat on a concentrated area. Temperature as high as 3500° F (1926° C) can be attained and the energy used to heat or cool large structures or clusters of buildings. Solar-furnace collectors must move with the sun.

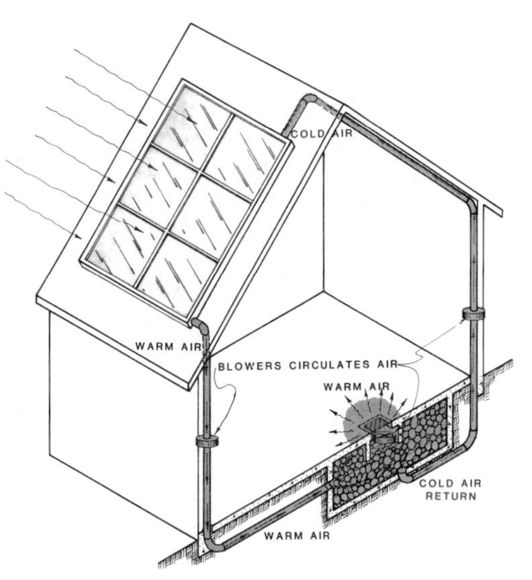

Fig. 63-1 Combination active and passive solar systems.

ELECTRICAL CONVERSION

This type of solar system uses the sun's heat for heating and cooling, and for converting sunlight into electricity to run home appliances. These tasks are accomplished by two large panels or collectors that consist of a number of solar cells. These solar cells are made of sandwiches of cadmium sulfide and copper sulfide between thin layers of glass. They produce electrical current upon exposure to sunlight. Part of the current produced in this manner is fed immediately into the home's electrical system, to operate lights and appliances. The remainder is used to charge a series of batteries that provide energy when the sun is not shining on the panels. As solar cells are improved, this form of electricity may become less expensive than turbine-generated electricity.

LIQUID COLLECTOR SYSTEMS

These systems, which are most popular for residences, use solar collectors to collect and trap the heat of the sun with liquid. The heated liquid is then pumped or blown to an insulated storage container and then, on demand, pumped or

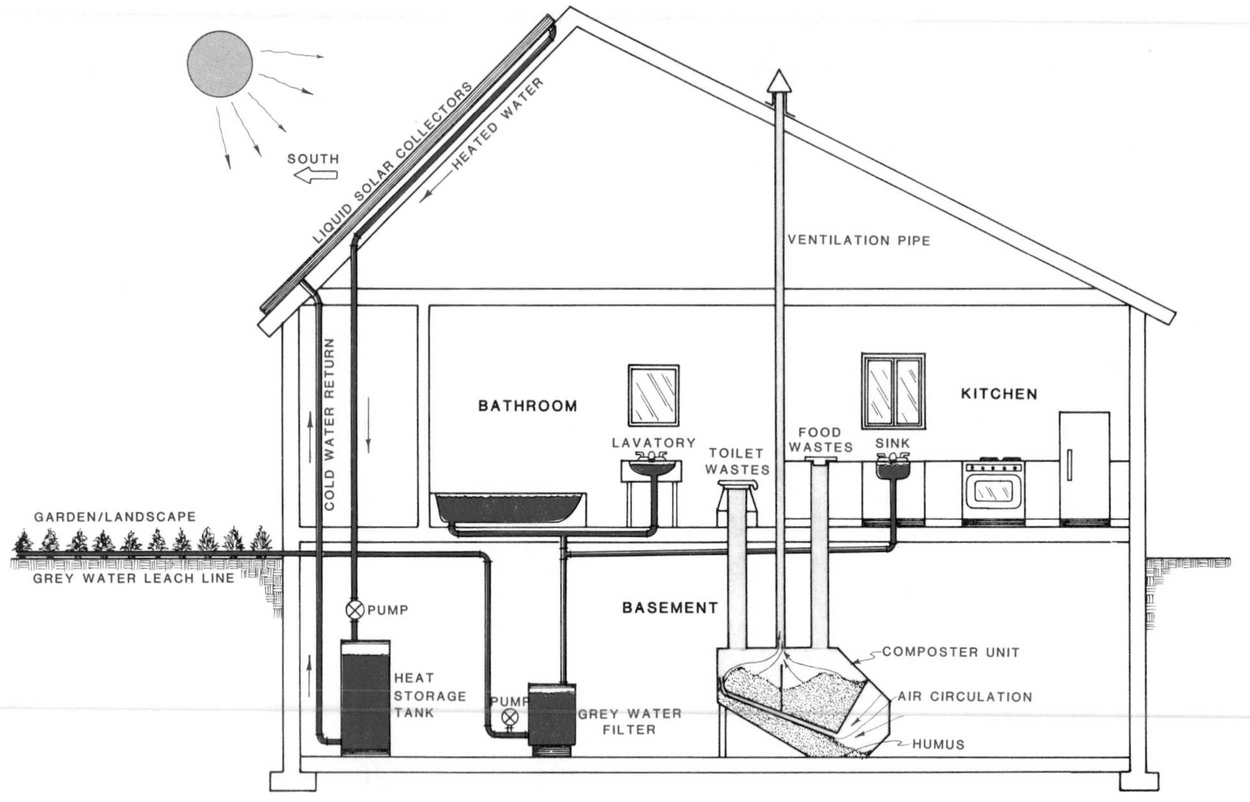

Fig. 63-2 Hybrid ecological design.

blown to appropriate parts of the structure. These solar systems are divided into three parts: collection, storage, and distribution.

In designing or choosing an individual active solar-heating system, the fluid used to transport the heat must first be determined. Water, antifreeze solutions, or oils may be used. The heated liquid is stored in large tanks and distributed through radiators, radiant panels, or liquid-to-air heat exchangers as shown in Fig. 63-3A.

Active liquid systems use collectors that consist of an *absorber* placed under a sheet of glass or plastic used as a *cover plate.* Under the absorber is a layer of *insulation* to help prevent heat loss. Collectors may be attached directly to a roof or installed in close proximity to the structure.

Each solar collector panel acts as a small greenhouse. Sunlight enters through the glass and warms the liquid circulating in pipes. The heat is trapped by the water and then pumped

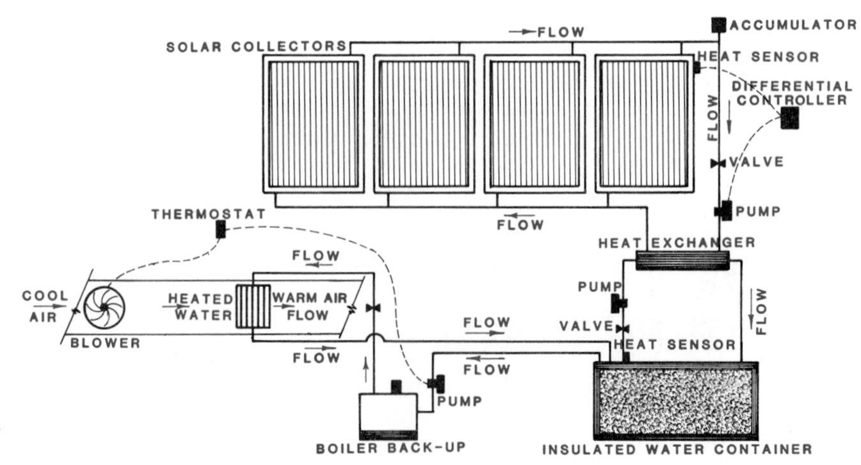

Fig. 63-3A Active solar liquid system.

into storage (Fig. 63-3B). Even on a very cold day with bright or filtered sunlight, these panels can be heated to 200° F.

In this system sunlight enters through a transparent cover as short-wave radiation and the heat is trapped as long-wave heat; then pumped into storage.

Liquid system absorbers are either flat metal or plastic sheets over which a liquid flows, or a network of pipes containing the liquid. Heat from the sun strikes the absorber and heats the liquid. The hot liquid flows to an insulated storage container to be held until needed.

Absorber plates are constructed of steel, copper, aluminum, or plastic, because of their heat conductivity. Absorbers are designed to retain a maximum amount of heat with low emittance. The amount of heat retained is called *absorptance,* and the amount of reflected heat is called *emittance.* Thus an efficient absorber should have high absorptance and low emittance.

Transparent glass or plastic absorber cover plates allow heat to penetrate the absorber while helping to hold the heat in. Insulation behind the absorber retards heat loss from the back side of the absorber.

COLLECTOR ORIENTATION

Active solar systems must have south-facing solar collectors set at an angle perpendicular to the sun. The angle can vary somewhat in either direction without a significant loss of efficiency. But for maximum efficiency, the sun's southern rays should strike the solar panels as perpendicularly as possible, as shown in Fig. 63-4A. Step angles location design increases the amount of area of the sun's exposure. Figure 63-4B shows optional locations for solar collectors and appropriate orientation to coincide with different site positioning restrictions.

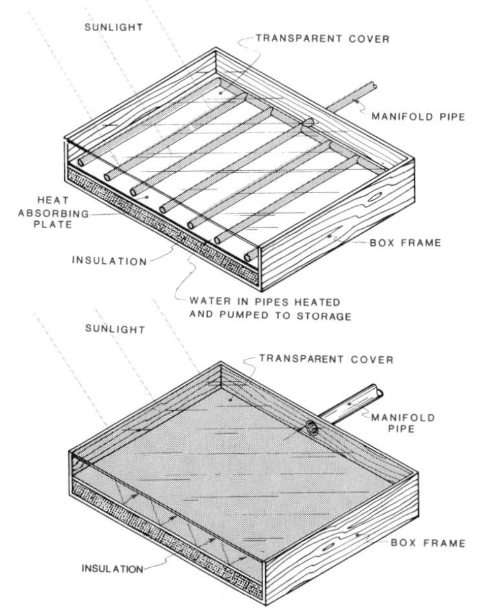

Fig. 63-3B Trapping of solar heat in air or liquid.

Ideally, heating collectors should face directly into the sun (at right angles) for the maximum number of hours each day. Unless rotating collectors are used, this ideal position is possible for only a short time each day. Therefore, fixed collectors are usually positioned to face the mid-afternoon sun, since air temperatures are usually higher at that hour. In most North American areas, this means the collector will face slightly west of south.

In addition to the horizontal orientation of the collectors, the vertical tilt must be considered. The tilt angle should be the same as the local latitude for maximum year-round effect; however, a tilt angle of 15° greater than the local latitude is best for winter heating because of the lower path of the sun.

HEAT STORAGE AND DISTRIBUTION

Because heat is needed when the sun is not shining, storage of the absorbed heat is necessary. Stored heat is limited to the capacity of the storage unit. The larger the storage unit, the longer the solar system will operate without the use of auxiliary devices. From 1 to 10 gallons (4 to 38 liters) of liquid storage are needed for each square foot of collector surface with liquid systems.

Once the solar-heated water is delivered to the storage facility, it remains there until thermostats call for it to be moved to various parts of the structure. When that occurs, the heated liquid is pumped through a series of pipes to radiant heat panels or liquid-air heat exchangers that feed warm-air systems.

Active liquid systems can be used with small collector and storage units to provide hot water and to heat water for boilers and/or swimming pools, thus reducing the cost of conventional heating. Figure 63-5 shows a typical active liquid solar system used for heating hot water.

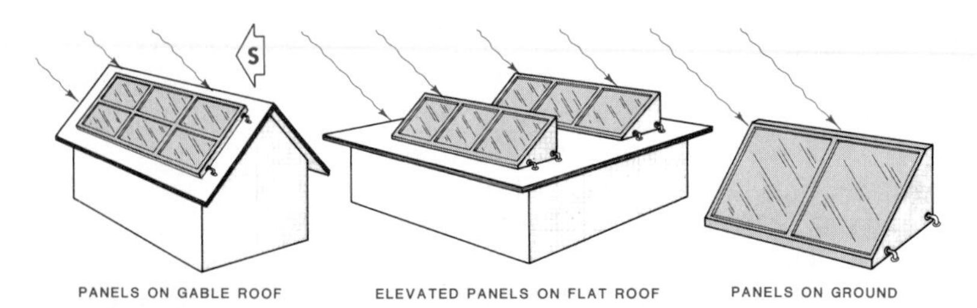

Fig. 63-4A Sun rays should be perpendicular to solar panel surfaces.

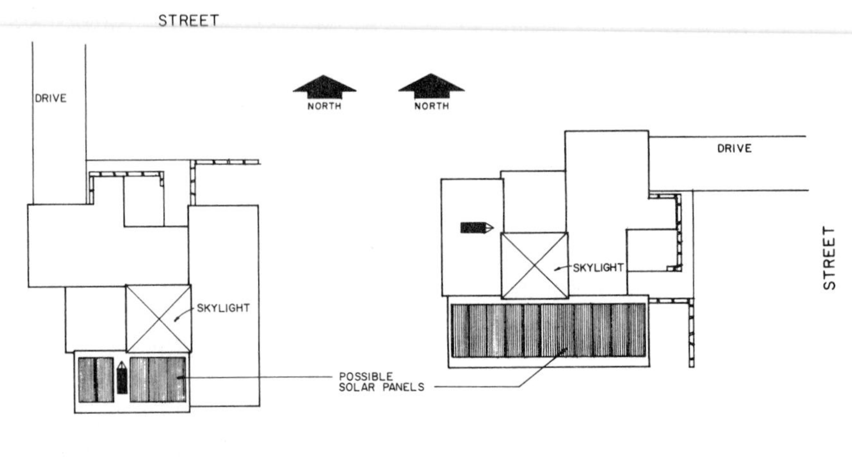

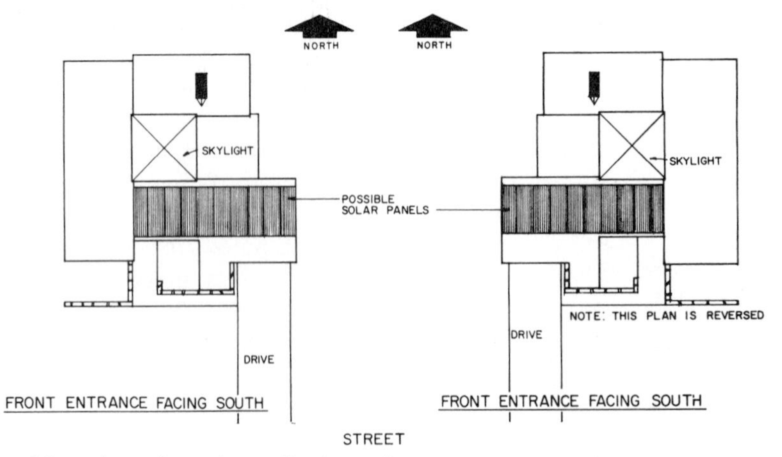

Fig. 63-4B Optional locations for solar collectors. *(Home Planners, Inc.)*

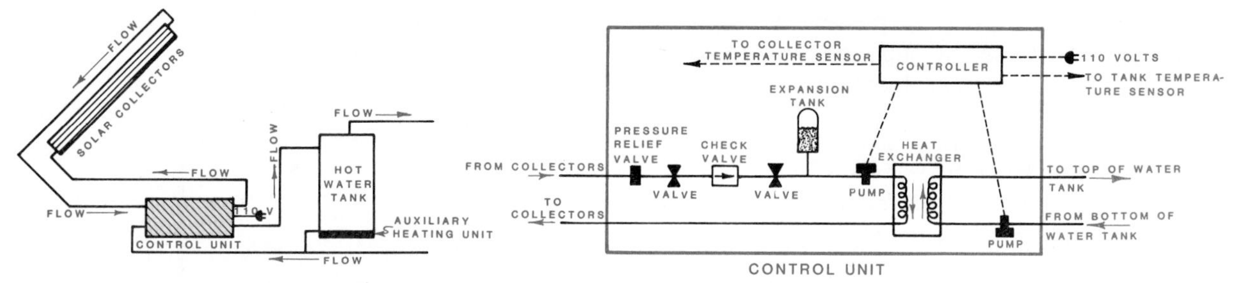

Fig. 63-5 Solar liquid system for water heating.

AIR SYSTEMS

Active solar systems using air as the heating element are more simple to construct and operate than liquid systems. They are effective for space heating; however, their inability to provide hot water is a serious drawback. Air-system collectors use sheet-metal absorber plates to heat air trapped between a cover plate and the absorber. This heated air is then blown to the storage facility. From the storage area, the heated air can then be blown directly through a duct system to appropriate rooms in a conventional warm-air system. Because air systems have low heat capacity, large ducts (up to 6″) must be used in this collector.

With air systems, heat can also be stored in rocks, gravel, or small containers of water and can be distributed through convection or radiant panels. 80 to 400 lb (36 to 181 kg) of rock are required for each square foot of collector for heat storage. Rock storage requires 2½ times as much volume as water to store the same amount of heat over the same temperature rise.

AUXILIARY HEATING SYSTEMS

Except in very mild climates, most solar-heat storage facilities cannot keep constant pace with peak demand. For this reason, an auxiliary heating system is usually recommended, especially for hot-water production. Figure 63-6 shows a typical active solar-heating unit with an integrated auxiliary heating unit. In the United

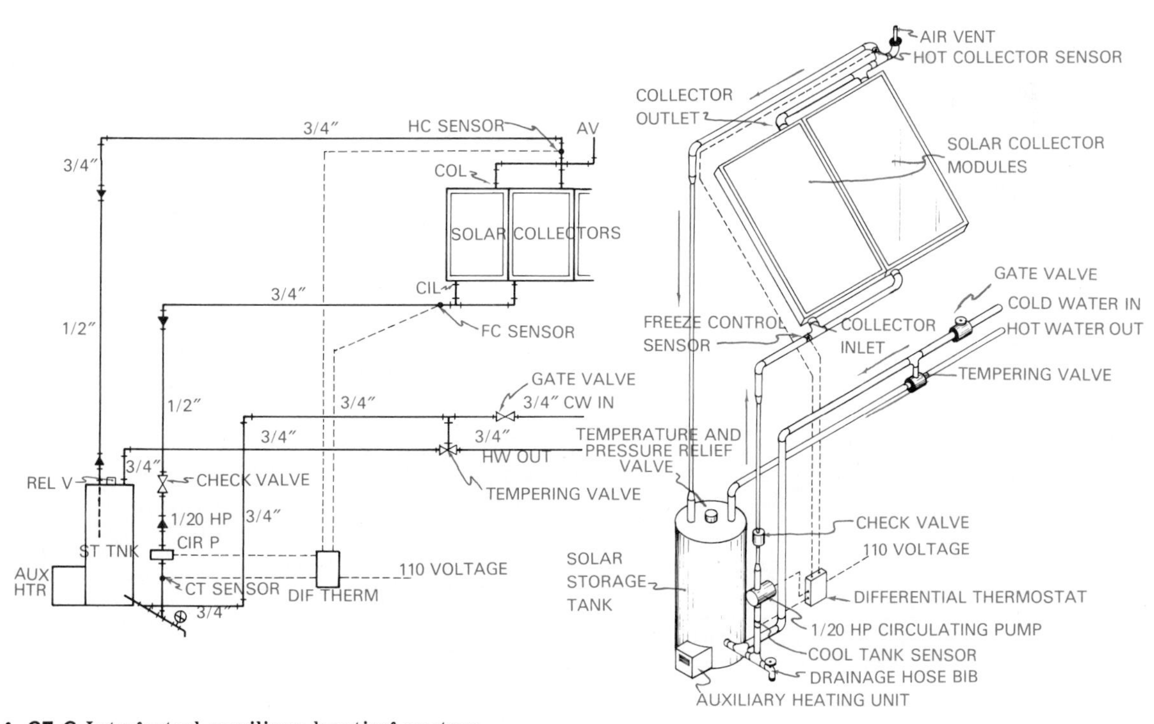

Fig. 63-6 Integrated auxiliary heating system.

States, between 50 and 90 percent of residential heating needs can be supplied through solar systems, depending upon location.

SOLAR COOLING

Solar cooling is possible through the same absorption-cooling method used in gas refrigerators. However, present equipment is very expensive. Figure 63-7A shows an active solar-heating and cooling system.

Collectors can be used minimally to help cool buildings by exposing them to cooler night air and closing them to daytime exposure. The cooler night air cools the liquid or air that returns to the storage area and is then released the next day to augment passive or conventional cooling systems. Figure 63-7B shows a solar system designed to remove warm air during warm days and hold heated air during cooler nights. This is done by either holding or expelling the solar heated air through the use of turbine ventilator control.

ACTIVE SOLAR APPLICATIONS

Figure 63-8 shows a residence designed for active solar heat utilization. Remember, a house

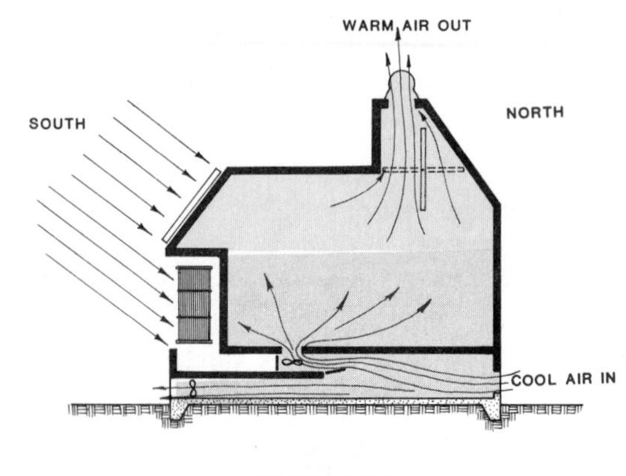

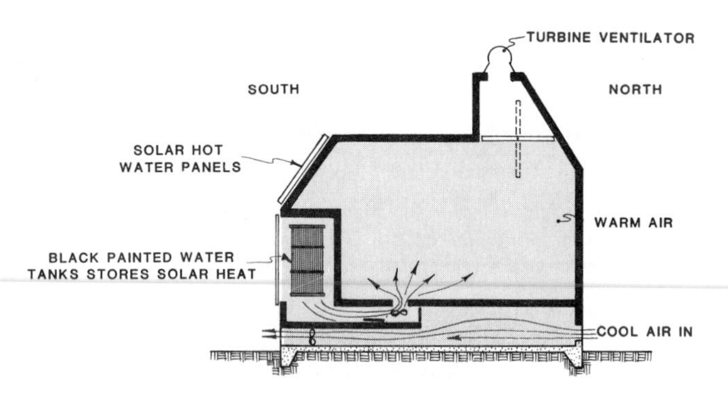

Fig. 63-7B Use of ventilator to adjust solar gained heat.

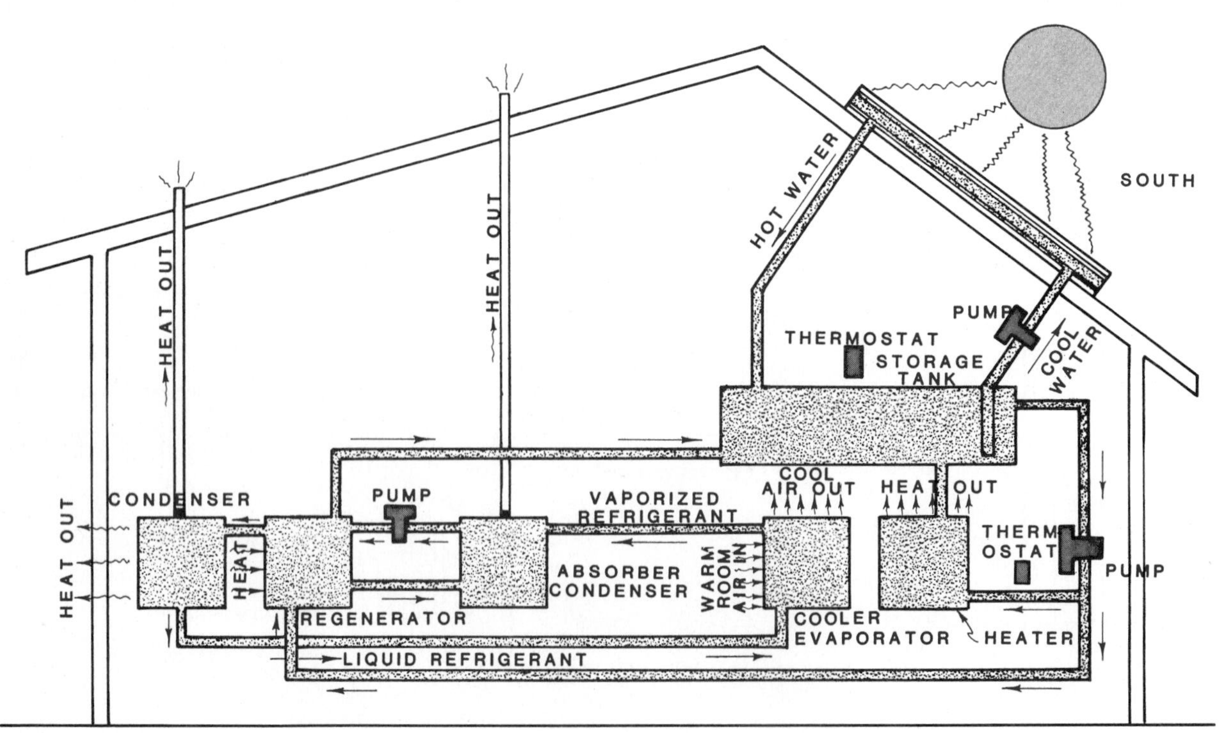

Fig. 63-7A Solar heating and cooling system.

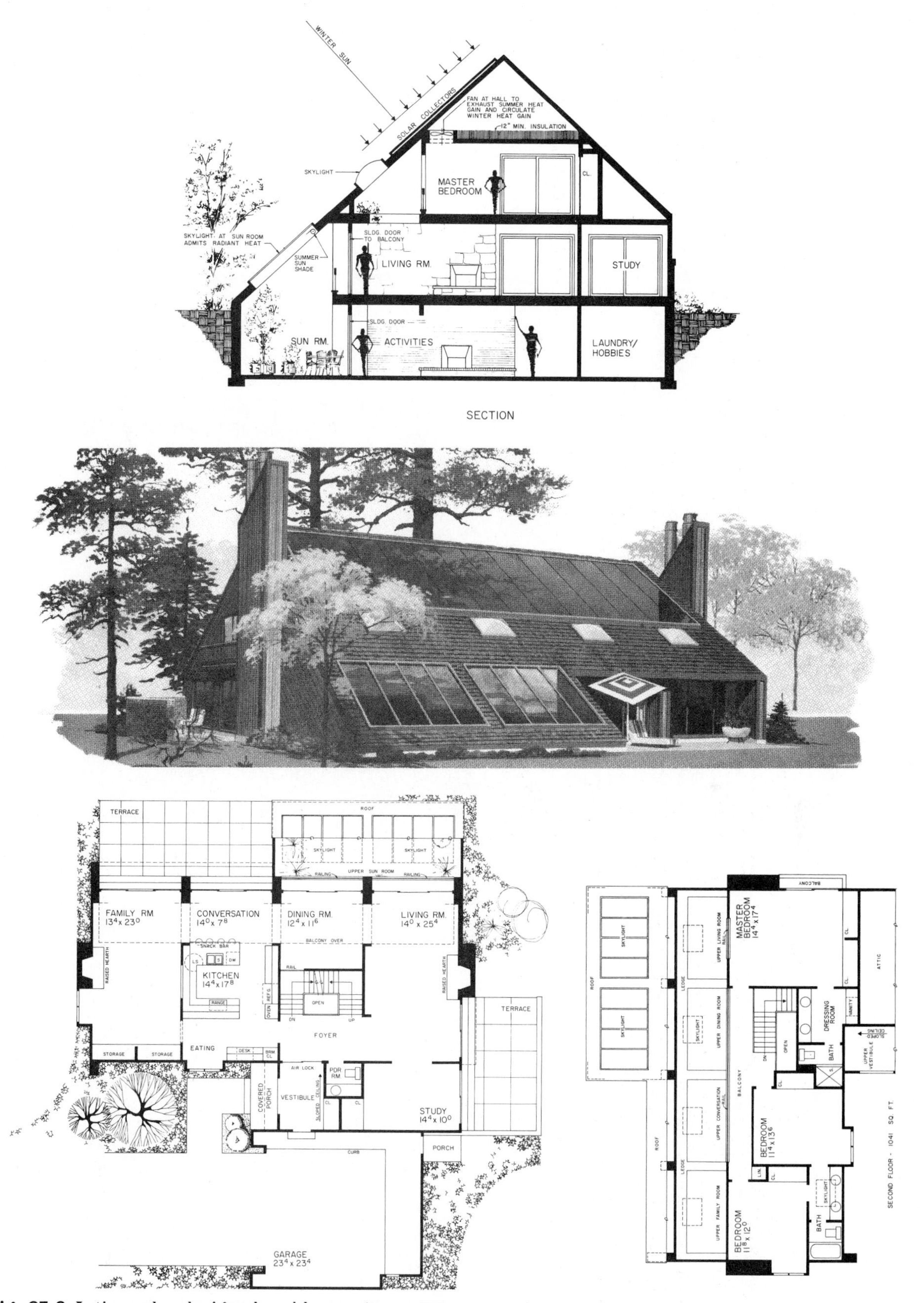

Fig. 63-8 Active solar-designed residence. *(Home Planners, Inc.)*

designed for an active solar system must be oriented so that the solar panels face south and the roof angle is perpendicular to the sun's rays at midday. In this plan, the solar equipment area houses the storage tanks for the hot water created in the solar panels. This residence is designed to utilize active solar devices and also incorporates many passive solar features to increase the efficiency of the active devices. In this plan the solarium creates a greenhouse effect which produces heat that can be circulated out or recirculated back into the house as the needs change.

Exercises

1. **On the plan shown in Fig. 69-1 indicate the best location of solar collection, storage, distribution, and control devices.**
2. **Add active solar features to a plan of your own design.**
3. **Define the following terms: *passive solar system, indirect solar system, solar furnace, absorbers, emittance, collectors, liquid system, air system.***
4. **List design features that will make any HVAC system more effective.**
5. **Design an active and passive solar system for the plan shown in Fig. 63-9.**

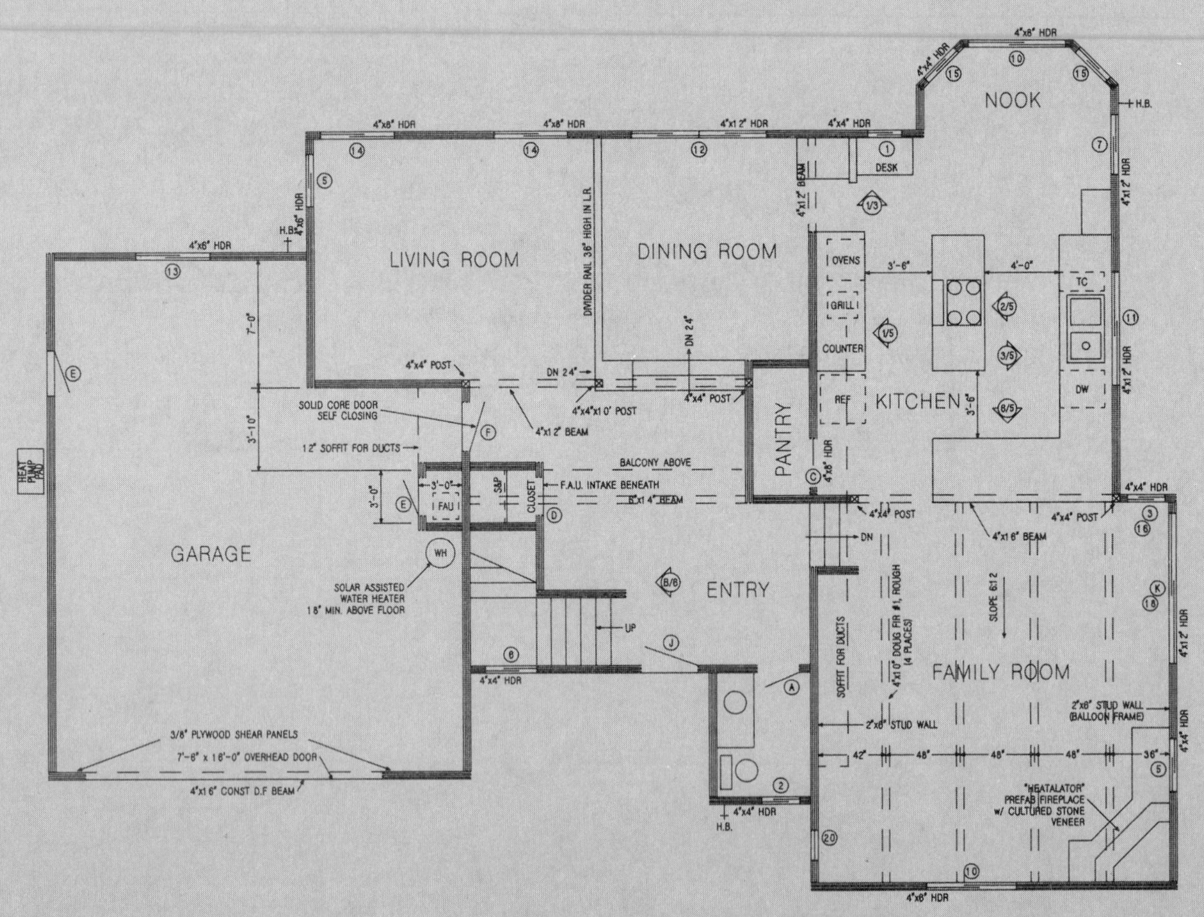

Fig. 63-9

SECTION 16

Plumbing Diagrams

Plumbing refers to the water supply and drainage of wastewater and sewage. A plumbing system consists of supply pipes that carry fresh water under pressure from a public water supply or individual wells to fixtures. It also includes a disposal system through which pipes carry wastes to a disposal area by gravity drainage. The design of plumbing systems also includes the selection and location of all fixtures that require water to function.

UNIT 64

PLUMBING CONVENTIONS

Like most architectural drawings, plumbing drawings must be prepared to a very small scale. Therefore schematic symbols are used as a substitute for drawing plumbing lines, components, and fixtures as they actually appear. Figures 64-1A and 64-1B show a detailed orthographic view of various plumbing lines and devices compared to a schematic drawing of the same items. These schematic symbols are used to show the type and location of fixtures, joints, valves, and other plumbing devices used to control the flow of liquids. Figures 64-2 and 64-3 show common plumbing symbols used on architectural drawings.

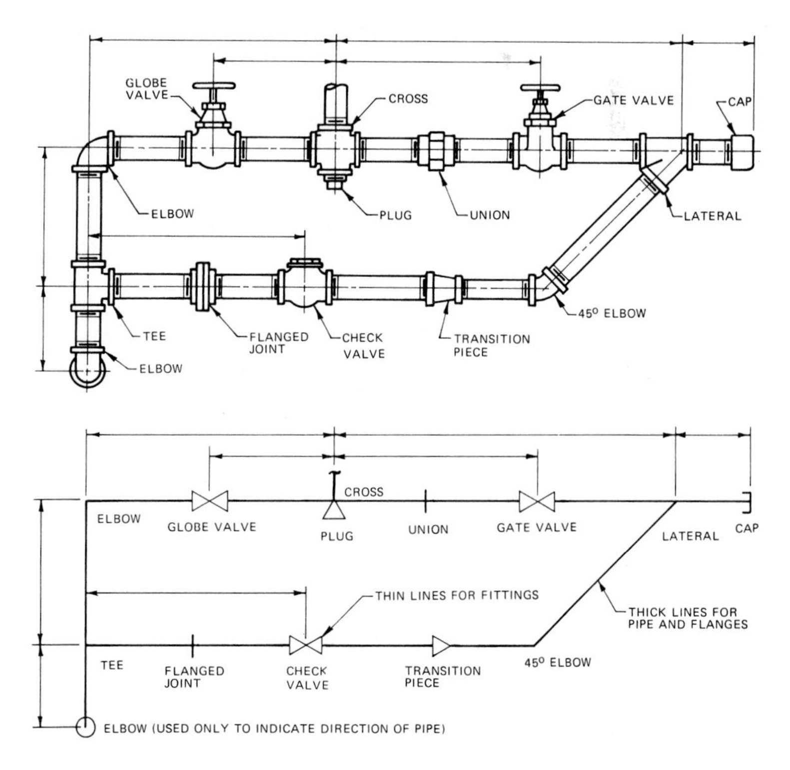

Fig. 64-1A Orthographic appearance of line components compared to schematic piping symbols.

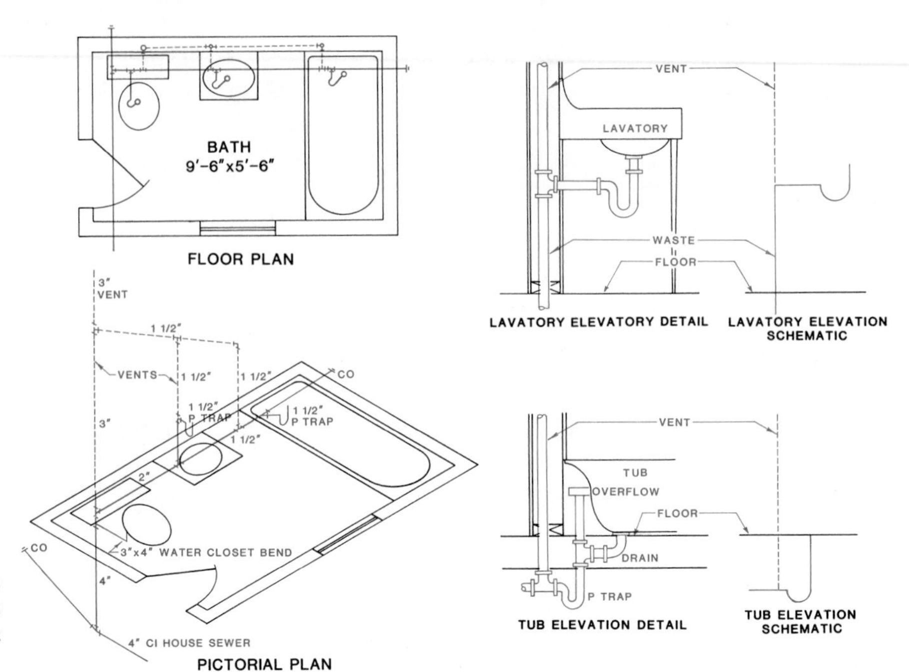

Fig. 64-1B Orthographic views of plumbing fixtures compared to schematic symbols.

Plumbing fixtures are available in a variety of sizes, colors, and materials. Since they are used continually for many years, durability is extremely important. High quality fixtures are a better investment than low quality fixtures. The following fixtures represent the most common types available for the bathroom and the kitchen.

Bathroom fixtures are divided into four types, as follows: *Water closets* may be tank and bowl in one piece, separate tank and bowl, or wall-

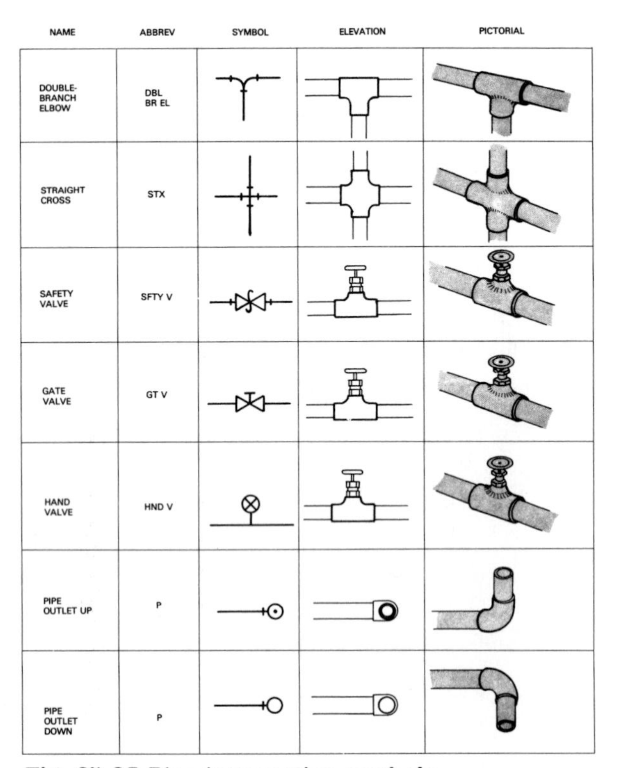

NAME	ABBREV	SYMBOL	ELEVATION	PICTORIAL
DOUBLE-BRANCH ELBOW	DBL BR EL			
STRAIGHT CROSS	STX			
SAFETY VALVE	SFTY V			
GATE VALVE	GT V			
HAND VALVE	HND V			
PIPE OUTLET UP	P			
PIPE OUTLET DOWN	P			

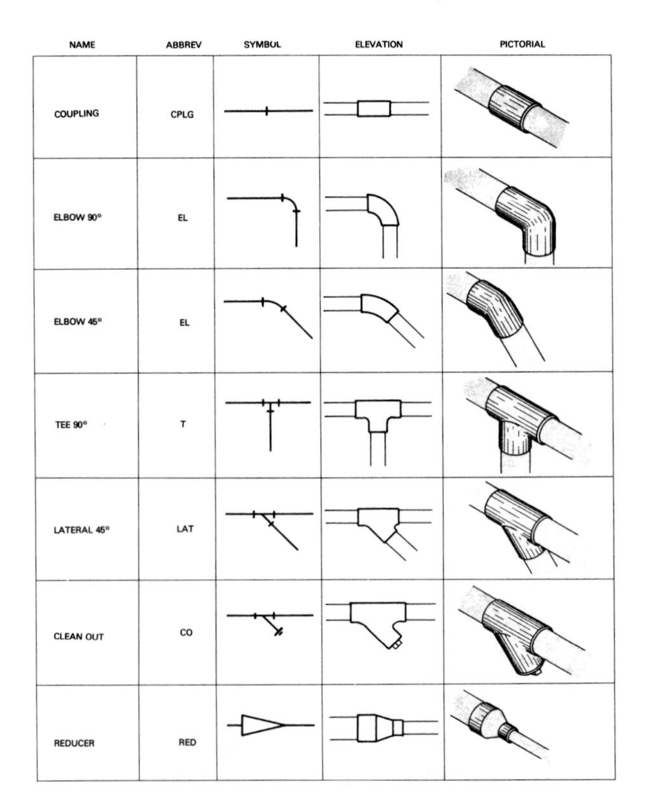

NAME	ABBREV	SYMBOL	ELEVATION	PICTORIAL
COUPLING	CPLG			
ELBOW 90°	EL			
ELBOW 45°	EL			
TEE 90°	T			
LATERAL 45°	LAT			
CLEAN OUT	CO			
REDUCER	RED			

Fig. 64-2A Pipe intersection symbols.

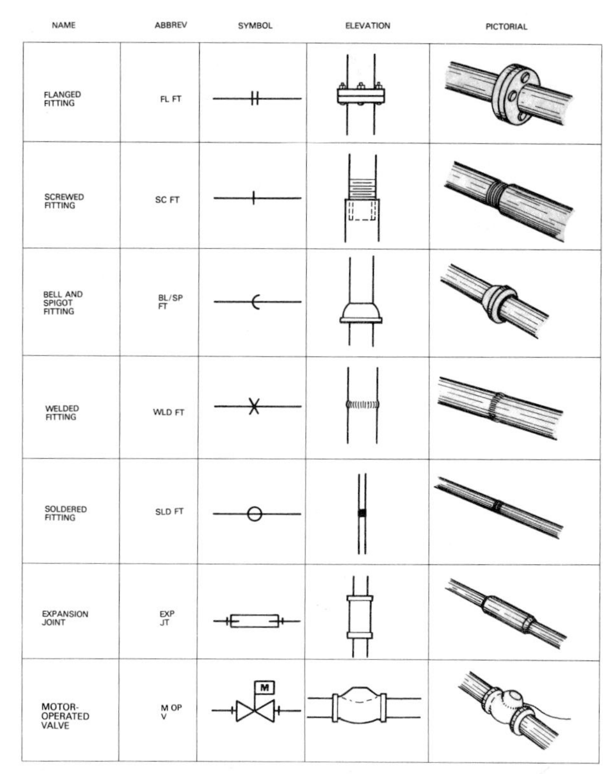

NAME	ABBREV	SYMBOL	ELEVATION	PICTORIAL
FLANGED FITTING	FL FT			
SCREWED FITTING	SC FT			
BELL AND SPIGOT FITTING	BL/SP FT			
WELDED FITTING	WLD FT			
SOLDERED FITTING	SLD FT			
EXPANSION JOINT	EXP JT			
MOTOR-OPERATED VALVE	M OP V	M		

Fig. 64-2B Pipe joint symbols.

NAME	ABBREV	SYMBOL	ELEVATION	PICTORIAL
METER	M	M		
FLOOR DRAIN	FD	FD		
CESS POOL	CP			
DRY WELL	DW			
SEPTIC TANK	SEP TNK			
SEPTIC-TANK DISTRIBUTION BOX	SEP TANK DIS BX			
SUMP PIT	SP	S P		

Fig. 64-2C Sanitary facilities symbols.

hung tank and bowl. *Showers* are prefabricated, built on the job, or placed over the bathtub. *Bathtubs* are recessed, square, freestanding, or sunken. *Lavoratories* may be wall-hung, cabinet-mounted, built-in countertop, or corner.

Kitchen fixtures are divided into five types, as follows: *Sinks* are available as a sink and drain-board unit, single sink, or double sink. *Laundry tubs* may be single, double, or triple. *Dishwashers* are either built-in or freestanding. *Hot-water heaters* are electric or gas.

For details relating to these fixtures, see Unit 16, Kitchens, Unit 17, Utility Rooms, and Unit 22, Baths.

NAME	ABBREV	SYMBOL	NAME	ABBRV	SYMBOL
COLD-WATER LINE	CW		AIR-PRESSURE RETURN LINE	APR	
HOT-WATER LINE	HW		ICE-WATER LINE	IW	IW
GAS LINE	G	—G—G—	DRAIN LINE	D	—D—D—
VENT	V		FUEL-OIL FLOW LINE	FOF	—FOF—
SOIL STACK PLAN VIEW	SS		FUEL-OIL RETURN LINE	FOR	----FOR----
SOIL LINE ABOVE GRADE	SL		REFRIGERANT LINE	R	—+—+—+—+—
SOIL LINE BELOW GRADE	SL		STEAM LINE MEDIUM PRESSURE	SL	

NAME	ABBREV	SYMBOL	NAME	ABBRV	SYMBOL
CAST-IRON SEWER	S-CI	S-CI	STEAM RETURN LINE—MEDIUM PRESSURE	SRL	
CLAY-TILE SEWER	S-CT	S-CT	PNEUMATIC TUBE	PT	
LEACH LINE	LEA		INDUSTRIAL SEWAGE	IS	
SPRINKLER LINE	SPR	—S—S—	CHEMICAL WASTE LINE	CW	
VACUUM LINE	VAC	—V—V—	FIRE LINE	F	—F—F—
COMPRESSED AIR LINE	COMP	—A—A—	ACID WASTE LINE	AC WST	ACID
AIR-PRESSURE LINE FLOW	APF		HUMIDIFICATION LINE	HUM	— - —H— - —

Fig. 64-3 Plumbing line symbols.

Exercises

1. **Add the plumbing fixtures to a plan of your own design.**
2. **Match the symbols with the terms in Fig. 64-4.**
3. **Define these terms: *plumbing fixture, water closet, wall-hung water closet, drainboard.***

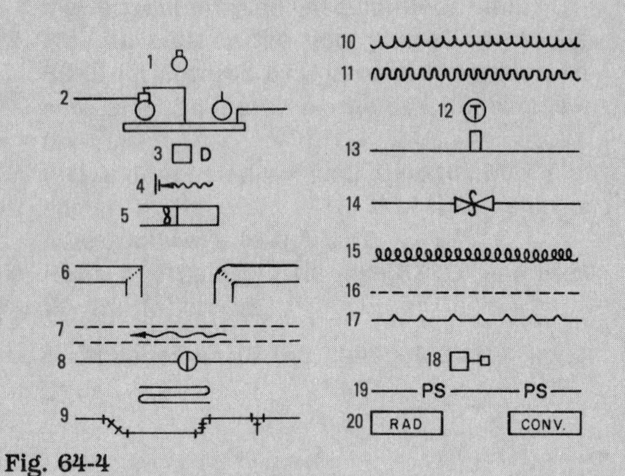

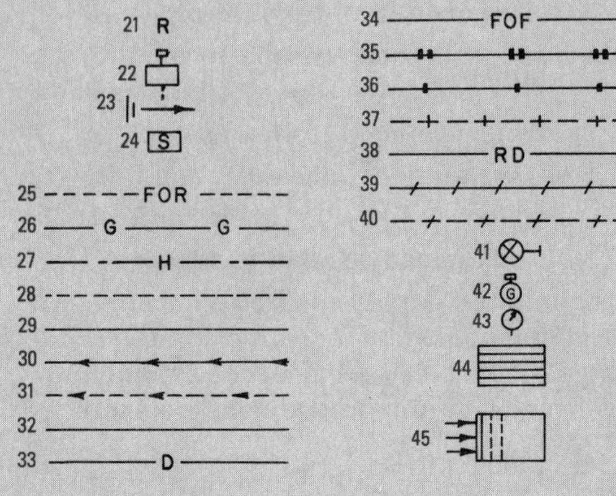

Fig. 64-4

UNIT 65

SCHEMATIC PLUMBING DRAWINGS

Plumbing diagrams are drawings that describe piping systems that supply water to, and drainage of waste material from, buildings. A plumbing system (Fig. 65-1) consists of supply pipes that carry fresh water under pressure from a public water supply or individual well to building fixtures and pipes, which, in turn, carry wastes to a disposal system by gravity drainage (Fig. 65-2). Plumbing diagrams are shown in both plan and elevation form.

SCHEMATIC PLUMBING PLANS

Plumbing symbols are indicated by one of two methods. They are added to an existing floor plan, or a separate plumbing plan is prepared as shown in Fig. 65-3. Sometimes plumbing symbols, electrical symbols, and heating and air-conditioning symbols are all combined in one plan. When that is done, dimensions and other construction notes are eliminated to facilitate reading the symbols without interference. Such procedure is possible because the plumbing contractor is concerned solely with the placement of plumbing fixtures and with the length of piping runs, and not with other construction details.

Since most plumbing fixtures are concentrated in the bathroom, the laundry, and kitchen areas, a detailed schematic plumbing plan is sometimes prepared for those rooms as shown in Figs. 65-4, 65-5, and 65-6.

WATER-SUPPLY SYSTEMS

Fresh water is brought to all plumbing fixtures under pressure. This water is supplied either from a public water supply or from private wells. Because this water is under pressure, the pipes may run in any convenient direction after leaving

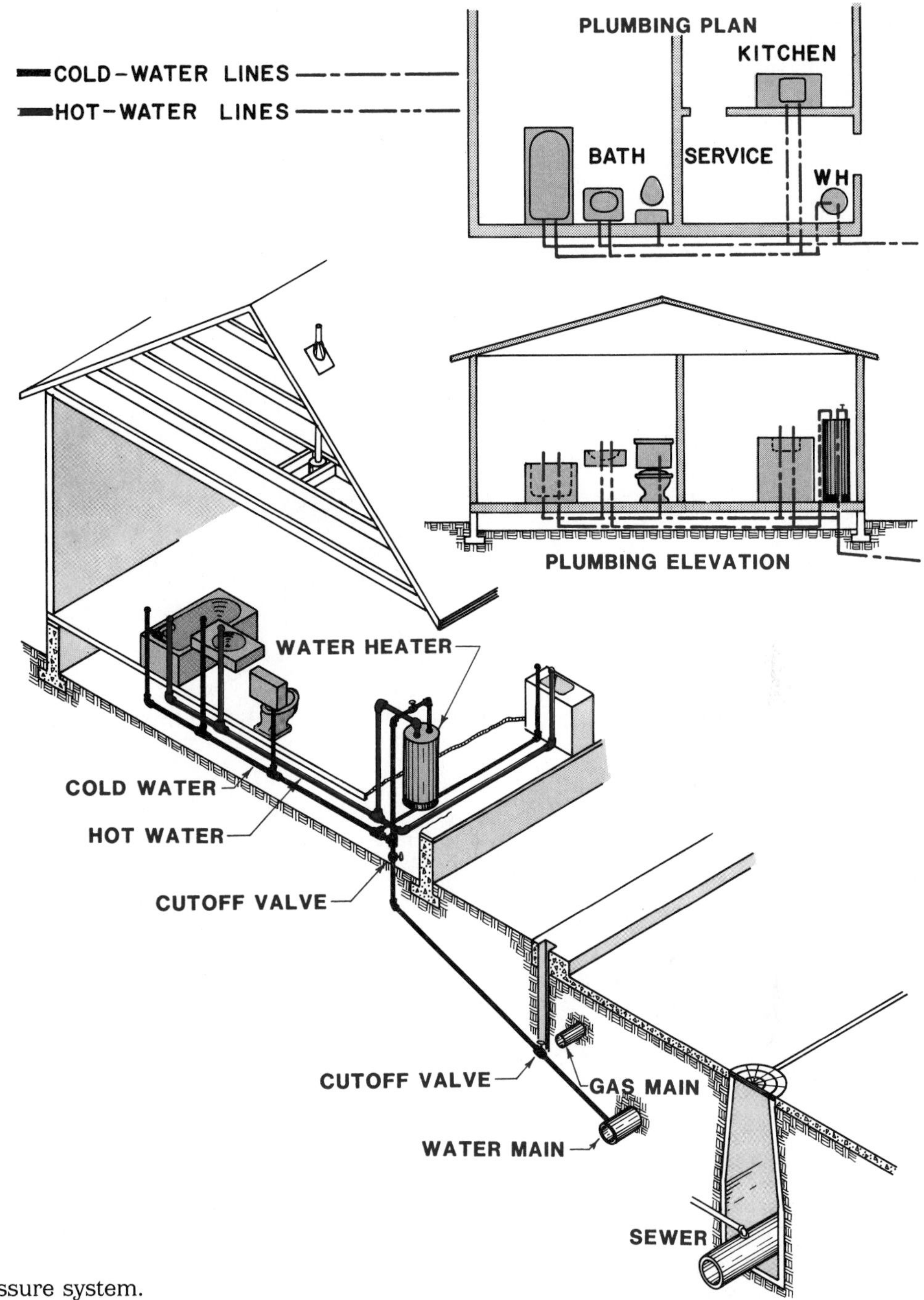

Fig. 65-1 A water pressure system.

the main control pipe, as shown in Fig. 65-7. Water lines require shutoff valves at the property line and at the entrance of the building. A water meter is located at the shutoff valve near the building.

Hot water is obtained by routing water through a hot-water heater. The hot water is then directed, under pressure, to appropriate fixtures. The hot-water valve is always on the left of each fixture as you face it. Placing insulation around hot-water lines conserves hot water and reduces the total cost of fuel for heating water. This is usually included in a note on the plumbing diagram, which specifies the type and thickness of the insulation material.

The size of all water-supply lines for a house ranges from 3/4″ to 1″ (19 to 25 mm). Each fixture should have a shutoff valve on the pipe to

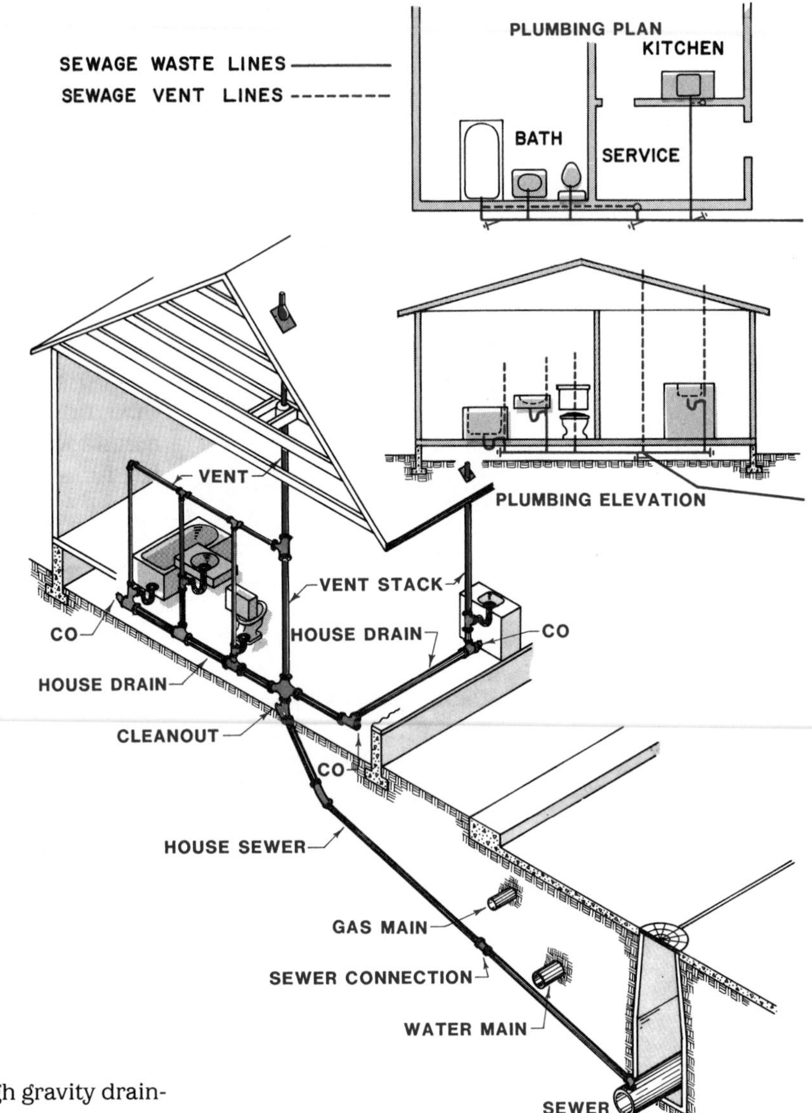

Fig. 65-2 Waste is discharged through gravity drainage.

allow repairs. All fixtures have a free-flowing supply of water if the lines are the correct size. Lines that are too small cause a whistling as the water flows through at high speeds. Air-cushion chambers stop hammering noises caused by closing valves. Too many changes of direction of pipe cause friction that reduces the water pressure.

WASTE-DISCHARGE SYSTEMS

Wastewater is discharged through the disposal system by gravity drainage, as shown in Fig. 65-2. All pipes in this system must slant in a downward direction so that the weight of the waste will cause it to move down toward the main disposal system and away from the structure. Because of this gravity flow, waste lines that connect to sewage systems are much larger than the water-supply lines, in which the water is under pressure, as shown in Fig. 65-8. Waste lines are concealed in walls and under floors. The vertical lines are called *stacks*, and the horizontal lines are called *branches*. There are also vents that provide for the circulation of air, permit sewer gases to escape through the roof to the outside, and equalize the air pressure in the drainage system. *Fixture traps* stop gases from entering the building. Each fixture has a separate trap (seal) to prevent backflow of sewer gas. Fixture traps are exposed for easy maintenance;

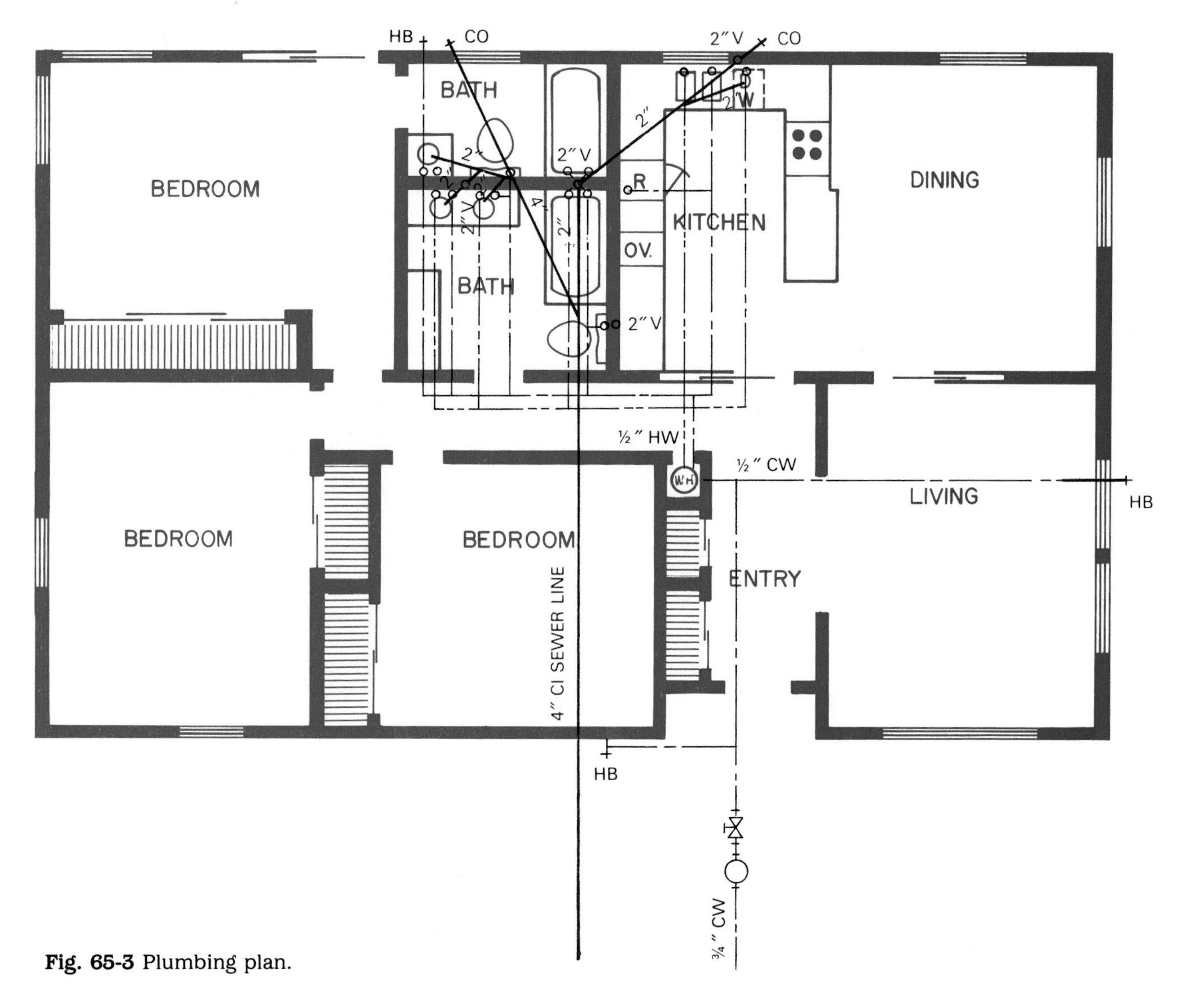

Fig. 65-3 Plumbing plan.

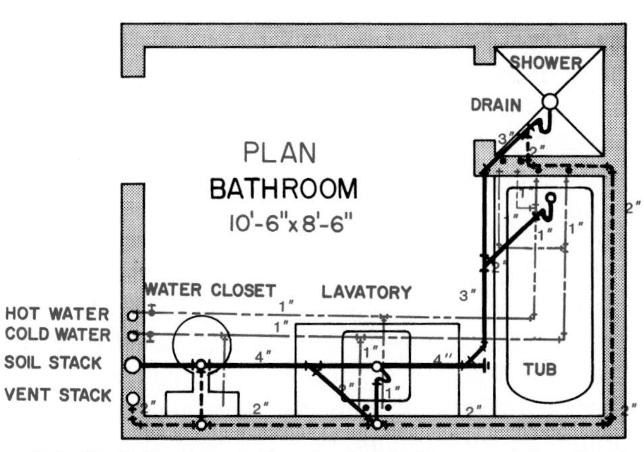

Fig. 65-4 A schematic plan of bathroom plumbing lines.

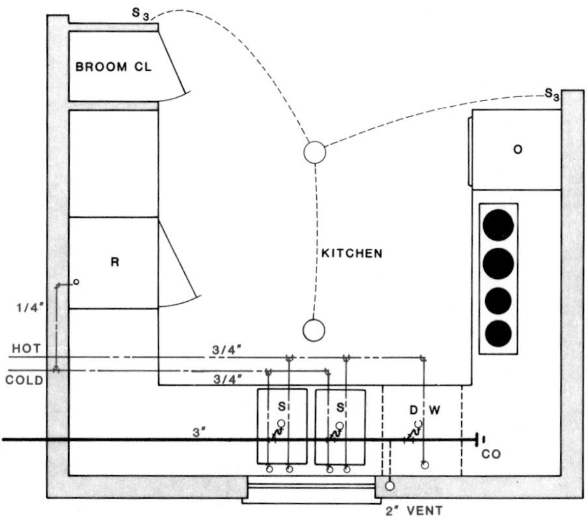

Fig. 65-5 A schematic plan of kitchen plumbing lines.

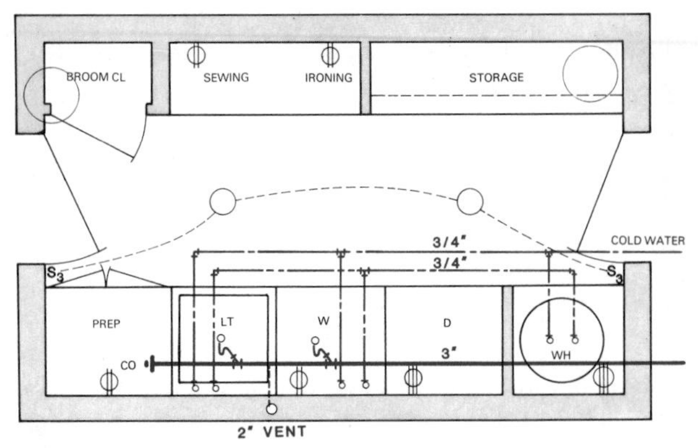

Fig. 65-6 A schematic plan of laundry plumbing lines.

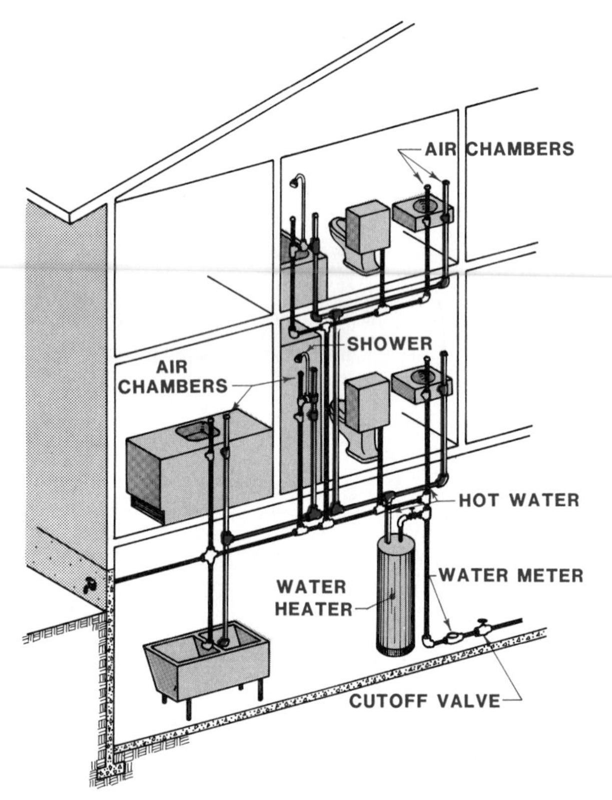

Fig. 65-7 A pressure plumbing supply system.

however, water-closet traps are built into the fixture.

The flow of wastewater starts at the fixture trap. It flows through the fixture branches to the soil stack. It then continues through the building drain to the sewer drain, and finally reaches the main sewer line. Waste stacks carry only wastewater. The lines that carry solid waste are called *soil lines*. Soil lines are the largest in the system and are flushed with water after each use.

Fresh-water-supply pipes are full of water *(wet pipes)* and under pressure all the time. The

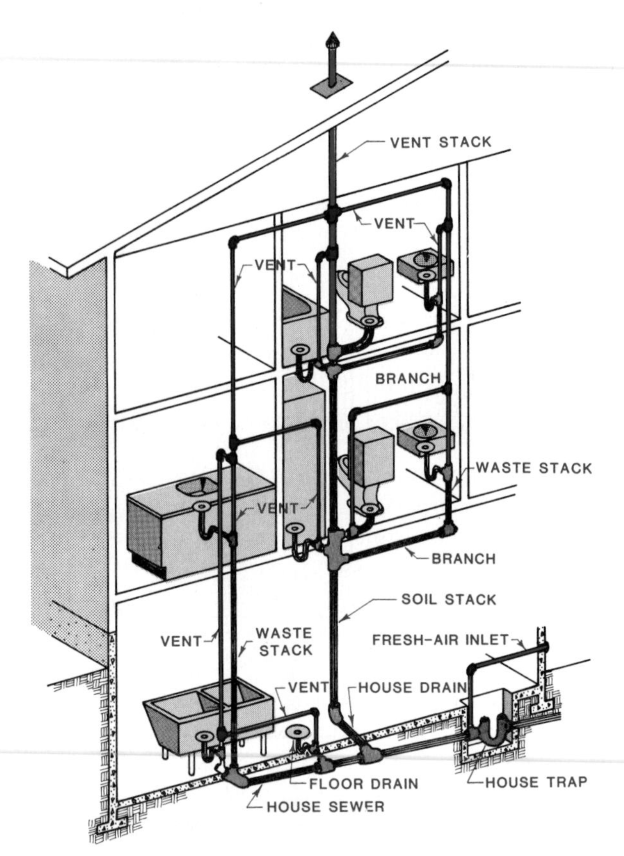

Fig. 65-8 Waste lines must be larger than water-supply lines.

waste and soil pipes are wet pipes that have water in them only when waste water is flushed through them. The vent-pipe system is composed of dry pipes that never contain water.

Outside Waste Systems

When municipal sewer-disposal facilities are available, connection of waste pipes to the public system are shown on the plot plan. However, when public systems are not available, a septic-tank system is used, and either a separate drawing is provided or the location of the system is indicated on the plot plan. This location drawing is usually required by the building code and supervised by the local board of health.

A *septic system* converts waste solids into liquid by bacterial action. The building wastes flow into a septic tank buried some distance from the building. The lighter part of the liquid flows out of the septic tank into drainage fields through porous pipes spread over an area to allow wide distribution of liquids.

The size and type of the septic system varies according to the number of occupants of a building, the contour of the terrain, and soil type. Figure 65-9 shows several types of drainage systems. The size of the lines and the distance of the septic tank and drainage fields from the building depend on the building codes of the community. Figure 65-10 shows the type of septic-system drawing required by most boards of health.

SCHEMATIC PLUMBING ELEVATIONS

Schematic plans show only the horizontal positioning of pipes. As a result, the amount of rise above floor level and the flow of fresh water and wastes between levels are difficult to read. Elevation drawings such as the water-distribution-system elevation, shown in Fig. 65-11, provide this vertical orientation. Elevation drawings are also used for some details such as the positioning of shutoff valves (Fig. 65-12).

Since most plumbing work is concentrated in the kitchen, the bathroom, and the utility room, separate schematic elevations are sometimes prepared for only those rooms, as shown in Fig. 65-13. Figure 65-14 shows only the waste-disposal system in elevation form from the second floor to the public sewer drain.

Prefabricated plumbing walls, installed at the factory, save the builder installation time because part of the labor is done on an assembly line. If fixtures are placed close together, many feet of pipe lines can be saved. The kitchen and bathrooms can also be back-to-back or over each other in a two-story house to eliminate long runs of pipe. However, it is more important to put the kitchen and bathrooms where they are most convenient, even if they are at opposite ends of the house. Figure 65-15 shows locations for storage tank, water softener and hot water tank.

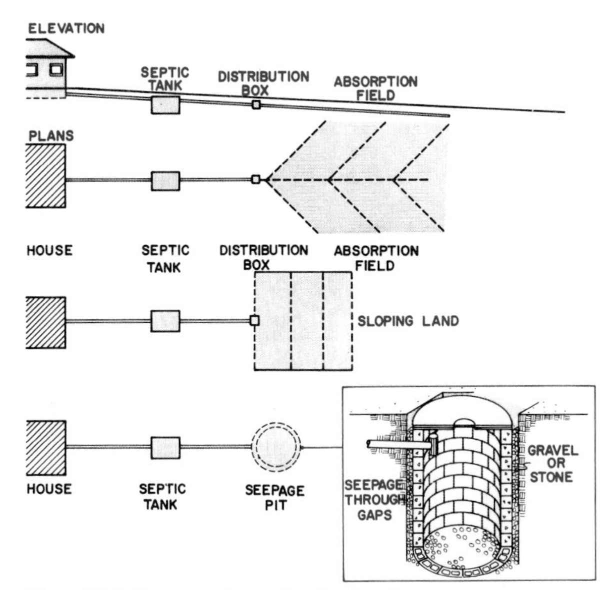

Fig. 65-9 Types of septic drainage systems.

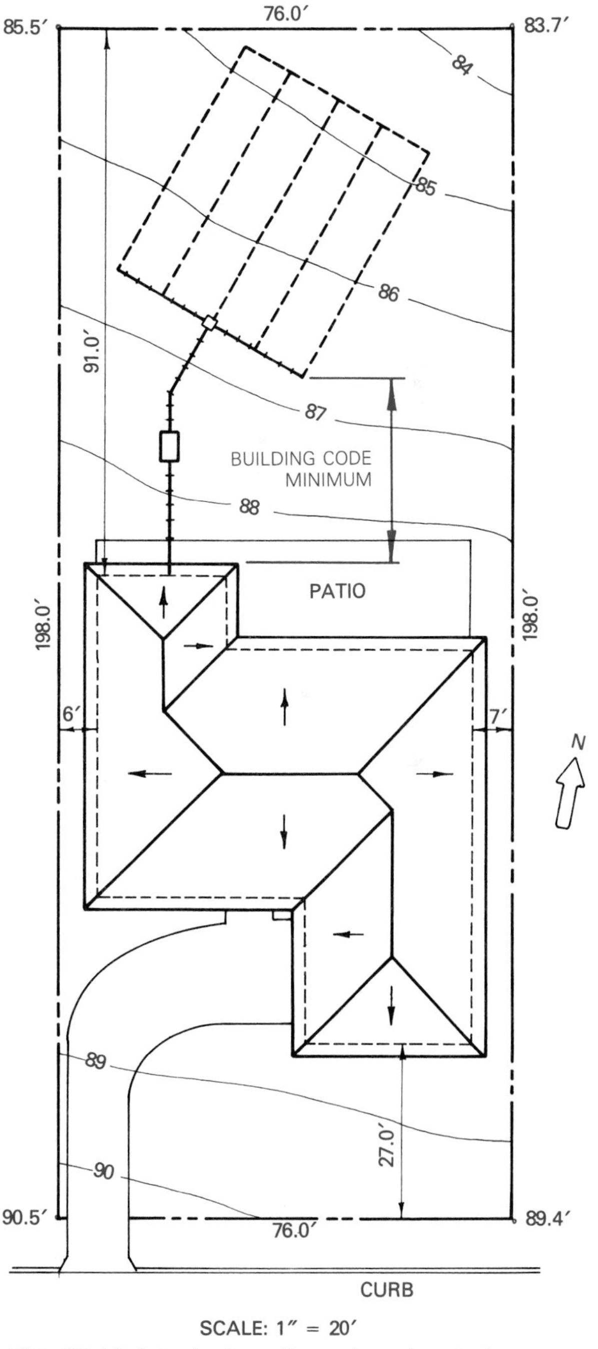

Fig. 65-10 A typical septic system drawing.

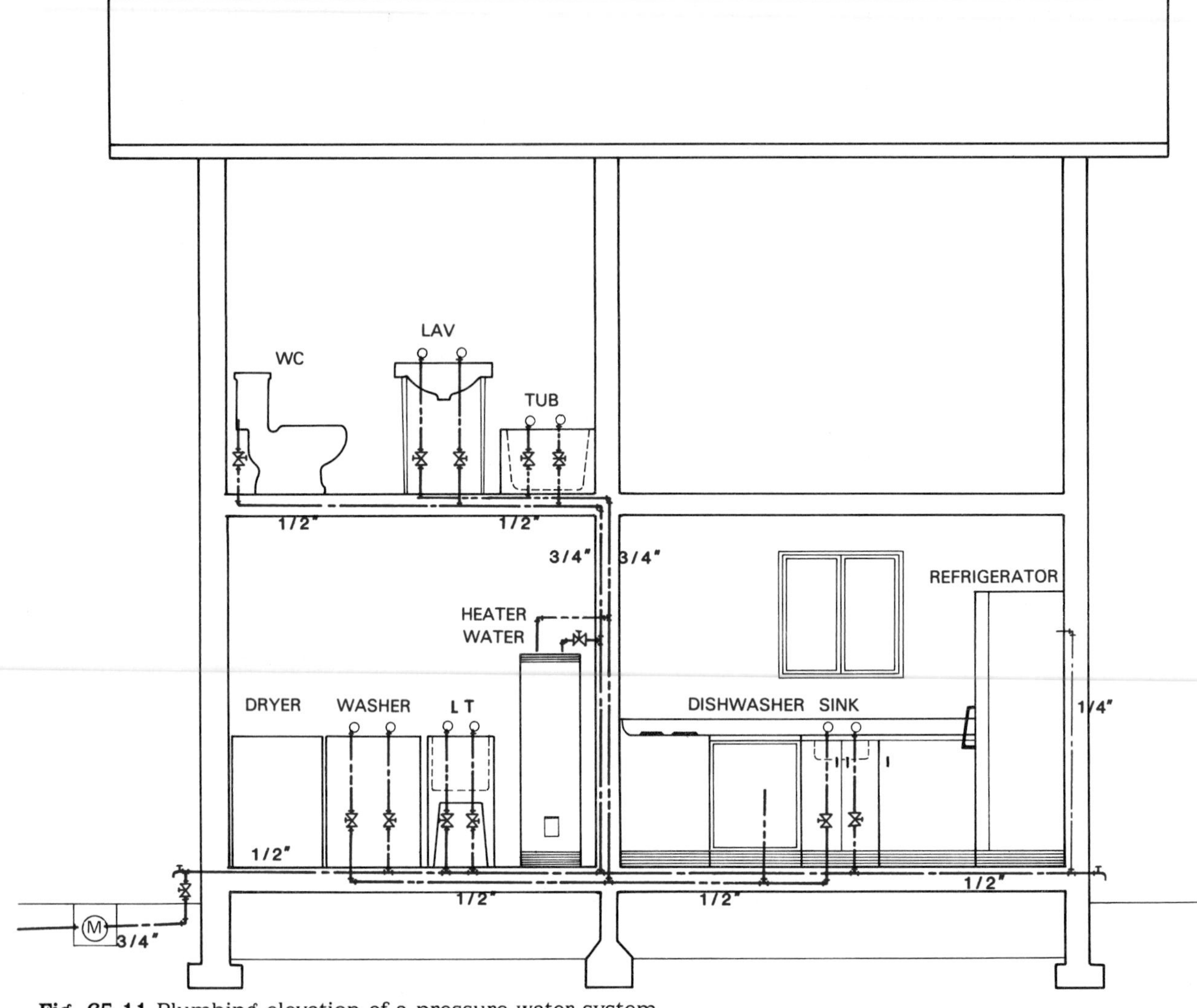

Fig. 65-11 Plumbing elevation of a pressure water system.

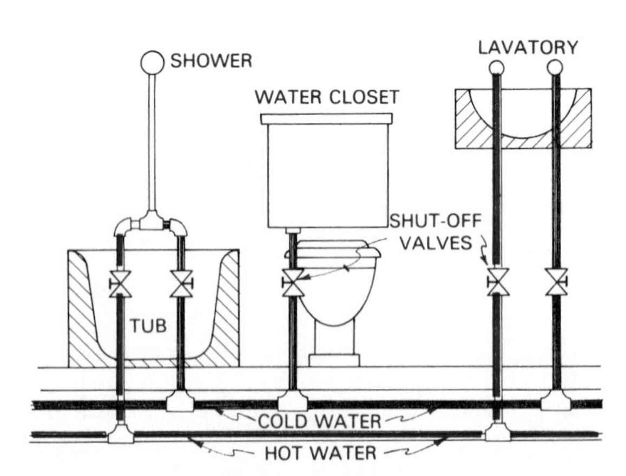

Fig. 65-12 Shutoff valve detail.

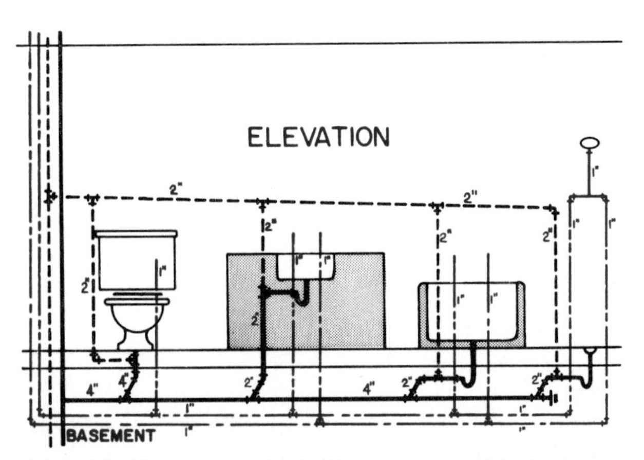

Fig. 65-13 Schematic bathroom plumbing elevation.

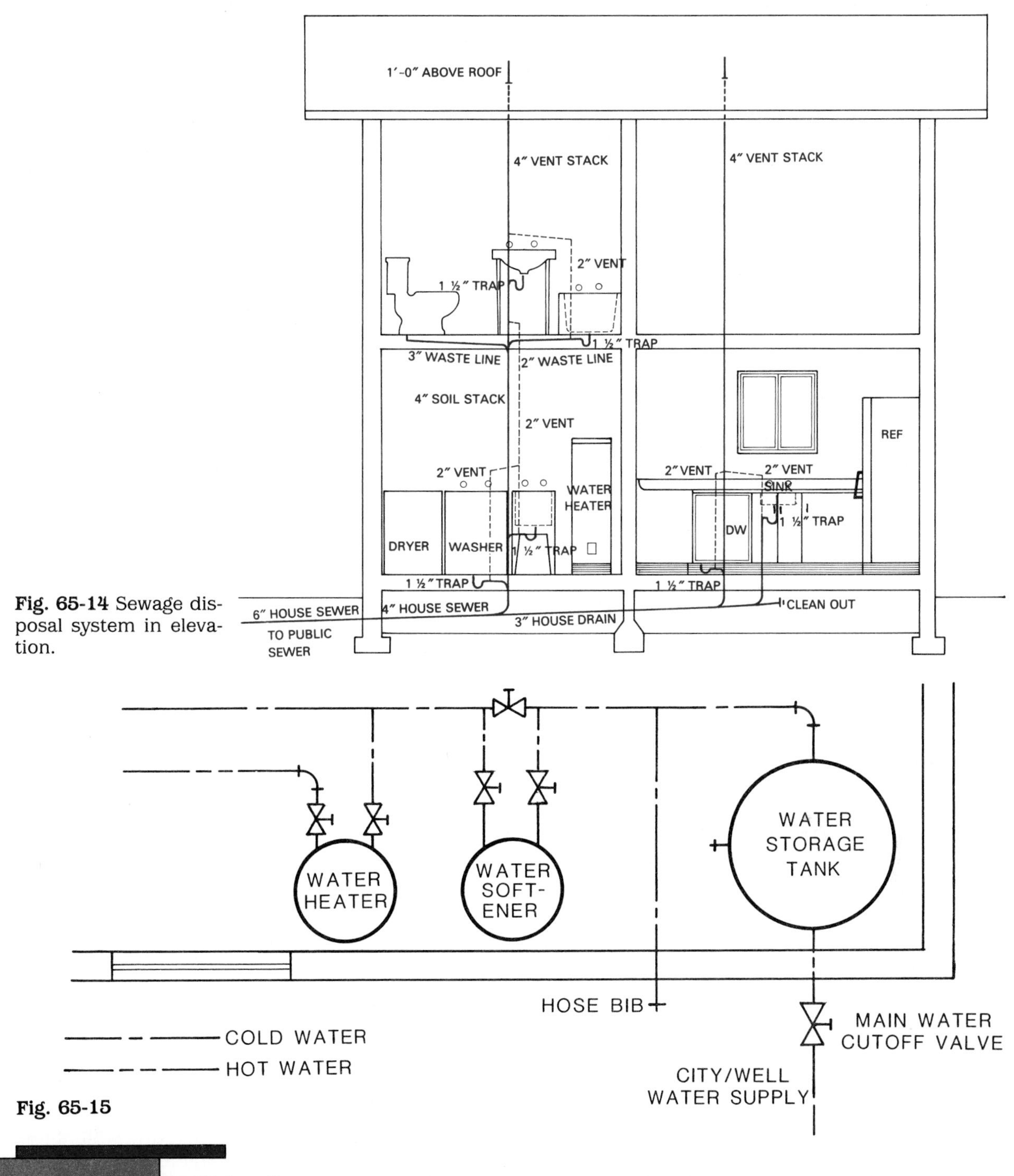

Fig. 65-14 Sewage disposal system in elevation.

Fig. 65-15

Exercises

1. **Prepare a plumbing plan for the house you are designing.**
2. **Define these terms: *gravity drainage, stack, branch, vent, fixture trap, soil stack, house drain, house sewer, waste stack, septic system, septic tank, drainage field, shutoff valve, air chambers, waste lines.***

SECTION 17

Modular Component Plans

Industrial automation methods have enabled manufacturers to produce high-quality and low-priced products more quickly than ever before. However, the construction industry has not industrialized or standardized its production methods to any great degree. Construction methods in the home-building industry continue to contribute to excessive waste of materials and time. However, to standardize and automate more fully the home-building industry, lumber manufacturers have developed a program of coordinating dimensional standards and components on standard structural parts such as windows, doors, and trusses. This standardization enables the part to fit into the plan for any home designed according to a standard modular component.

UNIT 66

MODULAR SYSTEMS OF DESIGN

The Unicom method of designing includes the use of modular components and the preparation of plans to a modular dimensional standard. *Unicom* means uniform manufacture of components. This modular system was developed by the National Lumber Manufacturers Association. Using this method, the designer must think of the home as a series of component parts. These component parts may be standard factory-made components or built on the job site. In either case, plans are prepared to be consistent with the size of these components. Plans must also be interchangeable and consistent with the Unicom method of dimensioning. Without this dimensional standard, the use of modular component parts is of no value.

ADVANTAGES

Buildings may be erected completely with factory-made parts (Fig. 66-1) or may be framed conventionally, or a combination method may be used. Faster planning and erection of the house benefit both builder and home buyer. This system makes possible the more efficient use of materials, thus reducing waste in construction.

Building inventory costs are cut because of the number of elements needed for maximum design flexibility are fewer. Perhaps the most important advantage of the Unicom system is that it makes possible the use of modern mass-production techniques. These techniques provide great accuracy in the construction of components and superior quality control in their fabrication.

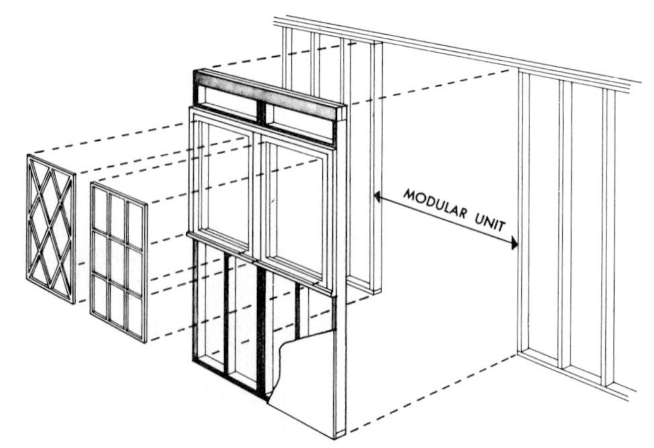

Fig. 66-1 Modular component doors and windows. *(National Forest Products Assoc.)*

DESIGNING WITH MODULAR GRIDS

The coordination of modular components with the system of modular dimensions utilizes all three dimensions—length, width, and height. The overall width and length dimensions are most critical in the planning process. The *modular planning grid* is a horizontal plane divided into equal spaces in length and width. It provides the basic control for the modular coordination system. The entire grid, shown in Fig. 66-2, is divided into equal spaces of 4″, 16″, 24″, and 48″. The Unicom system is not yet designed to metric standards. Composite dimensions are therefore all multiples of 4″. The 16″ unit is used in multiples for wall, window, and door panels to provide an increment small enough for flexible planning and optimum inventory of these components.

Increments of 24″ and 48″ are used for overall dimensions of the house. The 24″ module is called the *minor module.* The 48″ module is called the *major module.* Figure 66-3 shows the use of the 16″ module in design locations for window component panels. Since the Unicom method does not require the wall, floor, and roof elements to be tied to a large fixed panel size, there is no need for the designer to adhere to a fixed 4′ or larger increment in planning.

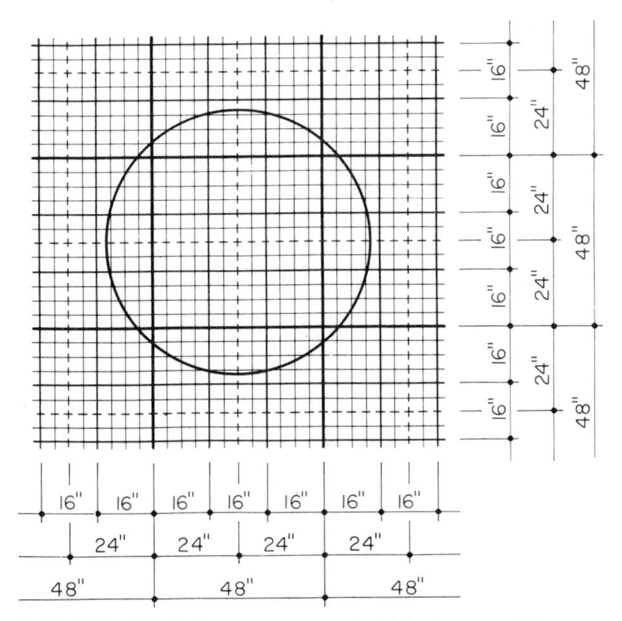

Fig. 66-2 Modular component grid. *(National Forest Products Assoc.)*

PREPARING THE MODULAR PLAN

When the basic floor-plan design is completed, the plan should be sketched or drawn on the Unicom modular grid and all nonmodular dimensions converted to the nearest modular size. Each square on the grid represents the basic 4″ module. Standard 16″ modules are represented by the intermediate heavy lines. Major 48″ module lines are indicated by the heaviest lines. Minor 24″ modules are represented by dotted lines. By employing modular dimensions in multiples of 2′ or 4′ for house exteriors, fractional spans for floor and roof framing are eliminated.

Variations in the thickness of exterior and partition walls interfere with true modular dimensioning; therefore the total thicknesses of all of these walls must be subtracted from the overall modular dimension to obtain the net inside dimension. In addition, nonmodular dimensions created by existing building laws or built-in equipment must be incorporated in the modular coordination system. This end is accomplished by dimensioning conventionally the nonmodular distances where they exist.

After the overall dimensions are established and nonmodular dimensions incorporated in the plan, the panels for exterior doors and windows and for exterior walls should be established on the 16″ module. The precise location of wall openings on the 16″ module also eliminates the extra wall framing commonly encountered in nonmodular home planning; and the 16″ spacing of structural members increases design flexibility by one-third, compared with spacing at the 24″ minor module.

Conventionally framed platform-floor systems provide the most flexible method of floor framing for the variety of design conditions encountered with the Unicom system. Modular sizes of floor-sheathing materials are economically applied to the modular spacing of floor joists.

As previously stated, the Unicom method is based on modular coordination of all three dimensions—width, length, and height. Although width and length are the most critical dimensions, standard heights are necessary to eliminate waste and to ensure the proper fitting of components. A standard height of 8′-1½″ for exterior-wall components allows for floor- and ceiling-finish applications with a combined

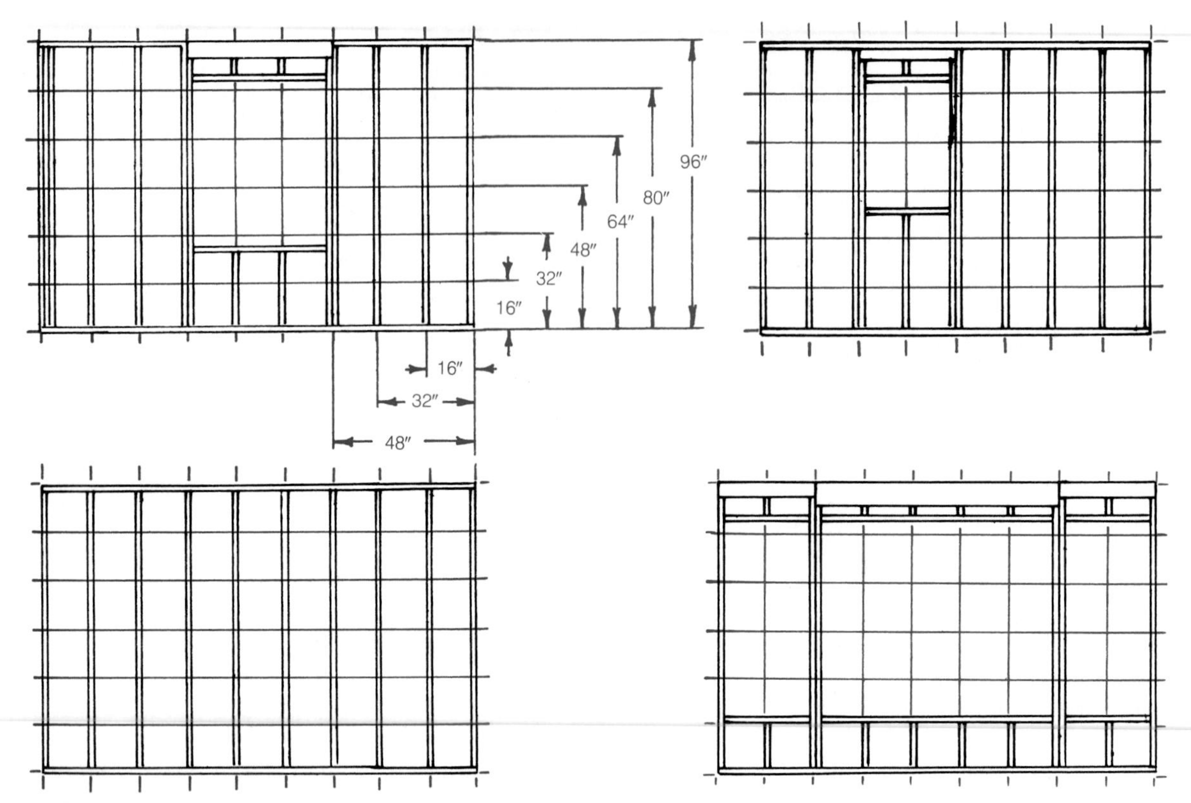

Fig. 66-3 Framing for modular components. *(National Forest Products Assoc.)*

thickness of 1½″. The result is a standard 8′ finished ceiling height.

IDENTIFICATION SYSTEM

A complete system of short-form identification for separate pieces and fabricated components simplifies the practical use of the modular coordination system of building. The shorthand has been extended to include complete descriptions of wall panels, including information pertaining to the slope, window style, and position, as shown in Fig. 66-4.

Modular construction in industry is important because it eliminates individually constructed items, thereby reducing costs. Portable modular units have been designed that can be transported to the site and assembled with many different floor-plan arrangements.

All structures are built as modules to some extent. That is, not all the materials or components are manufactured or put together on the site. Some structures are simply *precut*. This means that all the materials are cut to modular specification at the factory and then assembled on the site by conventional methods.

With the most common factory-built (prefabricated) homes, the major modular components, such as the walls, trusses, decks, and partitions, are assembled at the factory. The utility work, such as installation of electrical, plumbing, and heating systems, is completed on site. The final finishing work, such as installation of floors, roof coverings, and walls, is also done on site.

There are some factory-built homes, however, that are constructed in complete modules at the factory and require only final electrical-outlet, roof-overhang, and assembly-fastening work on site to complete the job. The time factor is a major advantage of this type of construction.

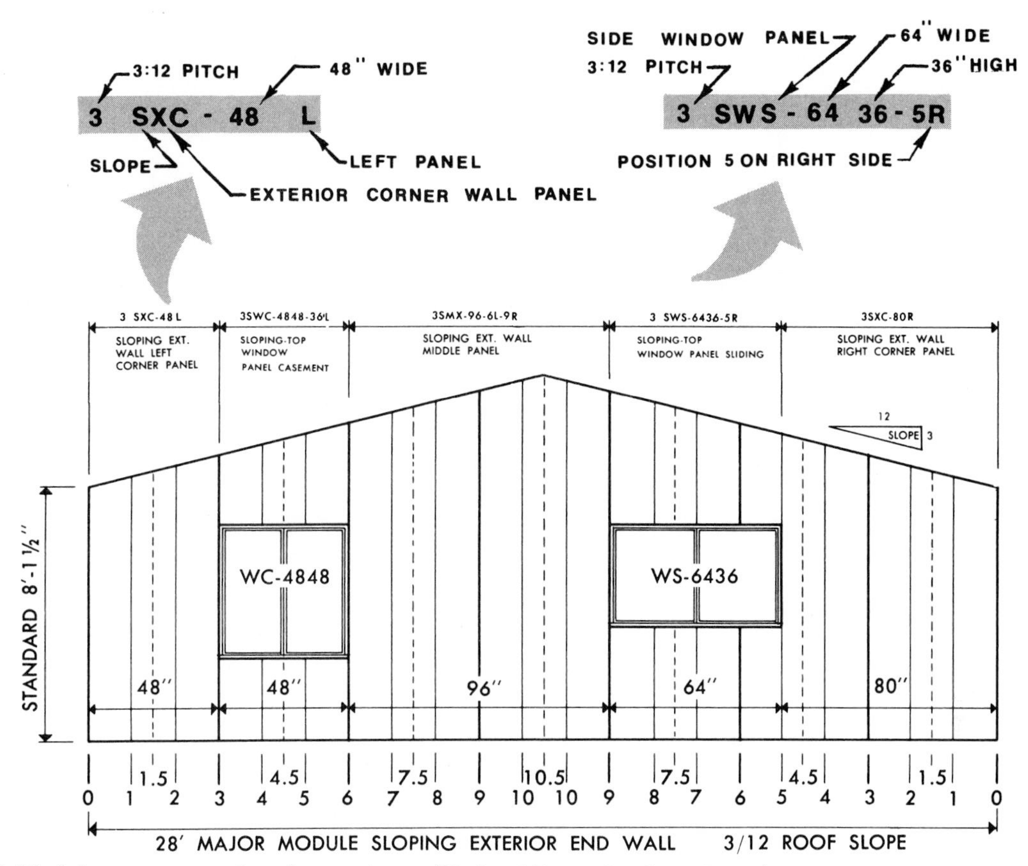

Fig. 66-4 Modular component code numbers. *(National Forest Products Assoc.)*

However, the size of the modules is limited to only 12′ widths. Twelve feet is the maximum width for a truck load on public roads.

Another advantage of modular design is the ability to design and build expandable plans in phases. If construction estimates exceed budget, first, the entire structure can be scaled down in size by selectively reducing the size of minimally used rooms or areas. Second, some expensive wants that were available on the first design can be eliminated, but none of the needs. Third, an expandable plan can be considered which will satisfy the needs for years to come but expand, in stages, into a more spacious home as the budget permits and/or as the family grows. In the expandable plan shown in Fig. 66-5, the family room functions as the bedroom and the powder room and utility room functions as the family bathroom by addition of a temporary bathtub or shower. The first stage is the addition of two bedrooms and a bathroom. The second stage is the expansion of the living area. Stage three in this plan involves the addition of the master bedroom suite. Stage four involves the addition of a garage. These stages, of course, can be interchanged, depending on the specific developing needs of the home owner.

When you use a CAD system to prepare modular architectural drawings, the grid pick mode is used in conjunction with the axis lock system. In the grid pick mode, stylus inputs are snapped to the nearest grid point. In the axis lock mode, stylus inputs are snapped to the closest *x* or *y* axis. Therefore, if the CAD grid lines are adjusted to coincide with modular grid dimensions, all line inputs will automatically fall on a modular increment. CAD grid lines can also be placed on a different level than object lines and later plotted either in a second color or in a narrower line to show the modular alignment. Nonmodular lines must be drawn in the non-axis lock and non-grid pick mode.

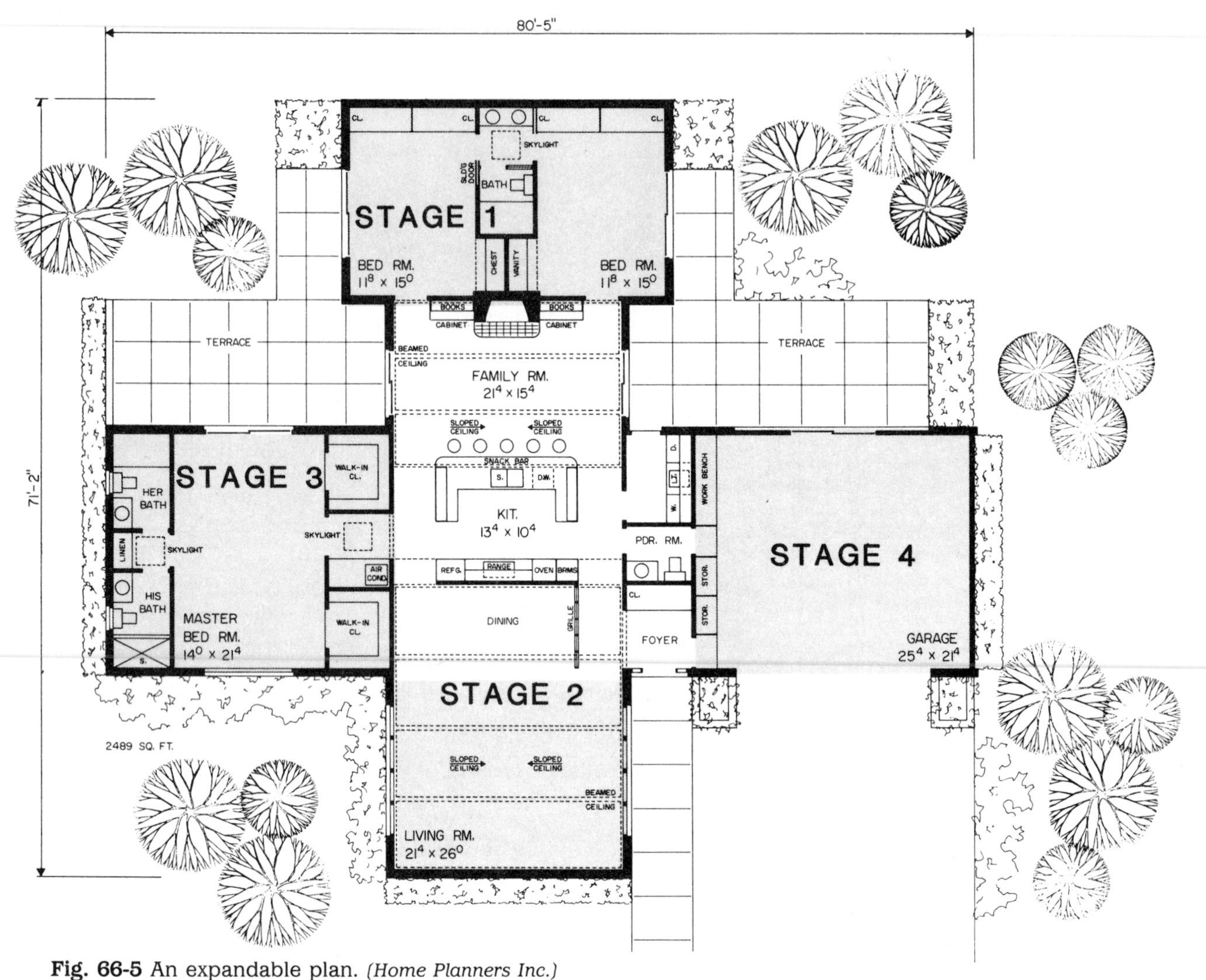

Fig. 66-5 An expandable plan. *(Home Planners Inc.)*

Exercises

1. **Develop an expandable plan to divide the construction of the plan shown in Fig. 69-1 into four stages.**
2. **Convert a floor plan of your own design to modular dimensions.**
3. **Define the following terms: *Unicom, modular components, modular planning grid, minor module, major module, nonmodular dimensions, modular dimensions, 16″ module.***

UNIT 67

DRAWING MODULAR PLANS

The floor plan shown in Fig. 67-1 has been fitted on the modular grid, and the designer has thus properly established the modular relationship of the foundation, floors, walls, windows, doors, partitions, and roof. Establishing the dimensions and components precisely on the 16″, 24″, and 48″ spaces of the modular grid assures accurate and less troublesome fitting of the components of the house at the time of its construction.

COMPONENT FLOOR PLANS

After completing the basic floor plan, a detailed component plan for fabrication and erection is prepared. Erection coding symbols are used to identify wall, window, door, and partition components by type and size (Fig. 67-2). The erection sequence for site assembly is also designated on the plan by ER 1, ER 2, and so on. This basic component floor plan for erection contains all basic dimensions necessary for layout work on the floor platform. Interior partitions may then be erected in varying sequences.

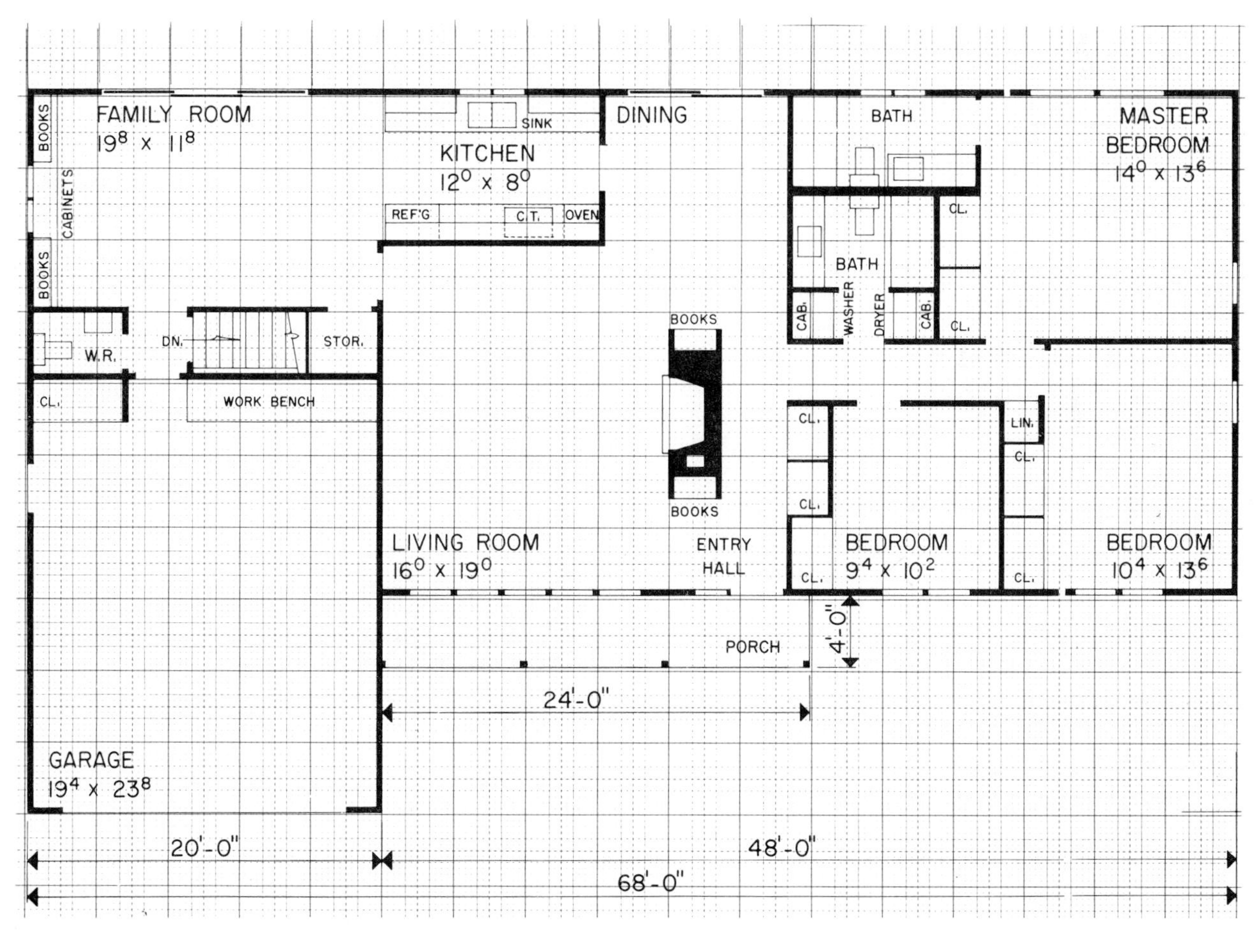

Fig. 67-1 Floor plan fitted on a modular grid. *(National Forest Products Assoc.)*

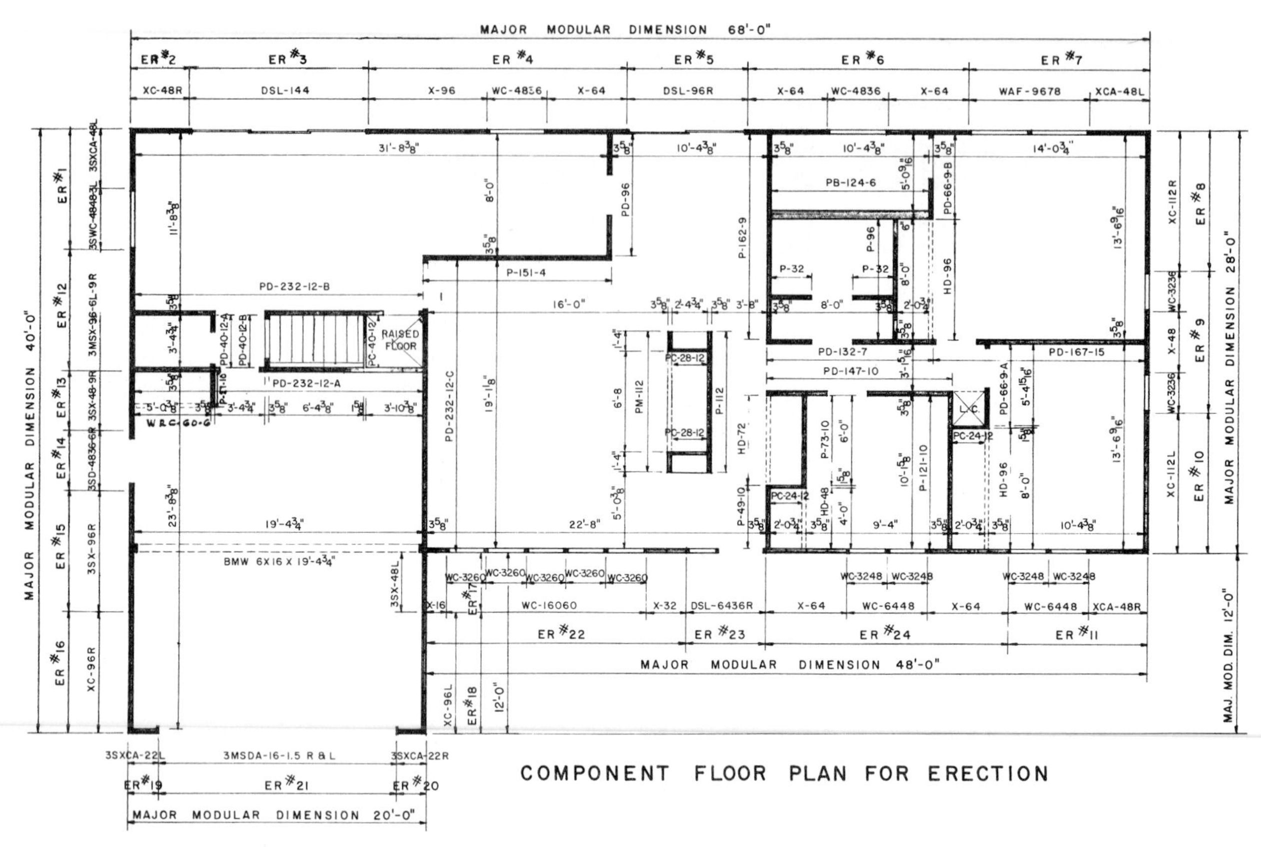

Fig. 67-2 Erection coding on floor plan. *(National Forest Products Assoc.)*

ELEVATION COMPONENT DRAWINGS

Just as the component floor plan for erection is prepared from the floor plan shown in Fig. 67-1, the elevation component drawings are prepared from standard elevations projected from the floor plan, as shown in Fig. 67-3. Elevation component drawings (Fig. 67-4) closely resemble conventional panel-framing elevations, except the wall, window, door, and roof components are identified and shown in their proper relationship, complete with standard Unicom nomenclature and erection sequences. The component elevation drawings are prepared from standard elevations shown in Fig. 67-3.

FRAMING PLANS

From the basic floor plan, a floor framing plan and stud layout, as shown in Fig. 67-5, are projected and dimensioned according to Unicom standards. Notice the precise rhythm of the 16", on-center wall studs and the 48" modular dimensions of the house perimeter. The advantage of designing to these standards is the reduction of wasted material. Waste occurs only in cutting at the stairwell and fireplace openings or other areas that are different from modular dimensions. Modular dimensions and panel code numbers are indicated on all plans.

INTERIOR PARTITION COMPONENTS

Partitions function as interior space separators for room privacy, traffic control, and storage. Their aesthetic value depends on decorative surface materials, doors, and trim designs. Partitions are a maze of intersecting planes that may carry the roof and ceiling loads, depending on the roof design. Partitions must fit between the basic exterior modular increments with allowance for exterior wall thicknesses. Space must be provided for intersecting partitions with backup members, proper door placement, vertical and horizontal plumbing runs, medicine cabi-

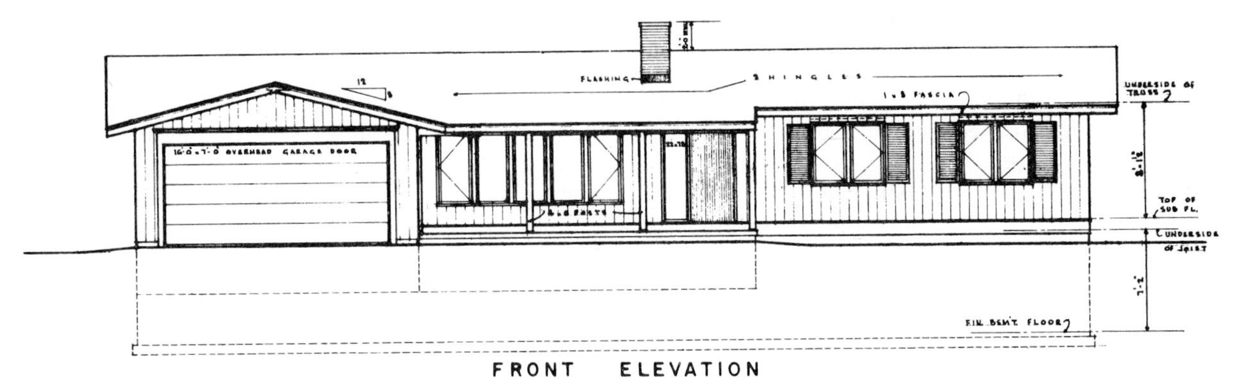

Fig. 67-3 Elevation component drawings are prepared from conventional elevations. *(National Forest Products Assoc.)*

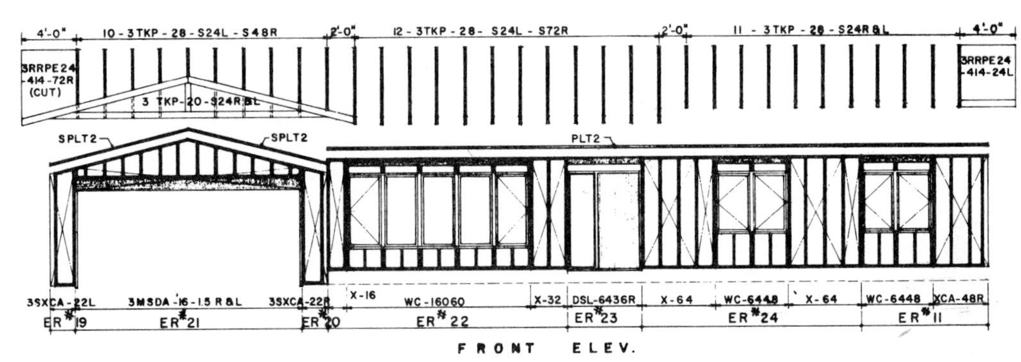

Fig. 67-4 Component elevation framing drawing. *(National Forest Products Assoc.)*

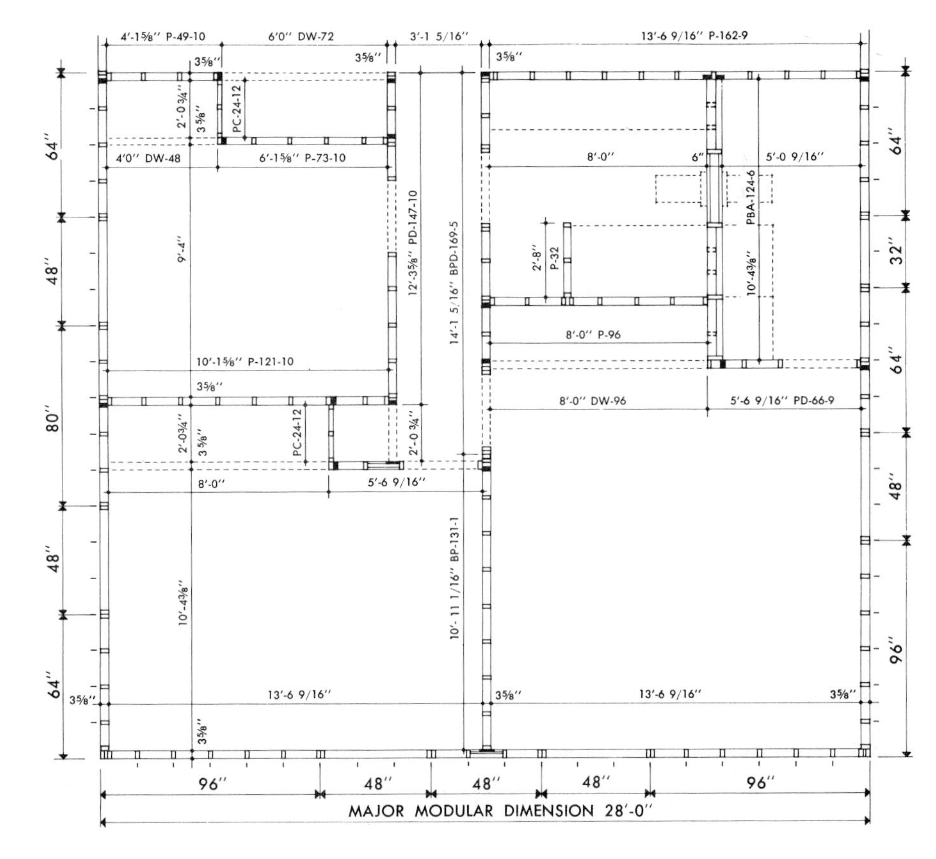

Fig. 67-5 Modular stud layout. *(National Forest Products Assoc.)*

nets, closets, fireplaces, and flexible room arrangements with varying designs.

Many more modular drawings are needed to describe completely a Unicom plan in every detail. Just as in structures of conventional design, the more detailed the drawings are, the better the chance of achieving the desired outcome.

Fig.67-6 Modular detail drawing.

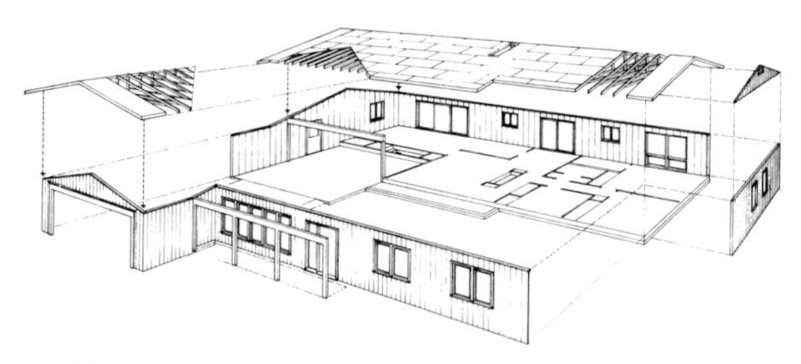

Fig. 67-7 Relationships of modular components. *(National Lumber Manufacturers Assoc.)*

Other plans that may become part of the complete Unicom design include transverse sections, truss and gable component drawings, roof-overhang details, and detailed drawings (Fig. 67-6) of many nonstandard components. Figure 67-7 shows the relationship of basic components in a house designed by the modular system. Figure 67-8 shows expansion possibilities through designing using modular systems.

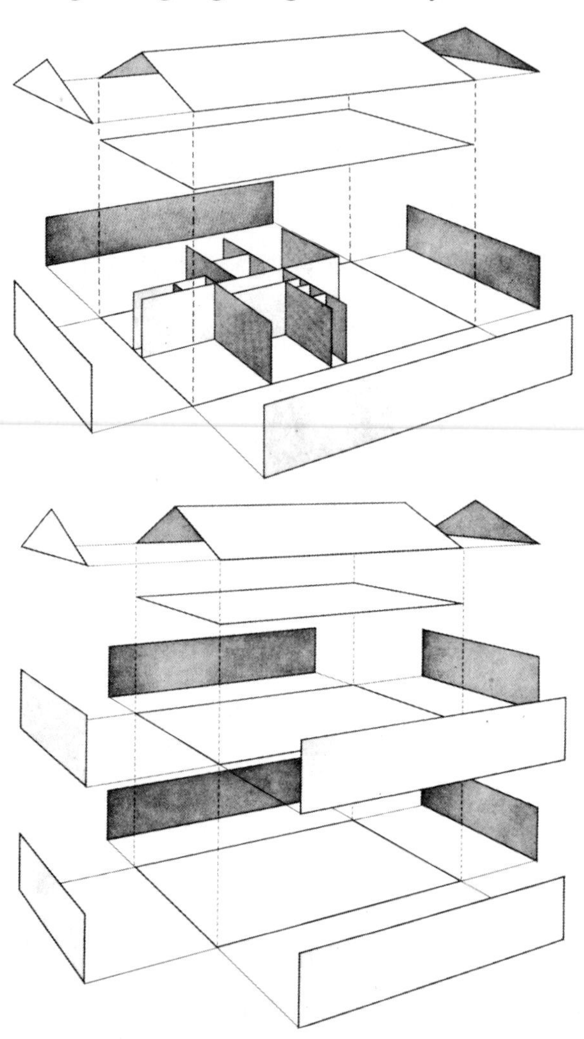

Fig. 67-8 Expansion options using modular units. *(National Lumber Manufacturers Assoc.)*

Exercises

1. Prepare a component floor plan for the floor plan shown in Fig. 69-9.

2. Prepare a component floor plan for the house you are designing.

3. Define the following terms: *component floor plan, elevation component drawing, modular increment.*

SECTION 18

Architectural Delineation Systems

Previous sections have covered the principles and practices involved in the preparation of each type of architectural drawing. In this section, the basic guidelines used in the preparation of a complete set of plans are developed. Special emphasis is placed on the relationship and consistency among the various plan features and dimensions.

UNIT 68

RELATIONSHIP OF DRAWINGS

Drawings used for construction vary from simple floor plans to comprehensive sets of plans complete with details, schedules, and specifications. The number of plans needed to construct a building depends on the complexity of the structure and on the degree to which the designer needs or wants to control the various methods and details of construction. For example, if only a floor plan is prepared, the builder must create the elevation designs. If only the floor plan and elevation drawings are prepared, but no details or specifications are provided, the builder assumes responsibility for many construction details, including selection of framing type, materials used, and many aspects of the interior design. Therefore, the more plans, details, and specifications developed for a structure, the closer the finished building will be to that conceived by the designer.

VOLUME

Figure 68-1 shows the types of construction drawings included in a minimum, average, and maximum set of plans. The preparation of only floor plans, elevations, surveys, and a section, as indicated in the minimum set of plans, provides the builder with great latitude in selection of materials and processes. The maximum set of plans will assure, to the greatest degree possible, agreement between the wishes of the designer and the final constructed building. Even though a building may be relatively simple, as many drawings as necessary should be prepared. Each detailed working drawing not prepared forces the builder into the role of the designer. For some buildings and some builders, this may be acceptable. For others, this is a highly unacceptable method of operation.

INDEXING SYSTEMS

The actual selection of drawings depends on the degree to which the designer wants to control various features of the building. Large sets of plans that include many different details require an indexing system. For large projects involving many buildings with many components, a much more detailed indexing system is needed to locate specific details on any building in a short period of time. In these cases, a master index is prepared that shows the sheet number where each drawing for each building is found. Figure 68-2 shows a title block with a simple indexing system. It includes a design number, sheet number, and the total number of drawings in the set.

Drawings	Size of set of plans		
	Min.	Aver.	Max.
FLOOR PLANS	X	X	X
FRONT ELEVATION	X	X	X
REAR ELEVATION		X	X
RIGHT ELEVATION	X	X	X
LEFT ELEVATION		X	X
AUXILIARY ELEVATIONS			X
INTERIOR ELEVATIONS		X	X
EXTERIOR PICTORIAL RENDERINGS		X	X
INTERIOR RENDERINGS			X
PLOT PLAN		X	X
LANDSCAPE PLAN			X
SURVEY PLAN	X	X	X
FULL SECTION	X	X	X
DETAIL SECTIONS		X	X
FLOOR-FRAMING PLANS			X
EXTERIOR-WALL FRAMING PLANS			X
INTERIOR-WALL FRAMING PLANS			X
STUD LAYOUTS			X
ROOF-FRAMING PLAN			X
ELECTRICAL PLAN		X	X
AIR-CONDITIONING PLAN			X
PLUMBING DIAGRAM			X
SCHEDULES			X
SPECIFICATIONS			X
COST ANALYSIS			X
SCALE MODEL			X

Fig. 68-1 Drawings needed for a set of plans.

The American Institute of Architects (AIA) recommends a numbering system that gives access to the information in a set of drawings. In this system, drawings are identified by letters and numbers for ease of referencing.

A readily identifiable alphabetical prefix is used to denote the specific discipline of work covered by a group of working drawings:

A Architectural
C Civil
D Interior design (color schemes, furniture, furnishings)
E Electrical
F Fire protection (sprinkler, standpipes, CO_2, and so forth)
G Graphics
K Dietary (food service)
L Landscape
M Mechanical (heating, ventilating, air-conditioning)
P Plumbing
S Structural
T Transportation/conveying systems

In this system, architectural drawings are divided into 10 specific groups, A0 through A9. The group number will always remain the same, no matter how large the project. Additional drawings may be added within groups without interrupting the alphanumerical order. Figure 68-3 explains the coding system as it would appear in use on a drawing. Figure 68-4 shows a listing of these groups and their corresponding code numbers.

If pin registration methods as described in Unit 28 have been used in the preparation of any of the drawings, the set should be aligned in this manner.

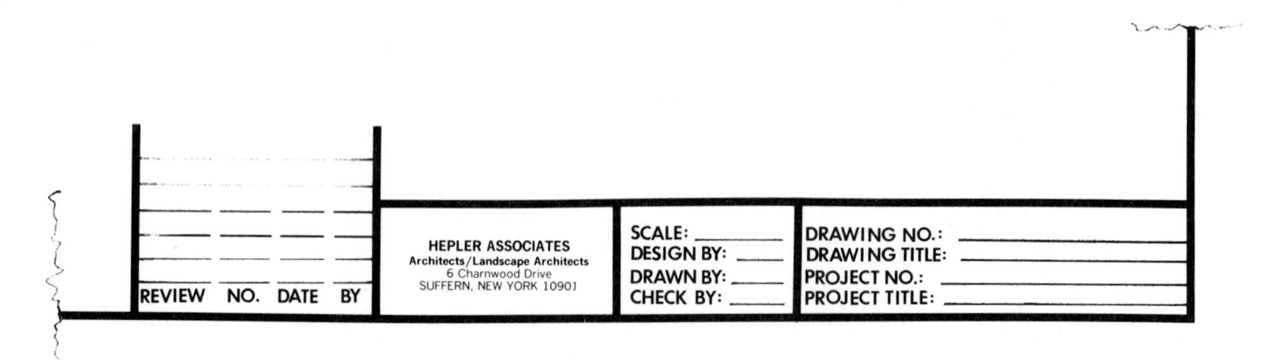

Fig. 68-2 Title block with indexing system labels.

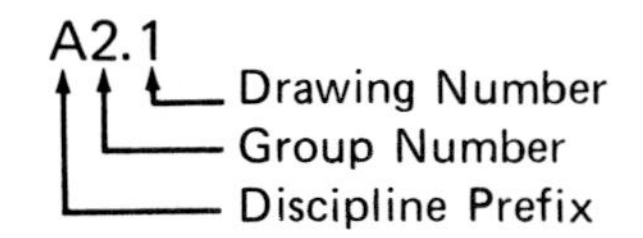

Fig. 68-3 Code numbers used on drawings.

ARCHITECTURAL DRAWINGS

A0.1,2,3 – General (Index, Symbols, Abbrev. notes, references)
A1.1,2,3 – Demolition, Site Plan, Temporary Work
A2.1,2,3 – Plans, Room Material Schedule, Door Schedule, Key Drawings
A3.1,2,3 – Sections, Exterior Elevations
A4.1,2,3 – Detailed Floor Plans
A5.1,2,3 – Interior Elevations
A6.1,2,3 – Reflected Ceiling Plans
A7.1,2,3 – Vertical Circulation, Stairs (Elevators, Escalators)
A8.1,2,3 – Exterior Details
A9.1,2,3 – Interior Details

STRUCTURAL DRAWINGS

S0.1,2,3 – General Notes
S1.1,2,3 – Site Work
S2.1,2,3 – Framing Plans
S3.1,2 – Elevations
S4.1,2 – Schedules
S5.1,2 – Concrete
S6.1,2 – Masonry
S7.1,2 – Structural Steel
S8.1,2 – Timber
S9.1,2 – Special Design

MECHANICAL DRAWINGS

M0.1,2 – General Notes
M1.1,2 – Site/Roof Plans
M2.1,2 – Floor Plans
M3.1,2 – Riser Diagrams
M4.1,2 – Piping Flow Diagram
M5.1,2 – Control Diagrams
M6.1,2 – Details

PLUMBING DRAWINGS

P0.1.2 – General Notes
P1.1,2 – Site Plan
P2.1,2 – Floor Plans
P3.1,2 – Riser Diagram
P4.1,2 – Piping Flow Diagram
P5.1,2 – Details

ELECTRICAL DRAWINGS

E0.1,2 – General Notes
E1.1,2 – Site Plan
E2.1,2 – Floor Plans, Lighting
E3.1,2 – Floor Plans, Power
E4.1,2 – Electrical Rooms
E5.1,2 – Riser Diagrams
E6.1,2 – Fixture/Panel Schedules
E7.1,2 – Details

Fig. 68-4 Coding system index.

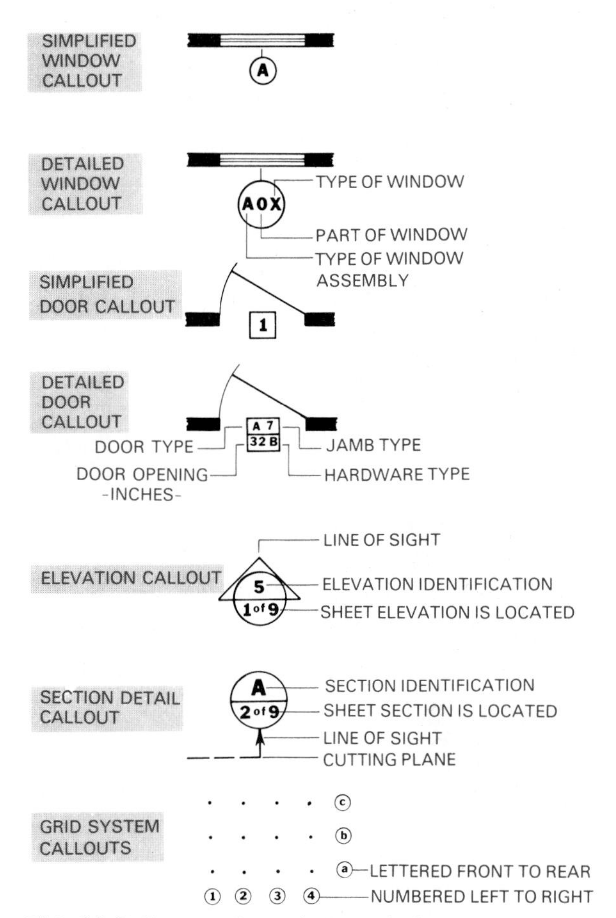

Fig. 68-5 Cross-referencing symbols.

REFERENCE AMONG DRAWINGS

The symbol reference system shown in Fig. 68-5 is typical of the type used in most architectural practice. The number on the top refers to the number of the detail. The number on the bottom refers to the drawing sheet where the symbol is located. Thus detail 6/4 is detail no. 6 located on page 4. When the detail represents a sectional view, the cutting-plane line and direction of sight is shown by connecting the key symbol to the arrowhead of the cutting-plane line.

COMBINATION PLANS

Sometimes, several plans in a complete set of plans are combined. For example, there may be no specific electrical, plumbing, or air-conditioning plan, but these symbols may be added to the basic floor plan. There may be no specific landscape plan, but the landscape features may be added to a survey plan. Details and sections are also quite often combined into one plan.

Combined plans are more difficult to read than separate specialized plans. For that reason, separate plans have been presented throughout this text for instructional purposes. Figure 68-6 shows part of a combination floor plan that includes not only information normally found on a floor plan, but also includes electrical, air-conditioning, plumbing, some landscape, survey, and plot plan symbols. CAD layering of these systems, in addition to interior design facilities and furniture, can eliminate much of the confusion of interpretation.

In studying this kind of plan, the specialized elements must be separated out of the plan through imagination to eliminate confusion. Study only one element at a time. If you are studying the electrical part, refer only to as much of the remainder of the plan as necessary to orient the position of switches, outlets, and so forth. Imagine that the plan is only an electrical plan and that all other features do not exist.

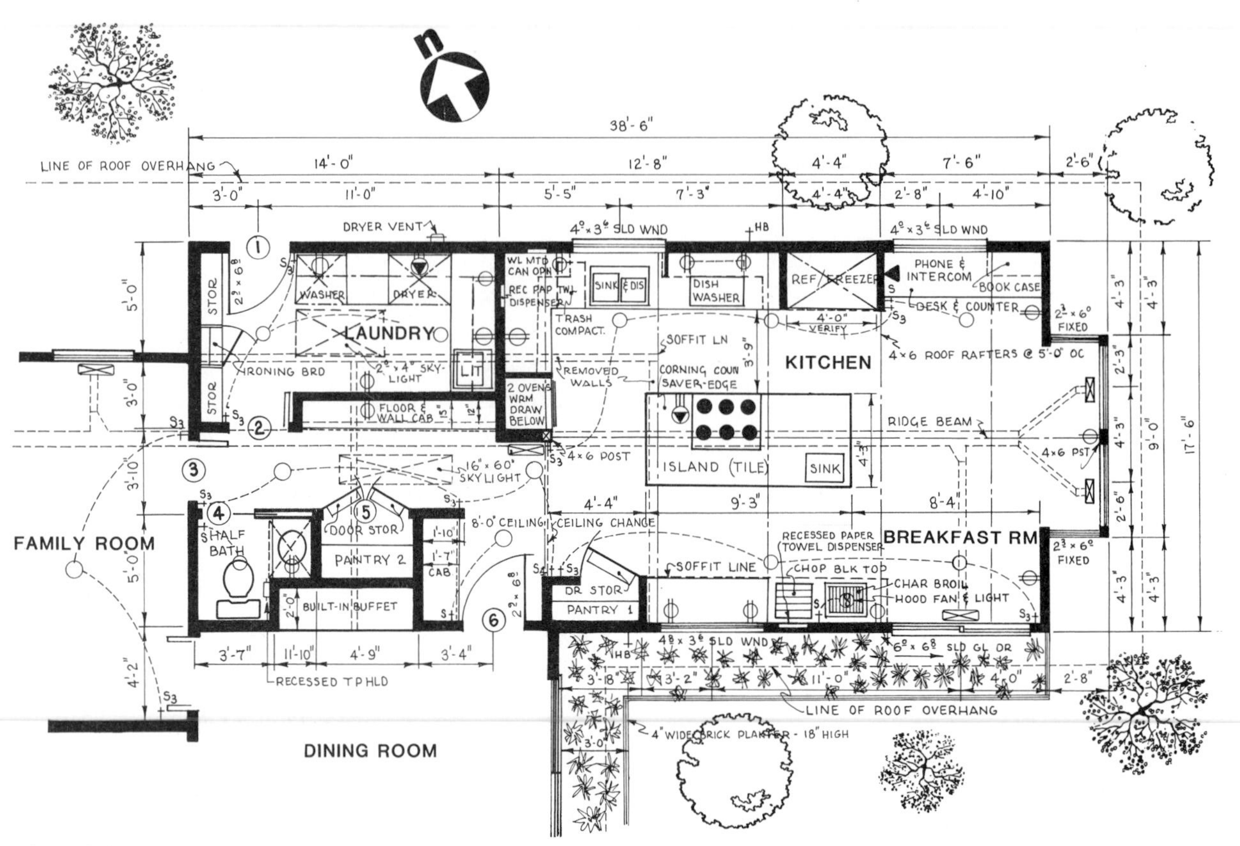

Fig. 68-6 Combination floor plan.

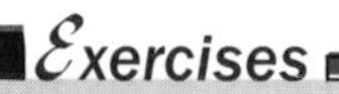

Exercises

1. Prepare an index for the set of plans you have developed.

2. List the minimum number of and the kind of drawings necessary to build the home shown in Fig. 69-1.

3. Identify these terms: *plan set, index, title block, master index, pin registration, key symbol.*

UNIT 69

ARCHITECTURAL DRAWING SET

In previous units, samples of architectural (construction) drawings were shown to illustrate various principles and practices related to the reading and interpretation of each specific type of drawing. However, these drawings did not necessarily relate to the same structure. Thus, there was no interrelationship between an electrical plan, a plumbing plan, and a survey plan. Drawings in this unit, however, are of the same building; therefore, the interpretation and agreement among plans can be studied more easily.

SEQUENCE

The floor plan that is prepared first relates to most other drawings. The second and successive floor plans are prepared by tracing the first-floor plan outline. Bearing partitions, plumbing wall, stairwells, fireplace and chimney openings, and other components can then be aligned vertically. The basement, floor, and roof framing plans are prepared in the same way. Stud layouts and the length of interior partition panel layouts are also derived from the floor plan. Horizontal distances on elevation drawings are also projected from the four sides of the floor plan. Because the floor plan functions as a base for so many other drawings, errors on the floor plan can easily be transferred to other drawings. For that reason, the floor plan must be carefully checked for accuracy before other drawings in the set are prepared. The usual sequence in preparing architectural drawings is as follows:

1. Floor plan
2. Foundation plans
3. Front elevation
4. Rear elevation
5. Right elevation
6. Left elevation
7. Auxiliary elevation
8. Plot plans
9. Landscape plans
10. Survey plans
11. Full sections
12. Detail sections
13. Electrical plans
14. Air-conditioning plans
15. Plumbing diagrams
16. Exterior pictorial rendering
17. Interior pictorial rendering
18. Landscape elevation
19. Floor framing plans
20. Exterior-wall framing plans
21. Interior-wall framing plans
22. Stud layouts
23. Roof framing plans
24. Fireplace details
25. Foundation details
26. Checking and checklists
27. Architectural models
28. Door and window schedules
29. Finish schedules
30. Specifications
31. Building cost and estimates

Arrangement

Although the above sequence is typically used in the preparation of drawings, often the arrangement of drawings in a set is different. Frequently drawings are positioned in a set in the order used on the job. For example, the above set may be arranged for the builder beginning with the survey plans followed by the plot plan, foundation plan, floor plan, floor framing plan, and so forth.

AGREEMENT AMONG DRAWINGS

It is important that elements of each drawing agree. For example, the dimensions on one side of a floor plan must agree with the total shown on the opposite side, of a rectangular building. It is also important that related features on different drawings agree. Thus checking the agreement of the same components on different drawings is extremely critical. If several drawings in a set contain conflicting information the builder cannot determine which is correct. In addition to the need for all drawings in the set to agree in every detail, each drawing in the set must also agree with the information found in all the related documents, including schedules, specifications, budget, and legal forms.

Handling Changes

Because of the importance of consistency among drawings the handling of changes to any drawing in the set is critical. Each change made on one drawing after the entire set is complete or partially complete has a chain reaction throughout the entire set. For example, if a stairwell opening is enlarged after the electrical, plumbing, and HVAC drawings are complete, wiring, pipe, and duct locations may need to be changed. In addition, changes are required on the floor plan, floor framing plan, stud layout, framing elevations, and specifications. Every drawing must be reviewed after each change to determine what other drawings need to be changed and what redesign is necessary. Each change must also be evaluated to determine if the improvement created by the change is worth the additional cost or secondary problems created.

Selected areas of each plan in this unit have been marked with geometric figures such as circles, rectangles, hexagons, crosses, squares, triangles, and diamonds. Each symbol identifies the identical area shown on each plan. By studying the position of these symbols on each drawing, you can observe how a specific area appears on each drawing in the set. For example, the position of the colored circle on the main-level floor plan represents the area covered with the same colored circle on the right elevation, lower-level plan, rear view, pictorial drawing, or any other plan so marked with a colored circle. The relationship between sectional views and basic drawings can be followed by tracing these geometric symbols and by locating the position

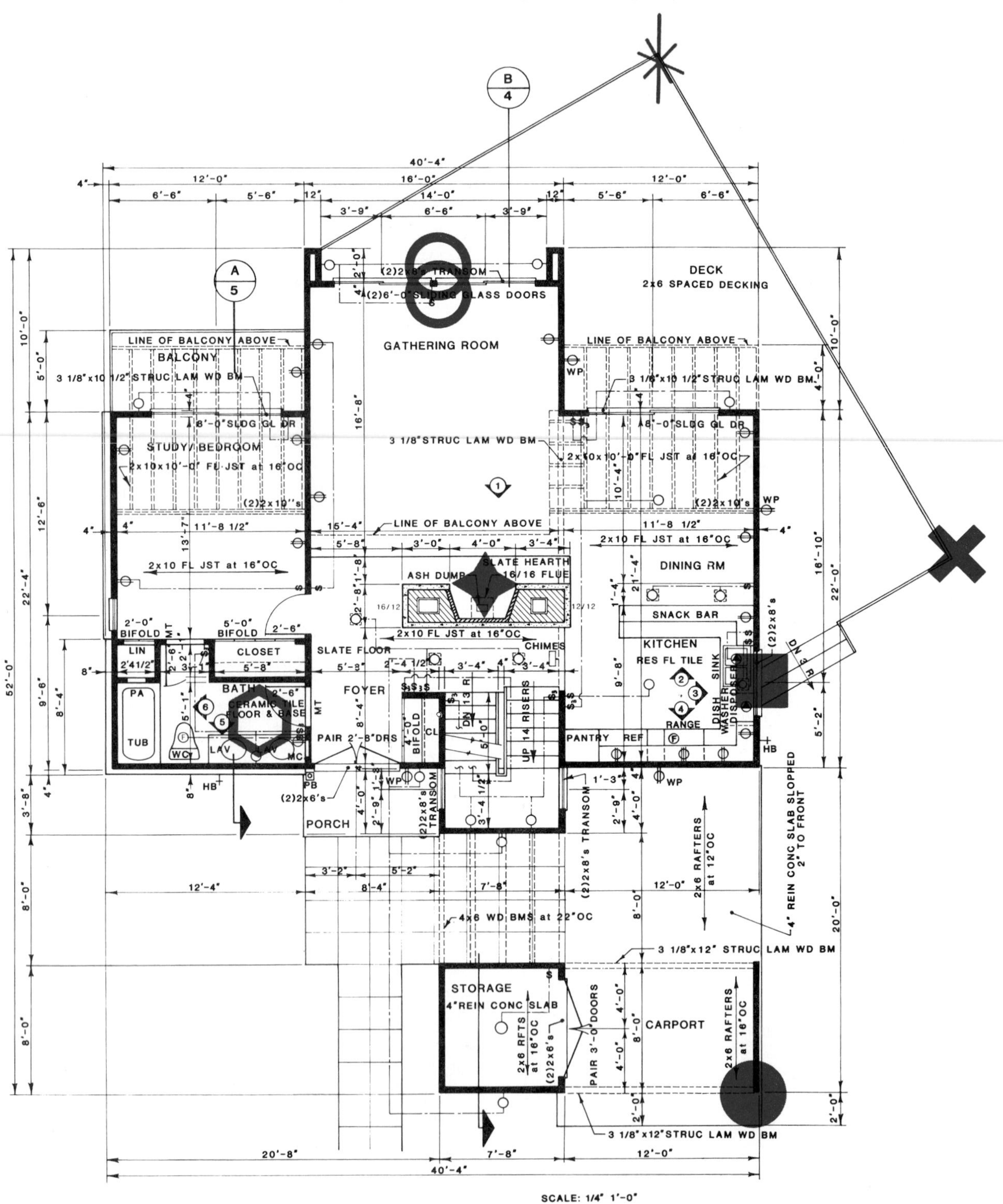

Fig. 69-1 Main-level floor plan. *(Home Planners, Inc.)*

of the appropriate cutting-plane line on the basic plans. Figures 69-1 through 69-12 show a related set of residential plans marked with geometric figures. Study the plans in the sequence listed above and trace the location of each of the geometric figures throughout the set of plans.

Commercial and industrial buildings normally have more floors, cover wider areas, use heavier construction members, and may be more interrelated with other buildings. Nevertheless, the

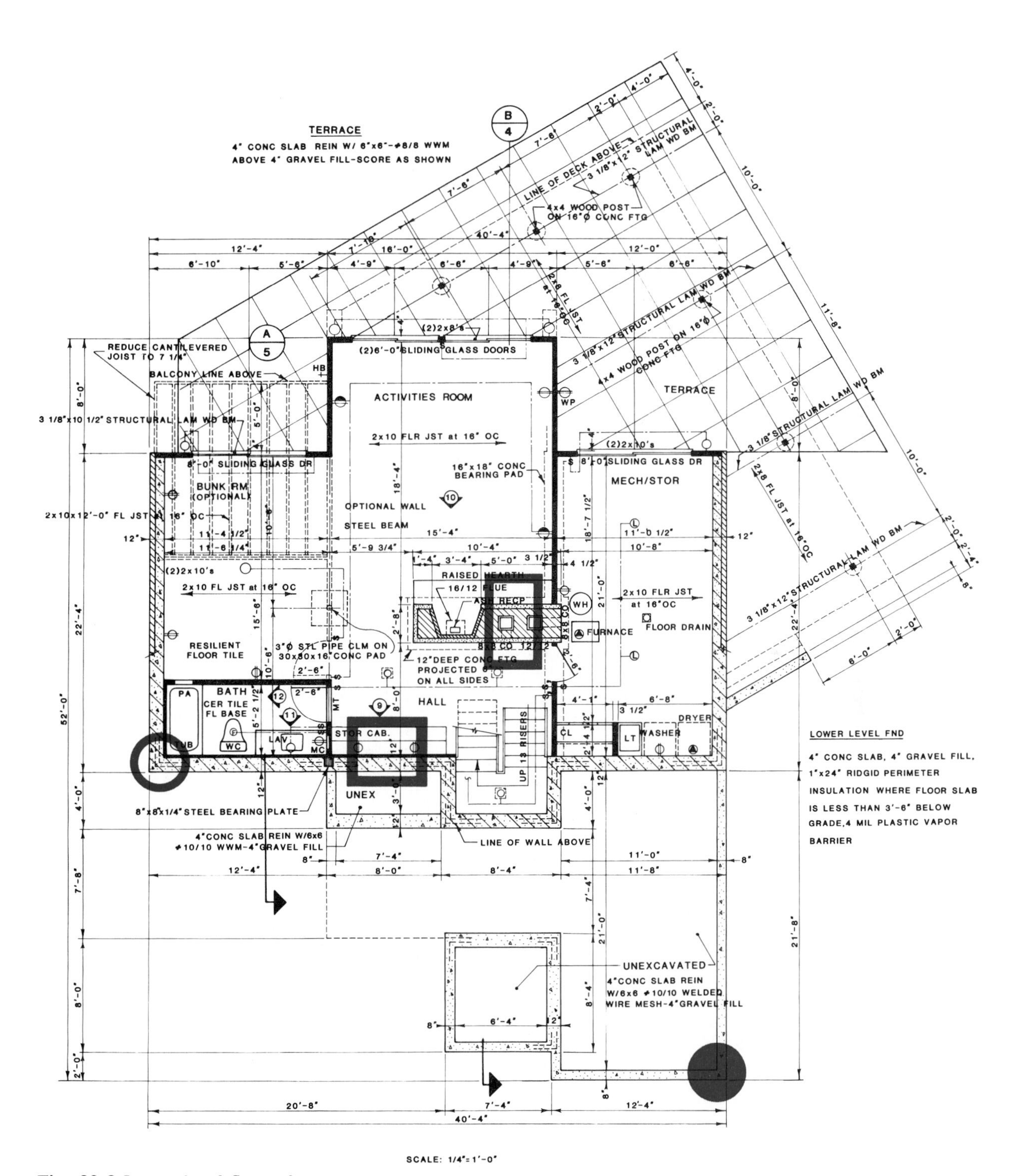

Fig. 69-2 Lower-level floor plan. *(Home Planners, Inc.)*

sequence of studying and understanding the relationship between drawings is identical.

DIMENSIONS

If a building is to be constructed as designed, it is extremely critical that dimensions describing the size of each component of a building agree on each drawing. If the dimensions of a basement plan do not match related dimensions on the floor plan, prefabricated wall panels may not fit, or the position of stairwell openings and fireplace footings may not align. Whether a dimension describes the overall length or width of a structure or only indicates the size of a subdimension of a detail, like dimensions must agree on each drawing in a set. For this reason, dimensional accuracy is verified by locating one dimension at a time throughout the entire set until each has been checked and agreement is determined.

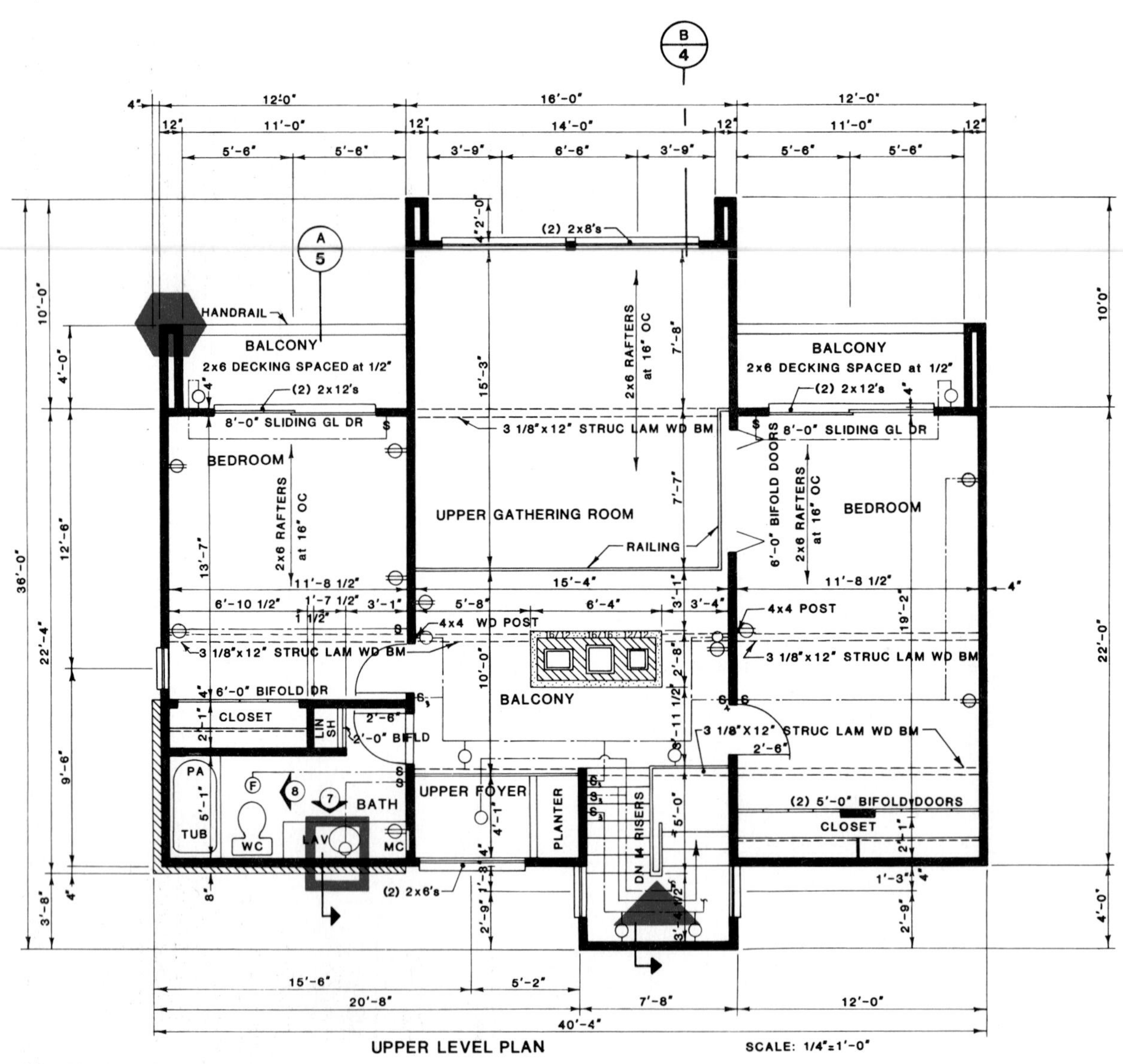

Fig. 69-3 Upper-level floor plan. *(Home Planners, Inc.)*

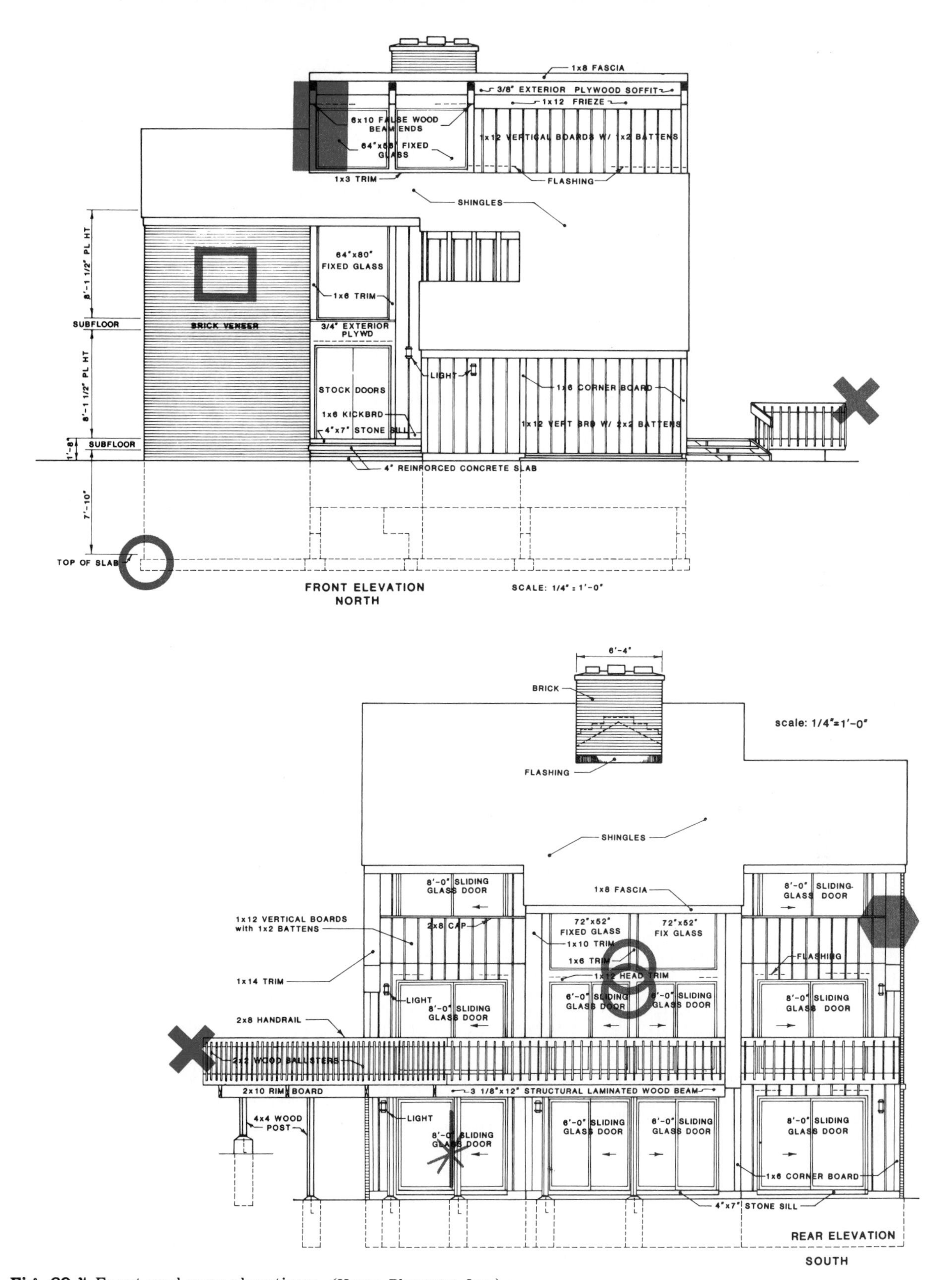

Fig. 69-4 Front and rear elevations. *(Home Planners, Inc.)*

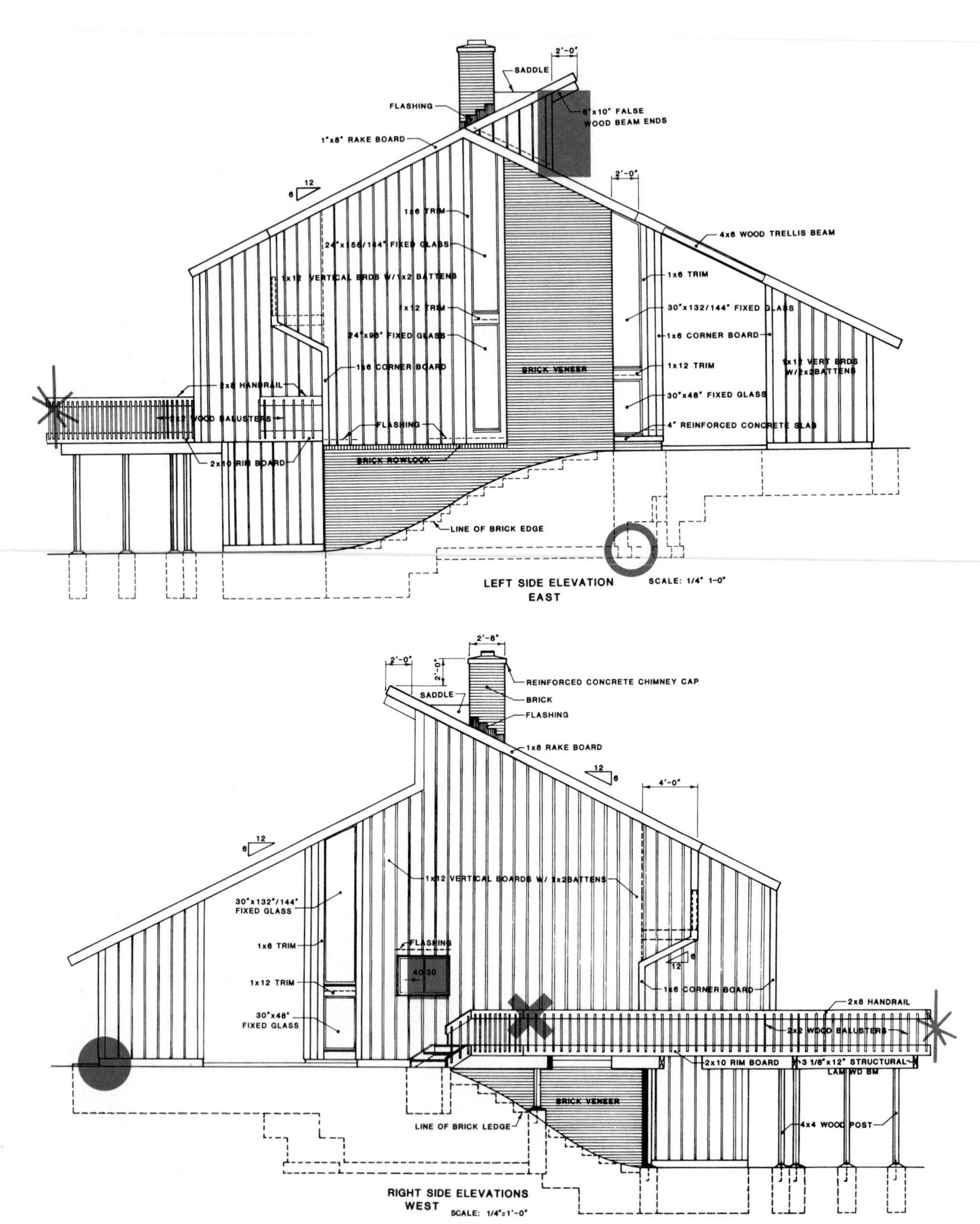

2'-0"
SADDLE
FLASHING
6"x10" FALSE WOOD BEAM ENDS
1"x8" RAKE BOARD
12
6
2'-0"
1x6 TRIM
24"x156/144" FIXED GLASS
4x6 WOOD TRELLIS BEAM
1x12 VERTICAL BRDS W/1x2 BATTENS
1x6 TRIM
1x12 TRIM
30"x132/144" FIXED GLASS
24"x96" FIXED GLASS
1x6 CORNER BOARD
1x6 CORNER BOARD
BRICK VENEER
1x12 TRIM
1x12 VERT BRDS W/2x2 BATTENS
2x8 HANDRAIL
30"x48" FIXED GLASS
2x2 WOOD BALUSTERS
FLASHING
4" REINFORCED CONCRETE SLAB
2x10 RIM BOARD
BRICK ROWLOCK
LINE OF BRICK EDGE
LEFT SIDE ELEVATION
EAST
SCALE: 1/4" 1-0"
2'-8"
2'-0"
2'-0"
REINFORCED CONCRETE CHIMNEY CAP
SADDLE
BRICK
FLASHING
1x8 RAKE BOARD
12
6
4'-0"
12
6
1x12 VERTICAL BOARDS W/ 1x2BATTENS
30"x132"/144"
FIXED GLASS
1x6 TRIM
FLASHING
40 30
1x12 TRIM
12
6
1x6 CORNER BOARD
2x8 HANDRAIL
30"x48"
FIXED GLASS
2x2 WOOD BALUSTERS
2x10 RIM BOARD
3 1/8"x12" STRUCTURAL
LAM WD BM
BRICK VENEER
LINE OF BRICK LEDGE
4x4 WOOD POST
RIGHT SIDE ELEVATIONS
WEST
SCALE: 1/4"=1'-0"

Fig. 69-5 Right and left elevations. *(Home Planners, Inc.)*

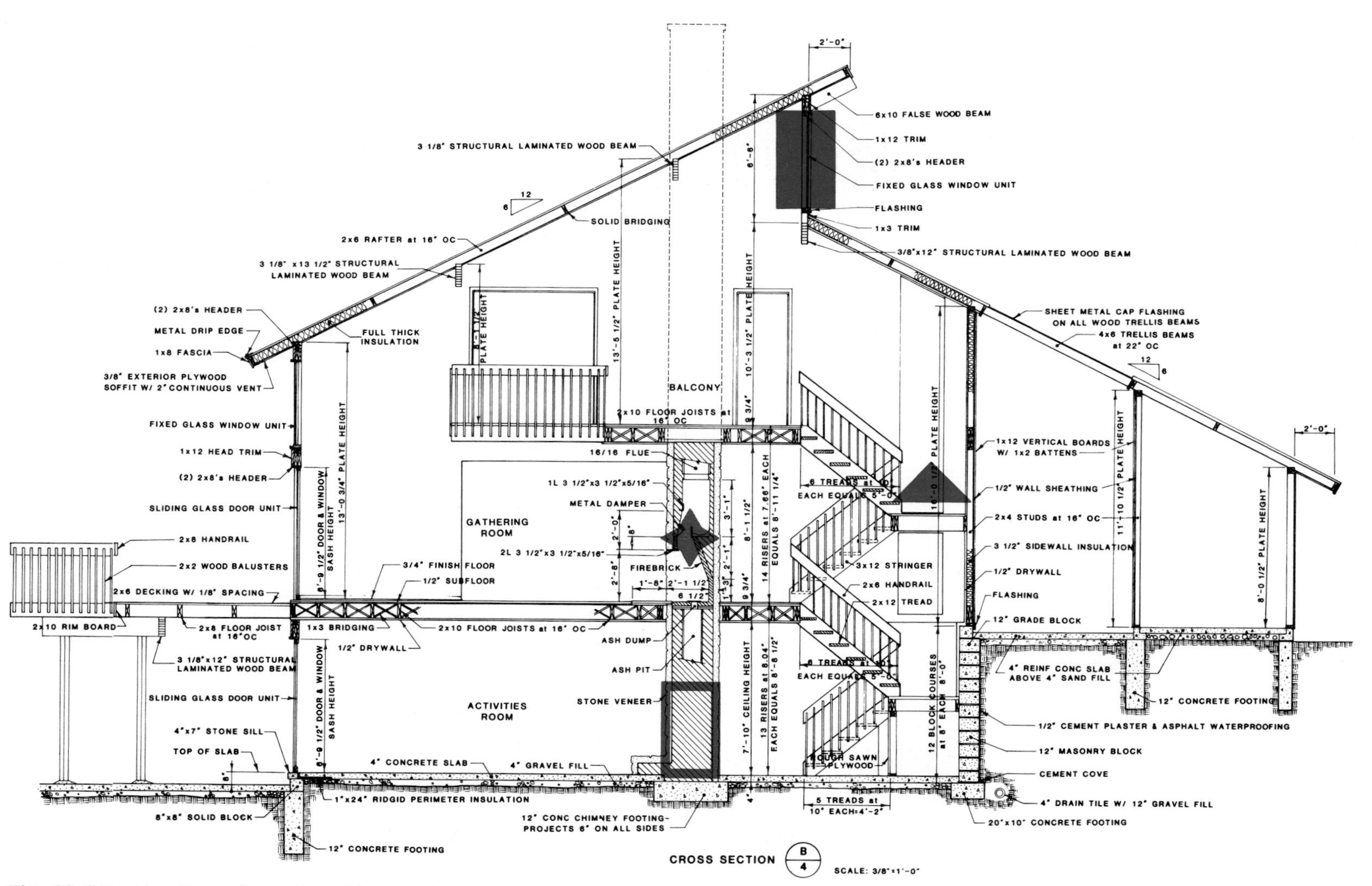

Fig. 69-6 Section through center of house. *(Home Planners, Inc.)*

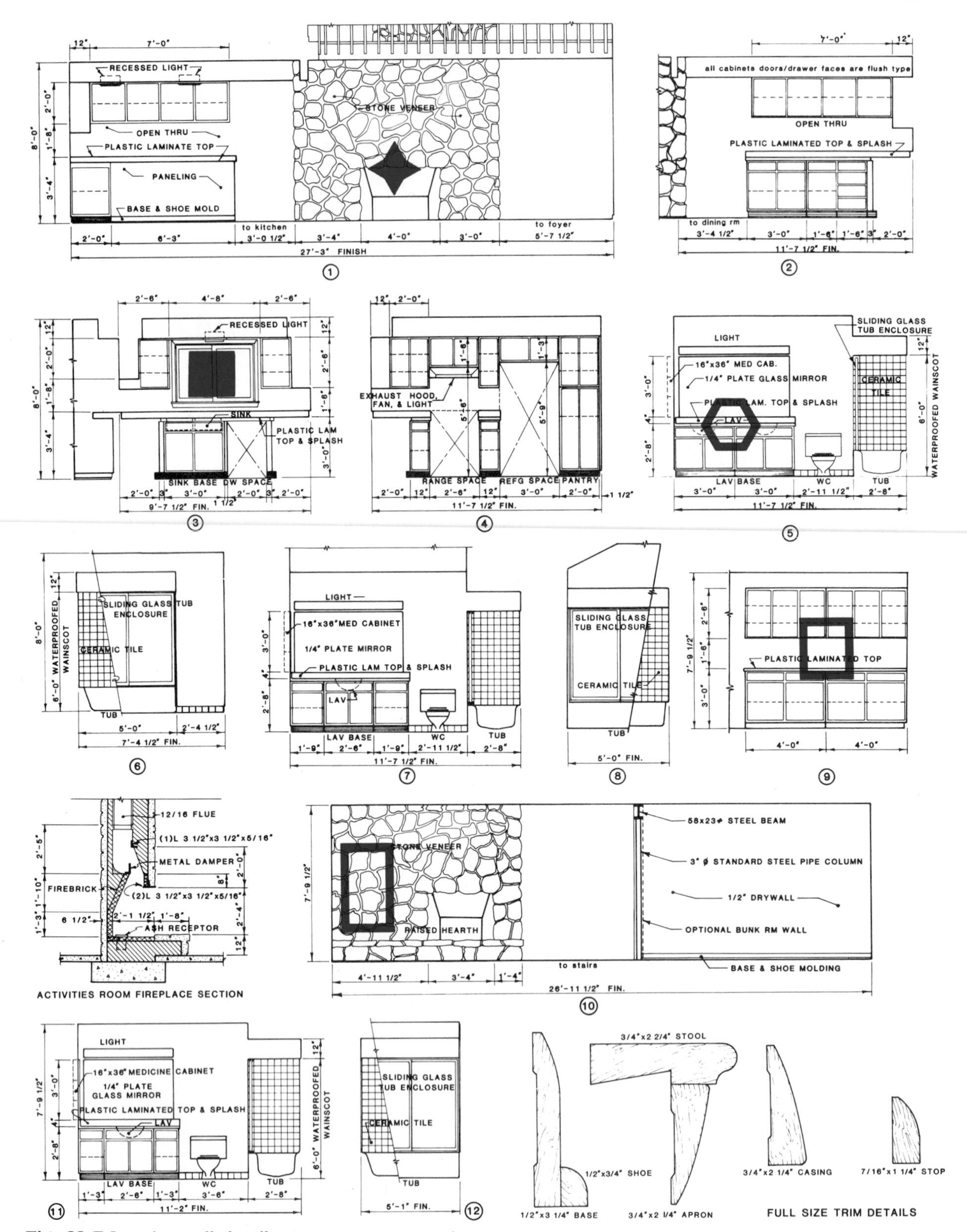

Fig. 69-7 Interior wall details. *(Home Planners, Inc.)*

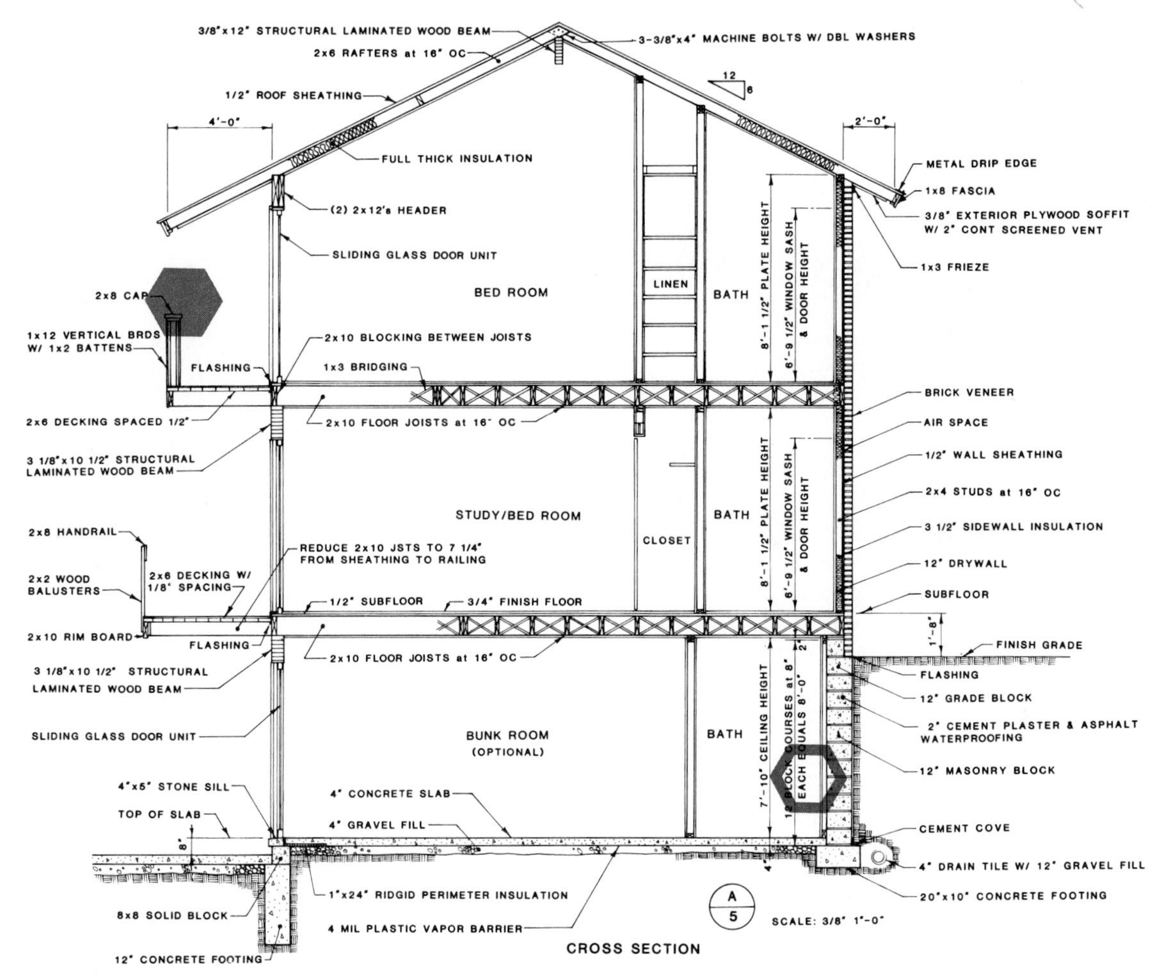

Fig. 69-8 Section through left portions of house. *(Home Planners, Inc.)*

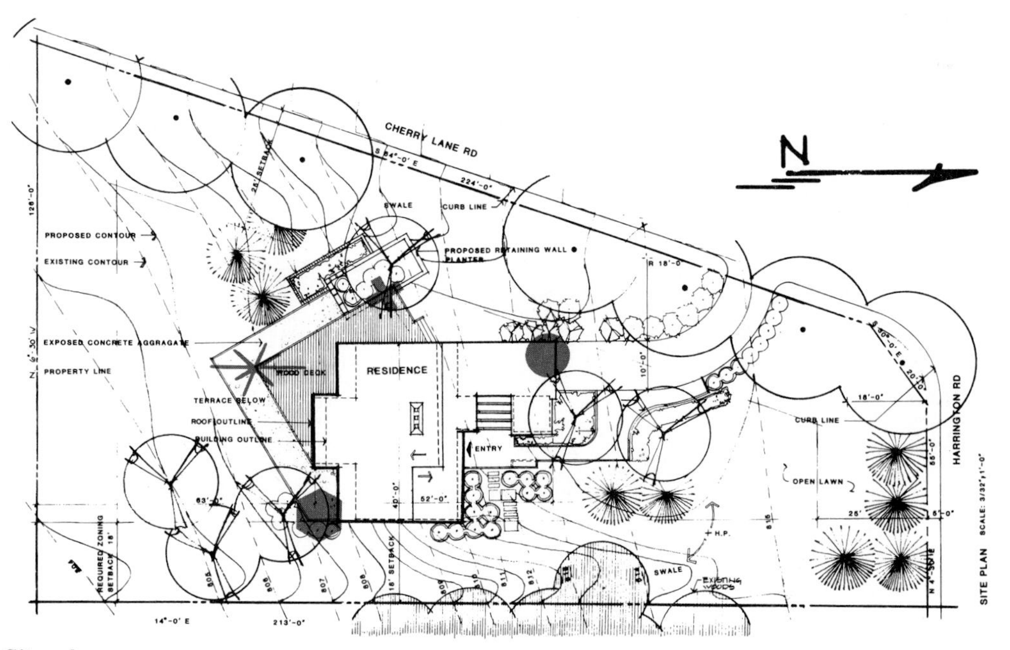

Fig. 69-9 Site plan.

Fig. 69-10 Exterior pictorials. *(Home Planners, Inc.)*

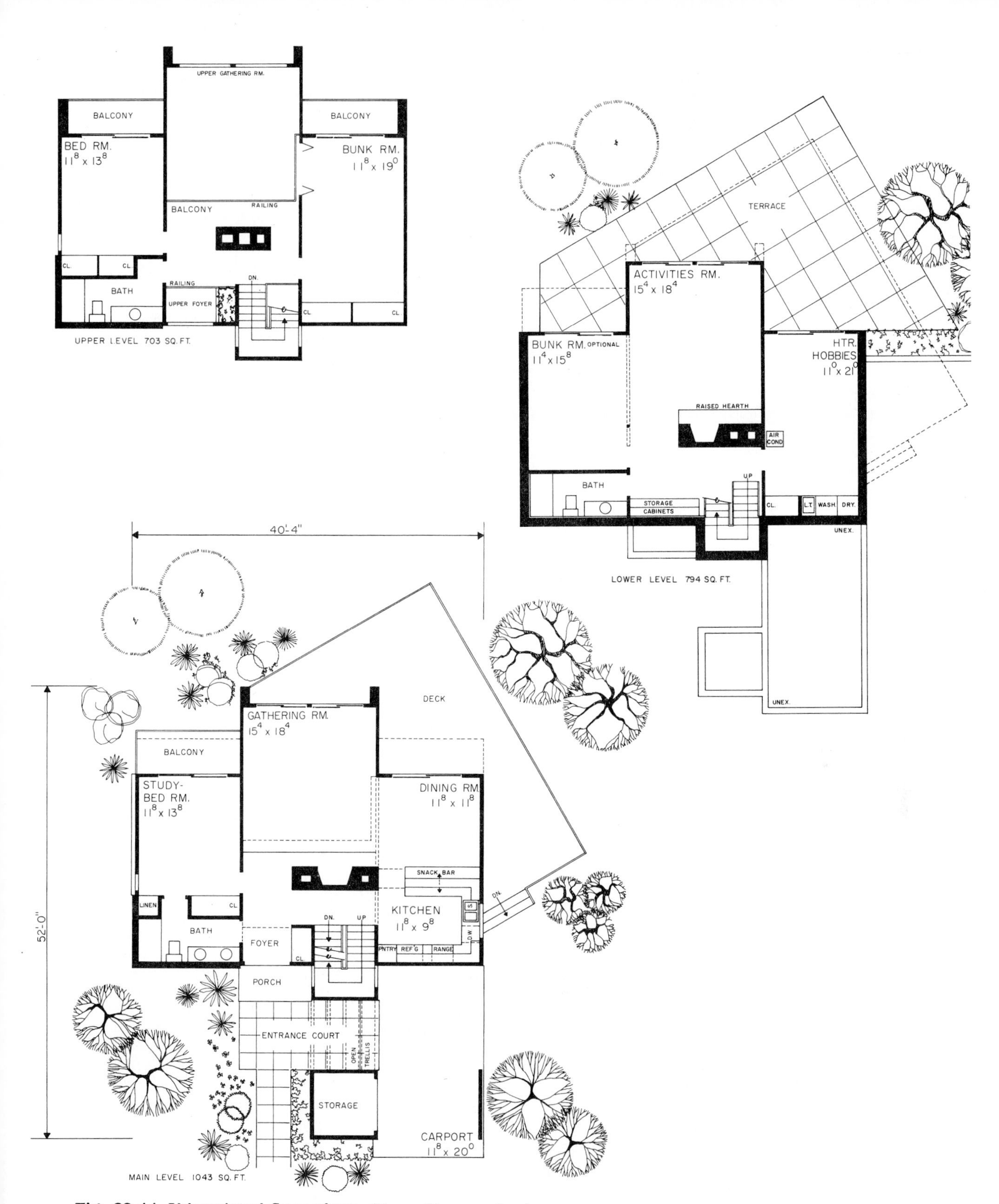
UPPER GATHERING RM.
BALCONY
BALCONY
BED RM.
11⁸ x 13⁸
BUNK RM.
11⁸ x 19⁰
BALCONY
RAILING
CL.
CL.
RAILING
BATH
UPPER FOYER
DN.
CL.
CL.
UPPER LEVEL 703 SQ. FT.
TERRACE
ACTIVITIES RM.
15⁴ x 18⁴
BUNK RM. OPTIONAL
11⁴ x 15⁸
HTR.
HOBBIES
11⁰ x 21⁰
RAISED HEARTH
AIR COND
BATH
UP
STORAGE CABINETS
CL.
L.T.
WASH.
DRY.
UNEX.
LOWER LEVEL 794 SQ. FT.
UNEX.
40'-4"
52'-0"
DECK
GATHERING RM.
15⁴ x 18⁴
BALCONY
STUDY-
BED RM.
11⁸ x 13⁸
DINING RM.
11⁸ x 11⁸
SNACK BAR
KITCHEN
11⁸ x 9⁸
DN.
LINEN
CL.
BATH
FOYER
DN.
UP
CL.
PNTRY
REF'G
RANGE
D.W.
PORCH
ENTRANCE COURT
OPEN TRELLIS
STORAGE
CARPORT
11⁸ x 20⁰
MAIN LEVEL 1043 SQ. FT.

Fig. 69-11 Abbreviated floor plans. *(Home Planners, Inc.)*

Exercises

1. List the order in which you would prepare each of the following drawings: stud layout, foundation plan, specifications, floor plan, front elevation, left elevation, exterior rendering, basement plan, second-floor plan.
2. Pick an area on the floor plan shown in Fig. 69-1 not covered by a geometric symbol. Locate that same area on the other drawings in this unit.
3. Pick a dimension not circled on the floor plan shown in Fig. 69-1. Find that dimension on other drawings in this unit.
4. Refer to Figs. 69-1 through 69-10 and identify the following:

   a. Dimensions of the chimney at the upper level

   b. Type of outside door used in activities room

   c. Plate height of gathering room

   d. Ceiling height of lower-level bath

   e. Width of lower-level east wall

   f. Type of construction used on lower-level west wall

   g. Terrace vertical support members

   h. Right-side siding materials

   i. Roof pitch

   j. Height of balcony handrails
5. Sketch the outline of Fig. 69-1 through 69-11 and place a number corresponding to the 22 numbers shown in Fig. 69-12 in related positions on each appropriate drawing.

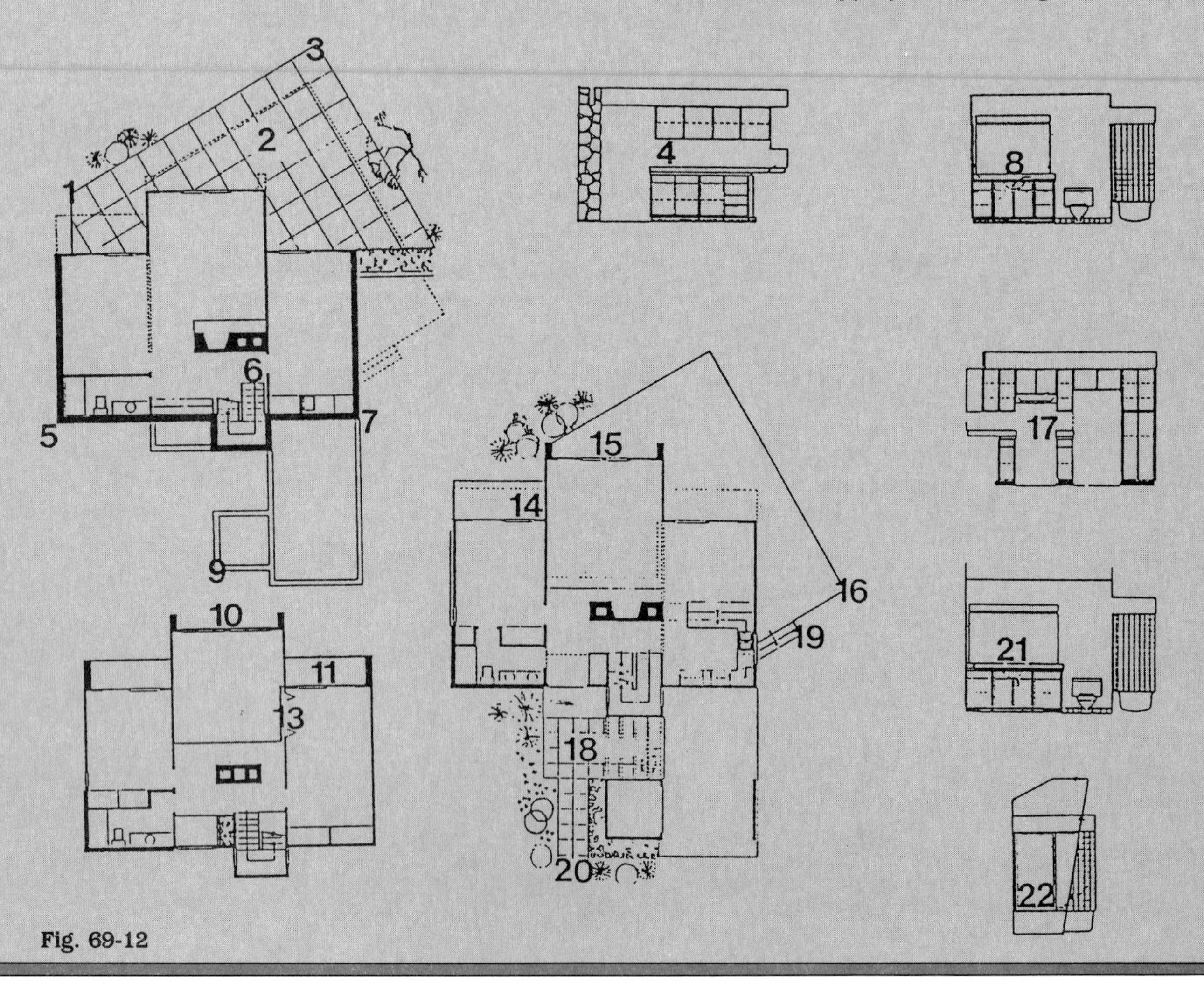

Fig. 69-12

PART FIVE

Architectural Support Services

(Southland)

The development of a total architectural design does not stop with the completion of a set of drawings. There is additional information needed by contractors, financial institutions, and governmental agencies that is not found on architectural drawings. The preparation and use of this information is presented in this part.

SECTION 19

Schedules and Specifications

The plans and drawings of a building are documents prepared to ensure that the building will be constructed as planned. It is sometimes difficult or impossible to show on the drawings all details pertaining to the construction of a building. All features not shown on a drawing should be listed in a schedule or in the specifications.

A schedule is a chart of materials and products. Most plans include a window, door, and interior-finish schedule. Schedules are also prepared for exterior finishes, electrical fixtures, and plumbing fixtures. Schedules can also be prepared for equipment and furnishings specified for the building.

Specifications are lists of details and building products to be included in the building. Specifications may be rather brief descriptions of the materials needed, or they may be complete specifications that list the size, manufacturer, grade, color, style, and price of each item of material to be ordered. Schedules and specifications are an extremely important part of the construction process. When conflicts occur relating to construction performance, the schedules or specifications hold legal preference over the information contained on the set of plans. All schedule and specification information must be consistent with the content of the drawings.

UNIT 70

DOOR AND WINDOW SCHEDULES

Door and window schedules conserve drawing time and space. Instead of including all of the information needed about a door or window directly on the drawing, a key number is placed next to each door or window symbol on the floor plan. Figure 70-1 shows the placement of key symbols on a floor plan. Letters are usually used for windows and numbers for doors. Also different geometric symbols are used to separate windows from doors and to keep the numbers or letters from being confused with labels or dimensions on the drawing.

Figure 70-2 shows a door and window schedule. Note that the key symbol for each door or

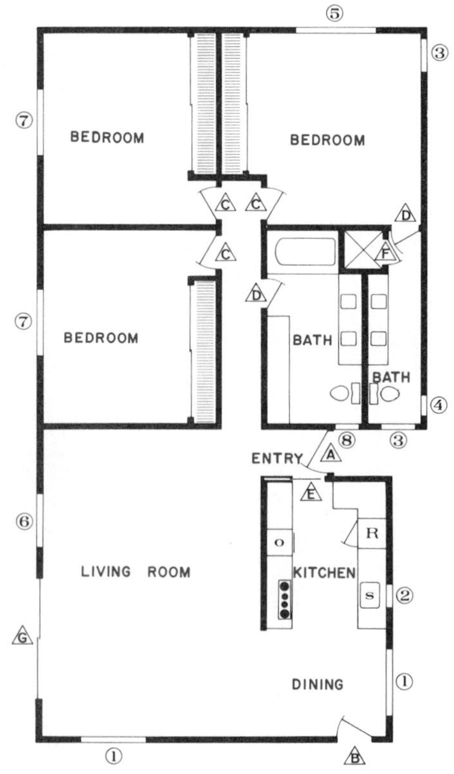

Fig. 70-1 Key symbols.

Door schedule

Symbol	Width	Height	Thickness	Material	Type	Screen	Quantity	Threshold	Remarks	Manufacturer
A	3′-0″	7′-0″	$1\frac{3}{4}$″	Wood—Ash	Slab core	No	1	Oak	Outdoor varnish	A. D. & D. Door, Inc.
B	2′-6″	7′-0″	$1\frac{3}{4}$″	Wood—Ash	Slab core	Yes	1	Oak	Oil stain	A. D. & D. Door, Inc.
C	2′-3″	6′-8″	$1\frac{3}{8}$″	Wood—Oak	Hollow core	No	3	None	Oil stain	A. D. & D. Door, Inc.
D	2′-0″	6′-8″	$1\frac{3}{8}$″	Wood—Ash	Hollow core	No	2	None	Oil stain	A. D. & D. Door, Inc.
E	2′-3″	6′-8″	$1\frac{1}{4}$″	Wood—Fir	Plywood	No	1	None	Sliding door	A. D. & D. Door, Inc.
F	1′-9″	5′-6″	$\frac{1}{2}$″	Glass & metal	Shower door	No	1	None	Frosted glass	A. D. & D. Door, Inc.
G	4′-6″	6′-6″	$\frac{1}{2}$″	Glass & metal	Sliding	Yes	2	Metal	1 sliding screen	A. D. & D. Door, Inc.

Window schedule

Symbol	Width	Height	Material	Type	Screen	Quantity	Remarks	Manufacturer	Catalog number
1	5′-0″	4′-0″	Aluminum	Stationary	No	2		A & B Glass Co.	18BW
2	2′-9″	3′-0″	Aluminum	Louver	Yes	1		A & B Glass Co.	23JW
3	2′-6″	3′-0″	Wood	Double Hung	Yes	2	4 Lites—2 High	A & B Glass Co.	141PW
4	1′-6″	1′-6″	Aluminum	Louver	Yes	1		Hampton Glass Co.	972BW
5	6′-0″	3′-6″	Aluminum	Louvered Sides	Yes	1		Hampton Glass Co.	417CW
6	4′-0″	6′-6″	Aluminum	Stationary	No	1		H & W Window Co.	57DH
7	5′-0″	3′-6″	Aluminum	Sliding	Yes	2	Frosted Glass	H & W Window Co.	22DH
8	1′-9″	3′-0″	Aluminum	Awning	Yes	1		H & W Window Co.	1711JB

Fig. 70-2 Key symbols indexed to a door and window schedule.

window is placed in the left column, followed by the detailed information relating to each. The data includes dimensions, specific characteristics, quantity, and often manufacturers' catalog numbers. If specific manufacturers' data is available and sufficiently detailed, the catalog or code number is often substituted for many of the schedule entries. Care must be taken to ensure that the style model and size specified exactly matches each door and window called for in the design. Figure 70-3 shows the basic minimum information required on door schedules. If all of this information is included on a floor plan there would be little space available for other dimensions and notes.

Door or window key symbols may also be included on elevation drawings. In some cases the outline of a door or window with a key symbol in the opening substitutes for drawing the detail of the door or window on the elevation. When this is done a separate elevation detail as shown in Figs. 37-9 and 70-4 must be used to show the style and form of the window. Many door details of this type also show surrounding details in a scale larger than possible on the elevation drawing. This procedure saves much time in preparing drawings. However, an elevation drawn in this manner lacks realism, so this type of drawing is not recommended for design or presentation purposes. Figure 70-4 shows door design drawings that are indexed to a drawing and schedule. Exterior door styles can be shown on elevation drawings. However, unless an eleva-

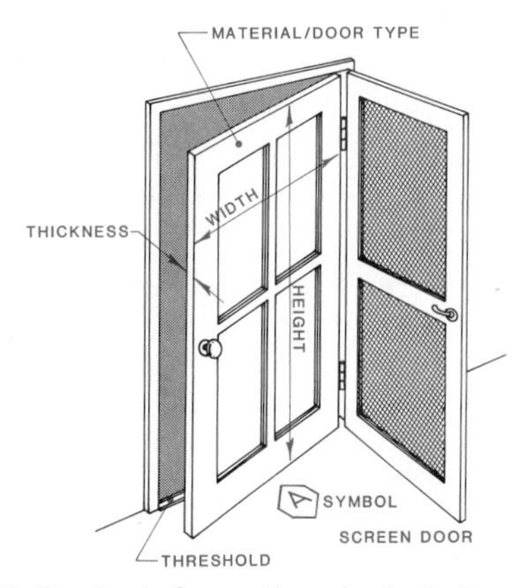

Fig. 70-3 Basic information included on a door schedule.

tion drawing is prepared for each room, interior door styles may never appear in a set of drawings. So drawing separate interior door designs and indexing these to the drawings and schedules is often the only way to see the features of the door style. Figure 70-5 shows a type of window pictorial representation that is often used to clarify details not possible on a plan, elevation, or schedule. In addition to better describing the design, drawings of this type are also helpful in eliminating errors in field installation.

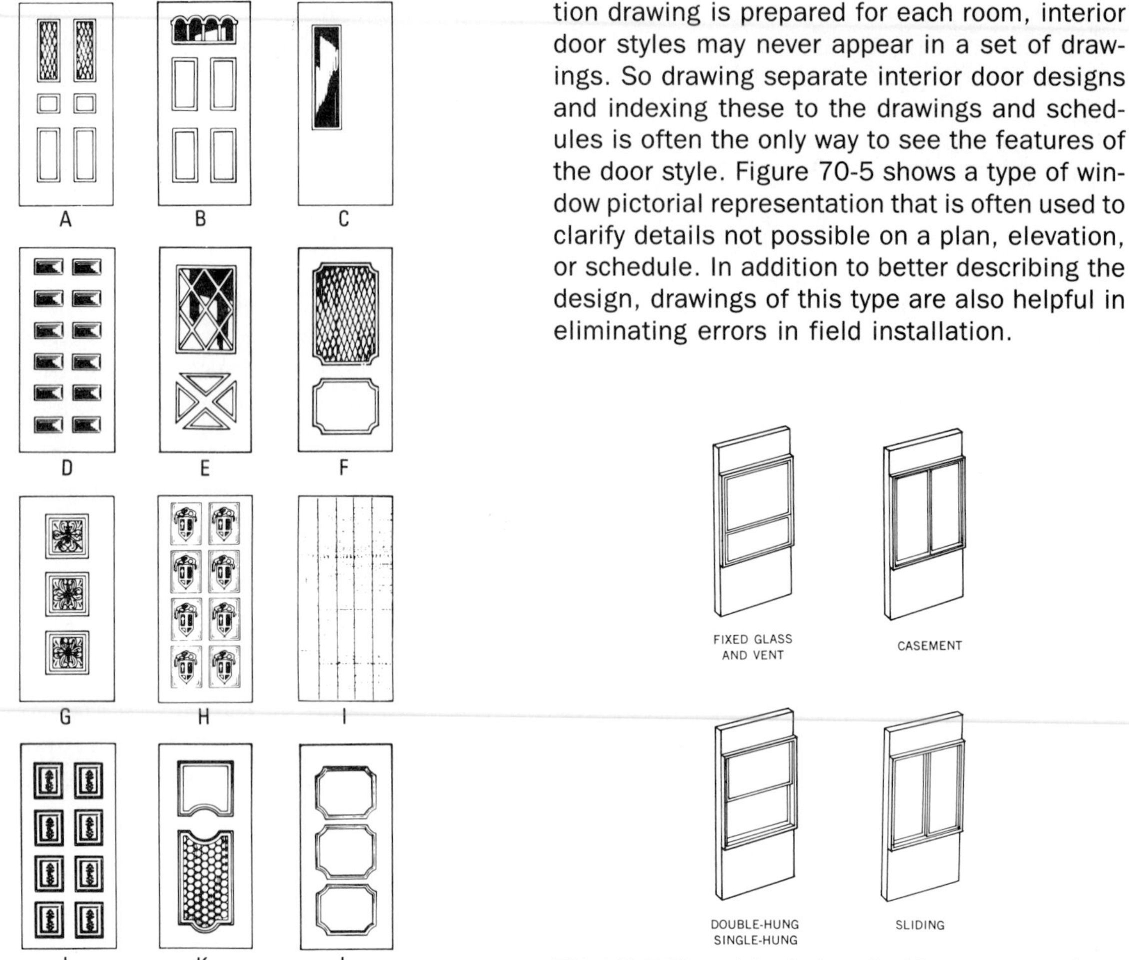

Fig. 70-4 Exterior door designs.

Fig. 70-5 Pictorial window designs are sometimes indexes to a schedule.

Exercises

1. **Prepare a window schedule for the windows shown in Fig. 70-6.**
2. **Prepare a door and window schedule for the windows and doors of the house you are designing.**
3. **Define these architectural terms: *schedule, specifications, sliding window, awning window, double-hung window, panel door, flush door, French door, Dutch door, key symbol.***

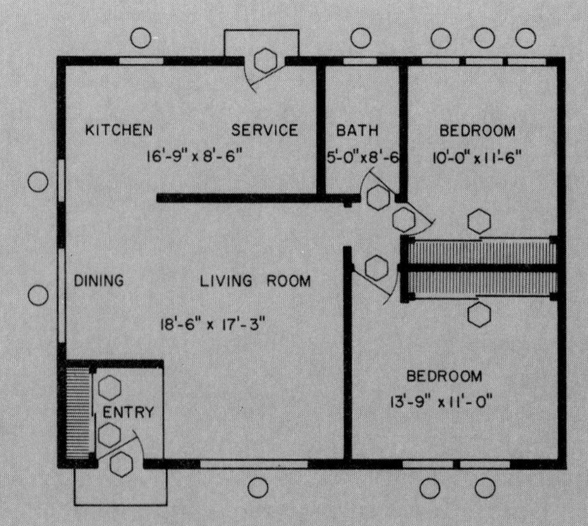

Fig. 70-6

UNIT 71

FINISH AND FIXTURE SCHEDULES

In addition to doors and windows which are a part of the structure, schedules can be developed for any category of item to be applied or installed in the building. However, schedules are usually only prepared for items that are ordered complete and not for materials to be used in constructing components on the site. These include schedules to describe the quantity and characteristics of fixtures, finishes, appliances, materials, and sometimes furniture and accessories.

FINISH SCHEDULE

Many types of finishes are used on the interior and the exterior of a building. To describe the type of finish enamel, paint, and stain, the amount of gloss and color of each finish for each room would require an exhaustive list with many duplications. A *finishing schedule* is a chart that enables the designer to condense all this information. The interior-finish schedule shown in Fig. 71-1 includes, in the horizontal column, the parts of each room and the type of finish to be applied. In the vertical column is information pertaining to the application of the finish and the room to which it is to be applied. The exact color classification has been noted in the appropriate intersecting block. The last column, headed "Remarks," is used for making notes about the finish application.

Schedules are not only useful in designing and ensuring that the finishing is completed as planned, but are also valuable as aids in ordering manufactured items, such as appliances and fixtures.

MATERIALS SCHEDULE

To ensure that all floor, wall, and ceiling coverings blend with the overall decor in each room, an interior-finish schedule can be prepared. All the possible materials for each part of the room should be listed in the horizontal column. The rooms are listed in the vertical column. The appropriate block can be checked for the suitable material for the ceiling, wall, wainscoting, base, and floor of each room (Fig. 71-2). When a schedule is prepared in this manner, it is easy to read and facilitates checking the color scheme of each room and of the overall decor. This kind of schedule condenses pages of unrelated material lists for each room into one chart, thus enabling the designer to see at a glance all the material that should be ordered for each room.

INTERIOR FINISHING SCHEDULE

ROOMS	FLOOR			CEILING					WALL					BASE					TRIM					REMARKS
	FLOOR VARNISH	UNFINISHED	WAXES	ENAMEL GLOSS	ENAMEL SEMIGLOSS	ENAMEL FLAT	FLAT LATEX	STAIN	ENAMEL GLASS	ENAMEL SEMIGLOSS	ENAMEL FLAT	FLAT LATEX	STAIN	ENAMEL GLOSS	ENAMEL SEMIGLOSS	ENAMEL FLAT	FLAT LATEX	STAIN	ENAMEL GLOSS	ENAMEL SEMIGLOSS	FLAT LATEX	ENAMEL FLAT	STAIN	
ENTRY			√				OFF WHT					OFF WHT			OFF WHT					OFF WHT				OIL STAIN
HALL			√					LT BRN		TAN					DRK BRN					DRK BRN				OIL STAIN
BEDROOM 1	√						OFF WHT					OFF WHT				GREY				GREY				ONE COAT PRIMER & SEALER —PAINTED SURFACE
BEDROOM 2	√						OFF WHT				LT YEL					YEL						YEL		ONE COAT PRIMER & SEALER —PAINTED SURFACE
BEDROOM 3		√					OFF WHT						LT BRN					DRK BRN					TAN	ONE COAT PRIMER & SEALER —PAINTED SURFACE
BATH 1				WHT					WHT					LT BLUE					LT BLUE					WATER-RESISTANT FINISHES
BATH 2				WHT					WHT					LT BLUE					LT BLUE					WATER-RESISTANT FINISHES
CLOSETS	√						BRN					BRN					BRN					BRN		
KITCHEN			√		WHT				YEL					YEL					YEL					
DINING			√					TAN		YEL					YEL				YEL					OIL STAIN
LIVING		√						TAN					LT BRN					LT BRN					LT BRN	OIL STAIN

Fig. 71-1 Interior finish schedule.

Rooms	Floor									Ceiling					Wall			Wainscot					Base					Remarks
	ASPHALT TILE	CERAMIC TILE	CORK TILE	LINOLEUM TILE	WOOD STRIP—OAK	WOOD SQS.—OAK	PLYWOOD PANEL	CARPETING	SLATE	TERAZZO	PLASTER	WOOD PANEL	ACOUSTICAL TILE	EXPOSED BEAM	PLASTER	WOOD PANEL	WALL PAPER	WOOD	CERAMIC TILE	PAPER	ASPHALT TILE	STONE VENEER	LINOLEUM	WOOD	RUBBER	TILE—CERAMIC	ASPHALT	
ENTRY									x	x		x				x								x				Terazzo step covering
HALL			x								x				x					x				x				
BEDROOM 1					x								x		x			x									x	Mahogany wainscot
BEDROOM 2					x								x		x		x	x									x	Mahogany wainscot
BEDROOM 3							x	x					x			x									x			See owner for grade carpet
BATH 1		x									x				x				x							x		Water seal tile edges
BATH 2	x										x				x				x							x		Water seal tile edges
KITCHEN				x									x		x						x		x					
DINING				x										x	x		x				x		x					
LIVING							x	x						x		x						x		x				See owner for grade carpet

Fig. 71-2 Materials schedule.

Appliance schedule

Room	Appliance	Type	Size	Color	Manufacturer	Model no.
KITCHEN	Electric stove	Cook top	4 burner	Yellow	Ideale Appliances	341 MG
KITCHEN	Electric oven	Built-in	30″ × 24″ × 24″	Yellow	Zeidler Oven Mfg.	27 Mg
SERVICE	Hot-water heater	Gas	50 gal	White	Oratz Water Htr.	249 KG

Fixture schedule

Room	Fixture	Type	Material	Manufacturer	Model number
LIVING	2 electric lights	Hanging	Brass reflectors	Hot Spark Ltd.	1037 IG
BEDROOM 1	2 spotlights	Wall bracket	Flexible neck—aluminum	Gurian & Barris Inc.	1426 SG
BATHS 1 & 2	2 electric lights	Wall bracket	Aluminum—water resistant	Marks Electrical Co.	2432 DG

Fig. 71-3 Fixture and appliance schedule.

FIXTURE SCHEDULE

To ensure that all appliances and fixtures are compatible with the plans, a fixture and appliance schedule is often prepared. These schedules control the type size, color, manufacturer, and model number of each item. Key symbols similar to the type used for door and window schedules are used to relate each item to the appropriate location on the floor plan. A partial fixture and appliance schedule is shown in Fig. 71-3.

Furniture and accessories are not always included as part of the basic architectural design. However, a totally integrated plan should involve all aspects of the design. When furniture schedules are developed, the information is keyed to the floor plan. Only the location and amount of space occupied can be shown conveniently for each piece of furniture or accessory on a floor plan. If additional information is added such as color, materials, and dimensions, the floor plan can become extremely cluttered and difficult to read. For this reason, furniture and accessory schedules are prepared. The schedule includes all pertinent information relating to each item, as shown in Fig. 71-4A. Each item is numbered and these numbers are placed on the appropriate item on the floor plan, as shown in Fig. 71-4B. This system allows the designer and client to record and find both the locations and significant specifications of each item. Pictures of major pieces are also sometimes keyed or attached to the schedule.

Often fabric, texture, color, or other features cannot be accurately described in writing on a schedule. In these cases, swatches of material or paint color chips are often attached to a sample collage. These are either keyed to the floor plan by number or related directly with leaders as shown in Fig. 71-5. The use of this method of describing materials eliminates much misunderstanding in the final execution of the design and selection of materials.

Sym.	Item	Room	Len.	Wid.	Ht.	Material	Color	Quan.	Manufac.	Cat. #	Cost	Remarks
1	Drapes	Bedroom	11′	—	7′	Cotton blend	Brown	1 Set	Sears	CD101	$75	Lined
2	Drapes	Mbr.	12′	—	7′	Cotton blend	Yellow	1 Set	Sears	CD107	$85	Lined
3	Drapes	Den	7′	—	7′	Cotton blend	Yellow	1 Set	Sears	CD106	$65	Lined
4	Drapes	Living/dining	24′	—	7′	Acrylic	Brown pat.	1 Set	Sears	CD203	$150	Lined
5	Chair	Mbr./den	18″	18″	18″	Plastic	Brown	4	ID Furn. Co.	X117	$45 ea.	
6	Chair	Din./kit.	18″	18″	18″	Oak	Natural	9	ID Furn. Co.	L217	$65 ea.	Oil finish
7	China cab.	Dining	6′	18″	5′-6″	Oak	Natural	1	Danish Furn.	13712	$650	Oil finish
8	Piano bench	Living	33″	10″	18″	Mahogany	Brn. stain	1	Music Co Inc.	23L19	$50	Piano finish
9	Piano	Living	5′-6″	2′-0″	5′-6″	Mahogany	Brn. stain	1	Music Co. Inc.	P17731	$1750	Piano finish
10	Up. wing ch.	Living	33″	30″	20″	Leather	Natural	1	Danish Furn. Co.	18979	$575	
11	Fl. lamp	Living	14″ Dia.	—	4′-6″	Metal/cloth	Tan	1	Danish Furn. Co.	37111	$85	
12	Stereo	Living	30″	11″	5′-0″	Teak	Brown	1	Danish Furn. Co.	60701	$450	
13	Sofa/sec.	Living	14′	30″	18″	Velveteen	Red	3 Pcs.	Danish Furn. Co.	42107	$1200	
14	Coffee tbl.	Living	30″ Dia.	—	15″	Teak	Natural	1	Danish Furn. Co.	77310	$110	Natural oil finish
15	Television	Den	21″	18″	30″	21″ Color	Brown	1	Sony	XL19	$675	
16	Coffee tbl.	Den	48″	15″	15″	Oak	Natural	1	Danish Furn. Co.	78325	$80	Natural oil finish
17	Sofa	Den	6′-6″	30″	18″	Cotton blend	Tan	1	Danish Furn. Co.	59781	$800	
18	Fl. lamp	Mbr./br./den	15″ Dia.	—	5′-0″	Wood/cloth	Brown	3	Danish Furn. Co.	66362	$75 ea.	
19	Fl. lamp	Den	12″ Dia.	—	4′-6″	Wood/plastic	Yellow	1	Danish Furn. Co.	65731	$50	
20	Desk	Mbr./den	39″	18″	29″	Oak	Natural	2	Danish Furn. Co.	47772	$225 ea.	Natural oil finish
21	Nightstand	Mbr./br.	18″	15″	24″	Oak	Natural	3	Danish Furn. Co.	64991	$45 ea.	Natural oil finish
22	Tbl. lamp	Mbr.	9″ Dia.	—	30″	Wood/plastic	Brown	2	Danish Furn. Co.	65820	$35 ea.	
23	Full bed	Mbr.	6′-9″	46″	20″	Standard	—	1	Acme Bed Co.	AC12	$235	Box spring/mat./frame
24	Dresser	Mbr.	39″	20″	48″	Oak	Natural	1	Danish Furn. Co.	37452	$125	Dbl. dresser
25	Dresser	Br.	48″	20″	52″	Oak	Natural	1	Danish Furn. Co.	37471	$200	Triple dresser
26	Twin bed	Br.	6′-9″	42″	20″	Standard	—	1	Acme Bed Co.	Ac08	$190	Box spring/mat./frame
27	Planter	Mbr./liv./den/por.	10″ Dia.	—	12″	Terra cotta	Brown	4	Flowers Inc.	23FP	$10 ea.	
28	Table	Kitchen	36″	22″	30″	Teak	Natural	1	Danish Furn. Co.	17832	$110	Natural oil finish
29	Table	Dining	5′-0″	3′-3″	30″	Teak	Natural	1	Danish Furn. Co.	17876	$235	Natural oil finish

Fig. 71-4A Furniture and accessory schedule.

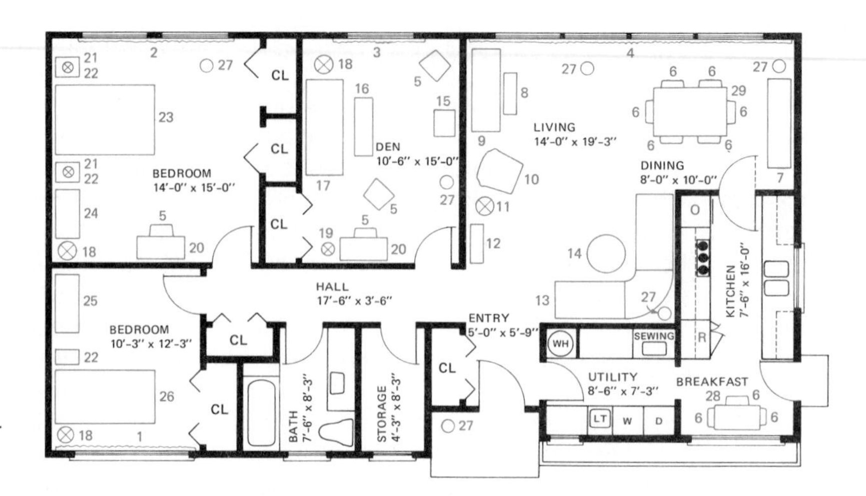

Fig. 71-4B Floor plan with furniture indexed to schedule.

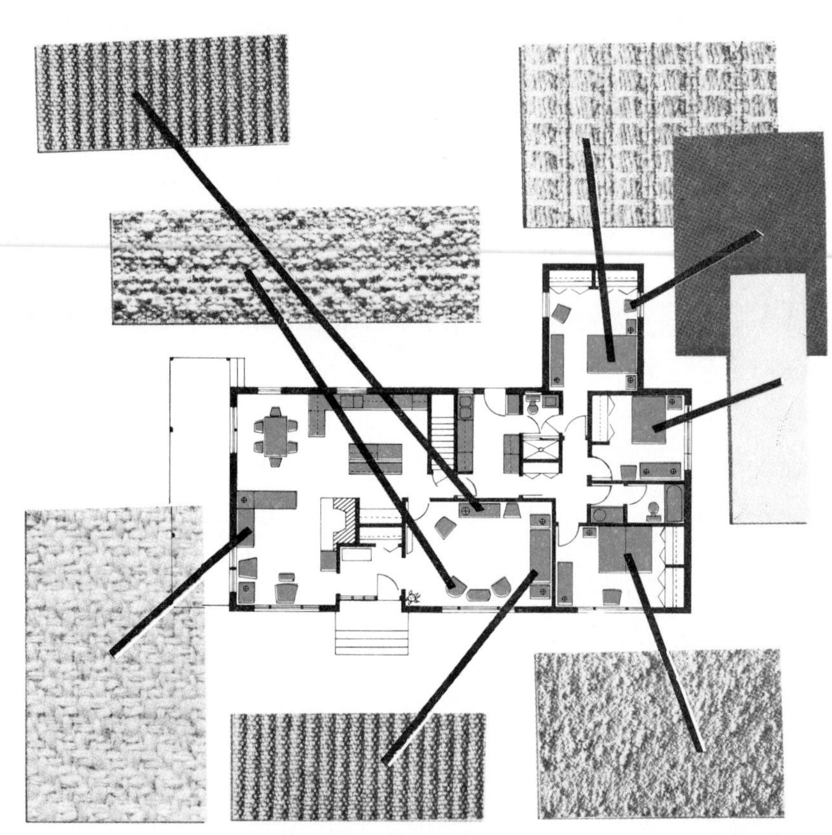

Fig. 71-5 Collage board indexed to floor plan.

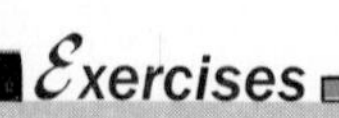

Exercises

1. Make a finish schedule for Fig. 69-1

2. Complete all the schedules for the home of your design. Include as much manufacturing information as you can obtain.

3. Know these architectural terms: *finish schedule, trim, base, wainscot, fixture schedule, appliance schedule, flat latex, enamel, primer, sealer, asphalt tile, slate, terrazzo, acoustical tile, veneer, ceramic tile.*

UNIT 72

SPECIFICATION WRITING

Schedules summarize information on items to be purchased for each building and installed as received. But specifications differ from schedules. *Specifications* are written instructions describing the basic requirements for constructing a building. Specifications describe sizes, kinds, and quality of building materials. The methods of construction, fabrication, or installation are also spelled out explicitly. Specifically, they tell the contractor, "These are the materials you must use, and this is how you must use them, and these are the conditions under which you undertake this job." Specifications guarantee the purchaser that the contractor will deliver the building when it is finished exactly as specified.

PURPOSE

Specifications are intended to simplify and clarify; not to make the construction process longer and more difficult. For this reason extreme care must be taken to ensure that the information on the last approved set of architectural working drawings agrees with the size, number, and material descriptions found in the specifications.

Accurate specifications are critical to the contractor. If the contractor's materials or methods are inferior to those specified, the project will not be approved. If the contractor uses materials and methods that exceed the limits described in the specifications, oversized materials and time will be wasted.

CONTENTS

Specifications contain information to be used for bidding requests, contract forms, ordering of materials, and budgeting of the project. Any information that cannot be included in the working drawings such as legal responsibilities, method of purchasing, insurance, and bond requirements may also be included in the specifications. However, the technical information sections covering the methods and materials required represent the major content of the specifications. This information legally supports and reinforces the information contained in the sets of working drawings. Specifications can be written several different ways: by describing each item by manufacturer's catalog number, by describing the minimum acceptable properties of each item, and/or by describing the required end result regardless of the product used or construction method employed.

USES

Specifications are used by contractors to make accurate construction estimates and to ensure that all contractors are bidding from the same data. Specifications help ensure that the building will be constructed according to the standards that the building laws require. Specifications are used frequently by banks and federal agencies in appraising the market value of a building.

ORGANIZATION

The type of information contained in the specifications includes the scope of the work, product descriptions, and methods of execution. The scope of the work describes the amount of work to be completed in each category of construction. Product descriptions detail the exact dimensional characteristics of every component and material that will become part of the finished structure. The methods of execution describe the approved methods of construction to be used. They may also include restrictions on the use of some equipment or devices, delivery access routes, or processes which may potentially interfere with public safety or convenience during the construction process.

To ensure consistency among specifications and to accelerate the specification writing and estimating process, the American Institute of Architects (AIA) and the Construction Specification Institute (CSI) have approved and recommended the use of the CSI format for construction specifications. This format organizes information under 16 major divisions with subdivisions included under each as shown in Fig.

Division section	Subsection	Detail
1. General requirements		
2. Site work		
3. Concrete	06100 Rough carpentry	06131 Timber trusses
4. Masonry	06130 Heavy timber	06132 Mill-framed structures
5. Metals	06150 Trestles	06133 Pole construction
6. Wood and plastics	06170 Prefab structural wood	
7. Thermal protection	06200 Finish carpentry	
8. Doors and windows	06300 Wood treatment	
9. Finishes	06400 Architectural woodworks	
10. Specialities	06500 Prefab structural plastics	
11. Equipment	06600 Plastics fabrications	
12. Furnishings		
13. Special construction		
14. Conveying systems		
15. Mechanical		
16. Electrical		

Fig. 72-1 Construction Specification Institute specification listings.

72-1. This organization isolates the information needed by each specialized subcontractor under one heading. This makes estimating by category extremely efficient.

INFORMATION SOURCES

Information to be included in sets of specifications (and schedules) can be found in manufacturers' literature, in trade association publications or through commercial sources such as Sweets Catalog files (Fig. 72-2). Sweets files are a compilation of information on a wide variety of architectural products from acoustical materials to wood products. Most major manufacturers are represented in these files, which are organized according to the Uniform Construction Index.

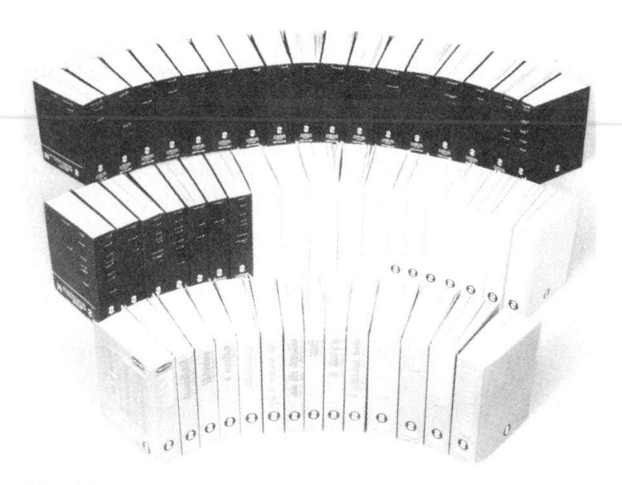

Fig. 72-2 Sweets Catalog file.

Exercises

1. **Make a specifications list for a home of your own design.**
2. **Make a specifications list for Fig. 70-6.**
3. **Obtain a specifications list from a set of plans.**
4. **Define these terms: *specifications*, *guarantee*, *documents*, *contractor*, *Sweets files*, *CSI*, *AIA*.**

SECTION 20

Building Codes

Building codes are collections of laws established to ensure that minimum building standards are achieved. These laws are enacted to safeguard life, health, property, and the public welfare. This is done through regulation and control of the design, construction, materials, location, use, and occupancy of structures within a municipality. Building codes also include information relating to building permits, fees, inspection, zoning, drawings, and legal documents required for approval.

Each municipality makes its own building-code requirements. Building codes are necessary to ensure that substandard, unsafe, and unattractive buildings are not built in the area. Building codes also help to regulate the kinds of structures that can be built in a specific area, by zoning. Zones are classified as residential, commercial, and industrial. Size restrictions are a vital part of every code.

Building-code information is presented in printed material, charts, sectional drawings, specifications, and detail drawings (Fig. 73-1). Building codes also contain regulations pertaining to building permits, fees, inspection requirements, drawings, property location, zoning, and general legal implications connected with building.

UNIT 73

CODE STANDARDS

Codes have been established for very good reasons. The enforcement of codes has lessened the loss of life and property from earthquakes, storms, fires, and floods. Codes also assure that buildings will be structurally sound, safe, secure, and free from preventable maintenance problems. Building codes are presented through printed materials, charts, detailed drawings, and specifications lists.

TYPES OF CODES

Some codes are prepared using end performance (finished product) criteria. Others use specific material and construction specifications. Performance-oriented codes do not limit or specify the use of most construction materials or methods. Performance codes establish only safety and performance requirements for the finished building. Most building codes are of the specification type because they are easier to control and provide a measure for compliance. Specification-type codes include very specific requirements for materials, location, and methods of construction. Options are included in many codes for some materials. However, material substitutions for items specified in this type of code must be equal to or better than the item required by the code.

Regardless of the type, most codes restrict materials, processes and locations that *cannot* be used, because of potential structural, environmental, or safety problems. Figure 73-1 illustrates some of the most common items controlled by building codes. Figure 73-2 shows an example of a few of the typical kinds of residential code requirements.

CODE CATEGORIES

Codes are generally divided by occupancy type, structural stability minimums, safety require-

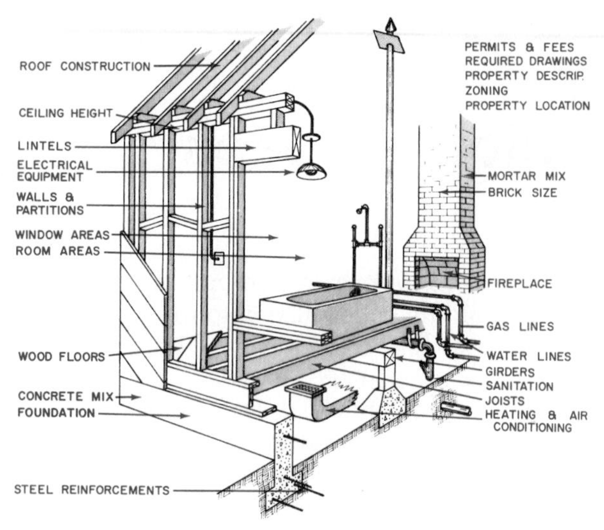

Fig. 73-1 Common items controlled by building codes.

ments and zoning restrictions. The types of occupancy defined in most codes are residential, commercial, industrial, school, and hospital. Other special occupancy categories often requiring code restrictions include garages, retail stores, hotels, and restaurants. For simplicity, many codes combine several of these categories if the same requirements apply for each. Each code category is specifically defined in the code, and then a unique set of code requirements is directed to each. These include such requirements and/or restrictions as: site size, number of exits, structure area, structure height, usage, parking facilities, lighting, ventilation, sanitation facilities, and setback distances.

UNIFORMITY

Building codes often differ among communities due to geographical, political, or social factors. Because of these inconsistencies, designers cannot assume a building planned for one location can be built in another. Codes, to be effective, must be updated frequently. If a code contains outdated information, the use of new construction methods and materials may be unintentionally restricted. For example, some codes still carry glass size restrictions because at one time large glass areas were unsafe. With new developments in glass technology this restriction is often unnecessary and may inhibit freedom in design. Likewise new developments in methods and materials never before used may be safe, but, if not covered in the code, cannot be used. In these cases the code should be updated. If the designer or builder will not wait for a code updating, which often takes a long time, variance to the code can be requested from the same authority that approves code changes.

Stairs for Single Family Residence		
A.	Minimum width	2′-6″
B.	Maximum stair rise	8″
C.	Minimum stair run	9″
D.	Minimum headroom	6′-6″
E.	Handrail height	30″ to 34″
Zoning for Single Family Residence		
A.	Minimum area of building site	10,000 sq. ft.
B.	Minimum width of site	80 ft.
C.	Minimum depth of site	100 ft.
D.	Maximum lot coverage by structure	35%
E.	Front yard setback	25 ft.
F.	Rear yard setback	15 ft.
G.	Side yard setback	8 ft.
H.	Building height	35 ft.
Doors for Single Family Residence		
A.	Minimum front door	3′-0″ × 6′-8″
B.	Minimum service door	2′-6″ × 6′-8″
C.	Minimum interior door	2′-6″ × 6′-8″
D.	Minimum bathroom door	2′-2″ × 6′-8″
E.	Exterior doors swing inside	
F.	Doors do not swing into halls	
G.	Door between connected garage and living area must be min 1½″ solid wood core faced with sheetmetal on garage side.	
Exterior Plastered Wall		
A.	Studs 16″ on center (oc)	
B.	Sheathing or #18 wire backing 6″ on center	
C.	Waterproof paper, horizontal lap 2″, vertical lap 6″	
D.	Metal reinforcement furred out ¼″ and nail with galvanized nails 6″ oc vertically and 16″ oc horizontally. Lap one full mesh at all joints. Vertical joints to be on sheathing or on studs.	
E.	First coat plater is ½″ thick—1 part cement to 4 parts sand	
F.	Second coat plaster is ¼″ thick—1 part cement to 5 parts sand	
G.	Finish coat is ⅛″ thick—1 part cement to 3 parts sand	

Fig. 73-2 Typical residential code requirements.

MODEL CODES

Rather than develop their own codes, many municipalities adopt the codes of other communities. However, to provide local authorities with more consistent source code material, several material and building organizations have prepared *model codes.* These codes are not intended to be adopted intact. They are designed to be used as a base from which local codes can be developed. Figure 73-3 shows a listing of model codes and the sponsoring organization. None of these codes are legal until passed into law by a municipality.

SAFETY CODES

In addition to the safety prevention implications of good structural, mechanical, and electrical design, codes also contain sections dealing specifically with personal and public safety. These cover such areas as electrical hazards, swimming pool enclosures, scaffolds, elevators, exit number and sizes, air and water pollution, health and disease prevention, and fire prevention and control. Fire codes define treatment of materials, building material size, sprinkler systems, escape routes, site security, and functioning of alarm systems.

Model code	**Sponsor**
BASIC BUILDING CODE	Building Officials Conference of America, Inc. 1313 East 60th Street Chicago, Illinois 60637
NATIONAL BUILDING CODE	American Insurance Association 85 John Street New York, New York 35203
NATIONAL PLUMBING CODE	American Standards Association, Inc. 10 East 40th Street New York, New York 10016
SOUTHERN STANDARD BUILDING CODE	Southern Building Code Congress Brown-Marx Building Birmingham, Alabama 35203
NATIONAL ELECTRIC CODE	National Fire Protection Association 60 Batterymarch Street Boston 10, Massachusetts
UNIFORM PLUMBING CODE	Western Plumbing Officials Association 520 Mission Road South Pasadena, California
UNIFORM BUILDING CODE—UBC	International Conference of Building Officials 5360 South Workman Road Whittier, CA 90601

Fig. 73-3 Model building codes and sponsors.

Building codes are extremely strict in stipulating the location, traffic patterns, and structural quality of buildings designed for public occupancy. Rigid code controls are placed on facilities such as factories and garages which may contain flammable substances or which may emit pollutants into the atmosphere. Structures in this risk category must also adhere to Environmental Protection Agency (EPA) standards in addition to local codes.

ZONING ORDINANCES

The use of specific geographical areas of a community is controlled through zoning ordinances. These laws specify the type of occupancy, population density, land use, and building type allowed in each zone of a community. Most codes divide municipalities into such zones as residential, commercial, or industrial. But often these are broken down further into such categories as single-family dwellings, multiple-family dwellings, and so forth.

Zoning ordinances may also include style and design requirements or restrictions. These are intended to maintain a degree of architectural consistency with a given area. Ordinances of this type may restrict or allow only specific styles, periods, materials, sizes, colors, and landscaping.

BUILDING PERMITS

Before a public structure or dwelling can be built, a building permit must be obtained from the local building department. Once the working drawings for a project are complete, a municipal building inspector carefully checks each area of the design for compliance with all existing codes and ordinances. Only after all aspects of the project are determined to adhere to the provisions of the code, as shown on the drawings and specifications, will a building permit be issued. Therefore much drawing revision work can be avoided by carefully checking all local and regional code requirements *before* finalizing the design and especially before beginning the preparation of working drawings for any building.

In addition to local building departments, the administration of local codes may be coregulated and/or controlled by other local or governmental agencies such as city planning commissions, air pollution control districts, fire departments, public health departments, water pollution control boards, and perhaps even art and design commissions or historical preservation societies. If federal building is involved, the department of Housing and Urban Development (HUD), Department of Health, Education, and Welfare (HEW), Federal Housing Authority (FHA), or other agencies may also be involved in the approval process.

STRUCTURAL STABILITY CODES

Structural codes deal with the loading capacity of materials specified and the structural integrity of the type of construction proposed. They involve detailed regulations on excavations, foundations, floor, roof, stairs, and bearing-wall construction. Specific coverage of loads and load-related factors in building codes is covered in Unit 74 and Section 24.

Exercises

1. **From study of your building code, what other parts of construction can be added to the list of building requirements in this unit?**
2. **Does your building code permit the use of prefabricated materials?**
3. **Obtain and read the residential building code for your community.**
4. **Compile a list of the major areas of construction covered by your local building code.**
5. **What are the zoning laws for a residential structure in your community?**

6. **Determine any code restrictions in building a structure of your design in your community.**
7. **Define these architectural terms: *building code, building permit, cutaway view, legal description, zoning, FHA, model code, EPA, HUD, public ordinance, building permit.***

UNIT 74

STRUCTURAL CODE REQUIREMENTS

The weight of all the materials used in the construction of a building, including all permanent structures and fixtures, constitutes the *dead load* of a building. All movable items, such as the occupants and furniture, are the *live load.* The total weight of the live load plus that of the dead load is called the *building load.*

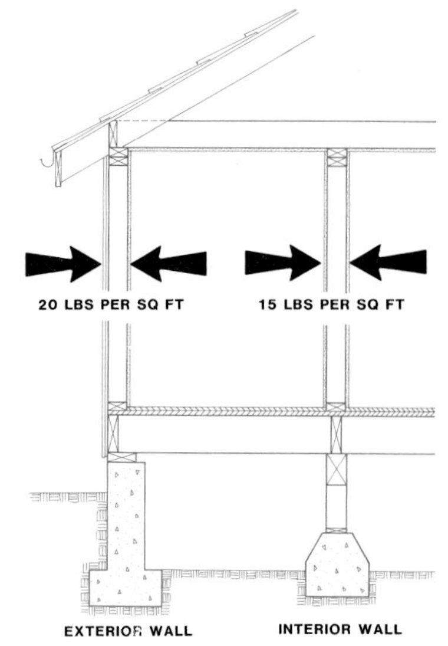

Fig. 74-1 The horizontal load acting on exterior walls is greater than the load acting on interior walls.

MAXIMUM ALLOWANCES

The maximum amount of load permissible for each kind of structure is always listed in the building code. The size of the various structural members to support the various loads is also included in the code.

When material-size regulations are compiled for the building codes, they are computed on the basis of maximum allowable loads. Engineers who are drawing up the code determine the correct size of construction members for carrying a maximum load. A safety factor is then added to the size of the materials to eliminate any possibility of building failure.

Structural sizes required by building codes not only provide for the support of all weight in a vertical direction but also allow for all possible horizontal loads, such as come from winds and earthquakes. Figure 74-1 shows horizontal loads on interior and exterior walls. Notice the difference between the loads exerted on the exterior walls compared to the loads on the inside walls. In addition to the support provided by the bearing wall support members, the wall coverings are also a factor in structural design. For example, Fig. 74-2 shows the ability of different types of sheathing to resist horizontal loads. The differences shown here illustrate why minimum sizes and types of sheathing are specified in most building codes.

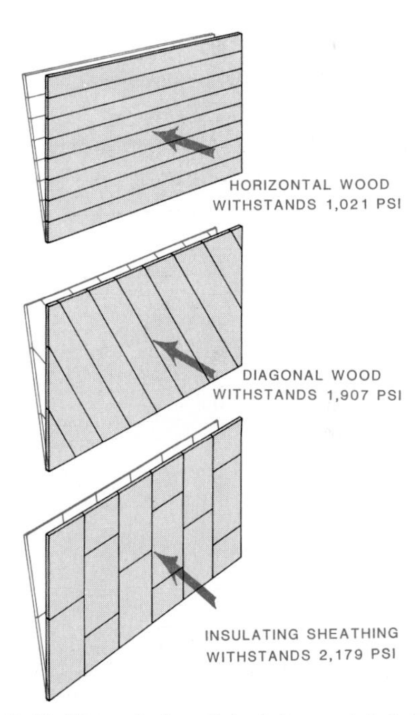

Fig. 74-2 Ability of sheathing to resist horizontal loads in pressure (pounds per square inch).

LIVE LOADS

Live loads include the weight of any movable object on the floors, roofs, or ceilings. Live loads acting on floors include persons and furniture. Live loads acting on roofs include wind and snow loads. Live loads acting on walls are wind loads.

Lateral loads from earthquakes are also considered.

DEAD LOADS

A building must be designed to support its own weight (dead load). Building codes specify the size and type of materials that are used in foundations to support maximum live loads. The size and spacing and type of materials used in walls that support the roof load is also specified. Loads become greater as the distance from the footing diminishes. For example, the load on the attic shown in Fig. 74-3 is only 25 lb/ft^2 or 119.7 pascals (Pa). (Pascal = Pound-force/ft^2 × 4.788). The load on the first floor is 45 lb/ft^2 (215.5 Pa). However, the typical floor load for an average room will vary from 30 to 40 lb/ft^2 (143.6 to 191.5 Pa).

Roof loads are comparatively light, but vary according to the pitch of the roof. Flat roofs offer more resistance to loads than do pitch roofs. Low-pitched roofs, those below 3/12 pitch, must often be designed to support wind and snow loads of 20 lb/ft^2. Note that the wind load will increase as the pitch increases. The snow load will increase as pitch decreases. High-pitched roofs, those over 3/12 pitch, need be designed to support live loads of 15 lb/ft^2. Since these loads, especially snow loads, vary greatly from one part of the country to another, local building codes establish the amount of load a roof must be designed to support at any given pitch (Fig. 74-4).

To eliminate excessive structural calculating, many building codes include engineering data tables for commonly used materials and components. Information in these charts includes minimum clearances, sizes, and composition of foundation components as shown in Fig. 74-5. Maximum joist, girder, and lintel spans are also frequently summarized on charts for different conditions as shown in Figs. 74-6, 74-7 and 74-8. In addition to stipulating spans and the spacing between rafters, building codes often chart ceiling joist spans as shown in Fig. 74-9.

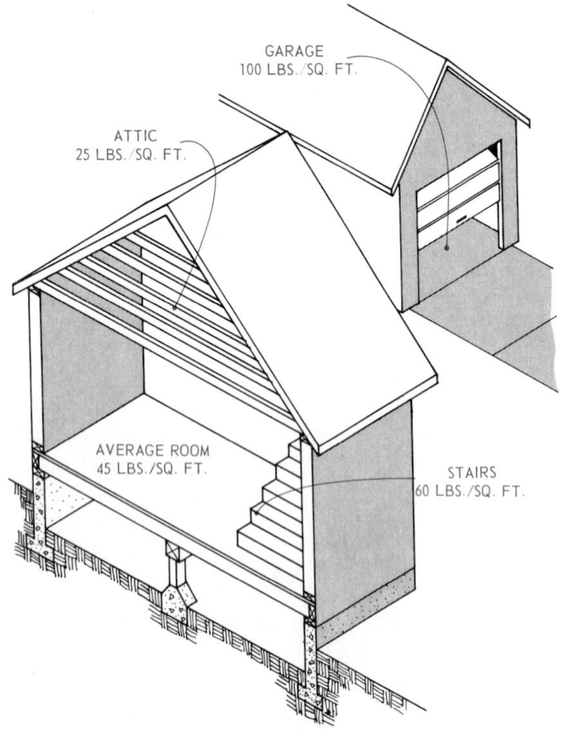

Fig. 74-3 Loads increase closer to the foundation.

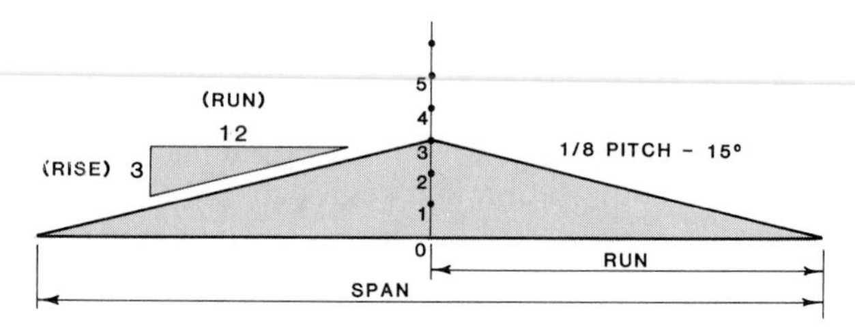

3:12 PITCH OR UNDER - 20 LBS/SQ FT

3:12 PITCH OR OVER - 15 LBS/SQ FT

Fig. 74-4 Roofs must be designed to support load limits established by local building codes.

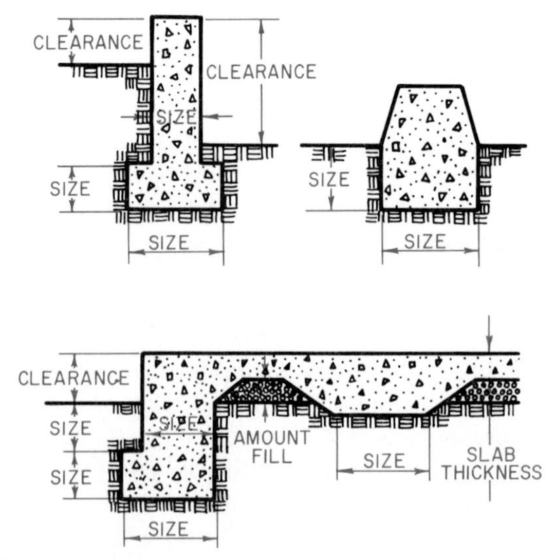

Fig. 74-5 Size, clearance, and composition of foundation determined by building codes.

TYPICAL SPANS FOR WOOD GIRDERS

Size	1-Story	2-Story
4″ × 6″	5′-0″	4′-0″
6″ × 6″	6′-0″	5′-0″
4″ × 8″	6′-6″	5′-6″
6″ × 8″	8′-0″	7′-0″
4″ × 10″	8′-0″	7′-0″
6″ × 10″	9′-0″	7′-0″

Fig. 74-6 Girder size, span, and spacing described in building codes.

LINTEL SPANS

Supporting roof & ceiling only		Supporting floor roof & ceiling only	
SIZE	SPAN	SIZE	SPAN
4 × 4	3′-0″	4 × 4	3′-6″
4 × 6	4′-0″	4 × 6	5′-0″
4 × 8	6′-0″	4 × 8	5′-6″
4 × 10	8′-0″	4 × 10	7′-0″
4 × 12	9′-0″	4 × 12	8′-0″
4 × 14	10′-0″	4 × 14	9′-0″
4 × 16	12′-0″	4 × 16	10′-0″

Fig. 74-7 Size and span of lintels outlined in building codes.

JOIST SPANS

Joist size	Joist spacing	Joist span
2 × 4	12″	10′-0″
	16″	9′-0″
	24″	7′-0″
2 × 6	12″	13′-0″
	16″	12′-0″
	24″	10′-6″
2 × 10	12″	16′-0″
	16″	15′-0″
	24″	12′-0″
2 × 12	12″	20′-0″
	16″	18′-0″
	24″	15′-0″
2 × 14	12″	23′-0″
	16″	21′-0″
	24″	17′-0″

Fig. 74-8 Size and spacing of joists outlined in building codes.

RAFTER SPANS

Size	Rafter spacing	Span-roof pitch less than 4:12	Span-roof pitch more than 4:12
2 × 4	12″	9′-0″	10′-0″
	16″	8′-0″	8′-6″
	24″	6′-6″	7′-0″
	32″	5′-6″	6′-0″
2 × 6	12″	14′-0″	16′-0″
	16″	12′-6″	13′-6″
	24″	10′-6″	11′-0″
	32″	9′-0″	9′-6″
2 × 8	12″	19′-0″	21′-6″
	16″	17′-0″	18′-6″
	24″	13′-6″	15′-0″
	32″	11′-6″	13′-0″
2 × 10	12″	23′-0″	25′-0″
	16″	21′-0″	22′-6″
	24″	17′-6″	19′-0″
	32″	15′-0″	16′-6″

CEILING JOIST SPANS

Size	Joist spacing	Span
2 × 4	12″	10′-0″
	16″	9′-0″
	24″	8′-0″
2 × 6	12″	16′-0″
	16″	14′-6″
	24″	12′-6″
2 × 8	12″	21′-6″
	16″	19′-6″
	24″	17′-0″

Fig. 74-9 Roof types, pitches, size of materials, and limits to spacing of rafters and joists in building codes.

By using this preapproved structural design data, the designer is assured of approval provided that the materials, sizes, and spacing are designed in accordance with the terms of the code. Where no such tables exist or where any of the variables change, the size, type, and spacing of members must be calculated and approved by the building inspector prior to construction. See Section 23 for information about structural calculations.

Exercises

1. **List 20 items used in the construction of a home that are part of the dead load.**
2. **List 20 objects that are part of the live load.**
3. **Why is load specified in pounds per square foot?**
4. **Why is a low-pitched roof constructed to support more weight per square foot than a roof of steeper pitch?**
5. **An exterior wall is 8′ high and 30′ long and is sheathed with diagonal sheathing. How much vertical load will the entire wall support?**

6. **Use your local building code to determine the minimum sizes and spacing of structural members required for a building of your own design.**
7. **Define these terms: *maximum allowance, live loads, dead loads, building load, wind load, snow load.***

SECTION 21

Checking Procedures

The appropriateness of the size and layout of a building should be completely checked by the architect, designer, builder, and occupant before a completed architectural plan is used for actual construction purposes. The technical authenticity of the drawing should first be checked by the drafter and/or by a checker. He or she should be certain that all dimensions and symbols are correct and that each detail drawing agrees with the basic plan.

People without technical training often find it difficult to interpret engineering and architectural drawings adequately. For this reason, it is sometimes advisable to construct an architectural model that represents the appearance of the finished building.

Another concept that is sometimes difficult for untrained people to grasp is the relationship of the size of rooms to the furniture and equipment that will actually be placed in these rooms. Checking the adequacy of room size can be done by placing templates of furniture, equipment, and even people on the drawings of the rooms. Checking methods include the use of three-dimensional, solid, structural, and cutaway models; templates; and computer-aided drafting (CAD) solid and wire-frame modeling.

UNIT 75

ARCHITECTURAL MODELS

Architectural models have been used for centuries to study the structural form and final appearance of buildings (Fig. 75-1). The use of an architectural model is the only way to actually see the finished design in three dimensions. A model is also the only representation of a building that can be viewed from any angle.

FUNCTION

Models may be used in planning cities or parts of city redevelopments. Models are often used to check the design of large commercial buildings. The appropriateness of size and layout can be seen better on a model than through any other device. The relationship to other objects, such as people, cars, trees, and other buildings, becomes more apparent when the structure is viewed in three-dimensional form. For these reasons, it is advisable to include within the overall model as many scale models of furniture, equipment, cars, and people as possible. Figure 75-2 shows a model with cutaway walls, revealing the relationship among furnishings, room size, and

Fig. 75-1 Architectural models have been used for centuries. *(Celotex Corp.)*

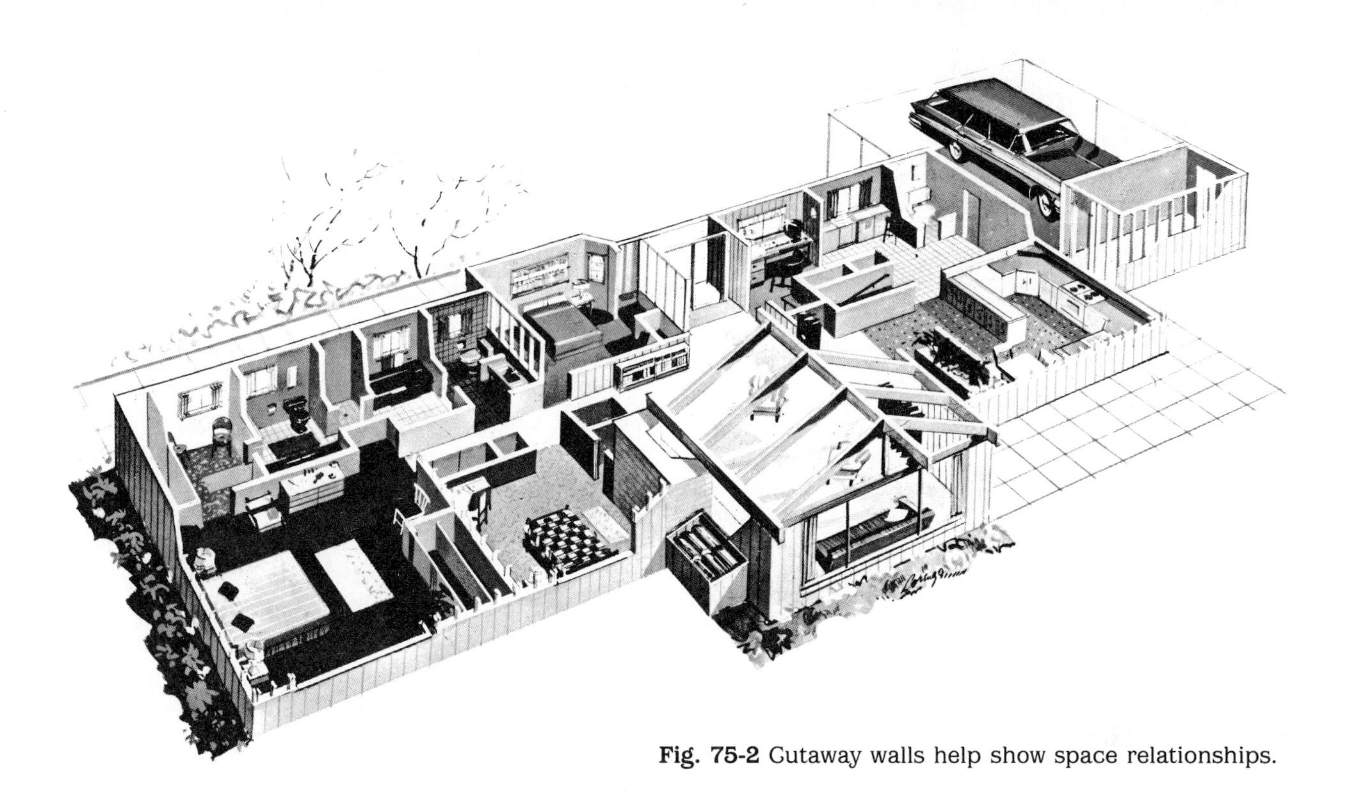

Fig. 75-2 Cutaway walls help show space relationships.

layout. It is sometimes advisable to prepare basic outline blocks representing adjacent buildings in order to compare their relative size and position with those of the building being designed. The model can also be used to design effectively or to check the color scheme of the entire house or of individual rooms since the entire house can be seen at a glance.

Some models are prepared to check only the structural qualities of the building. When a structural model of this type is prepared, soft wood strips are used to represent sills, studs, rafters, and beams. These structural members are prepared to the exact scale of the model, and the soft wood members are attached together with modelmakers' glue or with small pins, to approximate the methods used in nailing the full-sized house. Precut model lumber is often used.

COMPUTER MODELING

Three-dimensional architectural models can also be developed through the use of computer-aided-drafting systems. Employing a three-dimensional CAD system can eliminate the need for a physically constructed model during the design process. Since a computer-generated three-dimensional model can easily be changed, the computer model can be constantly updated to reflect design changes. Many 3D computer models are used in the design process to check the structural stability, solar relationship of structures, and pictorial appearance of the design from many different station points. The elements of design are often changed after the model is studied.

By selecting and inputing any combination of station-point location, vanishing-point distances and horizon-line levels, the computer can be used to calculate and display the appearance of the structure. To recreate the appearance of the structure from other angles, the compass position of the station point and/or elevation of the horizon line is simply entered into the computer and a new pictorial image will emerge on the computer's display screen. Three-dimensional computer programs allow the viewing of the interior of the structure from any point toward any direction by identifying the station point and direction of view on the floor plan displayed on the computer monitor.

Computer three-dimensional models are divided into two types: wire-frame (Fig. 75-3A) and solid models (Fig. 75-3B). Wire-frame models show the major outline of buildings so the viewer can "see through" the object without forefront solid objects, such as walls and roofs, obstruct-

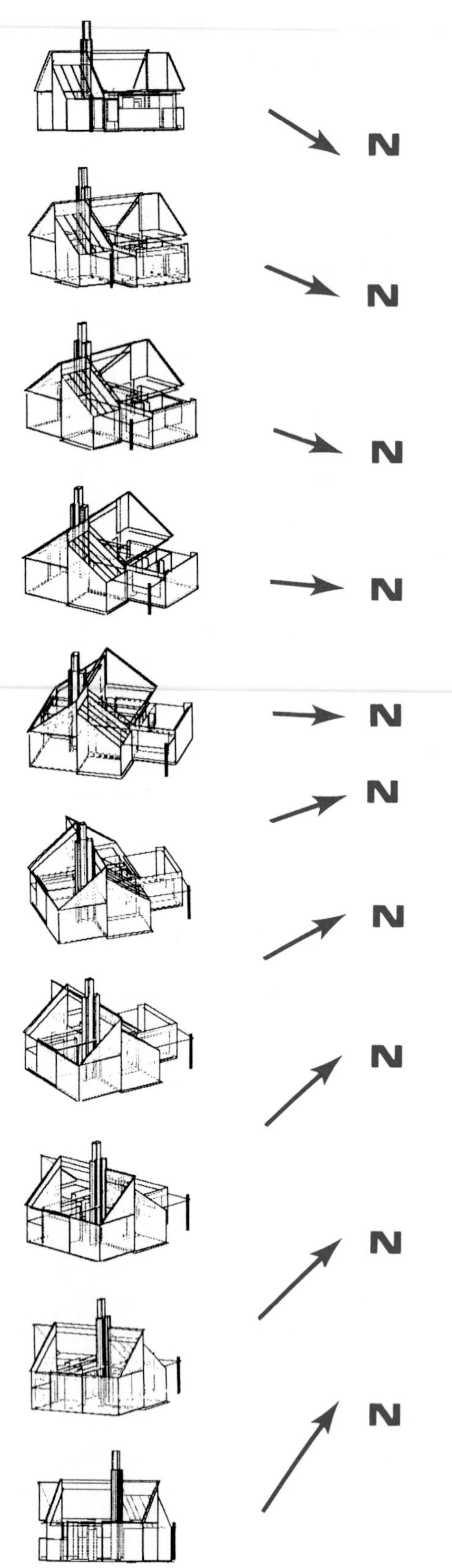

Fig. 75-3A Wire-frame computer model. *(Scribe International)*

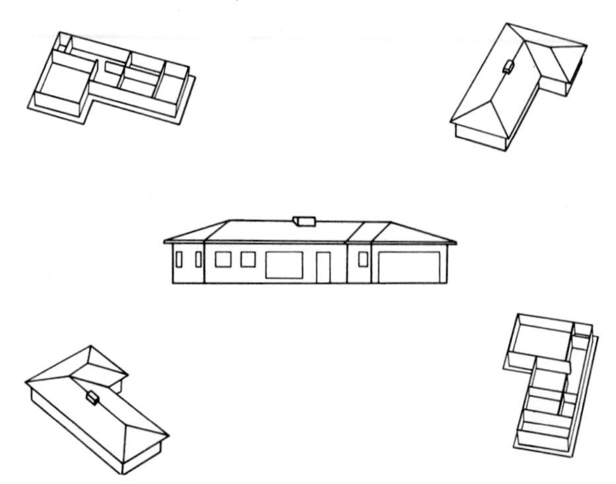
Fig. 75-3B Solid computer model.

ing the view. However, if too many building lines are used, the shape of the object can be difficult to interpret. Some computer programs have the capacity to show lines behind the front object as dotted (hidden) lines. This greatly helps to show the building more realistically and yet provides depth details.

Three-dimensional solid computer models show only the part of the building visible from the station point selected. Since no lines, solid or dotted, show behind solid areas, another station point must be selected if hidden areas are to be shown.

SOLID MATERIAL MODELS

At times it may be desirable to check only the proportions or relationship of one building to another. In these cases, a solid physical model as shown in Fig. 75-4 is used. Solid models of this type lack detail and can be constructed in a relatively short period of time. Styrofoam, balsa wood, acrylics, and sometimes bars of soap are used to create solid architectural models.

LAND FORM MODELS

Models are also frequently used to plan and check the land form of a site. Figure 75-5 shows the use of contour-interval modeling to represent the shape and slope of a site. In this type of model, the thickness of each layer is equal (in scale) to the contour interval. The shape of each layer is the same shape as the contour lines on a survey drawing.

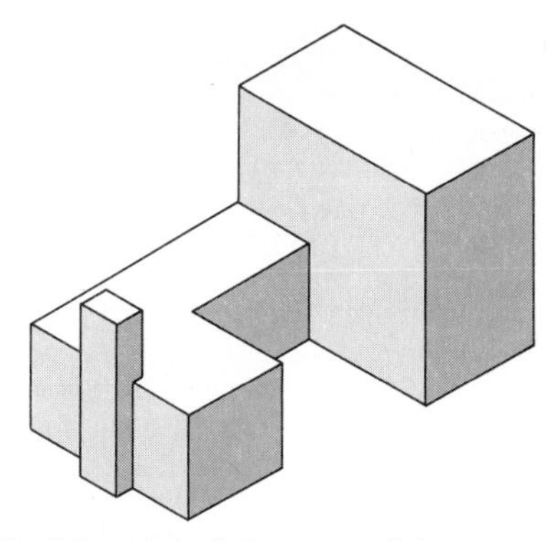

Fig. 75-4 Abbreviated form model.

Fig. 75-5 Site model with contour-interval layers.

MODEL CONSTRUCTION

Since the materials used in full-size buildings are too large for model construction, they must be simulated by other materials and other products. For example, coarse sandpaper may be used to simulate a built-up gravel roof. Sponges may be used to simulate trees, and green flocking may be used for grass. In addition to such substitute materials, a great variety of commercially prepared model materials is now available. Figure 75-6 shows some of the special materials used to make various parts of architectural models.

Methods of constructing models vary greatly, just as methods of constructing full-sized structures vary, according to the building materials used. Nevertheless, the following procedures represent the normal sequence of constructing models, even though some of the techniques may vary with the use of different materials.

1. A floor plan and elevation outline should be prepared to the size to which the model will be built. Small structures such as houses can be built conveniently at the scale ½″ = 1′-0″. Larger structures such as commercial office buildings should be built to the scale ¼″ = 1′-0″ or smaller.

2. The contour of the lot should be developed, as shown in Fig. 75-7.

3. Glue or trace the floor-plan outline on a pad ¼″ (6 mm) thick, and tack or glue this pad to a base, as shown in Fig. 75-8.

4. Cut exterior walls from balsa wood or heavy cardboard. Carefully cut out the openings for windows and doors with a razor blade, as shown in Fig. 75-9. When cutting side walls, cut then long to allow overlapping of the front and back walls.

5. Glue small strips of wood or paper around the windows to represent framing and trim, as shown in Fig. 75-10. Glue clear plastic inside the window openings to represent window glass and cover the joint with trim. Cut out doors and hang them on the door openings, using transparent tape, as shown in Fig. 75-11. Add siding materials to the exterior of the wall panels to simulate the materials that will be used on the house, as shown in Fig. 75-12.

6. Glue the exterior walls to the pad and to each other at the corners (Fig. 75-13).

7. Cut out openings in interior walls for doors, arches, and fireplaces, as shown in Fig. 75-14.

8. Glue interior walls to the partition lines of the floor plan (Fig. 75-15).

9. Add interior fixtures such as built-in cabinets, kitchen equipment, bathroom fixtures. Paint the interior surfaces to determine the color scheme and add furniture if desired.

10. Cut out roof parts, join roof parts together, and brace them so that the roof will lift

Part	Model materials	Methods of construction
WALLS	Soft wood; cardboard	Cut wall to exact dimensions of elevations. Allow for overlapping of joints at corners. Have wall thicknesses to scale.
ROOFS	Thin, stiff cardboard; paint-colored sand; sandpaper; wood pieces	Cut out roof patterns and assembly. For sand or gravel roof, paint with slow-drying enamel the color of roof. Sprinkle on sand. For shingle roof, cut sandpaper or thin wood pieces, and glue on as if laying shingle roof.
BRICK & STONE	Commercially printed paper	Glue paper in place; cut grooves in wood, and paint color of bricks or stones.
WOOD PANELING	Commercially printed paper; 1/32" veneer wood	Glue paper in place; with veneer wood, rule on black lines for strip effect, and glue in place. Mahogany veneer equals redwood.
STUCCO	Plaster of Paris	Mix and dab on with brush.
WINDOWS & DOORS	Preformed plastic; wood strips and clear plastic	Purchase ready-made windows to scale in model store; or frame openings with wood strips and glue in clear plastic for windows or wood panel for door.
FLOORS	Flocked carpet; commercially printed paper; 1/32" veneer wood	Paint area with slow-drying colored enamel, and apply flock, removing excess when dry. With paper, glue in place. With veneer, rule on black lines for strip effect, and glue in place.
FURNITURE	Cardboard, nails, flock, wood, clay	Fashion furniture to scale. Paint and flock to give effect of material.
SITE AREAS	Wood slab, wire screen, papier-mâché	Build up hilly areas with sticks and wire. Place papier-mâché over wire.
GRASS	Green enamel paint and flock	Paint grass area. Apply flock, removing excess when dry.
TREES & BUSHES	Sponge; lichen	Grind up sponges and paint different shades of green. Use small pieces for bushes. Glue small pieces to tree twigs for trees. Lichen may be purchased in model stores and used in the same manner as sponges.
AUTOS & PEOPLE	Toys and miniatures	If time permits, carve from soft wood.

Fig. 75-6 Materials used for architectural modeling.

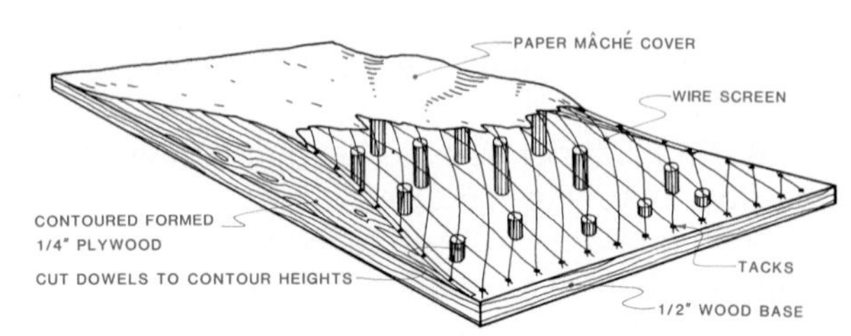

Fig. 75-7 Developing the contour of a site.

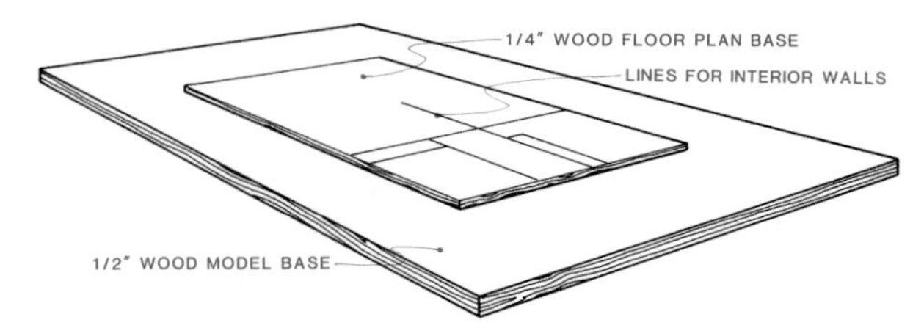

Fig. 75-8 A pad attached to a base.

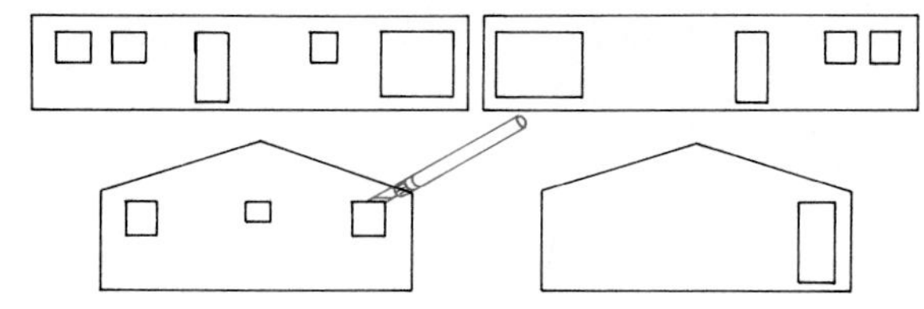

Fig. 75-9 Cutouts of exterior walls.

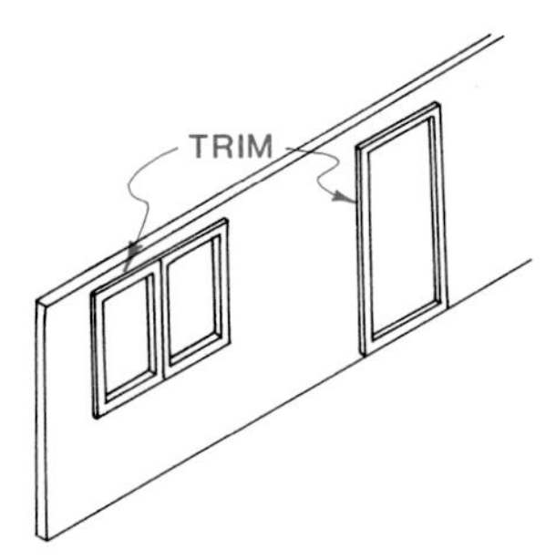

Fig. 75-10 Small strips of wood or paper can represent framing and trim around windows and doors.

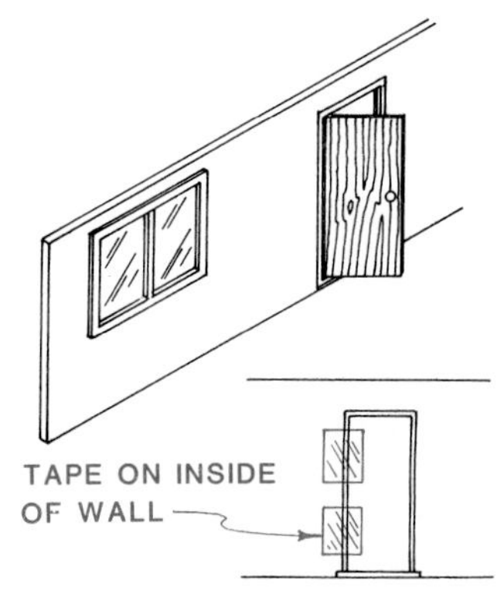

Fig. 75-11 Doors can be hung with transparent tape.

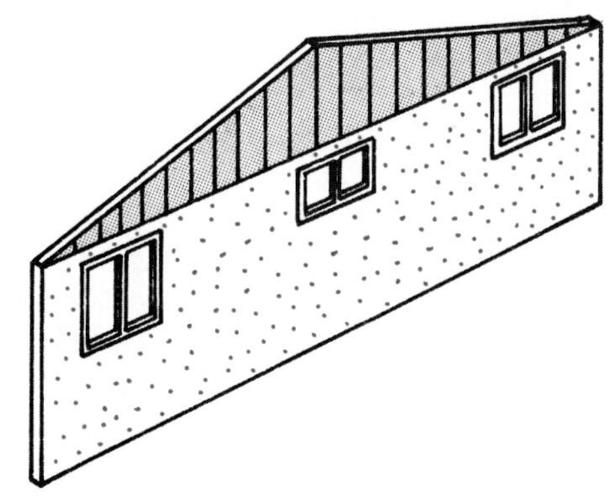

Fig. 75-12 Siding materials are added to the exterior panels.

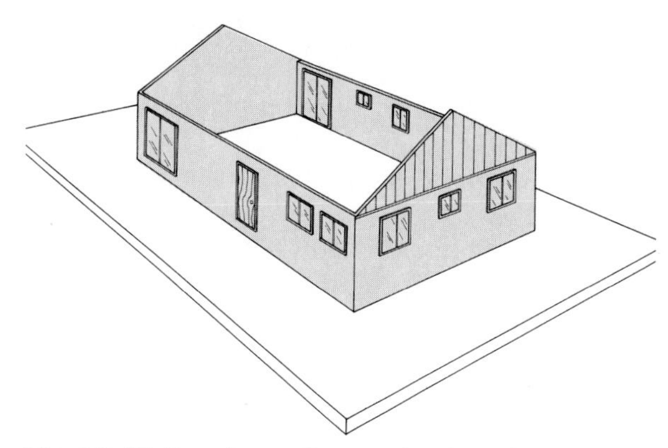

Fig. 75-13 Exterior walls are glued to the pad.

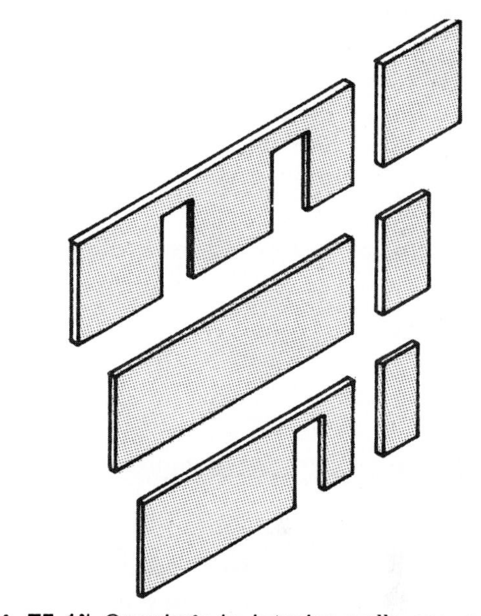

Fig. 75-14 Openings in interior walls are cut out.

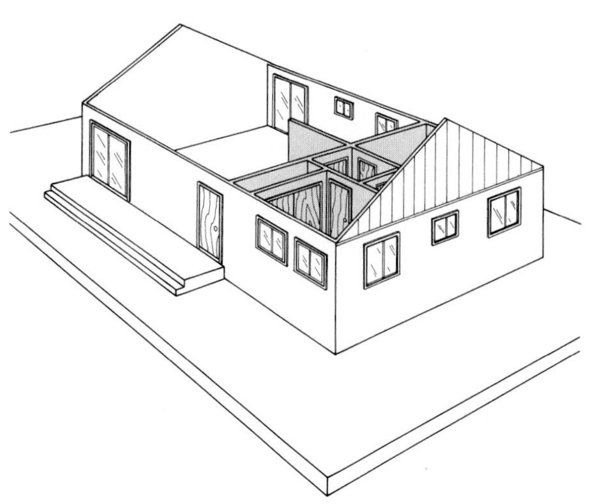

Fig. 75-15 Interior walls are glued to exterior walls and to the pad.

off. Add material to simulate the roof treatment on the exterior of the roof (Fig. 75-16).

11. Add landscaping features such as trees, shrubs, grass, patios, walkways, and outdoor furniture to add to the authentic appearance of the model.

12. Add entourage figures such as people, automobiles, trucks, and animals, where appropriate, to the site to create a comparison of the scale.

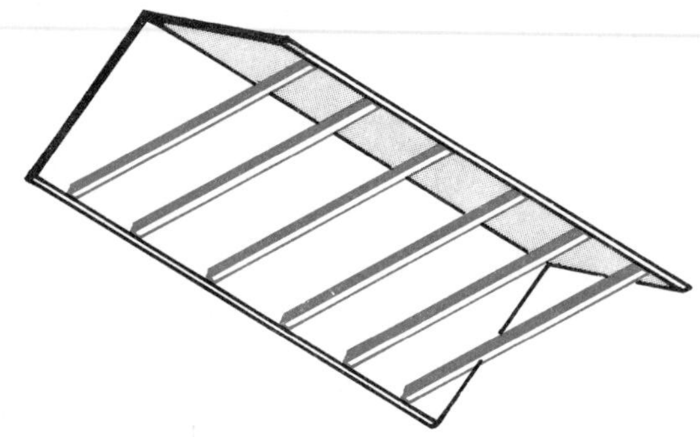

Fig. 75-16 Roof sections are joined and braced.

Exercises

1. **Construct a model similar to the one shown in Fig. 75-15.**

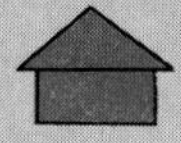

2. **Construct a model of the home you have designed.**
3. **Define these terms: *scale model, structural model, balsa wood, flocking, three-dimensional software.***

UNIT 76

TEMPLATE CHECKING

Room sizes often appear adequate on the floor plan, but when furniture is placed in the rooms, the occupants may find them too small or proportioned incorrectly. After the house is built, it is too expensive to change the sizes of most rooms. Therefore, extreme care should be taken in the planning stage to ensure that the plan as designed will accommodate the furniture, fixtures, and traffic anticipated for each area.

TEMPLATE PREPARATION

One method of determining the adequacy of room sizes and proportions is to prepare templates of each piece of furniture and equipment that will be placed in the room. These templates have the same scale as the floor plan. Placing the templates on the floor plan will show graphically how much floor space is occupied by each piece of furniture. The home planner can determine whether there is sufficient traffic space around the furniture, whether the room must be enlarged, whether the proportions should be changed, or whether, as a last resort, smaller or less furniture should be obtained.

The placement of the templates on the plan in Fig. 76-1 indicates that some rearrangement or adjustment should be made in the furniture placement in the master bedroom, and that the dining room might be inadequate if the table is expanded.

Checking by templates can be significant only if the templates are carefully prepared to the same scale as the floor plan and if the actual furniture dimensions are used in the preparation of the templates. The floor-plan design can be checked much more quickly by the template method than it can by a model. Furthermore, if templates are used while the plan is still in the sketching stage, adjustments can be made easily and rooms rearranged to produce a more desirable plan.

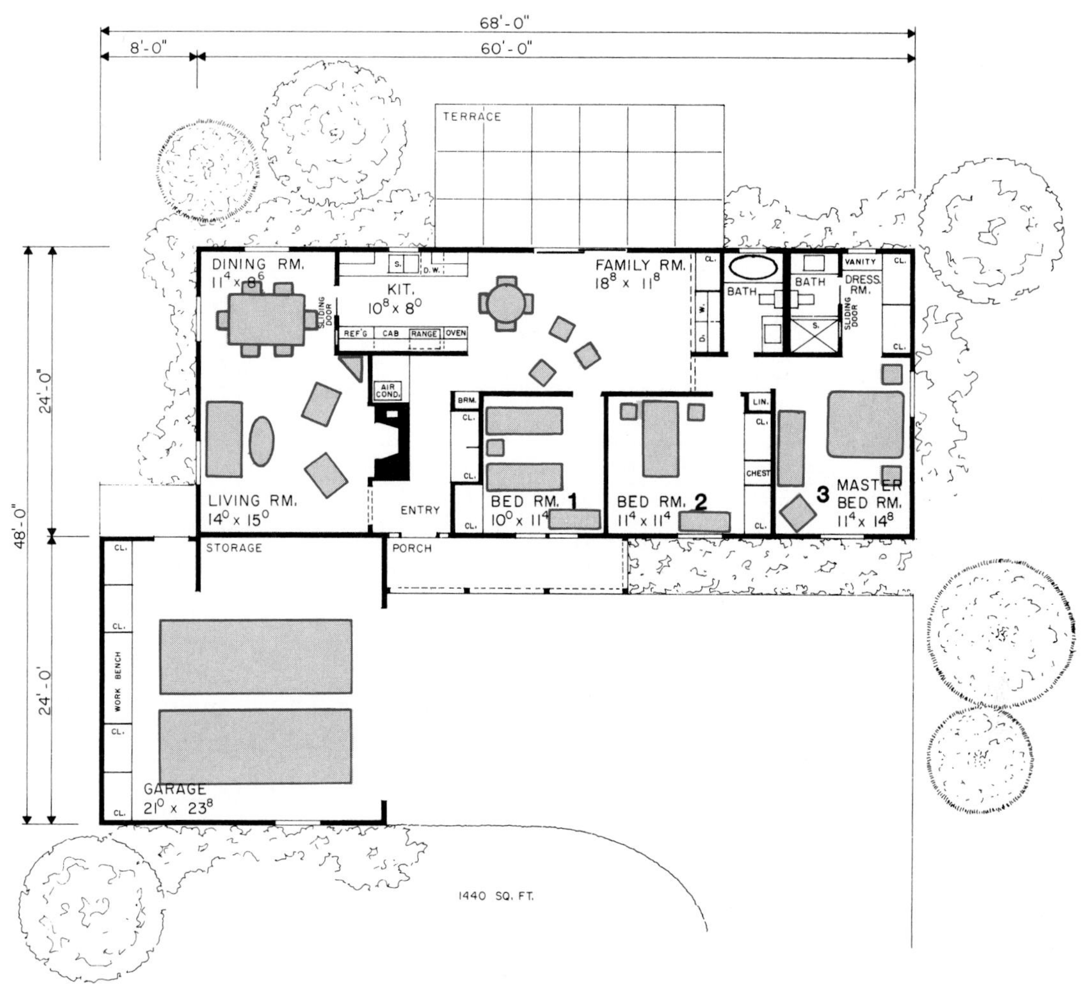

Fig. 76-1 Space requirements can be checked by placing furniture templates on the floor plans. *(Southern California Gas Co.)*

THREE-DIMENSIONAL TEMPLATES

The template method of checking room sizes does not reveal the three-dimensional aspect of space planning as a model does. One difference between the model method and the template method of checking drawings is the preparation of three-dimensional templates. Three-dimensional templates not only have width and length but also height, as shown in Fig. 76-2. This method of checking is essentially the same as placing furniture on a model, except that the walls do not exist and adjustments can be made much more easily. Figure 76-3 shows the height of typical pieces of furniture for use in preparing three-dimensional templates. These dimensions are typical and may vary slightly with different manufacturers.

TRAFFIC PATTERNS

A well-designed structure must provide for efficient and smooth circulation of traffic. Traffic patterns must be controlled, and yet sufficient space must be allowed for adequate passage and flexibility in the traffic pattern. Rooms

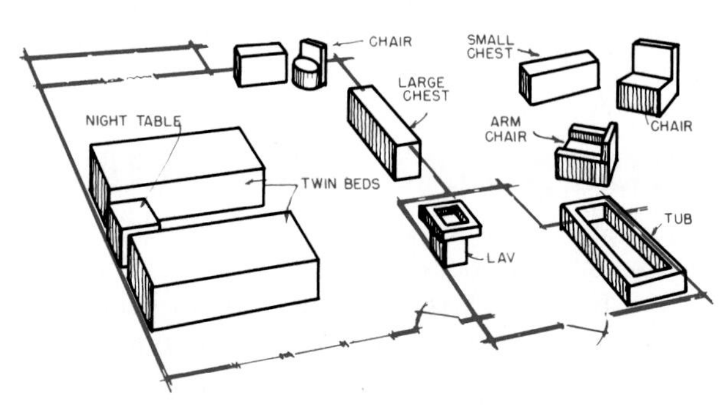

Fig. 76-2 Three-dimensional templates.

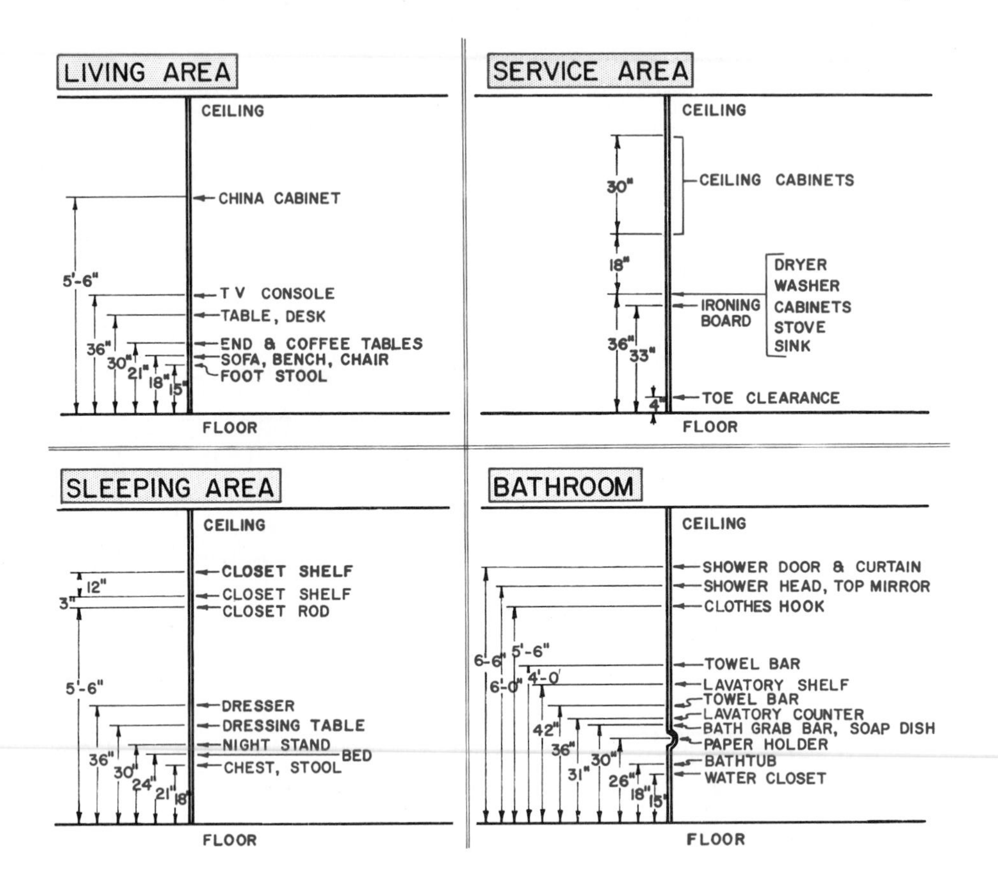

Fig. 76-3 Typical heights of cabinets, furniture, and fixtures.

should not be used as hallways or access areas to other rooms. Any room should be accessible from the front entrance or the service entrance without the necessity of passing through other rooms. The length of halls should be minimized because halls provide only traffic access and do not contribute livable space.

One method of checking the traffic pattern is to use a scale drawing and trace, with a pencil, the movements of your daily routine, as shown in Fig. 76-4. If you prepare a template of a person (an overhead view of yourself with arms outstretched), you will be able to determine the effectiveness of the traffic pattern as it relates to the size of each room and the layout of the entire plan.

TECHNICAL CHECKING

The drafter should always check dimensions, labels, and symbols before the drawing is removed from the board. A more formal check of these factors should be made on a print prepared for checking purposes, called a check print. After the drafter has made this check, another drafter, architect, or checker should also scrutinize the drawing for dimensional accuracy, proper labeling, and correctness of symbols.

One of the most effective methods of checking architectural drawings is the use of a colored pencil, as shown in Fig. 76-5. A checker draws a line through each dimension, label, and symbol as it is checked. Otherwise, the result is rechecking the same dimensions and missing many others. Many architectural offices use a color-coding system to indicate the checker's reaction to the drawing. In one system, a yellow pencil is used for checking items that are correct, a red pencil for checking errors, and a blue pencil for marking recommended changes.

Regardless of the checking method used, one of the most important items to check on architectural drawings is the correctness of dimensions. Dimensioning errors cause more difficulty on the construction job than any other single factor related to architectural drawings. Dimensions must be added to ensure that they total the overall dimension. Interior dimensions must be added to ensure that their total agrees with the exterior dimensions.

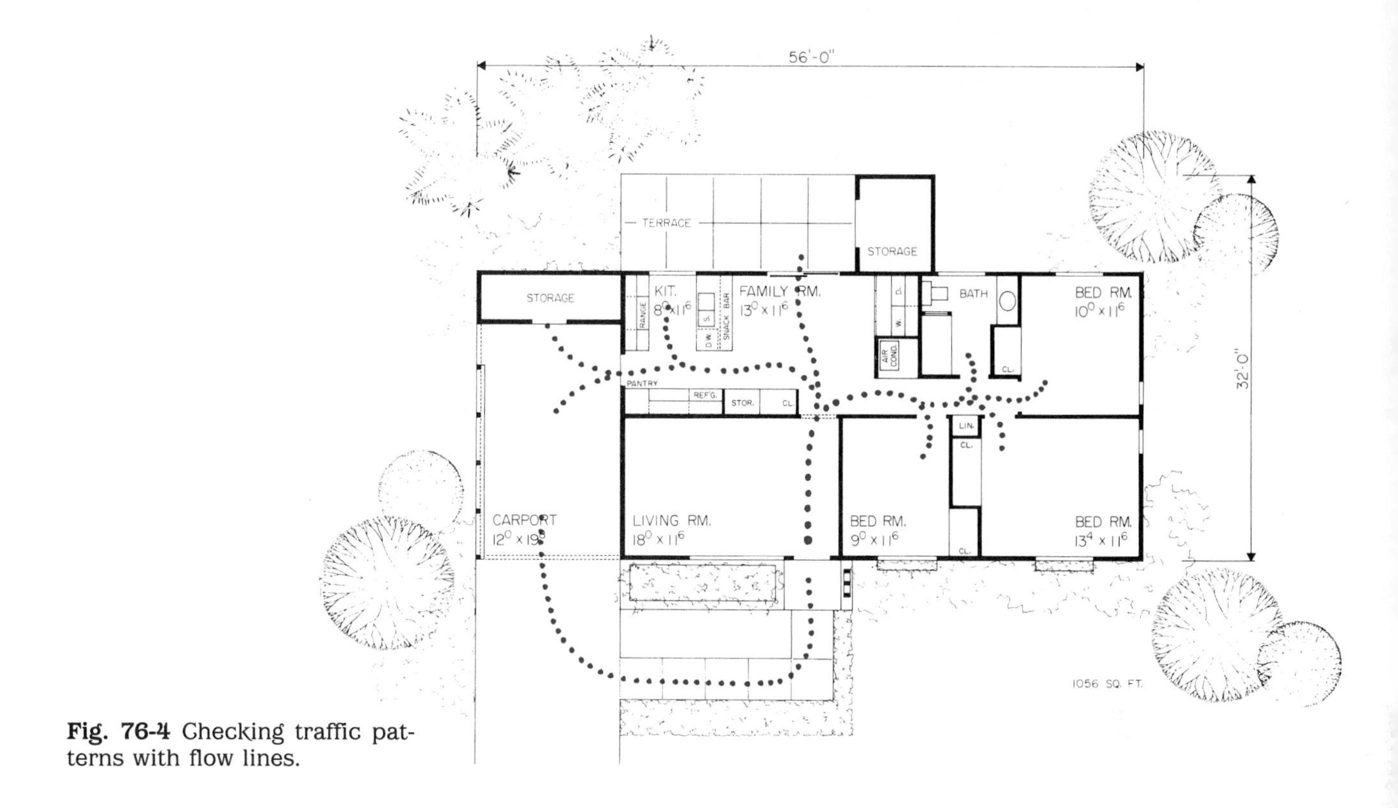

Fig. 76-4 Checking traffic patterns with flow lines.

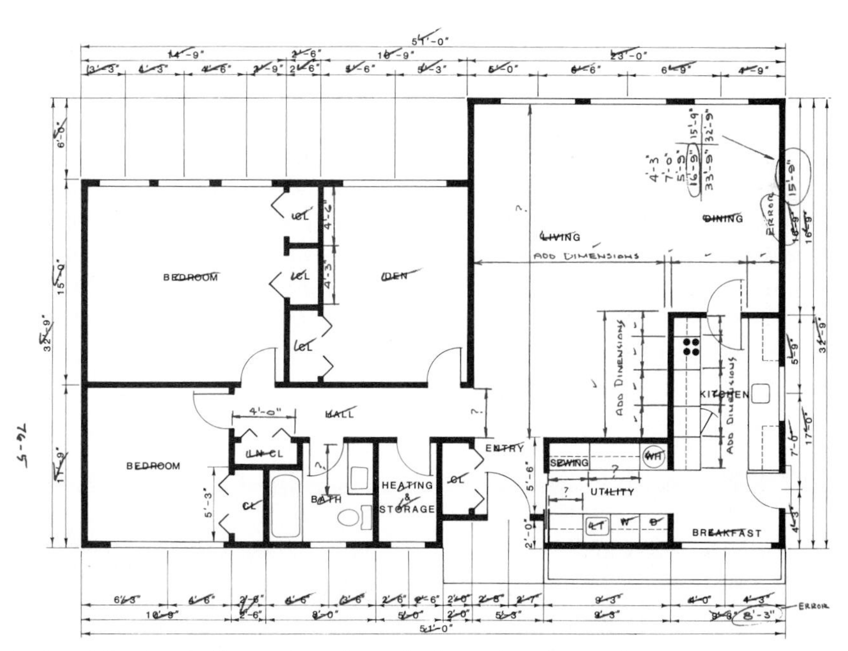

Fig. 76-5 Checking architectural dimensions by using a colored pencil.

Exercises

1. **Prepare a set of furniture templates. Sketch the floor plan of your own home and check its adequacy with these templates.**
2. **Use the templates you have prepared to check the floor plan of your own design.**
3. **Check the traffic pattern of the plan shown in Fig. 69-1. What recommendations would you make to improve the traffic pattern?**
4. **Define these architectural terms: *template, two-dimensional template, three-dimensional template, traffic pattern, checker, check print.***

UNIT 77

ARCHITECTURAL DRAWING CHECKLISTS

The number and type of architectural drawings prepared for any structure depend on the amount of control the designer demands. The more control the designer wants, the more drawings must be prepared. For example, if a designer prepares only a floor plan, the contractor is allowed to specify and build the exterior of the structure. Thus, the amount and type of drawings determine the degree of control the designer achieves. For maximum control, a complete set of architectural plans should include the drawings and schedules outlined in Unit 68.

Checklists, as provided in the teacher's manuals for this text, list items that may appear on each type of architectural drawing. Only by finding each item on each plan and checking for completeness of dimensioning and detail can a drafter be sure nothing was omitted from each drawing. Not all items apply to every design, but they are easily checked off as not applicable (N/A). By using checklists of this type, the drafter is reminded to determine whether the item was omitted by design (N/A) or by error.

Architectural drawings are prepared for a variety of purposes. They show dimensions of a lot, types of siding material, sizes and shapes of the floor plan, types and styles of exteriors, and so forth. Because each drawing must agree with all other drawings in a set, the designer must be sure that each drawing includes all the information for which it was intended. Checklists serve this purpose. Checklists should be used to ensure the completeness of each drawing and construction document.

Exercises

1. **Use the checklist provided in the teacher's manual to check the set of plans shown in Unit 69.**

2. **Use the checklist to check the completeness of a set of drawings of your own design.**

SECTION 22

Cost Analysis and Legal Considerations

Designers would welcome the opportunity to design a structure free from financial limitations. But this condition rarely exists. Budgets are the necessary framework within which designers must create most buildings, ranging from the smallest residence to the largest office building. There is, of course, more flexibility in some budgets than in others, but in every design problem the designer must strive to create architectural plans that will provide optimum facilities within an established budget.

UNIT 78

BUILDING COSTS

Approximately 40 percent of the cost of the average home is for materials. Labor costs account for another 40 percent. The remaining 20 percent is taken up by the price of the lot. As labor, material costs, and land values rise, the cost of homes increases proportionately.

TOTAL COST

Many factors such as site cost, labor costs, and material costs influence the total cost of the house. The location of the site is extremely important. An identical house built on an identical lot can vary several thousand dollars in cost, depending on whether it is located in a city, in a suburb, or in the country. Labor costs also vary greatly from one part of the country to another and from urban to rural areas. Normally, labor costs are lower in rural areas. The third important variable contributing to the difference in housing costs is the cost of materials. Material costs vary greatly, depending upon whether materials native to the region are used for the structure. In some areas, brick is a relatively inexpensive building material. In other parts of the country, a brick home may be one of the most expensive. Climate also has some effect on the cost of building. In moderate climates, many costs can be eliminated by excluding large heating plants and frost-deep foundations. In other climates, air-conditioning and maximum installation is mandatory.

ESTIMATING COSTS

Major-Project Estimating

Major construction projects are usually subject to a bidding system. Architects and contractors estimate how much they would charge to design and/or erect a building. That estimate must be based upon estimates of their costs in designing or constructing the building.

The owner of the proposed building may select an architect through referral or design competition or by reputation and/or past experience.

When the time comes for contractors to bid on public construction, they too find out about projects from advertisements in trade journals, or they may subscribe to Dodge Reports. (General contractors for private construction are invited directly to bid on the project.) Dodge Reports inform contractors when the kinds of jobs they are interested in are first conceived, who the owner is, when an architect is selected, who the architect is, who the general contractors bidding on the job are, and where and when the plans will be available. The Dodge Report also tells a contractor what kind of job it will be, where the building will be constructed, and how much it should cost. Each time a project moves to the next construction stage, another Dodge Report (Fig. 78-1) is issued.

A subcontractor must be able to see the plans and specifications of the building in order to estimate the cost of construction. These plans may be available from the architect in the form of blueprints, or the subcontractor may subscribe to the Dodge/SCAN microfilm system. With this system, bidding documents are reproduced on microfilm and mailed to subscribers. The subcontractor receiving the microfilm uses a patented SCAN viewing table (Fig. 78-2) to project the plans back to original size and accurate scale. From the accurate original sizes, contractors can make their estimates for the construction project. The SCAN system saves time and expense of travel and cuts costs of reproducing and distributing the blueprints and other bidding documents on a project.

Dodge Reports

```
• NY 777 444b 14-03x          CB          NEGOTIATING
Last rept 4-15-xx                         12-15-xx

OFFICE BLDG $510,000
Yourcity NY (Douglas Co) 123 Main St
WKG DWGS COMP-OWNER NEG CONTRACT-possible awd in
    2 weeks
Owner-Fahey & Co Frank Jones (Proj Mgr) 456 First
    Ave Yourcity NY 10011 (212/997-6184)
Arch-Robert Miller 3301 Elm St Yourcity NY 10015
    (212/443-1245)
Engr(str)-Evans Assoc 123 Market St Yourcity NY
    10015 (212/586-1165)
Engr(mech)-H L Brown & Co 201 W 89th St
    Yourcity NY 10018 (212/684-8446)
Engr(elec)-Patterson & Ross 1407 Altadena Dr
    Yourcity NY 10021 (212/783-3873)
    brk ext-1 sty-no bsmt-10,000 sq ft-wlbg-conc
    slab on grade flr-stl bar jst rf-mtl rf dk-
    incls pkg lot for 20 cars-prd conc fdns-elec
    htg-a/c-carpet vat & cer tile fin flrg-al
    sash-bu & pre-formed mtl rf dk-ptd drywl int
    wl fin-acoust clgs-fire spklrs-demol of small
    shed
PLANS ON SCAN FILM-#05-7777
NEG GC's
ABC Corp P O Box 123 Evansville NY 10203 (212/502-
    6100)
Riceclark Contrg Corp 1130 Park St Portland NY
    10116 (212/853-7772)
```


Dodge Reports

Fig. 78-1 An example of a Dodge Report. *(McGraw-Hill Information Systems Co.)*

Residential Estimating

There are two basic methods of determining the cost of a house. One is adding the total cost of all the materials to the hourly rate for labor multiplied by the anticipated number of hours it will take to build the home. The cost of the lot, landscaping, and various architect's and surveyor's fees must also be added to this figure. This process requires much computation and adequate techniques for estimating construction costs. Figure 78-3 shows a breakdown of the cost for an average residence.

Two quicker, rule-of-thumb methods for estimating the cost of the house are the square-foot method and the cubic-foot method. These methods are not as accurate as itemizing the cost of

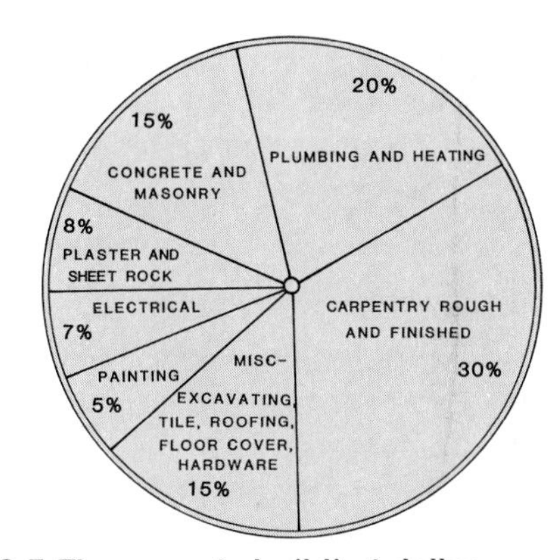

Fig. 78-3 The average building dollar.

Fig. 78-2 Using Dodge/SCAN microfilm to estimate building costs.

all materials, labor, and other items. However, they do provide a quick estimate for speculative purposes.

Building Materials Method Approximately 40 percent of the cost of an average house in this country is spent on building materials. However, the cost of construction materials varies greatly from one part of the country to another, depending on whether materials are native to the region or must be imported. For example, redwood is relatively inexpensive in California but very expensive on the east coast. Probably the greatest variable affecting the cost of building materials is the use of standard sizes of components versus the custom construction of components. For example, if kitchen cabinets are built on the site, the cost could easily be double the price of preconstructed factory units. Also, designing the residence in modular units will enable the builder to use standard size framing materials with a minimum of waste. So there is a direct relationship between the amount of on-site construction and the cost of construction. Also, specifying unique or exotic materials, such as rare stone or paneling, may not be worth the additional cost, depending on the budget restrictions of the builder. Regional climate conditions may also have a serious effect on the cost of building. For example, in moderate climates many costs can be eliminated by reducing heating equipment and frost-deep foundations. Figure 78-4A shows the cost of materials and labor for the residence shown in Fig. 78-4B.

Square-Foot Method In general, the cost of the average home ranges from \$60 to \$80 per square foot of floor space, depending on the geographical location. Each local office of the Federal Housing Administration can supply current estimating information. Figure 78-5 shows the computation of the cost of a one-level house at \$30 per square foot. This house is 60′ × 40′ or 1200 ft^2. At \$60 per square foot the house is estimated to cost approximately \$72,000, by the square foot method.

Principal items of construction	Labor costs	Material costs	Total cost	Cost per square foot
1. Foundation—excavations & footings	$ 2,547	$ 2,241	$ 4,788	$ 2.43
2. Slab foundation—fill, concrete, trowel fin.	1,381	1,572	2,953	1.50
3. Roof system—framing, sheathing, insulation	4,301	5,125	9,426	4.79
4. Roofing—shingles, flashing, gutters, downspouts	1,452	1,826	3,278	1.66
5. Exterior walls—framing, sheathing, wood siding, insulation, paint, wood doors, aluminum windows, insulating glass	5,903	9,084	14,987	7.61
6. Partitions—studs, drywall, doors	2,368	3,842	6,210	3.15
7. Wall finishes—paint, ceramic tile	1,417	697	2,114	1.07
8. Floor finishes—ceramic tile, vinyl tile, carpeting	725	2,688	3,413	1.73
9. Ceiling finishes—drywall, paint	1,634	745	2,379	1.21
10. Fixed equipment—range, range hood, oven exhaust fan, refrigerator, counters, cabinets, vanities, medicine cabinets, bathroom fittings	1,074	5,671	6,745	3.43
11. HVAC—forced hot air heating with air conditioning	2,505	2,409	4,914	2.50
12. Plumbing—water heater, bathroom fixtures, kitchen sink, laundry tub, pipes	3,017	2,335	5,352	2.72
13. Electrical—light fixtures, switches and outlets, connections for major appliances, service panel, wiring	1,591	1,409	3,000	1.52
Totals	$29,915	$39,644	$69,559	$35.32

Fig. 78-4A Materials and labor costs for the residence shown in Fig. 78-4B.

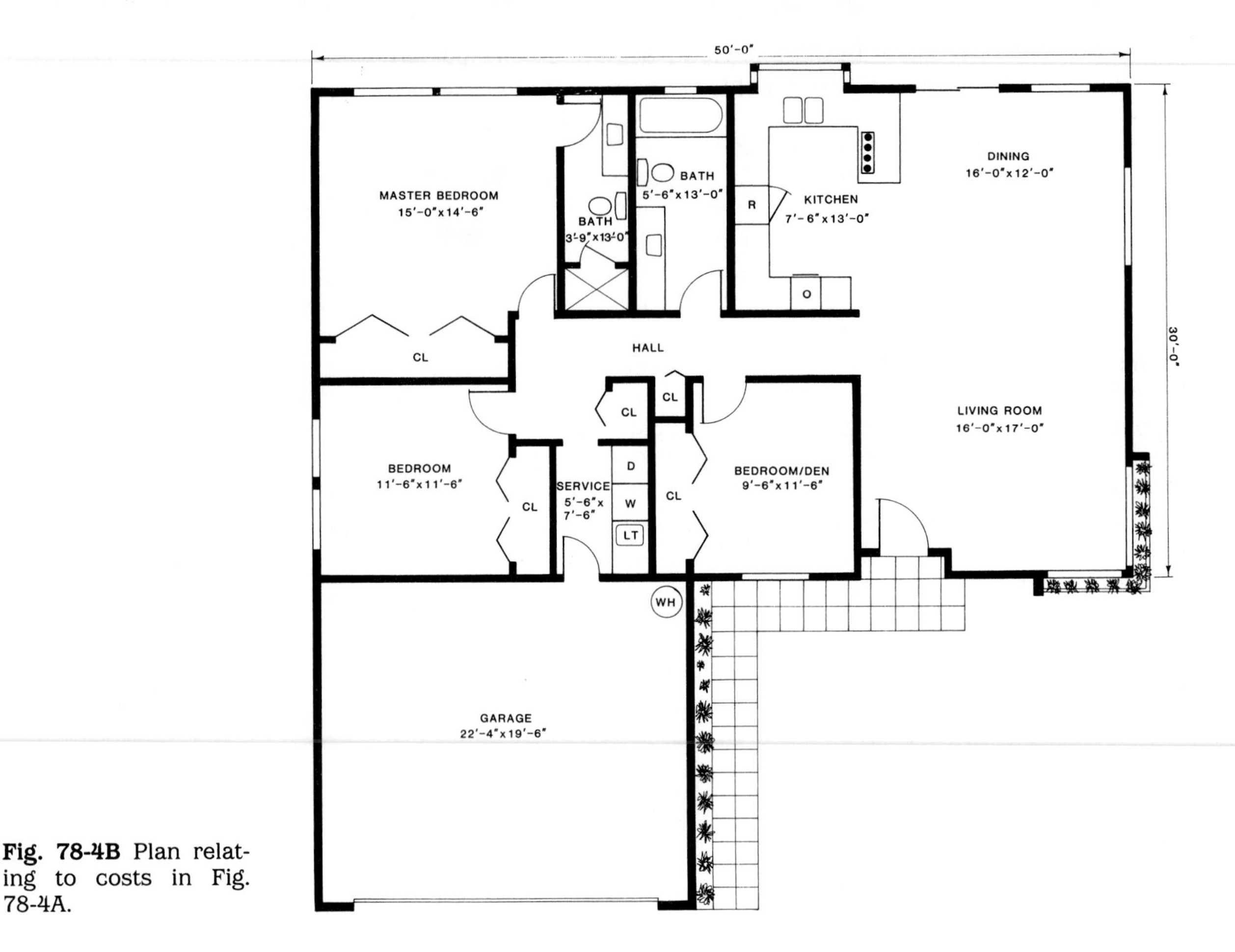

Fig. 78-4B Plan relating to costs in Fig. 78-4A.

Square Foot Method
Construction cost: \$60 per square foot
Square footage: 30′ × 40″ = 1,200 square feet
Cost: 1,200 × \$60 = \$72,000

Cubic Foot Method
Construction cost: \$5.00 per cubic foot
Cubic volume: floor area (square feet) × height
Cubic volume: 1,200 × 12 = 14,400 cubic feet
Total cost: cubic volume × cost per cubic foot
Total costs: \$14,400 × \$5.00 = \$72,000

Fig. 78-5 Square-foot and cubic-foot methods of estimating costs.

Cubic-Foot Method Using the cubic-foot method of estimating, this house would cost \$73,440. This estimate is determined by the cubic space (14,400 cubic feet) by multiplying the cost of cubic construction (\$5.82). The cubic foot method is a more accurate method for multilevel structures. The square foot and cubic foot methods are at best rough estimates, but they are very fast to calculate. However, the actual final cost of a building depends on the quality (unit cost) and quantity of the building materials used, plus the cost of financing and the cost of real estate.

Labor Costs

Labor costs also vary greatly from one part of the country to another and from urban to rural areas. Normally, labor costs are lower in rural areas. However, the amount of customized construction affects the labor cost of a building most dramatically. Not only does customizing take additional time, but it requires the services of more highly paid technicians. The labor cost factor can be reduced to whatever extent the owner may wish and is capable of becoming involved in some phase of the construction process.

Combination Costs

Rarely are final building costs based on one type of estimate. The estimated cost of most buildings is usually calculated by a combination of methods and cross-checking these for accuracy. Figure 78-6 shows one example of a combination of factors used to arrive at an estimated building cost.

MINIMIZING COSTS

Some construction methods and material utilization that can greatly reduce the ultimate cost of the home are listed as follows:

1. Square or rectangular homes are less expensive to build than irregular-shaped homes.

2. It is less expensive to build on a flat lot than on a sloping or hillside lot.

3. Using locally manufactured or produced materials cuts costs greatly.

4. Using stock materials and stock sizes of components takes advantage of mass-production cost reductions.

5. Using materials that can be quickly installed cuts labor costs. Prefabricating large sections or panels eliminates much time on the site.

6. Using prefinished materials saves labor costs.

7. Using prehung doors cuts considerable time from on-site construction.

8. Designing the home with a minimum amount of hall space increases the usable square footage and provides more living space for the cost.

9. Using prefabricated fireboxes for fireplaces cuts installation costs.

10. Investigating existing building codes before beginning construction eliminates unnecessary changes as construction proceeds.

11. Refraining from changing the design or any aspect of the plans after construction begins will keep costs from increasing.

12. Minimizing special jobs or custom-built items keeps costs from increasing.

13. Designing the house for short plumbing lines saves on materials.

14. Proper insulation will save heating and cooling costs.

15. Using passive solar features such as correct orientation reduces future utility costs.

1. ______ sq. ft. $70/sq. ft. =	$______
2. Basement—add $8.00/sq. ft.	$______
3. Extra bath (over one) add $5,000.	$______
4. Extra half-bath—add $3,000.	$______
5. One-car garage—add $3,000.	$______
6. Two-car garage—add $4,500.	$______
7. Covered porches—add $10.00/sq. ft.	$______
8. Each fireplace—add $2,000.	$______
9. Veneer masonry exterior—add $5.00/sq. ft.	$______
10. Finished attic—add $10.00/sq. ft.	$______
11. Cost of a lot in your community.	$______
12. Add 5% to 10% of item #1 for irregular design, hillside lot, special construction features.	$______
13. Cost of second story level $40.00/sq. ft.	$______
14. Cost of converting attic area to living space $15.00/sq. ft.	$______
15. Total estimated cost of home (within 10%)	$______

Fig. 78-6 Combination method of estimating.

DESIGN ESTIMATOR

Different types of cost estimates are developed during various design phases. Very rough estimates are needed during the conceptual design phase. Very accurate and precise estimates are required upon completion of working drawings.

Rough estimates are usually developed using the square foot or cubic foot method, although the costs of an existing similar building are often used. In this case the cost of unique materials or components, inflationary costs and/or geographical labor and material cost differences are added or subtracted. Building costs by region are available from a variety of sources such as Architectural Record.

A more accurate and much faster method of developing preliminary estimates is through the use of computer estimating software such as the Dodge Estimator. The Dodge Estimator software stores over 6000 material costs which can be combined with wage rates and other factors to produce preliminary construction estimates. These labor and material costs are updated for every zip code area every three months. The educational version (Design Estimator II) contains the average costs of labor and materials in the United States.

In estimating with the Dodge Estimator each building is divided into rectangular shapes as shown in Fig. 78-7. Then the basic dimensions of each area of a building are inputted using the information shown in Fig. 78-8. Next the type and size of building materials is selected and recorded on a materials selection form as shown in Fig. 78-9. If any labor costs are atypical a wage cost override can be used as shown in Fig. 78-10. If a building contains extremely unique materials or if some labor costs are very atypi-

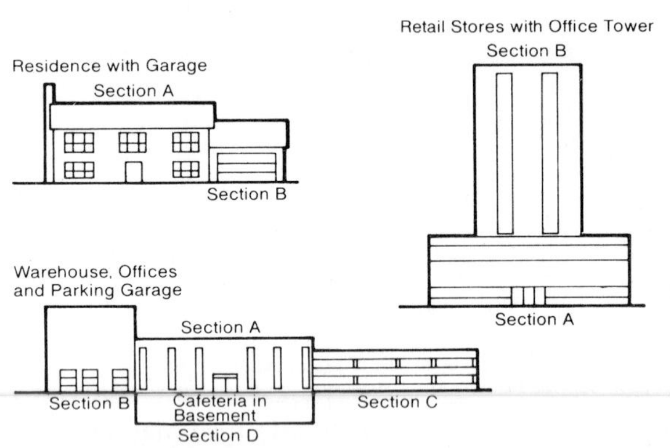

Fig. 78-7 Geometric breakdown of building areas. *(Dodge Microsystems)*

DESIGN ESTIMATOR II
WORKSHEET

McGraw-Hill

GENERAL DATA

Estimate Number	001—	BLDG 2
Client Name	006—	CLIENT NAME
Address	007—	CLIENT STREET
	008—	CLIENT CITY AND STATE
• Building Name	003—	BUILDING TWO
• Building Location	004—	CITY AND STATE
• Building Zip Code	016—	00001

OH & P/WAGE/AREA DATA

Overhead & Profit	030—	15 %
Wage Rates	026—	1 ☒ Union 2 ☐ Non-Union
• Gross Area	017—	7450

FEES/TAXES DATA

Architect's Fees	019—	%
Sales Tax	022—	5 %
Labor Taxes & Insurance	029—	27 %
Escalation Labor	024—	%
Material	025—	%
Inflation	027—	6 %
General Conditions	031—	6 %

BUILDING DIMENSIONS

Enter Section Descriptions ➡
† Do not duplicate common walls
†† Do not duplicate fixtures included in Baths

			A	B	C	D
Perimeter at Ground—Ln. Ft. †	•	001	200	105		
Ground Area—Sq. Ft.	•	002	2400	1050		
Floor to Floor Height or Eave Height	•	003	12	20		
Number of Floors (except Basement)	•	004	2	1		
Basement—%		005	67			
Depth—Ln. Ft.		006	12			
Number of Levels		027	1			
Crawl Space—%		007	33			
Grade Slab—%		008		100		
Piers—% (at Crawl Space)		009				
Total Floor Area Incl. Bsmt.	•	010	6400	1050		
Window Area—Sq. Ft.		011	630	105		
Partition Area—Sq. Ft.		026	2568			
Kitchens—No.		013				
Half Baths—No.		014				
Full Baths—No.		015				
Fireplaces—No.		016				
Outside Chimneys—No. (Residential)		017				
Plumbing Fixtures—No. ††		018	6			

• Must be filled in

Continued

Dodge MicroSystems & Cost Information Systems Division
Page 1 of 10

4-86

Fig. 78-8 Building dimensions input form. *(Dodge Microsystems)*

MATERIAL SELECTIONS

Floor Finish (continued)		A	B	C	D
Brick	211				
Marble	212				
Slate	213				
Flagstone	214				
Hardener	215				
Quarry tile	216				
Ceiling Finish					
Paint	221				
Plaster & paint	222				
Gypsum board & paint	223	X	200		
Metal (linear)	224				
Metal tile	225				
Wood & paint	226				
Acoustical-splined	228				
-lay-in 2'x2'	230	50			
-lay-in 2'x4'	231				
-lay-in high qual.	232				

Outside Chinmeys-Residential		A	B	C	D
Custom	261				
Average	262				
Simple	263				
Heating Only					
Residential-Radiation	271				
-Forced air	272				
-Electric	273				
Commercial-Radiation	275				
-Unit heaters	276				
Heating-By Area Use (omit if covered above)					
Apartments	277				
Manufacturing	278		X		
Warehouse	279	50			
Schools	280				
Supermarkets	281				
Residences	282				
*SF Cost	283				

*Enter SF Costs (In Cents-No Decimal)

Elevators-Equipment) (do not duplicate shafts)		A	B	C	D
Hydraulic-psgr	241				
-freight	242				
Electric-psgr.-low sp	243				
psgr.-med sp	244				
psgr.-high sp	245				
freight	246				
dumbwaiter	247				
residential	248				
Elevators					
Doors	249				
Kitchens (incl. plbg. conn) (omit if covered by write-ins or Special Features)					
Elaborate	400				
Average	401				
Minimal	402				
Fireplaces					
Custom	256				
Average	257				
Simple	258				

Cooling Only		A	B	C	D
Window units	284				
Residential—Central AC					
Via heat ducts	285				
Independent ducts	286				
HVAC By Area Use (omit if covered above)					
Apartments	295				
Laboratory	296				
Library, etc	297				
Office-commercial	298	50			
-corporate	299				
Residential	307				
Schools	300				
Supermarkets, etc.	301				
*SF Cost	306				
HVAC-Commercial					
Central Systems-Low	302				
-Med	303				
-High	304				

*Enter SF Cost (In Cents—No Decimal)

Fig. 78-9 Materials selection form. *(Dodge Microsystems)*

DESIGN ESTIMATOR II

WAGE/COST OVERRIDES

FIELD NUMBER	$/HR	TRADE
011		Asbestos Worker
012		Bricklayer
013		Carpenter
014		Cement Mason
015		Electrician
016		Glazier
017		Laborer
018		Lather
019		Oiler
020		Oper. Eng.—Hoisting
021		Oper. Eng.—Excavation
022		Painter
023		Pipefitter
024		Plasterer
025		Plumber
026		Reinf. Ironworker
027		Roofer
028		Sheet Metal Worker
029		Struct. Ironworker

WAGE/COST OVERRIDES (Cont'd)

FIELD NUMBER	$/HR	TRADE
030		Teamster
031		Tile Setter
032		Waterproofer
051		Cost of Concrete—Per Cu Yd
098		Material Cost Factor

Fig. 78-10 Wage-cost override form. *(Dodge Microsystems)*

cal, this can be entered using the write-in form as shown in Fig. 78-11. Once all the information is entered into the system, the computer can calculate and display the labor, material and total cost of each system as shown in Fig. 78-12. Each of the system detail totals can then be computed for the entire project as shown in Fig. 78-13. On CAD generated drawings precise take-off estimates can also be completed through digitizing directly from the drawing on a digitizing surface.

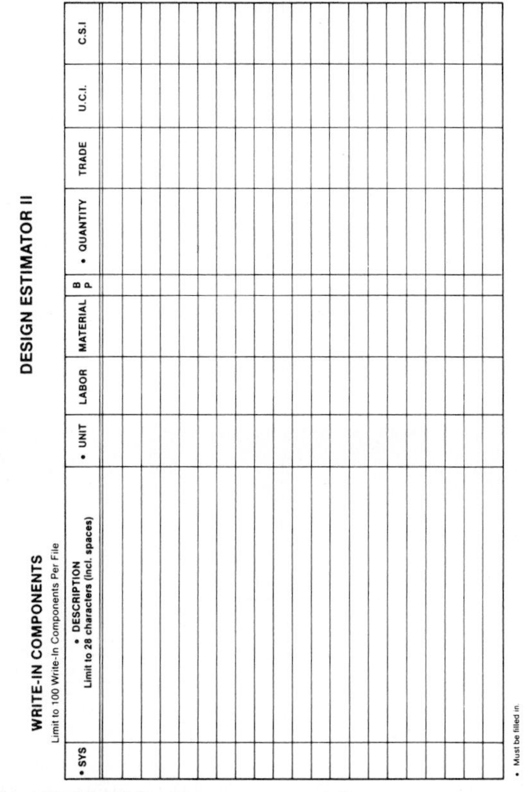

Fig. 78-11 Write-in component form.

SYSTEM DETAIL

DESCRIPTION	QUANTITY	UNIT	LABOR	MATERIAL	TOTAL
CEILING FINISHES					
GYPSUM BOARD	600	SQ FT	397	184	581
PAINT CEILING	600	SQ FT	497	88	585
			894	272	1,166
SPECIALTIES					
SCALE	1	EACH	1,475	7,300	8,775
			1,475	7,300	8,775
FIXED EQUIPMENT					
FIXED EQUIPMENT	600	SQ FT	465	1,644	2,109
FIXED EQUIPMENT	2400	SQ FT	698	2,522	3,220
			1,163	4,166	5,329
HVAC					
HEATING ONLY	3000	SQ FT	2,220	2,754	4,974
			2,220	2,754	4,974
PLUMBING					
FIXTURES & PIPING	4	EACH	2,027	2,570	4,597
			2,027	2,570	4,597
ELECTRICAL					
LIGHTING AND WIRING	600	SQ FT	1,642	2,051	3,693
LIGHTING AND WIRING	2400	SQ FT	3,158	3,927	7,085
			4,800	5,978	10,778
CONSTRUCTION TOTAL			63,241	72,277	135,518

Fig. 78-12 Systems costs.

SYSTEM SUMMARY

DESCRIPTION	LABOR	MATERIAL	TOTAL	SQ FT
FOUNDATIONS	7,219	4,833	12,052	4.01
FLOORS ON GRADE	5,449	5,730	11,179	3.72
SUPERSTRUCTURE	7,691	16,849	24,540	8.18
ROOFING	3,134	3,178	6,312	2.10
EXTERIOR WALLS	20,480	13,290	33,770	11.25
PARTITIONS	2,989	4,019	7,008	2.33
WALL FINISHES	1,543	372	1,915	0.63
FLOOR FINISHES	2,157	966	3,123	1.04
CEILING FINISHES	894	272	1,166	0.38
SPECIALITIES	1,475	7,300	8,775	2.92
FIXED EQUIPMENT	1,163	4,166	5,329	1.77
HVAC	2,220	2,754	4,974	1.65
PLUMBING	2,027	2,570	4,597	1.53
ELECTRICAL	4,800	5,978	10,778	3.59
CONSTRUCTION TOTAL	63,241	72,277	135,518	45.10

Fig. 78-13 Total costs.

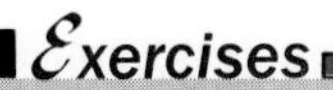

Exercises

1. At $70.00 a square foot, how much will the home shown in Fig. 69-1 cost?

2. At $6.00 per cubic foot, how much will the home shown in Fig. 69-1 cost?

3. Resketch the elevation of the building shown in Fig. 37-1. Substitute building materials that will reduce the cost of this home.

4. Resketch the plan shown in Fig. 74-1b to reduce the cost.

5. Compute the cost of the house you designed, based on the cost per square or cubic foot in your area.

6. Compute the cost of your present home based on the existing cost per square foot or cubic foot in your area.

7. Define these architectural terms: *real-estate survey, square footage, cubic footage, site costs, labor costs, materials costs.*

UNIT 79

FINANCIAL PLANNING

Few people can accumulate sufficient funds to pay for the entire cost of the home at one time. Therefore, most home buyers pay a percentage—approximately 9 to 15 percent—at the time of purchase, and arrange a loan (*mortgage*) for the balance. After analyzing the design and construction costs for a home, the prospective builder cannot determine the financial feasibility of the venture until the costs of financing are carefully considered.

MORTGAGE

A mortgage is a contract for a loan made to purchase a parcel of real estate and existing structures.

A mortgage is obtained from a mortgage company, bank, savings and loan association, or insurance company. The lending institution pays for the house, and through the mortgage loan agreement collects this amount from the home buyer over a long period of time: 10, 20, 30 years, or longer.

INTEREST

In addition to paying back the lending institution the exact cost of the house, the owner must also pay for the services of the institution. Payment is in the form of a percentage of the total cost of the home and is known as *interest*. Normal interest rates range from 9 to 15 percent. The interest payments increase with the length of time needed to pay back the loan. On a long-term loan—for example, 30 years—the monthly payments will be smaller but the overall cost will be much greater because the total interest is higher. A long-term loan can, over the life of the loan, accumulate 300 to 500 percent interest.

There are great variations in loan agreements. But remember low down payments usually result in higher interest payments over the life of the loan. One-time charges, called *points*, are also charged by some loan institutions to issue a loan. Each point represents 1 percent of the total loan amount.

A comparison of the time factor, amount of payment, and number of payments on a $60,000 loan at 13 percent interest is shown in Fig. 79-1. Figure 79-2 shows similar information for a $100,000 loan, and Fig. 79-3 shows the same information for a $150,000 loan.

Time in years	Monthly payments	Number of payments	Total cost
10	$447.94	120	$ 53,752.80
15	379.58	180	68,324.40
20	351.48	240	84,355.20
25	338.36	300	101,508.00
30	331.86	360	119,469.60

Fig. 79-1 Payments on a $60,000 mortgage at 13 percent interest.

Time in years	Monthly payments	Number of payments	Total cost
10	$746.56	120	$ 89,587.20
15	632.63	180	113,873.40
20	585.79	240	140,589.60
25	563.92	300	169,179.00
30	553.10	360	199,116.00

Fig. 79-2 Payments on a $100,000 mortgage at 13 percent interest.

Time in years	Monthly payments	Number of payments	Total cost
10	$1,194.49	120	$143,338.80
15	1,012.20	180	182,196.00
20	937.27	240	224,944.80
25	902.27	300	270,681.00
30	884.96	360	318,585.60

Fig. 79-3 Payment on a $150,000 mortgage at 13 percent interest.

TAXES

In addition to the *principal* (amount paid back that is credited to the payment for the house) and the interest, the taxes on the house must be added to the total cost of the home. Taxes on residential property vary greatly. Taxes on a $100,000 home in many residential suburban communities range between $2000 and $5000 per year.

INSURANCE

The purchase of a home is a large investment and must be insured for the protection of the home buyer and for the protection of the lending institution. Insurance rates vary greatly, depending on the cost of the home, location, type of construction, and availability of fire-fighting equipment (Fig. 79-4). The home should be insured against fire, public liability, property damage, vandalism, natural destruction, and accidents to trespassers and workers.

SERVICE COSTS

Other costs, in addition to the cost of the lot, structure, taxes, insurance, and financing, must

HOME VALUE	$75,000	$125,000	$187,500
COVERAGE INCLUDES: FIRE, PUBLIC LIABILITY, PROPERTY DAMAGE, THEFT, MEDICAL, PERSONAL LIABILITY, LANDSCAPE INS., WIND, FLOOD, EARTHQUAKE	$262.50	$437.50	$656.25

Fig. 79-4 Home insurance rates based on $3.50 per $1,000 of home value

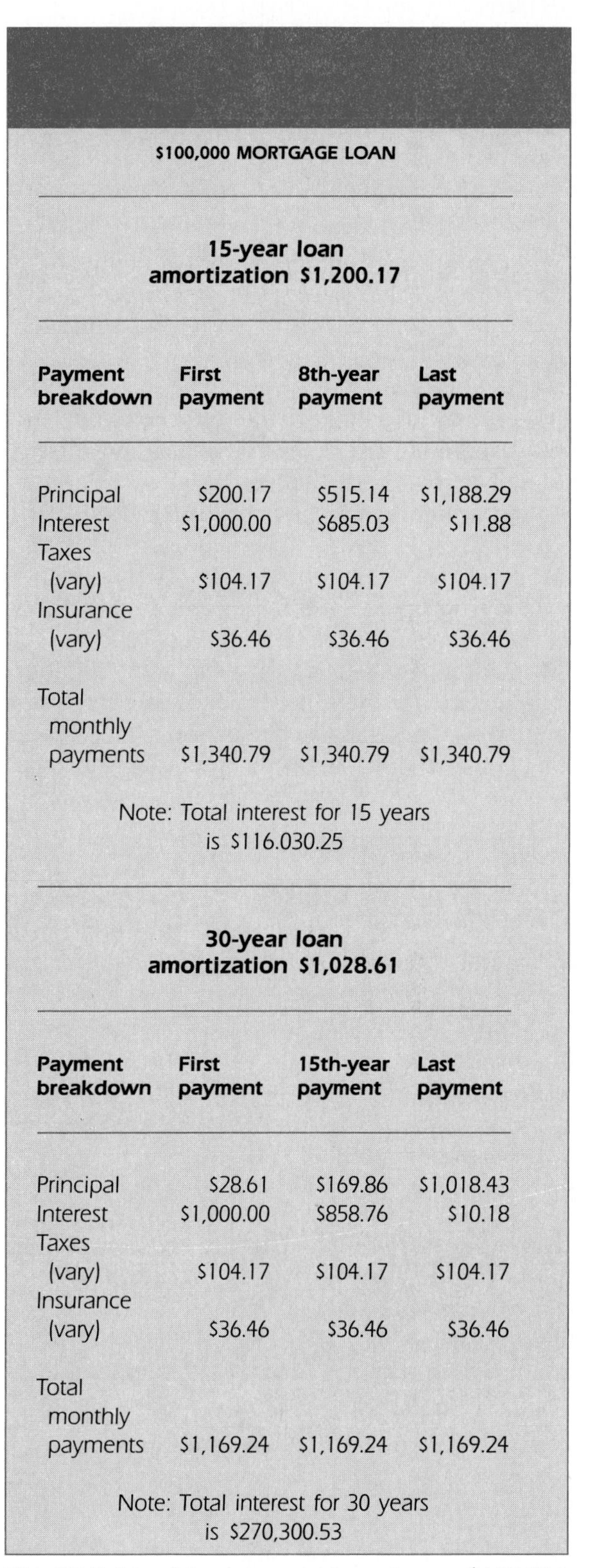

$100,000 MORTGAGE LOAN

15-year loan amortization $1,200.17

Payment breakdown	First payment	8th-year payment	Last payment
Principal	$200.17	$515.14	$1,188.29
Interest	$1,000.00	$685.03	$11.88
Taxes (vary)	$104.17	$104.17	$104.17
Insurance (vary)	$36.46	$36.46	$36.46
Total monthly payments	$1,340.79	$1,340.79	$1,340.79

Note: Total interest for 15 years is $116.030.25

30-year loan amortization $1,028.61

Payment breakdown	First payment	15th-year payment	Last payment
Principal	$28.61	$169.86	$1,018.43
Interest	$1,000.00	$858.76	$10.18
Taxes (vary)	$104.17	$104.17	$104.17
Insurance (vary)	$36.46	$36.46	$36.46
Total monthly payments	$1,169.24	$1,169.24	$1,169.24

Note: Total interest for 30 years is $270,300.53

Fig. 79-5 Monthly payments based on a $125,000 home with a $100,000 mortgage at 12 percent.

also be considered. These include service charges, title search, and transfer taxes. These are called "closing costs." The lawyer's, architect's, and surveyor's fees are sometimes included in the closing costs. Lawyer's fees range between $200 and $1000, and surveys cost between $300 and $500. Architects usually work on an 8 percent commission basis. If they supervise construction in addition to doing the designing, they may receive 10 percent of the cost of the project.

BUDGETS

Since most household budgets are established on a monthly basis, the monthly payments needed to purchase and maintain a residence are more significant than the total cost of the home. Monthly payments are broken into four categories: principal, interest, taxes, and insurance (PITI). Figure 79-5 shows the amortization for a 10-year, 20-year, and 30-year loan.

The prospective home buyer and builder should consider the following factors before selecting a particular institution for a mortgage. He or she should know the interest rate, the number of years needed to repay, prepayment penalties, total amount of monthly payment, conditions of approval, placement fees, amount of down payment required, service fees, and closing fees.

SALARY AND HOME COSTS

If the home buyer considers the purchase or the building of a home as an investment, he or she should take steps to ensure the maximum return on investment. If the home buyer purchases a home that costs considerably less than he or she can afford, the buyer is not investing adequately. On the other hand, if the home buyer attempts to buy a home that is more expensive than he or she can afford, the payments will become a drain on the family budget and undue sacrifices will have to be made to compensate.

Family budgets vary greatly, and a house that may be a burden for one person to purchase may be quite suitable for another, even if the two owners are earning the same relative salary. In general, the total cost of the mortgage should not exceed 4 times the annual income.

Exercises

1. **What is the cost of the home you could afford if you were earning $20,000 a year? $30,000 a year? $60,000 a year?**

2. **You purchase a home valued at $100,000 and make a 15 percent down payment. Your interest rate is 13 percent. What amount of interest will you pay per month?**

3. **What will be your total monthly payment on a $90,000 home if you have made a 10 percent down payment and are paying an interest rate of 12 percent? Your yearly taxes are $2000 and your insurance is $750.**

4. **What will your monthly mortgage payment be if you buy the house shown in Fig. 69-1 for $70 per square foot? Your interest is 14 percent for 25 years. You make a down payment of 9 percent of the total price. Your closing costs total $2000.**

5. **Compute the financing cost for a house of your own design using local mortgage information.**

6. **Define these terms: *interest, principal, escrow, mortgage, closing costs, taxes, insurance.***

UNIT 80

LEGAL DOCUMENTS

Legal documents (Fig. 80-1) define the agreement reached between the architect, builder, and owner of the building. This agreement indicates the fees to be paid the architect and the builder and the general conditions under which the project is undertaken. The legal document usually takes the form of a contract.

The following subsection of AIA document A201 describes the areas covered in general agreements between architects and builders.

A. Definitions.
B. Architect's supervision.
C. Architect's decision.
D. Notice.
E. Separate contracts.
F. Intent of plans and specifications.
G. Errors and discrepancies.
H. Drawings and specifications furnished to contractors.
I. Approved drawings.
J. Patents.
K. Permits, licenses, and certificates.
L. Supervision and labor.
M. Public safety and watchmen.
N. Order of completion.
O. Substitution of materials for those called for by specifications.
P. Materials, equipment, and labor.
Q. Inspection.
R. Defective work and materials.
S. Failure to comply with orders of architect.
T. Use of completed parts.
U. Rights of various interests.
V. Suspension of work due to unfavorable conditions.
W. Suspension of work due to fault of contractor.
X. Suspension of work due to unforeseen causes.
Y. Request for extension.
Z. Stoppage of work by architect.
AA. Default on part of contractor.
BB. Removal of equipment.
CC. Monthly estimates and payments.
DD. Acceptance and final payment.
EE. Deviations from contract requirements.
FF. Estoppel and waiver of legal rights.
GG. Approval of subcontractors and sources of material.
HH. Approval of material samples requiring laboratory tests.
II. Arbitration.
JJ. Bonds.
KK. Additional or substitute bonds.
LL. Public liability and property damage insurance.
MM. Workmen's Compensation Act.
NN. Fire insurance and damage due to other hazards.
OO. Explosives and blasting.
PP. Damages to property.
QQ. Mutual responsibility of contractors.
RR. Contractor's liability.
SS. Familiarity with contract documents.
TT. Shop drawings.
UU. Guarantee of work.
VV. Clean up.
WW. Competent workmen (state law).
XX. Prevailing wage act (state law).
YY. Residence of employees.
ZZ. Nondiscrimination in hiring employees (state law).
AAA. Preference to employment of war veterans (state law).
BBB. Hiring and conditions of employment (state law).

A contract includes fees and fee schedules, performance bond, labor and materials bonds, payments, time schedules, estimates, general conditions, and supplementary conditions. Schedules, specifications, and working drawings are also attached to the contracts so that they become legal documents too. Contracts describe the responsibilities for any financial changes that may be affected by time schedules or unavoidable delays caused by such things as acts of God and strikes.

A *deed* is a legal certificate of property ownership. Building code requirements are sometimes repeated in the contents of the property deed. These restrictions often describe the minimum setbacks, utility easements, building areas, and building types permitted. However, when building codes are updated usually the deeds are not changed accordingly. For this reason many deeds contain requirements which are no longer in existence. The designer must therefore check the deed for any restrictions and have the deed updated before proceeding with the design process if necessary.

A *performance bond* is offered by the contractor and guarantees that the performance of responsibilities as builder will be in accordance with the conditions of the contract.

Labor and materials bonds posted by the contractor guarantee that invoices for materials, supplies, and services of subcontractors will be paid by the prime contractor according to the terms of the contract.

Payment schedules are an important part of any contract. Payments are directly related to completion of various phases of the work, such as acceptance of the bid, beginning of the work, completion of phases of construction, and approval by the building inspector.

Just as most sets of plans must contain an architect's seal, licensed subcontractors are specified and required on most construction jobs. On larger projects, the seal of a licensed landscape architect may be required on all site design drawings. All plot plans or surveys registered with the local municipality must also contain the seal of a licensed surveyor. Indications of licensing requirements are usually included in construction contracts. Supervision of the labor force by journeymen in each specific area and the use of licensed electricians and licensed plumbers is also specified in contracts.

CONTRACT BIDS

Contractors receive invitations to bid by mail, through newspaper advertisements, or through private resources such as McGraw-Hill Dodge Reports. Construction bid forms are very specific in indicating the availability of documents, and when the documents can be examined. They also provide for the resolution of questions, approval for submission of materials, specific dates for bid submission, and the form for preparing bids. The bid form includes specific instructions to the bidders, the price of the bid, substitutions, restrictions, and the involvement of subcontractors.

The bid form is a letter that is sent from the owner or architect to the bidder. The letter covers the following points: verification of receipt of all drawings and documents, specific length of time the bid will be held open, price quotation for the bid, and a listing of substitute materials or components if any item varies from specified requirements. When the bidder signs this bid form, he or she agrees to abide by all conditions of the bid, including the price, time, quality of work, materials as specified in the contract documents, and drawings.

Sample instructions to the bidder, outlined by the Construction Specification Institute, are as follows:

Documents Bona fide prime bidders may obtain 3 sets of drawings and specifications from the architect upon deposit of $100 per set. Those who submit prime bids may obtain refund of deposits by returning sets in good condition no more than 10 days after bids have been opened. Those who do not submit prime bids will forfeit deposits unless sets are returned in good condition at least 2 days before bids are opened. No partial sets will be issued; no sets will be issued to sub-bidders by the architect. Prime bidders may obtain additional copies upon deposit of $100 per set.

Examination Bidders shall carefully examine the documents and the construction site to obtain firsthand knowledge of existing conditions. Contractors will not be given extra payments for conditions that can be determined by examining the site and documents.

Questions Submit all questions about the drawings and specifications to the architects, in writing. Replies will be issued to all prime bidders of record as addenda to the drawings and specifications and will become part of the contract. The architect and owner will not be responsible for oral clarification. Questions received less than 4 hours before the bid opening cannot be answered.

Substitutions To obtain approval to use unspecified products, bidders shall submit written requests at least 10 days before the bid date and hour. Requests received after this time will not be considered. Requests shall clearly describe the product for which approval is asked, including all data necessary to demonstrate acceptability. If the product is acceptable, the architect will approve it in an addendum issued to all prime bidders on record.

Basis of Bid The bidder must include all unit cost items and all alternatives shown on the bid forms; failure to comply may be cause for rejection. No segregated bids or assignments will be considered.

Preparation of Bids Bids shall be made on unaltered bid forms furnished by the architect. Fill in all blank spaces and submit two copies. Bids shall be signed with name typed below signature. Where bidder is a corporation, bids must be signed with the legal name of the corporation followed by the name of the state of incorporation and legal signatures of an officer authorized to bind the corporation to a contract.

Bid Security Bid security shall be made payable to the Irving Trust Co., in the amount of 10 percent of the bid sum. Security shall be either certified check or bid bond issued by surety licensed to conduct business in the State of NY. The successful bidder's security will be retained until bidder has signed the contract and furnished the required payment and performance bonds. The owner reserves the right to retain the security of the next 3 bidders until the lowest bidder enters into contract or until 2 days after bid opening, whichever is the shorter. All other bid security will be returned as soon as practicable. If any bidder refuses to enter into a contract the owner will retain his bid security as liquidated damages, but not as a penalty. The bid security is to be submitted 2 day(s) prior to the submission of bids.

Performance Bond and Labor and Material Payment Bond Furnish and pay for bonds covering faithful performance of the contract and payment of all obligations arising thereunder. Furnish bonds in such form as the owner may prescribe and with a surety company acceptable to the owner. The bidder shall deliver said bonds to the owner not later than the date of execution of the contract. Failure or neglecting to deliver said bonds, as specified, shall be considered as having abandoned the contract and the bid security will be retained as liquidated damages.

Subcontractors Names of principal subcontractors must be listed and attached to the bid. There shall be only one subcontractor named for each classification listed.

Submittal Submit bid and subcontractor listing in an opaque, sealed envelope. Identify the envelope with: (1)project name, (2)name of bidder. Submit bids in accord with the invitation to bid.

Modification and Withdrawal Bids may not be modified after submittal. Bidders may withdraw bids at any time before bid opening, but may not resubmit them. No bid may be withdrawn or modified after the bid opening except where the award of contract has been delayed for 2 days.

Disqualification The owner reserves the right to disqualify bids, before or after opening, upon evidence of collusion with intent to defraud or other illegal practices upon the part of the bidder.

Governing Laws and Regulations: Nondiscriminatory Practices Contracts for work under the bid will obligate the contractor and subcontractors not to discriminate in employment practices. Bidders must submit a compliance report in conformity with the President's Executive Order No. 11246.

U.S. Government Requirements This contract is federally assisted. The contractor must comply with the Davis-Bacon Act, the Anti-Kickback Act, and the Contract Work Hours Standards.

State Excise Tax Bidders should be aware of any state laws as they relate to tax assessments on construction equipment.

Opening Bids will be opened as announced in the invitation to bid.

Award The contract will be awarded on the basis of low bid, including full consideration of unit prices and alternatives.

Execution of Contract The owner reserves the right to accept any bid, and to reject any and all bids, or to negotiate contract terms with the various bidders, when such is deemed by the owner to be in his best interest.

Each bidder shall be prepared, if so requested by the owner, to present evidence of experience, qualifications, and financial ability to carry out the terms of the contract.

Notwithstanding any delay in the preparation and execution of the formal contract agreement, each bidder shall be prepared, upon written notice of bid acceptance, to commence work within 30 days following receipt of official written order of the owner to proceed, or on date stipulated in such order.

The accepted bidder shall assist and cooperate with the owner in preparing the formal contract agreement, and within 10 days following its presentation shall execute same and return it.

THE AMERICAN INSTITUTE OF ARCHITECTS

AIA Document A310

Bid Bond

AIA Document A311/CM

CONSTRUCTION MANAGEMENT EDITION

Performance Bond

AIA Document B161/CM

CONSTRUCTION MANAGEMENT EDITION

Standard Form of Agreement Between Owner and Architect for Designated Services

THIS DOCUMENT HAS IMPORTANT LEGAL CONSEQUENCES; CONSULTATION WITH AN ATTORNEY IS ENCOURAGED WITH RESPECT TO ITS COMPLETION OR MODIFICATION

This document is intended to be used in conjunction with AIA Document B162, Scope of Designated Services.

AIA Document C141

Standard Form of Agreement Between Architect and Engineer

Recommended for use with the current editions of standard AIA agreement forms and documents.

THIS DOCUMENT HAS IMPORTANT LEGAL CONSEQUENCES; CONSULTATION WITH AN ATTORNEY IS ENCOURAGED WITH RESPECT TO ITS COMPLETION OR MODIFICATION

AIA Document B801

Standard Form of Agreement Between Owner and Construction Manager

AIA Document C431

Standard Form of Agreement Between Architect and Consultant

THIS DOCUMENT HAS IMPORTANT LEGAL CONSEQUENCES. CONSULTATION WITH AN ATTORNEY IS ENCOURAGED. Recommended for use with the current editions of standard AIA agreement forms and documents.

AIA Document A177

Abbreviated Form of Agreement Between Owner and Contractor

for FURNITURE, FURNISHINGS AND EQUIPMENT where the Basis of Payment is a STIPULATED SUM

THIS DOCUMENT HAS IMPORTANT LEGAL CONSEQUENCES; CONSULTATION WITH AN ATTORNEY IS ENCOURAGED. This document includes abbreviated General Conditions and should not be used with other General Conditions. It has been approved and endorsed by the Contract Furnishings Council, the Business and Institutional Furniture Manufacturers Association, and the National Congress of Floor Covering Associations.

GEOTECHNICAL SERVICES AGREEMENT
AIA DOCUMENT G602
Geotechnical Engineer ☐ Owner ☐ Architect ☐

LAND SURVEY AGREEMENT
AIA DOCUMENT G601
Surveyor ☐ Owner ☐ Architect ☐

CONSTRUCTION CHANGE AUTHORIZATION
AIA DOCUMENT G713 (Instructions on reverse side)
Owner ☐ Architect ☐ Consultant ☐ Contractor ☐ Field ☐ Other ☐

OWNER'S INSTRUCTIONS FOR BONDS AND INSURANCE
AIA DOCUMENT G610

CHANGE ORDER
AIA DOCUMENT G701
Distribution to: OWNER ☐ ARCHITECT ☐ CONTRACTOR ☐ FIELD ☐ OTHER ☐

PROPOSAL REQUEST
AIA DOCUMENT G709
OWNER ☐ ARCHITECT ☐ CONTRACTOR ☐ FIELD ☐ OTHER ☐

CONSENT OF SURETY COMPANY TO FINAL PAYMENT
AIA DOCUMENT G707
OWNER ☐ ARCHITECT ☐ CONTRACTOR ☐ SURETY ☐ OTHER ☐

CONTRACTOR'S AFFIDAVIT OF PAYMENT OF DEBTS AND CLAIMS
AIA Document G706
OWNER ☐ ARCHITECT ☐ CONTRACTOR ☐ SURETY ☐ OTHER ☐

CONTRACTOR'S AFFIDAVIT OF RELEASE OF LIENS
AIA DOCUMENT G706A
OWNER ☐ ARCHITECT ☐ CONTRACTOR ☐ SURETY ☐ OTHER ☐

CONSENT OF SURETY TO REDUCTION IN OR PARTIAL RELEASE OF RETAINAGE
AIA DOCUMENT G707 A
OWNER ☐ ARCHITECT ☐ CONTRACTOR ☐ SURETY ☐ OTHER ☐

PROJECT: (name, address)

TO (Owner)

ARCHITECT'S PROJECT NO:

CONTRACT FOR:

CONTRACT DATE:

Fig. 80-1 Typical legal forms used by architects and builders.

Exercises

1. **What is the main purpose of a legal document?**
2. **What is meant by an "act of God"?**
3. **How do architects display their licensing as architects on the sets of plans?**
4. **Define the following terms: *contract, performance bond, materials bond, professional license, deed, contract bid.***

PART SIX

Appendix

Reference information most frequently needed in the preparation of architectural working drawings and documents is included in the appendix. Mathematical formulas for arithmetic and geometric calculations, plus coverage of structural design and calculations are included in Section 23, Basic Engineering Calculations.

A complete annotated glossary of architectural and construction terms is provided and architectural abbreviations used most frequently on architectural drawings and documents are defined. The appendix also includes a unit on major careers for which a background in architectural drafting is either mandatory or helpful.

SECTION 23

Basic Engineering Calculations

The field of architecture involves both art and science. Buildings *should* be aesthetically pleasing; but buildings *must* be structurally sound. Consequently a designer who deals with architecture only as an art form is extremely limited. A functional working knowledge of engineering mechanics is a necessity in architectural design and drafting.

The field of engineering mechanics is divided into two main parts: statics and dynamics. Statics deals with objects at rest. Dynamics deals with objects in motion or potential motion. The principles of statics and dynamics are combined with information on the strength of construction materials in the design of structures that will withstand the forces of nature and humanity.

In this section calculations most commonly used in the preparation of architectural working drawings and documents are presented first. This is followed by a coverage of the most frequently used geometric calculations. Then the principles of structural design and calculations necessary to determine the size, shape, and configuration of materials specified for an architectural design are presented.

UNIT 81

BASIC ARCHITECTURAL CALCULATIONS

The majority of errors found in architectural drawings are usually due to mathematical mistakes. Yet the majority of calculations performed in the process of preparing architectural drawings involves the basic arithmetic functions of adding, subtracting, multiplying and dividing.

CONVERSIONS

Part of the construction industry uses customary foot, inch and fractional dimensions. Other parts use decimal or metric dimensions. It is, therefore, important to convert dimensions from one system to another.

To convert inch fractions to decimals, divide the numerator by the denominator; for example, 7/8″ = .875, or 7 ÷ 8 = .875.

To convert decimals to fractions use the decimal number as the numerator and place a number 1 in the denominator, followed by as many zeros as there are to the right of the nominator. For example, .3 = 3/10, .45 = 45/100, and .675 = 675/1000.

To convert inches to millimeters (mm) multiply inches by 25.4, since 1″ = 25.4 mm. For example, 1/2″ = .5″, therefore, 1/2″ = 12.7 mm (.5 × 25.4). Likewise 6′-6″ = 6.5′ (78″); therefore, 6′-6″ = 1981.2 mm. Figure 81-1 shows conversion equivalents for common inch fractions and decimals.

ADDING DIMENSIONS

Most rows of dimensions include both feet and inches and may include fractional inches. In adding rows of mixed numbers such as these, add the inches separately, convert the inch total to feet and inches and then re-add the foot total. For example, in the plan shown in Fig. 81-2, three dimensions combine to equal 13′-3″:

$\frac{1}{64}$	0.015625	$\frac{33}{64}$	0.515625
$\frac{1}{32}$	0.03125	$\frac{17}{32}$	0.53125
$\frac{3}{64}$	0.046875	$\frac{35}{64}$	0.546875
$\frac{1}{16}$	0.0625	$\frac{9}{16}$	0.5625
$\frac{5}{64}$	0.078125	$\frac{37}{64}$	0.578125
$\frac{3}{32}$	0.09375	$\frac{19}{32}$	0.59375
$\frac{7}{64}$	0.109375	$\frac{39}{64}$	0.609375
$\frac{1}{8}$	0.1250	$\frac{5}{8}$	0.6250
$\frac{9}{64}$	0.140625	$\frac{41}{64}$	0.640625
$\frac{5}{32}$	0.15625	$\frac{21}{32}$	0.65625
$\frac{11}{64}$	0.171875	$\frac{43}{64}$	0.671875
$\frac{3}{16}$	0.1875	$\frac{11}{16}$	0.6875
$\frac{13}{64}$	0.203125	$\frac{45}{64}$	0.703125
$\frac{7}{32}$	0.21875	$\frac{23}{32}$	0.71875
$\frac{15}{64}$	0.234375	$\frac{47}{64}$	0.734375
$\frac{1}{4}$	0.2500	$\frac{3}{4}$	0.7500
$\frac{17}{64}$	0.265625	$\frac{49}{64}$	0.765625
$\frac{9}{32}$	0.28125	$\frac{25}{32}$	0.78125
$\frac{19}{64}$	0.296875	$\frac{51}{64}$	0.796875
$\frac{5}{16}$	0.3125	$\frac{13}{16}$	0.8125
$\frac{21}{64}$	0.328125	$\frac{53}{64}$	0.828125
$\frac{11}{32}$	0.34375	$\frac{27}{32}$	0.84375
$\frac{23}{64}$	0.359375	$\frac{55}{64}$	0.859375
$\frac{3}{8}$	0.3750	$\frac{7}{8}$	0.8750
$\frac{25}{64}$	0.390625	$\frac{57}{64}$	0.890625
$\frac{13}{32}$	0.40625	$\frac{29}{32}$	0.90625
$\frac{27}{64}$	0.421875	$\frac{59}{64}$	0.921875
$\frac{7}{16}$	0.4375	$\frac{15}{16}$	0.9375
$\frac{29}{64}$	0.453125	$\frac{61}{64}$	0.953125
$\frac{15}{32}$	0.46875	$\frac{31}{32}$	0.96875
$\frac{31}{64}$	0.484375	$\frac{63}{64}$	0.984375
$\frac{1}{2}$	0.5000	1	1.0000

Fig. 81-1 Dimensional equivalents.

these are 5′-3″, 5′-3″, and 2′-9″. To add these:

5′-3″
5′-3″
2′-9″
12′-15″, 15″ = 1′-3″
12′ + 1′-3″ = 13′-3″

If fractions are involved add the feet, inches, and fractions separately. Find the lowest common denominator in adding the fractions column. For example:

1′-7⅞″
2′-8¼″
6′-10⁹⁄₁₆″
11′-2¹¹⁄₁₆″

Step 1: 1′-7¹⁴⁄₁₆″
2′-8⁴⁄₁₆″
6′-10⁹⁄₁₆″
9′-25²⁷⁄₁₆″
Step 2: 9′-26¹¹⁄₁₆″
Step 3: 11′-2¹¹⁄₁₆″

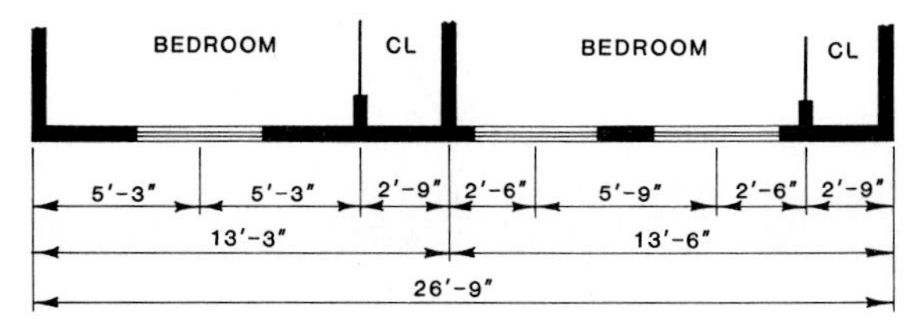

Fig. 81-2 Adding dimensions.

MODULAR MEASUREMENTS

It is often necessary in designing a structure to establish modular units. To determine the number of modular units needed for a given area divide the overall size by the modular unit size. For example, if a window opening dimension is planned to be 36″ × 66″ and the modular unit is 16″, proceed as follows:

36″ (height) ÷ 16″ = 2.25 units
Use two 16″ units to cover 32″ height.
66″ (width) ÷ 16″ = 4.125 units
Use four 16″ units to cover 64″ as shown in Fig. 81-3. A 32″ × 64″ modular window is the smallest size which will fit in the opening.

CONSTRUCTION MATERIAL CALCULATIONS

Construction materials are packaged or sold in a wide variety of quantities. It is, therefore, important to convert the total amount of material needed into the standard packaged measure. This often requires converting small volume units into larger units as shown in Fig. 81-4.

BOARD-FOOT MEASURE

Since lumber is purchased in bulk by the board foot, normal column measures such as cubic foot or yard do not apply. One board foot is 1″ ×

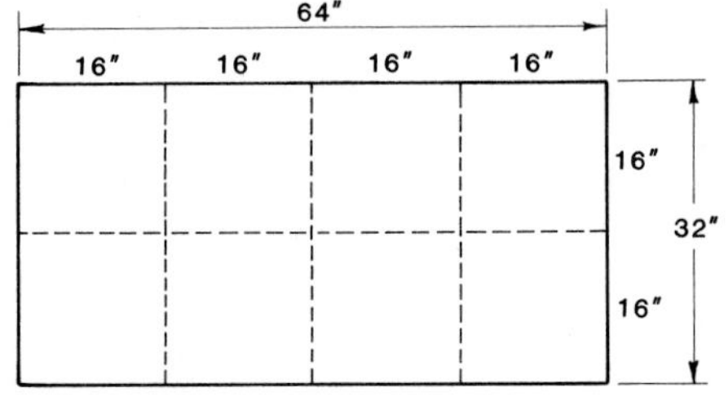

Fig. 81-3 Modular fitting.

Material	Measurement	Packaged or sold by
Cement	Bag 1 cubic ft.	Bag
Concrete	Cubic ft.	Cubic yard
Sand	Ton/lbs.	Ton/Cubic yds.
Blacktop	Sq. yds. after installation.	Cubic yd.
Gravel	Size of stone ¼″ to 3″	Ton/cubic yd.
Concrete block	Standard ht. 7⅝″ Standard length 15⅝″. Width 2″ to 14″.	Pallet—100 block Piece
Mortar	Bag 1 cubic ft.	Bag. Cu. ft. per bag.
Reinforcing rods	Width of bar ¼″ to 1⅝″	lbs.
Welded wire mesh	Size of wire 1/16″ to ¼″ Size of grid	Roll 5 × 100 ft. Sheet 5 × 20 ft.
Asphalt, static	lbs.	Bucket or barrel
Plywood	Thickness of sheet ¼ to 1″	Sheet 4′ × 8′
Paneling	Thickness of sheet ¼ to ⅜″	Sheet 4′ × 8′
2 × 4	Board ft./piece	Board ft./piece 6′ to 20′ lengths
2 × 6	Board ft./piece	6′ to 20′ lengths
2 × 8	Board ft./piece	6′ to 20′ lengths
2 × 10	Board ft./piece	6′ to 20′ lengths
2 × 12	Board ft./piece	6′ to 20′ lengths
Particleboard panels	Thickness of sheet 3/16 to 1″	Sheet—4′ × 8′
Hardboard panels	Thickness of sheet ⅛″ to ⅜″	Sheet—4′ × 8′
Roof shingles	Square–100 sq. ft.	⅓ sq. ft. per bundle. 33.3 sq. ft. per bundle.
Tar paper	Square feet	Rolls 3′ × 100′
Rain gutters	Depth 4″–5″	Lineal ft.
Aluminum siding	Square ft.	Square 100 sq. ft. = 1 sq.
Cedar siding	Square ft.	Square ft.
Drywall wallboard	Square ft.	Sheet 4′ × 8′
Sheathing panels	Square ft.	Sheet 4′ × 8′
Nails	Penny weight	Box or keg 1 lb.–25 lbs.
Pipe copper	¼–2½ dia.	Lin. ft. 20′ length.
Pipe plastic	¼–4″	Lin. ft.
Pipe iron	¼–4″ dia.	Lin. ft.
Pipe Galvanized	¼–4″ domestic ¼–36″ industrial	Lin. ft. Lin. ft.
Pipe cast iron	4″ dia.	Lin. ft.
Wire	Wire dia./gauge	Roll 50–100′
Conduit	Dia. ½″–4″	Lin. ft. 10′ lengths.

Fig. 81-4 Construction material quantities. *(Leo Kwolek)*

$12'' \times 12''$. To determine the number of board feet in a given piece of lumber multiply the length (in feet) × width (in inches) × thickness (in inches) ÷ 12:

$$BF = \frac{L' \times W'' \times T''}{12}$$

The board feet (BF) in the top piece of lumber in Fig. 81-5 is

$$\frac{3' \times 10'' \times 2''}{12} = 5 \text{ BF}$$

There are 1.75 board feet in the center piece. When dealing with multiple pieces of the same size (Fig. 81-5) either compute the BF for one piece and multiply it by the number of pieces or treat the entire package as one piece as follows:

$$\frac{5' \times 12'' \times 1''}{12} = 5 \text{ BF each} \times 3 \text{ pieces} = 15 \text{ BF}$$

or

$$\frac{3 \times 5' \times 12'' \times 1''}{12} = 15 \text{ BF}$$

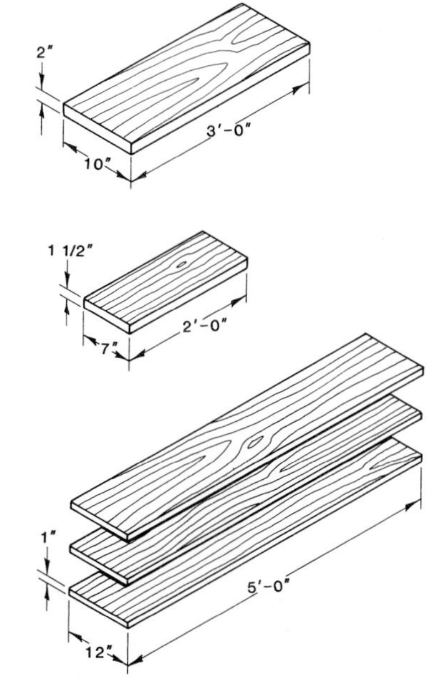

Fig. 81-5 Board-foot measure.

NUMBER OF RISERS CALCULATIONS

To determine the number of risers in a stair system (Fig. 81-6), divide the total height of the stair system by the height of each riser:

$$\frac{\text{Height of stairs}}{\text{Height of each riser}} = \text{riser number}$$

For example, to determine how many 7″ risers are needed for a stair system 9′-4″ high:

$$\begin{array}{lcr} 9'\text{-}4'' & & \\ 9' & = & 108'' \\ & & +\ \ 4'' \\ \hline & & 112'' \end{array}$$

$$\frac{112''}{7''} = 16 \text{ risers}$$

If the number is a fraction of a riser, round to the closest number of risers and adjust the riser height. For example:

$$\frac{115''}{7''} = 16.4 \text{ risers}$$

Adjust to 16 risers at 7.19″ each.

RISER HEIGHT CALCULATIONS

To determine the height of each riser in a stair system, divide the height of the stair assembly by the number of risers. For example, if a stair assembly is 9′-0″ high and has 15 risers, then:

$$9'\text{-}0'' = 108''$$

$$\frac{108''}{15} = 7.2'' \quad \text{(riser height)}$$

TREAD WIDTH CALCULATIONS

To calculate the tread width in a stair system, divide the stair run by the number of treads. For example, if a stair assembly has a run of 12′-0″ and 15 treads then:

$$12' = 144''$$

$$\frac{144''}{15} = 9.6'' \qquad \text{(width of each tread)}$$

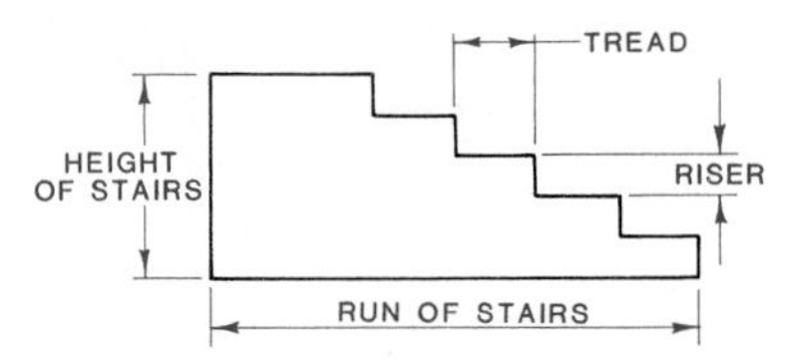

Fig. 81-6 Tread and riser calculations.

UNIT 82

GEOMETRIC CALCULATIONS

Architectural drawings contain a wide variety of geometric shapes. Some are two-dimensional flat surfaces and others are three-dimensional volume-containing areas. In estimating amounts of construction materials, labor, and costs, the ability to compute distances, areas, and volumes is important. The majority of geometric calculations in architectural work falls into these categories:

PERIMETER OF A POLYGON

Formula: Perimeter = sum of all sides

Examples: See Fig. 82-1.

Top drawing: 10 + 20 + 10 + 20 = 60′

Bottom drawing:
25′-0″
60′-6″
40′-0″
20′-0″
15′-0″
40′-6″
200′-12″ = 201′

CIRCUMFERENCE OF A CIRCLE

Formula: Circumference = $\pi \times$ diameter

$$C = \pi D$$

Example: See Fig. 82-2. $\pi = 22/7$ or 3.14.

Left drawing: $C = 3.14 \times 7 = 22'$

Right drawing: Inside-diameter
$C = 3.14 \times 14' = 43.96'$
Outside-diameter
$C = 3.14 \times 22' = 69.08'$

AREA OF A SQUARE OR RECTANGLE

Formula: Area = side × side

$$A = SS$$

Example: See Fig. 82-3.

Top drawing: $50' \times 25' = 1{,}250 \text{ ft}^2$

Center drawing: $175 \times 60 = 10{,}500 \text{ ft}^2$

Bottom drawing:
$60 \times 60 = 3{,}600$
$20 \times 20 = 400$
$4{,}000 \text{ ft}^2$

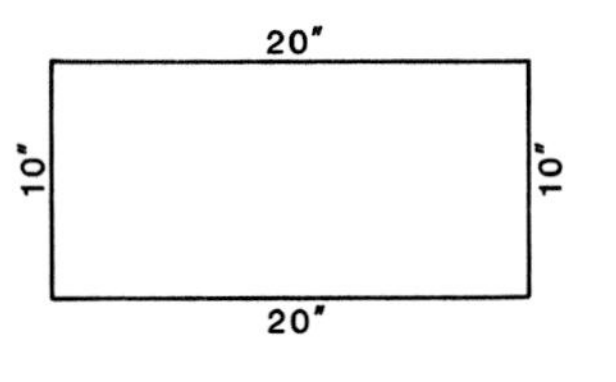

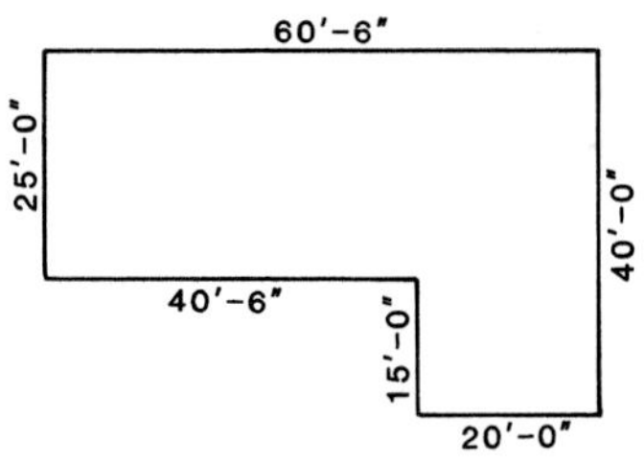

Fig. 82-1 Perimeter of a polygon.

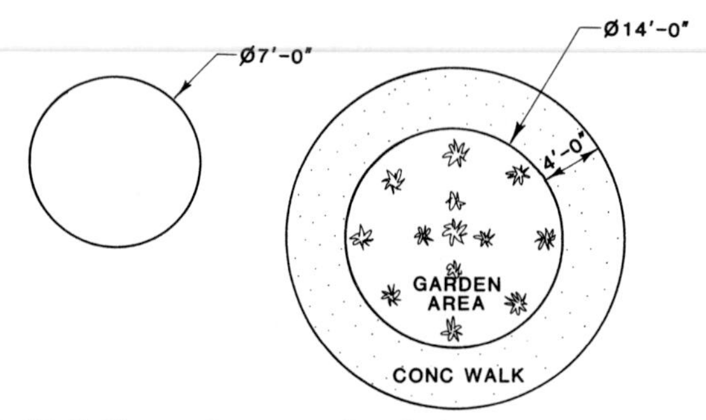

Fig. 82-2 Circumference of a circle.

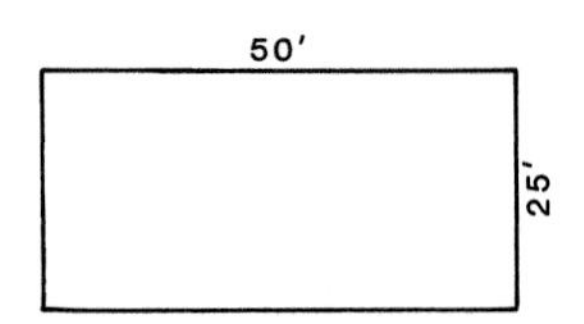

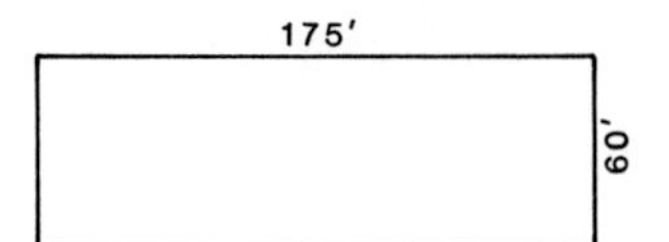

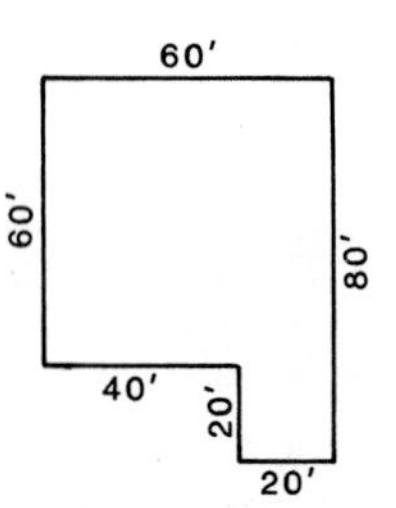

Fig. 82-3 Area of a square or rectangle.

AREA OF A TRIANGLE

Formula: Area = ½ base × altitude

$A = \frac{1}{2} BA$

Example: See Fig. 82-4.

Top drawing: $\frac{20'' \times 10''}{2} = 100 \text{ in}^2$

Center drawing: $\frac{15' \times 30'}{2} = 225 \text{ ft}^2$

Bottom drawing: $\frac{100 \times 70}{2} = 3500 \text{ ft}^2$

$\frac{100 \times 40}{2} = 2000$

$3500 + 2000 = 5500 \text{ ft}^2$

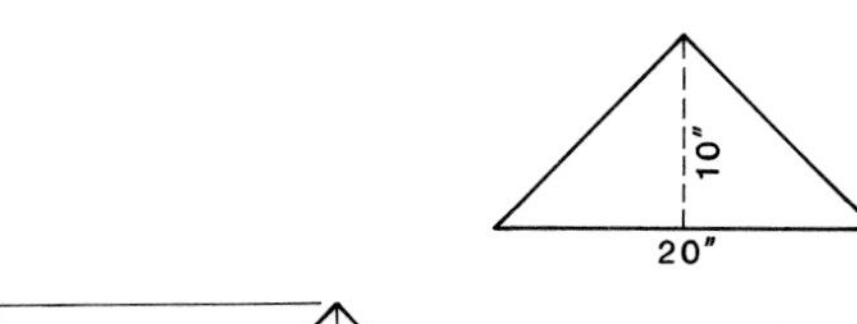

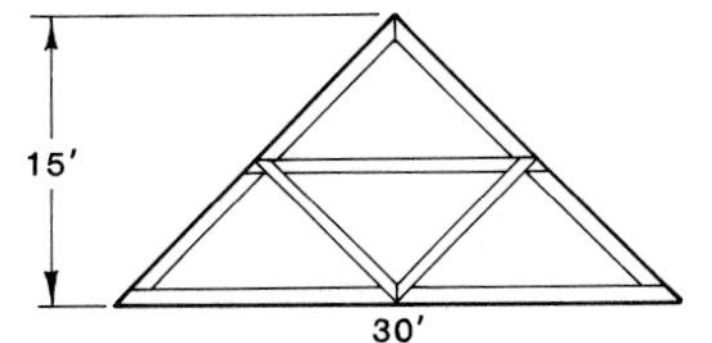

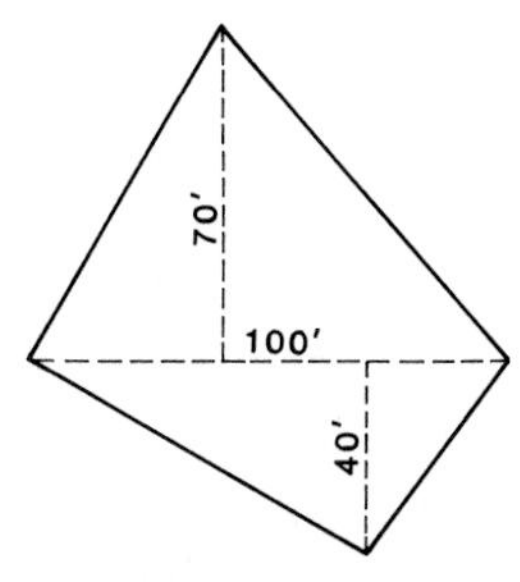

Fig. 82-4 Area of a triangle.

AREA OF A CIRCLE

Formula: Area = π × radius squared

$A = \pi R^2$

Example: See Fig. 82-5.

Top drawing: 3.14×7^2

$3.14 \times 49 = 153.86 \text{ in}^2$

Center drawing: $10 \div 2 = 5$

3.14×5^2

$3.14 \times 25 = 78.5 \text{ in}^2$

Bottom drawing: $20 \div 2 = 10$

$3.14 \times 10^2 = 314$

$314 \div 2 = 157 \text{ in}^2$

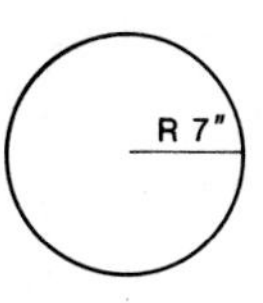

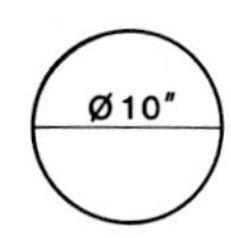

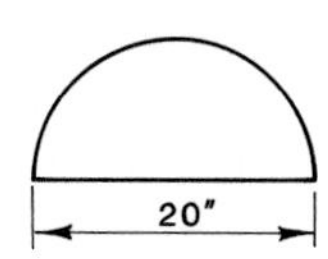

Fig. 82-5 Area of a circle.

SURFACE AREA OF A CYLINDER

Formula: Surface area = π × diameter × height

$A = \pi DH$

Example: See Fig. 82-6.

Top: $3.14 \times 10'' \times 20'' = 628 \text{ in}^2$

Center: $3.14 \times 2'' \times 100'' = 628 \text{ in}^2$

Bottom: Well is 10'-0" deep and 5'-0" wide

$3.14 \times 10' \times 5' = 157 \text{ ft}^2$

VOLUME OF A CUBE

Formula: Volume = length × height × width

$V = LHW$

Example: See Fig. 82-7.

Top: $V = 10'' \times 5'' \times 12''$

$= 600 \text{ in}^3$

Center: $V = 6 \times 50 \times 60$

$= 18{,}000 \text{ ft}^3$

$= 18{,}000 \div 27 = 666 \text{ yd}^3$

Bottom: $40' \times 24'' \times 6''$

$40' \times 2' \times 0.5' = 40 \text{ ft}^3$

$40' \times 6'' \times 12$

$40' \times 0.5' \times 1' = \underline{20} \text{ ft}^3$

60 ft^3

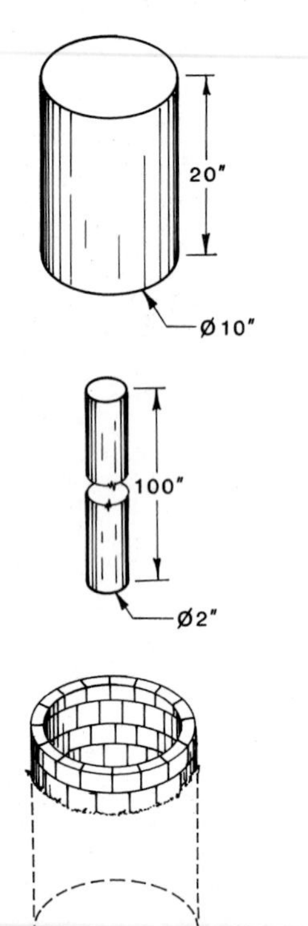

Fig. 82-6 Surface area of a cylinder.

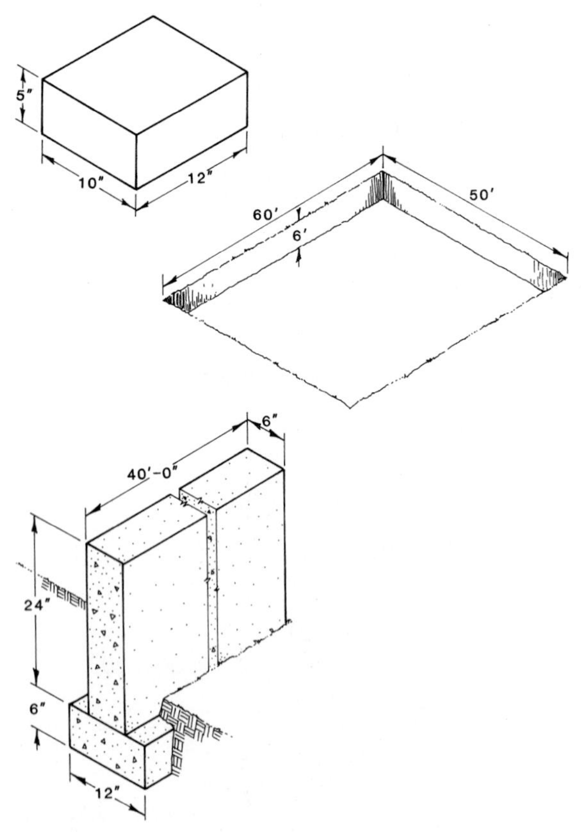

Fig. 82-7 Volume of a cube.

VOLUME OF A CYLINDER

Formula: Volume = π × radius squared × height

$$V = \pi R^2 H$$

Example: See Fig. 82-8.

Top: $V = 3.14 \times 5^2 \times 7$
$= 3.14 \times 25 \times 7$
$= 550 \text{ in}^3$

Center: $V = 3.14 \times 10^2 \times 14$
$= 3.14 \times 100 \times 14$
$= 4396 \text{ ft}^3$

Bottom: $V = 3.14 \times 3^2 \times 7$
$= 3.14 \times 9 \times 7$
$= 197.8 \text{ yd}^3$

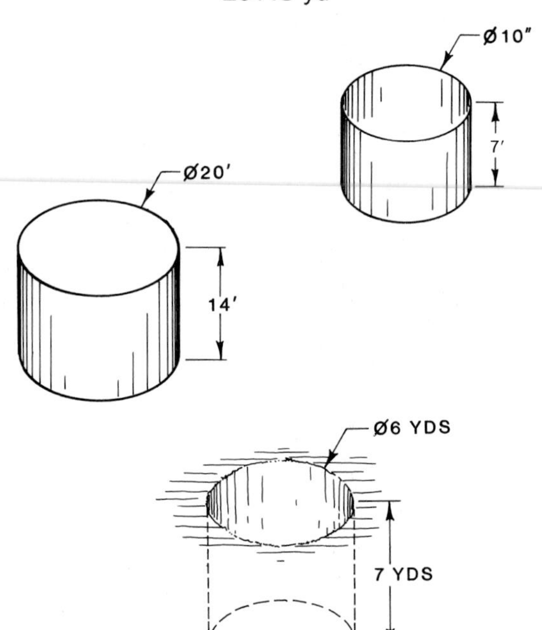

Fig. 82-8 Volume of a cylinder.

VOLUME OF A SQUARE PYRAMID

Formula: Volume = ⅓ × width of base × depth of base × height

$$V = \tfrac{1}{3} WDH$$

Example: See Fig. 82-9.

Top: $V = 0.33 \times 25'' \times 30'' \times 20''$
$= 4950 \text{ in}^3$

Bottom: $V = 0.33 \times 10' \times 12' \times 10'$
$= 396 \text{ ft}^3$

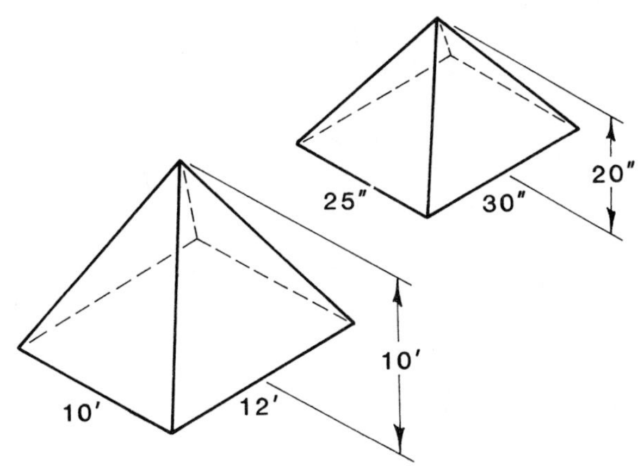

Fig. 82-9 Volume of a square pyramid.

VOLUME OF A CONE

Formula: Volume = $\frac{1}{3} \times \pi \times$ radius squared $\times$ height

$$V = \tfrac{1}{3}\pi R^2 H$$

Example: See Fig. 82-10.

Top: $V = .33 \times 3.14 \times 10^2 \times 30$
$= .33 \times 3.14 \times 100 \times 30 = 3108.6 \text{ in}^3$

Bottom: $V = .33 \times 3.14 \times 20^2 \times 20 \text{ ft}^2$
$= .33 \times 3.14 \times 400 \times 20 \text{ ft}^2$
$= 8289.6 \text{ ft}^3$

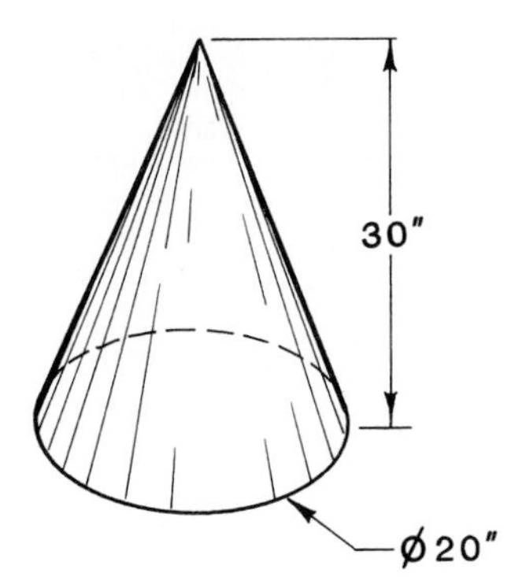

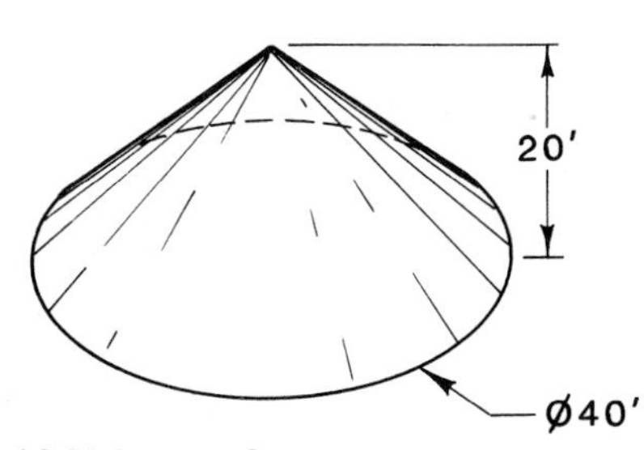

Fig. 82-10 Volume of a cone.

VOLUME OF A SPHERE

Formula: Volume = $\frac{1}{6} \times \pi \times$ diameter cubed

$$V = \tfrac{1}{6}\pi D^3$$

Example: See Fig. 82-11. $\frac{1}{6} = .166$.

$V = .166 \times 3.14 \times 3^3$
$= .166 \times 3.14 \times 27$
$= 14.07 \text{ in}^3$

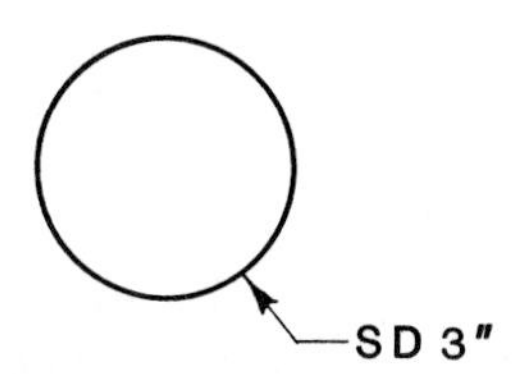

Fig. 82-11 Volume of a sphere.

RIGHT TRIANGLE LAW (PYTHAGOREAN THEOREM)

Formula: Square of the hypotenuse = the sum of the square of the two sides

$C^2 = A^2 + B^2$

Example: See Fig. 82-12.

$C^2 = A^2 + B^2$
$= \sqrt{A^2 + B^2}$
$= \sqrt{32^2 + 24^2}$
$= \sqrt{1024 + 576}$
$= \sqrt{1600}$
$= 40''$

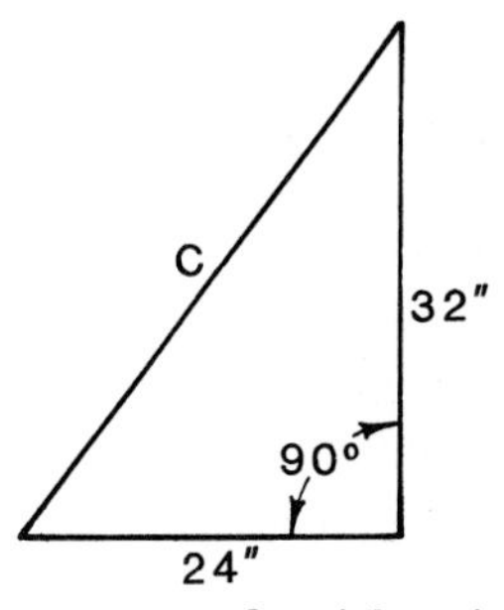

Fig. 82-12 Hypotenuse of a right triangle.

UNIT 83

STRUCTURAL DESIGN AND CALCULATIONS

Information in this unit is provided to aid the designer and drafter in designing and determining the most appropriate material, sizes, and construction methods. Some building codes specify materials and sizes for each specific situation, while other codes simply require the structural integrity be held to exact limits. The designer must use either precalculated tables or established formulas, or both, in arriving at the structural design. This unit includes reference information for this purpose.

STRUCTURAL DESIGN

An effective structural design is neither overdesigned nor underdesigned. When structures are underdesigned, systems fail and buildings sag or collapse. However, if a building is overdesigned, materials are wasted and costs greatly escalate. Thus the primary task of the structural designer is to design all components to be above, but not too far above, acceptable safe standards.

BUILDING LOADS

The structural effects of gravity on weights are divided into load categories as shown in Fig. 83-1: dead loads and live loads. Dead loads are permanent structural or unmovable parts of a building. Live loads are variable, movable bodies such as people, furniture, or cars. Lateral loads such as horizontal wind forces and earth movement must also be considered in the development of the design.

Loads vary greatly depending on the type of structure. Figure 83-2A shows typical loads for a two-level frame-construction building. Note loads are given in pounds per square foot. A more complete list of building material loads is shown in Fig. 83-2B.

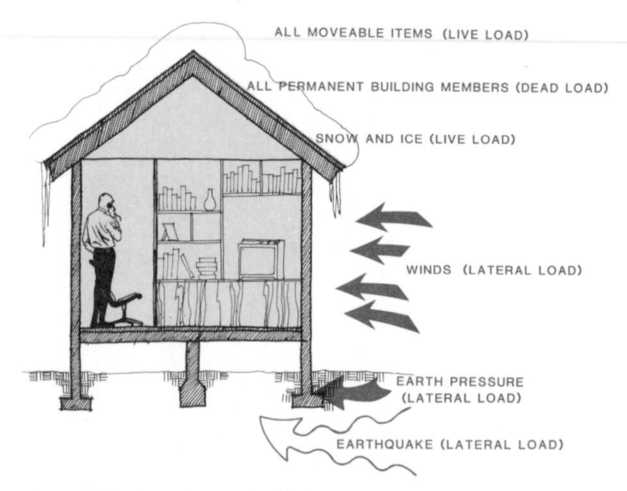

Fig. 83-1 Load categories.

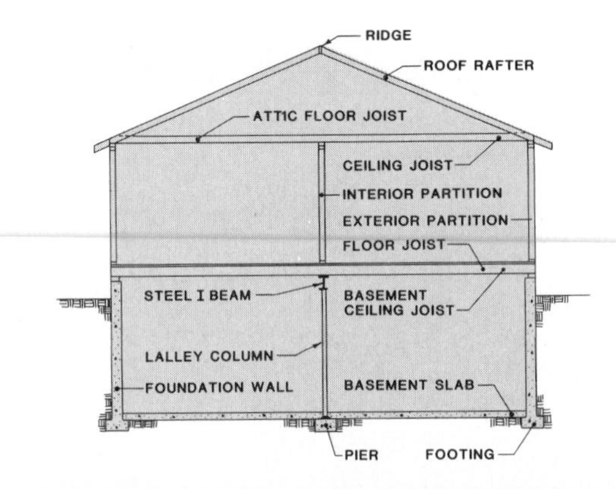

Construction area	Live load PSF	Dead load PSF
Roof	30	20
Ceiling joists attic/heavy storage	30	20
Ceiling joists attic no floor	0	10
Ceiling joists attic habitable rooms	30	10
Floors of rooms	40	20
Floors of bedroom	30	20
Partitions	0	20

Fig. 83-2A Typical loads.

TRIBUTARY AREAS

The tributary area of a structural member is the area of weight created by the member on a surface. The total load of the building is transmitted into the ground from footings and piers through

Roofing materials	Weight
built-up roofing, 3 ply and gravel	6 PSF
rafters, 2 × 4 at 16″ oc	2 PSF
rafters, 2 × 6 at 16″ oc	2.5 PSF
rafters, 2 × 8 at 16″ oc	3.5 PSF
sheathing, ½″ fiberboard	0.75 PSF
sheathing, ½″ gypsum	2 PSF
sheathing, 1″ wood	3 PSF
shingles, asbestos	4 PSF
shingles, asphalt	2 PSF
shingles, wood	2.5 PSF
skylights, glass 7 frame	11 PSF
tile, cement	15 PSF
tile, mission	13 PSF
Ceilings	
acoustical tile ½″	0.8 PSF
plaster on wood lath	8 PSF
suspended metal lath and cement plaster	15 PSF
suspended metal lath and gypsum plaster	10 PSF
Walls	
brick wall, 4″	35 PSF
brick wall, 8″	74 PSF
brick (4″) on 6″ concrete block	80 PSF
brick (4″) on wood frame with sheathing & plaster	45 PSF
building board, ½″	0.8 PSF
concrete block wall, 6″	40 PSF
concrete block wall, 8″	55 PSF
concrete wall, 8″	100 PSF
concrete wall, 10″	125 PSF
gypsum block, 2″	9.5 PSF
gypsum block, 4″	12.5 PSF
gypsum wallboard	2.5 PSF
plaster, ½″	4.5 PSF
plywood, ½″	1.5 PSF
tile, facing, 2″	15 PSF
tile, glazed ⅜″	3 PSF
wood siding, 1″	3 PSF
wood stud wall	5 PSF
wood stud wall, plastered one side	12 PSF
wood stud wall, plastered two sides	20 PSF
Floors	
cement finish 1″	12 PSF
clay tile on 1″ mortar base	23 PSF
concrete slab, 4″	48 PSF
hardwood flooring, 25/32″	4 PSF
floor joist, 2″ × 8″, 16″ oc/subflooring	6 PSF
floor joist, 2″ × 10″, 16″ oc/subflooring	6.5 PSF
floor joist, 2″ × 12″, 16″ oc/subflooring	7.0 PSF
marble on 1″ mortar base	28 PSF
plywood subflooring, ½″	1.5 PSF
quarry tile, ½″	6 PSF
terrazzo, 2″	25 PSF
vinyl asbestos tile, ⅛″	1.3 PSF
4 × 6 girder/post	4 PSF

Fig. 83-2B Building material loads.

General Building Materials	Weight
Woods at 12% moisture content, lbs. per cubic foot	
cedar	22 PCF
douglas fir	34 PCF
maple	42 PCF
oak	47 PCF
pine	27 PCF
poplar	28 PCF
redwood	28 PCF
Insulation	
bats, blankets, 3″	0.5 PSF
boards, vegetable fiber	1.7 PSF
cork board, 1″	0.6 PSF
foam board, 1″	0.1 PSF
Glass	
double strength, 1-8″	1.5 PSF
insulating plate, ⅛″ with air space	3.25 PSF
glass block, 4″	20 PSF
plate glass, ¼″	3.25 PSF
plastic, ¼″ acrylic	1.5 PSF
Masonry	
brickwork, 4″	35 PSF
concrete wall 6″	75 PSF
concrete wall 8″	100 PSF
poured concrete	150 lbs./cu. ft.
concrete block, lightweight 4″	22 PSF
concrete block, lightweight 6″	31 PSF
concrete block, stone 6″	50 PSF
concrete block, stone 8″	58 PSF
facing tile, 2″	16 PSF
facing tile, 4″	30 PSF
facing tile, 6″	41 PSF
marble, 1″	13 PSF
slate, 1″	14 PSF
stone, 1″	12 PSF
tile, structural clay, 4″ hollow	23 PSF
tile, structural clay, 6″ hollow	33 PSF
tile, structural clay, 8″ hollow	42 PSF
Metals, lbs./cu. ft.	
aluminum, cast	165 PCF
brass, yellow	528 PCF
bronze, commercial	552 PCF
copper, cast or rolled	556 PCF
iron, cast	450 PCF
iron, wrought	485 PCF
lead	710 PCF
steel rolled	490 PCF
tin, cast or hammered	459 PCF
steel beam S7 × 15.3	15.3 lbs./lineal foot
Soil, Sand & Gravel, lbs./cu. ft.	
clay, damp	110 PCF
clay, dry	63 PCF
clay and gravel, dry	100 PCF

Fig. 83-2B (continued)

Soil, Sand, & Gravel, lbs./cu. ft. (continued)	Weight
earth, loose, dry	76 PCF
earth, packed, dry	95 PCF
earth, moist, loose	78 PCF
earth, moist, packed	96 PCF
earth, mud, packed	115 PCF
sand/gravel, dry, loose	110 PCF
sand/gravel, dry, packed	120 PCF
sand/gravel, wet	120 PCF

Fig. 83-2B (continued)

a series of tributary areas. Since footings are continuous, load distribution is calculated as one lineal foot as shown in Fig. 83-3.

The tributary area under each pier extends one-half the distance to the next structural support members. For example, pier B in Fig. 83-4 is 9′-0″ from the nearest pier on both sides. The tributary area for pier B, therefore, extends one-half the distance to piers A and C (4′-6″) and one-half the distance to each foundation wall

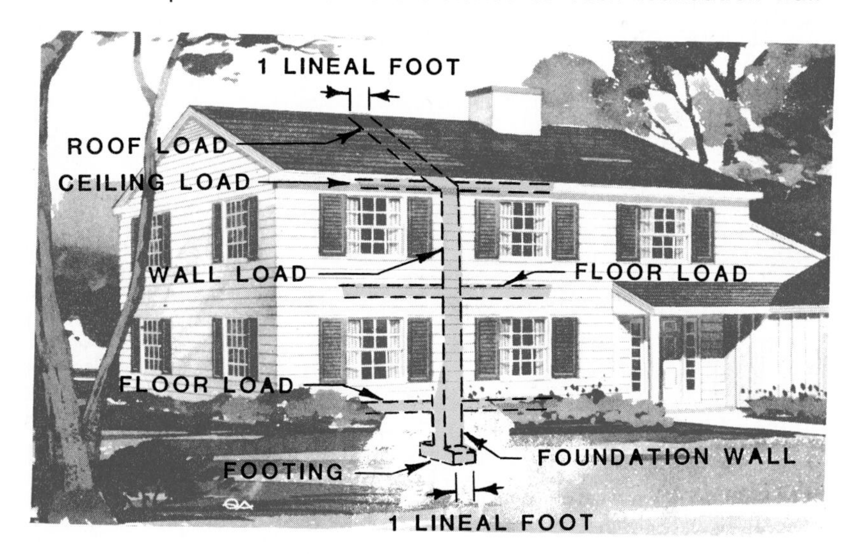

Fig. 83-3 Loads calculated on one foot of footing.

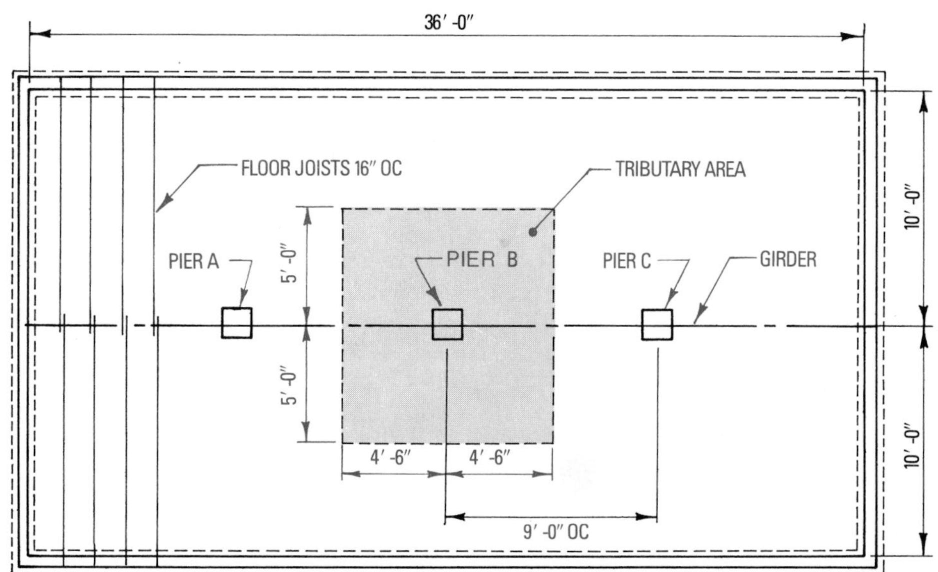

Fig. 83-4 Foundation tributary area.

(5′-0″). The tributary area is, therefore, 4′-6″ + 4′-6″ by 5′0″ + 5′-0″, or 9′-0″ × 10′-0″, or 90 ft^2. The tributary area in Fig. 83-5 is 12′-0″ × 7′-0″, or 84 ft^2.

SOIL CONDITIONS

The safety of a structure also depends on the type of soil and safe loading. Fig. 83-6 lists typical safe structure loads for different types of soils. It is important that the weight of the structure not exceed the safe load capacity of the soil. Footing sizes are also critical since they spread the weight of the structure on the soil. Footing area (FA) is calculated as follows (Fig. 83-7):

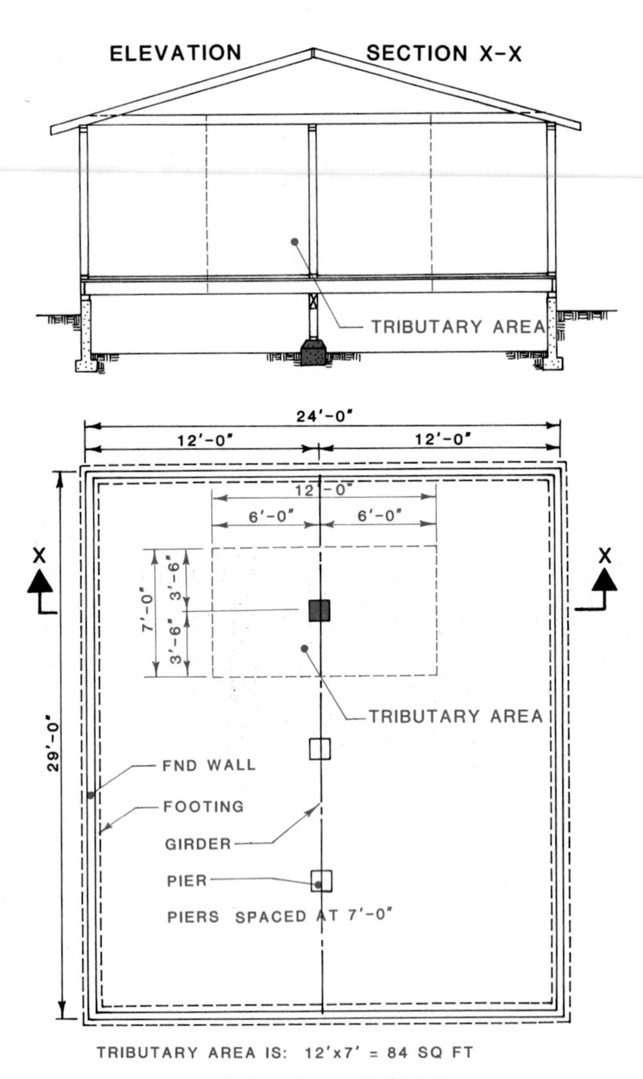

Fig. 83-5 Wall and foundation tributary areas.

Soil type	Safe load PSF
Soft clay; sandy loam	1,000
Firm clay; sand and clay mix; fine sand, loose	2,000
Hard dry clay; fine sand, compact; sand and gravel mixtures; coarse sand, loose	3,000
Coarse sand, compact; stiff clay; gravel, loose	4,000
Gravel; sand and gravel mixtures, compact	6,000
Soft rock	8,000
Exceptionally compacted gravels and sands	10,000
Hardpan or hard shale; sillstones; sandstones	15,000
Medium hard rock	25,000
Hard, sound rock	40,000
Bedrocks, such as granite, gneiss, traprock	100,000

Fig. 83-6 Safe soil loads.

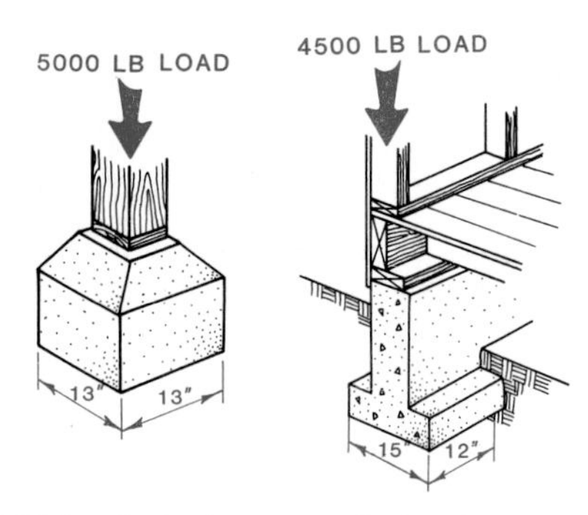

Fig. 83-7 Footing size spreads loads.

Pier footing calculation:

Structure load = 5000 lb/ft^2

Soil-bearing capacity = 6000 lb/ft^2 (given)

$$\text{Footing Area} = \frac{\text{structure load}}{\text{soil-bearing capacity}}$$

$$= \frac{5000}{6000} = 0.83$$

$$= 0.83 \text{ ft}^2 \times 144^* = 119.52 \text{ in}^2$$

$$= 11'' \times 11'' = 121 \text{ in}^2$$

(closest square to 119.52 in^2)

*144 sq. in. = 1 sq. ft.

Perimeter foundation wall footing calculation:

Structure load = 4500 lb

Soil-bearing capacity = 6000 lb/ft^2 (given)

$$
\begin{aligned}
\text{Footing Area} &= \frac{\text{structure load}}{\text{soil-bearing capacity}} \\
&= \frac{4500}{6000} \\
&= 0.75 \text{ ft}^2 \\
&= 0.75 \text{ ft}^2 \times 144 = 108 \text{ in}^2 \\
&= \frac{108 \text{ in}^2}{12} (\text{lineal }') \\
&= 9'' \text{ wide} \times 1 \text{ lineal foot}
\end{aligned}
$$

Use next standard size, 12″.

ROOF MEMBERS

The size of all structural members used for roof framing depends on the combined loads bearing on the member and the spacing and span of each member. If the load is increased, either the spacing or span must be decreased or the size of the member increased to compensate for the increased load. Consequently, if the size of a member is decreased, the members must be spaced more closely or the span must be decreased. If the length of the span is increased, the size of the members must be increased or the spacing made closer. To compute the most appropriate size of roof rafter for a given load, spacing, and span, follow these steps:

1. To determine the total load per square foot of the roof space, add the live load and the dead load (Fig. 83-8). Figure 83-9 shows the components of a roof system that are used in the calculation of roof loads as follows:

Component	*Dead load, lb/ft²*
2 × 6 roof rafter	3
3 layers felt, tar, and gravel	5
1″ insulation	1
½″ plywood sheathing	2
⅝″ gypsum board ceiling	2
2 × 6 ceiling joist	3
Dead load	16
Live load	20
Total load	36

In roof design, the live load is the safety factor allowing for weather conditions and movable objects the structure must support.

2. To determine the load per lineal foot on each rafter, multiply the load per square foot by the spacing of the rafters (Fig. 83-10). If rafters

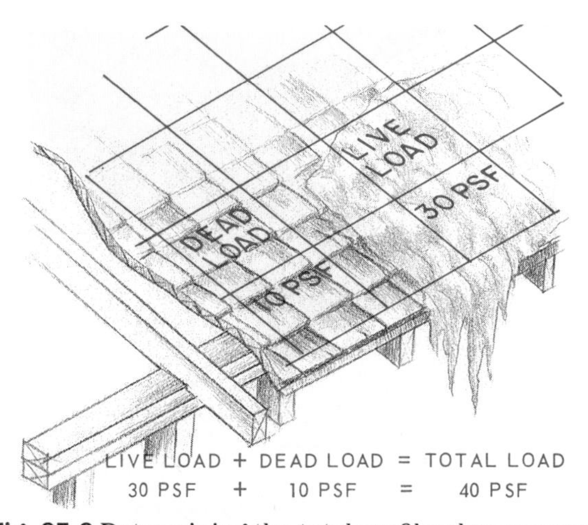

Fig. 83-8 Determining the total roof load per square foot.

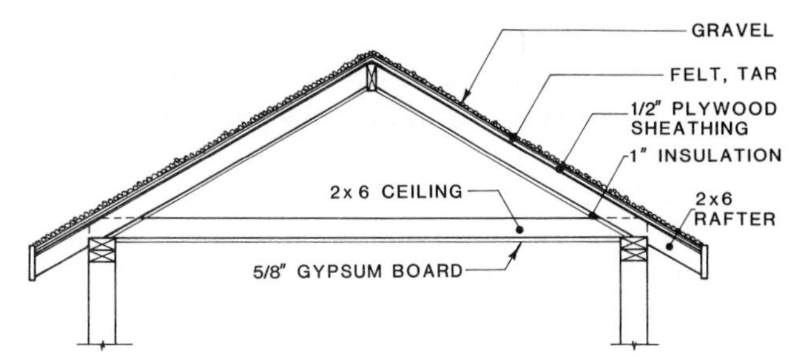

Fig. 83-9 Roof materials create loads.

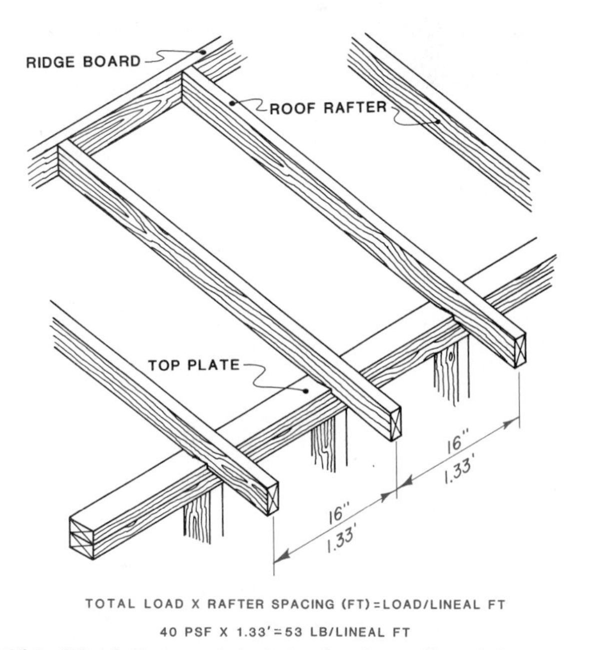

Fig. 83-10 Determining the load per lineal foot on each rafter.

are spaced at 12″ intervals, then the load per square foot and the load per lineal foot will be the same. However, if the rafters are spaced at 16″ intervals, then each lineal foot of rafter must support 1⅓ of the load per square foot.

3. To find the total load each rafter must support, multiply the load per lineal foot (Fig. 83-11) by the length of the span in feet (Fig. 83-12).

4. To compute the bending moment in inch-pounds, multiply total load × rafter span × 12. Divide this figure by 8. The bending moment (BM) is the force needed to bend or break the rafter. When the length of the span in pounds is multiplied by the length of the span in feet, the result is expressed in foot-pounds. The span must be multiplied by 12 to convert the bending moment into inch-pounds (Fig. 83-13). The bending moment (in inch-pounds) of a structural member (in pounds) equals the load (in pounds) times the length (in feet) times 12 divided by 8:

$$\text{Bending Moment} = \frac{L \times D \times 12}{8}$$

For example, the bending moment of the beam shown in Fig. 83-14 is calculated as follows:

$$\text{Bending Moment} = \frac{2000 \text{ lb} \times 10' \times 12}{8} = \frac{240{,}000}{8} = 30000 \text{ in-lb}$$

5. Set up the equation to determine the resisting moment. The resisting moment is the strength or resistance the rafter must possess to withstand the force of the bending moment of the rafter (Fig. 83-15), otherwise the member would bend or break. The resisting moment is determined by multiplying the fiber stress by the

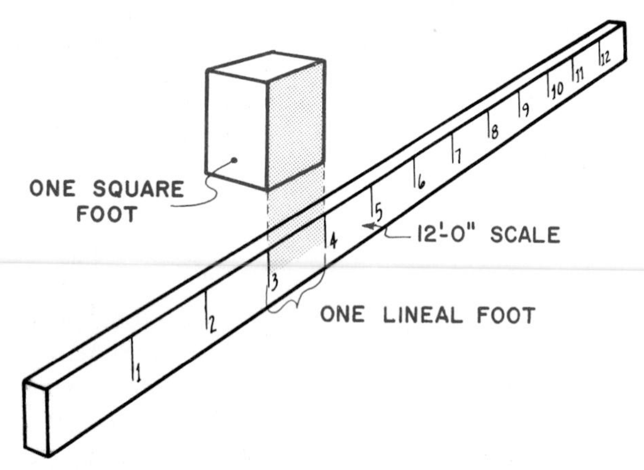

Fig. 83-11 Loads are expressed in lineal feet.

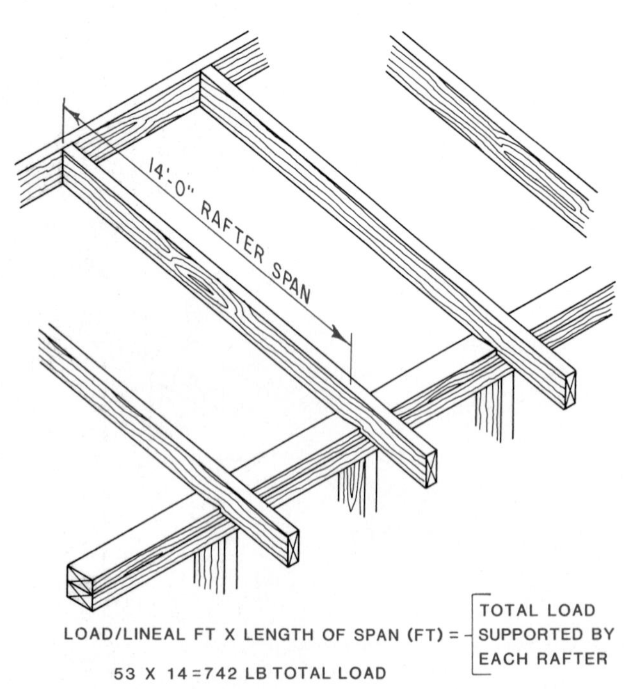

Fig. 83-12 Finding the total load each rafter must support.

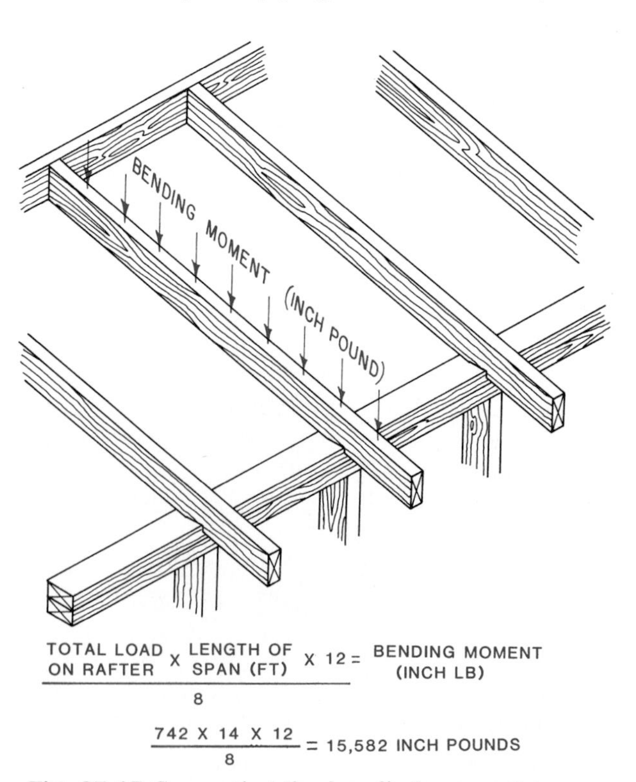

Fig. 83-13 Computing the bending moment.

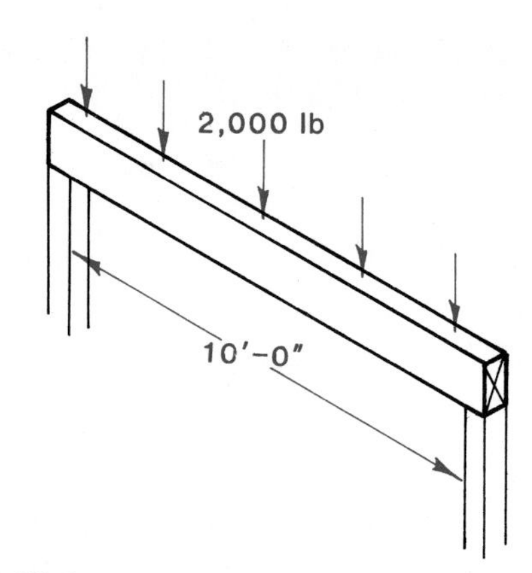

Fig. 83-14 Bending forces on a beam.

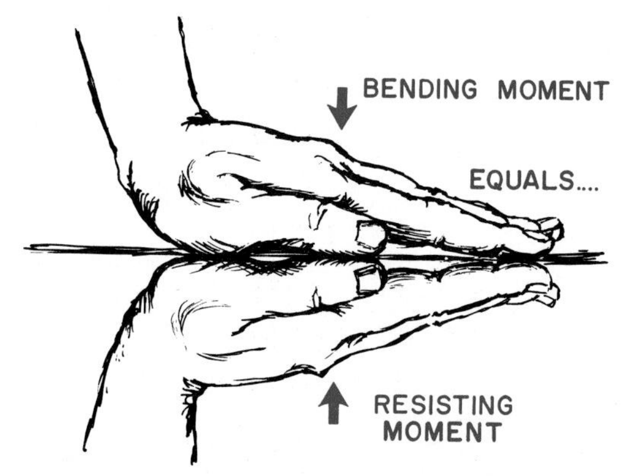

Fig. 83-15 The resisting moment must be equal or greater to the bending moment.

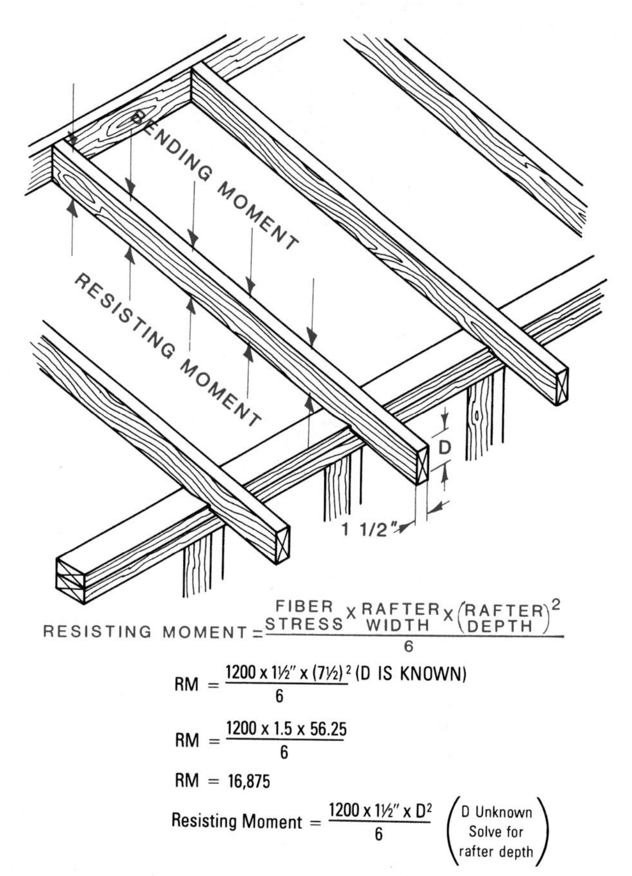

Fig. 83-16 Determining the resistance moment.

rafter width by the rafter depth squared. This figure is divided by 6. Rafter widths should be expressed in the exact dimensions of the finished lumber (Fig. 83-16). The fiber stress is the tendency of the fibers of the wood to bend and become stressed as the member is loaded. Fiber stresses range from 1750 lb/in^2 for southern dense pine select to 600 lb/in^2 for red spruce. Dense Douglas fir and southern pine have average fiber stresses of 1200 lb/in^2. The rafter depth is squared, since the strength of the member increases by squares. For example, a rafter 12″ deep is not 3 times as strong as a rafter 4″ deep. It is 9 times as strong.

6. Since the bending moment equals the resistance moment, the formulas can be combined, as shown in Fig. 83-17. The formula can then be followed for any of the variables, preferably for the depth of the rafter, since varying the depth will alter the resisting moment more than any other single factor. Another example of combining the bending moment and the resistance

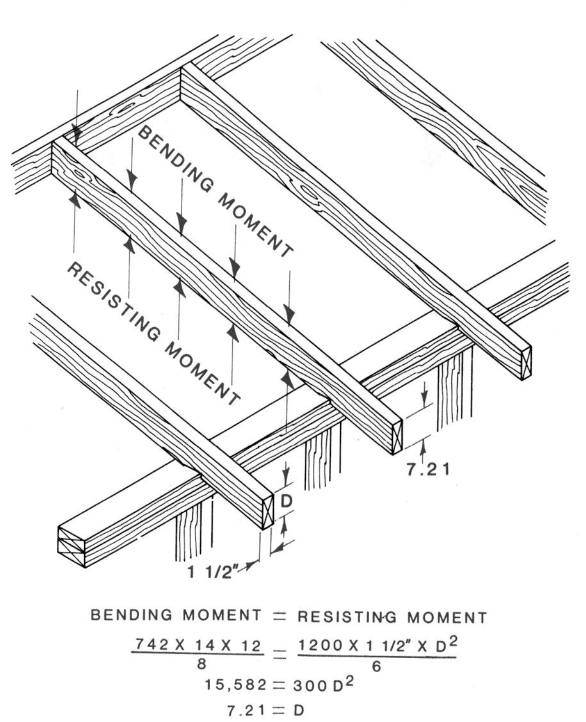

Fig. 83-17 Combining resistance moment and bending moment formulas.

moment as applied to Fig. 83-18 is summarized as follows:

$$\text{Bending Moment} = \text{Resisting Moment}$$

$$\frac{3000 \times 10 \times 12}{8} = \frac{1500 \times 1.5 \times 5.5^2}{6}$$

$$\frac{360{,}000}{8} = \frac{68{,}062.5}{6}$$

$$45{,}000 = 11{,}343.75$$

(Bending moment too great, increase beam size to 4″ × 8″.)

$$45{,}000 = \frac{1500 \times 3\tfrac{1}{2} \times 7\tfrac{1}{2}^2}{6}$$

$$45{,}000 = 49{,}218.75$$

(Beam is now in equilibrium—resisting moment is now greater than the bending moment.)

Care should be taken in establishing all sizes to ensure that the sizes of materials conform to manufacturers' standards and building-code allowances. Figure 83-19 shows a typical rafter space-span chart based on common lumber sizes and spacing. Figure 83-20 shows common truss specifications based on normal loading for residential work. The amount of load for wind and snow will vary considerably, depending on geographical location and the style of roof as shown in Fig. 83-21.

CEILING JOISTS

Tables for the selections of a standard ceiling joist are shown in Figure 83-22. If there is to be an attic above the ceiling for storage or living area, then the ceiling joists must be treated as floor joists using a floor joist table to calculate the joists for a heavier load. For a flat or very low-pitched roof, where the roof rafters are also the ceiling joists, then the table in Fig. 83-23 must be used.

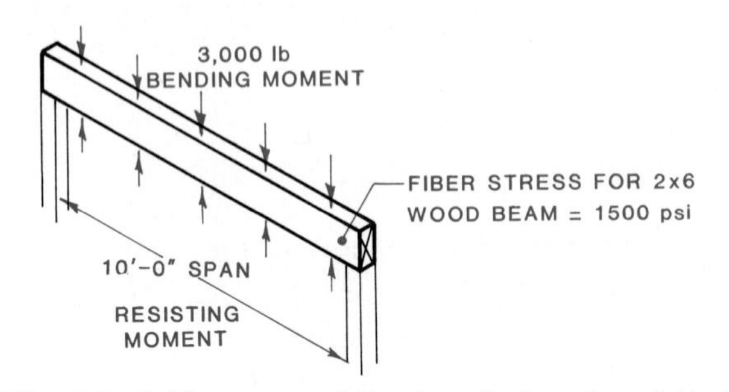

Fig. 83-18 Elements of the bending and resisting moment calculation.

PITCH: 5/12 LOAD: 40 PSF

Lumber size	Spacing, in Inches	Fiber stress, 1200 pounds, for douglas fir and southern yellow pine
2 × 4	24	6′—6″
	20	7′—3″
	16	8′—1″
	12	9′—4″
2 × 6	24	10′—4″
	20	11′—4″
	16	12′—6″
	12	14′—2″
2 × 8	24	13′—8″
	20	15′—2″
	16	16′—6″
	12	18′—4″

Fig. 83-19 Maximum rafter spans.

FLOORS

The major structural parts of the floor system are the floor joists (Fig. 83-24). The other parts of the floor system are the subfloor, finish floor, and supporting structural members. When the live load is determined, the size and spacing of joists can be established by referring to Fig. 83-25. This table is based on no. 1 southern white pine with a fiber stress of 1200 lb/in^2 and a modulus of elasticity (ratio of stress and strain) of 1,600,000 lb/in^2. For other materials, such as redwood or Douglas fir lumber, with a different fiber stress and different modulus of elasticity, a different table must be used.

An example of the use of Fig. 83-25 is as follows: If the live loads are approximately 40 lb/ft^2 and 16″ spaces are desired between joists, you can see that a 2″ × 8″ is good for a span of only 12′-1″. For a span larger than 12′-1″, a joist with larger cross section is necessary.

HEADERS AND LINTELS

The header (structural wood member) and the lintel (structural steel member) are horizontal

FOR DRY WALL CONSTRUCTION

Dead Load	12 lb/ft²
Live Load	33 lb/ft²
Total Load	45 lb/ft²
Trussed Rafter Spaced	2′ 0″ O.C.

FOR PLASTER FINISH

Dead Load	17 lb/ft²
Live Load	28 lb/ft²
Total Load	45 lb/ft²
Trussed Rafter Spaced	2′ 0″ O.C.

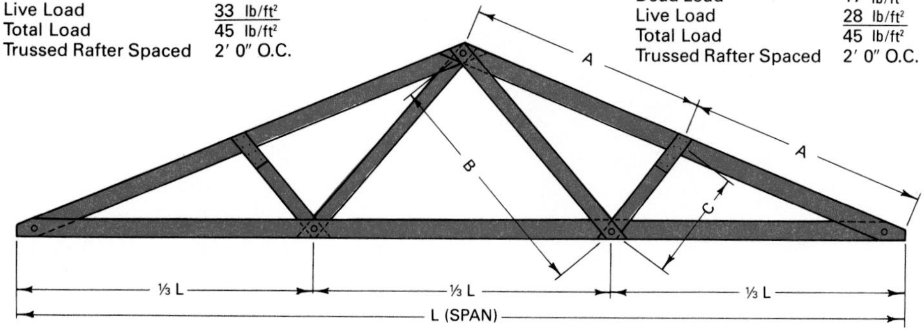

SLOPE	SPAN "L"	A	B	C
	20′-0″	5′-3¼″	4′-8 3/16″	2′-3 15/16″
	22′-0″	5′-9 9/16″	5′-1⅞″	2′-6¾″
4/12	24′-0″	6′-3⅞″	5′-7½″	2′-9 9/16″
	26′-0″	6′-10 3/16″	6′-1 3/16″	3′-0 7/16″
⅙ PITCH	28′-0″	7′-4 9/16″	6′-6 13/16″	3′-3¼″
	30′-0″	7′-10⅞″	7′-0½″	3′-6 1/16″
	32′-0″	8′-5 3/16″	7′-6 3/16″	3′-8⅞″
	20′-0″	5′-5″	5′-3⅜″	2′-7⅝″
	22′-0″	5′-11½″	5′-10 1/16″	2′-10 13/16″
5/12	24′-0″	6′-6″	6′-4 7/16″	3′-2″
	26′-0″	7′-0½″	6′-10⅞″	3′-5¼″
5/24 PITCH	28′-0″	7′-7″	7′-5¼″	3′-8 7/16″
	30′-0″	8′-1½″	7′-11 11/16″	3′-11⅝″
	332′-0″	8′-8″	8′-6 1/16″	4′-2 13/16″

SLOPE	SPAN "L"	A	B	C
	20′-0″	5′-7 1/16″	5′-11 11/16″	2′-11⅝″
	22′-0″	6′-1 13/16″	6′-6⅞″	3′-3¼″
6/12	24′-0″	6′-8½″	7′-2⅛″	3′-6⅞″
	26′-0″	7′-3 3/16″	7′-9 5/16″	3′-10 7/16″
¼ PITCH	28′-0″	7′-9 15/16″	8′-4 9/16″	4′-2 1/16″
	30′-0″	8′-4⅝″	8′-11¾″	4′-5 11/16″
	32′-0″	8′-11 15/16″	9′-6 15/16″	4′-9¼″
	20′-0″	5′-9 7/16″	6′-8 3/16″	3′-3⅞″
	22′-0″	6′-4 7/16″	7′-4¼″	3′-7 15/16″
7/12	24′-0″	6′-11⅜″	8′-0 5/16″	3′-11 15/16″
	26′-0″	7′-6 5/16″	9′-8⅜″	4′-4″
7/24 PITCH	28′-0″	8′-1¼″	9′-4 7/16″	4′-8″
	30′-0″	8′-8 3/16″	10′-0½″	5′-0 1/16″
	32′-0″	9′-3⅛″	10′-8 9/16″	5′-4 1/16″

Fig. 83-20 Common truss specifications.

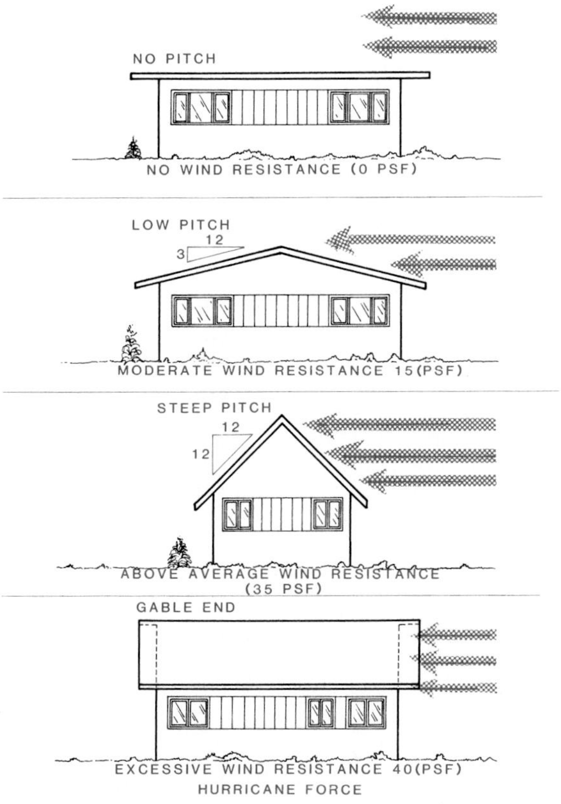

Fig. 83-21 Wind and snow load levels.

Size of ceiling joists	Spacing of ceiling joists (oc)	Maximum allowable span (feet and inches)			
		Group I	Group II	Group III	Group IV
2″ × 4″	12″	11′-6″	11′-0″	9′-6″	5′-6″
	16″	10′-6″	10′-0″	8′-6″	5′-0″
2″ × 6″	12″	18′-0″	16′-6″	15′-6″	12′-6″
	16″	16′-0″	15′-0″	14′-6″	11′-0″
2″ × 8″	12″	24′-0″	22′-6″	21′-0″	19′-0″
	16″	21′-6″	20′-6″	19′-0″	16′-6″

Fig. 83-22 Ceiling joist spans.

Size of roof joists	Spacing of roof joists (oc)	Maximum span			
		Group IV	Group III	Group II	Group I
2″ × 6″	24″	6′-6″	7′-2″	7′-10″	9′-4″
	16″	8′-0″	8′-8″	9′-8″	11′-4″
2″ × 8″	24″	9′-4″	10′	11′-2″	13′-0″
	16″	11′-6″	12′-4″	13′-8″	15′-10″
2″ × 10″	24″	12′-4″	13′-8″	14′-10″	17′-4″
	16″	15′-0″	16′-8″	18′-2″	19′-2″

Fig. 83-23 Low-slope joist sizes.

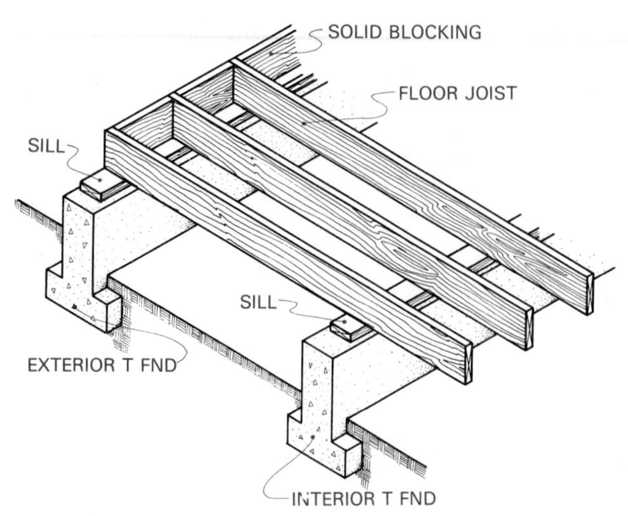

Fig. 83-24 Floor joist loads are transmitted to intermediate supports.

structural supports. They support the openings of windows and doors. Engineering tables for their selection are shown in Fig. 83-26.

GIRDERS AND BEAMS

All the weight of the floor system, including live loads and dead loads, is transmitted to bearing partitions. These loads are then transmitted either to the foundation wall, to intermediate supports, or to horizontal supports known as girders or beams.

To determine the exact spacing, size, and type of girder to support the structure, follow these steps:

1. Determine the total load acting on the entire floor system in pounds per square foot. Divide the total live and dead loads by the number of square feet of floor space. For example, if the combined load for the floor system shown in Fig. 83-27 is 48,000 lb, then there are 50 lb/ft^2 of load acting on the floor (48,000 lb divided by 960 ft^2 equals 50 lb/ft^2).

2. Lay out the proposed position of all columns and beams. It will also help to sketch the position of the joists to be sure that the joist spans are correct.

Live load—pounds per square foot	Spacing	2 × 6	2 × 8	2 × 10	2 × 12	2 × 14
10	12	12- 9	16- 9	21- 1	24- 0	—
	16	11- 8	15- 4	19- 4	23- 4	24- 0
	24	10- 3	14- 6	17- 3	20- 7	24- 0
20	12	11- 6	15- 3	19- 2	23- 0	24- 0
	16	10- 5	13-11	17- 6	21- 1	24- 0
	24	9- 2	12- 3	15- 6	18- 7	21- 9
30	12	10- 8	14- 0	17- 9	21- 4	24- 9
	16	9- 9	12-11	16- 3	19- 6	22- 9
	24	8- 6	11- 4	14- 4	17- 3	20- 2
40	12	10- 0	13- 3	16- 8	20- 1	23- 5
	16	9- 1	12- 1	15- 3	18- 5	21- 5
	24	7-10	10- 4	13- 1	15- 9	18- 5
50	12	9- 6	12- 7	15-10	19- 1	22- 4
	16	8- 7	11- 6	14- 7	17- 6	20- 5
	24	7- 3	9- 6	12- 1	14- 7	17- 0
60	12	9- 0	12- 0	15- 2	18- 3	21- 4
	16	8- 1	10-10	13- 8	16- 6	19- 3
	24	6- 8	8-11	11- 3	13- 7	15-11
70	12	8- 7	11- 6	14- 6	17- 6	20- 6
	16	7- 8	10- 2	12-10	15- 6	18- 3
	24	6- 5	8- 5	10- 7	12- 9	15- 0

Fig. 83-25 Maximum joist spans.

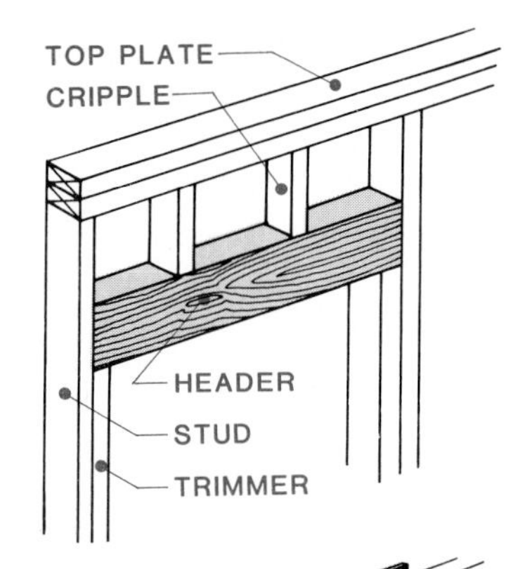

Safe spans for wood headers	
Size	Span
4″ × 4″	3′-6″
4″ × 6″	4′-6″
4″ × 8″	6′-0″
4″ × 10″	7′-6″
4″ × 12″	9′-0″

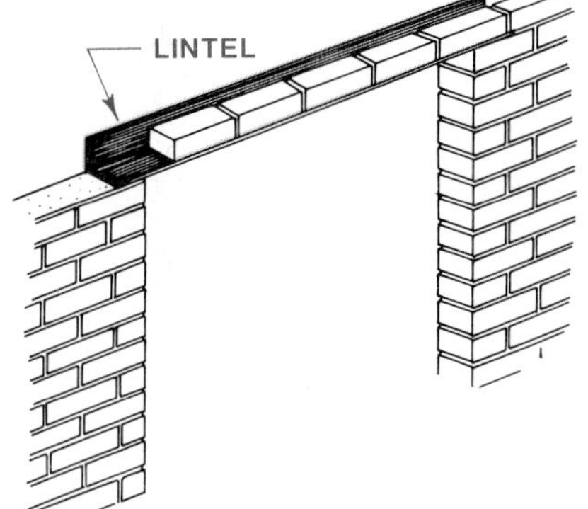

Safe spans for steel lintels 4″ masonry wall	
Size	Span
L-3½″ × 3½″ × ¼″	3′-0″
L-3½″ × 3½″ × 5/16″	4′-0″
L-4″ × 3½″ × 5/16″	6′-0″
L-5″ × 3½″ × 5/16″	8′-0″
L-6″ × 3½″ × 3/8″	10′-0″

Fig. 83-26 Lintel and header spans.

3. Determine the number of square feet supported by the girder (girder load area). The girder load area is determined by multiplying the length of a girder, from column to column, by the girder load width. The girder load width is the distance extending on both sides of the center line of the girder, halfway to the nearest support, as shown in Fig. 83-27. The remaining distance from a girder load area to the outside wall is supported by the outside wall.

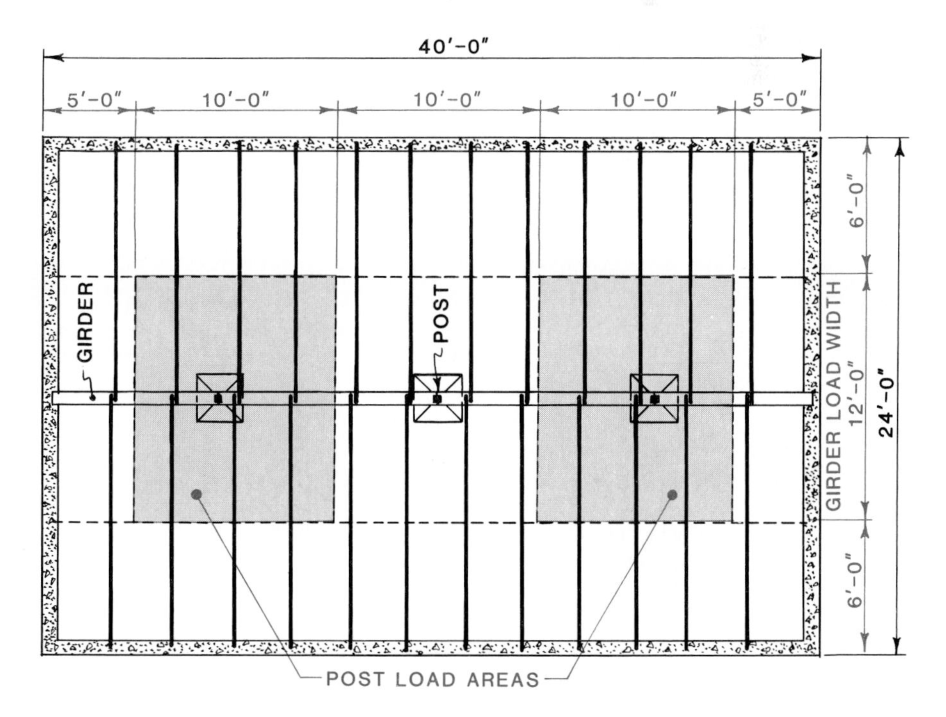

Fig. 83-27 Girder tributary area.

4. To find the load supported by the girder load area, multiply the girder load area by the load per square foot. For example, the girder load in Fig. 83-27 is 6000 lb (120 $ft^2 \times 50$ lb/ft^2).

5. Select the most suitable material to carry the load at the span desired. Built-up wood girders will span a greater length than a solid wood girder. However, I beams will span a greater length without intervening support. To compute the minimum cross-section area of a girder or beam, divide the load (in pounds) by the material's coefficient of elasticity:

$$CS = \frac{l}{E}$$

For example, if the combined live and dead loads on a wood member is 50,000 lb and the coefficient of elasticity is 1600 the cross section area is:

$$CS = \frac{50{,}000}{16{,}000} = 31.25 \text{ in}^2$$

The cross section of a 4″ × 10″ member surfaced to 3.5″ × 9.5″ is 33.25 in^2. A 4″ × 10″ beam is therefore the smallest member possible above the minimum 31.25 in^2.

6. Select the exact size and classification of the beam or girder. Use Fig. 83-28 to select the most appropriate wood girder. For example, to support 6000 lb over a 10′ span, either an 8″ × 8″ solid girder or a 6″ × 10″ built-up girder would suffice. The girder should be strong enough to support the load, but any size larger is a waste of materials. The only alternative to increasing the size of the girder is to decrease the size of span.

Calculating the tributary area that is supported by a girder or beam is shown in Fig. 83-29. The method used to calculate the safe load that is evenly distributed on a girder is shown in Fig. 83-30. The allowable fiber stress of the girder must be known to complete these calculations. For example, if a 6″ × 12″ girder (5½″ × 11¼″) spans 18′-0″ and the girder has a stress rated at 1800 lb/ft^2 the safe, evenly distributed, weight is calculated as follows:

$$W = \frac{f \times b \times d^2}{9 \times L}$$

where W = weight evenly distributed, lb
f = allowable fiber stress, lb/in^2
b = width of beam, in
d = depth of beam, in
L = Span, ft

$$W = \frac{1800 \times 5.5 \times (11.25)^2}{9 \times 18} = 7734 \text{ lb}$$

The stiffness of a girder, or any structural member, is the ability of the member to resist bending. The bending force that changes a straight member to a curved member is the force that causes deflection. The calculation of the deflection of a structural member uses Hooke's law. Data for a free support beam are shown in Fig. 83-31. For example, the deflection of a 2″ × 6″ (1½″ × 5½″) joist which has a 20′ span and supports a central load of 4000 lb is calculated as follows:

Girder size	Safe load in pounds for spans from 6 to 10 feet				
	6 ft	7 ft	8 ft	9 ft	10 ft
6 × 8 BUILT-UP	8 306	7 118	6 220	5 539	4 583
6 × 8 SOLID	7 359	6 306	5 511	4 908	4 062
6 × 10 BUILT-UP	11 357	10 804	9 980	8 887	7 997
6 × 10 SOLID	10 068	9 576	8 844	7 878	7 086
8 × 8 BUILT-UP	11 326	9 706	8 482	7 553	6 250
8 × 8 SOLID	9 812	8 408	7 348	6 544	5 416
8 × 10 BUILT-UP	15 487	14 732	13 608	12 116	10 902
8 × 10 SOLID	13 424	12 768	11 792	10 504	9 448

Fig. 83-28 Wood girder safe loads.

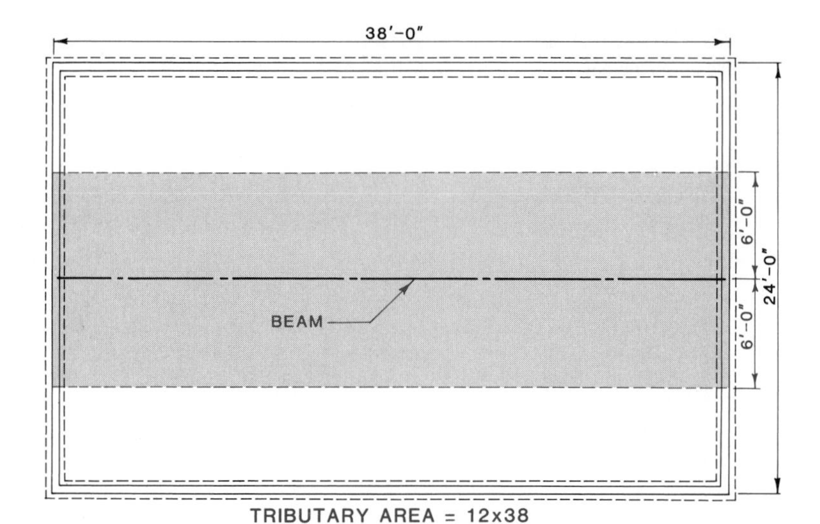

Fig. 83-29 Calculating beam tributary area.

$$D = \frac{PL^3}{48EI}$$

where D = deflection
P = force in center of span, lb
L = length of beam span, ft
E = modulus of elasticity (given in table as 1,000,000)
I = moment of inertia (for rectangular girders)
$I = bd/12$; b = width, d = height

$$D = \frac{4000 \times 20^3}{48 \times 1{,}000{,}000 \left(\dfrac{15 \times 5.5^3}{12}\right)}$$

$$= \frac{4000 \times 8000}{48 \times 1{,}000{,}000 \times 208}$$

$$= \frac{32{,}000{,}000}{9{,}984{,}000{,}000}$$

$$= .003''$$

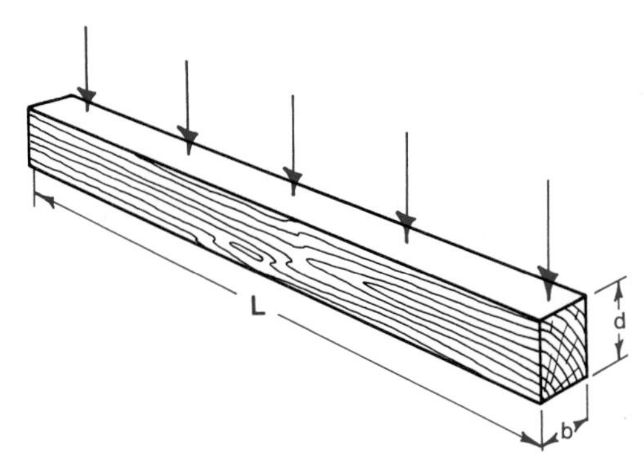

Fig. 83-30 Evenly distributed load.

STEEL BEAMS

The method of determining the size of steel beams is the same as for determining the size of

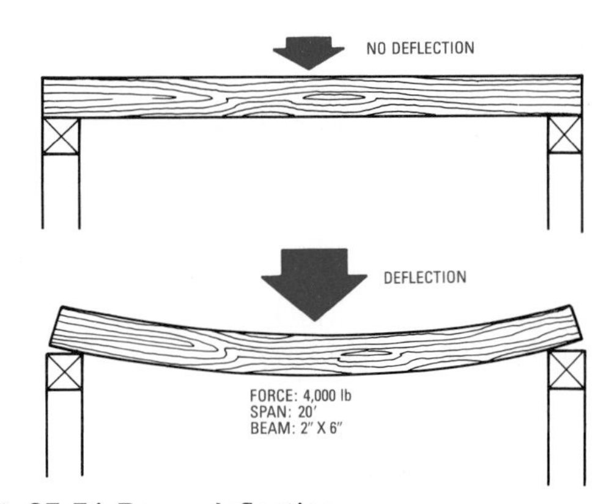

Fig. 83-31 Beam deflection.

wood beams. As wood beams vary in width for a given depth, steel beams vary in weight, depth, and thickness of webs and flanges. Classifications vary accordingly. Figure 83-32 shows the relationship of the span, load, depth, and weight of standard I beams and channels. A steel beam may be selected by referring to the desirable span and load and then choosing the most appropriate size (depth and weight) for the I beam. For example, a 5″ × 12.25-pound I beam will support 6.5 kips per given span of 10 ft. A kip is equal to 1000 lb.

COLUMNS AND POSTS

When girders or beams do not completely span the distance between foundation walls, then wood posts, steel-pipe columns, masonry columns, or steel beam columns must be used for intervening support. To determine the most appropriate size and classification of posts or col-

Size of beam	2⅝" × 4"	2¾" × 4"	3" × 5"	3¼" × 5"	3⅜" × 6"	3⅝" × 6"	3⅝" × 7"	3⅞" × 7"	4" × 8"	4⅛" × 8"	4⅝" × 10"	5" × 10"
Weight per foot	7.7	9.5	10.0	14.75	12.5	17.25	15.3	20.0	18.4	23.0	25.4	35.0
4	9.0	10.1	14.5	18.0	21.8	26.0	31.0	36.0	42.7	48.2	73.3	87.5
5	7.2	8.0	11.6	14.4	17.4	20.8	24.8	28.7	34.1	38.5	58.6	70.0
6	6.0	6.7	9.7	12.0	14.5	17.3	20.7	24.0	28.5	32.1	48.8	58.3
7	5.1	5.7	8.3	10.3	12.5	14.9	17.7	20.5	24.4	27.5	41.9	50.0
8	4.5	5.0	7.3	9.0	10.9	13.0	15.5	18.0	21.3	24.1	36.6	43.7
9	4.0	4.5	6.5	8.0	9.7	11.6	13.8	16.0	19.0	21.4	32.6	38.9
10	3.6	4.0	5.8	7.2	8.7	10.4	12.4	14.4	17.1	19.3	29.3	35.0
11	—	—	5.3	6.5	7.9	9.5	11.3	13.1	15.5	17.5	26.6	31.8
12	—	—	—	—	7.3	8.7	10.3	12.0	14.2	16.1	24.4	29.2
13	—	—	—	—	6.7	8.0	9.5	11.1	13.1	14.8	22.5	26.9
14	—	—	—	—	6.2	7.4	8.9	10.3	12.2	13.8	20.9	25.0
15	—	—	—	—	—	—	8.3	9.6	11.4	12.8	19.5	23.3
16	—	—	—	—	—	—	7.7	9.0	10.7	12.0	18.3	21.9
17	—	—	—	—	—	—	—	—	10.0	11.3	17.2	20.6
18	—	—	—	—	—	—	—	—	9.5	10.7	16.3	19.4
19	—	—	—	—	—	—	—	—	9.0	10.1	15.4	18.4
20	—	—	—	—	—	—	—	—	8.5	9.6	14.7	17.5

Fig. 83-32 Safe loads for I beams.

umns to support girders or beams, follow these steps:

1. Determine the total load in pounds per square foot for the entire floor area. Calculate this amount by multiplying the total load by the number of square feet of floor space.

2. Determine the spacing of posts necessary to support the ends of each girder. Great distances between posts should be avoided because excessive weight would concentrate on one footing. Long spans also require extremely large girders. For example, it is possible to span a distance of 30′, but to do so, a 15″ I beam would be needed. The extreme weight and cost of this beam would be prohibitive. On the other hand, if only a 6′ span is used, the close spacing might greatly restrict the flexibility of the internal design. As a rule, use the shortest span that will not interfere with the design function of the area.

3. Find the number of square feet supported by each post. A post will carry the load on a girder to the midpoint of the span on both sides. The post also carries half the load to the nearest support wall on either side of the post.

4. Find the load supported by the post support area. Multiply the number of square feet by the load per square foot (120 × 50).

5. Determine the height of a post. The height of the post is related to the span of a beam. The 4″ × 4″ post shown in Fig. 83-33 may be more than adequate to support a given weight if the height of the post is 6′. However, the same 4″ × 4″ post may be totally inadequate to support the same weight when the length is increased to 20′.

6. Determine the thickness and width of the post needed to support the load at the given height by referring to Fig. 83-34 for timber posts, Fig. 83-35 for I-beam columns, or Fig. 83-36 for the diameter of steel-pipe columns.

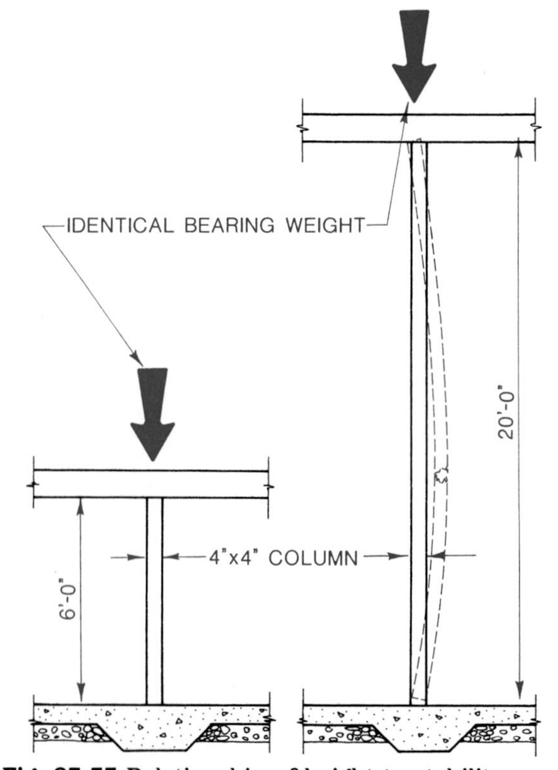

Fig. 83-33 Relationship of height to stability.

Nominal size, inches	3 × 4	4 × 4	4 × 6	6 × 6	6 × 8	8 × 8
Actual size, inches	$2\frac{1}{2} \times 3\frac{1}{2}$	$3\frac{1}{2} \times 3\frac{1}{2}$	$3\frac{1}{2} \times 5\frac{1}{2}$	$5\frac{1}{2} \times 5\frac{1}{2}$	$5\frac{1}{2} \times 7\frac{1}{4}$	$7\frac{1}{4} \times 7\frac{1}{4}$
Area in square inches	9.51	13.14	20.39	30.25	41.25	56.25
HEIGHT OF COLUMN:						
4 FEET	8 720	12 920	19 850	30 250	41 250	56 250
5 FEET	7 430	12 400	19 200	30 050	41 000	56 250
6 FEET	5 630	11 600	17 950	29 500	40 260	56 250
6 FEET 6 INCHES	4 750	10 880	16 850	29 300	39 950	56 000
7 FEET	4 130	10 040	15 550	29 000	39 600	55 650
7 FEET 6 INCHES	—	9 300	14 400	28 800	39 000	55 300
8 FEET	—	8 350	12 950	28 150	38 300	55 000
9 FEET	—	6 500	10 100	26 850	36 600	54 340
10 FEET	—	—	—	24 670	33 600	53 400
11 FEET	—	—	—	22 280	30 380	52 100
12 FEET	—	—	—	19 630	26 800	50 400

Fig. 83-34 Maximum loads for lumber posts.

Depth in inches	10	9	8	7	6	5	4	3
Weight per pound per foot	25.4	21.8	18.4	15.3	12.5	10.0	7.7	5.7
EFFECTIVE LENGTH:								
3 FEET	110.7	94.8	80.1	66.5	54.2	43.1	33.0	23.5
4 FEET	110.7	94.8	80.1	65.9	52.1	39.7	29.1	20.3
5 FEET	109.5	91.2	74.9	60.0	46.9	35.1	25.3	17.2
6 FEET	101.7	83.9	68.3	54.1	41.8	30.7	21.8	14.6
7 FEET	93.8	76.7	61.8	48.5	37.0	26.8	18.7	12.3
8 FEET	86.0	69.7	55.7	43.3	32.7	23.4	16.1	10.5
9 FEET	78.7	63.2	50.1	38.6	28.9	20.4	13.9	—
10 FEET	71.8	57.2	45.0	34.5	25.5	17.9	—	—
11 FEET	65.5	51.8	40.5	30.8	22.6	—	—	—
12 FEET	59.7	47.0	36.5	27.6	20.2	—	—	—
AREA IN SQUARE INCHES	7.38	6.32	5.34	4.43	3.61	2.87	2.21	1.64

Fig. 83-35 Safe loads for I-beam columns.

Nominal size, inches	6	5	$4\frac{1}{2}$	4	$3\frac{1}{2}$	3	$2\frac{1}{2}$	2	$1\frac{1}{2}$
External diameter, inches	6.625	5.563	5.000	4.500	4.000	3.500	2.875	2.375	1.900
Thickness, inches	.280	.258	.247	.237	.226	.216	.203	.154	.145
EFFECTIVE LENGTH:									
5 FEET	72.5	55.9	48.0	41.2	34.8	29.0	21.6	12.2	7.5
6 FEET	72.5	55.9	48.0	41.2	34.8	28.6	19.4	10.6	6.0
7 FEET	72.5	55.9	48.0	41.2	34.1	26.3	17.3	9.0	5.0
8 FEET	72.5	55.9	48.0	40.1	31.7	24.0	15.1	7.4	4.2
9 FEET	72.5	55.9	46.4	37.6	29.3	21.7	12.9	6.6	3.5
10 FEET	72.5	54.2	43.8	35.1	26.9	19.4	11.4	5.8	2.7
11 FEET	72.5	51.5	41.2	32.6	24.5	17.1	10.3	5.0	—
12 FEET	70.2	48.7	38.5	30.0	22.1	15.2	9.2	4.1	—
AREA IN SQUARE INCHES	5.58	4.30	3.69	3.17	2.68	2.23	1.70	1.08	0.80
WEIGHT PER POUND PER FOOT	18.97	14.62	12.54	10.79	9.11	7.58	5.79	3.65	2.72

Fig. 83-36 Safe loads for steel-pipe columns.

To check the compressive stress and deformations on posts apply the following formulas:

Compressive stress:

$$F = \frac{P}{A}$$

where F = compressive stress
P = compressive force
A = Cross section area

For example, in Fig. 83-37 the compressive force on the 4″ × 4″ post (16 in^2) is 32,000 lb; therefore

$$F = \frac{32,000}{16}$$
$$= 2000 \text{ lb/in}^2 \text{ or } 2 \text{ kips}$$

Deformation is the bending of structural members of a building caused by load stress. The amount of deformation on a member is calculated as follows:

$$\delta = \frac{PL}{AE}$$

where δ = deformation, in
P = force, lb
L = length, in
A = cross-section area
E = modulus of elasticity, lb/in^2 (for structural steel, E = 29,000,000 lb/in^2; for wood, E = 1,000,000 to 2,000,000 lb/in^2)

For example, a tensile force P of 50,000 lb is acting on the structural column shown in Fig. 83-38. The deformation is therefore

$$e = \frac{50,000 \text{ lb} \times 100''}{1'' \times 29,000,000} = \frac{5,000,000}{29,000,000}$$
$$= 0.172''$$

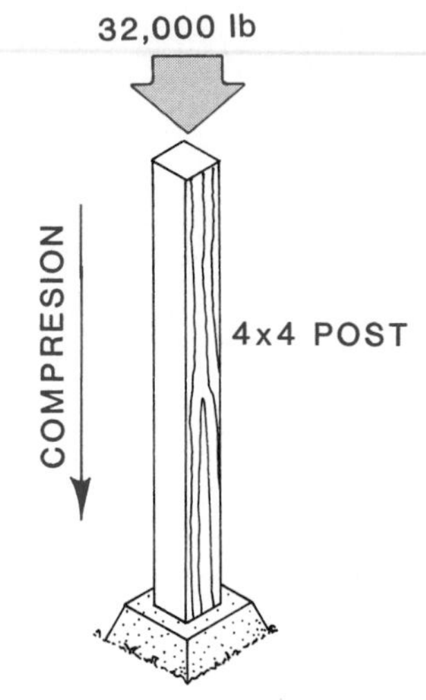

Fig. 83-37 Compression-force stress.

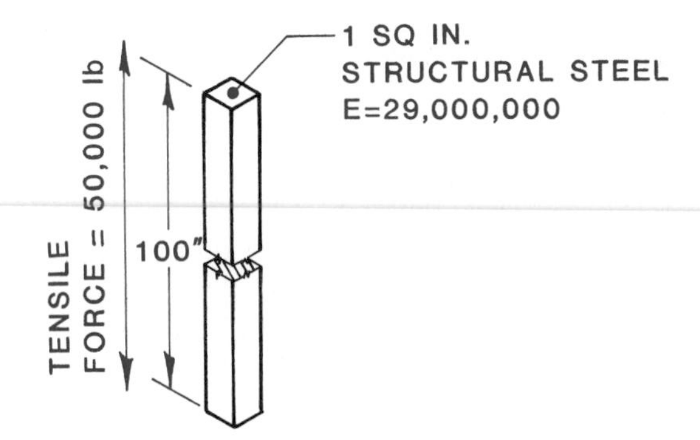

Fig. 83-38 Tensile-force deformation.

FOUNDATION FOOTINGS AND PIERS

The basic T foundation has an exterior concrete footing and a wall forming an inverted T. Generally, builders follow the sizes for foundation walls and footings as recommended in Fig. 83-39. The piers support the interior floor system. The piers for residential construction are usually 12″ high and 12″ to 18″ square at the base. To compute the bearing area of the pier or the T foundation's footing, divide the total load of the bearing on one square foot by the soil-bearing capacity.

FOOTING SIZE CALCULATIONS

To calculate the minimum bearing size of a lineal foot of footing refer to Fig. 83-40 (A through F) and Fig. 83-41. First find the total load on the footing as follows:

A. Roof live load	30 lb/ft^2
Roof dead load	
Rafters 2″ × 8″, 16″ OC	3.5 lb/ft^2
Wood sheathing 1″	3 lb/ft^2
Wood shingles	3 lb/ft^2
Batt insulation 3″	0.5 lb/ft^2
	40 lb/ft^2

Rafter length = 20′ (1 side) × 40 lb/ft^2
= 800 lb

Height description	Wood frame house: Minimum foundation wall thickness	Wood frame house: Footing projection each side of foundation wall	Masonry house: Minimum foundation wall thickness	Masonry house: Footing projection each side of foundation wall
ONE STORY-NO BASEMENT	6″	2″	6″	3″
ONE STORY-WITH BASEMENT	6″	3″	8″	4″
TWO STORY-NO BASEMENT	6″	3″	6″	4″
TWO STORY-WITH BASEMENT	8″	4″	8″	5″

Fig. 83-39 Footing sizes.

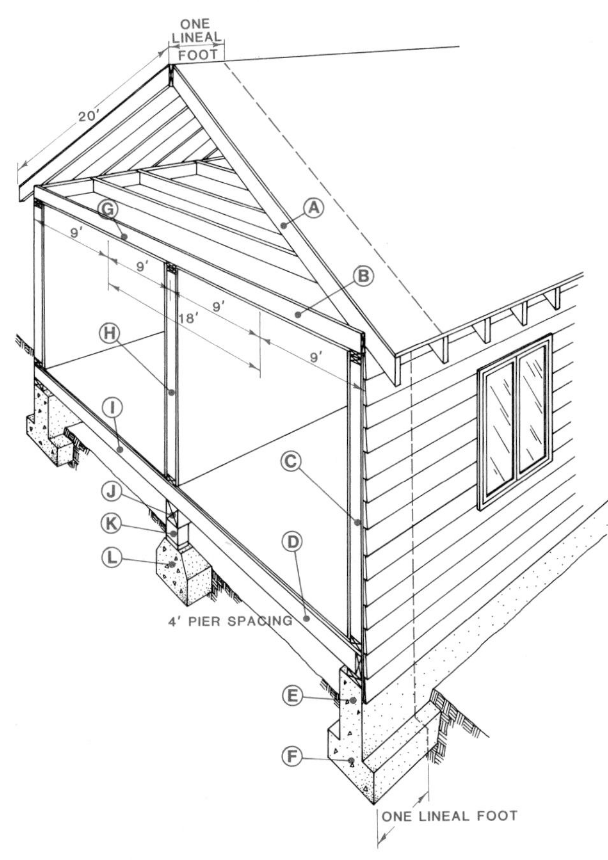

Fig. 83-40 Structural members used to calculate dead loads.

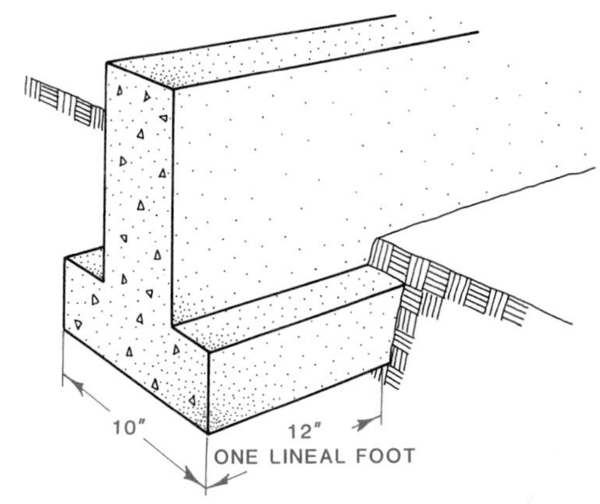

Fig. 83-41 Loads calculated on one lineal foot of footing.

B. Attic floor/ceiling live load ...... 20 lb/ft²
Attic floor/ceiling dead load
Joists 2″ × 10″, 16″ OC....... 6.5 lb/ft²
Lath and plaster (one side) ... 10 lb/ft²
Batt insulation 3″ .... 0.5 lb/ft²
37 lb/ft²
Tributary length 9′ × 37 = 333 lb
(See Fig. 83-42)

C. Exterior wall live load 0
Exterior wall dead load
2″ × 4″ wood stud/plaster
one side .... 12 lb/ft²
Exterior wood siding 1″ .... 3 lb/ft²
Batt insulation 3″ .... 0.5 lb/ft²
15.5 lb/ft²
Wall height 8′ × 15.5 = 124 lb

D. Floor live load .... 40 lb/ft²
Floor dead load
2″ × 10″ floor joists 16″
OC/subfloor .... 6.5 lb/ft²
Hardwood floor 25/32″ .... 4 lb/ft²
50.5 lb/ft²
Tributary length 9′ × 50.5 = 454.5 lb
(See Fig. 83-42)

E. Foundation wall live load 0
Foundation wall dead load
6″ concrete .... 75 lb/ft²
Approximately 3′ height × 75 = 225 lb

F. Footing live load 0
Footing dead load
6″ concrete . . . . 75 lb/ft^2
Approximately 1′-0″ wide × 1 lineal foot ×75 = 75 lb

Total load per lineal foot of footing 2011.5 lb.

To calculate the minimum footing size apply the following formula:

$$\frac{\text{Total load on footing:}\quad 2011.5\text{ lb.}}{\text{Soil bearing capacity (Given):}\quad 2500\text{ lb. ft}} = .80\text{ ft}^2\ (115.8\text{ in}^2)$$

A lineal foot is 12″, therefore the width of the footing should be 10″ (120 in^2) minimum. A 12″ standard footing width can therefore be used.

PIER SIZE CALCULATION

To calculate the size of a pier needed to support the structure refer to items G through L in Fig. 83-40 and calculate the load as follows:

G. Attic floor/ceiling live load . . . . . . 20 lb/ft^2
Attic floor/ceiling dead load
Floor joists 2″ × 10″, 16″ OC . . 6.5 lb/ft^2
Lath and plaster, one side . . . . 10 lb/ft^2
3″ batt insulation . . . . .5 lb/ft^2
37.0 lb/ft^2
See Fig. 83-42.
Tributary area (72 ft^2); 72 × 37 = 2664 lb

H. Wall live load 0
Wall dead load
2″ × 4″ wood stud/plaster
2 sides . . . . 20 lb/ft^2
Wall height 8′; 8 × 20 = 160 lb

I. Floor live load . . . . 40 lb/ft^2
Floor dead load
2″ × 10″ floor joist, 16″
OC/subfloor . . . . 6.5 lb/ft^2
Hardwood floor 25/32″ . . . . 4 lb/ft^2
50.5 lb/ft^2
Tributary area 4′ × 18′ = 72 ft^2;
72 × 50.5 = 3636 lb

J. Girder 4″ × 6″, dead load . . . . 4 lb/ft^2
4′ length × 4 = 16 lb

K. Post 4″ × 6″, dead load . . . . 4 lb/ft^2
Approximately 1.5′ × 4 = 6 lb

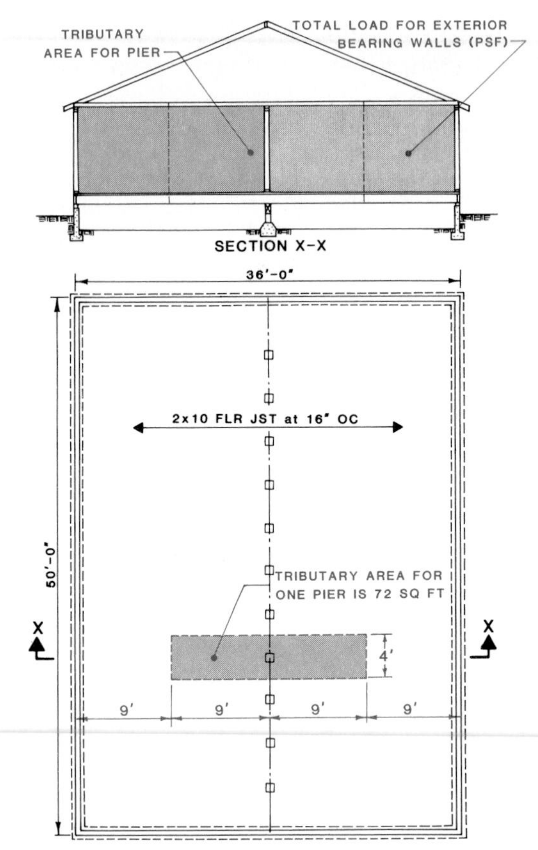

Fig. 83-42 Tributary areas for exterior walls and piers.

L. Concrete pier . . . . 150 lb/ft^2
Approx. 24″ × 24″ × 12″ = 4 cubic feet
4 ft^3 × 150 = 600 lb

Total load per pier = 7082 lb
(Total load on one pier)

Now use the total load and apply the following formula to compute the minimum size of pier:

$$\frac{\text{Total load on pier:}\quad 7082}{\text{Soil bearing capacity (Given):}\quad 2500} = 2.8\text{ ft.}^2$$

$$2.8\text{ ft.}^2 \times 144 = 403.2\text{ in}^2$$

21″ × 21″ (441 in^2) is next largest square size.

STRESS AND DYNAMICS FORMULAS

Some structural designs include other than standard construction methods, materials, and sizes which cannot be determined via precalculated charts. The following formulas relate to these aspects of structural design.

Direct Stress Calculations

To calculate the unit of direct stress on a structural member, divide the load by the cross section area of the member:

$$F = \frac{P}{A}$$

where F = unit of stress
P = load/force
A = Cross-section area

For example, in Fig. 83-43 a 40,000 lb I beam is supported by a 2″ × 2″ (4 in^2) support rod. The stress on the rod is

$$F = \frac{40{,}000}{4} = 10{,}000 \text{ lb/in}^2 \text{ or } 10 \text{ kips}$$

Shearing Stress

A unit of shearing stress equals the shearing force (in pounds) divided by the cross-section area (in square inches) of the structural member:

$$S = \frac{P}{A}$$

where S = unit of shearing stress
P = shearing force
A = Cross-section area of stress object

For example, in Fig. 83-44 a shearing force of 10,000 lb is acting on a 1″ pin. Therefore:

$$S = \frac{10{,}000}{\pi r^2} = \frac{10{,}000}{3.14 \times .5^2} = \frac{10{,}000}{.785}$$
$$= 12{,}739 \text{ lb/in}^2 \text{ or } 12.7 \text{ kips}$$

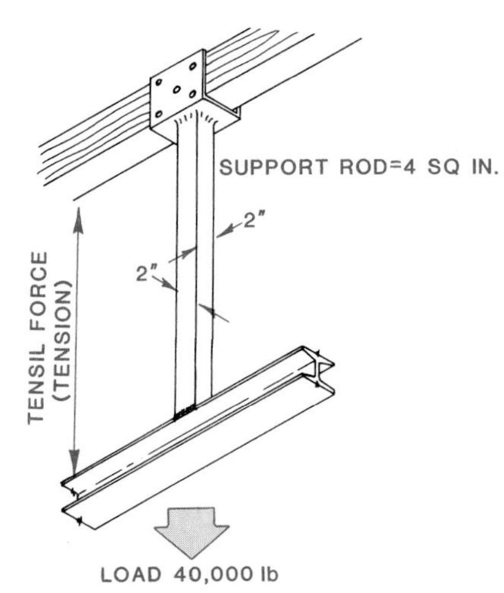

Fig. 83-43 Direct stress calculation.

Moments of Force

To calculate the moment of force (in foot-pounds) on a structional member multiply the force (in pounds) by the length of the moment arm (in feet):

Moment of force = force × arm length

For example, the left side of Fig. 83-45 shows a force of 200 lb acting on the end of a 5′ arm. The moment of force is, therefore, 1000 lb (200 lb × 5′). The moments of force on the 4′ arm on the right is 1200 ft-lb.

Forces in Equilibrium

The positive and negative moment values are always equal when there is no movement, that is, when the system is in equilibrium. The beam in Fig. 83-46 is in balance (equilibrium), therefore the moment of force on the right of the center of moment is 200 × 5 = 1000 ft-lb. The moment of force to the left of center is 100 lb × 10 = 1000 ft-lb.

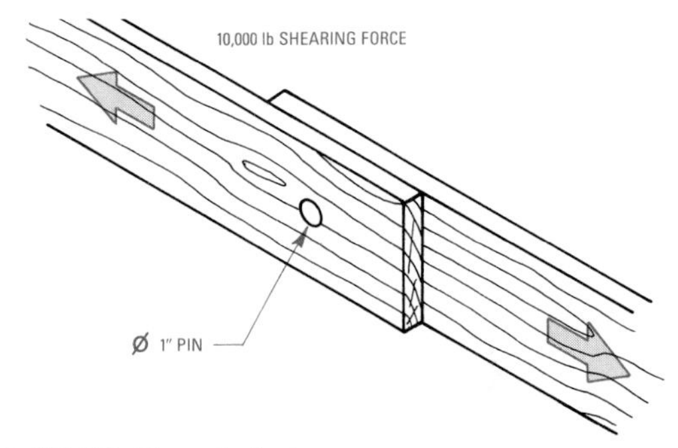

Fig. 83-44 Shearing stress.

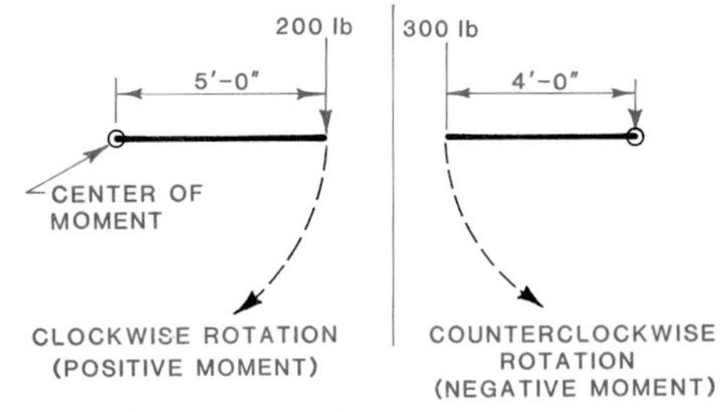

Fig. 83-45 Moment of force.

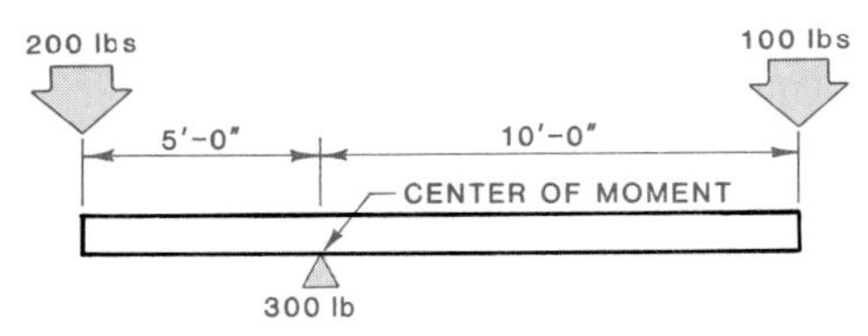

Fig. 83-46 Forces in equilibrium.

SECTION 24

Architectural-Related Careers

A career in one of the many architecture-related fields is rewarding, involves hard work, and offers excellent financial returns. The greatest satisfaction, regardless of the specific field, is the reward of seeing your creations and efforts take form in structures that become part of our physical environment. Architecture-related professional and technical careers are described in this section.

The *architect* must be an artist, engineer, and executive. Special qualities are required of the architect. An architect must understand people's needs, have a talent for creative design, and possess skill in math and science. And the architect must be able to communicate ideas and designs graphically.

The *architectural drafter* must be able to prepare all types of architectural drawings and documents. He or she must be able to take criticism and follow instructions carefully and must be able to work as a member of a team. An architectural drafter may become an architect by gaining architectural drafting and design experience, by obtaining letters of recommendation, by additional education, and by passing a state examination.

The *city planner* studies and plans the development or redevelopment of large areas such as cities, communities, housing projects, commercial projects, and so forth. The planning takes into account the utilities and necessities required for today's living. After designs are completed, the individual buildings may be designed by other architects. The other facets of the planner's overall design are handled and completed by other engineering specialists. With the growth of our country, the construction of new cities and the renovation of slum areas in older cities will require many new personnel in this area.

The *landscape architect* controls the design of the site, which includes earthwork, planting, layout of streets and walks, and the orientation of the structure. She or he should have an understanding of earth movement, drainage, and plant life and a background in math, art, architectural drafting, and rendering.

The *structural engineer*, through the use of calculations, designs the structural part of buildings. He or she is usually a civil engineer who specializes in structures. Of all the professional areas in the building trades, this is considered one of the most complex, because of the high competence it requires in physics and math.

The *civil engineer* does the calculating and designing that are also done by the structural engineer. In addition, he or she may survey, or may conduct large-scale planning of utilities, roads, structures, harbors, airfields, tunnels, bridges, and sewage plants. The field of civil engineering is so broad that a civil engineer usually specializes in one area.

Electrical engineers form the largest group of engineers. The electrical engineer in the construction field designs the electrical components of structures.

The *heating, ventilating, and air-conditioning engineer* designs the heating, ventilation, air-conditioning, and refrigeration systems for structures. This person's college degree is in mechanical engineering, and he or she will specialize in air-conditioning.

The *acoustics engineer* is responsible for controlling sound in the structure. However, this work is not confined to buildings; it can also be applied to noise suppression in machines, industrial factories, aircraft, and rockets; anywhere there is loud noise. This field is very technical. The acoustics engineer needs a broad background in math and physics.

The *mechanical engineer* is the engineer who does not specialize in one area. He or she works in production, the use of power, and machines which use power. The mechanical engineer who works in the building trades designs for operational parts of a structure.

The *estimator* prepares estimates of the cost

of building projects by figuring material requirements and labor costs. Her or his work must be accurate, because mistakes are expensive.

The *specification writer* prepares specifications (a written description of exact materials, methods of construction, finishes, and tests and performance of everything required for the structure). A knowledge of all types of construction is needed, as is a technical background and experience in building.

The *surveyor* defines in both words and pictures (usually maps) the specific space, position, and topography of a piece of land. The accuracy of the work is essential for proper foundations and construction. This work is the first step in the construction of roads, airfields, bridges, dams, and other structures.

The *architectural designer* designs and plans homes and other small buildings. She or he is usually an outstanding architectural drafter but does not have a degree in architecture. The engineering for their structures is done by architects or structural engineers.

The *contractor* is a business person who is responsible for the construction and finishing of a structure. He or she can do this work either with his or her own employees or by subcontracting other workers. The contractor can work from his or her own plans, but usually builds from an architect's or a designer's plans. Several years' experience in the building trades and a state license are required to become a contractor.

Carpenters form the largest group of skilled building tradespeople. The carpenter is a skilled worker who constructs the wooden parts of a building. A rough carpenter works on the framing, floor, and roof system. The finish carpenter does the trim, cabinets, hardware, and floor cover. Carpenters must be skilled with all hand and power tools used for wood and must be able to use all types of fasteners.

The *mason* works with stone of all kinds. The stone can be used either for the basic structure or as a veneer cover over another building structure. The mason must be highly skilled in the use of specialized masonry tools, and must be familiar with the characteristics of natural stone, fired clay products, artificial stone, and concrete masonry products.

Electricians install the total electrical system in a building including service and branch circuits. Electricians may also install building control and security systems.

Interior designers design the surfaces of all interior walls, ceilings and floors including window treatments, furniture selection and placement and often complete kitchen and bath layout and design.

Architectural model makers construct scale models of buildings and sites to show the actual appearance of the final design in three dimensions.

Architectural illustrators prepare renderings used for presentations and need a knowledge of both architectural drawing, commercial art techniques and often photography.

Schedulers prepare timetables for the delivery of materials to a building site. *Expeditors* insure that materials and components, as listed in the set of specifications, are delivered to the site at the scheduled time.

Estimators compute the cost of construction projects by determining the amount, number and cost of each building material or component specified in the set of drawings and specifications.

Detailers are drafters who specialize in the preparation of detail drawings such as sections and framing drawings.

Structural drafters prepare structural framework drawings and often compute and specify structural member types and sizes.

Professionals who are licensed by a state must first work under the direction of a licensed professional for several years and pass a state board examination. Once licensed, the professional can design and certify the safety of each design by affixing a license stamp on each approved drawing.

SECTION 25

Abbreviations

Architects and drafters print many words on a drawing. Often they use abbreviations to conserve space. By using the standard abbreviations listed below, they ensure that their drawings are accurately interpreted.

Here are five points to remember:

1. Most abbreviations are in capitals.

2. A period is used only when the abbreviation may be confused with a whole word.

3. The same abbreviation can be used for both the singular and the plural.

4. Sometimes several terms use the same abbreviation.

5. Many abbreviations are very similar.

Term	Abbreviation
Access panel	AP
Acoustic	ACST
Actual	ACT.
Addition	ADD.
Adhesive	ADH
Aggregate	AGGR
Air condition	AIR COND
Alternating current	AC
Aluminum	AL
American Institute of Architects	AIA
American National Standards Institute	ANSI
Ampere	AMP
Anchor bolt	AB
Apartment	APT.
Approved	APPD
Approximate	APPROX
Architectural	ARCH
Area	A
Asbestos	ASB
Asphalt	ASPH
At	@
Automatic	AUTO
Avenue	AVE
Average	AVG

Term	Abbreviation
Balcony	BALC
Basement	BSMT
Base	B
Bathroom	B
Bathtub	BT
Beam	BM
Bearing	BRG
Bedroom	BR
Bench mark	BM
Bending moment	BM
Between	BET.
Blocking	BLKG
Blower	BLO
Blueprint	BP
Board	BD
Board feet	BF
Boller	BLR
Both sides	BS
Brick	BRK
British thermal units	BTU
Bronze	BRZ
Broom closet	BC
Building	BLDG
Building line	BL

Term	Abbreviation
Cabinet	CAB.
Caulking	CLKG
Cast concrete	C CONC
Cast iron	CI
Catalog	CAT.
Ceiling	CLG
Cement	CEM
Center	CTR
Centerline	CL
Center to center	C to C
Centimeter	CM
Ceramic	CER
Circle	CIR
Circuit	CKT
Circuit breaker	CIR BKR
Circumference	CIRC
Cleanout	CO
Clear	CLR
Closet	CL
Coated	CTD
Column	COL
Combination	COMB.
Common	COM
Composition	COMP
Computer aided drafting	CAD
Concrete	CONC
Conduit	CND
Construction	CONST
Construction Standards Institute	CSI
Continue	CONT
Contractor	CONTR
Corrugate	CORR
Courses	C
Cross section	X-SECT
Cubic foot	CU FT
Cubic inch	CU IN.
Cubic yard	CU YD

Term	Abbreviation
Damper	DMPR
Dampproofing	DP
Dead load	DL
Deflection	D
Degree	(°) DEG
Design	DSGN
Detail	DET
Diagonal	DIAG
Diagram	DIAG
Diameter	ϕ
Dimension	DIM
Dining room	DR
Dishwasher	DW
Ditto	DO.
Division	DIV
Door	DR
Double	DBL
Double-hung	DH
Down	DN
Downspout	DS
Drain	DR
Drawing	DWG
Dryer	D

Term	Abbreviation
East	E
Electric	ELEC
Elevation	EL
Enamel	ENAM
Entrance	ENT
Environmental Protection Agency	EPA

Equal .... EQ
Equipment .... EQUIP.
Estimate .... EST
Excavate .... EXC
Existing .... EXIST.
Exterior .... EXT

Fabricate .... FAB
Feet .... (') FT
Feet board measure .... FBM
Finish .... FIN.
Fireproof .... FPRF
Fixture .... FIX.
Flashing .... FL
Floor .... FL
Floor drain .... FD
Flooring .... FLG
Fluorescent .... FLUOR
Foot .... (') FT
Foot pounds .... FT LB
Footcandle .... FC
Footing .... FTG
Footing area .... FA
Foundation .... FDN
Full size .... FS
Furred ceiling .... FC

Galvanize .... GALV
Galvanized iron .... GI
Garage .... GAR
Gas .... G
Gage .... GA
Girder .... G
Glass .... GL
Grade .... GR
Grade line .... GL
Ground-fault circuit interrupter .... GFCI
Gypsum .... GYP

Hall .... H
Hardware .... HDW
Head .... HD
Heater .... HTR
Height .... HT
Horizontal .... HOR
Hose bib .... HB
Hot water .... HW
House .... HSE
Hundred C
Hypotenuse .... H

I beam I
Impregnate .... IMPG
Inch .... (") IN.
Inch pounds .... IN.LB
Incinerator .... INCIN
Insulate .... INS
Intercommunication .... INTERCOM
Interior .... IINT
Iron I

Joint .... JT
Joist .... JST

Kilowatt .... kW
Kilowatt hour .... kWh
Kip (1000 lb.) .... K
Kitchen .... KIT

Laminate .... LAM
Laundry .... LAU
Lavatory .... LAV
Left L
Length .... L, LG
Length overall .... LOA
Light .... LT
Linear .... LIN
Linen closet .... L CL
Live load LL
Living room .... LR
Long .... LG
Louver LV
Lumber .... LBR

Main .... MN
Manhole .... MH
Manual .... MAN.
Manufacturing .... MFG
Material .... MATL
Maximum .... MAX
Medicine cabinet .... MC
Membrane .... MEMB
Metal .... MET.
Meter .... M
Millimeter .... MM
Minimum .... MIN
Minute .... (') MIN
Miscellaneous .... MISC
Mixture .... MIX.
Model .... MOD
Modular .... MOD
Molding .... MLDG
Motor .... MOT

Natural .... NAT
Nominal .... NOM
North .... N
Not to scale .... NTS
Number .... NO.

Obscure .... OB
On center .... OC
Opening .... OPNG
Opposite .... OPP
Overall .... OA
Overhead .... OVHD

Panel .... PNL
Parallel .... PAR.
Part .... PT
Partition .... PTN
Penny (nails) .... d
Permanent .... PERM
Perpendicular .... PERP
Pi (3.1416) .... π
Piece .... PC
Plaster .... PL
Plate .... PL
Plumbing .... PLMG
Pound .... LB
Pounds per square foot .... PSF
Pounds per square inch .... PSI
Precast .... PRCST
Prefabricated .... PREFAB
Preferred .... PFD

Quality .... QUAL
Quantity .... QTY

Radiator .... RAD
Radius .... R
Range .... R
Receptacle .... RECP
Reference .... REF
Refrigerate .... REF
Refrigerator .... REF
Register .... REG
Reinforce .... REINF
Reproduce .... REPRO
Required .... REQD
Resistance moment .... RM
Return .... RET
Riser .... R
Roof .... RF
Room .... RM
Round .... RD

Safety .... SAF
Sanitary .... SAN
Scale .... SC
Schedule .... SCH
Second .... (") SEC
Section .... SECT
Select .... SEL
Service .... SERV
Sewer .... SEW.
Sheet .... SH
Sheathing .... SHTHG

Term	Abbreviation
Shower	SH
Side	S
Siding	SDG
Similar	SIM
Sink	S
Soil pipe	SP
South	S
Specification	SPEC
Square	SQ
Square foot	FT^2; SQ FT
Square inch	IN^2; SQ IN
Cross section	CS
Stairs	ST
Standard	STD
Steel	STL
Stock	STK
Street	ST
Storage	STG
Structural	STR
Supply	SUP
Surface	SUR
Switch	SW
Symmetrical	SYM
System	SYS
Tar and gravel	T & G
Tangent	TAN.
Tarpaulin	TARP
Tee	T
Telephone	TEL
Television	TV
Temperature	TEMP
Terra-cotta	TC
Terrazzo	TER
Thermostat	THERMO
Thick	THK
Thousand	M
Through	THRU
Toilet	T
Tongue and groove	T & G
Total	TOT.
Tread	TR
Tubing	TUB.
Typical	TYP
Unfinished	UNFIN
Urinal	UR
Valve	V
Vaporproof	VAP PRF
Vent pipe	VP
Ventilate	VENT.
Vertical	VERT
Vitreous	VIT
Volt	V
Volume	V, VOL
Washing machine	WM
Water closet	WC
Water heater	WH
Waterproofing	WP
Watt	W
Weather stripping	WS
Weatherproof	WP
Weep hole	WH
Weight	WT
West	W
Width	W
Window	WDW
With	W/
Without	W/O
Wood	WD
Wrought iron	WI
Yard	YD

SYMBOL ABBREVIATIONS

Although most symbols consist of several letters of an abbreviated word, many abbreviations use a symbol as a substitute for a word, situation or material. The most commonly used symbol abbreviations used on architectural drawings are shown below.

SYMBOL ABBREVIATIONS					
AND	¢, &	FOOT	′	POUNDS	#
AT	@	I BEAM	S	RADIUS	R
ANGLE	L	INCH	″	SLOPE	◺
ARC LENGTH	⌒	LESS THAN	<	SPOTFACE	⌴
BY	X	MORE THAN	>	SQUARE	□
CENTERLINE	℄	NUMBER	#	SQUARE FOOT	□′
CHANNEL	[	ONE THOUSAND	K	SQUARE INCH	□″
COUNTERBORE	⌴	PARALLEL	//	STRAIGHTNESS	–
DEGREE	°	PENNY	d	TEE	T
DEPTH	↧	PER	/	TIMES/PLACES	X
DIAMETER	Ø	PERCENT	%	WIDE FLANGE	W
DITTO	DO	PERPENDICULAR	⊥	WITHOUT	WO
FIBER STRESS	f	PLATE	PL	ZERO	Φ

Symbol abbreviations used on architectural drawings.

SECTION 26

Glossary

Abstract of title A summary of all deeds, wills, and legal actions to show ownership.
Acoustics The science of sound. In housing, acoustical materials used to keep down noise within a room or to prevent it from passing through walls.
Adobe construction Construction using sun-dried units of adobe soil for walls; usually found in the southwestern United States.
Air conditioner An apparatus that can heat, cool, clean, and circulate air.
Air-dried lumber Lumber that is left in the open to dry rather than being dried by a kiln.
Air duct A pipe, usually made of sheet metal, that conducts air to rooms from a central source.
Air trap A U-shaped pipe filled with water and located beneath plumbing fixtures to form a seal against the passage of gases and odors.
Alcove A recessed space connected at the side of a larger room.
Alteration A change in, or addition to, an existing building.
Amortization An installment payment of a loan, usually monthly for a home loan.
Ampere The unit used in the measure of the rate of flow of electricity.
Anchor bolt A threaded rod inserted in masonry construction for anchoring the sill plate to the foundation.
Angle iron A structural piece of rolled steel shaped to form a 90° angle.
Apartment Living unit in a multifamily residential building.
Appraisal The estimated evaluation of property.
Apron The finish board immediately below a window sill. Also the part of the driveway that leads directly into the garage.

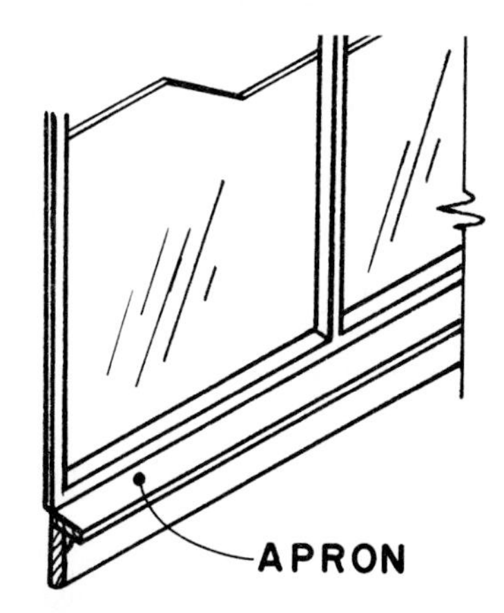

Arcade A series of arches supported by a row of columns.
Arch A curved structure that will support itself by mutual pressure and the weight above its curved opening.
Architect A person who plans and designs buildings and oversees their construction.
Area wall A wall surrounding an areaway.
Areaway A recessed area below grade around the foundation to allow light and ventilation into a basement window or doorway.
Ashlar A facing of squared stones.
Ashpit The area below the hearth of a fireplace which collects the ashes.
Asphalt Bituminous sandstones used for paving streets and waterproofing flat roofs.
Asphalt shingles Composition roof shingles made from asphalt-impregnated felt covered with mineral granules.
Assessed value A value set by governmental assessors to determine tax assessments.
Atrium An open court within a building.
Attic The space between the roof and the ceiling.
Awning window An out-swinging window hinged at the top.

Backfill Earth used to fill in areas around exterior foundation walls.
Backhearth The part of the hearth inside the fireplace.
Baffle A partial blocking against a flow of wind or sound.
Balcony A deck projecting from the wall of a building above the ground.
Balloon framing The building-frame construction in which each of the studs is one piece from the foundation to the roof of a two-story house.
Balustrade A series of balusters or posts connected by a rail, generally used adjacent to stairs.
Banister A handrail.
Base The finish of a room at the junction of the walls and floor.
Baseboard The finish board covering the interior wall where the wall and floor meet.
Base course The lowest part of masonry construction.
Base line A located line for reference control purposes.
Basement The lowest story of a building, partially or entirely below ground.
Base plate A plate, usually of steel, upon which a column rests.
Base shoe A molding used next to the floor in interior baseboards.
Batt A blanket insulation material usually made of mineral fibers and designed to be installed between framing members.

Batten A narrow strip of board, used to cover cracks between the boards in board-and-batten siding.

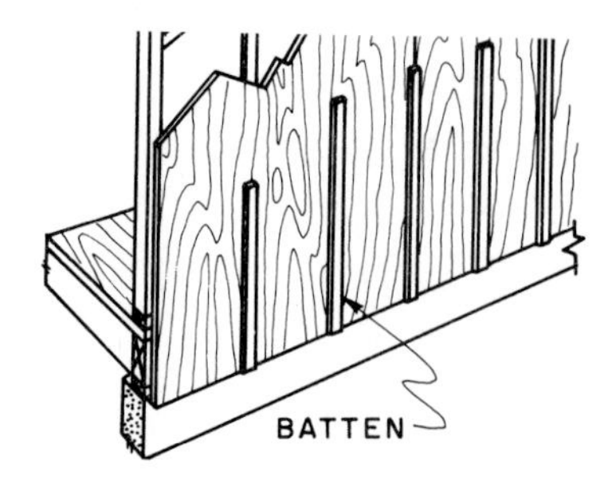

Batter A masonry or concrete wall which slopes backward from the perpendicular.

Batter boards Boards at exact elevations nailed to posts just outside the corners of a proposed building. Strings are stretched across the boards to locate the outline of the foundation.

Bay window A window projecting out from the wall of a building to form a recess in the room.

Beam A horizontal structural member that carries a load.

Beam ceiling A ceiling in which the ceiling beams are exposed to view.

Bearing plate A plate that provides support for a structural member.

Bearing wall or partition A wall supporting any vertical load other than its own weight.

Bench mark A metal or stone marker placed in the ground by a surveyor with the elevation on it. This is the reference point to determine lines, grades, and elevations in the area.

Bending moment A measure of the forces that break a beam by bending.

Bent A frame consisting of two supporting columns and a girder or truss used in vertical position in framing a structure.

Bevel siding Shingles or other siding board thicker on one edge than the other. The thick edge overlaps the thin edge of the next board.

Bib A threaded faucet allowing a hose to be attached.

Bill of material A parts list of material accompanying a structural drawing.

Blanket insulation Insulation in rolled-sheet form, often backed by treated paper that forms a vapor barrier.

Blocking Small wood framing members that fill in the open space between the floor and ceiling joists to add stiffness to the floors and ceiling.

Blueprint An architectural drawing used by workers to build from. The original drawing is transferred to a sensitized paper that turns blue with white lines when printed. Also, prints of blue lines on white paper.

Board measure A system of lumber measurement having as a unit a board foot. One board foot is the equivalent of 1 foot square by 1 inch thick.

Brace Any stiffening member of a framework.

Braced framing Frame construction with posts and braces used for stiffening.

Breezeway A roofed walkway with open sides. It connects the house and garage.

Broker An agent in buying and selling property.

Btu Abbreviation for British thermal unit; a standard unit for measuring heat gain or loss.

Buck Frame for a door, usually made of metal, into which the finished door fits.

Building code A collection of legal requirements for buildings designed to protect the safety, health, and general welfare of people who work and live in them.

Building line An imaginary line on a plot beyond which the building cannot extend.

Building paper A heavy, waterproof paper used over sheathing and subfloors to prevent passage of air and water.

Building permit A permit issued by a municipal government authorizing the construction of a building or structure.

Built-up beam A beam constructed of smaller members fastened together.

Built-up roof A roofing material composed of several layers of felt and asphalt.

Butterfly roof A roof with two sides sloping down toward the interior of the house.

Butt joint A joint formed by placing the end of one member against another member.

Buttress A projection from a wall to create additional strength and support.

BX cable Armored electric cable wrapped in plastic and protected by a flexible steel covering.

Cabinet work The finish interior woodwork.

Canopy A projection over windows and doors to protect them from the weather.

Cantilever A projecting member supported only at one end.

Cant strip An angular board used to eliminate a sharp right angle on roofs or flashing.

Carport An automobile shelter not fully enclosed.

Carriage The horizontal part of the stringers of a stair that supports the treads.

Casement window A hinged window that opens out, usually made of metal.

Casing A metal or wooden member around door and window openings to give a finished appearance.

Catch basin An underground structure for drainage into which the water from a roof or floor will drain. It is connected with a sewer drain or sump pump.

Caulking Soft waterproof material used to seal cracks.

Cavity wall A hollow wall usually made up of two brick walls built a few inches apart and joined together with brick or metal ties.

Cedar shingles Roofing and siding shingles made from western red cedar.

Cement A masonry adhesive material purchased in the form of pulverized powder.

Central heating A single source of heat that is distributed by pipes or ducts.

Certificate of title A document given to the home buyer with the deed, stating that the title to the property named in the deed is clearly established.

Cesspool A pit or cistern to hold sewage.

Chalk line A string that is heavily chalked, held tight, then plucked

to make a straight guideline against boards or other surfaces.
Chase A vertical space within a building for ducts, pipes, or wires.
Checks Splits or cracks in a board, ordinarily caused by seasoning.
Check valve A valve that permits passage through a pipe in only one direction.
Chimney A vertical flue for passing smoke and gases outside a building.
Chimney stack A group of flues in the same chimney.
Chord The principal members of a roof or bridge truss. The upper members are indicated by the term *upper chord*. The lower members are identified by the term *lower chord*.
Cinder block A building block made of cement and cinder.
Circuit The path of an electric current. The closed loop of wire in which an electric current can flow.
Circuit breaker A safety device used to open and close an electrical circuit.
Cistern A tank or other reservoir to store rainwater run off.
Clapboard A board, thicker on one side than the other, used to overlap an adjacent board.
Clearance A clear space to allow passage.
Clerestory A set of high windows often above a roof line.
Clinch To bend over the protruding end of a nail.
Clip A small connecting angle used for fastening various members of a structure.
Collar beam A horizontal member fastening opposing rafters below the ridge in roof framing.
Column In architecture: a perpendicular supporting member, circular in section; in engineering: a vertical structural member supporting loads.
Common wall A single wall that serves two dwelling units.
Compression A force that tends to make a member fail because of crushing.
Concrete A mixture of cement, sand, gravel, and water.
Concrete block Precast hollow or solid blocks of concrete.
Condemn To legally declare unfit for use.
Condensation The formation of frost or drops of water on inside walls when warm vapor inside a room meets a cold wall or window.
Conductor In architecture: a drain pipe leading from the roof; in electricity: anything that permits the passage of an electric current.
Conductor pipe A pipe used to lead water from the roof to the sewer.
Conduit A channel built to convey water or other fluids; a drain or sewer. In electrical work, a channel that carries wires for protection and for safety.
Construction loan A mortgage loan to be used to pay for labor and materials going into the house. Money is usually advanced to the builder as construction progresses and is repaid when the house is completed and sold.
Continuous beam A beam that has no intermediate supports.
Contractor The manager of a construction project.
Convector A heat-transfer surface that uses convection currents to transfer heat.
Coping The top course of a masonry wall that projects to protect the wall from the weather.

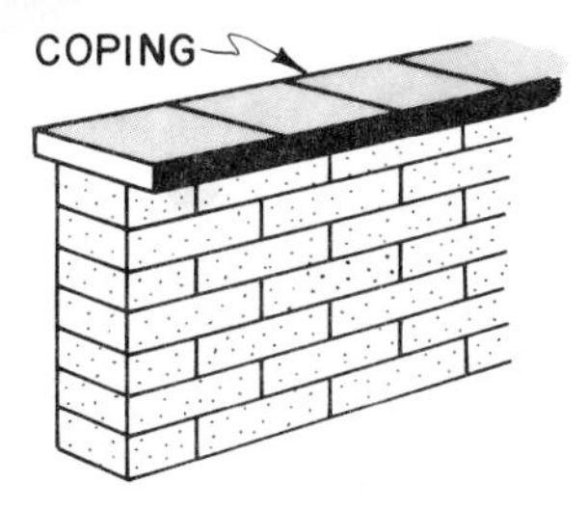

Corbel A projection in a masonry wall made by setting courses beyond the lower ones.
Corner bead A metal molding built into plaster corners to prevent the accidental breaking off of the plaster.
Cornice The part of a roof that projects out from the wall.
Counterflashing A flashing used under the regular flashing.
Course A continuous row of stone or brick of uniform height.
Court An open space surrounded partly or entirely by a building.
Crawl space The shallow space below the floor of a house built above the ground. It is surrounded by the foundation walls.
Cricket A roof device used at intersections to divert water.

Cripple A structural member that is cut less than full length, such as a studding piece above a window or door.
Cross bracing Boards nailed diagonally across studs or other boards to make framework rigid.
Cross bridging Bracing between floor joists to add stiffness to the floors.

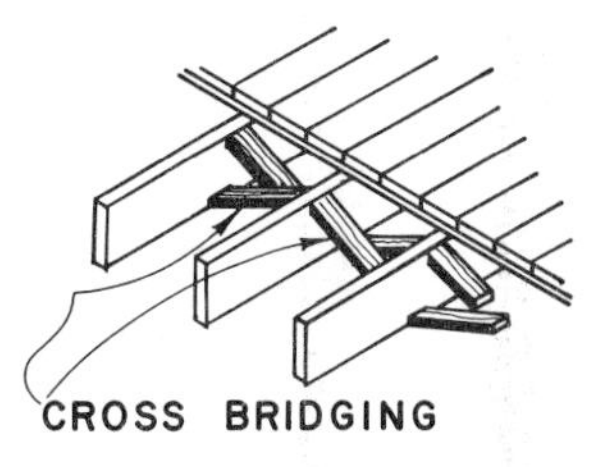

Crosshatch Lines drawn closely together at an angle to show a section cut.
Cull Building material rejected as below standard grade.
Culvert A passage for water below ground level.
Cupola A small structure built on top of a roof to provide ventilation.
Curb A very low wall.
Cure To allow concrete to dry slowly by keeping it moist to allow maximum strength.
Curtain wall An exterior wall that provides no structural support.

Damp course A layer of waterproof material.
Damper A movable plate that regulates the draft of a stove, fireplace, or furnace.
Datum A reference point of starting elevations used in mapping and surveying.
Deadening Construction intended to prevent the passage of sound.
Dead load All the weight in a structure made up of unmovable materials. See also Loads.

Decay The disintegration of wood through the action of fungi.
Dehumidify To reduce the moisture content in the air.
Density The number of people living in a calculated area of land such as a square mile or square kilometer.
Depreciation Loss of value.
Designer A person who designs houses but is not a registered architect.
Detail Information added to a drawing to provide specific instruction with a drawing, dimensions, notes, or specifications.
Dimension building material Building material that has been precut to specific sizes.
Dimension line A line with arrowheads at either end to show the distance between two points.
Dome A hemispherical roof form.
Doorstop The strips on the doorjambs against which the door closes.
Dormer A structure projecting from a sloping roof to accommodate a window.
Double glazing A pane made of two pieces of glass with air space between and sealed to provide insulation.
Double header Two or more timbers joined for strength.
Double-hung A window having top and bottom sashes each capable of movement up and down.
Downspout A pipe for carrying rainwater from roof to ground.

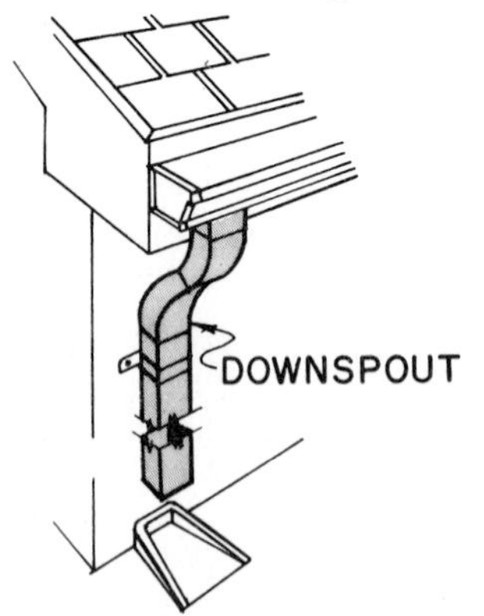

Drain A pipe for carrying wastewater.
Dressed lumber Lumber machined and smoothed at the mill. Usually ½ inch less than nominal (rough) size.
Drip A projecting construction member or groove below the member to prevent rainwater from running down the face of a wall or to protect the bottom of a door or window from leakage.
Dry rot A term applied to many types of decay, especially an advanced stage when the wood can be easily crushed to a dry powder.
Dry-wall construction Interior wall covering that uses precast sheets.
Dry well A pit located in porous ground and lined with rock that allows water to seep through the pit. Used for the disposal of rainwater or the effluent from a septic tank.
Ducts Sheet-metal conductors for warm- and cold-air distribution.

Easement The right to use land owned by another, such as a utility company's right-of-way.
Eave That part of a roof that projects over a wall.

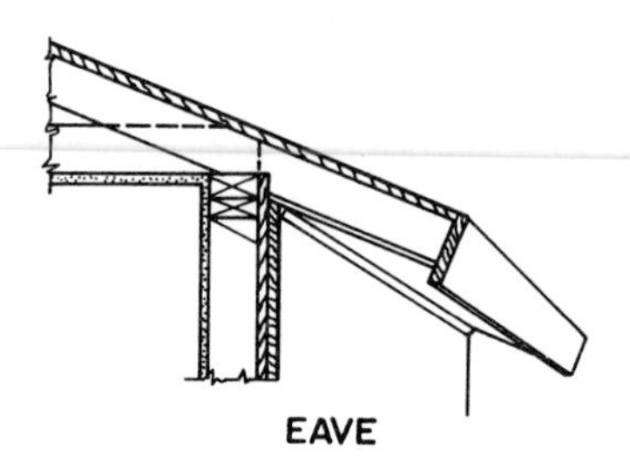

EAVE

Efflorescence Whitish powder that forms on the surface of bricks or stone walls due to evaporation of moisture containing salts.
Effluent The liquid discharge from a septic tank after bacterial treatment.
Elastic limit The limit to which a material can be bent or pulled out of shape and still return to its former shape and dimensions.
Elbow An L-shaped pipe fitting.

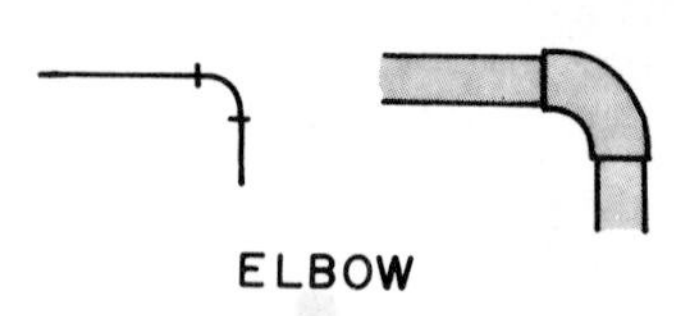

ELBOW

Elevation The drawings of the front, side, or rear face of a building.
Ell An extension or wing of a building at right angles to the main section.
Embellish To add decoration.
Eminent domain The right of the local government to condemn for public use.
Enamel Paint with a considerable amount of varnish. It produces a hard, glossy surface.
Equity The interest in or value of real estate the owner has in excess of the mortgage indebtedness.
Ergonomics The study of human space and movement needs.
Escutcheon The hardware on a door to accommodate the knob and keyhole.
Excavation A cavity or pit produced by digging the earth in preparation for construction.

Fabrication Work done on parts of a structure at the factory before delivery to the building site.
Facade The exterior face of a building.
Face brick A brick used on the outside face of a wall.
Facing A surface finish material used to cover another surface.
Fascia A vertical board nailed on the ends of the rafters at the eave line. It is part of the cornice.
Fatigue A weakening of structural members.
Federal Housing Administration (FHA) A government agency that insures loans made by regular lending institutions.
Felt papers Papers, sometimes tar-impregnated, used on roofs and sidewalls to give protection against dampness and leaks.
Fenestration The arrangement of windows.
Fiberboard A building board made with fibrous material—used as an insulating board.
Filled insulation A loose insulating material poured from bags or blown by machines into walls.
Finish lumber Dressed wood used for building trim and cabinet work.
Firebrick A brick that is especially hard and heat-resistant. Used in fireplaces.
Fireclay A grade of clay that can withstand a large quantity of heat. Used for firebrick.
Fire cut The angular cut at the end of a joist designed to rest in a brick wall.
Fire door A door that will resist fire.
Fire partition A partition designed to restrict the spread of fire.
Fire stop Obstruction across air passages in buildings to prevent

the spread of hot gases and flames. A horizontal blocking between wall studs.
Fished A splice strengthened by metal pieces on the sides.
Fixed light A permanently sealed window.
Fixture A piece of electric or plumbing equipment that is part of the structure.
Flagging Cut stone, slate, or marble used on floors.
Flagstone Flat stone used for floors, steps, walks, or walls.
Flashing The material used for and the process of making watertight the intersections on exposed places on the outside of the building.
Flat roof A roof with minimum pitch for drainage.
Flitch beam A built-up beam formed by a metal plate sandwiched between two wood members and bolted together for additional strength.
Floating Spreading plaster, stucco, or cement on walls or floors with use of a tool called a float.
Floor plan The top view of a building at a specified floor level. A floor plan includes all vertical details at or above windowsill levels.
Floor plug An electrical outlet flush with the floor.
Flue The opening in a chimney through which smoke passes.
Flue lining Terra-cotta pipe used for the inner lining of chimneys.
Flush surface A continuous surface without an angle.
Footing An enlargement at the lower end of a wall, pier, or column, to distribute the load into the ground.
Footing form A wooden or steel form used to hold concrete to the desired shape and size until it hardens.
Framing Wood skeleton of a building constructed one level on top of another.
Frieze The flat board of cornice trim that is fastened to the wall.

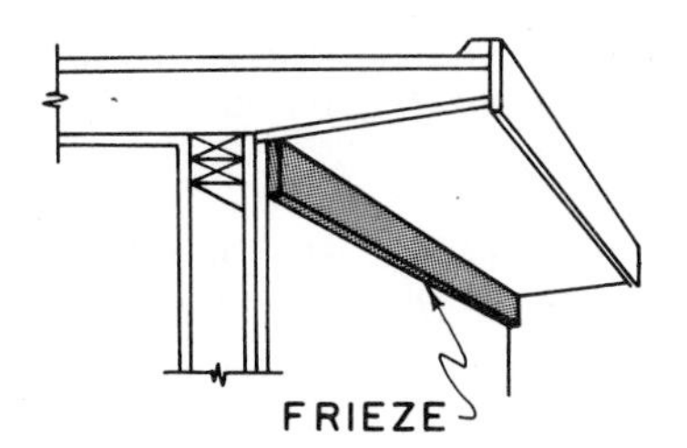

Frost line The depth of frost penetration into the soil.
Fumigate To destroy harmful insect or animal life with fumes.
Furring Narrow strips of board nailed upon walls and ceilings to form a straight surface for the purpose of attaching wallboards or ceiling tile.
Fuse A strip of soft metal inserted in an electric circuit and designed to melt and open the circuit should the current exceed a predetermined value.

Gable The triangular end of an exterior wall above the eaves.
Gable roof A roof that slopes from two sides only.
Galvanize A lead and zinc bath treatment to prevent rusting.
Gambrel roof A symmetrical roof with two different pitches or slopes on each side.
Garret An attic.
Girder A horizontal beam supporting the floor joists.

Glazing Placing of glass in windows or doors.
Grade The level of the ground around a building.
Gradient The slant of a surface expressed in percent.
Graphic symbols Symbolic representations used in drawing that simplify presentations of complicated items.
Gravel stop A strip of metal with a vertical lip used to retain the gravel around the edge of a built-in roof.
Green lumber Lumber that still contains moisture or sap.
Ground-fault circuit interrupter (GFCI) An electrical device that breaks an electric circuit when an excessive leakage current is detected. Intended to eliminate shock hazards to people.
Grout A thin cement mortar used for leveling and filling masonry holes.
Gusset A plywood or metal plate used to strengthen the joints of a truss.
Gutter A trough for carrying off water.
Gypsum board A board made of gypsum rock and fiber glass.

Half timber construction A frame construction of heavy timbers in which the spaces are filled in with masonry.
Hanger An iron strap used to support a joist, beam, or pipe.
Hardpan A compacted layer of soils.
Head The upper frame on a door or window.
Header The horizontal supporting member above openings that serves as a lintel. Also one or more pieces of lumber supporting ends of joists. Used in framing openings of stairs and chimneys.
Headroom The clear space between floor line and ceiling, as in a stairway.
Hearth That part of the floor directly in front of the fireplace, and the floor inside the fireplace on which the fire is built. It is made of fire-resistant masonry.
Heel plate A plate at the ends of a truss.
Hip rafter The diagonal rafter that extends from the plate to the ridge to form the hip.
Hip roof A roof with four sloping sides.
House drain Horizontal sewer piping within a building that receives wastes from the soil stacks.
House sewer The watertight soil pipe extending from the exterior of the foundation wall to the public sewer.
Humidifier A mechanical device that controls the amount of water vapor to be added to the atmosphere.
Humidistat An instrument used for measuring and controlling moisture in the air.

I beam A steel beam with an I-shaped cross section.

Indirect lighting Artificial light that is reflected from a surface before reaching source.
Insulating board Any board suitable for insulating purposes, usually manufactured board made from vegetable fibers, such as fiberboard.
Insulation Materials for obstructing the passage of sound, heat, or cold from one surface to another.
Interior trim General trim for all the finish molding, casing, baseboard, etc.

Jack rafter A short rafter, usually used on hip roofs.
Jalousie A type of window consisting of a number of long, thin, hinged panels.
Jamb The sides of a doorway or window opening.
Jerry-built Poorly constructed.
Joints The meeting of two separate pieces of material for a common bond.
Joist A horizontal structural member that supports the floor system or ceiling system.

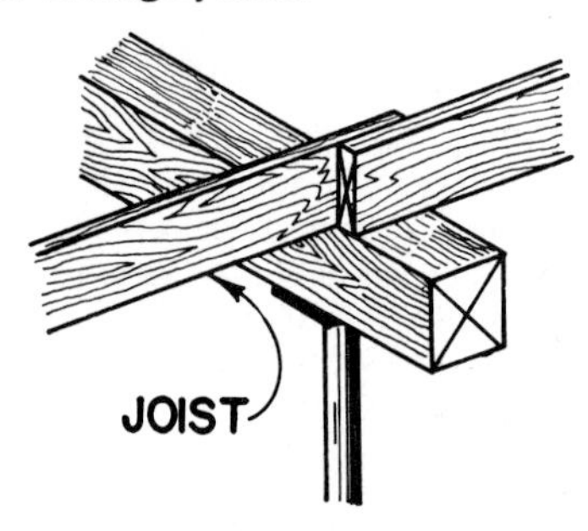

Kalamein door A fireproof door with a metal covering.
Keystone The top, wedge-shaped stone of an arch.
Kiln A heating chamber for drying lumber.
King post In a roof truss, the central upright piece.
Knee brace A corner brace, fastened at an angle from wall stud to rafter, stiffening a wood or steel frame to prevent angular movement.
Knee wall Low wall resulting from one-and-one-half-story construction.
Knob and tube Electric wiring through walls where insulated wires are supported with porcelain knobs and tubes when passing through wood construction members.

Lally column A steel column used as a support for girders and beams.

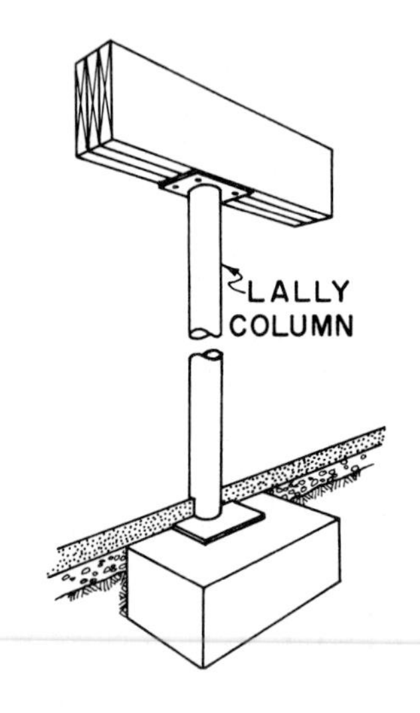

Laminated beam A beam made by bonding together several layers of material.
Landing A platform in a flight of steps.
Landscape architect A professional person who designs and adapts sites for people's use.
Lap joint A joint produced by lapping two pieces of material.
Lath (metal) Sheet-metal screening used as a base for plastering.
Lath (wood) A wooden strip nailed to studding and joists to which plaster is applied.
Lattice A grille or openwork made by crossing strips of wood or metal.
Lavatory A washbasin or a room equipped with a washbasin.
Leaching bed A system of trenches that carries wastes from sewers. It is constructed in sandy soils or in earth filled with stones or gravel.
Leader A vertical pipe or downspout that carries rainwater from the gutter to the ground also a line connecting a note to a part of a drawing.
Lean-to A shed whose rafters lean against another building or other part of the same building.
Ledger A wood strip nailed to the lower side of a girder to provide a bearing surface for joists.

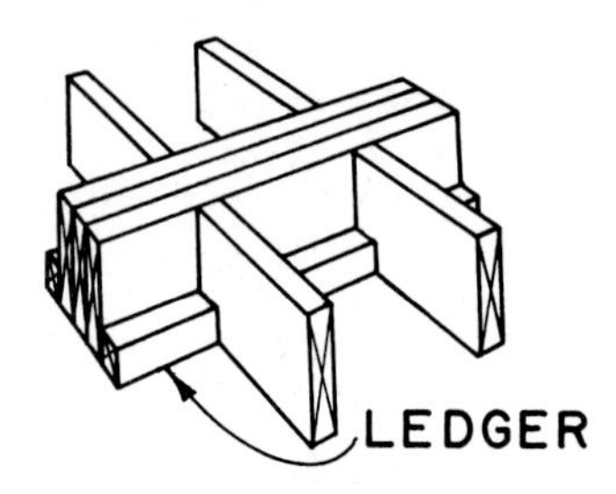

Lessee The tenant who holds a lease.
Lessor The owner of leased property.
Lien A legal claim on a property that may be exercised in default of payment of a debt.
Lineal foot A measurement of 1 foot along a straight line.
Lintel A horizontal piece of wood, stone, or steel across the top of door and window openings to bear the weight of the walls above the opening.
Loads Live load: the total of all moving and variable loads that may be placed upon a building. Dead load: the weight of all permanent, stationary construction included in a building.
Load-bearing walls Walls that support weight from above.
Loggia A roofed, open passage along the front or side of a building. It is often at an upper level, and it often has a series of columns on either or both sides.
Lookout A horizontal framing member extending from studs out to the end of rafters.
Lot line The line forming the legal boundary of a piece of property.
Louver A set of fixed or movable slats adjusted to provide both shelter and ventilation.

Mansard roof A roof with two slopes on each side, with the lower slope much steeper than the upper.
Mantel A shelf over a fireplace.
Market price The amount that property can be sold for at a given time.
Market value The amount that property is worth at a given time.

Masonry Stone, brick, tiles, or concrete.

Meeting rail The horizontal rails of a double-hung sash that fit together when the window is closed.

Member A single piece of material used in a structure.

Metal tie A strip of metal used to fasten construction members together.

Metal wall ties Strips of corrugated metal used to tie a brick veneer wall to framework.

Mildew A mold on wood caused by fungi.

Meter Unit of metric measure equal to 1.1 yard.

Millwork The finish woodwork in a building, such as cabinets and trim.

Mineral wool An insulating material made into a fibrous form from mineral slag.

Modular construction Construction in which the size of the building and the building materials are based on a common unit of measure.

Moisture barrier A material such as specially treated paper that retards the passage of vapor or moisture into walls and prevents condensation within the walls.

Monolithic Concrete construction poured and cast in one piece without joints.

Monument A boundary marker set by surveyors to locate property lines.

Mortar A mixture of cement, sand, and water, used as a bonding agent by the mason for binding bricks and stone.

Mortgage A pledging of property, conditional on payment of the debt in full.

Mortgagee The lender of money to the mortgagor.

Mortgagor The owner who mortgages property in return for a loan.

Mosaic Small colored tile, glass, stone, or similar material arranged on an adhesive ground to produce a decorative surface.

Mud room A small room or entranceway where muddy overshoes and wet garments can be removed before entering other rooms.

Mullion A vertical bar in a window that separates the window into sections.

Muntin A small bar separating the glass lights in a window.

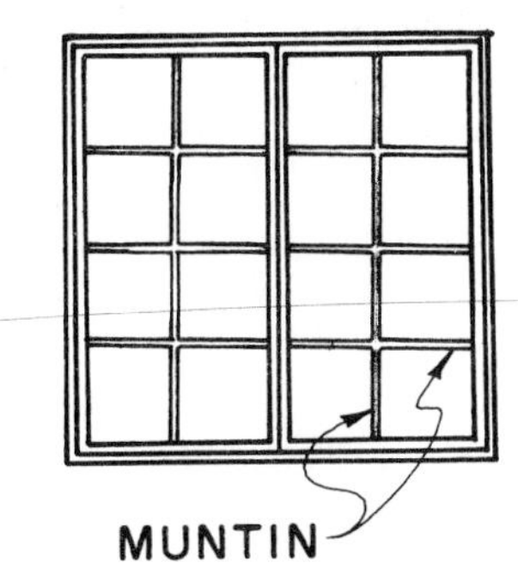

Newel A post supporting the handrail at the top or bottom of a stairway.

Nominal dimension Dimensions for finished lumber in which the stated dimension is usually larger than the actual dimension.

Nonbearing wall A dividing wall that does not support a vertical load.

Nonferrous metal Metal containing no iron, such as copper, brass, or aluminum.

Nosing The rounded edge of a stair tread.

Obscure glass Sheet glass that is made translucent instead of transparent.

On center Measurement from the center of one member to the center of another (noted OC).

Open-end mortgage A mortgage that permits the remaining amount of the loan to be increased, as for improvements, by mutual agreement of the lender and borrower, without rewriting the mortgage.

Orientation The positioning of a house on a lot in relation to the sun, wind, view, and noise.

Outlet Any kind of electrical box allowing current to be drawn from the electrical system for lighting or appliances.

Overhang The horizontal distance that a roof projects beyond a wall.

Panelboard The center for controlling electrical circuits.

Parapet A low wall or railing around the edge of a roof.

Parging A thin coat of plaster applied to masonry surfaces for smoothing purposes.

Parquet flooring Flooring, usually of wood, laid in an alternating or inlaid pattern to form various designs.

Partition An interior wall that separates two rooms.

Party wall A wall between two adjoining buildings in which both owners share, such as a common wall between row houses.

Patio An open court.

Pediment The triangular space forming the gable end of a low-pitched roof. A similar form is often used as a decoration over doors in classic architecture.

Penny A term for the length of a nail, abbreviated *d*.

Periphery The entire outside edge of an object.

Perspective A drawing of an object in a three-dimensional form on a plane surface. An object drawn as it would appear to the eye.

Pier A block of concrete supporting the floor of a building.

Pilaster A portion of a square column, usually set within or against a wall for the purpose of strengthening the wall. Also a decorative column attached to a wall.

Piles Long posts driven into the soil in swampy locations, or whenever it is difficult to secure a firm foundation, upon which the foundation footing is laid.

Pillar A column used for supporting parts of a structure.

Pinnacle Projecting or ornamental cap on the high point of a roof.

Plan A horizontal, graphic representational section of a building.

Planks Material 2 or 3″ (50 or 75 mm) thick and more than 4″ (100 mm) wide, such as joists, flooring, and the like.

Plaster A mortarlike composition used for covering walls and ceilings. Usually made of portland cement mixed with sand and water.

Plasterboard A board made of plastering material covered on both sides with heavy paper.

Plaster ground A nailer strip included in plaster walls to act as a

gage for thickness of plaster and to give a nailing support for finish trim around openings and near the base of the wall.
Plat A map or chart of an area showing boundaries of lots and other parcels of property.
Plate The top horizontal member of a row of studs in a frame wall to carry the rafters directly. Also a shoe or base member, as of a partition or other frame.
Plate cut The cut in a rafter that rests upon the plate. It is also called the *seat cut* or *birdmouth*.
Plate glass A high-quality sheet of glass used in large windows.
Platform Framing in which each story is built upon the other.
Plenum system A system of heating or air-conditioning in which the air is forced through a chamber connected to distributing ducts.
Plot The land on which a building stands.
Plow To cut a groove running in the same direction as the grain of the wood.
Plumb True vertical position perpendicular to the earth surface, as determined by a plumb bob.
Plywood A piece of wood made of three or more layers of veneer joined with glue and usually laid with the grain of adjoining piles at right angles.
Porch A covered area attached to a house at an entrance.
Portico A roof supported by columns, whether attached to a building or wholly by itself.
Portland cement A hydraulic cement, extremely hard, formed by burning silica, lime, and alumina together and then grinding up the mixture.
Post A perpendicular supporting member.
Post-and-beam construction Wall construction consisting of large, widely spaced posts to support horizontal beams.
Precast Concrete shapes made separately before being used in a structure.
Prefabricated buildings Buildings that are built in sections or component parts in a factory, and then assembled at the site.
Primary coat The first coat of paint.
Principal The original amount of money loaned.
Purlin A horizontal structural member which hold rafters together.

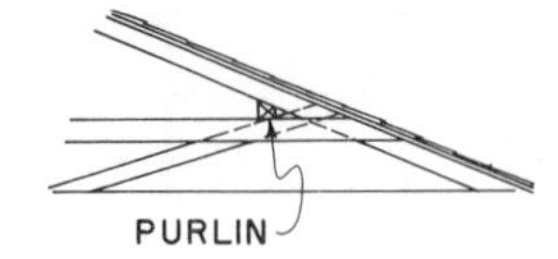

Quad An enclosed court.
Quarry tile A machine-made, unglazed tile.
Quoins Large squared stones set in the corners of a masonry building for appearance.

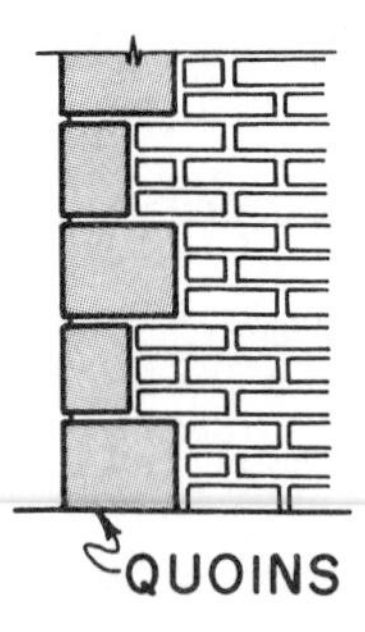

Radiant heating A system using heating elements in the floors, ceilings, or walls to radiate heat into the room.
Rafters Structural members used to frame a roof.
Raglin The open joint in masonry to receive flashing.
Realtor A real-estate broker who is a member of a local chapter of the National Association of Real Estate Boards.
Register The open end of a duct in a room for warm or cool air.
Reinforced concrete Concrete in which steel bars or webbing has been embedded for strength.
Rendering The art of shading or coloring a drawing.
Restoration Rebuilding a structure so it will appear in its original form.
Restrictions Limitations on the use of real estate building materials, size, or design styles.
Retaining wall A wall to hold back an earth embankment.
Rheostat An instrument for regulating electric current.
Ribbon A support for joists. A board set into studs that are cut to support joists.
Ridge The top edge of the roof where rafters meet.
Ridge cap A wood or metal cap used over roofing at the ridge.
Riprap Stones placed on a slope to prevent erosion. Also broken stone used for foundation fill.
Rise The vertical height of a roof.
Riser The vertical board in a stairway between two treads.
Rock wool An insulating material that looks like wool but is composed of such substances as granite or silica.
Rodding Stirring freshly poured concrete with a vibrator to remove air pockets.
Roll roofing Roofing material of fiber and asphalt manufactured in rolls.
Rough floor The subfloor on which the finished floor is laid.
Rough hardware All the hardware used in a house, such as nails and bolts, that cannot be seen in the completed house.
Roughing in Putting up the skeleton of the building.
Rough lumber Lumber as it comes from the saw.
Rough opening Any unfinished opening in the framing of a building.
Run The horizontal distance covered by a flight of stairs.

Saddle The ridge covering of a roof designed to carry water from the back of chimneys. Also called a *cricket*. A threshold.
Safety factor The ultimate strength of the material divided by the allowable working load. The element of safety needed to make certain that there will be no structural failures.
Sand finish A final plaster coat; a skim coat.
Sap All the fluids in a tree.
Sash The movable framework in which window panes are set.
Scab A small wood member, used to join other members, which is fastened on the outside face.

Scarfing A joint between two pieces of wood that allows them to be spliced lengthwise.
Schedule A list of parts or details.
Scratch coat The first coat of plaster. It is scratched to provide a good bond for the next coat.
Screed A guide for the correct thickness of plaster or concrete being placed on surfaces.
Scuttle A small opening in a ceiling to provide access to an attic or roof.
Seasoning Drying out of green lumber, either in an oven or kiln or by exposing it to air.
Second mortgage A mortgage made by a home buyer to raise money for a down payment required under the first mortgage.
Section The drawing of an object that is cut to show the interior. Also, a panel construction used in walls, floors, ceilings, or roofs.
Seepage pit A pit or cesspool into which sewage drains from a septic tank, and which is so constructed that the liquid waste seeps through the sides of the pit into the ground.
Septic tank A concrete or steel tank where sewage is reduced to liquid and gases by bacterial action. About half the sewage solids become gases that escape back through the vent stack in the house. The liquids flow from the tank into the ground through a leaching field tile bed.

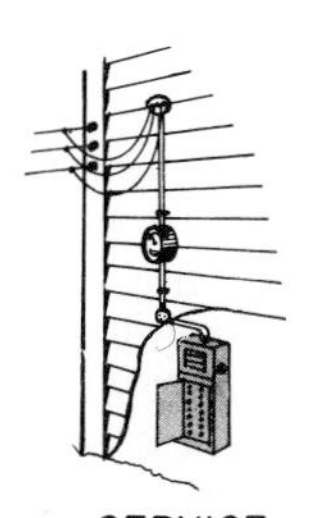

SERVICE CONNECTION

Service connection The electric wires to the building from the outside power lines.
Set The hardening of cement or plaster.
Setback A zoning restriction on the location of the home on a lot.
Settlement Compression of the soil or the members in a structure.
Shakes Thick hand-cut shingles.
Sheathing The structural covering of boards or wallboards, placed over exterior studding or rafters of a structure.
Sheathing paper A paper barrier against wind and moisture applied between sheathing and outer wall covering.
Shed roof A flat roof slanting in one direction.

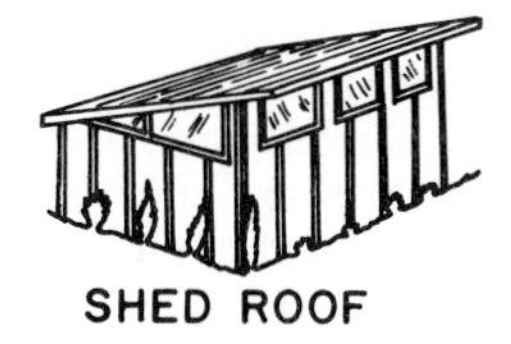

SHED ROOF

Shim A piece of material used to level or fill in the space between two surfaces.

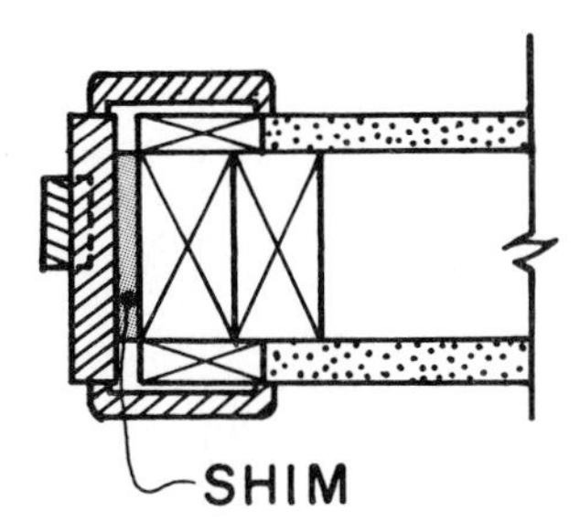

Shingles Thin pieces of wood or other materials that overlap each other in covering a roof. The number and kind needed depend on the steepness of the roof slope.
Shiplap Boards with lapped joints along their edges.
Shoe mold The small mold against the baseboard at the floor.
Shoring Lumber placed in a slanted position to support the structure of a building temporarily.
Siding The outside boards of an exterior wall.
Sill The horizontal exterior member below a window or door opening. Also the wood member placed directly on top of the foundation wall in wood-frame construction.

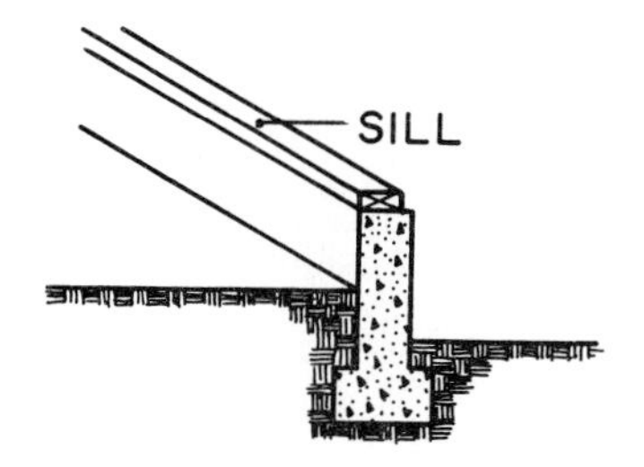

Skeleton construction Construction where the frame carries all the weight.
Skylight An opening in the roof for admitting light.
Slab foundation A reinforced concrete floor and foundation system.
Sleepers Strips of wood, usually 2×2's, laid over a slab floor to which finished wood flooring is nailed.

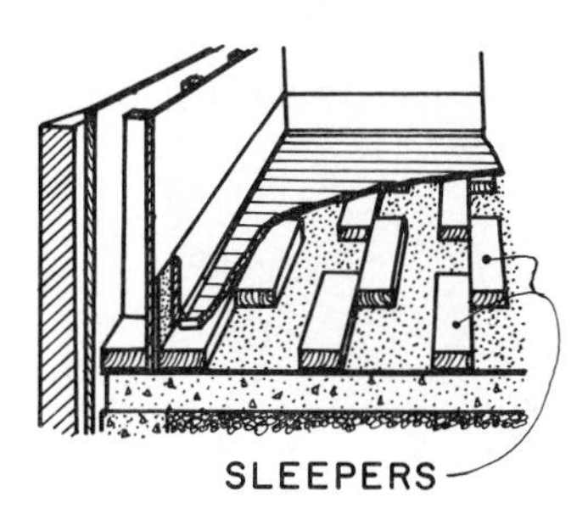

Smoke chamber The portion of a chimney flue located directly over the fireplace.
Soffit The undersurface of a projecting structure.

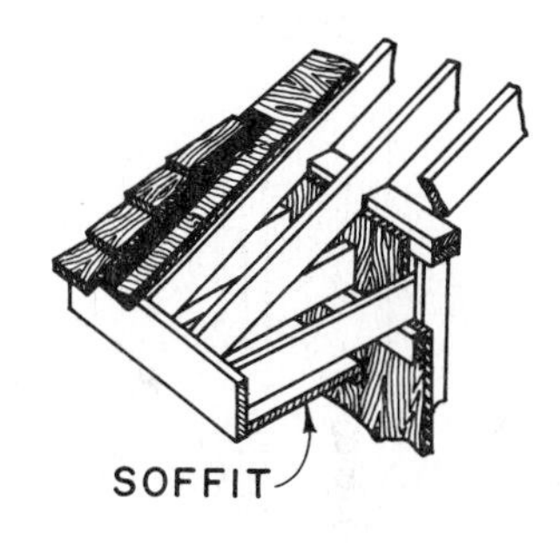

Softwood Wood from trees having needles rather than broad leaves. The term does not necessarily refer to the softness of the wood.
Soil stack The main vertical pipe that receives waste from all fixtures.
Solar heat Heat from the sun.
Sole The horizontal framing member directly under the studs.
Spacing The distance between structural members.
Spackle To cover wallboard joints with plaster.
Span The distance between structural supports.
Specification The written or printed direction regarding the details of a building or other construction not included in the set of working drawings.

Spike A large, heavy nail.
Splice Joining of two similar members in a straight line.
Stack A vertical pipe.
Stakeout Marking the foundation layout with stakes.
Steel framing Skeleton framing with structural steel beams.
Steening Brickwork without mortar.
Stile A vertical member of a door, window, or panel.
Stirrup A metal U-shaped strap used to support framing members and pipes.

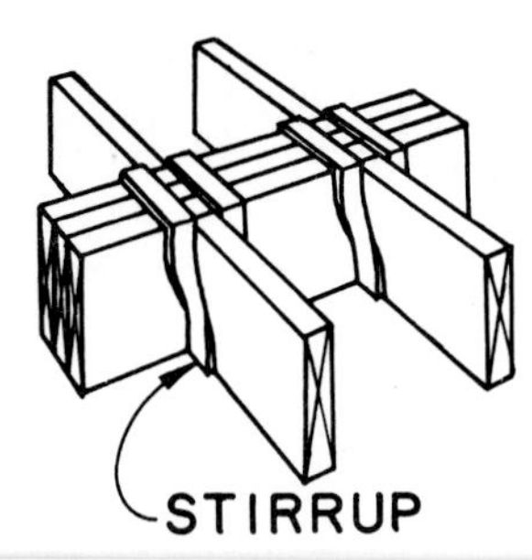

Stock Common sizes of building materials and equipment available from most commercial industries.
Stool An inside bottom part of a windowsill.
Stop A small strip to hold a door or window sash in place.
Storm door or window An extra door or extra window placed outside an ordinary door or window for added protection against cold.
Storm sewer A sewer that is designed to carry away water from storms, but not sewage.
Stress Any force acting upon a part or member.
Stress-cover construction Construction consisting of panels or sections with wood frameworks to which plywood or other sheet material is bonded with glue so that the covering carries a large part of the loads.
Stretcher course A row of masonry in a wall with the long side of the units exposed to the exterior.
Stringer One of the sides of a flight of stairs. The supporting member cut to receive the treads and risers.
Stripping Removal of concrete forms from the hardened concrete.
Stucco Any of various plasters used for covering walls, especially an exterior wall covering in which cement is used.
Stud Upright beams in the framework of a building. Usually referred to as 2×4's, and spaced at 16 inches from center to center.
Subfloor The rough flooring under the finish floor that rests on the floor joists.
Sump A pit in a basement floor to collect water, into which a sump pump is placed to remove water.
Surfaced lumber Lumber that is dressed by running it through a planer.
Surveyor A person skilled in land measurement.
Swale A drainage channel formed where two slopes meet.

Tamp To ram and compact soil.
Tar A dark heavy oil used in roofing and roof surfacing.
Tempered Thoroughly mixed cement or mortar.
Tensile strength The greatest stretching stress a structural member can bear without breaking or cracking.
Termite shield Sheet metal used to block the passage of termites.
Thermal conductor A substance capable of transmitting heat.
Thermostat A device for automatically controlling the supply of heat.
Threshold The beveled piece of stone, wood, or metal over which the door swings. It is sometimes called a carpet strip, or a saddle.
Throat A passage directly above the fireplace opening where a damper is set.
Tie A structural member used to bind others together.
Timber Lumber with a cross section larger than 4″×6″ (100 by 150 mm), for posts, sills, and girders.
Title insurance An agreement to pay the buyer for losses in title of ownership.
Toe nail To drive nails at an angle.
Tolerance The acceptable variance of dimensions from a standard size.
Tongue A projection on the edge of wood that joins with a similarly shaped groove.

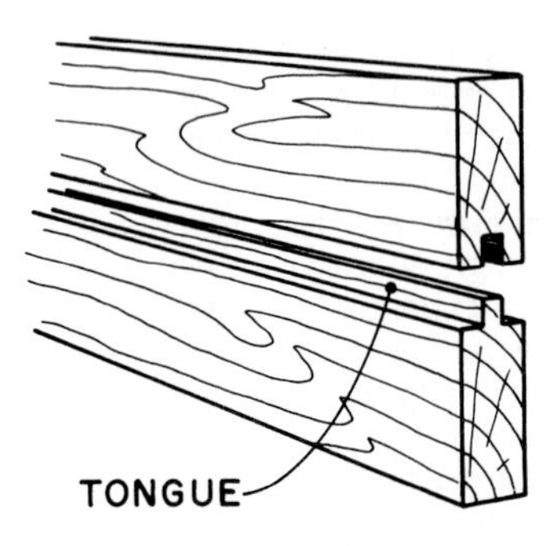

Total run The total of all the tread widths in a stair.
Transom A small window over a door.
Tread The step or horizontal member of a stair.
Trimmers Single or double joists or rafters that run around an opening in framing construction.
Truss A prefabricated triangular-shaped unit for supporting roof loads over long spans.

Underpinning A foundation replacement or reinforcement for temporary braced supports.
Undressed lumber Lumber that is not squared or finished smooth.
Unit construction Construction that includes two or more preassembled walls, together with floor and ceiling construction, for shipment to the building site.

Valley The internal angle formed by the two slopes of a roof.
Valley jacks Rafters that run from a ridgeboard to a valley rafter.
Valley rafter The diagonal rafter forming the intersection of two sloping roofs.
Valve A device that regulates the flow of material in a pipe.

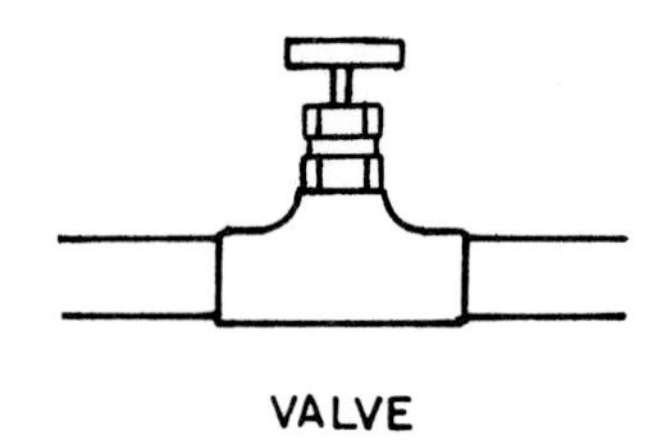

Vapor barrier A watertight material used to prevent the passage of moisture or water vapor into and through walls.
Veneer A thin covering of material over a core material.
Vent A screened opening for ventilation.
Ventilation The process of supplying and removing air by natural or mechanical means to or from any space.
Vent pipes Small ventilating pipes extending from each fixture of a plumbing system to the vent stack.
Vent stack The upper portion of a soil or waste stack above the highest fixture.
Vergeboard The board that serves as the eaves finish on the gable end of a building.
Vestibule A small lobby or entrance room.
Vitreous Pertaining to a composition of materials that resemble glass.
Volume The amount of space occupied by an object. Measured in cubic units.

Wainscot Facing for the lower part of an interior wall.
Wallboard Wood pulp, gypsum, or similar materials made into large rigid sheets that may be fastened to the frame of a building to provide a surface finish.
Warp Any change from a true or plane surface. Warping includes bow, crook, cup, and twist.
Warranty deed A guarantee that the property is as promised.
Wash The slant upon a sill, capping, etc., to allow the water to run off.
Waste stack A vertical pipe in a plumbing system that carries the discharge from any fixture.
Waterproof Material or construction that prevents the passage of water.
Water table A projecting mold near the base on the outside of a building to turn the rainwater outward. Also the level of subterranean water.
Watt A unit of electrical energy.
Weathering The mechanical or chemical disintegration and discoloration of the surface of exterior building materials.
Weather strip A strip of metal or fabric fastened along the edges of windows and doors to reduce drafts and heat loss.
Weep hole An opening at the bottom of a wall to allow the drainage of water.
Well opening A floor opening for a stairway.

Zoning Building restrictions as to size, location, and type of structures to be built in specific areas.

GLOSSARY OF CAD TERMS

Alphanumeric keyboard Letters and numerals combined on the same keyboard.
Bit A unit of computer information equal to one binary decision.
Byte A sequence of computer binary digits that function as a unit.
Cartesian coordinates Zero based rectangular grid system used as a guide to drawing width, height and depth.
Commands Computer operator's instructions to a computer via an input device.
Coordinates Intersecting points on a grid system.
Central processing unit (CPU) The heart of a computer system where operator input commands are transformed into pulses and currents which the computer can modify or store for future use.
Cathode ray tube (CRT) The display screen on which computer images are displayed.
Cursor A device used to locate points on a graphic table, touchscreen, or menu.
Database Information stored in a computer's memory.
Digitize The selection of computer commands or points with an input device.
Disk drive A device used to read information from a storage disk.
Enter The input of information into a computer.
Elements (entities) A related group of shapes or text which can be manipulated by the computer as a single shape or area.
Hard copy Computer output printed or plotted on paper or film.
Hard disk A storage device which is part of a CPU and holds, stores and retrieves more data, faster than a microdisk.
Hardware The physical components of a computer system which includes its CPU, input devices, memory and storage and output devices.
Joystick A lever device that controls the position of a cursor on a display screen.
Layering The separation of elements of the same drawing through the use of a computer command.
Microprocessor Electronic circuit embedded on a small silicon chip.
Modum Device used to transmit computer information over phone lines.
Mouse An input device which moves on a graphic table and moves the position of the cursor when activated by touch.
Network Multiple computers connected to a minicomputer or mainframe computer which enables each computer to use the minicomputer or mainframe data base.
Master menu (primary) A listing of the major headings of tasks available in a software program.
Software Computer program containing the data necessary to perform specific functions with a computer.
Solid modeling Three dimensional drawings showing only the exterior surfaces of an object.
Stylus An electronic pen used on a graphic tablet to locate the cursor on a screen or identify menu commands.
Wire frame Three dimensional drawings which reveals all hidden lines.
x Axis Horizontal axis in a cartesian coordinate system.
y Axis Vertical axis in a cartesian coordinate system.
z Axis Perpendicular axis in a three dimensional cartesian coordinate system.

ARCHITECTURAL SYNONYMS

Architectural terms are standard. Nevertheless, architects, draftsmen, and builders often use different terms for the same object. Geographic location can influence a person's word choice. For instance, what is referred to as a faucet in one area of the country is call a tap in another area. What one person calls an attic, another calls a garret, and still another a loft. Each entry below is followed by a word or words which someone—somewhere—uses to refer to the entry.

Anchor bolt securing bolt, sill bolt
Arcade corridor
Attic garret, loft, half story
Awning overhang, canopy
Back plaster parget
Baffle screen
Baseboard mopboard, finish board, skirting
Basement cellar
Base mold shoe mold
Batten cleat
Bearing plate sill, load plate
Bearing soil compact soil
Bibs faucets, taps
Birdsmouth seat of a rafter
Blanket insulation sheet insulation
Buck doorframe
Building area setback, building lines
Building code building regulations
BX conduit, tubing, metal casing, armored able
Caps coping
Carriage stringer
Casement window hinged window
Casing window frame
Catch basin cistern, dry well, reservoir
Caulking compound grout
Ceiling clearance headroom
Cesspool sewage basin, seepage pit
Chimney pot flue cap
Cleat batten
Colonnade portico
Coping caps
Domicile home, dwelling, residence
Doorsill saddle, threshold, cricket
Double hung double sashed
Drainage hole weep hole
Dry wall gypsum board, Sheetrock, wallboard, building board, rocklath, plasterboard, compo board, insulating board
Duct pipeline, vent, raceway, plenum
Easement right of way
Egress exit, outlet
Escalator motor stairs
Exit egress, outlet
Eyebrow dormer
Fillers shims
Filler stud trimmer
Finish work trim, millwork
Flush Plate switch plate
Footer anchorage, footing
Foundation sill mudsill
Gallery balcony, ledge, platform
Gazebo pavilion, belvedere
Glazing Bar muntin, pane frames, sash bars
Grade ground level, ground line, grade line
Ground line grade, grade line, ground level
Handrail newel
Hatchway opening, trapdoor, scuttle
Header lintel
Hoist lift, elevator, dumbwaiter
Hung ceiling drop ceiling, clipped ceiling
Jalousies louvers
Lacing lattice bars
Lavatory sink
Lintel header
Load weight
Load plate bearing plate, sill
Lobby vestibule, stoop, porch, portal
Lot plot, property, site
Mantel shelf
Millwork trim, finish work
Modern contemporary
Moving load live load
Newel handrail
Particle board composition board, fiberboard
Partition wall
Party wall common wall
Pilaster wall column
Pitch slant, slope
Plank and beam post and beam, post and lintel
Plate cut birdmouth, seat cut, seat of a rafter
Platform framing western framing
Plenum pipeline, vent, raceway, duct
Pressure stress
Rough floor subfloor
Rough lumber undressed lumber
Sill bearing plate, load plate
Sill bolt securing bolt, anchor bolt
Soffit underside
Spar lumber, beam, wood, timber
Spiral stairs screw stairs, winding stairs
Standard unit module
Step tread
Strutting bridging
Threshold saddle, doorsill, cricket
Tower turret
Underside soffit
Undressed lumber rough lumber
Veranda passageway, balcony, porch
Water closet toilet, W.C.
Water table water level
Weight load
Western framing platform framing

Index